THE REPLICATION OF
NEGATIVE STRAND VIRUSES

DEVELOPMENTS IN CELL BIOLOGY

Volume 1—Development and Differentiation in the Cellular Slime Moulds, P. Cappuccinelli and J.M. Ashworth (1977)

Volume 2—Biomathematics and Cell Kinetics, A.J. Valleron, P.D.M. Macdonald (1978)

Volume 3—Developmental Biology of Acetabularia, S. Bonotto, V. Kefeli, S. Puiseux-Dao (1979)

Volume 4—Physical and Chemical Aspects of Cell Surface Events in Cellular Regulation, Charles DeLisi, Robert Blumenthal (1979)

Volume 5—Structure and Variation in Influenza Virus, Graeme Laver and Gillian Air (1980)

Volume 6—Novel ADP-Ribosylations of Regulatory Enzymes and Proteins, Mark E. Smulson, Takashi Sugimura (1980)

Volume 7—The Replication of Negative Strand Viruses, David H.L. Bishop and Richard W. Compans (1981)

THE REPLICATION OF NEGATIVE STRAND VIRUSES

Proceedings of the 4th International Symposium on Negative Strand Viruses
held October 26–November 1, 1980 at Frenchman's Reef,
Saint Thomas, U.S. Virgin Islands

Editors:
DAVID H. L. BISHOP, Ph.D.
Professor of Microbiology
University of Alabama Medical Center
Birmingham, Alabama U.S.A.

and

RICHARD W. COMPANS, Ph.D.
Professor of Microbiology
University of Alabama Medical Center
Birmingham, Alabama U.S.A.

ELSEVIER/NORTH-HOLLAND
NEW YORK • AMSTERDAM • OXFORD

Published by:

Elsevier North Holland, Inc.
52 Vanderbilt Avenue, New York, New York 10017

Sole distributors outside USA and Canada:

Elsevier/North Holland Biomedical Press
335 Jan van Galenstraat, P.O. Box 211
Amsterdam, The Netherlands

Library of Congress Cataloging in Publication Data

International Symposium of Negative Strand Viruses
(4th : 1980 : St. Thomas, V.I.)
The replication of negative strand viruses.
(Developments in cell biology ; v. 7)
Bibliography: p.
Includes index.
1. Viruses, RNA—Reproduction—Congresses. I. Bishop, David H.L. II. Compans, Richard W. III. Title.
IV. Series.
QR395.I57 1980 616'.0194 81-3167
ISBN 0-444-00606-0 AACR2
Series SBN: 0-444-41607-2

Manufactured in the United States of America

Contents

PARAMYXOVIRUSES

Preface

The Negative Strand Viruses are responsible for many important common, or exotic, human and animal diseases, for example, influenza, mumps, measles, rabies, Rift Valley fever, certain human encephalitis and hemorrhagic fevers. Three previous Symposia concerning Negative Strand Viruses, organized by Drs. R.D. Barry and B.W.J. Mahy, were held in Cambridge, England, in 1969, 1973, and 1977. The proceedings of those meetings were published in the following books: "The Biology of Large RNA Viruses" (R.D. Barry and B.W.J. Mahy, eds.), Academic Press, 1970; "Negative Strand Viruses" (B.W.J. Mahy and R.D. Barry, eds.), Academic Press, 1975; and "Negative Strand Viruses and the Host Cell" (B.W.J. Mahy and R.D. Barry, eds.), Academic Press, 1978.

Since 1977, our understanding of the molecular biology and genetics of this group of viruses has expanded enormously. The fourth Negative Strand Virus Symposium, held at Frenchman's Reef, St. Thomas, U.S. Virgin Islands, October 27–November 1, 1980, brought together virologists and molecular biologists to discuss the recent advances made with these viruses, and present their research data on the following topics: the composition and function of the viral structural components; virus gene cloning and sequence analyses; virus infection processes; the mechanisms of viral RNA transcription and replication; virus genetics; the nature of antigenic determinants (as studied through hybridoma analyses), and the mechanism of antigenic variation; virus pathogenic and transmission capabilities; the molecular basis of virus virulence; interferon; viral defective particles; and studies of persistent virus infections in cell culture and *in vivo*.

This book, entitled "The Replication of Negative Strand Viruses" represents the collection of some 115 papers presented at the fourth Negative Strand Virus Symposium. The papers have been grouped into sections dealing with the studies concerning the various negative stranded virus families (Arenaviridae, Bunyaviridae, Orthomyxoviridae, Paramyxoviridae, Rhabdoviridae, and the unclassified Marburg-Ebola viruses). Within each section, papers are organized in the order of those concerning virus structure, replication and assembly, genetics, defective viruses, antigenic analyses and antigenic variation, virus persistence, transmission and the pathogenesis of virus infections.

We would like to acknowledge the sponsorship by the University of Alabama in Birmingham, and the financial support which was provided by the following organizations: Accurate Chemical and Scientific Corp.; American Cyanamid Co.; Bellco Glassware; Biocell Laboratories; Boehringer-Mannheim; Connaught Laboratories Inc.; Fogarty International Center-NIH; M.A. Bioproducts; Merck Sharp and Dohme Research Laboratories; National Institute of Allergy and Infectious Diseases; Rheem Manufacturing Co.; Sandoz Forschungsinstitut; Searle Research and Development; Vangard International Inc.; Warner-Lambert Co.

Finally, we would like to personally thank Mrs. Denice Lee-Montgomery and Mrs. Betty Jeffrey, both of whom worked cheerfully, extraordinarily hard and unselfishly in organizing the Symposium.

D.H.L. Bishop R.W. Compans

Department of Microbiology
The Medical Center
University of Alabama in Birmingham
Birmingham, Alabama 35294, U.S.A.
November 24, 1980

List of Those Attending Conference

GILLIAN M. AIR John Curtin School Medical Research, Canberra ACT 2601, Australia

THOMAS ALBRECHT Bureau of Biologics, FDA and NIAID, NIH, Bethesda, Maryland 20205

PASCU ATANASIU Pasteur Institut, Paris, Cedex 15, France

THOMAS BACHI University of Zurich, Zurich, Switzerland

MELVYN BAEZ Mt. Sinai School of Medicine of CUNY, New York, New York 10029

AMIYA K. BANERJEE Roche Institute of Molecular Biology, Nutley, New Jersey 07110

PAULINE H.S. BAY University of Illinois, Urbana, Illinois 61801

WILLIAM J. BEAN St. Jude Children's Research Hospital, Memphis, Tennessee 38101

BARRY J. BEATY Yale University, New Haven, Connecticut 06510

DAVID H.L. BISHOP University of Alabama in Birmingham, Birmingham, Alabama 35294

DAVID P. BOERSMA University of Alabama in Birmingham, Birmingham, Alabama 35294

MICHELE BOULOY Memorial Sloan-Kettering Cancer Center, New York, New York 10021

MICHAEL A. BRATT University of Massachusetts Medical Center, Worcester, Massachusetts 01605

GILBERT BRUN CNRS, Gif-sur-Yvette, France

DORIS J. BUCHER Mount Sinai Medical Center, New York, New York 10029

MICHAEL J. BUCHMEIER Scripps Clinic and Research Foundation, La Jolla, California 92037

BOYCE W. BURGE Christ Hospital Institute of Medical Research, Cincinnati, Ohio 45215

JAMES C. CHAN University of Texas System Cancer Center, Houston, Texas 77030

V.G. CHINCHAR St. Jude Children's Research Hospital, Memphis, Tennessee 38101

J.C.S. CLEGG PHLS Centre for Applied Microbiology and Research, Wiltshire, England

JOHN M.P. CLERX University of Alabama in Birmingham, Birmingham, Alabama 35294

CORRIE CLERX-VAN HAASTER University of Alabama in Birmingham, Birmingham, Alabama 35294

GAIL M. CLINTON Children's Hospital Medical Center, Boston, Massachusetts 02115

PETER L. COLLINS University of Connecticut, Storrs, Connecticut 06268

RICHARD W. COMPANS University of Alabama in Birmingham, Birmingham, Alabama 35294

RICHARD J. COLONNO Dupont Experimental Station, Wilmington, Delaware 19898

JON CONDRA NIH, Bethesda, Maryland 20205

CELIA E. COTO University of Buenos Aires, Buenos Aires, Argentina

JAMES H. COX Fed. Res. Inst. Animal Virus Diseases, Tubingen, Federal German Republic

NANCY J. COX Center for Disease Control, Atlanta, Georgia 30333

DONNA CRECELIUS St. Louis School of Medicine, St. Louis, Missouri 63104

JOAN C. CRICK Animal Virus Research Institute, Pirbright, United Kingdom

JOEL M. DALRYMPLE USAMRIID, Fort Detrick, Frederick, Maryland 21701

ELSA B. DAMONTE University of Buenos Aires, Buenos Aires, Argentina

JOHN ALUN DAVIES Searle Research and Development, High Wycombe, England

NANCY L. DAVIS University of North Carolina at Chapel Hill, Chapel Hill, North Carolina 27514

BERNARD DIETZSCHOLD The Wistar Institute, Philadelphia, Pennsylvania 19104

NIGEL J. DIMMOCK University of Warwick, Coventry, Warwickshire, CV4 7AL, England

KEN DIMOCK McMaster University, Hamilton, Ontario, L8S 4J9, Canada

PATRICIA DOWLING University of Pittsburgh, Pennsylvania 15261

EDWARD DUBOVI University of North Carolina at Chapel Hill, Chapel Hill, North Carolina 27514

FRANK J. DUTKO Scripps Clinic and Research Foundation, La Jolla, California 92037

ELLIE EHRENFELD University of Utah, Salt Lake City, Utah 84132

RICHARD M. ELLIOT Mt. Sinai School of Medicine of CUNY, New York, New York 10029

SUZANNE U. EMERSON University of Virginia, Charlottesville, Virginia 22908

JAMES R, ETCHISON University of California, Davis, California 95616

LINDA E. FISHER University of Michigan-Dearborn, Dearborn, Michigan 48128

ROBERT Z. FLORKIEWICZ University of Arizona, Tucson, Arizona 85721

TERYL K. FREY University of Pittsburgh, Pittsburgh, Pennsylvania 15261

PATRICIA N. FULTZ University of Texas Science Center, Dallas, Texas 75235

MARY JANE GETHING ICRF, Lincoln's Inn Fields, London WC2A 3PX, United Kingdom

HARA P. GHOSH McMaster Univ., Hamilton, Ontario L8S 4J9, Canada

OSCAR GRAU Univ. Nacional de la Plata, La Plata, Argentina

ANASTASIA GREGORIADES Public Health Research Institute City of New York, New York, New York 10016

KAILASH GUPTA St. Jude Children's Hospital, Memphis, Tennessee 38101

FREDERICK S. HAGEN Children's Hospital Medical Center, Boston, Massachusetts 02115

DELSWORTH HARNISH McMaster University, Hamilton, Ontario, L8S 4J9, Canada
ALAN J. HAY Natl. Inst. Med. Res., London NW7 1AA, England
RONALD C. HERMAN New York State Department of Health, Albany, New York 12205
GOERG HERRLER University of Alabama in Birmingham, Birmingham, Alabama 35294
MARTINEZ HEWLETT University of Arizona, Tucson, Arizona 85721
LAWRENCE E. HIGHTOWER University of Connecticut, Storrs, Connecticut 06268
ALAN L. HITI UCLA School of Medicine, Los Angeles, California 90024
FRANK M. HORODYSKI University of California, San Diego, La Jolla, California 92093
COLIN R. HOWARD London Sch. Hyg. and Trop. Med., London WC1E 7HT, England
ALICE S. HUANG Children's Hospital Medical Center, Boston, Massachusetts 02115
DIANE HUANG University of Pittsburgh, Pittsburgh, Pennsylvania 15261
STEPHEN C. INGLIS University of Cambridge, Cambridge CB2 2QQ, England
STEPHEN JACOBSON Rensselaer Polytechnic Institute, Troy, New York 12181
C. YONG KANG University of Texas Health Science Center, Dallas, Texas 75235
JACK D. KEENE Duke University Medical Center, Durham, North Carolina 27710
EDWIN D. KILBOURNE Mt. Sinai School of Medicine of CUNY, New York, New York 10029
MICHAEL P. KILEY Center for Disease Control, Atlanta, Georgia 30333
DAVID W. KINGSBURY St. Jude Children's Hospital, Memphis, Tennessee 38101
LAURA KINGSFORD California State University, Long Beach, California 90840
HANS-D. KLENK Justus Liebig, Univ., Giessen, Federal German Republic
DANIEL KOLAKOFSKY University of Geneve, Geneve, Switzerland
ROBERT M. KRUG Memorial Sloan-Kettering Cancer Center, New York, New York 10021
CAROL A. KRUSE UCLA, Los Angeles, California 90024
FLORENCE LAFAY Universite de Paris Sud, 91405 Orsay, France
CHING-JUH-LAI NIH, Bethesda, Maryland 20014
ROBERT A. LAMB Rockefeller University, New York, New York 10021
DENNIS M. LAMBERT Christ Hospital Institute of Medical Research, Cincinnati, Ohio 45215
FRANK R. LANDSBERGER Rockefeller University, New York, New York 10021
W. GRAEME LAVER John Curtin School Med. Research, Canberra ACT 2601, Australia
ROBERT A. LAZZARINI NIH, Bethesda, Maryland 20014
JOHN D. LEAVITT FDA and NIAID, NIH, Bethesda Maryland 20014
JACQUELINE LECOMTE NIBSC, Hampstead, London, NW3 6RB, England
FRITZ LEHMANN-GRUBE University of Hamburg, Hamburg, Federal German Republic
JOHN LENARD CMDNJ-Rutgers Medical School, Piscataway, New Jersey 08854
JO-ANN LEONG Oregon State University, Corvallis, Oregon 97331
WAI-CHOI LEUNG McMaster University, Hamilton, Ontario, L8S 4J9, Canada
DOUGLAS S. LYLES Bowman Gray School of Medicine, Winston-Salem, North Carolina 27103
H.F. MAASSAB University of Michigan, Ann Arbor, Michigan 48109
BRIAN W.J. MAHY University of Cambridge, Cambridge CB2 2QQ, England

JAMES C. MAO Abbott Labs, North Chicago, Illinois 60064
MARY ANN K. MARKWELL UCLA, Los Angeles, California 90024
LORRAINE MARNELL University of Utah, Salt Lake City, Utah 84132
GUSTAVE N. MBUY Christ Hospital Institute of Medical Research, Cincinnati, Ohio 45215
MICHELINE MCCARTHY NIH, Bethesda, Maryland 20205
MARCELLA MCCLURE Washington University School of Medicine, St. Louis, Missouri 63110
DALE MCPHEE CSIRO Division of Animal Health, Parkville, Victoria 3052, Australia
JAMES J. MCSHARRY Albany Medical College of Union University, Albany, New York 12208
HERBERT MEIER-EWERT Technische University, Munchen 40, Federal German Republic
CODY MEISSNER Harvard Medical School, Boston, Massachusetts 02115
DOUGLAS K. MILLER CMDNJ-Rutgers Medical School, Piscataway, New Jersey 08854
JULIE B. MILSTIEN Bureau of Biologics, Bethesda, Maryland 20205
EXEEN M. MORGAN St. Jude Children's Hospital, Memphis, Tennessee 38101
TRUDY MORRISON University of Massachusetts Medical Center, Worcester, Massachusetts 01605
SOLOMON L. MOWSHOWITZ Mt. Sinai School of Medicine of CUNY, New York, New York 10029
RICHARD MOYER Vanderbilt University School of Medicine, Nashville, Tennessee 37232
SUE A. MOYER Vanderbilt University School of Medicine, Nashville, Tennessee 37232
BRIAN MURPHY NIH, Bethesda, Maryland 20014
CLAYTON W. NAEVE University of Utah, Salt Lake City, Utah 84132
SETSUKO NAKAJIMA Center for Disease Control, Atlanta, Georgia 30333
DEBI P. NAYAK U.C.L.A. School of Medicine, Los Angeles, California 90024
JOSEPH ONGRADI Institute of Virology, Glasgow G11 5JR, Scotland
JOHN S. OXFORD NIBSCS, Hampstead, London, NW3 England
PETER PALESE Mt. Sinai School of Medicine of CUNY, New York, New York 10029
MICHAEL PARKER University of Arizona, Tucson, Arizona 85721
MARK E. PEEPLES University of Massachusetts Medical Center, Worcester, Massachusetts 01605
JACQUES PERRAULT Washington University School of Medicine, St. Louis, Missouri 63110
C.J. PETERS USAMRIID, Fort Detrick, Frederick, Maryland 21701
RALF PETTERSSON University of Helsinki, 00290 Helsinki 29, Finland
WILLIAM A. PETRI, JR. University of Virginia, Charlottesville, Virginia 22908
CHARLES J. PFAU Rensselaer Polytechnic Institute, Troy, New York 12181
STEPHEN K. PLOTCH Memorial Sloan-Kettering Cancer Center, New York, New York 10021
MARCEL PONS Christ Hospital Institute of Medical Research, Cincinnati, Ohio 45219
MIRÇEA POPESCU University of Illinois, Chicago, Illinois 61801
A.G. PORTER Searle Research and Development, High Wycombe, Bucks HP12 4HL, England

ALLEN PORTNER St. Jude Children's Research Hospital, Memphis, Tennessee 38101

OLIVIA T. PREBLE NIAMDD, Bethesda, Maryland 20205

CRAIG R. PRINGLE Institute of Virology, Glasgow, Scotland

DONALD RAO Washington University, St. Louis, Missouri 63110

ARLENE RAMSINGH McMaster University, Hamilton, Ontario, L8S 4J9, Canada

WILLIAM E. RAWLS McMaster University, Hamilton, Ontario, L8S 4J9, Canada

RUSSELL L. REGNERY Center for Disease Control, Atlanta, Georgia 30333

M.E. REICHMANN University of Illinois, Urbana, Illinois 61801

BERNARD RENTIER NIH, Bethesda, Maryland 20014

CHRIS RICHARDSON Rockefeller University, New York, New York 10021

STEVEN J. ROBBINS Pennsylvania State University; Hershey, Pennsylvania 17033

BETTY H. ROBERTSON University of Virginia, Charlottesville, Virginia 22908

JAMES S. ROBERTSON University of Cambridge, Cambridge CB2 2QQ, England

OLGA M. ROCHOVANSKY Christ Hospital Institute of Medical Research, Cincinnati, Ohio 45215

JOHN K. ROSE Salk Institute, San Diego, California 92138

MICHAEL G. ROTH University of Alabama in Birmingham, Birmingham, Alabama 35294

RUDOLF ROTT Justus Liebig University, Giessen, Federal German Republic

LAURENT ROUX Hopitol Cantonal, Infectious Disease Division, Geneve, 4 Switzerland

DAVE ROWLANDS Animal Virus Res. Institute, Pirbright, Surrey GU24 ONF, England

POLLY ROY University of Alabama in Birmingham, Birmingham, Alabama 35294

MILTON J. SCHLESINGER Washington University School of Medicine, St. Louis, Missouri 63110

SONDRA SCHLESINGER Washington University School of Medicine, St. Louis, Missouri 63110

LOTHAR SCHNEIDER Federal Research Institute Animal Virus Diseases, Tubingen, Federal German Republic

CHRISTOPH SCHOLTISSEK Justus Liebig-Univ., Giessen, Federal German Republic

MICHAEL SCHROM Albany Medical College, Albany, New York 12208

MANFRED SCHUBERT NIH, Bethesda, Maryland 20014

JEROME L. SCHULMAN Mount Sinai School of Medicine, New York, New York 10029

IRENE T. SCHULZE St. Louis School of Medicine, St. Louis, Missouri 63104

MARGARET J. SEKELLICK University of Connecticut, Storrs, Connecticut 06268

MICHAEL W. SHAW University of Alabama in Birmingham, Birmingham, Alabama 35294

KAZUFUMI SHIMIZU Nagasaki University School of Medicine, Nagasaki 852, Japan

ROBERT E. SHOPE Yale University, New Haven, Connecticut 06510

C.P. STANNERS University of Toronto, Toronto, Canada M4X 1K9

JOHN STEPHENSON University of Wurzburg, Wurzburg, Federal German Republic

HENRY O. STONE University of Kansas, Lawrence, Kansas 60045

KAZUO SUGIYAMA University of Alabama in Birmingham, Birmingham, Alabama 35294

DONALD F. SUMMERS University of Utah, Salt Lake City, Utah 84132

JOSEPH F. SZILAGYI Institute of Virology, Glasgow, G11 5JR, Scotland

OYEWALE TOMORI University of Ibadim, Nigeria

HIROSHI USHIJIMA University of Alabama in Birmingham, Birmingham, Alabama 35294

STEVEN WECHSLER Christ Hospital Institute of Medical Research, Cincinnati, Ohio 45215

PHILIP WECK Genentech Research Labs, South San Francisco, California 94080

GAIL T.W. WERTZ University of North Carolina at Chapel Hill, Chapel Hill, North Carolina 27514

TADEUSZ J. WIKTOR Wistar Institute of Anatomy and Physiology, Philadelphia, Pennsylvania 19104

JERRY S. WOLINSKY Johns Hopkins University, Baltimore, Maryland 21205

WILLIAM H. WUNNER Wistar Institute of Anatomy and Physiology, Philadelphia, Pennsylvania 19104

FUNMEI YANG NIH, Bethesda, Maryland 20014

JONATHAN YEWDELL Wistar Institute of Anatomy and Physiology, Philadelphia, Pennsylvania 19104

JAMES F. YOUNG Mt: Sinai School of Medicine of CUNY, New York, New York 10029

JULIUS S. YOUNGNER University of Pittsburgh, Pittsburgh, Pennsylvania 15261

JAMES J. ZAZRA Mt. Sinai School of Medicine of CUNY, New York, New York 10029

ARENAVIRUSES

David H.L. Bishop and Richard W. Compans, eds.
The Republication of Negative Strand Viruses

Characterization of the Arenaviruses Lassa and Mozambique

Michael P. Kiley,[a] Oyewale Tomori,[b] Russell L. Regnery,[a] and Karl M. Johnson[a]

The arenaviruses are a family of enveloped viruses containing single-stranded RNA of negative sense.[1,2] LCM virus has been chosen as the prototype virus for the family in which there are presently 11 members.[3,4] The group is divided into the LCM complex ("Old World") and the Tacaribe complex ("New World") on the basis of immunological cross-reactivity as determined by complement-fixation[5] or indirect fluorescent antibody staining.[6]

Besides LCM, the other members of the "Old World" group are Lassa virus and Mozambique virus. Lassa fever is an acute febrile disease of western Africa which may evoke hemorrhagic manifestations and is often fatal.[7,8] The disease is endemic in some areas of western Africa where man is infected by close contact with a single murine species, *Mastomys natalensis,* in which the virus is transmitted enzootically.[9] Mozambique virus, in contrast, while isolated from the same rodent species is found in the southeastern African country of Mozambique, where a hemorrhagic fever disease like that produced by Lassa fever virus has not been reported.[4]

We are currently investigating several aspects of Lassa fever and its etiologic agent and our facilities include a field station in Sierra Leone. We are studying viral pathogenesis, possible treatment regimens, the ecology of the rodent host, and the characterization of the Lassa and Mozambique viruses.

[a] Special Pathogens Branch, Bureau of Laboratories, Center for Disease Control, Atlanta, Georgia.
[b] Permanent address: University of Ibadan, Ibadan Nigeria.

We here present data concerning preliminary characterization of Lassa and Mozambique virions with emphasis on the viral proteins and antigenic differences.

Virus Growth and Purification

The Lassa virus used in this study was isolated from a patient in Sierra Leone and was passaged 3 times in Vero cells. The origin of the Mozambique virus used has been previously described.[4] We used the AN 20410 strain which had undergone 5 passages in suckling mouse brain and 3 in Vero cells. When these agents were used to infect Vero cells at a multiplicity of infection of $\approx$ 0.1 PFU/cell the growth curve depicted in Figure 1 was obtained. Both viruses grow well in Vero cells with Lassa reaching a peak titer of 5×10^7 PFU/ml on day 2 and Mozambique a peak titer of 5×10^6 on day 4. To obtain concentrated preparations of purified virions, roller bottle cultures were inoculated with dilute virus and at an appropriate time, virus in the supernatant fluid was concentrated and purified according to the method described by Obijeski et al.[10] The density of Lassa fever virus in a sucrose gradient was approximately 1.17 g/cc (data not shown) which is in agreement with previously determined values for other arenaviruses.[11] Purification of Lassa virus produced pools with titers of $> 10^9$ PFU/ml and 10% recovery of infectivity. When this concentrated virus was titrated by plaque assay significant interference was observed. In a pool with a titer of $> 10^9$ no plaques or cell destruction were seen until the 10^{-3} dilution (Kiley, unpublished). This phenomenon of autointerference has been previously reported for LCM and other arenaviruses.[12,13]

Virion Structural Proteins

Both Lassa and Mozambique virus were grown in Vero cells in the presence of either a ^{3}H-amino acid mixture or ^{35}S-methionine. When progeny virus from such cells was purified and analyzed on an SDS-phosphate polyacrylamide gell system, the electrophoretic profile presented in Figure 2 was obtained. Both viruses exhibited 3 protein peaks. To determine their molecular weights, virion proteins were compared to those of VSV proteins using the calculation technique of Weber and Osborn.[14] The VSV molecular weights used in our calculations were those reported by Obijeski et al.,[10] namely, 160,000, 65,000, 54,000, 42,000, and 27,000, respectively, for the L, G, N, NS, and M-proteins. The 3 Lassa virus proteins had molecular weights of 72,000, 52,000, and 39,000 and were similar, if not identical, to the corresponding Mozambique proteins.

To determine which of the virion proteins was glycosylated, virus was harvested and purified following growth in the presence of ^{3}H-glucosamine. Polyacrylamide gel profiles of glucosamine labeled proteins are presented in

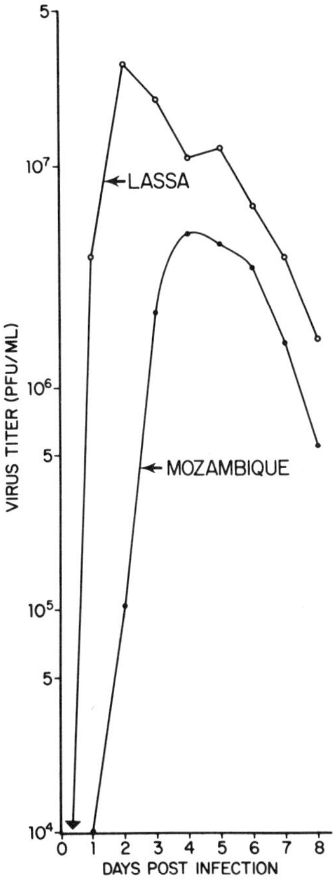

Figure 1. Growth curves of Lassa fever and Mozambique viruses. Monolayers of Vero cells in 150 cm^2 plastic flasks were infected with either virus at a multiplicity of infection of 0.1 PFU/cell. Virus was adsorbed for 30 min, cells were rinsed with PBS and 50 ml of Eagles minimal essential medium (MEM) containing 2% fetal calf serum was added to each flask. Duplicate 1 ml samples were collected at daily intervals and assayed by plaque titration in Vero cells.

Figure 3. Glucosamine labeled Lassa virus contained 2 major glycoproteins corresponding to the virion proteins with molecular weights of 52,000 and 39,000. Compared to the nucleocapsid peak, the glycoprotein peaks are quite broad. This appears to be due to varying degrees of glycosylation of the virion proteins.[15] An additional peak was seen at fraction 16 which corresponds to a molecular weight of approximately 115,000. The latter peak is not always present and its significance has not been determined. When ^3H-glucosamine labeled Mozambique proteins were electrophoresed, peaks similar in size to those described for Lassa were present and an additional

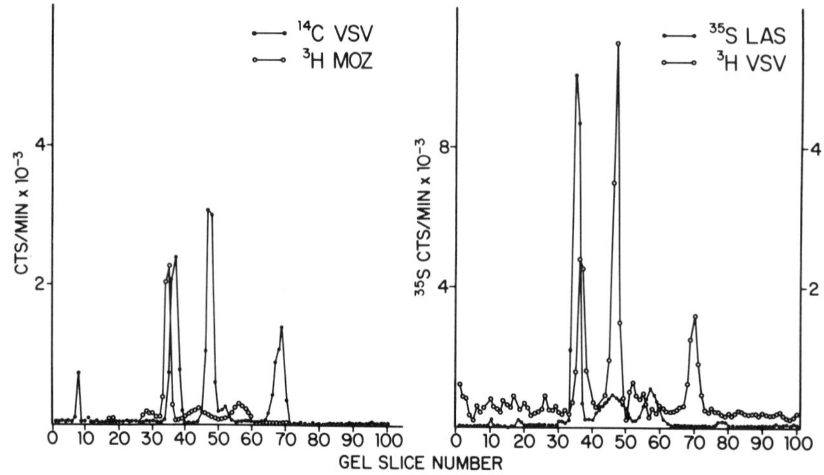

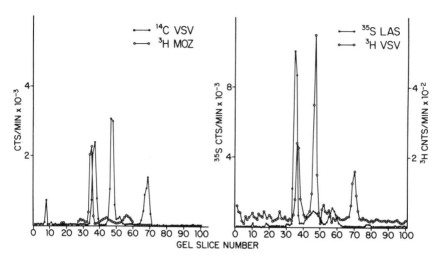

Figure 2. Analysis of Lassa and Mozambique virion proteins by polyacrylamide gel electrophoresis. Roller bottle cultures of Vero cells were infected with either Lassa or Mozambique virus at a multiplicity of $\approx 10^{-3}$ PFU/cell. MEM containing 2% FCS was added to each bottle before incubation at 36°. At 3 days postinfection, the media was replaced by one containing either a ^{3}H-amino acid mixture (10 μCi/ml and 10% of the normal amino acid concentration of ^{35}S-methionine (20 μCi/ml) in methionine tree medium. On day 4, supernatant was harvested, concentrated, and purified on KT-glycerol and sucrose gradients. The purification and gel analyses were done according to the procedures we have recently described for Ebola virus.[28] Labeled VSV virion proteins were included in each gel as molecular weight markers.

Figure 3. Polyacrylamide gel analyses of Lassa and Mozambique virion glycoproteins. Virus was grown purified and analyzed as described in Figure 2, only the medium for labeling contained ^{3}H-glucosamine (5 μCi/ml) in MEM with 50% of the standard glucose concentration.

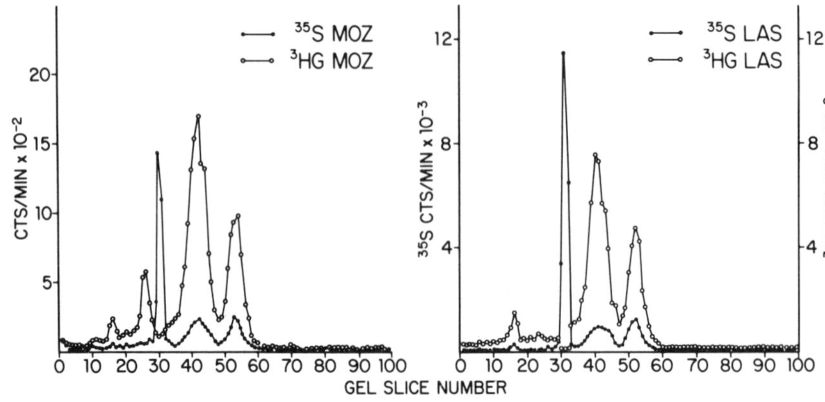

peak with an estimated molecular weight of approximately 84,000 also was detected. This latter polypeptide has appeared in the 4 Mozambique preparations examined to date.

Table 1 presents a summary of the Lassa and Mozambique virion proteins and compares them with the proteins of the other members of the arenavirus group. It is evident that the 3 major proteins of both Lassa and Mozambique are similar in size and distribution to most other arenaviruses and when the glucosamine data are considered, it is apparent that the Lassa and Mozambique proteins represent the N (72,000), G_1 (52,000), and G_2 (39,000) arenavirus proteins. The putative polymerase (P) is apparently absent in our Lassa and Mozambique preparations, unless in our system it is glycosylated with a molecular weight of either 84,000 or 115,000. Though this may be the case, a more likely role for either or both of the large glycoproteins is that of precursor for either the G1 or G2 virion proteins. Buchmeier and Oldstone[15] have described a large glycoprotein precursor in LCM infected BHK-21 cells and a similar uncleaved protein may be present in the Mozambique virion.

Antigenic Differences Between Lassa and Mozambique Viruses

A close antigenic relationship between Lassa and Mozambique viruses has been reported. Using both complement fixation and indirect fluorescent antibody staining with guinea pig, mouse, and hamster sera, it was demonstrated that antibody made to either virus reacts strongly with both viruses in either assay system.[4] Convalescent sera from Lassa fever patients also reacts with both viruses but do not react with other arenaviruses, with the possible exception of LCM.[6] In these reports and in recent results we have obtained using monkey sera (Kiley et al., unpublished), some quantitative difference in antigenicity was recognized.

To determine where the antigenic differences might reside, monoclonal antibodies react against one or both viruses were used as probes. Since we do not yet have monoclonal antibodies to Lassa or Mozambique viruses, we employed monoclonal antibody produced against LCM virus which reacted with Lassa and/or Mozambique viruses (LCM monoclonal antibody was kindly supplied by Dr. Michael Buchmeier). Figure 4 presents the results of an experiment in which LCM monoclonal antibodies reactive to either (1) Mozambique (9-7-5), (2) Lassa and Mozambique (24-A-21), or (3) neither virus (1-16-7) were reacted with detergent disrupted ^{35}S-methionine labeled Mozambique virus. The antigen-antibody complexes were precipitated with staphylococcal protein A and pellet and supernatant were analyzed on polyacrylamide gels. Some nucleocapsid was precipitated by protein A even in the absence of specific antibody (1-16-7), but the pellet-supernatant differences disclosed that monoclonal antibody 24-A-21 which reacts with all 3 viruses was directed against the virus nucleoprotein and that antibody 9-7-5 which reacts only with LCM and Mozambique virus was directed against the

Table 1. Estimated Molecular Weights ($\times 10^{-3}$) of Arenavirus Virion Proteins.

	Protein							
	X[a]	X[a]	P	N	G1	G2	X[a]	X[a]
Lassa	115G	—	—	72	52	39	—	—
Mozambique	115G	84(G)	—	72	54	40	—	—
LCM[21]				62	54	35	—	12
LCM[22]		91	75	67	55	39	25	12
Pichinde[23]				72	72	34		12
Pichinde[24]			77	66	64	38		12
Pichinde[25]			77	68	65	38	50	15
Tacaribe[26]			79	68	—	42		
Tacaribe[25]			77	68	—	44		
Junin[27]		91(G)	71(G)	64	52	38	25	
Machupo[25]		84	74	68	50	41		15

[a]Unknown function.
G: Glycoprotein.

virion G2 protein. When mixed with labeled Lassa virus, 24-A-21 precipitated the nucleoprotein and 9-7-5 did not react (data not shown). These results demonstrated that Mozambique shares with LCM an antigenic site on the G2 glycoprotein and that this site is not present on the Lassa virion.

Summary

These data demonstrate that both Lassa and Mozambique viruses possess the physicochemical characteristics of arenaviruses in the parameters measured. The density of 1.17 g/cc in sucrose is identical or similar to the values reported for other members of the group.[11] The autointerference detected in preparations of concentrated Lassa virus has been well documented for LCM[16,17] as well as for many other viral systems.[18]

Lassa virus contains 3 major proteins—1 nonglycosylated (presumably the nucleoprotein) and 2 smaller glycosylated proteins. While the molecular weights of the Lassa and Mozambique virus N-proteins (72,000) are somewhat larger than the 62,000-68,000 N-protein molecular weight range reported for other arenaviruses,[3] the pattern of the 3 Lassa and Mozambique proteins is generally very similar to the other arenaviruses. The 84,000 molecular weight glycoprotein found in the Mozambique virion may be the precursor for the G1 and/or G2 proteins. The presence of uncleaved glycoprotein on the Mozambique virion may play a role in the differential pathogenicity of the 2 viruses for monkeys.[19] A precursor glycoprotein similar to the glycoprotein present on the Mozambique virion has recently been demonstrated in LCM infected BHK-21 cells.[15]

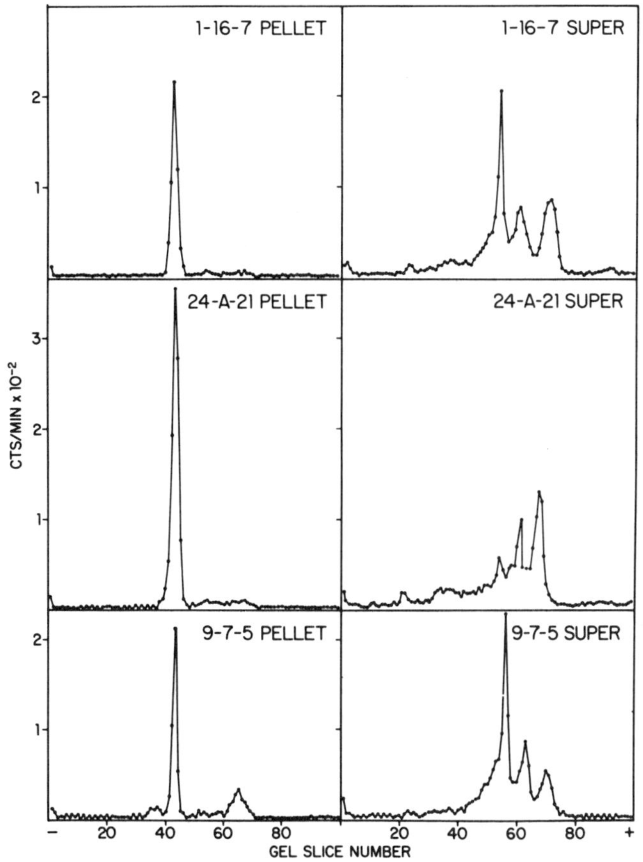

Figure 4. Immune precipitation of Mozambique virion proteins by LCM monoclonal antibodies. The method used was a modification of that described by Kessler.[29] ^{35}S-methionine labeled virions were disrupted in PBS containing 0.5% NP40, 0.5% sodium dodecyl sulfate. To 25 μl of the disrupted virus solution ($\approx$ 6000 CPM) was added 50 μl of undiluted antisera and the mixture was placed at 4° for 2 hr at which time 400 μl of a 10% solution of formalin-fixed *Staphylococcus* (Pansorbin, Calbiochem-Behring, LaJolla, CA) was added. This mixture was placed at 4° for 10 min and *Staphylococcus*-antibody complexes were pelleted at 8,700 $\times$ g for 2 min. The first supernatant was saved for protein analyses and the complexes were washed 3 times with PBS containing 0.5% NP40. Finally, 200 μl of a dissociating buffer (67.5 mm Tris-Hcl, pH 6.7, 4% SDS, 6M urea) was added to the pelleted complexes. After boiling for 1 minute, the mixture was repelleted and this supernatant was analyzed by gel electrophoresis. Electrophoresis conditions are described in Figure 2. By IFA staining monoclone 24-A-21 reacted with both Lassa and Mozambique, 9-7-5 reacted only with Mozambique and 1-16-7 reacted with neither virus.

When comparing arenavirus protein data (Table 1), it is clear that while there are similarities, especially regarding the N, G1, and G2 proteins, there are also many specific differences in the protein composition of these viruses. There seems to be no particular pattern that differentiates the "Old World" from the "New World" virus group or the human pathogens from the non-pathogens. It is also obvious that different laboratories can obtain different results when using the same virus. To resolve this problem it will be necessary to conduct a study comparing the virion proteins of all the arenaviruses in a single laboratory employing the same methodology. This laboratory must be able to work with all of the arenaviruses.

A recent report has demonstrated that some monoclonal antibodies raised against LCM virus will react not only with LCM but with Mozambique or Mozambique and Lassa.[20] We have demonstrated that the monoclonal antibody which reacts with only LCM and Mozambique binds to the G2 protein of Mozambique (9-7-5) while the antibody that reacts with all 3 viruses (24-A-21) binds to the N-protein of Lassa and Mozambique. LCM monoclonal antibody 9-7-5 also reacts with the LCM G2 protein[20] indicating that the LCM and Mozambique cross reactive sites reside on the G2 protein of both viruses.

In order to further delineate the antigenic differences among Lassa, Mozambique, and other arenaviruses and to aid in the identification of field isolates, we are currently producing monoclonal antibodies to both Lassa and Mozambique viruses.

ACKNOWLEDGMENTS

Any work performed in the Maximum Containment Laboratory requires the cooperation of all Special Pathogens Branch members. We especially want to thank Ms. Donna Sasso for her excellent technical assistance.

References

1. Rowe, W.P., Murphy, F.A., Bergold, G.H., Casals, J., Hotchin, J., Johnson, K.M., Lehmann-Grube, F., Mims, C.A., Traub, E., and Webb, P.A. (1970): *J. Virol.* 5:651.
2. Pfau, C.J., Bergold, G.H., Casals, J., Johnson, K.M., Murphy, F.A., Pedersen, I.R., Rawls, W.E., Rowe, W.P., Webb, P.A., Weissenbacher, M.D. (1974): *Intervirology* 4:207.
3. Pedersen, I.R. (1979): *Adv. Virus Res.* 24:277.
4. Wulff, H., McIntosh, B.M., Hamner, D.B., and Johnson, K.M. (1977): *Bull. WHO* 4:441–444.
5. Casals, J., Buckley, S.M., and Cedeno, R. (1975): *Bull. WHO* 52:421.
6. Wulff, H., Lange, J.V., and Webb, P.A. (1978): *Intervirology* 9:344–350.
7. Frame, J.D., Baldwin Jr., J.M., Gocke, D.J., and Troup, J.M. (1970): *Amer. J. Trop. Med. Hyg.* 19:670.
8. Buckley, S.M. and Casals, J. (1970): *Amer. J. Trop. Med. Hyg.* 19:680.
9. Monath, T.P. (1975): *Bull. WHO* 52:577.

10. Obijeski, J.F., Marchenko, A.T., Bishop, D.H.L., Cann, B.W., and Murphy, F.A. (1974): *J. Gen. Virol.* 22:21–33.

11. Rawls, W.E. and Buchmeier, M. (1975): *Bull. WHO* 52:393.

12. Welsh, R.M., Burner, P.A., Holland, J.J., Oldstone, M.B.A., Thompson, H.A., and Villarreal, L.P. (1975): *Bull. WHO* 52:403.

13. Dutko, F.J., Wright, E.A., and Pfau, C.J. (1976): *J. Gen. Virol.* 31:417–427.

14. Weber, K. and Osborn, M. (1969): *J. Biol. Chem.* 244:4406.

15. Buchmeier, M.J. and Oldstone, M.B.A. (1979): *Virology* 99:111–120.

16. Lehman-Grube, F. (1971): Lymphocytic Choriomeningitis Virus. *Virology Monographs,* 10, Springer-Verlag, New York.

17. Welsh, R.M., O'Connell, C.M., and Pfau, C.J. (1972): *J. Gen. Virol.* 17:355.

18. Huang, A.S. (1973): *Annual Reviews of Microbiology* 27:101.

19. Kiley, M.P., Lange, J.V., and Johnson, K.M. (1979): *Lancet* 2:738.

20. Buchmeier, M.J., Lewicki, H.A., Tomori, O., and Johnson, K.M. (1980): *Nature,* Vol 288:486.

21. Buchmeier, M.J., Elder, J.H., and Oldstone, M.B.A. (1978): *Virology* 89:133.

22. Pedersen, I.R. (1973): *J. Virol.* 11:416.

23. Ramos, B.A., Courtney, R.J., and Rawls, W.E. (1972): *J. Virol.* 10:661.

24. Vezza, A.C., Gard, G.P., Compans, R.W., and Bishop, D.H.L. (1977): *J. Virol.* 23:776.

25. Gangemi, J.D., Rosato, R.R., Connell, E.V., Johnson, E.M., and Eddy, G.A. (1978): *J. Gen. Virol.* 41:183–188.

26. Gard, G.P., Vezza, A.C., Bishop, D.H.L., and Compans, R.W. (1977): *Virology* 83:84.

27. Martinez-Segovia, Z.M. and DeMitri, M.I. (1977): *J. Virol.* 1977:579–583.

28. Kiley, M.P., Regnery, R.L., and Johnson, K.M. (1980): *J. Gen. Virol.* 49:333.

29. Kessler, S.W. (1975): *J. Immunol.* 115:1617.

David H.L. Bishop and Richard W. Compans, eds.
The Republication of Negative Strand Viruses

Junin Virus Structure

O. Grau,[a] M.T. Franze-Fernandez,[b]
V. Romanowski,[a] S.M. Rustici,[a] and M.F. Rosas[a]

Junin virus, the etiological agent of Argentine hemorrhagic fever, is a member of the arenavirus group.[1] Similarly to other arenaviruses, its genome consists of two single-stranded segments of RNA of 33S and 25S. In addition, the purified virions yield the ribosomal RNA species 28S and 18S, and 4S and 5S RNA.[2]

Analysis of the protein composition of several arenaviruses has established that a feature common to all the viruses studied is the presence of 2 or 3 major proteins: a nonglycosylated nucleocapsid polypeptide and either one or two glycoproteins. Only one glycoprotein has been found in Tacaribe and Tamiami viruses and two in LCM, Pichindé and Machupo viruses.[3] For Junin virus, six structural proteins, four of them glycosylated, have been reported.[4]

In order to clarify whether the protein composition of this pathogenic arenavirus differs from that of the other members of the group, we analyzed the polypeptides present in extensively purified preparations of two strains of Junin virus. One of them, MC_2, was isolated from a rodent from the endemic region;[5] the other, $XJCl_3$, is an attenuated derivative of the XJ prototype strain[6] that was used as an experimental vaccine.

Viruses from the supernatant medium of BHK-21 infected cells labeled with ^{3}H-leucine were pelleted through a sucrose cushion and purified through glycerol-tartrate and CsCl gradients.

[a] Cátedra de Química Biológica II, Facultad de Ciencias Exactas, Universidad Nacional de La Plata, 47 y 115, 1900 La Plata, Argentina.
[b] Facultad de Farmacia y Bioquímica, Universidad de Buenos Aires, Argentina.

Analysis of the labeled viral proteins by SDS-PAGE showed three peaks of radioactivity in both the MC_2 and $XJCl_3$ strains (Figure 1). The electrophoretic mobilities correspond to molecular weights of 60,000, 44,000 and 35,000. Similar patterns of radioactivity were obtained when the purification of the viruses was performed only up to the glycerol-tartrate step, or was carried further through the CsCl gradient. This last step removed the unlabeled serum proteins from the culture medium that contaminate the virus preparation, thus rendering coincident patterns of bands in the strained and labeled gels (results not shown).

Viral glycosylated polypeptides were analyzed by growing Junin virus in the presence of ^{3}H-glucosamine. After purification, the virions were dissociated with SDS and the proteins resolved by PAGE and localized by fluorography. Two intense bands with mol wt of about 39,000 and 34,000 and a faint band with a mol wt of 44,000 and observed (Figure 2). In some virus preparations instead of the two bands, a single broad band in the region of mol wt 39,000-34,000 was observed. This could be the result of different degrees of glycosylation as previously reported for other arenaviruses.[7]

To isolate Junin virus nucleocapsids, a preparation of ^{3}H-leucine-labeled virus was dissociated with NP40 and then fractionated by centrifugation in a 15-30% sucrose gradient. The radioactivity was distributed in two fractions, one at the top of the gradient and the other sedimenting with a sedimentation coefficient of about 170S. Analysis of the proteins in both fractions by gel electrophoresis revealed that the fast sedimenting fraction contained the virion 60,000 dalton polypeptide while the glycoprotein remained at the top of the gradient (Figure 3). As in other arenaviruses, the largest virion polypeptide is associated with the nucleocapsid.[8]

Figure 1. Gel electrophoresis analysis of the polypeptides of Junín virus MC_2 and $XJCl_3$. Junín viruses labeled with ^{3}H-leucine, purified up to the glycerol-tartrate step (Panels A and B) or up to the CsCl step (Panel C), were dissociated with SDS and electrophoresed on a 15% polyacrylamide gel.[9] Two millimeter slices were disolved with H_2O_2 at 60°C and counted in a 10 ml mixture of 2 volumes of 5% Omnifluor-toluene + 1 volume Triton X-100.

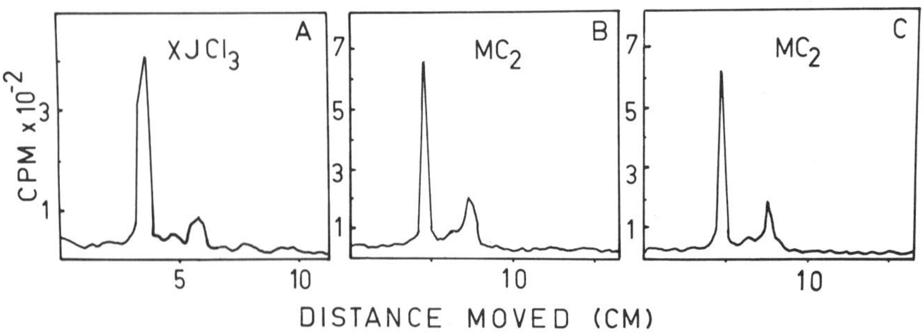

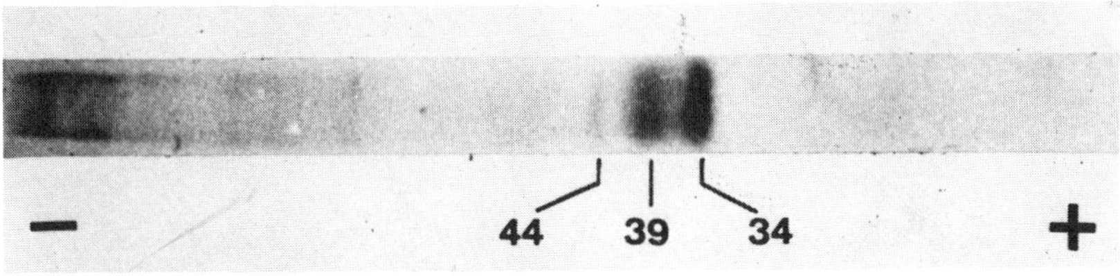

Figure 2. Glycoproteins of Junín virus. Junín virus MC_2 strain was labeled with ^{3}H-glucosamine and purified through a sucrose cushion. The viruses were dissociated and the proteins analyzed by gel electrophoresis as in Figure 1. Radioactive proteins were detected by fluorography.[10]

When viruses are labeled with ^{3}H-leucine, the amount of radioactivity incorporated into the 44,000 dalton protein is consistently low and in some cases it cannot be detected, as in the preparation analyzed in Figure 3. This protein is better labeled with ^{35}S-methionine as shown in the fluorography depicted in Figure 4. At present, we do not know whether this protein is a precursor of that of 39-34,000 mol wt, that was incorporated into the virus particle without being completely processed.

In conclusion, three proteins were resolved in purified preparation of Junin virus MC_2 and $XJCl_3$ strains. The largest, with a molecular weight of 60,000, is associated with the nucleocapsid. The other two, of molecular weights 44,000 and 39-35,000 respectively, are glycosylated. These data

Figure 3. Dissociation of Junín virus. Junín virus MC_2 was labeled with ^{3}H-leucine and purified up to the glycerol-tartrate step. The virus dissociated with 0.12% NP40 was layered on top of a 15-30% sucrose gradient and centrifuged for 150 min at 37,000 rpm in a Spinco SW41-Ti rotor. Radioactive proteins from the top and the fast sedimenting component of the gradient were analyzed by SDS-PAGE. A: Control virus (untreated with NP40); B: Protein associated with heavy structures; C: Glycoprotein solubilized by NP40.

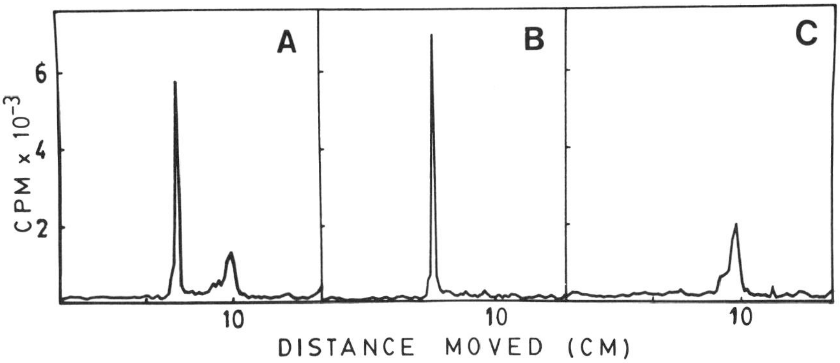

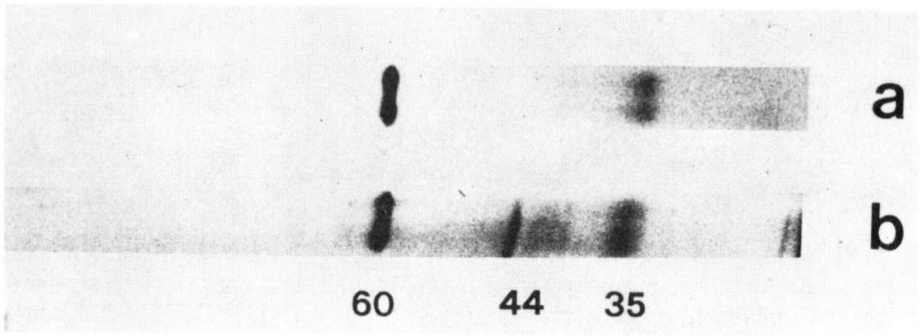

Figure 4. Gel electrophoresis analysis of the polypeptides of Junín virus MC_2 labeled with ^3H-leucine or ^{35}S-methionine. Purified virus labeled with ^3H-leucine (a) or ^{35}S-methionine (b) was dissociated and the proteins analyzed as in Figure 2.

show that Junin virus has a protein composition similar to that of the other members of the arenavirus group.

ACKNOWLEDGMENTS

This research was supported by grants from the Secretaría de Ciencia y Technología, Comisión de Investigaciones Científicas de la Provincia de Buenos Aires, Fundación Emilio Ocampo, Universidad Nacional de La Plata and a personal contribution of Mrs. J.G. de Blanco. V. Romanowski, S.M. Rustici and M.F. Rosas are fellows of the Comisión de Investigaciones Científicas de la Provincia de Buenos Aires. M.T. Franze-Fernández is a recipient of a research career award from the Consejo Nacional de Investigaciones Científicas y Técnicas. O. Grau is a recipient of a research career award from the Comisión de Investigaciones Científicas de la Provincia de Buenos Aires. We thank N. Sierra and L. Valenti for technical assistance and S.A. Moya for typing of the manuscript.

References

1. Pfau, C.J., Bergold, G.H., Casals, J., Johnson, K.M., Murphy, F.A., Pedersen, I.R., Rawls, W.E., Rowe, W.P., Webb, P.A. and Weissenbacher, M.C. (1974): *Intervirology* 4:207–213.
2. Añón, M.C., Grau, O., Martınez Segovia, Z. and Fernández, M.T. (1976): *J. Virol.* 18:833–838.
3. Rawls, W.E. and Leung, W.C. (1979): In: *Comprehensive Virology,* H. Fraenkel-Conrat and R. Wagner (eds.), Vol 14, pp. 157–192.
4. Martínez Segovia, Z.M. and De Mitri, M.I. (1977): *J. Virol.* 21:579–583.
5. Vilches, A.M., Barrera Oro, J.G., and Gutman Frugone, L.F. (1965): Segundas Jornadas Epidemiológicas Argentinas (Salta).
6. Guerrero, L.B. de, Weissenbacher, M.C., and Parodi, A.S. (1969): *Medicina* 29:1–8.
7. Buchmeier, M.J. and Oldstone, M.B.A. (1979): *Virology* 99:111–120.
8. Vezza, A.C., Gard, G.P., Compans, R.W., and Bishop, D.H.L. (1977): *J. Virol.* 23:776–786.
9. Laskey, R.A., Gurdon, J.B., and Crawford, L.V. (1972): *Proc. Natl. Acad. Sci. U.S.A.* 69:3665–3669.
10. Bonner, W.M. and Laskey, R.A. (1974): *Eur. J. Biochem.* 46:83–88.

David H.L. Bishop and Richard W. Compans, eds.
The Replication of Negative Strand Viruses

Analysis of the Structure and Function of Pichinde Virus Polypeptides

P.R. Young, A.C. Chanas, and C.R. Howard

Nearly all members of the Arenaviridae cause acute or persistent infections in their natural host. A number of arenaviruses also induce severe hemorrhagic disease in man, notably Lassa, Machupo and Junin and variously cross-react by complement fixation (CF) with other members that do not cause overt human illness. Of the latter group, Pichinde virus also displays a similar pathology to Lassa in experimentally infected animals and is therefore a particularly suitable model for those viruses pathogenic for man.

Pichinde has been shown to contain at least two RNA genome segments and three major polypeptide species, two of which are glycoproteins associated with the viral envelope. Although antisera prepared against Pichinde and Lassa viruses contain high titres of CF antibodies, they have been shown to exhibit minimal neutralizing capacity both *in vivo*[1] and *in vitro*.[2,3] A plaque size reduction test has been evaluated as an alternative method for the detection of Pichinde virus antibody.[3] In this technique, infectious aggregates formed after the reaction of Pichinde virus with the homologous antiserum result in the restriction of plaque development.

In order to analyze further the immunochemical properties of this virus, its morphological and biochemical properties have been re-examined and limited studies performed using antisera prepared against selected viral components.

Results and Discussion

Several studies have shown that all arenaviruses contain a major 54-68,000 mol wt nucleocapsid protein (N) together with one (G) or two (G1, G2) major glycoprotein species in the outer envelope.[4] It is likely that one or both glycoproteins induce a protective antibody response *in vivo;* sera containing

Department of Medical Microbiology, London School of Hygiene and Tropical Medicine, Keppel Street, London WC1E 7HT.

neutralizing antibodies to LCM virus have been shown to immunoprecipitate the two surface glycoproteins of the homologous virus.[5] In addition, a number nm long and appear club-shaped with a hollow central axis (Figures 2a and Pichinde virus (Figure 1). The radioiodination technique employing the enzyme lactoperoxidase has clearly shown that at least part of the 22,000 mol wt polypeptide is exposed on the virion surface and may contribute to the antigenic composition of the intact virus. The surface specificity of this reaction was confirmed by the finding that both the N and G2 proteins were metabolically labeled with ^{3}H-tyrosine whereas G2 was the only additional component labeled by the lactoperoxidase procedure which is specific for tyrosine residues.

The morphology and morphogenesis of all arenaviruses are remarkably similar. Electron microscopy of negatively stained preparations has shown the virus particles to be pleomorphic, enveloped structures between 80 and 150 nm in diameter. The viral envelope, which is formed from the plasma membrane of the infected cell, contains surface projections that are 5 to 10 nm long and appear clubshaped with a hollow central axis (Figures 2a and

Figure 1. SDS-PAGE analysis of purified Pichinde (a) unlabeled viral proteins separated on a 7.5–15% linear gradient slab gel and stained with Coomassie Blue. Migration of marker proteins is represented to the right of the profile. (b) viral proteins radiolabeled with ^{35}S-methionine (●—●) and ^{3}H-glucosamine (○ . . . ○) separated on a 10% cylindrical gel. Molecular weights indicated × 10^{-3}.

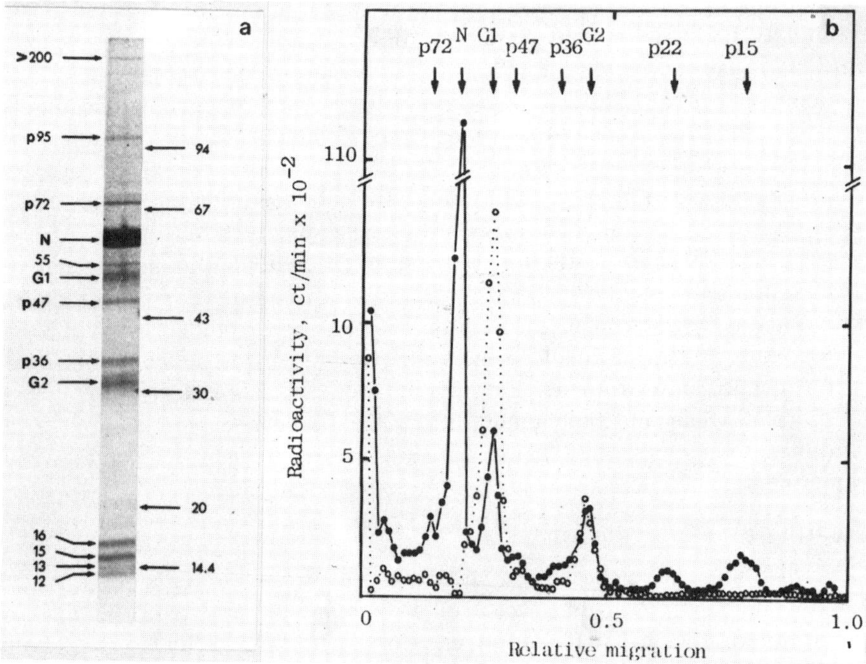

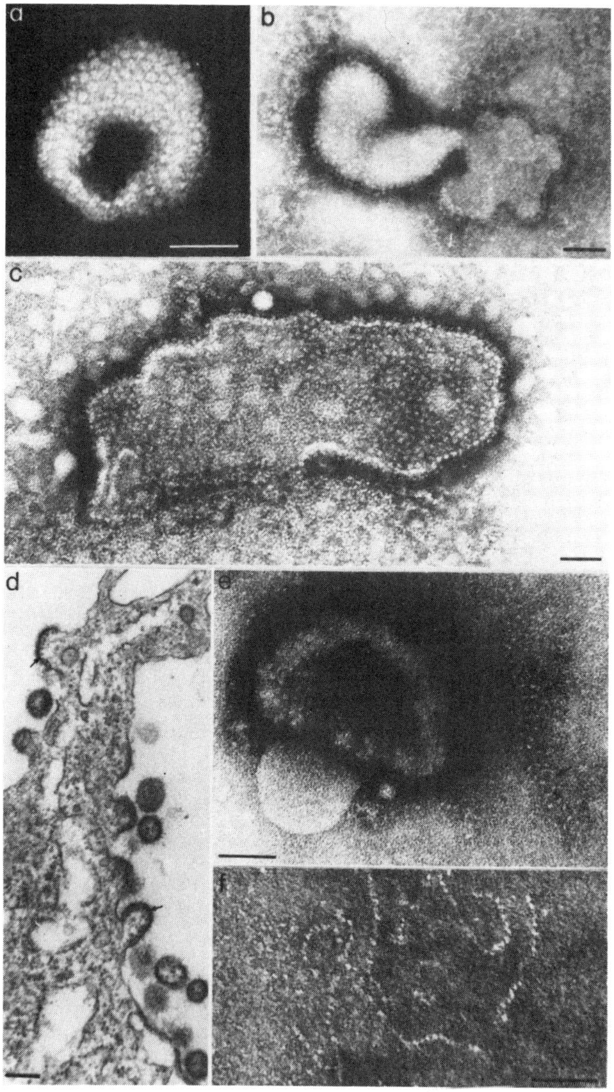

Figure 2. Electron micrographs of both (a, b) purified Pichinde and (c) virus collected from a tissue culture harvest, negatively stained with 2% potassium phosphotungstate (PTA) pH 7.3; (d) ultrathin section of infected Vero cells; (e) virus stained with 2% PTA pH 6.2 showing a liberated helical structure (arrowed) and (f) purified nucleocapsids stained with 2% PTA pH 7.3 demonstrating the "beader-on-a-string" appearance of nucleosomes. Bar represents 50 nm (a, b, c, e, f) or 200 nm (d).

2b). Variation in the spacing of these spikes as seen in different preparations suggests limited specific interaction between them and may simply reflect differing degrees of osmotic swelling (Figures 2a and 2c). Examination of ultrathin sections of infected cell cultures reveals the characteristic presence of host ribosomes within mature and budding arenavirus particles. In addition, an alignment of what appear to be parallel strands of nucleoprotein of 10 to 12 nm in diameter (arrowed) along the inner surface of the differentiating viral envelope was seen (Figure 2d).

Isolated structures thought to represent the nucleocapsids of Tacaribe and Pichinde viruses have been described.[6,7,8] In the present study, the internal nucleocapsid fraction was prepared by treatment of Pichinde virus with 1% Nonidet P40 in the presence of 0.2 M NaCl followed by isopycnic centrifugation in a linear Urografin gradient. Separated nucleocapsids were recovered at a density of 1.25 g/ml and subsequently found to contain the major nonglycosylated N protein and a small band migrating with an average mol wt of 15,000 (Figure 3). The latter band remained associated with the nuc-

Figure 3. Electron micrographs of purified nucleocapsids rotary shadowed with gold/palladium at an angle of 10° showing both (a) "spider-forms" and (b) isolated circles. Bar represents 100 nm.

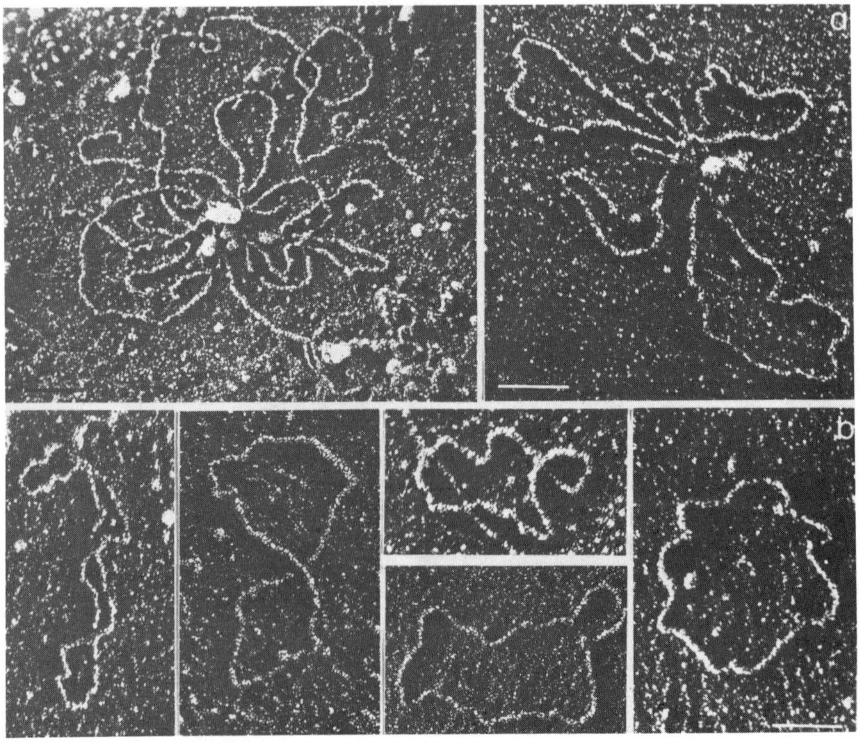

leocapsid fraction after solubilization in the presence of 0.8 M NaCl; these conditions removed trace amounts of other components with the exception of the minor 72,000 mol wt protein and high mol wt material. Direct visualization of the nucleocapsid fraction has revealed strands with a beaded appearance (Figure 2f). These beads, or nucleosomes, are 3 to 4 nm in diameter and are spaced on average 5 to 7 nm apart. Examination of virus gently disrupted with negative stain at low pH has shown that these nucleosome strands are organized as helical structures 10 to 12 nm in diameter (Figure 2e). When the purified nucleocapsid fraction was rotary shadowed, both circles and "spider-forms" were observed (Figure 4). The presence of the latter suggests the packaging of the virion nucleocapsid into a discrete core structure.

In a previous study, it was shown that rabbit antiserum containing antibodies directed against all the major polypeptides of Pichinde virus failed to neutralize infectivity despite extensive sensitization and aggregation of the virus particles.[3] Earlier studies have suggested that G2 is the most exposed component of the virion surface[8,9] and a monospecific antiserum to this glycoprotein was therefore prepared following separation by SDS-PAGE. Incorporation of anti-G2 serum in the agar overlay of infected Vero cells resulted in a significant difference in the average plaque diameter eight days after infection (Table 1). In contrast, an antiserum raised against the protein band of average mol wt 15,000 failed to sensitize infectious virus, a finding in accord with the location of this material within the virus particle. Thin-section electron microscopy showed the reaction of the anti-G2 serum with Pichinde virus particles budding from infected Vero cells and the formation of immune aggregates (Figure 5). No effect was seen in the presence of normal rabbit serum.

The specificity of the anti-G2 serum was confirmed by the removal of the G2 protein from infected cells. The SDS-PAGE profile of the fraction of an infected BHK-21 cell cytoplasmic extract retained on a protein A-Sepharose column bound with this antiserum is shown in Figure 6. The separation of the G2 and G1 glycoproteins under these experimental conditions provides indirect evidence that they may exist as separate structures in the viral membrane.

Preliminary examination of acetone fixed infected Vero cells by immunofluorescence with antiserum to the 15,000 mol wt protein band showed specific perinuclear fluorescence, a pattern also seen in the sera of patients convalescing from Junin infection and positive by the complement fixation test (J. Maiztegui, personal communication). Unfixed infected cells which showed specific surface immunofluorescence when stained with anti-G2 or hyperimmune sera to whole virus were negative when tested with anti-p15 serum. Although the technique of SDS-PAGE has enabled the separation of certain Pichinde viral proteins for serological studies, the difficulties of adequately separating other components together with possible denaturation following SDS treatment make this approach unsuitable for defining the total

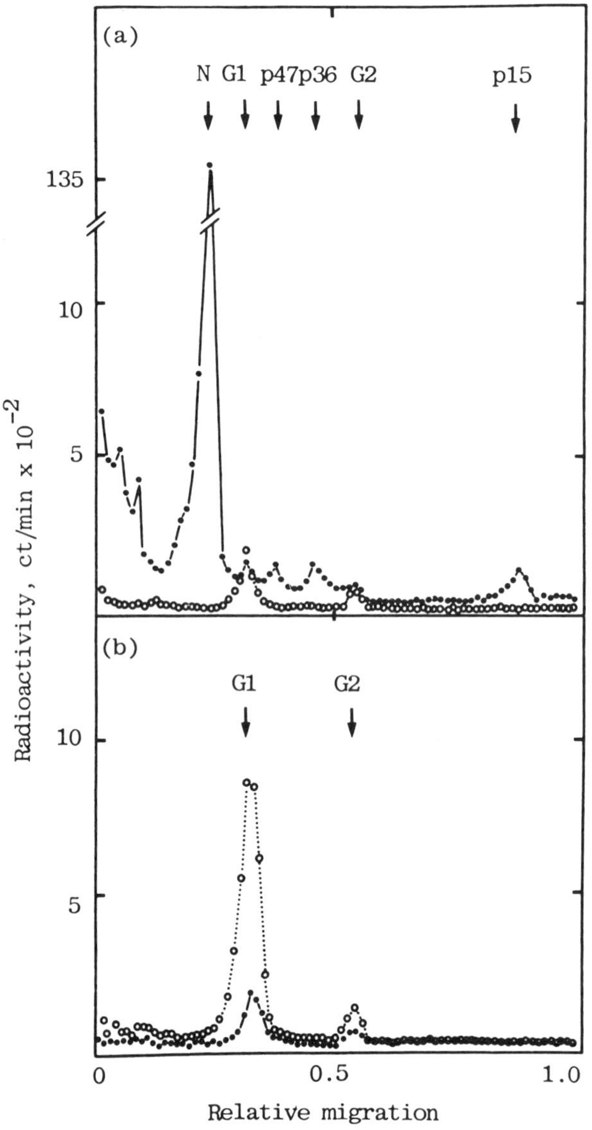

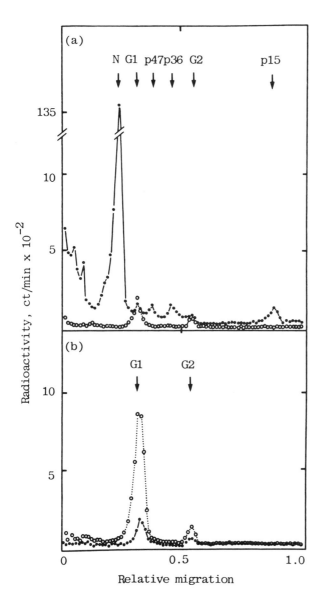

Figure 4. SDS-PAGE profiles of the disrupted fractions of purified Pichinde virus labeled with ^{35}S-methionine (●—●) and ^{3}H-glucosamine (○ . . . ○) and separated on a 10-60% linear Urografin gradient. (a) Material recovered at a density of 1.25 g/ml and (b) solubilized fraction remaining at the top of the gradient.

Table 1. Plaque Size Reduction in Infected Vero Cells in the Presence of Antiserum to Viral Components.

	Av diameter, mm ± SD
Anti-Pichinde serum	1.29 ± 0.22
Anti-G2 serum	1.40 ± 0.24
Anti-p15 serum	2.35 ± 0.37
Normal serum	2.32 ± 0.31

Test sera assayed at a final dilution of 1:40 and measurements made by projection 8 days after infection.

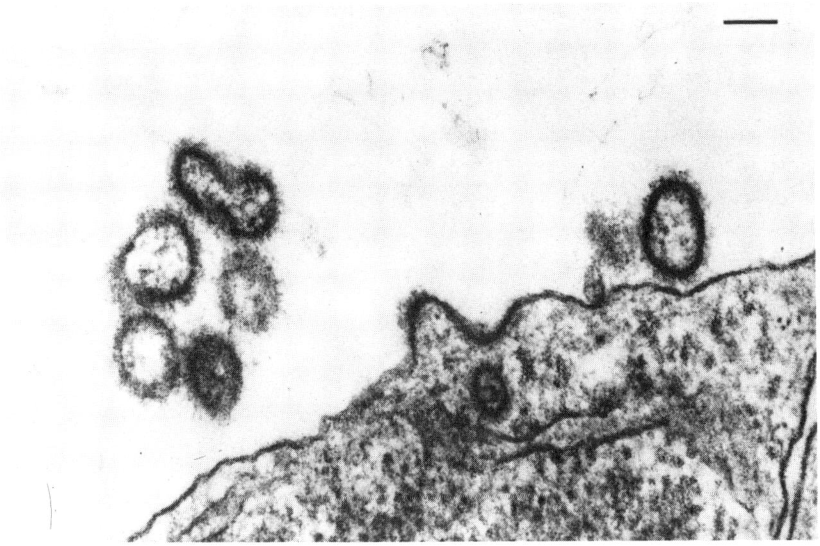

Figure 5. Pichinde virus particles budding from Vero cell plasma membrane in the presence of anti-G2 serum. Immune adsorption is restricted to the surface of mature particles, leading to the formation of aggregates, and to the plasma membrane at the sites of virus maturation. Bar represents 100 nm.

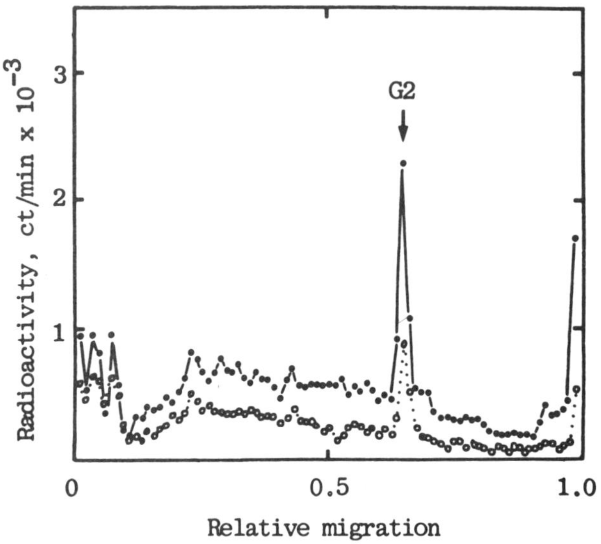

Figure 6. SDS-PAGE profile of protein removed from a cytoplasmic extract of infected BHK-21 cells labeled with ^{35}S-methionine (●—●) and ^{3}H-glucosamine (○ . . . ○) by an immunoadsorbent column containing bound anti-G2 serum. Only the G2 viral polypeptide was recovered following elution with 1 M acetic acid.

antigenic composition of this arenavirus. The production of monoclonal antibodies currently under development in this laboratory is therefore required for the complete immunochemical analysis of Pichinde and other arenaviruses.

References

1. Trapido, H. and Sanmartin, C. (1971): *Amer. J. Trop. Med. Hyg.* 20:631–641.
2. Monath, T.P. (1973): *Trop. Doc.* 4:155–161.
3. Chanas, A.C., Young, P.R., Ellis, D.S., Mann, G., Stamford, S., and Howard, C.R. (1980): *Arch. Virol.* 65:157–167.
4. Rawls, W.E. and Leung, W.-C. (1979): In: *Comprehensive Virology,* H. Fraenkel-Conrat and R.R. Wagner, (Eds.) Plenum, New York, Vol 14, pp. 157–192.
5. Buchmeier, M.J., Welsh, R.M., Dutko, F.J., and Oldstone, M.B.A. (1980): *Adv. Immunol.,* Vol 30:275–331.
6. Vezza, A.C., Gard, G.P., Compans, R.W., and Bishop, D.H.L. (1977): *J. Virol.* 23:776–786.
7. Palmer, E.L., Obijeski, J.F., Webb,P.A., and Johnson, K.M. (1977): *J. Gen. Virol.* 36:541–545.
8. Vezza, A.C., Clewley, J.P., Gard, J.P., Abraham, N.Z., Compans, R.W., and Bishop, D.H.L. (1978): *J. Virol.* 26:485–497.
9. Ramos, B.A., Courtney, R.J., and Rawls, W.E. (1972): *J. Virol.* 10:661–667.

Copyright 1981 by Elsevier North Holland, Inc.
David H.L. Bishop and Richard W. Compans, eds.
The Replication of Negative Strand Viruses

Immunoprecipitable Polypeptides in Pichinde Virus Infected BHK-21 Cells

D. Harnish, K. Dimock, W.-C. Leung, and W. Rawls

Department of Pathology, McMaster University, Hamilton, Ontario, Canada.

Pichinde virus, a member of the arenavirus group, is enveloped and contains single stranded genomic RNA of negative polarity.[1] Although purified virus preparations contain five distinct species of RNA (31S, 28S, 22S, 18S, 4-6S), only L (31S, 2.8×10^6 daltons) and S (22S, 1.3×10^6 daltons) RNAs are virus specific. The L and S RNAs contain unique sequences[2] and theoretically encode approximately 280,000 and 130,000 daltons of protein respectively.[3] However, the virion proteins which have been previously characterized, account for less than half of this coding potential and include the nucleoprotein (NP, 64,000 daltons) and two glycoproteins (GP1, 64-72,000 daltons; GP2, 34-38,000 daltons).[4,5] RNA transcriptase activity has been described[6] but the activity has not been assigned to a specific polypeptide.

We have characterized polypeptides which are immunoprecipitable from Pichinde virus infected BHK cells and herein report the existence of a previously unidentified high molecular weight protein (L) and several minor polypeptides. Interrelationships between polypeptides have been determined by two dimensional tryptic peptide analysis. We have also identified a 79,000 dalton glycoprotein (GPC) in infected cells and its nonglycosylated precursor (pGPC). In lymphocytic choriomeningitis virus infected cells GPC has been suggested to give rise to GP1 and GP2.[7]

Results

The number and kinetics of synthesis of polypeptides immunoprecipitable with immune sera from hamsters were assessed by examining the polypeptide profile obtained at 12, 24 and 48 hr after infection of BHK-21 cells at multiplicities of infection (moi) of 0.1, 1.0, 10 and 50. After a 1-hr pulse with L-^{35}S-methionine (50 μCi/ml; 900-1200 Ci/mmol), seven immune serum specific polypeptides with molecular weights of 28,000, 38,000, 48,000, 52,000, 64,000, 79,000 and approximately 200,000 were evident. All seven exhibited a time and moi-dependent appearance. The profile obtained at 24 hr postinfection is shown in Figure 1. The same profile was obtained by labeling with a mixture of ^{3}H-amino acids (data not shown).

Figure 1. Identification of immunoprecipitable polypeptides in Pichinde virus infected cells. BHK cells, infected at multiplicities of infection as indicated, were labeled with L-^{35}S-methionine for 1 hr at 24 hr postinfection. Cells were lysed in 50 mM Tris-HCl, pH 7.2 containing 0.15 M NaCl, 0.1% SDS, 1% DOC, 1% Triton X-100, 1 mM phenylmethylsulfonylfluoride and 1 mM benzamidine hydrochloride, and the supernatant was treated with hamster anti-Pichinde sera and protein A-Sepharose CL-4B. Polypeptides were analyzed on 7.5–15% gradient SDS-polyacrylamide gels. An moi of 0 refers to mock infected cells. IS and NS refer to hamster immune sera and normal sera, respectively. Polypeptides were visualized by fluorography.

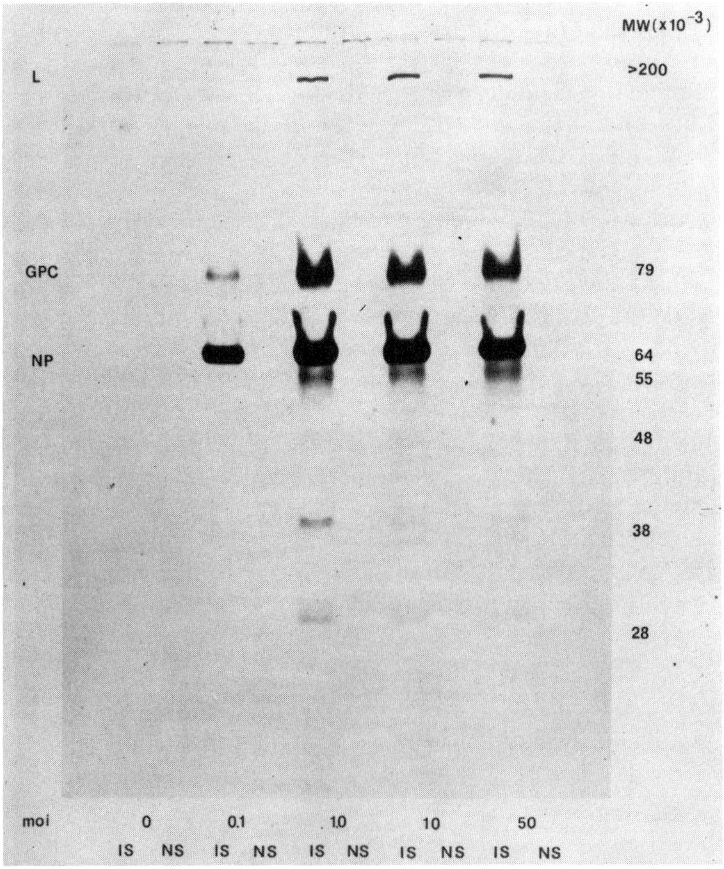

A series of pulse and pulse-chase experiments were performed to examine processing of the seven polypeptides within the infected cell. Polypeptides labeled with ^{35}S-methionine during a 10-min pulse were identical to those obtained during a 1-hr labeling period (data not shown). Following a 3-hr chase of radiolabel in the presence of at least a 100-fold excess of unlabeled methionine, polypeptides of 52,000, (GP1), 36,000, (GP2), 17,000, 16,500 and 14,000 daltons were apparent. Label in GPC shifted into GP1 and GP2 after a 9-hr chase (data not shown). All three components labeled with D-6-^{3}H-glucosamine hydrochloride although GP2 labeled poorly (Figure 2). These results suggest that GP1 and GP2 are derived from GPC.

Figure 2. Pulse and pulse-chase experiments to determine polypeptide interrelatedness. BHK cells were preincubated in methionine free medium for 1 hr and then pulse labeled for 10 min with ^{35}S-methionine. Label was chased for the indicated times in the presence of 100-fold excess unlabeled methionine. Immunoprecipitation was as described in Figure 1. M refers to mock infected cells; NS and IS as in Figure 1. GPC and GP1 were labeled with ^{3}H-glucosamine hydrochloride for 3 hr. Polypeptides were visualized by fluorography.

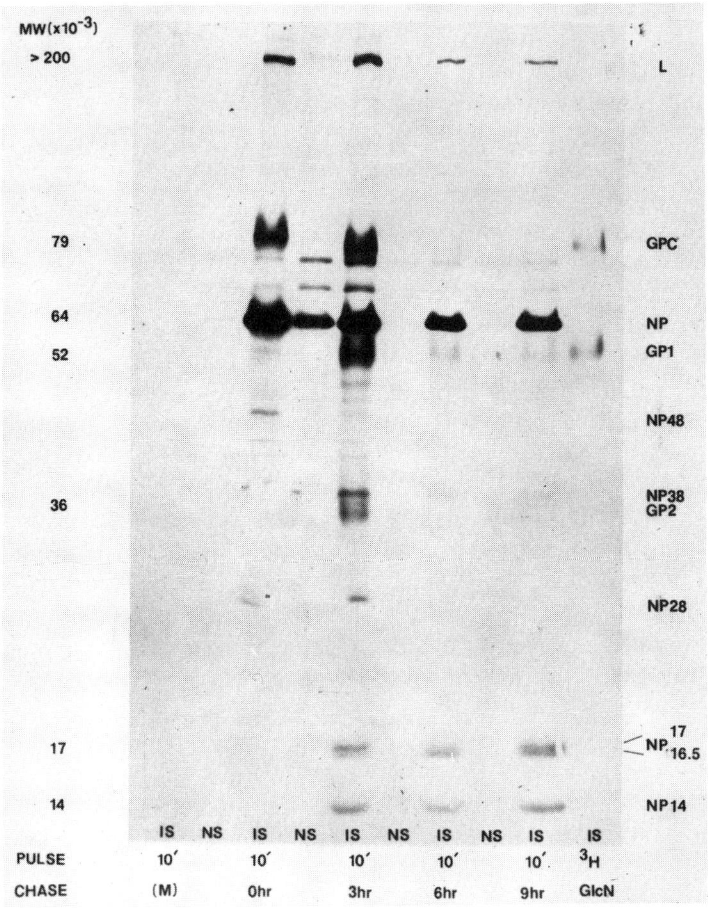

The nature of the 17,000, 16,500, and 14,000 dalton polypeptides was determined by two dimensional tryptic analysis of methionine-containing peptides. Similarities in the peptide maps indicated that these three polypeptides are proteolytic cleavage products of NP and therefore can be designated NP17, NP16.5, and NP14 (data not shown). The 48,000, 38,000 and 28,000 dalton polypeptides were also related to NP as determined by peptide mapping (data not shown). NP48 chased completely in 9 hr, but NP38 and NP28 were still evident at a reduced intensity. Two additional polypeptides, with mobilities between those of GPC and NP, were observed in immunoprecipitates using normal hamster sera or mock infected cells.

The apparent molecular weight of the glycoprotein precursor (GPC) was 79,000, including an undetermined proportion of carbohydrate. Experiments were performed to assess the molecular weight of the polypeptide portion of this molecule. To inhibit glycosylation, infected cells were labeled in the presence of 5 to 10 μg per tunicamycin ml. A polypeptide of 42,000 daltons was observed in immunoprecipitates. Identical peptide maps of the 42,000 dalton polypeptide and GPC suggest that the 42,000 dalton polypeptide is the nonglycosylated precursor to GPC and should be designated pGPC.

The number of primary translation products immunoprecipitable from infected cells was determined by tryptic peptide analysis. Although every polypeptide smaller than 64,000 daltons was found to be related to either NP or GPC, the methionine-containing tryptic peptides of L, NP, and GPC were distinct, which suggests that the three largest proteins are unrelated (Figure 3).

In order to ascertain the relationship of polypeptides immunoprecipitable from infected cells to polypeptides which are incorporated into mature virions, polypeptide profiles of purified virus and immunoprecipitates of infected cells were compared. A summary of the data is presented in Table 1. Preparations of purified virus labeled with either ^{3}H-amino acids or ^{35}S-methionine contained L, NP, GP1, NP48, NP38, GP2 and polypeptides of 20,000 and 15,000 daltons. GPC was found only in infected cells as was NP28. The relationship of the viral 15,000 dalton polypeptide to NP17, NP16.5 or NP14 has not been determined. Further proteolytic cleavage or phosphorylation could be responsible for changes in mobility of this polypeptide. The nature of the viral 20,000 dalton polypeptide is unknown. Other polypeptides present in immunoprecipitates and purified virus preparations were not consistently observed. Two polypeptides (40,000 and 30,000 daltons) with this property have been analyzed by peptide mapping and appear to be proteolytic cleavage products of NP.

Discussion

We have characterized polypeptides in Pichinde virus infected cells which are immunoprecipitable with hamster immune sera and have compared these

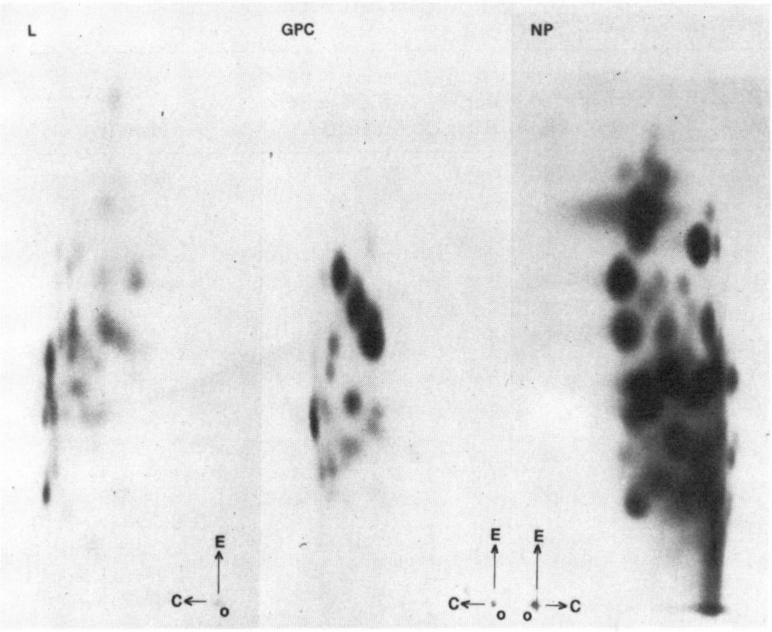

Figure 3. Two dimensional tryptic peptide analysis. Infected cells were labeled with ^{35}S-methionine (200 μCi/ml) and polypeptides were immunoprecipitated and electrophoresed as in Figure 1. Polypeptides were located by autoradiography. Protein bands were excised and digested with trypsin without prior elution from the gel. Peptides were separated in the first dimension by electrophoresis on silica gel thin layer sheets (Sil N-HR; Brinkman) and in the second dimension by ascending chromatography according to Dobos and Row.[10] Each peptide digest was separated with an NP standard for direct comparison between experiments. Spots were visualized by autoradiography. O: origin, C: chromatography, E: electrophoresis.

to polypeptides associated with purified virus preparations. The sera were obtained from hamsters which had been infected with 2000 pfu of virus and boosted six weeks later with about 10^7 pfu of virus. Sera collected two weeks later from six different animals precipitated the same polypeptides from infected cells. NP, GP1 and GP2 have been previously described, although GP1 was not clearly resolved from NP.[4,5] We have identified a 79,000 dalton glycoprotein (GPC) which is present in infected cells but not in mature virus. Pulse chase experiments suggested that GPC gives rise to GP1 and GP2. The glycoproteins are presently being compared by peptide mapping, and a similar result to that reported for lymphocyte choriomeningitis virus[7] is expected.

Tryptic peptide analysis has identified several polypeptides as cleavage products of NP (Table 1). The significance of consistent proteolytic processing of NP remains to be determined. NP17, NP16.5 and NP14 of infected

Table 1. Pichinde Virus Polypeptides.

Polypeptide M.W. ($\times 10^{-3}$)	Presence in virus preparation (V) or infected cells (IC)	Presence in pulse (P) of pulse chase (PC)	Identity
200	V, IC	P, PC	unassigned
79	IC	P	GPC
64	V, IC	P, PC	NP
52	V, IC	PC	GP1
48	V, IC	P	NP derived (NP48)
42	IC (in presence of tunicamycin)	P	pGPC
38	V, IC	P, PC	NP derived (NP38)
36	V, IC	PC	GP2
28	IC	P, PC	NP derived (NP28)
20	V		unassigned
17	IC	PC	NP derived (NP17)
16.5	IC	PC	NP derived (NP16.5)
15	V		Possibly related to NP17, 16.5 or 14
14	IC	PC	NP derived (NP14)

cells and the 15,000 dalton polypeptide of virus may represent the complement-fixing antigen described previously, which is composed of two polypeptides (15,000 and 20,000 daltons).[8] Other polypeptides smaller than NP have been observed but these were not consistent between preparations, and they likely represent degradation products of NP (as examples, the 30,000 and 40,000 dalton polypeptides). Although we have sometimes observed two polypeptides in virus and infected cells which have electrophoretic mobilities between those of NP and GPC, they appear to be of cellular origin. One of these resembles the 77,000 dalton polypeptide previously described in virions.[5]

The most interesting polypeptide evident in immunoprecipitates of infected cells and in purified virus preparations is L. This polypeptide is clearly not related to NP or GPC and its kinetics of synthesis suggest that it is viral encoded although we have no direct evidence for this. It is tempting to speculate an RNA polymerase role for this polypeptide.

L, NP and pGPC can account for most of the theoretical coding capacity of L and S viral RNA. NP and GPC are encoded by S RNA.[9] If L protein is viral specific it must be encoded by L RNA.[9] Since only three distinct polypeptides have been characterized, the mechanisms of regulation of Pichinde virus replication must be fairly integrate. These mechanisms are currently under investigation.

ACKNOWLEDGMENTS
These studies were supported by grants from The Medical Research Council of Canada. W-CL is an MRC Research Scholar and KD holds an MRC Postdoctoral Fellowship. DH is the recipient of a NSERC scholarship.

References

1. Rawls, W.E. and Leung, W.-C. (1979): Arenavirus. In: *Comprehensive Virology,* vol. 14. H. Fraenkel-Conrat and R.R. Wagner, (Eds.), New York, Plenum Publishing Corp., pp. 157–192.
2. Leung, W.-C., Ramsingh, A., Dimock, K., Rawls, W.E., Petrovich, J., and Leung, M. (1981): *J. Viron.,* Vol 37:48–54. Vol 14:31–46.
3. Ramsingh, A., Dimock, K., Rawls, W.E., and Leung, W. -C. (1980): *Intervirology:*
4. Ramos, B.A., Courtney, R.J., and Rawls, W.E. (1972): *J. Virol.* 10:661–667.
5. Vezza, A.C., Gard, G.P., Compans, R.W., and Bishop, D.H.L. (1977): *J. Virol.* 23:776–786.
6. Leung, W.-C., Leung, M.F.K.L., and Rawls, W.E. (1979): *J. Virol.* 30:98–107.
7. Buchmeier, M.J. and Oldstone, M.B.A. (1979): *Virology* 99:111–120.
8. Buchmeier, M.J., Gee, S.R., and Rawls, W.E. (1977): *J. Virol.* 22:175–186.
9. Leung, W.-C., Harnish, D., Ramsingh, A., Dimock, K., and Rawls, W.E.: This volume, p. 51–57.
10. Dobos, P. and Rowe, D. (1977): *J. Virol.* 24:805–820.

David H.L. Bishop and Richard W. Compans, eds.
The Replication of Negative Strand Viruses

Molecular and Genetic Studies of Tacaribe, Pichinde, and Lymphocytic Choriomeningitis Viruses

R.W. Compans,[a] D.P. Boersma,[a] P. Cash,[a,c] J.P.M. Clerx,[a] H.B. Gimenez,[a] W.E. Kirk,[a,d] C.J. Peters,[b] A.C. Vezza,[a,e] and D.H.L. Bishop[a]

Introduction

The Arenaviridae include the Old World species Lassa, Mozambique and lymphocytic choriomeningitis (LCM) viruses and the New World species, Pichinde (PIC), Tacaribe (TAC), Tamiami (TAM), Junin (the etiologic agent of Argentine hemorrhagic fever), Amapari, Latino, Machupo (the etiologic agent of Bolivian hemorrhagic fever) and Parana viruses. Arenaviruses have two viral single stranded RNA species, designated L (mol wt 3.2×10^6) and S (1.6×10^6), in addition to ribosomal 28 S and 18 S RNA species that are often, but not always, present. The viral RNA species are organized in strand-like nucleocapsids and ribosomes (when present) are encapsulated by an envelope in which are embedded surface glycoproteins. A schematic diagram of the structure of an arenavirus particle is shown in Figure 1.

Over the last few years, our laboratories have been investigating various aspects of the structure, replication and genetics of PIC, TAC, TAM and LCM viruses.[1-8] In the present paper, we present new information on the structural components of TAC virions, and review some recent observations on the construction of intertypic arenavirus recombinants and their use in determining coding assignments and viral pathogenic potentials.

[a] Department of Microbiology, University of Alabama in Birmingham, Birmingham, Alabama.
[b] U.S. Army Medical Research Institute of Infectious Diseases, Fort Detrick, Frederick, Maryland.
[c] Present address: Department of Bacteriology, University of Aberdeen, Foresterhill, Aberdeen, Scotland.
[d] Present address: Department of Microbiology, Medical Center, University of West Virginia, Morgantown, West Virginia.
[e] Present address: Laboratory of Parasitology, Rockefeller University, New York, New York.

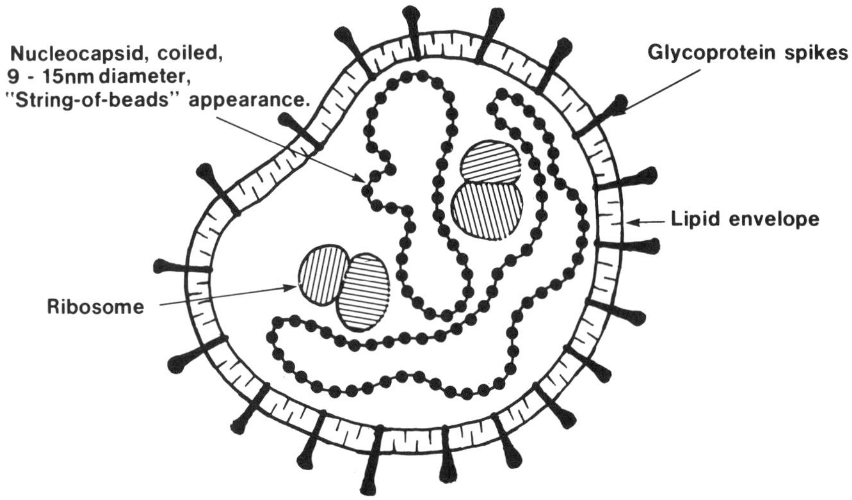

Figure 1. Schematic diagram of an arenavirus particle. The nucleocapsid is depicted in the form of two circular strands corresponding to the L and S RNA segments, in association with the N and P proteins. Ribosomes are usually present in the virion. The glycoprotein spikes are composed of either one or two glycosylated polypeptides; their arrangement and the extent of their penetration through the viral lipid layer are uncertain.[14]

Tacaribe Virion RNA Species

A comparison of the RNA species present in PIC and TAC virions is shown in Figure 2. Each virus has both L and S RNA species; the preparation of PIC virions shown in lane b contains 28 S RNA but no detectable 18 S ribosomal RNA. The S RNA of TAC virions (lane c) is clearly resolved from 28 S ribosomal RNA, and low levels of 18 S RNA are also observed.

To determine if the L and S RNA species of TAC virions contain distinct genetic information, the isolated RNA species were compared by oligonucleotide fingerprint analyses (Figure 3). The fingerprints of the L RNA (Figure 3A) and S RNA species (Figure 3B) are clearly distinguishable. In the fingerprint obtained from unfractionated TAC virion RNA (Figure 3C), it is possible to identify both L and S oligonucleotides; their identification is shown schematically in Figure 3D. The observation that the L and S RNA species of TAC virions have unique oligonucleotides indicates that these RNA species possess different genetic information, as has also been reported for the L and S RNA species of Pichinde and LCM viruses.[4,8]

Characterization of Tacaribe Virion Polypeptides

Whereas the available information indicates that all of the members of the arenavirus family possess genomic RNA species with similar sizes, some differences have been observed with respect to the polypeptide composition

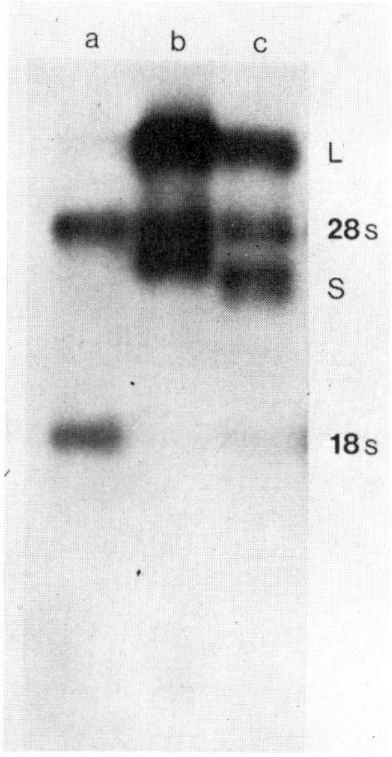

Figure 2. Analysis of virion RNA species of Pichinde and Tacaribe virions on a polyacrylamide slab gel. The virions were radiolabeled with $^{32}PO_4$, purified, and the RNA species extracted and analyzed by electrophoresis on a 2.4% polyacrylamide slab gel as described previously.[4,6] (a) 28 S and 18 S ribosomal marker RNA; (b) Pichinde virion RNA species; (c) Tacaribe virion RNA species.

SOURCE: Gimenez and Compans, submitted for publication.

of the virions. All members of the family have a major nucleocapsid protein N (mol wt 62,000-68,000).[1,2] When electrophoresis of the TAC viral polypeptides is carried out with a buffer change every 3 hr, the N protein is resolved into two distinct bands (Figure 4b). The two bands have been isolated and further characterized by limited proteolytic digestion with chymotrypsin (Figure 4, lanes e and f). The cleavage patterns obtained have been found to be essentially identical, indicating that the two polypeptide bands are related in primary sequence. It is not known whether proteolytic cleavage or other secondary modifications are responsible for the difference in the electrophoretic mobilities of the two forms of TAC N protein.

In the case of PIC, LCM, Lassa, and Machupo virions, two major surface glycoproteins have been reported (G1 and G2, mol wt 50,000-65,000 and 35,000-41,000, respectively).[1,9-12] However, only a single major glycoprotein size class has been observed in the case of TAC, and TAM viruses.[2] Arenaviruses have also been reported to contain minor amounts of a nonglycosylated polypeptide designated P (mol wt 77,000-79,000) associated with the nucleocapsid, although this polypeptide has not been consistently observed with all members of the group. The significance of these differences in virion polypeptide composition is not known. In order to resolve possible additional virion proteins, we have separated the TAC virion polypeptides by a 2-dimensional gel electrophoresis procedure (Figure 5).

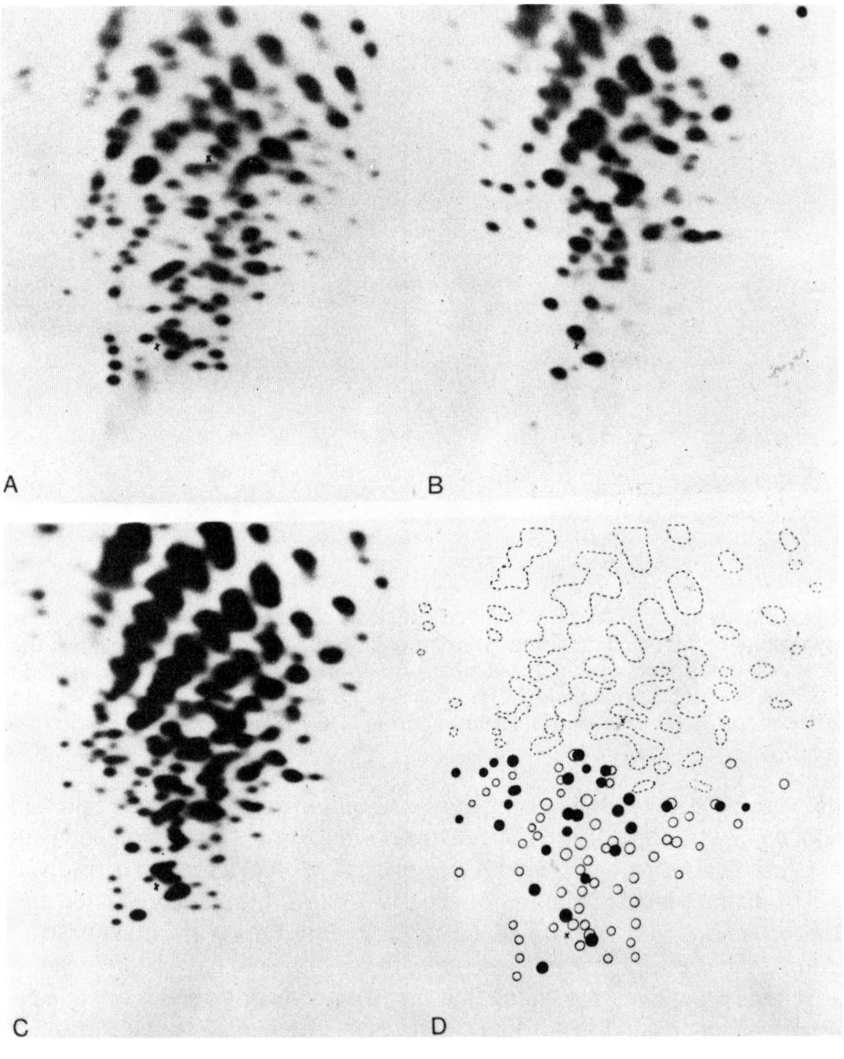

Figure 3. Oligonucleotide fingerprint analyses of Tacaribe virion RNA species. Virions were labeled with $^{32}PO_4$, purified, and RNA extracted as described previously.[6] The RNA species were separated by electrophoresis on 0.7% agarose as described by Wieslander,[15] and the RNA species located by autoradiography. The pieces of agarose containing each RNA species were heated for 20 min at 70°C in sterile electrophoresis buffer, and the RNA extracted with phenol-chloroform, ethanol-precipitated, pelleted, and resuspended in 2mM EDTA, 20mM Tris-HCl, pH 7.4. Oligonucleotide fingerprinting was carried out after digestion with RNase T1 as described by Clewley et al.[16] (A) L RNA; (B) S RNA; (C) Unfractionated TAC virion RNA. Panel D schematically depicts the L (o) and S (o) oligonucleotides in panel C.

SOURCE: Gimenez and Compans, submitted for publication.

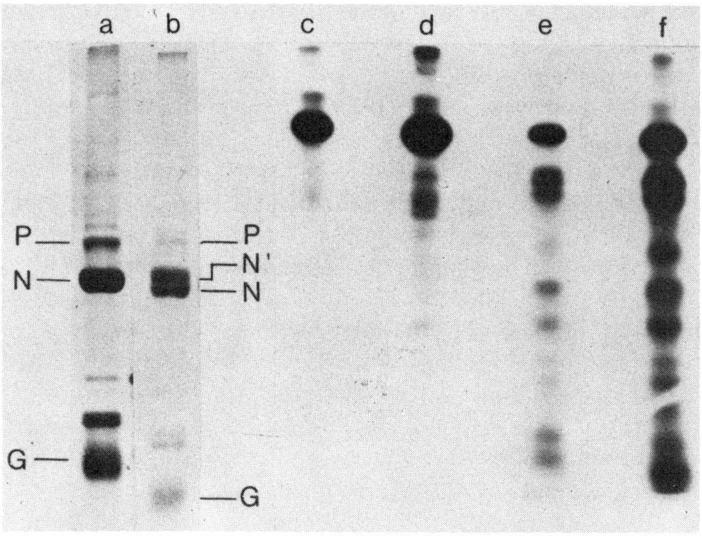

Figure 4. Demonstration of two forms of N proteins in *wt* Tacaribe virions. The polypeptides of *wt* Tacaribe virions were analyzed by SDS-polyacrylamide gel electrophoresis in a 1 mm slab gel at 5.5 mA. Electrophoresis time: (a) 16 hr (marker dye at the bottom of the gel); (b) 19.5 hr. Electrophoresis buffer in (b) was changed every 3 hr. On the right, the two forms of Tacaribe N protein seen in lane b are compared by partial proteolytic digestion. Bands were cut from the SDS gels and (c,d) reelectrophoresed without digestion, or treated (e,f) with 25 μg of chymotrypsin as described by Cleveland et al.[17] N′ protein: (c,e); N protein: (d,f). Occasional partial degradation of N protein produces the 2 new bands that migrate slower than the G protein.

Figure 5. Analysis of the polypeptides of purified ^{35}S-methionine labeled Tacaribe virions by two-dimensional polyacrylamide gel electrophoresis. First dimension was run on a cylindrical isoelectric focusing gel containing 4% acrylamide, 8 M urea, 2% ampholines pH 3.5-10, 2% NP-40, and 0.03% ammonium persulfate. After electrophoresis for 20 hr, the tube gel was equilibrated in 0.5 M Tris-HCl, pH 6.25, placed over a 1.5 mm slab gel, and anchored in 1% agarose. The second dimension slab gel contained 12% acrylamide, 0.3% bisacrylamide, 0.375 M Tris-HCl, pH 8.3 and 0.1% SDS. The stacking gel contained 4.75% acrylamide, 0.125 M Tris-HCl, pH 6.25 and 0.1% SDS. The cathode of the first dimension gel is to the left, the anode to the right.

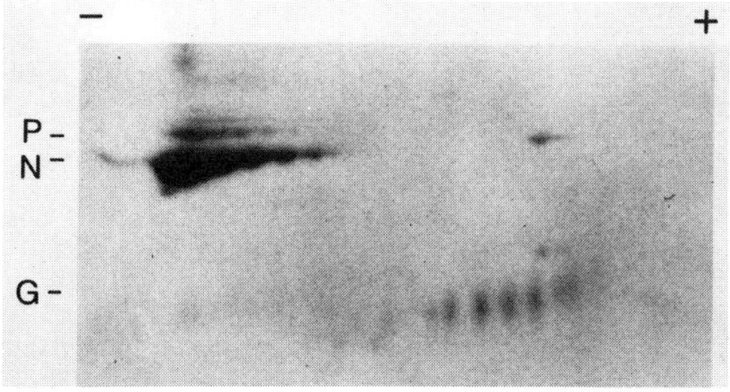

The patterns obtained indicate that each of the major structural proteins can be separated into multiple components with differing isoelectric points, with as many as ten distinct electrophoretic species being observed for the TAC G protein. The TAC P and N proteins appear to have similar isoelectric points, and each was partially resolved into several species. To analyze the sequence relationships among the separated virion polypeptides, they have been further characterized by limited proteolytic digestion and gel electrophoresis (Figure 6). The results indicate that treatment of the P and N polypeptides with *Staphylococcus* V8 protease produced a number of cleavage products of similar electrophoretic mobility suggesting a possible sequence relationship between P and N; however, some differences were found in the cleavage patterns. Each of the ten electrophoretic variants of the G polypeptide yielded a similar cleavage pattern, with three major cleavage products; thus the various forms of G protein are related in primary sequence. It is likely that differences in glycosylation are responsible for the differences is isoelectric points of the various forms of G protein.

Figure 6. Peptide mapping of Tacaribe virion proteins. Proteins labeled with ^{35}S-methionine were separated in two-dimensional polyacrylamide gels as described in Fig. 5. Proteins were located by autoradiography without PPO, excised, equilibrated to pH 6.25 in a solution containing 0.05 M Tris HCl, pH 6.25, 0.1% SDS, 10 mM EDTA, and 15% glycerol. The gel pieces containing protein were loaded into the sample wells of another polyacrylamide gel. Over each sample, 20 μl of *Staphylococcus* V8 protease at 10 μg/ml was layered. The fragments cleaved during the coelectrophoresis of the polypeptides and the protease were resolved in a 15% acrylamide gel. The P protein, N protein, and ten components of G separated by two dimensional polyacrylamide gels are compared.

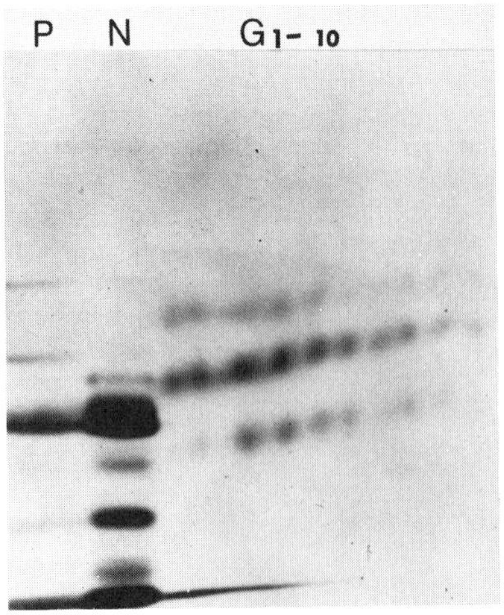

The Genetic Capacity of Arenaviruses

Using prototype PIC virus and an alternate PIC virus isolate (Pichinde Munchique, designated MUC), which can be distinguished from prototype PIC at the genome level by oligonucleotide fingerprinting, it has been possible to construct PIC-MUC reassortant viruses.[7] Two procedures have been used to obtain such reassortants, dual heterologous wild-type virus infections, and dual heterologous *ts* mutant virus infections.[7] Oligonucleotide fingerprint analyses have shown that the reassortant viruses formed by either procedure have genomes with one viral RNA species (L) derived from one parent virus, and the other viral RNA segment (S) derived from the other. So far, only one of the two possible reassortant virus genotypes has been obtained; it is designated according to the origins of its L/S RNA species, i.e., PIC-MUC (see Figure 7). No MUC-PIC reassortant virus has yet been isolated.

The plaque sizes of the PIC-MUC reassortant viruses resemble those of PIC virus and not those of MUC viruses, which are significantly smaller. Thus it appears that the plaque phenotype of Pichinde may be determined by its L RNA gene products.

Tryptic peptide analyses of the virion N polypeptides of PIC and MUC have shown that the N proteins of the two viruses can be distinguished, although they are very similar to each other (Figure 8). Tryptic peptide analyses of the PIC-MUC recombinant virus have demonstrated that it has a MUC type N polypeptide (Figure 8), proving that the S RNA of these viruses codes for the N polypeptide. Which RNA segment codes for the virion glycoproteins is under investigation.

Intertypic reassortant lymphocytic choriomeningitis (LCM) viruses have been obtained by coinfection of BHK-21 cells with two different LCM virus strains, WE and Armstrong (ARM).[8] The oligonucleotide fingerprints of the individual RNA segments of LCM-WE (Figure 9, top left and center panels) have shown that its L and S RNA species each contain unique oligonucleotides, some of which are indicated by arrows. The top right-hand panel of Figure 9 shows the oligonucleotide fingerprint of a mixture of LCM-WE L and S RNA species; most of the largest unique oligonucleotides of the viral L and S RNA segments can be easily identified in the composite fingerprint, some of which are also indicated by arrows. Similar analyses have been carried out on the virion RNA species of LCM-ARM virus (Figure 9, bottom panels). A comparison of the individual L RNA fingerprints (or those of the S RNA, or L and S RNA species), of each LCM strain showed that they could be easily distinguished.

To investigate the possibility of forming reassortant viruses between LCM-WE and LCM-ARM viruses, monolayers of BHK-21 cells were infected with both LCM-ARM (moi = 1) and LCM-WE (moi = 0.2) viruses,

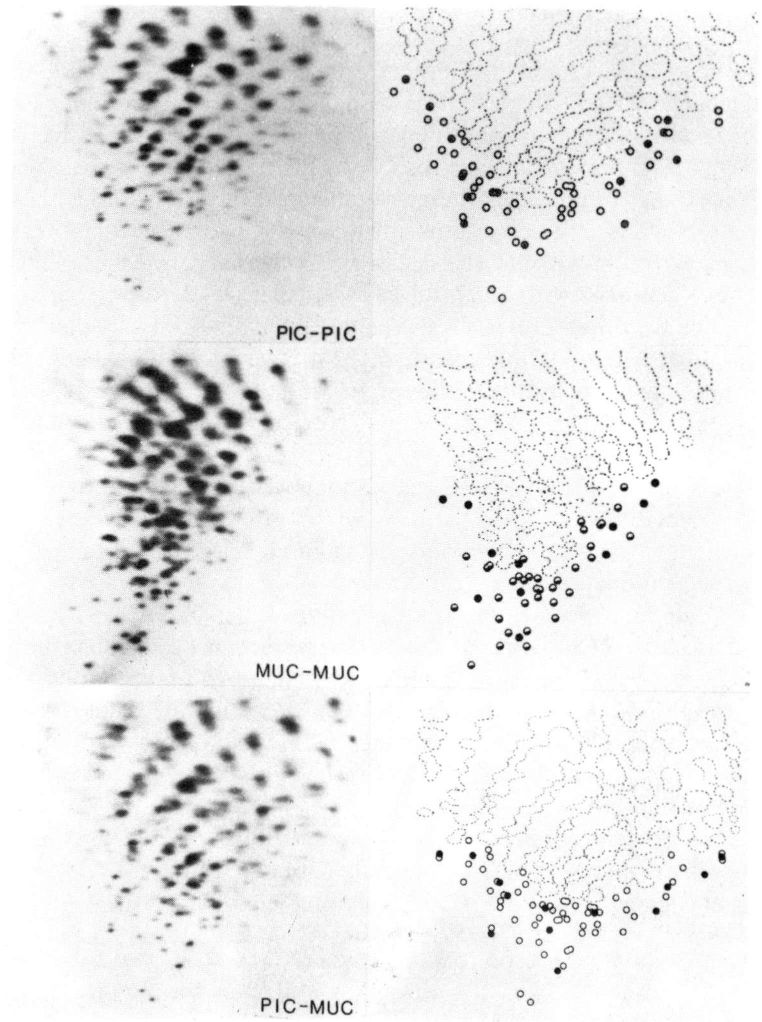

Figure 7. Composite RNA fingerprints of representative progeny virus clones derived from dual infections with wild-type PIC and MUC viruses. The RNase T1 derived oligonucleotides of the total ^{32}P-labeled viral RNA recovered from three cloned progeny viruses obtained from dual PIC and MUC wildtype virus infections were resolved by two-dimensional polyacrylamide gel electrophoresis.[16] In the left-hand panels are shown: top: PIC progeny virus RNA (PIC-PIC); middle: MUC progeny virus RNA (MUC-MUC); bottom: reassortant PIC-MUC type progeny virus RNA (PIC-MUC). In the right-hand panels are shown schematics of the three viruses with the origins indicated for oligonucleotides derived from PIC L (○), PIC S (⊗), MUC L (◒) and MUC S (●) viral RNA species.[7]

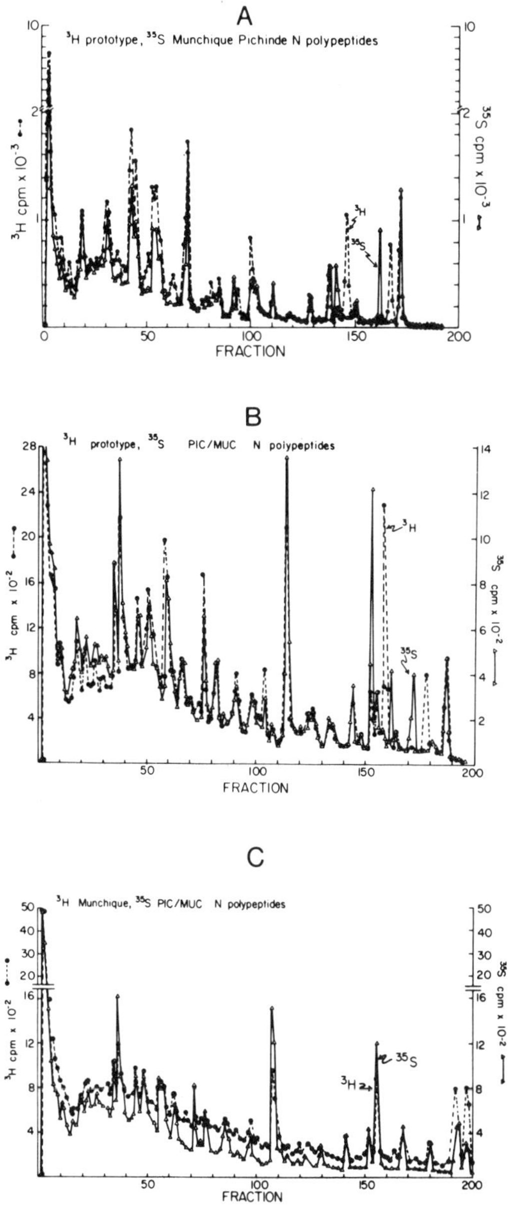

Figure 8. Tryptic peptide analyses of the viral N polypeptides of PIC, MUC, and a PIC-MUC reassortant virus. Preparations of (A) ^{3}H-methionine prototype PIC and ^{35}S-methionine MUC N polypeptides, (B) ^{3}H-methionine prototype PIC and ^{35}S-methionine PIC-MUC N polypeptides, and (C) ^{3}H-methionine MUC and ^{35}S-methionine PIC-MUC N polypeptides were digested with TPCK trypsin and the products resolved by high pressure ion-exchange column chromatography as described previously.[18] The results indicate that the PIC-MUC reassortant has a MUC type N polypeptide.[7]

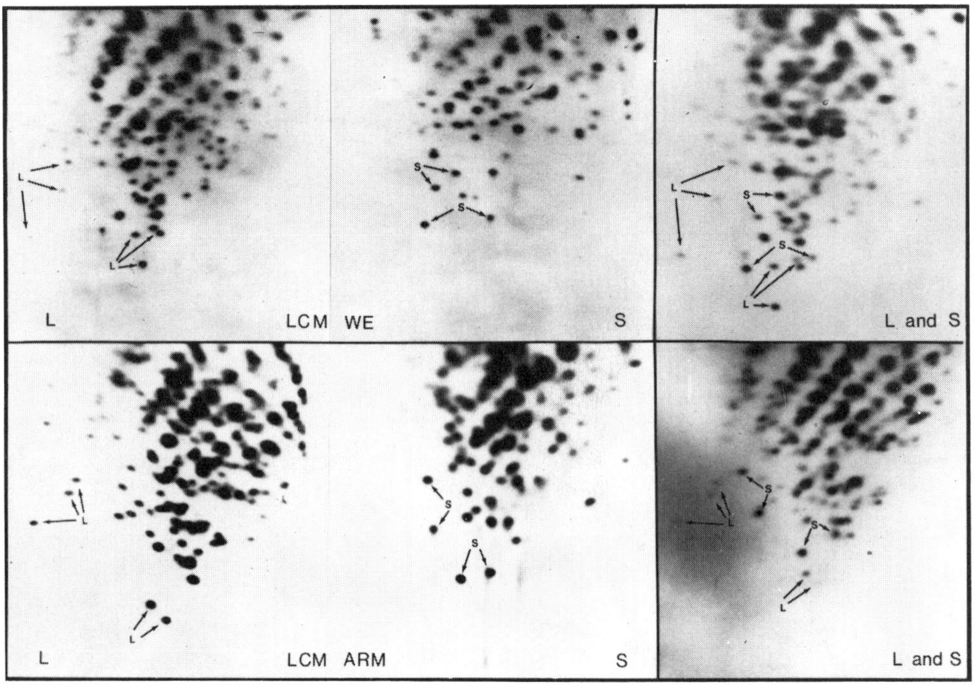

Figure 9. Oligonucleotide fingerprints of LCM-WE and LCM-ARM genome RNA species. The RNase T1 oligonucleotide fingerprints of either the combined L and S RNA species (right-hand panels), or the individual L and S RNA segments (left-hand and center panels, respectively) were obtained as described by Clewley et al.[16] Certain large unique oligonucleotides for the L and S RNA segments of each virus are indicated by arrows in the individual and composite fingerprints. In the ARM individual patterns the right-hand side oligonucleotides are somewhat lower than in the composite pattern due to a technical problem in the preparation of the second dimension gels. The upper center X in each pattern marks the position of the bromophenal blue due marker, the lower X in each pattern marks the position of the xylene cyanol FF dye marker.[8]

and incubated at 35°C for 2 days, at which time the culture fluids were harvested. The culture fluids from the mixed infection were plated on Vero cells using a semi-solid overlay. Several well isolated plaques were recovered. The viruses in these plaques were grown into stocks using BHK-21 cells and their virion RNA genotypes determined by oligonucleotide fingerprinting of the total RNA extracted from ^{32}P-labeled virus preparations. By comparison with the individual L,S and composite fingerprints of LCM-WE and LCM-ARM viruses, it was determined that of four progeny viruses analyzed, two had the L and S RNA genotypes of LCM-ARM, one had the L and S RNA genotype of LCM-WE, and one, as indicated in Figure 10, had a genotype consisting of the L RNA of WE and the S RNA of ARM (i.e., WE/ARM).

The pathogenic potential of the WE and ARM LCM virus stocks, as well as the WE/ARM reassortant, have been investigated in experimental animals (hamsters and guinea pigs). When 10^4 pfu of each virus were inoculated subcutaneously into groups of five 6-week-old female hamsters, none of the animals died that received the ARM or WE/ARM viruses, and indirect immunofluorescence studies indicated that all had seroconverted by six weeks postinoculation. Of the five hamsters that received 10^4 pfu of LCM-WE virus, 4 died (mean day of death 21.8 ± 2.6). Subcutaneous inoculation of Hartley or Strain 13 guinea pigs with 10^4 pfu of LCM-WE virus resulted in death of all of the animals by 2 weeks postinoculation, whereas no deaths were observed in animals similarly inoculated with LCM-ARM or WE/ARM. These results indicate that the S RNA segment of LCM virus codes for gene products that are responsible for its pathogenic potential in hamster and guinea pigs.

Figure 10. Oligonucleotide fingerprints of the individual species and total virion RNA of a WE/ARM LCM recombinant virus. The RNase T1 oligonucleotide fingerprint of the L (left panel), S (center panel) and total (right panel) virion RNA of a cloned virus recovered from a LCM-WE × LCM-ARM dual wild-type virus infection were obtained as described by Clewley et al.[16] The dye markers and certain unique oligonucleotides representing the LCM-WE L and LCM-ARM S RNA segments are indicated by arrows (see Figure 9).[8]

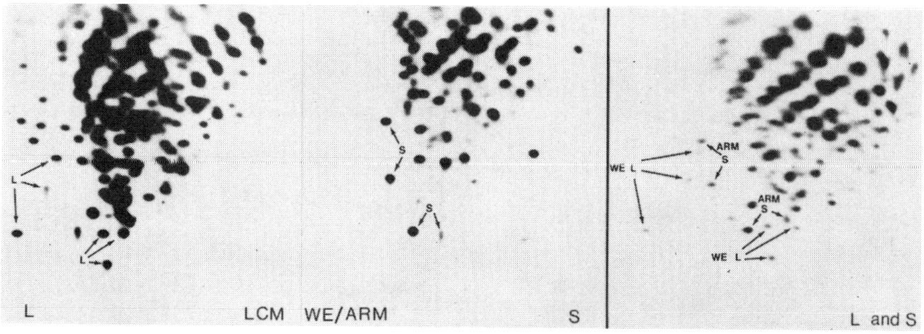

ACKNOWLEDGMENTS
This was supported by Grants CA 18611 from the National Cancer Institute and AI 14183 from the National Institute of Allergy and Infectious Diseases. D.P.B. and A.C.V. were supported by postdoctoral traineeships on NIAID training Grant T32 AI 07150.

References

1. Vezza, A.C., Gard, G.P., Compans, R.W., and Bishop, D.H.L. (1977): *J. Virol.* 23:776–786.
2. Gard, G.P., Vezza, A.C., Bishop, D.H.L., and Compans, R.W. (1977): *Virology* 83:84–95.
3. Vezza, A.C. and Bishop, D.H.L. (1977): *J. Virol.* 24:712–715.
4. Vezza, A.C., Clewley, J.P., Gard, G.P., Abraham, N.Z., Compans, R.W., and Bishop, D.H.L. (1978): *J. Virol.* 26:485–497.
5. Saleh, F., Gard, G.P., and Compans, R.W. (1979): *Virology* 93:369–376.
6. Gimenez, H.B. and Compans, R.W. (1980): *Virology* 107:229–239.
7. Vezza, A.C., Cash, P., Jahrling, P., Eddy, G., and Bishop, D.H.L. (1980): *Virology* 105:250–260.
8. Kirk, W.E., Cash, P., Peters, C.J., and Bishop, D.H.L. (1980): *J. Gen. Virol.,* 51:213–218.
9. Ramos, B.A., Courtney, R.J., and Rawls, W.E. (1972): *J. Virol.* 10:661–667.
10. Buchmeier, M.J., Elder, J.H., and Oldstone, M.B.A. (1978): *Virology* 89:133–145.
11. Kiley, M.P., Regnery, R.L., and Johnson, K.M. (1981): In: *Replication of Negative Strand Viruses.* (D.H.L. Bishop and R.W. Compans (Eds.). Elsevier, New York.
12. Gangemi, J.D., Rosato, R.R., Connell, E.V., Johnson, E.M., and Eddy, G.A. (1978): *J. Gen. Virol.* 41:183–188.
13. Grau, O., Romanowski, V., and Rustici, S.M. (1981): In: *Replication of Negative Strand Viruses.* D.H.L. Bishop and R.W. Compans (Eds.). Elsevier, New York.
14. Bishop, D.H.L. (1980): *Medicina (Buenos Aires)* 40:275–288.
15. Wieslander, L. (1979): *Anal. Biochem.* 98:305–309.
16. Clewley, J., Gentsch, J., and Bishop, D.H.L. (1977): *J. Virol.* 22:459–468.
17. Cleveland, D.N., Fischer, S.G., Kirschner, M.W., and Laemmli, U.K. (1977): *J. Biol. Chem.* 252:1102–1106.
18. Gentsch, J. and Bishop, D.H.L. (1978): *J. Virol.* 28:417–419.

Copyright 1981 by Elsevier Noroh Holland, Inc.
David H. L. Bishop and Richard W. Compans, eds.
The Replication of Negative Strand Viruses

Genome Structure of Lymphocytic Choriomeningitis Virus: Cohesive Complementary Termini?

F.J. Dutko,[a] S.I.T. Kennedy,[b] and M.B.A. Oldstone[a]

Introduction

The genetic information of several arenaviruses (lymphocytic choriomeningitis virus (LCMV),[1] Pichinde,[2] Junin,[3] Parana,[4] Tacaribe,[5] and Tamiami[5] viruses) has been found to consist of two RNA segments. By an examination of the RNA from several arenaviruses by sucrose gradient centrifugation or by polyacrylamide gel electrophoresis (PAGE), the apparent molecular weight of the L RNA varied from 2.1×10^6 to 3.2×10^6 daltons, and that of the S RNA varied from 1.1×10^6 to 1.6×10^6 daltons. For LCMV, the molecular weights determined by PAGE of L and S RNAs were reported to be 2.1×10^6 and 1.1×10^6 daltons, respectively.[1,2] Because all of these observations were obtained under nondenaturing conditions and because arenaviruses RNAs contain considerable secondary structure (resistance to RNase),[5] and as a prelude to an examination of recombinant arenaviruses and to gene cloning, we have studied the RNAs of LCMV under denaturing conditions using glyoxal.[7]

Results

Four LCMV RNAs

To examine the effect of glyoxal denaturation of LCMV (Armstrong) RNA, we added ^{32}P to infected BHK cells, purified the virus particles, extracted

[a] Department of Immunopathology, Scripps Clinic and Research Foundation, La Jolla, California.
[b] Department of Biology, University of California, San Diego, California.

the ^{32}P-labeled RNA, and analyzed the LCMV RNA by agarose gel electrophoresis with and without glyoxal treatment (Figure 1). Only two RNA bands (L and S) were found without glyoxal treatment. However, four RNA bands of lower electrophoretic mobility (L-1, L-2, S-1 and S-2) were observed after glyoxal denaturation. Uninfected BHK cell ribosomal 28S and 18S RNAs, and Semliki Forest virus 42S and 26S RNAs were the marker RNAs. Four LCMV RNAs were observed whether the virus was propagated in L929 or BHK cells. In addition, four RNAs were found after glyoxal denaturation of LCMV (strain WE) or Pichinde virus (data not shown).

Oligonucleotide Fingerprinting of LCMV RNAs

To investigate the sequence relationships among the four LCMV RNAs resolved by glyoxal denaturation and gel electrophoresis, we digested the isolated LCMV RNAs with RNase T_1 and fingerprinted the resulting oligonucleotides by two-dimensional gel electrophoresis. The fingerprints in Figure 2 show that the patterns and presumably the nucleic acid sequences of L-1 and L-2 RNAs were indistinguishable. Similarly those of S-1 and S-2 RNAs were indistinguishable from each other. However, the fingerprints of L-1 and L-2 RNAs were quite different from those of S-1 and S-2 RNAs.

Figure 1. Agarose gel electrophoresis of LCMV RNA. Two 175 cm^2 flasks of BHK-21 cells were infected with LCMV at a multiplicity of infection of 0.2. Viral RNA was labeled by adding $^{32}P_i$ to infected cells. ^{32}P-labeled RNA was extracted from purified virus particles by digestion with proteinase K and SDS, and extraction with phenol-chloroform. The LCMV RNAs were either not denatured (-G) or denatured with 1M glyoxal and 50% DMSO at 45°C for two minutes and left at room temperature for 15 minutes (+G). The positions of L-1, L-2, S-1 and S-2 RNAs are marked by arrows. ^{32}P-labeled intracellular RNAs from SFV-infected BHK-21 cells (rRNA; upper band = 28S rRNA, lower band = 18S rRNA) were glyoxal denatured. About 50,000 cpm of each RNA sample was applied to the 1.5% (w/v) agarose gel. After electrophoresis, the gel was dried and autoradiographed with an intensifying screen for 24 hours at room temperature (submitted to *J. Virol.*, 1980).

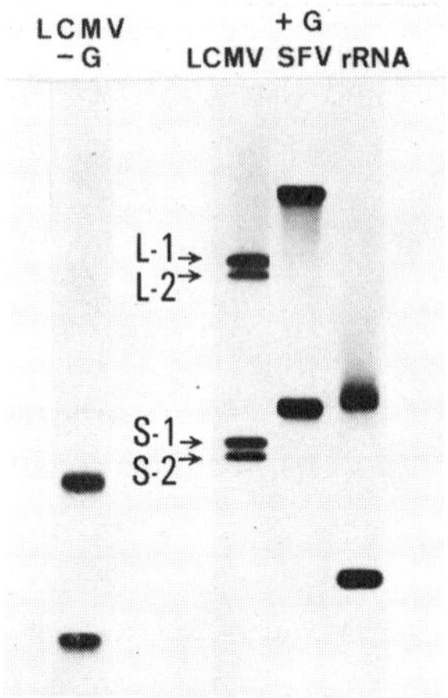

Variation in Temperature of Glyoxal Denaturation and Agarose Gel Concentrations

Since the nucleotide sequences of L-1 compared to L-2 (or S-1 compared to S-2) RNAs were similar to each other, we determined whether these RNAs might be conformational variants. We varied the temperature of the glyoxal denaturation and the concentration of the agarose gel used for electrophoresis. The results in Figure 3 show that standard temperatures of

Figure 2. RNase T_1 fingerprints of LCMV RNAs. ^{32}P-labeled LCMV (Armstrong) RNA was prepared as described in the legend for Figure 1. About 10^6 CPM of purified viral RNA was electrophoresed either in a 1.5% (w/v) or 0.7% (w/v) Seaplaque agarose gel to separate the L-1 and L-2 RNAs, or the S-1 and S-2 RNAs respectively. The isolated L-1 RNA, L-2 RNA, S-1 RNA, and S-2 RNA, together with 150 μg of carrier tRNA, were digested with RNaseT-l. The resulting oligonucleotides were fractionated by two-dimensional polyacrylamide gel electrophoresis. Electrophoresis in the first dimension (10% acrylamide − 0.3% bisacrylamide, buffer = 0.025 M citric acid, pH 3.5) was from left to right, and in the second dimension (20% acrylamide − 0.3% bisacrylamide, buffer = 0.2 M Tris; pH 8.3, 2.5 mM EDTA) was from top to bottom. Autoradiography was for two weeks at − 70°C with an intensifying screen. X and B indicate the positions of xylene cyanol FF and bromophenol blue marker dyes, respectively (submitted to *J. Virol.*; 1980).

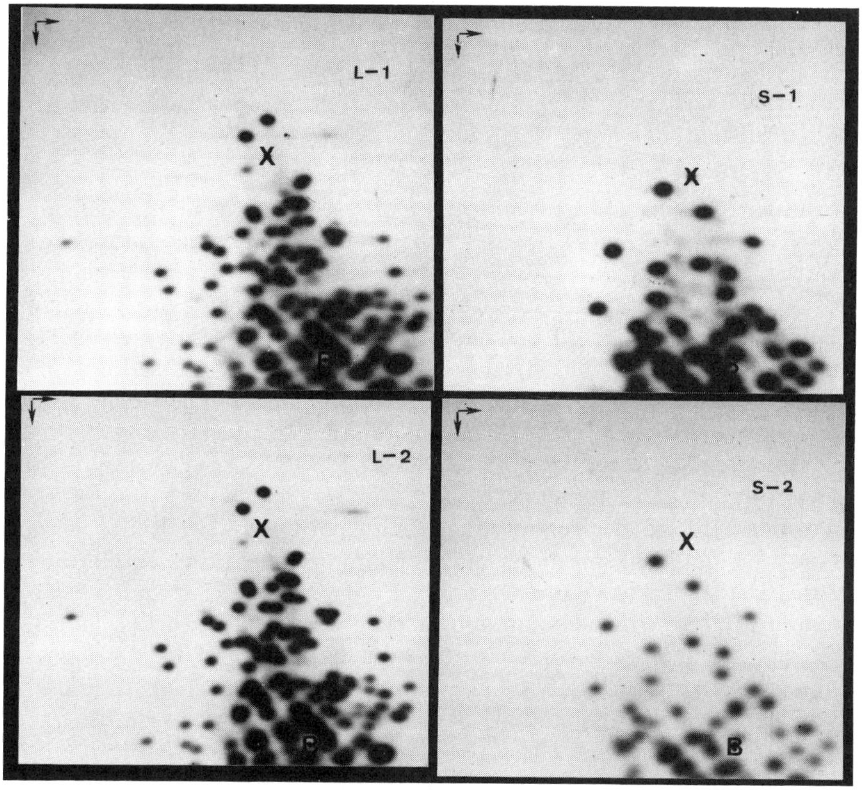

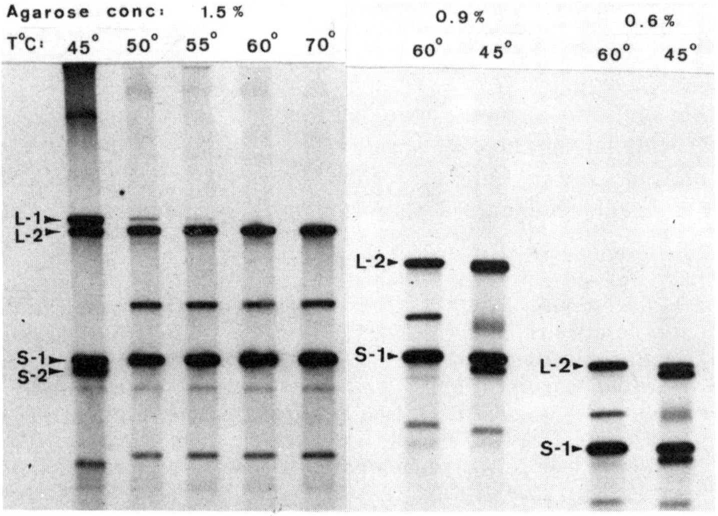

Figure 3. Denaturation of LCMV RNAs with glyoxal at different temperatures and electrophoresis in various concentrations of agarose gels. ^{32}P-labeled LCMV (strain Armstrong) RNA was isolated from purified virus as described in the legend to Figure 1. The ^{32}P-LCMV RNAs were denatured with glyoxal at 45°C for two minutes and then left at room temperature for 15 minutes (45°C), 50°C for three minutes followed by 50°C for 15 minutes (50°C, 55°C for three minutes followed by 50° for 15 minutes (55°C), 60°C for three minutes followed by 50°C for 15 minutes (60°C), or 70°C for three minutes followed by 50°C for 15 minutes (70°C). About 50,000 cpm of each dried and autoradiographed with an intensifying screen for 24 hours at room temperature (submitted to *J. Virol.*, 1980).

glyoxal denaturation (45°C) resolved LCMV RNA into the expected four RNAs (L-1, L-2, S-1 and S-2). After denaturation at any temperature above 45°C (i.e., 50°C, 55°C, 60°C, 70°C), the S-1, but not the S-2, RNA band was detected. After denaturation of LCMV RNA at 50°C or 55°C, trace acounts of L-1 RNA were detected but most of the L RNA migrated as the L-2 conformation. Only the L-2 RNA band was detected after denaturation of LCMV RNA at 60°C or 70°C. The RNA band that migrated between L-2 and S-1 after denaturation at 50°C to 70°C was 28S ribosomal RNA, and the RNA band that migrated faster than S-2 RNA was 18S RNA as shown by RNase T_1 fingerprinting (data not shown).

We compared the electrophoretic mobilities of the LCMV RNAs in various agarose gel concentrations after denaturation at 45°C or 60°C with glyoxal. The results in Figure 3 show that only 2 RNAs (L-2 and S-1) were observed after glyoxal denaturation at 60°C and electrophoresis in 1.5%, 0.9%, or 0.6% agarose gels. After glyoxal denaturation of LCMV RNA at 45°C, four RNAs (L-1, L-2, S-1 and S-2) were observed after electrophoresis in 1.5% agarose gels. The mobility of S-2 RNA was always greater than that of S-1 RNA in all three concentrations of agarose gels. However, the order of migration of L-1 and L-2 RNAs depended on the gel concentration. L-2 RNA migrated farther than L-1 RNA in 1.5% agarose gels, but L-1 migrated farther than L-2 RNA in 0.6% agarose gels. L-1 and L-2 RNAs comigrated in 0.9% agarose gels.

Glyoxal Denaturation of an RNA with Cohesive Complementary Termini

We examined an RNA known to have complementary ends by glyoxal denaturation at different temperatures and compared its electrophoretic behavior to that of LCMV RNA. For these studies, we used a defective interfering (DI) VSV particle whose RNA contains about 50 complementary nucleotides at the termini.[8] The results in Figure 4 show that glyoxal denaturation at 45°C resolved the DI VSV RNA into two bands. As the temperature of glyoxal denaturation was increased from 50°C to 55°C, increasing amounts of the DI VSV RNA was found as the more slowly migrating RNA. At the highest temperatures of gloxal denaturation (60°C and 70°C), all of the DI VSV RNA was found as the slowly migrating band. These results show that the electrophoretic behavior of glyoxal denatured DI VSV RNA is similar to that of LCMV RNAs—two RNA bands after denaturation at 45°C, but only a single RNA band after denaturation at 60°C or 70°C.

Figure 4. Agarose gel electrophoresis of VSV DI RNA denatured with glyoxal at various temperatures. ^{32}P-labeled VSV DI (VSV ts^{+} ATCC *DI 0.22* (5′, 20%) RNA was isolated from DI virions purified by rate zonal and equilibrium centrifugation. The ^{32}P-VSV DI RNA was denatured with glyoxal as described in the legend for Figure 3. About 25,000 cpm of each RNA sample was applied to a 1.5% (w/v) agarose gel. After electrophoresis, the gel was dried and autoradiographed with an intensifying screen for 24 hours at room temperature (submitted to *J. Virol.*, 1980).

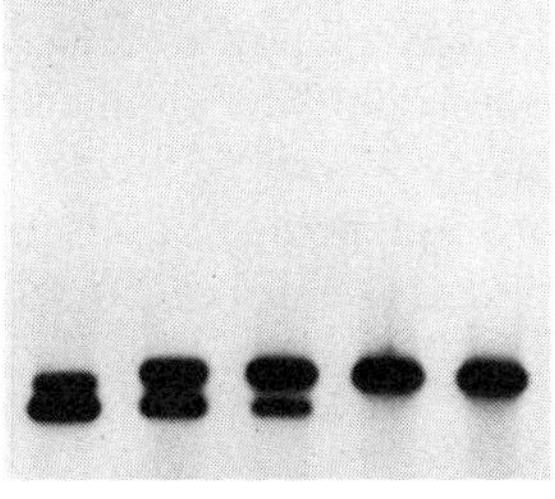

Discussion

We have demonstrated that glyoxal denaturation of LCMV RNA at less stringent (45°C) temperatures reveals the presence of four RNAs (L-1, L-2, S-1 and S-2). Fingerprinting of the oligonucleotides resistant to RNase T_1 revealed that the nucleic acid sequences of L-1 and L-2 and S-1 and S-2) RNAs are similar. However, the L-1 and L-2 RNAs are unique from S-1 and S-2 RNAs. RNase T_1 fingerprints of Pichinde virus L and S RNAs have shown each of these RNAs to be unique.[5] The LCMV L RNA fingerprint reported here differs markedly from that reported for Pichinde L RNA. Likewise, the LCMV S RNA fingerprint differs from that of Pichinde virus S RNA. In addition, preliminary results show that both the L and the S RNA fingerprints of LCMV strain WE differs significantly from that of LCMV strain Armstrong. These results show that there are significant nucleic acid sequence differences among these arenaviruses despite their antigenic relatedness.[9]

The molecular basis for the difference in electrophoretic mobility between LCMV L-1 and L-2 (and S-1 and S-2) RNAs is a change in the conformational structure. This conclusion is suggested by the results of increasing the temperature of the glyoxal denaturation which varied the relative amounts of L-1 to L-2 (and S-1 to S-2) RNAs. Only the L-2 and S-1 LCMV RNAs were observed after complete glyoxal denaturation. Furthermore, the electrophoretic mobility of L-1 and L-2 RNAs varied in different agarose gel concentrations such that L-2 RNA migrated faster than L-1 RNA in high concentrations (1.5%) agarose gels, but L-2 RNA migrated slower than L-1 RNA in lower concentration (0.6%) agarose gels. This electrophoretic behavior is similar to that of supercoiled DNA and linear DNA.[10] If this analogy is correct, then the L-1 RNA contains more secondary structure than the fully denatured L-2 RNA. In addition, by comparing the relative electrophoretic migration of LCMV RNAs completely denatured with glyoxal at 60°C to SFV 42S and ribosomal (28S and 18S) RNAs, we precisely determined the molecular weight of LCMV (strain Armstrong) L and S RNAs to be 2.85×10^6 and 1.3×10^6 daltons, respectively.

What is the nature of the conformational difference between L-1 and L-2 (and S-1 and S-2) RNAs? A reasonable explanation is that LCMV L and S RNAs have complementary termini. Circular nucleocapsids have been observed in preparations of another arenavirus.[11] Furthermore, the electrophoretic behavior of a glyoxal denatured DI VSV RNA known to have approximately 50 complementary nucleotides at the termini was similar to that of LCMV RNA. Formal evidence could be obtained from an electron microscopic examination of isolated LCMV RNAs or sequencing of both the 3′ and 5′ ends to show the location of the complementary region(s). It will also be interesting to examine defective interfering LCMV RNAs for their content of cohesive complementary termini.

ACKNOWLEDGMENTS
This is publication number 2295 from the Department of Immunopathology. The research was supported by U.S. Public Health Grants AI 09484, NS 12428, AI 15087, NSF Grant PCM 77-19383, and NIH Postdoctoral Fellowship 5 T32 GM 07437-03 to FJD. The authors gratefully thank Muriel Caruana and Cathy Booher for technical assistance, and Susan Edwards for preparation of the manuscript.

References

1. Pedersen, I.R. (1971): *Nature (New Biol.)* 234:112–114.
2. Carter, M.F., Biswal, N., and Rawls, W.E. (1973): *J. Virol.* 61–68.
3. Anon, M.C., Grau, O., Martinez-Segovia, Z., and Franze-Fernandez, M.T. (1976): *J. Virol.* 18:833–838.
4. Dutko, F.J., Helfand, J., and Pfau, C.J., unpublished observations.
5. Vezza, A.C., Clewley, J.P., Gard, G.P., Abraham, N.Z., Compans, R.W., and Bishop, D.H.L. (1978): *J. Virol.* 26:485–497.
6. Pederson, I.R. (1973): *J. Virol.* 11:416–423.
7. McMaster, G.K. and Carmichael, G.G. (1977): *Proc. Natl. Acad. Sci.* 74:4835–4838.
8. Reichmann, M.E., Villarreal, L.P., Kohne, D., Lesnaw, J., and Holland, J.J. (1974): *Virology* 58:240–249.
9. Buchmeier, M.J. and Oldstone, M.B.A. (1977): In: *Proceeds of conference on Negative Strand Viruses and the Host Cell.* Mahy, B.W.J. and Barry, R.D. (Eds.), Academic Press, New York.
10. Johnson, P.H. and Grossman, L.I. (1977): *Biochemistry* 16:4217–4224.
11. Palmer, E.L., Obijeski, J.F., Webb, P.A., and Johnson, K.M. (1977): *J. Gen. Virol.* 36:541–545.

Copyright 1981 by Elsevier North Holland, Inc.
David H. L. Bishop and Richard W. Compans, eds.
The Replication of Negative Strand Viruses

Gene Mapping in Pichinde Virus

W.-C. Leung, D. Harnish, A. Ramsingh, K. Dimock and W.E. Rawls[a]

Definition of the Pichinde viral genes and their gene products are essential as a basis for future studies of arenavirus replication and pathogenesis. The genome of Pichinde virus consists of two RNA species of negative polarity.[1-3] Previous estimations of the RNA size were performed under nondenaturing conditions and therefore might be inaccurate due to the presence of considerable secondary structure in the viral RNA.[2,3] In order to ascertain the RNA size, the viral RNAs were analyzed by gel electrophoresis in agarose gels under denaturation by methylmercury hydroxide (Figure 1a) or in agarose-polyacrylamide composite gels following denaturation by dimethylsulfoxide-glyoxal (Figure 1b).[4] As shown in Figure 1c, four major RNA species were found. The 28S RNA and 18S RNA were derived from host cell ribosomes incorporated into Pichinde virions.[2,3] The RNA species unique for the virus have been designated L and S. The virus specificity of the L and S RNA was supported by the findings that different oligonucleotide fingerprint patterns were found for the L and S RNAs from two different strains of Pichinde virus.[5] Measurements of the RNA sizes were taken following electrophoresis in gels of different concentrations to minimize the effect of gel concentrations on electrophoretic mobility of RNA. The average molecular weights of L and S Pichinde viral RNA are $2.63\text{-}2.83 \times 10^6$ and $1.26\text{-}1.31 \times 10^6$, respectively.[4]

The gene coding capacity of Pichinde virus RNA genome will be different if both RNAs contain unique sequences or if the S RNA is a subset of the L

[a] Department of Pathology, McMaster University, Hamilton, Ontario, L8N 3Z5, Canada.

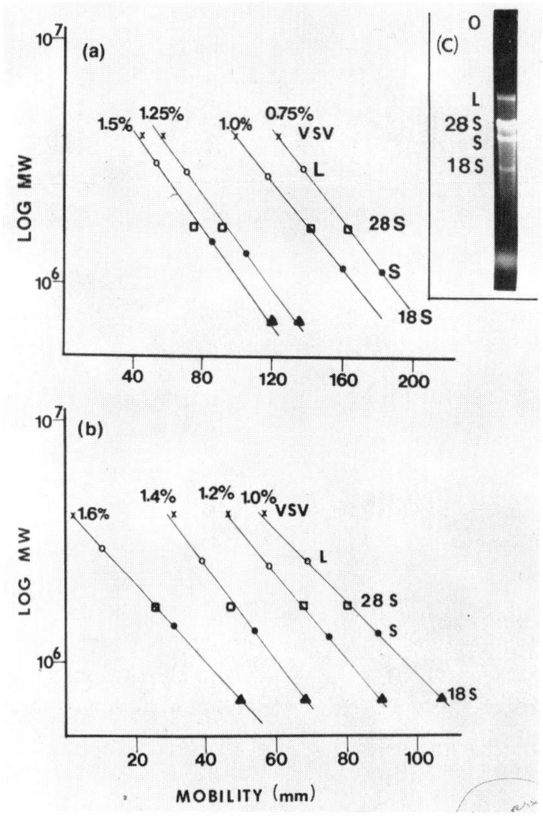

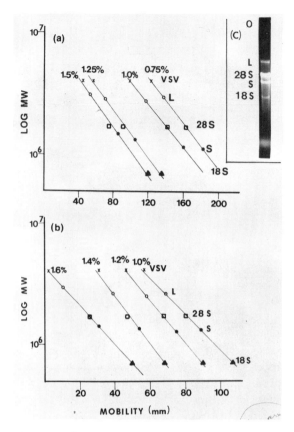

Figure 1. Size estimation of Pichinde viral RNA. Pichinde viral RNA was electrophoresed in (a) 10 mM methylmercury hydroxide in 4 different agarose concentrations (0.75%, 1.00%, 1.25%, 1.50%); (b) 4 different composite gels (polyacrylamide concentrations 1.0%, 1.2%, 1.4%, 1.6%; agarose concentration, 0.5%) after denaturing the RNA in glyoxal and DMSO. The Pichinde viral RNA bands were visualized by staining with ethidium bromide (c) and their log molecular weight versus mobility plots are presented in (a), (b).

RNA. To examine the nucleotide sequence relationship between L and S RNA, we have synthesized complementary DNA from individual L or S RNA for use as molecular probes. We have also investigated the optimal conditions for cDNA synthesis by reverse transcriptase and using oligodeoxyribonucleotides as random primers. We have achieved cDNA transcription from about 90% of either L or S RNA.[6] The cDNA molecules synthesized were then used as probes for sequence homology between L and S RNAs; about 90% hybridization was observed when the L cDNA reannealed with L RNA (Figure 2A) or when S cDNA reannealed with S RNA (Figure 2B). However, only background reannealing of about 3 to 5% hyb-

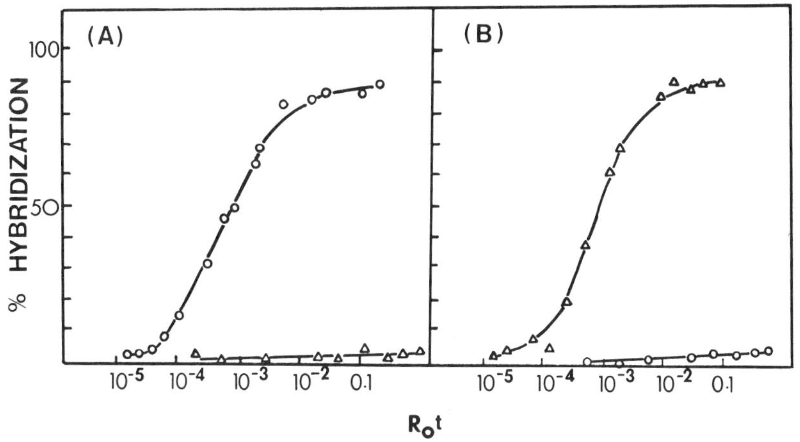

Figure 2. Hybridization of ^{3}H-cDNA with Pichinde viral L or S RNA. ^{3}H-cDNA synthesized from L RNA (Figure A) and S RNA (Figure B) were hybridized with the L RNA (o-o-o) or the S RNA (△-△-△). The degree of reannealing was monitored as S_1 nuclease resistant radioactivity and expressed as Rot (concentration of RNA in mol/L × time in S).

ridization was observed when either L cDNA was hybridized to S RNA (Figure 2A) or S cDNA was hybridized to L RNA (Figure 2B). These results suggested that L and S RNA sequences are predominantly unique and were in agreement with the high recombination rates between temperature-sensitive mutants of Pichinde virus[7] and unique oligonucleotide fingerprints for L and S RNAs.[8]

The L and S RNAs are single-stranded as determined by their sensitivity to RNase digestion and by buoyant density analyses.[2,3] RNA sequencing analysis on the 5′ and 3′ termini have demonstrated that the L and S RNAs contained unique sequences. It suggests that the L and S RNA each consists of only one single-stranded RNA species. No significant amount of positive stranded RNA was detected in the virion RNA. However, L and S RNAs do contain similar termini of about 30 nucleotides at both the 5′ and 3′ termini (Dimock, Rawls and Leung, unpublished data), a feature characteristic of negative stranded RNA viruses.

Having established that L and S RNAs sequences are unique, the coding capacity for each RNA can be calculated from its apparent molecular weight. The L RNA is calculated to code for polypeptides of 2.8×10^5 dalton and the S RNA for polypeptides of 1.3×10^5 dalton.[4]

Three primary gene products for Pichinde virus were found in virus infected cells, i.e., polypeptides L (mol wt about 200,000), pGPC (mol wt 42,000) and NP (mol wt 64,000). The nonglycosylated precursor pGPC gives rise to the fully glycosylated GPC (mol wt 79,000) which in turn is cleaved to yield the virion envelope glycopeptides GP-1 (mol wt 55,000) and GP-2 (mol wt 36,000). The nucleoprotein NP also gives rise, presumably by proteolytic cleavage to a series of lower molecular weight polypeptides.[9] The virus specificity of the 3 primary gene products was established by immune specificity, by their presence (or the presence of their derivatives) in purified virus preparations,[9] and by the demonstration that different strains of Pichinde virus contain different tryptic peptide maps for corresponding polypeptides (see below).

The gene assignments for these three polypeptides can be deduced from the following. The size of the L polypeptide exceeds the coding capacity of the S RNA, and therefore must be coded for by the L RNA. However, because of uncertainty in estimating the size of the L polypeptide, the L RNA conceivably could be large enough to code for all of L, pGPC and NP. A recombinant virus, Re-2, which contains the L RNA of Pichinde virus and the S RNA of Munchique virus, was used to locate pGPC and NP.[5] The virus specific polypeptides of Re-2 were examined by immune precipitation of infected cell extracts. Figure 3 demonstrates that the NP of Re-2 had an electrophoretic mobility similar to that of wt Munchique virus but different from that of wt Pichinde virus. Since Re-2 and wt Munchique virus share a common S RNA, this indicates that NP is coded for by S RNA. Since no apparent difference in electrophoretic mobility exists between the GPCs of wt Pichinde and wt Munchique viruses, the ^{35}S-methionine tryptic peptides of GP-C and NP of these viruses were analyzed. The peptide profile of NP of Re-2 (Figure 4A, C) was similar to that of wt Munchique (Figure 4D) but differed from that of wt Plichinde (Figure 4B). Similarly, the peptide profile of GPC of Re-2 (Figure 4F) was similar that of wt Munchique (Figure 4H) but different from that of wt Pichinde virus (Figure 4E, G). These results indicated that both NP and GPC, and therefore pGPC are coded for by the S RNA.

The model in which the L RNA codes for L-protein and the S RNA codes for polypeptides NP and GPC has the following implications. The NP and derivatives of GPC could provide the structural polypeptide components for the virion, however, a RNA transcription—replication function is required for the propagation of the viral genome. A RNA transcriptase activity has been found associated with Pichinde virus ribonucleoprotein complex.[10] It is tempting to propose that the transcriptase and/or replicase activity resides on the L protein. The above hypothesis could be tested by using ts mutants. The ts mutants of Pichinde virus have been grouped into two recombination groups.[7] It is conceivable that the ts lesions of one mutant group reside on the L RNA while the other resides on the S RNA. It is predicted that the ts mutants from the former group should possess temperature sensitive transcription and/or replication functions. Moreover, the very existence of Re-2 which contains the Pichinde L RNA and the Munchique S RNA suggests

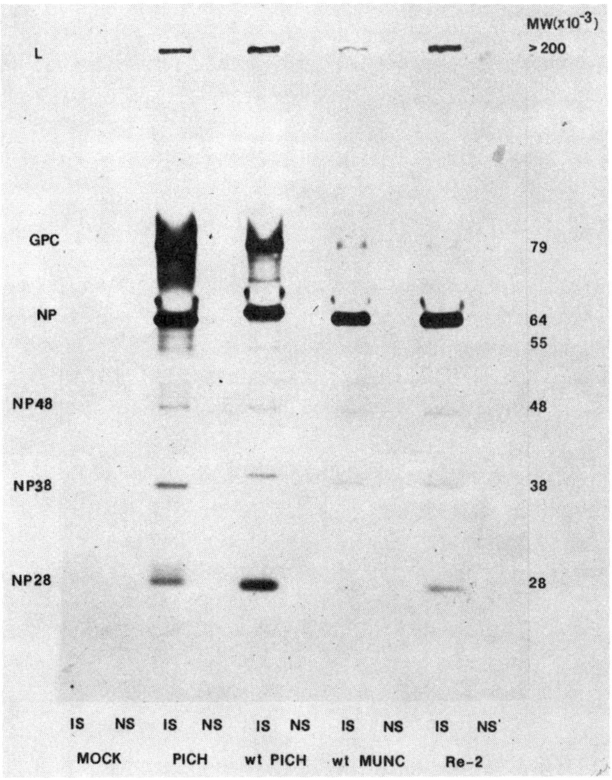

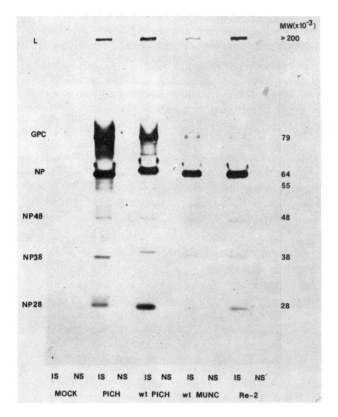

Figure 3. Electrophoresis of immune precipitates of virus infected cell polypeptides. The virus infected cell polypeptides were precipitated using hamster immune antiserum against Pichinde virus (from the author's laboratory). Wt Pichinde denotes the Pichinde virus strain (from Dr. D.H.L. Bishop laboratory) used to generate the recombinant. The electrophoresis was performed on a 7.5 to 15% gradient polyacrylamide gel. Polypeptides were visualized by fluorography.

that the hypothetical transcription and/or replication function coded for by L RNA can function in the transcription and/or replication of S RNA from a different strain.

On the other hand, in cells persistently infected by arenaviruses, weak immunofluorescence in the cytoplasm has been commonly observed.[2,3] The fluorescence is thought to represent the expression of NP polypeptide. Indeed this observation is substantiated by immune precipitation studies which demonstrates the presence of NP polypeptide in persistently infected cells (D. Harnish, unpublished observation). Since the available evidence suggests that in persistent infections, the viral genome probably does not

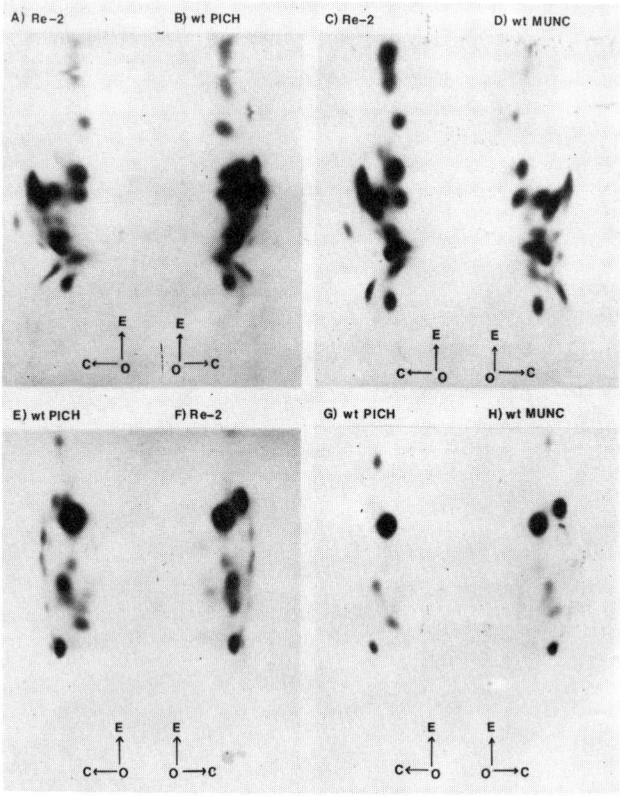

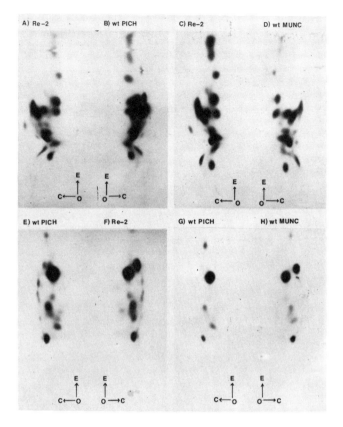

Figure 4. Two dimensional tryptic ^{35}S-methionine containing peptide analysis Figures A, B, C, D show the profile for NP and Figures E, F, G, H show the profile for GPC. E: electrophoresis C: chromatography O: origin.

exist in the form of proviral DNA, then the presence of L RNA and its expression may be required for the maintenance and expression of persisting S RNA genome. It follows that the expression of the L RNA gene function is an obligatory requirement for maintenance of Pichinde virus persistence. This hypothesis can be tested when highly sensitive molecular probes for L RNA and L polypeptides are available for examining their presence in persistently infected cells.

ACKNOWLEDGMENTS

The studies carried out in the authors' laboratories were supported by grants from Medical Research Council of Canada. WCL is a MRC scholar, DH, AR, KD are recipients of NSERC postgraduate scholarship, Ontario Government scholarship, and MRC postdoctoral scholarship, respectively.

References

1. Leung, W.-C. (1978): In: *Negative Strand Viruses and the Host Cell.* Mahy, B.W.J. and Barry, D. (eds.), Academic Press, New York, pp. 415–426
2. Rawls, W.E. and Leung, W.-C (1979): In: *Comprehensive Virology, Vol. 14.* Fraenkel Conrat, H. and Wagner, R.R. (Eds.), Plenum Publishing Corp., New York, pp. 157–192.
3. Pedersen, I.R. (1979): In: *Advances in Cancer Research, Vol. 24.* Academic Press, New York,
4. Ramsingh, A., Dimock, K., Rawls, W.E. and Leung, W.-C. (1980): Intervirology, Vol 14:31–36.
5. Bishop, D.H.L., personal communication.
6. Leung, W.-C., Ramsingh, A., Dimock, K., Rawls, W.E., Petrovich, J., and Leung, M. (1981): *J. Virol.* Vol 37:48–54.
7. Vezza, A.C. and Bishop, D.H.L. (1977): *J. Virol.* 24:712.
8. Vezza, A.C., Clewley, J.P., Gard, G.P., Abraham, N.Z., Compans, R.W., and Bishop, D.H.L. (1978): *J. Virol.* 26:485–497.
9. Harnish, D., Dimock, D., Leung, W.-C., and Rawls, W.E.: This volume.
10. Leung, W.-C., Leung, M., and Rawls, W.E. (1977): *J. Virol.* 30:98.

David H. L. Bishop and Richard W. Compans, eds.
The Replication of Negative Strand Viruses

Selection of Spontaneous ts Mutants of Junin and Tacaribe Viruses in Persistent Infections

Celia Esther Coto, Maria del Carmen Vidal, Ana Cristina D'Aiutolo and Elsa Beatriz Damonte[a]

The survival of arenaviruses in nature depends on their ability to persist in specific rodent hosts. The host-virus relationship established in *in vivo* chronic infections can be adequately studied by the development of appropriate *in vitro* systems.

Interaction of Junin virus (JV) with monkey kidney cells has been thoroughly studied in our laboratory. Vero cell lines persistently infected with JV (Vero-J cells) are readily produced despite virus cytopathogenicity. Characteristically, Vero-J cells released virus in a cyclical pattern during the early stage of persistence but later the cultures were no longer productive.[1] Independently of the presence of infectious virus, the cells always remained refractory to homologous virus superinfection.[1] Virus interfering particles were isolated during the non-producer stage [2] and small amounts of infectious virus have been rescued by a combined treatment of actinomycin D and cocultivation with normal Vero cells.[1] Studies on the properties of virus released from Vero-J cells during the productive period showed that as the number of cell transfers increased, the virus population became enriched in thermosensitive (ts) variants.[3] The role of ts mutants in virus persistence has been well documented for different virus families [4] but not for arenaviruses. Other authors, working with LCM virus, have reported DI particles,[5] slow growth virus variants [6] and recently a virus variant not affected by the concomitant presence of DI particles.[7]

[a] Laboratory of Virology, Department of Biochemistry, Faculty of Science, University of Buenos Aires, 1428, Argentina.

Since most of these studies were performed in long-term persistently infected cultures, we decided to concentrate our investigations on the first productive stage of persistent infection and to work with Tacaribe virus (TACV), which resembles JV in its interaction with Vero cells. Our purpose was to determine if selection of ts mutants is a regular event in persistent infection with arenaviruses. Simultaneously, we began to study the interaction of JV with mouse cells, an as yet unexplored system.

Infection of Vero Cells with TACV and Initiation of Persistent Infection

Stocks of wild type (wt) TACV were prepared in baby mouse brain or in Vero cells. In both cases, the virus obtained replicated with the same efficiency at 37 and 40°C. To originate persistently infected cultures, Vero cells were infected at a multiplicity of 0.01 or 0.001. After inoculation, the cultures were incubated at 37°C. Within 7 days, more than 99% of the cells had been destroyed. The few cells which remained repopulated the flasks giving rise to persistent infection. Three persistently infected cell lines designated Vero-T_2, Vero-T_6 and Vero-T_8 were developed in this way. Properties of Vero-T_8 cells will be fully described elsewhere. That the cultures were persistently infected was indicated by an absolute refractoriness of the cells to homologous virus superinfection during all their *in-vitro* life. Monolayers of Vero-T cells looked like normal Vero cells and divided at the same rate. Figure 1 illustrates the cyclic pattern of virus released from Vero-T_2 and Vero-T_8 cells. Although, it is not shown here, Vero-T_6 behaved like Vero-T_8 cells. Two stages can be clearly seen, a productive and a non-productive one. The extent of the first stage depends on the source of virus inoculum; when it is grown in mouse brain, the productive period is longer. On the contrary, virus passed in Vero cells originated cultures where infectious TACV is lost after a few transfers. Attempts to rescue virus by cocultivation of Vero-T_8 cells with normal Vero cells, at non-producer passage levels 18 and 20, failed. However, at later producer passages (13 and 15) cocultivation increased virus titres, indicating that disappearance of infectious virus during persistent infection is related to the loss of susceptible cells.

Properties of Virus Released

Results summarized in Table 1 show that virus released from Vero-T cell lines is thermosensitive. Data were selected to demonstrate that the ts viral population increased with time in culture. Both TACV released from Vero-T_8 cells and wt virus grow better at 37 than at 33°C. A certain degree of thermolability is shown by ts TACV from Vero-T_2 and Vero-T_6 cells in comparison with TAC wt virus. On the contrary, ts virus from Vero-T_8 cells displays a definite thermolabile character. None of the supernatants from Vero-T cell lines (between passages 2 and 22) have interfering activity. Table 1 also shows that ts mutants were more virulent for mice than the wt virus. Interestingly, the ts isolates could not initiate persistently infected cultures

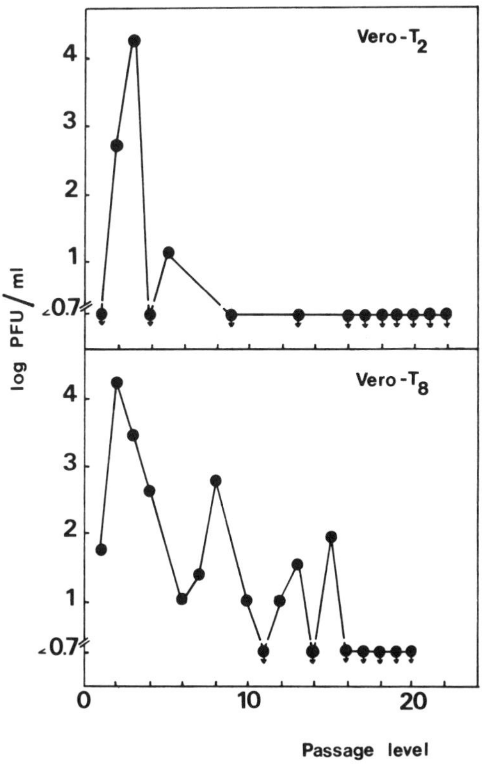

Figure 1. Virus production in Vero-T_2 and Vero-T_8 cell lines. Supernatants aliquots were taken from the cultures at 3-4 days after each cell transfer and titrated by plaque formation in RK_{13} cells at 37°C.

of Vero cells because infected cells were completely destroyed after 14 days of incubation at 37°C. An interpretation of these results is that another type of virus particles participate in the initiation of persistent infection. Interfering particles are the best candidates, since we know that they are present in our virus stocks, mainly when these are prepared in tissue culture.

Interaction of JV with Mouse Cells

Infection of L_{929} cells with JV, strain XJ-Cl_3 does not produce CPE if the virus was grown in baby mouse brain. However, after one passage through L cells, virus progeny caused extensive cytolysis starting on day 4 after infection. By day 8 most of the cells detached from the glass, but some isolated clones remained. Regrowth of the cells occurred 18 days after infection. Three independent L cell lines persistently infected with JV (LJ-A, LJ-B and LJ-C) were obtained after infection at a multiplicity of 0.01. In each case, the monolayers were fully recovered by days 38, 45 and 52, respectively. After

Table 1. Characteristics of Virus Released from Different Vero-T Cell Lines.

Virus	Days *in vitro*	Efficiency of[a] replication 40°C/37	37°C/33	Heat stability[b] (control/treated)	Ability to induce persistence	Mice virulence PFU/DL_{50}
wt	3	1.6	2.2	7.7×10^{-2}	yes	1.45
Vero-T_2	26	5×10^{-2}	ND[c]	2.7×10^{-2}	ND	0.87
Vero-T_6	50	3×10^{-3}	ND	2.1×10^{-2}	no	0.17
Vero-T_8	70	5×10^{-4}	2.1	$<8.6 \times 10^{-3}$	no	0.09

[a]Virus growth for 96 hr at 33, 37 and 40°C was compared, PFU assay was done at 37°C.
[b]Sample virus containing approximately 10^4 PFU was heated at 50°C for 30 min. Residual infectivity was determined by PFU assay at 37°C.
[c]ND: not done.

Table 2. Thermosensitivity of JV Released from LJ Cells.

Virus	Cell passage	Days *in vitro*	Virus yield[a] 37°C	40°C	40°C/37 ratio	Plaque[b] type
wt	—	2	$27. \times 10^3$	2.0×10^3	7.4×10^{-1}	t, m
	1	44	1.0×10^5	8.9×10^3	8.9×10^{-2}	t, l
LJ-A	3	69	6.2×10^4	2.9×10^3	4.7×10^{-2}	t, l
	4	73	1.3×10^3	< 5	$< 3.8 \times 10^{-3}$	t, l
	3	69	1.0×10^3	< 5	$< 4.7 \times 10^{-3}$	t, s
LJ-B	4	73	8.8×10^3	< 5	$< 5.7 \times 10^{-4}$	t, s
	5	76	3.0×10^3	< 5	$< 1.7 \times 10^{-3}$	t, s
LJ-C	2	69	3.5×10^3	3.0×10^1	8.7×10^{-3}	c, m
	3	78	7.2×10^3	< 5	$< 6.9 \times 10^{-4}$	c, m

[a]Virus growth for 72 hr at 37 and 40° was compared. Infectious virus was assayed in Vero cells (PFU/ml) at 37°C.

[b]Plaque type: t = turbid, c = clear. Size: small (1 mm), medium (1.5–2.0 mm), large (2–3 mm).

that, the cells were regularly transferred and cultivated at 37°C for a period of approximately 75 days.

Properties of Virus Released

Results recorded in Table 2 show that the virus liberated from the three lines, tested at different passage levels, is thermosensitive. In addition, a change is observed in the characteristics of the plaques produced by each ts mutant with respect to that produced by JV wt. LJ-A and LJ-B ts viruses produced turbid plaques but smaller (LJ-B) or larger (LJ-A) than those originated by the wt virus. On the contrary, LJ-C ts virus gave medium clear plaques very different from the others and easy to recognize. Based on these data, it is evident that a mixed population of ts mutants is already present in JV wt stock and that the destruction and regrowth of the monolayer from randomly surviving clones operate as selective pressure yielding one population in particular.

Discussion

Our previous data [3] and the results reported here strongly suggest that selection of spontaneous ts mutants is a regular event in the establishment of persistent infection with JV and TACV and that the process is independent of the host cell. However, the simultaneous presence of interfering particles which contribute to the state of persistence cannot be completely ruled out. Although not tested,another factor that could regulate the infection in L cells is interferon since very recently, we have been able to demonstrate a participation of interferon in mouse embryo cells persistently infected with JV.[8]

ACKNOWLEDGMENTS
This investigation was supported by Grants from SECYT (8641/79-4) and Secretaría de Estado de Salud Pública (Argentina).

References

1. Coto, C.E., León, M.E., Peralta, L.M., Help, G., and Laguens, R.P. (1979): In: *Humoral Immunity in Neurological Diseases*. Karcher, D., Lowenthal, A., and Strosberg, A.D. (Eds.), Plenum Press, New York and London, pp. 405–415.
2. Help, G.I, León, M.E., and Coto, C.E. (1976): *Rev. Soc. Arg. Microbiol.* 8:45–53.
3. Damonte, E.B. and Coto, C.E. (1979): *Intervirology* 11:282–287.
4. Preble, O.T. and Youngner, J.S. (1975): *J. Infect. Dis.* 131:467–473.
5. Welsh, R.M., O'Connell, C.M., and Pfau, C.J. (1972): *J. Gen. Virol.* 17:355–359.
6. Hotchin, J. and Sikora, E. (1973): *Infect. Immun.* 7:825–826.
7. Jacobson, S., Dutko, F.J., and Pfau, C. J. (1979): *J. Gen. Virol.* 44:113–121.
8. Vidal, M. del C. and Coto, C.E. (1980): *Medicina (Bs.As.)* 40:541–546.

David H. L. Bishop and Richard W. Compans, eds.
The Replication of negative Strand Viruses

Lymphocytic Choriomeningitis Virus Persistent Infection in EL4 Lymphoblastoid Cells

Mircea Popescu and Andrew Turek[a]

Mice infected with the virus of lymphocytic choriomeningitis (LCMV) congenitally or shortly after birth carry the virus through their entire life. This persistent infection is possible because the specific thymus dependent lymphocytes (T-Cells) response is impaired (reviewed in reference 1).

In the blood and lymphoid organs of mice, about 0.5% of lymphocytes were productively infected[2,3] with a non-cytolytic variant of virus.[4] The infected cells might represent the clone of lymphocytes responsible for virus elimination, but inactive because of the ongoing process of infection.[4] Details of virus-lymphocyte interactions in the carrier mouse are unknown. To provide more insight into the mechanism of LCMV persistence, we investigated the ability of LCMV to establish and maintain a persistent infection in mouse EL4 lymphoblastoid T-cell cultures.

Virus Multiplication During Persistent Infection

Multiplication in EL4 lymphoblastoid cells of parental cytolytic virus which forms clear plaques (C-virus) was characterized by an initial high production of both C-virus and defective interfering particles (DIP) able to form foci of interference (IFU) in L-cell monolayers [5] (Figure 1). Shortly after the peak multiplication was reached, approximately 60% of cells were killed, and in the cultures of surviving persistently infected cells (EL4-pi), the amount of both LCMV components declined. Later in persistence, the DIP were below

[a]Department of Microbiology and Immunology, University of Illinois, Chicago, Illinois.

the level of detectability and the parental C-virus was replaced by a turbid plaque forming variant of the virus. Strikingly similar virus kinetics were observed in the spleen and other tissues of neonatally infected mice (Figure 1).[6]

Characterization of Interfering Particles

DIP generated in EL4-pi cell cultures during the peak of virus multiplication had the same properties described for DIP which multiplied in non-lymphoid cell cultures.[5,7] In addition to their interfering ability (Table 1), the most notable characteristic was their resistance to UV irradiation under conditions which inactivate more than 1,000 times the PFU of C-virus (Table 2). DIP migrated slightly behind C-virus through density gradients of Urographin [8] and multiplied only together with C-virus, indicating that they were defective.

Figure 1. Multiplication of LCMV during persistent infection in EL4 lymphoblastoid cell cultures and in NMRI mice. 2×10^6 EL4 cells were infected (moi = 0.2) with LCMV strain WE (C-virus) and then cultivated in suspension at 37° by transferring 1×10^6 cells in 5 ml RPMI 1640 medium plus 15% fetal calf serum at intervals of 3–4 days. At indicated intervals, media were collected and after centrifugation, stored at −70°C. Virus units were assayed in L-cell monolayers under 0.5% agarose containing Eagle's medium. Determination of DIP as interference focus forming units (IFU) was done as described elsewhere.[5] Interference foci are islands of DIP protected cells on a C-virus cytolytic background with each focus being generated by a single DIP. PFU (—●— —▲—), IFU (—○—). Virus was titrated from tissue homogenates of combined carcasses of 3 suckling mice or of spleens from 2 virus-carrier mice.

SOURCE: *Virology 77:78, 1977, with permission Academic Press, New York.*

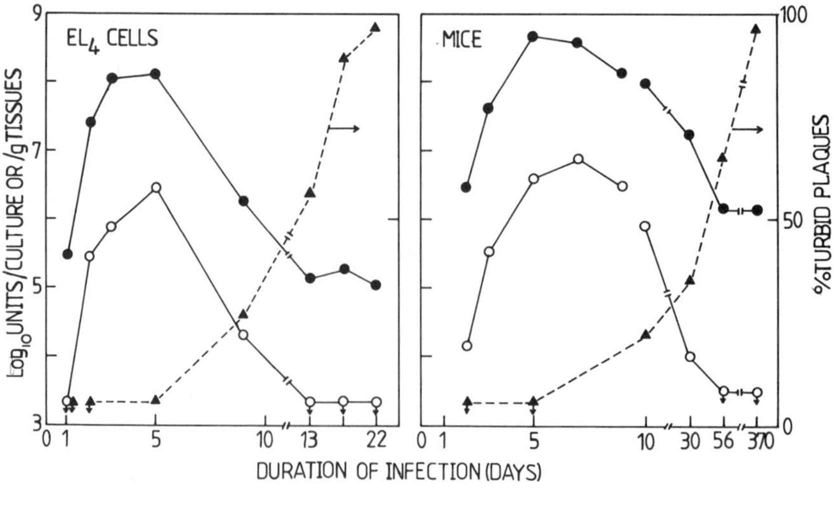

Table 1. Interference of Cytolytic LCMV by the Virus Generated in EL4 Lymphoblastoid Cell Cultures.

Virus added before challenge[a]	% Dead cells	C-virus[b] (PFU/culture)
IIP^-	4.9	6.82×10^4
IIP^+	6.8	$< 1.00 \times 10^6$
DIP	5.2	1.45×10^6
Control C-virus challenge	52.0	1.01×10^8

[a]Samples of 2×10^6 cells were infected for 2 hr with 0.1 ml of virus from E14-pi cells 96 weeks in culture (IIP^-), EL4-pi cells 5 weeks in culture (IIP^+), or from a pool of parental C-virus which was multiplied in EL4 cells for 96 hr and then UV irradiated 60 sec (DIP). Infected cells were incubated for 24 hr at 37°C. Cells were then wash centrifuged, challenged with C-virus (moi = 0.2), and further cultivated for 72 hr. As a control, non-infected EL4 cells were challenged in parallel. Virus yields were plaque titrated in L-cell monolayers.

[b]The titer of turbid plaques in the IIP^+ treated cultures was 2.91×10^7 PFU/culture and about 2% of plaques were of the cytolytic type.

In addition to DIP, EL4-pi cultures generated infectious non-cytolytic particles which were able to interfere with the parental C-virus. These infectious interfering particles (IIP) were not homogeneous. One class of IIP was composed of different plaque forming mutants, two of which are presented in Figure 2. Although the plaque characteristic was not changed during multiple passages of cloned mutants, a certain degree (0.1–2%) of back mutation to the parental C-virus plaque morphology was always observed. Later (more than 40–50 weeks in culture), IIP^+ virus was replaced entirely by an infectious non-cytolytic interfering virus unable to form plaques in L-cell monolayers (IIP^-). This virus killed or immunized mice depending on the

Table 2. UV Sensitivity of Interfering Particles Generated by LCMV in EL4 Lymphoblastoid Cells.

	Log_{10} titer/ml[b]			
	Before UV		After UV	
Virus[a]	PFU	IF units	PFU	IF units
IIP^-	0	3.70	0	1.00
IIP^+	7.21	4.00	4.32	1.70
DIP + C-virus	8.35	3.00	5.09	2.70

[a]Virus was purified from media of EL4-pi cells 90 weeks in culture (IIP^-), EL4-pi cells 40 weeks in culture (IIP^+), or EL4 cells in which C-virus and DIP multiplied for 120 hr. Virus suspensions were irradiated with an R-52 Mineralight UV lamp delivering 5×10^3 ergs/cm²/sec for 60 sec.

[b]The titer of IF units was indicated by the dilution of interfering virus which induced a slight, but recognizable inhibition of plaques produced by 10^2 PFU of C-virus added concomitantly with the interfering virus to L-cell monolayers.

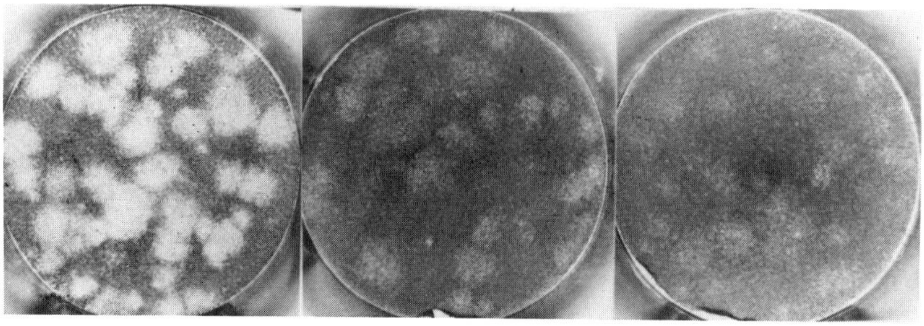

Figure 2. Plaque morphology of parental and persistent LCMV in L-cell monolayers. A: parental C-virus; B, C: clones of persistent IIP^+ virus. Petri dish 0 = 35 mm.

amount inoculated intracerebrally. Quantal titration of IIP-virus was based on interference with the cytolytic plaque formation of parental virus. The interference of IIP^+ and IIP^- with the yields of C-virus in EL4 cells is shown in Table 1. The data presented in Table 2 shows that IIP^+ and IIP^-, in contrast to DIP, were inactivated by UV irradiation at 258 nm indicating that the size of the target genome responsible for interference was different in the defective and infectious IP.

EL4-pi cells did not support C-virus multiplication, but were killed with virus production by heterologous Vesicular Stomatitis Virus.

Viral Antigens in EL4-pi Cells

Cell-associated lcmv antigens were visualized by the indirect immunofluorescence method using rabbit-antivirus antiserum, followed by fitc-conjugated goat anti-rabbit IgG antiserum. During 58 weeks of virus persistence, the proportion of EL4-pi cells with cytoplasmic fluorescence increased from 20 to 78%, while the proportion of cells with membrane fluorescence decreased from 20 to 3%. The intensity of fluorescence in comparison with the control C-virus infected cells gradually became weaker over the period of observations.

EL4-pi cells were cloned between 27 and 52 weeks of cultivation by spreading diluted monocell suspensions on a medium containing agar. Isolated colonies of cells were collected and grown further. The virus in the resulting cultures was detected by intracerebral inoculation of mice and by immunofluorescence microscopy. From 24 cloned cultures, 3 did not contain virus and were fully susceptible to C-virus challenge, suggesting that few EL4-pi cells were unable to transmit the virus to progeny cells. Inoculum from 3 other cultures did not kill or immunize the mice even though 10 to 34% of these cells contained cytoplasmic virus antigens, suggesting that a proportion of EL4-pi cells might be infected with a defective mutant unable to multiply in mice. In 18 cultures, the virus was present in media and 15 to 47%

of cells showed cytoplasmic immunofluorescence, suggesting that in more than half of the progeny cells derived from an infected parent cell, the virus was below detectability. These results were in partial agreement with similar observatuons made in LCMV persistently infected non-lymphoid cell cultures.[9,11] Our inability to obtain cultures in which all cells contained viral antigens might be the consequence of a selection imposed by the procedure or alternatively, may reflect the "shut off" mechanism of viral antigen synthesis described for LCMV infected L-cells.[12]

Mechanism of Virus Persistence—Concluding Remarks

The description of LCM persistence depicted in Figure 3 is based on the indication that DIP were mainly involved in the establishment of persistence, while its maintenance was supported by a continuous selection toward IIP variants devoid of any detectable cytolytic potential. Apparently, the process of selection is fueled by the competition for susceptible cells generated from those which were protected by DIP. Transition from resistance to susceptibility was inferred from the results of the cell cloning experiment presented here and from data of others.[13,14] It is also apparent that the role of DIP was diminished in IIP⁻ infected cells when the parasitism of cells by the virus reached perfection. The proposed model is in agreement with the previous view regarding homologous interference by LCMV.[15] In this system, interfering particles can be described as a range of mutants with various kinds and degrees of defectiveness, grouped here into three

Figure 3. Hypothetical mechanism proposed for LCMV persistent infection of EL4 lymphoblastoid cells.

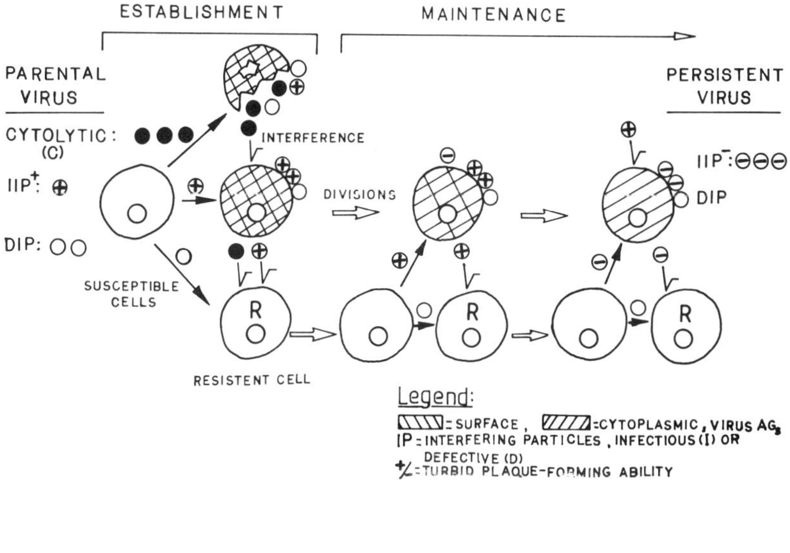

categories: IIP^+, IIP^- and DIP. The persistence seems to be achieved by reciprocal interference between infectious variants, resulting in selection of virus which does not disturb the essential cell functions.

Persistent infection with LCMV was shown to alter the "luxury" specialized function of neuroblastoma cells.[16] Whether the specialized function(s) of infected T-cells in LCMV carrier mice is also altered remains to be established.

ACKNOWLEDGMENTS

This work was supported by PHS Grant 5R01 A1 15138-02 from the National Institute of Allergy and Infectious Diseases and by a grant from the Arthritis Society, Illinois Chapter. A. Turek is a Predoctoral Trainee.

References

1. Lehmann-Grube, F. (1971): Lymphocytic choriomeningitis virus. In: *Virology Monographs*, Springer-Verlag, Wien, Vol. 10.
2. Popescu, M., Lohler, J., and Lehmann-Grube, F. (1977): *Z. Naturforsh.* 32c:1026–1028.
3. Doyle, M.V. and Oldstone, M.B.A. (1978): *J. Immunol.* 121:1262–1269.
4. Popescu, M., Lohler, J., and Lehmann-Grube, F. (1976): *J. Gen. Virol.* 42:481–492.
5. Popescu, M., Schaefer, H., and Lehmann-Grube, F. (1976): *J. Virol.* 20:1–8.
6. Popescu, M. and Lehmann-Grube, F. (1977): *Virology* 77:78–83.
7. Welsh, R.M., O'Connel, C.M., and Pfau, C.J. (1972): *J. Gen. Virol.* 17:355–359.
8. Gschwender, H.H. and Popescu, M. (1975): In: *Biological Separations in Iodinated Density Gradients*. D. Rickwood (Ed.), Information Retrieval Ltd., London, pp. 145–157.
9. Lehmann-Grube, F., Slenczka, W., and Tees, R. (1969): *J. Gen. Virol.* 5:63–81.
10. Stanwick, T.L. and Kirk, B.E. (1976): *J. Gen. Virol.* 32:361–367.
11. Staneck, T.L., Trowbridge R.S., Welsh, R.M., Wright, E.A., and Pfau, C.J. (1972): *Inf. and Immun.* 6:444–450.
12. Hotchin, J. (1973): *Nature New Biology* 241:270–272.
13. Jacobson, S., Dutko, F.J., and Pfau, C.J. (1979): *J. Gen. Virol.* 44:113–121.
14. Holland, J.J., Villarreal, L.P., Welsh, R.M., Oldstone, M.B.A., Kohne, D., Lazzarini, R., and Scolnick, E. (1976): *J. Gen. Virol.* 33:193–211.
15. Popescu, M. and Lehmann-Grube, F. (1973): *International Virology 3 Abstracts*. Bedson, H.S. (Ed.), IAMS, Madrid, pp. 190.
16. Oldstone, M.B.A., Holmstoen, J., and Welsh, R.M. (1977): *J. Cell. Physiol.* 91:459–472.

Copyright 1981 by Elsevier North Holland, Inc.
David H.L. Bishop and Richard W. Compans, eds.
The Replication of Negative Strand Viruses

Molecular Studies of LCM Virus Induced Immunopathology: Development and Characterization of Monoclonal Antibodies to LCM Virus

Michael J. Buchmeier and Michael B.A. Oldstone[a]

Introduction

Lymphocytic choriomeningitis virus (LCMV) infection of the mouse provides a unique model system for studying host-virus interaction. By appropriate manipulation of the route and age of the mouse, the outcome of infection can range from acute, rapidly lethal disease to a lifelong persistent infection. Examinations of both of these seemingly divergent courses of infection have yielded information of basic importance to our understanding of virus induced immunopathology. Studies of lymphocyte function during acute LCMV infection by Lundstedt,[1] and later by Cole and coworkers,[2] defined the lytic function of cytoxic T-lymphocytes induced during infection. Later experiments by Zinkernagel and Doherty[3] led to the definition of H-2 restricted killing of virus infected target cells by cytotoxic T-lymphocytes. Subsequent work by Welsh and Zinkernagel[4] provided the initial evidence for activation of natural killer (NK) cells early in virus infection. Investigation of persistent LCMV infection in neonatally infected mice led Oldstone and Dixon[5] to formulate the concept of virus induced immune complex disease resulting from the union of viral antigen and antibody leading to chronic glomerulonephritis. Despite a wealth of information that has been derived from studies of the immunobiology of LCMV infection, we have only a rudimentary understanding of the biochemical events involved in virus replication and persistence. Further understanding of complex pro-

[a]Department of Immunopathology, Scripps Clinic and Research Foundation, La Jolla, California

cesses such as immune mediate recognition and tissue injury, mechanisms of virus replication and persistence, and generation of defective interfering particles will require a more complete understanding of the molecular nature of the virus and its replication, as well as the regulation of viral gene expression in infected cells. To better understand these events, we have developed a series of monoclonal antibodies to utilize as specific probes to study these and other facets of the immunobiology and biochemistry of LCMV and other arenaviruses.

Results and Discussion

Generation and Characterization of Monoclonal Antibodies to LCMV

Mice were immunized by ip infection with a sublethal dose of LCMV, then boosted 4–6 weeks later with 3–7 μg purified virus. Fusion with P3 $\times$ 63Ag8 murine plasmacytoma cells was as described[6] using PEG, and hybrid cultures were selected in HAT medium. Proliferating cultures were screened for antibody to LCMV by indirect immunofluorescence on acetone fixed and viable target cells. Cultures producing antibody were cloned and often grown in ascites form to yield mouse ascites fluid of high antibody titer.

LCMV contains three major polypeptides,[7] a 63,000 d nucleocapsid protein (NP) and two glycopeptides, of 44,000 d (GP-1) and 35,000 d (GP-2) respectively. In addition, the infected cell contains a 75,000 d glycopeptide precursor (GP-C) which is proteolytically cleaved to yield the two virus structural glycopeptides.[8] To define specificity of the monoclonal antibodies, we immunoprecipitated radioactively labeled polypeptides from virions, and from cytosols prepared from infected cells. The precipitates were then analyzed by SDS-PAGE.

Fifty-four hybridoma cultures producing antibody to LCMV have been established. By immunofluorescent staining (Figure 1), 45 of these antibodies reacted with determinants expressed only in the cytoplasm of infected cells, while nine reacted with antigens expressed on the surface of infected cells. Analysis of polypeptide specificity of 29 of these monoclonals by immunoprecipitation demonstrated that 20, all of which reacted with cytoplasmic antigens, precipitated NP. The nine antibodies which reacted at the cell surface precipitated glycoproteins. Using these methods, we have identified antibodies to NP, GP1 and GP2 (Figure 2). Monoclonal antibodies to either GP1 or GP2 also precipitated the cell associated precursor glycopeptide, GP-C.

Methods are being developed utilizing monoclonal antibodies to study virus growth and antigen expression in tissues, and early results have been encouraging. Figure 3 illustrates immunofluorescent staining of NP antigen in the meninges and in ependymal cells lining the ventricles of the brain of a mouse infected six days earlier with LCMV. Antigens concentrated in these

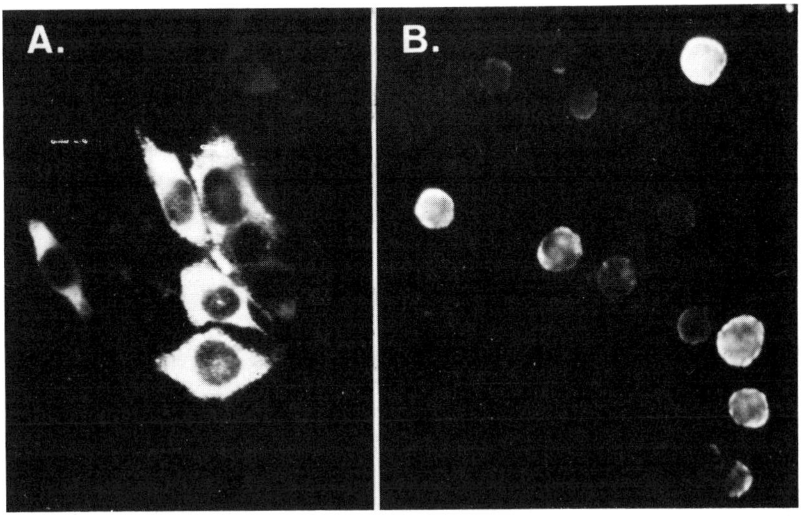

Figure 1. Immunofluorescent staining of LCMV antigens in the cytoplasm and on the surface of infected L-929 cells. (A) Monoclonal antibody to LCMV NP (clone 1-1.3) stained punctuate granular deposits of antigen in the cytoplasm, but did not stain antigens at the cell surface. (B) Monoclonal antibody to LCMV GP-1 (clone 2-11.10) stained antigens on the surface of infected cells. For both studies, uninfected cells were mixed with infected cells to insure an internal control.

regions may serve as targets for immune attack. We have obtained similar localization of LCMV GP1 antigen in ependymal cells of brains of infected mice, and of NP antigen in spleen, liver and lung of LCMV infected guinea pigs. It will be of interest to localize and to quantitate the concentrations of specific viral antigens over the course of acute and persistent infections and to relate this information to the immunopathologic events occurring in disease.

Use of Monoclonal Antibodies to Define Type Specific and Group Specific Antigens Among Arenaviruses

Arenaviruses are an interesting group taxonomically. [9,10] Although the members of the group appear similar in biology and structure,[11,12] serological analyses break the group into two distinct subdivisions. These are the so-called "New World" viruses, a group consisting of the Tacaribe complex, and the "Old World" viruses composed of LCM, Lassa and Mozambique viruses. We were interested in first determining whether any monoclonal antibodies could resolve fine antigenic differences between LCMV substrains. To accomplish this, we analyzed two commonly used LCMV strains, the Armstrong strain (ARM; CA1371)[13] which is benign in guinea

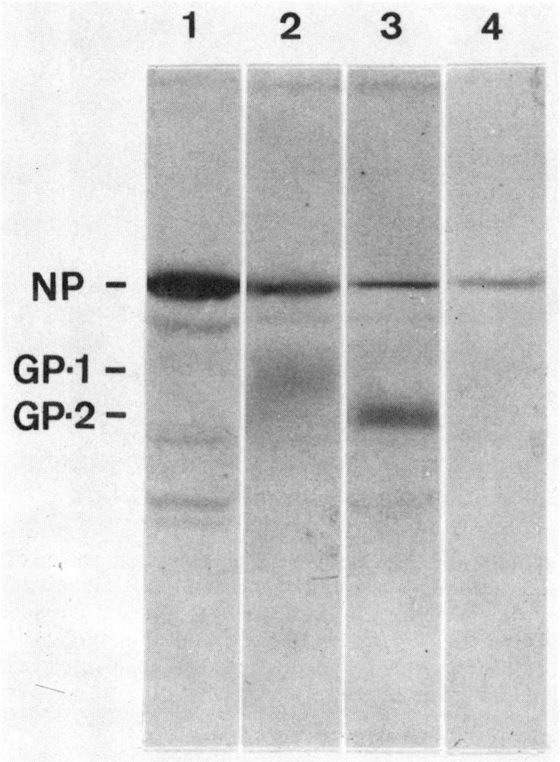

Figure 2. Immune precipitation of ^{35} S-methionine labeled LCMV polypeptides by monoclonal hybridoma antibodies. Lane 1, monoclonal antibody to NP (clone 24A-21.4); lane 2, anti-GP1 (clone 2-11.15); lane 3, anti-GP2 (clone 9-7.5); and as a control, lane 4 monoclonal antibody to Pichinde virus NP (clone P1-1.3). The faint bands observed in lane 1 are degradation products of NP.

pigs, and the WE strain,[14,15] which is lethal for guinea pigs.[16] We tested 22 antibodies to the nucleocapsid protein and found no antigens unique to either virus. In contrast, three of nine antibodies to the viral glycoproteins showed specificity for a single strain of LCMV. Of these three, two reacted in surface immunofluorescence only with LCMV-ARM infected cells, and one reacted only with LCMV-WE. The remaining six reacted with cells infected by either virus. Thus, monoclonal antibodies can resolve fine antigenic differences between LCMV strains.

In collaboration with Dr. Karl Johnson and Dr. Oyewale Tomori at CDC, Atlanta, we have extended our analyses to investigate the relationship between LCMV and the "Old World" viruses, Lassa and Mozambique. Forty-six culture fluids containing antibody to LCMV were analyzed by immunofluorescence for cross-reactivity with the African viruses. The results (Table 1) demonstrated six cross-reactive antibodies. Five of these, all antibodies to the viral NP, reacted with both Lassa and Mozambique vir-

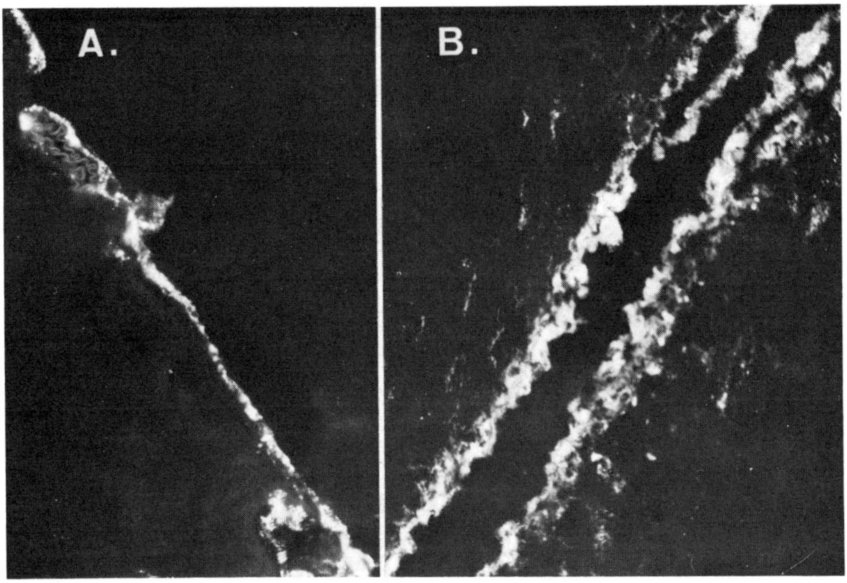

Figure 3. Immunofluorescent staining of LCMV NP antigen (clone 1-1.3) in the brain of a C_3H/St mouse infected six days earlier. Viral antigen is present in cells of the meninges (A) and in ependymal cells lining the ventricular system (B). The monoclonal antibody was conjugated directly with fluorescein isothiocyanate.

uses. A sixth antibody, directed against the LCMV GP2 glycopeptide, reacted only with Mozambique virus. These results have now been confirmed in analyses of five Lassa strains and nine Mozambique strains. None of the cross-reactive LCMV antibodies reacted with any of seven "New World" arenaviruses of the Tacaribe group.

Table 1. Cross-Reactivity of Hybridoma Antibodies to LCMV with Lassa and Mozambique Viruses.

Hybridoma antibody	Eliciting virus	Polypeptide specificity	Immunofluorescence reaction vs targets infected with		
			LCM	Lassa	Mozambique
1-1.3	LCMV	NP	+	+	+
24A-21	LCMV	NP	+	+	+
25-39	LCMV	NP	+	+	+
24B-2	LCMV	NP	+	+	+
3-3	LCMV	NP	+	+	+
9-7.5	LCMV	GP2	+	0	+
1-3.2	LCMV	NP	+	0	0
P1-1.3	Pichinde	NP	0	0	0
P3 × 53Ag8	None	—	0	0	0

Table 2. Cross-Reactivity of Hybridoma Antibodies to Pichinde Virus (PV) with Tacaribe Group Viruses.

Hybridoma antibody	Eliciting virus	Immunofluorescence reaction vs target infected with				
		Pichinde	Junin	Parana	Tamiami	Other[a]
P1-1.3	P.V.	+	0	0	+	0
P2-14.19	P.V.	+	0	+	+	0
P1-3	P.V.	+	+	0	0	0
P1-5	P.V.	+	0	0	0	0
24A-21	LCMV	0	0	0	0	ND[b]
P3 × 63Ag8	None	0	0	0	0	ND

[a]Other viruses tested included Amapari, Latino, Machupo, Tacaribe, LCM, Lassa and Mozambique.
[b]ND: not done.

Analyses of antigenic cross-reactivity among the viruses of the Tacaribe complex with a preliminary panel of four monoclonal antibodies to Pichinde virus illustrate similar results (Table 2). In this case, three of the four antibodies were cross-reactive with one or more than one agent. Each of the four antibodies cross-reacted with different heterologous viruses suggesting that each recognized a different determinant. We are now in the process of expanding our series of monoclonal reagents to LCMV and to Pichinde virus, and of extending these studies to other Tacaribe group agents to facilitate the use of these reagents as precisely defined molecular probes for studying virus replication, host-virus interactions, and the antigenic relationships that exist among the arenaviruses.

ACKNOWLEDGMENTS
We thank Hanna Lewicki and Ricarda DeFries for excellent technical assistance, and Susan Edwards for typing this manuscript. Work described here was supported by NIH Grants A1 16102, A1 09484 and NS 12428. MJB is the recipient of an Established Investigator Award from the American Heart Association.

This is publication number 2294 from the Department of Immunopathology, Scripps Clinic and Research Foundation, La Jolla, California.

References

1. Lundstedt, C. (1969): *Acta. Path. Microb. Scand.* 75:139.
2. Cole, G., Prendergast, P., and Henney, C. (1973): In: *Lymphocytic Choriomeningitis and Other Arenaviruses.* Lehmann-Grube, F. (Ed.), Springer, Berlin, pp. 60.
3. Zinkernagel, R. and Doherty, P. (1974): *Nature* (*London*) 251:547.
4. Welsh, R. and Zinkernagel, R. (1977): *Nature* 268:646.

5. Oldstone, M. and Dixon, F.J. (1969): *J. Exp. Med.* 129:483.
6. Kohler, G. and Milstein, C. (1975): *Nature* 256:495.
7. Buchmeier, M., Elder, J., and Oldstone, M. (1978): *Virology* 89:133.
8. Buchmeier, M. and Oldstone, M. (1979): *Virology* 99:111.
9. Rowe, W., Pugh, W., Webb, P., and Peters, C. (1970): *J. Virology* 5:289.
10. Casals, J. (1976): In: *Viral Infections of Humans—Epidemiology and Control.* Evans, A. (Ed.), Plenum, New York, pp. 103.
11. Johnson, K., Webb, P., and Justines, G. (1973): In: *Lymphocytic Choriomeningitis and Other Arenaviruses.* Lehmann-Grube, F. (Ed.), Springer, Berlin, pp. 241.
12. Rawls, W.E. and Leung, W.C. (1980): *Comprehensive Virology* 14:157.
13. Rowe, W., Black, P., and Levey, R. (1963): *Proc. Soc. Exp. Biol. Med.* 114:248.
14. Rivers, T. and Scott, T. (1936): *J. Exp. Med.* 63:415.
15. Lehmann-Grube, F. (1971): *Virology Monographs* 10:78.
16. Buchmeier, M. and Dutko, F.J. (1980): *Fed. Proc.* 39:675.

Published 1981 by Elsevier North Holland, Inc.
David H.L. Bishop and Richard W. Compans, eds.
The Replication of Negative Strand Viruses

Mechanisms of Pathogenesis in Lymphocytic Choriomeningitis Virus Infected Mice

Charles J. Pfau and Steven Jacobson[a]

Classic virus diseases such as smallpox, poliomyelitis and measles are being gradually vaccinated out of existence. Attention is now turning to those infections in which virus persists for months and even decades, often in spite of vigorous defense mechanisms of the host. How have viruses evolved to overcome these potentially hostile environments? Alternately, how has the host failed to halt virus invasion and proliferation? Lymphocytic choriomeningitis (LCM) virus infection of the mouse has gradually been recognized, since its discovery over 45 years ago,[1] as a model without equal for studying persistent infections. Transplacental or neonatal infection of the mouse leads to a life long, rather benign, infection. On the other hand intracerebral infection of the adult mouse causes a fatal convulsive central nervous system disease, whereas peripheral injection leads to an abortive immunizing infection.

Why are the outcomes to infection with LCM virus so varied? It has become increasingly evident that the answer to this question involves the understanding of an intricate and fascinating weave of virologic and immunologic phenomena.[2-6] Consider that after neonatal infection the virus: (a) largely or entirely prevents the appearance of subsets of T-lymphocytes which can recognize and destroy LCM infected tissue culture target cells;[7] (b) alters the humoral response so that only antibodies lacking virus-neutralizing activity are made; [8-9] (c) provokes a short-lived interferon response which transiently depresses the level of circulating virus; [10-11] but this interferon may incite acute disease in suckling mice[12] or late-onset glomerulonephritis;[11] (d) induces the rapid genesis of defective interfering (DI) particles[13] which can mask the cytolytic potential of the standard virus

[a] Department of Biology, Rensselaer Polytechnic Institute, Troy, New York

in tissue culture.[14] Attention will be focused here on the microevolution of viruses in persistently infected mice. The properties of these changing viruses will be discussed with regard to their growth in tissue culture at various temperatures, and in the presence of DI particles. Comparison will also be made of the disease syndromes and cytotoxic T-cell responses these viruses elicit after intracerebral (ic) infection of the adult mouse.

Shortly after Traub[15] discovered murine lymphocytic choriomeningitis, he observed that naturally passed virus isolated from the thoracic and abdominal organs differed markedly in pathogenicity for adult mice than the same virus which had been modified by multiple serial mouse brain-to-mouse brain transfers.[16] More recently, Hotchin and colleagues[17] found that virus repeatedly passed by intraperitoneal injection and harvested from the liver produced few, if any, deaths when injected into neonates (with all survivors being persistently infected). On the other hand, virus passed by ic injection and harvested from the brain invariably killed when injected into mice less than 24 hours old. In fact, they routinely observed that the brain of a persistently infected mouse contained a different substrain of virus from its liver and that these two populations coexisted at the same time in the same mouse.[18]

With the availability of a reliable plaque assay,[19] Hotchin and colleagues[20] showed that liver-passed LCM contained mostly turbid plaque types, while brain-passed LCM virus contained mostly clear plaque types. The cloned clear plaque virus killed neonatal mice and produced early convulsive death in adult mice, while the cloned turbid plaque virus induced persistent infection in neonates and either late wasting death or persistent infection in adult mice.[21] Popescu and Lehmann-Grube[22] confirmed that different organs of acutely as well as persistently-infected mice favored the replication of different plaque variants. Specifically, they also found that clear plaque types predominated in the brain while the replication of turbid plaque types was favored in the spleen (the turbid plaque types were subdivided into light and intense categories). Although Hotchin's and Lehmann-Grube's group were using a virus strain (UBC or WE) with the same origin, the latter group could not confirm that plaque morphology types reflected differences in pathogenic properties, i.e., all plaque types produced early convulsive death after ic injection of adult mice and induced persistent infection in neonates. As discussed later it seems likely that a strain of mouse was used that responded to the infections in ways that masked any difference in the disease-producing potential of the two plaque types. Nevertheless, Popescu and Lehmann-Grube[13] went on to follow the kinetics of appearance of these viruses. When clear plaque type virus was injected into neonates, it replicated rapidly and reached peak titer above seven days postinfection. Within the next three weeks, total virus titer dropped to a maximum of 1% of the former peak level, and at the same time a dramatic transition took place in plaque morphologies—most changed from clear to turbid in all organs except the brain.

With this knowledge that LCM virus somehow changes during the course of life-long infection in the mouse, and that the replication of LCM virus resistant to DI particles is a requisite for long-term infection in tissue culture,[23] we began to search for the *in vivo* equivalent to DI-resistant viruses. Using an MDCK monolayer assay[23] we examined the plaque morphologies of virus taken from the blood of a mouse that 10 months previously had been infected with an LCM strain (UBC) having a long history of tissue culture passage. Two plaque types were easily recognized. One was large (3-4 mm diameter) and clear and identical to the original input virus, while the other had a pinpoint (1-2 mm diameter) morphology.[24] These plaque types were easily cloned and two representing the clear and pinpoint plaque morphologies were chosen for further studies. First, the growth kinetics of the two types were determined in MDCK cells infected individually with these viruses at a multiplicity (moi) of 0.1. Peak progeny virus yields were obtained between 40 and 44 hours after infection but the clear plaque type consistently gave titers (1-2 $\times$ 10^8 pfu/ml) 0.5–1.0 $\log_{10}$ units higher than the pinpoint plaque type.[25] Both virus types caused almost complete destruction of the monolayers 60 hours after infection (MDCK cells are one of the few lines that are killed by LCM virus infection).[14] Each of these viruses, as well as the original virus used to initiate the persistent infection, were injected ic using a dose of 100 LD_{50} units into adult mice. In this way, the two clear plaque types could be easily distinguished from one another. After about 5 days, mice injected with the pinpoint plaque virus isolated from the 10-month-old persistently infected mouse, as well as the clear plaque virus used to initiate the infection, developed a nondescript illness with ruffled fur, hunched posture, and inactivity. If disturbed, particularly when spun by the tail, animals would go into clonic convulsions which frequently terminated in tonic extension of the hind legs and death. All mice eventually showed what Hotchin and colleagues[17] called the aggressive syndrome and virtually all died between 6 and 8 days after virus inoculation. In contrast, the most striking feature of the disease induced by the other clear plaque morphology virus (isolated from the 10-month-old persistently infected mouse) was a non-convulsive death with an unimpressive gross pathology. Although most mice infected with this docile[17] strain eventually died within 2–4 weeks, it appeared to be a consequence of progressive exhaustion and deterioration. Since these viruses from the persistently infected mouse were obviously different from the original input virus, we began to search for other characteristics which might indicate how the viruses were improving themselves with respect to the requirements for persistence. The experience of many workers has been that clones of virus isolated at various times after the initiation of persistent tissue culture infections differ from the original virus. Many of these viruses have been shown to be clearly temperature sensitive (TS), or small plaque mutants, or both.[26]

MDCK monolayer plaque assays were used to examine the possible TS characteristics of these three viruses. After an adsorption period at 37°C replicate plates containing serial 10-fold dilutions of virus were incubated at 35, 37, 39.5 and 40°C (below 35°C plaque formation, but not virus growth, was impossible to detect and above 40°C monolayers could not be maintained under agar). None of the viruses showed any decreased plating efficiency at 35°C but a 70% reduction in plaque number was consistently noted with the aggressive (pinpoint plaque) virus at 39.5 and 40°C.[25] This decreased efficiency of plating was not considered sufficient to classify, or work with these viruses as TS mutants. Since this slight temperature sensitivity was nearly identical to our experience with the SP strain of LCM—a virus resistant to DI particles and necessary for persistent tissue culture infection,[23] we determined the sensitivity of aggressive virus to *in vivo* generated DI particles.[24] Since tissue culture generated LCM DI particles were shown to prevent standard virus replication and induction of cytopathic effects (CPE) in MDCK cells,[14] this system was chosen to determine the DI particle sensitivity of the three LCM strains. Using DI particles generated in suckling mouse kidneys, we found that aggressive virus was relatively resistant to DI particles.[24] Furthermore, the same phenomenon could be reproduced in Vero and mouse neuroblastoma cells[25] although the Vero cells under normal circumstances never showed CPE after LCM infection.

We also began another line of experimentation to determine other characteristics that could be used to distinguish between the two viruses isolated from the persistently infected mouse. Riviere and colleagues[10] had shown that neonatal infection of outbred Swiss mice with the Pasteur Institute strain of LCM led to a high mortality rate (up to 80%). However, if these mice received anti-interferon serum, most could be protected against the infection. None of our three strains caused death in neonatal ICR outbred Swiss mice obtained from Blue Spruce Farms (Altamont, New York), but 100% of neonatal outbred Swiss mice (NYLAR) from the New York State Department of Health died when injected with the newly cloned aggressive and docile virus[25] (mice receiving docile virus did live twice as long—26 days—as mice injected with aggressive virus). We would predict from the studies of Riviere and colleagues[12] that the neonatal NYLAR mice produced much more interferon than the ICR mice. Using adult ICR mice, where clear differences could be seen between the newly cloned viruses, we found that the aggressive virus induced interferon and the docile did not[27] (the same relationship also held true for adult ic infected C3H and C57BL mice). Furthermore, induction of interferon by poly I:poly C in mice infected with docile virus caused them to die with the aggressive syndrome.[27] Our observations may have relevance for the entire Arenaviridae family.[28] A limited trial using poly-L-lysine stabilized poly I:poly C showed that such treatment was contraindicated in Machupo virus infected monkeys.[29] These animals not only had higher viremias but also died earlier than untreated virus infected controls. Furthermore, it would seem of interest to determine if the arenavirus Mozambique induced interferon in monkeys; and if it does not, what effect poly-L-lysine stabilized poly I:poly C would have on pathogenesis. This virus is closely related (by complement fixation and immunofluorescence serology) to the human pathogen Lassa fever virus. It does not kill rhesus monkeys but it will protect them against later challenge with Lassa.[30]

Death of mice following acute infection with LCM appears to be mediated by a population of virus specific T effector lymphocytes that possess lytic activity *in vitro* against virus infected syngeneic target cells.[7] Our initial experiments indicate that the cytolytic T-cell response in docile virus infected mice is as intense as in mice infected with aggressive virus (Jacobson and Pfau, unpublished observations). Why are the virus specific killer T-cells unable to kill mice infected with docile virus? Is interferon mediating the expression of histocompatibility antigens which are all important in effector cell-target recognition?[6,31] Could interferon be holding virus on the surface of productively infected cells thereby making them better targets?[27] Do T-cells go directly to the brain or in the presence of interferon do they recruit other cells (macrophages for example) to this area? The grand design of nature is still hidden but perhaps in these continuing studies we will see a little more fabric!

References

1. Armstrong, C. and Lillie, R.D. (1934): *Public Health Reports* 49:1019–1027.
2. Bro-Jørgensen, K. (1978): *Adv. Virus Research* 22:327–369.
3. Oldstone, M.B.A. (1975): *Prog. Med. Virol.* 19:84–119.
4. Pedersen, I.R. (1970): *Adv. Virus Research* 24:277–330.
5. Rawls, W.E. and Leung, W.C. (1970): In: *Comprehensive Virology.* Fraenkel-Conrat, H. and Wagner, R.R. (Eds.). Plenum Press, New York, pp. 157–192.
6. Zinkernagel, R.M. and Doherty, P.C. (1977): In: *Contemporary Topics in Immunobiology.* Stutman, O. (Ed.), Plenum Press, New York, pp. 179–220.
7. Cole, G.A., Prendergast, R.A., and Henney, C.S. (1973): In: *Lymphocytic Choriomeningitis Virus and Other Arenaviruses.* Lehmann-Grube, F. (Ed.), Springer Verlag, Berlin, pp. 61–71.
8. Oldstone, M.B.A. and Dixon, F.J. (1967): *Science* 158:1193–1195.
9. Oldstone, M.B.A. and Dixon, F.J. (1969): *J. Exptl. Med.* 129:483–505.
10. Riviere, Y., Gresser, I., Guillon, J.-C., and Tovey, M.G. (1977): *Proc. Nat. Acad. Sci. U.S.A.* 74:2135–2139.
11. Gresser, I., Morel-Maroger, L., Verroust, P., Riviere, Y., and Guillon, J.-C. (1978): *Proc. Nat. Acad. Sci. U.S.A.* 75:3413–3416.
12. Riviere, Y., Gresser, I., Guillon, J.-C., Bandu, M.-T., Ronco, P., Morel-Maroger, L., and Verroust, P. (1980): *J. Exptl. Med.* 152:633–640. I-A949.
13. Popescu, M. and Lehmann-Grube, F. (1977): *Virology* 77:78–83.
14. Dutko, F.J. and Pfau, C.J. (1978): *J. Gen. Virology* 38:195–208.
15. Traub, E. (1935): *Science* 81:298–299.
16. Traub, E. (1938): *J. Exptl. Med.* 68:229–250.

17. Hotchin, J., Benson, L., and Seamer, J. (1962): *Virology* 18:71–78.
18. Hotchin, J. (1972): *Monographs in Human Genetics* 6:172–181.
19. Pulkkinen, A.J. and Pfau, C.J. (1970): *Applied Microbiol.* 20:123–128.
20. Hotchin, J., Kinch, W., and Benson, L. (1971): *Infection & Immunity* 4:281–286.
21. Suzuki, S. and Hotchin, J. (1971): *J. Infectious Diseases* 123:603–610.
22. Popescu, M. and Lehmann-Grube, F. (1976): *J. Gen. Virol.* 30:113–122.
23. Jacobson, S., Dutko, F.J., and Pfau, C.J. (1979): *J. Gen. Virol.* 44:113–211.
24. Jacobson, S. and Pfau, C.J. (1980): *Nature* 283:311–313.
25. Jacobson, S. (1980): Ph.D. thesis, Rensselaer Polytechnic Institute, Troy, New York.
26. Youngner, J.S. (1977): In: *Microbiology-1977.* Schlessinger, D. (Ed.), American Society for Microbiology, Washington, D.C.
27. Jacobson, S., Friedman, R.M., and Pfau, C.J. (1981): *Nature,* submitted.
28. Pfau, C.J., Bergold, G.H., Casals, J., Johnson, K.M., Murphy, F.A., Pedersen, I.R., Rawls, W.E., Rowe, W.P., Webb, P.A., and Weissenbacher, M.C. (1974): *Intervirol.* 4:207–213.
29. Stephen, E.L., Scott, S.K., Eddy, G.A., and Levy, H.B. (1977): *Texas Reports Biol. Med.* 35:449–454.
30. Johnson, K.M., Lange, J.V., and Kiley, M.P. (1979): *Lancet* ii:738.
31. Vignaux, F. and Gresser, I. (1977): *J. Immunol.* 118:721–723.

Copyright 1981 by Elsevier North Holland, Inc.
David H. L. Bishop and Richard W. Compans, eds.
The Replication of Negative Strand Viruses

Reciprocal Patterns of Humoral and Cell-Mediated Anti-Viral Immune Responses of Mice Infected with High and Low Doses of Lymphocytic Choriomeningitis Virus (LCM)

Fritz Lehmann-Grube, Marita Varho, and Josef Cihak[a]

As a rule, mice die when inoculated intracerebrally (ic) with LCM virus. One exception is the paradoxical survival after infection with high virus doses. First described in 1936,[1] this phenomenon has been repeatedly documented,[2-4] but a satisfactory explanation cannot be given. We and others have found that the LCM virus-specific cytotoxic T-lymphocyte (CTL) activity of mice is high when low quantities of virus are inoculated but low when the infectious inoculum is increased,[5-7] indicating that reduced lethality is but one aspect of T-cell suppression.

While virus-specific T-cell suppression in itself is interesting, it became the more provoking when it was found that antibody titers attained higher levels after infection with high virus doses,[8] suggesting the possibility that some regulatory relationship exists between the two compartments of LCM virus-specific immune responses.

Materials and Methods

Female CBA/J mice, 8 to 12 weeks old, were used. The WE strain virus [9] was a clear plaque type variant;[10] E-350 ("Armstrong") had not been cloned. Virus was prepared in L-cells. Infectivity was quantitated in mice [11] and has been expressed as infectious units (IU). The ^{51}Cr-release assay for measuring the activity of CTL has been described.[7] For determination of the footpad reaction (FPR),[12] the right hind feet of mice were inoculated with 10^4 IU contained in 0.03 ml, and the degree of swelling is expressed as the ratio obtained by dividing the mean of thicknesses of the right feet by the mean of thicknesses of the left feet.

[a] Henrich-Pette-Institut für Experimentelle Virologie und Immuologie an der Universität Hamburg, Martinistrasse 52, 2000 Hamburg 20, Federal Republic of Germany.[2-4]

Results and Discussion

Mice of several strains infected by ic inoculation with the WE strain virus have exhibited high virus dose survival [224] and the same was found here for CBA/J mice (data not shown). Inasmuch as the LCM death of an acutely infected adult mouse is T-cell-mediated,[13] unexpected survival indicates that T-cell activity is impeded. Infection with high doses also resulted in depressed CTL activity (Figure 1) and all but abolished the FPR (data not shown), and we conclude that high doses of WE strain virus suppress LCM virus-specific cell-mediated immunity (CMI) in CBA/J mice.

In outbred albino mice infected ic with the E-350 strain of LCM virus, lethality had been 100% even after inoculation of high virus doses.[3] In a first experiment with CBA/J mice, 10% survived after infection with 10^7 IU of E-350, which was more than had been expected. The results of a dose-response experiment show (Figure 2) that mice were spared when inoculated with more than 10^{6-7} IU. Apparently, 10^7 is the quantity of this agent at which high virus dose survival just becomes demonstrable. The pattern of CTL activity in mice infected with E-350 differed markedly from the one

Figure 1. Cytoxic T-lymphocyte responses of CBA/J mice infected by intravenous inoculation with high (10^7 IU) or low (10^2 IU) doses of strain WE LCM virus. Data points are means of two values obtained from individual mice.

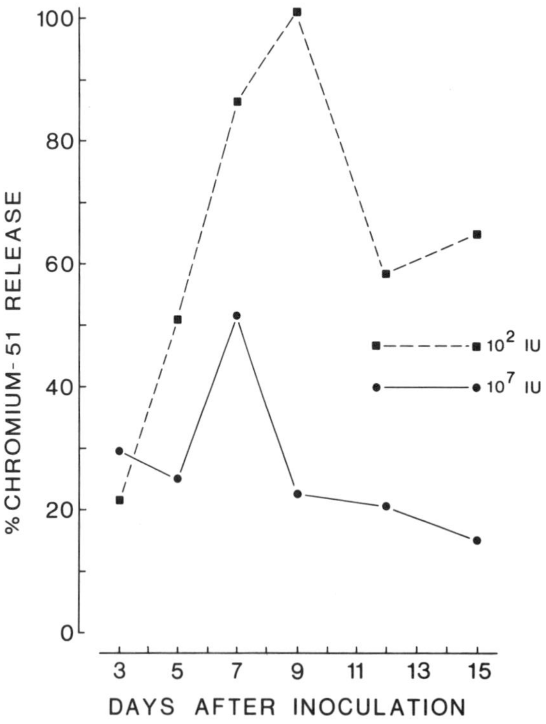

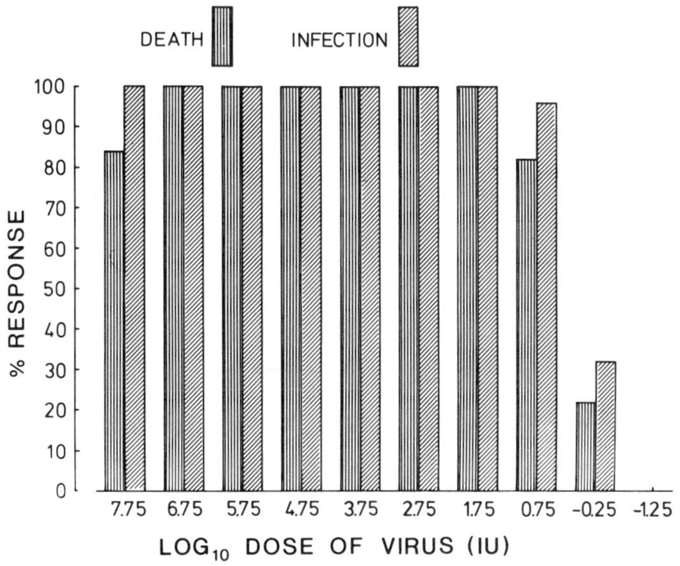

Figure 2. Dose-response relationship between CBA/J mice and strain E-350 LCM virus. Groups of 50 mice per decimal dilution step were inoculated intracerebrally and deaths were recorded between days 5 and 14. Survivors were again inoculated with a potentially lethal dose on day 28, and mice which had died after the first inoculation together with mice which had survived challenge were considered infected.

seen after infection with WE: chromium release was relatively low but was always higher after infection with 10^7 IU than with 10^2 IU (Figure 3). In contrast, the FPR closely resembled the one seen in CBA/J mice infected with WE: the parallel intravenous inoculation of 10^7 IU greatly reduced the footpad swelling caused by 10^4 IU (Figure 4). Whether these dissimilarities can be explained quantitatively or indicate true differences between the various LCM virus-specific CMI responses, we do not yet know.

Humoral responses of CBA/J mice to LCM virus have been described.[8] Both complement-fixing (CF) and neutralizing (N) antibodies attained higher titers after infection with high doses of the WE strain virus than after infection with low doses. In CBA/J mice infected with E-350 a similar pattern was detected for CF antibody, but N antibody could not be demonstrated if the inoculum had contained 10^2 IU; if 10^7 IU had been inoculated, low concentrations of N antibody were revealed more than 10 weeks later.

Mice infected with low or high doses of strain WE virus never expressed CMI when later challenged with 10^2 IU. When the second dose was raised to 10^7 IU, marked responses were seen in mice previously infected with 10^2 IU, but neither CTL activity (Figure 5) nor FPR (data not shown) could be elicited when the first inoculum had contained 10^7 IU. Surprisingly, in mice previously infected with high doses of E-350 not only the FPR was suppres-

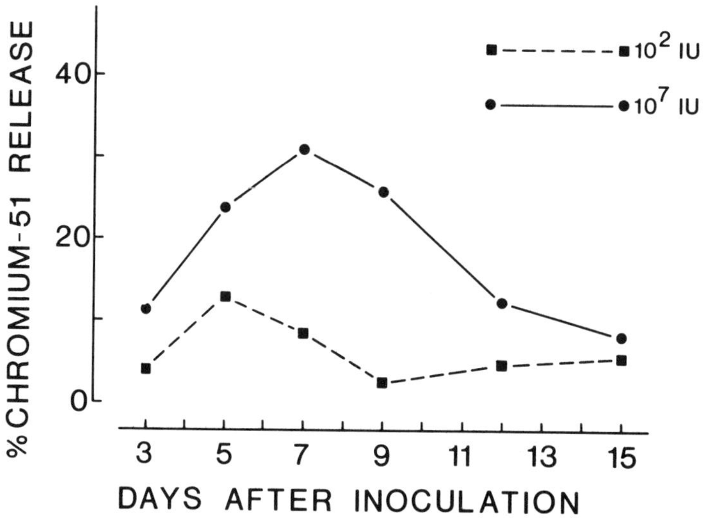

Figure 3. Cytotoxic T-lymphocyte responses of CBA/J mice infected by intravenous inoculation with 10^7 IU or 10^2 IU of strain E-350 LCM virus. Data points are means of two values obtained from individual mice.

Figure 4. Footpad reactions of CBA/J mice which had received 10^2 or 10^7 IU of strain E-350 LCM virus intravenously and immediately thereafter 10^4 IU into one hind foot. Data points are means of measurements from 15 mice.

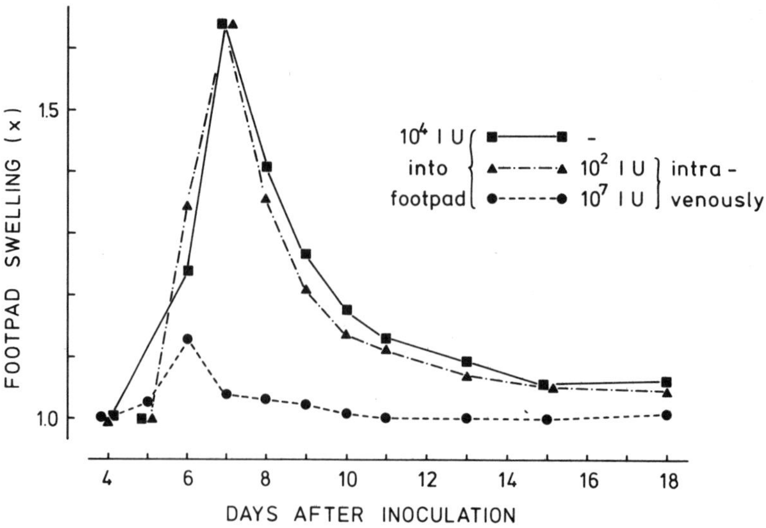

sed (Figure 6) but also CTL (Figure 7), although the latter had been shown to be higher after primary infection with 10^7 (Figure 3). When spleen cells from mice were taken at intervals up to 3 months after infection with 10^7 IU and were inoculated, together with virus, to lethally X-irradiated recipients, these developed CTL activities as fast and almost as high as recipients of spleen cells from mice previously infected with 10^2 IU (data not shown). Thus, in suppressed mice, memory CTL are present but are prevented from being activated. In further experiments suppressed mice (previously infected with 10^7 IU) received spleen cells from immune mice (previously infected with 10^2 IU): cells from immune mice were blocked in suppressed recipients but they were readily activated when transferred to normal control mice (Table 1).

While our observations show that suppression of LCM virus-specific CMI in mice by high doses of LCM virus is active and long lasting, they do not allow drawing conclusions as to the underlying mechanism, and further work

Figure 5. Secondary cytotoxic T-lymphocyte responses to the intravenous inoculation of 10^7 IU of strain WE LCM virus of CBA/J mice which had been infected by the same route 40 days previously with 10^2 or 10^7 IU.

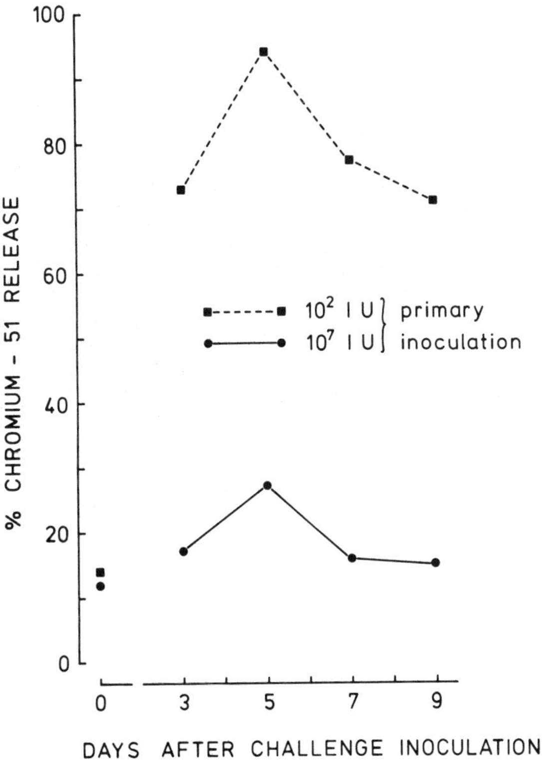

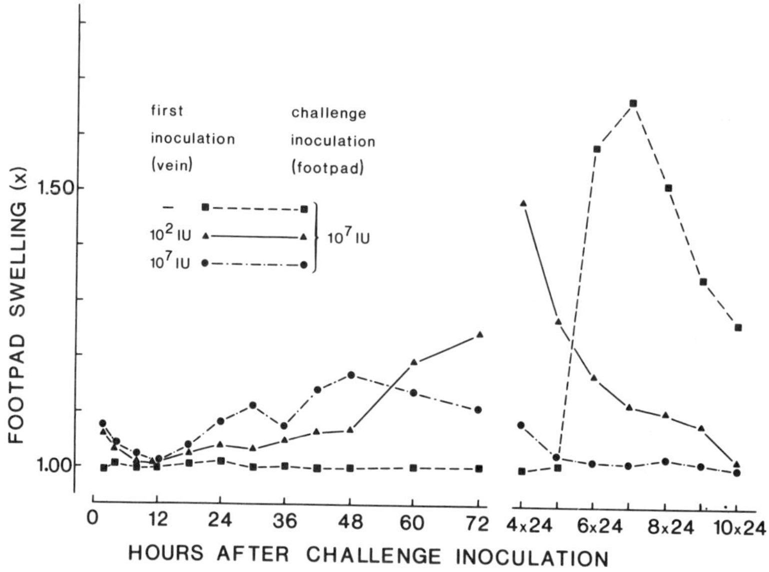

Figure 6. Secondary footpad reactions to 10^7 IU of E-350 strain LCM virus in CBA/J mice which had been infected 60 days previously be intravenous inoculation of 10^2 or 10^7 IU or had been left uninoculated. Data points are means of measurements from 15 mice.

Table 1. Effect of Transfer of Spleen Cells from Immune Mice on Cytotoxic T-Cell Response of Recipients whose Cell-Mediated Immunity had been Suppressed by Prior Infection with High Doses of WE Strain Virus.

	% ^{51}Cr-release[b]	
Recipient mice[a]	**Exp. 1**	**Exp. 2**
Normal	100	94
Suppressed	38	14

[a]Mice were infected by intravenous inoculation with 10^7 IU of WE strain LCM virus (suppressed) or were left untreated (normal). Two (Exp. 2) or 4 (Exp. 1) months later, both groups received by intravenous inoculation 5×10^7 spleen cells from mice which had been immunized 3 (Exp. 2) or 4 (Exp. 1) months previously by intravenous infection with 10^2 IU. The next day these mice were intravenously challenged with 10^7 IU of WE virus.

[b]Spleen cells from two recipient mice were pooled for the test, which was performed 5 days (Exp. 1) or 7 days (Exp. 2) after challenge.

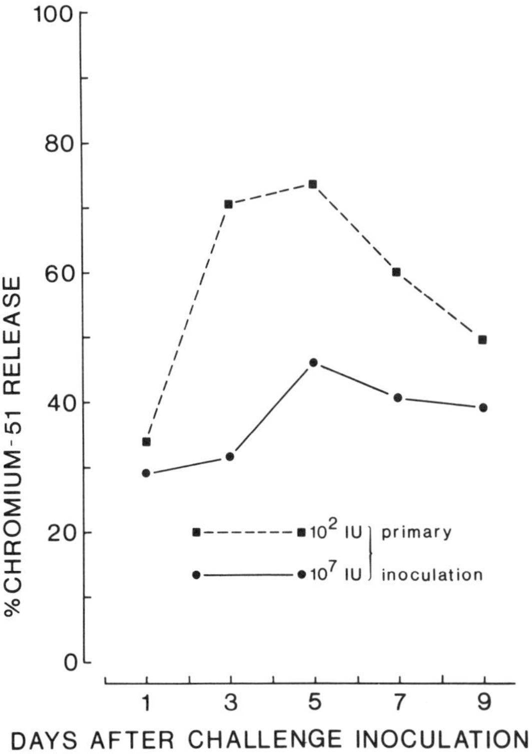

Figure 7. Secondary cytotoxic T-lymphocyte responses to the intravenous inoculation of 10^7 IU of strain E-350 LCM virus of CBA/J mice which had been infected by the same route 60 days previously with 10^2 or 10^7 IU. Data points are means of two values obtained from individual mice.

is required to reveal the roles cells and antibodies play. One explanation does not account for our findings, and that is neutralization of challenge virus by antibody.[14] Secondary responses cannot be induced in mice which had been infected before with high doses of E-350, although the level of N antibody is lower than in mice infected with low doses of WE, and in these latter animals secondary CMI responses are readily elicited.

ACKNOWLEDGMENTS
The Henrich-Pette-Institut is financially supported by Freie und Hansestadt Hamburg and Bundesministerium für Jugend, Familie und Gesundheit. This work was carried out with the help of a grant from the Deutsche Forschungsgemeinschaft.

References

1. Bengtson, I.A. and Wooley, J.G. (1936): *Public Health Rept. (Wash.)* 51:29–41.
2. Hotchin, J. and Benson, L. (1963): *J. Immunol.* 91:460–468.
3. Lehmann-Grube, F. (1969): *J. Hyg. (Camb.)* 67:269–278.
4. Suzuki, S. and Hotchin, J. (1971): *J. Infect. Dis.* 123:603–610.
5. Dohert, P.C., Zinkernagel, R.M., and Ramshaw, I.A. (1974): *J. Immunol.* 112:1548–1552.
6. Dunlop, M.B.C. and Blanden, R.V. (1977): *J. Exp. Med.* 145:1131–1143.
7. Cihak, J. and Lehmann-Grube, F. (1978): *Immunology* 34:265–275.
8. Kimmig, W. and Lehmann-Grube, F. (1979): *J. Gen. Virol.* 45:703–710.
9. Rivers, T.M. and Scott, T.F.M. (1936): *J. Exp. Med.* 63:415–432.
10. Popescu, M. and Lehmann-Grube, F. (1976): *J. Gen. Virol.* 30:113–122.
11. Lehmann-Grube, F. (1964): *Arch. Virusforsch.* 14:344–350.
12. Hotchin, J. (1962): *Virology* 17:214–216.
13. Cole, G.A. and Nathanson, N. (1974): *Progr. Med. Virol.* 18:94–110.
14. Yap, K.L. and Ada, G.L. (1979): *Scand. J. Immunol.* 10:325–332.

BUNYAVIRUSES

Copyright 1981 by Elsevier North Holland, Inc.
David H. L. Bishop and Richard W. Compans, eds.
The Replication of Negative Strand Viruses

Biochemical and Serological Comparisons of Australian Bunyaviruses Belonging to the Simbu Serogroup

D.A. McPhee and A.J. Della-Porta[a]

Several bunyaviruses belonging to the Simbu serogroup have been isolated in Australia over the past few years (see Table 1).[1,2] Akabane virus is known to cause congenital deformities in cattle, sheep and goats;[3-5] there is serological evidence implicating Aino virus as a possible causal agent of congenital deformities in calves,[6] and Tinaroo virus may be associated with congenital deformities in sheep (T.D. St. George, personal communication). Table 1 shows the possible invertebrate vectors of these viruses and their major vertebrate hosts. Thimiri virus is of interest as it differs from the other Australian Simbu viruses in having an avian host. This study was carried out primarily to determine the serological and biochemical relationships between the Australian Simbu group viruses. The results from these studies should aid in the identification of new Simbu group isolates and the interpretation of field serological results. Secondly these serological and biochemical comparisons could provide information useful in relation to their varied pathogenicity.

Serological Studies

The Simbu serogroup virus isolates in this study were principally assigned to the group by relationships demonstrated by the complement fixation (CF) test (T.D. St. George and D.H. Cybinski, personal communication).[1,2,7] In

[a] CSIRO Division of Animal Health, Australian National Animal Health Laboratory, Geelong at present located at Animal Health Research Laboratory, Private Bag No. 1, Parkville, Victoria, 3052, Australia.

Table 1. Australian Bunyaviruses of the Simbu Serogroup Used in This Study.

Virus	Isolate No.	Isolation details		Probable invertebrate host	Major vertebrate hosts	References
		Source	Location[e]			
Akabane[a]	B8935	*Culicoides brevitarsis*	QLD	*Culicoides brevitarsis*	cattle, sheep, goats	Doherty et al.[1]
Aino[b]	B7974	*Culicoides brevitarsis*	QLD	*Culicoides brevitarsis*	cattle, sheep, goats	Doherty et al.[1]
Tinaroo[c]	CSIR0153	*Culicoides brevitarsis*	QLD	*Culicoides brevitarsis*	cattle, sheep, goats	St. George et al.[2]
Peaton[d]	CSIR0110	*Culicoides brevitarsis*	QLD	*Culicoides brevitarsis*	cattle, sheep, goats	St. George et al.[2]
Douglas	CSIR0150	Cattle blood	NT	?	cattle, sheep, goats	St. George et al.[2]
Facey's Paddock	CSIR010	*Culicoides spp.*	NT	*Culicoides austropalpalis*	indigenous Australian fauna	Unpublished[f]
Thimiri	CSIR01	*Culicoides histrio*	NT	*Culicoides histrio*	migratory birds	Unpublished[f]

[a]Causes congenital deformities in cattle, sheep and goats.

[b]Serological evidence as the cause of congenital deformities in cattle.

[c]Suspected from limited serological evidence of causing congenital deformities in sheep.

[d]Serological evidence suggests no association with congenital deformities in cattle or sheep. (T.D. St George, personal communication).

[e]State from which original Australian isolate was made, Queensland (QLD) and Northern Territory (NT).

[f]H.A. Standfast, T.D. St. George and D.H. Cybinski, personal communication.

order to compare group-relationships, we used the agar gel diffusion precipitin (AGDP) test. There were significant cross-reactions between Akabane, Aino, Tinaroo and Peaton viruses (Table 2). Tinaroo virus, as antigen, appears most cross-reactive, reacting with most antisera in this test. Antiserum against Aino virus reacts more broadly. Facey's Paddock and Thimiri viruses show little cross-reactivity with the other viruses or their antisera. There are also some interesting one-way cross-reactions. Most notable are the ones between Akabane and Aino, and between Aino and Douglas viruses. These results suggest that there should be some interesting relationships between the N proteins of these viruses (the antigen probably associated with the AGDP test) [1] and that there could also be problems in interpreting field results using the AGDP test. It should be noted that the cross-reactivity of cattle or sheep sera in the AGDP test might be different to the hyperimmune mouse ascites fluids used in this test.

Cross-neutralization, using the microtitre neutralization test,[8] with the seven Australian Simbu serogroup viruses have shown them all to be distinct (T.D. St. George and D.H. Cybinski, personal communication).[2] Using a more sensitive serum neutralization test, a plaque-inhibition test,[9] we have found the seven Simbu viruses to be easily distinguishable with the exception of a few cross-reactions (Table 3). Tinaroo virus was found to be very closely related to Akabane virus. Tinaroo antiserum could not distinguish between these two viruses but there was a significant difference seen when Akabane antiserum was used. Tinaroo virus, as an antigen, showed low level cross-reactions detected by all antisera expect Thimiri virus which appeared to be distinct from all viruses in this study. It is interesting to speculate that Tinaroo virus surface antigens may represent more senior antigenic determinants, whereas the other members may have evolved from it to have modified junior determinants, as postulated by Fazekas de St. Groth for influenza virus.[10]

Biochemical Comparisons

For all bunyaviruses studied thus far virus particles have been shown to contain three major structural proteins, two external glycoproteins G1 and G2, and an internal nucleoprotein N.[7] A high molecular weight but minor protein L has also been observed for some of the viruses studied.[7] Analyses of purified virion proteins for Akabane (B8935), Aino (B7974), Tinaroo (CSIR0153), and Peaton (CSIR0110) viruses, labeled with ^{3}H-leucine and/or ^{35}S-methionine, in 10% Laemmli slab gels, revealed all members to have typical L, G1, G2 and N profiles; they were of similar relative amounts and molecular weights to those for the prototype species Bunyamwera virus (Figure 1). The molecular weights were determined using ^{14}C-methylated protein markers electrophoresed in the same slab gel (Figure 1 and Table 4). We have not completed virion protein analyses for all Australian Simbu serogroup viruses due to inadequate virus titres for some members.

Table 2. Agar Gel Diffusion Precipitin (AGDP) Test Comparison of Australian Simbu Group Viruses.

Antigen[a] in test	AGDP reaction[b] for antisera[c] against						
	Akabane	**Aino**	**Tinaroo**	**Peaton**	**Douglas**	**Facey's Paddock**	**Thimiri**
Akabane	3+	3+	3+	3+	—	—	—
Aino	?	3+	2+	3+	—	—	—
Tinaroo	2+	3+	3+	2+	1+	?	?
Peaton	—	3+	2+	3+	—	—	—
Douglas	—	3+	3+	—	3+	—	—
Facey's Paddock	—	—	—	—	—	3+	—
Thimiri	—	—	—	—	?	?	3+

[a]Antigen prepared from clarified infectious culture fluids harvested at 3 to 4+ CPE from virus infected BHK121 cells in roller bottles. The antigen was concentrated using an Amicon (Lexington, Mass.) model 402 stirred cell with a PM10 membrane.

[b]The AGDP test was performed after the method of Littlejohns (I.R. Littlejohns, personal communication) with antigen in the center well surrounded by 6 wells, each alternate well containing reference antiserum and the other wells the test sera. The reading was as follows:

— no significant turn in reference line (negative).
? an equivocal turn in the end of the reference line (doubtful).
1 a definite turn on the end of the reference line (weak positive).
2 a continuous line, showing identity with the reference line (positive).
3 equivalent to the reference line (strong positive).

[c]Hyperimmune mouse ascites fluids prepared against mouse brain grown virus.

Table 3. Serum Neutralization Plaque-Inhibition (PI) Test[a] Comparison of Australian Simbu Group Viruses.

Virus in test	PI zones (mean ± SE in mm)[b] for antisera against						
	Akabane	**Aino**	**Tinaroo**	**Peaton**	**Douglas**	**Facey's Paddock**	**Thimiri**
Akabane	*20.5 ± 1*	0	18.5 ± 1.3	0	0	0	4.7 ± 5.5
Aino	0	*19.7 ± 1.5*	0	0	13.5 ± 1.3	0	0
Tinaroo	13.7 ± 0.6	11.2 ± 1.3	*20.7 ± 1*	11.7 ± 2.4	7.2 ± 4	10 ± 0	0
Peaton	0	0	0	*24.7 ± 0.5*	0	0	0
Douglas	0	0	0	0	*26 ± 0.8*	0	0
Facey's Paddock	0	0	0	0	0	*22.7 ± 1.5*	0
Thimiri	0	0	0	0	0	0	*23.2 ± 1*

[a]The PI test was performed using the method described by Della-Porta et al.[9] Briefly, confluent monolayers of SVP cells in 90 mm petri dishes were infected with 50,000 PFU of virus and then overlaid with nutrient agar overlay. 7 mm sterile blotting paper discs were soaked in the serum under test and then placed on top of the agar overlay (6 discs per plate). The discs were removed and the plates stained, 3 days after infection, with 5 ml of nutrient agar containing 1/10,000 neutral red.

[b]Mean of diameters of zones measured on two separate tests, with duplicate samples per test (total fo 4 zones of inhibition). Antisera were hyperimmune mouse ascites fluids produced against mouse brain grown virus.

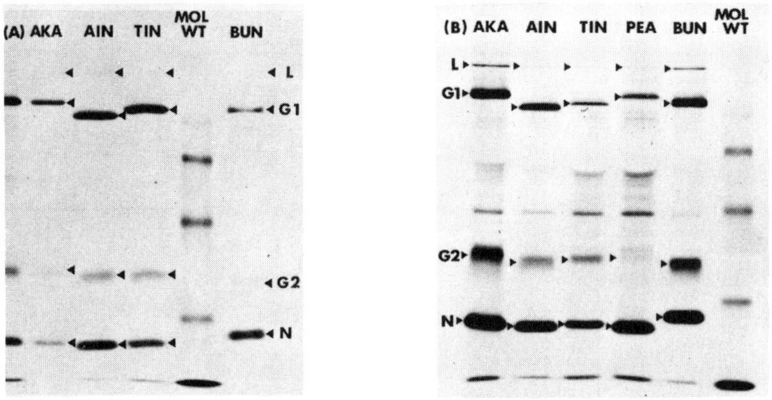

Figure 1. Definition by electrophoresis in 10% Laemmli[7] slab gels of the proteins of purified Akabane B8935 (AKA), Aino B7974 (AIN), Tinaroo CSIR0153 (TIN), Peaton CSIR0110 (PEA) and Bunyamwera (BUN) viruses. Each virus was grown in BHK-21/13 cells in the presence of ^{3}H-leucine (A) and/or ^{35}S-methionine (B) and twice purified in sucrose gradients. The proteins are designated L, G1, G2 and N according to published bunyavirus studies. Their molecular weights were estimated by reference to co-migration in 10% Laemmli slab gels of the following ^{14}C-methylated protein markers (MOL WT) (mol wt in parentheses: myosin (200,000), phosphorylase a and b (100,000 and 92,500), bovine serum albumin (69,000), ovalbumin (46,000) and carbonic anhydrase (30,000).

Table 4. Molecular Weights of Australian Simbu Group Virus Proteins.[a]

	Molecular weight (mean ± SE × 10^{-3})[b]							
Virus	**L**	**G1**	**G2**	**ns1**	**ns2**	**N**	**ns3**	**ns4**
Akabane	121±2	101±3	37±1	ND[c]	27	24±1	20±1	20±1
Aino	122±2	92±3	36±1	ND	ND	24±1	ND	ND
Tinaroo	121±2	98±0	37±1	31	28	24±0	20	19
Peaton	123	101±2	38±1	33	ND	24±1	21	20
Douglas	ND	ND	ND	ND	ND	ND	ND	ND
Facey's Paddock	123	102	ND	ND	ND	ND	ND	ND
Thimiri	123	101±2	37	ND	ND	24	20	19
Bunyamwera	120±0	98±0	36±1	ND	ND	26±0	ND	ND

[a]Molecular weights were determined by comparison with the following standards in 10% Laemmli gels: phosphorylase a and b (100 and 92.5 × 10^3), bovine serum albumin (69 × 10^3), ovalbumin (46 × 10^3) and carbonic anhydrase (30 × 10^3). The migration of the standards versus the log of their molecular weights were subjected to linear regression analysis and molecular weights were subjected to linear regression analysis and molecular weights of the virus proteins calculated by fitting to the curve. Analysis revealed that myosin (200 × 10^3) did not fit on this curve and was therefore omitted as a molecular weight standard.

[b]Values are for 2 to 5 determinations, except where the standard error (SE) is omitted; then only one determination was made.

[c]ND, not determined due to lack of definition of protein band.

Host cell protein synthesis, relative to the virus-specified proteins, was switched off during infection in cells infected with the Simbu serogroup viruses at a multiplicity of infection of 1 PFU/cell (Figure 2). In analyses of the proteins synthesized in Vero cells infected with Akabane, Aino, Tinaroo, Peaton, Thimiri and Facey's Paddock viruses we were able to identify the virus-specified proteins G1, G2, N and possibly L for all virus infected cells except for Facey's Paccock virus. Douglas virus infected cells were not analyzed as we were unable to obtain satisfactory titres of virus. The

Figure 2. Electrophoresis in a 10% Laemmli[7] slab gel of ^{3}H-leucine (A) and/or ^{35}S-methionine (B) labeled proteins synthesized in mock-infected (MOCK), Akabane B8935 (AKA), Aino B7974 (AIN), Tinaroo CSIR0153 (TIN), Peaton CSIR0110 (PEA), Facey's Paddock CSIR010 (FP) Thimiri CSIR01 (THI) and Bunyamwera (BUN)-infected Vero cells. Cells were labeled with ^{3}H-leucine and/or ^{35}S-methionine between 24–28, 36–40, 48–52 or 60–64 hr postinfection depending on when host cell protein synthesis was considered to be switched off.

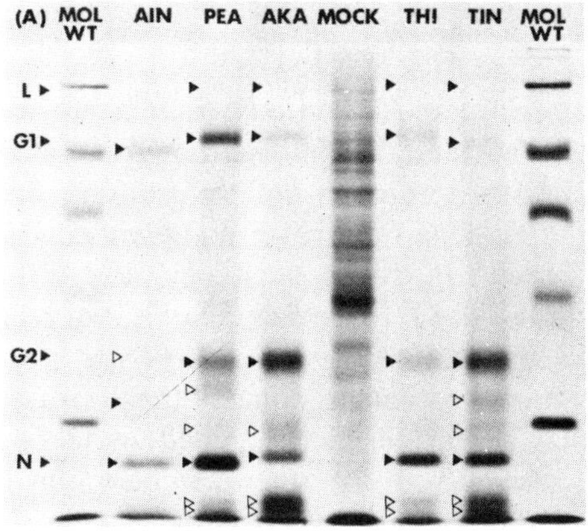

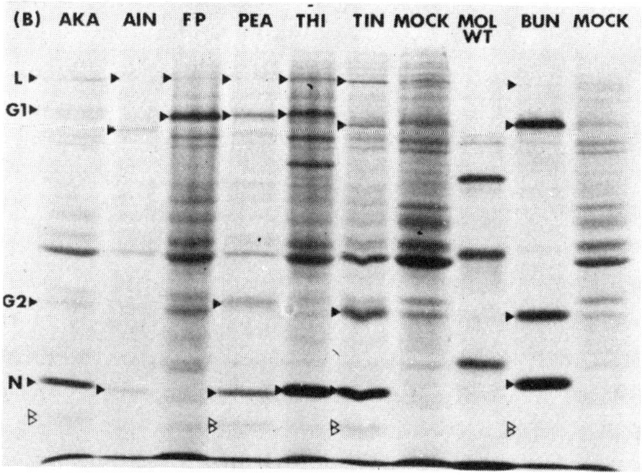

molecular weights, calculated relative to the molecular weight markers, are presented in Table 4. The molecular weight of L may be low, as myosin (220 $\times$ 10 ^{3} daltons) did not migrate in proportion to its log molecular weight in 10% Laemmli gels.

Virus-specified protein analyses for bunyavirus members are very limited with results being mainly for the prototype species Bunyamwera virus.[7,11,12] The three major structural proteins G1, G2 and N were readily identified in Bunyamwera virus infected cells with L possibly being detected in small amounts.[11,12] No non-structural proteins have been reported for any bunyavirus studied thus far.[7] In addition to the four structural proteins seen in purified virus preparations, several other proteins, migrating between G2 and N and below N, were detected in Akabane, Tinaroo and Peaton virus infected cells (Figure 2 and Table 4). Comparison of these profiles with those of mock-infected cells suggests that they may be virus-specified, but this needs to be confirmed by further biochemical studies.

Conclusions

The serological comparisons (Tables 2 and 3) of the Australian Simbu serogroup members indicate many interesting cross-reactions. The group test (AGDP; Table 2) results suggest that Akabane, Aino, Tinaroo and Peaton are related (although not identical), probably through their N protein.[7] The neutralization test (Table 3) suggests that Tinaroo virus has antigens on its surface which cross-react significantly with antisera against a number of the Simbu serogroup viruses. Further studies are required to define these relationships in more detail, but they do suggest that there may be some problems in serologically diagnosing previous infections by some of these viruses in animals. It is interesting to note that Thimiri virus, associated with a bird host cycle (Table 1), appeared very distinct from the other members; this is consistent with the observations of Bishop and Shope.[7]

In our analyses of the virion and virus-specified proteins of the Australian Simbu serogroup viruses we found all, except Douglas and Facey's Paddock, to be typical bunyaviruses. However, comparative slab gel analyses clearly shows the small differences in the migration of the major structural proteins G1, G2 and N. The molecular weight of the N protein for Aino and Peaton was slightly lower than for Akabane, Tinaroo and Thimiri viruses. The other obvious differences were in the migration of G1 and G2. G1 and G2 for Tinaroo and Aino were lower in molecular weight than for the other viruses. The molecular weight of G2 for all these viruses (about 37 $\times$ 10^3 daltons) was not found to be as high (75 $\times$ 10^3 daltons) as reported by D.W. Trent (cited in Obijeski and Murphy)[13] for Simbu virus. The role of the extra possible non-structural proteins seen in virus infected cells in unknown. If they are precursors, then their identity should be established by tryptic mapping and translation studies of mRNA. If they are not precursors, and still virus-specified, then the coding of the three RNA segments would extent beyond L, G1, G2 and N, as at present envisaged.[7,13]

References

1. Doherty, R.L., Carley, J.G., Standfast, H.A., Dyce, A.L., and Snowdon, W.A. (1972) *Aust. Vet. J.* 48:81–86.
2. St. George, T.D., Cybinski, D.H., Filippich, C., and Carley, J.G. (1979): *Aust. J. Exp. Biol. Med. Sci.* 57:581–582.
3. Omori, T., Inaba, Y., Kurogi, H., Miura, Y., Nobuto, K., Ohashi, Y., and Matsumoto, M. (1974): *Bull. Off. Int. Epizoot.* 81:447–458.
4. Parsonson, I.M., Della-Porta, A.J., and Snowdon, W.A. (1977): *Infect. Immun.* 15:254–262.
5. Porterfield, J.S. and Della-Porta, A.J. (1981): In: *Comparative Diagnosis of Viral Diseases, Volume 3.* Kurstak, E. and Kurstak, C. (Eds.), Academic Press, New York, in press.
6. Coverdale, O.R., Cybinski, D.H., and St. George, T.D. (1978): *Aust. Vet. J.* 54:151–152.
7. Bishop, D.H.L. and Shope, R.E. (1979): In: *Comprehensive Virology, Volume 14.* Fraenkel-Conrat, H. and Wagner, R.R. (Eds.), Plenum Press, New York, pp. 1–156.
8. Cybinski, D.H., St. George, T.D., and Paull, N.I. (1978): *Aust. Vet. J.* 54:1–3.
9. Della-Porta, A.J., Herniman, K.A.J., and Sellers, R.F. (1980): *Vet. Microbiol.,* in press.
10. Fazekas de St. Groth, S. (1969): *J. Immunol.* 193:1107–1115.
11. Lazdins, I. and Holmes, I.H. (1979): *J. Gen. Virol.* 44:123–133.
12. Pennington, T.H., Pringle, C.R., and McCrae, M.A. (1977): *J. Virol.* 24:397–400.
13. Obijeski, J.F. and Murphy, F.A. (1977); *J. Gen. Virol.* 37:1–14.
14. Laemmli, U.K. (1970): *Nature (Lond.)* 227:680–685.

David H. L. Bishop and Richard W. Compans, eds.
The Replication of Negative Strand Viruses

Biochemical Studies of Australian Akabane Virus Isolates and a Possible Structural Relationship to Virulence

A.J. Della-Porta,[a] I.M. Parsonson,[a] D.A. McPhee,[a] W.A. Snowdon,[a] H.A. Standfast,[b] T.D. St. George,[b] and D.H. Cybinski[b]

Akabane virus, a member of the Simbu serogroup in the family Bunyaviridae has been associated with congenital deformities in cattle, sheep, and goats.[1,4] The deformities most often observed are hydranencephaly (with complete or partial replacement of the cerebrum with fluid), arthrogryposis (the limbs locked in a fixed position) and porencephaly (with fluid filled cavities in the cerebrum). In 1972–1974, over 40,000 calves were lost in Japan from Akabane disease.[3] In Australia, during an epizootic in 1974, it is estimated that in excess of 5,000 calves were lost.[5] In Australia, the location of the major outbreaks of Akabane disease has been in southeastern New South Wales and the epizootics have been linked with the southward extension of the probably vector, *Culicoides brevitarsis,* to areas where there are susceptible pregnant animals.[1]

There is some data suggesting that there may be strain differences between isolates of Akabane virus. Examination of the pathogenicity of a number of isolates of Akabane virus for mice has shown that there was a significant difference between many of the isolates.[6] Comparison of the field data from the Japanese with data from the Australian epizootics of Akabane disease indicates that there may have been a higher level of abortions and stillbirths in Japan than in Australia.[3,5] This study was undertaken to determine whether Akabane virus isolates, made from diverse geographic locations, from a number of different species and over a 12-year-span (Table 1,

[a] CSIRO Division of Animal Health, Australian National Animal Health Laboratory, Geelong, at present located at Animal Health Research Laboratory, Private Bag No. 1, Parkville, Victoria 3052, Australia.
[b] CSIRO Division of Animal Health, Long Pocket Laboratories, Private Bag No. 3, Indooroopilly, Queensland, 4068, Australia.

Table 1. Details of the Australian Isolates of Akabane Virus Used in This Study.

Isolate No.	Ref.[a] No.	Isolation details: Source	Location[d]	Year	Reference
B8935	C1Q	*Culicoides brevitarsis*	Amberley, Qld.	1968	Doherty et al.[10]
R7949	C2Q	*C. brevitarsis*	Belmont, Qld.	1968	Doherty et al.[10]
CSIR016	C3Qa[b] C3Qb[c]	*C. brevitarsis*	Kairi, Qld.	1975	St. George et al.[11]
CSIR0240	C4Q	*C. brevitarsis*	Peachester, Qld.	1976	St. George et al.[11]
CSIR0241	C5N	*C. brevitarsis*	Hunter Valley, NSW	1978	Unpublished
CSIR0501	O1N	ovine fetus (D385)	Badgery's Creek, NSW	1976	Della-Porta et al.[2]
CSIR0502	O2N	ovine fetus (D410)	Badgery's Creek, NSW	1976	Della-Porta et al.[2]
CSIR0217	B1Q	bovine blood	Peachester, Qld.	1976	St. George et al.[11]
CSIR0223	B2Q	bovine blood	Peachester, Qld.	1976	St. George et al.[11]
CSIR0233	B3N	bovine blood	Tamworth, NSW	1976	Unpublished
CISR0235	B4T	bovine blood	Victoria River, NT	1979	Unpublished
CSIR0236	B5T	bovine blood	Douglas/Daly, NT	1979	Unpublished

[a]The reference number refers to the source (C is *C. brevitarsis*; O is ovine fetus; B is bovine blood), an arbitrary number for that isolate and location (Q is Queensland; N is New South Wales; T is Northern Territory) respectively.

[b]Tissue cultured passaged virus, pool B, figure 3.

[c]Insect passaged virus, pool C, figure 3.

[d]See Figure 1.

Figure 1), would show any obvious biochemical differences. Further, the original Australian isolate used in pathogenesis studies (B8935)[4] was compared with a more recent isolate to determine whether isolate or passage history might affect the pathogenicity.

Australian Akabane Virus Isolates—Comparison of Cytoplasmic Proteins

The virus-specified cytoplasmic proteins of 12 Akabane virus isolates (Table 1) were examined (Figure 2). No gross differences in the migration of any of the virus-specified proteins could be seen. In many cases complete switch off of the host cell protein synthesis was not achieved (possibly because most of the virus seeds had not been plaque-purified and there may have been a high level of defective-interfering particles in the seed stocks). However, the virus-specified protein patterns could be identified. Unlike El Said et al.,[7] we could identify differences in protein patterns of bunyaviruses in the same serogroup,[8] probably because of the use of slab gels which allow greater resolution than slicing cylindrical gels. It is possible

Figure 1. Map of Australia indicating locations of sites (⊡) where Akabane isolations were made (Table 1). Solid line indicates the normal southern limit of the distribution of the probably vector for Akabane virus, *Culicoides brevitarsis.*[1]

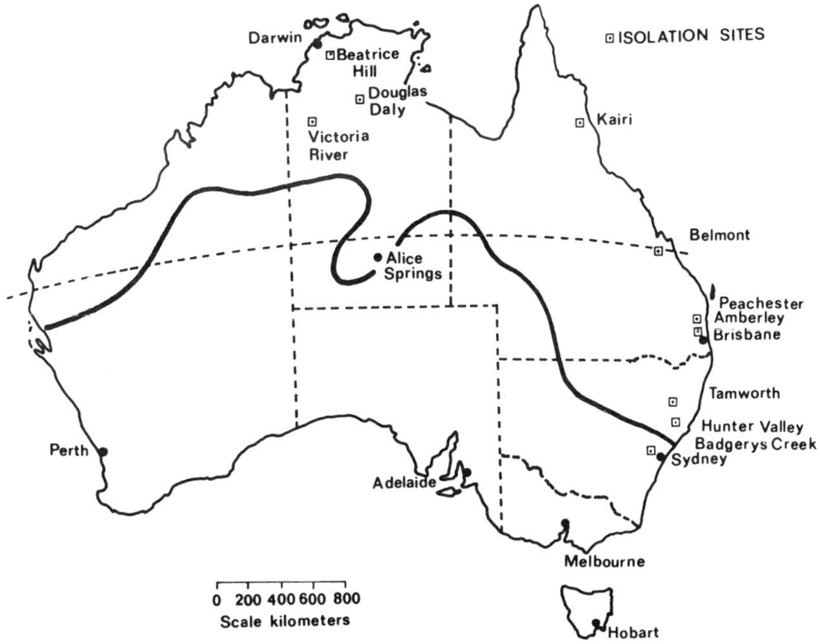

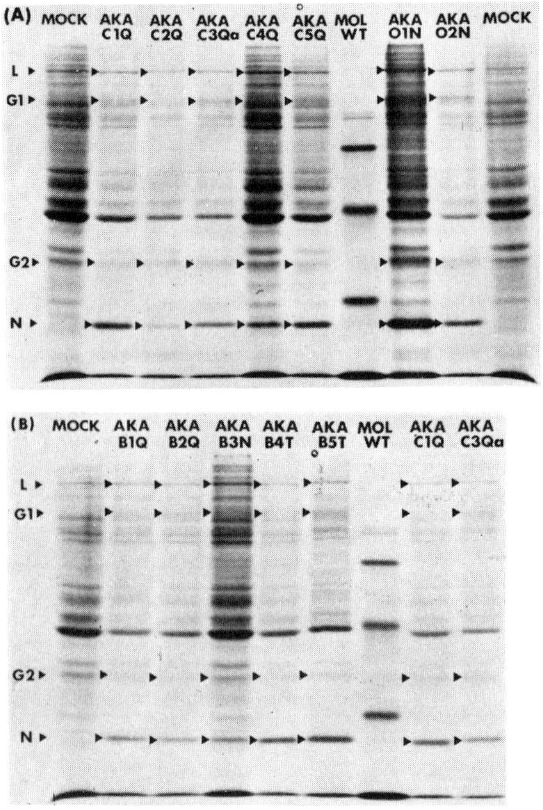

Figure 2. Virus-specified protein profiles of Akabane virus isolates, coded as described in Table 1. Vero cells were infected with virus at a multiplicity of infection of about 1 PFU/cell. The virus-specified proteins were labeled with ^{35}S-methionine (SJ.204, The Radiochemical Centre, Amersham, Bucks) between 36 and 40 or 60 to 64 hours postinfection, 3 μg/ml actinomycin D being added 2 hours before labeling and left there during the labeling period. The cells were solubilized and the protein dissociated in Laemmli dissociation buffer, boiled for 1 min and layered onto 10% Laemmli gels.[12] The same amount of radioactivity was loaded onto each well with the sample volumes being made equivalent by the addition of a cell extract. Gels were fixed and treated for fluorography.[13] Molecular weights for the virus proteins and virus-specified proteins were calculated and are designed as designated by McPhee and Della-Porta.[8] (A) *Culicoides brevitarsis,* C, and ovine, O, isolates. (B) Bovine, B, isolates.

that examination of oligonucleotide maps of the RNA species may have revealed differences between these isolates, as was reported for 12 isolates of La Crosse virus.[7]

Pathogenicity Studies—Effect of Isolate and Passage History

Pathogenicity Studies

Two isolates of Akabane virus (B8935 and CSIRO16) made from *C. brevitarsis* and receiving different passage histories (Figure 3) were examined for their ability to produce congenital deformities in ovine fetuses. The Akabane virus (B8935) passage level was identical to that used by Parsonson et al.[4] who found that they could experimentally produce deformed offspring in about 30% of the lambs, when about 10^6 to $10^{7.4}$ MLD_{50} doses of virus were inoculated intravenously into pregnant ewes at 30 to 36 days of gestation. In the experiments reported in this paper, less virus (about $10^{3.5}$) was used to infect ewes at the same stage of pregnancy. The reduced level of infectious virus, and the intravenous inoculation route, was used in order to better approximate the natural insect vector-vertebrate host situation.

The results of this study (Table 2) showed that although all ewes became infected (as observed by serological conversion to Akabane virus), only fetuses in group C, where the virus had received mosquito passages, showed congenital deformities. Virus was isolated from the placenta and fetal membranes, with all these fetuses in this group but from only 2 fetuses in group A and 1 in group B. No virus was isolated from any of the fetal tissues, probably because the ovine fetus produces neutralizing antibodies against Akabane virus at about 70 days gestation and fetuses were examined at 90 days gestation. The placentome and fetal membranes appear to be immunologically privileged sites isolated both from the maternal and from the fetal humoral immune system (I.M. Parsonson, A.J. Della-Porta, M.L. O'Halloran, W.A. Snowdon, K.J. Fahey and H.A. Standfast, *Vet. Microbiol.*, submitted for publication).

Figure 3. Passage histories of Akabane virus (isolates B8935 and CSIR016, Table 1) pools used in the pathogenesis studies.

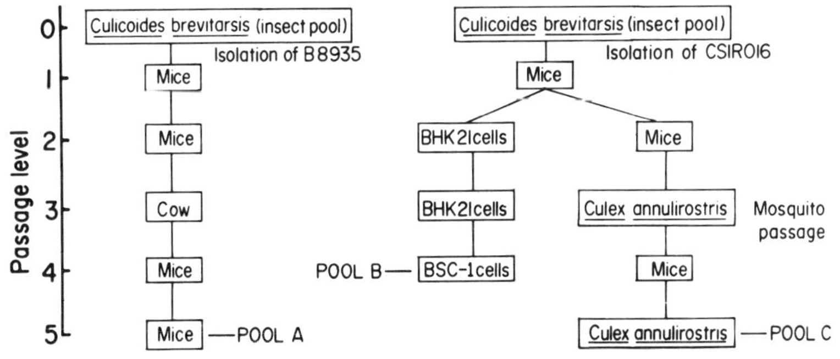

Table 2. Pathogenicity Studies in Pregnant Sheep with Different Isolates of Akabane Virus[a] with Different Passage Histories.

Virus group[b]	Designation	Fetus No.	Virus isolated	Gross pathology: Details	Group results: No.	%
A	B8935/M2/	TG77F1	—	—		
	BOVI/M2	TG96F1	—	—		
		SA38F1	—	—		
		SA40F1	—	—		
		SA49F1	—	—		
		SA81F1	—	—		
		SA98F1	+P, FM[c]	—		
		SA98F2	+P,FM			
					0/7	0
B	CSIR016/M1/	SB3F1	—	—		
	BHK2/BSC1	SB3F2	+P,FM	—		
		SB12F1	—	—		
		SB54F1	—	—		
		TG26F1	—	—		
		TG30F1	—	—		
					0/6	0
C	CSIR016/M2/	SB1F1	+P,FM	+PE[d]		
	MOS1/M1/	SB34F1	+P,FM	+AG/HE		
	MOS1	SB42F1	+P,FM	+HE		
		SB42F2	+P,FM	+AG/HE		
		SB45F1	+P,FM	+HE		
		SB45F2	+P,FM	+HE		
					6/6	100

[a]Ewes were infected intravenously with about $10^{3.5}$ MLD_{50} of virus at 30-36 days of pregnancy and the fetuses examined at 90 days of pregnancy.

[b]See figure 3..

[c]Virus only isolated from the placetome (P) and fetal membranes (FM) but not from other fetal tissues.

[d]Gross pathology of the fetuses included porencephaly (PE), arthrogryposis (AG) and hydranencephaly (HE).

Biochemical Studies

In order to determine whether there were any obvious biochemical differences of the viruses in the pools used in the pathogenicity study (Figure 3, Table 2), plaque-purified clones were prepared. These were initially compared by analysis of the virus-specified cytoplasmic proteins (Figure 4). All clones of Akabane (B8935) and Akabane (CSIR016), except one derived from the mosquito passaged material (pool C) had similar cytoplasmic protein profiles. One clone from the mosquito pool had an apparently lower molecular weight G2 glycoprotein than any other Akabane virus isolate (Figures 2 and 4). Although plaque-purified three times, this clone did not

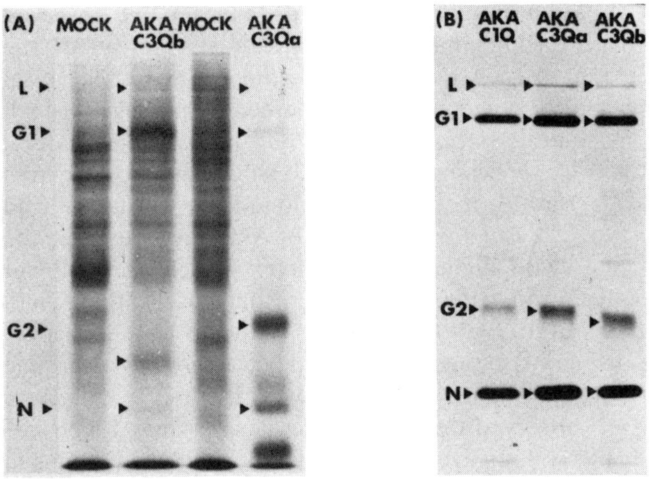

Figure 4. (A) Virus-specified protein profiles of clones of Akabane/B8935 (CIQ) (Table 1) and Akabane/CSIR016 (C3Qa and C3Qb) derived from the pools used in the pathogenicity study (Table 2; Figure 2). The proteins were labeled with ^{35}S-methionine and analyzed as described for Figure 3.
(B) Virus proteins of isolates in (A), labeled with ^{35}S-methionine, twice purfied by ultracentrifugation in sucrose density gradients and analyzed in 10% Laemmli gels.[8] Designation of virus proteins is as for Figure 2.

produce significant switch-off of host cell protein synthesis, relative to virus-specified protein synthesis late during infection. Typical bunyavirus virus-specified protein synthesis kinetics are the appearance of N first, with G1 and G2 seen later in large amounts, when host cell protein synthesis switch-off is evident. For this clone G1 was detected in large amounts relative to G2 and N.

Cloned virus from these three pools was further examined by labeling the virion proteins. This confirmed the difference of G2 in the clone from the mosquito passaged material (Figure 4). This virus was checked by serum neutralization tests and was confirmed to be an Akabane virus isolate. The proteins of a large number of clones of Akabane virus from a number of isolates have been examined (including other clones from the mosquito passaged material) and they have not shown any variation in migration of the virus-specified cytoplasmic proteins.

Conclusions

In this study, we observed that Akabane virus isolates from very diverse origins had virus-specified cytoplasmic protein profiles that could not be differentiated. These results suggested that the proteins of Akabane virus do not show significant variation in migration patterns. Further, we observed that alternating passage through insect/vertebrate hosts can affect the pathogenicity of Akabane virus. A more detailed study of this effect is required to confirm it as a general phenomenon. Whether the passage through an invertebrate, the mosquito *Culex annulirostris,* not the suspected natural midge (gnat) vector *Culicoides brevitarsis,* resulted in modification of the virus biochemically or a selection of a more virulent clone from within the wild population, is not known. The isolation of a clone of Akabane virus from the mosquito passaged pool (with higher pathogenicity) which had a lower molecular weight G2 glycoprotein was a very interesting observation, especially in view of the stability of the protein patterns noted earlier. Pathogenicity studies using this cloned virus would be required to determine whether it is more pathogenic than other clones with the higher molecular weight G2. Nagai et al.[9] observed that Newcastle disease virus which had both glycoproteins, F and HN, cleaved was more pathogenic than strains in which one or none of these proteins were cleaved. Whether G2 is cleaved in Akabane virus and whether this is a more general occurrence is not known.

The observations reported in this study present many questions concerning the nature of virulence in Akabane virus. They are far from conclusive but do emphasize the need to look in detail at vertebrate/invertebrate host effects on arboviruses and the likely importance of virion surface proteins in pathogenicity. Further detailed studies are required to clarify the factors involved in virulence of Akabane virus.

References

1. Della-Porta, A.J., Murray, M.D., and Cybinski, D.H. (1976): *Aust. Vet. J.* 52:496–502.
2. Della-Porta, A.J., O'Halloran, M.L., Parsonson, I.M., Snowdon, W.A., Murray, M.D., Hartley, W.J., and Haughey, K.J. (1977): *Aust. Vet. J.* 53:51–52.
3. Omori, T., Inaba, Y., Kurogi, H., Miura, Y., Nobuto, K., Ohashi, Y., and Matsumoto, M. (1974): *Bull. Off. Int. Epizoot.* 81:447–458.
4. Parsonson, I.M., Della-Porta, A.J., and Snowdon, W.A. (1977): *Infect. Immun.* 15:254–262.
5. Shepherd, N.C., Gee, C.D., Jessep, T., Timmins, G., Carroll, S.N., and Bonner, R.B. (1978): *Aust. Vet. J.* 54:171–177.
6. Kurogi, H., Inaba, Y., Takahashi, E.,Sato,K., Akashi, H., Satoda, K., and Omori, T. (1978): *Natl. Inst. Anim. Health Q. (Tokyo)* 18:1–7.
7. El Said, L.H., Vorndam, V., Gentsch, J.R., Clewley, J.P., Calisher, C.H., Kilmas, R.A., Thompson, W.H., Grayson, M., Trent, D.W., and Bishop, D.H.L. (1979): *Am. J. Trop. Med. Hyg.* 28:364–386.
8. McPhee, D.A. and Della-Porta, A.J. (1981): This volume.
9. Nagai, Y., Klenk, H.D., and Rott, R. (1976): *Virology* 72:494–508.
10. Doherty, R.L., Carley, J.G., Standfast, H.A., Dyce, A.L., and Snowdon, W.A. (1972): *Aust. Vet. J.* 48:81–86.
11. St. George, T.D., Standfast, H.A., and Cybinski, D.H. (1978): *Aust. Vet. J* 54:558–561.
12. Laemmli, U.K. (1970): *Nature (Lond.)* 227:680–685.
13. Bonner, W.M. and Laskey, R.A. (1974): *Eur. J. Biochem.* 46:83–88.

David H. L. Bishop and Richard W. Compans, eds.
The Replication of Negative Strand Viruses

The Effects of Proteolytic Enzymes on Structure and Function of La Crosse G1 and G2 Glycoproteins

Laura Kingsford[a] and Douglas W. Hill[b]

La Crosse (LAC) virus is a member of the California serogroup of viruses belonging to the family Bunyaviridae.[1] It is a negative strand virus composed ot three segments of single strand RNA,[2,3] four proteins (L, N, G1 and G2), and an envelope.[4] Two of the viral-coded proteins (G1 and G2) are embedded in the liquid bilayer of the envelope and are glycosylated.[4]

Studies with other viruses indicate that the glycoproteins in the viral envelope interact with host cell surfaces in the initial steps of virus replication, i.e., attachment.[5] When reacted with specific antibody made to these surface glycoproteins, the virus is rendered non-infectious (neutralized).[6] By contrast, little is known about the structure and function of LAC glycoproteins. Antibody made to the medium RNA gene products (G1 and G2) does have neutralizing capacity when reacted with LAC virions.[7] However, it is not known if antibody specific for each one of the glycoproteins alone can still neutralize the virus.

Reported here are a few experiments designed to learn more about the structure of G1 and G2 using proteolytic enzymes, to determine whether one or both of these glycoproteins are involved with viral-cell interaction, and whether specific antibody to each one will result in neutralization of the virion.

Effects of Trypsin and Chymotrypsin on LAC Glycoprotein Structure

As an initial experiment to determine what effect proteolytic enzymes might have on the infectivity and structure of LAC virus, ^{35}S-methionine-labeled

[a] Department of Microbiology, California State University, Long Beach, California.
[b] Department of Cellular, Viral and Molecular Biology, University of Utah School of Medicine, Salt Lake City, Utah.

virions were incubated with 10 μg/ml of trypsin or 20 μg/ml of chymotrypsin for 90 minutes at 35°C. This protease treatment resulted in a consistent loss of one log of viral infectivity.

Untreated and protease treated viruses were pelleted and electrophoresed in 10% reducing, SDS-polyacrylamide gels (Figure.1). LAC virus has four viral proteins: L (170,000 daltons), G1 (120,000 daltons), G2 (39,000 daltons) and N (23,000 daltons). With trypsin treatment, the G1 disappeared and new polypeptides of 95,000 and 67,000 daltons were apparent. The same was observed for virus treated with chymotrypsin, except that the cleavage products were 100,000 and 70,000 daltons. The other structural proteins, L, N and G2, were not affected by these two proteases. It thus apeared that tryspin and chymotrypsin cleaved only the G1 glycoprotein *in situ* and this cleavage resulted in two smaller components.

Effects of Protease Treatment on Neutralization Kinetics of LAC Virus

When antibody was prepared in rabbits by injecting G1 or G2 cut from polyacrylamide gels, only the antiserum containing antibody to G1 resulted

Figure 1. Polyacrylamide gel electrophoresis of LAC virions after treatment with trypsin or chymotrypsin. LAC virus was grown in baby hamster kidney cells and radiolabeled by the addition of 8–10 μCi of ^{35}S-methionine per milliliter of methionine-free Joklik medium. After purification by sucrose gradient centrifugation, virus was incubated with 10 μg/ml trypsin or 20 μg/ml chymotrypsin, pelleted and then electrophoresed in a 10%, reducing SDS-polyacrylamide gel. The autoradiograph shown is of a gel containing (a) structural proteins of LAC virions or cleavage products of virus incubated with (b) trypsin or (c) chymotrypsin.

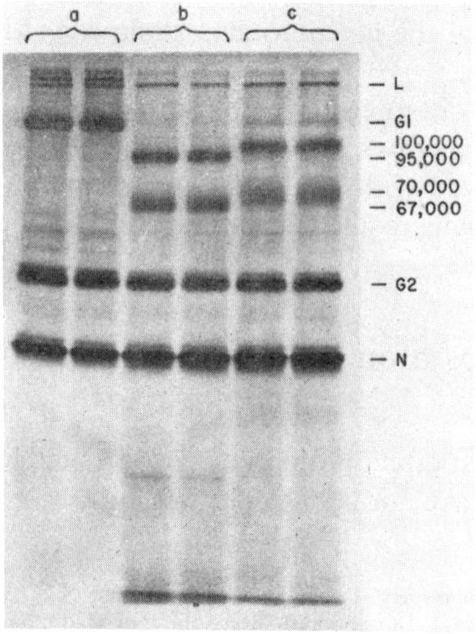

in neutralization of the virus. Because protease treatment removed a portion of G1 but left a significant number of infectious virions, it was desirable to know how this would effect the neutralization of LAC virus using specific antibody.

To insure that all enzyme could be removed from the virus so that it would not interfere with the neutralization reaction, ^{3}H-acetylated trypsin was incubated with the LAC virus. The enzyme-virus mixture was layered onto a sucrose gradient and centrifuged to equilibrium. All of the ^{3}H-trypsin remained at the top of the gradient while the virus banded at its density. Analysis of this sample indicated that G1 had been cleaved as usual.

Normal or protease treated LAC virus from peak fractions of the sucrose gradients were mixed with appropriate dilutions of antisera and incubated at 37°C. Samples were taken at time intervals and assayed for residual infectivity. Results shown in Figure 2 indicate that both trypsin and chymotrypsin treatment of LAC virus greatly decreased the rate of neutralization resulting in a much larger persistent fraction.

Trypsin inactivated with phenylmethylsulfonyl fluoride (PMSF) was used as a control to determine if proteolytic cleavage of G1 was directly related to the decreased neutralization kinetics. An equivalent of 10 μg/ml of PMSF-

Figure 2. Neutralization kinetics of LAC virus preincubated with trypsin, phenylmethylsulfonyl fluoride (PMSF) inactivated trypsin or chymotrypsin. Virus was grown and purified as in Figure 1 and incubated for 90 minutes at 35°C with 10 μg/ml trypsin (▲——▲), 20 μg/ml chymotrypsin (○——○), or 10 μg/ml of PMSF-trypsin (equal in activity to 7.5 × 10 $^{-5}$ units/ml) (△——△). After addition of 10% calf serum, the virus was layered onto 20–65% sucrose gradients and centrifuged to equilibrium. Virus from peak fractions were assayed for infectivity and 5 × 10^6 PFU's were mixed with an equal voluem of diluted anti-LAC antiserum for kinetics of neutralization studies. LAC virus that had not been incubated with proteases was used in the neutralization reaction as a control (●——●).

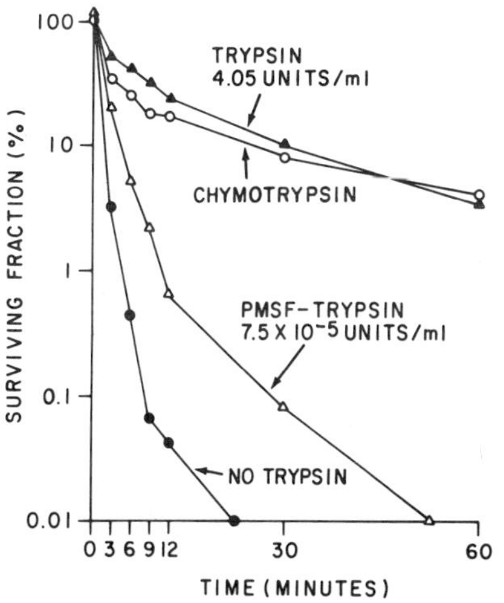

trypsin contained 7.5×10^{-5} units/ml as compared with 4.0 units/ml of untreated trypsin. When virus was treated with the PMSF-trypsin and then electrophoresed in polyacrylamide gels, some of the G1 glyoprotein still remained intact, although cleavage products of 100,000, 95,000 and 75,000 daltons could be seen in the gels. When incubated with PMSF-trypsin prior to admixture with antiserum, LAC virus was neutralized more like control virus (see Figure 2). It was observed that any time some G1 glycoprotein was left intact, the amount of virus neutralized was greater. This indicates that the degree of neutralization of LAC virus is directly related to the amount of intact G1 present in the envelope.

Proteolytic Cleavage of LAC Virus by Other Proteases

To determine what effect other proteolytic enzymes might have on LAC glycoproteins, ^{35}S-methionine-labeled virus was incubated with various proteases and electrophoresed in SDS-polyacrlamide gels. Virus was also assayed for reduction in infectivity. The results fall into three groups as shown in Table 1. Bromelain and pronase completely digested G1 rendering the virions non-infectious. Thermolysin, plasminogen activator, and streptokinase had no visible proteolytic effect on the virus glycoproteins or infectivity, although these enzymes were enzymatically active as determined by appropriate assays. Plasminogen, likewise, had no effect. However, plasmin and trypsin cleaved G1 to give two polypeptides of 95,000 and 67,000 daltons. The proteolytic treatment with trypsin, chymotrypsin and plasmin resulted in relatively little loss of viral infectivity. These three enzymes did not reduce infectivity by more than one log, but did decrease the subsequent neutralization by antibody.

The G2 glycoprotein was resistant to degradation by all proteolytic enzymes except bromelain. Of interest is the observation that pronase completely cleaved G1 but left G2 intact. Yet the infectivity of LAC was completely destroyed by pronase. This indicates that G2 alone is not capable of attaching LAC virus to cell receptors and implicates G1 in that function.

Discussion

The studies with the proteolytic enzymes raise questions about how G1 and G2 are oriented in the lipid membrane. The sum of the 67,000 and 95,000 trypsin cleavage products is 175,000 daltons, much larger than that of G1. The molecular weight of these two cleavage products were the same when electrophoresed in reducing or non-reducing gels indicating they are not disulfide linked. Thus, cleavage of G1 does not appear to be analogous to the cleavage of glycoproteins of viruses such as influenza,[8] Sendai,[9] etc. In addition, the two chymotrypsin digestion products were both larger than the trypsin cleaved polypeptides. These data indicate that cleavage of two dif-

Table 1. Effects of Proteolytic Enzymes on La Crosse Virus.

	La Crosse glycoproteins cleaved by protease[b]		Molecular weights of G1 cleavage products[b]	Reduction in infectivity of 10^8 PFU's	Effect on neutralization kinetics
Enzyme[a]	**G1**	**G2**			
Streptokinase	—	—	—[c]	None	None
Plasminogen Activator	—	—	—[c]	None	None
Thermolysin	—	—	—[c]	None	None
Trypsin	+	—	95,000 & 67,000	1 log	Less inactivation[d]
Plasmin	+	—	95,000 & 67,000	1 log	Less inactivation[d]
Chymotrypsin	+	—	100,000 & 70,000	1 log	Less inactivation[d]
Pronase	+	—	—[e]	>6 logs	—[f]
Bromelain	+	+	—[e]	>6 logs	—[f]

[a]LAC virus was incubated 90 minutes at 35°C with 500 units/ml streptokinase, 500 units/ml plasminogen activator, 10 μg/ml thermolysin, 10 μg/ml trypsin, 20 μg/ml chymotrypsin, 10 μg/ml bromelain, 10 μg/ml pronase, or 1:2 with plasmin prepared by mixing 1.1 ml of plasminogen (0.1 mg/ml) with 1.1 ml of streptokinase (1000 units/ml in 0.01 percent gelatin in saline).

[b]As determined by polyacrylamide gel electrophoresis.

[c]No cleavage products produced.

[d]Approximately 1.5 logs of protease treated virus neutralized in 15 minutes as compared with greater than 4 logs of control virus.

[e]G1 completely degraded.

[f]Neutralization could not be done.

ferent G1 molecules may be occurring. G2 is relatively resistant to the proteolytic enzymes which degrade G1 indicating that susceptible sites are not readily available. Perhaps an interaction between G1 and G2 not only protects G2 from enzymatic action but orients G1 in such a way that one of two tryptin (or chymotrypsin) susceptible sites are exposed on different G1 molecules.

The G2 glycoprotein alone does not seem to be required for virus-cell interaction, because virions with only G2 intact are non-infectious. In addition, antibody made to G2 does not result in neutralization of normal LAC virus. The observation that all surface glycoproteins are not involved in antibody neutralization has been described for other viruses such as Sindbis.[10]

The data presented indicate that the G1 glycoprotein is the surface component of the virion involved in attachment to cells and in antibody neutralization. When part of the G1 glycoprotein was removed by proteolytic cleavage, only limited neutralization occurred. One interpretation of this observation is that LAC virus requires the binding of at least two antibody molecules to each G1 glycoprotein. If this is correct, G1 would have at least two antibody binding sites or domains. When one of these antibody-binding sites is removed by proteases, antibody bound to the remaining site of the G1 polypeptide does not completely block virus-cell interaction. Thus, a critical site in terms of neutralization is on that part of the G1 molecule which can be removed by trypsin, chymotrypsin or plasmin.

References

1. Porterfield, J.S., Casals, J., Chumakov, M.P., Gaidamovich, S.Ya., Hannoun, C., Holmes, I.H., Horzinek, M.C., Mussgay, M., Oker-Blom, N., and Russell, P.K. (1975/1976): *Intervirology* 6:13–24.
2. Obijeski, J.F., Bishop, D.H.L., Palmer, E.L., and Murphy, F.A. (1976): *J. Virol.* 20:664–675.
3. Clewley, J., Gentsch, J., and Bishop, D.H.L. (1977): *J.Virol.* 22:459–468.
4. Obijeski, J., Bishop, D.H.L., Murphy, F.A., and Palmer, E.L. (1976): *J. Virol.* 19:985–997.
5. Fries, E. and Helenius, A. (1979): *Eur. J. Biochem.* 97:213–220.
6. Kelley, J.M., Emerson, S.U., and Wagner, R.R. (1972): *J. Virol.* 10:1231–1235.
7. Gentsch, J.R., Rozhon, E.J., Klimas, R.A., El Said, L.H., Shope, R.E., and Bishop, D.H.L. (1980): *Virology* 102:190–204.
8. Skehel, J.J. and Waterfield, M.D. (1975): *Proc. Nat. Acad. Sci. U.S.A.* 72:93–97.
9. Homma, M. and Ohuchi, M. (1973): *J. Virol* 12:1457–1465.
10. Dalrymple, J.M., Schlesinger, S., and Russell, P.K. (1976): *Virology* 69:93–103.

Copyright 1981 by Elsevier North Holland, Inc.
David H. L. Bishop and Richard W. Compans, eds.
The Replication of Negative Strand Viruses

In Vitro Translation of Uukuniemi Virus-Specific RNAs

Ralf F. Pettersson, Ismo Ulmanen, Esa Kuismanen and Päivi Seppälä[a]

Introduction

The bunyaviruses belong to the negative-strand viruses [1] and possess a segmented single-stranded RNA genome. Three segments have been identified from several members belonging to the California encephalitis complex serogroup,[1] as well as from Uukuniemi virus,[2] which is serologically unrelated to the members of the *Bunyavirus* genus. Each of the segments contains a unique nucleotide sequence and thus encodes different polypeptides.[2,3] In case of Uukuniemi virus, the RNA segments have been estimated to be 2.4 $\times$ 10^6 (L), 1.1 $\times$ 10^6 (M) and 0.5 $\times$ 10^6 (S) daltons in size.[2] Four structural proteins have been identified in purified virions of several bunyaviruses: [1] two membrane glycoproteins, G1 and G2; a nucleocapsid protein, N; and a minor large nucleocapsid-associated protein, L. The size of these polypeptides varies from one virus to another. In the case of Uukuniemi virus, the apparent molecular weights of the proteins are 180,000–200,000 (L) (Pettersson, unpublished results), 75,000 (G1), 65,000 (G2), and 25,000 (N).[4] Thus, there are four structural proteins to be encoded by only three RNA segments. One of the segments must therefore code for more than one protein.

To study the coding potential of the RNA segments, we have isolated from infected chick embryo cells the Uukuniemi virus-specific RNAs and translated them *in vitro* in a cell-free reticulocyte extract. An mRNA cosediment-

[a]Department of Virology, University of Helsinki, Haartmaninkatu 3, SF-00290, Helsinki 29, Finland.

ing with the virion M RNA directed the synthesis of a novel 110,000-dalton polypeptide, p110. This polypeptide was cleaved to G1/G2 during translation, when dog pancreas microsomes were added to the lysate. An mRNA(s) sedimenting at 12S and transcribed from the virion S RNA[5] encoded the N protein and a novel 30,000-dalton protein (NS), which may represent a nonstructural protein.

Isolation of Virus-Specific RNAs from Infected Cells

Total Cytoplasmic RNA

To isolate cytoplasmic virus-specific RNAs, chick embryo cells were infected with Uukuniemi virus. The RNA was labeled with ^{3}H-uridine at 4 to 8 hours after infection in the presence of actinomycin D, and the RNA fractionated on a sucrose gradient. Four peaks of radioactivity, sedimenting at about 29S, 23S, 17S and 12S were recovered, in addition to radioactivity left at the top of the gradient (Figure 1). No radioactivity corresponding to

Figure 1. Fractionation of Uukuniemi virus-specific RNAs by sucrose gradient centrifugation. Uukuniemi virus-specific RNAs, labeled with ^{3}H-uridine in the presence of actinomycin D were isolated from infected chick embryo cells at 8 hr postinfection. The cytoplasmic extract was analyzed on a 15–30% sucrose gradient (25.000 rpm, 17 hr, 23°C, SW27 rotor). An extract from mock-infected cells was analyzed on a separate gradient. The positions of Uukuniemi virion RNA segments run in a separate gradient are indicated as L, M and Sl. Peak fractions were pooled as indicated (I–IV) and used in subsequent translation assays. (●) RNA from infected cells; (○) RNA from uninfected cells.
SOURCE: Ulmanen et al., J. Virol., 37:72–79.

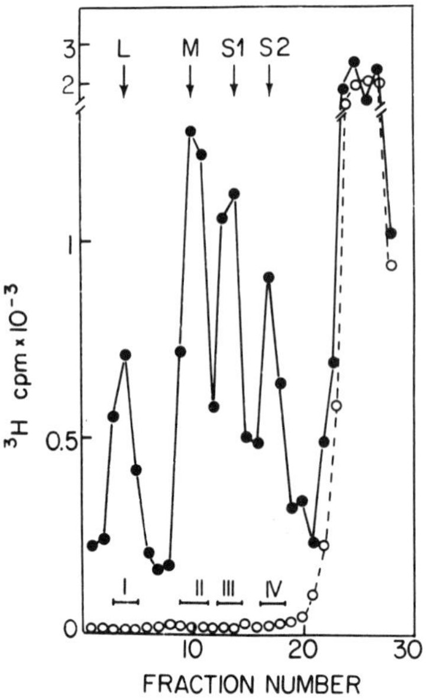

the four peaks was found from mock-infected cells. The virion L, M and S RNAs cosedimented with the three fastest sedimenting RNA species. We call the four RNA species L, M, S1 and S2. The S2 RNA was not found in virions. To enrich for mRNA species, each RNA was passed over an oligo (dT)-cellulose column. Under conditions where 90% of a poly(A)-containing Semliki Forest virus defective interfering RNA bound to the column [6] less than 5% of each of the Uukuniemi virus-specifics RNAs bound, suggesting that they do not contain a poly(A) tract long enough to allow binding to the column. Oligo(dT)-cellulose could therefore not be used to purify Uukuniemi virus mRNAs. Instead, all *in vitro* translation studies described below were done with total cytoplasmic RNA fractionated on sucrose gradients.

Isolation of Polysomal RNA

To study which of the four RNA species were associated with polyribosomes, a cytoplasmic extract was prepared in the absence of EDTA, and the polysomes collected onto a 60% sucrose cushion through a 15-40% sucrose gradient (Figure 2A). The radioactive material close to the bottom of the gradient was considered to represent virus-specific polysomes, since it was absent from extracts prepared from mock-infected cells or infected extracts treated with EDTA (Figure 2A). In both untreated and EDTA-treated extracts, three peaks of radioactivity sedimenting at 140S, 105S and 85S were found. These peaks were shown to contain RNA species cosedimenting with the virion L, M and S RNA species, respectively, and thus they probably represent the virion ribonucleoproteins.[4] The polysomes were extracted with phenol and the RNA analyzed on a sucrose gradient. Three peaks, sedimenting at 29S, 23S and 12S were recovered (Figure 2B). The RNA sedimenting at 17S (virion S RNA) was absent from polysomes. We thus conclude that there are at least three Uukuniemi virus-specific mRNAs, two of which correspond in size to the virion L and M RNAs, and one which is considerably smaller than the virion S RNA. We have recently shown by nucleic acid hybridization that the S2 mRNA is transcribed from the virion S segment.[5]

In Vitro Translation of Virus-Specific mRNAs

The peak fractions of each RNA species (Figure 1) were pooled individually (infected pools I-IV) as were corresponding fractions from mock-infected cells (mock-infected pools I-IV). The RNA species were then incubated in a cell-free micrococcal nuclease-treated reticulocyte lysate[7] and the products analyzed by SDS-polyacrylamide gel electrophoresis before and after immunoprecipitation with virion antiserum (Figure 3A-D). The L RNA (infected pool I) did not direct the synthesis of any detectable virus-specific product; the pattern of polypeptides made from infected and mock-infected pool I RNA was very similar (Figure 3A, lanes a and b) and no polypeptides were precipitated with virion antiserum (lane c).

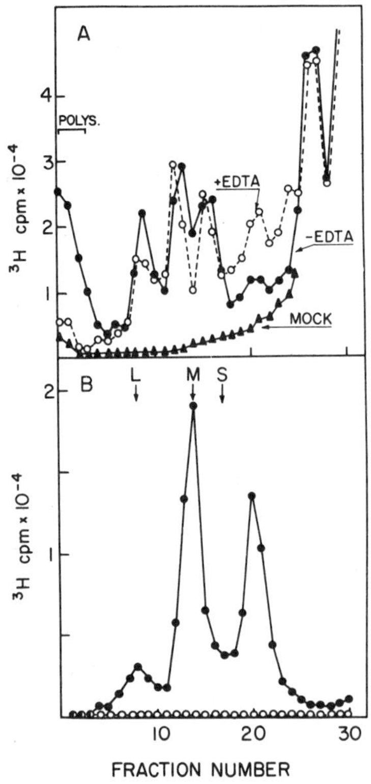

Figure 2. Identification of Uukuniemi virus-specific mRNAs associated with polyribosomes. (A) A cytoplasmic extract was prepared in the absence of EDTA from cells infected with Uukuniemi virus and layered on a 15–40% sucrose gradient containing at the bottom a 60% sucrose cushion. The polysomes were collected onto the cushion by centrifuging for 16 hr at 21,000 rpm and 4°C. Mock-infected (▲) and EDTA-treated (○) extracts were analyzed on separate gradients. (B) The RNA in the polysomes from infected (●) and mock-infected (○) cells was extracted with phenol and fractionated on a 15–30° sucrose gradient.

The RNA from infected pool II (M) directed the synthesis of a polypeptide with an apparent molecular weight of 110,000 (p110) (Figure 3B, lane b, arrow), which was immunoprecipitated with virion antiserum (lane c, arrow). p110 was absent from lysates programmed with the corresponding mock-infected RNA (lanes a and d). Therefore p110 is a virus-specific polypeptide. It has been shown that microsomal membranes isolated from dog pancreas are able to process glycoprotein precursors *in vitro* during translation.[8,9] To study whether p110 could be processed during translation, infected pool II RNA was incubated in a reticulocyte lysate together with dog pancreas microsomes (kindly prepared by Dr. B. Dobberstein and obtained from Dr. S. Kvist). The product after such an incubation was im-

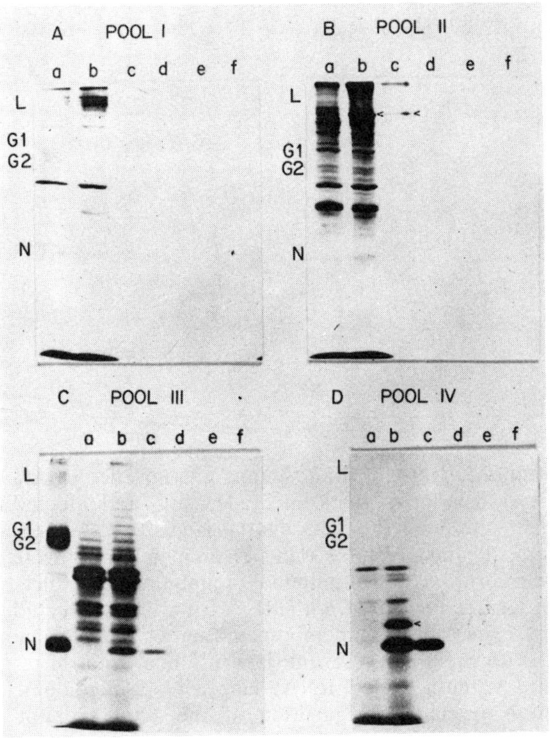

Figure 3. Cell-free translation of Uukuniemi virus-specific RNAs isolated from infected and mock-infected cells. Pool I–IV RNAs (panels A–D) from Figure 1 were translated in cell-free micrococcal nuclease-treated reticulocyte lysates. Translation of RNA from mock-infected cells (lane a) and immunoprecipitation of the products with antiserum (lane e) and preimmune serum (lane f). Translation of RNA from infected cells (lane b) and precipitation with antiserum (lane c) and preimmune serum (lane d). The positions of the virion protein markers are indicated on the left of each panel. Arrows in panel B (lanes b and c) indicate the position of the 110,000-dalton polypeptide (p110) and in panel D (lane b) that of the 30,000-dalton polypeptide (NS).
SOURCE: Ulmanen et al., J. Virol., 37:72–79.

munoprecipitated with virion antiserum and analyzed by SDS-gel electrophoresis. As shown in Figure 4, p110 was present in the lysate incubated without microsomes (lanes c and d, arrow). In the presence of microsomes, p110 was, however, absent. Instead an immunoprecipitable band migrating between the G1/G2 markers was found (lane e, arrow). We interpret the disappearance of p110 and the appearance of the faster migrating band to mean that p110 is cleaved by the microsomes roughly in the middle to yield G1/G2.

The infected pool III (S1) RNA did not direct the synthesis of any major virus-specific polypeptide, although small amounts of an immunoprecipita-

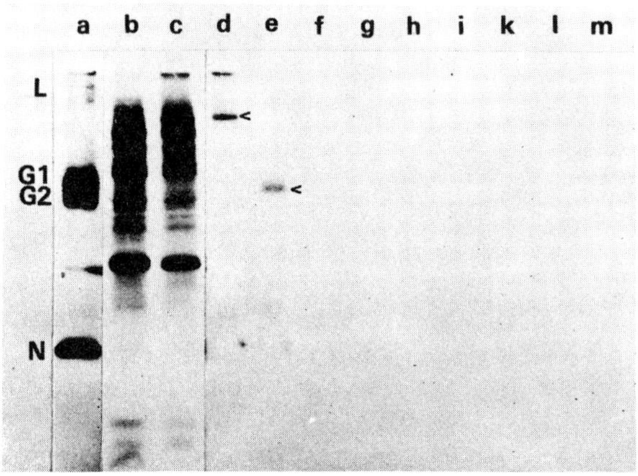

Figure 4. Cell-free translation of pool II RNA in the presence and absence of dog pancreas microsomal membranes. Translation of mock-infected (lane b) and infected (lane c) RNA in the absence of membranes. Products synthesized by RNA from infected cells in the absence (lane d) and presence (lane e) of membranes were immunoprecipitated with antivirion serum. Control immunoprecipitations: products precipitated with antiserum from lysates incubated with RNA from mock-infected cells in the absence (lane f) and presence (lane g) of membranes. Lanes h–l, the same as in lanes d–g, but precipitated with preimmune serum; lane m, incubation of reticulocyte lysate with membranes without added RNA; lane a, ^{35}S-methionine labeled Uukuniemi virion. Arrows indicate the position of the 110,000-dalton polypeptide (p110) (lane d), and its cleavage products (lane e).

SOURCE: Ulmanen et al., J. Virol., 37:72–79.

ble protein co-migrating with the N protein was seen (Figure 3C, lanes b and c). This protein was probably made from a contaminating mRNA present in pool IV, since infected pool IV RNA (S2) directed the synthesis of large amounts of the N protein (Figure 3D, lanes b and c). In addition, a polypeptide with an apparent molecular weight of about 30,000 was made from infected pool IV RNA (lane b, arrow), but not from mock-infected pool IV RNA (lane a). This polypeptide was not immunoprecipitated with virion antiserum (lane c). We therefore consider it to represent a non-structural protein and have designated it NS. This polypeptide has also been found in extracts from infected cells.[5]

Discussion

We report here the isolation, fractionation and *in vitro* translation of Uukuniemi virus-specific mRNAs from infected cells. Two of the four RNA species recovered directed the synthesis of virus-specific polypeptides; an mRNA cosedimenting with the virion M RNA (23S) was translated into a 110,000-dalton polypeptide (p110) and one (or two) 12S mRNA species directed the synthesis of the virion N protein and a novel 30,000-dalton (NS) protein not found in virions

None of the RNA pools from infected cells directed the synthesis of the L protein (180–200,000 daltons), the tentative RNA polymerase. The reason for this is unclear, but could be due to too small amounts of the mRNA or to degradation of the mRNA in the lysate or during the isolation procedure. Another possibility is that the mRNA and virion RNA could have formed hybrids, thus inactivating the mRNA as a template for translation.

p110 was considered to be virus-specific since it was immunoprecipitated with virion antiserum and was absent from lysates programmed with mock-infected RNA. The fact that p110 was cleaved in the presence of dog pancreas microsomes during translation to a polypeptide(s) co-migrating with G1/G2 suggested that p110 is a precursor to the glycoproteins. Similar *in vitro* processing by microsomes has also been described for other viral glycoproteins.[8,9] Final proof for the precursor-product relationship has to await comparison of tryptic peptides or partial proteolysis. We conclude from the results that the glycoproteins are encoded by the virion M RNA. This is in agreement with the results obtained by Gentsch and Bishop who showed that the M RNA of snowshoe hare virus also encodes both glycoproteins.[10]

The S2 (12S) mRNA gave rise to two proteins: the N (25K) and NS (30K). Since the N protein was efficiently immunoprecipitated with virion antiserum, whereas the NS was not precipitated at all, it is unlikely that NS is a precursor of N. The S2 mRNA must therefore consist of two separate mRNA species, unless one assumes that N and NS are translated from the same mRNA in different reading frames. We have recently shown that the S2 RNA specifically hybridizes to the virion S RNA segment.[5] There is barely enough genetic information in the S segment (0.5×10^6 daltons) to encode both N and NS. It will be of interest to see what the exact structure of the N and NS mRNAs is. It has recently been shown that the 8th segment of influenza virus is transcribed into two mRNA species, one of which is generated by a splicing event.[11] The two mRNAs are translated partially in different reading frames to give rise to the NS1 and NS2 proteins. The possibility that a similar situation might exist in the case of Uukuniemi virus S RNA segment should be kept in mind.

ACKNOWLEDGMENTS

We wish to thank Ms Tuula Rusi for her skillful technical assistance. This study was supported by the Sigrid Jusélius Foundation.

References

1. Obijeski, J.F. and Murphy, F.A. (1977): *J. Gen. Virol.* 37:1–14.
2. Pettersson, R.F., Hewlett, M.J., Baltimore, D., and Coffin, J.M. (1977): *Cell* 11:51–63.
3. Clewley, J., Gentsch, J., and Bishop, D.H.L. (1977): *J. Virol.* 22:459–468.

4. von Bonsdorff, C.-H. and Pettersson, R.F. (1975): *J. Virol.* 16:1296–1307.
5. Ulmanen, I., Seppälä, P., and Pettersson, R.F. (1981): *J. Virol.*, Jan. 1981, 37:72–79.
6. Pettersson, R.F. (1981): *Proc. Nat. Acad. Sci. U.S.A.*, Jan. 1981, in press.
7. Pelham, H.R.B. and Jackson, R.J. (1976): *Eur J. Biochem.* 247–256.
8. Garoff, H., Simons, K., and Dobberstein, B. (1978): *J. Mol. Biol.* 124:587–600.
9. Rothman, J.E. and Lodish, H.F. (1977): *Nature* 269:775–780.
10. Gentsch, J.R. and Bishop, D.H.L. (1979): *J. Virol.* 30:767–770.
11. Lamb, R.A. and Lai, C.-J. (1980): *Cell* 21:475–485.

David H. L. Bishop and Richard W. Compans, eds.
The Replication of Negative Strand Viruses

The 3′-Terminal Sequences of Uukuniemi and Inkoo Virus RNA Genome Segments

Michael D. Parker and Martinez J. Hewlett[a]

The 3′-terminal sequences of two bunyaviruses, Uukuniemi and Inkoo viruses have been determined by chemical sequencing techniques. The 3′-terminal sequences of the two viruses are totally unrelated but the genome segments of each virus are characterized by stretches of homology. The conserved sequence in the Uukuniemi genome is 13 nucleotides long, and 11 nucleotides long in the Inkoo genome. Sequence divergence first appears on the S segment of both viruses while homology continues on the M and L segments for 3 nucleotides in Uukuniemi and 5 nucleotides in Inkoo vRNA.

Introduction

The family Bunyaviridae consists of a group of viruses 100 nm in diameter characterized by a lipoprotein envelope surrounding a helical ribonucleoprotein. The genome is a tripartite, negative sense, single-stranded RNA.[1] The lipoprotein envelope consists of lipids derived from host cellular internal membranes[2] and two glycoproteins specified by the virus.[3] The ribonucleoprotein is composed of one major and one minor protein associated with three unique segments of tNA.[4,5]

Pettersson and von Bonsdorff[6] reported that Uukuniemi virus ribonucleoproteins were circular when spread for electron microscopy. Further, the virion RNAs were also circular when spread under relatively non-denaturing

[a] Department of Cellular and Developmental Biology, University of Arizona, Tucson, Arizona.

conditions and the RNAs were reversibly denatured under proper conditions.[7]

We report here, the sequence of the 3′-terminal nucleotides of Uukuniemi and Inkoo viruses. The sequences are conserved among the segments of an individual virus but differ substantially between members of different serological groups.

Materials and Methods

Virus and Cells

Uukuniemi virus is the same strain utilized in previous studies.[4] Inkoo virus was supplied by Dr. N. Karabatsos, CDC, Fort Collins, Colorado as a 10% suckling mouse brain suspension, and was plaque purified three times before use. Uukuniemi virus was grown in either secondary chicken embryo fibroblasts in Dulbecco's Modified Essential Medium (DMEM) with 3% calf serum or in the Wistar strain of BHK cells with 8% fetal calf serum. Inkoo virus was grown in BHK cells.

Growth and Purification of Virus

Confluent roller bottles (490 cm^2) were infected at an moi of approximately 0.1. After one hour, the inoculum was removed and replaced with 50 ml of complete growth medium. ^{32}P-labeled virus was prepared by the same protocol using phosphate-free DMEM containing 8% dialyzed fetal calf serum and 50 μCi/ml carrier free $H_3{}^{32}PO_4$. Uukuniemi virus was harvested at 24-hour intervals beginning at 72 hours postinfection. Inkoo virus was harvested at 48 hours postinfection. The medium was removed and clarified by centrifugation for 20 minutes at 10,000 $\times$ g at 4°. The virus was precipitated by adjusting the medium to 8% polyethylene glycol and 0.5 M NaCl as previously described.[8] The virus was pelleted by centrifugation at 4°C at 10,000 $\times$ g for 20 minutes. The pellet was resuspended in TNE buffer (0.1 M NaCl, 0.01 M tris-HCl pH 7.4, 0.001 M EDTA) and loaded onto 12.5–50% Renograffin (Squibb) gradients in TNE and spun at 28,000 rpm for 2 hours at 4°C in an SW 28 rotor. The visible virus band was collected and pelleted in an SW 28 rotor for 2 hours at 28,000 rpm. Some preparations were spun on an additional 30% glycerol-41% potassium tartrate gradient for 2 hours at 28,000 rpm. The virus band was then pelleted.

RNA Extraction and Ligation with Cytidine 3′, 5′-Bis (Phosphate)

Pelleted virus was disrupted with TNE buffer containing 0.5% SDS. The RNA was extracted three times with phenol:chloroform:isoamyl alcohol (50:48:2) and once with chloroform:isoamyl alcohol (96:4). The RNA was precipitated with 0.3 M sodium acetate and 2.5 volumes of ethanol.

Virion RNAs were tagged with pCp (NEN, sp. act > 2000Ci/mMole) according to Peattie.[9] The reaction (30 μl) typically contained 12 to 15 units of RNA ligase (P.L. Biochemicals) and 60 to 80 pMoles of pCp with a 20 molar excess of ATP. The reaction was held at 0°C for 16 hours. The RNAs were separated on 1.5% agarose gels containing 5 mM methylmercuric hydroxide.[10] The labeled RNAs were located by autoradiography and eluted from the crushed gel in 0.5 M ammonium acetate, 1% mercaptoethanal, 5 mM $MgCl_2$ and 1% SDS and carrier tRNA was added to 100 μg/ml. The RNAs were pelleted with ethanol and sequenced according to the methods of Peattie.[9] The cleaved RNAs were electrophoresed on 30% acrylamide gels in Tris-borate buffer,[11] pH 8.3 at 1000 volts and autoradiographed at $-70°$.

The 3′-terminal nucleotide was identified by nearest neighbor analysis. ^{32}pCp-labeled RNA or oligonucleotides were eluted from the gels and digested with 50 U/ml of RNAse T_1. 50 U/ml of RNAse T_2, 100 μg/ml RNAse A and 100 μg/ml of tRNA in 50 mM ammonium acetate buffer pH5.5. The samples were electrophoresed on Whatman 3 MM paper [12] at 1500 volts. The labeled spots were located by autoradiography and identified by mobility relative to mononucleotide standards.

Results

Virion RNA

Agarose gel electrophoresis under denaturing conditions of Uukuniemi virion RNA demonstrates the presence of 3 RNA species (Figure 1). Using BHK cell 28S and 18S rRNA as standards, the molecular weights of Uukuniemi virus RNAs are 2.95, 1.45 and 0.64 $\times$ 10 ^{6} daltons for L, M, and S, respectively. These values are slightly higher than those reported by Pettersson et al.[4] based upon complexity measurements. The Inkoo RNA preparation shown in Figure 1 has four components of: 2.85, 2.15, 0.8 and 0.35 $\times$ 10 ^{6} daltons for L, M, S_1 and S_2, respectively. The origin of the fourth RNA, S_1, is presently unknown as it has been previously reported that the California group of bunyaviruses contain only 3 RNA segments.[13] It is unlikely that S_1 is due to the presence of DI particles because these virus preparations were grown from low passages stocks prepared from plaque purified Inkoo virus.

Ligation and Sequencing of Virion RNA with pCp and T4 RNA Ligase

Total virion RNA was extracted and labeled according to Peattie.[9] The electrophoretic profile on preparative agarose gels of RNA obtained from the viruses after labeling with pCp was identical to that shown in Figure 1 (data not shown). The S_1 RNA of Inkoo was present but labeled to a much lesser extent than the other RNAs suggesting that it is present in much less than equimolar amounts.

It has been previously demonstrated that the segments of bunyavirus RNA are unique.[4,13] It was of interest to determine if the terminal sequences are conserved among the individual segments as reported for influenza vir-

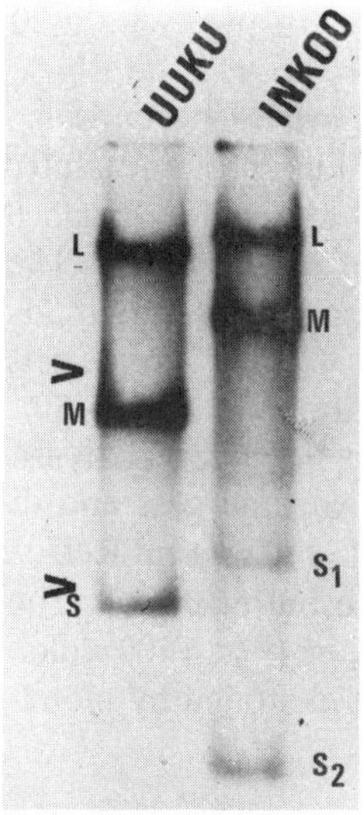

Figure 1. Agarose gel profile of ^{32}P-labeled RNA from UUKUNIEMI virus (Uuku) and Inkoo viruses. BHK cells were infected and labeled with 50 μCi/ml of carrier free H_3 $^{32}PO_4$. Virus was pelleted and purified on potassium tartrate-glycerol gradients. Extracted RNA was electrophoresed on 1.5% agarose slab gels containing 5 mM methylmercuric hydroxide. The dried gel was autoradiographed on Kodak X-OMAT film. The arrows denoted the position of HeLa cell 28S and 18S rRNA.

uses.[14] Figure 2 shows an autoradiograph of a sequence gel on which the fragments of the 3 segments of Uukuniemi RNA have been electrophoresed. It shows that the 3′-terminal base on each of the segments is U, as indicated by the band in the position of free pCp. Nearest neighbor analyses of Uukuniemi and Inkoo viral RNAs labeled with pCp and of fragments eluted from the sequencing gels showed that > 80% of the label was found in U. Figure 2 also shows that the 3′-terminal sequences of Uukuniemi viral RNA segments are identical for the first 13 nucleotides. However, the positions of G residues cannot be determined accurately from Figure 2 because RNAse T_1 was used for the G specific cleavage reactions while the remaining cleavages were done by the chemical method of Peattie.[9] RNAse T^1 yields products with 5′-hydroxyls while the chemical sequencing reactions yield products with 5-monophosphates. The differences in charge between these products have a significant effect on the relative electrophoretic mobility of small oligonucleotides.

Figure 3 shows an autoradiograph of a sequencing gel representative of those used to determine the 3′ sequence of the Uukuniemi RNA segments. The known sequences for all three segments arè shown in Figure 5. With the exception of a G at position 14 on the S segment which is missing on M and

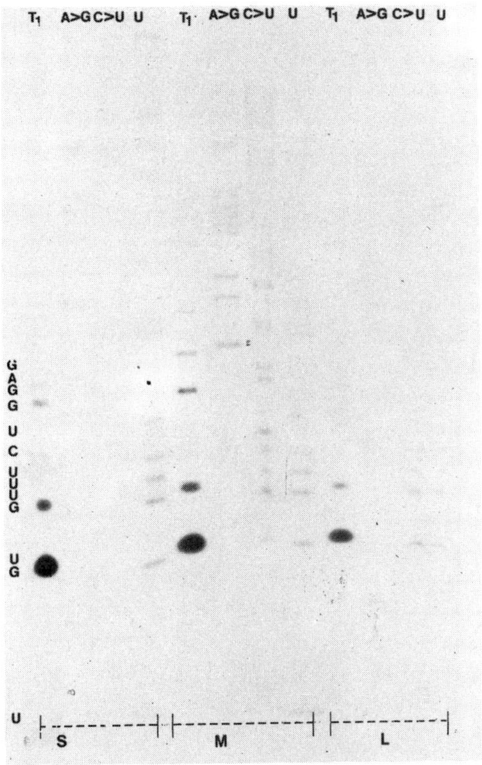

Figure 2. Autoradiograph of sequencing gel of 3′ ends of Uukuniemi virion RNA segments. Total virion RNA was ligated with pCp and T4 RNA ligase as described above. The segments were resolved on 1.5% agarose gels containing 5 mM methylmercuric hydroxide. The eluted segments were then sequenced according to Peattie[9] except the G specific reaction was done with RNAse T_1 in 20 mM sodium citrate, pH 5.0, 1 mM EDTA at an enzyme:substrate ratio of 10^{-4} at 55°C for 15 minutes. The fragments were electrophoresed on a 30% acrylamide gel in 50 mM tris-borate, pH 8.3, 1 mM EDTA at 1000 volts.

L, there is complete homology of sequence between the RNA segments to position 17. The known sequence of the L segment stops at position 17, but the remaining two segments show homology to position 20 after which the sequences diverge rapidly.

Figure 4 is an autoradiograph of a sequencing gel of the S segment of Inkoo RNA. The known sequence of the Inkoo RNAs are shown in Figure 5. The terminal nucleotide of Inkoo RNAs is also uridine. As shown for the Uukuniemi vRNAs, the terminal sequences of Inkoo are also conserved to position 11. At this point, the sequence of S diverges, while the L and M sequences are homologous to position 16.

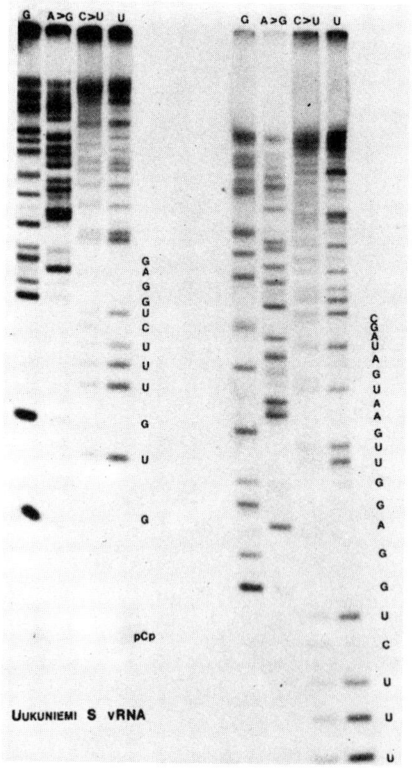

Figure 3. Autoradiograph of sequencing gel of Uukuniemi virus S RNA. Sequencing reactions and gel electrophoresis were performed as described in the legend to Figure 2.

Discussion

We have demonstrated extensive 3′-terminal sequence homology between the L, M and S segments of Uukuniemi virion RNA. Sequence homology has also been demonstrated at the 3′-termini of Inkoo virion RNAs. 3′-terminal sequence homology for other California group viruses and for Nairodirus genus viruses has been reported by Bishop and his colleagues (this volume). Sequence homology has also been reported for the termini of influenza virus, another segmented, negative stranded virus.[14,15]

The conserved terminal sequences of Uukuniemi and Inkoo vRNAs are totally unrelated. They differ not only in sequence but also in length, 13 nucleotides in Uukeniemi and 11 nucleotides in Inkoo virus. In addition, the region of homology in the Uukuniemi S segment is terminated by one G at position 14, while in Inkoo there are two G residues at the corresponding position of sequence interruption. In both viruses, limited sequence homol-

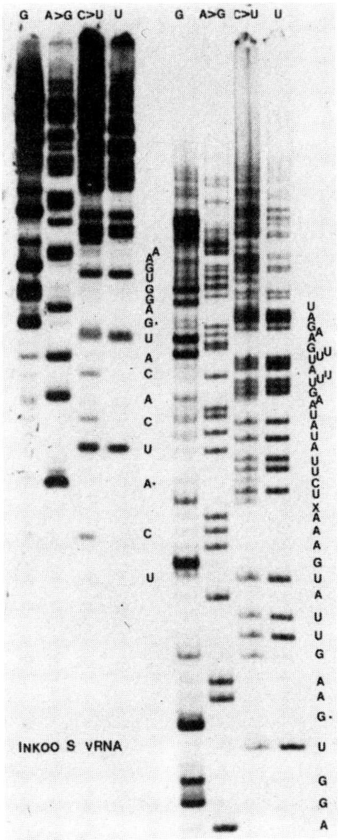

Figure 4. Autoradiograph of sequencing gel of Inkoo S RNA. The RNA was ligated and sequenced according to the techniques outlined in the legend to Fig. 2* marks segments eluted from the gel for nearest neighbor analysis.

ogy begins again after the interruption. The second conserved sequence begins at position 15 in Uukuniemi virus and continues to position 20. The known sequence of the Uukuniemi L segment ends at position 17, so it is not known if further homology is present in this segment. Inkoo RNA segments M and L show uninterrupted homology through position 16. However, the conserved sequence on the S segment ends at position 11.

The two viruses characterized here are serologically unrelated, were isolated from different sources [16] and exhibit different growth characteristics in mammalian, avian and invertebrate cells (Parker, Hewlett and Florkiewicz, unpublished data). The substantial differences in the conserved terminal nucleotide sequences between the two viruses may be another indicator of the significant differences which exist between the Uukuniemi and California serogroups of bunyaviruses.

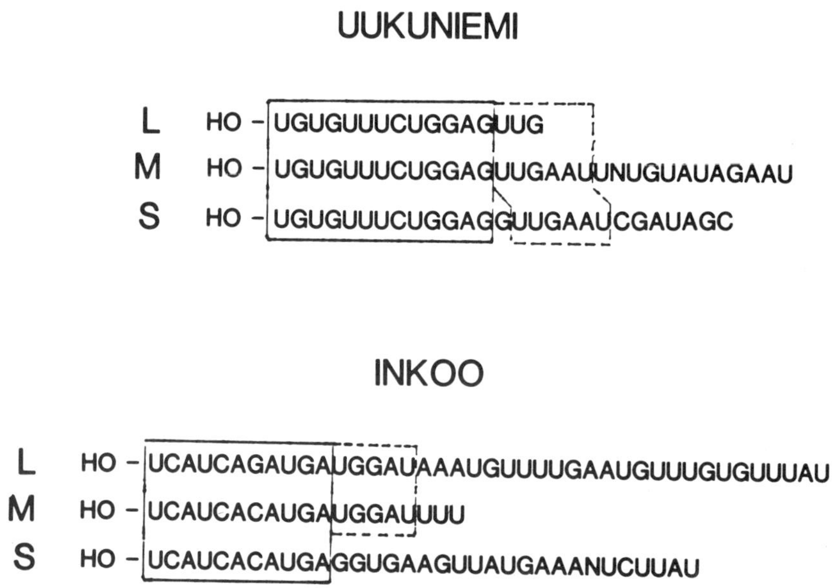

Figure 5. The known sequences of Uukuniemi and Inkoo viral RNA segments. The boxes denote regions of sequence homology.

We have not been able to efficiently label the 5′ termini of the Uukuniemi or Inkoo RNAs. The 5′ ends of the molecules are pppA[4,17] and more than 90% of these phosphates are removed by bacterial alkaline phosphatase. However, the product is not labeled with polynucleotide kinase and ATP under conditions which efficiently label control tRNA or HeLa rRNA. As a result we do not know the sequences of the 5′-terminal regions of either Uukuniemi or Inkoo viruses. However, Pettersson has analyzed the 5′-terminal T_1 fragment of Uukuniemi virus and found the sequence 5′-pppACACAAAG-3′ (personal communication) which is exactly complementary to the 3′ sequence. This may explain the finding that Uukuniemi vRNA is circular when spread under non-denaturing conditions for the electron microscope.[7] It will be of interest to determine if intermittent sequence homology also exists at the 5′ end of the Uukuniemi and Inkoo RNA segments and the length of the terminal complementarity.

Transcription of influenza mRNA terminates at a point 20–40 nucleotides short of the 5′ terminus of the template.[18] A similar situation may exist with the bunyaviruses, since it has been shown that S mRNAs of snowshoe hare virus have a higher mobility on gels than the corresponding genome templates.[18] It is tempting to speculate that base-pairing at the termini is involved in regulation of transcriptional termination. However, verification of this hypothesis requires the identification of specific sites at which transcription initiates and terminates on the vRNA.

Our sequencing data has not indicated the position of complements of possible translation initiation codons. Fowl plague virus genome segments have the complement of numerous potential initiation codons within 20–40 nucleotides of the 3′ end.[14] Our data indicates that bunyavirus initiation codons should be present beyond position 30 in the sequence. In fact, Bishop et al. (this volume) report that such codon complements can be found near position 80 in the California group sequences.

The break in sequence homology occurs first in the S strands of both viruses. The homology of sequence among the strands of influenza viruses is also interrupted by a short series of nucleotides and then resumes just as it does in Uukuniemi and Inkoo viruses. Whether this acts as a signal during virus assembly or during nucleic acid synthesis is unknown.

Several functions of the homologous sequences may be postulated. The terminal sequences may serve an important function in the initiation and/or specificity of the virus specific RNA polymerase. Regardless of whether or not circular structures actually play a role *in vivo,* the inverted complementarity of the terminal sequences dictate that the 3′ terminus of the vRNA and the 3′ terminus of the positive stranded templates from which they are replicated are identical. This indicates that potential polymerase binding sites or intiation sites are identical and would minimize the necessity for the polymerase to recognize multiple initiation sites for transcription and replication.

References

1. Bishop, D.H.L. and Shope, R. (1979): *Comp. Virol.* 14:1–156.
2. Murphy, F.A., Harrison, A.K., and Whitfield, S.G. (1973): *Intervirology* 1:297–316.
3. Gentsch, J. and Bishop, D.H.L. (1979): *J. Virol.* 30:767–776.
4. Pettersson, R.F., Hewlett, M.J., Baltimore, D., and Coffin, J.M. (1977): *Cell* 11:51–63.
5. Obijeski, J.F., Bishop, D.H.L., Murphy, F.A., and Palmer, E.L. (1976): *J. Virol.* 19:985–997.
6. Pettersson, R.F. and von Bonsdorff, C.H. (1975): *J. Virol.* 15:386–392.
7. Hewlett, M.J., Pettersson, R.F., and Baltimore, D. (1977): *J. Virol.* 21:1085–1093.
8. Della-Porta, A.J. and Westaway, E.G. (1972): *Appl. Micro.* 23:158–160.
9. Peattie, D. (1979): *Proc. Nat. Acad. Sci.* 76:1760–1764.
10. Bailey, J.M. and Davidson, N. (1976): *Anal. Biochem.* 70:75–85.
11. Donis-Keller, H., Maxam, A.M., and Gilbert, W. (1977): *Nuc. Acids Res.* 4:2527–2538.
12. Barrell, D. (1971): In: *Procedures in Nucleic Acid Research.* Vol. II. Contoni, G. L. and Davis, D.R. (Eds.), Harper and Row, New York, pp. 775–779.
13. Clewley, J., Gentsch, J. and Bishop, D.H.L. (1977): *J. Virol.* 22:459–468.
14. Roberts, J. (1979): *Nucleic Acid Res.* 6:3745–3752.
15. Both, G.W. and Air, G.M. (1979): *Eur. J. Biochem.* 96:363–372.
16. Berge, T.O. (1975): In: *International Catalogue of Arboviruses.* U.S. Dept. Health, Education and Welfare.
17. Obijeski, J., Bishop, D.H.L., Palmer, E.L., and Murphy, F.A. (1976): *J. Virol.* 20:664–675.
18. Cash, P., Vezza, A.C., Gentsch, J.R., and Bishop, D.H.L. (1979): *J. Virol.* 31:685–694.

David H.L. Bishop and Richard W. Compans, eds.
The Replication of Negative Strand Viruses

Molecular and Genetic Properties of Members of the Bunyaviridae

David H.L. Bishop, John P.M. Clerx, Corrie M. Clerx-van Haaster, Gloria Robeson, Edward J. Rozhon, Hiroshi Ushijima and Venkat Veerisetty[a]

Introduction

The Bunyaviridae include more than 200 serologically distinguishable virus isolates.[1,2] Some 124 viruses in the family have been assigned to the *Bunyavirus* genus (serogroups: Anopheles A (10), Bunyamwera (20), Bwamba (2), C Group (13), California (14), Capim (10), Gamboa (4), Guama (12), Koongol (2), Olifantsvlei (3), Patois (4), Simbu (22) and Tete (5); and 3 unassigned viruses).[2] Another 30 viruses are assigned to the *Phlebovirus* genus (e.g., Karimabad, KAR, Punta Toro, PT and Rift Valley fever viruses).[2] Some 18 viruses have been placed in the *Nairovirus* genus (serogroups: Crimean-Congo hemorrhagic fever (2), Nairobi sheep disease (3), Hughes (4), Dera Ghazi Khan (6) and Qalyub (2); and 1 unassigned virus).[2] Another 7 viruses constitute the *Uukuvirus genus*.[2] In addition to these 179 viruses there are more than 30 viruses that have not yet been assigned to a genus.[2]

Most of the bunyaviruses, as well as a few of the phleboviruses, have been isolated from mosquitoes.[3] Some have also been shown to replicate in, and be transmitted by, mosquitoes.[1] The nairoviruses and uukuviruses have come from ticks, while many of the phleboviruses have been obtained from phlebotomines.[3] As disease causing entities, although infection of vertebrates by many of the viruses in the family leads to the development of inapparent infections, some cause an occasionally fatal meningoencephalitis, while others produce a hemorrhagic fever.[1] In this paper we discuss

[a] *Department of Microbiology, The Medical Center, University of Alabama in Birmingham, Birmingham, Alabama*

the structural properties of members of different genera of the Bunyaviridae and the genetic and coding attributes of the bunyaviruses.

Structural Analyses of Members of the Bunyaviridae

Tables 1 and 2 give the results we have obtained from analyses of the structural components (RNA and polypeptides) of 22 bunyaviruses. The average sizes of their 3 major virion polypeptides are: glycoproteins, G1 and G2, 115 and 37 $\times 10^3$ daltons, respectively; nucleocapsid polypeptide, N, 22 $\times 10^3$ daltons (Table 1). The average sizes of their 3 unique viral RNA species are: large size RNA, L, 2.9 $\times 10^6$ daltons, medium size RNA, M, 2.0 $\times 10^6$ daltons, small size RNA, S, 0.44 $\times 10^6$ daltons (Table 2).[4]

Table 1. Molecular Weight ($\times 10^{-3}$) Estimations of the Major Structural Polypeptides of Several Bunyaviruses.

Virus	Virus polypeptide		
	G1	G2	N
A. Anopheles A serogroup			
Anopheles A	118	39	22
B. Bunyamwera serogroup			
Bunyamwera	115	38	19
Main Drain	115	38	21
Guaroa	115	32	21
C. Group C			
Itaqui	111	41	23
Oriboca	110	41	22
D. California serogroup			
California encephalitis BFS 283	115	39	21
Jamestown Canyon	118	38	21
Keystone	118	38	23
La Crosse	115	38	24
Lumbo	115	38	25
Melao	117	38	24
Snowshoe hare	115	38	21
South River	120	38	22
Tahyna	115	38	22
Trivitattus	118	38	23
E. Capim serogroup			
Capim	114	36	21
F. Guama serogroup			
Guama	114	41	21
G. Patois serogroup			
Shark river	113	35	22
H. Simbu serogroup			
Aino	109	29	19
Mermet	120	34	21
Simbu	108	32	20

Molecular weights were normalized to the reported values for snowshoe hare virus[1] by comparing the electrophoretic mobility of ^{3}H-labeled virus proteins with the structural polypeptides of ^{3}H-labeled snowshoe hare virus.

In Figure 1 are shown the electrophoretic separations of the viral RNA species of the phleboviruses KAR (mol wt: L, 2.6×10^6, M, 2.2×10^6 and S, 0.8×10^6) and PT (mol wt: L, 2.8×10^6, M, 1.8×10^6, S, 0.75×10^6) by comparison to BHK cellular ribosomal RNA and the viral RNA species of the bunyavirus snowshoe hare (SSH).[5] Figure 2 presents analyses of the 3 major virion polypeptides of KAR virus (mol wt: G1, 62×10^3, G2, 50×10^3 and N, 21×10^3).[5] Essentially similar results for their structural components have been obtained from analyses of the virion RNA and polypeptide species of the phleboviruses, Chagres, Icoraci, sandfly fever viruses Sicilian and Naples serotypes, Rift Valley fever and Buenaventura viruses (unpublished data).

Table 2. Molecular Weight ($\times 10^{-6}$) Estimations of the Three Virion RNA Species of Several Bunyaviruses.

Virus	Virus RNA Species		
	L	M	S
A. Anopheles A serogroup			
Anopheles A	2.7	2.1	0.43
B. Bunyamwera serogroup			
Bunyamwera	3.0	1.9	0.34
Guaroa	2.8	2.0	0.50
Main Drain	3.1	2.0	0.40
C. Group C			
Itaqui	3.0	1.9	0.50
Oriboca	2.7	1.8	0.43
D. California serogroup			
California encephalitis BFS 283	3.0	2.1	0.44
La Crosse	2.9	1.8	0.40
Lumbo	2.9	2.0	0.50
Snowshoe hare	2.9	1.9	0.45
Tahyna	3.0	1.9	0.45
Trivittatus	2.9	1.9	0.36
E. Capim serogroup			
Capim	3.1	2.3	0.43
F. Guama serogroup			
Guama	2.9	1.9	0.49
G. Patois serogroup			
Shark river	3.1	2.3	0.48
H. Simbu serogroup			
Aino	2.8	1.9	0.28
Mermet	3.0	2.1	0.46

Molecular weights were normalized to the reported values for snowshoe hare virus[1] by comparing the electrophoretic mobilities of ^{32}P- (or ^{3}H-) labeled virus RNA species with the ^{3}H- (or ^{32}P-) labeled virus or infected cell RNA species of snowshoe hare virus.

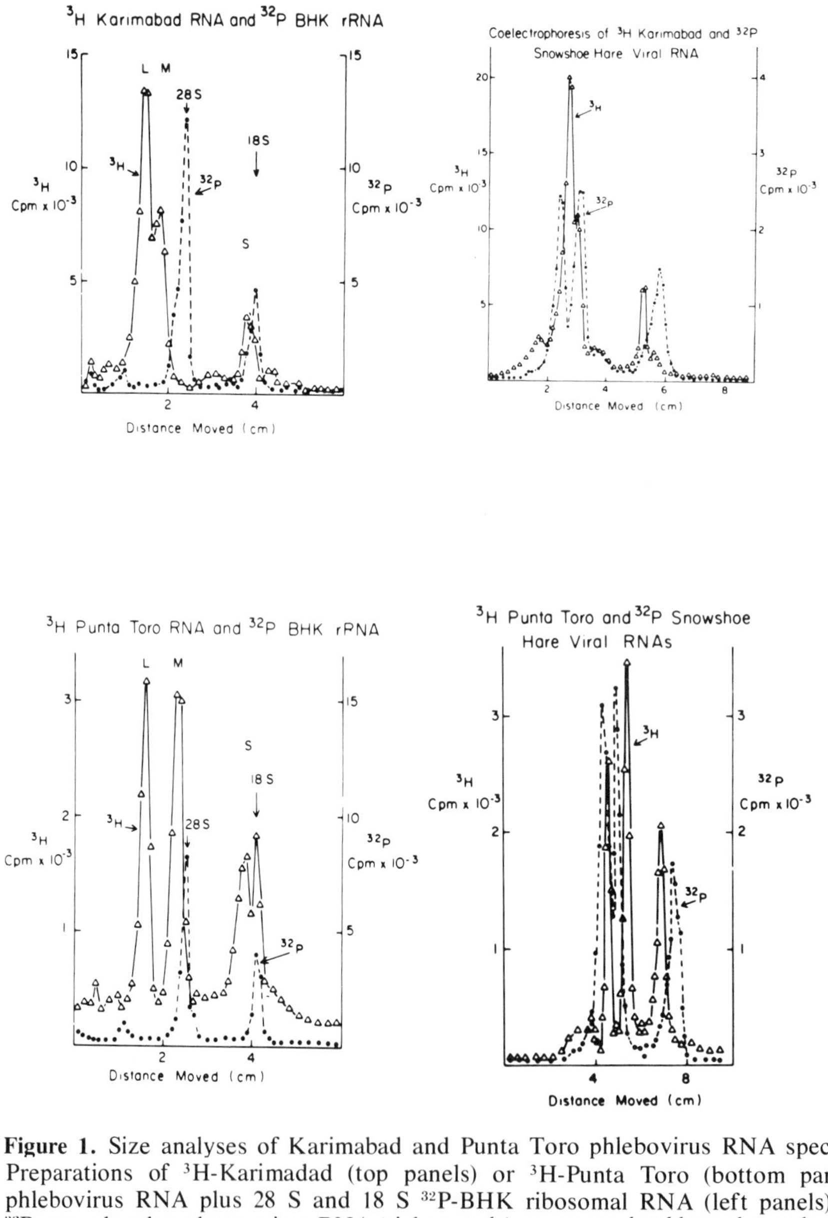

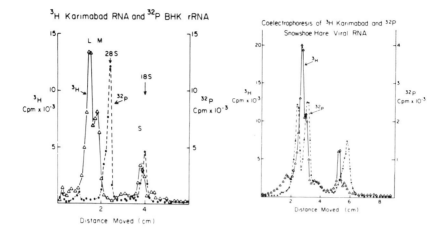

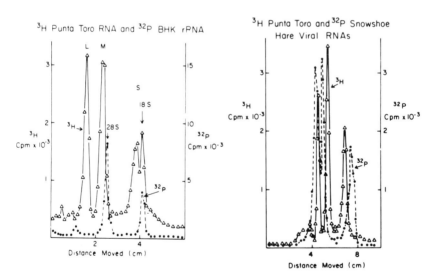

Figure 1. Size analyses of Karimabad and Punta Toro phlebovirus RNA species, Preparations of ^{3}H-Karimadad (top panels) or ^{3}H-Punta Toro (bottom panels) phlebovirus RNA plus 28 S and 18 S ^{32}P-BHK ribosomal RNA (left panels), or ^{32}P-snowshoe hare bunyavirus RNA (right panels), were resolved by polyacrylamide gel electrophoresis. The L, M, and S Karimibad viral RNA species are identified in the left hand panels.

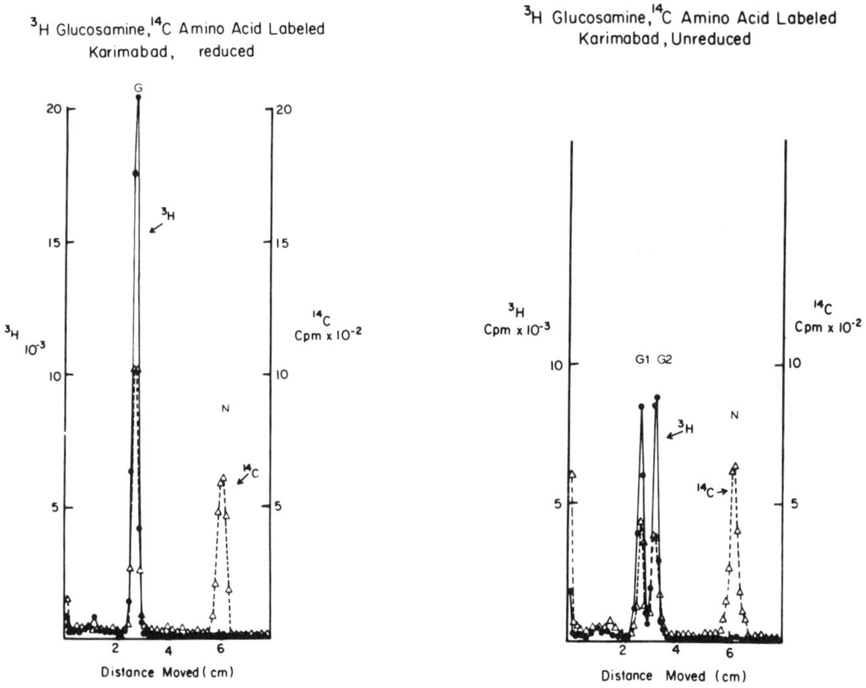

Figure 2. Analyses of Karimabad phlebovirus polypeptides. Preparations of ^{14}C-amino acid and ^{3}H-glucosamine labeled, reduced (left panel), or unreduced (right panel), Karimabad polypeptides were resolved by polyacrylamide gel electrophoresis.

The published studies of Uukuniemi virus indicate that it also has 3 unique virion RNA species (mol wt: L, 2.4×10^6, M, 1.1×10^6 and S, 0.5×10^6),[6] and 3 major virion polypeptides (mol wt: G1, 75×10^3, G2, 65×10^3 and N, 25×10^3).[7,8]

Figure 3 shows the electrophoretic separations of the viral RNA species of Hazara (HAZ), Qalyub (QYB),[9] Bandia (BDA) and Dugbe (DUG) nairoviruses by comparison with SSH viral RNA. The data obtained for the sizes of these nairovirus viral RNA species, by comparison to those of selected members of the other 3 Bunyaviridae genera, are given in Table 3. Figure 4 gives the results of virion and intracellular polypeptide analyses of QYB nairovirus.[9] The 3 major virion polypeptides of QYB (mol wt: G1, 75×10^3, G2, 40×10^3 and N, 53×10^3) are different from those of bunyaviruses, phleboviruses and uukuviruses. The QYB G1 (gp75) and G2 (gp40) polypeptides can be selectively removed from virions by protease treatment.[9] QYB N polypeptide (p53) is the major component of the viral nucleocapsids.[9] The intracellular polypeptides induced in QYB infected cells include N and the glycoprotein precursors gp115 and gp85, these precursors can be pulse

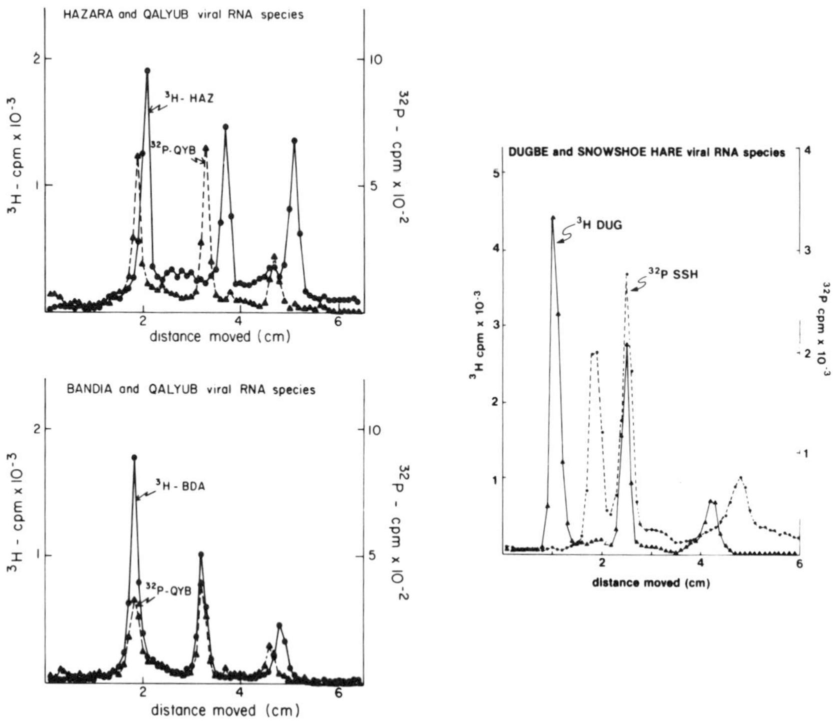

Figure 3. Size analyses of nairovirus RNA species. Preparations of ^{32}P-Qalyub (QYB) nairoviral RNA (left panels) plus ^{3}H-Hazara (HAZ), or ^{3}H-Bandia (BDA) nairoviral RNA, were resolved by argorose gel electrophoresis. In the right-hand panel, a preparation of ^{3}H-Dugbe (DUG) nairovirus RNA and ^{32}P-snowshoe hare (SSH) bunyavirus RNA were also analyzed by argorose gel electrophoresis.

labeled with radiolabeled mannose or leucine (etc.) and the label specifically chased into the virion G1 (gp75) and G2 (gp40) species.[9]

It is evident from the above that nairoviruses have virion structural components (e.g., their L RNA species and N, G1 and G2 polypeptides) that differ significantly in terms of their sizes from those of the bunyaviruses, phleboviruses and uukuviruses (see Table 3).

The structural analyses of members of the Bunyavirus and Nairovirus genera have been extended to include analyses of the 3′ end sequences of their virion RNA species. Shown in Figure 5 are the results obtained for the L, M and S RNA 3′ end sequences of the bunyavirus SSH[10] and La Crosse[10] (LAC, both California serogroup bunyaviruses), Main Drain (MD, Bunyamwera serogroup), Pahayokee (PAH, Patois serogroup), Mermet (MER, Simbu serogroup) and Boraceia viruses (BOR, unassigned Anopheles B serogroup). By reference to the LAC RNA sequences, identi-

Table 3. Molecular Weight ($\times 10^{-6}$) Estimations of the Three Virion RNA Species of Selected Bunyaviruses, Phleboviruses, Uukuviruses and Nairoviruses.

Virus	Virus RNA Species		
	L	M	S
A. Bunyavirus: snowshoe hare virus	2.9	1.9	0.45
B. Phlebovirus: Karimabad virus	2.6	2.2	0.8
C. Uukuvirus: Uukuniemi virus	2.4	1.1	0.5
D. Nairoviruses:			
Hazara	4.3	1.5	0.6
Dugbe	4.5	1.9	0.6
Qalyub	4.8	1.9	0.7
Bandia	4.8	1.9	0.6
Hughes	5.0	1.4	0.6

Molecular weights were normalized to the reported values for snowshoe hare virus[1] by comparing the electrophoretic mobilities of ^{32}P- (or ^{3}H-) labeled virus RNA species with the ^{3}H- (or ^{32}P-) labeled RNA species of snowshoe hare virus.

cal nucleotides in the sequences of the other viruses are indicated by filled-in circles, unknown nucleotides in a sequence by X and missing nucleotides by spaces in the sequences. UAC triplets in the sequences are underlined.

By comparison with the LAC RNA sequences, the L, M and S RNA sequences of the serologically closely related SSH California serogroup bunyavirus are almost identical. Single nucleotide differences between the LAC and SSH S RNA sequences occur from LAC residues 17 through 65, thereafter there is a stretch of 10 nucleotides with only 1 homologous nucleotide between the two sequences, this is followed by a region of almost complete sequence identity through LAC residue 122. This last region has a UAC triplet at LAC S positions 82–84. In the complementary S mRNA this UAC is transcribed into an AUG which is probably the translation initiation site for the synthesis of S RNA gene products.[10] Presumably the conservation of sequences after the first UAC triplet reflects the restrictions imposed against sequence evolution by the encoded gene products. The fact that the RNA sequences of the other bunyaviruses we have analyzed, have similar 3′ $_{HO}$UCAUCACAUG . . . sequences for their S, M and L RNA species (except for a PAH S RNA position 7 A residue, see Figure 5), indicates that the 3′-terminal sequences of these viruses are highly conserved. Following these 10 3′ nucleotides, the RNA species of the different viruses exhibit less sequence conservation (see Figure 5). The positions of the first UAC triplets in the S, M or L RNA species of the negative stranded viral genomes of these

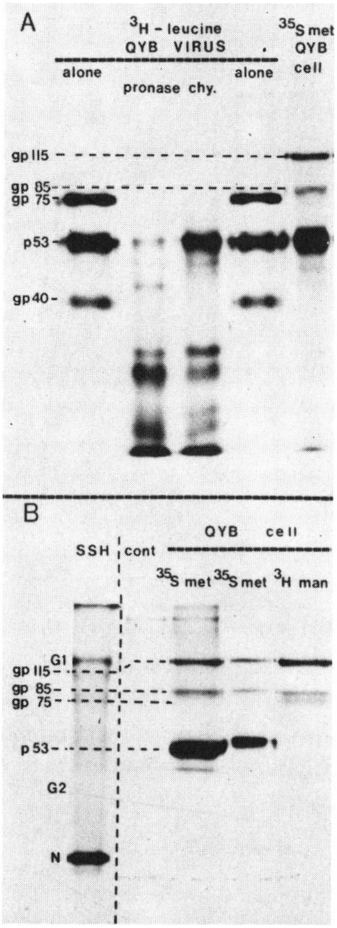

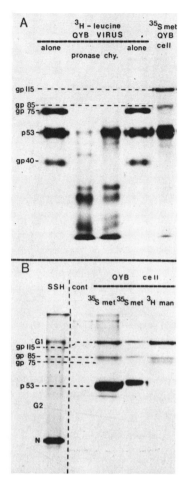

Figure 4. The virion and intracellular polypeptides of Qalyub (QYB) nairovirus. In (A) are shown the gel profiles of ^{3}H-leucine labeled virion polypeptides of QYB nairovirus either untreated (alone, first and fourth lanes), or after digestion with pronase (second lane), or chymotrypsin (third lane).[9] In the right-hand (fifth lane) QYB immune precipitated, ^{35}S-methionine labeled, polypeptides are displayed. The intracellular polypeptides (gp115, gp85, gp75) and nucleocapsid polypeptide (p53) as well as the virion gp75, gp40 and p53 polypeptides are indicated. In (B) are shown the virion polypeptides of snowshoe hare (SSH) viral polypeptides (G1, G2, N), and QYB infected (third-fifth lanes), or control (second lane), uninfected cell polypeptides labeled with ^{35}S-methionine (^{35}S met, third and fourth lanes), or ^{3}H-mannose (^{3}H man, fifth lane) and precipitated with QYB immune serum. The QYB induced gp115, gp85, gp75 and p53 intracellular polypeptides are indicated.

```
                              BUNYAVIRUS S RNA SPECIES

               5    10   15   20   25   30   35   40   45   50   55   60   65   70   75
LAC-S 3'   HO UCAUCACAUGAGGUGAACUUAUGAAACUUUUAUUUAACAACAACUGACAAAAAAUGGAUUCCCCUUUAAUAGUUCUCAC
SSH-S 3'   HO ●●●●●●●●●●●●●●●●G●●●●●●●●●●●●●●●GA●●●●●●●G●U●A●●●●●●● ●●●●●●●●● ●AACUG●ACCU●●●●
BOR-S 3'   HO ●●●●●●●●●●●●●●●●●G●●UCAUU●U●●CAG●G●U●
MD -S 3'   HO ●●●●●●●●●●●●●●UUUA●●U●UUG●GACGGU●GA●●UU●UU●AG
PAH-S 3'   HO ●●●●●●A●●●●A●AUUU●●●XXUUG●G●AACUAGC
MER-S 3'   HO ●●●●●●●●●●●●●●●UUAAGU●UUUGAA●●●CAG

              80   85   90   95   100  105  110  115  120
LAC-S cont.   ACUACAGCCUAAACCACAAAAUACUACAGCGUAGUUGUCCACG
SSH-S cont.   ●●●●●●●G●●●●●●●●●●●●●●●●●●●●●●●●●●●●●●●●●●●UUUACCUAAACUAGGACUACGUCCAUAUACCU

                              BUNYAVIRUS M RNA SPECIES

               5    10   15   20   25   30   35   40   45   50
LAC-M 3'   HO UCAUCACAUGAUGGUUCAUAUCUAUUGCAAACUUAUAAUUUCAAAAUUAG
SSH-M 3'   HO ●●●●●●●●●●●●●●●●●●●●●●●●●●●●●●●U●●●●G●●●●●●●G●●●●●ACUUCGGUUUCUACUAAACGUAUAACUA
BOR-M 3'   HO ●●●●●●●●●●●●●AACAU●U●●●GA●AGUUUUACUAGC●●
MD -M 3'   HO ●●●●●●●●●●●●●●G●UUAU●UXX●●●A●GUU●G●A

                              BUNYAVIRUS L RNA SPECIES

               5    10   15   20   25   30   35   40   45
LAC-L 3'   HO UCAUCACAUGAGGAUAGAUGUUUUGAAUGUCUUUUAAGUCAGUAUAGUG
SSH-L 3'   HO ●●●●●●●●●●●●●●●●●●●A●●●●●●●●●●U●●●●●●●●●A●●A●●●●●UAAUAUACUCAUUACCUGGUAGUUCUUAUG
BOR-L 3'   HO ●●●●●●●●●●G●●●●●●●●U●C●●U●U●UCU●●●AGUUAU●C

              80   85   90   95   100  105
SSH-L cont.   GUUGUUAAGUAUCGAUCUAAUUACAGACAG
```

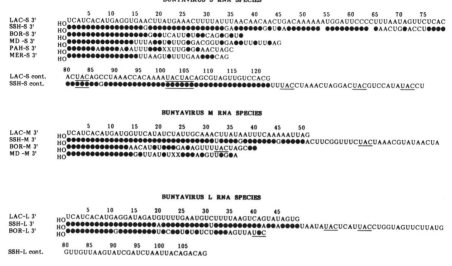

Figure 5. Sequence analyses of snowshoe hare (SSH), La Crosse (LAC), Main Drain (MD) Pahayokee (PAH) and Mermet (MER) bunyaviruses and the unassigned Anopheles B virus, Boraceia (BOR). The 3′ end sequences of the L, M and S viral RNA species were determined as described elsewhere and are presented as the best possible alignment.[10] Unknown nucleotides are indicated by an X, missing nucleotides from the aligned sequences are indicated by spaces, and identical nucleotides (by reference to the LAC sequence) by filled circles. UAC triplet positions are underlined in the sequences, the first UAC triplet in each sequence is probably equivalent to the respective mRNA initiation AUG for the encoded polypeptides.[10] Whether the other UAC triplets correspond to alternate translation initiation sites is under investigation.

serologically different bunyaviruses vary; none, however, is closer than 30 nucleotide residues from the 3′ end (see Figure 5).

By contrast to the results obtained for the bunyaviruses (and the unassigned BOR virus), analyses of the 3′ RNA sequences of the nairovirus QYB indicate that this virus has a different 3′ end sequence ($_{HO}$AGAGAUUCU. . .) on each of its 3 viral RNA species (Figure 6). The 3′-terminal sequences of the serologically related BDA virus (Qalyub serogroup), or more distantly related Hughes nairovirus (Hughes serogroup), are currently being analyzed. Preliminary results indicate that these viruses have almost identical 3′ end sequences (for the first 6–9 nucleotides) to those shown in Figure 3 for QYB virus (data not shown).

In conclusion, the results we (and others) have obtained from structural analyses of members of the Bunyaviridae, indicate that the members of the four recently defined genera (*Bunyavirus, Nairovirus, Phlebovirus* and *Uukuvirus*) have distinct structural components.[2] Whether the viruses share a common strategy in their replication processes remains to be determined.

NAIROVIRUS QALYUB L, M AND S RNA SEQUENCES

5 10 15 20 25 30 35 40 45 50 55 60 65 70 75

QYB-S 3' HO AGAGAUUCUGCCUGCACGGCGACGGCAACUUGAUUAUCAAGCAAACGUGUUACAGACGUAAUCUGAAAXCCCUGUUUG

QYB-M 3' HO ●●●●●●●●●CUA●●A●●●C●●UAUUUCGU●●C●CA●GG●●UACCCUUG●●●UUUAC

QYB-L 3' HO ●●●●●●●●●CUAXUU●G●●G●UAA' AU

Figure 6. Sequence analyses of Qalyub L, M and S viral RNA species.

Genetic Studies of Bunyaviruses

The genetic studies we have instituted with bunyaviruses belonging to the California serogroup (*Bunyavirus* genus) have had the purpose of defining (1) their reassortment potential, (2) their RNA-gene coding assignments and, in conjunction with Drs. R.E. Shope and B.J. Beaty of the Yale Arbovirus Research Unit, (3) the pathogenic and vector transmission potentials of the ressortant viruses by comparison to their parental strains, thereby determining which viral gene products are the principal determinants of viral pathogenesis and transmission. The latter studies are reported elsewhere in this volume.

We have described genetic recombination between SSH, LAC, Tahyna (TAH) and Trivittatus (TVT) viruses.[11–14] In unpublished studies, we have also obtained reassortant viruses involving Lumbo (LUM) and California encephalitis (CAL) viruses. The following 22 reassortants have been produced (L/M/S RNA genotypes): SSH/LAC/SSH, SSH/LAC/LAC, SSH/SSH/LAC, LAC/SSH/LAC, LAC/SSH/SSH, LAC/LAC/SSH, SSH/TAH/SSH, SSH/TAH/TAH, SSH/LUM/SSH, TVT/SSH/TVT, LUM/LAC/LAC, LAC/LUM/LAC, LAC/TAH/LAC, LAC/TAH/TAH, TAH/LAC/TAH, TAH/LAC/LAC, TVT/LAC/TVT, TVT/TAH/TVT, TAH/LUM/LUM, TAH/LUM/TAH, LUM/TAH/TAH/ and LAC/CAL/LAC. No genetic recombination has been detected, despite extensive experimentation involving a variety of protocols, between Guaroa virus and SSH, LAC, TVT, LUM or TAH viruses.[14] The reason for this apparent genetic incompatibility is not known. We have previously reported the difficulty in obtaining certain predicted reassortants, notably LAC/SSH/LAC and LAC/SSH/SSH.[14] In recent studies these two reassortants have been recovered, but at unexpectedly low frequencies (0.05% of a dually infected virus harvest), even though alternate reassortants (e.g., LAC/LAC/SSH) are recovered at the expected frequencies (1–40%). The reason for the difficulty in producing certain bunyavirus genotype combinations is not known.

Coding assignment analyses have shown that the M RNA species of bunyaviruses codes for the 2 viral glycoproteins (G1 and G2), also that the S RNA codes for the nucleocapsid polypeptide, N.[13,10] Presumably the viral L RNA codes for the minor virion polypeptide L that has been described.[1]

Conclusions

Studies have shown that the Bunyaviridae contain a diverse group of viruses that differ in their hosts, vectors, structural and genetic attributes. The possibility to produce reassortant viruses from parental viruses with different pathogenic potentials and different vector preferences, allows studies to be undertaken of the gene products that are involved. Such studies are currently underway.

ACKNOWLEDGMENTS
Support for these studies is acknowledged from the National Institutes of Health (Grant Al 15400), the National Sciences Foundation (Grant PCM 80-12782) and U.S. Army Medical Research and Development Command, Washington, D.C. (Contract DAMD 17-78-C-8017).

References

1. Bishop, D.H.L. and Shope, R.E. (1979): In: *Comprehensive Virology, vol. 14,* Fraenkel-Conrat H. and Wagner, R.R. (Eds.), Plenum Press, New York.
2. Bishop, D.H.L., Calisher, C., Casals, J., Chumakov, M.P., Gaidomovich, S.Ya., Hannoun, C., Lvov, D.K., Marshall, I.D., Oker-Blom, N., Pettersson, R.F., Porterfield, J.S., Russell, P.K., Shope, R.E., and Westerway, E.G. (1980): *Intervirology,* in press.
3. Berge, T.O. (Ed.): International Catalogue of Arboviruses Including Certain Other Viruses of Vertebrates. (1975) U.S. Dept. Health Education and Welfare, Pub. No. CDC 75-8301.
4. Clewley, Jr., Gentsch, J., and Bishop, D.H.L. (1977): *J. Virol.* 22:459–468.
5. Robeson, G., El Said, L.H., Brandt, W., Dalrymple, J., and Bishop, D.H.L. (1979): *J. Virol.* 30:339–350.
6. Pettersson, R.F., Hewlett, M.J., Baltimore, D., and Coffin, J.M. (1977): *Cell* 11:51–63.
7. von Bonsdorff, C.-H. and Pettersson, R.F. (1975): *J. Virol.* 16:1296–1307.
8. Pettersson, R.F., Kääriänen, L., von Bonsdorff, C.-H., and Oker-Blom, N. (1971): *Virology* 46:721–729.
9. Clerx, J.P.M. and Bishop, D.H.L. (1981): *Virology,* 108:361–372.
10. Clerx-van Haaster, C. and Bishop, D.H.L. (1980): *Virology* 105:564–574.
11. Gentsch, J. and Bishop, D.H.L. (1976): *J. Virol.* 20:157–169.
12. Gentsch, J., Wynne, L.R., Clewley, J.P., Shope, R.E., and Bishop, D.H.L. (1977): *J. Virol.* 24:893–902.
13. Gentsch, J.R., Rozhon, E.J., Klimas, R.A., El Said, L.H., Shope, R.E., and Bishop, D.H.L. (1980): *Virology* 102:190–204.
14. Gentsch, J.R., Robeson, G., and Bishop, D.H.L. (1979): *J. Virol.* 31:707–717.
15. Gentsch, J. and Bishop, D.H.L. (1978): *J. Virol.* 28:417–419.
16. Gentsch, J.R. and Bishop, D.H.L. (1979): *J. Virol.* 30:767–776.

Copyright 1981 by Elsevier North Holland, Inc.
David H.L. Bishop and Richard W. Compans, eds.
The Replication of Negative Strand Viruses

The Association of the Bunyavirus Middle-Sized RNA Segment with Mouse Pathogenicity

Robert E. Shope,[a] Gregory H. Tignor,[a]
Edward J. Rozhon,[b] and David H.L. Bishop[b]

Introduction

Bunyaviruses in the California serogroup have 3-segmented RNA genomes designated large (L), middle-sized (M), and small (S).[1,2] Reassortant viruses may be formed by reassortment of RNA segments following co-infection of cells with 2 related viruses.[3-8] The bunyavirus M RNA codes for the 2 viral glycoproteins and for the neutralization reaction, while the S RNA codes for the nucleocapsid protein.[6,9,10] It has been suggested although not proved, that the L RNA codes for the virion transcriptase.[11]

The pathogenicity of bunyaviruses in mice has been analyzed in this study using reassortant viruses derived from La Crosse (LAC), Tahyna (TAH), snowshoe hare (SSH), and trivittatus (TVT) cloned wild-type viruses. Analyses of segment composition indicate that the M RNA determines to a major extent the disease pattern in 4-week-old mice.

Materials and Methods

The origins of prototype LAC, SSH, TAH, and TVT viruses and their reassortants have been described.[3-5,12] Reassortant viruses resulted from coinfection of BHK-21 cells with ts mutants selected either as spontaneously occurring in wild-type virus or in progeny grown in the presence of 5-fluorouracil.[6] Viruses were identified on the basis of oligonucleotide fingerprint analyses by the method of Wachter and Fiers modified by Clewley et al.[13]

[a] Yale Arbovirus Research Unit, Yale School of Medicine, Box 3333 New Haven, Connecticut
[b] Department of Microbiology, The Medical Center, University of Alabama in Birmingham, Birmingham, Alabama

Four-week-old outbred Swiss Webster mice (Taconic Farms, New York) were inoculated in groups of 5 either by the intraperitoneal (ip) or intracerebral (ic) route. Mice were inoculated with serial 10-fold or 100-fold dilutions of supernatant fluid from infected BHK-21 cells and were observed for death for 21 days or for a week after the last observed death. Titers were calculated by the Reed-Muench formula.[14] Mouse brains for histopathologic examination were fixed in 10% formalin or Bouin's solution and stained by hematoxylin and eosin.

Results

Pathogenicity was measured by the ability of the viruses to kill mice following ic and ip inoculation. The cloned wild-type LAC and SSH viruses killed 4-week-old mice by both the ic and ip routes of inoculation. TAH and TVT viruses killed mice by the ic route but not by the ip route of inoculation. Most reassortant viruses with LAC or SSH M RNA, like the parent LAC and SSH viruses, killed mice after ip inoculation (Table 1). All reassortant viruses with TAH M RNA, like the parent TAH virus, failed to kill mice after ip inoculation (Table 2). None of the reassortant viruses were of increased ip virulence, i.e., none killed following ip inoculation in cases where neither parent killed mice.

Pathogenicity was also measured by survival time after ic inoculation. Mice inoculated with LAC cloned wild-type parent virus survived approxi-

Table 1. Intracerebral (ic) and Intraperitoneal (ip) Mouse Pathogenicity of Viruses with LAC and SSH M RNA.

	Log LD_{50}/ml[a]		
Virus	**ic**	**ip**	**Difference[b]**
LAC/LAC/LAC	7.5	4.4	3.1
SSH/LAC/SSH	6.8	4.5	2.3
SSH/LAC/LAC	7.7	4.0	3.7
TAH/LAC/TAH	7.4	4.8	2.6
LAC/LAC/SSH[c]	7.0	2.5	4.5
TAH/LAC/LAC	7.6	3.0	4.6
SSH/SSH/SSH	7.7	5.7	2.0
SSH/SSH/LAC	7.2	3.5	3.7
LAC/SSH/SSH	7.5	>1.5[d]	<6.0

[a]Five 4-week-old mice were inoculated per dilution.

[b]Difference between ic and ip titers.

[c]LAC/LAC/SSH derived from LAC/LAC/LAC I-20 × SSH/LAC/SSH II-13.[12]

[d]All mice died when inoculated undiluted, only dilution tested.

(Source: Adapted in part from Shope et al.[15])

Table 2. Intracerebral (ic) and Intraperitoneal (ip) Mouse Pathogenicity of Viruses with TAH M RNA.

	Log LD_{50}/ml[a]		
Virus	**ic**	**ip**	**Difference[b]**
TAH/TAH/TAH	8.6	<1.5	>7.1
LAC/TAH/LAC	6.5	<1.5	>5.0
SSH/TAH/SSH	5.9	<1.5	>4.4
LAC/TAH/TAH	8.9	<1.6	>7.3
SSH/TAH/TAH	7.1	<1.6	>5.5

[a]Five 4-week-old mice were inoculated per dilution.
[b]Difference between ic and ip titers.
(Source: Data from Shope et al.[15])

mately twice as long as mice inoculated with parent TAH virus. Similarly, as shown in Table 3, most reassortant viruses containing the LAC M RNA survived significantly longer than those bearing the TAH M RNA. The difference was more pronounced with doses of virus from 1.0 to 1.9 log LD_{50} than with larger doses.

Four reassortant viruses so far tested had a significantly reduced pathogenicity for mice when compared to the wild-type parent donor of the M RNA. The four viruses did not kill mice following ip inoculation and each had a markedly prolonged survival time after ic inoculation (Table 4). The LAC/LAC/SSH, as reported elsewhere,[15] has a non-ts L RNA defect which

Table 3. Survival Time of Mice Inoculated Intracerebrally with Viruses Containing LAC and TAH M RNA.

	Average survival in days		
Virus	**1–1.9 log LD_{50}**	**3–3.9 log LD_{50}**	**Difference[b]**
LAC/LAC/LAC	6.6	5.0	1.6
TAH/LAC/LAC	6.8	4.0	2.8
TAH/LAC/TAH	6.0	3.9	2.1
SSH/LAC/SSH	6.8	4.8	2.0
SSH/LAC/LAC	3.6	2.8	0.8
TAH/TAH/TAH	2.6	1.8	0.8
LAC/TAH/LAC	2.2	1.6	0.6
LAC/TAH/TAH	3.6	2.8	0.8
SSH/TAH/SSH	3.0	2.2	0.8
SSH/TAH/TAH	4.6	3.8	0.8

[a]Five 4-week-old mice were inoculated per dilution.
[b]Difference in survival times at 2 log LD_{50} interval.
(Source: Adapted in part from Shope et al.[15])

Table 4. Intracerebral (ic) and Intraperitoneal (ip) Pathogenicity of Viruses with Possible Variant RNA and of Wild-Type Parent Viruses.

	Log LD_{50}/ml[a]		AST[b] at log LD_{50}/ml of				
Virus	**ic**	**ip**	**0–.9**	**1–1.9**	**2–2.9**	**3–3.9**	**>3.9**
LAC/LAC/SSH[c]	6.6	<1.5	13.4	11.4	9.2	10.3	
TVT/SSH/TVT	6.5	<1.5	9.3	8.2		8.8	
TVT/LAC/TVT	6.3	<1.5	12.3		8.4		
TVT/TAH/TVT	5.7	<1.5		8.6		7.2	
LAC/LAC/LAC	7.5	4.4	5.0	6.6	5.4	5.0	3.6
SSH/SSH/SSH	7.7	5.7		4.8			
TAH/TAH/TAH	8.6	<1.5		2.6		1.8	1.6
TVT/TVT/TVT	6.9	<1.5	10.8				8.8

[a]Five 4-week-old mice were inoculated per dilution.
[b]AST: average survival in days following ic inoculation.
[c]LAC/LAC/SSH derived from LAC/LAC/LAC I-16 × SSH/LAC/SSH II-13.[12]
(Source: Adapted in part from Shope et al.[15])

accounts for its attenuation. The basis for modified virulence of TVT/LAC/TVT, TVT/SSH/TVT, and TVT/TAH/TVT has not yet been determined.

Histopathologic examination of brains of mice inoculated ic with LAC and TAH viruses was undertaken to try to determine the basis for the more rapid killing by TAH virus. Infection with each virus induced generalized encephalitis. The lesions associated with LAC infection were lytic with some inflammatory cells and granule cell destruction with microglial proliferation in the olfactory cortex and necrosis of the pyramidal cells of the hippocampus. The lesions associated with TAH infection differed. In addition to necrosis of the olfactory and hippocampal regions, there were necrotic foci in the granular layer of the cerebellum, lesions not found in LAC-virus infected mice.

Discussion

The bunyavirus M RNA is a major determinant of pathogenicity irrespective of the origin of the L and S RNA. This is true both for ability to kill mice by the ip route of inoculation and for rapidity of killing after ic inoculation. It is, however, possible for the pathogenicity of a bunyavirus to be altered by a non-ts lesion in the L RNA as has already been demonstrated for the LAC/LAC/SSH virus derived from LAC/LAC/LAC I-16 × SSH/LAC/SSH II-13 (Table 4).[15] A non-ts defect in either the L or S RNA probably also explains the attenuation of TVT/SSH/TVT, TVT/LAC/TVT, and TVT/TAH/TVT reassortant viruses, since the attenuation effect in the 3 viruses occurred irregardless of the type of the M RNA.

None of the bunyavirus reassortant viruses examined were found to have enhanced pathogenicity for mice. Rubin and Fields showed increased virulence for mice with reassortant reoviruses.[16] In the case of reoviruses, different determinants of virulence occur on at least 2 RNA segments, each of which codes for a different surface polypeptide. It may well be that since both bunyavirus surface glycoproteins are products of a single gene (M RNA) it is not possible to enhance bunyavirus virulence by segment reassortment.

TAH virus kills mice faster than LAC virus and there is an associated difference in the type of brain lesion found by histopathologic examination. It is premature to conclude that the differences in lesions between LAC- and TAH-infected brains account for the difference in survival time after infection. The differences do, however, offer the possibility of correlating the RNA segment with a specific CNS lesion.

Conclusions

Bunyavirus reassortants which were more pathogenic to mice than parent viruses were not found. The M RNA was a major determinant of mouse pathogenicity as measured by the difference between virus dose needed to kill ic and ip, and by average survival times after ic inoculation. Some reassortant bunyaviruses had mouse pathogenicity patterns which did not correspond to that predicted by the M RNA; these viruses may have non-ts mutations of one or more RNA segments. Tahyna and La Crosse wild-type clones had different average survival times afer ic inoculation and also had different histologic lesions in the CNS.

ACKNOWLEDGMENTS
This study was supported by Public Health Service Grant AI-15400 from the National Institute of Allergy and Infectious Diseases, and by National Science Foundation Grant PCM 7813701.

References

1. Gentsch, J.R., Bishop, D.H.L., and Obijeski, J.F. (1977): *J. Gen. Virol.* 34:257–268.
2. Obijeski, J.F., Bishop, D.H.L., Palmer, E.L., and Murphy, F.A. (1976): *J. Virol.* 20:664–675.
3. Gentsch, J.R. and Bishop, D.H.L. (1976): *J. Virol.* 20:351–354.
4. Gentsch, J., Wynne, L.R., Clewley, J.P., Shope, R.E., and Bishop, D.H.L. (1977): *J. Virol.* 24:893–902.
5. Gentsch, J.R., Robeson, G., and Bishop, D.H.L. (1979): *J. Virol.* 31:707–717.
6. Gentsch, J.R., Rozhon, E.J. Klimas, R.A., El Said, L.H., Shope, R.E., and Bishop, D.H.L. (1980): *Virology* 102:190–204.
7. Ozden, S. and Hannoun, C. (1978): *Virology* 84:210–212.
8. Ozden, S. and Hannoun, C. (1980): *Virology* 103:232–234.

9. Gentsch, J.R. and Bishop, D.H.L. (1978): *J. Virol.* 28:417–419.
10. Gentsch, J.R. and Bishop, D.H.L. (1979): *J. Virol.* 30:767–770.
11. Bishop, D.H.L. and Shope, R.E. (1979): In: *Comprehensive Virology, Vol. 14*. Fraenkel-Conrat, H. and Wagner, R.R. (Eds.), Plenum Press, New York and London, pp. 1–156.
12. Rozhon, E.J., Gensemer, P., Shope, R.E., and Bishop, D.H.L. (1981): *Virology,* in press.
13. Clewley, J., Gentsch, J., and Bishop, D.H.L. (1977): *J. Virol.* 22:459–468.
14. Reed, L.J. and Muench, H. (1938): *Am. J. Hyg.* 27:493–497.
15. Shope, R.E., Rozhon, E.J., and Bishop, D.H.L. (1980): *Virology,* submitted.
16. Rubin, D.H. and Fields, B.N. (1980): *J. Exp. Med.* 152, October, in press.

David H.L. Bishop and Richard W. Compans, eds.
The Replication of Negative Strand Viruses

Pathogenesis of Bunyavirus Reassortants in *Aedes triseriatus* Mosquitoes

B.J. Beaty,[a] B.R. Miller,[a] M. Holterman,[a]
W.J. Tabachnick,[a] R.E. Shope,[a] E.J. Rozhon,[b] and
D.H.L. Bishop[b]

Arboviruses of the family Bunyaviridae provide a unique opportunity to examine viral gene contributions to vector-virus interactions. The bunyavirus genome is composed of three RNA segments: a large (L), a middle-sized (M), and a small (S) segment.[1] Viruses of the California group of the Bunyaviridae can reassort segments during dual-infection of tissue cultures.[2,3] The new reassortant viruses contain two RNA segments from one parent virus and one from the other. Such reassortants of known genome composition were used to define bunyavirus gene coding assignments. The S RNA segment codes for the nucleocapsid protein,[4] the M RNA for the two glycoproteins,[5] and although not proven, the L RNA is thought to code for the polymerase.[1]

Two bunyaviruses of the California serogroup, La Crosse (LAC) and snowshoe hare (SSH) viruses and LAC-SSH reassortant viruses were used to determine the molecular basis of bunyavirus development in mosquitoes. Although serologically closely related and capable of segment reassortment *in vitro,*[2] the two viruses are maintained in distinct arbovirus cycles in nature, involving different vectors and vertebrate hosts.[6] LAC virus, a major cause of arthropod-borne encephalitis in the United States, is vectored by the mosquito *Aedes triseriatus*. The virus can be transovarially,[7] venereally,[8] and orally transmitted by the mosquito. Chipmunks and tree squirrels are the principal vertebrates hosts. The ecology of SSH virus is less well known;

[a] Yale Arbovirus Research Unit, Yale University School of Medicine, Department of Epidemiology and Public Health, 60 College Street, P.O. Box 3333, New Haven, Connecticut.
[b] The Medical Center, Department of Microbiology, University of Alabama in Birmingham, Birmingham, Alabama.

other *Aedes* spp have been incriminated as vectors, and the snowshoe hare is the principal vertebrate host.[6]

LAC, SSH, and their reassortant progeny viruses were used in studies to determine the efficiency of infection, transmission potential, and the comparative pathogenesis of each virus in *Aedes triseriatus* mosquitoes.

Efficiency of Infection Studies[9]

The most basic of vector-virus interactions involves the midgut. Upon ingestion of a viremic blood-meal, the virus must infect and replicate in midgut epithelial cells.[10] Subsequently, virus must escape from midgut cells into the hemocoele in order to infect organs involved in virus transmission.

To determine the molecular basis for bunyavirus infection of vector midgut cells, *Aedes triseriatus* mosquitoes were permitted to engorge upon suspensions of defibrinated rabbit blood and the respective parent or reassortant virus (Table 1). After 14 days extrinsic incubation, mosquitoes were killed by freezing. Heads and abdomens were severed and smeared on slides. After fixation, the tissue preparations were examined for the presence of viral antigen using immunofluorescence.[11] Detection of viral antigen in abdominal tissues indicated that the animal had become infected. Detection of antigen in tissues of the head indicated the virus had infected midgut cells and subsequently disseminated to infect secondary target organ systems. Not surprisingly, LAC/LAC/LAC was the most efficient virus in infection of *Aedes triseriatus* midgut cells (Table 1). This is what would be expected because *Aedes triseriatus* is its natural vector. However, there seemed to be little difference, with one exception, in the ability of the reassortant viruses of SSH virus to infect the mosquito. The virus genotypes and infection rates were: SSH/LAC/LAC—76%, SSH/LAC/SSH—83%, and SSH/SSH/SSH—

Table 1. Infection of *Aedes triseriatus* Midgut Cells by Reassortant Bunyaviruses.[a]

	Infected (%)		
Genome	**7.5–6.6**[b]	**6.5–5.6**	**Total**
LAC/LAC/LAC	25/25 (100)[c]	21/22(96)	46/47(98)
SSH/LAC/LAC	24/28(86)	8/14(57)	32/42(76)
SSH/LAC/SSH	11/12(92)	13/17(76)	24/29(83)
SSH/SSH/SSH	21/23(91)	9/17(53)	30/40(70)
LAC/LAC/SSH[d]	7/31(23)	0/28(0)	7/59(12)

[a]Adapted in part from Beaty et al. (1980a).

[b]Log_{10} pfu per ml in blood-virus suspension.

[c]Numerator-number of mosquitoes with viral antigen in midgut tissues. Denominator-number of mosquitoes engorged.

[d]This virus was subsequently found to have silent mutations in the L and possibly the other RNA segments which have resulted in attenuated biologic activity.

70%. The exception at 12% was LAC/LAC/SSH virus. This virus was subsequently found to have silent mutations in the L and possibly other RNA segments. These mutations also resulted in attenuated virulence in mice (unpublished data).

Although the viruses seemed equally capable of midgut infection, there was a difference in the ability of the viruses to escape from midgut cells and to infect secondary target organs (Table 2). Excluding the LAC/LAC/SSH virus (the virus with attenuated biologic activity), those viruses with the LAC M RNA segment efficiently established disseminated infection. The dissemination rates were: LAC/LAC/LAC—91%, SSH/LAC/LAC—53%, and SSH/LAC/SSH—50%. Only 6/30 (20%) of the mosquitoes fed SSH virus developed disseminated infections (Table 2).

Since *Aedes triseriatus* is its natural vector, it is logical that LAC virus was more efficient than the other viruses in midgut infection and virus dissemination. However, it was surprising that SSH virus was as capable of infection of midgut cells as the reassortant viruses containing the LAC M RNA. Evidently, if infection of midgut cells is a reception function, the receptor is incapable of discriminating between LAC and SSH M RNA gene products—the glycoproteins.

In contrast, those viruses with the LAC M RNA segment were more efficient in establishing a disseminated infection. Those mosquitoes infected with SSH virus that did not develop disseminated infections frequently contained substantial amounts of viral antigen in abdominal squash preparations. It seems reasonable to speculate that the SSH M RNA gene products either inefficiently interact with *Aedes triseriatus* cell membranes or that the glycoproteins are poorly recognized by the S RNA gene products of the parent viruses.

Table 2. Dissemination of Bunyavirus Infection from *Aedes triseriatus* Midgut Cells.[a]

	Disseminated infection (%)		
Genome	**7.5–6.6**[b]	**6.5–5.6**	**Total**
LAC/LAC/LAC	23/25(92)[c]	19/21(90)	42/46(91)
SSH/LAC/LAC	13/24(54)	4/8 (50)	17/32(53)
SSH/LAC/SSH	6/11(55)	6/13(46)	12/24(50)
SSH/SSH/SSH	6/21(29)	0/9 (0)	6/30(20)
LAC/LAC/SSH[d]	0/7 (0)	0/0 (0)	0/7 (0)

[a]Adapted in part from Beaty et al. (1980a).

[b]Log_{10} pfu per ml in blood-virus suspension.

[c]Numerator-number of mosquitoes with antigen in head tissues. Denominator-number of mosquitoes with antigen in midgut.

[d]This virus was subsequently found to have silent mutations in the L and possibly the other RNA segments which resulted in attentuated biologic activity.

Transmission Potential Studies.[12]

To determine transmission potential, parental and reassortant viruses were compared for their ability to be transmitted by *Aedes triseriatus,* the natural vector of LAC virus. Mosquitoes were infected by intrathoracic inoculation of 1,000 pfu of the respective virus. This route of inoculation was chosen in order to circumvent variables associated with midgut passage. After extrinsic incubation periods of 10 to 14 days, each mosquito was permitted to feed upon an individual suckling mouse. After the mosquito engorged, the mouse was identified and observed for morbidity and mortality for 10 days. Specificity of mouse mortality was confirmed by detection of viral antigen in brain impression smears by immunofluorescence. Engorged mosquitoes were also examined by immunofluorescence for the presence of viral antigen.

There was an extraordinary correlation between the presence of the LAC M RNA segment and transmission of virus to suckling mice (Table 3). Those viruses containing the LAC M RNA were transmitted by 96% of the engorging mosquitoes; however, those viruses containing the SSH M RNA were transmitted by only 35% of the mosquitoes. The LAC/LAC/SSH virus was again the exception; there was no successful transmission by 30 engorging mosquitoes. However, as previously described, this reassortant virus was found to have silent mutations in the L and M RNA segments resulting in attenuated biologic activity. This may be a factor in the unsuccessful transmission attempts or, alternatively, the virus may not be infectious to mice.

Table 3. Transmission of Bunyavirus Reassortants by *Aedes triseriatus* mosquitoes.[a]

Genome RNA segment composition	Number transmissions	Number infected	Percent transmission
LAC/LAC/LAC	60	60	100
SSH/LAC/LAC	22	23	96
SSH/LAC/SSH	35	39	90
Total LAC M RNA	117	122	96
LAC/SSH/LAC	8	19	42
LAC/SSH/SSH	5	14	36
SSH/SSH/LAC	11	35	31
SSH/SSH/SSH	12	36	33
Total SSH M RNA	36	104	35
LAC/LAC/SSH[b]	0	30	0

[a]Adapted in part from Beaty et al. (1980b). Four separate trials were conducted. The results of the trials are combined for ease of presentation.

[b]LAC/LAC/SSH virus was subsequently found to have silent mutations in the L and possibly the other RNA segments. A newly constructed LAC/LAC/SSH virus with a new L RNA but the same M RNA segment was transmitted by 9/14 (64%) of the mosquitoes.

To determine the anatomic basis for this phenomenon, salivary glands were dissected from selected engorging mosquitoes and examined for the presence of viral antigen. When the mosquitoes were infected with viruses containing thc LAC M RNA, there was uniform agreement between the presence of antigen in the salivary glands and virus transmission. Salivary glands were dissected from 13 mosquitoes infected with LAC/LAC/LAC. All 13 contained viral antigens and all 13 transmitted virus to mice. For SSH/LAC/LAC and SSH/LAC/SSH, 3/3 and 4/4 mosquitoes transmitted and contained viral antigen in the glands respectively. In contrast, less than 40% of those containing demonstrable SSH/SSH/LAC (2/8) and SSH/SSH/SSH (4/8) antigen in the glands transmitted the virus to mice.

Since virus antigen was detected in the salivary glands of nontransmitters, it seems that, as with dissemination of virus from midgut cells, those viruses with the SSH M RNA segment were less efficient in escaping from the infected cell. Alternatively, there may be major differences in the ability of the viruses containing the SSH M RNA to replicate in the salivary gland and midgut cells. If viruses containing the SSH M RNA replicate more slowly than those with the LAC M RNA, then longer extrinsic incubation periods may result in increased transmission and dissemination rates.

Summary

A genetic approach comparing wild type and reassortant LAC and SSH viruses was used to define at least partially the molecular basis for dissemination of bunyaviruses from the midgut and subsequent *per os* transmission by the vector. The M RNA segment through its gene products appears to be the major determinant of the phenomena.

ACKNOWLEDGMENTS

This research was supported by Public Health Service Research Grants AI-15426 and AI-15400 from the National Institute of Allergy and Infectious Diseases and by National Science Foundation Grant PCM-7813701. B.R.M. and E.J.R. were supported by Public Health Service postdoctoral fellowships (Training Grants 1-T32-AI-07098 and 1-T32-AI-07041 from NIAID).

References

1. Bishop, D.H.L. and Shope, R.E. (1979): *Bunyaviridae in Comprehensive Virology*. Fraenkel-Conrat, H. and Wagner, R.R. (Eds.), Plenum Press, New York.
2. Gentsch, J. and Bishop, D.H.L. (1976): Recombination and complementation between temperature sensitive mutants of the bunyavirus, snowshoe hare virus. *J. Virol.* 20:351–354.
3. Gentsch, J., Wynne, L.R., Clewley, J.P., Shope, R.E., and Bishop, D.H.L. (1977): Formation of recombinants between snowshoe hare and La Crosse bunyaviruses. *J. Virol.* 24:893–902.
4. Gentsch, J. and Bishop, D.H.L. (1978): The small viral RNA segment of bunyaviruses codes for the viral nucleocapsid (N) protein. *J. Virol.* 28:417–419.
5. Gentsch, J. and Bishop, D.H.L. (1979): The M RNA segment of bunyaviruses codes for two glycoproteins, G1 and G2. *J. Virol.* 30:767–770.

6. Le Duc, J. (1979): The ecology of California group viruses. *J. Med. Entomol.* 16:1–17.
7. Watts, D.M., Pantuwatana, S., De Foliart, G.E., Yuill, T.M., and Thompson, W.H. (1973): Transovarial transmission of La Crosse virus (California encephalitis group) in the mosquito *Aedes triseriatus*. *Science* 182:1140–1141.
8. Thompson, W.T. and Beaty, B.J. (1977): Venereal transmission of La Crosse (California encephalitis) arbovirus in *Aedes triseriatus* mosquitoes. *Science* 196:530–531.
9. Beaty, B.J., Miller, B.R., Shope, R.E., Rozhon, E.J., and Bishop, D.H.L. (1980a): Molecular basis of bunyavirus *per os* infection of mosquitoes: the role of the M RNA segment. *Proc. Natl. Acad. Science*, submitted.
10. Chamberlain, R.W. and Sudia (1961): Mechanisms of transmission of viruses in mosquitoes. *Ann. Rev. Entomol.* 6:371–390.
11. Beaty, B.J. and Thompson, W.H. (1976): Delineation of La Crosse virus in developmental stages of transovarially infected *Aedes triseriatus*. *Am. J. Trop. Med. Hyg.* 25:505–512.
12. Beaty, B.J., Holterman, M., Tabachnick, W., Shope, R.E., Rozhon, E.J., and Bishop, D.H.L. (1980b): Molecular basis of bunyavirus transmission by mosquitoes: the role of the middle-sized RNA segment. *Science*, submitted.

Copyright 1981 by Elsevier North Holland, Inc.
David H.L. Bishop and Richard W. Compans, eds.
The Replication of Negative Strand Viruses

Genetics of the Bunyamwera Complex

C.U. Iroegbu and C.R. Pringle[a]

Introduction

The genus *Bunyavirus* of the family Bunyaviridae contains 105 viruses arranged in 13 serological groups.[1] One of these groups, the Bunyamwera group, contains 19 viruses, 12 of which form the Bunyamwera Complex. Three of these viruses—Batai virus (BAT), Bunyamwera virus (BUN) and Maguari virus (MAG)—were chosen as representative of the span of antigenic variation in the complex. Batai virus and Maguari virus are more closely related to each other than either is to Bunyamwera virus.[2] Each virus originates from a different continent and there is no known overlap in their distribution. Batai virus was isolated initially from mosquitos in Malaysia, and subsequently from other regions of southeastern Asia, India, Japan, Czechoslovakia, Yugoslavia and the Ukranian SSR. There was serological evidence of infection of the human population throughout its range. Bunyamwera virus was isolated from mosquitos and man in Uganda, and 80% of the population of central and southern Africa showed serological evidence of past infection with this virus. Maguari virus was isolated from mosquitos in Brazil and appeared to be prevalent in human and animal populations of central and southern American.

One conclusion derived from this genetic study is that the Bunyamwera Complex can be regarded as a single gene pool, despite the antigenic diversity and ecological and geographical isolation of the individual viruses of the complex.[3]

[a] M.R.C. Virology Unit, Institute of Virology, Church Street, Glasgow G11 5JR, Scotland.

Genetic Interactions Between Temperature-Sensitive Mutants

The bunyaviruses are negative-stranded RNA viruses with single-stranded segmented genomes of 3 unique sub-units (L, M, and S) of 4–6 $\times$ 10^6 total mol wt and it has been established for the California encephalitis (CE) group that recombination occurred by reassortment of sub-units.[1] In the CE group, however, only 2 of the 3 potential recombination groups have been identified.[4]

High frequency recombination has been observed also between temperature-sensitive (*ts*) mutants of Batai, Bunyamwera and Maguari viruses obtained by mutagenesis with 5-fluorouracil. Recombination was an early event and the maximum frequency was observed at a multiplicity of 10 pfu/cell.[3] Recombination has been observed in both mammalian (BHK-21) and insect (*Aedes albopictus*) cells, despite lack of cytopathogenicity and poor growth in mosquito cells (Pringle and Iroegbu, unpublished data).

Mixed infections were carried out with all combinations of *ts* mutants of each virus and recombination frequencies (RF%) were calculated by a standard method.[3,4] The mutants fell into discrete groups with RF values at least 100-fold above the self-infection background and ranging between 1.2% and 98% for between group crosses. Table 1 gives the number of *ts* mutants of each virus isolated and their disposition by recombination group. One of the 46 *ts* mutants of Maguari virus was a double mutant, and one remains to be classified. Initially, as in the CE group, only two recombination groups (I and II) were identified, but recent experiments have clearly established that mutant MAG *ts* 23 represents the third group predicted by the existence of 3 genome sub-units (Iroegbu and Pringle, in preparation). The reason for the infrequency of group III mutants remains to be established.

Analogous crosses were carried out using heterologous pairs of *ts* mutants from groups I and II of each virus. Table 2 shows that significant recombination was observed in 9 of the 15 possible combinations of mutants, and with a single exception the RF values for homologous combinations were higher than for heterologous combinations. The data in Table 2 shows that recombination groups I and II of each virus were homologous, and consequently represent mutation of the same 2 sub-units of the tripartite genome.

Table 1. Classification of 5-Fluorouracil-Induced *ts* Mutants of Bunyamwera Complex Viruses into Recombination Groups.

		Recombination Group				
Virus	**Number isolated**	**I**	**II**	**III**	**I + II**	**Unclassified**
Batai	5	1	4	0	0	0
Bunyamwera	8	5	3	0	0	0
Maguari	46	12	31	1	1	1

Table 2. Homologous and Heterologous Recombination (RF%) Between *ts* Mutants of Groups I and II of Batai (BAT), Bunyamwera (BUN) and Maguari (MAG) Viruses.

Virus	**Batai**		**Bunyamwera**		**Maguari**	
Mutant	**BAT *ts* 1(I)**	**BAT *ts* 2(II)**	**BUN *ts* 5(I)**	**BUN *ts* 8(II)**	**MAG *ts* 7(I)**	**MAG *ts* 8(II)**
BAT *ts* 1 (I)	0.07	54.5	0.002*	53.6	0.006	0.2
BAT *ts* 2 (II)		0.05	6.5	0.05	0.3	0.0002
BUN *ts* 5 (I)			0.0003	27.9	0.05	0.66
BUN *ts* 8 (II)				0.005	0.79	0.003
MAG *ts* 7 (I)					0.06	27.1
MAG *ts* 8 (II)						0.001

*RF value from the cross BUN *ts* 2 (I) × BAT *ts* 1 (I).

The Polypeptides of Heterologous Recombinants

The bunyavirus virion has four major proteins, the L (large) and N (nucleoprotein) of the core, and the glycosylated G1 and G2 of the envelope. Of these the G1 and N polypeptides of the 3 Bunyamwera Complex viruses could be differentiated by slab gel electrophoresis.[3] Figure 1a shows that a recombinant from the cross MAG *ts* 7 (I) × BUN *ts* 8 (II) possessed a MAG G1 polypeptide and a BUN N polypeptide (track 3), whereas a recombinant from the cross BUN *ts* 1 (I) × MAG *ts* 8 (II) possessed a BUN G1 polypeptide and a MAG N polypeptide (track 4). Figure 1b shows that a recombinant

Figure 1. Electrophoresis in 7.5% polyacrylamide gel of the ^{35}S-methionine-labeled polypeptides of parental mutants and recombinant clones. (a) Bunyamwera virus × Maguari virus. Track 1, mutant MAG *ts* 7 (I); track 2, mutant MAG *ts* 8 (II); track 3, recombinant from MAG *ts* 7 (I) × BUN *ts* 8 (II); track 4, recombinant from MAG *ts* 8 (II) × BUN *ts* 1 (I); track 5, mutant BUN *ts* 1 (I); track 6, mutant BUN *ts* 8 (II). (b) Batai virus × Maguari virus. Track 1, mutant MAG *ts* 7 (I); track 2, mutant MAG *ts* 8 (II); track 3, recombinant from MAG *ts* 8 (II) × BAT *ts* 1 (I); track 4, recombinant from MAG *ts* 7 (I) × BAT *ts* 2 (II); track 5, mutant BAT *ts* 1 (I); track 6, mutant BAT *ts* 2 (II).

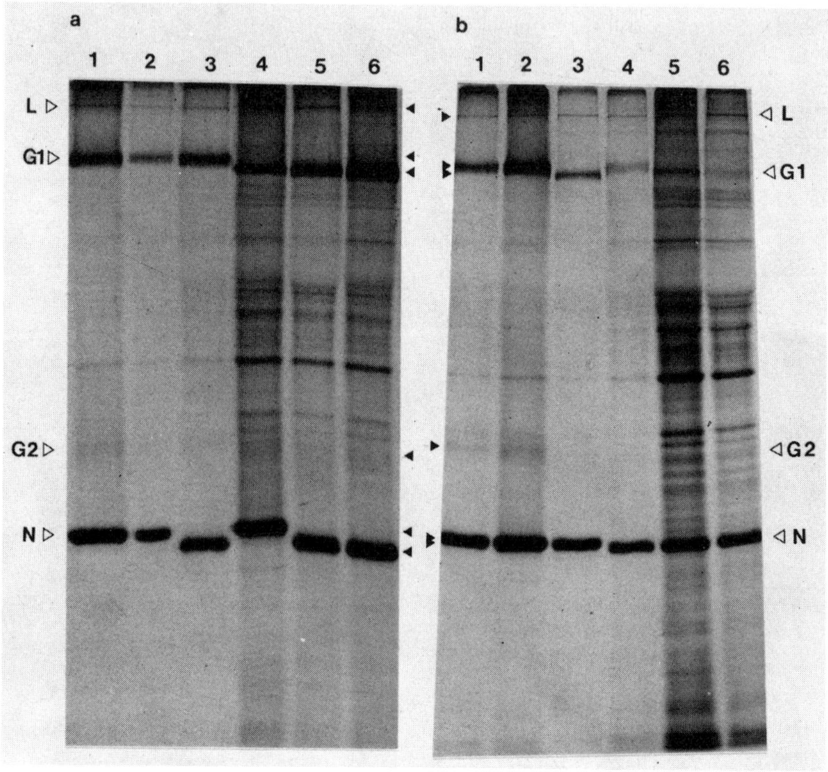

from the cross BAT *ts* 1 (I) × MAG *ts* 8 (II) possessed a BAT G1 and a MAG N (track 3), whereas a recombinant from the cross MAG *ts* 7 (I) × BAT *ts* 2 (II) possessed a MAG G1 and a BAT N (track 4). Likewise in crosses of Batai and Bunyamwera virus mutants (not shown), the group I parent donated the G1 (and G2, since both are encoded in the M-RNA[5]) polypeptide, and the group II parent the N polypeptide. Several recombinants from each cross had these properties, and it was concluded, therefore, that the group I mutation affected the N protein and the group II mutation the G1 (or G2) protein.

Neutralization Specificity of Heterologous Recombinants

The pattern of neutralization of recombinant clones by antisera against the 3 parental viruses confirmed that neutralization specificity was donated by the group I parent (Figure 2). This supports the conclusion that the group I parent donated the envelope proteins G1 and G2, and that the group II mutation affected the G1/G2 gene (the M RNA sub-unit).

Determination of Plaque Morphology

Maguari virus produced large opaque plaques on BS-C-1 monolayers. Figure 3 shows that recombinants from crosses of mutants of Batai and Maguari viruses, and of mutants of Bunyamwera and Maguari viruses, produced plaques which resembled the group I parent in size and the group II parent in

Figure 2. The neutralization characteristics of Batai, Bunyamwera and Maguari viruses and their heterologous recombinants. BAT: anti-Batai virus serum; BUN: anti-Bunyamwera virus serum; MAG: anti-Maguari virus serum.

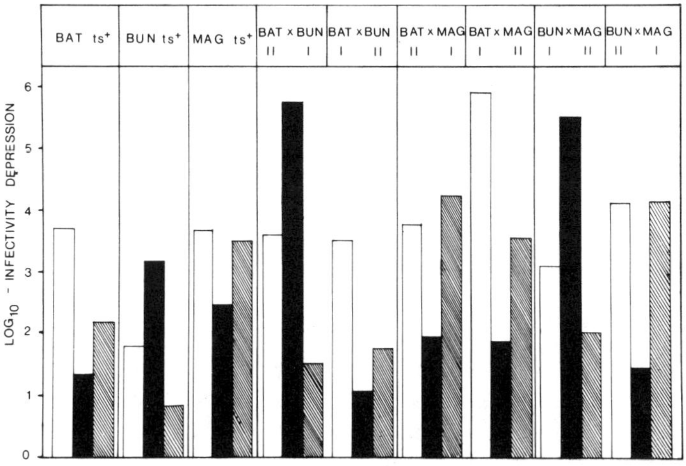

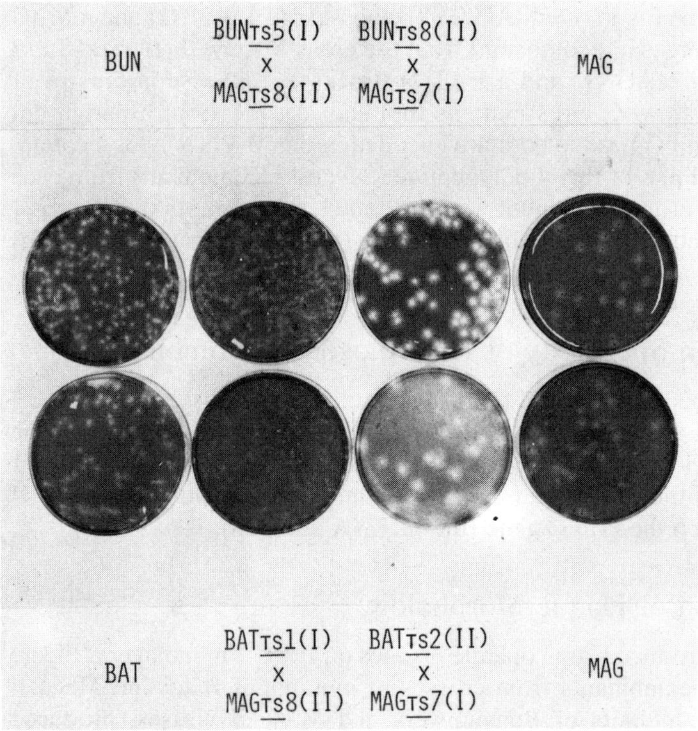

Figure 3. Plague morphology of parental and recombinant virus.

opacity. Thus it appeared that the G1 (and/or G2) proteins were important in determining rapidity of spread and the N protein lytic potential.

Intragenic Complementation

The high frequency and early occurrence of recombination obscured intergenic complementation, but complementation was detected between two non-recombining mutants of Maguari virus (Table 3). Complementation between mutants assigned to group I suggested that more than one gene may be encoded in the genome sub-unit encoding the N protein,[3] just as the M RNA segment encodes both the G1 and G2 proteins.[5]

Summary and Conclusions

(1) Recombination between *ts* mutants occurred at high frequency, probably by reassortment of sub-units, in both mammalian and insect cells. There appeared to be no genetic barrier to exchange of 2 at least of the 3 sub-units of the genome within the Bunyamwera Complex.

Table 3. Complementation and Absence of Recombination in a Mixed Infection with a Pair of Group I Mutants of Maguari Virus.

Complementation:

Virus	**Yield Assayed at 31°C**	**Yield Assayed at 38.5°C**	**Complementation index**
MAG *ts* 6	10	10	—
MAG *ts* 7	25	10	—
MAG *ts* 6 × MAG *ts* 7	1225	10	35

Recombination

Virus	**Yield assayed at 31°C**	**Yield assayed at 38.5°C**	**Recombination frequency (%)**
MAG *ts* 6	3.9×10^7	6.2×10^2	(0.003)
MAG *ts* 7	2.4×10^7	7.5×10^3	(0.06)
MAG *ts* 6 × MAG *ts* 7	3.4×10^7	9×10^3	0.005

(2) Three recombination groups have been identified by analysis of 46 *ts* mutants of Maguari virus. Analysis of the proteins of heterologous recombinants suggested that recombination group I corresponded to the N gene sub-unit, and group II to the G1/G2 gene sub-unit.
(3) The G1 (and/or G2) protein was associated with neutralization specificity and plaque size, and the N protein with plaque opacity.
(4) Intragenic complementation in group I may indicate that the N gene sub-unit encodes a second protein.

ACKNOWLEDGMENTS
C.U. Iroegbu was supported by a Junior Fellowship of the University of Nigeria, Nsukka. Valeria Devine and Margaret Wilkie provided valuable technical assistance. Figures 2 and 3 and Tables 2 and 3 are reproduced from the *Journal of Virology*.

References

1. Bishop, D.H.L. and Shope, R.E. (1979): In: *Comprehensive Virology,* Volume 14. Fraenkel-Conrat, H. and Wagner, R.R. (Eds.), Plenum Press, New York, pp. 1–56.
2. Hunt, A.R. and Calisher, C.H. (1979): *Am. J. Trop. Med. Hyg.* 17:78–89.
3. Iroegbu, C.U. and Pringle, C.R. (1981): *J. Virol.* 37:383–394.
4. Bishop, D.H.L. (1979): *Curr. Topics in Microbiol. Immunol.* 86: 1–33.
5. Gentsch, J.R. and Bishop D.H.L. (1979): *J. Virol.* 30:767–770.

Copyright 1981 by Elsevier North Holland, Inc.
David H.L. Bishop and Richard W. Compans, eds.
The Replication of Negative Strand Viruses

Antigenic Components of Punta Toro Virus

J.M. Dalrymple,[a] C.J. Peters,[a]
J.F. Smith[b] and M.K. Gentry[a]

The sandfly fever agents or Phlebotomus Fever Serogroup viruses have been recently proposed as members of a *Phlebovirus* genus of the family Bunyaviridae. Together with the most prominent member of this proposed genus, Rift Valley Fever virus, these agents share antigenic characteristics and common molecular characteristics. All members so far studied contain a negative-stranded, tripartite, RNA genome, two surface envelope glycoproteins and a smaller, nonglycosylated nucleocapsid or core protein. The participation of these structural proteins in the serological reactivities of these viruses has not yet been described.

Our research has been primarily concerned with developing models that will help to explain the diversity that exists between members of this group in their biology, antigenicity and molecular characteristics such as oligonucleotide fingerprints. We suspect that genetic change in nature may be a continual process and that dual infection of arthropod vectors resulting in natural reassortants is a real possibility. If these genetic changes are phenotypically reflected as changes in the antigens of these viruses, it becomes difficult to pursue vaccine development, interpret serological responses or predict immunity without defining specific antigenic determinants as markers. Thus, describing some of these antigens for a single virus was the purpose of this preliminary study.

Punta Toro virus was selected as a starting point because of its antigenic relationship to Rift Valley Fever virus, a major pathogen that we plan to

[a] Virology Division, USAMRIID, Frederick, Maryland.
[b] University of Maryland School of Medicine, Baltimore, Maryland.

ultimately investigate; however, Punta Toro virus does not require the strict laboratory containment of the Rift Valley Fever agent. Other reasons for the selection of this agent included the recent observation that isolates of Punta Toro virus from different geographic regions of Panama differ markedly in their virulence for laboratory animals (Table 1). Such variation appears quite common among this group of agents. An Egyptian strain of Rift Valley Fever virus produces a benign infection of Lewis rats, a fulminant infection of Wistar-Furth animals and a late encephalitis of ACI strain rats. Different strains of Rift Valley Fever virus can be readily distinguished by the disease produced in the Wistar-Furth rat. Lunyo, an agent antigenically indistinguishable from Rift Valley Fever virus, produces no detectable disease in mice. Avirulent variants can be selected from virulent isolates of RVFV that produce no disease in rat, hamster, or mouse and, as mentioned previously, different geographic isolates of Punta Toro virus differ markedly in virulence for the hamster.

Lymphocyte hybridomas producing monoclonal antibody were investigated as a possible method that would provide: (1) identification of the major virus antigens that contribute to immunity or protection from disease, (2) an adjunct to the classical biochemical and biophysical antigen separation procedures that classically result in a certain amount of degradation, and (3) the serological tools necessary to investigate the antigenic relationships that these viruses share. Infected suckling mouse brain suspensions were selected for animal immunization in anticipation of a preparation containing all virus-specified proteins and a minimum number of foreign host cell contaminants. Young adult BALB/C mice were given a primary immunizing injection and a booster intravenous injection at eight weeks. Three days after the second injection, spleens were harvested from the immunized mice and fused with the P3 plasmacytoma IgG_1 secreting cell line using polyethyleneglycol. Cultures of the fused products or lymphocyte hybridomas were screened for Punta Toro virus antibody with a solid phase, radioimmune assay (RIA) using nonionic detergent disrupted infected cell lysates as anti-

Table 1. Evidence of Genetic Variation in Various Phlebotomous Fever Serogroup Viruses as Measured by the Disease Produced in Laboratory Animals.

Genetic variable	Constant	Disease		
		Benign	Fulminant	Late encephalitis
Inbred rats	Egyptian RVF virus	Lewis	Wistar-Furth	ACI
RVF virus	Wistar-Furth rat	South African	Egyptian	—
RVF virus	Mouse	Lunyo	Other	—
RVF virus	Mouse, hamster, rat	Selected clones	Egyptian	
Punta Toro	Hamster	Western Panama	Eastern Panama	

gen which were attached directly to polyvinyl chloride microtiter plates (Table 2). Of the 864 cultures inoculated with the products of the cell fusion, 675 or 78% produced actively growing hybridomas. A total of these growing cultures yielded RIA detectable antibody in supernatant fluids. Over 300 of these culture fluids were also examined by fluorescent antibody (FA) procedures and obtained good correlation with the preceding RIA data. Because of the physical limitations of the laboratory, only 105 of these positive cultures were subsequently cloned on soft agar over feeder layers. Representative clones, presumably originating from a single hybridoma cell, were selected, again on the basis of an RIA positive supernatant, passed to larger cultures and frozen in liquid nitrogen. Only 19 of these have been removed from nitrogen, passed in culture to produce antibody containing cell culture fluids and injected into Pristane primed BALB/C mice for the production of ascitic fluids. These 19 clones are presented in some detail and include all cultures of current serological interest.

To determine if any of these monoclonal antibodies were directed at nonstructural virus proteins, the 105 positive cell supernatants were examined by RIA using both infected cell lysate antigens and purified virion antigens. All of these 105 hybridomas reacted with both infected cells and purified virions.

Having selected 105 clones on the basis of their RIA and FA reactions, it was next of interest to determine the specific viral proteins that contained the antigenic determinants recognized by these monoclonal antibodies. This was accomplished by an immunoprecipitation assay in which Punta Toro virus infected cells served as antigen. Infected cells were labeled for 5 hours with ^{3}H-leucine 16 hours after infection. These cells were subsequently solubilized in buffer containing detergents, and aliquots of this lysate were reacted with media supernatants of each of the selected clones. The resulting immune complexes were precipitated with solid phase protein A, washed repeatedly in lysis buffer and prepared for discontinuous electrophoresis and fluorography. The precipitates generated by anti-Punta Toro hyperimmune mouse ascitic fluids served to establish the electrophoretic migration of each of the known viral proteins.

Table 2. Fusion Efficiency in Lymphocyte Hybridoma Production and Selection of Representative Clones.

Production of lymphocyte hybridomas producing monoclonal antibody to Punta Toro virus	
Hybridomas produced	675/864—78%
Positive for Punta Toro antibody	435/675—64%
Selected for cloning and freezing	105/435—24%
Selected for intensive study	19/105—18%

Using these procedures, the majority of the selected clones (83) were shown to react specifically with the nucleocapsid protein (27K), four reacted with GP2 (56K) and one with GP1 (66K). Unreacting clones were further tested in a modified indirect assay utilizing rabbit antisera reactive against mouse IgM, IgA and IgG_1. This procedure demonstrated that seven additional clones contained activity against the nucleocapsid protein. The specificity of the remaining clones could not be determined and presumably represents clones that no longer produce antibody or do so at undetectable levels.

Each of the monoclonal antibody containing ascitic fluids were examined for their ability to inhibit virus hemagglutination (Table 3). The anti-GP1 and all four of the anti-GP2 clones reacted to high titer with the homologous Punta Toro antigen. None of the anti-nucleocapsid reactors exhibited any evidence of inhibition. All ascitic fluids were similarly tested for inhibition of Rift Valley Fever virus hemagglutinin but were uniformly negative.

Neutralizing activity of these antibodies was examined using the plaque reduction neutralization test (Table 4). Antibody to the GP1 determinant neutralized Punta Toro virus to high titer as did one of the reactors to GP2. The other three anti-GP2 reactors must be considered suspect because these low titers do not reflect the higher hemagglutination-inhibition titers of these same clones. There were surviving plaques in the titration of these three ascitic fluids, making the precise estimation of a titer difficult and perhaps suggesting a different antibody avidity or a separate determinant. Again the anti-nucleocapsid ascitic fluids were negative and all monoclonal antibodies failed to neutralize Rift Valley Fever virus.

Even though the ability of serum neutralizing antibody is often associated with immunity, the ultimate test for protection involves challenge of an

Table 3. Reaction of Ascitic Fluids Containing Monoclonal Antibody in the Hemagglutination-Inhibition Test.

	Antigen	
Antibody	**Punta Toro**	**Rift Valley Fever**
Anti-GP1	320	<10
Anti-GP2		
(A)	640	<10
(B)	320	<10
(C)	160	<10
(D)	1280	<10
Anti-nucleocapsid (N = 14)	all < 10	all < 10
Punta Toro HMAF	320	40
Rift Valley Fever HMAF	10	160

Table 4. Evidence of Plaque Reduction Neutralizing Antibody in Ascitic Fluid from Hybridomas Reactive to the Envelope Glycoproteins.

	Virus	
Antibody	**Punta Toro**	**Rift Valley Fever**
Anti-GP1	2560	<10
Anti-GP2		
(A)	40	<10
(B)	40	<10
(C)	±10	<10
(D)	2560	<10
Anti-nucleocapsid (N = 14)	<10	—
Punta Toro HMAF	>2560	40
Rift Valley Fever HMAF	±10	2560
Karimabad HMAF	10	40

animal with the infectious agent. We attempted to determine the protective capacity of these monoclonal antibodies by injecting hamsters with ascitic fluids and challenging with virulent Punta Toro virus (Table 5). The prototype Punta Toro virus used to immunize the mice for the preparation of the hybridomas does not normally kill hamsters. Consequently, the challenge virus consisted of the virulent isolate from eastern Panama previously described. The single anti-GP1 and an additional anti-GP2 were capable of protecting hamsters from a lethal challenge. It is not surprising that these are the same antibody preparations that contained high titered neutralizing antibody. One of the anti-GP2 ascitic fluids remains to be examined.

Table 5. Passive Protection of Hamsters Using Monoclonal Antibodies to Punta Toro Envelope Glycoproteins.

Hybridoma	**Protection**
Anti-GP1	Pos
Anti-GP2	
(A)	Neg
(B)	Neg
(C)	NT
(D)	Pos
Anti-nucleocapsid (N = 11)	Neg
Punta Toro HMAF	Pos

In summary, we have produced a number of lymphocyte hybridomas producing monoclonal antibodies to Punta Toro virus. The majority of these antibodies reacted exclusively with the nucleocapsid protein; however, four clones produced antibody to envelope glycoprotein GP2 and a single hybridoma produced antibody to GP1. All five of the monoclonal antibodies to envelope glycoproteins inhibited Punta Toro virus hemagglutination. The anti-GP1 and a single anti-GP2 clone had high virus neutralization titers and each of these were capable of passively protecting hamsters from lethal virus challenge. None of these monoclonal antibodies cross reacted with Rift Valley Fever virus in the limited number of experiments conducted. It is anticipated that further study of these viruses with monoclonal antibody will ultimately lead to a better picture of their antigenic composition and the biological function of these proteins.

ORTHOMYXOVIRUSES

David H.L. Bishop and Richard W. Compans, eds.
The Replication of Negative Strand Viruses

Analysis of Influenza C Virus Structural Proteins and Identification of a Virion RNA Polymerase

Herbert Meier-Ewert,[a] Arno Nagele,[a] Georg Herrler,[b] Sukla Basak,[b] and Richard W. Compans[b]

Influenza C virions possess a multisegmented, single-stranded RNA genome with at least seven segments, ranging from 0.94 to 0.23 $\times$ 10^6 in molecular weight.[1-3] Analysis of the viral proteins by polyacrylamide gel electrophoresis has revealed three major structural polypeptides: a glycoprotein (gp88) of about 88,000 mol wt which constitutes the spikes on the viral surface, a nucleoprotein (NP) of 66,000 mol wt and a membrane protein (M) of 26,000 mol wt.[4,5] In permissive host cells, gp88 is posttranslationally cleaved to yield two smaller glycoproteins, gp65 and gp30, of 65,000 and 30,000 mol wt, respectively. These smaller glycoproteins resemble HA1 and HA2 of influenza A viruses in that they are linked together by disulfide bonds to form a glycoprotein with an electrophoretic mobility similar to gp88.[6,7] Like the HA of influenza A and B viruses, cleavage of the influenza C viral glycoprotein is necessary for full viral infectivity.[6] However, influenza C viruses differ from other orthomyxoviruses in possessing only a single type of glycoprotein gene product, and in having a receptor-destroying enzyme that is not an α-neuraminidase.[4,8]

The coding capacity of the influenza C genome indicates that further gene products should exist, in addition to the gp88, NP and M proteins described previously. In this study, we have further characterized the virion proteins, and identified additional gene products as well as a virion RNA polymerase activity.

[a] Virologische Abteilung, Institut für Med. Mikrobiologie, Technische Universität, München, Biedersteiner Str. 29, 8000 München 40/West Germany.
[b] Department of Microbiology, University of Alabama in Birmingham, Birmingham, Alabama.

Fractionation of Virions with Triton X-100

Since the gp88 glycoprotein in influenza C virions always appears as a broad band on polyacrylamide gels, the possibility existed that minor viral protein components might be also present in this region. The nonionic detergent Triton X-100 has been used to selectively remove the glycoproteins from a number of enveloped viruses.[9] After treatment of purified influenza C virions with 2% Triton and separation of the viral glycoproteins from the internal components by centrifugation, the supernatant and pellet fractions were analyzed on polyacrylamide gels. Virus preparations grown in chick embryo fibroblast (CEF) and chicken kidney (CK) cells are compared in Figure 1. In addition to the main protein components NP and M in the pellet fraction, three discrete protein bands are resolved in the region of the gp88 glycoprotein. By analogy with the polypeptides of other orthomyxoviruses, these three polypeptides have been designated P_1, P_2 and P_3, and may represent

Figure 1. Fractionation of the polypeptides of influenza C virions with Triton X-100. Purified ^{3}H-leucine-labeled virions of influenza C/JHB/166 grown in CK or in CEF cells were disrupted with 2% Triton X-100 and centrifuged at 100,000 × g for 30 min. The polypeptides of the supernatant were precipitated with n-butanol. The polypeptides of the supernatant and pellet fractions were analyzed by SDS-polyacrylamide gel electrophoresis and detected by fluorography.[6] (a) pellet fraction, CEF-grown virus; (b) pellet fraction, CK grown virus. (c) and (f) supernatant fraction, CEF-grown virus; (d) and (e) supernatant fraction, CK grown virus. The samples in lanes c and d were electrophoresed under non-reducing conditions.

SOURCE: Herrler et al., submitted for publication.

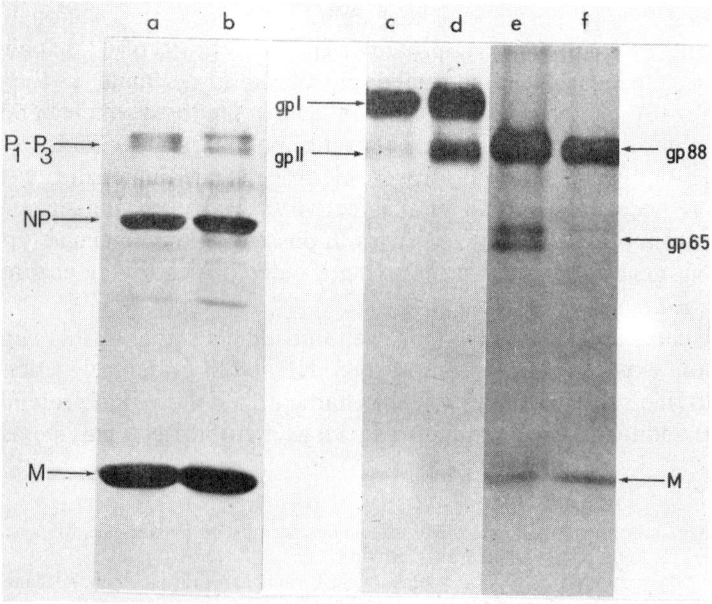

viral RNA polymerase proteins (see below). Based on the positions of NP and M proteins as markers, the mol wt of the P proteins were estimated to be 90,000, 86,000 and 85,000, respectively. Similar polypeptides have also been observed after the viral glycoproteins were digested by treatment with bromelain or trypsin.[10]

The supernatant fractions contained the purified glycoproteins of influenza C virus. Virus preparations grown in CEF and CK cells are compared in Figure 1, lanes c-f. Under non-reducing conditions, CEF grown virus contains only a single glycoprotein designated gpI, whereas two species of glycoprotein, gpI and gpII, are present in CK-grown virus.[6] When the samples were analyzed under reducing conditions, gpI migrates as a single glycoprotein, previously designated gp88,[5] whereas gpII appears as two sub-units, gp65 and gp30.

Identification of a Virion Polymerase

Infectious influenza A and B virions have been shown to contain an RNA-dependent RNA polymerase which catalyses *in vitro* RNA synthesis.[11–12] When Triton X-100 solubilized influenza C virions were tested for polymerase activity in a standard *in vitro* assay containing the four ribonucleoside triphosphates and ApG as primer,[13,14] incorporation of ^{3}H-UTP into a TCA-insoluble form could be detected (Figure 2). In the absence of the primer dinucleotide, the incorporation of ^{3}H-UTP was reduced by about 70% (not shown). Of the three temperatures at which this enzyme activity was measured, the highest incorporation of ^{3}H-UTP was observed at 33°.

Analysis of Tryptic Peptides and Glycopeptides of Viral Glycoproteins

We have further characterized the influenza C viral glycoproteins by analyses of their tryptic peptides by ion exchange chromatography. As seen in Figure 1, the large cleavage product gp65 regularly appears as a doublet, and if appropriate amounts of isolated glycoproteins are analyzed under reducing conditions on polyacrylamide gels, two bands can also be resolved for the smallest subunit gp30 (data not shown). Since the difference in the two bands of the glycoprotein subunit could be due to a different extent of glycosylation or to differences in the polypeptide portions, we compared their tryptic peptide patterns by ion exchange chromatography. Figure 3 shows that virtually identical patterns of peptides were obtained for each band in the doublet of gp65 when the tryptic digests were compared. Therefore, gp65 exists in two forms which contain polypeptides with similar amino acid sequences. When the extent of glycosylation of the two gp65 bands was estimated from the ^{3}H/^{14}C ratios in virus preparations doubly labeled with ^{3}H-glucosamine and ^{14}C-amino acids, it was found that the upper band had

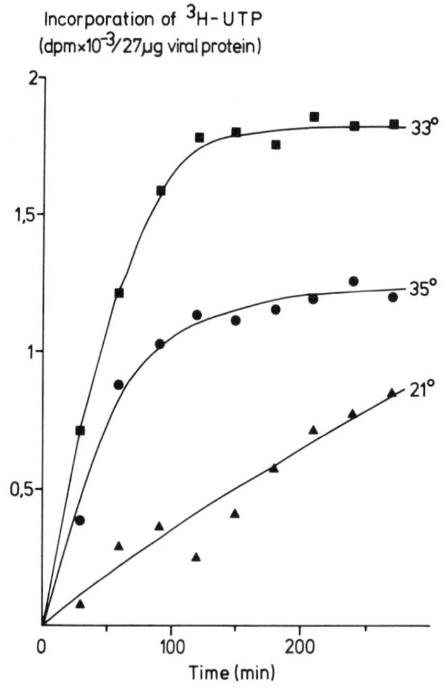

Figure 2. Assay of influenza C virion RNA polymerase activity at various temperatures. Influenza C JHB/1/66 virions at a protein concentration of 548 μg/ml were disrupted by Triton X-100 at a concentration of 1.64 μg/ml and incubated in a standard reaction mixture containing 1.37 mM ATP, CTP and GTP and 0.27 mM ^{3}H-UTP (specific activity, 99.8 Ci/mol.), 41 mM NaC1, 10 mM $MgC1_2$, 0.8 mM ApG, and 82 μM Tris HC1, pH 8.2. Aliquots of 50 μl were removed at intervals and the TCA insoluble radioactivity was determined. Incorporation of ^{3}H-UTP is expressed as dpm/μg of viral protein.

a ratio of 3.5 compared to 2.6 for the lower band, indicating the presence of additional glucosamine residues in the upper band. The glycopeptides derived from each band of gp65 after extensive digestion with pronase were compared by gel filtration on Bio-Gel P-6 (Figure 4). Three species of glycopeptides were resolved from each glycoprotein band, with estimated molecular weights of 5100, 2900 and 1600–2200. Three glycopeptides of similar size distribution were previously observed in unfractionated gp65 glycoproteins.[15] Further analysis will be required to determine whether the number of oligosaccharide side chains is different in the two bands, or whether differences in the composition of the individual oligosaccharide chains are responsible for the difference in migration. The possibility also exists that the difference in molecular weight is due in part to differences in the precise sites at which proteolytic cleavage occurred, which might not have been observed in the tryptic peptide patterns.

In order to confirm the relationship between gpII and the disulfide bonded subunits gp65 and gp30, the three species of glycoproteins were isolated

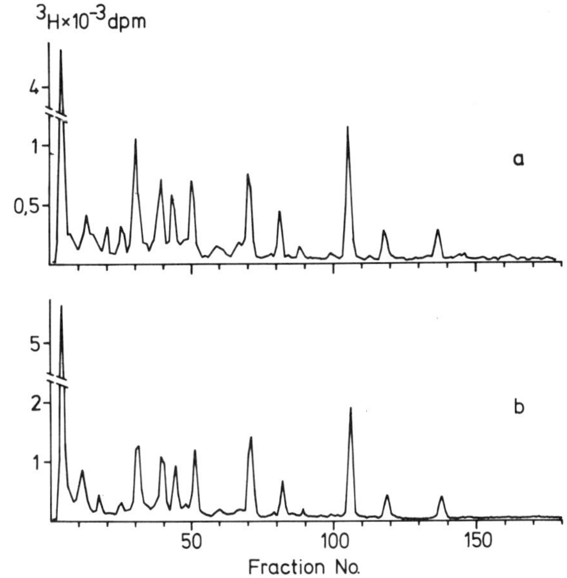

Figure 3. Tryptic peptide analysis of the doublet band of influenza C JHB/1/66 gp65 glycoprotein. The virus was grown in CK cells and labeled with ^{3}H-leucine. The upper (a) and the lower (b) band of gp65 were cut from polyacrylamide slab gels. The protein was eluted from the gel in 5 ml of 0.05 M ammonium carbonate, 0.1% SDS at 37° for 3 days. The samples were lyophylized and precipitated two times in TCA, resuspended in 0.5 ml 1 M NH_4OH, and alcohol precipitated. The samples were oxidized with performic acid according to the method of Crawford and Gesteland,[17] followed by extensive digestion with 25 μg/ml TPCK-trypsin (Merck). For ion-exchange chromatography, a procedure adapted from that described by Gentsch and Bishop[18] was used. (Source: Herrler et al., submitted for publication.)

from polyacrylamide gels and analyzed by ion exchange chromatography after extensive trypsin treatment (Figure 5). The patterns obtained from the glycoprotein cleavage products gp65 and gp30 show no peptides in common. However, if the peaks resolved for the two subunit glycoproteins gp65 and gp30 are superimposed, a pattern corresponding to that of gpII is obtained. These results demonstrate the interrelationship of the different glycoprotein species found in influenza C virions. Also, though the posttranslational cleavage of gpI to gpII is accompanied by a decrease of about 20,000 in apparent mol wt, the tryptic peptide patterns obtained for gpI and gpII were virtually indistinguishable (data not shown).

Discussion

The present observations indicate that three minor polypeptides are present as internal components in influenza C virions, analogous to the three P proteins of influenza A and B virions. These results, as well as the identifica-

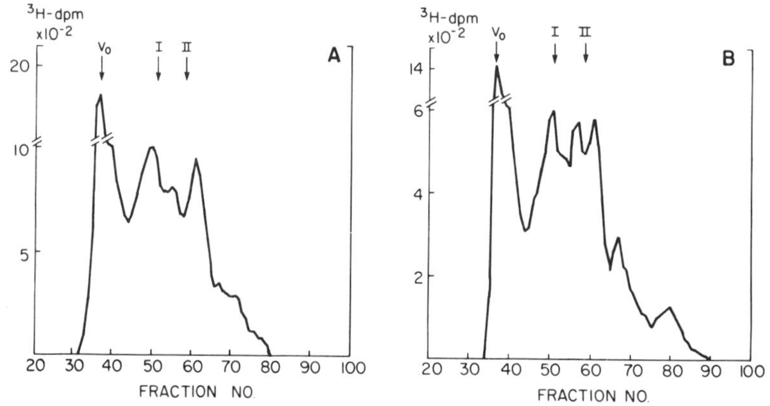

Figure 4. Glycopeptides obtained from the upper and lower band of gp65 glycoprotein. Influenza C/JHB/1/66 virus was grown in CK cells and labeled with ^{3}H-glucosamine. The upper (A) and lower (B) bands of gp65 were cut from polyacrylamide slab gels, minced and subjected to complete pronase digestion in 0.1 M Tris-HC1, pH 8.0, 0.01 M $CaC1_2$.[15] Glycopeptides were mixed with ^{14}C-labeled glycopeptides of A/WSN virus and chromatographed on a column of Bio-Gel P-6 (200/400 mesh). Arrows indicate the positions of the type I and type II glycopeptides of the A/WSN marker.

tion of RNA polymerase activity in influenza C virions, further indicates the close similarity in the molecular biology of these viruses to other orthomyxoviruses. We have also obtained evidence for the existence of two nonstructural influenza C proteins NS_1 and NS_2 with mol wt of 24,000 and 14,000,[10] which may correspond to the two products of the smallest RNA segment of influenza A viruses.[16] The entire coding capacity of the influenza C viral genome may be accounted for by these nonstructural proteins and the six structural polypeptides that have been identified. The exact coding relationships between influenza C virion RNA segments and gene products is yet to be determined.

Posttranslation cleavage of a precursor glycoprotein of influenza C virus into two subunits, linked by disulfide bonds, is a requirement for full infectivity that is shared with other myxo- and paramyxoviruses. Influenza C virus, however, seems to possess only a single glycoprotein gene product, which is unique for this family of viruses. The molecular basis of the observation that the two glycoprotein subunits gp65 and gp30 each exist in two molecular forms is not understood; however, the present results indicate that the polypeptide backbone does not differ detectably in the two bands observed with gp65. Further information on the glycoproteins of influenza C virus is required to determine whether the active sites for receptor binding and receptor-destroying enzyme activity can be distinguished.

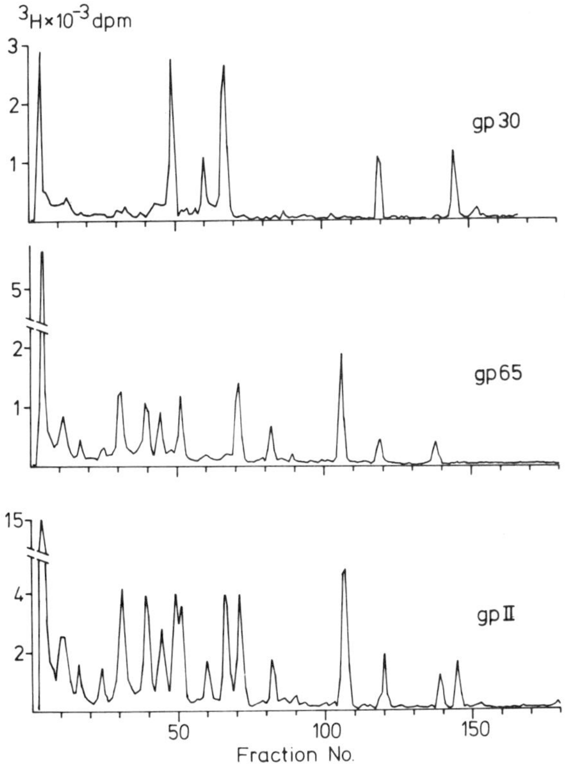

Figure 5. Tryptic peptide analysis of influenza C/JHB/1/66 virus glycoproteins gp30, gp65 and gpII. The isolation of the proteins and the subsequent treatment was performed as described in the legend to Figure 3. (Herrler et al., submitted for publication.)

ACKNOWLEDGMENTS

This research was supported by the Deutsche Forschungsgemeinschaft, USPHS Grant No. AI 12680 and NSF Grant PCM80-06498.

References

1. Cox, N.J. and Kendal, A.P. (1976): *Virology* 74:239–241.
2. Petri, T., Meier-Ewert, H., Crumpton, W.M., and Dimmock, N.J. (1979): *Arch. Virol.* 61:239–243.
3. Palese, P., Racaniello, V.R., Desselberger, U., Young, T., and Baez, M. (1980): *Philosophical Transaction of the Royal Society of London, Series B, Vol. 288* 1029:287–460.
4. Kendal, A.P. (1975): *Virology* 65:87–99.
5. Compans, R.W., Bishop, D.H.L., and Meier-Ewert, H. (1977): *J. Virol.* 21:239–241.
6. Herrler, G., Compans. R.W., and Meier-Ewert, H. (1979): *Virology* 99:49–56.

7. Meier-Ewert, H., Compans, R.W., Bishop, D.H.L., and Herrler, G. (1978): In: *Negative Strand Viruses and the Host Cell.* R.D. Barry and B.W.J. Mahy, (Eds.), Academic Press, London, pp. 127–133.
8. Hirst, G.K. (1950): *J. Exp. Med.* 91:177–184.
9. Scheid, A., Caliguiri, L.A., Compans, R.W., and Choppin, P.W. (1972): *Virology* 50:640.
10. Petri, T., Herrler, G., Compans, R.W., and Meier-Ewert H. (1980): *FEMS Microbiol. Lett.* 9:43–47.
11. Chow, N. and Simpson, R.W. (1971): *Proc. Natl Acad. Sci. U.S.A.* 68:752–756.
12. Oxford, J.S. (1973): *J. Virol.* 12:827–835.
13. McGeoch, D. and Kitron, N. (1975): *J. Virol.* 15:686–695.
14. Plotch, S.J. and Krug, R.M. (1978): *J. Virol.* 25:579–586.
15. Nakamura, K., Herrler, G., Petri, T., Meier-Ewert, H., and Compans, R.W. (1979): *J. Virol.* 29:997–1005.
16. Lamb, R.A. and Choppin, P.W. (1979): *Proc. Natl. Acad. Sci. U.S.A.* 76:4908–4912.
17. Crawford, L.V. and Gesteland, R.F. (1973): *J. Mol. Biol.* 74:627–634.
18. Gentsch, J. and Bishop, D.H.L. (1978): *J. Virol.* 28:417.

Copyright 1981 by Elsevier North Holland, Inc.
David H.L. Bishop and Richard W. Compans, eds.
The Replication of Negative Strand Viruses

Changes in Conformation and Charge Paralleling Proteolytic Activation of Myxovirus Glycoproteins

H.-D. Klenk, W. Garten, and T. Kohama[a]

Posttranslational proteolytic cleavage is involved in the biosynthesis of the hemagglutinin of influenza virus and of both envelope glycoproteins of paramyxoviruses. In the case of the influenza hemagglutinin, the precursor HA is cleaved into the amino-terminal fragment HA_1 and the carboxy-terminal fragment HA_2 which are both present in the spike as a disulfide-bonded complex.[1] Similarly, 2 disulfide-bonded fragments, the amino-terminal F_2 and the carboxy-terminal F_1, are derived from the precursor F_0 of the fusion protein of paramyxoviruses.[2] In contrast, only one large fragment (HN) is found in the virion, when the precursor HN_0 of the paramyxovirus hemagglutinin-neuraminidase is cleaved which has been observed only with some strains of Newcastle disease virus (NDV). A glycopeptide is lost in this cleavage reaction.[3] Cleavage which depends on the presence of an appropriate protease in the host cell is not necessary for particle formation, but is essential for the biological activities of the glycoproteins and for their role in initiation of infection. This has been clearly demonstrated by experiments in which virions containing the uncleaved precursors have been activated by *in vitro* protease treatment. Differences in the susceptibility of the glycoproteins to proteolytic cleavage have been found to account for variations in important biological properties of the viruses, such as host range, tissue tropism, and pathogenicity.[1]

To learn more about the molecular details of proteolytic activation we have now carried out isoelectric focusing studies on the influenza

[a] Institut für Virologie, Fachbereich Humanmedizin der Justus-Liebig-Universität, D-6300 Giessen, Germany.

hemagglutinin and on the NDV glycoproteins. Evidence will be presented for similarities in the cleavage mechanism of the influenza hemagglutinin and of the F protein of NDV. We present also CD spectra of the cleaved and uncleaved NDV glycoproteins which demonstrate that cleavage is paralleled by a conformational change of the glycoproteins.

Charge Shift Paralleling Cleavage of the Influenza Virus Hemagglutinin

The elucidation of the complete amino acid sequence of the hemagglutinin gives detailed insight into the structure of the cleavage site. Figure 1 shows a comparison of the sequences at the cleavage sites of 3 different hemagglutinins. The data show that after *in vivo* cleavage a basic peptide containing several arginine and lysine residues is eliminated from the hemagglutinin of FPV (Hav 1), whereas with serotypes H2 and H3 only a single arginine is removed. It is conceivable that such differences in the structure of the cleavage site account for the above mentioned differences in the susceptibility of the various hemagglutinins to proteolytic activation (F.X. Bosch, unpublished results).

Elimination of a basic connecting piece should result in a charge shift of the hemagglutinin towards a more acidic isoelectric point. Figure 2 demonstrates that this is indeed the case, even when the hemagglutinin is cleaved *in vitro* with trypsin as the only enzyme added to the purified virus. Thus, it appears that after the initial action of trypsin or a trypsin-like endoprotease, a proteolytic enzyme of different specificity, perhaps an exopeptidase of the carboxypeptidase B type, is involved in the cleavage reaction. Our data suggest also that the latter enzyme activity is present in the virus particle. It is not clear, however, if it is a virus-coded protein or a tightly bound host component.

Figure 1. The cleavage sites of 3 different influenza A hemagglutinins. Sections of the sequence of the uncleaved precursor HA of serotypes H2,[18] H3,[19,20] and Hav1 (FPV) [21] are shown. Arrows indicate the carboxy-termini of HA_1 and the amino-termini of HA_2 determined on the cleaved forms of H2,[22] H3,[23] and FPV (W. Garten, unpublished results) hemagglutinins.

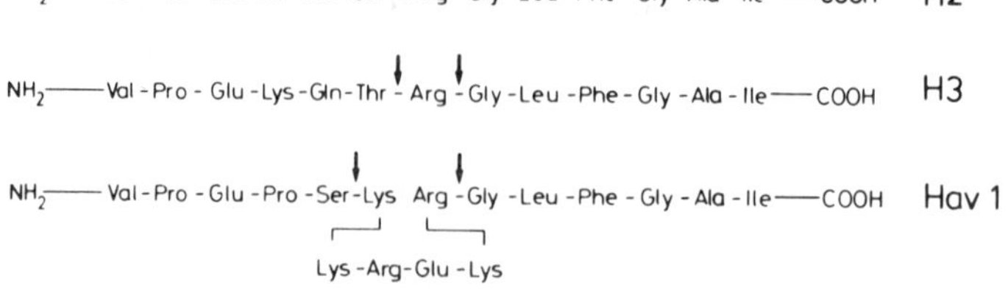

What HA is cleaved by the non-activating enzyme thermolysin, there is a shift of the cleavage site by a single amino acid with leucine instead of glycine forming the amino terminus of HA_2.[4,5] Figure 2 shows that, under these conditions, the cleaved and the uncleaved hemagglutinin have very similar isoelectric points. Thus, the basic intervening sequence is not elimi-

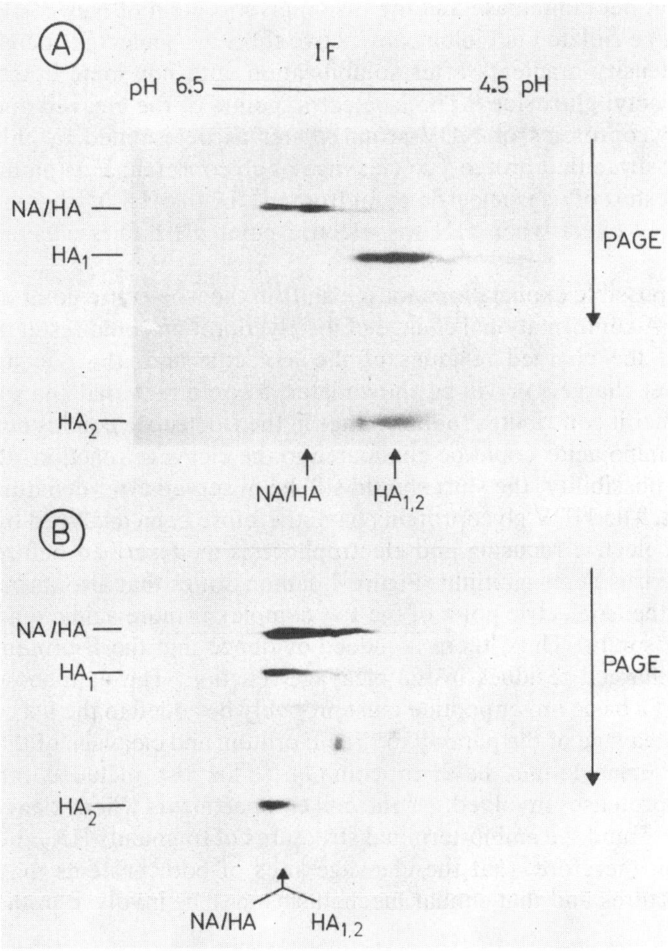

Figure 2. Two-dimensional isoelectric focusing and electrophoresis of the glycoproteins of influenza A virus. Virions of strain MRC11 (H3N2) labeled with ^{3}H-amino acids and grown in chick embryo fibroblasts have been used. The hemagglutinin was cleaved by *in vitro* treatment with trypsin (A) and thermolysin (B). Both samples were mixed with virus containing the uncleaved hemagglutinin. After disintegration in 2% NP40 and 9 M urea at 25°C, viral glycoproteins were analyzed on polyacrylamide gels by electrofocusing under non-reducing conditions in horizontal direction followed by electrophoresis under reducing conditions in vertical direction.[6] The origin is in the upper right corner. Identical results were obtained, when virus was pretreated with 1.5% SDS.

nated, when glycine is present at its carboxy-terminal end. This is further support for the involvement of a carboxypeptidase-B-like activity, since this enzyme cleaves only arginine and lysine in carboxy-terminal position.

Charge Shift Paralleling Cleavage of the F Protein of NDV

The hemagglutinin-neuraminidase and the fusion glycoprotein of Newcastle disease virus can be isolated in biologically active form by isoelectric focusing on sucrose density gradients after solubilization with non-ionic detergents, such as n-octyl-glucoside.[6] The isoelectric points of the cleaved and the uncleaved glycoproteins of NDV strain Ulster as determined by this procedure demonstrate that proteolytic cleavage of glycoprotein F_0 is paralleled by a distinct shift of its isoelectric point from pH 6.3 to pH 5.0, whereas there is hardly any effect when HN_0 (isoelectric point pH 5.6) is cleaved (Figure 3).

There are two possible explanations for the shift in the isoelectric point of the F protein: (1) A conformational change of the glycoprotein could result in rearrangement of the charged residues of the molecule and, thus, in an alteration of its net charge. As will be shown later, a conformational change occurs, but whether it contributes to the change in the isoelectric point is not clear. (2) Basic amino acids could be eliminated in the cleavage reaction. In case of the latter possibility, the shift should still be observed after denaturing of the protein. The NDV glycoproteins have therefore been analyzed by 2-dimensional isoelectric focusing and electrophoresis as described before for the influenza virus hemagglutinin. Figure 4 demonstrates that also under these conditions the isoelectric point of the $F_{1,2}$ complex is more acidic than that of the precursor F_0. Thus, there is indeed evidence that the F protein loses positively charged residues in the cleavage reaction. The data show that elimination of a basic linker peptide must probably be added to the list of parameters that cleavage of the paramyxovirus F protein and cleavage of the influenza virus hemagglutinin have in common. This list includes the specificity of the proteases involved,[7–10] the cell compartments where cleavage takes place.[11–13] and the amino-terminal structures of fragments HA_2 and F_1.[14,15] It appears, therefore, that the cleavage sites of both proteins may have similar structures and that similar mechanisms may be involved in the cleavage reaction.

Changes in Conformation Paralleling Cleavage of NDV Glycoproteins

In order to obtain information on the conformation of the viral glycoproteins before and after cleavage, circular dichroism (CD) spectra of the isolated

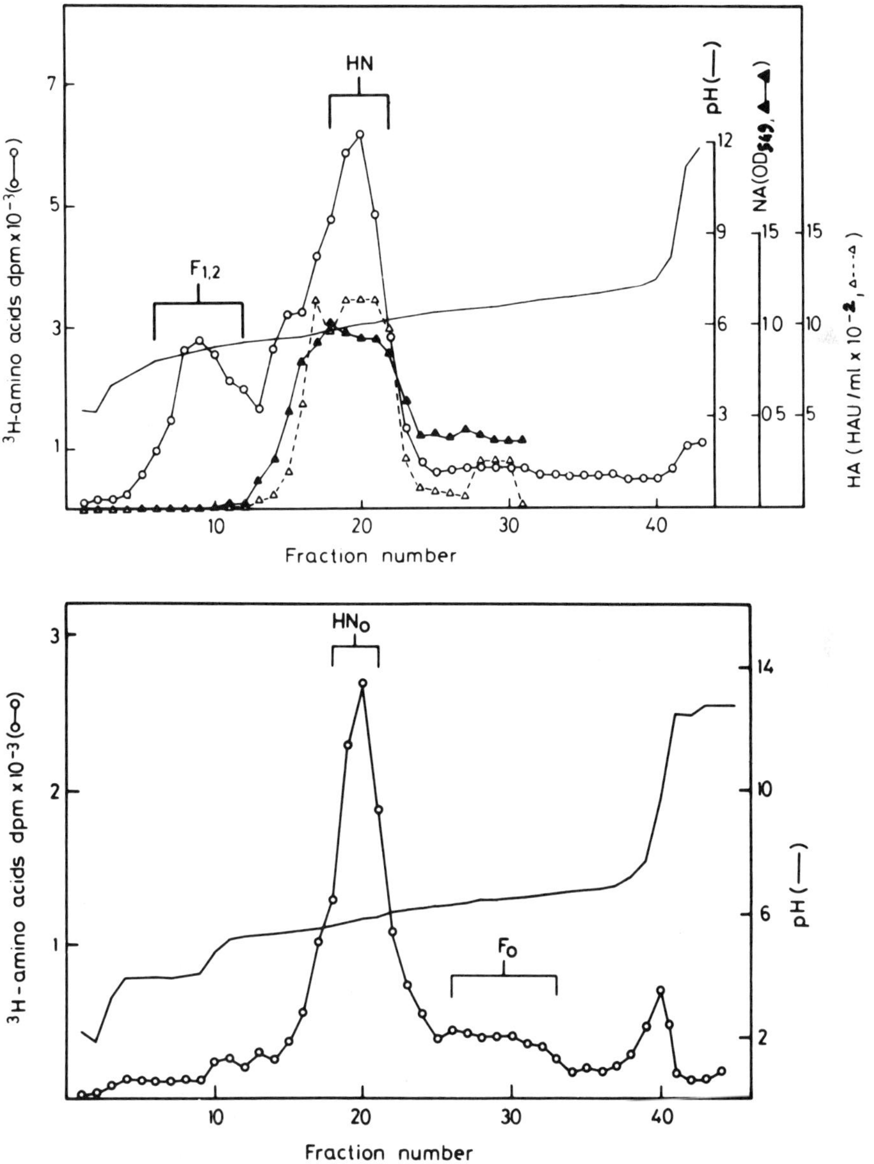

Figure 3. Separation of the glycoproteins of NDV strain Ulster by isoelectric focusing on density gradients. Virus was grown in MDBK cells and labeled with ^{3}H-amino acids. The glycoproteins were cleaved by trypsin (upper panel), or they were present in the uncleaved form (lower panel). After solubilization in 2% n-octyl-glucoside, the glycoproteins were electrofocused in a LKB 8101 column.[6]

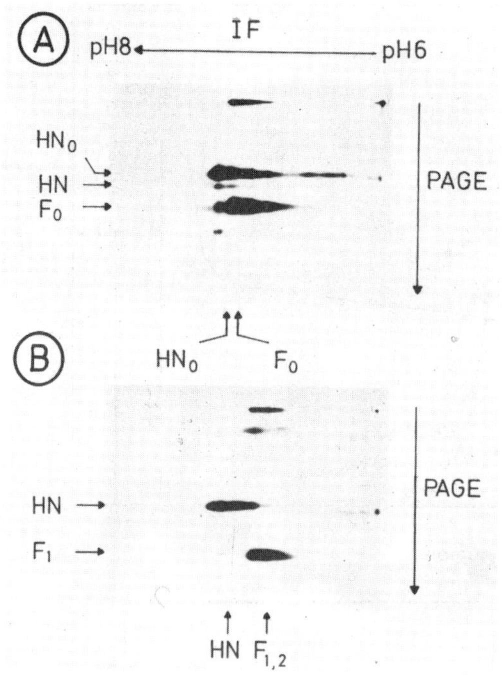

Figure 4. Two-dimensional isoelectric focusing and elctrophoresis of the glycoproteins of NDV strain Ulster. ^{3}H-glucosamine labeled virions grown in MDBK cells were analyzed as described in Figure 2. (A) Virions contain uncleaved glycoproteins. (B) Virions contain glycoproteins that are cleaved by trypsin treatment.

glycoproteins were determined in the far UV region. As shown in Figure 5, there are differences between F_0 and $F_{1,2}$ as well as between HN_0 and HN. When compared to theoretical curves derived from CD spectra of proteins whose secondary structure had been eludicated by X-ray diffraction studies,[16] it can be calculated that cleavage of F_0 results in a significant decrease of alpha-helix and of β-structure and in a concomitant increase in unordered structure. The CD spectra of HN_0 and HN are more difficult to interpret, but again it appears that cleavage is paralleled by a decrease in helicity.[6] Thus, cleavage causes a distinct conformational change of both glycoproteins.

As pointed out above a glycopeptide (MW 8,000) is removed at the amino-terminal end, when HN_0 is converted into HN. This glycopeptide appears to be rich in alpha-helix structure, and its elimination may account for the decrease in helicity observed in this cleavage reaction. It is reasonable to assume that in the precursor the glycopeptide masks the active site of the molecule and that it is therefore responsible for the biological inactivity of glycoprotein HN_0. Two possible explanations have to be considered for

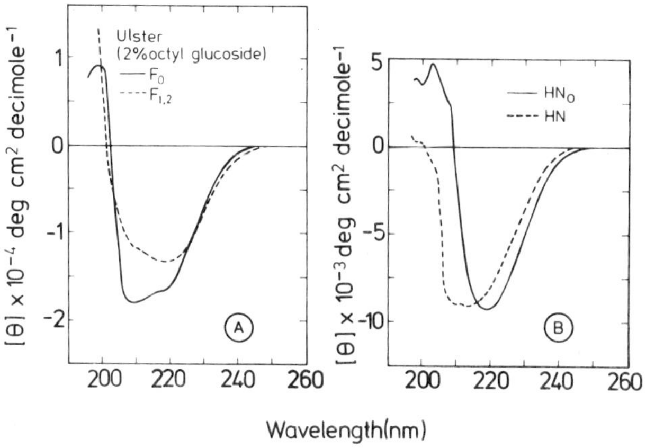

Figure 5. CD spectra of the glycoproteins of NDV strain Ulster before and after proteolytic cleavage. Glycoproteins were isolated by isoelectric focusing as described in Figure 3 and analyzed in a Jasco model J-41A spectropolarimeter.

the change in the secondary structure of glycoprotein F: (1) Again a large peptide with a relatively high content of alpha-helix and β-pleated sheet may be lost in the cleavage. The eliminated peptide would have to be the presumptive segment connecting F_1 and F_2. (2) Alternatively, as pointed out above there is good evidence to believe that almost the entire polypeptide complement of the precursor F_0 is preserved in the cleaved glycoprotein $F_{1,2}$ and that the linking peptide is too small to contribute by itself to the overall secondary structure of the molecule. It is therefore conceivable that the polypeptide is rearranged after cleavage in such a way that areas can interact that are far apart from each other in the uncleaved molecule. As a result of these new interactions, the areas may undergo changes in secondary structure. CD spectra of the influenza virus hemagglutinin have also presented evidence for a conformational change taking place when this glycoprotein is cleaved.[17] Thus, there is another feature common to proteolytic activation of both the influenza virus hemagglutinin and the F protein of paramyxoviruses.

ACKNOWLEDGMENT

This work was supported by the Deutsche Forschungsgemeinschaft (Sonderforschungsbereich 47).

References

1. Klenk, H.-D. and Rott, R. (1980): *Curr. Top Microbiol. Immunol.* 90:19–48.
2. Scheid, A. and Choppin, P.W. (1977): *Virology* 80:54–66.
3. Garten, W., Kohama, T., and Klenk, H.-D. (1980): *J. Gen. Virology*, 50:135–147.
4. Garten, W. and Klenk, H.-D. (1980); *Z. Physiol. Chem.* 361:248.

5. Klenk, H.-D. (1980): In: *Structure and Variation in Influenza Virus.* Laver, G. and Air, G. (Eds.), Elsevier/North-Holland, New York, pp. 213–222.
6. Kohama, T., Garten, W., and Klenk, H.-D. *Virology,* submitted.
7. Lazarowitz, S.G. and Choppin, P.W. (1975): *Virology* 68:440–455.
8. Klenk, H.-D., Rott, R., and Orlich, M. (1977): *J. Gen. Virol.* 36:151–161.
9. Scheid, A. and Choppin, P.W. (1976): *Virology* 69:265–277.
10. Nagai, Y. and Klenk, H.-D. (1977): *Virology* 77:125–134.
11. Klenk, H.-D., Wöllert, W., Rott, R., and Scholtissek, C. (1974): *Virology* 57:28–41.
12. Hay, A.J. (1974): *Virology* 60:398–418.
13. Nagai, Y., Ogura, H., and Klenk, H.-D. (1976): *Virology* 69:523–538.
14. Scheid, A., Graves, M.C., Silver, S.M., and Choppin, P.W. (1978): In: *Negative Strand Viruses and the Host Cell.* Mahy, B.W.J. and Barry, R.D. (Eds.), Academic Press, London, pp. 181–193.
15. Gething, M.J., White. J.M., and Waterfield, M.D. (1978): *Proc. Natl. Acad. Sci. U.S.A.* 75:2737–2740.
16. Chen, Y.-H., Yang, J.T., and Martinez, H.M. (1972): *Biochemistry* 11:4120–4131.
17. Flanagan, M.T. and Skehel, J.J. (1977): *FEBS Letters* 80:57–60.
18. Gething, M.J., Bye, J., Skehel, J.J., and Waterfield, M.D. (1980): In: *Structure and Variation in Influenza Virus.* Laver, G. and Air, G. (Eds.), Elsevier/North-Holland, New York, pp. 1–10.
19. Min Jou, W., Verhoeyen, M., Devos, R., Saman, E., Fang, R., Huylebroeck, D., Fiers, W., Threlfall, G., Barber, C., Carey, N., and Emtage, S. (1980): *Cell* 19:683–696.
20. Sleigh, M.J., Both, G.W., Brownlee, E.E., Bender, V.J., and Moss, B. (1980): In: *Structure and Variation in Influenza Virus.* Laver, G. and Air, G. (Eds.), Elsevier/North Holland, New York, pp. 69–80.
21. Porter, A.G., Barber, C., Carey, N.H., Hallewell, R.A., Threlfall, G., and Emtage, J.S. (1979): *Nature* 282:471–477.
22. Waterfield, M.D., Gething, M.J., Scrace, G., and Skehel, J.J. (1980): In: *Structure and Variation in Influenza Virus.* Laver, G. and Air, G. (Eds.), Elsevier/North Holland, New York, New York, pp. 11–20.
23. Ward, C.W. and Dopheide, T.A.A. (1980): *Virology* 103:37–53.

Published 1981 by Elsevier North Holland, Inc.
David H.L. Bishop and Richard W. Compans, eds.
The Replication of Negative Strand Viruses

Morphological and Immunological Studies of Influenza Virosomes

J.S. Oxford,[a] D.J. Hockley,[a] T.D. Heath,[b] nd S. and S. Patterson[c]

Introduction

Liposomes are phospholipid vesicles[1] and certain virus proteins such as influenza virus hemagglutinin (HA) and neuraminidase (NA)[2-4] and the spike glycoprotein of Semliki Forest virus[5] can interact with liposomes and assume an orientation similar to that observed in their native state (virosomes). Virosomes constitute a model system for studies on the organization of virus membrane proteins and lipids and interaction between virus particles and membranes.[6,7] We have investigated in detail the morphology of influenza virosomes prepared using different phospholipids and particularly the location and insertion of HA spikes. We report an increased immunogenicity of HA incorporated in virosome preparations compared to the immunogenicity of free HA sub-units.

Materials and Methods

Purification of Influenza Virus and Preparation of HA Antigen

Approximately 20 mg of purified virus (X-47 recombinant) was disrupted by the addition of Triton X-100 detergent to a final concentration of 2% v/v for 15 mins. The disrupted virus was fractionated on 20–50% w/v sucrose gra-

[a] Division of Virology, National Institute for Biological, Standards and Control, Holly Hill, London NW3 6RB, England.
[b] Chester Beatty Research Institute, Fulham Road, London SW3, England.
[c] Clinical Research Center, Northwick Park Hospital, Watford Road, Harrow, Middlesex, England.

dients in the SW-41 rotor of a Beckman ultracentrifuge by centrifugation at 35,000 rpm for 16 hours. Gradient fractions were examined for the presence of virus antigens by rocket immunoelectrophoresis.[8]

Virosomes

For the preparation of small unilamellar liposomes, 30 mg lipid in a 1:1 chloroform methanol solution was dried in a quickfit B24 boiling tube by rotary evaporation. 3 ml of PBS was added to the lipid, which, after suspension in the agueous medium, was kept under a stream of nitrogen and placed in a BTL bath type sonicator until clear (usually 20 min). Virosomes were prepared by mixing 10 mg/ml sonicated suspensions of phospholipid with an equal volume of 200 μg/ml influenza HA, with further sonication for 20 min.

Results

Comparative Morphology of Virosomes

The morphology of the virosomes prepared by sonication or by the REV technique[9] was similar to those described by Almeida et al. (1975) but some additional features were noted. The HA sub-units, which were triangular or wedge-shaped, were always attached to the liposome by their narrower end, so that they had the same orientation as on the virus particle (Figure 1). Occasionally, HA sub-units were observed both on the outside and inside of virosomes. Influenza HA attached more readily to negatively charged or neutral liposomes than to positively charged liposomes. In addition, there were differences in the average number of sub-units attached to different types of virosomes. PI virosomes consistently carried more HA sub-units than PC:DCP virosomes, even though the two types of preparations contained a similar percentage of virosomes. The number of HA sub-units on the virosomes was always less than the number found on intact virus particles. A mean of 27 HA sub-units were counted around the periphery of 50 nm diameter intact virus particles, compared to 13 peripheral HA sub-units on PI virosomes. Scanning electron microscopy of PC and PC/SA virosomes clearly showed their spherical nature and the considerable size variation of PC virosomes (Figure 1). Sections of a pellet of PC:SA virosomes revealed unilamellar vesicles ranging in diameter from 70 to 500 nm. The HA sub-units and the smaller virosomes (25–60 nm) were not detectable by these latter techniques. The width of the lipid bilayer in sections of virosomes and virus particles was identical and approximated to 7.5 nm.

Depth of Insertion of HA spikes on Virus Particles and Virosomes

Measurements were taken from 14 electron micrographs of the length of 500 HA spikes on virosomes, on virus particles and as single free spikes. The mean lengths and standard deviations were 13.3 $\pm$ 0.7 nm, 12.2 $\pm$ 0.7 nm and 14.2 $\pm$ 0.9 nm respectively. Thus there was no evidence from the measure-

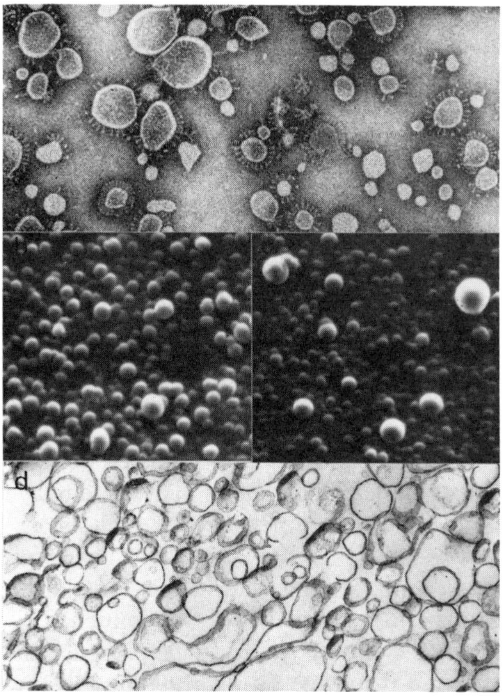

Figure 1. Virosome size and morphology. (a) small, unilamelar PI virosomes prepared by ultrasonication. Negative stained preparation ×130,000. (b) Scanning electron micrograph of PC: SA virosomes ×18,000. (c) Scanning electron micrograph of PC virosomes ×18,000. (d) Sectioned PC: SA virosomes—predominantly unilamellar vesicles are present with a 7.5 nm lipid bilayer.

ments of deep insertion of HA sub-units in either virus or virosome membranes. Furthermore, freeze-fracture preparations of virosomes indicated that the HA spikes did not penetrate far into the lipid bilayer.[4]

Interaction of Virosomes and cells

Numerous PC: DCP virosomes were found on the surfacc of Vero cells after 30-min incubation at 37°C, although the majority of the virosomes were observed in cytoplasmic vacuoles. There was no evidence of fusion between the virosomes and the plasma membrane in contrast to the findings of Rott[11] for virosomes which contained NA and cleaved HA.

Immunogenicity of Virosomes

Virosomes, virus particles or HA sub-units were used to immunize guinea pigs and the post-immunization sera tested for antibody to virus HA by the quantitative single radial hemolysis test.[10] Following primary immunization with 2 μg of HA, antibodies were detected by SRH tests in a high proportion of animals immunized with PC: DCP or PI virosomes (Table 1). A less frequent antibody response was detected in animals immunized with HA sub-units alone or with a mixture of HA sub-units and PC: SA liposomes where HA was not attached to the liposome. In addition, the virosome preparations induced higher levels of antibody than the HA sub-unit preparations. In an antigen extinction test 2.5 μg of HA sub-units was required to produce a detectable primary immune response (mean SRH zone area of 1.5 mm^2), whereas a ten-fold lower concentration of HA incorporated into virosomes (0.25 μg) produced a greater immune response measured as 9.6 mm^2 mean zone area in the SRH test.

Discussion

Earlier studies on the morphology of influenza A virus[12-14] have not established the precise depth of attachment of the HA sub-units to the lipid bilayer of the virus. There is evidence with other enveloped viruses such as VSV derived from experiments with cross-linking agents[15] and spin label probe analysis[16] that the glycoprotein may penetrate the lipid bilayer envelope to contact the underlying M protein. The measurements of free influenza HA sub-units and sub-units attached to virosomes or virions in the present study indicated that the HA sub-units did not penetrate through the lipid bilayer. The difference in length between free monomer HA or HA rosettes and HA attached to virion or virosomes was no greater than 2 nm, whilst the distance between the two leaflets of the lipid bilayer was approximately 7 nm. It is unlikely that the mild non-ionic detergent used to release HA from virions caused contraction or elongation in the HA sub-unit. However, the present results do not exclude the possibility of a hydrophobic tail which may partially or completely penetrate the lipid membrane but which is not detectable by electron microscopy.

Table 1. Comparison of Immunogenicity of Influenza Virus HA Sub-Units and Virosomes.

		No. of Animals with antibody post-immunization / No. of animals immunized		Anti-HA antibody (mm^2 zone area ±SD) measured by SRH	
Immunogen	**μg/HA dose**	**Primary**	**Secondary**	**Primary**	**Secondary**
HA sub-units	2	3/10	9/10	3.9 ± 1.9	22.9 ± 5.8
PC:DCP virosomes	2	10/10	10/10	39.1 ± 6.2	74.6 ± 8.6
PI virosomes	2	7/10	10/10	12.1 ± 3.4	41.8 ± 6.4
PC:SA virosomes	2	1/10	10/10	0.8	64.3 ± 8.0

The immunogenicity studies indicated that the attachment of HA onto a PI or PC:DCP virosome increased its ability to stimulate an antibody response *in vivo,* compared to free HA sub-units. Virosomes have potential value in enhancing the immunogenicity of viral glycoprotein sub-units used as vaccines. The immunostimulating effect of liposomes on influenza HA is in keeping with previous observations of the adjuvant effects of liposomes and many lipid substances with bovine serum albumin,[17] diphtheria toxoid,[18] hepatitis B antigen[19] and malarial membrane antigens.[20] The major effect of lipid substances on the immune response may concern the presentation of the antigen and the insertion of the influenza HA into the lipid bilayer would appear to be critical for its increased immunogenicity.

References

1. Bangham, A.D., Hill, M.W., and Miller, N.G.A. (1974): *Methods in Membrane Biology* 1:1–68.
2. Almeida, J.D., Brand, C.M., Edwards, D.C., and Heath, T.D. (1975): *Lancet* 2:899–901.
3. Huang, R.T.C., Wahn, K., Klenk, H-D., and Rott, R. (1979): *Virology* 97:212–217.
4. Oxford, J.S., Hockley, D.J., Heath, T.D., and Patterson, S. (1980): *J. Gen. Virol.,* in press.
5. Helenius, A., Fries, E., and Martenbech, J. (1977): *J. Cell Biol.* 75:866–880.
6. Haywood, A.M. (1974): *J. Mol. Biol.* 87:625–628.
7. Hsu, M-C., Scheid, A., and Choppin, P.W. (1979): *Virology* 95:476–491.
8. Oxford, J.S. and Schild, G.C. (1977): *J. Gen. Virol.* 38:187–193.
9. Szoka, F. and Papahadjopoulos, D. (1978): *Proc. Nat. Acad. Sci.* 75:4194–1498.
10. Oxford, J.S., Schild, G.C., Potter, C.W., and Jennings, R. (1979): *J. Hyg.* 82:51–61.
11. Rott, R. (1980): In: *Structure and Variation in Influenza Virus.* Laver, W.G. (Ed.), pp. 201–209.
12. Nermut, M.V. (1972): *J. Gen. Virol.* 17:317–331.
13. Schulze, I.T. (1973): In: *Advances in Virus Research,* 18. Academic Press, New York and London, pp. 1–56.
14. Wrigley, N.G. (1979): *Brit. Med. Bull.* 35:35–38.
15. Dubovi, E.J. and Wagner, R.R. (1977): *J. Virol.* 22:500–509.
16. Lenard, J. (1978): *Ann. Rev. Biophys. Bioengineer.* 7:139–165.
17. Heath, T.D., Edwards, D.C., and Ryman, B.E. (1976): *Biochem. Soc. Transact.* 4:129–133.
18. Allison, A.C. and Gregoriadis, G. (1974): *Nature* 252:252.
19. Manesis, E.K., Cameron, C.H., and Gregoriadis, G. (1979): *FEBS Letters* 102:107–111.
20. Siddiqui, W.A., Taylor, D.W., Kan, S-C, Kramer, K., Richmond-Crum, S.M., Kotani, S., Shiba, T., and Kusimoto, S. (1978): *Science* 201:1237–1239.

David H.L. Bishop and Richard W. Compans, eds.
The Replication of Negative Strand Viruses

Interaction of Influenza M Protein with Lipid

Anastasia Gregoriades[a]

The membrane (M) protein of influenza virus is localized directly beneath the lipid bilayer and has been visualized as a tightly adherent sac surrounding the nucleoprotein.[1-3] Although this protein is the major constituent of the virus,[4,5] its structural relationship to the lipid or other viral proteins is unknown. It is not clear whether M is an integral part of the viral membrane by insertion into the lipid bilayer, or whether it associates with the ribonucleoprotein to be part of a nucleocapsid structure. A slight space has been reported between M and the lipid bilayer suggesting that it is not part of the virus membrane,[6] while other data using fluorescent dyes embedded into the lipid bilayer suggest that M may be penetrating into the lipid.[7]

The question as to whether M is an integral part of the membrane was approached by determining if M can interact *in vitro* with lipid extracted from the virus, or with other lipids of defined composition. Artificial membrane vesicles have been widely used to study biological membranes and *in vitro* reconstitution experiments of membrane components indicate that the proteins insert themselves into the liposomes in an identical orientation to that found in the original host.[8-13] The hemagglutinin and neuramidase incorporate into preformed vesicles when they are mixed and sonicated[14] or when the protein and lipid are mixed together and vesicles formed by the detergent dialysis method.[15] The resulting spike structures resemble those of the virus with either method. The hemagglutinin is stably attached to the virus lipid bilayer by a sequence of hydrophobic amino acids at the C-terminus of the

[a] Department of Virology, The Public Health Research Institute of the City of New York, New York, New York.

protein[16,17] and the same has been discerned for the neuraminidase.[18,19] The work reported here was undertaken to determine if M can interact *in vitro* with viral or other types of lipid; a high affinity for lipid based on insertion of M into vesicles, would suggest that M could be interacting with the virus lipid to be an integral part of the membrane.

Association of M with Viral Lipid and Phosphatidylcholine

The ability of M to interact with lipid was determined by mixing purified M from the WSN strain of influenza virus with viral lipid or egg phosphatidylcholine. Vesicles were allowed to form and these were then analyzed for M content by floating these upward through sucrose gradients. Alternatively, M was mixed with preformed vesicles and these then analyzed for M content after flotation.[20] This type of analysis indicated that M protein alone remained at the bottom of gradient tubes (Figure 1A) and lipid alone floated to the top of such gradients (Figure 1B). When M was mixed with viral lipid or egg phosphatidylcholine, M and the lipid floated upward (Figure 1C). Between 90 and 100% of the M floated upwards after centrifugation, irrespective of the method used to purify M. M also became tightly associated with preformed vesicles when these were mixed with dry films of M and agitated 5 minutes on a Vortex mixer. Under these conditions, approximately 60% of the M attached to the preformed vesicles to float upwards. The ability of M to float upwards through 35% sucrose even in the presence of 0.5M NaC1 where electrostatic forces would be eliminated indicated that the protein was tightly bound to the lipid. The ribonucleoprotein of the virus, on the other hand, did not interact with lipid.

Proteolytic Digestion of Vesicles Containing M

The manner in which M was interacting with lipid vesicles was investigated . Vesicles containing radioactively labeled M were exposed to trypsin or thermolysin and the lipid vesicles were then floated upward through 35% sucrose. The radioactivity associated with lipid vesicles which were untreated with proteolytic enzymes floated to the top of the gradient (Figure 2A). After treating vesicles with 1% trypsin or thermolysin, approximately 80% of the radioactivity remained at the bottom of the tube, while the rest remained associated with the vesicles which floated to the top of the gradients (Figure 2B). Those vesicles which floated to the top of the gradients after proteolytic digestion were collected and analyzed on SDS polyacrylamide gels. Undigested vesicles contained M only (Figure 3A). Vesicles containing M which were treated with trypsin or thermolysin showed a decrease in the M content with the concomitant appearance of a fragment of about 5,000 daltons. With excess enzyme (50%, only the fragment remained associated with the vesicles (Figure 3B). Two-dimensional analysis of the

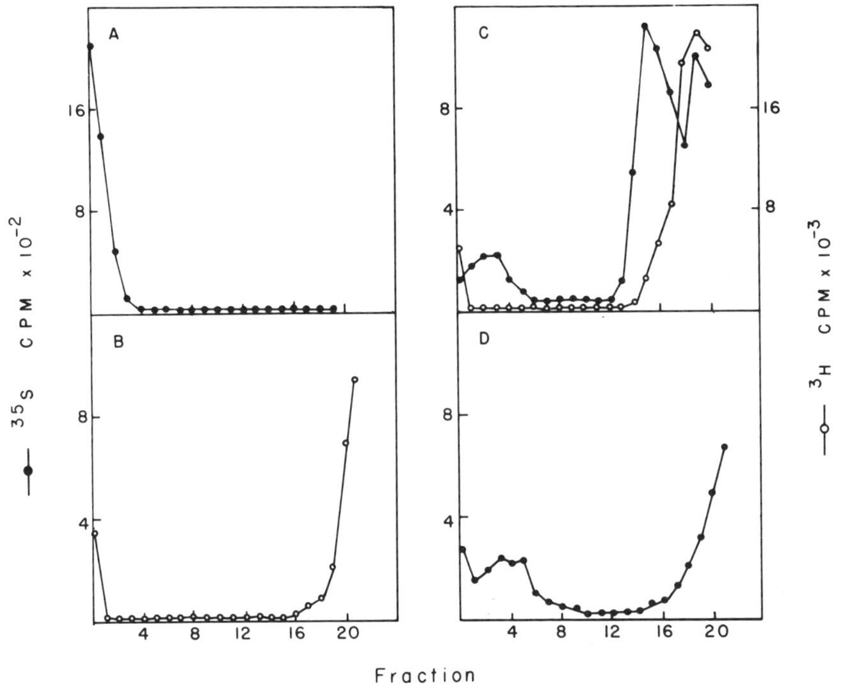

Figure 1. Flotation analysis of M mixed with lipid, through discontinuous gradients. ^{35}S-labeled M was mixed with lipid and deoxycholate as previously described.[20] The detergent was removed by dialysis, the samples were made 45% with respect to sucrose and floated through 35% sucrose. Trace amounts of ^{3}H-phosphatidylcholine were added as marker. -●- ^{35}S-methionine labeled M; -○- ^{3}H-phosphatidylcholine. (A) M alone; (B) Virus lipid alone; (C) M mixed with viral lipid; (D) Dried film of M agitated 5 minutes on a Vortex mixer with preformed phosphatidylcholine vesicles. Vesicles were prepared by the detergent dialysis method.[20]

vesicles remaining after extensive trypsinization indicated that only one peptide was associated with the vesicles, that this represented a specific part of the M molecule and that its amino acid composition was different from that of the whole molecule.[20] The fragment was very insoluble and aggregated in aqueous buffers once the lipid was removed, so that trypsin or chymotrypsin were ineffective in digesting it further.

The complete nucleotide sequence of the RNA coding for the M protein has been published and the amino acid sequence deduced in its entirety.[21,22] The data above indicated that a portion of the M was embedded into the lipid bilayer. An effort was made to determine the amino acid sequence of that part of the protein in order to localize the area in the whole molcule. This however, proved to be unproductive since an N-terminal was unavailable. Therefore, a second approach was used to determine the location on the molecule where insertion into the bilayer might be occurring.

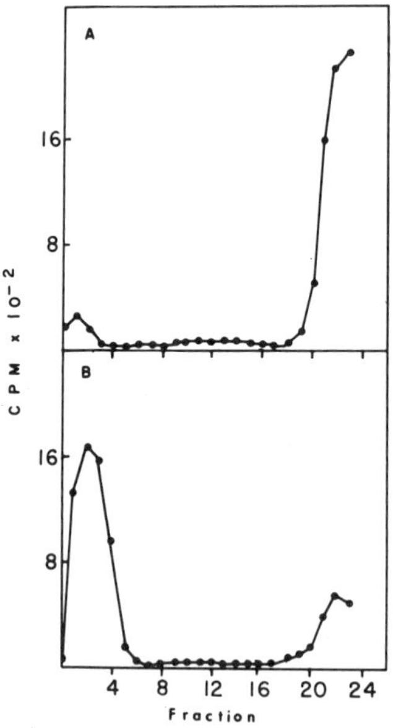

Figure 2. Discontinuous gradient centrifugation of trypsin treated vesicles containing M. ^3H-amino acid labeled M was extracted with chloroform-methanol[23] and incorporated into vesicles by the detergent dialysis method. Samples were made dense with sucrose and analyzed by flotation through 35% sucrose. (A) Untreated vesicles containing M; (B) Trypsin treated vesicles containing M.

Association of Cyanogen Bromide Fragments of M with Lipid

The ability of a specific part of the M protein to interact with lipid vesicles *in vitro* was further tested by determining whether specific fragments of M could also interact with lipid. The M protein was therefore cleaved at its methionine residues with cyanogen bromide and lipid was added to these fragments; the lipid vesicles formed were then floated upwards through sucrose gradients, harvested and analyzed to determine if any M fragments associated with them. Two-dimensional analysis on silica thin layer plates indicated that 12–13 fragments were produced after cleavage with cyanogen bromide (Figure 4A) and of these, three specifically associated with lipid vesicles to float upwards (Figure 4B). Thus, the results using fragments of M also indicated that a specific part of the molecule bound lipid tightly.

The three cyanogen bromide fragments that bound to lipid vesicles have been separated on Biogel P-10 with 50% formic acid. These have been analyzed for amino acid composition and sequence. The data indicate that one fragment represents the N-terminus of the M protein and that the other

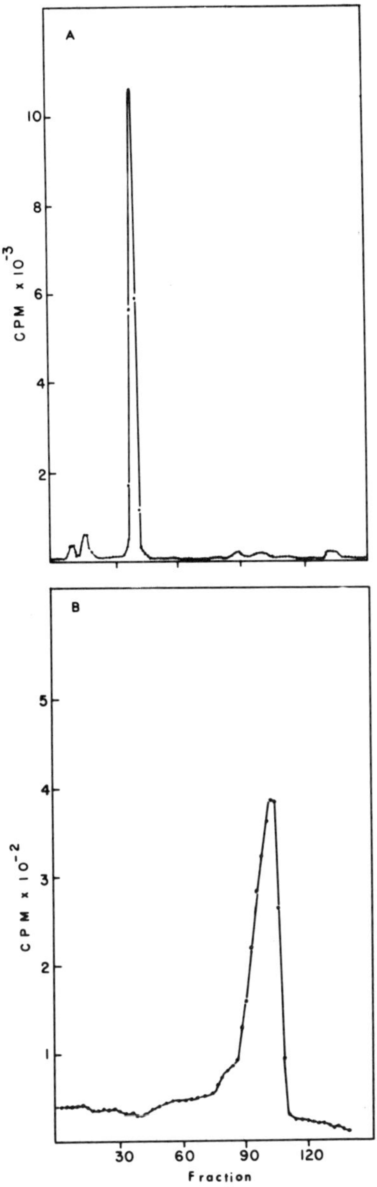

Figure 3. Polyacrylamide gel electrophoresis[24] of vesicles containing M after incubation with excess concentrations of trypsin. M was mixed with phosphatidylcholine and vesicles allowed to form by the detergent dialysis method. Vesicles were trypsinized, isolated by flotation and analyzed on 12% polyacrylamide gels. (A) Untreated vesicles containing M; (B) Trypsin treated vesicles.

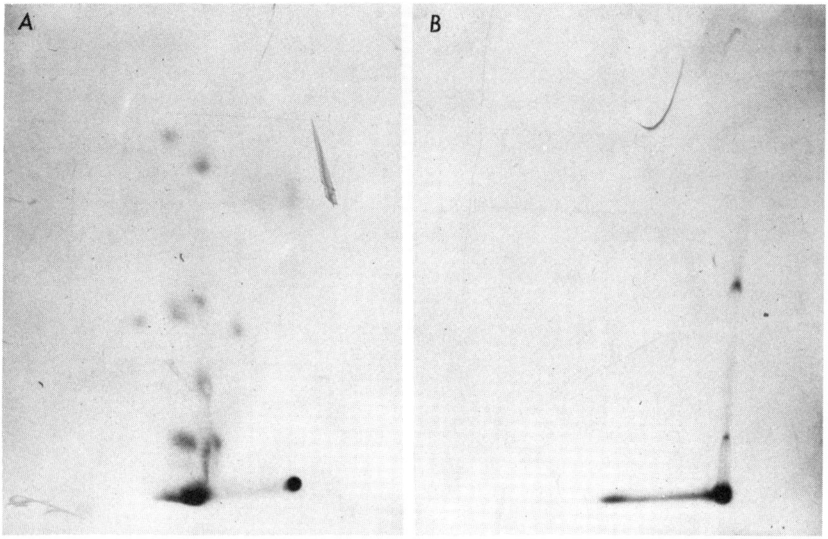

Figure 4. Two-dimensional analysis of cyanogen bromide fragments of M which floated upwards with lipid, through 35% sucrose. M was cleaved at the methionine residues with cyanogen bromide, mixed with phosphatidylcholine and the vesicles isolated by flotation to determine if fragments of M could associate with lipid. The conditions for two-dimensional analysis were as previously described.[20] (A) Total mixture of cyanogen bromide fragments with phosphatidylcholine; (B) Cyanogen bromide fragments of M which associated with lipid vesicles and floated to the top of sucrose gradients.

SOURCE: *Gregoriades, 1980.*

two fragments are contiguous to this. The evidence also indicates that the cyanogen bromide fragments represent part of the same area of the M molecule which remains associated with vesicles after extensive trypsinization discussed above (Gregoriades and Frangione, in preparation).

Conclusion

The work reported here was carried out to determine if protein-lipid interactions can occur between the influenza M protein and viral or other lipids. The data indicate that M has a very high affinity for lipid *in vitro* and associates with free lipid or with preformed vesicles. Aqueous buffers containing lipid vesicles have a marked ability to disaggregate insoluble M preparations where only acids or strong denaturants would do so. The M protein attaches to lipid vesicles at a specific area of the molecule, represented by a fragment of approximately 5,000 daltons. This fragment is resistant to further proteolytic digestion, which would suggest that it is embedded into the lipid bilayer. The primary structure of M can allow certain areas of the molecule to interact with lipid, as shown by the finding that three contiguous cyanogen bromide fragments, beginning at the N-terminus of the protein bound specifically to lipid vesicles. The very high affinity of M for viral or other types of lipid *in vitro* is strong suggestive evidence that the protein interacts with the virus lipid bilayer to be an integral part of the membrane. This would be in agreement with data suggesting that M may be penetrating into the bilayer[7] and would also be consistent with the solubility of M in chloroform-methanol[23] which classifies it as a proteolipid.[24]

ACKNOWLEDGMENT
This work was supported by Public Health Service Grant AI-12567 from the National Institute of Allergy and Infectious Diseases. The author thanks Theresa Christie for most capable technical assistance.

References

1. Apostolov, K. and Flewett, T.H. (1969): *J. Gen. Virol.* 4:365–370.
2. Bachi, T., Gerhard, W., Lindenmann, J., and Muhlethaler, K. (1969): *J. Virol.* 4:769–776.
3. Schulze, I.T. (1972): *Virology* 47:181–196.
4. Compans, R.W., Klenk, H-D., Caliguiri, L.A., and Choppin, P.W. (1970): *Virology* 42:880–889.
5. Schulze, I.T. (1970): *Virology* 42:890–904.
6. Nermut, M.V. (1972): *J. Gen. Virol.* 17:317–331.
7. Lenard, J., Wong, C.Y., and Compans, R.W. (1974): *Biochim. Biophys. Acta* 332:341–349.
8. Engelhard, V.H., Guild, B.C., Helenius, A., Terhorst, C., and Strominger, J.L. (1978): *Proc. Natl. Acad. Sci. U.S.A.* 75:3230–3234.
9. Helenius, A., Fries, E., and Kartenbeck, J. (1977): *J. Cell Biol.* 75:866–880.
10. Hosaka, Y. and Shimizu, Y.K. (1972): *Virology* 49:627–639.
11. Kagawa, Y. and Racker, E. (1966): *J. Biol. Chem.* 241:2475–2482.
12. Littman, D.R., Cullen, S.E., and Schwartz, B.D. (1979): *Proc. Natl. Acad. Sci. U.S.A.* 76:902–906.
13. Petri, W.A. and Wagner, R.R. (1979): *J. Biol. Chem.* 254:4313–4316.
14. Almeida, J.D., Edwards, D.C., Brand, C.M., and Heath, T.D. (1979): *Lancet* 2:899–901.
15. Huang, R.T.C., Wahn, K., Klenk, H-D., and Rott, R. (1979): *Virology* 97:212–217.
16. Bucher, D.J., Li, L. S-L., Kehoe, J.M., and Kilbourne, E.D. (1976): *Proc. Natl. Acad. Sci. U.S.A.* 73:238–242.
17. Skehel, J.J. and Waterfield, M.D. (1975): *Proc. Natl. Acad. Sci. U.S.A.* 72:93–97.
18. Lazdins, I., Haslam, W.A., and White, D.O. (1972): *Virology* 49:758–765.
19. Wrigley, N.G., Skehel, J.J., Charlwood, P.A., and Brand, C.M. (1973): *Virology* 51:525–529.
20. Gregoriades, A. (1980): *J. Virol.* 36:470–479.
21. Both, G.W. and Air, G.M. (1979): *Eur. J. Biochem.* 96:363–372.
22. Winter, G. and Fields, S. (1980): *Nucl. Acid Res.* 8:1965–1974.
23. Gregoriades, A. (1973): *Virology* 54:369–383.
24. Stern, W. and Dales, S. (1976): *Virology* 75:242–255.
25. Folch, J. and Lees, M. (1951): *J. Biol. Chem.* 191:807–817.

David H.L. Bishop and Richard W. Compans, eds.
The Replication of Negative Strand Viruses

Interaction of Neuraminidase with M-Protein in Liposomes

J.F. Davis and D.J. Bucher[a]

Purified viral protein antigens can be engineered to produce a multiplicity of structures. Glycoprotein surface antigens from enveloped viruses can be "forced" to aggregate upon removal of detergents by dialysis or they may be incorporated into liposomes.[1-5] Viral glycoproteins in liposomes may closely simulate the original conformation in the viral envelope. These liposomal structures may have enhanced immunogenicity.[4]

The availability of purified viral proteins also provides an opportunity for the study of interactions between different protein components of the virion. The mechanism by which these proteins interact with each other and the lipid bilayer may assist us in understanding the maturation of enveloped viruses.

We have recently shown the great avidity of purified M-(matrix or membrane) protein for the lipid bilayer.[6] M-protein readily associated with lipid to form liposomes which formed a band of heavy density (1.22 g/ml) on sucrose gradient centrifugation. The M-protein appeared to partition into the lipid bilayer by a co-operative process since only two discrete liposomal fractions were found on gradient analysis; a heavy density fraction containing M-protein or a light fraction which was devoid of M-protein. Freeze-fracture preparations of M-protein liposomes examined by electron microscopy showed the bilayers to be studded with 80-90-Å particles at a distribution of 3000/μm^2. The size and distribution was similar to that found for intramembranous particles of other biological membranes.

[a] Department of Microbiology, Mount Sinai School of Medicine of The City University of New York.

In studies of Sendai virus infected cells, Bächi has found crystalline patterns of intramembranous particles on the inner surface of the plasma membrane.[7] He has shown the association of Sendai specific glycoproteins and nucleocapsids with these domains. Bächi has suggested that the intramembranous particles may consist of aggregated M-protein or M-protein interacting with lipid.

A model system for studying the assembly of the viral envelope may consist of the surface glycoproteins and M-protein incorporated into liposomes. We are able to independently isolate M-protein and neuraminidase by chromatographic techniques [8-10] and insert them into the lipid bilayer. The enzymatic activity of the neuraminidase serves as a sensitive means of quantitating M-protein-neuraminidase interaction in liposomes.

Materials and Methods

Preparation of Liposomes

M-protein was purified by SDS gel chromatography from a batch of X-38 (Heq1N2) virus prepared for vaccine use (a gift of Lederle Laboratories, Pearl River, New York). SDS was removed as previously described.[6] Neuraminidase was isolated by affinity chromatography from the X-7 recombinant (HON2) as previously described.[10] Neuraminidase from such preparations is highly immunogenic if aggregated (by removal of Triton X-100) or incorporated into liposomes (Davies and Bucher, manuscript in preparation).

Liposomes were formed by dialysis following solubilization of the lipids or lipid-protein mixture with octyl glucoside. Control liposomes were prepared after solubilizing the mixture of lecithin : dicetyl phosphate : cholesterol (molar ratio 9:2:1) in 2% octyl glucoside. Liposomes containing viral proteins were formed by the addition of the purified protein(s) to the same ratio of lipids as used for control liposomes. The protein : lipid ratios (by weight) were respectively: M-protein liposomes, 1:5 (0.8 mg:4 mg); neuraminidase liposomes, 1:25 (0.16 mg:4 mg); and neuraminidase-M-protein liposomes, 1:5:25 (0.16 mg:0.8 mg:4mg) (NA:M:lipid). Liposomal preparations were dialyzed against 2 mM $CaCl_2$ for one week to remove detergent.

Liposomes were characterized by sucrose gradient centrifugation as previously described.[6] Absorbance at 260 nm was monitored in a Uvicord I (LKB) ultraviolet monitor on elution of the gradients. Protein was detected by the Lowry assay.[11] The neuraminidase assay was performed on individual aliquots as described previously.[10]

Results and Discussion

Control liposomes. Liposomes containing no added viral protein showed multiple forms which banded in an area of light density (1.04–1.07 g/ml) as shown in Figure 1. These multiple bands may represent liposomes containing varying numbers of lamellae in the liposome.

M-protein liposomes. The finding of two major liposomal forms in this case confirms our earlier observations with M-protein liposomes (Figure 2).[6] Sucrose gradient centrifugation showed liposomes of two distinctly different densities, a fraction rich in M-protein of heavy density and a second fraction of light density with little or no detectable protein. The density of the heavy band (1.15 g/ml) is lighter than the 1.22 g/ml found in our earlier studies. The protein found at the top of the gradient is not M-protein but appears to be a contaminant of the lecithin preparation; the protein co-migrates with ovalbumin.

Neuraminidase liposomes. Neuraminidase activity was found associated with the light density liposomal bands with respective densities of 1.03, 1.06 and 1.07 g/ml (Figure 3). Associated neuraminidase did not significantly alter the densities of the liposomal fractions, unlike the case with M-protein liposomes where the M-protein served to greatly increase the density of the M-protein liposomes. A substantial proportion of the neuraminidase aggregated and sedimented to the bottom; no substantial amount of lipid associated with this fraction as judged from the lack of turbidity of 260 nm.

Neuraminidase-M-protein liposomes. The presence of M-protein greatly enhances the incorporation of neuraminidase into the lipid bilayer (Figure 4). Three times as much neuraminidase activity is associated with liposomes in the presence of M-protein (see Figure 4) as compared to that found in the absence of M-protein (see Figure 3); the same quantity of neuraminidase had

Figure 1. Control liposomes. Turbidity of the liposomal fractions on the gradient was monitored by absorbance at 260 nm (--------). Density was measured by refractive index (——).

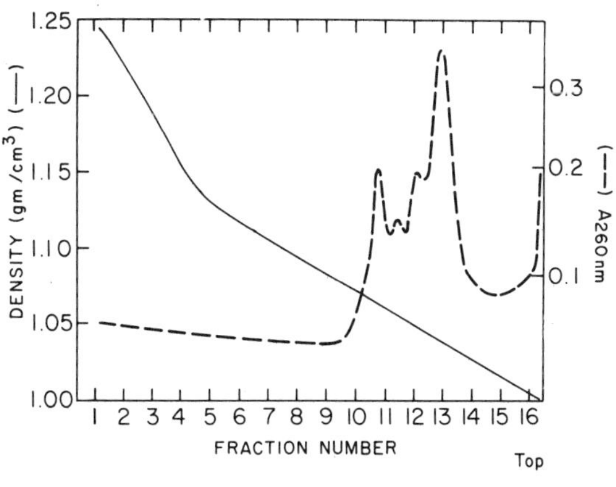

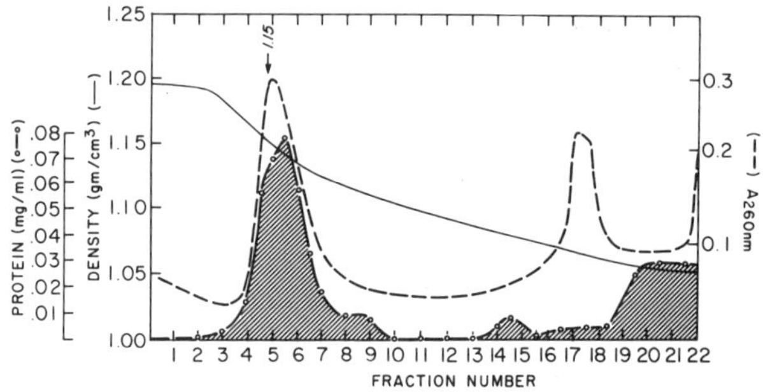

Figure 2. M-protein liposomes. Protein concentration is indicated by the shaded area (—O—O—O). M-protein is associated with the liposomal fraction of 1.15 g/cm³.

been added during formation of liposomes in both cases. Nearly two-thirds of the neuraminidase activity associates with the heavy density M-protein containing liposomes. A slight increase in density is detected from 1.15 g/ml as seen in Figure 2 for M-protein liposomes to 1.17 g/ml, an increase of 0.02 g/ml. A similar amount of neuraminidase activity is associated with the low density liposomes as seen above for the neuraminidase liposomes in Figure 3.

The enhanced incorporation of neuraminidase into liposomes in the presence of M-protein suggests two possibilities for the mechanism of formation of neuraminidase-M-protein liposomes. Neuraminidase may *directly interact* with M-protein through its hydrophobic tail resulting in incorporation of the enzyme into liposomes containing M-protein. A second explanation may be

Figure 3. Neuraminidase liposomes. Neuraminidase activity is represented by the stippled area (—●—●—●). Neuraminidase activity is associated with low density liposomal fractions (1.03 to 1.07 g/cm³). A significant portion of the neuraminidase activity sediments under these conditions.

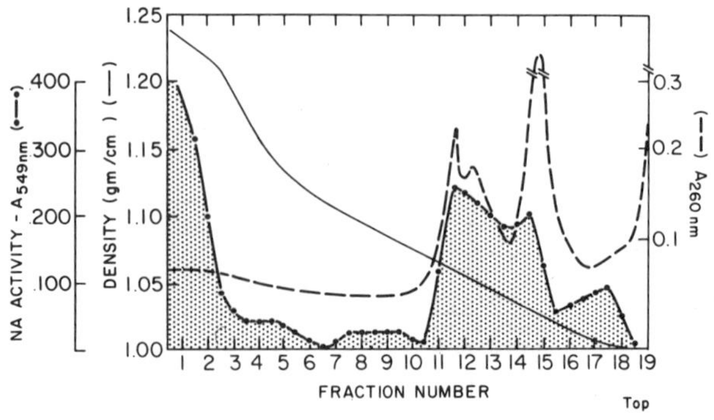

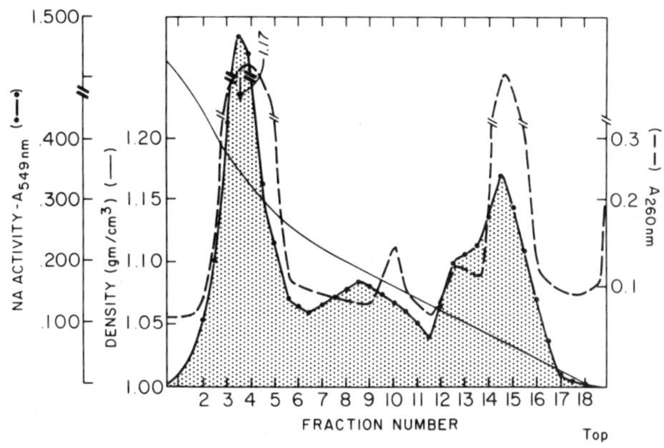

Figure 4. Neuraminidase-M-protein liposomes. Neuraminidase activity is represented by the stippled area (—●—●—●). A substantial portion of neuraminidase activity is associated with the heavy density liposomal fraction containing M-protein (1.17 g/cm^3). Neuraminidase activity of the M-protein associated fraction reaches nearly 1.50 optical density units as determined at 549 nm; the liposomal fraction at 1.04 and 1.06 g/cm ^{3} has neuraminidase activities represented by 0.35 and 0.20 respectively.

that M-protein *alters the lipid bilayer* in such a way that the lipid is more conducive to insertion of the neuraminidase.

We have previously shown the considerably affinity of M-protein for interaction with itself and the lipid bilayer.[6] We have now demonstrated the interaction of neuraminidase and M-protein in co-formation of liposomes. Interaction of M-protein with the lipid bilayer could alter membrane structure during viral infection of the host cell. This interaction could result in concentration of glycoprotein spikes in an area of "condensed" M-protein-lipid domains.[7,12,13] The M-protein in turn may then interact with ribonucleoprotein complexes causing the "budding" and maturation of viral particles.[7,13] Studies of inhibitors which will interact with M-protein and prevent its interaction with itself, other viral proteins, or the lipid bilayer in model liposomal systems could be of considerable assistance in the design of antiviral pharmacologic agents effective against influenza and other enveloped viruses.

ACKNOWLEDGMENTS

The expert assistance of Ms. Tatiana Tomko in preparation of M-protein is appreciated. The authors thank Ms. Ellen Pitler for preparation of the manuscript.

This work was monitored by The Commission on Influenza of The Armed Forces Epidemiological Board and was supported by the U.S. Army Medical Research and Development Command under Research Contract No. DADA 17-69-C-9137. Support for this work was also provided by The New York State Health Research Council.

References

1. Hosaka, Y. and Shimizu, Y.K. (1972): *Virology* 49:640–646.
2. Almeida, J.D., Brand, C.M., Edwards, D.C., and Heath, T.D. (1975): *Lancet* 2:899–901.
3. Helenius, A., Fries, E., and Kartenbeck, J. (1977): *J. Cell Biol.* 75:866–880.
4. Morein, B., Helenius, A., Simons, K., Pettersson, R., Kaariainin, L., and Schirmacher, V. (1978): *Nature* 276:715–718.
5. Huang, R.T.C., Wahn, K., Klenk, H.-D., and Rott, R. (1979): *Virology* 97:212–217.
6. Bucher, D.J., Kharitonenkov, I.G., Zakomirdin, Ju.A., Grigoriev, V.B., Klimenko, S.M., and Davis, J.F. (1980): *J. Virol.* 36:586–590.
7. Bächi, T. (1980): *Virology* 106:41–49.
8. Bucher, D.J. (1975): In: *The Negative Strand Viruses. Vol. 1.* Mahy, B.W.J. and Barry, R.D. (Eds.), Academic Press, London, pp. 133–143.
9. Bucher, D.J., Li, S.S.-L., Kehoe, J.M., and Kilbourne, E.D. (1976): *Proc. Nat. Acad. Sci. U.S.A.* 73:238–242.
10. Bucher, D.J. (1977): *Biochim. Biophys. Acta:* 482:393–399.
11. Lowry, O.H., Rosebrough, N.J., Farr, A.L., and Randall, R.J. (1951): *J. Biol. Chem.* 193:265–275.
12. Choppin, P.W. and Compans, R.W. (1975): In: *The Influenza Virus and Influenza.* Kilbourne, E.D. (Ed.), Academic Press, New York, pp. 15–51.
13. Robertson, B.H., Bhown, A.S., Compans, R.W., and Bennett, J.C. (1979): *J. Virol.* 30:759–766.

Published 1981 by Elsevier North Holland, Inc.
David H.L. Bishop and Richard W. Compans, eds.
The Replication of Negative Strand Viruses

Advantages and Limitations of the Oligonucleotide Mapping Technique for the Analysis of Viral RNAs

James F. Young, Ronald Taussig,
Robert P. Aaronson and Peter Palese[a]

Introduction

The technique of oligonucleotide mapping of RNAs has successfully been used by many investigators for the comparison of different RNA molecules.[1-3] The method is based on the two-dimensional separation of oligonucleotides which derive from a complete digest with a specific ribonuclease, in most instances ribonuclease T1. The resulting oligonucleotides are then separated either by homochromatography or by polyacrylamide gel electrophoresis.

We have been interested in the analysis of genetic variation of different influenza viruses and have used oligonucleotide mapping extensively for the characterization of the genomes of these viruses. With the advent of cloning and nucleic acid sequencing techniques, it has now become possible to precisely quantitate nucleotide differences among various influenza virus genes. In the present study, we have compared the sequence differences of several RNAs as estimated by oligonucleotide mapping techniques with the previously published sequence data for these genes. This comparison allows us to establish the range of sensitivity of oligonucleotide fingerprinting, so that it can be more effectively applied for the rapid characterization of different viral RNAs.

[a] Department of Microbiology, Mount Sinai School of Medicine of CUNY, New York, New York.

Materials and Methods

The influenza A-PR/8/34 (H0N1) (PR8) and A-Udorn/72 (H3N2) (Udorn) viruses were grown in embryonated eggs. Viral RNAs were extracted from purified viruses,[4] separated by electrophoresis on 2.6% or 2.8% polyacrylamide gels [4] and isolated as described.[5] Isolated RNA segments were ribonuclease T1-digested and the resulting oligonucleotides were 5′-end labeled and separated by two-dimensional polyacrylamide gel electrophoresis.[3,5] For the analysis of the nucleotide sequencing data we used DEC PDP11/45 digital computer.

Results and Discussion

Figure 1 shows the oligonucleotide maps of the NS genes of two human influenza A viruses, A/PR/8/34 and A-Udorn/72. Comparison of these two oligonucleotide patterns at first suggested that they were quite different. However, when both oligonucleotide preparations were co-electrophoresed on the same gel, it was found that six of the oligonucleotides from each strain co-migrated to the same positions on the oligonucleotide maps (indicated by solid circles in the schematic diagrams, Figures 1c and d). These "common" or unchanged oligonucleotides represent 46% of the oligonucleotides (6/13) in the map of the PR8 NS gene and 35% (6/17) of the oligonucleotides in the map of the Udorn NS gene. This analysis includes only those oligonucleotides more than eleven bases in length; oligonucleotides shorter than this may not represent unique sequences in the RNAs. The lengths of these oligonucleotides were determined by electrophoresis of the eluted spots on a 20% polyacrylamide gel together with an alkali generated "ladder" for sizing (data not shown). An estimate from this analysis suggests that the large oligonucleotides in the two RNA fingerprints have an average size of 16 nucleotides (with a range from 12 to 42 nucleotides).

In order to correlate the observed changes in the oligonucleotide maps with the changes determined by direct sequence analysis of the two RNAs, we proceeded to calculate the number of unchanged oligonucleotides in a random nucleotide sequence following the introduction of random sequence changes. Figure 2 shows the statistically predicted values of unchanged oligonucleotides 16 bases in length when sequence changes of 1 to 20% were introduced into an RNA sequence. According to this calculation, an average of only 50% of these oligonucleotides are conserved, when two RNAs with an overall sequence difference of 4.3% are compared (Figure 2). Thus, the observed conservation of 46% and 35% of the PR8 and Udorn NS gene-specific oligonucleotides, respectively, reflects a predicted 5–6% overall base sequence difference in their RNAs. It should be noted, however, that this calculation is based on the assumption that all changes can be distinguished, are randomly distributed, and that all oligonucleotides have a length of 16 bases.

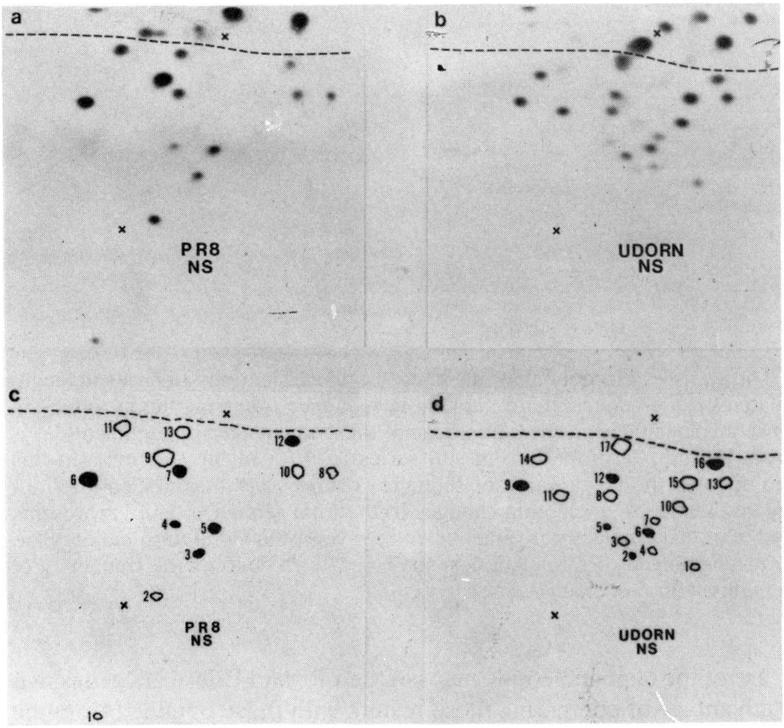

Figure 1. Oligonucleotide map analysis of the NS genes of influenza A/PR/8/34 and A/Udorn/72 viruses. Panels a and b show actual maps of the two RNAs. Xs represent xylene cyanol and bromophenol blue dye markers. Only the spots below the broken line were used for the analysis. Panels c and d represent schematic diagrams of these maps. A tentative assignment of these spots was made using the sequence information provided in Table 1 (see spot numbers). Six oligonucleotides in the map of the A/PR/8/34 NS gene and in the map of the A/Udorn/72 NS gene were found to comigrate when both oligonucleotide preparations were run on the same gel (solid circles).

Analysis of Figure 2 also suggests that few oligonucleotides of an average size of 16 nucleotides remain unchanged when two RNA sequences differ by more than 10–15% (assuming the changes are randomly distributed over the entire genome). Thus, oligonucleotide mapping does not appear to be a sensitive tool to determine the relatedness of RNAs whose overall sequence homology is less than 90%, since these RNAs should give rise to completely different fingerprints. However, should these RNAs share regions of homology greater than 90% (e.g., strains with conserved sequences or recombinant strains), oligonucleotide mapping becomes a valuable tool for the analysis of these portions of the RNAs.

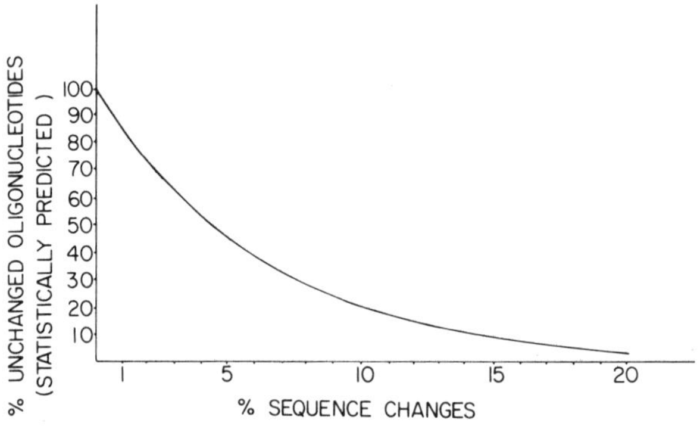

Figure 2. The statistical frequency of unchanged oligonucleotides 16 bases in length was calculated for a given (random) nucleotide sequence following the introduction of random sequence changes. The percentage of these unchanged oligonucleotides is derived from the first term of the Poisson series [e^{-m} (1, m, m^2, ····, m^i)]. In this formula, m denotes the mean value of sequence changes per oligonucleotide (i.e., size of oligonucleotide/% nucleotide changes of the total sequence) and i represents the number of actual changes per oligonucleotides (i.e., the $i = 0$ term denotes the frequency of unchanged oligonucleotides; the $i = 1$ term denotes the frequency of oligonucleotides with one change, etc.).

In the case of the oligonucleotide maps of the PR8 and Udorn NS genes, we have the advantage of comparing these results with those obtained from the complete nucleotide sequences of these RNAs. Based on the published sequences of the PR8[6] and Udorn NS [7] genes, a series of large oligonucleotides is expected to be generated by ribonuclease T1 treatment (Table 1). Attempts were made to correlate the expected oligonucleotides with specific spots in the two maps, taking into account the predicted size and composition of the oligonucleotides.[2] In the case of the Udorn NS, perfect agreement was found for the expected and the actual fingerprint (Figure 1d and Table 1). For the PR8 NS RNA there appears to be one exception. Oligonucleotide #2 (of 19 nucleotide length, Table 1) could not be found in the expected position and a larger spot was seen in the fingerprint. Since we have not attempted to sequence specific oligonucleotides obtained from the gels, we cannot determine whether this discrepancy is the result of a mutation on the viral RNA used for fingerprint analysis, whether a clonal variant of PR8 was used for sequencing,[6] or whether the secondary structure of the RNA is responsible for an incomplete RNase T1 digestion. Despite the one exception, the excellent agreement between the predicted and actual fingerprints of the two NS genes (Figure 1) provides further evidence for the high resolving power of two-dimensional polyacrylamide gels and the reliability of the T1 fingerprinting technique. In comparison with the estimated sequence homology of 5–6% between the PR8 and Udorn NS genes which was based on a comparison of the oligonucleotide maps (Figure 1), we find an actual value of 8.9%

Table 1. Large Ribonuclease T1 Resistant Oligonucleotides of A/PR/8/34 and A/Udorn/72 Viral RNAs.

A/Udorn/72				A/PR/8/34			
Spot	Size	Position[a]	Sequence	Spot	Size	Position	Sequence
1	19	17	TTTTTTATCATTAAATAAG[b]	1	42	141	TCTCACTTCTTCAATCAACCATC TTATTTCTTCAAACTTCTG
2	19	232	TCTCCCATTCTCATTACTG	2	19	17	TTTTTTATTACTAAATAAG
3	18	141	TCTCACTTCTTCAATCAG	3	19	232	TCTCCCATTCTCATTACTG
4	18	161	AATCTTCTCTACTATCTG	4	17	108	CTCAAAACTATTCTCTG
5	17	108	CTCAAAACTATTCTCTG	5	17	296	TTATCATTCCATTCAAG
6	17	296	TTATCATTCCATTCAAG	6	15	449	TCAAAAATCACACTG
7	16	159	CCATCTTATCTCTTCG	7	14	725	CCCCTTCCTCTTAG
8	15	63	TTCCACTTCAAATAG	8	13	50	TTCTTATCTCTTG
9	15	449	TCAAAAATCACACTG	9	13	570	ACATTTCCTCAAG
10	15	526	TCCTTCCACTTTCTG	10	13	648	ATTCTTCTTTCAG
11	15	570	ACAATTCCTCAATAG	11	12	63	CTCCACTTCAAG
12	14	725	CCTCTTCCCCTTAG	12	12	485	TTCTTATCCATG
13	13	50	TTCTTATCTCCTG	13[c]	12	667	CTCCACTATCTG
14	13	344	ACATCCTCAATAG				
15	13	648	ATTCTTCCTTCAG				
16	12	485	TTCTTATCCATG				
17[c]	12	679	CTTTCCAACATG				

[a] Indicates position of the 5′-end of the oligonucleotide; numbering is in the 5′ to 3′-direction along the virion RNAs. (Total length = 890 bases.) For this list the published sequences of the NS genes of A/PR/8/34 and A/Udorn/72 viruses were used.[6,7]

[b] Underlined sequences are found in both the A/PR/8/34 and A/Udorn/72 NS genes. The two sequences underlined with dashed lines are spots predicted by computer analysis to migrate identically; one of them (PR8 spot #2) was not observed in the actual fingerprint.

[c] The 13 and 17 oligonucleotides of the PR8 and Udorn NS genes represent 24.5% and 29.3% of the entire sequences, respectively.

overall sequence difference between the two genes.[6] The underestimation of the calculated sequence difference derived by T1 mapping may in part be due to the co-migration of PR8 spot #7 and Udorn spot #12, which contain counterbalancing C → T and T → C changes resulting in identical nucleotide compositions of the two spots and therefore identical mobilities (Table 1). However, the occurrence of such counterbalancing changes should be less frequent in more closely related genes, permitting a more accurate appraisal of their differences.

Since the "true" oligonucleotide maps appear to correlate well with the fingerprints predicted from sequence data, we performed a similar analysis on the computer generated oligonucleotide maps of several H3 hemagglutinin gene sequences.[8-11] The T1 map of the HA gene of A-Aichi/2/68 virus was compared to the fingerprints of the HAs of A/NT/60/68/29C, A/Memphis/102/72, and A/Victoria/3/75 viruses. Using this technique, these HAs were predicted to differ in sequence homology by 0.8% to 1.8%. This compares with the values of 0.5% to 3.8% observed in the actual sequence data of these genes. Although not strictly comparable, the predicted values are within 2% of the actual values and demonstrate the ability to discriminate closely related genes (0.5 to 4% different), making it possible to predict the derivations of very similar strains [5] and to elucidate the gene origin of naturally occurring recombinant strains.[12] Such analyses have greatly helped our understanding of the genetic variation of influenza virus and its epidemiology.

Thus, based on theoretical considerations and the analysis of available sequence data we were able to show that oligonucleotide mapping is a technique best suited to compare RNAs or regions of RNA molecules with overall base sequence homologies of more than 90%. These comparisons can be made rapidly and, in the absence of sequence data, the technique also helps to obtain an estimate of the overall sequence homology of two RNAs.

ACKNOWLEDGMENTS

Influenza A-Udorn/72 virus was kindly provided by Dr. Robert Webster. P.P. is a recipient of an I.T. Hirschl Career Research Award and R.P.A. is a recipient of a Research Career Development Award from the NIH (5K04GM0278). This work was supported by Grant AI-11823 from the NIH, Grant PCM78-07844 from the NSF, and Grant MV-23A from the American Cancer Society.

References

1. Szekeley, M. and Sanger, F. (1969): *J. Mol. Biol.* 43:607–617.
2. Frisby, D. (1977): *Nucl. Acids Res.* 4:2975–2996.
3. Pedersen, F.S. and Haseltine, W.A. (1980): In: *Methods in Enzymology*. Grossman, L. and Moldave, K. (Eds.), Academic Press, New York, pp. 680–687.
4. Palese, P. and Schulman, J.L. (1976): *Proc. Natl. Acad. Sci. U.S.A.* 73:2142–2146.
5. Young, J.F., Desselberger, U., and Palese, P. (1979): *Cell* 18:73–83.

6. Baez, M., Taussig, R., Zazra, J.J., Young, J.F., Palese, P., Reisfeld, A., and Skalka, A.M. (1980): *Nucl. Acids Res.,* in press.
7. Lamb, R.A. and Lai, C.-J (1980): *Cell* 21:475–485.
8. Min Jou, W., Verhoeyen, M., Devos, R., Saman, E., Fang, R., Huylebroeck, D., Fiers, W., Threlfall, G., Barber, C., Carey, N., and Emtage, S. (1980): *Cell* 19:683–696.
9. Verhoeyen, M., Fang, R., Min Jou, W., Devos, R., Huylebroeck, D., Saman, E., and Fiers, W. (1980): *Nature* 286:771–776.
10. Sleigh, M.J., Both, G.W., Brownlee, G.G., Bender, V.J., and Moss, B.A. (1980): In: *Structure and Variation in Influenza Virus.* Laver, G. and Air, G. (Eds.), Elsevier/North Holland, New York, pp. 69–79.
11. Both, G.W. and Sleigh, M.J. (1980): *Nucl. Acids Res.* 8:2561–2575.
12. Young, J.F. and Palese, P. (1979): *Proc. Natl. Acad. Sci. U.S.A.* 76:6547–6551.

David H. L. Bishop and Richard W. Compans, eds.
The Replication of Negative Strand Viruses

Complete Sequence Analysis of the Influenza A/WSN/33 (H0 Subtype) Hemagglutinin and its Relationship to the A/Japan/305/57 (H2 Subtype) Hemagglutinin

Alan L. Hiti, Alan R. Davis, and Debi P. Nayak[a]

Introduction

Human influenza viruses undergo extensive antigenic variation in both surface glycoproteins, the hemagglutinin (HA) and the neuraminidase (NA). The HA appears to play the more important role in both minor changes (antigenic drift) and major changes (antigenic shift) of the virus. To understand the cause and mechanism of antigenic variation, the HA genes of a number of influenza virus strains have recently been completely sequenced.[1-5] Here we report the complete HA gene sequence of the A/WSN/33 virus strain (H0 subtype) derived by similar recombinant DNA methods. Since the H0 and H1 subtypes can probably be grouped under one subtype,[6] at least one of each of the known human subtypes has now been sequenced. By a comparison to the known nucleotide and protein sequences, we have found a close relationship between the hemagglutinins of the H0 and H2 subtypes.

Materials and Methods

The construction of plasmid 2-29 and its characterization as a chimeric plasmid containing a nearly complete double-stranded copy of the A/WSN/33 RNA gene segment 4 (which codes for the HA protein) has been described in detail elsewhere.[7] Briefly, the HA gene DNA has been inserted into the Pst 1 site of pBR322 via G-C tailing, cloned in χ1776, and completely sequenced by established techniques.

[a] Department of Microbiology and Immunology, UCLA School of Medicine, Los Angeles, California.

Results and Discussion

The complete sequence of the A/WSN/33 HA gene (plus sense) is shown in Figure 1 along with the predicted amino acid sequence. A discussion of the common structural characteristics in relation to other human hemagglutinins has been presented elsewhere.[8] The sequence of the A/WSN HA gene predicts a structure for the H0 subtype HA that is consistent with the structural characteristics found for the HAs of the other human subtypes (Table 1). This discussion focuses on the interesting similarities of the HA sequence of the H0 and H2 subtypes.

A comparison of the HA amino acid sequences of the A/WSN (H0 subtype) and the A/Japan (H2 subtype) virus strains suggests a close relationship between them (Figure 1). Only two insertions are required to align the A/WSN sequence to the A/Japan sequence: a single amino acid at position 122 (Figure 1) and two amino acids at positions −1 and −2 (in the "signal" prepeptide, Figure 1). This close relationship becomes more evident when the H0-H2 sequence homology is compared to other intersubtypic comparisons (Table 2). The overall amino acid sequence homology between the H0 and H2 subtypes (67%) is 18% higher than the next most homologous intersubtypic pairing (Hav 1 and H3, 49%). When HA_1 regions are compared, the H0-H2 pairing reveals a 57% homology which is 20% greater than the next most homologous HA_1 pairing (Hav 1 and H3, 37%). The high degree of homology between the H0 and H2 amino acids occurs throughout the HA_1 sequence (Figure 1). The HA_1 region varies considerably more than the HA_2 region in field strains. However, the degree of homology in the HA_1 regions of the HO and H_2 subtypes is even greater than the HA_2 region homology in 4 out of 5 subtype pairings (Table 2).

The HA_2 regions of the H0 and H2 subtypes are $>$ 80% homologous (179/222, Figure 1) and, moreover, 40% of the changes (17/43) can be explained by single nucleotide changes. The amino acid sequence homology between the H0 and H2 subtypes appears throughout the HA_2 sequence, also. The largest stretch of nonhomology is only 6 residues long (pos. 526–531, Figure 1). Additionally the 17 terminal residues of the HA_2 are identical in the H0 and H2 strains.

The coding region nucleotide homologies (Table 3) show data consistent with the amino acid homologies (Table 2). The amino acid homologies of the H0-H3, H0-Hav 1, H2-H3, and H2-Hav 1 pairings are all similar. The same is true for the coding region nucleotide homologies (Table 3). The H0-H2 pairing shows considerably more homology in both the HA_1 and HA_2 regions in both the amino acid (Table 2) and the coding region nucleotide (Table 3) comparisons. Although the amino acid homology is the strongest argument for a close relationship between the H0 and H2 proteins, the nucleotide homology is also consistent with a close genetic relationship between the two. The H0-H2 codon comparison also showed that 27% of the codons in the HA_1 are completely conserved, which is over twice the percentage found in H0-H3 and H0-Hav 1 comparisons (13% and 12%, respectively, data not

Table 1. Comparison of Influenza Hemagglutinin Gene Sequences.

Virus strain and subtype	Total nucleotides	5′UT	"Signal"	HA_1	Connector	HA_2	3′UT
A/WSN/33 (HO)[8]	1775	32	51 (17)	975 (325)	3 (1)	666 (222)	48
A/Japan/305/57 (H2)[2]	1773	43	45 (15)	972 (324)	3 (1)	666 (222)	44
A/Victoria/3/75 (H3)[4]	1768	29	48 (16)	987 (329)	3 (1)	663 (221)	38
A/Memphis/102/72 (H3)[3]	1765	29	48 (16)	984 (328)	3 (1)	663 (221)	38
A/FPV/Rostock/34 (Hav1)[1]	1742	21	54 (18)	957 (319)	15 (5)	663 (221)	32

Number of coded amino acids listed in parentheses next to the nucleotide totals. UT: untranslated.

Plus strand sequence of A/WSN gene 4 5' AGCAAAAGCA GGGGAAAAUA AAAACAACCA AA 32

AUG AAG GCA AAA CUA CUG GUC CUG UUA UAU GCA UUU GUA GCU 74
A/WSN/33/(H0) Met Lys Ala Lys Leu Leu Val Leu Leu Tyr Ala Phe Val Ala -4
A/Japan/305/57(H2) Met Ala Ile Ile Tyr Leu Ile Leu Leu Phe Thr Ala Val Arg

ACA GAU GCA - - - - - - - - - - GAC ACA AUA UGU AUA GGC UAC 104
Thr Asp Ala - - - - - - - - - - ASP Thr Ile Cys Ile Gly Tyr 7
Gly - - - - - - - - - - - - ASP Gln Ile Cys Ile Gly Tyr

CAU GCG AAC AAC UCA ACC GAC ACU GUU GAC ACA AUA UUC GAG AAG AAU GUG GCA GUG ACA 164
His Ala Asn Asn Ser Thr Asp Thr Val Asp Thr Ile Phe Glu Lys Asn Val Ala Val Thr 27
His Ala Asn Asn Ser Thr Glu Lys Val Asp Thr Asn Leu Glu Arg Asn Val Thr Val Thr

CAU UCU GUU AAC CUG CUC GAA GAC AGA CAC AAC GGG AAA CUA UGU AAA UUA AAA GGA AUA 224
His Ser Val Asn Leu Leu Glu Asp Arg His Asn Gly Lys Leu Cys Lys Leu Lys Gly Ile 47
His Ala Lys Asp Ile Leu Glu Lys Thr His Asn Gly Lys Leu Cys Lys Leu Asn Gly Ile

GCC CCA CUA CAA UUG GGG AAA UGU AAC AUC ACC GGA UGG CUC UUG GGA AAU CCA GAA UGC 284
Ala Pro Leu Gln Leu Gly Lys Cys Asn Ile Thr Gly Trp Leu Leu Gly Asn Pro Glu Cys 67
Pro Pro Leu Glu Leu Gly Asp Cys Ser Ile Ala Gly Trp Leu Leu Gly Asn Pro Glu Cys

GAC UCA CUG CUU CCA GCG AGA UCA UGG UCC UAC AUU GUA GAA ACA CCA AAC UCU GAG AAU 344
Asp Ser Leu Leu Pro Ala Arg Ser Trp Ser Tyr Ile Val Glu Thr Pro Asn Ser Glu Asn 87
Asp Arg Leu Leu Ser Val Pro Glu Trp Ser Tyr Ile Met Glu Lys Glu Asn Pro Arg Asp

GGA GCA UGU UAU CCA GGA GAU UUC AUC GAC UAU GAG GAA CUG AGG GAG CAA UUG AGC UCA 404
Gly Ala Cys Tyr Pro Gly Asp Phe Ile Asp Tyr Glu Glu Leu Arg Glu Gln Leu Ser Ser 107
Gly Leu Cys Tyr Pro Gly Ser Phe Asn Asp Tyr Glu Glu Leu Lys His Leu Leu Ser Ser

GUA UCA UCA UUA GAA AGA UUC GAA AUA UUU CCC AAG GAA AGU UCA UGG CCC AAC CAC ACA 464
Val Ser Ser Leu Glu Arg Phe Glu Ile Phe Pro Lys Glu Ser Ser Trp Pro Asn His Thr 127
Val Lys His Phe Glu Lys Val Lys Ile Leu Pro Lys Asp Arg - Trp Thr Gln His Thr

UUC AAC GGA GUA ACA GUA UCA UGC UCC CAU AGG GGA AAA AGC AGU UUU UAC AGA AAU UUG 524
Phe Asn Gly Val Thr Val Ser Cys Ser His Arg Gly Lys Ser Ser Phe Tyr Arg Asn Leu 147
Thr Thr Gly Gly Ser Arg Ala Cys Ala Val Ser Gly Asn Pro Ser Phe Phe Arg Asn Met

CUA UGG CUG ACG AAG AAG GGG GAU - - UCA UAC CCA AAG CUG ACC AAU UCC UAU GUG 578
Leu Trp Leu Thr Lys Lys Gly Asp - - Ser Tyr Pro Lys Leu Thr Asn Ser Tyr Val 165
Val Trp Leu Thr Lys Glu Gly Ser - - Asp Tyr Pro Val Ala Lys Gly Ser Tyr Asn

AAC AAU AAA GGG AAA GAA GUC CUU GUA CUA UGG GGU GUU CAU CAC CCG UCC AGC AGU GAU 638
Asn Asn Lys Gly Lys Glu Val Leu Val Leu Trp Gly Val His His Pro Ser Ser Ser Asp 185
Asn Thr Ser Gly Glu Gln Met Leu Ile Ile Trp Gly Val His His Pro Ile Asp Glu Thr

GAG CAA CAG AGU CUC UAU AGU AAU GGA AAU GCU UAU GUC UCU GUA GCG UCU UCA AAU UAU 698
Glu Gln Gln Ser Leu Tyr Ser Asn Gly Asn Ala Tyr Val Ser Val Ala Ser Ser Asn Tyr 205
Glu Gln Arg Thr Leu Tyr Gln Asn Val Gly Thr Tyr Val Ser Val Gly Thr Ser Thr Leu

AAC AGG AGA UUC ACC CCG GAA AUA GCU GCA AGG CCC AAA GUA AAA GAU CAA CAU GGG AGG 758
Asn Arg Arg Phe Thr Pro Glu Ile Ala Ala Arg Pro Lys Val Lys Asp Gln His Gly Arg 225
Asn Lys Arg Ser Thr Pro Glu Ile Ala Thr Arg Pro Lys Val Asn Gly Gln Gly Gly Arg

AUG AAC UAU UAC UGG ACC UUG CUA GAA CCC GGA GAC ACA AUA AUA UUU GAG GCA ACU GGU 818
Met Asn Tyr Tyr Trp Thr Leu Leu Glu Pro Gly Asp Thr Ile Ile Phe Glu Ala Thr Gly 245
Met Glu Phe Ser Trp Thr Leu Leu Asp Met Trp Asp Thr Ile Asn Phe Glu Ser Thr Gly

AAU CUA AUA GCA CCA UGG UAU GCU UUC GCA CUG AGU AGA GGG UUU GAG UCC GGC AUC AUC 878
Asn Leu Ile Ala Pro Trp Tyr Ala Phe Ala Leu Ser Arg Gly Phe Glu Ser Gly Ile Ile 265
Asn Leu Ile Ala Pro Glu Tyr Gly Phe Lys Ile Ser Lys Arg Gly Ser Ser Gly Ile Met

ACC UCA AAC GCG UCA AUG CAU GAG UGU AAC ACG AAG UGU CAA ACA CCC CAG GGA UCU AUA 938
Thr Ser Asn Ala Ser Met His Glu Cys Asn Thr Lys Cys Gln Thr Pro Gln Gly Ser Ile 285
Lys Thr Glu Gly Thr Leu Glu Asn Cys Glu Thr Lys Cys Gln Thr Pro Leu Gly Ala Ile

```
AAC AGC AAU CUC CCU UUC CAG AAU AUA CAC CCA GUC ACA AUA GGA GAG UGC CCA AAA UAU   998
Asn Ser Asn Leu Pro Phe Gln Asn Ile His Pro Val Thr Ile Gly Glu Cys Pro Lys Tyr   305
Asn Thr Thr Leu Pro Phe His Asn Val His Pro Leu Thr Ile Gly Glu Cys Pro Lys Tyr

GUC AGG AGU ACC AAA UUG AGG AUG GUU ACA GGA CUA AGA AAC AUC CCA UCC AUU CAA UAC  1058
Val Arg Ser Thr Lys Leu Arg Met Val Thr Gly Leu Arg Asn Ile Pro Ser Ile Gln Tyr   325
Val Lys Ser Glu Lys Leu Val Leu Ala Thr Gly Leu Arg Asn Val Pro Gln Ile Glu Ser

 -   -   -   -   -  AGA GGU CUA UUU GGA GCC AUU GCU GGU UUU AUU GAG GGG GGA UGG ACU  1106
 -   -   -   -   -  Arg GLY Leu Phe Gly Ala Ile Ala Gly Phe Ile Glu Gly Gly Trp Thr   341
 -   -   -   -   -  Arg GLY Leu Phe Gly Ala Ile Ala Gly Phe Ile Glu Gly Gly Trp Gln

GGA AUG AUA GAU GGA UGG UAU GGU UAU CAU CAU CAG AAU GAA CAG GGA UCA GGC UAU GCA  1166
Gly Met Ile Asp Gly Trp Tyr Gly Tyr His His Gln Asn Glu Gln Gly Ser Gly Tyr Ala   361
Gly Met Val Asp Gly Trp Tyr Gly Tyr His His Ser Asn Asp Gln Gly Ser Gly Tyr Ala

GCG GAU CAA AAA AGC ACA CAG AAU GCC AUU AAC GGG AUU ACA AAC AAG GUG AAC UCU GUU  1226
Ala Asp Gln Lys Ser Thr Gln Asn Ala Ile Asn Gly Ile Thr Asn Lys Val Asn Ser Val   381
Ala Asp Lys Glu Ser Thr Gln Lys Ala Phe Asp Gly Ile Thr Asn Lys Val Asn Ser Val

AUC GAG AAA AUG AAC ACU CAA UUC ACA GCU GUG GGU AAA GAA UUC AAC AAC UUA GAA AAA  1286
Ile Glu Lys Met Asn Thr Gln Phe Thr Ala Val Gly Lys Glu Phe Asn Asn Leu Glu Lys   401
Ile Glu Lys Met Asn Thr Gln Phe Glu Ala Val Gly Lys Glu Phe Gly Asn Leu Glu Arg

AGG AUG GAA AAU UUA AAU AAA AAA GUU GAU GAU GGG UUU CUG GAC AUU UGG ACA UAU AAU  1346
Arg Met Glu Asn Leu Asn Lys Lys Val Asp Asp Gly Phe Leu Asp Ile Trp Thr Tyr Asn   421
Arg Leu Glu Asn Leu Asn Lys Arg Met Glu Asp Gly Phe Leu Asp Val Trp Thr Tyr Asn

GCA GAA UUG UUA GUU CUA CUG GAA AAU GAA AGA ACU UUG GAU UUC CAU GAC UUA AAU GUG  1406
Ala Glu Leu Leu Val Leu Leu Glu Asn Glu Arg Thr Leu Asp Phe His Asp Leu Asn Val   441
Ala Glu Leu Leu Val Leu Met Glu Asn Glu Arg Thr Leu Asp Phe His Asp Ser Asn Val

AAG AAU CUG UAC GAG AAA GUA AAA AGC CAA UUA AAG AAU AAU GCC AAA GAA AUC GGA AAU  1466
Lys Asn Leu Tyr Glu Lys Val Lys Ser Gln Leu Lys Asn Asn Ala Lys Glu Ile Gly Asn   461
Lys Asn Leu Tyr Asp Lys Val Arg Met Gln Leu Arg Asp Asn Val Lys Glu Leu Gly Asn

GGG UGU UUU GAG UUC UAC CAC AAG UGU GAC AAU GAA UGC AUG GAA AGU GUA AGA AAU GGG  1526
Gly Cys Phe Glu Phe Tyr His Lys Cys Asp Asn Glu Cys Met Glu Ser Val Arg Asn Gly   481
Gly Cys Phe Glu Phe Tyr His Lys Cys Asp Asp Glu Cys Met Asn Ser Val Lys Asn Gly

ACU UAU GAU UAU CCA AAA UAU UCA GAA GAA UCA AAG UUG AAC AGG GAA AAG AUA GAU GGA  1586
Thr Tyr Asp Tyr Pro Lys Tyr Ser Glu Glu Ser Lys Leu Asn Arg Glu Lys Ile Asp Gly   501
Thr Tyr Asp Tyr Pro Lys Tyr Glu Glu Glu Ser Lys Leu Asn Arg Asn Glu Ile Lys Gly

GUG AAA UUG GAA UCA AUG GGG GUG UAU CAG AUU CUG GCG AUC UAC UCA ACU GUC GCC AGU  1646
Val Lys Leu Glu Ser Met Gly Val Tyr Gln Ile Leu Ala Ile Tyr Ser Thr Val Ala Ser   521
Val Lys Leu Ser Ser Met Gly Val Tyr Gln Ile Leu Ala Ile Tyr Ala Thr Val Ala Gly

UCA CUG GUG CUU UUG GUC UCC CUG GGG GCA AUC AGU UUC UGG AUG UGU UCU  -  AAU GGG  1703
Ser Leu Val Leu Leu Val Ser Leu Gly Ala Ile Ser Phe Trp Met Cys Ser  -  Asn Gly   540
Ser Leu Ser Leu Ser Ile Met Met Ala Gly Ile Ser Phe Trp Met Cys Ser  -  Asn Gly

UCU UUG CAG UGC AGA AUA UGC AUC UGA GAUUACGAUU UCAGAAAUAU AAGGAAAAAC ACCCUUGUUU  1770
Ser Leu Gln Cys Arg Ile Cys Ile STOP                                               548
Ser Leu Gln Cys Arg Ile Cys Ile

CUACU      3'                                                                     1775
```

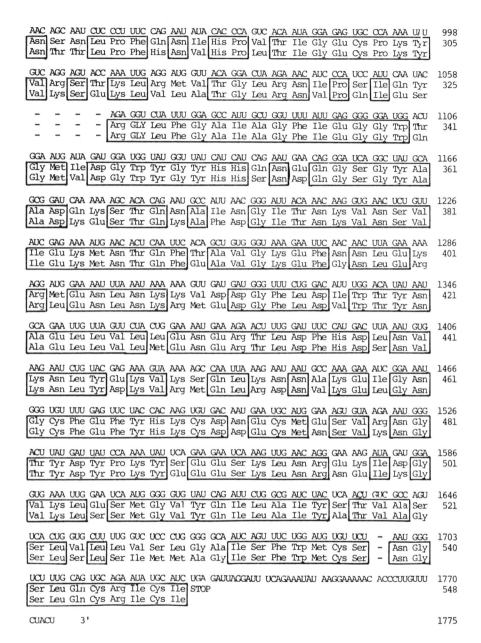

Figure 1. The A/WSN/33 nucleotide sequence of the "+" strand of gene 4 (top line) and amino acid sequences of the hemagglutinins of the H0 subtype (A/WSN/33, second line) and the H2 subtype (A/Japan/305/57,[2] third line). Boxed regions indicate homology between the two subtype sequences. Dashed lines are occasionally inserted for improved alignment with other subtype HAs (A/Memphis/102/72 (H3) [3] and A/Fowl plague/Rostock/34 (Hav 1) [1]). The N-terminal amino acids of the HA_1 and HA_2 chains are capitalized.

Table 2. Intersubtypic Amino Acid Homologies.

	H2(A/Japan/57)	H3(A/Memphis/72)	Hav 1 (A/FPV/Rostock/34)
H0 (A/WSN/33)	57, 81, 67[a]	35, 52, 42	35, 51, 41
Hav 1 (A/FPV/ Rostock/34)	36, 53, 43	37, 66, 49	—
H3 (A/Memphis/72)	39, 49, 42	—	—

[a]The numbers indicate the percent amino acid homology between indicated subtype pairings in the (1) HA_1 chain (pos. 1-325), (2) HA_2 chain (pos. 327-548), and (3) total (pos. 1-548) based on the alignment in reference 7. Homologies were calculated as number of homologous residues/ total number of pairings. Deletions were counted as mismatches unless they coincided. Coincident deletions were not counted.

shown). The amino acid sequence homology, the coding region nucleotide homology, and the similarity in sizes of the HA_1 and HA_2 regions all argue for a close relationship between the H0 and H2 HA molecules.

This evidence strengthens the argument that all HAs are derived from a single primordial ancestral gene, but, more importantly, poses questions of antigenic drift and shift. Two recent studies of drifted H3N2 field strains [5,9] indicate a rate of amino acid change in human populations that is too low to explain the direct "drift" of the H0 sequence to the H2 sequence in the 24 years that separate their isolation. Hybridization data also indicate that only four of the H2N2 virus genes (NP, M, NS, and one P) are almost completely homologous to the H0 or H1 genes.[6] Preliminary sequence data that the H1 subtype HA sequence is quite similar to that of the H0 subtype for at least the first 70 amino acids [10] may indicate that the H1 is not a direct intermediary to the H2. However, the exact relationship between the H1 and H2 subtype HA genes must await the complete sequence analysis of the H1 gene. These considerations argue against a direct "drift" from the H0 gene to the H2 gene in human populations, but in favor of gene reassortment as

Table 3. Instersubtypic Coding Region Nucleotide Homologies.

	H2 (A/Japan/57)	H3 (A/Memphis/72)	Hav 1 (A/FPV/Rostock/34)
H0 (A/WSN/33)	59, 72, 64[a]	45, 56, 49	43, 57, 49
Hav 1 (A/FPV/ Rostock/34)	46, 59, 51	45, 66, 53	—
H3 (A/Memphis/72)	45, 55, 49	—	—

[a]The numbers indicate the coding region percent nucleotide homology between indicated subtype pairings in the (1) HA_1 region (pos. 1-325), (2) HA_2 region (pos. 327-548), and (3) total (pos. 1-548) based on the alignment in reference 7. Homologies were calculated as numbers of homologous nucleotides/total of compared nucleotides. Deletions were counted as mismatches unless they coincided. Coincident deletions were not counted.

the cause in the emergence of the H2 subtype. However, the source of the H2 virus genes which, by hybridization studies, are not homologous to those of the H0 (HA, NA, and two P genes) has not been identified.

Our study indicates the possibility that the H2 is directly related to the H0, separated by 24 years of reassortment event(s) and different selective pressures in non-human hosts, but more probably, in the absence of evidence that would indicate a faster rate of variation in non-human hosts, the H2 and H0 HA genes share a close common ancestor.

ACKNOWLEDGMENTS
This work was supported by a Grant from Genentech, Inc. ALH was supported by USPHS-NRSA 07104.

References

1. Porter, A.G., Barber, C., Carey, N.H., Hallewell, R.A., Threlfall, G., and Emtage, J.S. (1979): *Nature* 282:471–477.
2. Gething, M.J., Bye, J., Skehel, J., and Waterfield, M. (1980): *Nature* 287:301–306.
3. Sleigh, M.J., Both, G.W., Brownlee, G.G., Bender, V.J., and Moss, B.A. (1980): In: *Structure and Variation in Influenza Virus.* Laver, G. and Air, G. (Eds.), Elsevier, Amsterdam, pp. 69–78.
4. Min Jou, W., Verhoeyen, M., Devos, R., Saman, E., Fang, R., Huylebroeck, D., Fiers, W., Threlfall, G., Barber, C., Carey, N., and Emtage, S. (1980): *Cell* 19:683–696.
5. Verhoeyen, M., Fang, R., Min Jou, W., Devos, R., Huylebroeck, D., Saman, E., and Fiers, W. (1980): *Nature* 286:771–775.
6. Scholtissek, C. (1978): *Curr. Top. Micobiol. Immunol.* 80:139–169.
7. Davis, A.R., Hiti, A.L., and Nayak, D.P. (1980): *Gene* 10:205–218.
8. Hiti, A.L., Davis, A.R., and Nayak, D.P. Manuscript submitted.
9. Laver, W.G., Air, G.M., Dopheide, T.A., and Ward, C.W. (1980): *Nature* 283:454–457.
10. Air, G.M. (1980): In: *Structure and Variation in Influenza Virus.* Laver, G. and Air, G. (Eds.), Elsevier, Amsterdam, pp. 135–146.

Copyright 1981 by Elsevier North Holland, Inc.
David H. L. Bishop and Richard W. Compans, eds.
The Replication of Negative Strand Viruses

Sequence Relationships in Influenza Viruses

G.M. Air, J. Blok and R.M. Hall[a]

Genetic variation in influenza viruses is most obviously manifest as antigenic variation, and until recently the only detailed assays of genetic differences and similarities were serological ones. The advances in nucleic acid sequence technologies and their assistance in protein sequence analysis means that it is now possible to study in detail the sequence diversity in influenza A virus genes.

Antigenic variation of the influenza A virus surface proteins, hemagglutinin and neuraminidase, is of two forms. Mutation in the genes coding for these antigens and selection for mutations which enable the proteins to escape (at least partially) antibody neutralization results in antigenic drift, as has been demonstrated in the Hong Kong subtype of viruses by both protein [1] and nucleic acid [2,3] sequence analysis. Antigenic shift is the term applied to sudden and dramatic changes in circulating viruses. The "new" virus does not arise from the "old" but can be the result of recombination between human and animal or bird strains [4] or is a reappearance after prolonged absence in humans of a particular human virus type.[5]

A new classification of influenza A viruses recently announced by W.H.O.[6] recognizes 12 subtypes of hemagglutinin (HA) and 9 subtypes of neuraminidase (NA), based on the absence of any common antigenic determinants detectable in double-diffusion tests. Many of these subtypes have been isolated from human and non-human hosts, and human epidemic strains can be readily isolated from birds or animals.

[a] Australian National University, Canberra.

```
H1  (H0,H1)              Met Lys Ala Asn Leu Leu Val Leu Leu Cys Ala Leu Ala Ala Ala Asp Ala↑Asp Thr Ile Cys Ile Gly Tyr His Ala Asn Asn Ser Thr Asp 31
H2  (H2)                         Met Ala Ile Ile Tyr Leu Ile Leu Leu Phe Thr Ala Val Arg Gly↑Asp Gln Ile Cys Ile Gly Tyr His Ala Asn Asn Ser Thr Glu 29
H5  (Hav5)                   Met Glu Arg Val Val Leu Leu Leu Ala Met Ile Ser Leu Val Lys Ser↑Asp Gln Ile Cys Ile Gly Tyr His Ala Asn Asn Ser Thr Glu 30
H11 (Hav3)                   Met Lys Lys Val Leu Leu Phe Ala Ala Ile Ile Ile Cys Ile Arg Ala↑Asp Glu Ile Cys Ile Gly Tyr Leu Ser Asn Asn Ser Thr Glu 30
H6  (Hav6)                   Met Ile Ala Ile Ile Val Val Ala Ile Leu Ala Thr Ala Gly Arg Ser↑Asp Lys Ile Cys Ile Gly Tyr His Ala Asn Asn Ser Thr Thr 30
H8  (Hav8)               Met Glu Lys Phe Ile Ala Ile Ala Met Leu Leu Ala Ser Thr Asn Ala Tyr↑Asp Arg Ile Cys Ile Gly Tyr Gln Ser Asn Asn Ser Thr Asp 31
H9  (Hav9)           Met Glu Thr Lys Ala Ile Ile Ala Ala Leu Leu Met Val Thr Ala Ala Asn Ala↑Asp Lys Ile Cys Ile Gly Tyr Gln Ser Thr Asn Ser Thr Glu 32
H12 (Hav10)              Met Glu Lys Phe Ile Ile Leu Ser Thr Val Leu Ala Ala Ser Phe Ala Tyr↑Asp Lys Ile Cys Ile Gly Tyr Gln Thr Asn Asn Ser Thr Glu 31
H7  (Hav1)           Met Asn Thr Gln Ile Leu Val Phe Ile Ala Cys Val Leu Ile Lys Ala Lys Gly↑Asp Lys Ile Cys Leu Gly His His Ala Val Ala Asn Gly Thr 32

H4 (Hav4)    Met Leu Ser Ile Thr Ile Leu Phe Leu Leu Ile Ala Glu Gly Ser Ser Gln Asn Tyr Thr Gly Asn Pro Val Ile Cys Leu Gly His His Ala Val Ser Asn Gly Thr 36
H3 (H3) Met Lys Thr Ile Ile Ala Leu Ser His Ile Phe Cys Leu Val Leu Gly↑Gln Tyr Leu Pro Gly Asn Asp Asn Ser Thr Ala Thr Leu Cys Leu Gly His His Ala Val Pro Asn Gly Thr 40
```

```
H1   Thr Val Asp Thr Val Leu Glu Lys Asn Val Thr Val Thr His Ser Val Asn Leu Leu Glu Asp Ser His Asn Gly Lys Leu Cys Arg Leu Lys Gly Ile Ala Pro Leu Gln Leu Gly Lys 71
H2   Lys Val Asp Thr Ile Leu Glu Arg Asn Val Thr Val Thr His Ala Lys Asp Ile Leu Glu Lys Thr His Asn Gly Lys Leu Cys Lys Leu Asn Gly Ile Pro Pro Leu Glu Leu Gly Asp 69
H5   Gln Val Asp Thr Ile Met Glu Lys Asn Val Thr Val Thr His Ala Gln Asp Ile Leu Glu Lys Thr His Asn Gly Lys Leu Cys Ser Leu Asn Gly Val Lys Pro Leu Ile Leu Arg Asp 70
H11  Lys Val Asp Thr Ile Ile Glu Ser Asn Val Thr Val Thr Ser Ser Val Glu Leu Val Glu Asn Glu Tyr Thr Gly Ser Phe Cys Ser Ile Asp Gly Lys Ala Pro Ile Ser Leu Gly Asp 70
H6   Gln Ile Asp Thr Ile Leu Glu Lys Asn Val Thr Val Thr His Ser Val Glu Leu Leu Glu Asn Gln Lys Glu Glu Arg Phe Cys Lys Ile Leu Lys Lys Ala Pro Leu Asp Leu Lys Gly 70
H8   Thr Val Asn Thr Leu Thr Glu Gln Asn Val Pro Val Thr Gln Thr Met Glu Leu Val Glu Thr Glu Lys His Pro Ala His Cys Asn Thr Asp Leu Gly Ala Pro Leu Glu Leu Arg Asp 71
H9   Thr Val Asp Thr Leu Thr Glu Ser Asn Val Pro Val Thr His Thr Lys Glu Leu Leu His Thr Glu His Asn Gly Met Leu Cys Ala Thr Asp Leu Gly His Pro Leu Val Leu Asp Thr 72
H12  Thr Val Asn Thr Leu Ser Glu Leu Asn Val Pro Val Thr Gln Val Glu Glu Leu Val His Gly Gly Ile Asp Pro Ile Leu Cys Gly Thr Glu Leu Gly Ser Pro Leu Val Leu Asp Asp 71
H7   Lys Val Asn Thr Leu Thr Glu Arg Gly Ile Glu Val Val Asn Ala Thr Glu Thr Val Glu Thr Ala Asn Ile Gly Lys Ile Cys Thr Gln Gly Lys Arg --- Pro Thr Asp Leu Gly Gln 71
H10
H4   Met Val Lys Thr Leu Thr Asp Asp Gln Val Glu Val Val Thr Ala Gln Glu Leu Val Glu Ser Gln His Leu Pro Glu Leu Cys Pro Ser Pro Leu Lys --- Leu Val Asp Gly Gln Thr 75
H3   Leu Val Lys Tnr Ile Thr Asn Asp Gln Ile Glu Val Thr Asn Ala Thr Glu Leu Val Gln Ser Ser Ser Thr Gly Lys Ile Cys Asn Asn Pro His Arg ---,Ile Leu Asp Gly Ile Asp 79
```

H1	Cys	Asn	Ile	Ala	Gly	Trp	Leu	Leu	Gly	Asn	Pro	Glu	Cys	Asp	Pro	Leu	Leu	Pro	Val	Arg	Ser	Trp	Ser	Tyr	Ile	Val	Glu	Thr	Pro	Asn	Ser	Glu	Asn								
H2	Cys	Ser	Ile	Ala	Gly	Trp	Leu	Leu	Gly	Asn	Pro	Glu	Cys	Asp	Arg	Leu	Leu	Ser	Val	Pro	Glu	Trp	Ser	Tyr	Ile	Met	Glu	Lys	Glu	Asn	Pro	Arg	Asp	Gly	Leu	Cys	Tyr	Pro	Gly	Ser	
H5	Cys	Ser	Val	Ala	Gly	Trp	Leu	Leu	Gly	Asn	Pro	Met	Cys	Asp	Glu	Phe	Leu	Thr	Val	Pro	Glu	Trp	Ser	Tyr	Ile	Val	Glu	Lys	Asp	Asn	Pro	Ile	Asn	Gly	Leu	Cys	Tyr	Pro	Gly	Ile	
H11	Cys	Ser	Phe	Ala	Gly	Trp	Ile	Leu	Gly	Asn	Pro	Met	Cys	Asp	Asp	Leu	Ile	Gly	Lys	Thr	Ser	Trp	Ser	Tyr	Ile	Val	Glu	Asn	Gln	Ser											
H6	Cys	Thr	Ile	Glu	Gly	Trp	Ile	Leu	Gly	Asn	Pro	Gln	Cys	Asp	Leu	Leu	Leu	Ser	Val	Pro	Glu	Trp	Ser	Tyr	Ile	Val	Glu														
H8	Cys	Lys	Ile	Glu	Ala	Glu	Ile	Tyr	Gly	Asn	Pro	Lys	Cys	Asp	Ile	His	Leu	Lys	Val	Asn	Gly	Trp	Ser	Tyr	Ile	Val	Glu	Arg	Pro												
H9	Cys	Thr	Ile	Glu	Gly	Leu	Ile	Tyr	Gly	Asn	Pro	Ser	Cys	Asp	Ile	Leu	Leu	Gly	Gly	Lys	Glu	Trp	Ser	Tyr	Ile	Val	Glu														
H12	Cys	Ser	Leu	Glu	Gly	Leu	Ile	Leu	Gly	Asn	Pro	Lys	Cys	Asp	Leu	Tyr	Leu	Asn	Gly	Arg	Glu	Trp																			
H7	Cys	Gly	Leu	Leu	Gly	Thr	Leu	Ile	Gly	Pro	Pro	Gln	Cys	Asp	Gln	Phe	Leu	Glu	Phe	---	Glu	Leu	Asp	Leu	Ile	Ile	Glu	Arg	Arg	Glu	Gly	Asn	Asn	---	Ile	Cys	Tyr	Pro	Gly	Lys	Phe
H10																																									
H4	Cys	Asp	Ile	Val	Asn	Gly	Ala	Leu	Gly	Ser	Pro	Ser	Cys	Asp	His	Leu	Asn	Gly	Ala	---	Glu	Trp	Asp	Val	Phe	Ile	Glu														
H3	Cys	Thr	Leu	Ile	Asp	Ala	Leu	Leu	Gly	Asp	Pro	His	Cys	Asp	Gly	Phe	Gln	Asn	Glu	---	Thr	Trp	Asp	Leu	Phe	Val	Glu	Arg	Ser	Lys	Ala	Phe	Ser	---	Asn	Cys	Tyr				

Figure 1. Amino acid sequences predicted from cDNA sequences transcribed from the 3′ end of the hemagglutinin gene (RNA segment 4) of 11 of the 12 known HA subtypes. The subtypes are designated by the new nomenclature [1] with the older form in parentheses. The viruses from which these sequences were derived were: – A/PR/8/34 (H1N1), A/RI/5⁻/57 (H2N2), A/Shearwater/Australia/75 (H5N3), A/duck/Mem/546/76 (H11N9), A/Shearwater/Australia/72 (H6N5), A/turkey/Ont/6118/68 (H8N4), A/turkey/Wis/1/66 (H9N2), A/duck/Alb/60/76 (H2N5), A/turkey/Oregon/71 (H7N3), A/duck/Alb/28/76 (H4N6), A/Mem/1/71 (H3N2).

The sites of cleavage of the signal peptide to generate the N-terminal amino acid of HA1 are shown by arrows; these are known for the H1, H2, H7 (Asp) [18,20] and H3 (Gln) [10] subtypes and inferred by homology for H5, H11, H6, H8, H9 and H12. The amino terminus of HA1 of H4 is not known; the presence of a "glycosylation sequence" (Asn-X-Ser/Thr) [23] preceding the Asp start of the other subtypes suggests that the N-terminus is extended, since the other signal peptides do not have glycosylation sequences.

The sequences have been aligned at cysteine residues, and these and other amino acids conserved through all subtypes are boxed. Potential glycosylation sequences (Asn-X-Ser/Thr) [23] are underlined.

The ease of isolating virus recombinants in the laboratory and of growing all known subtypes in chicken eggs suggests that there are no species barriers to any subtype of HA or NA.

Within a subtype, there is considerable antigenic variation (drift), although the sequence changes are rather few. It is also recognized that variation occurs in genes other than those coding for proteins under antibody pressure.[7]

We have therefore obtained nucleotide sequence data from several genes of many influenza viruses. We have been studying drift in several subtypes of HA and NA (the result of natural selection?) and in the matrix and NS genes (random drift?). Since the complete sequences of HA genes from three subtypes, H7 (Hav 1),[8] H2,[9] and H3[10,11] show rather dramatic differences, we have studied sequence relationships between 11 of the 12 subtypes of HA and 8 of the 9 subtypes of NA.

The sequence data is obtained by primed synthesis of cDNA from the 3′ end of the viral RNA and use of the Sanger dideoxy method.[12] We can obtain up to 350 nucleotides of sequence, which is 20% of the HA gene and 40% of the smallest gene (NS). The data is therefore limited, but as yet we have no reason to believe that the 3′ end region is not representative of the whole. Complete sequence information from several strains is so far available only for the HA, and in that case the relationships deduced from the 3′ 350 nucleotides are the same as from the whole sequences. The viruses used are described in the figure legends. The methods of virus purification, extraction of RNA, and dideoxy sequencing using a synthetic dodecanucleotide primer have been described previously, with preliminary results for the HA gene,[13,14] NA gene,[15] matrix gene,[16] and NS gene.[17]

The Hemagglutinin Gene

The hemagglutinin is known to be synthesized as a single polypeptide which is processed to the mature surface antigen by removal of a "signal" peptide, glycosylation, and cleavage into HA1 and HA2.[18] The cDNA sequences obtained therefore represent part of the 5′ non-coding sequence (the cap and extra nucleotides are derived from cellular mRNAs,[19] the sequence coding for the hydrophobic signal peptide, and the N-terminal 80 or so amino acids of HA1.

Figure 1 shows the amino acid sequences predicted from the HA gene sequences of viruses representing 11 of the 12 subtypes. The conservation of Cys residues noted in H2, H3 and H7[8-10] is extended to all these subtypes. Subtypes H3, H11, H6, H8, H9 and H12 all have Asp residues corresponding to the Asp N-terminus known for HA1 of H1, H2 and H7.[18,20]

As shown in Figure 1, there are very few amino acid residues conserved through all subtypes in this region, and the signal peptides show no relationships at all to each other. However, features can be seen which place some strains closer to each other than to others. Figure 2 shows these relation-

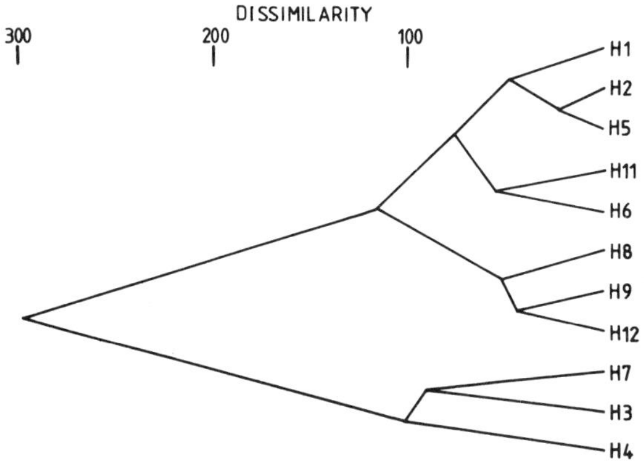

Figure 2. Relationship between 11 of the 12 HA subtypes from direct amino acid sequence comparisons.[21,22]

ships as obtained by direct comparison of amino acid sequences between strains by "diagnol" computer analysis [21] and "multclas" analysis.[22]

Drift in the Hemagglutinin Gene

Figure 3 shows the sequences of some human H1 (H0N1) strains isolated from 1933 to 1957, immediately before the H2 (Asian) epidemic, and two "Hsw" viruses. There are several silent base changes, but few amino acid substitutions in HA1. Although the "swine" strains are closely related to each other and somewhat different to the human HAs, the magnitude of difference is well below that seen between subtypes in Figure 1. Similarly the sequences of PR8/34 and Loyang/57 show so few differences that it is not possible to say if the H1 subtype was drifting towards any other.

A similar analysis of early and later strains of the H2 subtype (RI/5⁻/57, Tok/67, Ned/68, Berk/68) gives similar results. There are a few amino acid changes in the signal peptide, but in the N-terminal region of HA1 the only amino acid changes are Lys 15 → Thr in Ned/68 and Ser 72 → Arg in Tok/67, Ned/68 and Berk/68.

The Neuraminidase Gene

Figures 4 and 5 show the sequence variation in the neuraminidase gene within and between subtypes. In the N1 strains there is 1–20% variation in nucleotide sequences and up to 30% variation in predicted amino acid sequence. The N-terminus of the NA is not known for any strain, nor is it known if there is processing of the polypeptide. However, following the first ATG in the cDNA, there is a region where nucleotide and amino acid sequences are highly conserved; the first 6 predicted amino acids are conserved in all subtypes and the next 6 are conserved in most subtypes. After

					35 ATG Met	AAG Lys	GCA Ala	AAC Asn	CTA Leu	CTG Leu	GTC Val	CTG Leu
PR/8/34 (PR8)	AGCAAAAGCA	GGGGAAAATA	AAAACAACCA	AA	ATG Met	AAG Lys	GCA Ala	AAC Asn	CTA Leu	CTG Leu	GTC Val	CTG Leu
NWS/33 (NWS)								AGA Arg				
Bellamy/42 (Bel)								AGA Arg				
FW/1/50 (FW)				G		AAA		AAA Lys				
Loyang/4/57 (Loy)		T				AAA		AAA Lys			ATC Ile	
Swine/Wis/15/30 (SW)			G G					ATA Ile		TTA		TTG
NJ/11/76 (NJ)			G G					ATA Ile		TTA		TTG

PR8	ACT Thr	GTT Val	GAC Asp	ACA Thr	GTA Val	CTC Leu	GAG Glu	AAG Lys	AAT Asn	GTG Val	ACA Thr	GTG Val	ACA Thr	CAC His	TCT Ser	GTT Val	AAC Asn
NWS					CTA Leu									CAT			
Bel					ATA Ile		GAA									GTC	
FW							GAA		AAC							GTC	
Loy																GTA	
SW									AAC	GTA	ACC						
NJ			GAT			CTA	GAA			GTA		GTA					AAT

PR8	TGT Cys	AAC Asn	ATC Ile	GCC Ala	GGA Gly	TGG Trp	CTC Leu	TTG Leu	GGA Gly	AAC Asn	CCA Pro	GAA Glu	TGC Cys	GAC Asp	CCA Pro	CTG Leu	CTT Leu
NWS															TCA Ser		
Bel			ATT				ATC Ile			AAT				GAA Glu	TCA Ser		
FW	TGC		ATT				ATC Ile	TTA						GAA Glu	TCA Ser		TTT Phe
Loy			ATT				GTC Val	TTA				CAA Gln		GAA Glu	TCA Ser	TTG	CTA
SW		AAT	ATT											GAT	TTG Leu		CTC
NJ			ATT				CTT						TGT	GAA Glu	TTA Leu	CTA	CTC

Figure 3. Drift in hemagglutinin sequences from the H1 (H0, H1 and HSW) subtype. The strains shown are A/PR/8/34, A/NWS/33, A/Bellamy/42, A/FW/1/50, A/Loyang/4/57, A/Swine/Wis/15/30 and A/NJ/11/76. The nucleotide sequence and predicted amino acid sequence are given for cDNA transcribed from the 3′ end of RNA segment 4 of PR8. For the other strains, where there is a nucleotide change the

																						125
TTA	TGT	GCA	CTT	GCA	GCT	GCA	GAT	GCA	GAC	ACA	ATA	TGT	ATA	GGC	TAC	CAT	GCG	AAC	AAT	TCA	ACC	GAC
Leu	Cys	Ala	Leu	Ala	Ala	Ala	Asp	Ala	Asp	Thr	Ile	Cys	Ile	Gly	Tyr	His	Ala	Asn	Asn	Ser	Thr	Asp
									↑													
			ATT																			
			Ile																			
				TCA																		
				Ser																		
		GCG		TCC																		
				Ser																		
			TTT				AAT				CTA			GGT			GCA	AAT				
			Phe				Asn				Leu											
		ACA	TTT		GCC		AAT				CTA			GGT	TAT		GCA	AAT	AAT		ACT	
		Thr	Phe				Asn				Leu											

																						245
CTG	CTC	GAA	GAC	AGC	CAC	AAC	GGA	AAA	CTA	TGT	AGA	TTA	AAA	GGA	ATA	GCC	CCA	CTA	CAA	TTG	GGG	AAA
Leu	Leu	Glu	Asp	Ser	His	Asn	Gly	Lys	Leu	Cys	Arg	Leu	Lys	Gly	Ile	Ala	Pro	Leu	Gln	Leu	Gly	Lys
							GGG				AAA											
											Lys											
							GGG		TTA					GGG								
CTA		GAG		AGT						TGC		CTA										
CTA						AAT	GGG		TTA	TGC												AAC
																						Asn
	CTT											CTA	GGG					TTA		CTG		
													Gly									
CTT	CTT			AGA	CAT		GGG				AAA	CTG	GGG	GGG				TTG	CAC		GGT	
				Arg							Lys		Gly						His			

															344	
CCA	GTG	AGA	TCA	TGG	TCC	TAC	ATT	GTA	GAA	ACT	CCA	AAC	TCT	GAG	AAT	
Pro	Val	Arg	Ser	Trp	Ser	Tyr	Ile	Val	Glu	Thr	Pro	Asn	Ser	Glu	Asn	
	GCG									ACA			│			
	Ala												│			
TCA	GAG					│										
Ser	Glu					│										
TCT	AAG	AAA						GCA		ACA			TCC	│		
Ser	Lys	Lys						Ala						│		
TCC	AAA				TCT	│										
Ser	Lys					│										
ACA		AGC				TAT				ACA	TCG		│			
Thr		Ser									Ser		│			
ACA	GTA	AGC			│											
Thr		Ser			│											

(continued)

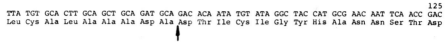

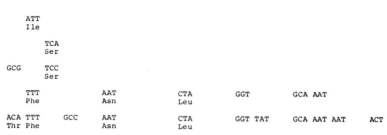

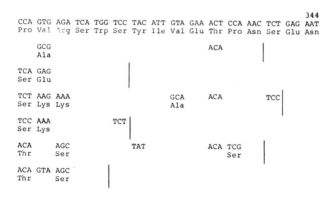

whole codon is printed; if this also involves an amino acid change the new amino acid is printed. Blank areas therefore indicate regions where nucleotide and amino acid sequences are identical. Vertical lines indicate the end of data for the strain, and the arrow shows the start of mature HA1.

												20									
A/NWS/33 (H0N1)	AGCAAAAGCA	G	G	A	G	T	T	T	-	A	A	ATG	AAT	CCA	AAC	CAG	AAA	ATA	ATA	ACC	ATT
												Met	Asn	Pro	Asn	Gln	Lys	Ile	Ile	Thr	Ile
												21									
A/PR/8/34 (H0N1)									A						AAT						
A/Bellamy/42 (H0N1)						A			A						AAT						
A/FW/1/50 (H1N1)									A						AAT						
A/Loyang/4/57 (H1N1)									A		C		AAC		AAT					AAC	
																				Asn	
A/USSR/90/77 (H1N1)				X					A						AAT						
A/Memphis/10/78 (H1N1)									A						AAT						
A/Swine/Wis/15/30 (HswN1) (SHOPE)				X					A					ACA	AAT	CAA					
														Thr							
A/New Jersey/11/76 (HswN1) (X53A)				G					A					ACA	AAT	CAA	AGA				
														Thr			Arg				

NWS	ATT	AGC	CAT	TCA	ATT	CAA	ACT	GGA	AAT	CAA	AAC	CAT	AAT	GGA	ATA	TGC
	Ile	Ser	His	Ser	Ile	Gln	Thr	Gly	Asn	Gln	Asn	His	Asn	Gly	Ile	Cys
PR8									AGT				ACT			
									Ser				Thr			
Bel									AGT				ACT			
									Ser				Thr			
FW	GTA		CAC						AGT				ACT			
	Val								Ser				Thr			
Loy			CAC						AGT				ACT	GGG		
									Ser				Thr			
USSR	GTT		CAC						AGT			CAC	ACA		ACA	
	Val								Ser				Thr		Thr	
Mem 78	GTT		CAC						AGT				ACA			
	Val								Ser				Thr			
SHOPE								AGA	GAT				GCT	GAA	ACA	
								Arg	Asp				Ala	Glu	Thr	
X53A	ATG					CAG			GAA	AAA	AGC		CCT	AAA	GGA	
	Met								Glu	Lys	Ser		Pro	Lys	Gly	

```
                                                                                        118
GGA TCA ATC TGT ATG GTA GTC GGA ATA ATC AGC CTA ATA TTG CAA ATA GGG AAT ATA ATC TCA ATA TGG
Gly Ser Ile Cys Met Val Val Gly Ile Ile Ser Leu Ile Leu Gln Ile Gly Asn Ile Ile Ser Ile Trp

                CTG             CTA ATT
                Leu             Leu

                                    ATT                                 ATT


                    GCA ATC         ATT AGT                             ATT
                    Ala Ile

                    GCA ATC     ACA     AGT         CTG                 ATT
                    Ala Ile     Thr

                    GCA ATC         ATT AGT                             ATT
                    Ala Ile

                    GCA ATC         ATT AGT                             ATT
                    Ala Ile



GGG             CTA ATA GTT         ACT             TTA
                Leu Ile             Thr

GGG ACA         CTA ATA GTT         ATT AGT     TTA     CAG     GGA             TTG TTA
    Thr         Leu Ile                         Leu                             Leu Leu




                            190
AAC CAA AGA ATC ATT ACC CAT AGG   *
Asn Gln Arg Ile Ile Thr His Lys
                                                206
    CAT AAC             TAT AAA AAT AGC ACC TGG GTA   *
    His Asn             Tyr     Asn Ser Thr Trp Val
                                                                                            239
    CAT AGC             TAT AAA                 GTT AAT CAA ACA TAT GTT AAT ATT AGC AAC ACA AAC
    His Ser             Tyr                         Asn Gln Thr Tyr Val Asn Ile Ser Asn Thr Asn

                        TAT GAA     *
                        Tyr Glu

                        TAT   *
                        Tyr

    CAT                 TAT GAA                         TAC         AGT         ACT
    His                 Tyr Glu                                     Ser

                        TAT GAA                 *
                        Tyr Glu


        AGC             TAT GAA AAC AAC ACA     GTG                     AAC         AAT GCT
        Ser             Tyr Glu     Asn                                                 Ala

        AGT GTC         TAT GAA AAC GAC ACA     GTG AAC     ACT *
        Ser Val         Tyr Glu     Asp
```

Figure 4. Drift in neuraminidase gene sequences from the N1 subtype. The strains shown are A/NWS/33, A/PR/8/34, A/Bellamy/42, A/FW/1/50, A/Loyang/4/57, A/USSR/90/77, A/Memphis/10/78, A/Swine/Wis/15/30 and A/NJ/11/76. The nucleotide and predicted amino acid sequence for cDNA transcribed from the 3′ end of segment 6 of NWS is shown and the nucleotide and amino acid changes are indicated as in Figure 3. *denotes end of sequence data for that particular strain.

```
N2
                                                        20
A/NWS/33HA-Tokyo/3/67NA (H0N2)      AGCAAAAGCA GGAGTGAAAATG AAT CCA AAT CAA AAG ATA ATA ACA ATT GGC
                                                        Met Asn Pro Asn Gln Lys Ile Ile Thr Ile Gly
A/Memphis/102/72 (H3N2)                                G

A/Turkey/Wis/1/66 (H9N2)                         X                      CAG
    (IVY)

N3 (Nav2)
                                                        19
A/Turkey/Ore/71 (H7N3)              AGCAAAAGCA  GGXCGAAGATG AAT CCA AAT CAG AAG ATA ATA ACA ATC GGG
    (GREG)                                              Met Asn Pro Asn Gln Lys Ile Ile Thr Ile Gly

A/Dk/Alb/77/77 (H2N3)                                                   ATC         ATT GGT
    (DAWN)
```

```
NWS/TOK                             TTG CAT TTT AAG CAA CAT GAT TGC GAC TCC ACC GCA CAG CTA CCA AGT
                                    Leu His Phe Lys Gln His Asp Cys Asp Ser Thr Ala Gln Leu Pro Ser
MEM 72                                      TTC         TAT GAG                 GGG AAT CAA
                                                        Tyr Glu                 Gly Asn Gln

IVY                                 CTG                 AAT GAA     AAC CCC CTT CGC GAA CAA TCA
                                                        Asn Glu     Asn Pro Leu Arg Glu Gln Ser

GREG                                ATA CAT GAG AAA ATA GGG AAT CAC CAA ACA GTG ATT CAC CCA ACA ATA
                                    Ile His Glu Lys Ile Gly Asn His Gln Thr Val Ile His Pro Thr Ile

DAWN                                    CAC                 GAT             GCG     CGC         GTA
                                                            Asp             Ala     Arg         Val
```

```
                                                                                         121
TCT GTC TCT CTC ACC ATT GCA ACA GTA TGC TTT CTC ATG CAG ATT GCC ATG TTG GTA ACT ACT GTA ACA
Ser Val Ser Leu Thr Ile Ala Thr Val Cys Phe Leu Met Gln Ile Ala Met Leu Val Thr Thr Val Thr

                                ATA     TTC                         CTG
                                Ile

                                        TTC                     ATC CTG GCA ACA     GTG
                                                                Ile     Ala
```

```
                                                                                         120
GTA GTG AAC ACT ACT CTA TCA ACA ATA GCC CTT CTC ATT GGA GAG GGA AAT CTG GTT TTC AAC ACA GTC
Val Val Asn Thr Thr Leu Ser Thr Ile Ala Leu Leu Ile Gly Glu Gly Asn Leu Val Phe Asn Thr Val

CTA     AAT                     AGA     CTA                 GGG     TTG
Leu                             Arg
```

```
                                                                                         238
AAT GCC GTG TTA ACC AAT AAT AAT ACA AAG GCA TAT AAC AGA GAT AGT GCA TTT CAA TTA CAC CAC CAA
Asn Ala Val Leu Thr Asn Asn Asn Thr Lys Ala Tyr Asn Arg Asp Ser Ala Phe Gln Leu His His Gln

AAC GCT                         AGA     GAA CAT                         GAC
                                Arg     Glu His                         Asp

AGT     AGG GGA         CAT     AGA     GAA CAT                 GTA TTC GAA CAA TAT
Ser     Arg Gly         His     Arg     Glu His                 Val     Glu Gln Tyr
```

```
                                                                                             240
ACG ACT CCT GCA GTA CCA AAC TGC AGT GAC ACT ATA ATA ACA TAC AAC AAC ACT GTG ATA AAC AAC ATA ACA
Thr Thr Pro Ala Val Pro Asn Cys Ser Asp Thr Ile Ile Thr Tyr Asn Asn Thr Val Ile Asn Asn Ile Thr

    ACC         ATA                                         AAT AAT                         ACG
                Ile
```

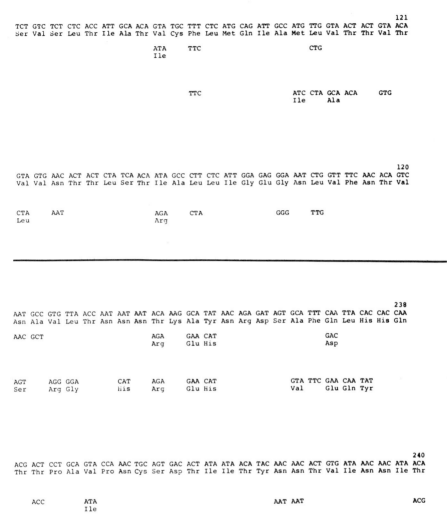

Figure 5. The strains sequenced from two NA subtypes, N2 and N3, are: A laboratory recombinant A/NWS/33_{HA}—Tokyo/67_{NA} (H1N2), A/Memphis/102/72 (H3N2), A/turkey/Wis/1/66 (H9N2), and A/turkey/Ore/71 (H7N3) and A/Duck/Alb/77/77 (H2N3). The nucleotide and predicted amino acid sequences obtained from the 3′ end of segment 6 of NWS/Tok and A/turkey/Ore/71 are shown while the differences in the other N2 and N3 strains are indicated as in Figures 3 and 4.

Table 1.

Nucleotide differences in 238 residues at the 3'-terminus of RNA 7 and amino acid substitutions in the predicted polypeptide.

Nucleotide Position / Strain	31	55	58	68	70	94	100	124	130	136	142	143	147	205	208	214
PR/8/34 H0N1	T	C	A	A	C	G	A	A	G	C	T	G	T	A	A	G
FW/4/50 H1N1	T	C	T	G	C	G	A	T	G	C	T	G	C	A	A	G
Loyang/57 H1N1	T	C	T	G	C	G	A	T	A	C	T	G	C	G	A	A
RI/5⁻/57 H2N2	T	C	T	G	C	G	G	T	G	C	T	G	C	G	A	A
CbrGr/77 H3N2	C	T	T	G	T	A	G	T	G	C	T	C	C	G	A	A
Black Duck/702/78 Hav7 Neq2	T	C	T	G	C	G	G	A	A	T	C	G	C	A	G	G
Amino acid Substitutions	-	-	-	Ile → Val	-	-	-	-	-	-	-	Glu → Gln	Val → Ala	-	-	-

Table 2.

Nucleotide differences in 221 residues at the 3'-terminus of RNA 8 and amino acid substitutions in the predicted polypeptide.

Nucleotide Position / Strain	21	33	35	50	62	68	81	83	88	90	107	110	128	131	134	149
PR/8/34 H0N1	A	C	A	C	T	T	C	C	G	G	A	C	T	G	T	A
RI/5⁻/57 H2N2	G	C	T	C	T	C	C	C	A	G	A	T	T	G	T	G
Cbr.Gr/77 H3N3	G	T	C	T	T	C	A	A	A	A	A	T	T	A	T	G
Black Duck/702/78 Hav7 Neq2	A	T	C	C	C	T	C	C	G	T	G	T	C	G	C	G
Amino acid Substitutions	Asn ↓ Asp	Pro ↓ Ser		-	-	-	-		Arg ↓ Gln	Val ↓ Ile Phe	-	-	-	-	-	-

Nucleotide Position / Strain	155	158	164	170	176	182	183	189	191	192	194	197	200	202	205	206
PR/8/34 H0N1	A	A	G	T	C	G	G	A	G	A	A	C	A	G	C	T
RI/5⁻/57 H2N2	A	G	A	T	C	G	A	G	A	A	A	C	C	G	T	T
Cbr.Gr/77 H3N3	A	G	A	C	C	A	G	G	A	G	A	C	C	A	T	T
Black Duck/702/78 Hav7 Neq2	G	A	A	C	T	G	G	G	A	A	G	T	T	G	C	A
Amino acid Substitutions	-	-	-	-	-	-	Asp ↓ Asn ↓ Asp	Lys ↓ Glu		Thr ↓ Ala	-	-	-	Arg ↓ His	Ala ↓ Val	-

this region both the nucleotide and amino acid sequences show remarkably little similarity among the different NA subtypes. There are no features such as cysteine residues or potential glycosylation sites which are conserved in the first 100 amino acids of the predicted protein sequence.
as cysteine residues or potential glycosylation sites which are conserved in the first 100 amino acids of the predicted protein sequence.

Drift in the Matrix (M) and NS Genes

Table 1 shows the sequence differences found in 230 nucleotides of the M gene from NWS/33 (H1N1) to Canberra Grammar/77 (H3N2) and for comparison, an avian virus matrix gene. A drift is clearly seen in the human viruses with time, while the avian sequence is somewhat different. Table 2 shows a similar analysis and similar results in the NS gene (RNA segment 8).

During the 23 years from NWS/33 to RI/5⁻/57 there are 7/230 nucleotide changes in the M gene and 13/220 in the NS gene, and from RI/5⁻/57 to Cbr/Gr/77 there are 5/230 nucleotide differences in the M gene and 12/220 in the NS gene. These nucleotide differences are of the same order of magnitude as those found during drift in the NA and HA genes (Figures 3 and 4).

Therefore the rate of drift in the influenza genes under antigenic selection (HA and NA) is not significantly greater than in the "non-selected" genes. Although there are varying relationships between the subtypes of HA, these relationships are very distant compared with the variation within a subtype, and neuraminidase subtypes show no clear relationships. From the present data there is no explanation of how the 12 HA and 9 NA subtypes have arisen.

ACKNOWLEDGMENTS
We thank Dr. Adrian Gibbs for computer facilities and analyses, and Anne Mackenzie for expert technical assistance. This work was supported in part by Grant No. AI 15343 for NIAID.

References

1. Laver, W.G., Air, G.M., and Ward, C.W. (1980): *Nature* 283:454–457.
2. Both, G.W., Sleigh, M.J., Bender, V.J., and Moss, B.A. (1980): In: *Structure and Variation in Influenza Virus*. Laver, W.G. and Air, G.M. (Eds.), Elsevier, pp. 81–90.
3. Verhoeyen, M., Fang, R.; Min Jou, W., Devos, R., Huylebroeck, D., Saman, E., and Fiers, W. (1980): *Nature* 286:771–776.
4. Laver, W.G. and Webster, R.G. (1973): *Virology* 51:383–391.
5. Nakajima, D., Desselberger, U., and Palese, P. (1978): *Nature* 334–339.
6. Bull. W.H.O. (1980): 58, in press.
7. Scholtissek, C. and von Hoynigen-Huene, V. (1980): *Virology* 102:13–20.
8. Porter, A.G., Barber, C., Carey, N.H., Hallewell, R.A., Threlfall, G., and Emtage, J.S. (1979): *Nature* 282:471–477.

9. Gething, M.J., Bye, J., Skehel, J., and Waterfield, M. (1980): *Nature* 287:301–306.
10. Ward, C.W. and Dopheide, T.A. (1980): In: *Structure and Variation in Influenza Virus*. Laver, W.G. and Air, G.M. (Eds.), Elsevier, pp. 27–38.
11. Sleigh, M.J., Both, G.W., Brownlee, G.G., Bender, V.J., and Moss, B.A. (1980): In: *Structure and Variation in Influenza Virus*. Laver, W.G. and Air, G.M. (Eds.), Elsevier, pp. 69–80.
12. Sanger, F., Nicklen, S., and Coulson, A.R. (1977): *Proc. Natl. Acad. Sci.* 74:5463–5467.
13. Air, G.M. (1979): *Virology* 97:468–472.
14. Air, G.M. (1980): In: *Structure and Variation in Influenza Virus*. Laver, W.G. and Air, G.M. (Eds.), Elsevier, pp. 135–146.
15. Blok, J. and Air, G.M. (1980): *Virology*, 107:50–60.
16. Both, G.W. and Air, G.M. (1979): *Europ. J. Biochem.* 96:363–372.
17. Air, G.M. and Hackett, J.A. (1980): *Virology* 103:291–298.
18. Waterfield, M.D., Espelie, K., Elder, K., and Skehel, J.J. (1979): *Brit. Med. Bull.* 35:57–64.
19. Plotch, S.J., Bouloy, M., and Krug, R.M. (1979): *Proc. Natl. Acad. Sci* 76:1618–1622.
20. Bucher, D.J., Li, S.S.L., Kehoe, J.M., and Kilbourne, E.D. (1976): *Proc. Natl. Acad. Sci.* 73:238–242.
21. Gibbs, A.J. and McIntyre, G.A. (1970): *Europ. J. Biochem.* 16:1–11.
22. Lance, G.N. and Williams, W.T. (1967): *Aust. Comput. J.* 1:15–20.
23. Neuberger, A., Gottscalk, A., Marshall, R.O., and Spiro, R.G. (1979): in: *The Glycoproteins: Their Composition, Structure and Function*. Gottschalk, A. (Ed.), Elsevier, pp. 450–490.

Published 1981 by Elsevier North Holland, Inc.
David H. L. Bishop and Richard W. Compans, eds.
The Replication of Negative Strand Viruses

Complete Nucleotide Sequences of Cloned Copies of the RNA Genes Coding for the Hemagglutinin and Matrix Proteins of a Human Influenza Virus

Mary-Jane Gething and Hamish Allen[a]

Introduction

The genome of influenza type A viruses consists of eight separate and unique RNA gene segments of negative polarity varying in length from 890 to 2500 nucleotides.[1] We have previously reported [2,3] the molecular cloning of the genes of a high yielding recombinant virus [4] which contains genome segments from strains A/Japan/305/57 (segments 1, 2, 4 and 5), A/Bel/42 (segments 6 and 8) and A/PR/8/34 (segments 3 and 7). This paper describes the complete nucleotide sequences of segments 4 and 7 which code for the hemagglutinin (HA) and matrix proteins of the virus.

Results and Discussion

A summary of the procedures used to clone influenza virus nucleotide sequences is presented in Figure 1. A complete description of this work and of the methods used has been published previously.[2,5] The cloning of the HA gene involved preparation of double-stranded cDNA copies of the gene before insertion at the Pst 1 site of plasmid pAT153 and transfection of *E. coli* χ1776. An alternative procedure in which vRNA/cDNA hybrids were inserted at the Pst 1 site of plasmid pBR322 was employed to clone the matrix gene. In both cases almost full length copies (90–95%) of the genes were obtained and these were mapped with restriction endonucleases and the DNA sequences determined. The missing sequences corresponded to the 5′ ends of the vRNA segments and these were obtained using restriction frag-

[a] Imperial Cancer Research Fund, Lincoln's Inn Fields, London, WC2A 3PX, United Kingdom.

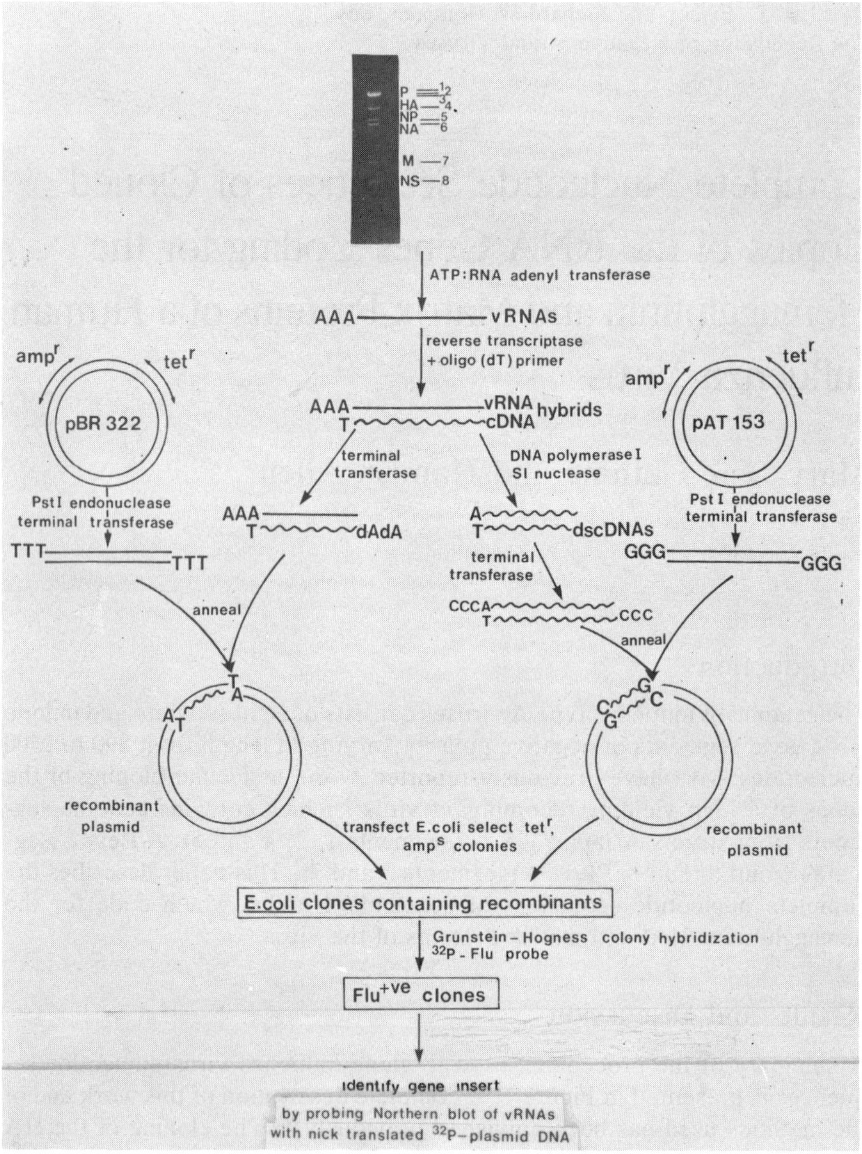

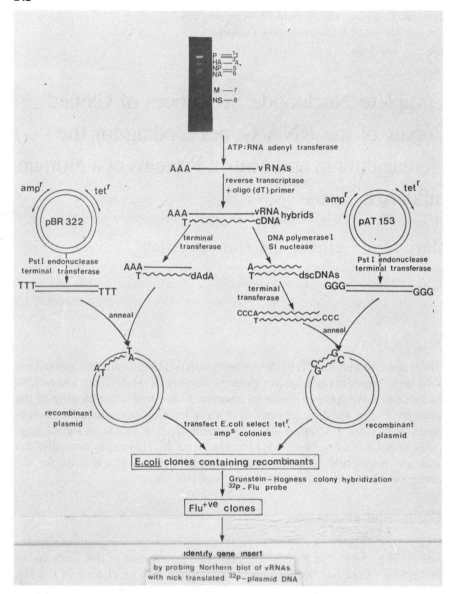

Figure 1. Strategy for cloning influenza nucleotide sequences.

ments from the distal ends of the cloned inserts to prime synthesis of cDNA copies from vRNA template. In the case of the HA gene, a clone containing the missing sequence was prepared and used together with the original clone to construct a plasmid containing the entire coding sequence of the HA protein.[2] This clone is being used for experiments on the expression of HA in prokaryotic and eukaryotic cells.

The Hemagglutinin

The HA glycoprotein is the major surface antigen of the influenza virus. It is responsible for binding the virus particle to the cell during infection [6] and possibly mediates virus-cell membrane fusion.[7-9] Antibodies which react with the HA neutralize virus infectivity [10] and, as a consequence, viruses which have the potential to cause new epidemics or pandemics of disease in an immune population have antigenically novel HAs.

The correspondence between the cloned HA gene insert, the cRNA (message strand) and the protein chain is shown schematically in Figure 2 and the complete nucleotide sequence and derived amino acid sequence of the HA is presented in Figure 3. The gene is 1773 nucleotides long and contains an uninterrupted coding sequence of 1686 nucleotides with short 5′ and 3′ non-translated sequences.

In infected cells, the HA glycoprotein is synthesized as a single polypeptide precursor which is proteolytically cleaved into 2 disulphide bonded subunits HA1 and HA2.[7,8] This cleavage is required for the formation of infectious virus. The mature Japan HA consists of 547 amino acids and a single Arg residue is removed during formation of HA1 (324 amino acids) and HA2 (222 amino acids). The Japan HA is a member of the H2 pandemic subtype. The sequences of HAs from a number of strains of the H3 subtype[2,11-15] and of an avian strain Hav 1 [16] have also been reported. Comparison of these sequences [2] reveals considerable homology between all 3 types (30%) and there is a striking conservation of the location of the cys-

Figure 2. Schematic correlation of the cloned Japan HA gene insert with the cRNA (message strand) and the protein chain.

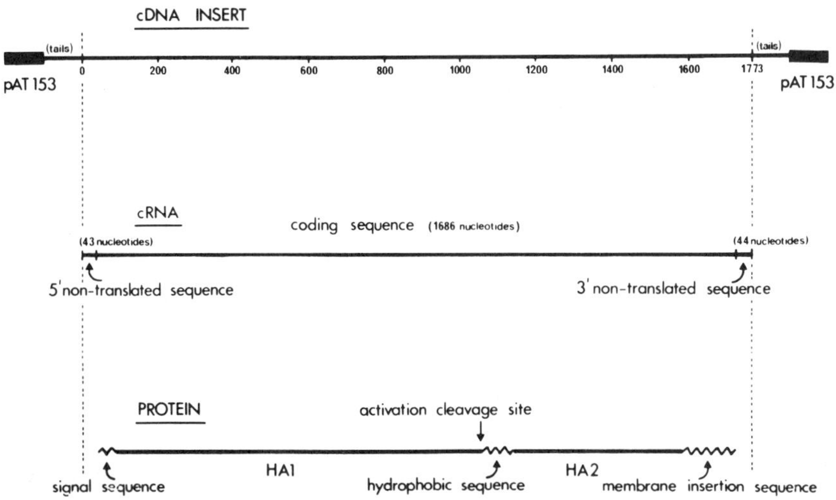

```
                                                                     Signal sequence
 5'-nontranslated sequence                                                                    61
                                                                 MET ALA ILE ILE TYR LEU
AGCA AAA GCA GGG GTT ATA CCA TAG ACA ACC AAA AGC AAA ACA ATG GCC ATC ATT TAT CUC

                                     HA1
                                                                                        121
ILE LEU LEU PHE THR ALA VAL ARG GLY ASP GLN ILE CYS ILE GLY TYR HIS ALA ASN ASN
ATT CTC CTG TTC ACA GCA GTG AGA GGG GAC CAG ATA TGC ATT GGA UAC CAT GCC AAT AAT

                                    151                                                 181
SER THR GLU LYS VAL ASP THR ASN LEU GLU ARG ASN VAL THR VAL THR HIS ALA LYS ASP
TCC ACA GAG AAG GTC GAC ACA AAT CTA GAG CGG AAC GTC ACT GTG ACT CAT GCC AAG GAC

                                    211                                                 241
ILE LEU GLU LYS THR HIS ASN GLY LYS LEU CYS LYS LEU ASN GLY ILE PRO PRO LEU GLU
ATT CTT GAG AAG ACC CAT AAC GGA AAG TTA TGC AAA CTA AAC GGA ATC CCT CCA CTT GAA

                                    271                                                 301
LEU GLY ASP CYS SER ILE ALA GLY TRP LEU LEU GLY ASN PRO GLU CYS ASP ARG LEU LEU
CTA GGG GAC TGT AGC ATT GCC GGA TGG CTC CTT GGA AAT CCA GAA TGT GAT AGG CTT CTA

                                    331                                                 361
SER VAL PRO GLU TRP SER TYR ILE MET GLU LYS GLU ASN PRO ARG ASP GLY LEU CYS TYR
AGT GTG CCA GAA TGG TCC TAT ATA ATG GAG AAA GAA AAC CCG AGA GAC GGT TTG TGT TAT

                                    391                                                 421
PRO GLY SER PHE ASN ASP TYR GLU GLU LEU LYS HIS LEU LEU SER SER VAL LYS HIS PHE
CCA GGC AGC TTC AAT GAT TAT GAA GAA TTG AAA CAT CTC CTC AGC AGC GTG AAA CAT TTC

                                    451                                                 481
GLU LYS VAL LYS ILE LEU PRO LYS ASP ARG TRP THR GLN HIS THR THR THR GLY GLY SER
GAG AAA GTA AAG ATT CTG CCC AAA GAT AGA TGG ACA CAG CAT ACA ACA ACT GGA GGT TCA

                                    511                                                 541
ARG ALA CYS ALA VAL SER GLY ASN PRO SER PHE PHE ARG ASN MET VAL TRP LEU THR LYS
CGG GCC TGC GCG GTG TCT GGT AAT CCA TCA TTT TTC AGG AAC ATG GTC TGG CTG ACA AAG

                                    571                                                 601
GLU GLY SER ASP TYR PRO VAL ALA LYS GLY SER TYR ASN ASN THR SER GLY GLU GLN MET
GAA GGA TCA GAT TAT CCG GTT GCC AAA GGA TCG TAC AAC AAT ACA AGC GGA GAA CAA ATG

                                    631                                                 661
LEU ILE ILE TRP GLY VAL HIS HIS PRO ILE ASP GLU THR GLU GLN ARG THR LEU TYR GLN
CTA ATA ATT TGG GGG GTG CAC CAT CCC ATT GAT GAG ACA GAA CAA AGA ACA TTG TAC CAG

                                    691                                                 721
ASN VAL GLY THR TYR VAL SER VAL GLY THR SER THR LEU ASN LYS ARG SER THR PRO GLU
AAT GTG GGA ACC TAT GTT TCC GTA GGC ACA TCA ACA TTG AAC AAA AGG TCA ACC CCA GAA

                                    751                                                 781
ILE ALA THR ARG PRO LYS VAL ASN GLY GLN GLY GLY ARG MET GLU PHE SER TRP THR LEU
ATA GCA ACA AGG CCT AAA GTG AAT GGA CAA GGA GGT AGA ATG GAA TTC TCT TGG ACC CTC

                                    811                                                 841
LEU ASP MET TRP ASP THR ILE ASN PHE GLU SER THR GLY ASN LEU ILE ALA PRO GLU TYR
TTG GAT ATG TGG GAC ACC ATA AAT TTT GAG AGT ACT GGT AAT CTA ATT GCA CCA GAG TAT

                                    871                                                 901
GLY PHE LYS ILE SER LYS ARG GLY SER SER GLY ILE MET LYS THR GLU GLY THR LEU GLU
GGA TTC AAA ATA TCG AAA AGA GGT AGT TCA GGG ATC ATG AAA ACA GAA GGA ACA CTT GAG

                                    931                                                 961
ASN CYS GLU THR LYS CYS GLN THR PRO LEU GLY ALA ILE ASN THR THR LEU PRO PHE HIS
AAC TGT GAG ACC AAA TGC CAA ACT CCT TTG GGA GCA ATA AAT ACA ACA TTG CCT TTT CAC

                                    991                                                1021
ASN VAL HIS PRO LEU THR ILE GLY GLU CYS PRO LYS TYR VAL LYS SER GLU LYS LEU VAL
AAT GTC CAC CCA CTG ACA ATA GGT GAG TGC CCC AAA TAT GTA AAA TCG GAG AAG TTG GTC

                                                      HA2
                                   1051                                                1081
LEU ALA THR GLY LEU ARG ASN VAL PRO GLN LEU GLU SER ARG GLY LEU PHE GLY ALA ILE
TTA GCA ACA GGA CTA AGG AAT GTT CCC CAG CTT GAA TCA AGA GGA TTG TTT GGG GCA ATA
```

```
                                            1111                                                      1141
ALA GLY PHE ILE GLU GLY GLY TRP GLN GLY MET VAL ASP GLY TRP TYR GLY TYR HIS HIS
GCT GGT TTT ATA GAA GGA GGA TGG CAA GGA ATG GTT GAT GGT TGG TAT GGA TAC CAT CAC

                                            1171                                                      1201
SER ASN ASP GLN GLY SER GLY TYR ALA ALA ASP LYS GLU SER THR GLN LYS ALA PHE ASP
AGC AAT GAC CAA GGA TCA GGG TAT GCA GCA GAC AAA GAA TCC ACT CAA AAG GCA TTT GAT

                                            1231                                                      1261
GLY ILE THR ASN LYS VAL ASN SER VAL ILE GLU LYS MET ASN THR GLN PHE GLU ALA VAL
GGA ATC ACC AAC AAG GTA AAT TCT GTG ATT GAA AAG ATG AAC ACC CAA TTT GAA GCT GTT

                                            1291                                                      1321
GLY LYS GLU PHE GLY ASN LEU GLU ARG ARG LEU GLU ASN LEU ASN LYS ARG MET GLU ASP
GGG AAG GAA TTC GGT AAC TTA GAG AGA AGA CTG GAG AAC TTG AAC AAA AGG ATG GAA GAC

                                            1351                                                      1381
GLY PHE LEU ASP VAL TRP THR TYR ASN ALA GLU LEU LEU VAL LEU MET GLU ASN GLU ARG
GGG TTT CTA GAT GTG TGG ACA TAC AAT GCT GAG CTT CTA GTT CTG ATG GAA AAT GAG AGG

                                            1411                                                      1441
THR LEU ASP PHE HIS ASP SER ASN VAL LYS ASN LEU TYR ASP LYS VAL ARG MET GLN LEU
ACA CTT GAC TTT CAT GAT TCT AAT GTC AAG AAT CTG TAT GAT AAA GTC AGA ATG CAG CTG

                                            1471                                                      1501
ARG ASP ASN VAL LYS GLU LEU GLY ASN GLY CYS PHE GLU PHE TYR HIS LYS CYS ASP ASP
AGA GAC AAC GTC AAA GAA CTA GGA AAT GGA TGT TTT GAA TTT TAT CAC AAA TGT GAT GAT

                                            1531                                                      1561
GLU CYS MET ASN SER VAL LYS ASN GLY THR TYR ASP TYR PRO LYS TYR GLU GLU GLU SER
GAA TGC ATG AAT AGT GTG AAA AAC GGG ACG TAT GAT TAT CCC AAG TAT GAA GAA GAG TCT

                                            1591                                                      1621
LYS LEU ASN ARG ASN GLU ILE LYS GLY VAL LYS LEU SER SER MET GLY VAL TYR GLN ILE
AAA CTA AAT AGA AAT GAA ATC AAA GGG GTA AAA TTG AGC AGC ATG GGG GTT TAT CAA ATC

                                            1651                                                      1681
LEU ALA ILE TYR ALA THR VAL ALA GLY SER LEU SER LEU ALA ILE MET MET ALA GLY ILE
CTT GCC ATT TAT GCT ACA GTA GCA GGT TCT CTG TCA CTG GCA ATC ATG ATG GCT GGG ATC

                                            1711                                                      1741
SER PHE TRP MET CYS SER ASN GLY SER LEU GLN CYS ARG ILE CYS ILE
TCT TTC TGG ATG TGC TCC AAC GGG TCT CTG CAG TGC AGG ATC TGC ATA TGA TTA TAA GTC

                                            1771
    3'-nontranslated sequence
ATT TTA TAA TTA AAA ACA CCC TTG TTT CTA CT
```

Figure 3. Nucleotide and amino acid sequence of the hemagglutinin from influenza virus strain A/Japan/305/57.

teine residues in the primary sequences, which is probably due to constraints on the folding of the molecule and the formation of disulphide bridges.[17] There is also conservation of the number and position of carbohydrate attachment sites,[2] although there is some latitude for variation in their location.

The HA molecules contain 3 hydrophobic regions which are conserved in all the virus strains.[2] These are: the signal sequence at the NH_2-terminus of the precursor[18,19] which is probably responsible for transport of the nascent polypeptide through the membrane of the endoplasmic reticulum; the NH_2-terminus of HA2 which may have a functional role in interaction with and penetration of the cell membrane during infection;[9,20] and a sequence near the COOH terminus of HA2 which is probably involved in anchoring the HA in the lipid envelope. Although hydrophobicity is conserved in these 3 regions in all the HAs, specific amino acid sequence conservation is only seen at the NH_2-terminus of HA2.

In general, there is much greater degree of amino acid conservation in HA2 (42%) than in HA1 (22%) which is consistent with the observation [21,22] that the major antigenic sites are located in HA1 and that it is variation in these sites which allows the virus to escape from neutralizing antibody. The results of the sequence comparisons suggest that antigenic drift occurs by a simple stepwise mutation of the HA nucleotide sequence. However such stepwise changes cannot explain the more dramatic differences between the HAs of the H2 and H3 pandemic subtypes. Recombination or reassortment of HA genes between influenza viruses of humans and those of domestic or wild animals has been suggested as the mechanism of antigenic shift [23] and the data from the comparison of shift strains are not inconsistent with such a theory.

The Matrix

The complete sequence of the matrix gene (segment 7) is shown in Figure 4. The gene is 1027 nucleotides long. The sequence reveals that there are two open reading frames which overlap by 68 bases and have a coding capacity for polypeptides of 27,000 and 11,000 daltons. The amino acid sequence predicted for the larger protein fits the available data of amino acid composition and limited sequence analysis of the matrix protein.[24] The retention of an extended second open reading frame argues for the existence of a second gene product. There is a precedent for an influenza RNA segment coding for two polypeptides since it has been shown that gene 8 codes for the two non-structural proteins NS1 and NS2.[25-27] NS2 mRNA is a spliced message with the splice junctions conforming to the consensus sequences found at the splicing sites of intervening regions in eukaryotic mRNAs.[28] Although there are two possible splice donor sites (AGGT) following closely after the AUG codon in the 5′ nucleotide sequence of gene 7, there are no consensus sequences in the second open reading frame indicating acceptor sites which would result in a spliced mRNA that could code for the putative M2 protein. An alternative possibility is that a second message for M2 is generated by an internal initiation from the in phase AUG codon at the beginning of the second open frame. However there is as yet no evidence that the coding capacity of gene 7 is used to synthesize a second protein.

The amino acid sequence of the matrix protein and of the putative M2 protein are shown in Figure 4. The matrix protein underlies the lipid membrane of the virus and it is presumed to interact with the lipid membrane, with the carboxyl portions of the viral glycoproteins and perhaps with the nucleocapsid protein. However the natur eof these interactions is unresolved. The primary structure of matrix derived from the nucleotide sequence shows that the protein does not have the typical hydrophobic signal sequence characteristic of transmembrane proteins,[29] which confirms predic-

```
                                     31                                     61
                                 MET SER LEU LEU THR GLU VAL GLU THR TYR VAL LEU
AGCG AAA GCA GGT AGA TAT TGA AAG ATG AGT CTT CTA ACC GAG GTC GAA ACG TAC GTA CTC

                                     91                                    121
SER ILE ILE PRO SER GLY PRO LEU LYS ALA GLU ILE ALA GLN ARG LEU GLU ASP VAL PHE
TCT ATC ATC CCG TCA GGC CCC CTC AAA GCC GAG ATC GCA CAG AGA CTT GAA GAT GTC TTT

                                    151                                    181
ALA GLY LYS ASN THR ASP LEU GLU VAL LEU MET GLU TRP LEU LYS THR ARG PRO ILE LEU
GCA GGG AAG AAC ACC GAT CTT GAG GTT CTC ATG GAA TGG CTA AAG ACA AGA CCA ATC CTG

                                    211                                    241
SER PRO LEU THR LYS GLY ILE LEU GLY PHE VAL PHE THR LEU THR VAL PRO SER GLU ARG
TCA CCT CTG ACT AAG GGG ATT TTA GGA TTT GTG TTC ACG CTC ACC GTG CCC AGT GAG CGA

                                    271                                    301
GLY LEU GLN ARG ARG ARG PHE VAL GLN ASN ALA LEU ASN GLY ASN GLY ASP PRO ASN ASN
GGA CTG CAG CGT AGA CGC TTT GTC CAA AAT GCC CTT AAT GGG AAC GGG GAT CCA AAT AAC

                                    331                                    361
MET ASP LYS ALA VAL LYS LEU TYR ARG LYS LEU LYS ARG GLU ILE THR PHE HIS GLY ALA
ATG GAC AAA GCA GTT AAA CTG TAT AGG AAG CTC AAG AGG GAG ATA ACA TTC CAT GGG GCC

                                    391                                    421
LYS GLU ILE SER LEU SER TYR SER ALA GLY ALA LEU ALA SER CYS MET GLY LEU ILE TYR
AAA GAA ATC TCA CTC AGT TAT TCT GCT GGT GCA CTT GCC AGT TGT ATG GGC CTC ATA TAC

                                    451                                    481
ASN ARG MET GLY ALA VAL THR THR GLU VAL ALA PHE GLY LEU VAL CYS ALA THR CYS GLU
AAC AGG ATG GGG GCT GTG ACC ACT GAA GTG GCA TTT GGC CTA GTA TGT GCA ACC TGT GAA

                                    511                                    541
GLN ILE ALA ASP SER GLN HIS ARG SER HIS ARG GLN MET VAL THR THR THR ASN PRO LEU
CAG ATT GCT GAC TCC CAG CAT CGG TCT CAT AGG CAA ATG GTG ACA ACA ACC AAT CCA CTA

                                    571                                    601
ILE ARG HIS GLU ASN ARG MET VAL LEU ALA SER THR THR ALA LYS ALA MET GLU GLN MET
ATC AGA CAT GAG AAC AGA ATG GTT TTA GCC AGC ACT ACA GCT AAG GCT ATG GAG CAA ATG

                                    631                                    661
ALA GLY SER SER GLU GLN ALA ALA GLU ALA MET GLU VAL ALA SER GLN ALA ARG GLN MET
GCT GGA TCG AGT GAG CAA GCA GCA GAG GCC ATG GAG GTT GCT AGT CAG GCT AGA CAA ATG

                                    691                                    721
VAL GLN ALA MET ARG THR ILE GLY THR HIS PRO SER SER SER ALA GLY LEU LYS ASN ASP
GTG CAA GCG ATG AGA ACC ATT GGG ACT CAT CCT AGC TCC AGT GCT GGT CTG AAA AAT GAT
                                                                     LYS MET ILE
                                    751                                    781
LEU LEU GLU ASN LEU GLN ALA TYR GLN LYS ARG MET GLY VAL GLN MET GLN ARG PHE LYS
CTT CTT GAA AAT TTG CAG GCC TAT CAG AAA CGA ATG GGG GTG CAG ATG CAA CGG TTC AAG
 PHE LEU LYS ILE CYS ARG PRO ILE ARG ASN GLU TRP GLY CYS ARG CYS ASN GLY SER SER
                                    811                                    841

TGA TCC TCT CAC TAT TGC CGC AAA TAT CAT TGG GAT CTT GCA CTT GAC ATT GTG GAT TCT
 ASP PRO LEU THR ILE ALA ALA ASN ILE ILE GLY ILE LEU HIS LEU THR LEU TRP ILE LEU
                                    871                                    901

TGA TCG TCT TTT TTT CAA ATG CAT TTA CCG TCG CTT TAA ATA CGG ACT GAA AGG AGG GCC
 ASP ARG LEU PHE PHE LYS CYS ILE TYR ARG ARG PHE LYS TYR GLY LEU LYS GLY GLY PRO
                                    931                                    961

TTC TAC GGA AGG AGT GCC AAA GTC TAT GAG GGA AGA ATA TCG AAA GGA ACA GCA GAG TGC
 SER THR GLU GLY VAL PRO LYS SER MET ARG GLU GLU TYR ARG LYS GLU GLN GLN SER ALA
                                    991                                   1021

TGT GGA TGC TGA CGA TGG TCA TTT TGT CAG CAT AGA GCT GGA GTA AAA AAC TAC CTT GTT
 VAL ASP ALA ASP ASP GLY HIS PHE VAL SER ILE GLU LEU GLU

TCT ACT
```

Figure 4. Nucleotide sequence of the matrix gene from influenza virus strain A/PR/8/34. The derived amino acid sequence for the matrix protein is shown above the nucleotide sequence while that for the putative M2 protein is shown below it.

tions from biosynthetic studies.[30] However the sequence contains several regions of 10 or more non-polar residues including a stretch of 37 amino acids (residues 116–153) of which only 2 (a glutamic acid and an arginine) are charged. These non-polar regions may be involved in hydrophobic interactions that may be protein-lipid or protein-protein and that contribute to the stability of the virus, and indeed to the unusual chloroform-methanol solubility of the protein.

References

1. Palese, P. and Schulman, J.L. (1976): *J. Virol.* 17:301–884.
2. Gething, M.J., Bye, J., Skehel, J.sand Waterfield, M. (1980): *Nature* 287:301–306.
3. Gething, M.J., Bye, J., Skehel, J., and Waterfield, M. (1980): In: *Structure and Variation in Influenza Virus.* Laver, G. and Air, G., (Eds.), Elsevier/North-Holland, pp. 1–10.
4. Hay. A.J., Bellamy, A.R., Abraham, G., Skehel, J.J., Brand, C.M., and Webster, R.G. (1977): *Devel. Biol. (Standard)* 39:15–24.
5. Allen, H., McCauley, J., Waterfield, M., and Gething, M.J. (1980): *Virology,* in press.
6. Hirst, G.K. (1942): *J. Exp. Med.* 75:49–64.
7. Klenk, H.D., Rott, R., Orlich, M., and Blodorn, J. (1975): *Virology* 68:426–439.
8. Lazarowitz, S.G. and Choppin, P.W. (1975): *Virology* 68:440–454.
9. Gething, M.J., White, J.M., and Waterfield, M.D. (1978): *Proc. Natl. Acad. Sci. U.S.A.* 75:2737–2740.
10. Laver, W.G. and Kilbourne, E.D. (1966): *Virology* 30:493–501.
11. Min Jou, W., Verhoeyen, M., Devos, R., Saman, E., Fang, R., Huylebroeck, D., Fiers, W., Threlfall, G., Barber, C., Carey, N., and Emtage, J.S. (1980): *Cell* 19:683–696.
12. Both, G.W. and Sleigh, M.J. (1980): *Nucleic Acids Res.* 8:2561–2575.
13. Verhoeyen, M., Fang, R., Min Jou, W., Devos, R, Huylebroeck, D., Saman, E., and Fiers, W. (1980): *Nature* 286:771–776.
14. Ward, C.W. and Dopheide, T.A. (1979): *Brit. Med. Bull.* 35:51–56.
15. Ward, C.W. and Dopheide, T.A. (1980): *Virology* 103:37–53.
16. Porter, A.G., Barber, C., Carey, N.H., Hallewell, R.A., Threlfall, G., and Emtage, J.S. (1979): *Nature* 282:471–477.
17. Waterfield, M., Gething, M.J., Scrace, G., and Skehel, J.J. (1980): In: *Structure and Variation in Influenza Virus.* Laver, G. and Air, G. (Eds.), Elsevier/North-Holland, pp. 11–20.
18. Elder, K.T., Bye, J.M., Skehel, J.J., Waterfield, M.D., and Smith, A.E. (1979): *Virology* 95:343–350.
19. McCauley, J., Bye, J., Elder, K., Gething, M.J., Skehel, J.J., Smith, A.E., and Waterfield, M.D. (1979): *FEBS Lett.* 108:422–426.
20. Waterfield, M.D., Espelie, K., Elder, K., and Skehel, J.J. (1979): *Brit. Med. Bull.* 35:57–64.
21. Brand, C.M. and Skehel, J.J. (1972): *Nature New Biol.* 238:145–147.
22. Jackson, D.C., Dopheide, T.A., Russell, R.J., White, D.O., and Ward, C.W. (1979): *Virology* 93:458–465.
23. Laver, W.G. and Webster, R.G. (1979): *Brit. Med. Bull.* 35:29–33.
24. Robertson, B.H., Bhown, A.S., Compans, R.W., and Bennett, J.C. (1979): *J. Virol.* 30:759–766.
25. Scholtissek, C. (1978): *Curr. Top. Microbiol. Immunol.* 80:139–169.

26. Porter, A.G., Smith, J.C., and Emtage, J.S. (1980): *Proc. Natl. Acad. Sci. U.S.A.* 77:5074–5078.

27. Lamb, R.A. and Lai, C-J. (1980): *Cell* 21:475–485.

28. Lerner, M.R., Boyle, J.A., Mount, S.M., Wolin, S.L., and Steitz, J.A. (1980): *Nature* 283:220–224.

29. Blobel, G., Walter, P., Chang, C.N., Goldman, B.M., Erikson, A.H., and Lingappa, V.R. (1979): *Symp. Soc. Exp. Biol. (G.B.)* 33:9–36.

30. Hay, A.J. (1974): *Virology* 60:398–418.

David H. L. Bishop and Richard W. Compans, eds.
The Replication of Negative Strand Viruses

Interrupted mRNA(s) and Overlapping Genes in Influenza Virus

Robert A. Lamb[a] and Ching-Juh Lai[b]

We showed previously in influenza virus infected cells that in addition to the eight recognized viral polypeptides which are encoded in the eight virion RNA segments, a ninth virus coded polypeptide of ~ 11,000 mol wt was synthesized.[1,2] This polypeptide designated NS_2 was shown to be unique from the other viral polypeptides by tryptic peptide mapping, and was translated from a separate mRNA.[2] We suggested that one of the eight virion RNA segments contained the genetic information for two polypeptides.[2] Subsequently it was shown by using genetic recombinants of influenza virus and by hybrid-arrested translation that segment 8 of the genome coded for both the NS_1 and NS_2 polypeptides.[3,4] As the size of RNA segment 8 was estimated to be 900 nucleotides[5] and as NS_1 (mol wt ~ 23,000) would need 615 nucleotides and NS_2 (mol wt ~ 11,000) 294 nucleotides to code for these two polypeptides, we suggested that different reading frames were being used to translate NS_1 and NS_2.[3] We mapped the mRNAs for NS_1 and NS_2 on RNA segment 8 using cloned full length DNA (NS DNA).[6] This was done by a modification of the S_1 nuclease technique [7] and showed that the body of the NS_1 mRNA (860 nucleotides) mapped from 0.05–0.95 units, and the body of the NS_2 mRNA (340 nucleotides) mapped from 0.59–0.95 units of the NS DNA (5′ to 3′ in the mRNA orientation) suggesting that the two mRNAs were 3′ coterminal and share the same site of polyadenylation. These data were confirmed in hybrid-arrested translation experiments by using restric-

[a] The Rockefeller University, New York, New York
[b] Laboratory of Infectious Diseases, National Institute of Allergy and Infectious Diseases, Bethesda, Maryland

tion fragments of the NS DNA and hybridizing to the mRNAs so that translation *in vitro* of the NS_1 and NS_2 polypeptides was inhibited. From the size of premature termination products of the NS_1 polypeptide, it could be estimated that the carboxyl terminus of NS_1 was ~ 0.76 map units. Therefore, we predicted that as NS_1 and NS_2 polypeptides did not share common ^{35}S-methionine or ^{3}H-leucine-containing tryptic or chymotryptic peptides, that NS_1 and NS_2 overlap by 50–60 amino acids that are translated from different reading frames.[6]

Although the above experiments mapped the positions of the bodies of the mRNAs for NS_1 and NS_2, it was possible that we had failed to detect the existence of small RNA segments, specifically any of less than 50 nucleotides. We therefore decided to examine the precise 5′-terminal nucleotides of the NS_2 mRNA and to demonstrate directly that translation of the NS_2 mRNA could occur in a reading frame different from that used for NS_1, by sequencing the cloned NS DNA and the NS_2 mRNA. This would also give us information as to how the mRNA for NS_2 was transcribed.

The complete sequence of the NS DNA was obtained for both strands of the DNA,[8] using 5′-uniquely-labeled fragments and base specific chemical cleavages.[9] The sequence is shown in Figure 1 and also included for comparison is the sequence of gene 8 of an avian influenza virus, fowl plague virus, obtained by Porter and co-workers.[10]

The NS DNA is a full length copy of influenza virus gene 8 as it contains both the 3′ and 5′-terminal sequences of the RNA segment.[11] In addition, as the NS DNA clone was constructed from cDNA copies of both the virion RNA segment 8 and the NS_1 mRNA,[12] the NS DNA contains 10–11 non-viral nucleotides which were derived from the 5′ end of the mRNA and are presumably derived from a cellular mRNA.[13] Sequence analysis of independently cloned DNA segments showed that these non-viral nucleotides are heterogeneous in length and sequence, providing further evidence that cellular RNA sequences are used to prime influenza viral mRNA transcription in infected cells.[14]

The sequence of the NS DNA indicates that the colinear NS_1 mRNA consists of at least 864 nucleotides depending on the exact site of termination of transcription and polyadenylation which occurs within the AAAAA region (869-874).[15] There is a 5′ non-coding region of 26 nucleotides before the first AUG, followed by an open reading frame that could code for a protein of 237 amino acids, which is compatible with the size of the NS_1 polypeptide. The + 1 reading frame is open from nucleotide 460 to 861, a region large enough to code for the NS_2 polypeptide, a conclusion supported by our previous data showing that the body of the NS_2 mRNA contained ~ 340 nucleotides.[6]

To investigate the nucleotide arrangement at the 5′ end of the NS_2 RNA, this mRNA was sequenced by the primer extension method. A DNA primer from the body region of the NS_2 RNA was hybridized to purified NS_2 mRNA and the primer extended, with reverse transcriptase, using the mRNA as a template, and then the single-stranded DNA was sequenced. These data showed that the 5′ end of the NS_2 mRNA is heterogeneous in sequence for 10–20 nucleotides. Following this heterogeneous sequence, there are ~ 56 nucleotides that are complementary to the 3′ end of the virion RNA segment 8 and are the same nucleotides that are found at the 5′ end of the NS_1 mRNA. After this ~ 56 nucleotide leader sequence, there is an interrupted region of 473 nucleotides, with the leader being covalently linked to the body of the NS_2 mRNA beginning at nucleotide 526–529. The exact nucleotide at which the 5′-terminal leader sequence is joined to the body of the NS_2 mRNA cannot be determined because of the repetition of CAGG at 54–57 and 526–529, but as described below, it is likely to be nucleotides 56 and 529. This mRNA arrangement creates an open reading frame from the initiation codon of protein synthesis at nucleotide 27–29, and continues until the termination codon at 862–864. Thus 9 amino acids coded by nucleotides 27–56 would be shared by NS_1 and NS_2, and then translation of NS_2 would occur in the + 1 reading frame after the interrupted region. The sequences indicate that NS_1 and NS_2 overlap by 70 amino acids that are translated from different reading frames. A schematic representation for the arrangement of the mRNAs for polypeptides NS_1 and NS_2 is shown in Figure 2.

In eukaryotic mRNAs, the nucleotides at both sides of intervening sequences have been found to follow a distinct pattern. The 5′ (donor) site has the preferred form AG↓GTA (↓ = cleavage), and the 3′ (acceptor) site has the preferred sequence PyPyNPyAG↓, with no AG in the preceeding 10 nucleotides.[16-18] These consensus sequences are found at the junctions of the intervening sequence in the NS_2 mRNA. In addition, it has been found that the small stable U-1 nuclear RNA is exactly complementary to the consensus sequences, and it has been suggested that the U-1 RNA is a recognition component in the cleavage and ligation mechanism.[18,19] In Figure 3, the NS_2 mRNA is aligned with the U-1 RNA sequence to show that the NS_2 mRNA can readily be fitted to this proposed model.

The synthesis of the NS_2 mRNA may occur by processing of its colinear transcript, NS_1 mRNA by a splicing mechanism. If the NS_1 mRNA is a substrate for splicing, the event is tightly controlled, because most of the mRNA in the cytoplasm is NS_1 mRNA.[6] In addition, we showed earlier that the synthesis of the NS_2 mRNA is dependent on early protein synthesis.[2] For eukaryotic mRNAs, splicing is a nuclear event, and thus this processing may be an important nuclear event in influenza virus replication. It is possible that a pool of the NS_1 mRNA in the nucleus is modified in a subtle way, e.g., by methylation, which makes it a substrate for splicing. Alternative, but less likely, possibilities for the formation of the NS_2 mRNA have been discussed by us previously.[8] These include transcriptional jumping exactly at the complement of the consensus sequence, or from formation of DI particles with internal deletions exactly at the complement of the consensus

Non-viral

Virion strand (-) 3' -TAGGAAAACGT TCGTTTTCGTCCCACTGTTTCTGTAT TAC CTA AGG TTG TGA CAC AGT TCA AAA GTC CAT CTG ACG
20 *Hinf I* 40 60

mRNA strand (+) 5' -ATCCTTTTGCA AGCAAAAGCAGGGTGACAAAGACATA ATG GAT TCC AAC ACT GTG TCA AGT TTT CAG GTA GAC TGC
A/Udorn/72 (H3N2) *N-terminus NS₁ & NS₂* *Met-Asp-Ser-Asn-Thr-Val-Ser-Ser-Phe-Gln-Val-Asp-Cys-(13)*

mRNA strand (+) 5' -AGCAAAAGCAGGGUGACAAAAACAUA AUG GAU UCC AAC ACU GUG UCA AGC UUU CAG GUA GAC UGC
A/FPV/Rostock/34 (Hav1N1) *N-terminus NS₁(& NS₂?)* *Met-Asp-Ser-Asn-Thr-Val-Ser-Ser-Phe-Gln-Val-Asp-Cys-*

AAG GAA ACC GTA CAG GCT TTT GTT CAA CAT CTG GTT CTT GAT CCA CTA CGG GGT AAG GAA CTA GCC GAA GCG GCT CTA GTC TTC
80 100 120 140

TTC CTT TGG CAT GTC CGA AAA CAA GTT GTA GAC CAA GAA CTA GGT GAT GCC CCA TTC CTT GAT CGG CTT CGC CGA GAT CAG AAG
Phe-Leu-Trp-His-Val-Arg-Lys-Gln-Val-Val-Asp-Gln-Glu-Leu-Gly-Asp-Ala-Pro-Phe-Leu-Asp-Arg-Leu-Arg-Arg-Asp-Gln-Lys-(41)

UUU CUU UGG CAU GUC CGC AAA CGA UUU GCA GAC CAA GAA AUG GGU GAU GCC CCA UUC CUU GAC CGA CUU CGC CGA GAU CAG AAG
Phe-Leu-Trp-His-Val-Arg-Lys-Arg-Phe-Ala-Asp-Gln-Glu-Met-Gly-Asp-Ala-Pro-Phe-Leu-Asp-Arg-Leu-Arg-Arg-Asp-Gln-Lys-

---

AGG GAT TCC CCT TCT CCG TCG TGA GAG CCA GAT TTG TAG CTT CGT CGG TGG GTA CAA CCT TTC GTC TAT CAT CTC TTC TAA GAC
160 *Mbo II* 180 *Taq I* 200 220 *Hinf I*

TCC CTA AGG GGA AGA GGC AGC ACT CTC GGT CTA AAC ATC GAA GCA GCC ACC CAT GTT GGA AAG CAG ATA GTA GAG AAG ATT CTG
Ser-Leu-Arg-Gly-Arg-Gly-Ser-Thr-Leu-Gly-Leu-Asn-Ile-Glu-Ala-Ala-Thr-His-Val-Gly-Lys-Gln-Ile-Val-Glu-Lys-Ile-Leu-(69)

UCC CUG AGG GGA AGA GGC AGC ACU CUU GGU CUG GAC AUC GAC ACA GCU ACU CGU GUU GGA AAG CAG AUA GUG GAG CGG AUU CUG
Ser-Leu-Arg-Gly-Arg-Gly-Ser-Thr-Leu-Gly-Leu-Asp-Ile-Asp-Thr-Ala-Thr-Arg-Val-Gly-Lys-Gln-Ile-Val-Glu-Arg-Ile-Leu-

---

TTC CTT CTT AGA CTA CTC CGT GAA TTT TAC TGG TAC CGG AGG TGT GGA CGA AGC GCT ATG TAT TGA CTG TAC TGA TAA CTC CTT
240 *Hinf I* *Mbo II* 260 *Hae III* 280 300

AAG GAA GAA TCT GAT GAG GCA CTT AAA ATG ACC ATG GCC TCC ACA CCT GCT TCG CGA TAC ATA ACT GAC ATG ACT ATT GAG GAA
Lys-Glu-Glu-Ser-Asp-Glu-Ala-Leu-Lys-Met-Thr-Met-Ala-Ser-Thr-Pro-Ala-Ser-Arg-Tyr-Ile-Thr-Asp-Met-Thr-Ile-Glu-Glu-(97)

GAG GAC GAA UCC GAU GAG GCA CUU AAA AUG ACC AUU GCC UCU GUA CCU GCU ACA CGC UAC CUA ACU GAC AUG ACU CUU GAA GAG
Glu-Asp-Glu-Ser-Asp-Glu-Ala-Leu-Lys-Met-Thr-Ile-Ala-Ser-Val-Pro-Ala-Thr-Arg-Tyr-Leu-Thr-Asp-Met-Thr-Leu-Glu-Glu-

---

AAC AGT TCC CTG ACC AAG TAC GAT TAC GGG TTC GTC TTT CAC CTT CCT GGA GAA ACG TAG TCT TAT CTG GTT CGT TAG TAC CTA
320 340 360 380 400

TTG TCA AGG GAC TGG TTC ATG CTA ATG CCC AAG CAG AAA GTG GAA GGA CCT CTT TGC ATC AGA ATA GAC CAA GCA ATC ATG GAT
Leu-Ser-Arg-Asp-Trp-Phe-Met-Leu-Met-Pro-Lys-Gln-Lys-Val-Glu-Gly-Pro-Leu-Cys-Ile-Arg-Ile-Asp-Gln-Ala-Ile-Met-Asp-(125)

AUG UCA AGG GAC UGG UUC AUG CUC AUG CCC AAA CAG AAA GUG GCA GGC UCC CUU UGC AUC AGA AUG GAC CAG GCG AUC AUG GGG
Met-Ser-Arg-Asp-Trp-Phe-Met-Leu-Met-Pro-Lys-Gln-Lys-Val-Ala-Gly-Ser-Leu-Cys-Ile-Arg-Met-Asp-Gln-Ala-Ile-Met-Gly-

---

TTC TTG TAG TAC AAC TTT CGC TTA AAG TCA CAC TAA AAA CTG GCC GAT CTC TGG GAT TAT AAT GAT TCC CGA AAG TGG CTT CTC
420 440 *Hpa II* 460 480

AAG AAC ATC ATG TTG AAA GCG AAT TTC AGT GTG ATT TTT GAC CGG CTA GAG ACC CTA ATA TTA CTA AGG GCT TTC ACC GAA GAG
Lys-Asn-Ile-Met-Leu-Lys-Ala-Asn-Phe-Ser-Val-Ile-Phe-Asp-Arg-Leu-Glu-Thr-Leu-Ile-Leu-Leu-Arg-Ala-Phe-Thr-Glu-Glu-(153)

AAG AAC AUC AUA CUG AAA GCA AAC UUC AGU GUG AUU UUC GAU CGG CUG GAA ACU CUC AUA CUA UUA AGG GCU UUA ACC GAU GAG
Lys-Asn-Ile-Ile-Leu-Lys-Ala-Asn-Phe-Ser-Val-Ile-Phe-Asp-Arg-Leu-Glu-Thr-Leu-Ile-Leu-Leu-Arg-Ala-Leu-Thr-Asp-Glu-

---

CCT CGT TAA CAA CCG CTT TAG AGT GGT AAC GGA AGA AAA GGT CCT GTA TGA TAA CTC CTA CAG TTT TTA CGT TAA CCC CAG GAG
Mbo II 500 520 *Bst NI* 540 560

GGA GCA ATT GTT GGC GAA ATC TCA CCA TTG CCT TCT TTT CCA GGA CAT ACT ATT GAG GAT GTC AAA AAT GCA ATT GGG GTC CTC
Gly-Ala-Ile-Val-Gly-Glu-Ile-Ser-Pro-Leu-Pro-Ser-Phe-Pro-Gly-His-Thr-Ile-Glu-Asp-Val-Lys-Asn-Ala-Ile-Gly-Val-Leu-(181)
-Asp-Ile-Leu-Leu-Arg-Met-Ser-Lys-Met-Gln-Leu-Gly-Ser-Ser-(24)

GGA GCA AUU GUC GGC GAA AUU UCA CCA UUG CCU UCU CUU CCA GGA CAU ACU GAU GAG GAU GUC AAA AAU GCA AUU GGG GUC CUC
Gly-Ala-Ile-Val-Gly-Glu-Ile-Ser-Pro-Leu-Pro-Ser-Leu-Pro-Gly-His-Thr-Asp-Glu-Asp-Val-Lys-Asn-Ala-Ile-Gly-Val-Leu-
-Asp-Ile-Leu-Met-Arg-Met-Ser-Lys-Met-Gln-Leu-Gly-Ser-Ser-

TAG CCT CCT GAA CTT ACC TTA CTA TTG TGT CAA GCT CAG AGA TTT TGA GAT GTC TCT AAG CGA ACC CCT TCG TCA TTA CTC TTA

580 600 Taq I 620 Hinf I 640

Hinf I

ATC GGA GGA CTT GAA TGG AAT GAT AAC ACA GTT CGA GTC TCT AAA ACT CTA CAG AGA TTC GCT TGG GGA AGC AGT AAT GAG AAT

Ile-Gly-Gly-Leu-Glu-Trp-Asn-Asp-Asn-Thr-Val-Arg-Val-Ser-Lys-Thr-Leu-Gln-Arg-Phe-Ala-Trp-Gly-Ser-Ser-Asn-Glu-Asn-(209)

-Ser-Glu-Asp-Leu-Asn-Gly-Met-Ile-Thr-Gln-Phe-Glu-Ser-Leu-Lys-Leu-Tyr-Arg-Asp-Ser-Leu-Gly-Glu-Ala-Val-Met-Arg-Met-(52)

AUC GGA GGA CUU GAA UGG AAU GAU AAC ACA GUU CGA GUC UCU GAA ACU AUA CAG AGA UUC GCU UGG AGA AGC AGU AAU GAG AAU

Ile-Gly-Gly-Leu-Glu-Trp-Asn-Asp-Asn-Thr-Val-Arg-Val-Ser-Glu-Thr-Ile-Gln-Arg-Phe-Ala-Trp-Arg-Ser-Ser-Asn-Glu-Asn-

-Ser-Glu-Asp-Leu-Asn-Gly-Met-Ile-Thr-Gln-Phe-Glu-Ser-Leu-Lys-Leu-Tyr-Arg-Asp-Ser-Leu-Gly-Glu-Ala-Val-Met-Arg-Met-

---

CCC TCT GGA GGT GAA TGA GGT TTT GTC TTT GCC TTT TAC CGC TCT TGT CGA TCC AGT TTT CAA GCT TCT CTA TTC TAC CGA CTA

660 680 700 Alu I 720 Mbo II

Taq I

GGG AGA CCT CCA CTT ACT CCA AAA CAG AAA CGG AAA ATG GCG AGA ACA GCT AGG TCA AAA GTT CGA AGA GAT AAG ATG GCT GAT

Gly-Arg-Pro-Pro-Leu-Thr-Pro-Lys-Gln-Lys-Arg-Lys-Met-Ala-Arg-Thr-Ala-Arg-Ser-Lys-Val-Arg-Arg-Asp-Lys-Met-Ala-Asp.(237)

-Gly-Asp-Leu-His-Leu-Leu-Gln-Asn-Arg-Asn-Gly-Lys-Trp-Arg-Glu-Gln-Leu-Gly-Gln-Lys-Phe-Glu-Glu-Ile-Arg-Trp-Leu-Ile-(80)

GGG GGA CCU CCA CUC CCU CCA AAG CAG AAA CGG AAA AUG GCG AGA ACA AUU GAG UCA GAA AUU UGA AGA AAU AAG AUG GUU AAU

Gly-Gly-Pro-Pro-Leu-Pro-Pro-Lys-Gln-Lys-Arg-Lys-Met-Ala-Arg-Thr-Ile-Glu-Ser-Glu-Ile. (**230**) *C-terminus NS_1*

-Gly-Asp-Leu-His-Ser-Leu-Gln-Ser-Arg-Asn-Gly-Lys-Trp-Arg-Glu-Gln-Leu-Ser-Gln-Lys-Phe-Glu-Glu-Ile-Arg-Trp-Leu-Ile-

---

A CTT CTT CAC TCT GTG TCT GAC TTC TGT TGT CTC TTA TCA AAA CTC GTT TAT TGT AAG TAC GTT CGG AAT GTC GAT GAT AAA CTT

740 Mbo II 760 780 800 Alu I 820

T GAA GAA GTG AGA CAC AGA CTG AAG ACA ACA GAG AAT AGT TTT GAG CAA ATA ACA TTC ATG CAA GCC TTA CAG CTA CTA TTT GAA

C-terminus NS_1

-Glu-Glu-Val-Arg-His-Arg-Leu-Lys-Thr-Thr-Glu-Asn-Ser-Phe-Glu-Gln-Ile-Thr-Phe-Met-Gln-Ala-Leu-Gln-Leu-Leu-Phe-Glu-(108)

U GAA GAA GUG AGA CAU AGA UUG AAG AUA ACA GAG AAU AGU UUU GAA CAA AUA ACA UUU AUG CAA GCC UUA CAA CUA CCG CUU GAA

-Glu-Glu-Val-Arg-His-Arg-Leu-Lys-Ile-Thr-Glu-Asn-Ser-Phe-Glu-Gln-Ile-Thr-Phe-Met-Gln-Ala-Leu-Gln-Leu-Pro-Leu-Glu-

---

CAC CTT GTC CTC TAT TCT TGA AAG AGC AAA GTC GAA TAA ATT ACTATTTTTTGTGGGAACAAAGATGA-5' Virion strand (-)

840 Alu I 860 880 890

GTG GAA CAG GAG ATA AGA ACT TTC TCG TTT CAG CTT ATT TAA TGATAAAAAACACCCTTGTTTCTACT-3' mRNA strand (+)

Val-Glu-Gln-Glu-Ile-Arg-Thr-Phe-Ser-Phe-Gln-Leu-Ile.(121) C-terminus NS_2

GUG GAG CAA GAG AUA AGA ACU UUC UCG UUU CAG CUU AUU UAA UGAUAAAAAACACCCUUGUUUCUACU-3' mRNA strand (+)

Val-Glu-Gln-Glu-Ile-Arg-Thr-Phe-Ser-Phe-Gln-Leu-Ile. C-terminus NS_2

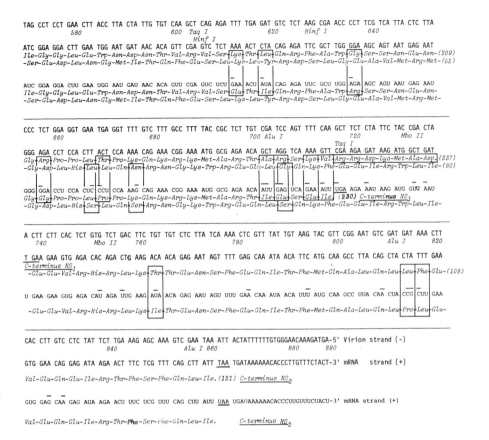

Figure 1. Complete nucleotide sequence of influenza virus gene 8 derived from cloned NS DNA. The top line shows the sequence of the virion strand (−), and the second line shows the sequence of the mRNA strand (+) of the A/Udorn/72 (H_3N_2) strain. The deduced amino acid sequence of polypeptides NS_1 and NS_2 is shown. The arrows at nucleotides 56 and 529 show the probable donor and acceptor nucleotides in the interrupted mRNA for NS_2. The third line of the nucleotide sequence is that of the mRNA strand (+) of influenza A/FPV/Rostock/34 (Hav1N1) obtained by Porter et al.[10] Nucleotides that differ between Udorn and FPV are indicated by a line over the FPV nucleotide and amino acid differences are denoted by rectangles. The sequence of the FPV NS_2 mRNA has not been obtained, but because the consensus sequences for the interrupted region are conserved, it is highly probably that NS_1 and NS_2 of FPV share a common N terminus.

SOURCE: *Lamb and Lai.*[8]

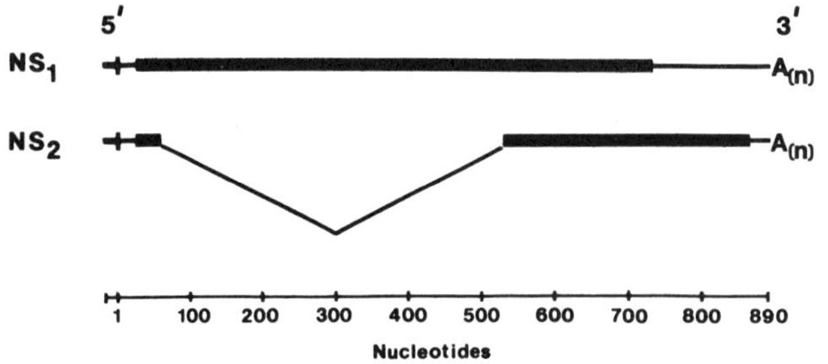

Figure 2. Schematic representation for the arrangement of the NS_1 and NS_2 mRNAs. The thin lines at the 5′ and 3′-termini of the NS_1 and NS_2 mRNAs represent non-coding regions. The thick lines represent the coding regions of the two mRNAs. In the region 529–861, the NS_2 mRNA is translated in a reading frame different from that used for NS_1. The V-shaped thin line in the NS_2 mRNA represents the interrupted sequence in the NS_2 mRNA. The medium line and bar before nucleotide 1 at the 5′-termini represent the heterogeneous nucleotides derived from cellular mRNAs that are donated to the NS_1 and NS_2 mRNAs.
SOURCE: *Lamb and Lai.*[8]

sequence. The nucleotide sequence of the NS DNA provides indirect evidence concerning the amino acid sequence of NS_1 and NS_2 in the absence of direct protein sequence information. NS_1 is predicted to have a mol wt of 26,815 and NS_2 a mol wt of 14,216. The amino acid composition of NS_1 obtained for the WSN strain,[20] compares well with that predicted for the Udorn strain.[8] In addition, the predicted tryptic or chymotryptic peptides containing methionine or leucine for NS_1 and NS_2 compare very well to those found previously.[2,3,6]

Figure 3. Base pairing of NS_1 mRNA to U-1 snRNA at the junction of the interrupted sequence found in the NS_2 mRNA. The alignment of the donor and acceptor sequences of the interrupted region in the NS_2 mRNA to the U-1 small stable nuclear RNA is adapted from the model of Lerner et al.[18] and Rogers and Wall.[19] These authors propose that the U-1 snRNA is the recognition component of the nuclear RNA splicing complex and forms base pairs with both ends of an intervening sequence (intron) so as to align them for cleavage and ligation (indicated by dots).

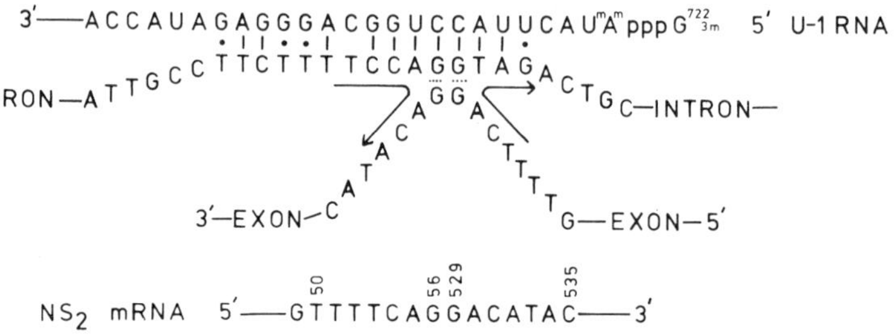

The sequence of gene 7 of influenza virus, which codes for the M-protein, has been obtained [21,22] for the PR8 strain and it has been observed that, as in the case of genome segment 8 there is a second open reading frame at the 3′ end of the mRNA that could code for a maximum of 97 amino acids. If a second mRNA is generated from the M gene by a splicing mechanism using the consensus sequences described for the NS_2 mRNA, the only donor site that could be used would be the AG↓ GTA at nucleotides 11–15 in the

Figure 4. Hybrid-arrested translation of influenza virus mRNAs with separate virion RNA segments. Cytoplasmic mRNAs from influenza virus infected cells were hybridized to individual genome RNA segments, and then the RNAs were translated *in vitro* using wheat germ extracts and the polypeptide products analyzed on polyacylamide gels. Lane C = control mRNA with no added virion RNA segment. Lanes 1–3, 4, 5, 6, 7, and 8 show the polypeptides synthesized after hybridization of the respective virion segments to mRNA. Conditions of hybridization were described previously.[3]

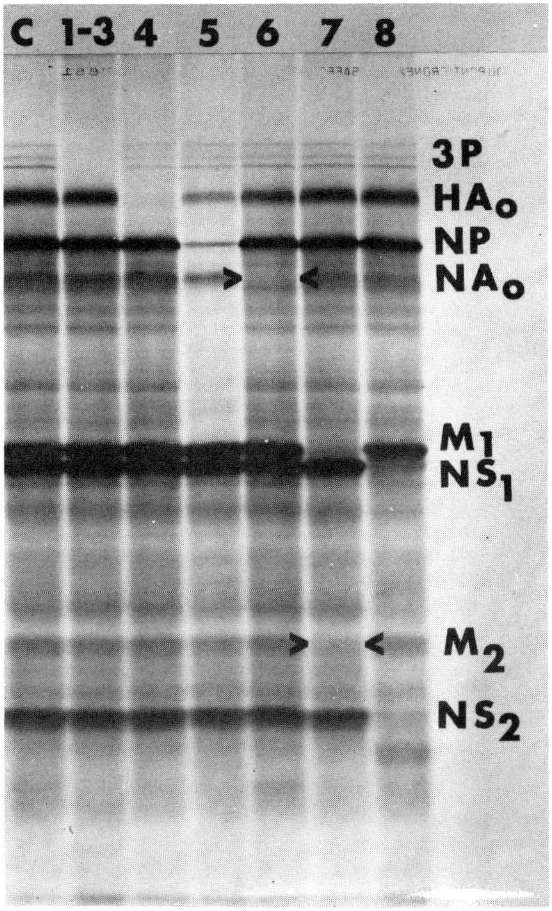

noncoding region of the M mRNA, which would match 3′ receptor site (e.g., at 691–696) so that translation could occur in the + 1 reading frame. If this is the case, the polypeptide product would consist of 96 amino acids initiating with the first AUG in the + 1 reading frame.

We have searched for a second gene 7 derived product. In infected cells, and after *in vitro* translation of mRNAs extracted from infected cells, a new polypeptide slightly larger than NS_2 could be seen. In hybrid-arrested translation experiments in which the virion RNA segments were hybridized to total infected cells mRNAs and then the mRNAs translated *in vitro,* RNA segment 7 specifically prevented the synthesis of M and the same small polypeptide (Figure 4—here we have used M_1 and M_2 to identify the products). This small polypeptide (M_2) has a tryptic peptide map unique from that of the membrane protein (M_1), and contains only one major methionine-containing peptide as predicted from the sequence analysis. Therefore, these data suggest that this (M_2) is a second polypeptide derived from gene 7. Although it can be predicted from the sequence that the second polypeptide derived from segment 7 would only have a mol wt of ~ 11,000, and the polypeptide (M_2) we have identified is larger than NS_2 (14,216), many small polypeptides migrate anomalously on polyacrylamide gels depending on their amino acid composition, e.g., NS_2 was originally estimated to be 11,000.

ACKNOWLEDGMENTS
This work was supported by Grant AI-05600 from the National Institute of Allergy and Infectious Diseases. R.A.L. is an Irma T. Hirschl Career Scientist Awardee.

References

1. Lamb, R.A. and Choppin, P.W. (1978): Effect of the host cell on early and late synthesis of influenza virus polypeptides. In: *Negative Strand Viruses and the Host Cell.* Mahy, B.W.J. and Barry, R.D. (Eds.), Academic Press, London, pp. 229–237.
2. Lamb, R.A. Etkind, P.R., and Choppin, P.W. (1978): Evidence for a ninth influenza virus polypeptide. *Virology* 91:60–78.
3. Lamb, R.A. and Choppin, P.W. (1979): Segment 8 of the influenza virus genome is unique in coding for two polypeptides. *Proc. Natl. Acad. Sci. U.S.A.* 76:4908–4912.
4. Inglis, S.C., Barrett, T., Brown, C.M., and Almond, J.W. (1979): The smallest genome RNA segment of influenza virus contains two genes that may overlap. *Proc. Natl. Acad. Sci. U.S.A.* 76:3790–3794.
5. Sleigh, M. J., Both, G.W., and Brownlee, G.G. (1979): A new method for the size estimation of the RNA genome segments of influenza virus. *Nucl. Acids Res.* 6:1309–1321.
6. Lamb, R.A., Choppin, P.W., Chanock, R.M., and Lai, C.-J. (1980): Mapping of the two overlapping genes for polypeptides NS_1 and NS_2 on RNA segment 8 of the influenza virus genome. *Proc. Natl. Acad. Sci. U.S.A.* 77:1857–1861.
7. Berk, A.J. and Sharp, P.A. (1978): Spliced early mRNAs of simian virus 40. *Proc. Natl. Acad. Sci. U.S.A.* 75:1274–1278.

8. Lamb, R.A. and Lai, C.-J. (1980): Sequence of interrupted and uninterrupted mRNAs and cloned DNA coding for the two overlapping nonstructural proteins of influenza virus. *Cell* 21:475–485.
9. Maxam, A.M. and Gilbert, W. (1977): A new method for sequencing DNA. *Proc. Natl. Acad. Sci. U.S.A.* 74:560–564.
10. Porter, A.G., Smith, J.C., and Emtage, J.S. (1980): The sequence of influenza virus RNA segment 8 indicates that the coding regions of the NS_1 and NS_2 proteins overlap. *Proc. Natl. Acad. Sci. U.S.A.* 77:5074–5078.
11. Robertson, J.S. (1979): 5′ and 3′-terminal nucleotide sequences of the RNA genome segments of influenza virus. *Nucl. Acids Res.* 6:3745–3757.
12. Lai, C.-J., Markoff, L.J., Zimmerman, S., Cohen, B., Berndt, J.A., and Chanock, R.M. (1980): Cloning DNA sequences from influenza viral RNA segments. *Proc. Natl. Acad. Sci. U.S.A.* 77:210–214.
13. Krug, R.M., Broni, B.A., and Bouloy, M. (1979): Are the 5′ ends of influenza viral mRNAs synthesized *in vivo* donated by host mRNAs? *Cell* 18:329–334.
14. Dhar, R., Chanock. R.M., and Lai, C.-J. (1980): Non-viral oligonucleotides at the 5′-terminus of cytoplasmic influenza virus mRNAs deduced from cloned complete genomic sequences. *Cell* 21:495–500.
15. Robertson, J.S.: Personal communication and see this volume.
16. Konkel, D.A., Tilghman, S.M., and Leder, P. (1978): The sequence of the chromosomal mouse β-globin major gene: homologies in capping, splicing and poly(A) sites. *Cell* 15:1125–1132.
17. Seif, I., Khoury, G., and Dhar, R. (1979): BKV splice sequences based on analysis of preferred donor and acceptor sites. *Nucl. Acids Res.* 6:3387–3398.
18. Lerner, M.R., Boyle, J.A., Mount S.M., Wolin, S.L., and Steitz, J.A. (1980): Are snRNPs involved in splicing? *Nature* 283:220–224.
19. Rogers, J. and Wall, R. (1980): A mechanism for RNA splicing. *Proc. Natl. Acad. Sci. U.S.A.* 77:1877–1879.
20. Shaw, M.W. and Compans, P.W. (1978): Isolation and characterization of cytoplasmic inclusions from influenza A virus-infected cells. *J. Virol.* 25:608–615.
21. Winter, G. and Fields, S. (1980): Cloning of influenza cDNA into M13: the sequence of the RNA segment encoding the A/PR8/34 matrix protein. *Nucl. Acids Res.* 8:1965–1974.
22. Allen, H., McCauley, J., Waterfield, M., and Gething, M.-J. (1980): Influenza virus RNA segment 7 has the coding capacity for two polypeptides. *Virology*, 107:548–551.

David H. L. Bishop and Richard W. Compans, eds.
The Replication of Negative Strand Viruses

Characterization of Subgenomic Polyadenylated mRNAs Encoded by Genome RNA Segment 7 of Influenza Virus

S.C. Inglis[a]

Introduction

Winter and Fields[1] have recently deduced the nucleotide sequence of virion RNA (vRNA) segment 7 (which encodes the virus matrix protein)[2] of influenza virus A/PR8. Their work suggests that vRNA segment 7 may, like vRNA segment 8, contain two overlapping genes. The gene encoding the matrix protein (M) begins near the 3′-terminus of the vRNA, and consists of an uninterrupted sequence which occupies about 75% of the molecule. A second potential gene, with the capacity to encode a polypeptide of about 11,000 mol wt, spans the remaining 25%, and overlaps the M-protein gene (in a different codon-reading frame) by about 60 nucleotides (see Figure 3a). A polypeptide specified by this nucleotide sequence has not yet been identified in virus infected cells, but its presence cannot be ruled out. In this report evidence is presented that infected cells contain, in addition to the mRNA encoding the M-protein, two small polyadenylated RNAs which are complementary to sequences of vRNA segment 7. Characterization of complementary DNA (cDNA) copies of these RNAs has indicated that both contain sequences complementary to the section of the vRNA segment which contains the putative second gene, and so either or both RNAs may specify the predicted polypeptide product. Furthermore, the data suggest that each RNA may derive its sequences from non-contiguous regions on the vRNA molecule, and so could arise through processing of a larger precursor.

[a] Division of Virology, Department of Pathology, University of Cambridge, Laboratories Block, Addenbrooke's Hospital, Hills Road, Cambridge, England.

Results

Chicken embryo fibroblasts (CEF) were infected with the avian influenza, fowl plague virus (FPV) in the presence of ^{32}P-orthophosphate, and poly(A)-containing RNA was prepared 4.5 hr later.[3] Poly(A) tracts were subsequently removed from the RNA by incubation with ribonuclease H in the presence of oligo (dT).[4] Messenger RNAs complementary to vRNA segment 7 were then isolated by hybridization with a bacterial plasmid (pBR322) which contained a cloned DNA copy of vRNA segment 7 of FPV;[5] the plasmid DNA was immobilized on a nitrocellulose membrane prior to annealing. Messenger RNAs which specifically hybridized were eluted from the immobilized DNA and analyzed by gel electrophoresis (Figure 1a).

The largest and most abundant RNA in the preparation (RNA7) was approximately the same length as vRNA segment 7, and was therefore presumably the matrix protein mRNA.[6] Two smaller RNAs (labeled 10 & 11) were also present, though in much smaller quantity. Both small RNA species were detected in three separate experiments. These RNAs were further characterized through analysis of cDNA copies. ^{32}P-labeled cDNA copies of virus mRNAs (cmDNAs) were synthesized using reverse transcriptase[4] and were hybridized with immobilized segment 7 DNA sequences as before. The hybridized cmDNAs were then eluted and analyzed by gel electrophoresis (Figure 1b). The two small cmDNA species (labeled cmDNA 10 and 11) were considered to be true reverse transcripts of RNAs 10 and 11 rather than short transcripts of mRNA 7 since when the small RNAs were isolated by sucrose density gradient sedimentation and then incubated with reverse transcriptase, cmDNAs 10 and 11 were the predominant products (Figure 2). The sizes of cmDNAs 10 and 11 were estimated to be approximately 350 and 310 nucleotides respectively based on their gel mobility relative to known DNA markers. Their RNA templates are likely to be about 15 nucleotides shorter [excluding poly(A)] since oligo $(dT)_{12\text{-}18}$ was used to prime reverse transcription.

10 and 11 are represented, plasmid DNA containing the cloned copy of vRNA segment 7 was digested separately with restriction endonucleases Hinf I and Alu I, and DNA fragments corresponding to different regions of the vRNA were isolated by two cycles of preparative gel electrophoresis (Figure 3a). These DNA fragments were immobilized separately on nitrocellulose membranes, and each was hybridized with ^{32}P-labeled reverse transcripts of infected cell mRNA as before. The cmDNAs, which hybridized specifically to each fragment, were analyzed by gel electrophoresis (Figure 3b). As expected, cmDNA 7, the reverse transcript of the mRNA encoding matrix protein, hybridized specifically to each of the Hinf I and Alu I frag-

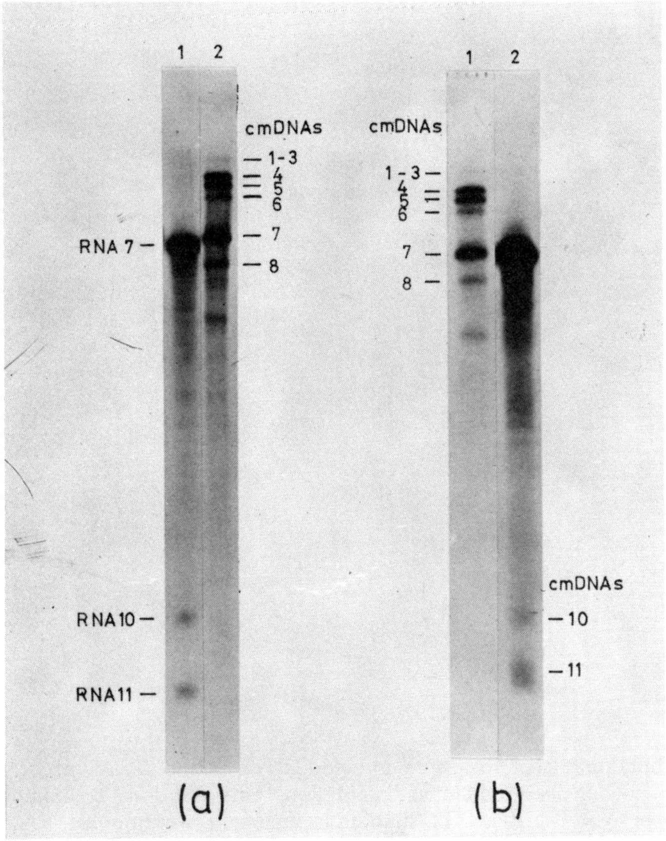

Figure 1. (a) Primary CEF cells were infected with FPV in the presence of ^{32}P-orthophosphate (2 mCi), and poly(A)-containing RNA was prepared 4.5 hr later.[3] Poly(A) was subsequently removed using RNase H.[4] Ten micrograms of bacterial plasmid pBR322 which contained a complete DNA copy of vRNA segment 7[5] was bound to a 1 cm diameter nitrocellulose disc,[7] and the disc was annealed with the ^{32}P-labeled RNA preparation. Sequences which hybridized specifically with the immobilized DNA were recovered and analyzed by polyacrylamide gel electrophoresis and autoradiography (track 1). Details of these procedures have been described elsewhere.[4] ^{32}P-labeled cmDNAs, prepared by reverse transcription of poly(A)-containing RNA from FPV-infected cells, were loaded in a parallel slot (track 2) to act as approximate size markers for virus-specific mRNAs.

(b) ^{32}P-labeled reverse transcripts of poly(A)-containing RNA from FPV-infected cells (cmDNAs) were prepared and annealed with immobilized segment 7-specific DNA as described in (a). DNA which specifically hybridized was recovered and analyzed by gel electrophoresis and autoradiography (track 2). A sample of the unfractionated cmDNA was loaded in a parallel slot (track 1).

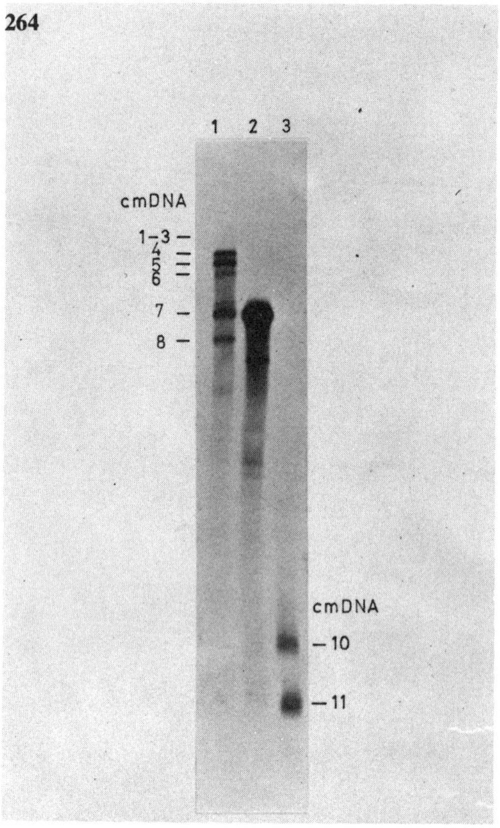

Figure 2. Poly(A)-containing RNA from FPV-infected cells was fractionated according to size by sucrose density gradient sedimentation,[8] and fractions containing RNAs of a similar size to RNAs 10 and 11 (marked in Figure 1a) were pooled. RNA was recovered by ethanol precipitation and incubated with reverse transcriptase.[4] The ^{32}P-labeled complementary DNA products were then annealed with immobilized segment 7-specific DNA as before, and those sequences which hybridized were analyzed by gel electrophoresis (track 3). Tracks 1 and 2 show cmDNAs prepared in the same way as for the equivalent-numbered track in Figure 1b).

Figure 3. (a) A restriction endonuclease map of the pBR322 plasmid which contained a DNA copy of vRNA segment 7. Plasmid DNA (50 μg) was digested with Hpa II and a 1280 base pair DNA fragment which contained the virus-specific sequences was isolated by preparative agarose gel electrophoresis. This DNA was redigested with Alu I and Hinf I, and the resulting fragments were purified by two cycles of preparative polyacrylamide gel electrophoresis. The relative orientation of these fragments was determined by the method of Smith and Birnsteil.[9]
(b) Alu I fragments A, B, C, and D, and Hinf I fragments A, B, C, D/E* and G [prepared as described in (a)] were immobilized separately on nitrocellulose membranes as before, and each was hybridized with ^{32}P-cmDNA as described for Figure 1b. Hybridized cmDNAs were recovered and analyzed by gel electrophoresis and autoradiography. Tracks 1-4 show cmDNAs which annealed to Alu I fragments A, D, E, and B respectively, and tracks 5-9 show cmDNAs which annealed to Hinf I fragments A, C, D/E, G, and B respectively. The order of the samples reflects the relative position of each fragment in the cloned DNA (see Figure 3a). Approximately equal amounts of radioactivity were loaded in each slot. *Hinf I fragments D and E were isolated together, since they did not separate during gel electrophoresis.

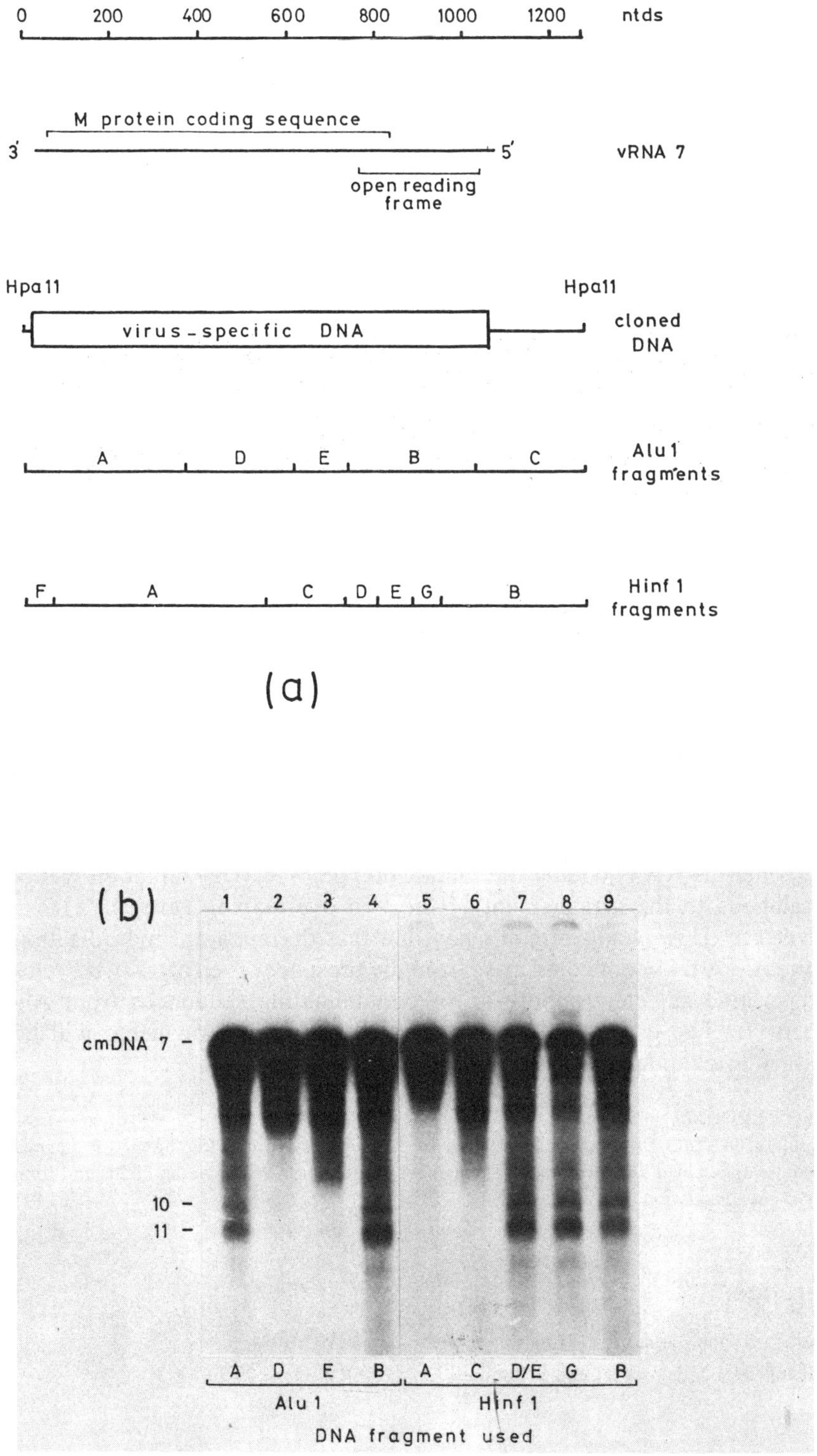
0
200
400
600
800
1000
1200
ntds
M protein coding sequence
3'
5'
vRNA 7
open reading frame
Hpa11
Hpa11
virus-specific DNA
cloned DNA
A
D
E
B
C
Alu 1 fragments
F
A
C
D
E
G
B
Hinf 1 fragments
(a)
(b)
1 2 3 4 5 6 7 8 9
cmDNA 7 -
10 -
11 -
A D E B A C D/E G B
Alu 1
Hinf 1
DNA fragment used

ments used. CmDNAs 10 and 11 however only hybridized efficiently to certain of the fragments. Of these, Hinf D/E, Hinf G, Hinf Band Alu B correspond to a region of about 300 nucleotides close to the 5′-terminus of vRNA segment 7 (see Figure 3a). The other restriction fragment which appeared to hybridize with both cmDNAs 10 and 11, Alu I fragment A represents the 3′-terminal sequences of the vRNA molecule.

Conclusions

Fowl plague virus infected cells contain at least three species of polyadenylated RNA which hybridize with sequences derived from vRNA segment 7. The largest of these presumably encodes the matrix polypeptide, since it is similar in size to vRNA segment 7. The two smaller RNAs are approximately 330 and 290 nucleotides in length, but their coding function, if any, is unknown. Most of the sequences from both these small mRNAs appear to correspond with the 5′-terminal region of vRNA segment 7, which implies that the two RNAs are related. Since it has been demonstrated that the nucleotide sequence of this part of vRNA segment 7 could encode a polypeptide of approximately 11,000 mol wt,[1] either or both of these small RNAs may represent a genuine virus-specific mRNA. Cell-free translation studies should establish whether this is indeed the case.

Both small RNAs appeared also to hybridize quite efficiently with the Alu I A fragment, which contained sequences from the 3′-terminus of vRNA segment 7 (though not with sequences from the Hinf I F fragment, which maps near the 3′-terminus). This suggests that the RNAs may derive their information from non-contiguous regions on the vRNA segment, which in turn suggests that they may arise through processing of a larger precursor, perhaps the mRNA encoding the matrix polypeptide. This situation would be analogous to the arrangement of the two overlapping genes in vRNA segment 8.[10] However it remains possible that the apparent hybridization with the Alu A fragment could arise from the presence, even after two cycles of preparative gel electrophoresis, of contaminating sequences from Alu fragment B. The use of individually cloned restriction fragments will be necessary to eliminate this possibility.

ACKNOWLEDGMENTS

I am grateful to Carol Brown for expert assistance, to Andrew Caton for the gift of plasmid DNA containing cloned vRNA segment 7 sequences, and to Greg Winter, Stan Fields and John McCauley for discussion.

References

1. Winter, G. and Fields, S. (1980): *Nucleic Acids Res.* 8;1965–1974.
2. Ritchey, M.B., Palese, P., and Schulman, J.L. (1976); *J. Virol.* 20:307–313.

3. Inglis, S.C. and Mahy, B.W.J. (1979): *Virology* 95:154–164.
4. Inglis, S.C., Gething, M.-J., and Brown, C.M. (1980): *Nucleic Acids Res.* 8:3575–3589.
5. Caton, A.J. and Robertson, J.S. (1980): *Nucleic Acids Res.* 8:2591–2603.
6. Hay, A.J., Abraham, G., Skehel, J.J., Smith, J.C., and Fellner, P. (1977): *Nucleic Acids Res.* 4:4197–4210.
7. Gillespie, D. and Spiegelman, S. (1965): *J. Molec. Biol.* 12:829–842.
8. Inglis, S.C., Barrett, T., Brown, C.M., and Almond, J.W. (1979): *Proc. Natl. Acad. Sci. U.S.A.* 76:3790–3794.
9. Smith, H.O. and Birnsteil, M. (1976): *Nucleic Acids Res.* 3:2387–2398.
10. Lamb, R.A. and Lai, C.-J. (1980): *Cell* 21:475–485.

David H. L. Bishop and Richard W. Compans, eds.
The Replication of Negative Strand Viruses

On the Initiation of Myxovirus Infection

R. Rott, R.T.C. Huang, K. Wahn and H.-D Klenk[a]

It is generally accepted that viral glycoproteins play important roles in the interaction between viral envelopes and cellular membranes and that their primary biological function is the initiation of infection. After adsorption of the virus particle to cellular receptors, the virus genome of enveloped viruses has to penetrate not only through the cellular membrane to enter the host cell, but also through its own membrane before the genome can be uncoated to initiate replication.

For paramyxoviruses it is widely believed that the common principle underlying these processes is insertion of the viral envelope into the cell membrane by fusion. It is also commonly accepted that in these viruses the proteolytically cleaved F-glycoprotein is the sole structure responsible for induction of membrane fusion, while the second glycoprotein, the HN-protein, is responsible for attachment of the virus particle to the cell receptor.[1]

Penetration of Orthomyxoviruses through the Cell Membrane

There is increasing evidence that penetration of orthomyxoviruses is also a result of fusion and that the mechanism underlying this process is principally similar in ortho- and paramyxoviruses. One of the two glycoproteins, the hemagglutinin, is suggested to play a role in penetration corresponding to that of the F-protein of paramyxoviruses. This concept is supported by the following observations:

[a] Institute für Virologie, Justus-Liebig-Universität Giessen, D-6300 Giessen, Germany.

1. As the F-glycoproteins of paramyxoviruses, the hemagglutinin of orthomyxoviruses has to be posttranslationally cleaved by host proteases to obtain its full biological activity necessary for viral infectivity.[2,3] One cleavage fragment, HA_2 of orthomyxoviruses, possesses an identical N-terminus in all influenza A virus strains examined, and there is a striking similarity in amino acid sequence and hydrophobicity between the N-termini of HA_2 and the F_1-glycoprotein of paramyxoviruses.[1] Studies with different proteases have shown that a specific cleavage site is required for biological activity and that removal of a single amino acid on the N-terminus of HA_2 results in inactivity of the virus.[4] Interestingly, oligopeptides with the amino acid sequence of the N-termini of HA_2 and F_1 were able to inhibit the replication of ortho- and paramyxoviruses.[5] These observations emphasize the biological significance of the N-termini in HA_2 and F_1 glycoproteins and suggest related functions of these compounds of the two different viruses.
2. Nuclear magnetic resonance spectra of chicken fibroblasts exposed to orthomyxoviruses indicate that viruses containing the cleaved hemagglutinin cause a change in fluidity of the lipid bilayer of the cellular membrane similar to paramyxoviruses. This alteration is not seen when cells are infected with virus containing the uncleaved HA.[6]
3. The integration of the orthomyxovirus glycoproteins into the plasma membrane in the initial phase of infection has been shown by cell-mediated cytotoxicity. Cytotoxic T-lymphocytes primed by influenza viruses can lyse the target cells only when they are infected with influenza virus containing cleaved hemagglutinin. Virus particles with uncleaved hemagglutinin are fully capable of adsorbing to the cell surface. Host cells which carry such virus particles are, however, not lysed by cytotoxic T-lymphocytes.[7] This means that the formation of target cells implies insertion of viral glycoproteins and that mere adsorption is insufficient.
4. The involvement of the orthomyxovirus hemagglutinin in fusion between the viral envelope and the host cell membrane could be demonstrated directly with viral glycoprotein-loaded liposomes. Electron microscopic studies showed (Figure 1) that liposomes containing both influenza virus glycoproteins[8] fused with cell membranes when the hemagglutinin was present in the proteolytically cleaved form. Liposomes containing the uncleaved hemagglutinin adsorbed to cells without causing membrane fusion.[9]

Function of the Viral Neuraminidase in Myxovirus Penetration

Besides the cleaved hemagglutinin of orthomyxoviruses or the active F protein of paramyxoviruses, neuraminidase activity is also required for myxovirus-induced membrane fusion. This became evident from experi-

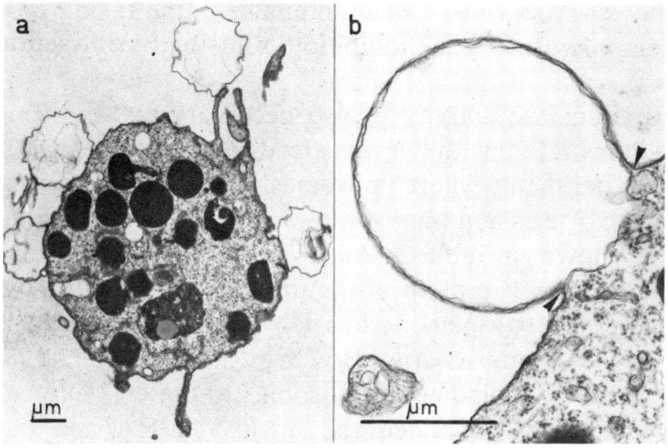

Figure 1. Interaction of chick embryo cells with liposomes containing neuraminidase and uncleaved (a) or cleaved (b) hemagglutinin of virus N (Hav 2Neg 1). (for details see Huang et al.).[8]

ments involving the interaction of cell membranes with liposomes containing ortho- and paramyxovirus glycoproteins, individually or in various combination.[10]

It could be shown electron microscopically that liposomes containing only the cleaved hemagglutinin of orthomyxoviruses adsorb massively to cells without causing fusion (Table 1). To induce fusion under these conditions, neuraminidase was necessary. This could be even soluble neuraminidase of viral or bacterial origin. The fact that neuraminidase which was not inserted into liposomes can induce fusion in the presence of cleaved hemagglutinin, indicates that the enzyme does not act as a linker between liposomal and

Table 1. Requirement for Neuraminidase for Fusion Activity of Liposomes Containing the Glycoproteins of Orthomyxoviruses.

Liposomes containing	Percentage fusion
$HA_{1,2}$[a] + NA	60
HA[b] + NA	4
$HA_{1,2}$	0
$HA_{1,2}$ + neuraminidase of *V. cholerae*[c]	75
$HA_{1,2}$ + soluble NA[c]	50
$HA_{1,2}$ + NA + anti-NA	5
NA	0

[a]Cleaved HA.

[b]Uncleaved HA.

[c]Neuraminidases containing 0.2 unit of enzyme activity/ml of liposome-cell mixture were used.

cellular membranes. The essential role of neuraminidase in the fusion process could be further demonstrated by the inhibition of fusion by antineuraminidase antibodies.

Correspondingly, neuraminidase activity is also necessary for fusion of paramyxoviruses as demonstrated with Newcastle disease virus (NDV). This could be shown by experiments where liposomes were used containing glycoproteins of orthomyxoviruses and the F-protein of NDV.[11] The results of such experiments are shown in Table 2. As to be expected liposomes containing both the HN and the F protein of paramyxoviruses could fuse with cells. Cell adsorption still took place, when HN was replaced by the cleaved or uncleaved HA of orthomyxovirus. However, no fusion could be observed and there was extensive binding of liposomes to the cell surface. The liposomes fused when they contained the uncleaved HA and the neuraminidase of orthomyxovirus and the F-glycoprotein of paramyxovirus. In this case, uncleaved hemagglutinin was required since cleaved HA in the presence of neuraminidase already induced membrane fusion as mentioned above. Liposomes containing only the HN or the F-glycoprotein are known to possess no fusion activity.[11]

The mechanism underlying the cooperative effect of neuraminidase in membrane fusion and thereby in virus penetration is not known. It might be that myxovirus infection is a two-step process. It has long been known that, in the initial phase of infection, myxoviruses react with the negatively charged neuraminic acid-containing cellular receptors. The active site of the viral surface involved in this reaction is probably located on the HN-glycoprotein of paramyxoviruses, or the HA_1 fragment of the hemagglutinin of influenza virus. It is possible that after adsorption, the N-termini of F_1 or HA_2 bring the viral envelope in closer contact with the host cell membrane. We propose that neuraminidase unmasks a new receptor which could react now with the hydrophobic N-termini of F_1 or HA_2, or the hemagglutinin has to unbind from neuraminic acid containing receptors so that HA_2 can react with the cell surface.

Table 2. Fusion Property of Liposomes Containing the Mixed Glycoproteins of Ortho- and Paramyxoviruses.

Liposomes containing	**Percentage fusion**
HN + F	80
$HA_{1,2}$ + F	0
HA + F	0
HA + NA + F	60

For details see Huang et al.[10]

Table 3. pH-Dependence of Hemolysis[a] Induced by Different Viruses.

Viruses	pH 5.00	5.25	5.50	5.75	6.00	6.25	6.50	6.75	7.00
	Extinction[b] measured at 540 nm								
Influenza viruses									
A/FPV/Rostock/34 (FPV; Hav 1N1)	0.45	0.60	0.60	1.00	0.91	0.05	0	0	0
A/chicken/Germany/49 (Virus N; Hav 2Neq1)	1.00	0.85	0.13	0	0	0	0	0	0
A/equine/Miami/1/68 (Equi-2; Heq2Neq2)	0.96	1.00	0.91	0.10	0	0	0	0	0
A/Hong Kong/1/68 (Hong Kong; H3N2)	1.00	0.47	0.20	0 0	0	0	0	0	0
A/swine/1976/31 (Swine; Hsw1N1)	0.85	0.85	1.00	0.21	0	0	0	0	0
A/PR/8/34 (PR8; HON1)	0.80	0.91	1.00	0.28	0.08	0	0	0	0
A/Asia/M/57 (Asia; H2N2)	0.94	1.00	0.52	0.11	0	0	0	0	0
Paramyxoviruses									
Newcastle disease virus (NDV, strain Ulster)	0.11	0.20	0.64	0.81	1.00	1.00	0.98	0.90	0.80
Sendai virus (Sendai, strain Z)	0	0	0	0.05	0.20	0.32	0.61	0.85	1.00

[a]Purified virus containing about 400 hemagglutinin units were added to 2.5 ml of 1% chick erythrocytes in saline buffered with 0.1 M sodium acetate at various pHs and incubated at 37°C for 15 minutes. The mixtures were then centrifuged and supernatants were measured for hemoglobin at 540 nm.

[b]For the ease of comparison, the maximal extinction value for each virus was set at 1.00, and other values obtained from the same virus were calculated proportionally.

Orthomyxovirus-Induced Hemolysis and Cell Fusion

All data briefly presented above support the idea that ortho- as well as paramyxoviruses gain entry into cells by fusion of the viral envelope with the host cell membrane. Recent results demonstrate in addition that orthomyxoviruses, like paramyxoviruses, can cause hemolysis and cell fusion under suitable conditions.[12]

All orthomyxoviruses listed in Table 3 induced hemolysis within a narrow pH range when their hemagglutinin was present in the cleaved form. The pH-optimum of hemolysis varied from 5.0 to 5.75 depending on the viruses and hemolysis decreased sharply from maximal values with an increase of 0.5 pH units. In comparison, paramyxoviruses hemolyzed maximally at pH 6.25 or 7.00 and hemolysis remained at a high level within a wide pH range. Orthomyxoviruses possessing uncleaved HA caused no significant hemolysis.

In Figure 2, it can be seen that various viruses hemolyze with different efficiencies when hemolysis was measured at optimal pH conditions for each virus after adjustment to the same hemagglutinating activity. Figure 2 also

Figure 2. Hemolysis induced by ortho- and paramyxoviruses. Purified virus containing 10 HA units per 10 μl were added in increasing amounts to 2.5 ml of 1% chicken erythrocytes in saline buffered with 0.1 M sodium acetate to pH values which would cause maximal hemolysis for the viruses used (see Table 3). The mixtures were incubated at 37°C for 15 min and then centrifuged. The supernatants were measured for hemoglobin at 540 nm.

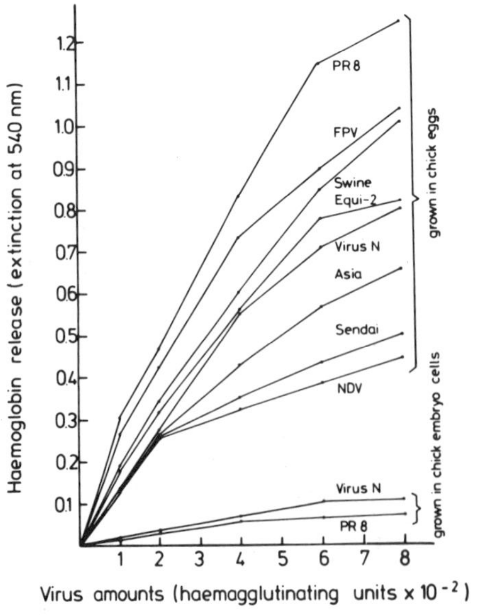

shows that, as a whole, influenza viruses hemolyzed more strongly than paramyxoviruses.

In addition to causing hemolysis, orthomyxoviruses also induce fusion of cells. This is shown in Figure 3 for fowl plague virus infected BHK cells which fused within 2 hours when exposed to the optimal pH for hemolysis at 6 hours after infection. Fusion was inhibited when cells were incubated 1 hour prior to the pH shift with specific anti-hemagglutinin or anti-neuraminidase antiserum. Fusion of other cells like chicken erythrocytes or fusion of heterologous cells, such as between chicken erythrocytes and chicken fibroblasts, also occurred readily in the presence of influenza viruses. Besides the critical pH, a proteolytically cleaved HA was also found to be a prerequisite for orthomyxovirus-induced cell fusion.

These results indicate that orthomyxoviruses are effective fusogens for cells and that for each of the ortho- and paramyxoviruses an optimal pH exists at which hemolysis and fusion of cells become most apparent. Recently it was reported that alphaviruses could also cause hemolysis and cell fusion at low pH.[13-15] The molecular basis of the pH-dependence of these processes is not clear. However, it appears reasonable to assume that it is virus-specific rather than cell-specific, since each virus has its specific pH optimum for hemolysis and cell fusion.

Figure 3. Fusion of BHK-21-F cells induced by orthomyxoviruses. Cells were infected with fowl plague virus (Hav1N1) (a) or with influenza virus strain PR8 (H0N1) (b) at a multiplicity of about 50 PFU per cell. After an adsorption period of 45 min, the inoculum was removed and replaced by Dulbecco's medium. After incubation at 37°C for 6 hours, the medium was replaced by citrate buffered saline (pH 5.75 or pH 5.50). The cultures were incubated for another 2 hours at 37°C and observed by phase contrast microscopy. Multiplication × 1000.

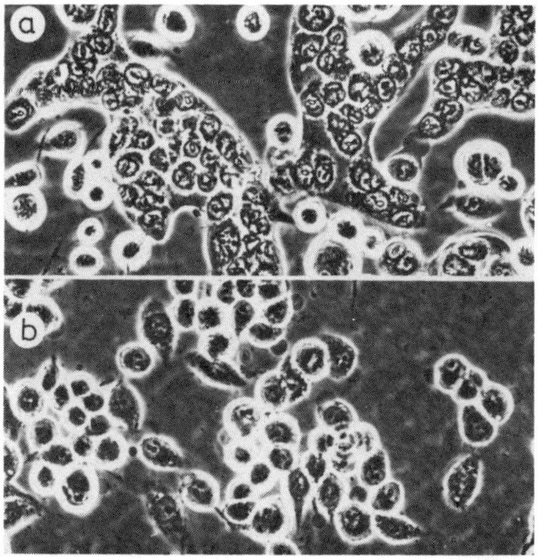

ACKNOWLEDGMENT
This work was supported by the Deutsche Forschungsgemeinschaft (Sonderforschungsbereich 47).

References

1. Klenk, H.-D. and Rott, R. (1980): *Curr. Top. Microbiol. Immunol.* 90:19–48.
2. Klenk, H.-D., Rott, R., Orlich, M., and Blodorn, J. (1975): *Virology* 68:426–439.
3. Lazarowitz, S.G. and Choppin, P.W. (1975): *Virology* 68:440–454.
4. Klenk, H.-D., Garten, W., Kohama, T., Huang, R.T.C., and Rott, R. (1980): *Biochem. Soc. Transact.* 8:419–422.
5. Richardson, C.D., Scheid, A., and Choppin, P.W. (1980): *Virology* 105:205–222.
6. Nicolau, C., Klenk, H.-D., Reimann, A., Hildenbrand, K., and Bauer, H. (1978): *Biochim. Biophys. Acta* 511:83–92.
7. Kurrle, R., Wagner, H., Rollinghoff, M., and Rott, R. (1980): *Eur. J. Immonol.* 9:107–111.
8. Huang, R.T.C., Wahn, K., Klenk, H.-D., and Rott, R. (1979): *Virology* 97:212–217.
9. Huang, R.T.C., Wahn, K., Klenk, H.-D., and Rott, R. (1980): *Virology* 101:294–302.
10. Huang, R.T.C., Rott, R., Wahn, K., Klenk, H.-D., and Kohama, T. (1980): *Virology* 107:313–319.
11. Hosaka, Y. (1975): In: *Negative Strand Viruses,* Mahy, B.W.J. and Barry, R.D. (Eds.), Academic Press, London, pp. 885–903.
12. Huang, R.T.C., Rott, R., and Klenk, H.-D. (1981): *Virology,* submitted.
13. Väänänen, P. and Kääriäinen, L. (1979): *J. Gen. Virol.* 43:553–601.
14. Väänänen, P. and Kääriäinen, L. (1980): *J. Gen. Virol.* 46:467–475.
15. White, J. and Helenius, A. (1980): *Proc. Natl. Acad. Sci. U.S.A.* 77:3273–3277.

David H. L. Bishop and Richard W. Compans, eds.
The Replication of Negative Strand Viruses

The Ratios of Influenza Virus Complementary RNA segments as a Function of Time after Infection

Marcel W. Pons[a]

A number of workers have reported that the intracellular synthesis of influenza virus polypeptides is under temporal control.[1-9] In most cases the evidence suggested that this control was mediated at the transcriptional level,[4, 10-12] and that the host cell had a modulating effect on this transcriptional control.[5, 13] In most cases, these analyses were done by determining the amounts of viral polypeptides synthesized as a function of time after infection. Inhibitors of polypeptide and/or RNA synthesis were used to estimate whether the temporal control of polypeptide synthesis was at the transcriptional or translational level. As stated above, it was concluded control was at the transcriptional level. Recently, however, Tekamp and Penhoet (1980), using quantitative hybridization techniques, showed that the variation in levels of individual mRNAs did not reflect the variation in levels of the viral polypeptides which they coded for.[14] These workers concluded control mechanisms operated at both transcriptional and translational levels.

From the experiments to be described here, we have concluded that (1) with the WSN strain of influenza grown in chick embryo fibroblast monolayers (CFM) there is no temporal control of mRNA synthesis, (2) the ratios of template (tp) RNA segments present within the cell are not those of the RNA within virions, possibly indicating a control or selective mechanism for packaging of RNA segments within mature virions, and (3) the polyacrylamide gel electrophoresis (PAGE) patterns of tp and mRNAs obtained from CFM and MDBK cells infected with the WND strain of virus are different, indicating again the influence of the host cell.

[a] The Christ Hospital Institute of Medical Research, 2141 Auburn Avenue, Cincinnati, Ohio.

Determination of the Amounts of m and tp RNAs Present in Virus Infected CFM

The basic protocol for these experiments was the same. Confluent primary CFM cultures were infected with the WSN strain of influenza virus at multiplicities of 0.5 to 1.0 pfu/cell. After a 30-minute adsorption period at room temperature, medium was added which contained 100 μCi/ml $^{32}PO_4$ or 50 μCi/ml 5,6-^{3}H-uridine and the cultures incubated at 37°. At the indicated intervals, usually 45 min, 1½, 2¼, 3, and 3¾ hr, cultures were removed and processed to obtain the polyadenylated and non-polyadenylated RNAs as described earlier.[15] These RNAs were annealed with approximately a 100-fold excess of purified vRNA;[15] after annealing, the hybrids were treated with 30 units S1 nuclease for 2 hours, after which they were reextracted with phenol, chromatographed on CF11 cellulose and analyzed by PAGE as described elsewhere.[16]

Figure 1 is a PAG electropherogram where aliquots of each sample were applied to the gel; consequently, since the amounts of material vary so greatly between the early and late intervals, one must choose between showing the early vs late samples by the length of film exposure. We chose the latter in this case. Tracks 1 to 5 are the polyadenylated RNAs and 6 to 10 are the non-polyadenylated RNAs. Another way to show the amounts of each segment present is to adjust the concentration of each sample so that the same amount of radioactivity is present in each. Figure 2 depicts such an electropherogram. Clearly, each segment is present as early as 45 minutes after infection and there is no evidence for any one or more segments being synthesized early versus others being synthesized later. This is more easily seen by determining the molar ratios of each segment at each time interval. Table 1 shows the results of such an analysis. Clearly, within the limits of experimental error, there is no evidence for any segment being synthesized in a greater or lesser amount at different times after infection. In addition, the ratios of segments relative to one another during any given interval is approximately the same at the different intervals. These values are close to those reported by Tekamp and Penhoet (1980)[14] with the exception of segment 7, which is higher in this system.

Included in Table 1 is the mean of 15 separate determinations of the ratios of RNA segments obtained from purified virions. The segments are not in the ratio of 1:1:1: etc. McGeoch et al. (1976)[17] showed relative molar ratios of segments of fowl plague virus to be closer to one than those shown here. Our determinations are quite close to those of McGeoch et. al.[17] for segments 4-8 but differ by approximately ½ for segments 1-3. This may reflect a greater proportion of incomplete virus particles in our preparations.

If one makes the admittedly large assumption that the ratio of template RNAs directly reflects the ratio of virion RNAs in the cell, one must conclude that the ratios of vRNA segments in virions is quite different from those in the cell. If the original assumption is correct, we can make the

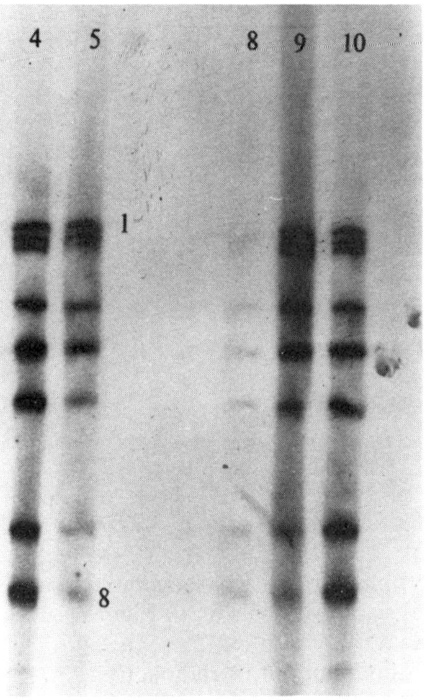

Figure 1. Polyacrylamide gel electrophoresis (PAGE) of influenza virus specific complementary RNAs isolated from chick fibroblast monolayers (CFM) at various times after infection. CFM were infected with the WSN strain of influenza virus and labeled with 5,6-^{3}H-uridine. At ¾, 1½, 2¼, 3, and 3¾ hours, cells were harvested, RNA extracted and chromatographed on oligo(dT) cellulose columns. The polyadenylated and non-polyadenylated RNAs were isolated, annealed to an excess of vRNA obtained from purified virions, treated with S_1 nuclease, chromatographed on CF_{11} cellulose and analyzed by PAGE[15] by taking 20 μl aliquots of the final yields. Lanes 1–5 are the hybrids formed by polyadenylated RNAs isolated at ¾, 1½, 2¼, 3, and 3¾ hours, respectively. Lanes 6–10 are the hybrids formed by non-polyadenylated RNAs isolated at the same intervals.

further assumption that there is a selective mechanism which determines the number of each segment packaged or there is a control at the replication level.

Another interesting aspect of the data presented in Table 1 can be seen by multiplying the molar ratios of the segments by their molecular weights or by the relative differences in their molecular weights. Table 2 depicts the results of such an analysis. In this case, the molecular weight of segment 8 was taken as unity. Again, segments 1-3 are present in diminished amounts due, as suggested above, to the presence of incomplete virus. Segments 4-8 are rather similar, however, indicating that the amount of each mRNA segment present is directly related to its molecular weight. That is, there are 3 copies of segment 8 for each single copy of segment 4, the former being 1/3 the length of the latter.

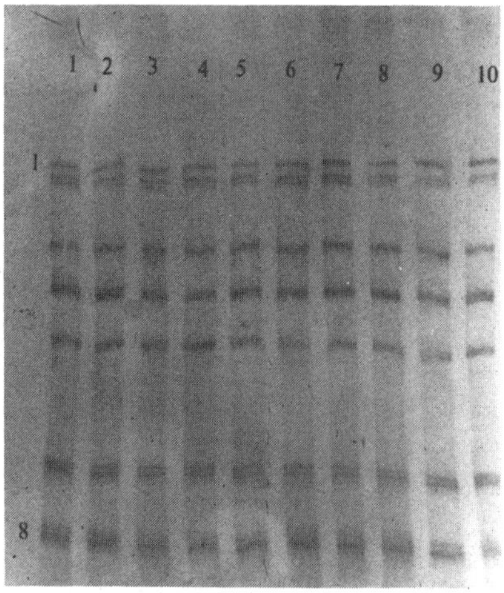

Figure 2. PAGE of influenza virus specific complementary RNAs isolated from CFM at various times after infection. These are the same samples as in Figure 1 with the same lane designations. The amount of radioactivity applied to each lane was adjusted by dilution so each lane received the same amount of radioactivity.

Table 1. Molar Ratios of Influenza Virus (WSN Strain) m and tp RNA segments at Various Times after Infection.

	Time after infection (hours)										
	3/4		1½		2¼		3		3-3/4		
Segment #	**m**	**tp**	**m**	**tp**	**m**	**tp**	**m**	**tp**	**m**	**tp**	**Virion**
1	.5	.6	.5	.4	.3	.3	.4	.4	.5	.4	.34
2	.5	.6	.5	.5	.3	.3	.4	.4	.5	.4	.80
3	.6	.6	.6	.5	.3	.3	.3	.4	.4	.3	
4	.8	.9	.8	.7	.8	.6	.8	.8	.9	.7	1.00
5	1.0	1.0	1.0	1.0	1.0	1.0	1.0	1.0	1.0	1.0	1.0
6	1.2	1.2	1.2	1.1	.8	1.1	1.0	1.0	.8	1.0	1.03
7	2.0	2.1	1.9	1.8	1.4	2.1	2.0	1.7	1.5	2.1	1.29
8	2.7	3.0	2.6	2.6	2.3	3.1	2.8	2.5	1.7	3.5	1.30

Molar ratios were calculated by determining the amount of radioactivity in each band by excision of the band and counting or by densitometer scanning and weighing the peaks. The values so obtained were divided by the molecular weights of each segment[18] and normalized to the NP segment (#5). Each value is the mean of values obtained from nine separate experiments, the values for virion RNA from 15 experiments.

Table 2. Relative Amounts of mRNA segments present.

Segment #	Mol wt[a] × 10^6	Mol wt Ratio	Molar ratio[b]	Product
1	1.00	3.7	.4	1.48
2	.97	3.59	.4	1.44
3	.95	3.52	.3	1.06
4	.76	2.81	.8	2.25
5	.65	2.41	1.0	2.41
6	.60	1.78	1.0	1.78
7	.37	1.37	2.0	2.74
8	.27	1.00	2.8	2.80

[a]From Desselberger and Palese (1978).[18]

[b]From Table 1—mRNA at 3 hours.

The m and tp RNAs in WSN Virus Grown in MDBK Cells

Since the results presented above are somewhat at variance with those reported by others, an attempt was made to determine the host cells' influence on complementary RNA synthesis. MDBK cells were infected with the WSN strain of influenza virus which had been passaged two times in MDBK cells. This same virus was also used to infect CFM. A third set was CFM infected with virus only passaged in CFM. After a two-hour incubation period in medium containing 25 μCi/ml 5,6-^{3}H-uridine the cells were harvested and mRNA and tp RNA isolated. The samples were annealed with an excess of vRNA obtained from CFM grown virus treated with S1 nuclease, purified by chromatography, and analyzed by PAGE.

The results are shown in Figure 3. Lanes 1-3 are respectively, mRNAs from MDBK virus in MDBK cells, MDBK virus in CFM, CFM virus in CFM. Lanes 4-6 are the tp RNA counterparts of the above while lanes 7-9 are identical to the samples in lanes 1-3 but treated with 2 μl of a mixture of 10 μg/ml RNase A-1 μg/ml RNase T1 for 5 min just prior to electrophoresis. Although we emphasize the preliminary nature of these results, it seems clear that the m and tp RNAs from MDBK virus in MDBK cells are different from those grown in CFM. The nature of this difference is unknown but could represent a decreased fidelity of transcription in MDBK cells. In any case, the host cell did influcnce the results obtained and this may explain partially why different laboratories have reported different results.

Conclusions

The data presented above have been interpreted to mean that there is no temporal control on transcription of influenza virus grown in primary CFM. Secondly, preliminary evidence indicates that the host cell influences transcription in some way which is still unclear.

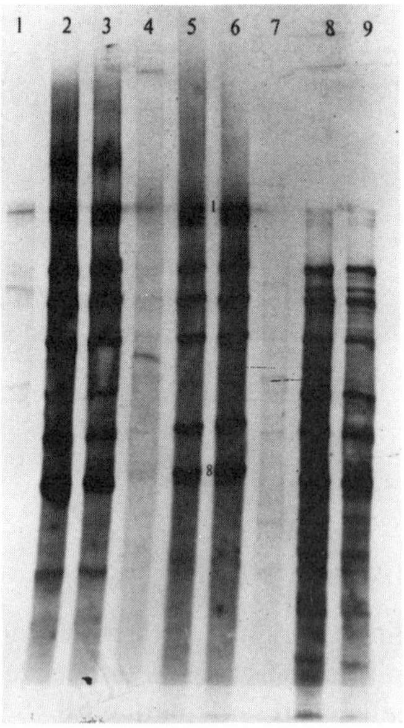

Figure 3. PAGE of influenza virus specific complementary RNAs isolated from CFM and MDBK cells. Cells were infected with influenza virus (WSN strain) and harvested at 2 hours after infection. Polyadenylated and non-polyadenylated RNA were isolated, annealed to vRNA obtained from virus grown in CFM and analyzed by PAGE as described in Figure 1. Virus grown for 2 passages in MDBK cells was used to infect MDBK cells, lane 1 is the polyadenylated RNA, lane 4 the non-polyadenylated RNA. The same virus as above used to infect CFM, lane 2 polyadenylated lane 5, non-polyadenylated RNAs. Virus passaged only in CFM was used to infect CFM. Lane 3 is polyadenylated, lane 6 non-polyadenylated RNAs. Lanes, 7, 8, 9 are the same samples as 1, 2, and 3 but treated with 2 μl of a 10 μg/ml RNase A-1 μl/ml RNase T_1 mixture for 5 minutes just prior to electrophoresis.

ACKNOWLEDGMENT
This work was supported in part by Grant AI-14225 of the National Institute of Allergy and Infectious Diseases U.S. Public Health Service.

References

1. Hay, A.J. (1964): *Virology* 60:398–418.
2. Inglis, S.C., Carroll, A.R., Lamb, R.A., and Mahy, B.W.J. (1976): *Virology* 74:489–503.
3. Klenk, H.-D. and Rott, R. (1973): Formation of influenza virus proteins. *J. Virol.* 11:823, 831.
4. Lamb, R.A. and Choppin, P.W. (1976): *Virology* 74:504–519.
5. Lamb, R.A. and Choppin, P.W. (1978): *Negative Strand Viruses and the Host Cell.* Mahy, B.W.J. and Barry, R.D. (Eds.), Academic Press, London, pp. 229–237.
6. Lazarowitz, S.G., Compans, R.W., and Choppin, P.W. (1971): *Virology* 46:830–843.

7. Meier-Ewert, H. and Compans, R.W. (1974): *J. Virol.* 14:1083–1091.

8. Skehel, J.J. (1972): *Virology* 49:23–26.

9. Skehel, J.J. (1973): *Virology* 56:394–399.

10. Etkind, P.R., Buchagen, D.L., Herz, C., Broni, B., and Krug, R.M. (1977): *J. Virol.* 22:346–352.

11. Hay, A.J., Lomniczi, B., Bellamy, A.R., and Skehel, J.J. (1977): *Virology* 83:337–355.

12. Inglis, S.C. and Mahy, B..J. (1979): *Virology* 95:154–164.

13. Bosch, F.X., Hay, A.J., and Skehel, J.J. (1978): *Negative Strand Viruses and the Host cell.* Mahy, B.W.J. and Barry, R.D. (Eds.), Academic Press,London, pp. 465–474..

14. Tekamp, P.A. and Penhoet, E.E. (1980): *J. Gen. Virol.* 47:449–459.

15. Pons, M.W. and Rochovansky, O.M. (1979): *Virology* 97:183–189.

16. Pons, M.W. (1980): *Virology* 100:43–52.

17. McGeoch, D., Fellner, P., and Newton, C. (1976): *Proc. Natl. Acad. Sci. U.S.A.* 73:3045–3049.

18. Desselberger, U. and Palese, P. (1978): *Virology* 88:394–399.

Copyright 1981 by Elsevier North Holland, Inc.
David H. L. Bishop and Richard W. Compans, eds.
The Replication of Negative Strand Viruses

Heterogeneity of Influenza Viral Polypeptides During Productive and Abortive Infections

Michael Schrom and Lawrence A. Caliguiri [a]

Introduction

Influenza virus produces an abortive infection in HeLa cells and the virus yield from these cells is less than 1% of that from permissive cell lines such as MDBK cells. Previous results from this laboratory have shown that in the absence of defective interfering (DI) particles in the inoculum, the restriction of influenza virus in HeLa cells is due to a defect in virus maturation resulting in the accumulation of budding virus particles on the plasma membrane and viral products in the cytoplasm.[1] The synthesis of viral components and transport of viral specific ribonucleoproteins (RNPs) are similar in abortively infected HeLa cells and productively infected MDBK cells.[2]

There are several possible explanations for the defect in virus maturation and the accumulation of aberrant elongated virions on the surface of HeLa cells. It is clear that the assembly of enveloped viruses requires involvement of host cell components, as well as interaction of viral RNPs with a modified cell membrane. The viral RNP appears to bind specifically to regions of the cell membrane modified by the insertion of viral glycoproteins and the presence of the nonglycosylated matrix (M) protein on the inner surface of the cell membrane. The restriction in viral maturation in HeLa cells may be the result of a defect in a viral or a host cell component required for the complex process of maturation to proceed.

The results we report in this paper show that there is a defect in the phosphorylation of viral proteins in HeLa cells which appears to be host

[a] Department of Microbiology and Immunology, Albany Medical College of Union University, Albany, New York.

dependent. NP and NS_1 proteins are phosphorylated in productively infected MDBK cells, but are not phosphorylated in HeLa cells. Since the NP protein is the major structural component of viral RNPs, the lack of phosphorylation of this protein may account for the aberrant maturation of influenza virus in HeLa cells.

Time Course of Synthesis and Phosphorylation of Viral Polypeptides During Productive Infection

The growth of influenza virus Ao/WSN (HON1) in MDBK cells has been reported to yield a population relatively free of DI particles.[3] Such virus was used as inoculum during a single-cycle infection of MDBK cells and viral protein synthesis was examined at various times after infection. Figure 1 shows the synthesis of NP and NS_1 by 4 hours postinfection in permissive cells. In cells labeled with $^{32}PO_4$, the phosphorylation of NP and NS_1 can be observed. The degree of phosphorylation of NP and NS_1 increased with

Figure 1. Time course of synthesis and phosphorylation of viral polypeptides during permissive infection. MDBK cells were infected with influenza virus Ao/WSN (HON1) at a multiplicity of 30-50 PFU/cell. The cells were labeled with ^{35}S-methionine for 15 min or ^{32}P-orthophosphate for 2 hr in the absence of either methionine or inorganic phosphate, respectively, ending at the indicated times in hours after infection. The cells were subjected to PAGE as described by Laemmli[11] using a gradient of acrylamide of 7-15% and autoradiographed. The patterns obtained by labeling uninfected cells with either ^{35}S-methionine or $^{32}PO_4$ are indicated in the lanes marked "U". The positions of the viral polypeptides are indicated.

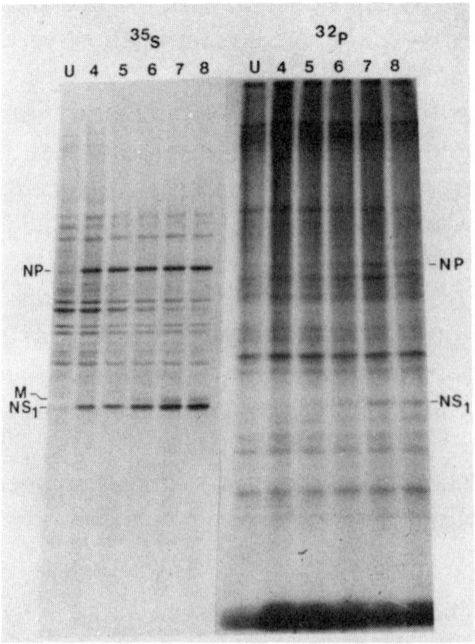

increasing time after infection (Figure 1). This pattern of phosphorylation of NP and NS_1 is consistent with that seen in productively infected cells as reported by others.[4,5]

Synthesis and Phosphorylation of Viral Polypeptides During Abortive Infection

When HeLa cells were infected in the same manner and viral protein synthesis examined at various times after infection, significant differences in phosphorylation of viral proteins were apparent (Figure 2). The viral NP and NS_1 proteins were first detected at 4 hours postinfection and increased during the 8 hours of observation. Inhibition of cellular polypeptide synthesis was apparent by 8 hours after infection in HeLa cells. The most striking difference in the synthesis of viral proteins in HeLa and MDBK cells was the absence of $^{32}PO_4$ incorporation into viral proteins in HeLa cells (Figure 2). This suggests that the phosphorylation of the viral proteins is host cell dependent and may be related to the maturation of the virus particle. As we have reported previously, the block in virus replication in HeLa cells appears to be at a late stage in virus maturation.[1] It was therefore of interest to determine whether the small quantity of virus produced by HeLa cells contained phosphorylated NP.

Figure 2. Synthesis and phosphorylation of viral polypeptides during abortive infection. HeLa cells were infected and labeled as described for MDBK cells in Figure 1 and analyzed by PAGE and autoradiography.

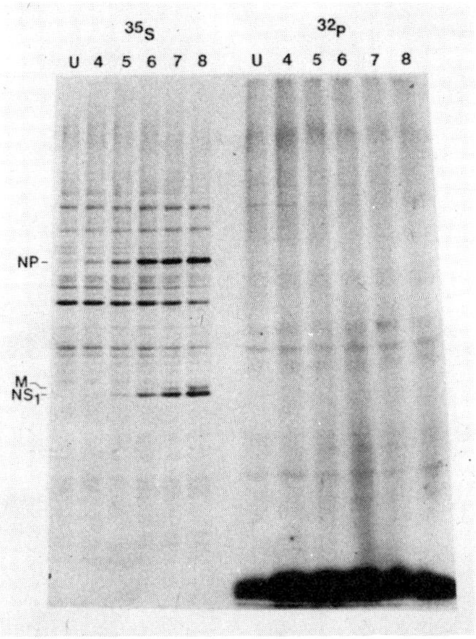

Phosphorylation of Viral Proteins in Virus Released During Productive and Abortive Infection

When the virus released from HeLa and MDBK cells 24 hours after infection is analyzed by polyacrylamide gel electrophoresis (PAGE), the viral proteins in released virus from MDBK cells were easily distinguished (Figure 3A). The NP polypeptide was the only virion protein observed to contain $^{32}PO_4$ label (Figure 3B). Although many cellular proteins were observed to contain $^{32}PO_4$, the small amount of virus released from HeLa cells did not contain any phosphoproteins which comigrated with the NP polypeptide (Figure 3C).

These results suggest that phosphorylation of viral proteins is important for the efficient maturation and release of influenza virus. However, small amounts of infectious virus may be released from HeLa cells despite the lack

Figure 3. Phosphorylation of viral proteins in virus released during productive and abortive infections. MDBK and HeLa cells were infected at a multiplicity of 30-50 PFU/cell. Two hours after infection replicate cultures were labeled with either ^{35}S-methionine or ^{32}P-orthophosphate as indicated. Twenty-four hours after infection the media were collected and clarified by low speed centrifugation. The released virus was pelleted at 70,000 rpm for 90 min in the Beckman SW50.1 rotor. The pelleted virus was subjected to PAGE and autoradiography as described in Figure 1. (A) ^{35}S-methionine-labeled MDBK released virus. (B) $^{32}PO_4$-labeled MDBK released virus. (C) $^{32}PO_4$-labeled HeLa released virus. The positions of the viral polypeptides are indicated.

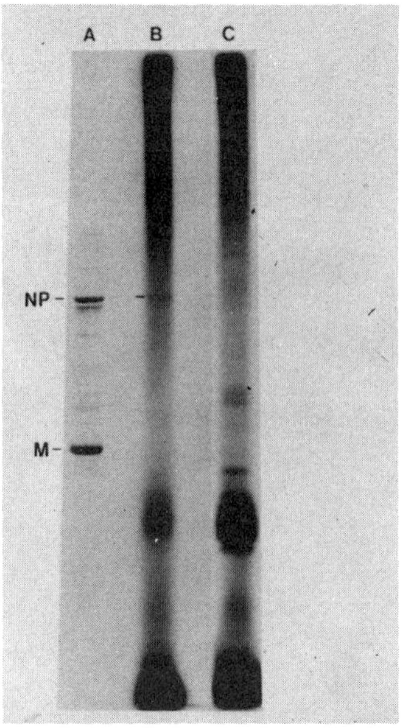

of phosphorylation. Thus, phosphorylation of NP is not required for infectivity.

It is conceivable that host-dependent modifications, in addition to the failure to phosphorylate NP, may account for the aberrant maturation of influenza virus in HeLa cells. Preliminary results of one-dimensional peptide mapping[6] (not shown) suggest that qualitatively similar methionine-containing NP peptides are generated from both HeLa and MDBK cells.

Conclusions

We have previously suggested that there is a host cell defect in the final stages of virus maturation which results in a block in the release of influenza virus from HeLa cells.[1] The present study has described the failure of HeLa cells to phosphorylate either of the two phosphoproteins (NP and NS_1) which are influenza gene products. Neither the cell-associated, nor the released NP polypeptide, appears to be phosphorylated. Since influenza viruses are not known to contain any protein kinase activities, it can be assumed that the failure to phosphorylate the viral proteins is the result of the lack of a specific cellular protein kinase. This is further emphasized by the fact that the defect is dependent upon the host cell. The NP polypeptide, as the major structural protein of the viral ribonucleoprotein (RNP) complex, is closely associated with the viral RNA and this association appears to occur normally as evidenced by the apparently normal synthesis and transport of the RNP complexes in HeLa cells.[1] However, an altered charge by virtue of the lack of a phosphate group may alter its association with the matrix protein (M) on the inner surface of the cytoplasmic membrane and thus affect the formation of the virus particle.

Phosphorylation of viral proteins is a posttranslational modification frequently observed in other virus-cell interactions.[7-10] Our results suggest a possible role for phosphorylated NP in the assembly process of influenza virus, but leave open the functional significance of the phosphorylation of NS_1. If NS_1 is a regulatory protein, then a charge alteration could have important implications for its activity. Further work is required to elucidate the biological activity of NS_1.

ACKNOWLEDGMENTS
This work was supported by a Grant from the New York State Health Research Council (No. 1737). During the course of this work, M.S. was supported by fellowships from the American Lung Association and the National Science Foundation (No. SPI-7914894). We thank Marilyn Schmidt for expert technical assistance and Kathleen Cavanagh for expert typing.

References

1. Caliguiri, L.A. and Holmes, K.V. (1979): *Virology* 92–15.

2. Caliguiri, L.A. and Gerstein, H. (1978): In: *Negative Strand Viruses and the Host Cell.* Barry, R.D. and Mahy, B.W.J. (Eds.), Academic Press, N.Y., pp. 483–492.
3. Choppin, P.W. (1969): *Virology* 39:130.
4. Privalsky, M.L. and Penhoet, E.E. (1977): *J. Virol.* 24:401.
5. Privalsky, M.L. and Penhoet, E.E. (1978): *Proc. Natl. Acad. Sci. U.S.A.,* 75:3625.
6. Cleveland, D.W., Fischer, S.G., Kirschner, M.W., and Laemmli, U.K. (1977): *J. Biol. Chem.* 252:1102.
7. Goldberg, A.R., Krueger, J.G., and Wang, E. (1980): *Cold Spring Harbor Symp. Quant. Biol.* 44:991.
8. Collett, M.S. and Erikson, R.L. (1978): *Proc. Natl. Acad. Sci. U.S.A.,* 75:2021.
9. Clinton, G.M., Burge, B.W., and Huang, A.S. (1978): *J. Virol.* 27:340.
10. Kingsford, L. and Emerson, S.U. (1980): *J. Virol.* 33:1097.
11. Laemmli, U.K. (1970): *Nature* 227:680.

David H. L. Bishop and Richard W. Compans, eds.
The Replication of Negative Strand Viruses

The Mechanism of Initiation of Influenza Viral RNA Transcription by Capped RNA Primers

Robert M. Krug,[a] Stephen J. Plotch,[a] Ismo Ulmanen,[a] Carolyn Herz[a] and Michele Bouloy[b]

Introduction

Capped eukaryotic RNAs act as primers for the synthesis of influenza messenger RNA (mRNA) *in vitro* and transfer their 5′-terminal methylated cap structure and a short stretch of nucleotides (about 10-15) to the viral mRNA.[1-4] A similar process apparently also occurs *in vivo,* because the viral mRNAs synthesized in the infected cell contain a short stretch of nucleotides at their 5′ end, including the cap, that are not viral-coded.[5-7]

As an initial approach for determining the mechanism for this priming reaction, we previously identified those bases of a representative primer, β-globin mRNA, that were transferred to the viral RNA transcripts. Using ^{125}I-labeled globin mRNA as primer for *in vitro* transcription, we found that the predominant sequence at the 5′ end of each viral mRNA segment was identical to the first 13 nucleotides (plus the cap) at the 5′-terminus of β-globin mRNA[4], which has the sequence: $m^7Gpppm^6{}_1AmC(m)ACUUG$-$CUUUU_{13}GAC$. . . .[8] Because only the C residues were labeled with ^{125}I, these results indicated that either the first 12, 13 or 14 5′-terminal bases of β-globin mRNA were transferred to the viral mRNAs. Analysis of the minor ^{125}I-labeled oligonucleotides found in the viral mRNAs indicated that shorter, 5′-terminal fragments of β-globin mRNA (8-11, or 2-3 bases in length) were sometimes transferred and that the transferred pieces were most likely linked to G as the first base incorporated by the transcriptase. Evidence for G being the first base added during viral RNA transcription *in vivo* has also been obtained.[7]

[a] Molecular Biology and Genetics Unit of the Graduate School, Memorial Sloan-Kettering Cancer Center, New York, N.Y.
[b]Present address: Institut Pasteur, 75015 Paris, France.

The priming mechanism must involve recognition of the 5′-terminal methylated cap structures ($m^7GpppXm$), because only RNAs containing a cap are active as primers,[1-3,9] Removal of the m^7G of the cap by chemical or enzymatic treatment eliminates all priming activity, and this activity can be restored by enzymatically recapping the RNA.[2] The absence of either methyl group from the cap of an RNA greatly reduces its priming activity,[3] and the absence of both methyl groups completely eliminates activity.[9]

These and other experiments (to be discussed here) suggested to us a possible mechanism for priming, for which we will provide support here.

Results and Discussion

Priming Activity of an RNA does not Require Hydrogen-bonding with the Template Virion RNA

In terms of understanding the mechanism of priming, it was necessary to determine whether the priming RNA must contain a sequence complementary to the common 3′ end (3′UCGU)[10,11] of the virion RNA (vRNA) templates. Two approaches to this problem were taken. In one, we determined whether 5′-terminal fragments of natural mRNAs lacking such a complementary sequence were active as primers.

When globin mRNA was partially digested with mild alkali and then subjected to sucrose density gradient centrifugation (Figure 1A), priming activity was found in two regions of the gradient: (1) a relatively small amount of activity associated with RNA sedimenting with intact globin mRNA; and (2) a large amount of activity associated with molecules sedimenting less than 4S. Thus, fragments of globin mRNA were effective primers. However, there was about a 30% loss of total priming activity after fragmentation with mild alkali. This was almost certainly due to ring-opening of the terminal m^7G by alkali,[12] because when globin mRNA was fragmented by partial ribonuclease T1 digestion, the resulting fragments sedimenting less than 4S were 4-8-fold more effective as primers on a molar basis than the intact globin mRNA. To determine the exact nucleotide length of the active 5′-terminal fragments, globin mRNA and 2′-0-methylated alfalfa mosaic virus (A1MV) RNA 4 (each labeled with ^{32}P in the cap) were fragmented by partial alkali digestion, and the fragments of various chain length from each of these RNAs were resolved by gel electrophoresis and tested for priming activity (Figures 1B and 1C). With both these RNAs, 5′ fragments as short as 14-23 nucleotides long were effective primers. In fact, with A1MV RNA 4, fragments of this size were more active per 5′-termini than larger-size fragments. Since 5′-terminal fragments 14-23 nucleotides long from either A1MV RNA 4 or β-globin mRNA do not contain a sequence complementary to the 3′ end of influenza vRNA,[8,13] the observed stimulation by these fragments cannot result from hydrogen-bonding between them and the 3′ end of vRNA.

The same conclusion was obtained using capped ribopolymers as primers.[14] Capped poly A and capped poly AU, neither of which contain a

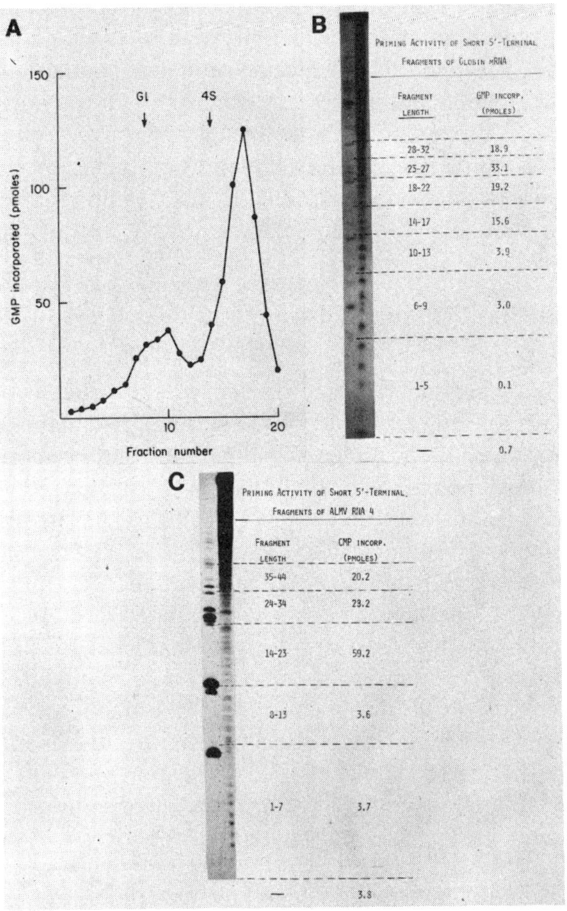

Figure 1. Priming activity associated with alkali-generated fragments of globin mRNA (A and B) and 2'-0-methylated A1MV RNA 4 (C). (A): Globin mRNA was subjected to mild alkali digestion (50°C mM Na_2CO_3, pH 8.5, 55 C 30 min) and sucrose density gradient centrifugation (SW 50.1 rotor for 14 hr). An aliquot of each gradient fraction was assayed for 1 hr at 31°C in a transcriptase reaction with (α-^{32}P) GTP as labeled precursor (●—●). Sedimentation markers: G1, intact globin mRNA; 4S, *E. coli* tRNA. (B): A mixture of unlabeled and ^{32}P-cap-labeled globin mRNA was partially digested with mild alkali and then subjected to electrophoresis on a 20% gel (right lane). The gel slices containing the indicated fragments were eluted, and each of the fragment preparations was assayed in a transcriptase reaction for 1 hr at 31°C with 8-^{3}H-GTP as labeled precursor. The left lane of the gel shows the partial T1 ribonuclease digest of the ^{32}P-labeled globin mRNA. (C): A mixture of unlabeled and ^{32}P-cap-labeled A1MV RNA 4 was partially digested with mild alkali and subjected to electrophoresis on a 20% gel (right lane). Each of the fragment preparations was assayed in a transcriptase reaction with (α-^{32}P)CTP as labeled precursor. The left lane shows the partial U2 ribonuclease digest of the ^{32}P-labeled A1MV RNA 4.

sequence complementary to the 3′ end of vRNA, were about as effective primers as globin mRNA. Some, though lower, activity was seen with capped poly U, even though the 3′-terminal 12-nucleotide sequence common to the eight influenza vRNA segments contain no A residues.[10,11] Of the various polymers tested as primers, capped poly C was the least effective. Thus, while there is some effect of sequence on the efficiency of transcriptase priming by an exogenous capped ribopolymer (see later) there is no requirement for the presence of a sequence complementary to the 3′ end of the vRNA template.

Postulated Mechanism for the Priming of Influenza Viral RNA Transcription

Based on our results, we can postulate a mechanism for the priming of influenza viral RNA transcription by capped RNAs (Figure 2). The capped RNA is cleaved by a virion-associated nuclease to generate one or more 5′-terminal fragments. In the example shown, β-globin mRNA is cleaved at the G residue 13 nucleotides from the 5′ end. Some, or all, of these fragment(s) are the actual primers initiating transcription. In the absence of hydrogen-bonding between the primer and vRNA, the stimulation of initiation would result from a specific interaction between the capped RNA fragment and one or more proteins in the transcriptase complex. This specific interaction presumably requires recognition of the 5′-terminal methylated cap structure of the primer. As a result of this interaction, the transcriptase acquires the ability to initiate transcription and links a G residue to the 3′ end of the 5′-terminal fragment generated by the nuclease. Elongation of the viral RNA transcripts would then follow. It is most likely that the incorporation of the initial G residue is directed by the 3′ penultimate C of the vRNA. As a

Figure 2. Postulated mechanism for the priming of influenza viral RNA transcription by β-globin mRNA and other capped RNAs.

SOURCE: *Plotch, Bouloy and Krug, submitted for publication.*

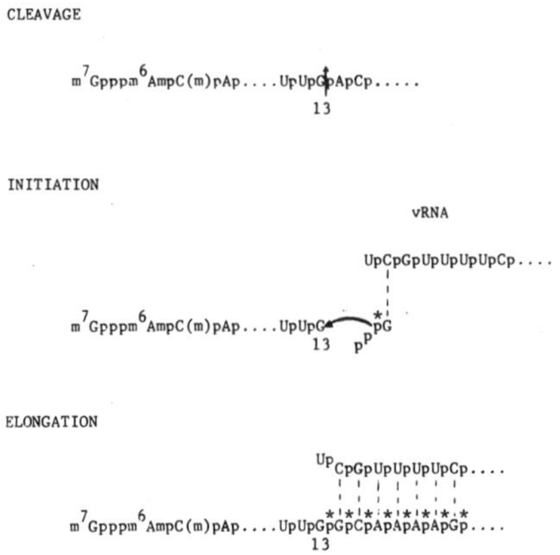

consequence, the viral RNA transcripts would not necessarily contain an A complementary to the 3′-terminal U of the template vRNA. In order for an A to be found opposite the 3′-terminal U of the vRNA template, the capped fragment that initiates transcription would need to have a 3′-terminal A.

This mechanism predicts the existence of an intermediate in the priming reaction: 5′-terminal fragment(s) cleaved from the mRNA primer. In addition, in a reaction in which the only triphosphate present is GTP, those 5′-terminal fragments that are actual primers initiating transcription should be linked to a G residue. Here, we will identify both these species, thereby providing strong support for the postulated mechanism of priming.

Identification of the 5′-Terminal Fragments that are the Actual Primers Initiating Transcription

Globin mRNA was incubated with detergent-treated virus in the presence of (α-^{32}P)GTP as the only ribonucleoside triphosphate, and the reaction products were analyzed by electrophoresis on 20% acrylamide gels containing 7M urea (Figure 3). Two major labeled bands were observed in the size range of about 15 nucleotides. Sequence analysis of band 2 indicated that it was the fragment resulting from cleavage at the G13 residue of β-globin mRNA[8] to which one labeled G residue was added, i.e., m^{7}Gpppm^{6}Am-CACUUGCUUUUGpG. This is the species predicted from the postulated mechanism. Thus, the 5′-terminal fragment of β-globin mRNA cleaved at G13 can be presumed to be the actual primer that initiates influenza viral RNA transcription. Band 1 was the same G13 fragment of β-globin mRNA with two G residues added, indicating that more than one G can be added to the primer fragment in the absence of the other ribonucleoside triphosphates. No fragments of α-globin mRNA with 3′-terminal labeled G residues were found, even though α-globin mRNA constituted 20% of the globin mRNA preparation.

Similar experiments were performed using either (α-^{32}P)ATP, (α-^{32}P)UTP, or (α-^{32}P)CTP as the only ribonucleoside triphosphate in the transcriptase reaction. With labeled ATP or UTP, no incorporation of label into fragments of globin mRNA was detected. With labeled CTP, however, terminally labeled fragments of both β- and α-globin mRNA were found. The β-globin mRNA species was the same G13 fragment with a single C residue added, and the α-globin mRNA species was the cleavage product at nucleotide G10[15] to which one C residue was added, i.e.,m^{7}Gpppm^{6}Am-CAUUCUGGpC. This C incorporation was most likely directed by the G residue at the third position from the 3′ end of vRNA (see Figure 2). To account for this, we presume that with β-globin mRNA, the G residue (G13) of the primer fragment lines up opposite the 3′ penultimate C of the vRNA template. With α-globin mRNA, the two G residues at the 3′ end of its G10 fragment would hydrogen-bond to the UC at the 3′ end of the vRNA (see Figure 2), thereby precluding G addition to the fragment and allowing only C addition.

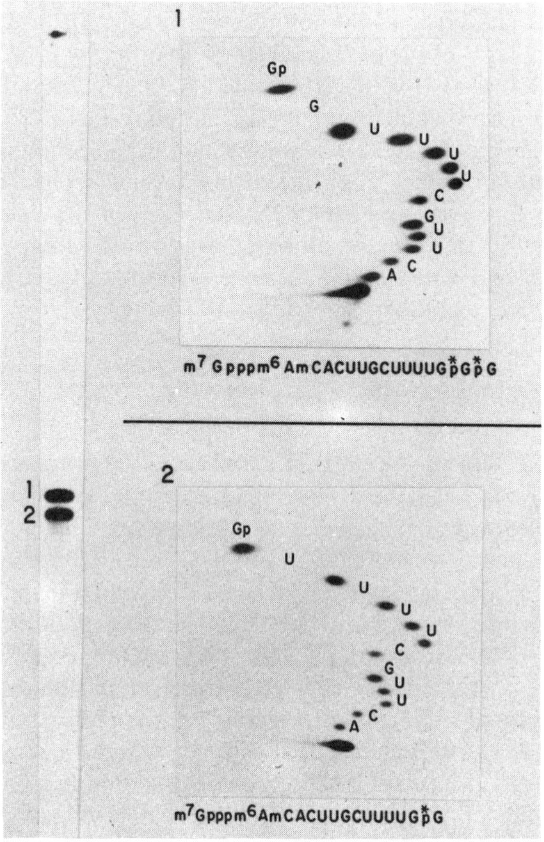

Figure 3. Identification and sequencing of the 5′-terminal fragments of β-globin mRNA that initiate influenza viral RNA transcription. Globin mRNA was incubated with detergent-treated virus in the presence of (α-^{32}P)GTP, and the RNA products were analyzed on a 20% acrylamide gel in 7M urea. The RNA in the two bands (1 and 2) was eluted, and after β-elimination,[2] was partially digested with sodium bicarbonate. The digest was analyzed by two-dimensional gel electrophoresis. The identity of the 3′-terminal nucleoside monophosphate (Gp in both RNAs 1 and 2) was determined by thin-layer chromatography of the RNase T2 digest of the RNA. The m^{7}Gpppm^{6}AmC sequence, which could not be directly determined, is based on the known cap structure of β-globin mRNA.[8]

SOURCE: *Plotch, Bouloy and Krug, submitted for publication.*

Consequently, we would predict that primer fragments of mRNAs which contain any nucleotide other than G at their 3′ ends should not be capable of lining up with the 3′ penultimate C of the vRNA template and hence should not be terminally labeled in the presence of only (α-^{32}P)CTP. These fragments should only be labeled in the presence of (α-^{32}P)GTP. This was shown to be the case, using as primers capped poly A, capped poly AU, and A1MV

RNA 4 (with the 5′-terminal sequence m^{7}GpppGmUUUUUAUUUUU-AAUUU... and no other G residue until position 39 from the 5′ end).[13] With these primers, the only base incorporated in single triphosphate reactions was G (and not C). In the presence of both unlabeled GTP and labeled CTP, however, C incorporation was detected. These results thus indicate that G and C incorporation are almost certainly directed by the second and third bases at the 3′ end of vRNA: C incorporation only occurs next to a G residue, which was either already present at the 3′ end of the primer fragment or was put there by prior incorporation.

The primer fragments derived from several capped RNAs were 10 to 13 nucleotides long (plus the cap) and almost always contained a purine (A or G) at their 3′ end. As noted above, the β-globin and α-globin mRNA primer fragments contained a 3′-terminal G residue. With the ribopolymer capped poly AU, all the primer fragments contained 3′-terminal A and not U, and with brome mosaic virus (BMV) RNA 4 the primer fragment was the cleavage product at the A10 residue,[16] m^{7}GpppGmUAUUAAUAA. Even with A1MV RNA 4, which has the extremely U-rich sequence at its 5′ end noted above, the predominant primer fragment (about 80% of the time) was the cleavage product at A13 (see above). In this case, a fraction (about 20%) of the primer fragments contained a 3′-terminal U residue. The ability to occasionally utilize U-terminated fragments would explain why capped poly U exhibits some priming activity. Nevertheless, with most cellular capped RNAs which contain a relatively even distribution of purines and pyrimidines near their 5′ end, primer fragments would contain almost exclusively purine (A or G) residues at their 3′ ends.

A Cap-Dependent Endonuclease Generates the 5′-Terminal RNA Fragments that Prime Transcription

In order to assay directly for an endonuclease generating 5′-terminal fragments from capped RNAs, we used as substrate several capped RNAs containing ^{32}P label in their 5′-terminal methylated cap structure, and incubated these RNAs with detergent-treated virus or purified viral cores in the absence of all four nucleoside triphosphates. When BMV RNA 4 containing a ^{32}P-labeled, cap 1 structure m^{7}G$\overset{*}{p}$ppGm was incubated in the absence of ribonucleoside triphosphates (Figure 4, lane 2), two major fragments were produced. The slower-moving of these two fragments was identified as the cleavage product at the G12 residue,[16] i.e., m^{7}GpppGmUAUUAAUA-AUG$_{12}$, and the faster-moving fragment was the cleavage product at A10. These cleavage products were shown to contain a 3′-hydroxyl group, as expected for primer molecules, and thus they migrate slower than the alkali-derived marker fragments of the same chain length, which contains a 2′ or 3′ phosphate group. No cleavage at U11 was detected, indicating a strong preference for cleavage at purines. Preferential cleavage at a purine

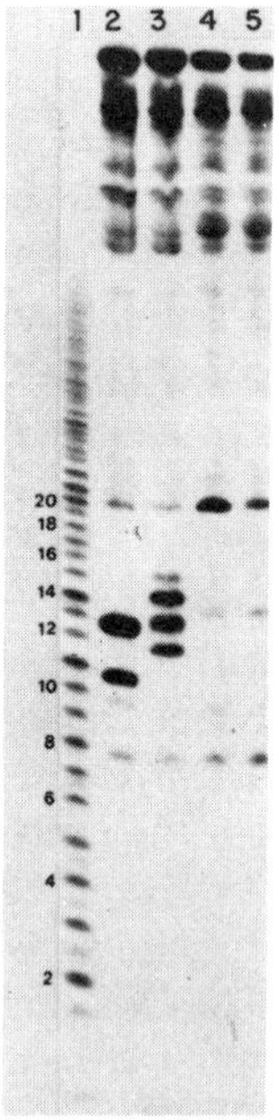

Figure 4. Identification of a cap-dependent endonuclease in influenza virions that cleaves BMV RNA 4 at specific positions near its 5′ end. BMV RNA 4 containing either a ^{32}P-labeled methylated cap structure $m^7G\overset{*}{p}ppGm$ (lanes 2 and 3) or a ^{32}P-labeled unmethylated cap structure $G\overset{*}{p}ppG$ (lanes 4 and 5) was incubated with detergent-treated virus in the absence of ribonucleoside triphosphates (lanes 2 and 4) or in the presence of 25 μM GTP (lanes 3 and 5). After incubation, the phenol-extracted RNA was analyzed by electrophoresis on a 20% acrylamide gel in 7M urea. Lane 1 is the partial alkali digestion product of BMV RNA 4 containing a m^7GpppGm cap, and the numbers on the left refer to the chain lengths of these products, counting the Gm residue as the first base.

SOURCE: *Bouloy, Plotch, Ulmanen and Krug, submitted for publication.*

was also seen when cap-labeled A1MV RNA 4 was incubated with virus in the absence of ribonucleoside triphosphates: in the extremely U-rich region at the 5′ end of this RNA, the predominant cleavage product was at the A13 residue. Thus, the primary reason that primer fragments contain almost exclusively a purine residue at their 3′ ends is that the endonuclease strongly prefers to cleave at purines.

It should be noted that with BMV RNA 4 (Figure 4, lane 2), the G12 fragment was generated in larger amounts than the A10 fragment, though as noted above, only the A10 fragment was utilized as primer. The utilization of the A10 fragments as primer can also be seen by adding unlabeled GTP to the reaction (Figure 4, lane 3). The A10 fragment disappeared, and three more slowly migrating bands were then predominently observed. These bands correspond to the A10 fragment with one, two or three G residues added.

However, these three bands contained more radioactivity than was present in the A10 fragment generated in the absence of GTP, and a band corresponding to the G12 fragment was not apparent in the presence of GTP. These results suggest that, in the presence of GTP, the G12 fragment was cleaved further to yield an additional A10 fragment that was then used as primer. This conclusion was verified by isolating the A10 and G12 fragments from a nuclease reaction (done in the absence of GTP) and then incubating each of these fragments in a subsequent reaction in the presence of GTP. The A10 fragment was used as a primer directly without further modification, indicating that the utilization of an RNA fragment as a primer does not have to be directly coupled to the nuclease reaction that generates that fragment. In contrast, in the presence of GTP, the G12 fragment was cleaved further to yield the A10 fragment which was then used as primer. This preferential utilization of the A10 fragment over the G12 fragment as primer reflects a generalized preference of A-terminated over G-terminated fragments as primers. Thus, we have found that the A-terminated fragments of AlMV RNA 4 and BMV RNA 4 were much more efficiently utilized for initiation (G-addition) than the G-terminated fragment of globin mRNA.

To determine whether the cleavage of the primer RNA is dependent on the presence of a methylated cap structure, we prepared BMV RNA 4 containing a 5′-terminal ^{32}P-labeled G$\overset{*}{p}$ppG blocking group. When this RNA was incubated with virus in the absence (Figure 4, lane 4) or presence (lane 5) of unlabeled GTP, the specific cleavages at nucleotides A10 and G12 did not occur. Instead, presumably non-specific cleavages, generating fragments that were not utilized to initiate transcription, occurred at nucleotides 7, 9, 13 and 20. Some of these cleavages were also seen in lanes 2 and 3 with the BMV RNA containing a methylated cap structure. Thus, a methylated cap structure in an RNA is required by the viral nuclease to generate those specific fragments that serve as primers for viral RNA transcription. This is the only known example of a cap-dependent endonuclease.

Conclusions

Our results have shown that the priming of influenza viral RNA transcription by capped RNAs occurs as shown in Figure 2. An endonuclease associated with the transcriptase complex cleaves the capped RNA predominantly, if not entirely, at a purine residue, A or G, 10 to 13 nucleotides from the 5′ cap. These specific cleavages require the presence of a methylated cap structure in the RNA substrate. After the specific cleavage, the transcriptase preferentially utilizes A-terminated fragments to initiate transcription via the initial incorporation of a G residue. The cleavage and initiation steps are not necessarily coupled, because an appropriate nuclease-generated fragment isolated from one reaction can without any apparent modification directly initiate transcription when added to a second reaction.

The initial G incorporated onto the capped primer fragment is most likely directed by the 3′ penultimate C of the vRNA template. The next base incorporated is a C residue, apparently directed by the third base (a G residue) at the 3′ end of vRNA. The G-linked fragments, therefore, can be presumed to be the intermediates which are subsequently elongated to form the viral mRNA. To verify this, we intend to sequence the 5′ ends of the complete viral mRNA molecules synthesized in the presence of all four ribonucleoside triphosphates using the primer RNAs discussed above. Actually, this has, in essence, already been done for β-globin mRNA in the experiment using ^{125}I-globin mRNA as primer in the presence of all four ribonucleoside triphosphates[4] (see Introduction).

As a consequence of the preference for the utilization of A-terminated over G-terminated fragments as primers, an AGC sequence would preferentially be generated in the viral mRNA, complementary to the UCG sequence at the 3′ end of the vRNA. We expect that this also occurs *in vivo*. At present, however, there are conflicting reports concerning whether the viral mRNA synthesized in the infected cell contains an A residue complementary to the 3′-terminal U of the vRNA.[7,17] It is important to resolve this issue because if an A residue is predominantly found at this position, then it is conceivable that the same initiation step could be used for the synthesis of both viral mRNA and the full-length transcripts that are the putative templates for vRNA synthesis.[18] Generation of the full-length transcripts would then involve removal of the host-derived sequence by nucleolytic cleavage at the 5′ side of the A residue, thereby leaving these transcripts with the proper 5′ AGC . . . sequence.

The 5′ methylated cap structure is recognized at the cleavage step of the reaction, but as mentioned at the outset it is likely that the cap is also recognized at the initiation step, where it would mediate the specific interaction with one or more transcriptase proteins that causes the stimulation of initiation. Because the cleavage and initiation steps are not necessarily coupled, it may be possible to determine whether the cap is indeed recognized at initiation, by using as primers 5′-terminal fragments of RNAs of the proper size and 3′-terminus, e.g., the A10 fragment of BMV RNA 4, but lacking methyl groups in the cap.

All the steps shown in Figure 5 are carried out by viral cores which contain four known viral proteins, the nucleocapsid (NP) protein and the three P proteins.[19,20] The NP protein, which comprises about 90 percent of the viral proteins in the core, probably has primarily a structural role, as it is situated along the vRNA chains at approximately 20 nucleotide intervals.[21] The P proteins are therefore most likely the viral proteins which catalyze transcription. It will be of great interest to establish which P protein(s) is the endonuclease and which is the transcriptase and which recognize methylated cap structures.

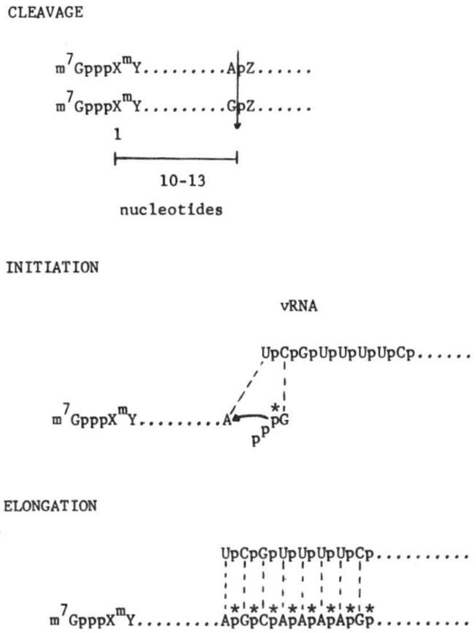

Figure 5. Mechanism for the priming of influenza viral RNA transcription by capped RNAs.

Because similar priming by capped RNAs also apparently occurs in the infected cell, we can explain the α-amanitin-sensitive (RNA polymerase II) step required for viral RNA transcription *in vivo,*[22-24] i.e., the host RNA polymerase II is required for the synthesis of capped RNA primers.[1-4] Recently, we have been able to demonstrate directly that only those capped cellular RNAs made after infection and not those pre-existing before infection serve as primers. Using a new assay for pulse-labeled viral RNA transcripts in the infected cell, we have shown that the cap which is donated by the host RNA primer is synthesized only after and not before infection. An important question then is why only newly synthesized capped cellular RNAs serve as primers. One possibility is that pre-existing, but not newly synthesized, capped RNAs are tied up in ribonucleoprotein structures (including polyribosomes) and cannot be used by the viral transcriptase. If so, the viral transcriptase would be expected to function near the site of synthesis of capped cellular RNAs in the nucleus. Several lines of evidence have suggested that some steps in viral RNA transcription occur in the nucleus,[17,24-26] but this has not yet been directly demonstrated. We are in the process of determining the site(s) of viral RNA transcription in the infected cell, using our new assay for pulse-labeled RNA. If, however, viral RNA transcription does not occur in the nucleus, then other explanations would be needed.

ACKNOWLEDGMENTS
We thank Barbara A. Broni for expert technical assistance. These investigations were supported by U.S. Public Health Service Grants AI 11772 and CA 08748; by U.S. Public Health International Fellowship TW 02590 to M.B., and by European Molecular Biology Fellowship ALTF 190-1979 to I.U.

References

1. Bouloy, M., Plotch, S.J., and Krug, R.M. (1978): *Proc. Natl. Acad. Sci. U.S.A.* 75:4886–4890.
2. Plotch, S.J., Bouloy, M., and Krug, R.M. (1979): *Proc. Natl. Acad. Sci. U.S.A.* 76:1618–1622.
3. Bouloy, M., Morgan, M.A., Shatkin, A.J., and Krug, R.M. (1979): *J. Virol.* 32:895–904.
4. Robertson, H.D., Dickson, E., Plotch, S.J., and Krug, R.M. (1980): *Nucleic Acids Res.* 8:925–942.
5. Krug, R.M., Broni, B.B., and Bouloy, M. (1979): *Cell* 18:329–334.
6. Dhar, R., Chanock, R.M., and Lai, C.-J. (1980): *Cell* 21:495–500.
7. Caton, A.J. and Robertson, J.S. (1980): *Nucleic Acids Res.* 8:2591–2603.
8. Lockard, R.E. and RajBhandary, U.L. (1976): *Cell* 9:747–760.
9. Bouloy, M., Plotch, S.J., and Krug, R.M. (1980): *Proc. Natl. Acad. Sci. U.S.A.* 77:3952–3956.
10. Skehel, J.J. and Hay, A.J. (1978): *Nucleic Acids Res.* 4:1207–1219.
11. Robertson, J.S. (1979): *Nucleic Acids Res.* 6:3745–3757.
12. Shatkin, A.J. (1976): *Cell* 9:645–653.
13. Koper-Zwarthoff, E.C., Lockard, R.E., Alzner-DeWeerd, B., RajBhandary, U.L., and Bol, J.F. (1977): *Proc. Natl. Acad. Sci. U.S.A.* 74:5504–5508.
14. Krug, R.M., Broni, B.B., La Fiandra, A.J., Morgan, M.A., and Shatkin, A.J. (1980).
15. Baralle, F.E. (1977): *Nature* 267:279–281.
16. Dasgupta, R., Harada, F., and Kaesberg, P. (1976): *J. Virol* 18:260–267.
17. Lamb, R.A. and Lai, C.-J. (1980): *Cell* 21:475–485.
18. Hay, A.J., Lomniczi, B., Bellamy, A.R., and Skehel, J.J. (1977): *Virology* 83:337–355.
19. Rochovansky, O. (1976): *Virology* 73:327–338.
20. Inglis, S.C., Carroll, A.R., Lamb, R.A., and Mahy, B.W.J. (1976): *Virology* 74:489–503.
21. Compans, R.W., Content, J., and Duesberg, P.H. (1972): *J. Virol.* 10:795–800.
22. Lamb, R.A. and Choppin, P.W. (1977): *J. Virol.* 23:816–819.
23. Spooner, L.L.R. and Barry, R.D. (1977): *Nature* 268:650–652.
24. Mark, G.E., Taylor, J.M., Broni, B., and Krug, R.M. (1979): *J. Virol.* 29:744–752.
25. Krug, R.M., Morgan, M.M., and Shatkin, A.J. (1976): *J. Virol.* 20:45–53.
26. Barrett, J., Wolstenholme, A.J., and Mahy, B.W.J. (1979): *Virology* 98:211–225.

David H.L. Bishop and Richard W. Compans, eds.
The Replication of Negative Strand Viruses

The Sites of Initiation and Termination of Influenza Virus Transcription

James S. Robertson,[a] Andrew J. Caton,[a]
Manfred Schubert[b] and Robert A. Lazzarini[b]

Introduction

The mRNA transcripts of the influenza virus genome are not exact copies of the individual genome segments. Their 5′-termini apparently contain a cap structure and additional nucleotides derived from the host cell. This phenomenon has been studied in detail *in vitro*[1-3] but little is known concerning the process *in vivo*. It has also been established that the 5′-terminal 20-30 nucleotides of the individual genome segments are not transcribed into mRNA.[4] Using a variety of sequencing techniques, we have analyzed in detail the structure of the 5′ and 3′-terminal regions of the influenza virus mRNA extracted from infected cells.

The 5′-Terminal Region of Influenza Virus mRNA

We have recently developed a method of cloning influenza virus mRNA which allows us to deduce the nucleotide sequence at the 5′-terminal region of individual mRNA molecules.[5,6] Sequence analysis of one such clone reveals the presence of 13 additional non-virus specific nucleotides at the 5′ end of the mRNA (Figure 1). These nucleotides are presumably host derived and their presence suggests that a priming mechanism of virus transcription

[a] University of Cambridge, Department of Pathology, Division of Virology, Addenbrooke's Hospital, Cambridge, England.
[b] Laboratory of Molecular Biology, NINCDS, NIH, Bethesda, Maryland.

primer sequence

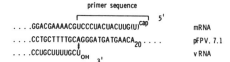

	. . . .GGACGAAAACGUCCCUACUACUUG(U)^{cap} 5'	mRNA
	. . . .CCTGCTTTTGCAGGGATGATGAACA$_{20}$	pFPV. 7.1
	. . . .CCUGCUUUUGCU$_{OH}$ 3'	vRNA

Figure 1. Sequence analysis of clone pFPV.7.1 corresponding to the 5′-terminal region of the mRNA from Caton and Robertson.[6] The mRNA of segment 7 of A/FPV/Rostock strain of influenza virus was cloned as previously described.[5,6] The sequence of the original mRNA as deduced from the clone, the sequence of the DNA of pFPV.7.1 and the 3′-terminal sequence of the vRNA are shown.

similar to that observed *in vitro* also exists *in vivo*. An additional feature of the structure at the 5′ end of the mRNA was noted. This was the absence in the mRNA of the nucleotide complementary to the 3′-terminal nucleotide of vRNA, i.e., the virus-specific sequence of the mRNA apparently begins opposite the penultimate nucleotide of the vRNA.

To investigate the extent of sequence diversity of the primer region, we directly examined the 5′ end of total infected cell mRNA by dideoxy chain-termination sequence analysis of the mRNA utilizing a restriction fragment derived from clone pFPV.7.1 as primer (Figure 2). This analysis avoided any sequence artifact which may have arisen through molecular cloning. The restriction fragment used as primer represents nucleotides 32-121 from the 3′ end of genome segment 7. In the sequencing gel (Figure 2), the control track

Figure 2. Dideoxy sequence analysis of the 5′-terminal region of segment 7 mRNA from Caton and Robertson.[6] A specific restriction fragment from clone pFPV.7.1 was used to prime the synthesis of DNA complementary to the 5′ end of segment 7 mRNA in the presence of dideoxytriphosphates as indicated and the products analyzed by polyacrylamide gel electrophoresis. Track N contained no ddNTPs. The numbering indicates nucleotide positions from the 3′ end of vRNA.[7]

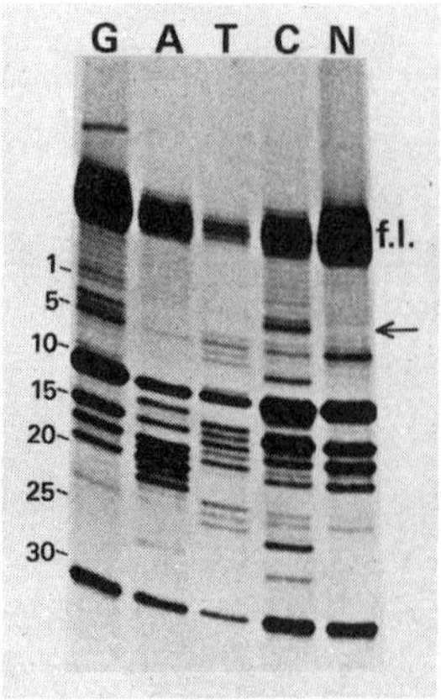

which contained no dideoxy triphosphates showed several artifacts and these were present in each of the other four tracks. Comparison of this sequencing gel with the established sequence of the 3′ end of genome segment 7 allowed us to deduce certain features regarding the structure of the 5′-termini of the mRNA. Firstly, the series of full length products in each track indicates size heterogeneity of the 5′-termini of the mRNA. In addition, beyond the region corresponding to the virus-specific sequence, there are no specific products in any of the tracks, indicating sequence heterogeneity. Thus the host-derived primer of the mRNA corresponding to segment 7 has a heterogeneous sequence which varies from 9-15 nucleotides in length.

Figure 2 also indicates that the last specific nucleotide is the C residue (indicated by the arrow) corresponding to position 2 in the vRNA, whilst position 1 can be occupied by any of the four nucleotides. This indicates that initiation of transcription *in vivo* of vRNA segment 7 occurs at the penultimate residue of the vRNA molecule.

By analogy with recent *in vitro* analyses of influenza virus transcription, we conclude that influenza virus mRNAs derive 9-15 nucleotides of heterogeneous sequence from the host cell during acquisition of a cap structure. The exact mechanism by which this is achieved remains unclear at present.

Polyadenylation Sites of Influenza Virus RNA

A specific feature of the 5′-terminal region of the vRNA which is common to all eight segments is a tract of 5-7 uridine residues approximately 20 nucleotides from the 5′ end (Figure 3).[7] The possibility that these uridine tracts are the sites of polyadenylation was tested by the following analysis. Individual vRNA segments were digested to completion with T_1 RNase and the products labeled at their 5′-termini with polynucleotide kinase and γ-^{32}P-ATP. These labeled oligonucleotides were annealed with an excess of infected cell mRNA and subsequently treated with T_1 and pancreatic RNases (T_1 RNase only with segment 5, see below). The RNase resistant hybrids which remain attached to the poly A tails of the mRNA were selected by oligo(dT) cellulose chromatography. Hybrids specifically selected in this way should contain the T_1 oligonucleotide which is complementary to the region of the mRNA adjacent to the poly A tail. If the U tracts common to each segment are involved in polyadenylation, then analysis of the genome segments as described above should specifically select the T_1 oligonucleotides underlined in Figure 3.

Segments 4, 5, 7, and 8 were studied in this analysis. (The U tract containing T_1 oligonucleotide of segment 6 is too small, whilst for segments 1, 2 and 3 an adequate quantity of mRNA was not readily available for this analysis). The procedure predicts that a specific 27 residue T_1 oligonucleotide will be

Segment	5′ 25 50
8	AGUAGAAACAAGGGUGUUUUUUAUCAUUAAAUAAGCUGAAACGAGAAAGUUCUUAUCUCU
7	AGUAGAAACAAGGUAGUUUUUUACUCCAGCUCUAUGUUGACAAAAUGACCAUCGUCAACA
6	AGUAGAAACAAGGAGUUUUUUUGAACAAACUACUUGUCAAUGGUGAAUGGCAACUCAGCA
5	AGUAGAAACAAGGGUAUUUUUCUUUAAUUGUCAUACUCCUCUGCAUUGUCUCCGAAGAAA
4	AGUAGAAACAAGGGUGUUUUUUUCCAAACUUAUAUACAAAUAGAGCACCGCAUGUUUCCG
3	AGUAGAAACAAGGUACUUUUUUGGACAGUAUGGAUAGCAAAUAGUAGCAUUGCCACAACU
2	AGUAGAAACAAGGCAUUUUUUCAUGAAGGACAAGCUAAAUUCACUAUUUCUGCCGUCUGA
1	AGUAGAAACAAGGUCGUUUUUAAACAAUUCGACAUUAAUUGAUGGCCAUCCGAAUUCUUU

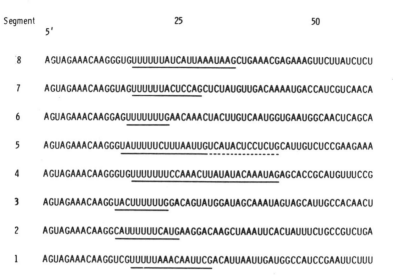

Figure 3. 5′-terminal nucleotide sequences of influenza virus (A/FPV/Rostock) genome RNA from Robertson.[7] The solid lines indicate the U tract containing T_1 oligonucleotides. The dashed line indicates the additional T_1 oligonucleotide specifically selected from segment 5.

selected from segment 4, a specific 16 residue product from segment 5, a specific 13 residue product from segment 7 and a specific 19 residue product from segment 8. The T_1 oligonucleotides which were specifically selected by oligo(dT) cellulose chromatography are shown in Figure 4. With segment 5, two major products were selected; with segment 4, several products were selected; with segment 7 and 8, only one major product was selected. Sequence analysis of the products indicated by the symbols in Figure 4 confirmed that they were the U tract containing T_1 oligonucleotides (Figure 5). Analysis of the smaller products selected from segment 4 indicated that they were derived from the U tract containing oligonucleotide presumably by degradation during the analysis. The 27 residue U tract containing oligonucleotide of segment 4 was found to be more stable during an analysis of total unfractionated genome RNA.

The specific selection of these T_1 oligonucleotides indicates that the common U tracts are the sites of polyadenylation of the mRNA corresponding to segments 4, 5, 7 and 8. The presence of this U tract in all eight segments suggests that it is the polyadenylation site for segments 1, 2, 3, and 6 also. A "chattering" or "stuttering" mechanism which allows repetitive copying of a tract of uridine residues by the transcriptase complex has been suggested for generation of poly A tails at such a sequence.[8]

Sequence analysis of the smaller additional product selected from segment 5 indicated that it is the T_1 oligonucleotide adjacent to the U tract containing oligonucleotide corresponding to positions 31-43 from the 5′ end of vRNA. Because of the nature of the sequence at the 5′ end of the U tract containing

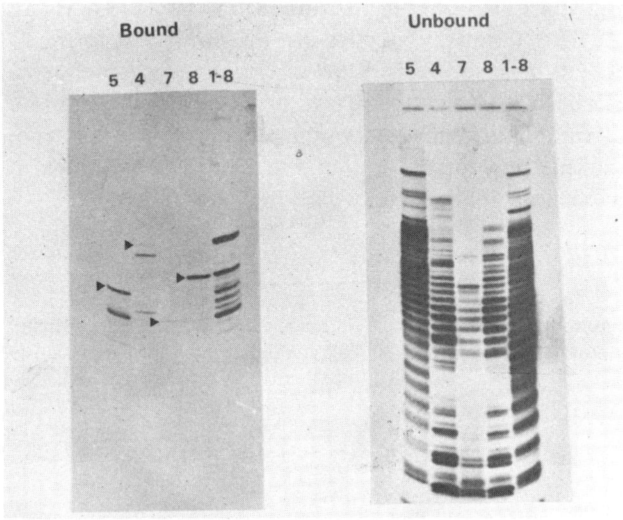

Figure 4. Polyacrylamide gel electrophoresis of the products specifically selected by oligo(dT) cellulose chromatography. Samples treated as described in the text were applied to oligo(dT) cellulose and unbound and bound fractions analyzed by gel electrophoresis. The triangles indicate the U tract containing oligonucleotides.

oligonucleotide, only T_1 RNase was used to digest the hybrids to avoid loss of the 5′-terminal label. Specific selection of the smaller 13 residue oligonucleotide indicates that the poly A tail remains attached to the region of the mRNA complementary to the 13 residue oligonucleotide. This could only occur if no G residues exist between the poly A tail and the double-stranded

Figure 5. Sequence analysis of the larger product specifically selected from segment 5. (a) Tracks from left to right are: control, Phy I, Phy I, ladder, U_2; (b) two-dimensional wandering spot analysis after formamide degradation.

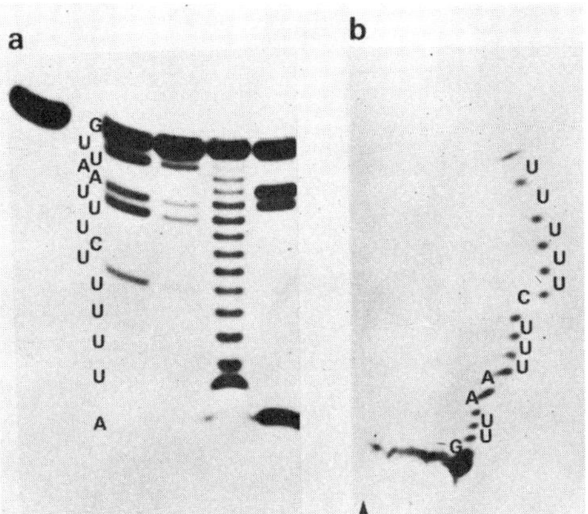

region and indicates that the C residue at position 22 in the vRNA is not transcribed into mRNA. This C residue occurs not only adjacent to the U tract of segment 5 but within a series of 8 uridine residues. Absence of transcription of this residue may indicate that synthesis of a poly A tail by polymerase slippage at a tract of uridine residues can occur, even if that tract is interrupted by a different nucleotide or that the 3 uridine residues at position 23-25 are sufficient to signal polyadenylation.

ACKNOWLEDGMENTS
We would like to thank Elaine Robertson for her excellent technical assistance, Stuart Nichol for assistance in preparation of this manuscript and Nucleic Acids Research for permission to reproduce Figures 1, 2 and 3.

References

1. Bouloy, M., Plotch, S.J., and Krug, R.M. (1978): *Proc. Natl. Acad. Sci. U.S.A.* 75:4886–4890.
2. Krug, R.M., Broni, B.A., and Bouloy, M. (1979): *Cell* 18: 329–334.
3. Robertson, H.D., Dickson, E., Plotch, S.J., and Krug, R.M. (1980): *Nucl. Acids Res.* 8:925–942.
4. Hay, A.J., Abraham, G., Skehel, J.J., Smith, J.C., and Fellner, P. (1977): *Nucl. Acids Res.* 4:4197–4209.
5. Caton, A.J. and Robertson, J.S. (1979): *Nucl. Acids Res.* 7:1445–1456.
6. Caton, A.J. and Robertson, J.S. (1980): *Nucl. Acids Res.* 8:2591–2603.
7. Robertson, J.S. (1979): *Nucl. Acids Res.* 6:3745–3757.
8. Schubert, M., Keene, J.D., Herman, R.C., and Lazzarini, R.A. (1980): *J. Virol.* 34:550–559.

David H.L. Bishop and Richard W. Compans, eds.
The Replication of Negative Strand Viruses

Comparison of Transcriptional Capabilities of Influenza Virions and Nucleocapsid Complexes

Olga M. Rochovansky[a]

Previous studies on the composition of functional transcriptional complexes isolated from the WSN strain of influenza virus showed that the eight RNA genomic segments, one or more of the P-proteins and the nucleoprotein (NP) were present. Very little membrane (M) protein was detected.[1] Under conditions where Mn^{++} was the divalent cation, comparisons of the polymerase activity to RNA concentration ratios of purified virions and complexes were similar suggesting that transcriptional capabilities were retained. Most of the cRNA synthesized by complexes was 4-12S and probably represented incomplete transcripts. Addition of a suitable dinucleoside phosphate, such as adenylyl 3′→5′ guanosine (ApG) or guanylyl 3′→5′ guanosine (GpG), shown earlier by McGeoch and Kitron[2] to stimulate the rate of transcription, increased the size of transcripts only slightly. It was, however, demonstrated that complexes, free of contaminating intact or only partially disrupted virus particles, were capable of establishing a productive infection in suitably treated host cells.[3] This finding indicated that at least some complexes did retain the viral activities required for the first biosynthetic step carried out under the direction of the virus, the formation of viral mRNA.

More recently, Plotch and Krug[4] demonstrated that detergent treated influenza virus was capable of *in vitro* synthesis of polyadenylated cRNAs. The requirements were Mg^{++} as divalent cation and the addition of ApG or GpG. Subsequent studies by Krug and co-workers[5-8] showed that eukaryotic

[a] The Christ Hospital Institute of Medical Research, 2141 Auburn Avenue, Cincinnati, Ohio.

mRNAs, notably globin mRNA, served as primers for the *in vitro* synthesis of polyadenylated transcripts. The primers donated their 5′-terminal cap structures and 12-14 nucleotides to the transcripts.

This paper compares the synthetic capabilities of complexes and unfractionated influenza virions. The relative abilities of the two systems to use either ApG or globin mRNA as primer for *in vitro* synthesis and the products formed were examined. Also, evidence which indicates that complexes in the absence of added primer were capable of significant RNA formation is presented.

Transcripts Formed in Presence of ApG

In order to determine whether complexes were capable of polyadenylated transcript synthesis, the products formed by unfractionated virions and complexes in the presence of Mg^{++} and ApG were compared. Affinity chromatography on oligo(dT) cellulose was used to separate transcripts into polyadenylated and unbound species. The unbound RNAs were freed from unreacted incubation components by chromatography on CF-11 cellulose.[9] Experimental details are provided in the legend to Figure 1. It was found that approximately 30% of the total RNA synthesized by either virions or complexes was bound by the affinity columns. Two conclusions were drawn from these data: (1) Complexes synthesized polyadenylated transcripts; (2) The two systems formed the same relative proportions of polyadenylated and unbound RNAs.

Both bound and unbound transcripts were analyzed by polyacrylamide gel electrophoresis after annealing to unlabeled vRNA and S1 nuclease treatment to remove single-stranded regions.[10] Lanes 4 and 5 of Figure 1 show the patterns given by hybrids obtained from polyadenylated transcripts synthesized by virions and complexes, respectively. As may be seen, the eight expected transcripts were formed by both systems and migrations of the hybrids were similar to that of marker influenza dsRNA (lane 1). Scans of the gel showed that the relative amounts of individual segments present within each complete set of transcripts were very similar. Thus with ApG as primer, complexes and virions synthesized polyadenylated transcripts that appeared identical in size when analyzed as dsRNAs and in the relative amounts of individual segments.

Electrophoretic analyses of the unbound RNAs formed by virions and complexes are shown in lanes 2 and 3, respectively. The presence of some apparently completed transcripts may have resulted either from loss of poly A during manipulation of the RNA, poor binding to affinity columns, or failure of the systems to add poly A sequences. Clearly the gel shows that unbound RNAs contained incomplete transcripts. This is in contrast to the analyses of the polyadenylated transcripts which indicated the presence of only small amounts of incomplete transcripts. The results suggest that only complete mRNA-length transcripts were polyadenylated.

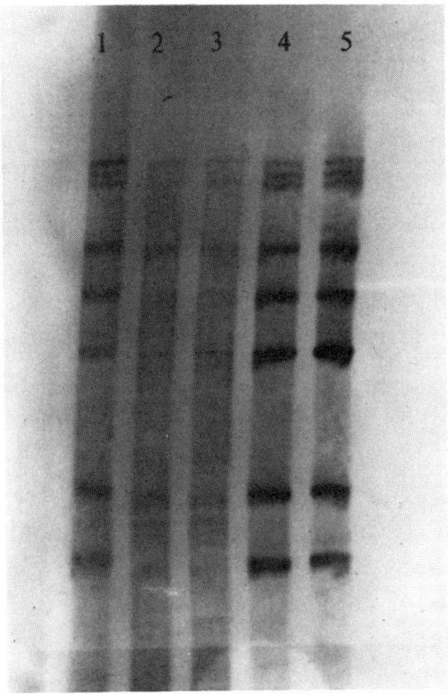

Figure 1. Polyacrylamide gel electrophoresis of polyadenylated and unbound cRNAs formed by unfractionated influenza virus and complexes. Incubation conditions were as described before[1] with the exception that Mg^{++} at 8mM was the divalent cation and ApG at 0.4 mM was present. ^{3}H-UTP at 0.1 mM containing 320 cpm per pmol was added as radioactive substrate. Incubations were at 31° for 75 min. Reaction mixtures were then adjusted to 0.5% SDS and 0.5 M NaCl and directly applied to oligo(dT) cellulose columns. Unbound and bound RNAs were eluted, phenol extracted and precipitated with 2.5 volumes of 95% ethanol.[10] Unreacted reaction components were removed from unbound RNA by CF-11 cellulose chromatography.[9] After annealing in the presence of excess vRNA, hybrids were treated with S1 nuclease to remove single-stranded regions[10] phenol extracted, precipitated and analyzed. (1) Marker influenza dsRNA; (2) Unbound RNA formed by virions; (3) Unbound RNA formed by complexes; (4) Polyadenylated RNA formed by virions; (5) Polyadenylated RNA formed by complexes.

Globin mRNA as Primer

Recent analyses of complexes prepared from ^{35}S-methionine-labeled virions showed that the 3 P-proteins, NP and a trace of M-protein, were present (unpublished results). The 3 P-proteins were in the same relative ratios to each other and to NP in both complexes and virions. The M-protein, relative to NP, represented less than 5% of the amount present in virions. No other viral proteins were detected. It was of interest to determine whether such complexes could carry out the modifications necessary to convert eukaryotic mRNAs into suitable primers.

Table 1 shows the activities of complexes and virions in the presence of either ApG or various amounts of globin mRNA. In order to compare the

Table 1. Activity of Complexes with Globin mRNA.[a]

Added primer	pmoles ^{3}H-UMP incorporated by	
	Complexes	Virions
none	2.74	0.45
ApG	8.92	8.46
mRNA (0.8 μg)	5.88	5.64
mRNA (1.2 μg)	8.90	8.46
mRNA (1.6 μg)	10.16	9.84

[a]Reaction conditions were described in the legend to Figure 1 except that incubation volumes were 0.05 ml and reaction time was 60 min. Complexes and virions were present at 0.8 μg and 4.5 μg, respectively. Product formation was assayed as TCA precipitable cpm which were corrected for an incubated blank (20 cpm).

results directly, conditions were selected so that the activities of isolated complexes and purified virus in the reactions would be equivalent. It was calculated from earlier data[1] that in terms of viral protein, 4.5 μg of virus contained approximately 0.8 μg of complexes. Accordingly, these amounts were used in the assays. Comparisons of the activities in Table 1 show that about the same synthesis of RNA was achieved by complexes and virions in the presence of either ApG or various amounts of globin mRNA. These data suggested that complexes were primed by globin mRNA as efficiently as was intact virus. This conclusion was supported by results to be presented elsewhere which showed that the same amounts of polyadenylated transcripts were formed by both complexes and unfractionated virions with globin mRNA as primer.

The stimulations given by ApG or globin mRNA in the two systems were, however, quite different. This was due to the synthesis of about 6-fold more RNA by complexes than virions in the absence of added primer. Activities of virions were increased 19-fold by ApG and 22-fold by 1.6 μg of globin mRNA. In contrast, complexes were stimulated only 3.3 and 3.7-fold, respectively.

Activity of Complexes in Absence of Added Primer

The results in Table 1 indicated that complexes formed substantial amounts of RNA when either ApG or globin mRNA was omitted. The first evidence for this synthetic capability was obtained from other studies to be presented elsewhere. The data in Table 1 do, however, illustrate one significant finding. Complexes synthesized 30% as much RNA in the absence as in the presence of added primer. In contrast, unfractionated virus was only about 5% as active when primer was omitted. In order to evaluate this synthesis by complexes, the formation of polyadenylated transcripts with and without ApG was investigated. It was found that 30% of the total RNA synthesized by complexes in the absence of ApG was bound by oligo(dT) cellulose. Although the total synthesis of RNA was only 1/3 of that observed with ApG added, the relative proportions of polyadenylated and unbound transcripts formed under both sets of conditions were very similar.

Electrophoretic analyses of the two species of transcripts after annealing to unlabeled vRNA and S1 nuclease treatment[10] are presented in Figure 2. Lanes 4 and 5 show the patterns given by hybrids obtained from polyadenylated transcripts synthesized with and without ApG, respectively. As may be seen, the eight expected transcripts were formed under both conditions. Incomplete transcripts were not detected. Although only 30% as much polyadenylated product was synthesized in the absence of ApG, comparisons of the relative amounts of individual transcripts in the two sets were similar. Omission of ApG did not, therefore, result in a disproportionate synthesis of certain segments.

The migrations of annealed unbound RNA formed by complexes in the presence and absence of ApG are shown in lanes 1 and 2, respectively. Again, the patterns were very similar. In addition to some full length trans-

Figure 2. Polyacrylamide gel electrophoreses of polyadenylated and unbound cRNAs formed by complexes in the presence and absence of ApG. Reaction conditions and isolation of the RNAs were as described in the legend to Figure 1. (1) Unbound RNA formed in presence of ApG; (2) Unbound RNA formed in absence of ApG; (3) Marker influenza dsRNA; (4) Polyadenylated RNA formed in presence of ApG; (5) Polyadenylated RNA formed in absence of ApG.

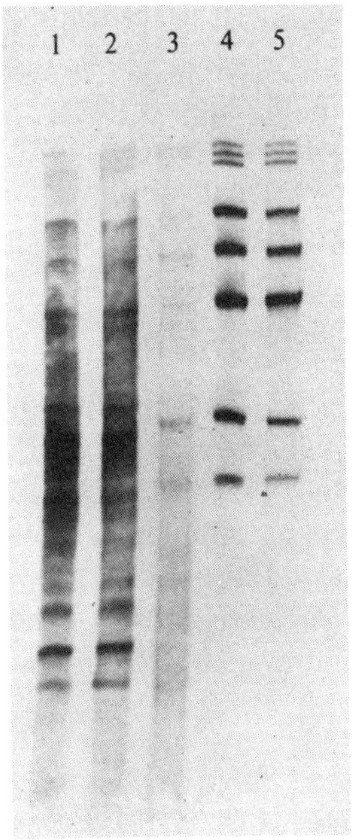

cripts, larger amounts of apparently incomplete transcripts were present. Except for the differences in total RNA synthesized, the products formed by complexes under the two conditions were virtually identical.

Discussion

Results presented here indicate that like purified unfractionated influenza virus[4] transcriptional complexes isolated from the virus synthesized eight polyadenylated transcripts when ApG was added as primer. Comparisons of the activities of virions and complexes showed no discernible differences in either the relative amounts of individual transcripts formed or the proportions of poly A^+ RNA in total cRNA. Furthermore, addition of globin mRNA to approximately equivalent activities of complexes and virions resulted in the formation of similar amounts of cRNA. These findings strongly suggest that complexes comprised of only the viral RNA genomic segments, NP, 3 P-proteins and a trace of M-protein contained functional RNA-dependent RNA and poly A polymerases as well as the activity necessary for conversion of globin mRNA to a suitable primer. Whether these activities are all viral-specified and catalyzed by distinct and separable proteins or by a multimeric protein is not presently known.

One significant difference was found, however, when the activities of complexes and virions in the absence of added primers were compared. Complexes synthesized approximately 30% as much RNA without as with primer. Except for the overall decrease in activity, RNA formed under both conditions appeared identical by the methods of analysis used. A substantial proportion of complexes was, therefore, capable of initiation of RNA synthesis, elongation and addition of poly A in the absence of added primer. In contrast, unfractionated virus was reported by Plotch and Krug[11] to be only 1 to 2% as active when primer was omitted. Although small amounts of polyadenylated transcripts were detected, virions were apparently much less able to initiate RNA synthesis and/or elongate under these conditions.

It is presently unclear whether the activity observed with complexes resulted from a primer-independent initiation or from the presence of an endogenous primer. The results suggest, however, that an inhibitory modulator of RNA synthesis is present in virions and at least partially absent in complexes. Zvonarjev and Ghendon[12] recently reported that addition of M-protein to complexes isolated from the A/PR/1/74 strain of influenza virus inhibited transcriptase activity by about 40%. Although the conditions of assay were somewhat different from those used here, investigations of the effects of M-protein on the primed and "unprimed" syntheses would appear warranted.

ACKNOWLEDGMENTS

The technical assistance of Mr. P.J. Ciraolo is appreciated. This work was partly supported by U.S. Public Health Service Research Grant AI 14733.

References

1. Rochovansky, O.M. (1976): *Virology* 73:327–338.
2. McGeoch, D. and Kitron, N. (1975): *J. Virol.* 15:686–695.
3. Rochovansky, O.M. and Hirst, G.K. (1976): *Virology* 73:339–349.
4. Plotch, S.J. and Krug, R.M. (1977): *J. Virol.* 21:24–34.
5. Bouloy, M., Plotch, S.J., and Krug, R.M. (1978): *Proc. Natl. Acad. Sci. U.S.A.* 75:4886–4890.
6. Plotch, S.J., Bouloy, M., and Krug, R.M. (1979): *Proc. Natl. Acad. Sci. U.S.A.* 76:1618–1622.
7. Robertson, H.D., Dickson, E., Plotch, S.J., and Krug, R.M. (1980): *Nucleic Acids Res.* 8:925–942.
8. Bouloy, M., Plotch, S.J., and Krug, R.M. (1980): *Proc. Natl. Acad. Sci. U.S.A.* 77:3952–3956.
9. Franklin, R.M. (1966): *Proc. Natl. Acad. Sci. U.S.A.* 55:1504–1511.
10. Pons, M.W. and Rochovansky, O.M. (1979): *Virology* 97:183–189.
11. Plotch, S.J. and Krug, R.M. (1978): *J. Virol.* 25: 579–586.
12. Zvonarjev, A.Y. and Ghendon, Y.Z. (1980): *J. Virol.* 33:583–586.

Published 1981 by Elsevier North Holland, Inc.
David H.L. Bishop and Richard W. Compans, eds.
The Replication of Negative Strand Viruses

RNA Synthesis of Temperature-Sensitive Mutants of WSN Influenza Virus

Solomon L. Mowshowitz[a]

Temperature-sensitive mutants of WSN influenza virus fall into at least seven complementation-recombination groups.[1-3] At the non-permissive temperature (39.5), mutants from groups I, II, III and V synthesize greatly reduced amounts of virus-specific RNA in infected cells.[1] For each group, the specific viral protein which bears the temperature-sensitive lesion has been identified.[4,5] Members of two complementation groups, I and III—corresponding to defects in viral proteins P3 and P1 respectively[4]—did not synthesize detectable cRNA in infected cells at the non-permissive temperature.[6] Mutants from groups II and V (now known to correspond to lesions in P2 and NP, respectively)[4] were able to synthesize cRNA at the non-permissive temperature and were presumed to have defects in vRNA synthesis.[6]

It is now appreciated, however, that there are at least two different species of cRNA within the infected cell, which have different putative functions:[7,8] (1) mRNA molecules which are capped and polyadenylated, and represent incomplete transcripts of the vRNA segments (transcription may terminate at a run of Us 17 nucleotides from the 5′ end of the vRNA segments)[9] and (2) template cRNA molecules, which most likely function as templates for the synthesis of vRNA, and which are apparently complete and faithful transcripts of the vRNA. It is entirely possible, therefore, that P2 and NP mutants are deficient with respect to the synthesis of one or the other of these species at the non-permissive temperature. Furthermore, the possible role of any of

[a] Mt. Sinai School of Medicine, Fifth Avenue & 100th St., New York, New York.

the gene products in vRNA synthesis is undefined. P3 and P1 may also participate directly in this process. On the other hand, P2 and/or NP mutants may be deficient in template cRNA synthesis, and may be able to synthesize vRNA at the non-permissive temperature if provided with template cRNA. It is clear that in order to fully characterize the mutants, the synthesis of all three species must be measured directly.

RNA Synthesis: Single Cycle Experiment

Two basically different approaches have been used to describe the natural history of RNA synthesis in cells infected with influenza virus. In the first approach, RNA is isolated from cells at different times following infection, and the abundances of the different virus-specific RNA species are measured by hybridization kinetics using appropriate radiolabeled probes. This approach is not optimal for shift-up experiments with ts mutants. For one thing, further accumulation of an RNA species following a shift-up in temperature must be measured against a background of material already accumulated. More serious is the fact that the amount of each species present at a given time is a balance between its synthesis and degradation. In fact, Taylor et al[10] and Mark et al,[11] studying WSN-infected MDCK cells at 37°, showed that from 2.5 hours postinfection on, degradation of cRNA far exceeds synthesis. The initial characterization of cRNA synthesis with the mutants of WSN influenza virus[12] was performed by following cRNA accumulation after a temperature shift *four hours* postinfection at 33°. In that report, cRNA continued to accumulate at least up to six hours postinfection at 39° and at least up to ten hours at 33°. The discrepancy may be due in part to the use of MDBK cells in the latter study. In any event, the inherent difficulties of this approach are evident.

The second approach, that of pulse-labeling cells at different times postinfection, is not without its own attendant difficulties. The problem of precursor pools is always present, and in fact has been invoked[13] to help explain major discrepancies between the picture of RNA-synthesis derived by pulse-labeling experiments and that derived by measuring accumulation with radiolabeled probes. Authors employing the latter method[10,11,13] report a major shift in predominance from cRNA (early in infection) to vRNA at or around two hours postinfection. Hay et al,[8] using pulse-labeling techniques, found approximately equal amounts of vRNA and cRNA synthesis throughout the infectious cycle, with some predominance of cRNA synthesis from three hours postinfection on. Differential rates of degradation of the various species may also partially account for the difference in results.

It is clear that to obtain a complete picture of the events of RNA synthesis in infected cells, both approaches are required.

Since I intend to characterize the ts mutants using pulse-labeling techniques, it was important to establish the characteristics of viral RNA synthesis through the infectious cycle by this approach to see if it could be reconciled with the data obtained by others measuring RNA accumulation with radiolabeled probes.

MDBK monolayers were pulsed with 100 μC_i of ^{3}H-uridine various times after infection with WSN influenza virus. Pulse-labeled RNA was isolated by phenol:chloroform extraction and ethanol precipitation, before being annealed with unlabeled probes.

To estimate total cRNA, unlabeled vRNA was used as a probe. Experimental details are in the legend to Figure 2.

To ascertain what portion of the total cRNA was template cRNA, a new approach was used, which took advantage of the ability of benzoylated DEAE-cellulose (BD-cellulose) to bind nucleic acids with appreciable single-stranded regions.[16] Figure 1 illustrates that only hybrids involving vRNA with complete template cRNA lack appreciable single-stranded regions and will not be retained by BD cellulose columns. Procedural details are in the legend to Figure 1.

Previous estimates of the relative proportions of mRNA and template cRNA have relied on the resolution of the total cRNA into two fractions by oligo(dT)-cellulose chromatography. The polyadenylated fraction of cRNA has been equated to mRNA while the unpolyadenylated fraction has been assumed to represent template cRNA. A disadvantage of this approach is that partially degraded mRNA will appear in the template cRNA fraction, and may result in an overestimation of the proportion of template cRNA. Conversely, since degraded template cRNA will be retained on the BD column, this latter method may result in an underestimation of the proportion of template cRNA.

Various approaches have been used to determine the amount of pulse-labeled vRNA. Hay et al.[8] determined total virus-specific RNA with a probe consisting of a mixture of vRNA and cRNA, which they generated by incubating purified virions under conditions appropriate for *in vitro* transcription, in the presence of unlabeled ribonucleoside triphosphates. Thus, the amount of vRNA could be computed by subtracting the amount of cRNA from the total. A different approach was used in an earlier study by Scholtissek and Rott[17] in which cRNA synthesized *in vitro* by a microsomal preparation from infected cells was used to detect pulse-labeled vRNA. In the present study, mRNA was isolated from cells three hours postinfection by oligo(dT)-cellulose chromatography[18] and used as a probe for vRNA. Further details are in the legend to Figure 2.

The results of the experiment are summarized in the same figure. For the first two hours after infection, little or no vRNA synthesis predominated. These results are harmonious with reports on the accumulation of cRNA and vRNA in MDCK cells infected with WSN.[10,11] Furthermore, they are in substantial agreement with the pulse-labeling studies of Scholtissek and Rott with the fowl-plague-CEF system.[17]

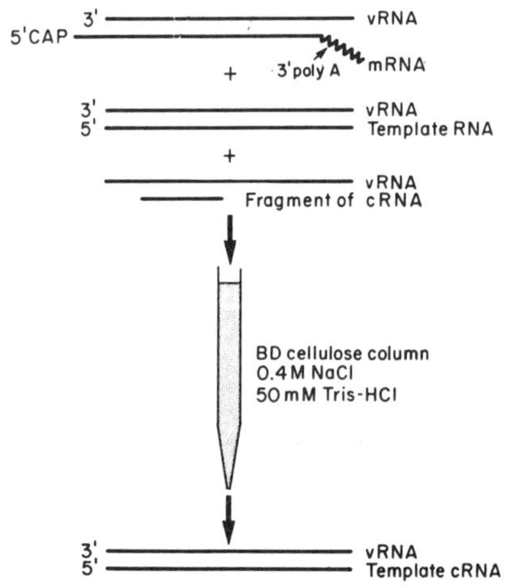

Figure 1. BD cellulose chromatography of hybrids. A portion of the hybridization reaction mixture was diluted into 3.5 ml of 50 mM Tris-HCl pH 8.2 containing 0.4 M NaCl. The hybrids were twice passed through a (1 ml) column of BD cellulose (Bio Rad), which has been equilibrated previously with the same buffer. The column was washed with 1 ml of buffer and the effluents were pooled, treated with RNase, and processed as described for the determination of total cRNA. The RNase-resistant counts were taken to be equivalent to template cRNA. The salt conditions chosen were such that over 95% of the label was retained when hybrids of ^{3}H-labeled vRNA with unlabeled mRNA isolated from infected cells were passed through the column.

Messenger RNA synthesis preceded that of template cRNA (as also found by Barrett et al.)[13] possibly reflecting the dependence of the latter on the synthesis of protein.[8,11,13] Once begun, template cRNA represented roughly a third of the total cRNA synthesized. Mark et al.[11] found template cRNA to be about half of the total cRNA accumulated in WSN-infected cells 2.5 hours postinfection. This minor difference may be due to differential turnover of the two species of cRNA. Alternatively, it may reflect the tendency of oligo(dT)-cellulose chromatography to overestimate (and of BD-cellulose chromatography to underestimate) the proportion of template cRNA.

cRNA and vRNA Synthesis by Temperature-Sensitive Mutants

In temperature-shift experiments, cRNA and vRNA synthesis by various ts mutants of WSN were measured from four to five hours postinfection at the permissive and non-permissive temperatures (Table 1).

The P1 and P3 mutants were grossly deficient in cRNA synthesis at the non-permissive temperature, while the P2 and NP mutants were able to synthesize some cRNA. This confirms the earlier findings of Krug et al. who measured accumulation of cRNA with the same mutants.[6]

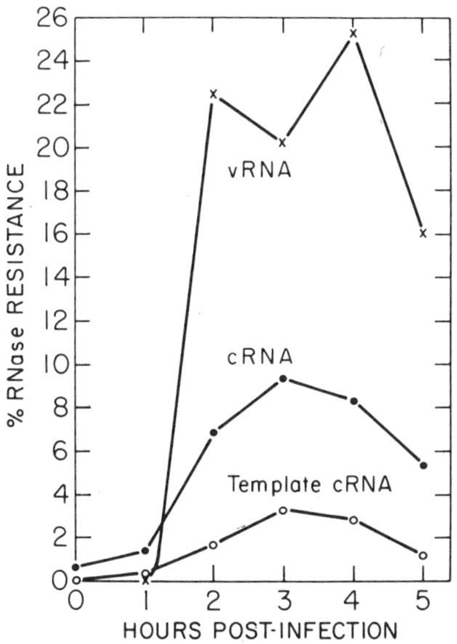

Figure 2. Synthesis of vRNA, total cRNA and template cRNA. To estimate total cRNA, approximately 100,000 cpm of pulse-labeled RNA were annealed with 1.0 μg of unlabeled vRNA isolated from purified virions of WSN virus grown in embryonated eggs. Annealing conditions were as described by Hay et al.,[8] after Ito and Joklik.[15] After annealing, a portion of the reaction mixture was removed, precipitated with 10% TCA and counted to determine total cpm. A second portion was removed, diluted into 1XSSC, and incubated 30 minutes at 39° with 10 μg of RNase A and 10 units of RNase T1. RNase-resistant material was then precipitated with 10% TCA and counted. Under these conditions, RNA from uninfected cells contained an RNase-resistant residue of approximately 0.1% of the total. This background was subtracted from the other values.

To determine vRNA, approximately 100,000 cpm of pulse-labeled RNA were hybridized with mRNA isolated from 20×10^6 infected cells. The annealing procedure and the processing of the hybrids was similar to the determination of total cRNA. Template cRNA was measured as described in Figure 1. x: vRNA; ●: total cRNA; O: template cRNA.

The P3 mutant was also ts$^-$ for vRNA synthesis. P3 mutants have a ts$^-$ phenotype *in vitro* with respect to the activity of the virion associated transcriptase.[14] It appears that P3 function is required for all virus-specific RNA synthesis. Since chain elongation is required for all viral RNA synthesis, P3 may participate in this function.

Although ts53 (P2) was presumed to be deficient for vRNA synthesis,[6] other alternatives were possible, such as a deficiency in template cRNA synthesis. Table 1 shows that P2 is indeed required directly for vRNA synthesis.

Table 1. RNA Synthesis by Temperature-Sensitive Mutants.

Experiment 1: Total cRNA				
Virus	**Lesion**	**%RNase Resistant: 33°**	**%RNase Resistant: 39°**	**39°/33°**
ts+	none	2.60	5.08	1.95
ts1	P3	4.59	<0.2	<0.04
ts15	P1	5.28	0.3	0.06
ts53	P2	6.93	6.11	0.88
ts56	NP	2.73	3.65	1.33

Experiment 2: Total vRNA				
Virus	**Lesion**	**%RNase Resistant: 33°**	**%RNase Resistant: 39°**	**39°/33°**
ts+	none	5.20	5.21	1.00
ts1	P3	8.08	0.54	0.07
ts53	P2	3.88	0.2	0.05

Duplicate MDBK cell monolayer cultures were infected with various ts mutants of WSN at an approximate MOI of 5.0. After 4 hours at 33°, one of each pair was shifted to 39° and all the cultures were pulsed for 1 hour with 100 μCi of ^{3}H-uridine. Determination of total cRNA and vRNA synthesis was as described in Figure 2.

Work is in progress to determine the ts phenotypes of P1 and NP mutants with respect to vRNA synthesis and those of P2 and NP for the synthesis of template cRNA.

ACKNOWLEDGMENTS
I wish to thank J.G. Wetmur for helpful discussions. T. Flanders provided expert technical assistance. This work was supported by Grant no. AI-09304 from the NIAID.

References

1. Sugiura, A., Tobita, K., and Kilbourne, E.D. (1972): *J. Virol.* 10:639–647.
2. Ueda, M. (1972): *Arch. Ges. Virusforch.* 39:360–368.
3. Sugiura, A., Ueda, M., Tobita, K., and Enomoto, C. (1975): *Virol.* 65:363–373.
4. Palese, P., Ritchey, M.B., and Schulman, J.L. (1977): *J. Virol.* 21:1187–1195.
5. Ritchey, M.B. and Palese, P. (1977): *J. Virol.* 21:1196–1204.
6. Krug, R.M., Veda, M., and Palese, P. (1975): *J. Virol.* 16:790–796.
7. Hay, A.J., Abraham, G., Skehel, J.J., Smith, J.C., and Fellner, P. (1977): *Nucleic Acids Res.* 4:4197–4209.
8. Hay, A.J., Lomniczi, B., Bellamy, A.R., and Skehel, J.J. (1977): *Virol.* 83:337–355.
9. Skehel, J.J. and Hay, A.J. (1978): *J. Gen. Virol.* 39:1–8.
10. Taylor, J.M., Illmensee, R., Litwin, S., Herring, L., Broni, B., and Krug, R.M. (1977): *J. Virol.* 21:530–540.
11. Mark, G.E., Taylor, J.M., Broni, B., and Krug, R.M. (1979): *J. Virol.* 29:744–752.

12. Krug, R.M., Ueda, M., and Palese, P. (1975): *J. Virol.* 16:790–796.
13. Barrett, T., Wolstenholme, A.J., and Mahy, B.W.J. (1979): *Virol.* 98:211–225.
14. Mowshowitz, S.L. and Ueda, M. (1976): *Archives of Virol.* 52:135–141.
15. Ito, Y. and Joklik, W.K. (1972): *Virol.* 50:189–201.
16. Sedat, J., Lyon, A., and Sinsheimer, R.L. (1967): *J. Mol. Biol.* 26:537–540.
17. Scholtissek, C. and Rott, R. (1970): *Virol.* 40:989–996.
18. Stephenson, J.R., Hay, A.J., and Skehel, J.J. (1977): 36:237–248.

David H.L. Bishop and Richard W. Compans, eds.
The Replication of Negative Strand Viruses

Studies on the Action of Nucleic Acid Inhibitors of the Influenza Virion Transcriptase

Phillip Weck,[a] Mark Jackson,[a]
Nowell Stebbing,[a] and Robert Raper[b]

Introduction

The development of *in vitro* assays for the transcriptase of influenza virions following detergent disruption[1] provide a means for evaluating inhibitors of this viral enzyme. The study of such systems could also lead to the discovery of nucleotide sequences that bind either to the viral enzyme or template thus preventing viral replication.[2]

The virion-associated transcriptase of influenza virus is known to be sensitive to chelating agents[3] and can also be inhibited by pyrophosphate analogs.[4] Previous work has demonstrated that certain polyribonucleotides such as poly(rA), poly(rI), and poly(rU) significantly inhibit the virion transcriptase.[5] The most effective of these against a range of influenza viruses are poly(rU), a thiolated derivative poly (4-thiouridylic acid) and a copolymer of cytidine and 4-thiouridine residues [poly(C,S^4U_{10})]. As with polynucleotides which inhibit the reverse transcriptases of RNA tumor viruses,[6] inhibition of the influenza virion transcriptase is dependent on the polymeric nature of the compounds.[5] In addition, the sugar phosphate backbone appears to be necessary for inhibition by polynucleotides because vinyl analogs of poly(rU) and poly(rA) have little or no effect on virus transcription *in vitro*.[5]

The studies described here were designed to examine the role of strandedness of polynucleotide inhibitors, the effects of polydeoxyribonucleotides

[a] Department of Biology, Genentech, Inc., 460 Point San Bruno Boulevard, South San Francisco, California.
[b] Searle Research Laboratories, Lane End Road, High Wycombe, Bucks, United Kingdom.

and analogs of ApG, which is known to stimulate the influenza virion transcriptase.[1]

Materials and Methods

The virion transcriptase activity of purified A/Victoria/75 (Glaxo Labs., Liverpool, United Kingdom) was determined at 31°C in the presence of .20 mM ApG by measuring the incorporation of ^{3}H-UMP into acid-insoluble products using the virion RNAs as templates in reaction mixtures containing either 8 mM $MgCl_2$ + 0.2 mM $MnCl_2$, or 8 mM $MgCl_2$ alone. Standard reactions were set up in a volume of 100 μl and contained 50 mM Tris-HCl (pH 7.8), 150 mM KCl, 2 mM ATP, 2 mM CTP, 0.4 mM GTP, 0.05 mM ^{3}H-UTP (2 Ci/mmol), 5 mM dithiothreitol, 0.5 percent Nonidet P-40, and 100 μg/ml viral protein. Reactions were routinely set up in 96-well microtiter plates, thus allowing simultaneous measurement of a number of inhibitors. Inhibitors were added to the reactions in the presence of ApG five minutes prior to the addition of the divalent cation to initiate the reaction and the reaction was terminated by the addition of ice-cold 8 percent TCA generally 20 minutes after initiation.

Poly- and oligo-ribonucleotides were obtained from P-L Biochemicals, Milwaukee, Wisconsin, d(AG) polymers were obtained from Collaborative Research, Waltham, Massachusetts. Other oligonucleotides containing deoxyribonucleotides were synthesized by Dr. Roberto Crea. The copolymer containing 9 percent cytidine residues and 91 percent 4-thiouridine residues, poly(C,S^4U_{10}) was prepared by Dr. L.D. Bell as described by Hochberg and Keren-Zur.[7] The ID_{50} values for poly(C,S^4U_{10}), poly(rU), and poly(rI) in this study were comparable to those reported previously.[5]

Results and Discussion

Because the *in vitro* inhibition of the reverse transcriptase of retroviruses by polynucleotides is known to be dependent on single-strandedness,[8-10] we determined the effect of annealing polynucleotides to the known polynucleotide inhibitors of the influenza transcriptase. The data in Figure 1 show that inhibition is reduced or absent when poly(rU) or poly(rI) is annealed to its complementary polyribonucleotide. The results shown refer to incubations containing Mg^{++} as the sole divalent cation but the same results were obtained in the presence of Mg^{++} and Mn^{++}.

The complex of poly(rA) with poly(rU) shows some inhibitory effect at concentrations over 10 μg/ml and this reaches about 50 percent at 200 μg/ml. However, there is a 200-fold difference in the ID_{50} for poly(rU) and the complex with poly(rA). In part, the inhibitory effect of the complex of poly(rA) and poly(rU), which occurs at high concentrations, may be due to

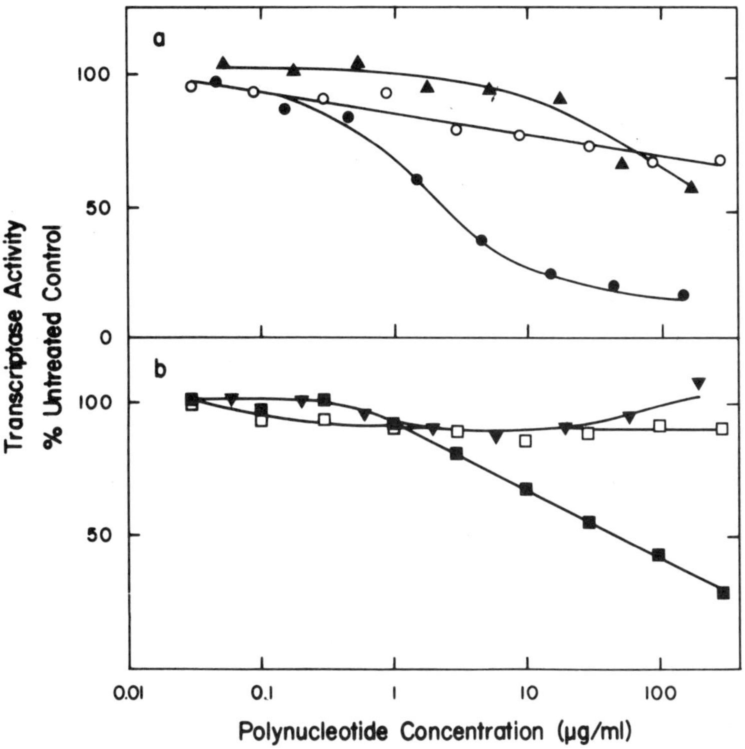

Figure 1. The effect of annealing poly(rA) to poly(rU) or poly(rC) to poly(rI) on the inhibitory effect of the polynucleotides against the transcriptase of influenza A/Victoria/75. (●) Poly(rU); (○) Poly(rA); (▲) Annealed mixture of poly(rU) and poly(rA); (■) Poly(rI); (□) Poly(rC); (▼) Poly(rI).Poly(rC).

Panel a: Solutions of poly(rA) and poly(rU) in 0.1 M NaCl, 0.01M Tris-HCl,pH7.4, were annealed at 21°C for 2 hr. Dilutions of the polynucleotides were added to the assay prior to addition of ApG and divalent cations.

Panel b: Dilutions of poly(rI), poly(rC), and poly(rI).poly(rC) were added to the assay as described for Panel a.

release of poly(rA) from the triple-stranded structure which can form between the component polynucleotides.[11] There are also double- and triple-stranded complexes formed between poly(rA) and poly(S^4U) but the lower Tm of the triple- and double-stranded complexes between these polymers made it difficult to assess the inhibitory effect of the complexes in the absence of the free polymers. However, at 10 μg/ml, poly(C,S^4U_{10}) causes over 90 percent inhibition of the transcriptase[5] and with increasing concentrations of poly(rA) the inhibitory effect of this concentration of poly (C,S^4U_{10}) was progressively decreased, so that at a concentration of 100 μg/ml the inhibition observed could be accounted for by the poly(rA).

It has been shown that substitution of the phosphodiester backbone by a polyvinyl eliminates the inhibitory activity of poly(rA) and greatly reduces the inhibitory activity of poly(rU.[5] The role of the sugar moiety in inhibition of the influenza transcriptase was further studied by comparing the inhibitory effects of poly(rU) with poly(dU) and poly(rI) with poly(dI). The results, in Figure 2, show that on a weight basis the ID_{50} for poly(dU) is approximately 7-fold higher than that of poly(rU) and that the inhibition achieved with poly(dI) is only 20 percent at 100 μg/ml and the ID_{50} for poly(rI) occurs at a slightly lower concentration. The results shown refer to incubations with Mg^{++} but similar results were obtained with Mg^{++} and Mn^{++}.

It is possible that the molar concentration of polynucleotides is of importance for inhibition, although this may not be so significant if the polynucleotide can bind to more than one transcriptase molecule. However, the $S_{20,w}$ values of the poly(rI) and poly(dI) were 9.43 and 7.65 respectively so that, on a molar basis, the difference in the inhibitory effects of these two polynucleotides above 10 μg/ml at least, will be greater than indicated in Figure 2. The $S_{20,w}$ values of the poly(rU) and poly(dU) were 6.10 and 8.60, respectively, so that on a molar basis the difference in the inhibitory effect of these two polynucleotides will be less than indicated by the data in Figure 2 and from estimates of the difference in molecular weight,[12,13] the difference in the ID_{50} may be very small. These results indicate that inhibition of the influenza transcriptase is in part dependent on the nature of the sugar moiety such that a 2′-hydroxyl tends to decrease inhibition although this effect is greater for poly(I) than for poly(U).

Figure 2. The effect of poly(rU) (○), poly(dU) (●), poly(rI) (□) and poly(dI) (■) on the transcriptase activity of influenza A/Victoria/75.

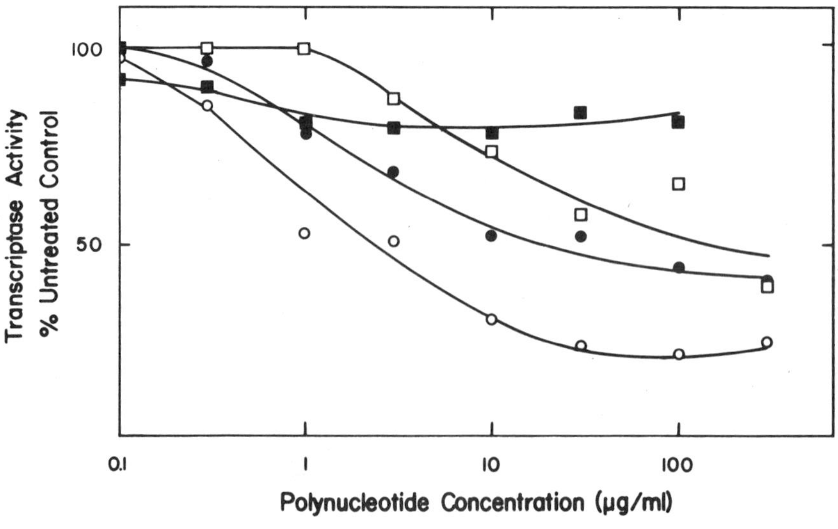

Significant although reduced inhibition by polydeoxyribonucleotides compared with polyribonucleotides indicated that nucleotide binding sites of the influenza transcriptase may not be specific for ribo-sugars. Because ribo-ApG stimulates the transcriptase,[14] we examined related deoxynucleotide compounds for inhibitory activity on the basis that they might bind but not promote transcription. The results, in Table 1, demonstrate that polymers of dp(AG) show some inhibitory activity, the most potent being $dp(AG)_2$ but relatively high concentrations of greater than 300 μg/ml were required to cause near total inhibition. Other deoxynucleotides, such as $dp(T)_4$ dp(GA), and dp(AC) and even several dideoxynucleotides, such as no significant reduction in transcriptase activity primed with ApG. In addition, ribonucleotides containing AG were examined for inhibition of the ApG stimulated reaction and of these only UAG showed significant activity (Table 1). Of the ribonucleotides tested, none showed inhibition of the reaction in the absence of AG and AGC caused a slight stimulation which is to be expected, because this trinucleotide is complementary to the 3′-termini of the influenza virion RNAs.[15] Because $dp(AG)_2$ and ribo-UAG each caused significant inhibition of the virion transcriptase, mixed ribo/deoxy-compounds were synthesized for testing in the standard assay system. Neither dAGrA nor dAGrU inhibited incorporation of ^{3}H-UTP into RNA products except at fairly high concentrations.

The presence of $MnCl_2$ causes random initiation and premature termination of influenza virion transcription *in vitro*. Because inhibitors of transcription could prevent initiation and/or elongation, different ionic conditions were employed to measure inhibition by $dp(AG)_2$, UAG, and poly(C,S^4U_{10}). All three inhibitors reduced transcription in the presence of Mg^{++}, Mn^{++}, or a combination of these divalent cations (Table 2). However, the degree of inhibition was reduced under conditions which favored random initiations,

Table 1. Inhibitory Doses of Various Nucleic Acids on Influenza Transcriptase.

	Material	Inhibitory dose, 50%
deoxy-compounds	dp (AG)	1,000 μg/ml
	dp $(AG)_2$	140 μg/ml
	dp $(AG)_4$	260 μg/ml
	dp $(AG)_{6\text{-}9}$	>1,000 μg/ml
ribo-compounds	AGC	>1,000 μg/ml
	AGU	>1,000 μg/ml
	UAG	150 μg/ml
mixed ribo/deoxy-compounds	$dAGA_r$	780 μg/ml
	$dAGU_r$	2,400 μg/ml

Increasing concentrations of nucleic acids were added to standard transcriptase assays prior to incubation at 31°C for 20 min. The data are presented as the final concentrations that caused a 50 percent reduction in incorporation of ^{3}H-UTP. 1000 μg indicates little or no inhibition was observed at this concentration of nucleic acid.

Table 2. Inhibition of Virion Transcriptase by Nucleic Acid in the Presence of Mg^{++} and Mn^{++}

Divalent cation	Control reactions	Inhibitor added		
		$dp(AG)_2$	UAG	poly(C, S^4U_{10})
		(dpmx 10^{-4}/mg viral protein)		
8 mM Mg^{++}	52.6	27.0 (49)	15.5 (71)	16.1 (69)
1 mM Mn^{++}	2.5	1.9 (24)	0.1 (96)	1.7 (32)
8 mM Mg^{++} + 0.2 mM Mn^{++}	17.1	11.2 (35)	8.3 (52)	11.1 (35)

Nucleic acids were added to reaction mixtures to give final concentrations of 340 μg/ml $dp(AG)_2$, 200 μg/ml UAG or 2 μg/ml poly(C,S^4U_{10}) in the presence of 2 mM ApG. All reactions were incubated at 31°C for 20 min. The numbers in parentheses indicate the percent inhibition in the reaction mixtures as compared to control reactions.

indicating that these molecules inhibit proper initiation of transcription but fail to prevent random initiations which occur in the presence of Mn^{++}. A more direct means for measuring inhibition of elongation involves addition of inhibitors after the transcription reaction has been initiated with divalent cations. The results in Figure 3 show that addition of poly(C,S^4U_{10}) either 10 min or 20 min post-initiation stops elongation of previously initiated RNA chains. In contrast, $dp(AG)_2$ prevents transcription only when present with ApG prior to initiating the reaction with Mg^{++} and Mn^{++}. The lack of total inhibition by prior addition of $dp(AG)_2$ probably reflects random initiations occurring during incubation, since the experiment in Figure 3 was performed with Mg^{++} and Mn^{++}.

These results demonstrate that two possible classes of inhibitors of influenza virion transcriptase exist and can function at the levels of initiation and/or elongation. Because essentially identical results were observed, with regard to strandedness and the different sugar moieties of polynucleotides, using Mg^{++}, or Mg^{++} and Mn^{++}, it appears that the inhibitory effect of poly (C,S^4U_{10}) probably involves elongation although effects on initiation cannot be ruled out. This conclusion is confirmed by inhibition of the non-primed reaction,[5] lacking AG, and inhibition even 20 minutes after initiation of the reaction. Also, the relatively greater inhibitory effect of poly(C,S^4U_{10}) in the absence of Mn^{++} implies some effect on initiation. In contrast, inhibition by $dp(AG)_2$ appears to be related to AG priming because low levels of transcription still occur when $dp(AG)_2$ is added, prior to initiation in the presence of Mn^{++} (Figure 3) and inhibition is insignificant after initiation of transcription. The action of these nucleic acid sequences is presently being investigated in virion transcriptase systems containing cellular mRNAs, which should offer further insight to the mode of action of these inhibitors.

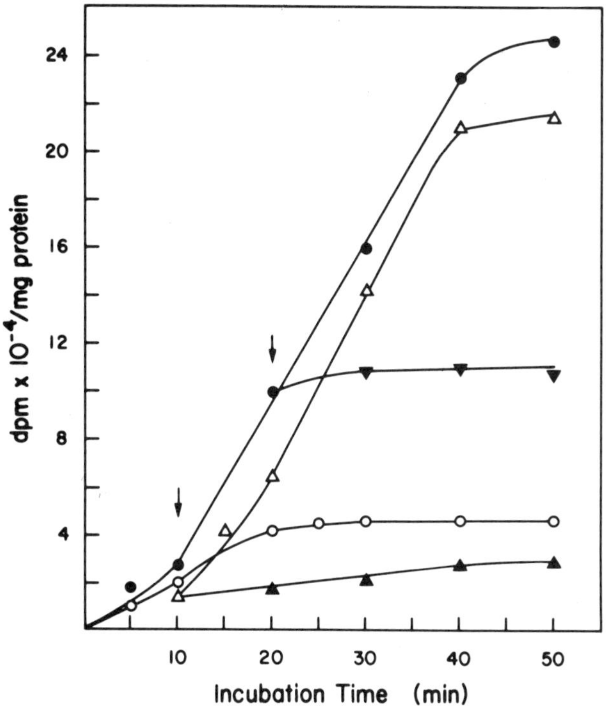

Figure 3. The effect of adding dp(AG)$_2$ or poly(C,S^{4}U$_{10}$) before or after initiating the transcriptase of influenza A/Victoria/75. dp(AG)$_2$ was added at a final concentration of 340 μg/ml before (○) or 10 min after (△) initiating the reaction with Mg^{++} and Mn^{++}. Poly (C,S^{4}U$_{10}$) at 2 μg/ml final concentrations was added at 10 min (▲) or 20 min (▼) after the reaction was started. The solid circles (●) indicate reactions which received no additions and addition of poly(C,S^{4}U$_{10}$) prior to initiation resulted in no incorporation of radioactive precursor.

ACKNOWLEDGMENTS

We thank Dr. L.D. Bell for providing poly(C,S^{4}U$_{10}$) and Dr. R. Crea for synthesis of the oligonucleotides used in these studies.

References

1. Plotch, S.J. and Krug, R.M. (1977): *J. Virol.* 24:24–34.
2. Stebbing, N. (1979): *Pharmac. Ther.* 6:291–332.
3. Oxford, J.S. and Perrin, D.D. (1975): In: *Negative Strand Viruses.* Mahy, B.W.J. and Barry, R.D. (Eds.), Academic Press, New York, pp. 433–444.
4. Stridh, S., Helgstrand, E., Lannero, B., Misiorny, A., Stening, G., and Oberg, B. (1979): *Arch. Virol.* 61:245–250.

5. Smith, J.C., Raper, R.H., Bell, L.D., Stebbing, N., and McGeoch, D. (1980): *Virology* 103:245–249.
6. Erickson, R.J., Janik, B., and Sommer, R.G. (1973): *Biochem. Biophys. Res. Commun.* 52:1475–1482.
7. Hochberg, A.A. and Keren-Zur, M. (1974): *Nucl. Acids Res.* 1:1619–1630.
8. Arya, S.K. and Chawda, R. (1977): *Mol. Pharmac.* 13:374–377.
9. De Clerg, E., Billiau, A., Hobbs, J., Torrence, P.F., and Witkop, B. (1975): *Proc. Nat. Acad. Sci. U.S.A.* 72:284–288.
10. Erickson, R.J. and Grosch, J.C. (1974): *Biochemistry* 13:1987–1993.
11. Blake, R.D., Massoulie, J., and Fresco, J.R. (1967): *J. Mol. Biol.* 30:291–308.
12. Boedtker, H. (1968): *Meth. Enzymology* 12:429–458.
13. Eigner, J. (1968): *Meth. Enzymology* 12:386–429.
14. Plotch, S.J. and Krug, R.M. (1978): *J. Virol.* 25:579–586.
15. Skehel, J.J. and Hay, A.J. (1978): *Nucleic Acids Res.* 4:1207–1219.

Copyright 1981 by Elsevier North Holland, Inc.
David H.L. Bishop and Richard W. Compans, eds.
The Replication of Negative Strand Viruses

Synthesis of Influenza Virus RNAs

Geoffrey L. Smith and Alan J. Hay[a]

During influenza infection, the virus genome is transcribed by two distinct mechanisms differing both in initiation and termination of synthesis.[1] The products of one of these, the A-cRNAs, are exact complements of the eight genome RNAs with 5′-terminal pppA and are presumed to function as templates for the synthesis of progeny virus RNAs (vRNAs). The virus messenger RNAs (mRNAs), on the other hand, are incomplete transcripts which lack sequences complementary to the 16 5′-terminal nucleotides of virus RNAs and have additional 3′-terminal polyadenylate sequences and non-viral coded extensions at their 5′-termini.[1-3] The latter are heterogeneous in length and sequence,[4,5] the different mRNAs having a similar spectrum of terminal sequences, and appear to be derived from cellular messenger RNAs or their precursors that are required in initiating synthesis of these molecules.[6] The synthesis of these mRNAs is regulated such that they are produced in differing relative amounts throughout infection—a characteristic feature of most influenza A virus infections being the preferential synthesis early in infection of mRNAs for the nucleoprotein and nonstructural NS_1 polypeptides,[7-9] resulting in the predominant synthesis of these polypeptides during early stages of infection. Although it is evident that the pattern of mRNA synthesis later in infection is influenced by the host cell,[10] the factors regulating transcription are as yet unknown. Investigations into the relationships between synthesis of vRNAs and their transcripts have recently included analyses of vRNAs synthesized during infection and this

[a] Division of Virology, National Institute for Medical Research, Mill Hill, London.

communication presents some of the results and conclusions deriving from this work.

The composition of vRNA synthesized during infection was examined in a manner analogous to that used in studying the synthesis of virus cRNAs.[7] RNA, extracted from virus infected cells incubated with ^{3}H-uridine, was annealed with an excess of unlabeled mRNA and the radioactive double-stranded RNA molecules were analyzed by polyacrylamide gel electrophoresis as described in the legend to Figure 1. The cRNA used in these experiments was a mixture of primary transcripts and mRNAs from infected cells incubated in the presence of cycloheximide between 2 and 8 hours after infection and purified by chromatography on oligo(dT)-cellulose and CF-11 cellulose. Control experiments showed that under the conditions used, competition of ^{3}H-cRNA was greater than 95% and the composition of vRNA as determined by analyses of double-stranded RNAs after hybridization

Figure 1. The synthesis of virus mRNAs and vRNAs during infection of chick cells by FPV. Chick cell monolayers (10^7 cells/culture) were infected with FPV at 100 pfu/cell and at the following times after infection pairs of cultures were labeled with ^{3}H-uridine (20 μCi/culture) for 30 mins: 0 hr (1), ½ hr (2), 1 hr (3), 1½ hr (4), 2 hr (5), 2½ hr (6), 3½ hr (7), 4½ hr (8), 6½ hr (9), 8½ hr (10). Equal aliquots of RNA extracted from infected cells were denatured and hybridized with an unlabeled excess of either vRNA, or single-stranded virus mRNA that had been extracted from infected cells incubated in the presence of cycloheximide (100 μg/ml) and selected by sequential oligo(dT)-cellulose and CF-11 cellulose chromatography. RNA samples were digested with S_1 nuclease before electrophoresis on a 4% polyacrylamide gel at 60 V for 16 hrs and fluorography as described elsewhere.[7]

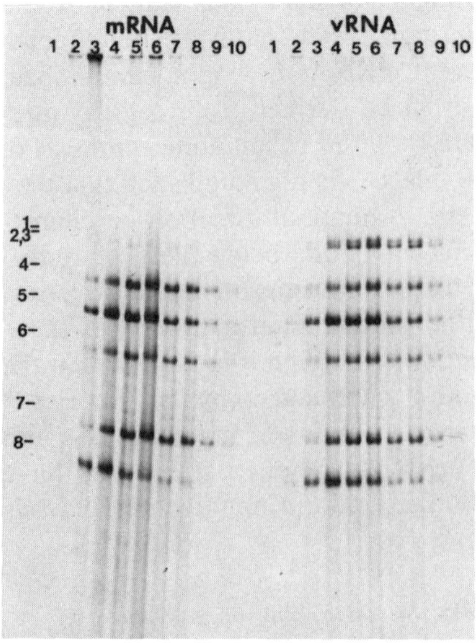

was similar to that indicated by direct analyses of the single-stranded vRNAs (Figure 4). Figure 1 shows the results of comparative analyses of vRNAs and mRNAs synthesized during infection of chick cells by fowl plague virus. The most striking features of this data are the similarities between syntheses of the two types of RNA in regard to the time course of their synthesis and the relative amounts of the individual RNAs produced at different times after infection. Thus vRNA and mRNA are synthesized in similar amounts from around 30 minutes of infection reaching a maximum at 2-2½ hours some 60 minutes later than peak production of A-cRNA.[7] The characteristic changes in relative synthesis of the individual RNAs, including the early predominance of RNAs 5 and 8 between 30 and 90 minutes, and the later diminished synthesis of RNA 8 are the same for both types of RNAs. With the exception of RNAs 1, 2 and 3 (the vRNAs of which are produced in 10-fold greater amounts than the mRNAs) the corresponding vRNA and mRNA molecules are produced in similar amounts.

This correlation between the regulation of synthesis of vRNAs and mRNAs was also apparent during abortive infections of several cell lines, e.g., L cells, HeLa cells and Vero cells, by fowl plague virus. In L cells (Figure 2) the different pattern of mRNA synthesis as compared to that in

Figure 2. The syntheses of virus mRNAs and vRNAs during infections of chick cells and L cells by FPV. Cell monolayers (10^7 cell/culture) were infected and labeled with ^{3}H-uridine (20 μCi/culture) for 30 mins at the following times after infection: 2 hr (1), 3 hr (2), 4 hr (3). Equal aliquots of RNA extracted from infected cells were denatured and hybridized with an excess of either vRNA or virus mRNA. ^{3}H RNAs were digested with S_1 nuclease, electrophoresed on a 4% polyacrylamide gel at 80 V for 16 hr and detected by fluorography as described in Figure 1.

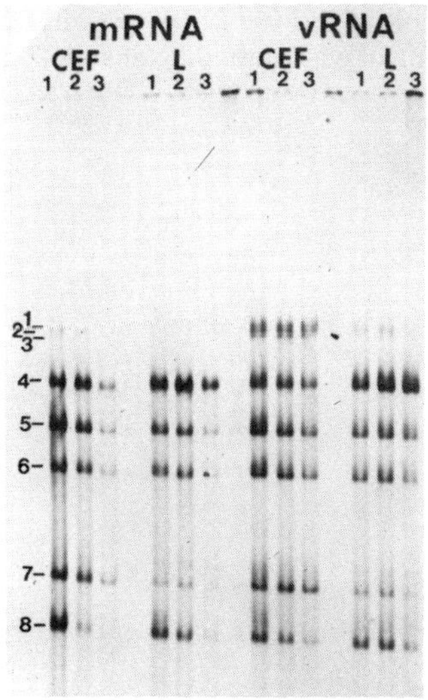

infected chick cells is reflected in the synthesis of vRNA, in particular with respect to the low level of synthesis of RNA 7 throughout infection and the predominance of RNA 4 at later times. Again the various mRNAs are produced in amounts similar to those of the corresponding vRNAs with the exception of RNAs 1, 2 and 3.

The changes in relative predominance of synthesis of the individual RNAs during infection is inconsistent with a mechanism involving sequential synthesis of either vRNAs or mRNAs and this conclusion has been confirmed by the results of analyses of RNA synthesis following ultra-violet irradiation of infected cells. From the results shown in Figure 3, it is apparent that the degree of inhibition of synthesis of the individual vRNAs and A-cRNAs by increasing doses of irradiation is proportional to the sizes of the RNAs, indicating that synthesis of these molecules is initiated independently, as shown previously for mRNAs.[11,12]

In contrast to the similarities between syntheses of vRNAs and mRNAs there is no obvious correspondence between synthesis of vRNAs and production of their presumed templates, the A-cRNAs. Although later in infec-

Figure 3. The sensitivities of A-cRNA and vRNA syntheses to irradiation of infected cells by ultraviolet light. At either 1 hr or 2 hr after infection, the culture medium was removed from chick cell monolayers (10^7 cells/culture) and duplicate cultures were UV-irradiated at 10 ergs/mm²/sec for 0, 5, 10, 20 or 40 secs. After 5 mins incubation at 37°C cells were labeled with ^{3}H-uridine (20 μCi/culture) for 30 mins. RNA extracted from cells irradiated at 1 hr after infection was hybridized with an excess of vRNA and the nonpolyadenylated and polyadenylated double-stranded RNAs were analyzed as described elsewhere.[7] RNA extracted from cells irradiated 2 hr after infection was denatured and hybridized with an excess of virus mRNA and the radioactive RNAs analyzed as described in Figure 2. Radioactivity was determined by counting bands cut from the gels after fluorography. The % of radioactivity remaining in each band compared with unirradiated control was plotted against dose of irradiation.

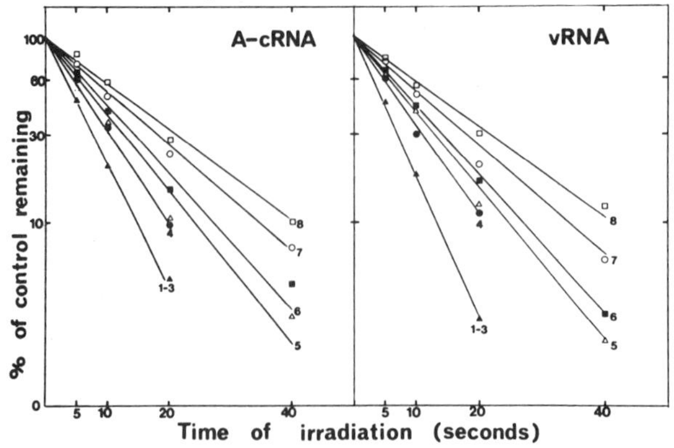

tion, some changes in the synthesis of the latter does occur, in particular the relative reduction in synthesis of RNA 8 (cf vRNA and mRNA synthesis), there is no preferential synthesis at early times.[7] This difference between relative syntheses of vRNAs and relative abundance of the corresponding A-cRNAs is further exemplified during infection of chick cells by a ts mutant of FPV, ts 166,[13] at permissive temperature. Whereas there is a specific 10-fold increase in the production of A-cRNA 6 in these cells, there is no difference in the syntheses of either vRNA or mRNA as compared with the wild-type virus infection.

Without further information on the template utilized for syntheses of vRNAs and mRNAs, evidence as to the regulating mechanisms is equivocal. The available data do, however, suggest that the preferential synthesis of certain vRNAs early in infection is the result of selective transcription of the template. The subsequent changes in the pattern of synthesis is clearly the result of a multiplicity of factors which may include, for example, changes in the relative availability of enzyme and template and host cell factors which are important in influencing synthesis later in infection. Dependence of vRNA replication on the host cell has also been indicated from studies of host-range mutants.[14] Changes in synthesis of vRNA also appear dependent on continuing synthesis of viral or host cell mRNA, since following addition of actinomycin D to infected cells the pattern of vRNA synthesis is invariant. Regulation of secondary mRNA synthesis, on the other hand, may be largely the consequence of alterations in the available template. The close correlation of the relative levels of synthesis of mRNAs with relative levels of synthesis of vRNAs, rather than with the relative abundance of these molecules in the infected cell, suggests an interdependence of the two synthetic processes. Analyses of the synthesis of vRNA and mRNA at nonpermissive temperature in cells infected with certain temperature-sensitive mutants have also suggested that mRNA synthesis is dependent on continuing synthesis of vRNA, although this dependence may be "uncoupled" in the absence of functional protein synthesis. For example, in infected cells incubated in the presence of ρ-fluorophenylalanine, viral mRNA synthesis continues in the absence of vRNA synthesis and since under these conditions the pattern of mRNA synthesis changes in a manner similar to that late in a normal infection, it is apparent that mRNA synthesis may also be modified by factors similar to those responsible for regulating vRNA synthesis.

Another question arising from the variable synthesis of the vRNAs during infection concerns the relationship between the synthesis of these molecules and their incorporation into the genome of progeny virus particles. In the absence of suitable pulse-chase conditions, comparisons between the compositions of vRNA synthesized in infected cells and in released virus particles were analyzed only during later stages of infection. From the results of experiments such as described in Figure 4 it is apparent that the composition

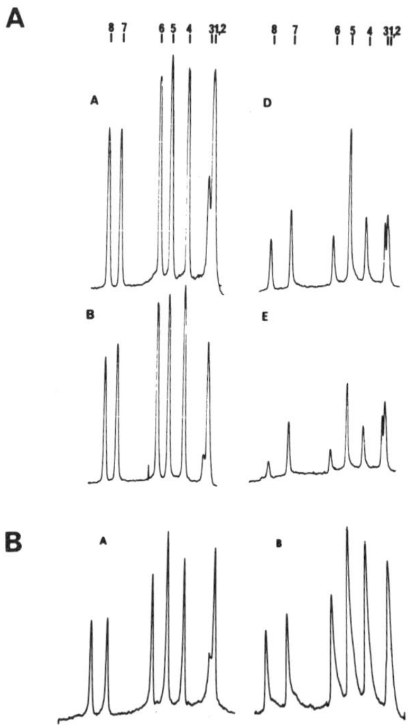

Figure 4. (A) Microdensitometer tracings of fluorographs showing the patterns of vRNAs present in infected cells (D and E) and in virus particles released from these cells (A and B). Infected chick cells (3×10^7 cells/culture) were labeled with ^{3}H-uridine (100 μCi/culture) from either 0-10 hr (D) or 4-10 hr (E) postinfection. An aliquot of RNA extracted from infected cells 10 hr postinfection was denatured and hybridized with an excess of virus mRNA. Samples were digested with S_1 nuclease and analyzed by electrophoresis on a 4% polyacrylamide gel at 80 V for 16 hr as described in Figure 1. Virus particles released from infected cells labeled 0-10 hr (A) or 4-10 hr (B) were purified as described elsewhere[7] and vRNA was extracted and analyzed on a 3% polyacrylamide gel at 80 V for 16 hr. Fluorographs were scanned with a Joyce-Loebl microdensitometer.

(B) Microdensitometer tracings of fluorographs showing the patterns of vRNA before (A) and after (B) hybridization with an excess of virus mRNA. Double-stranded RNAs (B) were treated with S_1 nuclease prior to electrophoresis. Electrophoresis of single and double stranded RNAs were as described in A.

of the genome of released virus particles is independent of the time at which vRNA is synthesized and does not reflect the composition of vRNA in infected cells. It is clear therefore, that following synthesis, a mechanism exists for selecting the genome complement for progeny virus particles.

References

1. Hay, A.J. Skehel, J.J., and McCauley, J. (1980): *Phil. Trans. R. Soc. Lond. B.* 288:341–348.
2. Skehel, J.J. and Hay, A.J. (1978): *Nucl. Acids Res.* 5:1207–1219.
3. Krug, R.M., Broni, B.A., and Bouloy, M. (1979): *Cell* 18:329–334.
4. Caton, A.J. and Robertson, J.S. (1980): *Nucl. Acids Res.* 8:2591–2603.
5. Dhar, R., Chanock, R.M., and Lai, C.-J. (1980): *Cell* 21:495–500.
6. Robertson, H.D., Dickson, E., Plotch, S.J., and Krug, R.M. (1980): *Nucl. Acids Res.* 8:925–942.
7. Hay, A.J., Lomniczi, B., Bellamy, A.R., and Skehel, J.J. (1977): *Virology* 83:337–355.
8. Etkind, P.R., Buchhagen, D.L., Hertz, C., Broni, B.B., and Krug, R.M. (1977): *J. Virol.* 22:346–352.
9. Barrett, T., Wolstenholme, A.J., and Mahy, B.W.J. (1979): *Virology* 98:211–225.
10. Bosch, F.X., Hay, A.J., and Skehel, J.J. (1978): In: *Negative Strand Viruses and the Host Cell.* Barry, R.D. and Mahy, B.W.J. (Ed.), Academic Press, New York, pp. 465–474.
11. Abraham, G. (1979): *Virology* 97:177–182.
12. Pons, M.W. and Rochovansky, O.M. (1979): *Virology* 97:183–189.
13. Ghendon, Y.Z., Markushin, S.G., Blagoveshenskaya, O.V., and Ghenkina, D.B. (1975): *Virology* 66:454–463.
14. Israel, A. (1980): *Virology* 105:1–12.

David H.L. Bishop and Richard W. Compans, eds.
The Replication of Negative Strand Viruses

Electrophoretic Analysis of Influenza Virus Ribonucleoproteins from Purified Virus and Infected Cells

P.J. Rees and N.J. Dimmock[a]

Influenza virus ribonucleoproteins (RNPs) have been separated into five major bands[1] by electrophoresis on a gradient of polyacrylamide gel and with high loads, a number of minor bands became apparent also (Figure 1a). The electrophoresis buffer solution contains 0.1% sodium deoxycholate (DOC) and it is necessary to recirculate the buffer to keep the DOC in solution. Previous analyses of RNPs have included (a) density gradient centrifugation[2,3] which resolved them as single band and (b) velocity gradient centrifugation[4,5] and gel electrophoresis[4] which separated RNPs into one to three broad bands. Separation in our system depends on size and not charge but despite using a variety of gel concentrations (Figure 1b) we could resolve only five species. However, on occasion RNPc could be seen as two separate bands. Resolution was equally good, using ^{35}S-methionine or ^{32}P-radiolabel. (RNA, in this system, migrates off the bottom of the gel). Quantitation of radioactivity in excised RNPs suggested that RNPa is a trimer and RNPc a dimer. Analysis of RNA from the major RNPs by electrophoresis on RNA slab gels demonstrated that RNPa contained RNA of the same mobility as virion RNAs 1, 2 and 3; RNPb: RNA 4; RNPc: RNAs 5 and 6; RNPd: RNA 7 and RNPe: RNA 8.[1] Thus our data support the generally held assumption that each segment of virion RNA is not covalently linked to another. However, we do not know whether or not RNAs 1, 2 and 3 are contained in a single RNP. The minor RNPs are also present in infected cells (see below) but they have not been studied further.

[a] Department of Biological Sciences, University of Warwick, Coventry, CV4 7AL, United Kingdom.

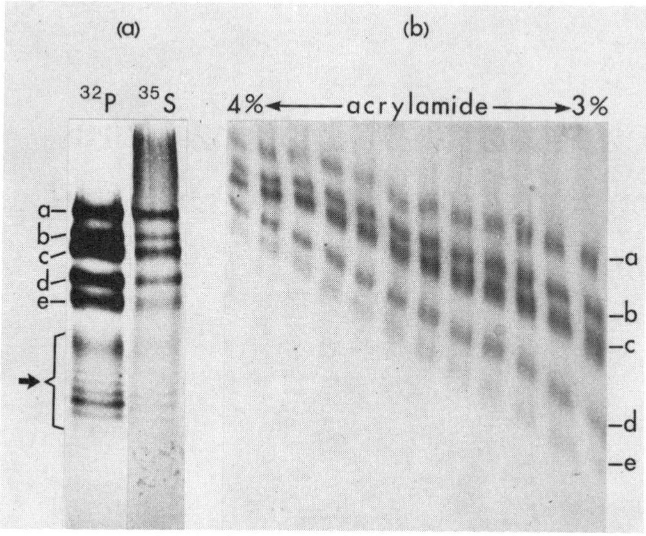

Figure 1. Separation of RNPs by polyacrylamide gel electrophoresis (PAGE) of ^{35}S-methionine or ^{32}P-labeled A/FPV/Rostock/34 (a) on a slab gradient gel of 3 to 5% acrylamide for 23 hr and (b) on a slab of 3 to 4% acrylamide which had been turned through 90° prior to electrophoresis for 45 hr. RNPs were extracted from purified virus contained in low salt buffer[4] (0.2 M NaCl, 0.02 M Tris, pH 7.4, 2 mM EDTA), by disruption with 0.2 vol of 5% nonidet P40 (BDH Chemicals Ltd., Poole, United Kingdom), 2.5% sodium deoxycholate (DOC; BDH Chemicals Ltd.) followed by incubation at 37°C for 4 min[7] before loading onto the gel. Gels were formed from the following solutions: 3% (w/v) acrylamide, 0.14% (w/v) bis-acrylamide, 10% (v/v) glycerol, 0.048% (v/v) N,N,N′,N′-tetramethylene diamine (TEMED), 0.48% (w/v) ammonium persulphate and 4% (or 5%) acrylamide, 0.19% (0.23% for 5% gel) bis-acrylamide, 30% glycerol, 0.2% TEMED, 0.2% ammonium persulphate. All solutions were prepared in electrophoresis buffer (0.02 M sodium acetate, 0.04 M Tris-acetic acid, pH 7.2, 1 mM EDTA).[4] The gel was poured and allowed to polymerize under water-saturated butanol. This was removed and replaced with about 7 ml starter gel[8] containing 3% (w/v) acrylamide, 0.14% (w/v) bis-acrylamide, 0.1% (v/v) TEMED and 0.1% (w/v) ammonium persulphate, prepared in one-fifth strength electrophoresis buffer. Samples were electrophoresed at a constant current of 30 mA in recirculating electrophoresis buffer containing 0.1% DOC. Major RNPs are labeled a to e; minor bands are arrowed. Virus was prepared by infecting CEF monolayers with 0.1 PFU/cell. The PFU:HA ratio of purified FPV was about $10^{4.7}$.

Analysis of ^{35}S-methionine labeled proteins from slab gel-separated RNPs showed that they contained predominantly NP and M-proteins.[1] Trace amounts of presumptive HA_1 were removed by chromatography on Concanavalin A-Sepharose. P-proteins were not associated with isolated RNPs and are probably removed by the deoxycholate.[6] Data presented elsewhere

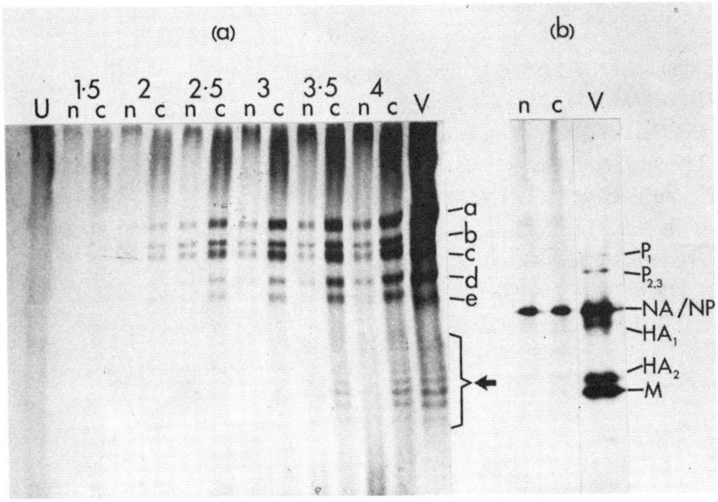

Figure 2. (a) Analysis of RNPs from FPV-infected chick embryo fibroblast cells which were labeled continuously with 25 μCi/ml ^{35}S-methionine from 1 hr postinfection. Cells were fractionated into nucleus (N) and cytoplasm (C) and harvested at the times indicated. No RNPs were resolved in non-infected cells (U). (b) PAGE of proteins, on a 10-30% linear gradient of acrylamide,[9] isolated from nuclear (N) or cytoplasmic (C) RNPs excised from an RNP gel. The marker track (V) is purified FPV virions.

show that each RNP has the same amount of M despite RNPs differing considerably in size,[1] and we suggest that this is indicative of a specific RNP: M association.

The slab gel analysis of RNPs can be used for samples from infected cells (Figure 2a). Non-infected chick embryo fibroblast cells have no recognizable RNPs. Five major bands and a number of minor bands are seen which increase with the time of labeling and these comigrate with RNPs from virus. RNPs appeared in both nucleus and cytoplasm and all species accumulate at about the same rate. A detailed study of the kinetics of appearance of RNPs is in progress. Analysis of proteins of intracellular viral RNPs isolated from a preparative gel (Figure 2b) showed that NP was the major component, thus confirming that they were virus-specified. Trace amounts of other proteins were also present. The proportion of M was less in intracellular RNPs, whether from cytoplasmic or nuclear compartments, than in RNPs obtained from virions.

ACKNOWLEDGMENTS

We are grateful to the Cancer Research Campaign and Medical Research Council for financial support.

References

1. Rees, P.J. and Dimmock, N.J. (1981): *J. Gen. Virol.,* in press.
2. Krug, R.M. (1971): *Virology* 44:125–136.
3. Caliguiri, L.A. and Gerstein, H. (1978): *Virology* 90:119–132.
4. Duesberg, P.H. (1969): *J. Mol. Biol.* 42:485–499.
5. Kingsbury, D.W. and Webster, R.G. (1969): *J. Virol.* 4:219–225.
6. Inglis, S.C., Carroll, A.R., Lamb, R.A., and Mahy, B.W.J. (1976): *Virology* 74:489–503.
7. Pons, M.W. (1971): *Virology* 46:149–160.
8. Jeppesen, P.G.N. (1974): *Anal. Biochem.* 58:195–207.
9. Cook, R.F., Avery, R.J., and Dimmock, N.J. (1979): *Infect. Immun.* 25:396–402.

Published 1981 by Elsevier North Holland, Inc.
David H.L. Bishop and Richard W. Compans, eds.
The Replication of Negative Strand Viruses

Cellular and Viral Control Processes Affect the Expression of Matrix Protein During Influenza Virus Infection of Avian Erythrocytes

N.J. Dimmock,[a] R.F. Cook,[a]
W.J. Bean[b] and Janis M. Wignall[a]

We recently reported the novel finding that avian erythrocytes could be directly infected by several human and avian strains of influenza type A virus.[1,2] Virus proteins are synthesized at about the same rate as in cells in culture and show the same early and late phases.[1-3] However, erythrocytes synthesized about 0.7% virus protein per cell compared with chick embryo fibroblast (CEF) cells. No infectious progeny have been detected despite rigorous efforts and the cause of this abortiveness is unknown. Synthesis of viral or cellular RNA species is too low to be detected by radiolabeling with ^{32}P or ^{3}H-uridine. Virus proteins are synthesized in erythrocytes from chickens of all ages, but in the following experiments we have used those from 13-day-old embryos as these are more convenient to handle and give a higher yield of radiolabeled viral proteins.

The avian erythrocyte has several special features which recommend it for morphological and biochemical studies of influenza virus multiplication: for instance, the nucleus has clearly defined pores, cytoplasm is uncluttered by endoplasmic reticulum (Figure 1a) and there is little macromolecular synthetic activity (see below). Virus particles attach to the cell (Figures 1b,c) and we have observed their uptake by pinocytosis.[4] Fusion between viral and cellular membranes has not been seen and neither are subvirus or virus particles apparent inside the erythrocyte.

Although avian blood can be easily depleted of white cells by centrifugation through Ficoll,[1] fractionation of the resulting "red cell" population by

[a] Department of Biological Sciences, University of Warwick, Coventry, United Kingdom.
[b] St. Jude Children's Research Hospital, Memphis, Tennessee.

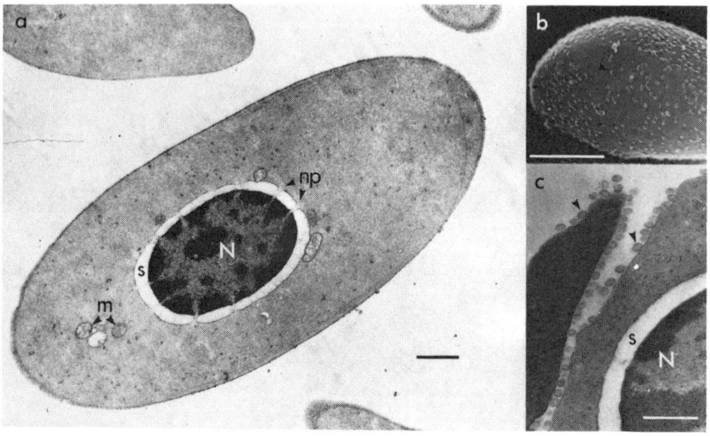

Figure 1. (a) Section of a non-infected erythrocyte from a 13-day-old chicken embryo. The bar is 1 μ. Cells were fixed in 2% glutaraldehyde in phosphate buffered saline, stained with 2% uranyl acetate and embedded by the Spurr resin procedure. Blocks were sectioned with a Reichert OMU2 ultramicrotome and examined with an A.E.I. Corinth 275 electron microscope. (b) Erythrocyte infected A/FPV/Rostock/34 (500 PFU/cell) for 37°C for 1 hr. Cells were fixed with glutaraldehyde, dehydrated, critical point dried and sputter coated with gold before being examined with a Jeol Temscan 100CX scanning electron microscope. The bar is 1 μ. (c) Section of an erythrocyte, infected at 4° for 15 min and processed as (a). The bar is 1 μ. Nucleus: N; nuclear pores: NP; degenerating mitochondria: M. The space between the nuclear membranes (S) is always observed. Virus particles are arrowed.

centrifugation on discontinuous gradients of BSA showed that this was heterogenous, although 85% cells were located in the densest fraction (number 5) of the gradient (Figure 2a). These cells contained hemoglobin, had the lowest ability to incorporate precursors into DNA, RNA or protein (Figure 3) and the characteristic ovoid morphology of mature erythrocytes.

Figure 2. (a) Fractionation of blood from 13-day-old chicken embryos by centrifugation on discontinuous gradients of bovine serum albumin (BSA)[7] at 1000 g for 30 min at 10°C. Cells from the top two interfaces were pooled to form fraction 1. Shown is the % distribution of cells from 13-day embryos and of white cell-free blood from adult laying birds, the proportion of cells with ovoid morphology (+++ is > 99%), synthesis of globin per cell as revealed by ^{35}S-methionine labeling (+++ is maximal, see Figure 4) and % cells containing hemoglobin (Hb) by histochemical staining. (b) Polyacrylamide gel electrophoresis (PAGE) of FPV/Rostock proteins synthesized by fractions 1-5 from the BSA gradients above. These were infected with about 30 PFU/cell and labeled for 30 min at 6 hr postinfection with 100 μCi/ml ^{35}S-methionine. Extracts of each cell fraction were loaded onto the gel in proportion to their distribution in blood. Note the synthesis of globin (g) by erythrocytes and the parallel infected (I) and non-infected (U) chick embryo fibroblast cells and purified virions (V). Experimental details are in reference 1.

a

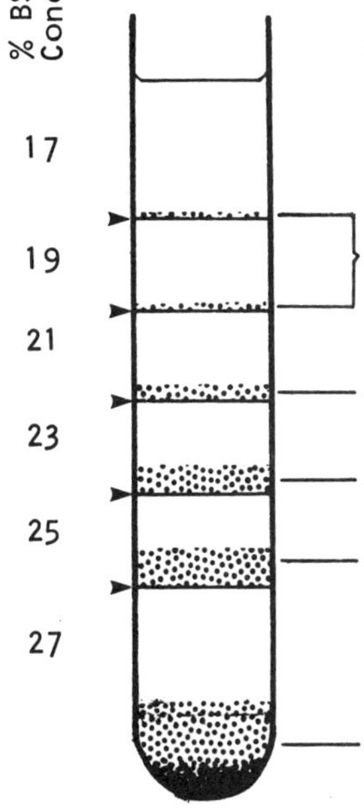

Fraction No.	% Distribution of cells: 13 Day	% Distribution of cells: Adult	Proportion with ovoid morphology	Synthesis of globin	% Containing haemoglobin
1	0.1	<0.1	-	tr	0
2	0.2	0.1	-	±	0
3	2.3	0.1	+	+	0
4	11.2	5.0	++	++	15
5	86.1	95.0	+++	+++	>99

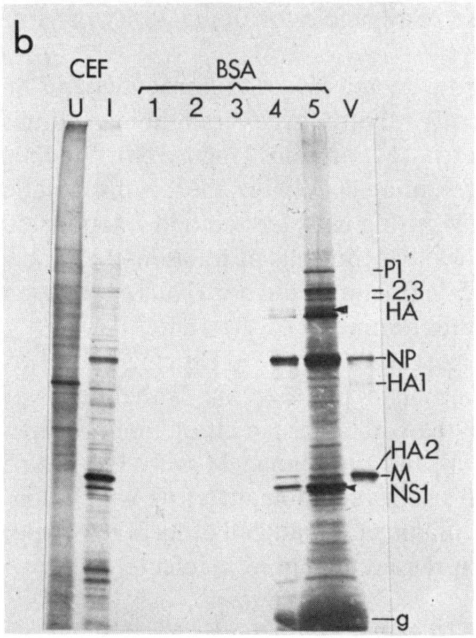

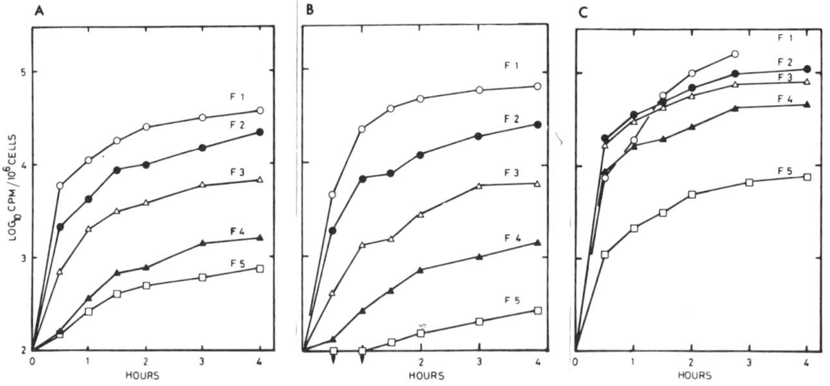

Figure 3. Normalized incorporation of radioactive precursors into TCA-insoluble material by cell fractions obtained by BSA gradient centrifugation of blood from 13-day-chicken embryos (Figure 2a). Cells were labeled continuously with 40 μCi/ml ^{3}H-tymidine (18-25 Ci-mmol.), 50 μCi/ml ^{3}H-uridine (25-30 μCi-mmol), or 10 μCi/ml ^{3}H-leucine (40-60 μCi/mmol).

Compared with cultured CEF cells, fraction 5 erythrocytes synthesize 1.6% DNA, 0.2% RNA and 7.0% protein. All the biochemical, histological and morphological data support the view that fractions 1-5 represent increasingly differentiated cells of the erythropoietic series.[5,8] All these cell fractions synthesized influenza virus proteins (unpublished data). However, when radiolabeled extracts from infected cell fractions were loaded onto a polyacrylamide slab gel in the same proportion as they occur in whole blood, it was clear that fraction 5 erythrocytes were responsible for the majority of virus protein synthesis (Figure 2b).

Although all influenza A strains so far tested infect erythrocytes and induce the synthesis of similar amounts of viral proteins overall, human strains such as A/NWS do not synthesize matrix (M) protein[2] (Figure 5a). We then investigated whether or not less differentiated cells of the erythropoietic series had the capacity to synthesize M, using cells from blood fractionated as described in Figure 2. Figure 4 shows that the cells of fractions 1, 2 and 3 expressed the M-protein of A/NWS while fraction 4 did not. Thus expression of M is dependent upon the state of differentiation of the cell.

Why should avian strains (A/FPV/Rostock/34, A/FPV/Dutch/27), A/Turkey/Ontario/7732/66) and not human strains synthesize M? To help answer this question we co-infected erythrocytes with an avian strain and a human strain and found that both the avian M and the human M were expressed (Figure 5).[2] Another approach was to use a recombinant virus which contained RNA 7 (encoding M) from the human parent and all other RNAs from the avian parent.[6] The recombinant expressed human M in infected erythrocytes (Figure 6).

To explain how the expression of M depends on both the virus strain (Figures 5, 6) and the infected cell (Figure 4), we suggest that there has to be

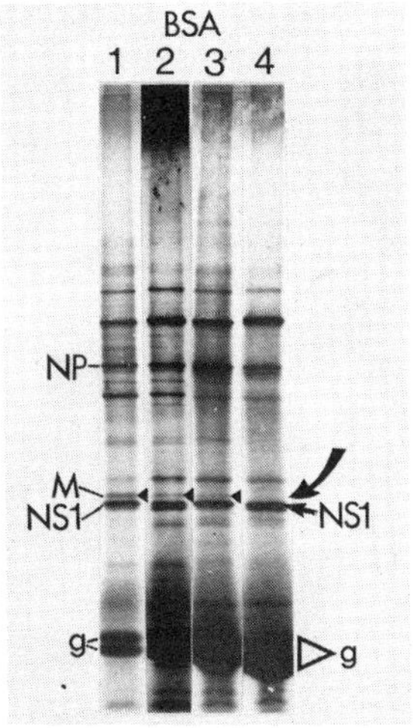

Figure 4. PAGE of proteins synthesized by cell fractions obtained by BSA gradient centrifugation of blood from 13-day chick embryos. Cells were infected with a human strain A/NWS. BSA fraction 5 is not shown. Radiolabeling was from 6 to 6.5 hr postinfection and approximately equal amounts of radioactivity were loaded onto each track. The large arrow indicates the absence of M from fraction 4 and the small arrow the presence of M in fractions 1-3. The amounts of globin (g) synthesized increase from fractions 1 to 4.

Figure 5. Synthesis of the M-protein of a human influenza virus strain (A/NWS) in erythrocytes co-infected with A/FPV/Dutch. (A) Erythrocyte infected (i) or not infected (n.i.) with A/FPV/Dutch or A/NWS at about 10 and 100 PFU/cell respectively. Radiolabeling was with 100 μCi/ml ^{35}S-methionine/10^7 cells from 4 to 4.5 hr after infection. (B) Mixed infection of erythrocytes by A/FPV/Dutch and A/NWS together. (C) Photographic enlargement of part of B. Outer tracks are markers of infected CEF cells; inner tracks are the mixed infection pulsed at 4 hr after infection for 10 min or chased for an additional 10 min (10 + 10). (This figure first appeared in reference 2).

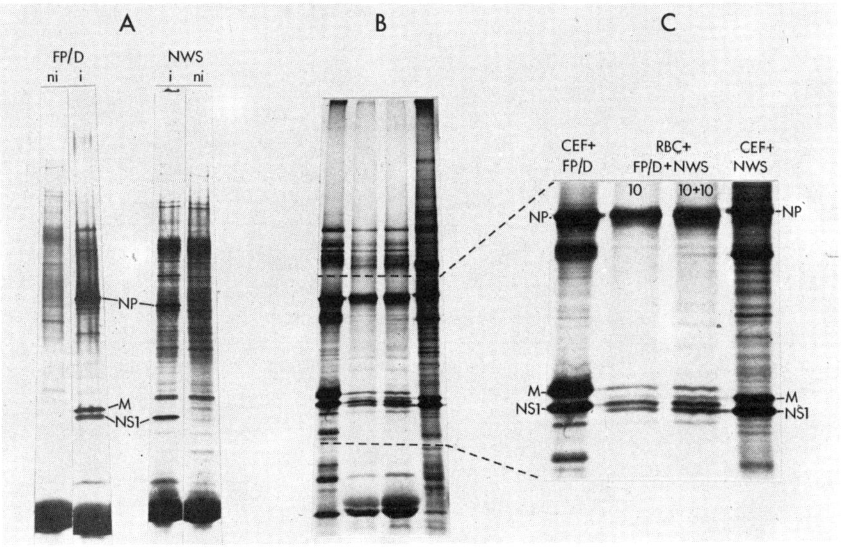

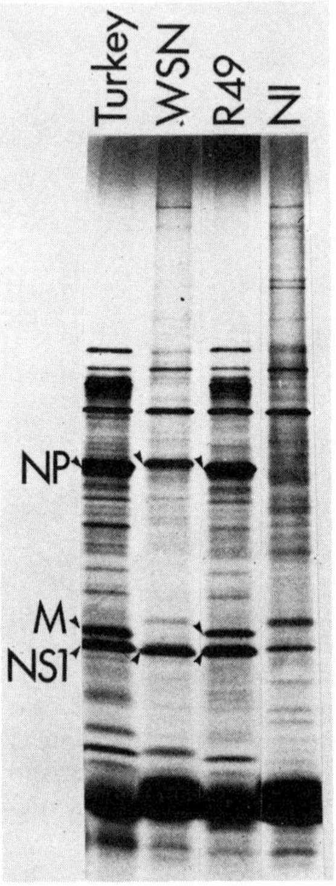

Figure 6. Infection of erythrocytes with a hybrid strain (R49) having the matrix gene (RNA 7) of the human parent (A/WSN) and RNAs 1-6 and 8 of the avian parent (A/Turkey/Ontario/7732/66). Cells were infected at a multiplicity of about 10 PFU/cell and radiolabeled from 6 to 6.5 hr postinfection. Conditions of infection, radiolabeling and analysis are described in the figure legends above. NP, M and NS1 proteins are arrowed. Other viral proteins can be distinguished by comparison with the non-infected erythrocytes (NI).

Table 1. Hypothetical Scheme of the Control of Expression of Matrix (M) Protein by Viral and Cellular Genes in Infected Avian Erythrocytes.

Avian strain →→ $X_{av} + Y_{cell} \rightarrow X_{av}Y_{cell}$	→→ M_{av}
	→→ M_{hu}
Human strain →→ $X_{hu} + Y_{cell}$ → incompatible, no M_{hu} expressed	

X: virus gene products(s); Y: cell gene product(s); →→: unknown number of steps in the pathway; av: avian; hu: human.

"compatibility" between a viral gene product (or products) and a cell gene product (or products) before M is expressed (Table 1). This compatibility may be of a quantitative nature (e.g., reflecting the affinity between viral X and cellular Y in Table 1) or qualitative (meaning that the appropriate form of Y is absent in mature erythrocytes).

While viruses are by definition dependent on host cell gene products, this report describes the dependence of a particular viral gene product upon a host cell function which is lost as the cell differentiates.

ACKNOWLEDGMENTS

We thank the Agricultural Research Council for financial support and Margaret K. Arnold (N.E.R.C., Unit of Invertebrate Virology, Oxford) for expert advice and use of the scanning electron microscope. We acknowledge the helpful discussions of part of this work with R.J. Avery and Grant AI 16841 from the NIAID.

References

1. Cook, R.F., Avery, R.J., and Dimmock, N.J. (1979): *Infect. Immun.* 25:396–402.
2. Cook, R.F., Avery, R.J., and Dimmock, N.J. (1980): *Archives Virol.* 65:319–324.
3. Skehel, J.J. (1977): *Virology* 56:394–399.
4. Bossart, W., Meyer, J., and Bienz, K. (1973): *Virology* 55:295–298.
5. Lucas, A.M. and Jamroz, C. (1961): *Atlas of Avian Hematology,* U.S. Department of Agriculture. Monograph No. 25.
6. Bean W.J. and Webster, R.G. (1978): In: *Negative Strand Viruses and the Host Cell.* Mahy, B.W.J. and Barry, R.D. (Eds.) Academic Press, London, pp. 685–692.
7. Williams, N., Kraft, N., and Shortman, K. (1972): *Immunology* 22:885–899.
8. Kabat, D. and Attardi, G. (1967): *Biochim. Biophys. Acta* 138:382–399.

David H.L. Bishop and Richard W. Compans, eds.
The Replication of Negative Strand Viruses

Interaction of the Structural Polypeptides of Influenza Virus with the Cellular Cytoskeleton During Productive Infection of Human Fibroblasts

John Leavitt,[a] Grace Bushar,[a] Nibedita Mohanty,[a] Ronald Mayner,[a] Takeo Kakunaga,[b] and Francis A. Ennis[a]

Influenza A virus infected cells synthesize eight viral polypeptides: P_1, P_2, P_3, HA_0, NP, NA, M and NS.[1-4] All viral polypeptides except NS are structural components of the infectious virion. The diploid human fibroblast strain known as KD,[5,6] although relatively unproductive as a host for influenza infection, provides a model system for comparison of the permissive and non-permissive host cell state. Clonal sublines of KD cells have been isolated which are elevated 100 to 1000-fold in ability to produce infectious influenza A virus. These clonal sublines revert to an unproductive antiviral state when preincubated with human fibroblast interferon. The clonal cell line HUT-14[6,7] exhibits an altered morphology which is indicative of a change in the ultrastructure of its cytoskeleton. Recently, in collaboration with others, we have shown that this KD substrain synthesizes a mutant form of β actin[6] which is diminished in ability to integrate into the cytoskeleton.[5,6]

The observation that cellular assembly complexes of pox virus are found attached to a reorganized cytoskeleton of the infected host cell[7] led us to investigate the possible relationship of the cytoskeleton to influenza A infection in KD and HUT-14 cells. We report here that newly synthesized influenza A virus polypeptides are bound almost exclusively to the cytoskeleton in nearly equal proportions in both infected cell types. Our observations suggest that the cytoskeleton may play an important role in the cellular infectious process.

[a] Department of Health and Human Services, Public Health Service, Food and Drug Administration, Bureau of Biologics, Division of Virology 8800 Rockville Pike, Bethesda, Maryland.
[b]National Cancer Institute, Bethesda, Maryland.

Materials and Methods

Infectious H3N2 influenza A virus, the Hong Kong strain (A/Aichi/2/68) used in this study, was provided by Dr. Akira Sugiura.[8] Purified ^{35}S-methionine labeled virus grown in MDBK cells was also provided by Dr. Sugiura. KD and HUT-14 cells were maintained as described previously.[5] Labeling of cellular proteins with ^{35}S-methionine and preparation of the soluble and cytoskeletal cell fractions with a solution of 1% Triton X-100 has been described.[5] Radioactive cell-free virus secreted into the supernatant culture medium was prepared for gel electrophoresis by centrifugation of the culture fluid at 30,000 rpm (Ti50 rotor) for two hours, followed by solubilization of the virus pellet with electrophoresis sample buffer.[5] The procedures for separation and identification of influenza A polypeptide by one- and two-dimensional polyacrylamide gel electrophoresis and quantitative densitometry of stained gels gels and autoradiographs have been described in detail previously.[5,9]

Results

Detection of influenza polypeptides in the cytoskeleton. Titering experiments show that HUT-14 cells produce 10^6 infectious units per 10^6 cells while KD cells produce a virus titer which is 100 to 1000-fold lower. HUT-14 cells are killed by the infection while KD cells exhibit no apparent cytopathic effect within the same time period.

Figure 1 shows a separation by SDS polyacrylamide gel electrophoresis of ^{35}S-methionine labeled polypeptides of subcellular fractions of KD and HUT-14 cells either uninfected or infected with a mouse adapted strain of A/Hong Kong/68 H3N2 virus. In each fraction, the influenza polypeptides have been identified by migration rate in SDS polyacrylamide gels.[9] The relative amount of each viral polypeptide (^{35}S-methionine radioactivity) in the soluble, cytoskeletal, or cell-free supernatant fraction of each cell type was measured by densitometry of unsaturated autoradiographs from one-dimensional gels. These measurements shown in Table 1 indicate that greater than 90 percent of the cell-associated viral polypeptides M, NP, and HA_0 are present in the cytoskeleton of both infected cell-types. A low percentage of each viral polypeptide is secreted in a particulate form into the cell-free supernatant.

In Figure 1, two novel influenza polypeptides, NP_2 and M_x, are indicated. The identity of these polypeptides as modified forms of NP and M, respectively, is based on their electrophoretic behavior during two-dimensional gel electrophoresis which separates polypeptides by both charge and molecular weight. The two-dimensional technique also allows separation of the influenza polypeptide NP_1 from vimentin, an abundant cytoskeletal polypeptide which shares the same migration rate (M_r = 56,000) in SDS gels.

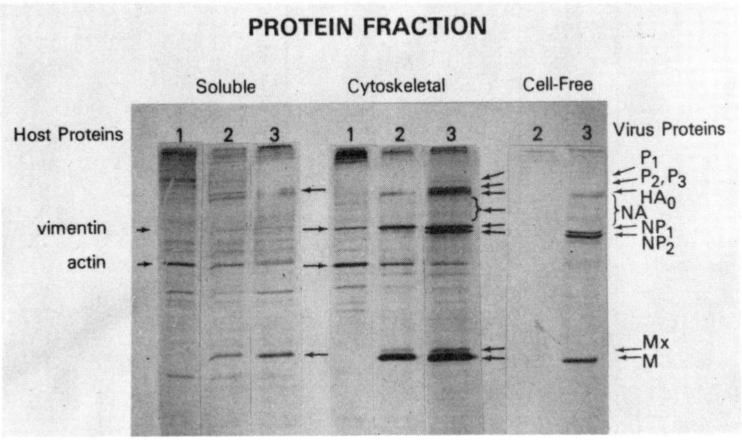

Figure 1. Autoradiography of one-dimensional SDS-polyacrylamide gel electrophoresis[5] of ^{35}S-methionine labeled polypeptides: Lane 1, uninfected KD cells; Lane 2, A/Hong Kong infected KD cells; and Lane 3, A/Hong Kong-infected HUT-14 cells. Vimentin is 10 micron filament protein;[5,7] influenza polypeptides have been identified previously[9] or are described in the text. Polypeptides were labeled for the final 12 hours of an 18-hour infection. Preparation of soluble proteins, the cytoskeletal fraction, and the cell-free fraction is described elsewhere[5] or in the Methods.

Figure 2 shows two-dimensional electrophoretic patterns of cytoskeletal polypeptides and secreted cell-free virus from infected cells. The HA_0, NP, and M are clearly visible in Coomassie blue stained gels indicating that the viral polypeptides amount to an abundant percentage of the total cytoskeletal protein. Autoradiographs of these gels indicate that viral polypeptides are synthesized preferentially with concomitant shut-off of host protein synthesis, since there is a dimunition of radioactivity in the actin and vimentin spots (seen by comparison to the same spots in uninfected cytoskeletons). The NP_2 polypeptide is absent in the KD cytoskeleton but is found to be a major component of the HUT-14 cytoskeleton; otherwise, KD cells are nearly as efficient as HUT-14 cells in synthesis of HA_0, NP, and M (Table 1).

Although NP_1 is the only form of NP found in the KD cytoskeleton, virus secreted by KD cells into the cell-free supernatant exhibited both NP_1 and NP_2. The ratio of NP_1 to NP_2 differed greatly between KD virus, HUT-14 virus, and virus yielded by the additional permissive cell line MDBK (Table 2). The ratio of NP ($NP_1 + NP_2$) to M in cell-free virus also differed greatly between KD virus (NP: M = 0.5), HUT-14 virus (NP: M = 1.4) and MDBK virus (NP: M = 1.1).

Localization of influenza antigens in the cytoskeleton. In addition to the cytoplasmic intermediate filament and microfilament systems, the cytoskeleton

Table 1. Relative Amounts of Matrix, Nucleocapsid, and Hemagglutinin Polypeptides in the Cell-free, Cytoskeletal, and Soluble Cellular Fractions.

Influenza polypeptide	KD cellular fractions						HUT-14 cellular fractions					
	Cell-free		Cytoskeletal		Soluble		Cell-free		Cytoskeletal		Soluble	
A/Hong Kong	O.D.	%	O.D.	%	O.D.	%	O.D.	%	O.D.	%	O.D.	%
Matrix (M + Mx)	0.5	<1	55.2	90	6.0	10	2.8	4	62.9	90	4.1	6
M	0.5		51.2		6.0		2.8		60.7		4.1	
Mx	ND	4.0			ND		ND		2.2		ND	
NP ($NP_1 + NP_2$)	trace		26.8	97	0.7	3	4.0	10	35.6	87	1.2	<3
NP_1	trace		26.8		0.7		1.3		18.7		0.7	
NP_2	trace		ND		ND		2.7		16.9		0.5	
HA_0	trace		10.8	>90	trace		1.4	10	13.9	90	trace	

Optical density measurements (O.D.) made on unsaturated audioradiographs were normalized to reflect a 56-hour autoradiography (Figure 1) of polypeptides from 6×10^4 cells which were infected 18 hours (moi 5) and labeled with ^{35}S-methionine[5] for the final 12 hours. ND indicates that no polypeptide band was detectable. The percentages refer to the relative distribution of a single polypeptide species between the three cell fractions. The O.D. measurements are a measure of the ^{35}S-methionine residue in each polypeptide.

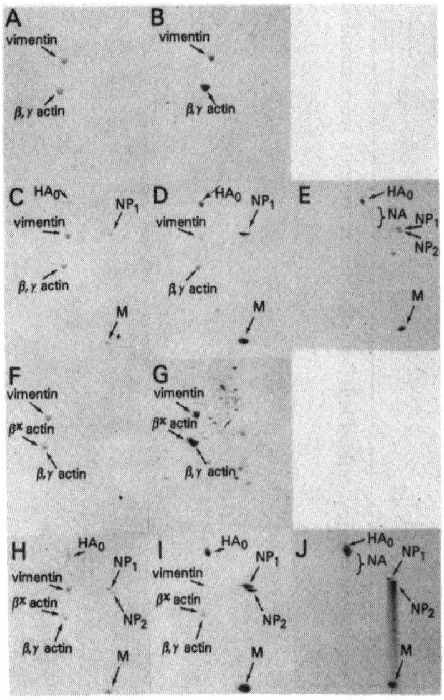

Figure 2. Two-dimensional gel electrophoresis[6] with non-equilibrium pH gradient electrophoresis in the first dimension.[10] Coomassie blue stained gels A, C, F, and H of cytoskeletal proteins are presented with the corresponding autoradiographs of ^{35}S-methionine-labeled polypeptides B, D, G, and I, respectively. Cytoskeletal patterns A, B, F, and G are from uninfected cells and C, D, H, and I are from infected cells. The KD polypeptides are A through E and the HUT-14 polypeptides are F through J. Autoradiographs of KD and HUT-14 cell-free viral polypeptides are shown in E and J respectively. β^x, a mutant actin polypeptide is described elsewhere.[6] Vimentin and β, γ actin are host cytoskeletal proteins.[5,7]

consists of a nuclear component which includes the nuclear matrix, the DNA genome, and DNA bound proteins such as histones.[7] A panel of 95 anti-influenza monoclonal antibodies which are specific for antigenic determinants of HA, NP, and additional unidentified influenza antigens were used in an indirect immunofluorescence microscopy experiment to determine the location of antigenic influenza polypeptides in the cytoskeleton. All influenza antigens recognized by this panel of monoclonal antibodies were found localized in the cytoskeleton of infected KD cells.

Four distinct cytoskeletal patterns of antibody-influenza antigen complexes shown in Figure 3 were observed with the 95 monoclonal antibodies. Table 3 summarizes the specificities of monoclonal antibodies of each immunofluorescence class (Figure 3) determined by (a) immunoprecipitation experiments using ^{35}S-methionine labeled A/Hong Kong influenza polypeptides and protein A-Sepharose and (b) functional assays including plaque

Table 2. Proportions of Nucleocapsid Polypeptide Subunits NP_1 and NP_2 in Cell-Free Virus Particles, Cytoskeleton, and Soluble Cellular Fractions.

		KD cells			HUT-14 cells		MDBK cells
	Cell-free virus %	Cytoskeleton %	Soluble %	Cell-free virus %	Cytoskeleton %	Soluble %	Cell-free virus %
NP_1	55	>99	>99	32	53	58	85
NP_2	45	ND	ND	68	47	42	15

The percentages reflect the ratio NP_1:NP_2 in a particular cell fraction. MDBK cells are bovine kidney cells. Densitometry measurements were made on autoradiographs of two-dimensional gels. ND indicates that the NP_2 spot was not detectable.

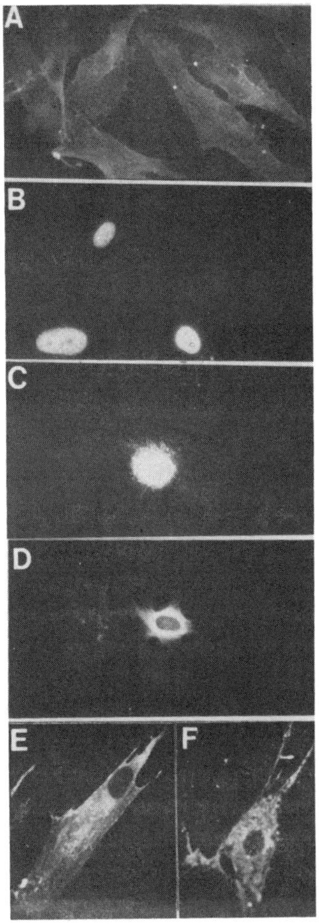

Figure 3. Detection of A/Port Chalmers H3N2 influenza antigens in the KD cellular cytoskeleton by indirect immunofluorescence microscopy with individual mouse monoclonal antibodies and anti-mouse IgG antibody conjugated with fluorescein: A, anti-human cytoskeletal monoclonal antibody 5; B, anti-NP monoclonal antibody 304; C, monoclonal antibody 335 specific for influenza infected cells but with unknown antigenic specificity; D, anti-NP monoclonal antibody 192; E, KD cell and F, KD cytoskeleton, anti-HA monoclonal antibody 107. Cells were infected at a low multiplicity so that less than 5 percent of the cells exhibited influenza antigens.

inhibition, hemagglutination inhibition, neuraminidase inhibition and complement-mediated hemolysis. Considering the large number of monoclonal antibodies that have been examined thus far, it has been disappointing that no anti-M or anti-NA monoclonal antibodies have been identified.

Table 3. Localization of Influenza Antigens in the Cytoskeleton with Monoclonal Antibodies and Indirect Immunofluorescence Microscopy.

Secondary immunization before hybridoma preparation	Cytoskeletal antigen distribution pattern B, C, D or E	Number of monoclonal antibodies tested	Antibody specificities included
Homologous H3N2			
A/Port Chalmers core particles	Cytoplasm (E)	50	Hemagglutinin HA_0 H3N2 Plaque inhibiting antibodies Hemagglutination inhibiting antibodies Hemolytic antibodies
	Nucleus (B)	17	NP Subtype specific Subtype cross-reactive
	Perinuclear cytoplasm (D)	8	NP Subtype specific Subtype cross-reactive
Heterologous H1N1			
A/USSR core particles	Nucleus (B)	15	NP Subtype cross-reactive
	Nucleus and perinuclear cytoplasm (C)	1	Unknown Subtype cross-reactive
	Perinuclear cytoplasm (D)	3	NP Subtype cross-reactive

Core particles used for immunizations were prepared by incubation of virus in 1% T X-100 detergent and sucrose gradient sedimentation to separate core particles from solubilized HA and NA.[9] Core particles still retain a low percentage of the viral HA and NA. The patterns of distribution of antigen B, C, D, or E refer to those shown in Figure 3.

Discussion

The results presented show clearly that influenza A polypeptides M, NP, and HA_0 are associated with the host cell cytoskeleton during vegetative infection. Even at the earliest stages of infection, when viral gene products are first detectable, the viral polypeptides are found associated exclusively with the cytoskeleton (unpublished result). The amount of M, NP, and HA_0 left with the cytoskeleton after release of the maximum yield of infectious virus represents about 90% of the total amount of these three polypeptides synthesized. The cytoskeletal-bound influenza polypeptides for the most part do not represent a cell-associated form of the assembled virus, since differing antigens are found located in different compartments of the cytoskeleton including the nucleus. Thus virus assembly, while possibly an important function of the cytoskeleton,[7] may be only one functional aspect of the interaction between viral and host structural polypeptides.

Recently, Fulton et al.[10] have shown that non-viral cytoskeletal proteins are synthesized on polyribosomes that are associated with the cytoskeleton, and that nascent cytoskeletal polypeptides are incorporated into the cytoskeleton framework at the site of synthesis. This observation may explain the compartmentalization of influenza polypeptide antigens, if indeed they are synthesized on cytoskeletal polyribosomes. Failed attempts to solubilize influenza polypeptides by treating cytoskeleton preparations of infected cells with either ribonuclease or deoxyribonuclease suggest that influenza polypeptides are bound to the cytoskeleton by association with host cytoskeletal proteins. The finding that HA is resistant to release from the cytoskeleton by Triton detergent, which solubilizes membrane phospholipids,[10] argues that cellular HA is not exclusively membrane-bound. The insolubility of cytoskeletal HA distinguishes this form of HA from virion HA which is solubilized from the envelope membrane of the virus particle by non-ionic detergents.[9] The highly acidic charge of the cytoskeletal and nuclear matrix structural polypeptides[5] make them suitable electrostatic receptors for the relatively basic domain of HA_0 (HA_1) and the highly basic M and NP polypeptides.[9]

The appearance of a second electrophoretic form of NP (NP_2) in the cell-free virus of KD and HUT-14 cells and the absence of this electrophoretic form on the KD cytoskeleton suggests that this modification of NP may reflect a maturation process of the influenza nucleocapsid. The contrasting abundance of NP_2 on the HUT-14 cytoskeleton is one distinguishable difference between the two host infections that might with future study provide an explanation for the differing yields of virus from the two cell-types.

In future experimentation, we hope to explore the relationship of the mutant β actin to the acquired behavior of HUT-14 cells as an influenza host. Future investigation of the effects of cytoskeletal variation on virus productivity in this model cell system and in differentiated cell-types which exhibit unique cytoskeletal architectures may unveil critical features of the host cell-virus relationship.

References

1. Lazarowitz, S.G., Compans, R.W., and Choppin, P.W. (1971): *Virology* 46:830–843.
2. Compans, R.W., Klenk, H-D., Caliguiri, L.A., and Choppin, P.W. (1970): *Virology* 42:880–889.
3. Inglis, S.C., Carroll, A.R., Lamb, R.A., and Mahy, B.W.J. (1976): *Virology* 74:489–503.
4. Ritchey, M.B., Palase, P., and Schulman, J.L. (1977): *Virology* 76:122–128.
5. Leavitt, J. and Kakunaga, T. (1980): *J. Biol. Chem.* 255:1650–1661.
6. Vandekerckhove, J., Leavitt, J., Kakunaga, T., and Weber, K. (1980): *Cell,* in press.
7. Hiller, G., Weber, K., Schneider, L., Parajsz, C., and Jungwirth, C. (1979): *Virology* 98:142–153.
8. Sugiura, A. and Ueda, M. (1980): *Virology* 101:440–449.
9. Leavitt, J.C., Phelan, M.A., Leavitt, A.H., Mayner, R.E., and Ennis, F.A. (1979): *Virology* 99:340–348.
10. Fulton, A.B., Wan, K.M., and Penman, S. (1980): *Cell* 20:849–857.

David H.L. Bishop and Richard W. Compans, eds.
The Replication of Negative Strand Viruses

Genetic Characterization of an Influenza Virus from Seals

William J. Bean, Jr., Virginia S. Hinshaw, and Robert G. Webster[a]

In December, 1979, an unusually large number of dead or moribund harbor seals (*Phoca vitulina*) were found on the beaches of Cape Cod, Massachusetts. Post-mortem examination of these animals revealed severe lung consolidation typical of primary viral pneumonia (Geraci et al., manuscript in preparation). The total mortality during the winter of 1979-1980 for this region was estimated at 500, or about 20% of the local seal population.

Analysis of tissue samples from these animals revealed high titers of influenza virus in the lungs (10^6-10^7 EID_{50}/gm) and lower titers in the brain ($10^{1.5}$-$10^{2.5}$ EID_{50}/gm).[1] All virus isolates were of the serotype, HavlNeql. The prototype virus with this antigenic combination is A/Fowl Plague/Dutch/27, a strain that had previously been isolated only from domestic fowl.[2] The host range of this virus was tested by experimental infection of several species of birds and mammals.[1] Replication in birds was sporadic and limited to the respiratory tract. There were no clinical symptoms and no intestinal replication of the virus as is seen with many avian influenza virus strains. In mammals, the virus replicated in the respiratory tract of swine, cats, ferrets, guinea pigs and mice.

Although the surface proteins of the seal isolates are clearly related to those of avian viruses, the host range of the virus suggested the possibility that the genes coding for the non-surface proteins might have been derived from a mammalian influenza strain. The study described here compared the genetic homology of a series of mammalian and avian influenza strains with the seal isolate to determine which virus strains contain gene segments most closely related to those of the seal virus.

[a] Division of Virology, St. Jude Children's Research Hospital, Memphis, Tennessee.

Methods

Genetic homologies were measured by competitive RNA:RNA reassociation as described recently.[3] The RNA gene segments of A/Seal/Mass/1/80 were isolated by polyacrylamide gel electrophoresis, labeled with ^{125}-iodine, and annealed to homologous complementary RNA in the presence of unlabeled RNA from the homologous seal virus or from other virus strains. Annealing reactions were run at 15° below the homologous melting temperature. This modest level of stringency was used to show overall levels of homology rather than small differences in base sequence which are amplified when the reaction is run at a higher temperature.[4]

Results and Discussion

The influenza virus strains chosen for comparison with the genes of the seal virus are listed at the top of Figure 1. These include representatives of all of the human, equine and swine serotypes and a series of avian influenza isolates from several species representing all of the avian hemagglutinin and neuraminidase types.

RNA from each of these strains was used in competitive reassociation assays with individual labeled seal virus RNA segments coding for the nonsurface proteins. With all seal RNA segments, the most closely related corresponding RNAs were found in various avian influenza strains. For example, with RNA segment 3 (Figure 1), the most closely related strain was A/Gull/Md/5/77, while with RNA segment 5 (Figure 2) the mostly related strain was A/duck/Alberta/60/76. None of the strains tested contained genes closely related to all of the seal RNAs and only one of the strains with a closely related gene (Ty/Oregon/71, gene 7) also had the appropriate hemagglutinin.

The data clearly indicate that all of the genes of A/Seal/Mass/1/80 are closely related to genes from different avian viruses. It is not clear, however, how or when this strain originated or why it is adapted to replicate in seals. Although we have not yet found an avian influenza virus containing all of the genes of the seal virus, such a strain presumably could have formed by reassortment since the exchange of genes among avian influenza viruses appears to be a frequent event.[5] We do not know if this reassortment occurred in a bird or in a seal following chance infection with two or more avian virus strains. Also uncertain is whether the virus was immediately capable of efficient replication in seals or if a prolonged period of adaptation preceded its recent association with epizootic pneumonia. The precise evolutionary history of this strain may not be traceable, if critical intermediates in this process are not found. Retrospective immunological analysis of seal sera, however, may tell us if undetected infections of seals with influenza viruses have occurred in the past.

This virus, A/Seal/Mass/1/80, provides the first evidence suggesting that an influenza strain deriving all of its genes from one or more avian influenza viruses can be associated with severe disease in a mammalian population.

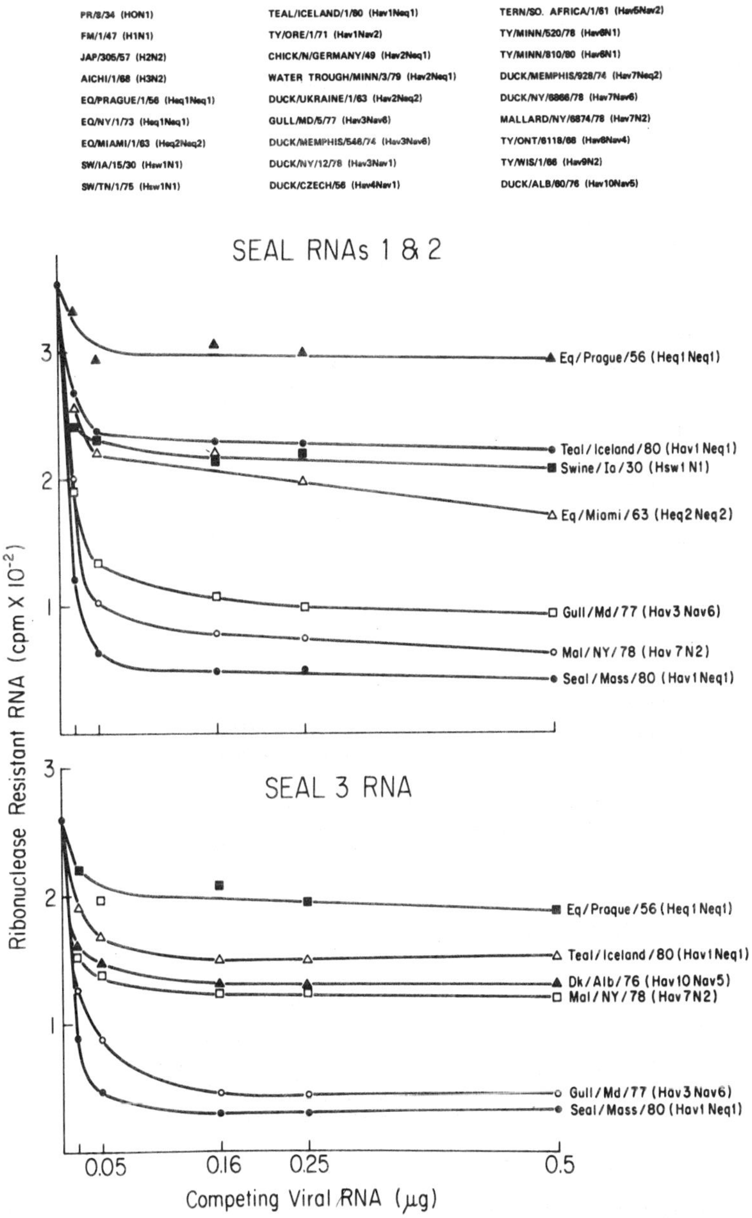

Figure 1. Competitive hybridization analysis of RNA gene segments of A/Seal/Mass/1/80 with influenza virus strains of avian and mammalian origin. Labeled RNA segments of A/Seal/Mass/1/80 were prepared as described in the text[3] and annealed with homologous complementary RNA in the presence of increasing concentrations of RNA from the virus strains listed at the top of the figure. Not all of the assays are shown, but those included show the full range of values obtained and include the strain or strains showing the closest homology with the seal RNA segment. The RNA segments used in this figure code for the three "P" proteins of the seal virus.

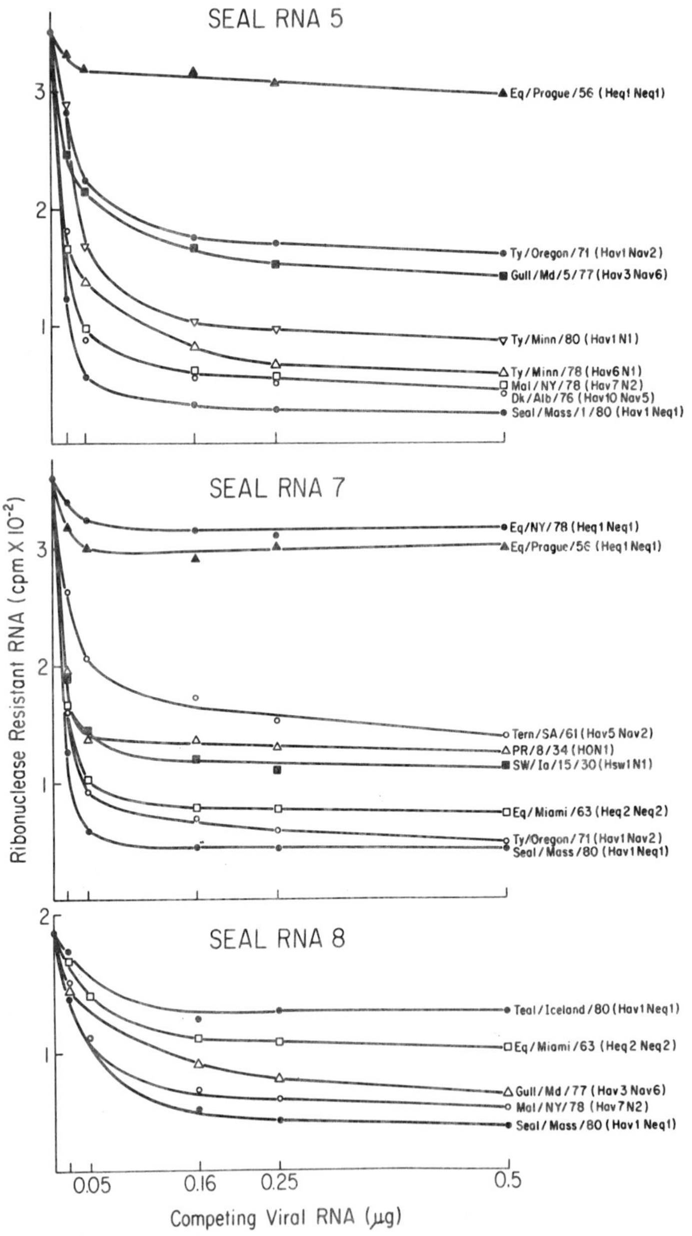

Figure 2. Competitive hybridization of RNAs 5, 7 and 8 of the seal virus. These RNAs code for the nucleoprotein, matrix and non-structural protein, respectively. Experimental details are given in the legend of Figure 1.

Whether this breach of species specificity represents a unique event in influenza evolution remains to be determined. The possibility that future human or animal influenza viruses may be directly derived from avian strains should be considered.

ACKNOWLEDGMENTS
We wish to acknowledge the excellent technical assistance of Raymond Wilson, Janet Ulm and Alford Pointer. This work was supported by Grants AI 16841, AI 08831 and AI 52524 from the National Institute of Allergy and Infectious Diseases, National Institutes of Health, by Cancer Center Support (CORE) Grant CA 21765 from the National Cancer Institute, and by ALSAC.

References

1. Lang, G., Webster, R.G., Hinshaw, V.S., Bean, W.J., van Wyke, K.L., Geraci, J.R., and Petursson, (1981), manuscript in preparation.
2. Hinshaw, V.S., Webster, R.G., and Rodriquez, R.J. (1979): *Arch. Virol.* 62:281–290.
3. Bean, W.J. Jr., Cox, N.J., and Kendal, A.P. (1980): *Nature* 284:638–640.
4. Sriram, G., Bean, W.J., Hinshaw, V.S., and Webster, R.G. (1980): *Virology* 105:592–599.
5. Hinshaw, V.S., Bean, W.J., Webster, R.G., and Sriram, G. (1980): *Virology* 102:412–419.

Copyright 1981 by Elsevier North Holland, Inc.
David H.L. Bishop and Richard W. Compans, eds.
The Replication of Negative Strand Viruses

Temperature-Sensitive Mutants of Influenza A/Udorn/72 (H3N2) Virus: Intrasegmental Complementation and Temperature-Dependent Host Range (*td-hr*) Mutation

Kazufumi Shimizu,[b] Brian R. Murphy,[a]
and Robert M. Chanock[a]

Temperature-sensitive (*ts*) mutants of influenza A virus have been useful in understanding the genetic organization of the influenza A virus genome and the mechanism of virus replication.[1-15] In addition, *ts* mutants are being evaluated for use in live virus vaccines as donors of attenuating *ts* genes to new variants of influenza A virus.[16-20] The *ts* mutants we have generated have been assigned to six complementation-recombination groups[21] which is less than the expected number of eight, i.e., one group for each of the eight segments of the influenza A virus genome.[22] We, therefore, sought to produce a large number of *ts* mutants of influenza A virus to obtain a complete set of complementation-recombination groups for several reasons: (a) to study the organization and function of the virus genome; (b) to study the relationship between the locus of *ts* mutation, its level of temperature sensitivity *in vitro,* and its effect upon replication *in vivo;* and (c) to aid us in genotyping live *ts* virus vaccines produced by recombination.

One of the most important problems still remaining for further investigation is the host range specificity of influenza A virus. Knowledge about the factors that control the host range of the virus is necessary for understanding the pathogenesis of a virus in different hosts and in different tissues within a host. Such information might also contribute to understanding the transmission of virus from animals to man which is considered to be one of the ways that new human pandemic strains are generated.[23] In order to identify the

[a] Laboratory of Infectious Diseases, National Institute of Allergy and Infectious Diseases, National Institutes of Health, Bethesda, Maryland.
[b] Present address: Department of Bacteriology, Nagasaki University School of Medicine, Nagasaki 852, Japan.

genes which might be involved in controlling host range, we attempted to produce host range mutants of influenza A virus.

The present study describes the isolation and genetic analysis of 136 *ts* mutants and 36 temperature-dependent host range (*td-hr*) mutants of influenza A/Udorn/72 (H3N2) virus.

Isolation of *ts* Mutants

Influenza A/Udorn/72 (H3N2) virus was mutagenized with ICR 191 (an acridine-based compound), nitrous acid or ultraviolet light, and 6185 clonal populations were examined for plaque formation on both primary rhesus monkey kidney (RMK) and Madin-Darby canine kidney (MDCK) monolayer cultures at 34°C and 40°C. Viruses which exhibited a 40°C/34°C plaque titer of 10^{-4} were further investigated as *ts* mutants and the viruses which exhibited a RMK/MDCK plaque titer of 10^{3} or 10^{-3} were further investigated as host range mutants. A summary of the isolation of the *ts* mutants is presented in Table 1. Significantly more clones exhibited the *ts* phenotype on MDCK than on RMK cells, i.e., 133 mutants were *ts* on MDCK and 83 on RMK cells. The total number of *ts* mutants was 136 since 80 mutants were scored as *ts* on both cells.

Host Dependency of Temperature Sensitivity of *ts* Mutants

The replication of a considerable number of mutants exhibited a significant host dependency of temperature sensitivity. The mutants were classified into three types depending on the host dependency. Type I mutants exhibited no distinct host dependency, type II mutants exhibited ≥1000-fold more restriction of plaque formation on RMK cells than MDCK cells at 38°C, 39°C and/or 40°C, and type III mutants conversely exhibited ≤1000-fold more restriction of plaque formation on MDCK cells than RMK cells at 38°C, 39°C and/or 40°C.

The distribution of mutants among the three types is shown in Table 2. Thirty-six of the 136 *ts* mutants exhibited distinct host dependent restriction of plaque formation at high temperature. Four of the 36 mutants exhibited RMK cell dependent restriction (Type II) and the other 32 mutants exhibited MDCK cell dependent restriction (Type III).

None of the clones among the 6,185 screened exhibited significant host dependent restriction of plaque formation at 34°C when tested on RMK and MDCK cells.

Complementation and Recombination Groups

After extensive complementation assays for plaque formation in pairwise matings of the 83 RMK *ts* mutants on RMK cells at 40°C, 13 *ts* mutants were selected as prototypes of complementation groups. These 13 prototypes

Table 1. Production of *ts* Mutants of Influenza A/Udorn/72 (H3N2) Virus.

Mutagenesis	Selection	No. of clones[a] tested	No. of *ts*[b] mutants on RMK cells	(% *ts*)	No. of *ts* mutants on MDCK cells	(% *ts*)	Designation of mutants
Ultraviolet irradiation of virions	none	2008	14	(0.7)	23	(1.1)	UV
Nitrous acid treatment of virions	none	303	6	(2.0)	8	(2.6)	NA
ICR191 treatment during replication	none	1509	27	(1.8)	41	(2.7)	ICR
ICR191 treatment of virions	none	603	6	(1.0)	11	(1.8)	ICR
	immune lysis[c]	305	8	(2.6)	12	(3.9)	ICRC
Spontaneous	none	1329	18	(1.4)	34	(2.6)	SP
	immune lysis[c]	128	4	(3.1)	4	(3.1)	SPC
Total		6185	83	(1.3)	133	(2.2)	

[a]All clones were picked from plaques of primary calf kidney monolayer cultures and tested for temperature sensitivity of plaque formation on both RMK (primary rhesus monkey kidney) and MDCK (Madin-Darby canine kidney) monolayer cultures.

[b]*ts* mutant is defined as a virus that exhibits a 40°C/34°C plaque titer of $\leq 10^{-4}$.

[c]Antiserum and complement were added to cell monolayer incubated at the non-permissive temperature of 40°C after infection with the mutagenized virus to kill cells expressing viral antigens on the cell surface by immune lysis; subsequently temperature of incubation was lowered to the permissive temperature of 34°C.

Table 2. Classification of Influenza A/Udorn/72 *ts* Mutants.

Type	Definition	No. of mutants
I	No. distinct host dependency	100
II	Restricted ≥1000-fold more on RMK cells than on MDCK cells at 38°C, 39°C and/or 40°C	4
III	Restricted ≥1000-fold more on MDCK cells than on RMK cells at 38°C, 39°C and/or 40°C	32
	Total	136

complemented each other in all of the 78 possible pairwise matings. The other 70 mutants shared *ts* lesions with at least one of the 13 prototypes. We were surprised that the number of complementation groups, 13, exceeded the number of the RNA genome segments, 8. This suggested the occurrence of intrasegmental or intracistronic complementation.

To determine if recombination had occurred in the pairwise matings of the 13 complementation group prototypes, the progeny produced by the matings were analyzed for *ts* phenotype (Table 3). The progeny from the crosses were predominantly ts^+ recombinants in all pairwise matings with several exceptions. The progeny from a mating of complementation groups 4 and 6 were predominantly *ts*, whereas the group 4 to 6 prototype yielded predominantly ts^+ virus in matings with prototypes of the other complementation groups. These data suggested that the group 4 and group 6 mutants possessed *ts* lesions on the same gene and were thus assigned to recombination group B. Complementation groups 1 and 11; 2 and 12; and 5, 7, 9 and 12 were similarly assigned to recombination groups C, F and H, respectively. Complementation groups 3, 8, 10 and 13 were assigned to recombination groups A, D, E, and G, respectively. The prototype of group 12 seemed to possess *ts* lesions on two genes corresponding to recombination groups F and H. Thus, we obtained 8 recombination groups, A to H. The number of recombination groups, 8, is in good agreement with the number of influenza A RNA gene segments.

Mapping *ts* Mutations on the RNA Gene Segments

The location of the *ts* mutations was studied by a method similar to that of Almond et al. (1977)[24] using urea-polyacrylamide gel electrophoresis of RNA. A set of 8 independently derived ts^+ recombinant viruses were produced at 40°C by mating a Udorn *ts* mutant with a WSN *ts* mutant. The locus of the *ts* lesion in the Udorn *ts* virus could be identified by determining the RNA segment uniformly replaced with the corresponding WSN gene since the viral recombinant must receive a WSN wild type gene to render it ts^+. It

Table 3. Progeny Analysis of Pair-Wise Crosses Between the Prototypes of 13 Complementation Groups on RMK Cells.

Complementation group	Prototype	% of wild type (ts^+) recombinants in cross-progeny[a]													Recombination group[b]
		UV 257	UV 1617	UV 1942	UV 186	UV 1958	NA 233	ICRC 282	UV 268	SP 712	SP 456	UV 1841	SPC 45	ICR 516	
3	UV 257	—													A
4	UV1617	92	—												B
6	UV 1942	88	49	—											B
1	UV 186	100	92	100	—										C
11	UV 1958	57	93	76	48	—									C
8	NA 233	96	93	92	82	78	—								D
10	ICRC 282	88	71	92	88	94	92	—							E
2	UV 268	92	96	92	83	85	79	92	—						F
12	SP 712	79	82	75	77	81	81	93	49	—					F,H
13	SP 456	91	96	82	100	91	96	89	83	88	—				F
5	UV 1841	91	96	79	83	56	96	83	92	44	58	—			H
7	SPC 45	75	82	88	92	57	88	92	83	41	96	28	—		H
9	ICR 516	92	83	88	93	80	88	92	96	45	91	32	16	—	H

[a]Three plaques from each cross at 40 °C on RMK monolayer cultures were picked and the progeny in the plaques were cloned at 34°C on RMK monolayer cultures. From 24 to 72 progeny clones of each cross were tested for temperature-sensitivity of plaque formation on RMK monolayer cultures. The clones which exhibited a 40°C/34°C plaque titer of ≥ 0.1 were scored as wild type (ts^+) recombinants.

[b]Recombination groups were made up of members whose cross progeny were predominantly ts^+ recombinants, i.e., the cross progeny contained greater than 50% ts^+ recombinants.

can be seen (Figure 1) that each of the eight RNA segments of the WSN virus migrated differently than Udorn RNA when electrophoresis was carried out at 30°C using 3.0% polyacrylamide gel. Under these conditions, Udorn RNA separated into 7 bands, but at 4°C 8 bands were distinguished. By comparing the migration rate of parental and *ts*$^+$ recombinant RNA, the parental origin of all genes in the *ts*$^+$ recombinants could be determined. RNA patterns of the 8 independent *ts*$^+$ recombinants derived from the cross between Udorn UV 1958 (complementation group 11 and recombination group C) and WSN *ts* 15 which was known to have a *ts* defect on RNA segment 2 (P1 polymerase gene)[9,12] are shown on Figure 1. Each of the 8 *ts*$^+$ recombinants had an RNA segment 3 derived from WSN; this gene codes for the P2 protein. This indicated that the Udorn *ts* mutant UV 1958 had a *ts* mutation on this RNA segment. The locus or loci of the *ts* mutation(s) of the other prototypes were similarly mapped and a summary of the results is presented in Table 4. It should be noted that intrasegmental complementation occurred with *ts* mutations that affected the genes coding for P2, NP and NS and probably P1 as well. Each of the 8 recombination groups corresponded to one of the 8 RNA gene segments. There was no apparent overlap, although the localization of group G on RNA 7 is not yet conclusive.

Temperature-Dependent Host Range (*td-hr*) Mutants

Type II and type III mutants exhibited more than a 1000-fold difference in efficiency of plaque formation on RMK and MDCK cells in the restrictive temperature range (38°-40°C) whereas they plaqued with the same efficiency on RMK and MDCK cells at 34°C. Therefore, they appeared to be temperature-dependent host range (*td-hr*) mutants. Sixteen of the 32 type III mutants had a 10,000-fold or greater reduction in plaque formation at 40°C on MDCK cells compared to that at 34°C, whereas there was less than a 100-fold reduction in plaque formation on RMK cells at 40°C compared to 34°C, i.e., these 16 mutants were O *ts* on MDCK cells but not *ts* on RMK cells. Therefore their *td-hr* phenotype reflected a host dependent temperature sensitive (*hd-ts*) mutation. The *hd-ts* lesions of the 16 mutants were analyzed by complementation with the prototype *ts* mutants of the 8 recombination groups on MDCK cells at 40°C. *hd-ts* mutations were identified in each of the 8 recombination groups. To be certain that the *td-hr* mutants did not share a common *ts* lesion, complementation between the *td-hr* mutants themselves was tested on MDCK cells at 40°C. If the *td-hr* mutants shared a common lesion, complementation would not be observed in any of the matings. However, this was not the case. Complementation did occur except when both mutants tested were assigned to the same recombination group (Table 5). It was clearly shown that *td-hr* mutations were not limited to any single gene. On the contrary, each influenza gene can undergo host range mutation.

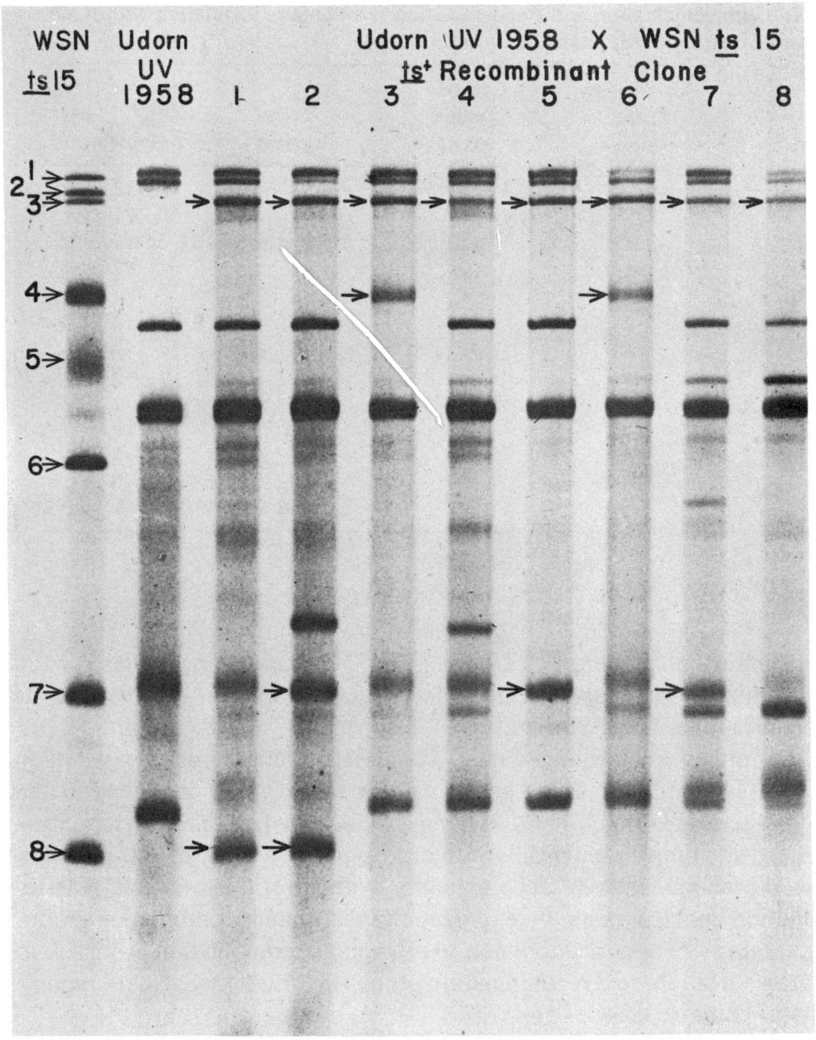

Figure 1. Determination of the parental origin of genes in *ts*$^+$ recombinant viruses derived from the cross between Udorn UV 1958 and WSN *ts*15. An Udorn *ts* mutant, UV 1958, and a WSN *ts* mutant, *ts*15,[9] were mated on RMK cells at 40°C. The cross progeny were plaqued on RMK cells at 40°C to select *ts*$^+$ recombinant clones. The parental viruses and 8 *ts*$^+$ recombinant viruses were propagated in embryonated chicken eggs and purified by two cycles of sucrose gradient centrifugation. RNA was extracted according to Palese et al. (1976)[25] and subjected to electrophoresis on 3% polyacrylamide gel containing 6M urea at 75 V for 8 hours at 30°C. RNA was stained with ethidium bromide (0.5 μg/ml) for 30 min and visualized under ultraviolet light. Photograph was taken through an orange filter. The RNA segments which originated from the WSN are marked by an arrow.

Table 4. Complementation and Recombination Groupings: Correlation with RNA Segment(s) Bearing *ts* Mutation(s).

Mutants	Complementation group	Recombination group	RNA segment	Protein
UV 257	3	A	1	P3
UV 1617	4	B	2	P1
UV 1942	6	B	2, 7 and/or 8	P1, M and/or NS
UV 186	1	C	3	P2
UV 1958	11	C	3	P2
NS 233	8	D	4	HA
ICRC 282	10	E	5	NA
UV 268	2	F	6	NP
SP 712	12	F and H	6 and 8	NP and NS
SP 456	13	G	2 or 7	P1 or M
SPC 45	7	H	8	NS
ICRC 516	9	H	8	NS
UV 1841	5	H	3, 6 and/or 8	P2, NP and/or NS

Summary

136 *ts* mutants of influenza A/Udorn/72 (H3N2) virus were isolated from unmutagenized virus suspensions or after mutagenesis with ICR191, nitrous acid or ultraviolet light.

The *ts* mutants were arranged into 13 complementation groups based upon plate complementation assays performed on RMK cell monolayers at the restrictive temperature of 40°C. Analysis of progeny from the pairwise matings of the 13 complementation group prototypes revealed that co-infection produced predominantly ts^+ recombinants, with several exceptions. The recombination deficient pairs were assigned to the same recombination group. The mutants were assigned to 8 non-overlapping recombination groups A to H. Mutants from these 8 recombination groups invariably yielded ts^+ recombinants in pairwise crosses.

It was shown that each of the 8 recombination groups corresponded to one of the eight RNA segments. The *ts* mutations of complementing pairs of mutants belonging to the same recombination group were mapped to the same RNA segment. Intrasegremental complementation was demonstrated for *ts* mutations that affected the gene coding for P1, P2, NP or NS protein.

Thirty-six of the 136 *ts* mutants exhibited distinct host dependent restriction of plaque formation at high temperature and were referred to as temperature-dependent host range (*td-hr*) mutants. The *td-hr* mutants did not share a common lesion; instead the *td-hr* mutations were shown to be distributed among the 8 recombination groups. These results suggested that each influenza gene can undergo host range mutation.

Table 5. Complementation Between *td-hr* Mutants of Influenza A/Udorn/72 Virus.

td-hr mutant	Plaque titer MDCK/RMK cells (log) at 34°C	Plaque titer MDCK/RMK cells (log) at 40°C	Recombination group	Complementation index[a] of the indicated pairs of mutants on MDCK cells at 40°C: SP 37	SP 392	SP 443	UV 1766	SP 185	ICR 800
SP 37	−0.69	<−3.3	A	—					
SP 392	0.30	<−3.0	B	3.2	—				
SP 443	−0.19	<−3.7	C,D	16	1.4	—			
UV 1766	0.35	−3.1	E,F	14	6.8	3.3	—		
SP 185	0.10	<−4.1	F,G	15	2.4	16	<0.013	—	
ICR 800	−0.22	<−3.6	H	13	11	91	13	12	—

[a]Complementation index was defined as the ratio of the number of plaques observed at 40°C to the expected number of cells dually infected with the two viruses. The expected number of dually infected cells was estimated assuming that the distribution of the virus to the cell fits the Poisson distribution using formula, $(1-e^{-m1})(1-e^{-m2})$(No. of cells) where m1 and m2 were the multiplicities of infection of the two viruses determined at permissive temperature of 34°C.

ACKNOWLEDGMENTS
We wish to acknowledge the excellent technical assistance of Mrs. Mary G. Mullinix. We also thank Dr. Tsutomu Miyamoto, Dr. Shigeo Hino and Miss Shizuko Murata for discussion and help in preparation of the manuscript.

References

1. Simpson, R.W. and Hirst, G.K. (1968): *Virology* 35:41–49.
2. Mackenzie, J.S. (1970): *J. Gen. Virol.* 6:63–75.
3. Sugiura, A., Tobita, K., and Kilbourne, E.D. (1972): *J. Virol.* 10:639–647.
4. Markushin, S.G. and Ghendon, Y.Z. (1973): *Acta Virol.* 17:369–376.
5. Ghendon, Y.Z., Markushin, S.G., Marchenko, A.T., Shitnikov, B.S., and Ginzburg, V.P. (1973): *Virology* 55:305–319.
6. Hirst, G.K. (1973): *Virology* 55:81–93.
7. Palese, P., Tobita, K., Ueda, M., and Compans, R.W. (1974): *Virology* 61:397-410.
8. Scholtissek, C., Kruczinna, R., Rott, R., and Klenk, H.-D. (1974): *Virology* 58:317–322.
9. Sugiura, A., Ueda, M., Tobita, K., and Enomoto, C. (1975): *Virology* 65:363–375.
10. Ghendon, Y.Z., Markushin, S.G., Blagovezhinskaya, O.V., and Genkina, D.B. (1975): *Virology* 66:454–463.
11. Scholtissek, C. and Bowles, A.L. (1975): *Virology* 67:576–587.
12. Krug, R.M., Ueda, M., and Palese, P. (1975): *J. Virol.* 16:790–796.
13. Scholtissek, C., Harms, E., Rohde, W., Orlich, M., and Rott, R. (1976): *Virology* 74:332–344.
14. Palese, P., Ritchey, M.B., and Schulman, J.L. (1977): *J. Virol.* 21:1187–1195.
15. Almond, J.W., McGeoch, D., and Barry, R.D. (1979): *Virology* 92:416–427.
16. Mills, J. and Chanock, R.M. (1971): *J. Inf. Dis.* 123:145–157.
17. Murphy, B:R., Chalhub, E.G., Nusinoff, S.R., and Chanock, R.M. (1972): *J. Inf. Dis.* 126:170–178.
18. Spring, S.B., Nusinoff, S.R., Mills, J., Richman, D.D., Tierney, E.L., Murphy, B.R., and Chanock, R.M. (1975): *Virology* 66:522–532.
19. Murphy, B.R., Tierney, E.L., Spring, S.B., and Chanock, R.M. (1976): *J. Inf. Dis.* 134:577–584.
20. Richman, D.D., Murphy, B.R., Chanock, R.M., Gwaltney, J.M., Douglas, R.G., Betts, R.F., Blacklow, N.R., Rose, F.B., Parrino, T.A., Levine, M.M., and Caplan, E.S. (1976): *J. Inf. Dis.* 134:585–594.
21. Spring, S.B., Nusinoff, S.R., Mills, J., Richman, D.D., Tierney, E.L., Murphy, B.R., and Chanock, R.M. (1975): *Virology* 66:542–550.
22. Palese, P. (1977): *Cell* 10:1–10.
23. Laver, W.G. and Webster, R.G. (1973): *Virology* 51:383–391.
24. Almond, J.W., McGeoch, D., and Barry, R.D. (1977): *Virology* 81:62–73.
25. Palese, P. and Schulman, J.L. (1976): *J. Virol.* 17:876–884.

David H.L. Bishop and Richard W. Compans, eds.
The Replication of Negative Strand Viruses

Analysis of the Functions of Influenza Virus Genome RNA Segments By Use of Temperature-Sensitive Mutants of Fowl Plague Virus

B.W.J. Mahy, T. Barrett, S.T. Nichol,
C.R. Penn and A.J. Wolstenholme[a]

Introduction

The influenza virus genome consists of eight negative single-stranded RNA segments ranging in size from 890 to approximately 2,400 nucleotides. The protein coding functions of each virion RNA segment have been determined for several influenza A viruses by hybrid-arrested translation, or by analysis of virus recombinants formed by genome reassortment between different influenza virus strains whose RNA segments and virus-specific polypeptides are distinguishable by polyacrylamide gel electrophoresis.[1] The major proteins encoded by RNA segments 4-7 are well-defined structural components of the virus, namely the hemagglutinin (HA), nucleoprotein (NP), neuraminidase (NA) and matrix protein (M).

Much less is known concerning the functional significance of those proteins encoded by RNA segments 1-3, 8, and the second reading frame present within 7.[2] The proteins encoded by RNA segments 1-3, which together form more than fifty percent of the influenza virus genome, are presently referred to as P_1, P_2 and P_3 on the basis of their relative mobilities during polyacrylamide gel electrophoresis. These proteins do not always migrate in the same order in different influenza virus strains. For example, comparisons between two avian influenza A (fowl plague) virus strains showed that although the P_1 proteins are functionally equivalent, the P_2 protein of the Rostock strain is equivalent to the P_3 protein of the Dobson strain and vice

[a] Division of Virology, Department of Pathology, University of Cambridge, Addenbrooke's Hospital, Hills Road, Cambridge CB2 200, England.

versa.[3] The P-proteins are known to associate with the NP protein and to be present in transcriptive complexes isolated from infected cells[4,5] but their precise functions during replication have not been defined.

One approach to elucidating these functions is to utilize temperature-sensitive (*ts*) mutants with a defined lesion in a specific genome RNA segment. Such studies carried out with groups of *ts* mutants isolated in various laboratories, have recently been reviewed.[6] We have begun to investigate the phenotypes of several *ts* mutants of fowl plague virus originally isolated and grouped in our laboratory by J.W. Almond.[7] Studies carried out with group IV *ts* mutants, having a lesion in RNA segment 8 coding for two non-structural proteins NS_1 and NS_2, have been described.[8] In this paper, we summarize our findings with *ts* mutants in groups I, II, III and V which affect the functions of proteins NP, P_1, P_2 and P_3 respectively.

Primary Transcription

Analysis of protein synthesis in fowl plague virus infected cells following release from a cycloheximide block indicated three stages in the production of virus-specific messenger RNA.[9] In the first, primary transcription stage, synthesis of all mRNAs (except that coding for the NS_2 polypeptide) occurs at similar rates. It was of interest to determine whether primary transcription could be detected at the non-permissive temperature during infection with P-protein mutants.

Chick embryo fibroblasts were treated for 1 hr prior to and during infection with 100 μg/ml cycloheximide at either 34° or 40.5°C.

At five hours postinfection, the total cellular RNA was extracted, and the amount of virus-specific polyadenylated cRNA was determined by hybridization to I^{125}-labeled virion RNA as described previously.[10] The results, shown in Table 1, indicate that although all the mutants studied were capable of some cRNA synthesis at the non-permissive temperature, only the P_3 protein mutant synthesized similar amounts of cRNA at either temperature. Those mutants studied which had a *ts* protein P_1 or P_2 synthesized less than 50% A(+) cRNA during primary transcription *in vivo* at the non-permissive temperature.

Secondary Transcription and vRNA Synthesis

We next studied the ability of the mutants to synthesize non-polyadenylated A(−) cRNA (template RNA) and vRNA at the non-permissive temperature. As shown in Table 2, all the mutants studied were defective in the synthesis of both these classes of virus-specific RNA at 40.5°C. However, the failure to synthesize vRNA could result either from a *ts* defect in the polymerase component(s) involved in vRNA synthesis, or a defect in template cRNA synthesis. Temperature shift-up experiments were therefore carried out to allow accumulation of template cRNA at 34°C for 2 hours before shifting cells to the non-permissive temperature from 2-4 hours (Table 3). These

Table 1. Primary Transcription by P-Protein Mutants of Influenza.

Mutant	Temperature	ng Virus-specific RNA per μg cell RNA	
*ts*15 (P_1)	34°C	0.244	
	40.5°C	0.113	(46%)[a]
mN5 (P_2)	34°C	0.231	
	40.5°C	0.078	(34%)
*ts*17 (P_2)	34°C	0.983	
	40.5°C	0.281	(29%)
*ts*45 (P_3)	34°C	0.456	
	40.5°C	0.495	(108%)

Confluent monolayers of CEF cells were infected with approximately 20 pfu per cell of virus at either 34°C or 40.5°C. The cells and virus were incubated for 1 hr prior to infection at the appropriate temperature. Cycloheximide at 100 μg/ml was present throughout. At 5 hr postinfection total cellular RNA was extracted and the amount of virus-specific cRNA was determined by hybridization to ^{125}I-labeled virion RNA (vRNA) as described by Barrett et al.[10]

[a]Percentage of the corresponding value at 34°C.

results showed that mutants *ts*15 (P_1), *ts*45 (P_3) and US1 (NP) are all defective in vRNA synthesis even in the presence of template cRNA. Only mutant mN5 (P_2) appeared capable of normal vRNA synthesis after temperature shift-up at 2 hours postinfection. These results implicate both the fowl plague virus Rostock proteins P_1 and P_3, as well as the NP protein, in the process of vRNA synthesis.

Table 2. Synthesis of Virus-Specific RNA in *ts* Mutant-Infected Cells.

Mutant	Temperature of infection	Time (hours postinfection)	Genome equivalents per cell: A (−) cRNA	Genome equivalents per cell: vRNA
*ts*15 (P_1)	34°C	2	1900	720
	40.5°C	2	314	91
mN5 (P_2)	34°C	4.5	5760	1580
	40.5°C	4.5	380	80
*ts*17 (P_2)	34°C	4.5	5140	2420
	40.5°C	4.5	40	120
*ts*45 (P_3)	34°C	4.5	2220	4010
	40.5°C	4.5	210	100
US1 (NP)	34°C	4.5	1590	3430
	40.5°C	4.5	70	20

Confluent monolayers of CEF cells were infected with 5-10 pfu per cell of virus at either 34°C or 40.5°C. Cells and virus were incubated for 1 hr prior to infection at the appropriate temperature. RNA was extracted at the times indicated and separated into polyadenylated and non-polyadenylated fractions by oligo(dT) cellulose chromatography. The amount of virus-specific RNA in the non-polyadenylated fraction was determined by hybridization to ^{3}H-labeled DNA complementary to virion RNA (cDNA) and ^{125}I-labeled vRNA as described by Barrett et al.[10]

Table 3. Temperature-Shift Experiments.

Mutant	Incubation temperature 2-4 hrs	Genome equivalents/cell at 4 hrs	
		A (−) cRNA	vRNA
*ts*15 (P_1)	34°C	1440	2370
	40.5°C	1630	550
mN5 (P_2)	34°C	5760	1580
	40.5°C	3790	1679
*ts*45 (P_3)	34°C	1210	1210
	40.5°C	890	350
US1 (NP)	34°C	2760	1480
	40.5°C	2840	600

Confluent monolayers of CEF cells were infected with 5-10 pfu per cell of virus at 34°C. At 2 hr postinfection, half the bottles from each mutant infection were shifted to the non-permissive temperature. At 4 hr postinfection, RNA was extracted from all samples and separated into polyadenylated and nonpolyadenylated fractions by oligo(dT) cellulose chromatography. The amount of virus-specific RNA in the non-polyadenylated fraction was determined by hybridization to ^{3}H-labeled DNA complementary to vRNA (cDNA) and ^{125}I-labeled vRNA as described by Barrett et al.[10]

Virus-Induced Protein Synthesis

We have previously reported that the synthesis of virus-induced proteins in cells infected with *ts* mutants having a defect in segment 8 is blocked at the non-permissive temperature before the onset of late protein synthesis.[8] A similar investigation of the P and NP protein mutants was carried out. Chick embryo fibroblast cells were infected at 34°C or 40.5°C with the various mutants, and subsequently pulse-labeled for 30 min with ^{35}S-methionine at 2 and 5 hours postinfection before analysis of the proteins by polyacrylamide gel electrophoresis as described previously.[4] The results (Figures 1 and 2) showed considerable variation, even between mutants from a single group, in the ability to induce virus-specific proteins. Of five group III (P_2) mutants, two (*ts*17 and *ts*33) gave only an early pattern of protein synthesis and failed to induce the synthesis of normal amounts of matrix protein at 40.5°C. With mutants mN5, *ts*5 and *ts*9, the pattern of protein synthesis was not significantly different at the two temperatures. Similar results were obtained with a group I mutant (US1) and a group II mutant (*ts*15) (Figure 3).

One mutant studied from group V (*ts*45, defective in the P_3 protein) induced the synthesis at a low rate, of similar amounts of all the virus-specific proteins except NS_2 (Figure 3). This pattern of protein synthesis is seen during the initial stages after release of normal virus infected cells from a cycloheximide block. It suggests that mutant *ts*45 is blocked at 40.5°C at the stage of primary transcription, and may be unable to initiate the synthesis of full-length template cRNA.

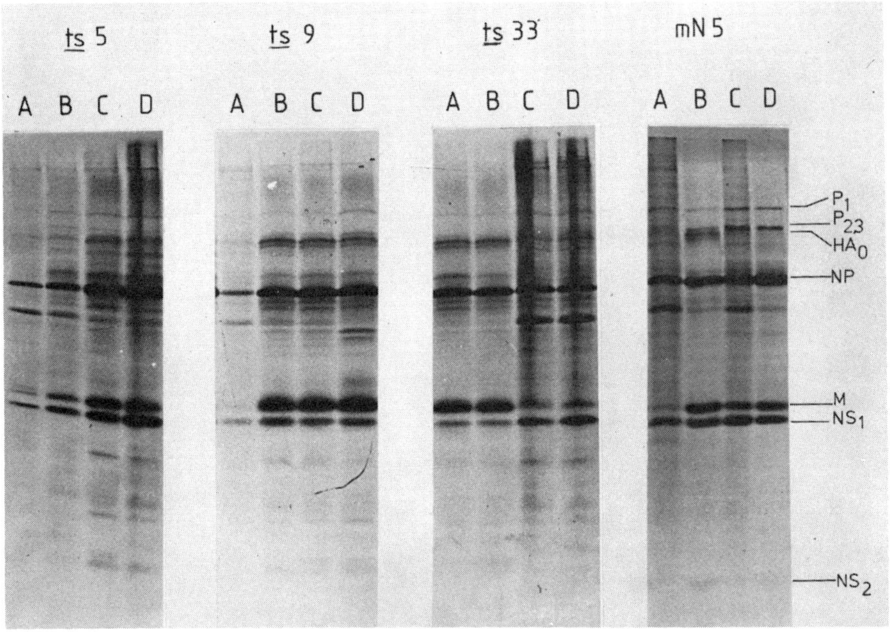

Figure 1. Polypeptides synthesized in cells infected with P_2 mutants of FPV at 34°C and 40.5°C. A, 2 hpi at 34°C. B, 5 hpi at 34°C. C, 2 hpi at 40.5°C. D, 5 hpi at 40.5°C. Cells were pulse-labeled with ^{35}S-methionine for 30 min before harvesting and the polypeptides separated by polyacrylamide gel electrophoresis as described by Inglis et al.[4]

RNA Transcription *in vitro*

Analysis of *in vitro* RNA synthesis by *ts* mutant virions is complicated by the fact that the optimal temperature for the assay is 31°C. Using the reaction conditions previously reported with ApG as a primer,[8] the activity of wild type fowl plague virus RNA transcriptase at 40.5°C was approximately 10% of the activity at 31°C. A large number of *ts* mutants were assayed for virion transcriptase activity at both 31°C and 40.5°C. In each case, a kinetic analysis of the reaction was performed with six time-points taken at intervals over a 90-minute incubation period. The ratios of the activities at 40.5°C compared to 31°C varied from 20% to 1%. We decided to take values of less than 2.5% as being negative and using this criterion the only *ts* mutant which was obviously negative for virion transcriptase activity at 40.5°C was *ts*17 (Table 4).

In an attempt to identify the virus-specific proteins involved in priming by capped mRNA, which appears to be essential for mRNA synthesis *in vivo,*[11-13] *ts* mutants were analyzed for the ability of globin mRNA to stimulate transcriptase activity at either 31°C or 40.5°C. Globin mRNA was an effec-

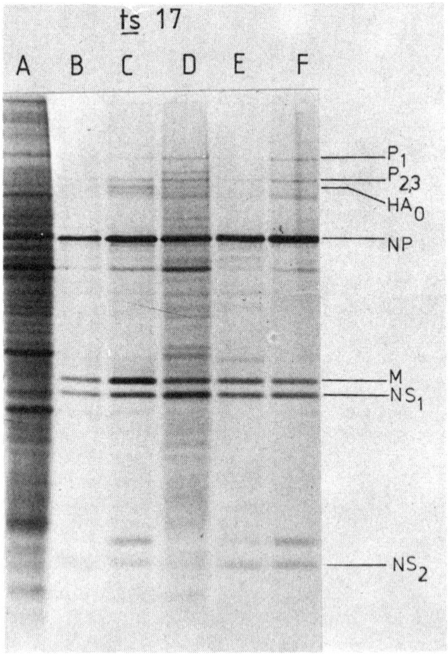

Figure 2. Polypeptides synthesized in cells infected with FPV mutant *ts*17 (P_2) at 34°C and 40.5°C. A, B, C, polypeptides synthesized at 34°C at 2, 5 & 8 hpi respectively. D, E, F, polypeptides synthesized at 40.5°C at 2, 5 & 8 hpi respectively. Cells were labeled with ^{35}S-methionine and the polypeptides separated by polyacrylamide gel electrophoresis as described by Inglis et al.[4]

tive primer for the wild type virion transcriptase activity at 40.5°C. The results of the study of globin mRNA priming of *ts* mutant virion transcriptases are summarized in Table 4. The only *ts* mutants found to be obviously negative for globin mRNA priming at the non-permissive temperature were *ts*17 and *ts*44 (group III) both of which had lesions in the RNA segment encoding the P_2 protein. These results suggest that the P_2 protein of this influenza virus strain is directly involved in mRNA priming as well as in the transcription reaction itself. However, mRNA synthesis can occur to a limited extent with *ts*17 *in vivo* at 40.5°C (see Figure 2 and Table 1). Clearly the *in vitro* transcriptase assay does not provide an exact indication of the transcription reaction as it occurs *in vivo,* where many primers other than globin mRNA are available.

Mutants of both positive and negative phenotypes with respect to *in vitro* transcriptase activity and mRNA priming at the restrictive temperature were found in group III. This may result from the defective protein (P_2) having more than one function, as has been suggested by intragenic complementation between influenza mutants.[14] Alternatively, there may be more than one

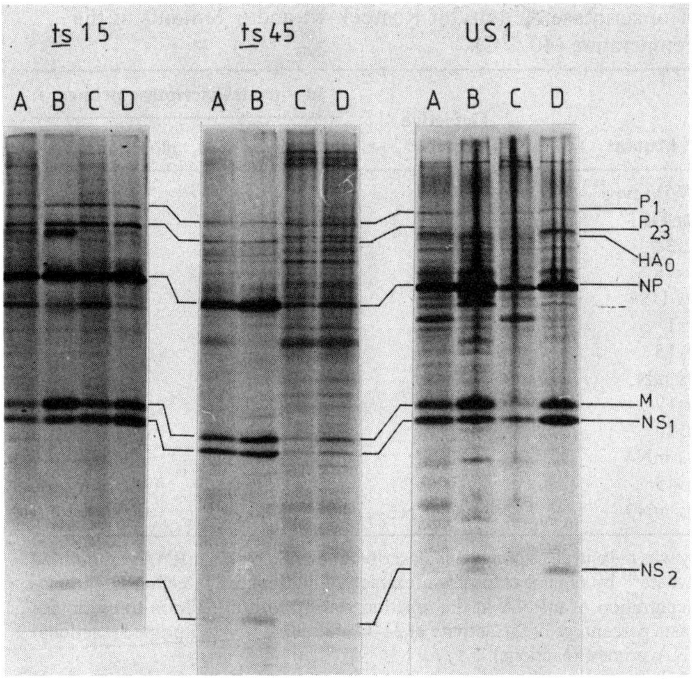

Figure 3. Polypeptides synthesized in cells infected with FPV mutants *ts*15 (P_1), *ts*45 (P_3) and US1 (NP) at 34°C and 40.5°C. A, 2 hpi at 34°C. B, 5 hpi at 34°C. C, 2 hpi at 40.5°C. D, 5 hpi at 40.5°C. Cells were labeled with ^{35}S-methionine and the polypeptides separated by polyacrylamide gel electrophoresis as described by Inglis et al.[4]

gene product or the defective protein may be functional at the nonpermissive temperature if synthesized at the permissive temperature.

Conclusions

It cannot be assumed that a *ts* lesion in a particular influenza RNA segment will affect all the functions of its gene product(s) equally. Consequently, it is necessary to examine the phenotype of a series of mutants within each recombination group in order to define completely the function of the gene product(s) of that RNA segment. However, since certain functions are *ts* in these mutants it is possible to assign minimum functions to each gene.

Our results indicate that all three P-proteins and the NP protein play a role in virus-specific RNA synthesis in infected cells. A functional P_1 protein appears to be required for the synthesis of all three classes of virus-specific RNA. The P_2 protein appears to be required for both primary transcription and synthesis of template cRNA, and both NP and P_3 appear to be involved in synthesis of template cRNA and vRNA.

Table 4. *In Vitro* Transcriptase Activity of Rostock wt and *ts* Mutants at the Non-Permissive Temperature (40.5°C).

Recombination group	Mutant	Defective protein[a]	In vitro transcriptase primer	
			ApG	globin mRNA
—	Wild type	—	9.4 (+)	11.6 (+)
I	*ts* US1	NP	22.4 (+)	19.1 (+)
	*ts*34	NP	10.8 (+)	6.4 (+)
VI	*ts*46	HA	6.1 (+)	7.4 (+)
	ts US4	HA	7.4 (+)	7.7 (+)
II	*ts*1	P_1	5.6 (+)	5.0 (+)
	*ts*15	P_1	4.8 (+)	10.2 (+)
III	*ts* mN5	P_2	7.4 (+)	7.0 (+)
	*ts*17	P_2	1.4 (−)	1.3 (−)
	*ts*44	P_2	6.2 (+)	2.2 (−)
V	*ts* mN4	P_3	20.2 (+)	26.6 (+)
	*ts*45	P_3	9.7 (+)	6.7 (+)
IV	*ts* mN3	NS_1 &/or NS_2	7.3 (+)	7.2 (+)

Transcriptase activity was measured as previously described.[8] Rabbit globin mRNA was purified from a reticulocyte lysate[20] by phenol-chloroform extraction and oligo(dT) cellulose chromatography.[10] The concentration of mRNA in the reaction was 100 μg/ml. Virion transcriptase activity is expressed as a percentage of the activity at 31°C after 90 min (ApG primed reactions) or 120 minutes (mRNA primed reactions).

[a]Almond and Barry;[3] Almond et al.[21]

Since it has been shown that RNA segment 1 of fowl plague virus Rostock is functionally equivalent to RNA segment 2 of WSN strain influenza virus,[6,15] there is a clear correlation between our results with fowl plague virus Rostock and previous studies with influenza WSN,[16,17] influenza A/Ann Arbor/6/60[18] and fowl plague virus Weybridge.[19] In each case a protein essential for the function of the virion transcriptase is encoded in the genome RNA segment equivalent to influenza WSN RNA segment 2.

ACKNOWLEDGMENTS
We thank Margaret Willcocks for excellent technical assistance. This work was supported by an MRC Program Grant (G979/140). S.T.N. was a SCR CASE student partly supported by Beecham Pharmaceuticals, CRP and AJW were in receipt of studentships from the Medical Research Council.

References

1. Barry, R.D. and Mahy, B.W.J. (1979): *Brit. Med. Bull.* 35:39–46.
2. Winter, G. and Fields, S. (1980): *Nucl. Acids Res.* 8:1965–1974.
3. Almond, J.W. and Barry, R.D. (1979): *Virology* 92:407–415.
4. Inglis, S.C., Carroll, A.R., Lamb, R.A., and Mahy, B.W.J. (1976): *Virology* 74:489–503.
5. Caliguiri, L.A. and Gerstein, H. (1978): *Virology* 90:119–132.

6. Scholtissek, C. (1979): *Adv. in Genetics* 20:1–36.
7. Almond, J.W., McGeoch, D. and Barry, R.D. (1979): *Virology* 92:416–427.
8. Wolstenholme, A.J., Barrett, T., Nichol, S.T., and Mahy, B.W.J. (1980): *J. Virol.* 35:1–7.
9. Inglis, S.C. and Mahy, B.W.J. (1979): *Virology* 95:154–164.
10. Barrett, T., Wolstenholme, A.J., and Mahy, B.W.J. (1979): *Virology* 98:211–225.
11. Krug, R.M., Broni, B.A., and Bouloy, M. (1979): *Cell* 18:329–334.
12. Caton, A.J. and Robertson, J.S. (1980): *Nucleic Acids Res.* 8:2591–2603.
13. Dhar, R., Chanock, R.M., and Lai, C.-J. (1980): *Cell* 21:495–500.
14. Heller, E. and Scholtissek, C. (1980): *J. Gen. Virol.* 49:133–139.
15. Scholtissek, C. (1978): *Curr. Topics Microbiol. Immunol.* 80:139–169.
16. Mowshowitz, S.L. (1978): *Virology* 91:493–495.
17. Palese, P. (1977): *Cell* 10:1–10.
18. Kendal, A.P., Cox, N.J., Galphin, J.C., and Maassab, H.F. (1979): *J. Gen. Virol.* 44:443–456.
19. Ghendon, Y.Z., Markushin, S.G., Blagovezhenskaya, O.V., and Genkina, D.B. (1975): *Virology* 66:454–463.
20. Hunt, T. and Jackson, R.J. (1974): In: *Modern Trends in Human Leukemia.* Neth, R., Gallo, R.C., Spiegelman, S., and Stohlman, F. (Eds.), Verlag, Munich, pp. 300–307.
21. Almond, J.W., McGeoch, D., and Barry, R.D. (1977): *Virology* 81:62–73.

Copyright 1981 by Elsevier North Holland, Inc.
David H. L. Bishop and Richard W. Compans, eds.
The Replication of Negative Strand Viruses

Suppressor Recombinants of an Influenza A Virus

Christoph Scholtissek and Susan B. Spring[a]

Introduction

Since influenza viruses have a segmented genome, the recombination frequency is extremely high due to reassortment of the 8 RNA segments. Properties like host range, tropism, and pathogenicity can be affected by this kind of recombination (for a review see reference 1). There are efforts underway to develop live influenza vaccines by replacing genes of a pathogenic strain by those with a genetic defect. So far, it has been found that these new recombinants carrying for example, a gene with a temperature-sensitive (*ts*) defect express the same phenotype as the parent strain, from which the defective gene was derived.[2,3] However, the possibility exists that a gene whose product cooperates with other viral proteins, in a new constellation, might change its phenotype. By rescuing a *ts*-mutant of a fowl plague virus (FPV, Hav1N1) recombinant carrying a *ts*-defect in segment 8 (non-structural (NS) protein gene) with the PR8 (H0N1) strain we have obtained recombinants with wild type phenotype (*ts*⁺). These recombinants have replaced segment 2 (Ptra gene), however, keep segment 8 with the original *ts*-defect. We propose to call these recombinants suppressor recombinants, since by replacement of another gene the original defect is phenotypically suppressed.

[a] Institut für Virologie, Justus-Liebig-Universität Giessen, D-6300 Giessen, Federal German Republic.

Methods and Results

As a parent for the isolation of *ts*-mutants, we used the partial heterocygote 19N, which was obtained by rescuing *ts*19 of FPV with virus N (Hav2Neq1).[4] This recombinant was mutagenized with 5-fluorouracil, and *ts*-mutants were isolated according to Simpson and Hirst.[5] One of them (*ts*526) was further studied. By using a set of standard *ts*-mutants covering 7 recombination groups,[6,7] 6 of these *ts*-mutants were able to rescue *ts*526. Only the *ts*-mutant (*ts*412) carrying a *ts*-defect in segment 8 was unable to do so (Table 1). According to these results the *ts*-defect should not reside in segments 1 to 6, but in segment 8. Next, marker rescue experiments were performed by double infection of chick embryo cells with *ts*526 and PR8 at 33° (T) or 40° (P). The plaques obtained at 40° were purified by 3 consecutive plaque passages at 40°. The gene constellation of 9 ts^{+}-isolates was determined by the hybridization technique.[8] The results are presented in Table 2. None of these isolates had replaced segment 8, however, all of them carry segment 2 of the rescuing virus PR8. Thus there is a clear discrepancy: according to the data presented in Table 1 the *ts*-defect should reside in segment 8 and not in segment 2; according to the results of Table 2 the opposite might be concluded. One way to reconcile these contradictory results would be to assume that in the recombinants the gene with the *ts*-genotype is still present, but by replacement of segment 2 by the corresponding segment of PR8 the *ts*-genotype is phenotypically suppressed.

In Figure 1, a polyacrylamide gel is shown, on which the various viral proteins were separated after a 2-hour pulse with ^{35}S-methionine. It can be seen that the NS1-protein of *ts*526 migrates a little faster than that of the parent 19N. It is assumed that this difference is caused by the mutation related with the *ts*-character of *ts*526. In the recombinants T6, T7, T10, and T23 the NS1-protein with the faster migration rate is retained. (The NS1-protein of PR8 migrates even faster, which is not shown here). This suggests that the NS1-protein gene with the *ts*-defect is still present in the recombin-

Table 1. Recombination Frequencies of *ts*526 after Double Infection of CEC with a Panel of Standard *ts*-Mutants of FPV.

Recombination frequencies[a]						
***ts*3** segm. 1	***ts*90** segm. 2	***ts*263** segm. 3	***ts*227** segm. 4	***ts*19** segm. 5	***ts*113** segm. 6	***ts*412** segm. 8
10	6	4	30	50	50	<0.1

[a]The recombination frequencies were calculated from the following formula: $\{[AB_{33}]_{40} - ([A_{33}]_{40} + [B_{33}]_{40}) \times 100 \times 2\}/[AB_{33}]_{33}$. $[AB_x]_y$ indicates that the cells were infected with viruses A and B at temperature x and the yield was subsequently assayed at temperature y.[9]

Table 2. Derivation of the Various Genes after Marker Rescue of *ts*526 with PR8.[a]

Strain	RNA segments 1	2	3	4	5	6	7	8
*ts*526	F	F	N	F	N/F[b]	F	F	F
T6	F	P	N	F	N/F	P	F	F
T7	F	P	P	F	N/F	F	F	F
T10	F	P	N	F	N/F	F	F	F
T23	F	P	N	F	N/F	F	F	F
T26	F	P	N	F	N/F	F	F	F
P2	F	P	P	F	N/F	F	F	F
P3	F	P	P	F	N/F	F	F	F
P4	F	P	P	F	N/F	F	F	F
P5	F	P	P	F	N/F	F	F	F

[a]CEC were doubly infected with *ts*526 and PR8 either at 33° (T) or 40° (P). The yield was passaged at 40°. The derivation of the RNA segments was determined by hybridization of ^{32}P-labeled vRNA segments of FPV with a surplus of non-labeled cRNA of 431 (a plaque derivative of 19N[4]), PR8, or of the recombinants.[8] F = FPV, P = PR8, N = virus N.

[b]N/F means crossover segment containing about 65% sequences of virus N and about 35% sequences of FPV.[4]

Figure 1. Labeling of viral proteins of repressor recombinants. CEC were infected with *ts* 526 or recombinants. After 3 hours incubation at 33°, a pulse with 5 μCi ^{35}S-methionine per culture was started. Two hours later, the cells were processed, and the proteins were separated by polyacrylamide gel electrophoresis according to Bosch et al.[10]

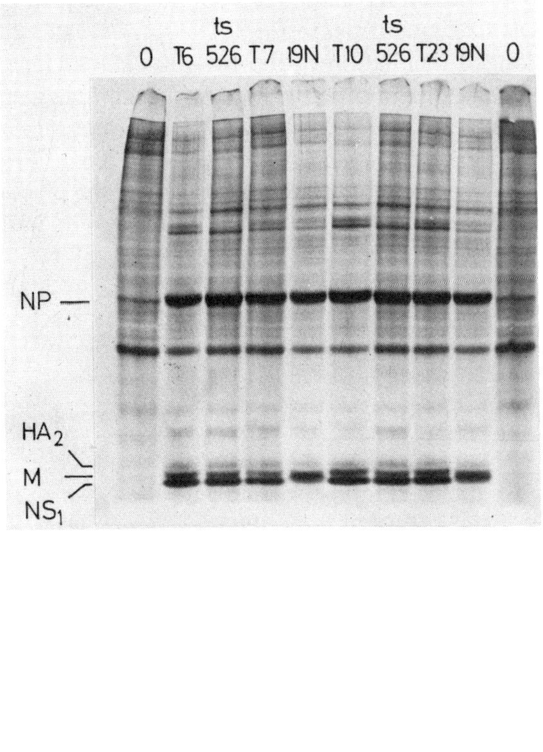

ants. Furthermore, if this is true it should be possibly by a backcross with the original parent strain 19N to recover again a certain percentage of *ts*-mutants, namely in those cases where segment 2 of PR8 is again replaced and segment 8 with the *ts*-genotype is retained. For this purpose we have doubly infected CEC with one of these recombinants and 19N (or a derivative of it) at 33°. A plaque test of the yield was started at 33°. Three days later, the size of the plaques was marked at the bottom of the dishes, and the cultures were transferred to 40°. Two days later, the cells were stained with neutral red, and both types of plaques were counted, those which grew further at 40° and those which did not grow at 40°. The latter ones by definition are *ts*-mutants. The results are shown in Table 3. It can be seen that in singly infected cells the number of plaques which did not enlarge at 40° was about 1%, while in doubly infected cells it was about 10%. Four plaques which were obtained after double infection with the T7 isolate the 19N and which did not enlarge at 40° were injected into embryonated chicken eggs without further purification. In the allantoic fluids of 3 eggs, only *ts*-viruses were found; the fourth contained virus with wild type character. One of these *ts*-isolates was analyzed further: by recombination with the standard *ts*-mutants it could be shown that the *ts*-mutation was in segment 8; and by hybridization we found that the gene constellation was identical with the parent *ts*526. These results are taken as evidence, that in the 9 recombinants (Table 2) derived from *ts* 526 and PR8 the gene with the *ts*-mutation is still present.

Next we studied how far the host cell influences the *ts*-character of these recombinants. Plaque tests were performed on CEC and MDCK cells at 33° and 40°. As shown in Table 4, the parent 19N does not form plaques on MDCK cells, either at 33° or at 40°, in contrast to FPV. The same was found with *ts*526. Only after a prolonged incubation of the MDCK cells at 33° (5 days) very tiny plaques were seen. T7 and other isolates with the same gene constellation readily formed plaques on MDCK cells at 33°, but not at 40°. Suppressor recombinants with a different gene constellation like T6 or T10, formed plaques on MDCK cells at both temperatures. (As a rule, the plaque number is higher on CEC compared with MDCK cells). These results indicate that the host plays an important role in the expression of the *ts* phenotype (and host range).

Discussion

The results presented here on suppressor recombination have several implications:

1. Since we are able to suppress the *ts*-genotype related with the NS-protein gene by replacing segment 2 (Ptra gene of the polymerase complex), we

Table 3. Backcross of Recombinants Obtained after Marker Rescue of *ts*526 with PR8 by Double Infection with 19N.[a]

CEC infected with	No. of plaques growing further at 40° / No. of plaques not growing further at 40°
19N	250/3
431	193/1
T6	159/2
T6 × 19N	83/15
T7	192/2
T7 × 19N	169/18
T10	115/1
T10 × 19N	116/13
T23	186/1
T23 × 431	82/13
T26	162/2
T26 × 431	132/17
P2	165/1
P2 × 19N	165/14
P5	209/2
P5 × 19N	180/14

[a]CEC were infected either singly or doubly. Fourteen hr after incubation at 33°, a plaque test was performed on CEC using the supernatant medium. After 3 days at 33°, the plaques were marked on the bottom of the dish without prior staining; and the cultures were incubated further at 40° for 2 days. After staining with neutral red the plaques were examined. 431 is a plaque derivative of 19N.[4]

Table 4. Plaque Tests of Recombinants on CEC and MDCK Cells.[a]

	CEC				MDCK			
	33°		40°		33°		40°	
Recombinant	PFU	ø[b]	PFU	ø	PFU	ø	PFU	ø
19N	1×10^8	2	4×10^7	5	$<10^3$	—	$<10^3$	—
*ts*526	2×10^8	2	$<10^3$	—	$<10^3$	—	$<10^3$	—
T6	2×10^7	2	8×10^6	4	5×10^6	2	2×10^6	4
T7	3×10^8	2	2×10^8	4	2×10^7	1	$<10^3$	—
T10	4×10^8	2	3×10^8	4	2×10^7	1	2×10^6	3
P2	4×10^8	2	2×10^8	4	4×10^7	1	$<10^3$	—
P4	3×10^8	2	1×10^8	4	1×10^7	1	$<10^3$	—
P5	1×10^8	2	3×10^7	4	8×10^6	1	$<10^3$	—

[a]The tissue cultures were stained 3 days after infection.

[b]Approximate diameters of plaques in mm.

have to assume that the products of these two genes somehow cooperate with each other. The preferential cotransfer of segments 2 and 8 during recombination between virus N and FPV is also in favor of this idea.[1] Thus we have here a tool at hand to study functional interactions of viral gene products.

2. The assignment of RNA segments to *ts*-lesions and corresponding functions by marker rescue of *ts*-mutants with other prototype influenza strains gives an unequivocal answer only in combination with backcrosses of the corresponding recombinants with the parent strain of the *ts*-mutants. We have checked recombinants obtained with our standard *ts*-mutants in this way. We found that the assignments as published [7,8] are correct.
3. Concerning the production of a live vaccine strain, the mere introduction of the RNA segment with a *ts*-lesion into another strain does not necessarily mean that this new recombinant also exerts the *ts*-phenotype. Or, if it exerts the *ts*-phenotype in one host cell, it might not do so in another cell type, and therefore might not have a reduced virulence for a certain host. Thus, the most urgent problem is to provide a reliable and fast test system for a potential human live vaccine.

ACKNOWLEDGMENTS

We thank V. von Hoyningen-Huene and H. Buerger for providing us with the ^{32}P-labeled vRNA segments of FPV. The work was supported by the Sonderforschungsbereich 47, Virologie, of the Deutsche Forschungsgemeinschaft.

References

1. Scholtissek, C. (1979): *Adv. Genetics* 20:1–36.
2. Murphy, B.R., Wood, F.T., Massicot, J.G., and Chanock, R.M. (1978): *Virology* 88:244–251.
3. Murphy, B.R., Wood, F.T., Massicot, J.G., and Chanock, R.M. (1980): *Arch. Virol.* 65:175–186.
4. Scholtissek, C., Rohde, W., Harms, E., Rott, R., Orlich, M., and Boschek, C.B. (1978): *Virology* 89:506–516.
5. Simpson, R.W. and Hirst, G.K. (1968): *Virology* 35:41–49.
6. Scholtissek, C. and Bowles, A.L. (1975): *Virology* 67:576–587.
7. Koennecke, I., Boschek, C.B., and Scholtissek, C. (1980): *Virology,* in press.
8. Scholtissek, C., Harms, E., Rohde, W., Orlich, M., and Rott, R. (1976): *Virology* 74:332–344.
9. Sugiura, A., Tobita, K., and Kilbourne, E.D. (1972): *J. Virol.* 10:633–647.
10. Bosch, F.X., Orlich, M., Klenk, H.-D., and Rott, R. (1979): *Virology* 95:197–207.

Copyright 1981 by Elsevier North Holland, Inc.
David H. L. Bishop and Richard W. Compans, eds.
The Replication of Negative Strand Viruses

Characterization of Influenza Virus "Cold" Recombinants Derived at the Non-Permissive Temperature (38°)

H.F. Maassab, C.W. Smitka, A.M. Donabedian, A.S. Monto, N.J. Cox, and A.P. Kendal[a]

Recombination-genetic reassortment at 25°C is being used to transfer genes from the "Master Strain," cold mutant (A/AA/6/60-H2N2) to new antigenic strains of type A influenza virus of epidemiological relevance. The genetic manipulation has provided recombinants which are genetically stable, grow to high titers and form plaques in primary chick kidney cells (PCKC) incubated at 25°. In addition, these recombinants were temperature-sensitive, shared lesions with *ts*-mutants derived by mutagenesis and were attenuated in animals and man.[1-6]

Dual infection of PCKC at 25° can be a selective process for isolation of recombinants with constant gene constellation and also in inducing mutations in the wild virus gene during recombination and cloning at 25°. In addition, genes from the cold mutant parent may predominate in recombinants simply because of greater efficiency of replication of the mutant than the wild type parent in cells mixedly infected and incubated at 25°. Thus, the process of recombination at 25° will preferentially transfer the cold gene(s) to the co-infecting wild type, hindering the attempt in identifying with certainty the gene(s) of the "Master Strain" which are responsible for attenuation and in determining the nature of the lesion(s) responsible for cold-adaptation (ca) and for temperature-sensitivity (*ts*).

In this report, recombinations at the non-permissive temperature of 38° where the ca donor strain exhibits a precipitous decrease in infectivity (> 5

[a] Department of Epidemiology, School of Public Health, University of Michigan, Ann Arbor, Michigan. World Health Organization Collaborating Center for Influenza, Bureau of Laboratories, Center for Disease Control, Public Health Service, U.S. Department of Health, Education, and Welfare, Atlanta, Georgia.

$\log_{10}$) might offer an approach for the derivation of recombinant virus having a more diverse mixture of cold and wild type genes. Biological and biochemical evaluation of these clones might offer a more definitive role of the ca and *ts* phenotypes in conferring attenuation. Analysis of clones with dissociation in the ca and *ts* markers and relating the acquisitions of these phenotypes to a gene or sets of genes responsible for the level of attenuation are paramount to our understanding of the determinants of attenuation of these recombinants for animals and humans.

Growth of the "Master Strain" (ca) and the Wild Type (wt) Lines of Influenza Viruses at the Non-permissive Temperature (NP) 38°

Experiments were performed to evaluate the growth characteristics of the wild type strain A/Alaska/6/77-H3N2 E111-Clone 3-A and the cold mutant, "Master Strain," A/AA/6/60-H2N2. Further insight from studies comparing the replication of the wt, and the cold mutant (ca) at different temperatures will help in defining the nature of lesion(s) in the cold mutant.

In earlier studies, we have reported that viral adsorption, uncoating or transcription and translation are not affected in primary kidney cells infected with cold mutant and incubated at the non-permissive temperature of 40°. However, a marked decrease in infectious virus of about 100,000-fold was evident.[7-9] Defects attributable to inhibition of RNA synthesis were not observed probably because cells were infected at high multiplicities. It is known that under this condition, significant levels of RNA synthesis can occur with mutants that otherwise exhibit an RNA^- phenotype when cells are infected at lower multiplicity.[10]

Figure 1 shows the growth characteristics of the wt and the cold mutant lines at 38°C. It is apparent that the maximum infectious yield of progeny virus of the wild type at 38° (2×10^7 PFU/ml.) was not different from the titer at 33° (5×10^7 PFU/ml.). As for the cold mutant (ca) with temperature-sensitive (*ts*) characteristics, some increase in titer was evident between 6 and 10 hours postinfection at the NP temperature (3×10^3 PFU/ml) followed by a decline in titer, with a yield of 2×10^2 PFU/ml, apparent 24 hours postinfection. No infectious titer was detected at 39°. Thus, a precipitous difference in infectivity titer of greater than 5 $\log_{10}$ is evident, when one compares the infectious yield at 24 hours in primary chick kidney cells of the wt and the ca lines. The cold mutant line exhibited similar growth at 25° and 33° both considered permissive to its growth, while the wt parent did not produce any plaques at 25°.

The input multiplicity used for the two lines at 38° was equivalent (3 PFU/cell). Input multiplicities at 38° influence the height and the slope of the curve. For instance, an input multiplicity of 10 PFU/ml or greater, the infectious yield of the cold-adapted line is greater, and progressively decreases as

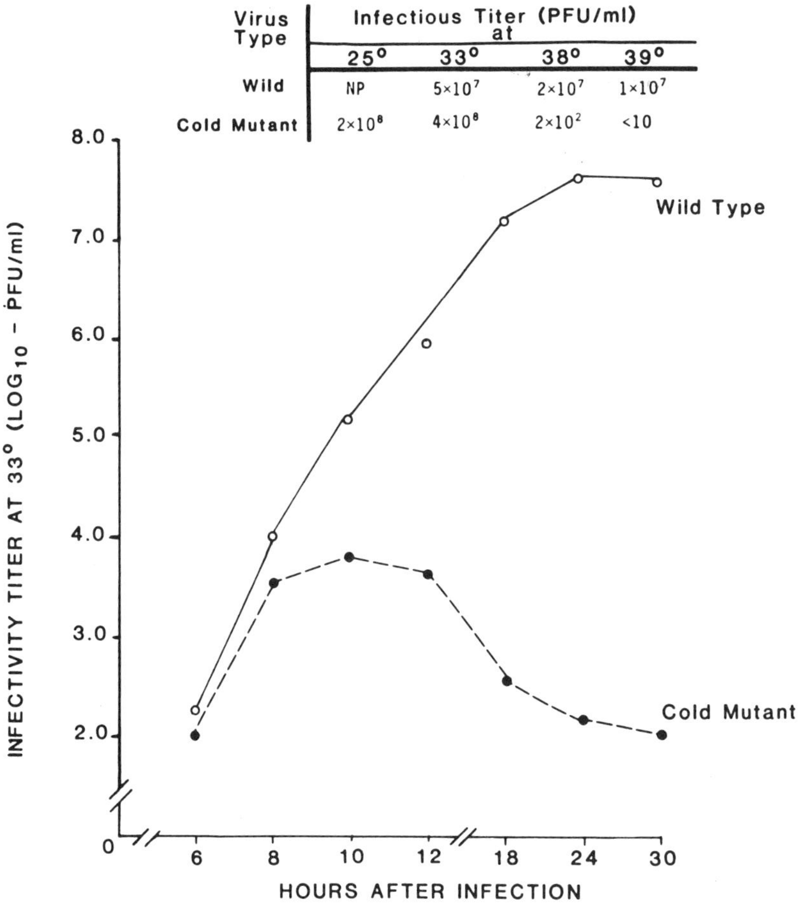

Virus Type	Infectious Titer (PFU/ml) at 25°	33°	38°	39°
Wild	NP	5×10^7	2×10^7	1×10^7
Cold Mutant	2×10^8	4×10^8	2×10^2	<10

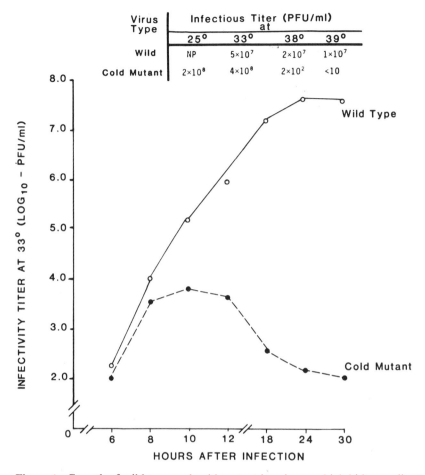

Figure 1. Growth of wild type and cold mutant in primary chick kidney cells at 38°.

the input is decreased. Thus, in cells mixedly infected and incubated at 38° RNA synthesis can occur, and where recombination-reassortment may proceed to furnish recombinants with different genealogy which is of major importance for defining the nature of the ca and *ts* lesion(s) responsible for attenuation in the cold-adapted lines.

Derivation of Clones of AA-CR43-H3N2 at the Non-Permissive Temperature 38°

Derivation of independent clones was accomplished using the individual "tube method." [4] The procedure illustrated in Figure 2 is designated to prepare independent recombinant clones and to minimize spontaneous mu-

INDEPENDENT DERIVATION OF RECOMBINANTS
AT THE NONPERMISSIVE TEMPERATURES (38°)

HISTORY "WILD PARENT"	PASSAGE OF THE RECOMBINANT	HISTORY "COLD VARIANT"
A/Alaska/6/77-H3N2 Lot-E111-Clone 3-A	1. Co-infect 20 culture tubes of CK cells with equal input of A/Ann Arbor/6/60 cold mutant and "Wild" parent. Incubate at 38° until CPE observed. (Muitiplicity= 3 PFU/cell) for each parent).	7PI-A/AA/6/60-H2N2 SE3-25°
↓	2. Harvest from each individual culture tube, and repassage separately in CK cells at 33° in the presence of A/Ann Arbor/6/60 serum.	↓
	3. Same as 2.	
	4. Same as 2, but omitting immune serum.	
	5. Grow each separate line of recombinant in CK cells at 33° under agar overlay, without immune serum. Pick one large, well defined plaque for each separate line of recombinant. 1st plaque purification.	
	6. Grow virus in the plaques from step 5 in separate cultures of CK cells at 33° under agar overlay. Again pick a single, large, well defined plaque for each line of recombinant. 2nd plaque purification.	
	7. Same as 6. 3rd plaque purification.	
Virus recorded was wild, ts+ and not cold-adapted (ca)	8. Prepare egg pools for each line of independent recombinant clone by inoculation into egg the virusin the plaques recovered in step 7.	No infectious virus or hemagglutinin recovered

tations. Each culture tube was passaged separately and was designated as a clone in contrast to the earlier procedure which calls for pooling individual cultures similarly treated before initiating the next passage.[2] The wild type parent of AA-CR43 was the A/Alaska/6/77-H3N2, Lot E111, Clone 3-A prepared by the Flow Laboratories for use in human volunteers. In general, mixed infection of primary chick kidney cells (PCKC) was initiated at 38° using 3 PFU/cell with the attenuated cold mutant A/AA/6/60-H2N2. After 24 hours, complete lysis of cells was evident. The second and third passages were made at 48-day intervals at 33° with antisera directed against the cold mutant. A 4th passage without antiserum was carried at 33° in PCKC or embryonated eggs. Each independent culture was cloned and each clone was plaque-purified 3 times. Biological characterization and the geneology of each clone followed. Parallel passage of the cold mutant, "Master Strain," failed to yield to virus, demonstrating that the incubation temperature of 38° and the immune serum have been effective in suppressing growth of the attenuated parent. Passage of the wild type parent under identical conditions did not alter the quality of the line, which is *ts*$^+$ and not cold-adapted (ca).

Phenotype of Recombinants Derived at Permissive (P-25°) and Non-Permissive (NP-38°) Temperatures

Dissociation of temperature-sensitive (*ts*) properties and virulence has been documented for certain animal viruses.[11,12] In recent studies, we have shown that in the cold recombinant AA-CR13-H3N2, such dissociation did exist. The line was not *ts* since it did not exhibit the 100-fold reduction in infectivity when compared to the wt, however, it was cold-adapted.[4] In addition, it was attenuated and immunogenic when administered to ferrets. Similarly, the observation was made for yet another cold recombinant the AA-CR12-H3N2. Upon infection of hamsters with AA-CR12-H3N2 line, *ts*$^+$ revertants were isolated and with retention of the ca property.[4]

In addition, such dissociation was also shown when recombination was done at the permissive temperature of 33°, a temperature which does not preferentially select for cold gene(s). In this study the derivation of AA-CR23-H3N2 was accomplished by acquisition of cold genes when cells were mixedly infected at 33° with cold mutant A/AA/6/60 and the wild type A/Victoria/3/75-H3N2 (Lot E-81). Clones from the cold recombinant AA-CR23-H3N2 exhibited different biological properties. All the clones were *ts*, however, some had both ca and *ts* markers, while the rest were not cold-adapted.[4]

In Table 1, using the same parental types, and with recombination done in the primary chick kidney cells, at 25° or 38°, it is evident that the 25° incubation (AA-CR31) did sevect for clones where both phenotypes, ca and *ts* were expressed. However, with the incubation at NP-38°, differences in phenotypes were expressed. Of ten clones so far analyzed, 60% had both ca and *ts* markers, 20% had ca alone and 20% were *ts*. The fact that in this selective environment, (NP-38°), the predominant clones isolated were still *ts* and ca, and wt type clones were not isolated provides further assurance about the potential safety of cold-adapted viruses as live influenza vaccines to be used in man.

Genotypes of Clones of the Recombinant AA-CR43-H3N2 Derived at the Non-Permissive Temperature (NP-38°)

Preliminary gene composition of 4 clones of the recombinant AA-CR43-H3N2 is illustrated in Table 2. It is evident that a different pattern has emerged as the result of this approach. In four clones the source of RNA 3 was not determined, while the source for the 7 genes was determined. In

Table 1. Evaluation of Recombinants of Influenza Virus Derived at Permissive (P-25°), and Non-Permissive Temperatures (NP-38°).

Parental types—"Cold mutant—Master Strain"—7PI-A/AA/6/60-H2N2 and the "wild type," A/Alaska/6/77-H3N2 Clone 3-A, E111

Recombinants designation	Antigenicity	Phenotype	Genotype
AA-CR31-H3N2 P-25°	7 Clones-$H3_{77}$-$N2_{77}$ 3 Clones-$H3_{77}$-$N2_{60}$	All clones are ca[a] and *ts*	−40% have 6 genes from CV[b] −50% have 5 genes from CV −10% have 7 genes from CV
AA-CR43-H3N2 NP-38°	All clones have $H3_{77}$-$N2_{77}$	6 Clones are *ts* and ca 2 Clones—ca only 2 Clones—*ts* only	Different—more WT genes

[a]ca: cold-adapted marker.
[b]CV: cold variant-master strain.

previous reports, complete assignments of genes in cold recombinants of influenza viruses used in vaccine trials has been resolved, although the gene(s) in the lines, responsible for cold-adaptation or temperature-sensitivity, were still matters of speculation.[13] The preliminary findings have suggested that the genetic composition of cold recombinants might be restricted so that RNA1 as well as NP and M-protein genes, were always derived from the cold mutant parent.[8,13] Thus, the recombinants derived at NP-38° show similar genealogy to the ones derived at 25°. Biological characteristics of these two types of recombinants were also similar. The data from the present study reinforce the concept that the cold mutant, A/AA/6/60-H2N2 is a genetically stable donor of attenuated genes which can furnish the necessary attributes to candidates of the influenza virus vaccines with predictable levels of attenuation in man.

Characterization *in vivo* of the 4 Clones of AA-CR43-H3N2

Evaluation of the reactogenicity and reversion of the 4 clones *in vivo,* along a comparison to the parental types is shown in Table 3. The infecting dose for each line was adjusted for equivalence and 4 ferrets were infected intranasally with each of the line.

Clinical signs such as fever, anorexia and runny nose, virus shedding and extent of viral growth in the turbinates and lungs were used to characterize the properties of these lines. Rectal temperatures were taken twice daily. Coryza was noted daily if present. At the 3rd and 8th day postinfection, the ferrets were sacrificed and viral contents of the lungs and turbinates were determined. The data shows that the growth in turbinates of the four clones

Table 2. Preliminary Gene Composition and Biological Characteristics of Clone of AA-CR43-H3N2 Derived at 38° Non-Permissive Temperature.

Virus clones	Antigenicity	Ca[d] and *ts*[e] phenotypes	Gene composition-Gene derivation							
			RNA 1	RNA 2	RNA 3	HA	NA	NP	M	NS
1-1	$H3_{77}$-$N2_{77}$	Ca and *ts* (38°)	M[a]	W[b]	?[c]	W	W	M	M	M
5-1	↓	Non Ca- and *ts* (38°)	?	W	?	W	W	M	M	W
7-1	↓	Ca and *ts*	M	W	?	W	W	M	M	M
8-1	↓	Ca and *ts*	M	W	?	W	W	M	M	W

[a]M: Mutant.

[b]W: Wild type.

[c]The RNA 3 was not determined.

[d]Ca: Cold adapted.

[e]*ts*: temperature sensitive when titrated in PCKC.

Table 3. Response of Ferrets to Infection with 4 Clones of AA-CR43-H3N2 with Differing Genotypes.

1. Infecting dose was approximately $10^{7.3}$ EID_{50}/ml.
2. Intranasal route of inoculation.
3. Nasopharyngeal swabs were taken daily for 8 days.
4. Four ferrets were used for each line.
5. Two ferrets sacrificed at the 3rd and again at the 8th day postinfection.

Virus line	Clinical signs	Virus shedding[a]	Virus Content (log_{10}-Eid_{50}/ml) T[b]		L[b]		Comments
			3 d[b]	8 d	3 d	8 d	
1-1	None	6 days	4.3	<1.0	<1.0	<1.0	ca, *ts* (38°)
5-1	Mild rhinitis	5 days	5.0	<1.0	<1.0	<1.0	Non ca,[c] *ts* +
7-1	None	5 days	5.5	<1.0	<1.0	<1.0	ca, *ts* (38°)
8-1	None	7 days	4.0	<1.0	<1.0	<1.0	ca, *ts* (38°)
Parental types "Master strain" cold mutant A/AA/6/60-H2N2	none		4.5	<1.0	<1.0	<1.0	*ts* (38°), Ca
A/Alaska/6/77-H3N2 "Wild Type"	Fever with rhinitis		6.3	2.0	2.5	1.0	Non ca, *ts* +

[a]Virus shed from nasopharyngeal swabs retained the markers (s).

[b]T: turbinates; L: Lungs; d: days.

[c]Revertants were isolated from the turbinates.

was similar to the cold-adapted "Master Strains" with no clinical signs evident. Virus growth was not detected in the lungs. In contrast, the infection with the parental wild type resulted in higher yield of virus in the turbinates and also evidence of growth in the lungs. Clinical manifestations were also present. Thus, the four clones are considered attenuated to the same extent as the parental cold mutant. Analysis of virus shed shows that clones 1-1, 7-1, 8-1, as expected, retained the ca and *ts* phenotypes, since their genealogy fits the concept advanced in previous studies. However, for clone 5-1, where at least 4 genes were derived from the wild type, reversion was apparent. It follows that in developing laboratory criteria for attenuated influenza virus vaccine, the RNA 1 gene (virus transcriptase) must be derived from the cold mutant.

Summary

For production of live influenza virus vaccines at the time of appearance of new antigenic variants, recombination-reassortment using a cold-adapted (ca) and temperature-sensitive (*ts*) "Master Strain" as a donor of ca and *ts* genes associated with attenuation has been carried out in primary chick kidney cells (PCKC) at 25°. To minimize the possible preferential transfer of "cold" gene(s) and or spontaneous mutations at 25°, recombination was made at the non-permissive temperature of 38°, using the same host-system. Independent clones were obtained, exhibiting differences in their gene composition. More wild type genes were transferred at 38°. Dissociation of the phenotypic characteristics in the recombinants was also possible at 38°. Some recombinants had similar genotypes as those derived at 25°. Analysis of the data shows that RNA2 is not associated with cold-adaptation, while RNA 1, NP and M of the cold mutant appear to be linked with cold-adaptation in recombinants with the proper level of attenuation.

ACKNOWLEDGMENT
This investigation was supported by Contract 1-AI-72521, National Institutes of Health, National Institute of Allergy and Infectious Diseases, Development and Applications Branch, Bethesda, Maryland.

References

1. Maassab, H.F. (1969): *J. Immunol.* 102:728–732.
2. Maassab, H.F. (1975): In: *Negative Strand Viruses.* Mahy, B.W.J. and Barry R.D. (Eds.), New York, Academic Press, pp. 755–761.
3. Maassab, H.F., Cox, N.J., Murphy, B.R., and Kendal, A.P. (1977): In: *International Symposium on Influenza Immunization.* Geneva Development of Biological Standardization, Basel, S. Krager, II, 39, pp. 25–31.
4. Maassab, H.F., Spring, S.B., Kendal, A.P., and Monto, A.S. (1978): In: *Negative Strand Viruses and the Host Cell.* Mahy, B.W.J. and Barry, R.D. (Eds.), New York, Academic Press, pp. 721–732.

5. Spring, S.B., Maassab, H.F., Kendal, A.P., Murphy, B.R., and Chanock, R.M. (1977): *Arch. Virol.* 55:233–243.
6. Spring, S.B., Maassab, H.F., Kendal, A.P., Murphy, B.R., and Chanock, R.M. (1977): *Arch. Virol.* 55:233–243.
6. Spring, S.B., Maassab, H.F., Kendal, A.P., Murphy, B.R., and Chanock, R.M. (1977): *Virology* 337–343.
7. Kendal, A.P., Kiley, M.P., and Maassab, H.F. (1973): *J. of Virol.* 12:1503–1511.
8. Kendal, A.P., Cox, N.J., Spring, S.B., and Maassab, H.F. (1978): In: *Negative Strand Viruses and the Host Cell.* Mahy, B.W.J. and Barry, R.D. (Eds.), New York, Academic Press, pp. 773–744.
9. Kendal, A.P., Cox, N.J., Galphin, J.C., and Maassab, H.F. (1979): *J. Gen. Virol.* 44:443–456.
10. Krug, R.M., Veda, M., and Palese, P. (1975): *J. Virol.* 16:790–795.
11. Preble, O.T. and Yaunger, J.S. (1973): *J. Virol.* 12:481–485.
12. Stanners, C.P. and Goldberg, V.J. (1975): *J. Gen. Virol.* 29:281–285.
13. Cox, N.J., Maassab, H.F., and Kendal, A.P. (1979): *Virology* 97:190–194.

Copyright 1981 by Elsevier North Holland, Inc.
David H. L. Bishop and Richard W. Compans, eds.
The Replication of Negative Strand Viruses

Genetic Synergism Between Matrix Protein and Polymerase Protein Required for Temperature-Sensitivity of the Cold-Adapted Influenza A/Ann Arbor/6/60 Mutant Virus

Nancy J. Cox,[a] Alan P. Kendal,[a] Hunein F. Maassab,[b] Christof Scholtissek,[c] and Susan B. Spring[d]

Introduction

The cold-adapted (ca) and temperature-sensitive (*ts*) mutant of influenza A/Ann Arbor/6/60 has been used to attenuate a series of wild type (wt) viruses by recombination (reassortment). Human volunteer studies in the U.S. and Europe have indicated that the ca recombinants are infectious for susceptible young adults and children and rarely cause symptoms greater than those acceptable for a live vaccine. Because of the promise these ca viruses have as a model to provide information about the usefulness of live, attenuated influenza vaccines, we are characterizing biochemically and genetically the ca mutant and its recombinants in order to define the nature of the lesions responsible for the ca, *ts* and ultimately the attenuation properties of this mutant. Previous studies showed that recombinants prepared at 25°C had a highly restricted gene composition with five or six non-hemagglutinin (HA), non-neuraminidase (NA) RNA segments derived from the A/Ann Arbor/6/60 ca parent.[1-3] This prevented us from determining which gene(s) contained the *ts* or ca lesions. Therefore, we prepared a separate series of recombinants at 33° and 39°, conditions less likely to favor selection of genes from the ca parent. In this report, we describe the genetic and biological properties of these recombinants, the biochemical detection of lesions in RNAs 1 and 7 in the A/Ann Arbor/6/60 mutant as well as the

[a]World Health Organization Collaborating Center for Influenza, Center for Disease Control, Atlanta, Georgia.
[b]University of Michigan, School of Public Health, Ann Arbor, Michigan.
[c]Institut für Virologie Justus-Liebig-Universität Giessen, 6300 Giessen, Germany,
[d]Laboratory of Infectious Diseases, National Institutes of Health, Bethesda, Maryland.

functional assignments for the three largest genes of A/Ann Arbor/6/60 which have not been reported previously. We conclude that the *ts* property of the A/Ann Arbor/6/60 mutant depends on synergism between a protein coded for by RNA 7 and a polymerase protein coded for by RNA 1.

Methods

Viruses. Production of the ca A/Ann Arbor/6/60 mutant has been described.[4] WSN *ts*-mutants (*ts*6, *ts*53, *ts*15, *ts*56 and *ts*51) were obtained from Dr. Simpson and Dr. Palese. Experiments using recombinants with fowl plague virus (FPV)[5,6] were carried out in the Institut für Virology, Universität Giessen, Germany. All virus seeds were propagated in eggs.

Preparation of recombinants. Recombinants were made to mixedly infecting MDCK cells with either a combination of the A/Ann Arbor/6/60 mutant and WSN *ts*-mutant or a combination of the A/Ann Arbor/6/60 mutant and the A/Ann Arbor/9/73 (H3N2) wt virus. Multiplicity of infection (moi) was 1–10 PFU/cell with each virus. Cells were maintained at 33°C for 18–20 hr. in the presence of 2 μg/ml TPCK trypsin. For experiments utilizing the WSN *ts*-mutants, recombination frequencies were determined by plaque titration of the virus yield at 33° and 39°.[7] Our WSN *ts*-mutant stocks had titer reductions of greater than 3 $\log_{10}$ PFU/ml at 39° compared with the titer at 33°. Clones of ts^+ virus were selected at 39° when dilutions of 10^{-4} or 10^{-5} produced plaques. Virus clones were grown in eggs once at 34°C, retitrated at 39°, recovered from 39° plaques and verified to be ts^+ virus. In order to reduce selective pressure, clones from the A/Ann Arbor/6/60 mutant X A/Ann Arbor/9/73 recombination mixtures were plaqued at 33° and then screened to determine their properties.

RNA analysis. Polyacrylamide gel electrophoresis was carried out as described previously.[2,3] Unequivocal identification of RNAs 2 and 3 of the Ann Arbor mutant X WSN mutant recombinants was not always possible with PAGE analysis. Therefore, when necessary the ts^+ recombinants were analyzed by RNA-RNA hybridization of ^{32}P-labeled vRNA segments 1, 2 and 3 of A/Ann Arbor/6/60 mutant with complementary RNA obtained from cells infected with each of the recombinants or their parents.[8,9] RNA-RNA hybridization experiments to determine functional gene assignments were carried out according to established procedures.[5]

Oligonucleotide analysis Total RNA extracted from purified egg-propagated virus or the isolated RNA segments were digested with T_1 ribonuclease. The resultant oligonucleotides were labeled at the 5′ ends using γ-^{32}P-ATP and polynucleotide kinase,[10,11] and the oligonucleotides were mapped by two-dimensional electrophoresis in acrylamide gels using standard procedures.[11,12]

Analysis of double-stranded RNAs. Monolayers of CEF cells were treated with 100 μg/ml cycloheximide for 1 hr before, during and after infection with either A/Ann Arbor/6/60 mutant or wt virus at a moi of approximately 100 PFU/cell. ^{3}H-uridine (300 μCi/ml) was added 1 hr postinfection and cells were harvested at 5 hr postinfection by scraping into ice-cold TSE. Nuclei were removed by centrifugation after treatment with 1% NP-40 for 20 min. at 4°. The cytoplasmic supernatant was extracted 3 times with phenol: chloroform: isoamylacohol mixture (50:48:2) after treatment at 37° for 30 min. with 1% SDS, 0.75 mg/ml pronase and 1% 2-mercaptoethanol. RNA was ethanol-precipitated, dissolved in 0.001 M EDTA and divided into aliquots to each of which was added 10 μg vRNA. RNA was denatured, hybridized and analyzed on 7.5% polyacrylamide gels containing 6 M urea as described.[8,13]

Results

Analysis of recombinants prepared at 39°. A/Ann Arbor/6/60 mutant was recombined at 39° with single-step *ts*-mutants of A/WSN/33 derived by Suguira and coworkers.[14] We chose viruses from each WSN *ts* group except groups IV and VI which carry lesions in the glycoproteins (because HA and NA of A/Ann Arbor/6/60 are known not to contain critical lesions required for ca and *ts* properties), and group VIII which was not available. This approach was designed to see which WSN mutant failed to produce ts^+ recombinants with the A/Ann Arbor/6/60 mutant and thus identify the *ts* gene in A/Ann Arbor/6/60. From 0.5% to 5.0% of the progeny in the yield from the original mixed infections of the A/Ann Arbor/6/60 mutant and WSN *ts* mutants of groups I (P_3), II (P_2), V (NP) and VII (M) were ts^+ (Table 1). WSN mutant group III (P_1) consistently failed to recombine with A/Ann Arbor/6/60, however, suggesting the *ts* lesion in the A/Ann Arbor/6/60 virus was in the gene coding for P_1. To verify that ts^+ virus obtained with WSN groups I, II, V and VII mutants were not revertants, but contained the predicted A/Ann Arbor/6/60 rescuing gene, representative isolates were genotyped. Gene assignments confirmed the presence of the appropriate rescuing A/Ann Arbor/6/60 genes in each case. However, we obtained the surprising result that no single gene of the A/Ann Arbor/6/60 mutant appeared to be responsible for the *ts* phenotype because each A/Ann Arbor/6/60 gene was represented several times among the recombinants (Table 2). Only RNAs 2 and 7 of A/Ann Arbor/6/60 are absent from IIg_1, suggesting the involvement of one or both of these genes in the *ts* phenotype; however recombinants $VIIb_1$ and $VIIc_1$ which contain the A/Ann Arbor/6/60 M-protein gene were also ts^+ and the A/Ann Arbor/6/60 RNA 2 was present in three ts^+ recombinants. This suggested a possible synergism between two or more A/Ann Arbor/6/60 genes for the *ts* phenotype. The combination of RNA 1 and RNA 7 was one combination missing in all the ts^+ recombinants, and therefore was a candidate for such a synergistic constellation.

Table 1. Frequency of Production of ts^+ Recombinants from Mixed Infections with Pairs of *ts*-Mutants.

MDCK cells mixedly infected with mutant A/Ann Arbor/6/60 and A/WSN/33 *ts*-mutants:	WSN gene with RNA lesion	WSN proteins with functional defect	Mean recombination frequency[a]
WSN group I (*ts*6)	RNA 1	P_3[b] (polymerase)	2.6%
WSN group II (*ts*53)	RNA 3	P_2 (polymerase)	3.1%
WSN group III (*ts*15)	RNA 2	P_1 (polymerase)	<0.0008%
WSN group V (*ts*56)	RNA 5	NP (nucleoprotein)	0.5%
WSN group VII (*ts*51)	RNA 7	M (matrix protein)	3.8%

[a]Recombination frequency calculated as (yield titrated at 39°C)/(yield titrated at 33°C) × 100; all results are the mean of at least two experiments.

[b]Terminology of Palese and coworkers.

Analysis of recombinants prepared at 34° To examine further the genetic basis for the *ts* phenotype of A/Ann Arbor/6/60, we produced a second set of recombinants with a different virus strain, wt A/Ann Arbor/9/73. The cloning of recombinants was done at 33° so not to exert selective pressure for either ca genes (as at 25°) or ts^+ revertant or suppressor genes (as at 39°). Identification and characterization of recombinants was done by genotyping using antigenic analysis, PAGE of virion RNA segments, and by plaque titration at 25°, 33° and 39°. This experiment yielded both *ts* and ts^+ recombinants (Table 3). All contained A/Ann Arbor/6/60 M-protein. However *ts* and ts^+ recombinants differed in the derivation of their RNA 1; like the recombinants derived with the WSN group VII mutant, ts^+ viruses did not possess RNA 1 of the A/Ann Arbor/6/60 RNA 1. A noteworthy observation was the presence of only a single wt gene, RNA 1 in the ts^+ clone S_{34}. We conclude, therefore, on the basis of the properties of these two sets of recombinants produced at 33° and 39° that the presence of both RNAs 1 and 7 of A/Ann Arbor/6/60 is necessary to confer the *ts* property. The ca property also appears to be dependent on interaction of two or more genes; however, biological analysis of the ca property of the recombinants is still in progress.

Oligonucleotide mapping of the RNA. Oligonucleotide maps were prepared for RNA segments 1 and 7 of the mutant and wt viruses to confirm the existence of mutations in these genes. RNA 1 of the mutant virus exhibited a one-spot difference from the corresponding wt RNA segment (Figure 1). Oligonucleotide differences were also seen in RNA 3 and the NP gene (data not shown). The detection of a difference in the oligonucleotides of RNA 1 in

Table 2. Properties of ts^+ Recombinants Derived from Mixed Infections of Mutant A/Ann Arbor/6/60 and WSN Mutants.

ts^+ recombinant	Gene derivation[a]								EOP[c]
Clone designations[b]	1 (P_3/TRA)[d]	2 (P_1/pol. 1)[e]	3 (P_2/pol. 2)[e]	4 (HA)[f]	5 (NP)[d]	6 (NA)[e]	7 (M)[g]	8 (NS)[d]	PFU 39°/PFU 33°
Ia_3	A	A	W	W	A	A	W	A	12%
Iu_2	A	W	W	W	W	A	W	W	67%
Im_3	A	W	W	W	A	A	W	A	25%
IIg_1	A	W	A	A	A	A	W	A	25%
IIh_2	W	A	A	A	W	A	W	A	40%
Vh_5	W	W	A	W	A	W	W	W	50%
$VIIb_4$	W	W	W	A	W	A	A	W	100%
$VIIc_1$	W	A	W	W	W	A	A	A	25%
No. recombinants with Ann Arbor gene	4/8	3/8	3/8	3/8	4/8	7/8	2/8	5/8	

[a] W or A indicates that the gene was derived from the WSN mutant or from the A/Ann Arbor/6/60 mutant. Functional assignments in parentheses include those by Palese/Scholtissek where different systems are in use.

[b] Roman numerals indicate the mutant group of the WSN parent used in recombination. WSN mutants were as follows: Group I, *ts*6 (defect in P_3); Group II, *ts*53 (defect in P_2); Group V, *ts*56 (defect in NP); Group VII, *ts*51 (defect in M). No ts^+ recombinants were obtained with *ts*15, a group III mutant (defect in P_1).

[c] Efficiency of plaquing calculated as (yield titrated at 39°)/(yield titrated at 33°) × 100%. Each value is the mean of at least three determinations.

[d] Determined by PAGE of vRNA.

[e] Determined by PAGE of vRNA, RNA-RNA hybridization and oligonucleotide mapping.

[f] Determined by vRNA and neuraminidase-inhibition or hemagglutination-inhibition tests.

[g] Determined by PAGE of vRNA and proteins.

Table 3. Properties of A/Ann Arbor/6/60 × A/Ann Arbor/9/73 Recombinants.

Recombinant clone	Gene derivation[a]								EOP
	1	2	3	4 (HA)	5 (NP)	6 (NA)	7 (M)	8 (NS)	PFU 39°/ PFU 33°
S_1	A	W	W	W	W	A	A	A	<.0001%
S_{10}	A	A	A	A	A	A	A	W	<.001%
M_4	A	W	A	W	A	W	A	A	<.002%
S_{32}	W	W	A	A	A	A	A	A	51%
S_{34}	W	A	A	A	A	A	A	A	19%

[a]W indicates that the gene was derived from the wild type virus; A indicates that the gene was derived from the A/Ann Arbor/6/60 mutant.

mutant and wt virus is consistent with a previous observation that the electrophoretic migrational rates differ for RNA 1 of the two viruses.[2] No oligonucleotide differences were observed for RNA 7 (Figure 1).

Analysis of RNA-RNA hybrids. Since no differences were observed in the RNA oligonucleotide maps of RNA 7 from the mutant and wt viruses, we applied other methods to look for evidence of a mutation in this gene during cold-adaption. Single-base changes can sometimes be detected by PAGE analysis of S_1 nuclease-treated double-stranded (ds) hybrids formed between virion RNA and cRNA transcripts from cells infected in the presence of cycloheximide.[13] Analysis on 7.5% acrylamide gels (Figure 2) revealed different electrophoretic mobilities for the heterologous deuplex RNAs 4, 5, 6, 7 and 8 compared to the homologous RNA obtained with wt and mutant A/Ann Arbor/6/60 viruses. This ds RNA analysis indicates that base changes have occurred in RNA 7 of the A/Ann Arbor/6/60 mutant.

Coding assignments of RNAs 1–3 The coding assignments for RNAs 4,5,6,7 and 8 of A/Ann Arbor/6/60 have been published;[1,2] however, functional gene assignments have not been reported for RNAs 1–3. These have now been determined by hybridization of isolated ^{32}P-labeled RNA segments of mutant A/Ann Arbor/6/60 with cRNA of recombinants containing only a single, known RNA of influenza A/Singapore/57 (that is closely related to A/Ann Arbor/6/60), but other large RNAs of the distantly related avian FPV.[5,6] Results demonstrated that RNAs 1 and 3 of A/Ann Arbor/6/60 code for products functionally similar to FPV "transport gene" and "polymerase 2" respectively (Table 4). Therefore, RNA 2 of A/Ann Arbor/6/60 must code for a product functionally similar to FPV "polymerase 1." Correlation of WSN and FPV functional gene assignments has been reported.[16] By inference, RNAs 1, 2 and 3 of the A/Ann Arbor/6/60 mutant correspond to RNAs 1,2 and 3 of WSN that code for "P_3" (cRNA synthesis), "P_1" (cRNA synthesis) and "P_2" (vRNA synthesis), respectively, in the terminology of Palese and coworkers.[17,18]

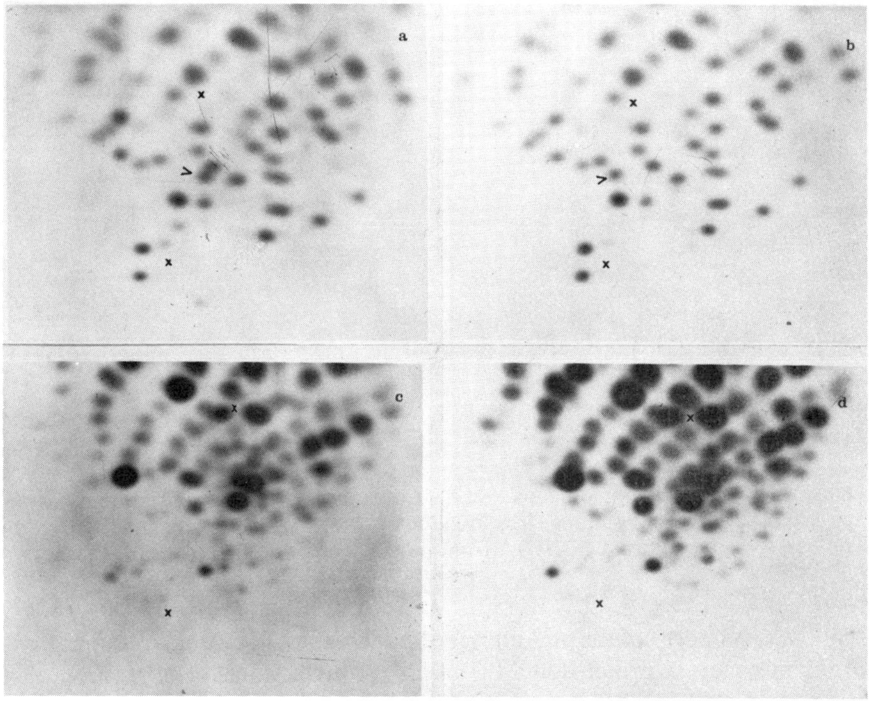

Figure 1. Oligonucleotide maps of the RNAs 1 and 7. (a) RNA 1 of A/Ann Arbor/6/60 wt virus. (b) RNA 1 of A/Ann Arbor/6/60 mutant virus. (c) RNA 7 of A/Ann Arbor/6/60 wt virus. (d) RNA 7 of A/Ann Arbor/6/60 mutant virus. Positions of dye markers xylene cyanol FF and bromophenol blue are indicates by Xs. Oligonucleotide differences noted by arrows.

Discusssion

The failure of the A/Ann Arbor/6/60 mutant to rescue a WSN group III *ts*-mutant in recombination experiments implied that the corresponding RNA of the A/Ann Arbor/6/60 mutant (RNA2) contained the *ts* lesion. However, RNA genome analysis revealed that this is not the case, because RNA 2 of the A/Ann Arbor/6/60 mutant is present in several ts^+ virus clones obtained by recombination with other WSN *ts*-mutants. Furthermore, several *ts* recombinants of A/Ann Arbor/6/60 which contain wt RNA 2 have been described previously.[2] These results, as well as those reported by others,[19] point out the inadequacies of complementation/recombination analysis without supporting gene analysis to identify the site of *ts* mutations.

It is expected that multi-step mutations occurred during stepwise adaptation of A/Ann Arbor/6/60 for maximum replication at 25°. The selection of mutations during this process would be influenced by the interdependence of genes and gene products as they operate in combination. Our results indicate

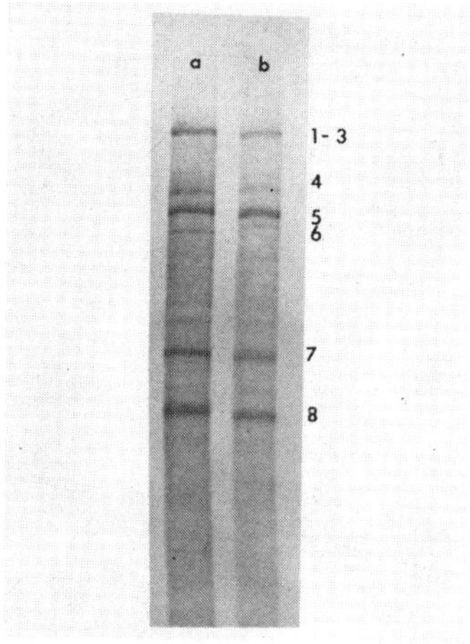

Figure 2. Analysis of homologous and heterologous ds RNAs. (a) The homologous mixture of cRNA from mutant virus infected cells and vRNA from mutant virus. (b) The heterologous mixture of cRNA from mutant virus and vRNA from wt virus.

that the *ts* property of the A/Ann Arbor/6/60 mutant is bigenic, requiring one of the polymerase proteins and the matrix protein. It has been reported that isolated M-protein of vesicular stomatitis virus [20] and influenza A virus [21] can inhibit the level of *in vitro* transcriptase activitiy of these negative strand viruses. Our conclusions are consistent with these findings in demonstrating that interactions between polymerase and matrix protein genes or gene products may occur during replication of influenza A virus.

Table 4. Homology Between Large RNAs of A/Ann Arbor/6/60 and Fowl Plague Virus Recombinants Possessing a Single A/Singapore/1/57 Gene.

	^{32}P vRNA segments of A/Ann Arbor/6/60[a]	
cRNA of	RNA 1	RNA 3
A/Singapore/1/57	1620[c]	1980
Fowl plague virus	720	1190
Recombinant FPV *ts*3 × Singapore (-RNA 1 of FPV)[b]	700	1010
Recombinant FPV *ts*90 × Singapore (-RNA 2 of FPV)	1510	1140
Recombinant FPV *ts*263 × Singapore (-RNA 3 of FPV)	670	1660

[a]Separated by PAGE at 39° to 41°C.

[b]RNA composition of recombinants reported previously by Scholtissek and coworkers.

[c]RNase resistant counts/5 min. After hybridization samples were heated for 10 min at 75° in 1% formaldehyde before digestion.

ACKNOWLEDGMENTS
We thank Eve Bingham and Judy Galphin for expert technical assistance. This work was supported, in part, by National Institutes of Health, Contract No. 1-Y01-AI-80001-00 (CDC IRA No. H900-06X).

References

1. Kendal, A.P., Cox, N.J., Murphy, B.R., Spring, S.B., and Maassab, H.F. (1977): *J. Gen. Virol.* 37:145–159.
2. Kendal. A.P., Cox, N.J., Galphin, J.C., and Maassab, H.F. (1979): *J. Gen. Virol.* 44:443–456.
3. Cox, N.J., Maassab, H.F., and Kendal, A.P. (1979): *Virology* 97:190–194.
4. Maassab, H.F. (1967): *Nature (London)* 213:612–614.
5. Scholtissek, C., Harms, E., Rohde, W., Orlich, M., and Rott, R. (1976): *Virology* 74:332–344.
6. Scholtissek, C., Rohde, W., Harms, E., and Rott, R. (1978): In: *Negative Strand Viruses and the Host Cell.* Mahy, B.W.J. and Barry, R.D. (Eds.), Academic Press, London, pp. 19–26.
7. Cox, N.J., O'Neill, M.C. and Kendal, A.P. (1977): *J. Gen. Virol.* 37:161–173.
8. Hay, A.J., Lomniczi, B., Bellamy, A.R., and Skehel, J.J. (1977): *Virology* 83:337–355.
9. Bean, W.J., Cox, N.J., and Kendal, A.P. (1980): *Nature (London)* 284:638–640.
10. Frisby, D. (1977): *Nucleic Acids Res.* 4:2975–2996.
11. Pederson, F.S. and Haseltine, W.A. (1980): In: *Methods in Enzymology, Vol. 65.* Grossman, L. and Moldane, K. (Eds.), Academic Press, New York, pp. 680–687.
12. deWachter, R. and Fiers, W. (1972): *Anal. Biochem.* 49:184–197.
13. Hay, A.J., Skehel, J.J., and Webster, R.G. (1979): *J. Gen. Virol.* 45:245–248.
14. Sugiura, A., Ueda, M., Tobita, K., and Enomoto, C. (1975): *Virology* 65:363–373.
15. Cox, N.J. and Kendal, A.P. (1978): *J. Gen. Virol.* 40:229–232.
16. Scholtissek, C. (1979): In: *Advances in Genetics.* Caspiri, E.W. (Eds.), Academic Press, New York, pp. 1–36.
17. Palese, P., Ritchey, M.B., and Schulman, J.L. (1977): *J. Virol.* 21:1187–1195.
18. Krug, R.M., Ueda, M., and Palese, P. (1975): *J. Virol.* 29:134–142.
19. Murphy, B.R., Wood, F.T., Massicot, J.G., Spring, S.B., and Chanock, R.M. (1978): *Virology* 88:231–243.
20. Carroll, A.R. and Wagner, R.R. (1979): *J. Virol.* 29:134–142.
21. Zvonarjev, A.Y. and Ghendon, Y.Z. (1980): *J. Virol.* 33:583–586.

Copyright 1981 by Elsevier North Holland, Inc.
David H. L. Bishop and Richard W. Compans, eds.
The Replication of Negative Strand Viruses

Molecular Organization of Defective Interfering Influenza Viral RNAs and Their Role in Persistent Infection

Debi P. Nayak, Alan R. Davis and Barun K. De[a]

von Magnus virus [1] or defective interfering (DI) virus particles [2] are produced by passing virus at high multiplicity and have been found in all cell-virus systems that have been carefully analyzed. Over the last few years, we have studied the biochemical and biological properties of DI influenza virus.[3] DI influenza virus can be readily produced by passing virus at high multiplicity in either embryonated chicken eggs, chick embryonic fibroblasts, MDBK, MDCK or HeLa cells. Like other DI viruses, influenza DI virus also requires the helper function of homologous infectious virus, but in turn interferes with the replication of infectious viruses. Interfering property of DI viruses can be directly quantitated as defective interfering unit (DIU)/ml using an infectious center reduction assay in which the same cell, co-infected with DI virus and infectious virus, does not produce a visible plaque.[4] DI influenza virus can not physically be separated from infectious virus by density or velocity gradient centrifugation and contains essentially the same structural proteins as infectious virus, although the amount of specific proteins such as nucleoprotein and hemagglutinin may vary. DIU and PFU content in different DI virus preparations produced from separate clones can vary considerably from 2 to 8×10^6 DIU/ml with a DIU/PFU ratio from 27 to 1500. Interfering property is 30–40 times more resistant to UV irradiation compared to infectivity.[5]

[a] Department of Microbiology and Immunology, UCLA School of Medicine, Los Angeles, California.

Nature of DI Viral Genome

When DI viral RNAs were analyzed by PAGE, they were shown to contain, in addition to 8 viral RNA segments, a set of smaller RNA segments (DI RNA) which are characteristic of a specific DI virus preparation.[3] Clonal stocks or even plaques may contain a small amount of DI viruses which later become amplified. Stock preparations of wild type viruses were contaminated with DI viruses even after repeated plaque purification. The characteristic RNA of DI virus appears to change after many passages during amplification with added standard virus.[6] Usually new smaller DI RNAs become predominant over larger DI RNA populations. Thus unlike VSV DI virus, influenza DI particles are rather difficult to propagate indefinitely. Different DI RNAs are not always mutually exclusive, although they are not found in equimolar ratios in a given preparation, and some may interfere with the replication of others.

DI virus from one subtype can interfere with the replication of other subtypes of influenza A viruses.[6] Equally DI viral RNAs can replicate with the helper function of other subtype viruses. These results would suggest that the gene products required for the replication of DI RNAs as well as those for the interference can function equally well among viral subtypes. Since genetic reassortment occurs freely among the different subtypes, gene products of one subtype must be functional genetic exchange among viral subtypes to the DI virus mediated interference.

The nature of host cells greatly influence the replication of DI influenza virus. Both DI and infectious viruses are produced most efficiently in MDBK cells and least in HeLa cells.[6] CEF cells produce an intermediate amount of both DI and infectious viruses. Such a result would suggest that the steps needed in the replication of DI viruses are essentially the same as for the replication of infectious virus. Similar DI RNA patterns in all 3 types of cells indicate that these cells do not exert a great selective pressure in the amplification of one DI RNA segment over another.

Composition and the nature of the RNA of individual DI and infectious virus particles remain unknown at present. Although occasionally specific vRNA segment is greatly reduced in a DI preparation,[4,7,9] the majority of DI viruses contain reduced amounts of larger RNA segments, suggesting that the loss of infectivity is probably due to a random rather than a specific loss of one or more RNA segments.[5] Partial complementation of some virus function at a high multiplicity of DI virus supports this idea.[10] The distribution of DI RNAs in the DI virus population is also not clear. It appears unlikely that DI virus contains only DI RNAs and no viral RNA segments. However, since all DI RNAs are not in equimolar ratio, all DI RNA segments may not be present in each DI particle producing a heterogeneity of DI RNA content among DI virus particles in a population. Although unlikely, we can not rule out that each DI particle contains only one kind of DI RNA.

Since DI virus preparation contains viral RNA segments and DI virus particles cannot be separated from contaminating infectious and non-infectious (non-interfering) particles, the role of DI RNA segments to interference is rather difficult to establish directly. We, therefore, isolated RNP preparation from infectious and DI viruses and demonstrated that fractions containing only DI RNP can interfere in infectious center reduction assay, while RNP preparations of infectious or DI virus which did not contain DI RNA, could not interfere indicating the etiological role of DI RNAs in interference.[11]

Origin of DI RNAs

Since DI RNAs appeared smaller and were responsible for interference, they possessed properties similar to the DI RNAs of VSV and Sendai viruses which have been extensively studied. However, it soon became apparent that the influenza DI RNAs possessed properties unique to influenza viruses.[12]

(1) The majority of the DI viruses originate from polymerase genes (P1, P2, P3) and possess the same polarity as vRNAs. Of the 14 DI RNAs analyzed, 7 arose from P3, 6 from P1 and 1 from P2. We have not yet detected any DI RNAs from the 5 other genes (HA, NP, NA, M or NS). The significance of this observation is not yet understood, although it is quite possible that DI RNAs are formed from other genes at a reduced frequency which we have not detected. Alternatively, only P genes may be capable of yielding DI RNAs, either because of their secondary structure, increased length, specific internal sequence recognition by polymerase or the involvement of a defective polymerase in interference. There is as yet no evidence to support one or the other possibility.

(2) Unlike the majority of VSV or Sendai DI RNAs, influenza DI RNAs appear to be internally deleted. A number of observations support this conclusion.[13]

(a) Oligonucleotide analyses showed that DI RNAs that arise from the same gene contain a number of identical oligonucleotide spots suggesting that they were not formed at random but some specific segments of the progenitor genes were preserved. However, in many cases oligonucleotides of smaller DI RNAs were not a complete subset of the larger DI RNAs. Each may have contained one or more spots present in the progenitor vRNA but not present in the othe rDI RNAs. This could be explained only by the formation of internal deletion.[12]

(b) Terminal sequence analysis showed that all DI RNAs contained the common 5′ and 3′ ends of progenitor RNAs. Therefore, internal deletion must have occurred in the formation of DI RNAs.[13]

(c) Cloning, restriction mapping, hybridization followed by nuclease treatment, heteroduplex mapping and complete sequence analyses of a number of DI and progenitor RNAs are currently being investigated. Preliminary data suggest that one or more deletion of polymerase genes is required for the formation of influenza DI RNAs.

Role of Influenza DI Viruses in Persistent Infection of Cell Cultures

Although DI viruses have been shown to be produced with all types of viruses, their role in the natural history of viruses and viral infection remains unclear. One attractive postulation, made by Huang and Baltimore,[2] implies that DI particles, because of their ability to reduce the virus growth and cytocidal activity of homologous lytic virus, may affect the course of acute and chronic viral infections. Over the years, a number of experimental systems involving viral infection have supported this hypothesis.[14] More importantly, the role of DI virus in the establishment of persistent infection in culture has been clearly established for a number of cell-virus systems.[3] We have also investigated the role of DI virus in establishing persistent infection of influenza virus in MDBK and HeLa cells.[6]

Persistently infected (Pi) cultures could be routinely established by coinfecting cells with both temperature-sensitive mutant and DI virus but not by using infectious ts^- or ts^+ virus alone or by a combination of ts^+ and DI virus. Pi cells showed a cyclical pattern of cell lysis and virus production at early passages (<p−7), but at later passages (<p−20) Pi cells became stable and behaved like normal cells. A number of experiments suggested that although very little virus was produced by Pi cells, the majority if not all of the Pi cells contained the viral genome and expressed viral antigens. (1) Pi cultures were resistant to superinfection by homologous virus but not by heterologous virus, e.g., NDV, VSV. Both cytopathic effect, PFU/ml and HAU/ml were greatly reduced in Pi cells after homologous challenge (Table 1). (2) Indirect immunofluorescent studies showed that influenza viral antigen(s) was expressed in the majority of cells. (3) Electron-microscopic analyses also confirmed that although very little virus was released in the supernatant, budding virus like particles were present on the surface of Pi cells. (4) Finally Pi cells produced plaques (50–80%) when plated on monolayers of normal MDBK cells.[6]

The nature of virus release by Pi cells was analyzed by a number of experiments. We found that the virus particles were highly temperature-sensitive, produced a lower yield, and smaller plaques at 34°C. However Pi cells did not release any DI virus as measured by interference assay or by RNA analysis in gels. We also concluded that interferon, either exogenous or endogenous, was not playing a significant role in these Pi cells. There was neither any detectable production of interferon, nor were Pi cells resistant to heterologous infection. Furthermore, persistent infection in these cultures

Table 1. Virus Production After Superinfection of Persistently Infected Cells by Homologous and Heterologous Viruses.

Cells (passage No.)	Superinfecting virus (moi)	Yield	
		PFU/ml	HAU/ml
MDBK-Control	WSN (2)	300×10^6	2048
Pi MDBK 9 (p-44)	WSN (2)	0.003×10^6	32
HeLa-Control	WSN (10)	4.0×10^6	1024
Pi HeLa (p-41)	WSN (10)	0.0015×10^6	16
MDBK-Control	NDV (2)	70×10^6	8192
Pi MDBK (p-30)	NDV (2)	60×10^6	4096
HeLa-Control	NDV (10)	0.5×10^6	1024
Pi HeLa (p-29)	NDV (10)	0.3×10^6	512
Pi MDBK	0	0	<2
Pi HeLa	0	0	<2

Normal or persistently infected (Pi) MDBK and HeLa cells were superinfected with WSN virus or NDV for 14 hr and supernatants were analyzed for hemaglutinin (HAU) and infectivity (PFU).

was not maintained at the population level (i.e., a small percentage of cells are infected at one time)—a characteristic of interferon mediated persistent infection.

We believe that DI virus is crucial in the initial phase of persistent infection because DI virus helps the survival of infected cells by suppressing the cytopathic effect of infectious virus. Furthermore, DI virus may aid in the selection of a variant with a reduced growth capacity and cell killing effect. Once an appropriate variant is selected and a stable cell-virus association is established, DI virus may not be required and is eliminated from the culture. Since DI viruses occur commonly in influenza virus replication, they may further provide additional selection pressure in the creation of diversity and evolution of influenza virus in nature.

ACKNOWLEDGMENTS
This work was supported by grants from the National Institute of Allergy and Infectious Diseases (AI12749, AI 16348) and the National Science Foundation (PCM 7823220).

References

1. von Magnus, P. 1954: *Advan. Virus. Res.* 2:59–78.
2. Huang, A.S. and Baltimore, D. 1970: *Nature (London)* 226:325–327.
3. Nayak, D.P. 1980: *Ann. Rev. Microbiol.* 34:619–644.
4. Janda, J.M., Davis, A.R., Nayak, D.P., and De., B.K. 1979: *Virology* 95:48–58.
5. Nayak, D.P., Tobita, K., Janda, J.M., Davis, A.R., and De, B.K. 1978: *J. Virology* 28:375–386.
6. De, B.K. and Nayak, D.P. 1980: *J. Virology*, in press.

7. Choppin, P.W. and Pons, M.W. 1970: *Virology* 42:603–610.
8. Crumpton, W.M., Dimmock, N.J., Minor, P.D., and Avery, R.J. 1978: *Virology* 90:370–373.
9. Nakajima, K., Ueda, M., and Sugiura, A. 1979: *J. Virology* 29:1142–1148.
10. Kavern, N., Kolomietz, L., and Rudneva, I. 1980: *J. Virology* 34:506–511.
11. Janda, J.M. and Nayak, D.P. 1979: *J. Virology* 32:697–702.
12. Davis, A.R. and Nayak, D.P. 1979: *Proc. Natl. Acad. Sci., U.S.A.* 76:3092–3096.
13. Davis, A.R., Hiti, A.L., and Nayak, D.P. 1980: *Proc. Natl. Acad. Sci. U.S.A.* 77:215–219.
14. Jones, C.L. and Holland, J.J. 1980: *J. Virology* 32:697–702.

Published 1981 by Elsevier North Holland, Inc.
David H. L. Bishop and Richard W. Compans, eds.
The Replication of Negative Strand Viruses

Towards a Universal Influenza Vaccine

W.G. Laver,[a] G.M. Air,[a] and R.G. Webster[b]

The aim of this work is to produce a vaccine capable of inducing antibodies which will neutralize the infectivity of type A influenza viruses of every subtype.

If this aim cannot be achieved, it may at least be possible to produce a vaccine which will neutralize the infectivity of all viruses within a subtype, thus overcoming the problem of antigenic drift, if not that of the major shifts.

We believe the best candidate for such a vaccine will be a modified hemagglutinin (HA) molecule, but a detailed analysis of the antigenic structure of the HA is first required.

The HA forms one of the surface "spikes" on particles of influenza virus. Antibody to the HA neutralizes virus infectivity and variation in its structure is the main reason we are unable to control influenza by vaccination.

Influenza virus HA is a triangular, rod-shaped molecule composed of three pairs of disulphide-linked polypeptide chains, HA1 and HA2.[1] The amino acid sequence of these is known for strains within three different subtypes of type A influenza, fowl plague (Hav1N1) virus,[2] Asian (H2N2) influenza virus[3] and Hong Kong (H3N2) influenza virus.[4-6] Incomplete, bromelain-released HA molecules of Hong Kong (H3N2) influenza virus have been crystallized[7] and the three-dimensional structure is being determined by X-ray diffraction methods.[8]

[a] Australian National University, Canberra.
[b] St. Jude Children's Research Hospital, Memphis, Tennessee.

We have begun to investigate the structure of the antigenic sites on the Hong Kong (H3) HA molecule and the way in which these change during antigenic drift. Others [9] have obtained antigenic fragments from both HA1 and HA2 but so far these have been of large size and no small antigenically active peptides have been isolated. We have therefore concentrated on examining the differences in amino acid sequence occurring in the HA during antigenic drift. We have also examined the effect of chemical modification of specific amino acids in the HA on its antigenic activity.

The Hong Kong (H3N2) subtype of type A influenza appeared in man 1968. Since then antigenic drift has occurred in the HA of the virus to the extent that recent H3N2 isolates (e.g., A/Texas/1/77) show hardly any cross-reactivity with the original 1968 strain. This considerable change in the antigenic properties of the HA is associated with approximately 18–20 changes in the amino acid sequence of HA1.[10] However, which of these cause the antigenic changes is not known and in order to relate sequence changes to antigenic changes, we have selected antigenic variants of Hong Kong (H3N2) influenza virus under pressure of monoclonal antibodies to different antigenic sites on the HA. The amino acid sequence changes associated with the altered capacity of the HA to combine with each monoclonal antibody were then determined.

Variants selected with different monoclonal antibodies showed dramatic changes in antigenicity (in tests with monoclonal antibodies) and single amino acid sequence changes in widely separated regions of HA1.[11] Many changes occurred in or near residues number 142–146 in HA1. In 10 variants of Hong Kong (H3N2) influenza virus selected with monoclonal antibodies, the proline at position 143 in HA1 changed to serine, threonine, leucine or histidine. In other variants, asparagine 133 changed to lysine, glycine 144 to aspartic acid and serine 145 to lysine. All these changes are possible by single base changes in the RNA except the last, which requires a double base change. Residues 142–146 also changed in field strains of Hong Kong influenza isolated between 1968 and 1977.[10] The single amino acid sequence changes in HA1 of the monoclonal variants were detected by comparing the compositions of the soluble tryptic peptides from the variants with the known sequences of these peptides from wild type virus. Two insoluble tryptic peptides, comprising residues 110–140 and 230–255 in the HA1 molecule, were not examined and we do not know if additional changes occurred in these regions.

It has been proposed [12-15] that the hemagglutinin molecules of influenza viruses possess a single antigenic site and that antigenic drift occurs by the sequential substitution of increasingly bulky hydrophobic amino acids at a unique locus. No direct experimental evidence for this theory has ever been presented and in order to see whether such sequential substitution did occur (since analysis of natural variants suggested it did not) antibody was prepared against the new antigenic site on the variants in which proline 143 changed to histidine or threonine. This was then used to select second generation variants of these variants. In the first case, the glycine residue (144) next to the histidine changed to aspartic acid and in the second, the threonine at position 143 reverted to proline and the virus regained the antigenicity of wild type.

Although monoclonal antibodies revealed dramatic antigenic differences between the variants and wild type virus, only those variants with changes at position 144 of glycine to aspartic acid or at position 145 of serine to lysine could be distinguished from wild type virus using heterogeneous rabbit or ferret antisera. The other variants, including those which showed sequence changes in widely separated positions of HA1, could not be distinguished from wild type with heterogeneous antisera.

These findings suggest that sequence changes in the region comprising residues 142–146 of HA1 affect an important antigenic site on the hemagglutinin molecule.

We do not know, however, if the amino acids which changed formed part of the antigenic determinants recognized by the monoclonal antibodies used to select the variants, or whether the sequence alterations we found caused conformational changes in the hemagglutinin, which in turn altered an antigenic site some distance away. We have tried to answer this question by examining the effects of chemical modification of certain amino acid residues on the antigenic properties of the HA. We have examined monoclonal variants of A/Mem/1/71 (H3N2) virus in which the change in sequence associated with the alteration of an antigenic site on the hemagglutinin was to an amino acid capable of reacting with 1-fluoro 2, 4-dinitrobenzene (FDNB), tetranitromethane (TNM) or diazotized sulphanilic acid (DSA) to form stable derivatives.

The following variants were used:

1. Variants where the asparagine at position 53 and the serine at position 145 in HA1 were replaced by lysine. These were reacted with FDNB.
2. A variant where the serine at position 205 changed to tyrosine was reaction with TNM and DSA.
3. A variant in which the proline at position 143 in HA1 was replaced by histidine. This was reacted with DSA and FDNB.

We thought that if the amino acids which changed formed part of an antigenic site and were capable of binding antibody, they and the "new" amino acids should also be accessible to the small organic reagents. Furthermore, if these amino acids did form part of the antigenic site, their chemical modification should interfere with the binding of antibody directed against the new site. The experiments should obviously have been done using monoclonal variants in which lysine, tyrosine or histidine residues changed to other amino acids, but so far such variants of Mem/71 virus have not been found.

DNP-substituted hemagglutinin molecules, isolated from FDNB treated A/Memphis/1/71$_H$-BEL$_N$(H3N1) virus particles, had up to 58% of lysines substituted with DNP. These molecules, nevertheless, retained hemagglutinin activity and, as far as could be measured, the same capacity as the unsubstituted hemagglutinin to react with heterogeneous antiserum or a panel of monoclonal antibodies.

These results suggest that those amino acid side chains able to react with FDNB (lysine, histidine, tyrosine and cysteine) are either not present in the antigenic sites on the HA, or if they are, then either the side chains which bind antibody do not react with DNP, or the presence of DNP in the site does not affect its ability to combine with antibody.

The results also suggest that substitution of more than half of the lysine in the hemagglutinin molecule does not cause any marked conformational changes, for such changes would be expected to affect the ability of the HA to combine with both cell receptors and antibody molecules.

Similar findings with TNM-treated HA suggest that tyrosine is not an essential part of any antigenic site on H3 type HA. HA treated with diazotized sulphanilic acid lost HA activity, but its antigenicity was similar to that of untreated HA when tested in double immunodiffusion tests with heterogeneous antisera, suggesting that histidine was not present in the antigenic sites.

Sequence studies on HA1 from HA molecules treated with FDNB, TNM or DSA showed that certain lysine, tyrosine or histidine residues were 100% substituted after the reaction, while others apparently did not react at all.

We were also surprised to find that when monoclonal variants, showing changes at positions 53 in HA1 of asparagine to lysine or at position 145 of serine to lysine, were treated with 2, 4-dinitro fluorobenzene, the new lysine residues did not react to form DNP derivatives, except under drastic conditions when Lys 53 reacted, but much more slowly than other lysine residues in the molecule. Treatment of the variant with a sequence change at position 205 in HA1 of serine to tyrosine with tetranitromethane also did not lead to any nitration of the new tyrosine residue, although some other tyrosine residues in the HA1 polypeptide were 100% nitrated by this reagent. We imagined that the "new" lysines and tyrosines would be located within antigenic sites and accessible to FDNB and tetranitromethane, but apparently this was not the case.

However, when HA from the monoclonal variant of A/Mem/1/71 (H3N2) virus with a sequence change from wild type in HA1 of proline (143) to histidine was reacted with diazotized sulphanilic acid. the histidine at position 143 in HA1 reacted completely and the HA lost the ability to bind antibody specific for the new antigenic site on this variant. However, treatment of this variant with FDNB did not lead to substitution of histidine 143.

ACKNOWLEDGMENTS
Jean Clark, Sally Campbell and Martha Sugg provided excellent technical assistance. This collaborative project was greatly helped by international direct dialing telephone facilities provided by the Australian Overseas Telecommunications Commission. This work was supported in part by Grants AI-15343 and AI-08831 from the National Institute of Allergy and Infectious Diseases, and by ALSAC.

References

1. Wiley, D.C., Skehel, J.J., and Waterfield, M. (1977): *Virology* 79:446–448.
2. Porter, A.G., Barber, C., Carey, N.H., Hallewell, R.A., Threlfall, G., and Emtage, J.S. (1979): *Nature* 282:471–477.
3. Gething, M.J., Bye, J., Skehel, J., and Waterfield, M.D. (1980): *Nature* 287:301–306.
4. Ward, C.W. and Dopheide, T.A. (1980): In: *Structure and Variation in Influenza Virus.* Laver, W.G. and Air, G.M. (Eds.), Elsevier, New York, pp. 27–38.
5. Sleigh, M.J., Both, G.W., Brownlee, G.G., Bender, V.J., and Moss, B.A. (1980): In: *Structure and Variation in Influenza Virus.* Laver, W.G. and Air, G.M. (Eds.), Elsevier, New York, pp. 69–79.
6. Verhoeyen, M., Fang, R., Min Jou, W., Devos, R., Huylebroeck, D., Saman, E., and Fiers, W. (1980): *Nature* 286:771–776.
7. Wiley, D.C. and Skehel, J.J. (1977): *J. Mol. Biol.* 112:343–347.
8. Wilson, I.A., Skehel, J.J., and Wiley, D.C. (1980): In: *Structure and Variation in Influenza Virus.* Laver, W.G. and Air, G.M. (Eds.), Elsevier, New York, pp. 339–349.
9. Brown, L.E., Dopheide, T.A.A., Ward, C.W., White, D.O., and Jackson, D.C. (1980): *J. Immunol.* 125:1583–1588.
10. Laver, W.G., Air, G.M., Dopheide, T.A., and Ward, C.W. (1980): *Nature* 283:454–457.
11. Laver, W.G., Air, G.M., Webster, R.G., Gerhard, W., Ward, C.W., and Dopheide, T.A. (1979): *Virology* 98:226–237.
12. Fazekas de St. Groth, S. (1969): *Bull. Who* 41:651–657.
13. Fazekas de St. Groth, S. (1970): *Arch. Environ. Health* 21:293–303.
14. Fazekas de St. Groth, S. (1973): In: *Negative Strand Viruses: Proceedings, Vol. 2.* Mahy, B.W. and Barry, R.D. (Eds.), pp. 741–75.
15. Underwood, P.A. (1980): *Infection and Immunity* 27:397–404.

David H. L. Bishop and Richard W. Compans, eds.
The Replication of Negative Strand Viruses

Studies on Antigenic Variation in Influenza A(H1N1) Virus: Significance of Certain Antigenic Determinants on the Hemagglutinin Molecule

Setsuko Nakajima[b] and Alan P. Kendal[a]

Introduction

Influenza A virus of a given subtype changes antigenic specificity several times during interpandemic periods ("antigenic drift"), contributing to the survival of the virus by circumventing the effects of immunity in the population. Analysis of this phenomenon is benefiting from the development of techniques for preparing monoclonal antibodies that can be used in fine antigenic analyses of individual antigenic determinants.[1-3] By such means the presence of three to five antigenic determinants on the hemagglutinin (HA) molecule of influenza viruses has been suggested.[4-7] Furthermore, by selecting for viruses that grow in the presence of monoclonal antibodies, variants have been isolated which have changes in any of the antigenic determinants of their HA molecules recognized by the monoclonal antibodies. With one exception, however, none of the variants selected by using a single-step procedure with monoclonal antibody to H3 or H0 hemagglutinin could be distinguished from parental virus by heterogeneous antisera,[5,6,8] so that it was difficult to estimate the relevance of the changes of antigenic determinants on the HA molecules of these *in vitro* variants compared to the variation detected in natural isolates. We are attempting to obtain further information about this by studying influenza A(H1N1) viruses, among which several antigenic variants were detected in a short period after their reappearance in 1977, (unpublished data).[9,10]

[a] World Health Organization Collaborating Center for Influenza, Center for Disease Control, Virology Division, Atlanta, Georgia.
[b] Present adress: Department of Microbiology, The Institute of Public Health, Minato-ku, Tokyo 108, Japan.

Materials and Methods

Viruses. Natural variants used in this study were selected on the basis that they exhibited antigenic drift from A/USSR/90/77 virus when tested with postinfection ferret sera to A/USSR/90/77 virus (i.e., were inhibited to titers at least four-fold lower than the homologous titer of A/USSR/90/77 serum). Some of the H1N1 influenza variant strains have been previously described.[9,10]

Hemagglutination-inhibition (HI) tests and preparation of antisera. These were done as previously described.[9-11] Ascitic fluids containing monoclonal antibodies specific for the HA molecule of A/USSR/90/77(H1N1) virus were kindly provided by Dr. R.G. Webster.

Selection of antigenic variants and safety precautions. A/USSR/90/77(H1N1) virus was plaque-purified twice in MDCK cells and several clones were grown in eggs. Antigenic variants were selected by neutralization of the egg-grown cloned virus for 60 min at room temperature with ascitic fluids containing mouse monoclonal antibody. Virus-antibody mixtures were then titrated for infectious virus in flasks of MDCK cells in the absence of monoclonal antibody. After several days of growth, well-isolated plaques were picked from each flask and inoculated into eggs to obtain antigens. To prevent infection of laboratory personnel with new antigenic variants that might be produced by this study, biosafety guidelines of the Center for Disease Control for Class III studies were adopted with infectious *in vitro* variants. Infectivity of virus in allantoic fluid harvests used for antigenic analysis was inactivated with γ-radiation under conditions previously shown not to affect antigenic specificity.[12] The name "monoclonal variant" was used to describe antigen variants selected *in vitro* with monoclonal antibodies.[6]

Results

Isolation of in vitro variants with mouse monoclonal antibodies to A/USSR/90/77 hemagglutinin. Variant clones resistant to each of six monoclonal antibodies to A/USSR/90/77 hemagglutinin were isolated from plaques developed in MDCK cells. Antigenic variants resistant to the IgG class monoclonal antibody 264 were obtained at relatively high frequencies ($10^{-3.7}$ to $10^{-4.2}$), while antigenic variants were selected at low frequencies ($< 10^{-7}$) by the IgM class monoclonal antibody 22 (Table 1). Frequencies of isolation of antigenic variants with the other four (IgG class) monoclonal antibodies were similar, at about $10^{-5.1}$ to $10^{-5.6}$.

Changes in epitopes of monoclonal variants. Preliminary analysis was undertaken with two to eight clones selected with each monoclonal antibody to determine their reaction patterns. Experiments A, B, D and F where monoclonal antibodies 70, 264, 110 and 22, respectively, were used in selection resulted in the isolation in each case of one apparently homogenous group of antigenic variants as judged by the reaction patterns with the six available monoclonal antibodies. When the other two monoclonal antibodies (W18 and 385) were used for selection in experiments C and E, however, two groups of variants were identified that had different reaction patterns. Representative clones for each group were selected for more extensive analysis (Table 2). All of the monoclonal variants failed to react in HI tests with the monoclonal antibodies used for their selection. Five of the eight groups of monoclonal variants (A-1 clone 42, C-1 clone 1, C-2 clone 14, E-1 clone 1 and F-1 clone 1) also had reduced reactivities to monoclonal antibodies not used in their selection. Several of the epitopes therefore appeared to be interactive with each other. Examples are epitopes 22 and 70, which interacted reciprocally, and epitopes W18 and 264 that exhibited a one-way interaction detected only when W18 was used as the selective antibody. Epitope 110 was independent of all others, but epitope 385 appeared to interact in an unpredictable manner with W18, 22 and 264.

Changes in antigenicity of monoclonal variants detected by heterogeneous ferret antisera. Monoclonal variants were tested by HI tests for changes in antigenic specificity recognizable with heterogeneous ferret antisera (Table 2). Monoclonal variant C-1-1 which had the changes in epitopes W18, 264 and 385 showed a reciprocal antigenic drift from parental A/USSR/90/77 virus and closely resembled a natural variant A/Lackland/3/78. Monoclonal variants B-1-23, C-2-14 and E-1-1 exhibited asymmetric drift from A/

Table 1. Frequencies of Antigenic Variants Detected in Cloned Populations of A/USSR/90/77(H1N1) Virus with Monoclonal Antibodies.[a]

Monoclonal antibody used for selection	Immunoglobin class	Frequency (log_{10}) of antigenic variants in: Clone 3-1	Clone 4-5
W18	IgG	=5.3	=5.1
22	IgM	=7.3	<=7.3
70	IgG	=5.1	=5.6
110	IgG	=5.2	=5.4
264	IgG	=4.2	=3.7
385	IgG	=5.3	<=5.3

[a] A/USSR/90/77 virus clones were neutralized with monoclonal antibody to A/USSR/90/77 HA molecule. Virus-antibody mixtures were titrated in MDCK cell monolayers. Multiple plaques from each experiment were picked and inoculated into eggs to obtain antigens for HI testing. Only those which were recovered and failed to react in HI tests with the monoclonal antibody used for neutralization were included in calculations of selection frequency.

Table 2. Hemagglutination-Inhibition Reactions of Ferret Sera and Monoclonal Antibodies with Influenza A(H1N1) Natural and Monoclonal Variants.

Antigen	Ferret antiserum[a]									Monoclonal antibody to A/USSR/90/77 hemagglutinin					
	A/USSR/90/77 cl. 3-1	A/Lackland/3/78	A/Arizona/14/78	A/Brazil/11/78	A/California/45/78	B-1-23	C-1-1	C-2-14	E-1-1	W18	22	70	110	264	385
Natural variants:															
A/USSR/90/77 cl. 3-1	640	640	160	240	120	160	160	120	480	+	+	+	+	+	+
A/Lackland/3/78	80	2560	120	240	480	60	320	80	80	–[b]	+	+	+	–	+
A/Arizona/14/78	80	640	640	240	100	80	60	80	160	+	+	+	+	–	+
A/Brazil/11/78	80	320	80	480	50	60	60	80	320	+	+	+	+	–	+
A/California/45/78	<20	320	40	80	320	≤20	20	20	40	–	+	+	+	–	–
A/Texas/7742/78	160	480	160	960	60	60	80	160	320	+	–	–	+	–	+
A/Texas/8820/79	160	480	160	320	120	60	160	160	320	+	+	+	–	–	+
A/Texas/23/79	80	320	160	240	120	60	60	40	160	–	+	+	+	–	+
A/USSR/50/79	80	120	60	60	20	40	30	<20	160	+	+	+	+	–	–
Monoclonal variants:															
A-1-42 (70)[c]	480	1280	240	480	30	160	160	160	960	+	–	–	+	+	+
B-1-23 (264)	80	800	240	80	60	120	40	40	120	+	+	+	+	–	+
C-1-1 (W18)	80	3840	60	200	240	30	640	60	80	–	+	+	+	–	–
C-2-14 (W18)	80	640	80	120	50	40	120	80	120	–	+	+	+	–	+
D-1-14 (110)	480	640	160	240	40	120	120	80	480	+	+	+	–	+	+
E-1-1 (385)	160	240	60	240	50	40	80	60	640	+	+	+	+	–	–
E-2-1 (385)	320	320	240	120	120	20	80	60	480	+	+	+	+	+	–
F-1-1 (22)	1920	2560	640	120	120	640	640	480	2560	+	–	–	+	+	–

[a]Results are the reciprocal of the highest two-fold dilutions causing complete inhibition of four hemagglutinating units of each virus. Each result is the mean of duplicate tests.

[b]—indicates HI titer ≥16-fold lower than that obtained with A/USSR/90/77 cl. 3-1 virus.

[c]monoclonal antibody used for selection of antigenic variant is shown in parentheses.

USSR/90/77 parental virus because they were inhibited 4- to 8-fold less than A/USSR/90/77 virus by antiserum to the parental virus, but themselves elicited antibodies that reacted well with A/USSR/90/77 virus. Other monoclonal variants, A-1-42, D-1-14, E-2-1 and F-1-1 did not show significant ($\geq$ 4-fold) differences in antigenicity from parental virus when tested with ferret sera in HI tests, despite the changes in some epitopes detected by monoclonal antibodies. All monoclonal variants exhibiting four-fold or greater drift from A/USSR/90/77 virus in HI tests with ferret sera had reduced reactivity with at least monoclonal antibody 264, whereas none of the monoclonal variants having reduced reactivity with antibodies 22, 70 or 110 exhibited drift from A/USSR/90/77 virus detectable with ferret antisera.

Antigenic analysis of influenza A(H1N1) natural isolates in 1977–1980. HI tests of natural variants with mouse monoclonal antibodies to A/USSR/90/77 HA molecule showed that every variant selected on the basis of reduced activity with ferret antiserum to A/USSR/90/77 virus also had reduced reaction with one or more of the available monoclonal antibodies (Table 2). HI tests with ferret antisera indicated that the selected natural variants studied were all distinguishable from each other even when they had similar monoclonal antibody reaction patterns (e.g., A/Brazil/11/78 and A/Arizona/14/78, A/Lackland/3/78 and A/Texas/23/79). All of these natural variants had at least a 32-fold reduction in reactivity with monoclonal antibody 264 compared to A/USSR/90/77 virus.

Discussion

Antigenic variants of H1N1 influenza virus which could be distinguished from the parental A/USSR/90/77 strain with one or more monoclonal antibodies were produced by single-step selection in a manner analogous to that described by others with H0N1 and H3N2 influenza A viruses.[4–6,8,13] Although the frequency of detection of most variants in the cloned population of viruses was of a similar order (about 10^{-5}) to that previously described for influenza or other viruses,[4-6,14] a much lower frequency of selection of variants (10^{-7}) was observed with IgM class monoclonal antibody. Lubeck et al.[13] also detected monoclonal variants of influenza viruses with a very low frequency on some occasions, but did not determine the class of antibody involved. Influenza viruses might have a low frequency of escaping neutralization by IgM class antibodies, presumably on account of that antibody's multivalency and consequently high avidity.

As has been observed with monoclonal antibodies to H0 and H3 hemagglutinins, it appeared possible to arrange the six monoclonal antibodies to A/USSR/90/77 hemagglutinin into subsets according to the interactions of the epitopes they recognize; at least three non-interactive determinants were identified in our studies; one detected by antibodies W18 and 264, a second detected by antibodies 22 and 70, and a third detected by antibody 110.

Antigenic variants detected *in vitro* with the highest frequency were obtained by selection in the presence of monoclonal antibody 264. Changes in this epitope was also observed in naturally occurring variants of A/USSR/90/77 virus detected in 1977–1978 and maintained since that time in prevalent strains and only changes in the epitopes were detected by monoclonal antibody 264 (with or without other changes) were consistently associated with significant antigenic drift detected with heterogeneous ferret antisera. The frequencies with which the monoclonal variants were selected indicate they probably arose by single point mutations in the amino acid sequence of the HA molecule. Thus our studies support the possibility for significant antigenic variation of influenza viruses to occur by means of a point mutation at a critical locus.

Summary

Antigenic variants of influenza A(H1N1) virus were selected with six monoclonal antibodies to A/USSR/90/77(H1N1) hemagglutinin by a single-step procedure. Antigenic variants were obtained with one of the monoclonal antibodies (264) at a frequency of 10^{-4} and the frequency was higher than those of 10^{-4} to $<10^{-7}$ obtained with other monoclonal antibodies used in the selection process. Antigenic analysis showed that only those variants which had changes in the antigenic determinant recognized by monoclonal antibody 264 could be distinguished from the parental strain with heterogeneous ferret antisera. The findings obtained corresponded well to the detection since 1977 of many natural variants of H1N1 influenza virus which also exhibited changes detected by monoclonal antibody 264. At least one variant selected *in vitro* with monoclonal antibody closely resembled a natural variant in HI tests with ferret antisera. The findings support the possibility for significant antigenic variation of H1N1 influenza viruses to occur by means of a point mutation at a critical locus.

References

1. Gerhard, W., Braciale, T.J., and Klinman, N.R. (1975): *Eur J. Immunol.* 5:720–725.
2. Gerhard, W. (1976): *J. Exp. Med.* 144:985–995.
3. Koprowski, H., Gerhard, W., and Croce, W.M. (1977): *Proc. Nat. Acad. Sci. U.S.A.* 74:2985–2988.
4. Yewdell, J.W., Webster, R.G., and Gerhard, W. (1979): *Nature* 279:246–248.
5. Laver, W.G., Air, G.M., Webster, R.G., Gerhard, W., Ward, C.W., and Dopheide, T.A.A. (1979): *Virology* 98:226–237.
6. Webster, R.G. and Laver, W.G. (1980): *Virology* 104:139–148.

7. Gerhard, W., Yewdell, J., and Frankel, M. (1980): In: *Structure and Variation in Influenza Virus.* Laver, W.G. and Air, G.M. (Eds.), Elsevier, Amsterdam, pp. 273–281.

8. Gerhard, W. and Webster, R.G. (1978): *J. Exp. Med.* 148:383–392.

9. Kendal, A.P., Joseph, J.M., Kobayashi, G., Nelson, D., Reyes, C.R., Ross, M.R., Sarandria, J.L., White, R., Woodall, D.F., Noble, G.R., and Dowdle, W.R. (1979): *Amer. J. Epidem.* 110:449–461.

10. Webster, R.G., Kendal, A.P., and Gerhard, W. (1979): *Virology* 96:258–264.

11. Dowdle, W.R., Kendal, A.P., and Noble, G.R. (1979): In: *Diagnostic Procedures for Viral Rickettsial and Chlamydial Infections.* Lennette, E.H. and Schmidt, N.J. (Eds.), Amer. Publ. Hlth Assoc., New York, pp. 585–609.

12. White, L.A., Hall, H.E., Chappell, W.A., and Kendal, A.P. (1979): *Abstract, Annual Meeting, Amer. Soc. Microbiol.* p. 272.

13. Lubeck, M.D., Schulman, J.L., and Palese, P. (1980): *Virology* 102:458–462.

14. Portner, A., Webster, R.G., and Bean, W.J. (1980): *Virology* 104:235–238.

David H. L. Bishop and Richard W. Compans, eds.
The Replication of Negative Strand Viruses

Evidence for More Than One Antigenic Determinant on the Matrix Protein of Influenza A Viruses

J. Lecomte[a] and J.S. Oxford[b]

The matrix protein (M) is one of the major internal polypeptides of influenza virus. Its antigenicity was first demonstrated by Schild [1] in agar gel precipitation. Although not identical to the nucleoprotein, it has an antigenic reactivity common to all human influenza A viruses. Because of this common reactivity, it was proposed that the M-protein was also a type specific antigen. This common antigenic reactivity was also found on M-proteins derived from influenza viruses isolated from other mammalian or avian species.[2] Different electrophoretic migration profiles of RNA segments coding for M-protein [3] the different tryptic peptide maps and amino acid compositions of the M-protein have indicated that mutations can occur on the genetic information for the M-protein.[4-7] No evidence so far has shown that these mutations are reflected in antigenic alterations in the M-protein.

A competitive radioimmunoassay was designed to examine the possible occurrence of more than one class of antigenic determinants in the M-protein of influenza A viruses and to reevaluate this strong interspecies reactivity. The results obtained demonstrate that at least two antigenic sites are found in M-protein of the A/PR/8/34 (H0N1) virus, one of which is shared with other influenza A virus subtypes, while another is lacking in the M-protein of the A/chick/Germ/N/49 (Hav2Neq1) virus. Morover, the cross-reactive determinants in the N virus M-protein are not equally shared by other influenza A viruses.

[a] Centre de recherche en virologie, Institut Armand-Frappier, Laval, Québec, Canada.
[b] National Institute for Biological Standards and Control, Holly Hill, Hampstead, London NW3 6RB.

Representative strains of influenza A viruses from four human subtypes were used together with strains occurring in swine, birds and horses as inhibitors in a competitive radioimmunoassay (Table 1). In this assay, unlabeled M-protein or unlabeled disrupted whole virus is incubated with a known amount of the radioactively labeled M-protein and a predetermined amount of anti-M-protein serum. Unlabeled M-protein then competes with labeled M-protein in the antigen-antibody reaction so that the greater the amount of unlabeled M-protein the less the binding of labeled M-protein to anti-M serum. The degree of inhibition of binding will then be an expression of the quantity and homology of the unlabeled M-protein with labeled M-protein.

Bindings Assays with Anti-M-Protein Sera

Binding assays carried out with N virus anti-M serum and ^{125}I-labeled M-protein from N virus and PR8 virus as well as PR8 virus anti-M serum and both ^{125}I-labeled M-proteins constantly showed a greater reactivity with the

Figure 1. Binding assays with ^{125}I-labeled M-proteins to anti-M-protein sera. (a) ^{125}I-labeled M-protein from N virus to: swine anti-M-protein (●); N virus anti-M-protein (○); PR8 virus anti-M-protein (▲) and (b) ^{125}I-labeled M-protein from PR8 virus to: PR8 virus anti-M-protein (●); N virus anti-M-protein (○). Both ^{125}I-labeled M-proteins had a specific activity of 1–4 × 10^{6} cpm/μg. Antisera were reacted with 10,000 cpm of ^{125}I-labeled M-protein. Arrows indicate the dilution used in the respective competitive radioimmunoassays.

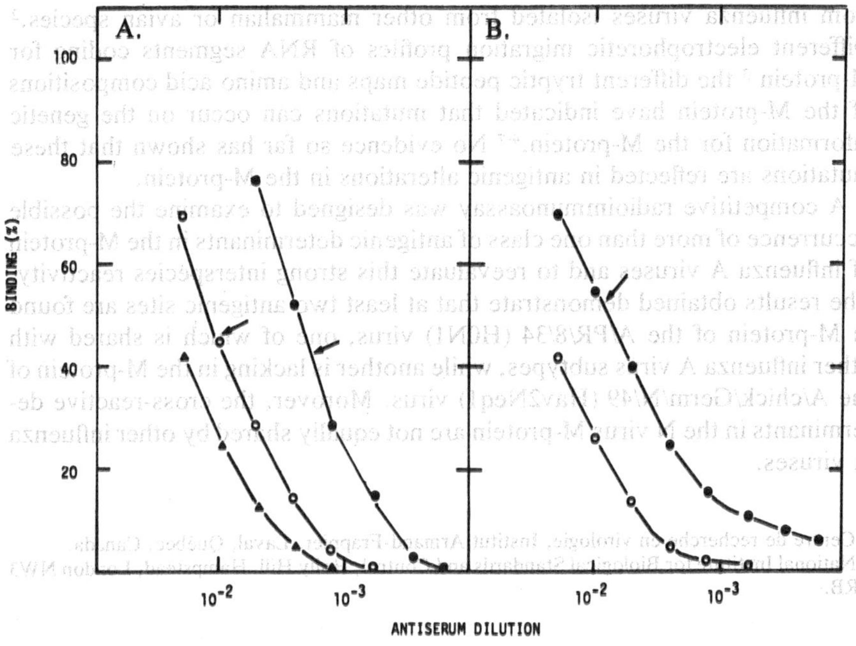

Table 1. Influenza Viruses or Recombinants Purified or from Infected Allantoic Fluid Used in the Competitive Radioimmunoassay.

Origin	Virus or recombinants		Abbreviation
Human	A/PR/8/34	(H0N1)	PR8
	A/FM/1/47	(H1N1)	FM
	A/Singapore/1/57	(H2N2)	Sing
	A/Aichi/2/68	(H3N2)	Aichi
	A/NJ/11/76	(Hsw1N1)	NJ
Recombinant	X-31 (PR8 × Aichi)	(H3N2)	X-31
	X-47 (PR8 × A/Vic/3/75	(H3N2)	X-47
	X-49 (PR8 × A/Texas/1/77	(H3N2)	X-49
	X-53a (PR8 × NJ)	(Hsw1N1)	X-53a
Swine	A/Swine/15/31	(Hsw1N1)	Sw-1
	A/Sw/Wisc./79/76	(Hsw1N1)	Sw-W
	A/Sw/Man/674/67	(Hsw1N1)	Sw-M
Avian	A/chick/Germ/N/49	(Hav2Neq1)	N
	A/Tern/SA/61	(Hav5Nav2)	Tern
	A/Duck/NB/373/77	(Hsw1N1)	DK373
	A/Duck/Que/426/77	(Hsw1N1)	DK426
Equine	A/Eq/Prague/1/56	(Heq1N1)	Eq1
	A/Eq/Miami/63	(Heq2N1)	Eq2
Type B	B/Lee/40		Lee
	B/Hong Kong/3/75		B/H.K.

homologous ^{125}I-labeled M-protein (Figure 1). The data also show that an antiserum prepared against a swine virus (heterologous antiserm) had a higher titer with the ^{125}I-labeled M-protein (from N virus), thus indicating strong reactions of this antiserum toward the cross-reactive determinants on this M-protein.

Inhibition of Binding of ^{125}I-Labeled M-Protein from N Virus to the N Virus Anti-M Serum (Homologous System)

The antigenic determinants in M-protein of N virus were examined by the extent to which M-proteins of different viruses competed with ^{125}I-labeled homologous M-protein in its reaction with N virus, anti-M serum.

Figure 2 shows that about 40 ng of unlabeled M-protein from N virus gave 50% inhibition and that three-fold more protein was required to achieve the same degree of inhibition with purified N virus. This is in accordance with the amount of M-protein present in N virus.[8] Heating the N virus at 100° for 2 min did not affect the antigenic reactivity of the M-protein in its capability to compete with the labeled M-protein. Influenza B viruses, B/Lee and B/HK, showed no inhibition. On a weight basis, 50% inhibition could not be obtained with M-proteins of the purified representative human, swine or other avian strains. (See Table 1 for a complete list). Similar results were obtained with viruses obtained from infected allantoic fluid (Figure 2b).

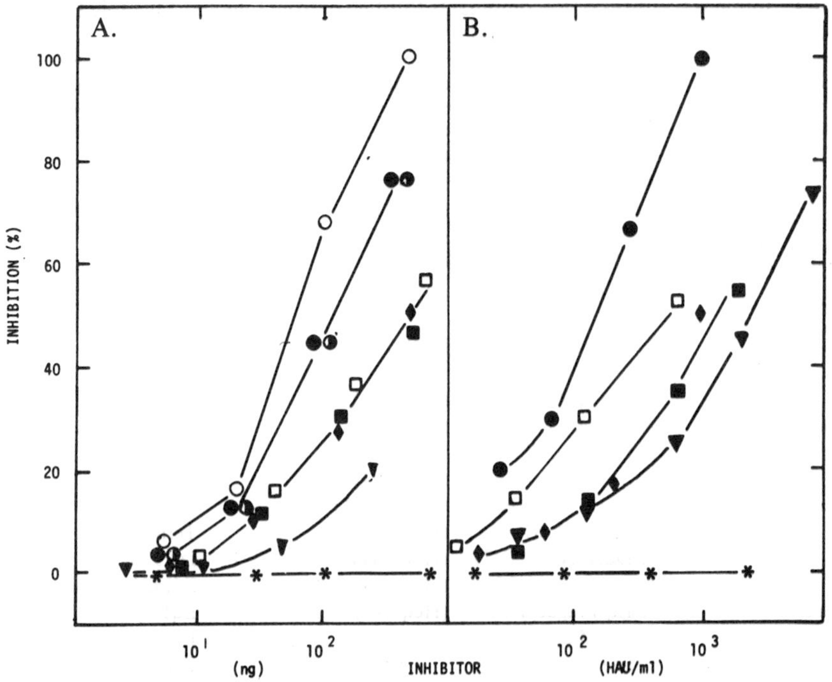

Figure 2. Inhibition of binding of ^{125}I-labeled M-protein from N virus to N virus anti-M-protein serum (homologous system). Inhibitors are (a) unlabeled purified whole viruses disrupted with 1% Sarcosyl, (b) unlabeled viruses from infected allantoic fluid. Isolated M-protein from N virus (○): N virus (●) and N virus heated at 100°C for 2 min (○); PR8 (◆); Sw (■) FM1 (□); NJ (▲); B/HK (*).

Inhibition of Binding of ^{125}I-Labeled M-Protein from N Virus to a Swine Anti-M Serum (Heterologous System)

These small differences observed in a homologous system could be due to differences in the concentration of M-protein in the different viruses. To test this possibility, the competing unlabeled viruses were analyzed with the same labeled M-protein from N virus but with the anti-M (A/Sw/Cambridge/39) serum. In this system, only the cross-reactive determinants of the M-protein of N virus would be bound. All disrupted purified viruses and viruses from infected allantoic fluid competed equally well (Figure 3). Furthermore, it is of interest to note that as little as 10–50 ng were sufficient to achieve 50% inhibition, whereas in the homologous system 50% inhibition could not be obtained with as much as 500 ng of virus proteins. It thus appears that the cross-reactive antigenic determinants present on the M-protein from N virus are not equally shared by other human and animal strains.

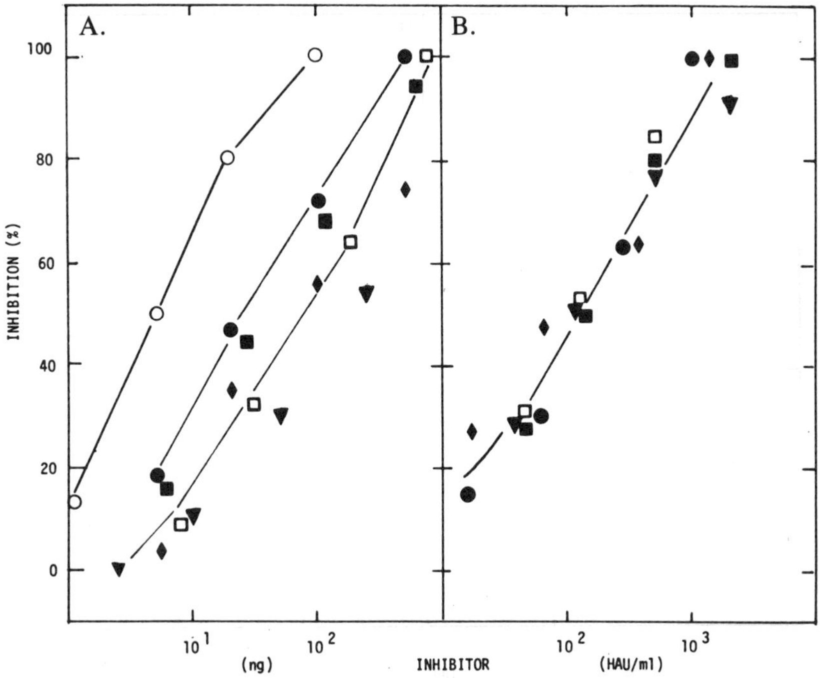

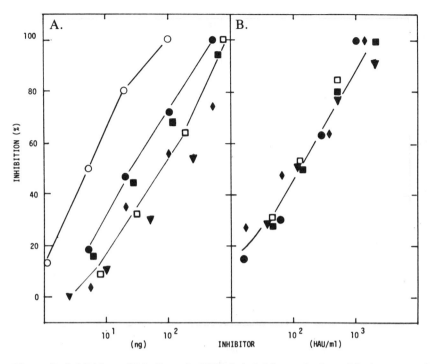

Figure 3. Inhibition of binding of ^{125}I-labeled M-protein from N virus to anti-M-protein (A/Sw/Cambridge/39) serum (heterologous system). Inhibitors are (a) unlabeled purified whole virus disrupted with 1% Sarcosyl, (b) unlabeled virus from infected allantoic fluid. Symobols as in Figure 2.

Inhibition of Binding of ^{125}I-Labeled M-Protein from PR8 Virus to the PR8 Anti-M Serum (Homologous System)

These findings were further extended by comparing the M-proteins of PR8 and N virus in another homologous system, using anti-M-protein serum prepared to M-protein from PR8 virus [9] and ^{125}I-labeled M-protein from PR8 virus. As shown in Figure 4, the PR8 virus M ^{125}I-labeled protein could be efficiently inhibited by all human and animal strains but the M-protein from N virus competed poorly. Thus, it becomes evident that M-protein from N virus lacks some determinants shared by all the other strains tested.

Conclusions

These findings suggest that the majority of the viruses analyzed, whether of human, avian, equine or swine origin have an M-protein with a unique specificity which is lacking in the N virus M-protein. Furthermore, the cross-reactive determinants in the M-protein from N virus are not equally

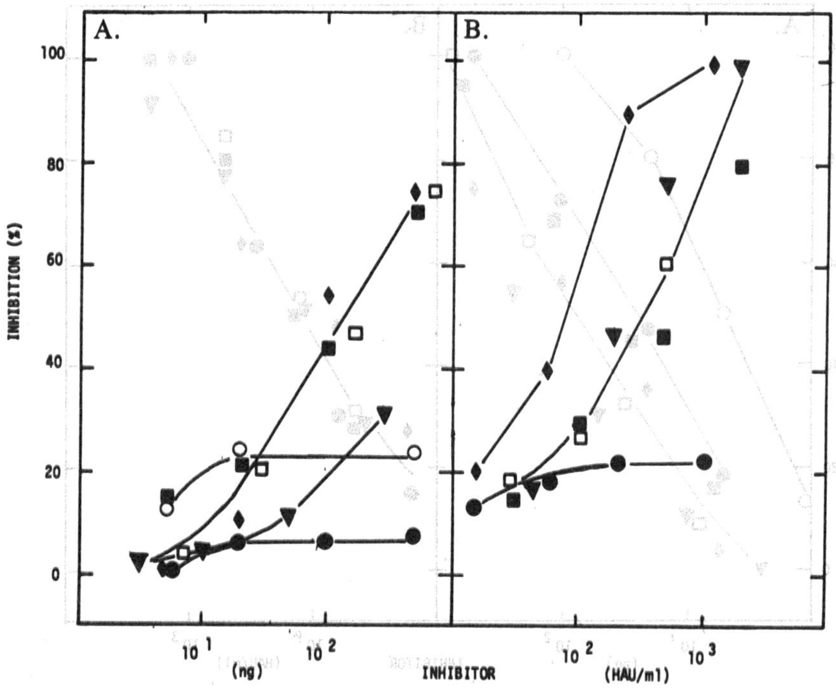

Figure 4. Inhibition of binding of ^{125}I-labeled M-protein from PR8 virus to PR8 anti-M-protein serum (homologous system). Inhibitors are (a) unlabeled purified virus disrupted with 1% Sarcosyl, (b) unlabeled virus from infected allantoic fluid. Symbols as in Figure 2.

shared among other influenza A strains. The data presented here are the first to clearly show antigenic differences among the M-proteins of influenza A virus strains. It may be predicted that with the use of other antisera and monoclonal antibodies with a more restricted specificity multiple antigenic determinants will be found on the M-protein of influenza A strains.

ACKNOWLEDGMENT

This work was supported by the Canadian Medical Research Council No MA-6014. We wish to thank Dr. M. Reginster for kindly providing the anti-M PR8 serum. Mr. G. Croteau provided excellent technical assistance.

References

1. Schild, G.C. (1972): *J. Gen. Virol* 15:99
2. Oxford, J.S. and Schild, G.C. (1975): In: *Negative Strand Viruses.* Mahy, W.J. and Barry, P.D. (Eds.), Academic Press, New York, pp. 611–620.
3. Young, J.F. and Palese, P. (1979): *Proc. Nat. Acad. Sci. U.S.A.* 76:6547.

4. Laver, W.G. and Downie, J.C. (1976): *Virology* 70:105.
5. Brand, C.M., Stealy, V.M., and Rowe, J. (1977): *J. Gen. Virol.* 36:385.
6. Dimmock, N.J., Carver, A.S., Kennedy, S.I.T., Lu, M.R., and Luscombe, S. (1977): *J. Gen. Virol.* 36:503.
7. Erickson, A.H. and Kilbourne, E.D. (1980): *Virology* 100:34.
8. Lecomte, J. and Croteau, G. (1981): *J. Virol Methods,* in press.
9. Reginster, M., Joassin, L., and Fontaine-Delcambe, P. (1979): *J. Gen. Virol.* 45:283.

David H. L. Bishop and Richard W. Compans, eds.
The Replication of Negative Strand Viruses

Cytotoxic Cellular Immune Responses During Influenza A Infection in Human Volunteers

J.A. Daisy,[a] M.D. Tolpin,[b] G.V. Quinnan,[a] A.H. Rook,[a]
B.R. Murphy,[b] K. Mittal,[a] M.L. Clements,[c]
M.G. Mullinix,[b] S.C. Kiley [a] and F.A. Ennis [a]

A cytotoxic T-lymphocyte response that is uniformly H-2 restricted and virus specific is now well established in the mouse model for both influenza A and B infection.[1-5] Such a response has been detected *in vivo* during infection in tests of freshly isolated lymphocytes [1-4] and *in vitro* after secondary stimulation with virus of splenic lymphocytes from previously primed mice.[5] Depending on the experimental conditions, this response is either broadly cross-reactive for influenza A subtypes or is hemagglutinin specific.[3,4] Also, an increase in natural killer cell activity *in vivo* has recently been observed in mice following influenza infection (J. Djeu, unpublished observations).

McMichael et al.[6] established that a human cytotoxic T lymphocyte response can be generated by stimulation with virus (influenza) *in vitro*. This response of peripheral blood lymphocytes from normal, uninfected adults was HLA restricted, and, although broadly cross-reactive for influenza A subtypes, could distinguish between influenza A and B virus infected target cells (virus-specific). Biddison et al.[7-9] have elaborated on the immunogenetics of this *in vitro* response. Similarly, *in vitro* augmentation of natural killer cell activity has been demonstrated in human lymphocytes stimulated with influenza A virus.[10] The only reported well-defined *in vivo* cytotoxic activity to influenza has been an antibody dependent cellular response detectable in the peripheral blood lymphocytes of antibody positive, uninfected adults.[11]

[a] Department of Health and Human Services, Public Health Service, Food and Drug Administration, Bureau of Biologics, Division of Virology, Bethesda, Maryland.
[b] National Institute of Allergy and Infectious Diseases, National Institutes of Health, Bethesda, Maryland.
[c] Center for Vaccine Development, University of Maryland, School of Medicine, Baltimore, Maryland.

We have been interested in the specific cellular immune responses to influenza infection in humans and, in particular, the study of the cytotoxic T-lymphocyte responses in infected individuals. To this end, we have studied volunteers who received live influenza A viruses as part of ongoing studies by the National Institute of Allergy and Infectious Diseases to develop live, attenuated vaccines. We report, here, the results of natural killer cell and cytotoxic T-lymphocyte studies performed on peripheral blood lymphocytes from these volunteers.

Methods

Informed, healthy adult volunteers were inoculated intranasally with live attenuated or wild type influenza A virus. Peripheral blood lymphocytes were assessed prior to inoculation and at several points postinoculation for natural killer (NK) cell activity and cytotoxic T-lymphocyte (CTL) activity. NK cell activity was determined in an 18-hour ^{51}Cr release assay using fresh peripheral blood lymphocytes as effectors and ^{51}Cr-labeled K562 cells as targets.

In assays for cytotoxic T-lymphocytes, peripheral blood lymphocytes were used both as targets and effector cells in two different ways. In both types of assays, lymphocytes to be used as targets were infected with influenza A virus and labeled with ^{51}Cr. Influenza B infected targets and uninfected lymphocyte targets were used as controls. In one type of assay, fresh lymphocytes were used as effector cells against those target cells and, in the other type, lymphocyte effectors were first stimulated for 5 days *in vitro* with homologous influenza A virus. To test HLA restriction, each effector cell population was tested against autologous targets and target lymphocytes of an HLA matched (namely, sharing at least one HLA A or B locus) and an HLA mismatched donor. Cytotoxic T-lymphocyte activity was defined as activity that was both HL-A restricted and virus-specific. Peripheral blood lymphocytes of uninfected humans do not possess such cytotoxic T-cell activity unless they are first stimulated *in vitro* with virus.

We have now looked at two separate groups of volunteers. One group of volunteers was inoculated with a pseudorevertant of a temperature-sensitive recombinant strain of A/Alaska/6/77 (H3N2) virus, the second group received wild type A/California/10/78 (H1N1) virus.

Results

In the first study, 8 volunteers were inoculated with the A/Alaska virus and we evaluated responses in 3 of those volunteers (Table 1). Two of the volunteers had mild upper respiratory tract symptoms and shed virus. The third showed only a rise in antibody titer. Cytotoxic T-cell activity, using fresh effector cells collected on days 0, 3, 6 and 28 postinoculation, was not

Table 1. Clinical and Laboratory Findings in Volunteers After Intranasal Inoculation with an Influenza A/Alaska/6/77 (H_3N_2) Virus.

Volunteer	Signs and symptoms	Viral shedding	Seroconversion
FH	−	−	+
JW	+	+	+
CP	+	+	+

detected. When the volunteers' peripheral blood lymphocytes were stimulated *in vitro*, HLA restricted, virus-specific cytotoxic activity was generated from day 0 lymphocytes, and no increase was discernible on subsequent study days in the 2 volunteers on whom we have data. The reason for failing to detect a cytotoxic T-lymphocyte response *in vivo* in these volunteers is unclear, but may, in part, have been related to some residual attenuation of the virus as evidenced by the mild infection caused by this strain.

In a second study, 5 volunteers were studied after inoculation with the wild type A/California (H1N1) virus (Table 2). Three had systemic clinical manifestations which included fever and malaise as well as upper respiratory tract symptoms; they also shed the virus in their nasal secretions. All volunteers had antibody responses, although in one case this was only detected by a sensitive ELISA test.

We also examined the peripheral blood lymphocytes of these volunteers in a natural killer cell assay. Figure 1 shows the NK cell response in two different effector to target cell ratios. Using this assay, all 5 volunteers responded initially with significant decreases in natural killer cell activity on day 2. By day 6, however, the average NK cell activity was above baseline and on day 10, it was significantly increased. This activity returned to baseline values by day 31.

Table 2. Clinical and Laboratory Findings in Volunteers Following Inoculation With Wild Type Influenza A/California/10/78 (H_1N_1) Virus.

Volunteer	Symptoms and signs	Viral shedding	Seroconversion by HI	Seroconversion by NT	Seroconversion by ELISA
JR	+	+	+	+	+
RS	+	+	+	+	+
IT	+	+	+	+	+
SV	−	−	−	+/−	+
SW	−	−	+	+	+

HI: Hemagglutinating inhibition test; NT: Neutralization test; ELISA: Enzyme-linked immunosorbent assay.

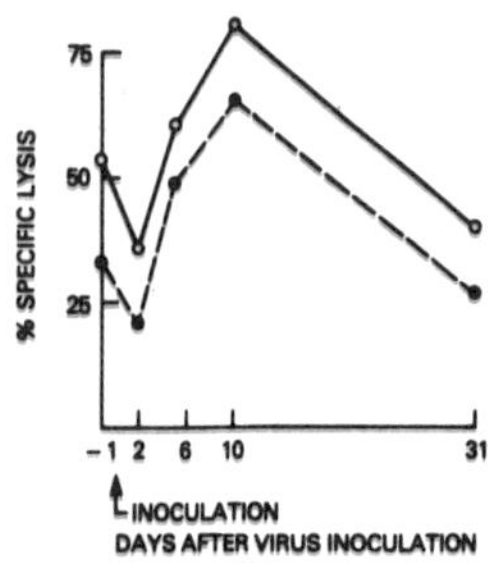

Figure 1. Natural killer cell activity following intranasal inoculation with influenza A/California (H_1N_1) virus. The results shown are the mean values from 5 volunteers measured at 50:1 (○—○) and 10:1 (●—●) E:T ratios, and at baseline (day -1) were not different from the mean values of normals measured in our laboratory. Individual values among the 5 volunteers also fell within the normal range at baseline.

The results of cytotoxic T-lymphocyte assays using fresh effector lymphocytes by volunteers are presented in Figure 2. In 4 volunteers (JR, RS, SW and IT) significant lysis was detected on day 6 and/or day 10. This lysis was restricted to targets sharing HLA antigens with the effectors, except that lymphocytes from JR lysed unmatched targets on day 6. This level of lysis was low but statistically significant. In this case, the unmatched target cell possessed an HLA-B7 antigen which cross-reacts strongly in serological testing with an antigen present on the effector cells, HLA-B27. Thus, the reason that HLS restriction was only partial could be explained by either cross-reactive T-cell recognition of the B7 antigen or by a mixed T-cell and non-T-cell-mediated cytotoxic response. In all other instances, specific lysis values for non-matched target cells were not statistically significant. Lymphocytes from SV displayed no activity on any of the assay days. Target cells infected with influenza B virus were not lysed in any case (results not shown). Thus, an HLA restricted, virus-specific response typical of a cytotoxic T-cell response developed during influenza A infection in four of the five volunteers.

As was observed in the first group of volunteers, HLA-restricted virus-specific cytotoxic activity could be stimulated *in vitro* both prior to and after virus challenge (results now shown).

Discussion

These studies involved a small number of volunteers, but the data indicated that human adults infected with influenza A virus transiently augmented their natural killer cell activity in response to the infection. Additionally, an HLA-restricted, virus-specific cytotoxic response, typical of cytotoxic T-lymphocytes, occurred.

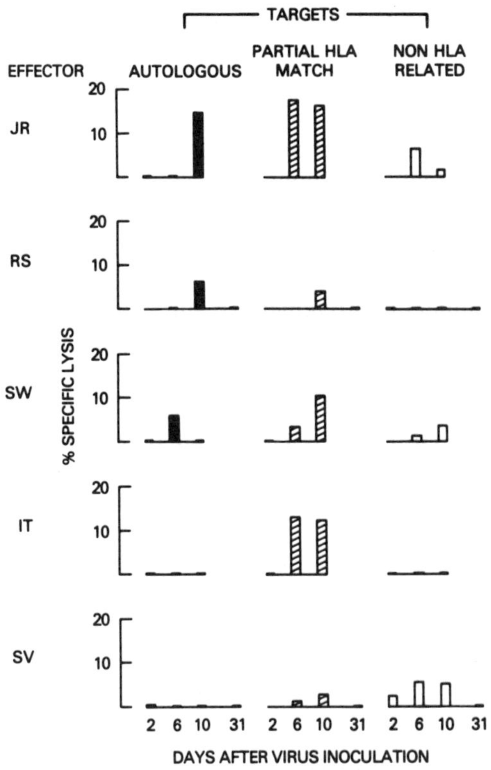

Figure 2. Virus-specific cytotoxic activity of fresh peripheral blood lymphocytes from human volunteers following intranasal inoculation with influenza A/California/10/78 (H_1N_1) virus. Lymphocytes from IT were not used as targets for effectors other than the autologous effectors; therefore, their reliability as targets could not be satisfactorily assessed.

Previously published studies of *in vivo* cytotoxic cellular responses to influenza infection in humans have centered on a model employing a xenogeneic target cell line (baby hamster kidney cells) that, in consequence, has been capable of only demonstrating non-T-cell activity. Greenberg et al., using this assay, demonstrated antibody dependent cellular activity in the fresh, unstimulated peripheral blood lymphocytes of normal adults.[11] These investigators were also able to show a transient response, following inoculation with either killed influenza A vaccine or live virus, that was not clearly antibody dependent.[12] In this xenogeneic system, however, the cytotoxicity was not T-cell-mediated. Quinnan et al.[13] demonstrated virus-specific cytotoxicity in the peripheral blood lymphocytes of volunteers inoculated with killed influenza A vaccine, using autologous lymphocyte targets, however, HLA restriction was not documented. Herberman et al.[14] refer, in passing, to the augmentation after 2 days of NK cell activity in human volunteers following the use of swine influenza vaccine, no further detail is given.

The work reported here is the first documentation of both a CTL and NK cell *in vivo* response to influenza A infection in humans. A similar response has now been shown for cytomegalovirus infection in bone marrow transplant recipients.[15] These transient responses occur before the antibody response to influenza virus is detectable and correlate temporally with both the decline of virus shedding and clinical improvement (results not shown). One can speculate, therefore, about the role these cytotoxic cellular responses play in recovery from infection. In the mouse, CTL [4] and NK cells (J. Djeu, unpublished observations) are generated in the lungs early in the course of acute influenza infection. Although it is difficult to accurately assess the role of these cytotoxic responses to influenza infection, it is reasonable to argue that they are important for the recovery from infection in both man and mouse.

References

1. Ennis, F.A., Martin, W.J., Verbonitz, M.W., and Butchko, G.M. (1977): *Proc. Natl. Acad. Sci. U.S.A.* 74:3006–3010.
2. Ennis, F.A., Martin, W.J., and Verbonitz, M.W. (1977): *J. Exp. Med.* 146:893–898.
3. Effros, R.B., Doherty, P.C., Gerhard, W.E., and Bennink, J. (1977): *J. Exp. Med.* 145:557–568.
4. Ennis, F.A., Wells, M.A., Butchko, G.M., and Albrecht, P. (1978): *J. Exp. Med.* 148:1241–1250.
5. Braciale, T.J. (1979): *J. Exp. Med.* 149:856–869.
6. McMichael, A.J., Ting, A., Zweerink, H.J., and Askonas, B.A. (1977): *Nature* 270:524–526.
7. Biddison, W.E. and Shaw, S. (1979): *J. Immunol.* 122:1705–1709.
8. Shaw, S. and Biddison, W.E. (1979): *J. Exp. Med.* 149:565–575.
9. Biddison, W.E., Ward, F.E., Shearer, G.M., and Shaw, S. (1980): *J. Immunol.* 124:548–552.
10. Trinchieri, G. and Santoli, D. (1978): *J. Exp. Med.* 147:1314–1321.
11. Greenberg, S.B., Criswell, B.S., Six, H.R., and Couch, R.B. (1975): *J. Immunol.* 119:2100–2106.
12. Greenberg, S.B., Criswell, B.S., Six, H.R., and Couch, R.B. (1978): *Infect. Immun.* 20:640–645.
13. Quinnan, G.V., Ennis, F.A., Tuazon, C.U., Wells, M.A., Butchko, G.M., Armstrong, R., McLaren, C., Manischewitz, J.F., and Kiley, S. (1980): *Infect. Immun.* 30:362–369.
14. Herberman, R.B., Djeu, J.Y., Kay, D.G., Ortaldo, J.R., Riccardi, C., Bonnard, G.D., Holden, H.T., Fagnani, R., Santoni, A., and Puccetti, P. *Immun. Rev.* 44:43–70.
15. Quinnan, G.V., Kirmani, N., Esber, E., Saral, R., Manischewitz, J.F., Rogers, J.L., Rook, A.H., Santos, G.W., and Burns, W.H.: *J. Immunol.*, in press.

Published 1981 by Elsevier North Holland, Inc.
David H. L. Bishop and Richard W. Compans, eds.
The Replication of Negative Strand Viruses

Transmission in Swine of Hemagglutinin Mutants of Swine Influenza Virus

Edwin D. Kilbourne,[a] Sandy McGregor,[b] and Bernard C. Easterday[b]

The occurrence of dimorphic hemagglutinin variants of swine influenza virus (A/NJ/11/76 [H1N1]) was first recognized during the cloning of high-yield vaccine recombinant viruses in 1976 [1,2] and was also noted in field isolates.[2,3] L mutants grow to low titer in chicken embryos and MDCK cells, react in hemagglutination inhibition (HI) and neutralization tests with heterotypic A/sw/Cam/39 antiserum, and are more infective for swine, the natural host, than are H mutants, which have converse characteristics. Segregation of the L or H hemagglutinin (HA) genes in recombinants containing all other genes from A/PR/8/34(H1N1) virus has confirmed that the HA phenotype is responsible for the differing replication characteristics of the mutants.

Although virus of L phenotype prevails in most contemporary isolates,[2,3] earlier strains appear to have been of H phenotype by serologic analysis.[3] Thus, it is unlikely that the H mutant (less infective for swine) is merely an artifact of selection in the chicken embryo, in which it has a replicative advantage. Furthermore, H virus has been recovered directly from swine using antiserum suppressive to the L mutant.[2] Under laboratory conditions with cloned virus, mutation of L to H phenotype has been demonstrated.[1] Therefore, the persistence of H mutants is explicable as allelic variation arising from point mutations in L as the currently predominant form. We have tried to ascertain whether or not H mutants are fortuitous or offer the virus potential survival advantage. Therefore, we have undertaken the present experiment to determine whether differences in transmission or viral

[a] Department of Microbiology, Mount Sinai School of Medicine, New York, New York.
[b] Department of Veterinary Science, University of Wisconsin, Madison, Wisconsin.

persistence occurred when the mutants were introduced simultaneously into a herd of susceptible swine. The experimental design is summarized in Figure 1.

Results

Twenty-four hours following inoculation with L or H virus as indicated, inoculated (infector) pigs were reintroduced into the herd. Infector and contact animals were observed under conditions in which they mixed freely. Nasal swabbings were obtained from infector animals 3 days after infection and from infector and contact animals 6, 8, 10 and 13 days after inoculation of infectors. Serum was obtained from all animals 4 weeks after inoculation. Nasal swabbings were collected in transport medium containing stabilizing agents and antibiotics and were stored at −70°C prior to their injection into chicken embryos in serial dilutions, with or without antiserum selective against L virus.

Virus Excretion Of Infector Swine

The early virus shedding patterns of infector swine are shown in Figure 2. First, it is clear that all L inoculated but not all H inoculated pigs were infected (i.e., shed virus) although it had been calculated that they would receive the same swine ID_{50}. Furthermore, L virus predominated in 80% of both L and H inoculated pigs on day 8 postinfection, despite clear evidence of replication of H to high titer in 3 of 5 animals 2 days earlier (Table 1). Indeed, as early as day 6 postinfection, the only H inoculated pig that had not shed H virus, excreted L virus.

Figure 1. Experimental Design. Timetable of inoculation of infector pigs and exposure of contact animals.

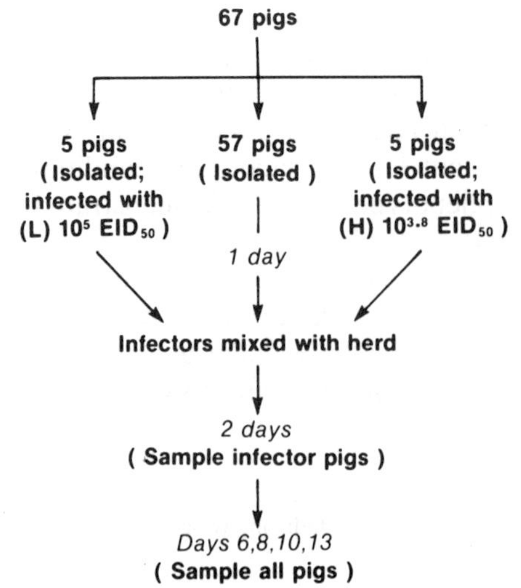

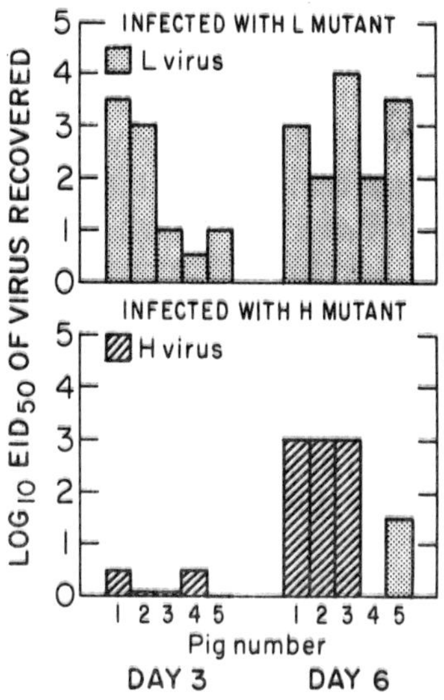

Figure 2. Virus shedding of infector swine.

Virus Excretion By Contact Swine As Evidence Of Infection

The overwhelming prevalence of infection of contact animals with the L mutant is illustrated in Figure 3, the incidence reaching a peak on the 7th day of contact and declining rapidly with the exhaustion of susceptibles. On the basis of HI antibody responses (100%) as well as cumulative virus shedding data, we conclude that all animals in the herd became infected within the 13-day observation period.

Table 1. Virus Excretion from Initially Inoculated (Infector) Swine.

		L		H	
Day[a]	**Infected with**	**No.**	**%**	**No.**	**%**[b]
3	L mutant	5/5	100	0/5	0
6	L mutant	5/5	100	0/5	0
8	L mutant	4/5	80	0/5	0
10	L mutant	0/5	0	0/5	0
13	L mutant	0/5	0	0/5	0
3	H mutant	0/5	0	2/5	40
6	H mutant	1/5	20	3/5	60
8	H mutant	4/5	80	1/5	20
10	H mutant	1/5	20	0/5	0
13	H mutant	0/5	0	0/5	0

[a]Postinfection.

[b]80% of H infectors shed virus on 1 or more days.

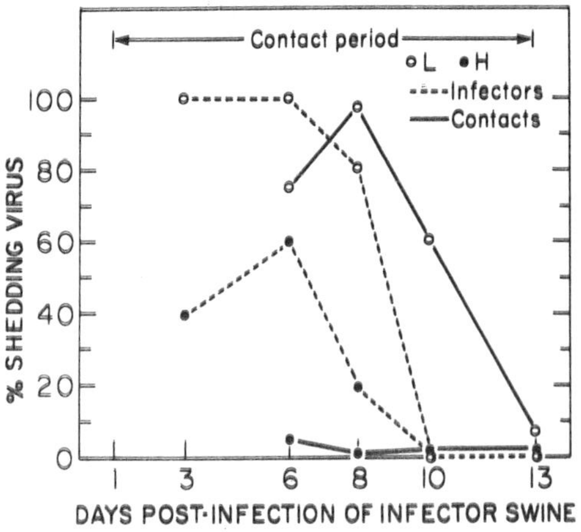

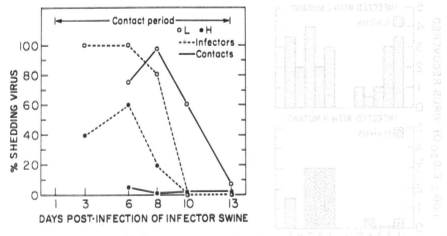

Figure 3. Percentage of infector or contact pigs shedding L or H virus during interval sampling of nasal secretions during contact period.

Only 4 of 56 contact pigs shed H virus following exposure (Table 2)). In two instances, H virus excreted on day 6 was rapidly supplanted by shedding of the L mutant but in two others virus was isolated on 2 different days, attesting to the fact that transmitted infection with H did occur. In 2 cases, viruses were titrated and found to be present in concentrations of 10^{5} or $10^{3.5}$ EID_{50}/0.2 ml on the first day on which virus was detected. Shedding patterns in H infected contact pigs are shown in Table 3.

Discussion and Summary

This experiment has established conclusivley that both L and H mutants of A/NJ/11/76 virus can be transmitted independently from swine to swine

Table 2. Virus Excretion of Contact Swine as Evidence of Infection.

	Virus isolated			
	L		H[a]	
Day of exposure	No.	%	No.	%
5	42/56	75	3/56	5
7	55/56	98	1/56	2
9	34/56	61	2/56	4
12	4/56	7	0/56	0

[a] 4/56 contact pigs acquired H infection; 2/4 excreted virus on 2 days; 3/4 excreted L virus also on one or more days.

Table 3. Virus Shedding Patterns in Contact Pigs Acquiring H Infection.

Pig	Day Post contact			
	5	7	9	12
14	—	H[a]	H	—
48	H	L[b]	H	—
52	H	L	L	L
58	H	L	L	L

[a]H mutant isolated.
[b]L mutant isolated.

under conditions of barnyard contact. Unfortunately, the relative degree of transmissibility of the 2 variants has not been ascertained because of demonstrably lower levels of virus excretion in H infector animals, probably related to their inadvertant inoculation with a lesser *swine* infective dose of virus, although identical EID_{50} were administered. Therefore, the L variant had a significant head start in competing for the respiratory tracts of contact animals, producing a brisk epidemic and saturating the herd by day 7 post contact. The fact that no virus was detected in infector pigs after 8 days (with one exception) suggested that transmission must have occurred within the first week. So pervasive was the L variant that prior H infection was insufficient to interfere with subsequent L colonization. On day 6, when the peak of shedding from H infectors occurred, most contact animals had already been infected with the L mutant.

This experiment adds to prior evidence that each mutant can infect swine, demonstrating that each can induce transmissible infection. It remains to be determined which virus in fact has a competitive advantage in nature when introduced in similar swine ID_{50}, or whether H virus has other attributes conducive to its continuing survival. We now have antisera selective against either L or H mutant that will facilitate our qualitative and quantitative study of field isolates in the future.

References

1. Kilbourne, E.D. (1978): *Proc. Natl. Acad. Sci. U.S.A.* 75:6258.
2. Kilbourne, E.D., McGregor, S., and Easterday, B.C. (1979): *Infection and Immunity* 26:197.
3. Kendal, A.P., Noble, G.R., and Dowdle, W.R. (1977): *Virology* 82:111.

David H. L. Bishop and Richard W. Compans, eds.
The Replication of Negative Strand Viruses

Characteristics of an Influenza Virus Resistant MDBK Cell Variant

I.T. Schulze, K.J. Whitlow, D.M. Crecelius, and M.V. Lakshmi [a]

Introduction

Within recent years, continuous cell lines have been persistently infected with a number of different viruses. The mechanisms involved in the establishment of such cultures have been reviewed by Walker [1,2] and by Joklik.[3] One important factor in maintaining the balance between viral growth and cell survival is the genetically determined permissiveness of the host cell for the virus in question. Very little is presently known about these determinants, largely because virus resistant cell variants which differ from the parent cell in only a few characteristics are not easy to obtain. In the course of trying to establish persistent infection in MDBK cells with influenza virus, we have obtained a cell line which we consider to be such a variant. Since these cells are specifically and stringently resistant to influenza virus, we expect that they will be valuable for studying the replication of this virus. The process by which the cell line was obtained, along with some of its properties, is described here.

Results

Persistent Infection of MDBK Cells with Plaque Type Variants of Influenza A Virus

Establishment of Persistently Infected Cultures

We have previously reported that stocks of influenza A (HON1), strain WSN, which were grown in chick embryo fibroblasts contained viruses of

[a] Department of Microbiology, St. Louis University School of Medicine, 1402 South Grand Boulevard, St. Louis, Missouri.

two plaque types, designated F and C.[4] In our first attempts at establishing persistently infected MDBK cells, we used F plaque type virus since it produced less severe cytopathogenic effects (CPE) in these cells than did the C variant. When confluent cultures of MDBK cells were infected with F plaque type virus at a multiplicity of 0.005 PFU per cell, the monolayers were still intact after 48 hours at 37°, although the culture medium contains a high titer ($10^7 \times 10^8$ PFU per ml) of virus. When the culture medium was replaced and again harvested 48 hours later, drastic reduction in virus yield was observed but most of the cells in the culture remained intact. Periodic refeeding of those cultures produced monolayers which could be subcultured by the same procedure as that used for the growth of uninfected MDBK cells. After about three weeks, these monolayers were free of CPE. We have monitored virus yields from such cultures over a period of approximately three months. As shown in Figure 1, they produced virus in varying amounts for an extended period after which the yields dropped below that detectable by hemagglutination.

Experiments were then carried out to determine whether persistently infected cultures could be established with the C plaque variant which produced extensive CPE by 48 hours. Except that most of the cells were initially destroyed, those cultures went through the same course of events as that just described. Thus, establishment of influenza virus persistence in MDBK cells did not require that the virus in the original inoculum be only mildly cytopathogenic.

Properties of Virus Produced after Prolonged Culture of Cells

We have previously reported that, during serial passage in MDBK cells, C plaque type virus appeared in the F progeny and eliminated the parental virus by non-reciprical interference.[4,5] On day forty-five of the experiment presented in Figure 1, plaque assays were therefore performed in order to determine the plaque types of the virus. As expected from our previous work, only C plaque type virus was present. Titrations were carried out at 33° and 39° and it was found that the virus had not changed in temperature-sensitivity. We also found that infected cultures continued to yield low levels of virus when put back into culture after they had been frozen and stored at −70°C.

Viral Susceptibility of Persistently Infected Cultures

Since the cell cultures showed no CPE after their recovery from the initial cycles of cell destruction, these monolayers could be used for plaque assays despite the fact that they were producing low levels of virus as measured by plaque production on MDBK cells. We were therefore able to challenge the persistently infected cultures with the C variant of influenza virus as well as with an unrelated virus. As shown in Table 1, when five persistently infected lines were examined, two produced no plaques when exposed to 10^5 PFU of

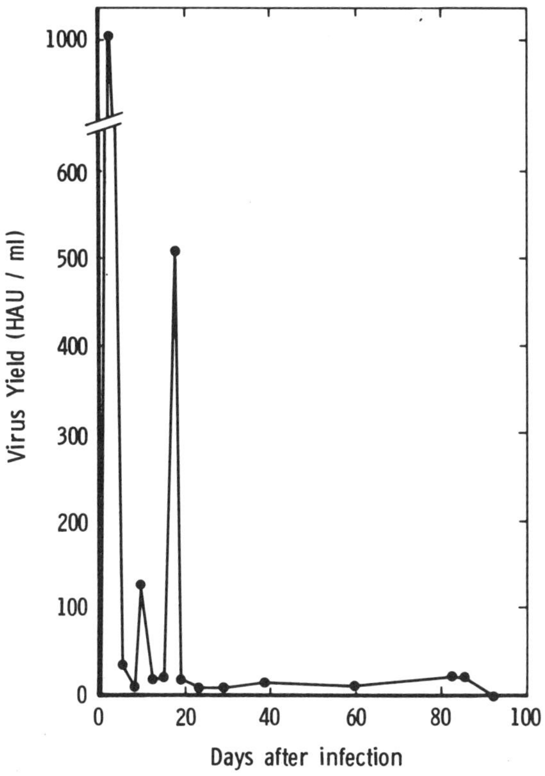

Figure 1. Influenza virus yields from persistently infected MDBK cells. Confluent monolayers of MDBK cells were infected with F plaque type variant of the WSN strain of influenza A virus at a multiplicity of 0.005 PFU per cell as previously described.[4] Culture medium was removed and replaced at 43 and 96 hours and at two to five-day intervals thereafter. Cells were subcultured at 6 days and then at weekly intervals. Virus concentrations in the culture medium, measured as hemagglutinin units (HAU) per ml, were determined as previously described.[4]

influenza virus, whereas three others produced plaques at 3, 6 and 21% the efficiency of the control MDBK cells. When challenged with vesicular stomatitis virus (VSV), the two lines which were most resistant to superinfection with influenza virus produced more plaques than did the control MDBK cells. We concluded that these two cell lines were metabolically capable of supporting the growth of virus, but that they were specifically resistant to infection by influenza virus.

As shown in Figure 1, after 92 days of culture, virus was no longer detectable in the medium by the hemagglutination assay. Plaque assays revealed low levels of virus which decreased with passage and then disappeared completely. Similar results were obtained with a number of lines from cells originally infected with either F or C plaque type virus. With line 20, suscep-

Table 1. Susceptibility of Persistently Infected Cell Lines to Influenza Virus and to VSV.

Cells used for plaque assays	Virus titer (PFU per ml)	
	Influenza virus	VSV
MDBK	1.0×10^7 (100)	4.3×10^6 (100)
Persistently infected MDBK lines:		
2	$<1.0 \times 10^2$ (<0.001)	5.2×10^6 (122)
11	3.2×10^5 (3)	2.3×10^6 (54)
13	$<1.0 \times 10^2$ (<0.001)	5.1×10^6 (119)
14	6.1×10^5 (6)	3.4×10^6 (86)
20	2.1×10^6 (21)	1.2×10^6 (30)

Plaque assays were carried out as previously described[4] using confluent monolayers of MDBK cells and the persistently infected cell lines. Preparations of the WSN strain of influenza A virus containing 10^7 PFU per ml and VSV containing 4.5×10^6 PFU per ml were plated at appropriate dilutions to permit counting of the plaques. Virus titers were calculated from duplicate plates. The numbers in parenthesis represent plaque titers as percentages of MDBK controls. Cell lines 2 through 14 were derived from cells intially infected with the F plaque variant, line 20 from cells originally infected with C plaque variant.

tibility to influenza virus was substantially restored when the cultures stopped producing virus. However, with line 13, resistance to superinfection by influenza virus was retained even after virus production had completely ceased, suggesting that maintenance of resistance did not depend on the production of infectious particles. Other more sensitive techniques aimed at detecting virus, such as cocultivation of the resistant cells with MDBK cells, also failed to show any virus production.

Conclusions

Taken together, the experiments described above indicate that persistently infected MDBK cell cultures can be rather easily established with either plaque type variant and can be maintained over a period of months. In addition, we have shown that when spontaneous cessation of virus production occurs, the cultures may either regain susceptibility to influenza virus or they may retain complete resistance. Thus far, those which have remained resistant have been obtained from cultures originally infected with the F plaque type virus, suggesting that a low level of cell destruction during the initial cycles of virus growth may be involved in producing resistant cells. If an initial period of concomitant rather than alternating growth of virus and cells is required, we may be unable to obtain resistant cells from cultures initially infected with the C variant.

So far we have chosen a single resistant line (line 13) for further investigation. These cells appear to be stably resistant to influenza virus and have been grown in continuous culture for about two years. Since they are not resistant to other viruses, we consider them to be of potential value for studying influenza virus replication. The remainder of the work presented here pertains to these cells.

Properties of an Influenza Virus Resistant MDBK Cell Line

Karyotype of the Influenza Virus Resistant Cells

In order to insure that the influenza virus resistant cells were indeed variants of MDBK cells rather than a contaminating cell, our current MDBK cell cultures (grown from cells originally obtained in 1969 from Dr. Purnell Choppin of the Rockefeller University) and line 13 of the virus resistant cells were karyotyped in the laboratory of Dr. Patricia Monteleone in the Department of Pediatrics of St. Louis University School of Medicine. The modal number of chromosomes, determined from 100 cells of each type, was found to be 51 for each cell line. In addition, the chromosomes from the two lines were virtually identical in morphology. Thus we have concluded that our virus resistant cells are indeed MDBK cells and have designated them MDBK/IV to denote both their origin and their resistance to influenza virus.

Morphology and Growth Properties of MDBK/IV Cells

Although MDBK and MDBK/IV cells are indistinguishable in morphology when the cultures are sparse, extensive vacuolation occurs when MDBK/IV cultures are kept after they have reached confluency (Figure 2). In addition, MDBK/IV cultures contain low numbers of very large cells, some of which contain multiple nuclei. Such cells are not seen in MDBK cultures. Neither the basis for these morphological differences nor their significance for virus production is known at this time. Whatever the explanation, the growth rate of the cells is not affected. During logarithmic growth, the doubling time of both cell lines is approximately fifteen hours.

Virus Susceptibility of MDBK/IV Cells

Thus far we have tested the susceptibility of MDBK/IV cells to four viruses in an attempt to determine to what extent the block in virus production is restricted to influenza virus replication. Based on growth of virus under liquid medium followed by assay of virus yields by plaque formation on MDBK cells, MDBK/IV cells made approximately 2.5 times as much VSV as did MDBK cells. Yields of herpes simplex virus, type 1, and Sindbis virus were approximately one-fourth that obtained from MDBK cells. These values are in striking contrast to those obtained with influenza virus; under liquid medium MDBK/IV cells infected at multiplicities from 0.005 to 2 PFU per cell produced virus titers which were five or more logs below those obtained from MDBK cells. Thus most, if not all, cells in MDBK/IV cultures were capable of replicating both RNA and DNA viruses, but only a very small minority of the cells made influenza virus.

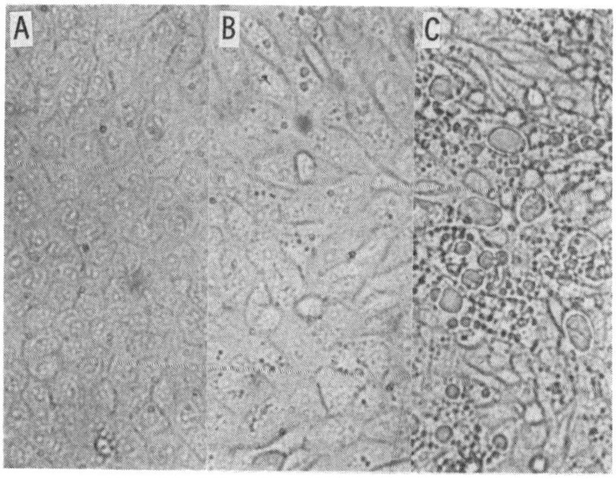

Figure 2. Light micrographs of MDBK and MDBK/IV cells. (A) Confluent monolayers of MDBK cells, showing typical cell morphology. (B) Newly confluent monolayers of MDBK/IV cells showing cells highly similar in morphology to the parent cells. A few cells contain small vacuoles. (C) Heavy monolayers of MDBK/IV cells two days after the monolayer had reached confluence. A majority of the cells contain large vacuoles.

Membrane Properties of MDBK/IV Cells

In an attempt to locate the block in virus synthesis in MDBK/IV cells we have initiated three separate experimental approaches. In the first, the initial events of virus infection (i.e., adsorption and penetration) are being investigated. Assuming that the vacuolation observed when MDBK/IV cultures become confluent may result from changes in membrane structure, we are exploring the chemical composition of their plasma membranes. Thus far we have observed that MDBK cell membranes contain twice the amount of glycosphingolipid found in MDBK/IV cells. Since other lines of investigation in our laboratory indicate that glycolipids are involved in binding and/or penetration of influenza virus to MDBK cells,[6] we have investigated the binding of radioactive influenza virus to the virus resistant cells. As shown in Table 2, when the two cultures were exposed to the same concentrations of virus, MDBK/IV cells bound significantly less virus per cell than did MDBK cells. This reduction in binding correlated in a general way with the reduction in total glycolipids. However, since it did not adequately account for the resistance of MDBK/IV cells to influenza virus, it prompted an investigation of the intracellular steps in influenza virus replication in MDBK/IV cells.

Attempts to Detect Viral Antigens in MDBK/IV Cells

Using the method of indirect immunofluorescence, we have attempted to detect influenza virus proteins in the resistant cells after exposure to 10 PFU per cell. As shown in Figure 3, under conditions in which viral antigens were

Table 2. Binding of Influenza Virus to MDBK and MDBK/IV Cells.-

Cells	Virus added (HAU/monolayer)	Particles bound per cell	MDBK/IV / MDBK
MDBK/IV	20	36	0.30
MDBK	20	120	
MDBK/IV	50	45	0.15
MDBK	50	300	
MDBK/IV	100	180	0.33
MDBK	100	540	

Binding of purified ^{32}P-labeled influenza virus was measured as previously described,[7] using confluent cell cultures with 1.0×10^6 to 1.2×10^6 cells per monolayers. Virus preparations contained 22 ^{32}P cpm/HAU. One HAU equals 3×10^7 particles.

detected in virtually all MDBK cells, the resistant cells were essentially devoid of detectable viral antigens. MDBK/IV cell cultures incubated for long periods (up to 30 hours) after exposure to virus also showed only occasional cells which contained viral antigens. The antisera used in these studies contained antibodies against the matrix, nucleoprotein, and hemagglutinin. We have therefore concluded that these three major virion proteins are either not synthesized at all by MDBK/IV cells or are present in insufficient amounts to be detected by the techniques used. Experiments employing pulses of ^{35}S-methionine are underway to determine whether viral proteins, including the non-structural proteins and the P-proteins which are associated with RNA polymerase activity,[9-11] are synthesized by the resistant cells.

Viral RNA Synthesis in MDBK/IV Cells

Based on the experiments described thus far, we were unable to distinguish between a block in virus penetration or uncoating and a block in any of a number of steps leading to virion protein synthesis. For this reason, we have begun to investigate viral RNA synthesis in MDBK/IV cells following exposure to high multiplicities of influenza virus. Current data indicate that the only viral RNA species synthesized by MDBK/IV cells is polyadenylated RNA which is complementary to the virion RNA and that the synthesis of this species is not amplified at about three hours as it is in MDBK cells. The experiments suggest that MDBK/IV cells are fully capable of supporting primary transcription and that no other viral RNA synthesis occurs in the cells. Experiments are underway to confirm and expand this data and to determine whether the primary block involves aberrant transcription or inability to support viral protein synthesis.

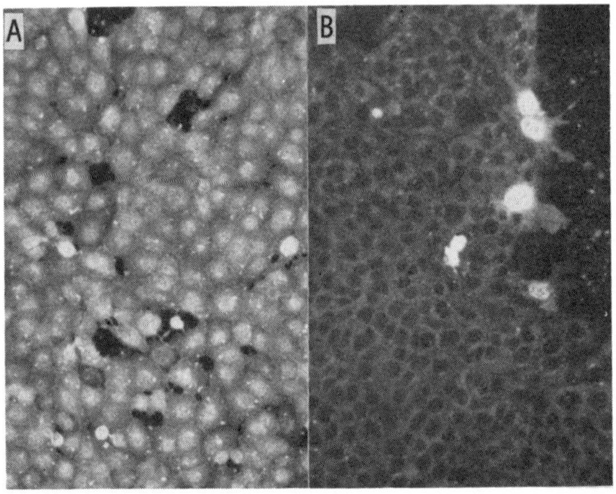

Figure 3. Indirect immunofluorescent staining of MDBK and MDBK/IV cells after exposure to high multiplicities of influenza virus. The method of Emmons and Riggs[8] was used. Cell monolayers on coverslips were fixed with acetone and stained 5.5 hours after exposing the cells to virus at a multiplicity of 10 PFU per cell. The virus-specific antisera used was prepared in rabbits. It contained antibodies against the matrix, nucleoprotein and hemagglutinin of the WSN strain of influenza virus. Goat anti-rabbit IgG, conjugated to fluorescein isothiocyanate, was used to detect bound antibodies. Both nuclear and cytoplasmic fluorescence is seen in virtually all MDBK cells (A), whereas only a few MDBK/IV cells are stained (B).

Discussion

Based on the information presented above, there appear to be two mechanisms by which these influenza virus resistant cells could have arisen. Either they are cell variants which have been selected by long-term exposure to influenza virus or they are genetically altered cells into which viral genetic material has been inserted. Other possibilities, such as control of virus production by interferon or defective interfering particles, have been essentially eliminated by our existing data. Protection by interferon should prevent the growth of heterologous challenge viruses and MDBK/IV cells are even more sensitive to VSV than are the parent cells. Secondly, we have observed that mixed monolayers, obtained by growing the two cells together, produce 75% as many influenza virus plaques as do MDBK cells grown alone. Thus, MDBK/IV cells can be neither constitutive producers of interferon nor inducible high level producers, since, if they were, they would protect the MDBK cells from infection and thereby prevent plaque development. Their inability to prevent plaque formation under these conditions also argues that they are not making defective interfering particles. In addition, all direct attempts at detecting non-infectious virus like particles have failed. Thirdly, prolonged replication of defective particles would require that some infectious virus be available to provide the missing gene products, and no virus has been detected in these cells for months.

Since the MDBK/IV cells have grown out of cultures in which there was concomitant replication of host and viral genes, opportunities for virus-host genome interactions were certainly available. However, the conditions were also appropriate for repeated selection of resistant cells. Experiments presently underway should let us distinguish between these possibilities, i.e., between influenza virus resistant variants of MDBK cells and cells which have unexpressed viral genes. There is no way at present to predict what kind of alteration in cell function should be observed, were the latter the case. Certainly, there is no *a priori* reason to expect that integration of influenza virus genetic information would render the recipient cell resistant to superinfection.

Independent of how these cells were derived, their high level of resistance to influenza virus along with the sensitivity of the parent cells to this virus should make this an excellent system for studying the role of the host in influenza virus replication.

ACKNOWLEDGMENTS

The authors wish to acknowledge the excellent technical assistance of Ms. Barbara Macon who is responsible for the growth of our cell cultures. We also wish to thank Dr. Max Arens for data on the growth of herpes simplex virus in the cells described here and for his advice during the preparation of this manuscript.

References

1. Walker, D.L. (1964): *Prog. Med. Virol.* 6:111–148.
2. Walker, D.L. (1968): In: *Medical and Applied Virology.* Sanders, M. and Lennette, E.H. (Eds.), Warren H. Green, Inc., St. Louis, pp. 99–110.
3. Joklik, W.K. (1977): In: *Microbiology-1977.* Schlessinger, D. (Ed.), *Amer. Soc. Microbiol.*, Washington, D.C. pp. 434–439.
4. Noronha-Blob, L. and Schulze, I.T. (1976): *Virology* 69:314–322.
5. Noronha-Blob, L. and Schulze, I.T. (1976): *Virology* 70:509–516.
6. Lakshmi, M.V. and Schulze, I.T. (1978): *Fed. Proc.* 37:1747.
7. Lakshmi, M.V. and Schulze, I.T. (1978): *Virology* 88:314–324.
8. Emmons, R.W. and Riggs, J.L. (1977): In: *Methods in Virology, Vol. VI.* Maramorosch, K. and Koprowski, H. (Eds.), Academic Press, New York, pp. 1–28.
9. Lazarowitz, S.G., Compans, R.W., and Choppin, P.W. (1971): *Virology* 46:830–843.
10. Skehel, J.J. (1972): *Virology* 49:23–26.
11. Lamb, R.A., Etkind, P.R., and Choppin, P.W. (1978): *Virology* 91:60–78.

PARAMYXOVIRUSES

David H. L. Bishop and Richard W. Compans, eds.
The Replication of Negative Strand Viruses

Evidence for Two Different Sites on the HN Glycoprotein Involved in Neuraminidase and Hemagglutinating Activities

A. Portner [a]

The glycoproteins which form the spike-like projections on the surfaces of paramyxoviruses play essential roles in the initiation of infection.[1-4] The HN glycoprotein is responsible for host cell receptor binding and neuraminidase activities,[3,5-7] whereas the other virus glycoprotein F is involved in virus penetration and has hemolysis and cell fusion activities.[1,8] Although the role of the HN glycoprotein in adsorption to neuraminic acid receptors on erythrocytes and host cells is well accepted,[9-12] little is known about the biological significance of the neuraminidase activity. By analogy to studies with influenza viruses,[13] it has been suggested that the enzyme may be involved in virus entry or release but evidence with paramyxoviruses for either of these concepts is lacking.

An important point to be clearly established is whether the active sites on the HN molecule responsible for hemagglutination and neuraminidase activities are the same or whether two different sites are involved. In this report, evidence is presented that supports the concept that different sites are involved in virus attachment and neuraminidase activity. This is based on the analysis of a Sendai virus temperature-sensitive (*ts*) mutant with a defect in the HN glycoprotein that prevents attachment, and the use of antigenic variants and monoclonal antibodies to analyze hemagglutination and neuraminidase activities.

[a] Division of Virology, St. Jude Children's Research Hospital, Memphis, Tennessee.

A *ts* Mutant of Sendai Virus is Temperature-Sensitive for Hemagglutination, But Not For Neuraminidase Activity

We previously reported that Sendai virus mutant *ts*271, grown at the permissive temperature (30°), is *ts* for hemagglutination and attachment to host cells.[14,15] This mutant is able to agglutinate erythrocytes at 30° but not at the non-permissive temperature (38°). The same temperature-sensitive behavior was exhibited by the HN glycoprotein fraction isolated from the mutant virions, which almost certainly places the defect in the HN polypeptide.[14,15] To determine whether the neuraminidase activity of the mutant also showed the same temperature-sensitive behavior, the *ts*271 enzyme was compared to wild type virus at 30°C and 38°C using the soluble substrate neuraminlactose. The results in Figure 1 show that the kinetic behavior of the wild type and *ts*271 neuraminidases at 30° or 38° are the same, indicating that in the mutant, enzyme activity was not temperature-sensitive. These observations that the *ts* lesion has no effect on neuraminidase activity, but dramatically reduces the ability of the mutant to agglutinate erythrocytes or adsorb to host cells, provided evidence for different active sites on the HN molecule involved in these activities.

Figure 1. The effect of 30 and 38°C on neuraminidase activities of wild type Sendai virus and *ts*271. Replicate tubes containing 10 hemagglutinating units were placed at 30 or 38°C and after 10 minutes 200 μg of neuraminlactose was added to start the reaction. At intervals, one tube of each was removed, rapidly chilled, and assayed for sialic acid.[17]

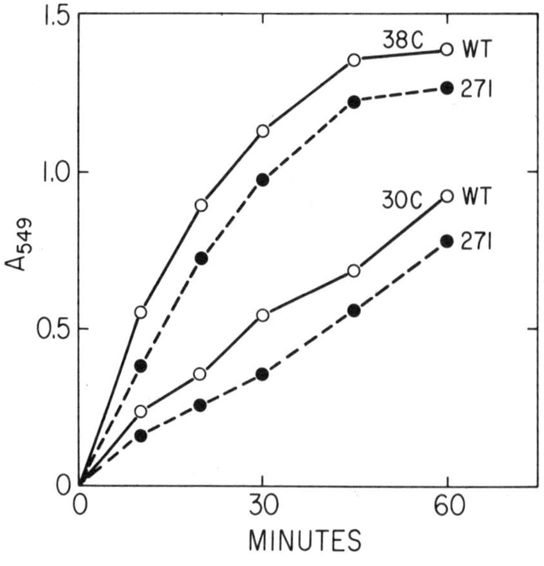

Monoclonal Antibodies Discern Differences in the Hemagglutinating and Neuraminidase Activities of the HN Molecule

To support the conclusions reached with the *ts*-mutant, we isolated monoclonal antibodies directed against the HN molecule and used these antibodies to isolate antigenic variants. The variants were then analyzed for their reactivity in neuraminidase (NI) and hemagglutination inhibition (HI) tests. Ideally, antigenic mutants could be isolated that were inhibited by monoclonal antibodies directed at one active site but not the other. Previously, we showed that fluids containing monoclonal antibodies were specific for the Sendai virus HN molecule.[16] Three different antibodies designated S15/15, S16/27, and S18/17 were used to select antigenic variants.[16] To demonstrate that the Sendai viruses selected with monoclonal antibodies were variants, we analyzed their reactivities by HI (Table 1). Variants selected after a single passage of cloned virus showed a marked reduction in their capacity to react with the monoclonal antibodies used in their selection. Monoclonal antibodies S15/15 and S16/27 showed a 6400-fold reduction in HI titer when reacted with variants selected with S15/15 or S16/27 antibodies, while the S18 antibody showed a 800-fold reduction in titer with the S18 variants. The variants selected with one monoclonal antibody had different reactivity patterns with the other monoclonal antibodies, and thus could be separated into three antigenic groups. These results indicate that these three Sendai virus monoclonal antibodies were directed against different antigenic sites on the HN molecule, and as shown previously by double neutralization tests,[15] these sites must overlap.

To determine whether the antigenic sites involved in HI were also involved in neuraminidase activity, monoclonal antibodies were reacted with antigenic variants and parental virus in a NI test (Table 2). The striking difference between the HI and NI reactivity patterns was that some variants that retained their ability to bind antibody at an antigenic site that was responsible for inhibiting hemagglutination, *were not* inhibited in neuraminidase activity by the same monoclonal antibody. For example, the neuraminidase activities of variants V3 and V4 selected with S16/27 antibody and V1, V2, and V3 selected with S18/17 antibody were completely unreactive in the NI test with S15/15 antibody (Table 2). In contrast, the HI results (Table 1) showed that the S15/15 antibody clearly inhibited the hemagglutinating activity of these variants. The finding that some variants were inhibited to high titer in HI tests but not inhibited in the NI tests with the same monoclonal antibody suggests that the antigenic sites involved in hemagglutinating and neuraminidase activities may not be identical.

The parental virus reacted to high titer in the HI and NI tests with all three monoclonal antibodies suggesting that the antigenic sites involved in

Table 1. Reactivity of Antigenic Variants with Monoclonal Antibodies Used for Their Selections.

Monoclonal antibody to Sendai virus (HN)	Parental virus	HI titer with antigenic variants of Sendai virus: Variants selected with S15/15 V1	V2	V3	V4	Variants selected with S16/27 V1	V2	V3	V4	Variants selected with S18/17 V1	V2	V3
S15/15	6400	–[a]	–	–	–	–	–	400	1600	800	1200	3200
S16/27	6400	–	–	–	–	–	–	–	–	–	–	6400
S18/17	800	200	200	100	100	3200	1600	–	–	–	–	–

[a]Less than 25.

Table 2. Reactions in Neuraminidase Inhibition Tests Between Sendai Virus Antigenic Variants and Monoclonal Antibodies Used for Their Selection.

Monoclonal antibody to Sendai virus (HN)	Parental virus	NI titer with antigenic variants: S15/15 variants V1	V2	V3	V4	S16/27 variants V1	V2	V3	V4	S18/17 variants V1	V2	V3
S15/15	2290	–[a]	–	–	–	–	–	–	–	–	–	–
S16/27	4786	–	–	–	–	–	–	–	–	–	–	402
S18/17	479	66	57	191	–	–	68	–	–	–	–	–

[a]Less than 25.

hemagglutinating and neuraminidase inhibition fall within the overlapping epitopes recognized on the HN molecule by these antibodies.[16] Therefore, it is possible that the antigenic sites involved in the HI and NI activities, share some of the same amino acids.

It is not known whether the antigenic sites and the hemagglutinating and neuraminidase functional sites are the same. Indeed, it is possible that antibody binding to an antigenic site removed from the functional sites causes a conformational change in the HN molecule effecting hemagglutinating and neuraminidase activities. However, this would not change our view for two separate sites. It seems unlikely that an antibody-induced conformational change would affect one activity and not the other, if the same site on the HN molecule was responsible for attachment and neuraminidase activity, using the same neuraminic acid substrate. Therefore, these data provide evidence that suggests independent sites are involved in adsorption and neuraminidase activity.

Summary

A Sendai virus mutant *ts* for hemagglutination and adsorption to host cells was not *ts* for neuraminidase activity. Monoclonal antibodies specific for the HN molecule were able to inhibit hemagglutinating activity but not neuraminidase activity of Sendai virus antigenic variants. These results suggest that two different sites are involved in virus adsorption and neuraminidase activity.

ACKNOWLEDGMENTS
This work was supported by Research Grant AI 11949 from the National Institute of Allergy and Infectious Disease, by Childhood Cancer Center Support Grant CA 21765 from the National Cancer Institute, and by ALSAC.

References

1. Homma, M. (1975): In: *Negative Strand Viruses.* Mahy, B.W.J. and Barry, R.D. (Eds.), Academic Press, London, p. 685.
2. Nagai, Y. and Klenk, H-D. (1977): *Virology* 77:125.
3. Scheid, A., Caliguiri, L.A., Compans, R.W., and Choppin, P.W. (1972): *Virology* 50:640.
4. Scheid, A. and Choppin, P.W. (1975): In: *Negative Strand Viruses.* Mahy, B.W.J. and Barry, R.D. (Eds.), Academic Press, London, p. 177.
5. Scheid, A. and Choppin, P.W. (1973): *J. Virol.* 11:263.
6. Shimizu, K., Shimizu, Y.K., Kohama, T., and Ishida, N. (1974): *Virology* 62:90.
7. Tozawa, H., Watanabe, M., and Ishida, N. (1973): *Viology* 55:242.
8. Scheid, A. and Choppin, P.W. (1974): *Virology* 57:475.
9. Holmgren, J., Svennerholm, L., Elwing, H., Fredman, P., and Strannegard, O. (1980): *Proc. Nat. Acad. Sci. U.S.A.* 77:1947.
10. Meindl, P., Boda, G., Palese, P., Schulman, J., and Tuppy, H. (1974): *Virology* 58:457.
11. Paulson, J.C., Sadler, J.E., and Hill, R.L. (1979): *J. Biol. Chem.* 254:2120.

12. Scheid, A. and Choppin, P.W. (1974): *Virology* 62:125.
13. Compans, R.W. and Klenk, H-D. (1979): In: *Comprehensive Virology, Vol. 13.* Fraenkel-Conrat, H. and Wagner, R.R. (Eds.), Plenum Press, New York, p. 324–325.
14. Portner, A., Marx, P.A., and Kingsbury, D.W. (1974): *J. Virol.* 13:298.
15. Portner, A., Scroggs, R.A., Marx, P.A., and Kingsbury, D.W. (1975): *Virology* 67:179.
16. Portner, A., Webster, R.G., and Bean, J.W. (1980): *Virology* 103:235.
17. Aymard-Henry, M., Coleman, M.T., Dowdle, W.R., Laver, W.G., Schild, G.C., and Webster, R.G. (1973): *Bull. Wld. Hlth. Org.* 48:199.

David H. L. Bishop and Richard W. Compans, eds.
The Replication of Negative Strand Viruses

Conformation and Activity of the Newcastle Disease Virus HN Protein in the Absence of Glycosylation

T.G. Morrison, P.A. Chatis, and D. Simpson [a]

Introduction

Enveloped viruses such as vesicular stomatitis virus (VSV) and Newcastle disease virus (NDV) contain on their external surface spike structures composed of glycoproteins.[1] It has been proposed that the carbohydrate found on the glycoprotein may play a role in: (1) the stability of the protein;[2,3] (2) the intracellular transport of the glycoprotein to the appropriate membrane;[4] and/or (3) the biological activity of the protein.[3]

Studies of the role of carbohydrate found on glycoproteins have been facilitated by the use of tunicamycin, a glucosamine-containing antibiotic which prevents glycosylation of nascent glycoproteins.[5,6] The role of carbohydrate in the life cycle of VSV has been extensively studied using this antibiotic.

Role of Carbohydrate in the VSV Life Cycle

Initial studies with VSV (San Juan) suggested that carbohydrate addition played a role in the intracellular transport of G to the cell surface.[6,7] Subsequent studies[8] with another strain of VSV, the prototype strain, and G mutants derived from the strain[9] revealed that proteins with different amino acid sequences have different requirements for carbohydrate.[9] Thus, as suggested by Schlesinger et al., carbohydrate may be required by some G proteins to assume a conformation necessary for migration to the cell surface.[8]

[a] Department of Molecular Genetics and Microbiology, University of Massachusetts Medical School, Worcester, Massachusetts.

Surprisingly, carbohydrate does not play a role in viral infectivity. Leavitt et al.[7] showed that virus released from tunicamycin treated cells was fully infectious. Similarly, mutant virions including *ts*044R (a pseudorevertant of *ts*044)[9] released from tunicamycin treated cells are fully infectious (Table 1).

Thus, there is no primary, obligatory role for carbohydrate in the formation or infectivity of VSV in tissue culture. To explore these conclusions we examined another viral system—NDV. We have found that, like VSV (San Juan) carbohydrate plays a primary role in the conformation of the NDV glycoproteins. However, the ramifications of an altered conformation of the glycoproteins are different than those found in the VSV life cycle.

Synthesis and Stability of NDV Glycoproteins in the Absence of Carbohydrate

NDV, a paramyxovirus, is structurally similar to VSV and, like VSV, matures at the host cell plasma membrane.[10] NDV, however, has two glycoproteins, HN and F.[10] We have shown that NDV infected cells treated with tunicamycin synthesize all viral encoded proteins including the unglycosylated forms of the viral glycoprotein HN (hemagglutinin-neuraminidase protein) and the F_0 protein (fusion protein).[11] During maturation, the F_0 glycoprotein is cleaved to generate F_1 and F_2. This cleavage is necessary to the infectivity of the virion.[12] We have found that unglycosylated F_0 is similarly cleaved.[11]

Table 1. Infectivity of Unglycosylated Virions.

Virus	**Temperature**	**TM**	**CPM × 10^{-4} ^{35}S-methionine**	**% of control**	**pfu/ml × 10^{-7}**	**% of control**
VSV						
(prototype)	30°	−	6.40		5.10	
		+	1.90	29%	1.40	27%
	39.5°	−	3.00		0.80	
		+	0.13	4%	0.00	0%
VSV *ts*044R	30°	−	2.90		2.20	
		+	2.20	75%	1.20	60%
	39.5°	−	2.20		0.96	
		+	1.00	45%	0.40	41%
NDV-AV	37°	−	3.00		0.40	
(Australia-Victoria Strain)		+	1.70	57%	0.01	3%

Confluent monolayers (2 × 10^6 cells) of CHO cells pretreated with tunicamycin (TM) (1 μg/ml) for 2 hours were infected with virus at a multiplicity of 5 pfu/cell. VSV was labeled and purified as previously described.[9] NDV infected cells (strain AV) were radioactively labeled with ^{35}S-methionine from 5-12 hours postinfection. Virions were purified on a discontinuous sucrose gradient follwed by an equilibrium density sucrose gradient. Infectivity was determined on monolayers of CHO cells.

Furthermore, the unglycosylated glycoproteins are relatively stable in infected cells.[11]

Release of Particles From Tunicamycin Treated NDV Infected Cells

Monolayers of NDV infected chinese hamster ovary cells (CHO) treated with tunicamycin release radioactively labeled particles which have a similar size and density (not shown) as virions released from untreated-infected CHO cells or from infected eggs. The amount of radioactivity in particles released from tunicamycin treated cells is 60% that released from untreated cells (Table 1). Thus, like VSV (prototype),[8] inhibition of glycosylation does not significantly block particle release.

Particles released from tunicamycin treated cells contained, in near normal amounts, all viral polypeptides including the unglycosylated HN protein and the cleaved unglycosylated form of the fusion protein, F_1 (Figure 1).

Iodination of the Surface of NDV Infected Cells

Release of particles containing the unglycosylated HN and F_1 proteins suggested that these molecules are inserted into the plasma membrane. To identify plasma membrane associated viral proteins, the surfaces of infected cells were labeled by lactoperoxidase-mediated iodination. Iodination of the proteins at the surface of untreated NDV infected cells clearly labels both the HN protein and the F_1 protein (Figure 2B). However, surprisingly, iodination of the surfaces of NDV infected tunicamycin treated cells did not label the unglyosylated HN protein (HN_{ug}) (Figure 2C). The unglycosylated forms of F_0 (F_{0ug} and F_1 (F_{1ug}) appeared to be labeled.

Antibody Binding to Cell Surfaces

The failure to label HN_{ug} at the cell surface could indicate that the molecule is not inserted into plasma membranes. Alternatively, the molecule may be in a conformation such that it cannot be labeled by lactoperoxidase mediated iodination. We, therefore, utilized another method to identify cell surface molecules. Anti-NDV antibody was bound to the surface of intact NDV infected cells at 4°C. Excess unbound antibody was carefully washed away leaving antibody molecules which were bound to exposed viral proteins. Since internal proteins should not be accessible to antibody binding, only cell surface proteins should be precipitated in antigen-antibody complexes following cell disruption. After such a procedure with NDV infected untreated cells, the radioactive material precipitated was resolved on polyacrylamide gels in the absence of β-mercaptoethanol. No labeled NP or M was present in the antigen-antibody complexes while both labeled HN and F_0

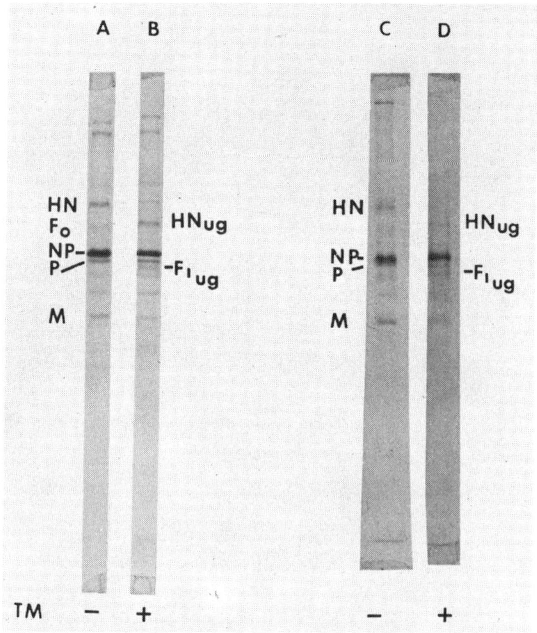

Figure 1. Polypeptides in particles released from tunicamycin treated cells. Figure shows autoradiograms of 10% polyacrylamide gels containing ^{35}S-methionine labeled proteins in purified particles released from untreated and tunicamycin treated cells (lanes C and D respectively). ^{35}S-methionine labeled proteins present in total cell extracts are shown for comparison (lanes A and B). HN_{ug}: unglycosylated hemagglutinin-neuraminidase protein; F_{1ug}: unglycosylated F_1 protein; NP: nucleocapsid protein; M: membrane protein; P: phosphorylated nucleocapsid associated protein; F_0: uncleaved fusion protein.

proteins were present (Figure 2G). No viral proteins were precipitated following the binding of preimmune sera to the surfaces of infected cells (Figure 2M). Thus, only those molecules expected to be at the surface of NDV infected cells were isolated by this procedure.

The antigen-antibody complexes isolated after binding antibody to cell surfaces were also resolved on polyacrylamide gels in the presence of β-mercaptoethanol. Again HN and some F_0 were detected. In addition the disulfide bond between F_1 and F_2, derived from cleavage of F_0, was broken[13] and F_1 was detected. There are two additional polypeptides, one of which has an apparent size between F_0 and F_1 and one which has a size similar to the M-protein. These molecules are of unknown origin and are currently being investigated.

Similarly, antibody was bound to the surfaces of tunicamycin treated infected cells. The resulting antigen-antibody complexes were isolated and the

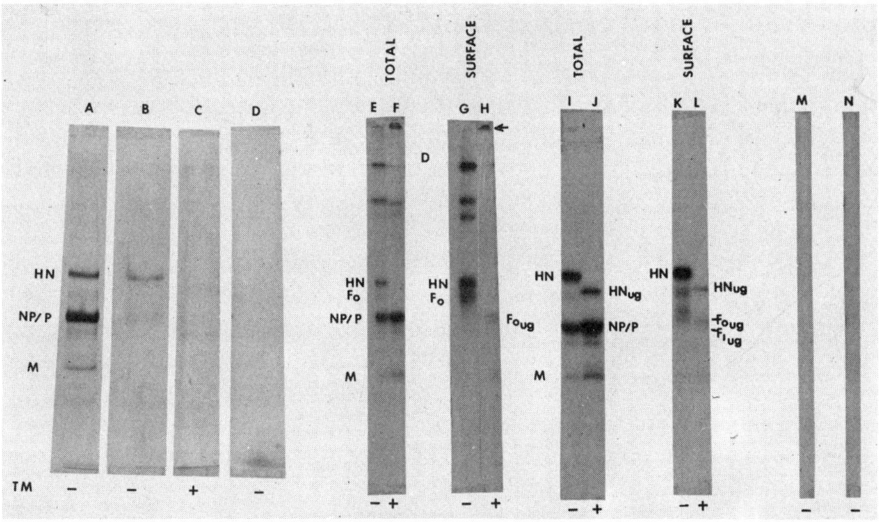

Figure 2. Cell surface NDV proteins. *Surface iodination:* Surfaces of NDV infected cells (B) and NDV infected tunicamycin treated cells (C) were radioactively labeled with ^{125}I at 6 hours postinfection, as described previously.[9] Mock infected cells (D) were labeled in parallel. Cells were lysed in 1% triton × 100 and viral protein precipitated with anti-NDV antibody.[9,11] Figure shows autoradiograms of the fixed dried 10% polyacrylamide gel. Lane A contains ^{35}S-methionine labeled NDV proteins as markers. *Cell surface proteins detected by antibody:* NDV infected cells were labeled with ^{35}S-methionine for 1.5 hours at 6 hours postinfection. NDV proteins were precipitated with anti-NDV antibody (raised in rabbits with UV inactivated egg grown virus). Total cellular NDV proteins precipitated after adding antibody to disrupted cells are shown in E, F, I and J. Cell surface NDV proteins precipitated with antibody bound only to the external surface of cells are shown in G, H, K, and L. Proteins precipitated were resolved on 10% polyacrylamide gels in the absence (E-H) or presence (I-L) of β-mercaptoethanol. Proteins precipitated from tunicamycin treated cells are shown in lanes F, H, J and L. Proteins precipitated from untreated cells are shown in lanes E, G, I and K. No proteins were precipitated when preimmune antisera was added to monolayers of infected cells (M) or when anti-NDV was added to monolayers of uninfected cells (N).

viral proteins present resolved on polyacrylamide gels. In the absence of a reducing agent, one polypeptide was detected. This polypeptide has the size of the unglycosylated F_0 protein (56,000 daltons). Unglycosylated HN does not enter the gel under these conditions [11] and, in fact, aggregates of HN can be seen at the top of the gel (Figure 2H arrow). In the presence of a reducing agent, the radioactive material migrates with sizes identical to the unglycosylated forms of the HN, F_0, and F_1 (52,000 daltons) proteins.[11] Thus, the unglycosylated glycoproteins are accessible to antibody at the cell surface. The amount of viral protein accessible by this procedure was approximately 20% the amount detected on the surface of untreated cells.

Biological Activity of Unglycosylated NDV Glycoproteins

Thus, like VSV (prototype), failure to add carbohydrate does not block the migration of glycoproteins to cell surfaces nor the formation of virus-like particles. In contrast to VSV (prototype),[7] the particles released from tunicamycin treated cells are non-infectious (Table 1). This finding suggested that one or both glycoproteins have altered activity in the absence of sugar side chains.

The HN protein has associated with it two activities—the ability to bind to red blood cells and a neuraminidase activity.[14] These activities can be assayed on monolayers of NDV infected cells. Early in infection, monolayers bind no detectable avian red blood cells. However, as infection proceeds significant numbers of red blood cells are bound (Table 2). Similarly, early in infection, monolayers have no detectable cell surface neuraminidase activity, but as infection proceeds, significant amounts of neuraminidase can be detected (Table 2).

In contrast, monolayers of tunicamycin treated infected cells have no detectable capacity to bind red blood cells (Table 2). Nor is there any detectable neuraminidase activity (Table 2). Thus, no activity associated with the HN protein can be detected at the surface of tunicamycin treated cells.

Conclusion

Results presented demonstrate that inhibition of glycosylation has no effect on the morphogenesis of NDV-AV. Near normal levels of virus-like particles are released hemoglobin in the presence of tunicamycin and these particles contain all viral polypeptides. However, in contrast to the glycosylated form

Table 2. Biological Activity of Cell Surface Unglycosylated HN Protein.

TM	Time after infection (hours)	RBC $\times 10^7$ Bound per 2×10^6 cells	Enzyme units $\times 10^3$ neuraminidase per 2×10^6 cells
−	1	0.01	0.05
	5	4.20	2.00
	8	9.10	5.50
+	1	0.01	0.05
	5	0.02	0.06
	8	0.01	0.04

Monolayers of infected cells, either untreated or treated with tunicamycin (1 μg/ml) from the beginning of infection, were assayed for their ability to bind avian red blood cells (RBC) or for neuraminidase activity. To assay hemadsorption a suspension of RBC (4%) was added to washed monolayers on ice. After 15 min, unbound RBC were washed away, and bound RBC lysed in 0.05M NH_4Cl. Released hemaglobin was measured at 560 *nm*. Neuraminidase activity was assayed as previously described.[15]

of the HN protein, HN_{ug} cannot be detected at the cell surface by lactoperoxidase-mediated iodination. HN_{ug} can, however, be detected at the cell surface by anti-NDV antibody. These results suggest that HN_{ug} is in a conformation that is different from the glycosylated form of the molecule.

An altered conformation may explain why this protein no longer exhibits the activities normally associated with the HN protein: hemadsorption and neuraminidase. Such a change in HN may account for the lack of infectivity of these particles.

Thus, like VSV (San Juan), the primary role of carbohydrate in NDV may be its influence on the conformation of the glycoproteins. However, the consequence of a change in conformation is different for the two viruses. The unglycosylated G of VSV (San Juan) is unable to migrate to the cell surface. The unglycosylated HN can reach the cell surface but the molecule is inactive in the biological activities associated with the protein.

ACKNOWLEDGMENT
This work was supported by National Institutes of Health Grant R01 AI 13847-03.

References

1. Lennard, J. and Compans, R. (1974): *Biochem. Biophys. Acta* 344:51–95.
2. Olden, K., Pratt, R.M., and Yamada, K.M. (1978): *Cell* 13:461–473.
3. Sharon, N. (1975): *Complex Carbohydrates.* Addison-Wesley Publishing Co., Reading, Massachusetts.
4. Eylar, E.H. (1966): *J. Theor. Biol.* 10:89–113.
5. Takatsoki, A., Kohro, A., and Tamura, G. (1975): *Agri. Biol. Chem.* 39:2089–2091.
6. Leavitt, R., Schlesinger, S., and Kornfeld, S. (1977): *J. Virol.* 21:375–385.
7. Gibson, R., Leavitt, R., Kornfeld, S., and Schlesinger, S. (1978): *Cell* 13:671–679.
8. Gibson, R., Schlesinger, S., and Kornfeld, S. (1979): *J. Biol. Chem.* 254:3600–3607.
9. Chatis, P.A. and Morrison, T.G. (1981): *J. Virol.* January.
10. Choppin, P.W. and Compans, R.W. (1975): In: *Comprehensive Virology IV.* Fraenkel-Conrat, H. and Wagner, R. (Eds.), Plenum Publishing Corp, New York, pp. 95–174.
11. Morrison, T. and Simpson, D. (1980): *J. Virol.* 36:171–180.
12. Scheid, A. and Choppin, P. (1974): *Virology* 57:475–490.
13. Scheid, A. and Choppin, P. (1974): *Virology* 80:54–66.
14. Scheid, A., Caliguiri, L., Compans, R., and Choppin, P. (1972): *Virology* 50:640–652.
15. Tallman, J. and Brady, R. (1972): *Methods in Enzymology XXVIII.* Ginsburg, V. (Ed.), Academic Press, New York and London, p. 826.

Copyright 1981 by Elsevier North Holland, Inc.
David H. L. Bishop and Richard W. Compans, eds.
The Replication of Negative Strand Viruses

Biochemical Properties of the NDV P Protein

Glenn W. Smith and Lawrence E. Hightower [a]

Introduction

A unique core-associated phosphoprotein designated P by analogy to a putative polymerase component of other paramyxoviruses was identified in Newcastle disease virus (NDV). Biochemical characteristics of the NDV P polypeptides including data comparing the kinetics of assembly of each of the viral proteins into mature virions are described.

Results

Characterization of P

The NDV counterpart of the paramyxoviral P-protein eluded identification in previous studies because it comigrated with other structural proteins of similar size during standard SDS-polyacrylamide gel electrophoresis (SDS-PAGE). A typical gel pattern is shown in Figure 1a. Radioactively labeled polypeptides extracted from purified virions of strain *AV* (Australia-Victoria, 1932) in the presence of mercaptoethanol separated into five size classes during SDS-PAGE. In order of increasing electrophoretic mobility, the proteins included: the large L polypeptide, the hemagglutinin-neuraminidase (HN) glycopolypeptide, the major nucleocapsid component NP, a band composed of both fusion glycopolypeptide F_1 and P polypeptide, and the small nonglycosylated membrane protein M.[1]

[a] Microbiology Section, Biological Sciences Group, University of Connecticut, Storrs, Connecticut

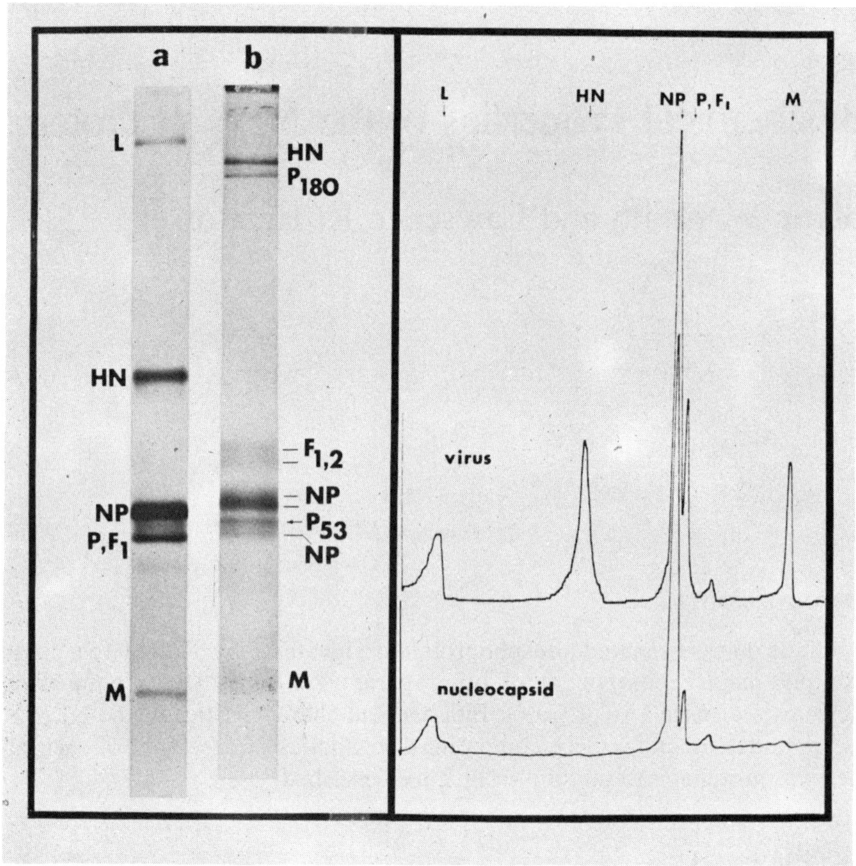

Figure 1. Structural proteins of NDV. ^{35}S-methionine labeled polypeptides were extracted from virions purified by equilibrium density ultracentrifugation and were separated by SDS-PAGE[19] under reducing (a) and nonreducing (b) conditions. Fluorograms of the gels were made according to the methods of Laskey and Mills.[12] Densitometer scans of ^{35}S-methionine labeled virion and nucleocapsid polypeptides separated on reducing gels are shown at right. Nucleocapsids were purified from 1% Triton X-100-0.4 M KCl solubilized virions as described previously.[1,20]

Separation of P and F_1 polypeptides was achieved by SDS-PAGE in the absence of mercaptoethanol (Figure 1b). Under these conditions F_1 (53 kilodaltons [K]) remains disulfide-linked to the F_2 glycopolypeptide (10K) and migrates as a 62-64K species ($F_{1,2}$). $F_{1,2}$ is the active form of the fusion glycoprotein produced by proteolytic cleavage of the precursor F_0.[2] Only the P-proteins remained in the 53K region of nonreducing gels. These nonglycosylated polypeptides were unrelated to any of the other known viral proteins as determined by peptide mapping.[1] Nonreducing gels also contained a 180K protein which was composed of three disulfide-linked 53-55K P monomers, as determined by peptide mapping and electrophoresis in a second dimension under reducing conditions. Sendai virus P-proteins are also found as disulfide-linked trimers suggesting that trimer formation may be a common feature of paramyxoviral P-proteins.[3]

Recently Collins and coworkers, using a cell-free system programmed with mRNA transcribed by virions, proved that P is encoded by the viral genome. They also determined that the gene specifying P was second in the transciptional order (NP-P-F_0-M-HN-L).[4]

In addition to the $F_{1,2}$ glycoprotein, which separated into two electrophoretically distinct forms in strain *AV* (Figure 1b), the HN glycoprotein had a lower electrophoretic mobility in nonreducing gels. The formation of HN dimers is strain-specific, but its biological significance is not known.[5] Three electrophoretic variants of NP, presumably formed by different patterns of intrachain disulfide bonds, were found on nonreducing gels. This result suggests that conformationally distinct subclasses of NP may exist in nucleocapsids.

As in other paramyxoviruses, P polypeptides were associated with ribonucleoprotein cores of NDV.[6-8] Radioactively labeled cores from Triton-salt solubilized virions were purified by equilibrium density ultracentrifugation. Densitometer scans of polypeptides extracted from whole virions and nucleocapsids and separated by reducing SDS-PAGE were compared in Figure 1. The identity of P in nucleocapsids was confirmed by peptide mapping. No 53K F-related peptides were detected indicating that F_1 was removed effectively by detergent-salt solubilization.[1]

Another method which effectively separated P from other viral proteins was isoelectric focusing according to the methods of O'Farrell (IEF-SDS).[9] Four distinct charge variants of P were separated into two size classes of polypeptides by this technique (Figure 2). We have found that P-proteins from virions, infected cells, and those synthesized in cell-free systems shared qualitatively similar two-dimensional gel patterns, although the relative amounts of each of the four spots differed in each sample. In virions, all four species of P are represented in P_{180} trimers.

Other investigators have shown that P-proteins of Sendai[10] and measles viruses[11] are phosphorylated. We analyzed ^{32}P-labeled virions of NDV by IEF-SDS and found that only the two most acidic forms of P (spots 3 and 4, Figure 2) were phosphopolypeptides.[1]

Assembly of Polypeptides into Virions

In the final stages of paramyxovirus maturation, the nucleocapsid is enveloped by the viral modified plasma membrane as the virus buds from the cell surface. We examined the kinetics with which newly synthesized polypeptides assembled into budding virions during the steady state phase of virus production.

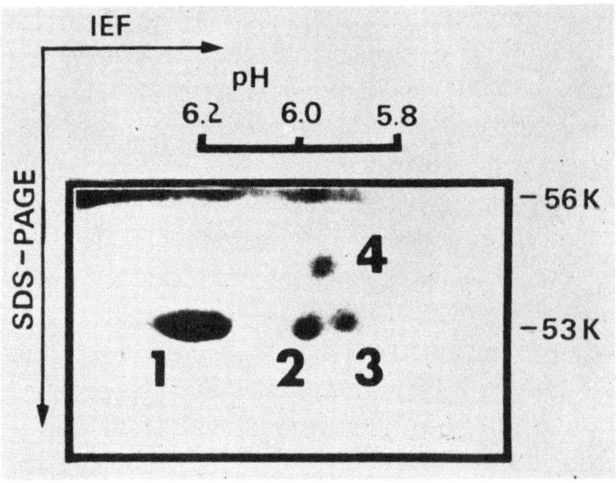

Figure 2. ^{35}S-methionine labeled polypeptides extracted from purified virions and analyzed by isoelectric focusing followed by SDS-PAGE under reducing conditions.[9] A fluorogram of the portion of the gel containing P polypeptides is shown.

Virus infected cell cultures were pulse-labeled with ^{35}S-methionine, washed, and then incubated in chase medium that contained excess non-labeled methionine. At various intervals, virions that were released into the chase medium were collected from replicate cultures and concentrated by differential ultracentrifugation. Viral polypeptides were separated on reducing gels to insure quantitative recovery of proteins (protein aggregates usually remained on the top of nonreducing gels). Individual protein bands located by fluorography[12] were excised from the gel, digested with H_2O_2, and the radioactivity quantified in a liquid scintillation spectrometer. Only 1% of the NDV polypeptides synthesized in infected cells ever appeared in virions.[13] Polypeptides made during a 10-minute pulse were available for incorporation into virions for at least 4 hours, except for P. As shown in the individual accumulation curves (Figure 3), the first radioactively labeled proteins detected in virions after a short chase period were two core proteins, P and NP, and the M-protein. During the first 10 minutes of chase the accumulation of both NP and P were especially rapid. In contrast, the incorporation of newly synthesized membrane glycoproteins HN and $F_{1,2}$ into virions was delayed for 20–30 minutes. Since P and F_1 comigrated on reducing gels, resulting in a composite accumulation curve, viral proteins were separated under nonreducing conditions to show the lag in the appearance of $F_{1,2}$ in virions (inset, Figure 3). The delay in the incorporation of HN and $F_{1,2}$ presumably indicated the time required for transport from sites of synthesis on internal membranes to the cellular surface. After a 60-minute chase period, little additional radioactive P entered virions and most of the radioactivity that accumulated in the 53K size class was contributed by F_1. Therefore, as originally shown for Sendai virus, P polypeptides of NDV were available for assembly into virions for only a relatively short time after

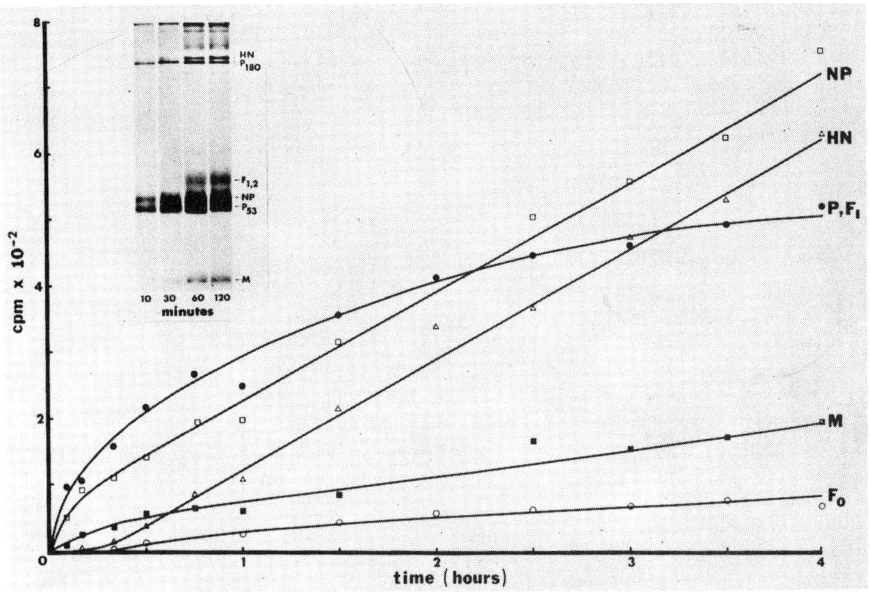

Figure 3. Kinetics of incorporation of ^{35}S-methionine labeled proteins into virions. Experimental procedure is described in the text. Inset: Radioactively labeled virions collected at the indicated chase intervals and analyzed by nonreducing SDS-PAGE.

synthesis.[13] A small amount of F_0 glycoprotein was incorporated into strain *AV* virions. However, following prolonged incubation in the culture medium the polypeptide disappeared, presumably due to the cleavage to $F_{1,2}$.

Discussion

The P-proteins of paramyxoviruses are thought to be components of the RNA dependent RNA polymerase of virions.[6,14,15] Although nucleocapsids contain only three types of viral proteins (NP, L and P), there are a variety of other enzymatic activities associated with these structures, including poly (A) adding,[16] methylating[17] and capping activities.[18] Different modifications to P polypeptides could provide NDV with a means of generating a variety of functions from a single gene product. However, recent studies by Chinchar and Portner (personal communication) indicate that nearly one-half of the Sendai P molecule can be eliminated by proteases without drastically reducing any of the enzymatic activities of nucleocapsids, raising doubts about the possible enzymatic roles for P. Alternatively, P may have a structural role in virions. It is relatively abundant in NDV virions and it is among the last proteins assembled into budding virus particles. Perhaps the various forms of P aid in the final folding of nucleocapsids into virions. Further studies of the P-proteins are required to test these possibilities.

ACKNOWLEDGMENTS
Dana Astheimer provided excellent technical assistance. We thank Michael Bratt, Peter Collins, Charles Madansky, Trudy Morrison and Mark Peeples for helpful discussions, and Jean Winters for help in preparing the manuscript. This work was supported by PHS Grant HL23588 and NSF Grant PCM78-08088. We benefited greatly from the use of a cell culture facility supported by PHS Grant CA14733.

References

1. Smith, G.W. and Hightower, L.E. (1981): *J. Virol.* 37, in press.
2. Scheid, A. and Choppin, P.W. (1977): *Virology* 80:54–66.
3. Markwell, M.A.K. and Fox, C.F. (1980): *J. Virol.* 33:152–166.
4. Collins, P.L., Hightower, L.E., and Ball, L.A. (1980): *J. Virol.* 35:682–693.
5. Moore, N.F. and Burke, D.C. (1974): *J. Gen. Virol.* 25:275–289.
6. Buetti, E. and Choppin, P.W. (1977): *Virology* 82:493–508.
7. Kingsbury, D.W., Hsu, C.H., and Murti, K.G. (1978): *Virology* 91:86–94.
8. Stallcup, K.C., Wechsler, S.L., and Fields, B.N. (1979): *J. Virol.* 30:166–176.
9. O'Farrell, P.H. (1975): *J. Biol. Chem.* 250:4007–4021.
10. Lamb, R.A. and Choppin, P.W. (1977): *Virology* 81:382–397.
11. Bissell, R.H., Waters, D.J., Seals, M.K., and Robinson, W.S. (1974): *Med. Microbiol. Immun.* 160:105–124.
12. Laskey, R.A. and Mills, A.D. (1975): *Eur. J. Biochem.* 56:335–341.
13. Portner, A. and Kingsbury, D.W. (1976): *Virology* 73:79–88.
14. Stone, H.O., Kingsbury, D.W., and Darlington, R.W. (1972): *J. Virol.* 10:1037–1043.
15. Scheid, A. and Choppin, P.W. (1974): *Virology* 57:475–490.
16. Weiss, S.R. and Bratt, M.A. (1974): *J. Virol.* 13:1220–1230.
17. Colonno, R.J. and Stone, H.O. (1975): *Proc. Natl. Acad. Sci. U.S.A.* 72:2611–2615.
18. Colonno, R.J. and Stone, H.O. (1976): *Nature* 261:611–614.
19. Laemmli, U.K. (1970): *Nature* 277:680–685.
20. Colonno, R.J. and Stone, H.O. (1976): *J. Virol.* 19:1080–1089.

David H. L. Bishop and Richard W. Compans, eds.
The Replication of Negative Strand Viruses

In Vitro Transcription of the Measles Virus Genome

Julie B. Milstien and Adele S. Seifried [a]

Transcription of the measles virus genome can be carried out *in vitro* using the appropriate precursors and nucleocapsids prepared from virions by detergent treatment.[1] The simplicity of this cell-free system compared to virus infected cells makes it an attractive alternative to study measles virus replication and transcription. However, the fidelity of *in vitro* cRNA synthesis to the events in the cell must be demonstrated before the cell-free system can be used to probe measles RNA synthesis.

RNA synthesized in measles virus infected cells was analyzed and compared to the *in vitro* transcription products with respect to size, polyadenylation, methylation and capping, and ability to be translated into proteins. The influence of reaction conditions on the *in vitro* product was studied as a basis for understanding the controls of transcription and replication.

Time Course of RNA Synthesis in Infected Cells

To find the optimum time to study RNA synthesis in measles virus infected cells, the kinetics were followed, and compared to those for expression of measles antigens as determined by immunofluorescence. In these experiments, replicate plates of Vero cells were infected at an moi of 3 with Edmonston measles virus, and labeled at regular intervals with ^{3}H-uridine in the presence of actinomycin D for 3 hours prior to harvesting or stained for immunofluorescence using measles-immune monkey serum. The labeled cell

[a] Bureau of Biologics, Bethesda, Maryland.

extracts were applied to CsCl gradients to separate measles nucleocapsids from measles mRNA,[2] the amount of RNA was measured and it was hybridized to measles virion RNA or measles mRNA to determine its polarity. Measles RNA is first detectable at 4 hours, and is maximum at 8 hours postinfection, but antigens are not detectable until 18 hours postinfection. At 4 hours postinfection, more than 50% of the RNA made is complementary to virion RNA and is associated with the mRNA pellet on CsCl. These data suggest that mRNA synthesis can be readily studied early after infection with measles virus.

Polyadenylation of RNA Transcripts

The degree of polyadenylation of RNA from infected cells was determined as the proportion sticking to oligo (dT) cellulose.[3] cRNA made using measles nucleocapsids was characterized as well. Table 1 summarizes the results, which lead to the following conclusions: (1) the mRNA fraction isolated on CsCl from infected cells is at least partially polyadenylated, while nucleocapsid and virion RNA is much less so. (However, the polyadenylated fraction of virion RNA reproducibly sediments as a 50S species in DMSO gradients); (2) the cRNA can be polyadenylated *in vitro* in reaction mixtures containing only purified viral nucleocapsids, S-adenosyl methionine (SAM), and RNA precursors. Manipulation of the SAM concentration can change the extent of polyadenylation.

Capping and Methylation of Measles RNA

Recent reports have suggested that the presence of a methylated "cap" plays an important function in viral mRNA translation.[4] For example, it has

Table 1. Binding of Measles RNA to Oligo(dT) Cellulose.

RNA source	0.5 M KCl wash, cpm	Low salt elution, cpm	% polyadenylated
Measles infected Vero cells			
mRNA fraction from CsCl	159,200	49,600	24
nucleocapsid-associated fraction from CsCl	275,200	1,160	0.4
virion RNA	36,800	2,970	6.6
Measles cRNA			
1 μM SAM	94,000	7,740	7.6
10 μM SAM	2,400	10,640	82

cRNA was synthesized by incubating viral nucleocapsids with 0.7 mM each ATP, CTP, GTP, and UTP, 0.2 mg/ml yeast RNA in 0.05 M Tris, pH 7.5, 0.5M KAo, 3 mM dithiothreitol, and 4 mM $MgAc_2$ for 2 hours at 28°. The reaction products were applied to a linear 20-40% CsCl gradient[2] and the pellet applied to an oligo(dT) cellulose column. Nucleocapsid-associated and messenger RNA fractions were isolated on similar gradients using proteinase-digested cytoplasmic extracts. Virion RNA was prepared by phenol extraction of purified virions.

been shown that VSV mRNA is "capped" prior to translation and that this reaction can occur as part of the *in vitro* polymerase reaction.[5] The presence of a methylated cap was tested for in two ways: directly, by characterizing the site of incorporation of methyl label, and indirectly, by assessing the effect of cap removal or cap binding chemicals on *in vitro* translation in a rabbit reticulocyte lysate.

Measles virus infected cells were labeled with 20 μCi/ml ^{3}H-methyl-methionine for 3 hours prior to harvesting, after treating cells with 10 mM sodium formate and 5 μg/ml actinomycin D for 1 hour. Messenger RNA selected on oligo (dT) cellulose chromatography was treated with RNase T2 and applied to a DEAE-cellulose column in 7 M urea which was eluted with a linear salt gradient.[6] Figure 1A shows the results of this procedure with measles mRNA. The arrows show the positions of cap 0, cap I, and cap II, T2 RNase products possessing the structures m ^{7}G(5′)ppp(5′)Ap; m ^{7}G(5′)ppp(5′)A^{m}pXp; and m ^{7}G(5′)ppA^{m}pX^{m}pYp respectively, determined from the elution positions of known markers. This elution pattern suggests that measles mRNA is indeed capped, with the majority of the methyl label, excluding that in mononucleotides (−2), incorporated into structures eluting in the positions of capped structures.

Figure 1B shows the results of a similar procedure applied to the polyadenylated cRNA product made by virus nucleocapsids *in vitro*. ^{3}H-SAM was used in the reaction mixture, the product was purified and selected on oligo (dT) cellulose (see Table 1), T2 RNase treated, and applied to a DE-urea column. In this case the predominant labeled structure coeluted with cap 0.

Thus direct analysis of both measles mRNA and cRNA suggests a capped and methylated 5′ end. To demonstrate that this cap structure was actually involved in translation, the mRNA was translated *in vitro* [3] under two sets of conditions: the cap was removed by treatment with $NaIO_4$ and aniline,[7] or translation was carried out in the presence of methyl glyoxal, a potent cap-dependent inhibitor of globin mRNA translation.[8] The results are shown in

Figure 1. Elution from a DE-urea column of T2 RNase-digested RNA. (A) ^{3}H-methyl-methionine labeled measles mRNA. (B) ^{3}H-SAM-labeled measles cRNA.

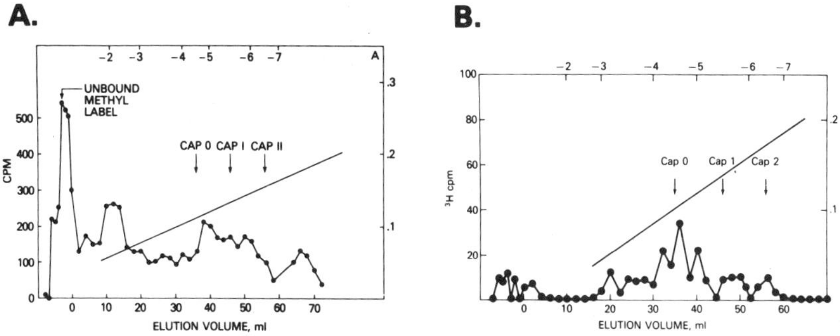

Figure 2. Chemical decapping of measles mRNA interferes with its ability to function in directing polypeptide synthesis, under conditions where polio RNA messenger activity is unaffected; methyl glyoxal abolishes the translation directing activity. Both of these results implicate the cap as necessary for translation of measles mRNA.

Measles cRNA has a cap, and is translated *in vitro* into a product resembling nucleocapsid, as shown in Figure 3. We have been unable to detect the synthesis of any other than the nucleocapsis protein from the cRNA product; however, at least a portion of the cRNA synthesized is authentic translatable RNA.

Separation of mRNAs by Size and Translation Product

Measles-specific RNA found in infected cells includes the 50 S virion RNA plus mRNAs sedimenting at 35 S and a heterogenous group at 16–20S.[9-10] To separate this broad band into its component mRNA species, ^{3}H-uridine labeled measles mRNA was selected on oligo (dT) cellulose, treated with an equal volume of formamide and sedimented on a 15–30% glycerol gradient (SW 60, 5.5 hours, 55,000 rpm). The gradient fractions were ethanol precipitated, and either glyoxal treated and run on agarose gels,[11] or translated in a reticulocyte lysate translation system.[3] Figure 4 shows the results obtained from this experiment. No translation was obtained from RNA sedimenting faster than 25 s. Discreet size classes of RNA, sedimenting at 19 s, 16 s, and

Figure 2. *In vitro* translation of RNA. (A) After chemical decapping of measles (M) and polio (P) mRNA: 0; no treatment; I: oxidation at 5′ end by $NaI0_4$; I + A: $NaI0_4$ and aniline treatment to remove cap. −; no exogenous RNA present; +; 0.8 μg measles mRNA added to reticulocyte lysate system. (B) In the presence of 0–4 mM methyl glyoxal. 8 μg measles mRNA was used in each experiment. ^{35}S-labeled purified virus was used as a measles marker.

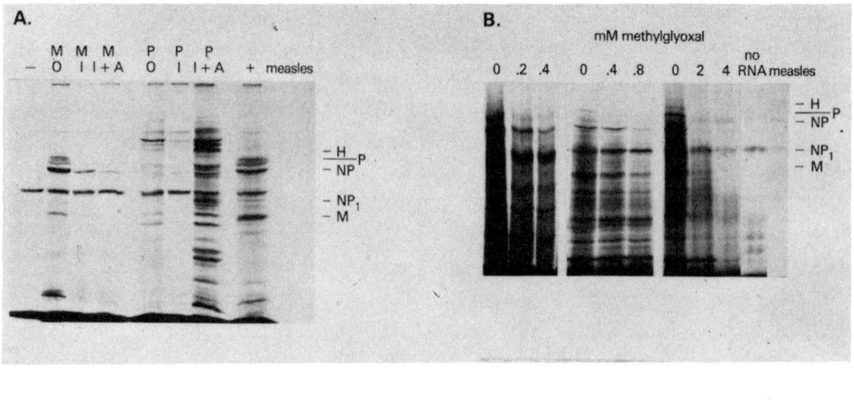

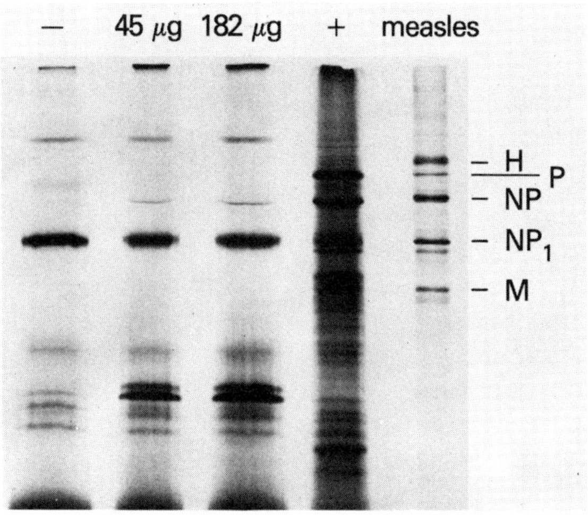

Figure 3. *In vitro* translation of 45 μg and 182 μg measles cRNA. –: no exogenous RNA added; +; 8 μg measles mRNA present in reaction. The measles marker was ^{35}S-labeled purified virus.

12 s respectively were the only species to be translated *in vitro*. The primary products were P and NP from the 16 s (fractions 27–29) class. The 12 s mRNA, the M message (fractions 31–33) is clearly separable from other mRNA species on gradient centrifugation or glyoxal gels. The largest *in vitro* polypeptide (which may be F) is also coded for by a readily separable mRNA species (fractions 24–25). It may be possible to use this information in designing specific nucleic acid probes.

Effect of Reaction Conditions on the cRNA Product

Table 1 shows that the amount of SAM in the reaction influences the polyadenylation of the cRNA products. The conditions used in this paper routinely give a significant degree of polyadenylation and mRNA-sized translatable cRNA products. Moreover, the extent of polyadenylation can be raised to roughly 100% by increasing the ATP concentration from 0.7 mM to 4 mM. In contrast, using the previously reported conditions,[1] a significant amount of the product was 50 s in size, there was no detectable polyadenylation, nor has this product been translated *in vitro*. The sense of the 50 s product has not been determined by the appropriate hybridization experiments. Nevertheless, the differential effects of reaction conditions on the characteristics of the *in vitro* product can be exploited to analyze the different functions of the measles RNA polymerase complex.

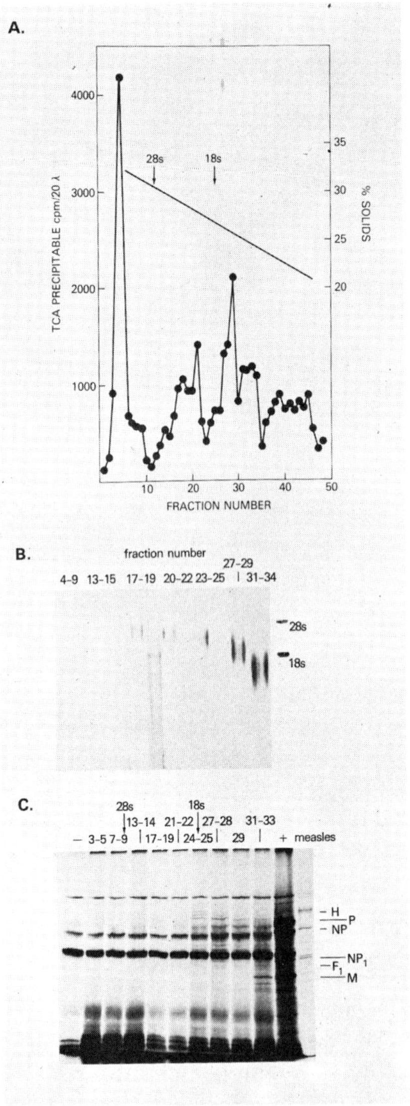

Figure 4. Separation and coding assignments of measles mRNAs. (A) Glycerol gradient centrifugation of measles mRNA. (B) Glyoxal gel electrophoresis of glycerol gradient fractions. 28 s and 18 s ribosomal RNA markers are in center track. (C) Translation in reticulocyte lysate system of glycerol gradient fractions.

References

1. Seifried, A.S., Albrecht, P., and Milstien, J.B. (1978): *J.Virol* 25:781–787.
2. Simonsen, C.C., Batt-Humphries, S., and Summers, D.F. (1979): *J. Virol.* 31:124–132.
3. Sprague, J., Eron, L.J., Seifried, A.S., and Milstien, J.B. (1979): *J. Virol.* 32:688–691.
4. Shatkin, A.J., Banerjee, A.K., and Both, G.W. (1977): In: *Comprehensive Virology,* Vol. 10. Fraenkel-Conrat, H. and Wagner, R.R. (Eds.), Plenum Press, New York, pp. 1–71.
5. Abraham, G., Rhodes, D.P., and Banerjee, A.K. (1975): *Cell* 5:51–58.
6. Moyer, S.A. and Banerjee, A.K. (1976): *Virol.* 70:339–351.
7. Rose, J.K. and Lodish, H.F. (1967): *Nature* 262:32–37.
8. Kozarich, J.W. and Deegan, J.L. (1979): *J. Biol. Chem.* 254:9345–9348.
9. Carter, C., Schluederberg, A., and Black, F.L. (1973): *Virol.* 53:379–383.
10. Hall, W.W. and ter Meulen, V. (1977): *J. Gen. Virol.* 35:497–510.
11. McMaster, G.K. and Carmichael, G.G. (1977): *Proc. Natl. Acad. Sci. U.S.A.* 74:4835–4838.

David H. L. Bishop and Richard W. Compans, eds.
The Replication of Negative Strand viruses

Transcription of the Newcastle Disease Virus Genome *in Vitro* in a Hepes Buffered System

Timothy J. Miller and Henry O. Stone [a]

Introduction

The RNA genome of negative strand viruses serves as a template for messenger RNA synthesis by a virion-associated RNA dependent RNA polymerase. Appropriate conditions have been described to assay *in vitro* the virion-associated RNA polymerase activity for a variety of rhabdo-, myxo-, and paramyxoviruses. However, the *in vitro* RNA polymerase activity of all paramyxoviruses[1-4] is significantly lower than that of the Indiana strain of vesicular stomatitis virus.[4,5] In addition, only a limited portion of the paramyxovirus genome is actually transcribed[1,2] into messenger RNA *in vitro*.

We have re-examined the optimal conditions for transcription *in vitro* by Newcastle disease virus (NDV). Transcription of the entire NDV genome was achieved *in vitro* using a Hepes rather than a Tris buffer, a higher pH (8.0) and a higher ionic strength than previously described in studies of this paramyxovirus. The total amount of RNA synthesized by NDV virions in the Hepes system is comparable to that reported for all strains of VSV except for the Indiana strain.[5] A small amount of full length (50S) plus strand was synthesized *in vitro* in the Hepes buffered system.

Materials and Methods

The Beaudette C strain of NDV was grown in embryonated eggs and purified as described previously.[3] Standard conditions for transcription assays (0.1

[a] Department of Microbiology, University of Kansas, Lawrence, Kansas.

ml) for the Hepes-KOH system contained 140 mM Hepes-KOH, pH 8.0, 120 mM NH_4Cl, 0.4 mM $MnCl_2$, 0.015% (v/v) Triton N101, 3 mM dithiothreitol (DTT), 0.7 mM ATP, 0.7 mM CTP, 0.7 mM GTP, and 0.23 mM UTP with ^{3}H-UTP or ^{32}P-UTP added to a final specific activity of 0.108 Ci/mM and 0.16 Ci/mM, respectively. The Km NTP for ATP, CTP, GPT and UTP were 82.7, 52.8, 56.9 and 28.9 M, respectively. The Tris system was modified from that described by Huang et al.[4] by use of manganese instead of magnesium. Standard conditions for the transcription assays (0.1 ml) for the Tris-HCl system contained 50 mM Tris-HCl, pH 7.5, 100 mM NH_4Cl, 0.4 mM $MnCl_2$, 0.015% (v/v) Triton N101, 3 mM DTT, 0.7 mM ATP, 0.7 mM CTP, 0.7 mM GTP, and 0.09 mM UTP with ^{3}H-UTP added to a final specific activity of 0.16 Ci/mM. All reactions contained 10–20 μg of NDV protein and were incubated at 32° for 3–6 hr. Labeled monophosphate incorporation was assayed by trichloracetic acid precipitable radioactivity as described previously.[3]

Procedures for phenol extraction, ethanol precipitation, chromatography on Sephadex G-50, and velocity sedimentation on sucrose gradients have been described previously.[3] For dimethyl sulfoxide (DMSO) treatment, the ethanol precipitated RNA was suspended in 0.1 ml TE buffer (5 mM Tris-KCl, 1 mM $EDTA_1$, pH 7.4) and 0.9 ml 99% DMSO. The RNA was incubated in DMSO for 1 hr at 37°, ethanol precipated, and resuspended in 0.4 ml of TE buffer containing 0.5% (w/v) sodium dodecyl sulfate (SDS). The RNA was centrifuged on sucrose gradients, the gradient was fractionated, and the radioactivity in the fractions was determined as previously described.[3]

For glyoxal treatment, the *in vitro* RNA product was incubated 10 min at 60° in 0.3 M glyoxal (Polysciences) and 0.2 M Hepes-KOH, pH 8.5. The glyoxal treated RNA was sedimented on glyoxal gradients as described by Hsu, Jung, and Davidson.[6]

The RNA in purified NDV virions was iodinated with ^{125}I by the Commerford [7] procedure and the reaction products were chromatographed on a Sephadex G-50 column. The iodinated RNA was then centrifuged on a 15–30% linear sucrose gradient and the 50S genome size RNA was pooled and ethanol precipitated. Using this technique, the specific activity of the virion RNA was 10^6 cpm/μg.

The ^{3}H-labeled RNA synthesized *in vitro* and ^{125}I-labeled genome RNA were annealed for 30 min at 70° in 2 × SSC. An aliquot was TCA precipitated and an additional aliquot was treated with ribonuclease A (100 μg/ml) for 1 hr at 37°, followed by TCA processing to determine the amount of ^{125}I-labeled genome resistant to ribonuclease.

Results

RNA products are virus-specific transcripts. The RNA synthesized *in vitro* in both systems annealed to unlabeled NDV genome RNA. The results of the hybridization indicated that 93.8% of the RNA synthesized in the Tris system and 98.7% of the RNA synthesized in the Hepes system could be hybridized to the virion RNA. However, three times the amount of unlabeled virion RNA was required to saturate all the RNA sequences synthesized in the Tris system than in the Hepes system, indicating that the RNA products made in the Tris system were more highly reiterated.

Sizes of RNA products synthesized in vitro. The ^{3}H-UMP labeled RNA products synthesized *in vitro* were analyzed by rate zonal centrifugation on sucrose gradients. The RNAs synthesized in the Hepes system have distinct 7.8S, 18S, and 22S peaks plus a larger RNA species (28–33S) present at the bottom of the gradient (Figure 1). This is strikingly different from the RNA synthesized in the Tris system which sedimented in the range of 12–24S. When the RNA from the Hepes system which accumulated on the cushion in the gradient in Figure 1 was denatured with DMSO and residimented at a lower speed, most of the RNA migrated in the 28–33S region of the gradient (Figure 2). This suggests that it was not an aggregate of smaller RNA species or nascent RNA chains still attached to genome RNA in a transcriptive complex.

Complexity of the RNA synthesized in vitro. Self annealing experiments were performed to determine if the RNA species synthesized *in vitro* from both systems were of a single polarity. The self annealing values ranged from 1.8 to 6.4 per cent. The low values for each fraction of RNA prove that they consist of one polarity. All hybridizations performed below were corrected for the appropriate self annealing value.

Complete transcription of the NDV genome. The total RNA synthesized *in vitro* from both systems was hybridized to ^{125}I-labeled genome RNA (Figure 3). The RNA synthesized in the Hepes system rendered 92–95% of the NDV genome resistant to ribonuclease, whereas the RNA synthesized in the Tris system protected only 34% of the genome from ribonuclease . The genome RNA has a 4.6% self annealing value. Thus, only 95.4% of the genome can be proven to be transcribed. This demonstrates that the RNA transcribed in the Hepes system represents sequences from essentially the entire NDV genome, while the transcripts synthesized in the Tris system are from a limited portion of the genome.

If the RNAs transcribed in the Hepes system (12–24S and 28–33S) are unique species, each should represent transcripts from different segments of the genome. The 12–24S and 28–33S RNAs (see Figures 1 and 2) were hybridized separately to ^{125}I-labeled genomes. After hybridization, 34% and 66% of the genome were protected from ribonuclease by the (28–33S) and (12–24S) RNA, respesctively (data not shown). When saturating amounts of 28–33S RNA (20 pmol rNTP) were mixed with increasing amounts of 12–24S RNA from the Hepes system and hybridized to ^{125}I-

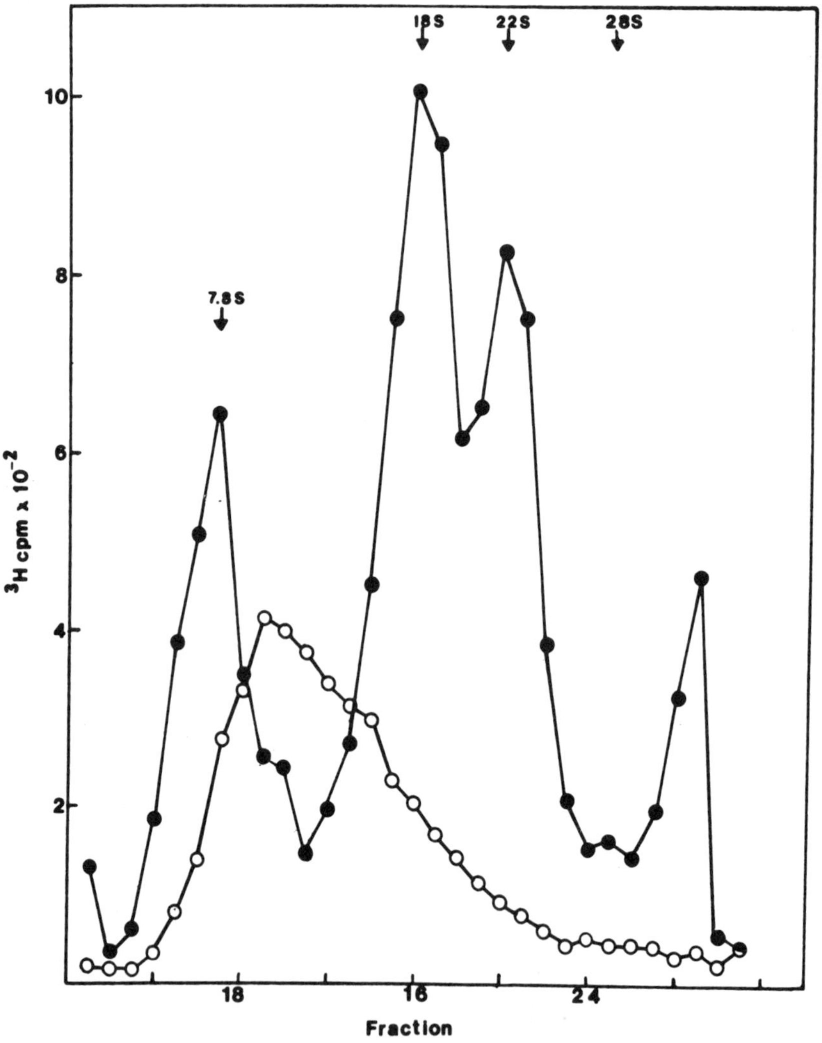

Figure 1. Sucrose gradient analysis of RNA synthesized *in vitro*. The ^3H labeled RNA products synthesized *in vitro* in the Hepes (●----●) and Tris (●----●) buffered transcription systems were phenol extracted, chromatographed on Sephadex G-50, and after denaturation in 90% dimethyl sulfoxide were centrifuged on 15–30% linear sucrose gradinets in a SW41 Beckman rotor (38,000 rev/min, 9 hr at 20°). Each fraction was assayed for ^3H-UMP labeled RNA using an aqueous scintillation cocktail. The migration is from left to right and the position of unlabeled 28S rRNA is indicated by an arrow.

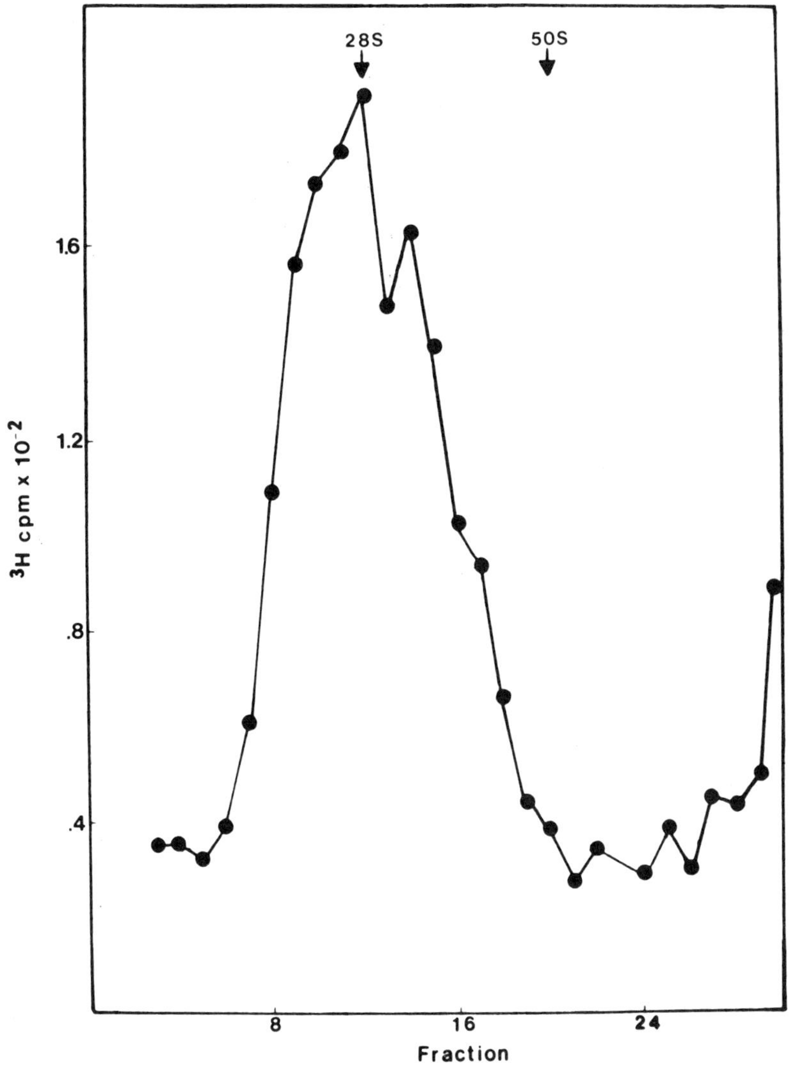

Figure 2. Sucrose gradient analysis of the 28–33S RNA synthesized *in vitro* in the Hepes system. The RNA present at the bottom of the gradient in Figure 1 was recentrifuged at a slower speed in order to prevent large RNAs from accumulating on the 60% sucrose cushion. The RNA was treated with 90% DMSO for 1 hr at 37°, ethanol precipitated and resuspended in TE buffer plus 0.5% (w/v) SDS. The sample was then centrifuged on a 15–30% linear sucrose gradient in a Beckman SW41 rotor (36,000 rev/min, 4 hr at 20°). Each fraction was assayed for ^{3}H-UMP labeled RNA using an aqueous scintillation cocktail.

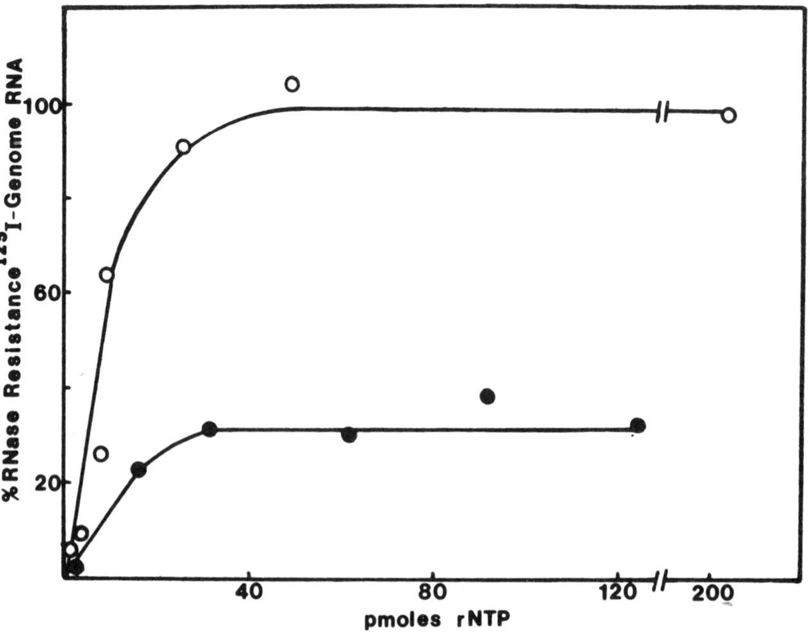

Figure 3. Annealing of virion RNA and of *in vitro* polymerase product RNA. The total RNA products symthesized *in vitro* in either the Hepes (0----0) or the Tris (●----●) polymerase systems were phenol extracted, ethanol precipitated and then were annealed to ^{125}I-labeled genome. Increasing amounts of *in vitro* synthesized ^{3}H-UMP labeled RNA were added to 1 ng of ^{125}I-labeled genome (1 ng, 1.6×10^{4} DPM) for each data point. Samples were then hybridized as described in Materials and Methods.

labeled genomes, 94% of the genome became resistent to ribonuclease treatment (Figure 4). Therefore, separately these two fractions of RNAs (12–24S and 28–33S) do not represent transcripts from the entire genome. Combined, their effect is additive which does represent transcripts from the entire genome. Thus, 12–24S and 28–33S RNAs must be transcribed from different portions of the genome, since they do not share common sequences between them. The total RNA synthesized in the Tris system does not contain any of the sequences found in the 28–33S RNA made in the Hepes system (data not shown).

Synthesis of large RNA products. Among the total RNA products in the Hepes system appear to be RNA molecules with sedimentation coefficients in excess of 33S. Some of the large RNA products sediment at 50S. To conclusively establish that RNA molecules sedimenting at 50S were intact single-stranded RNA molecules and were not aggregates of smaller molecules, the RNA products were denatured using two different procedures. One method (Figure 5a) employed dimethyl sulfoxide (DMSO) which

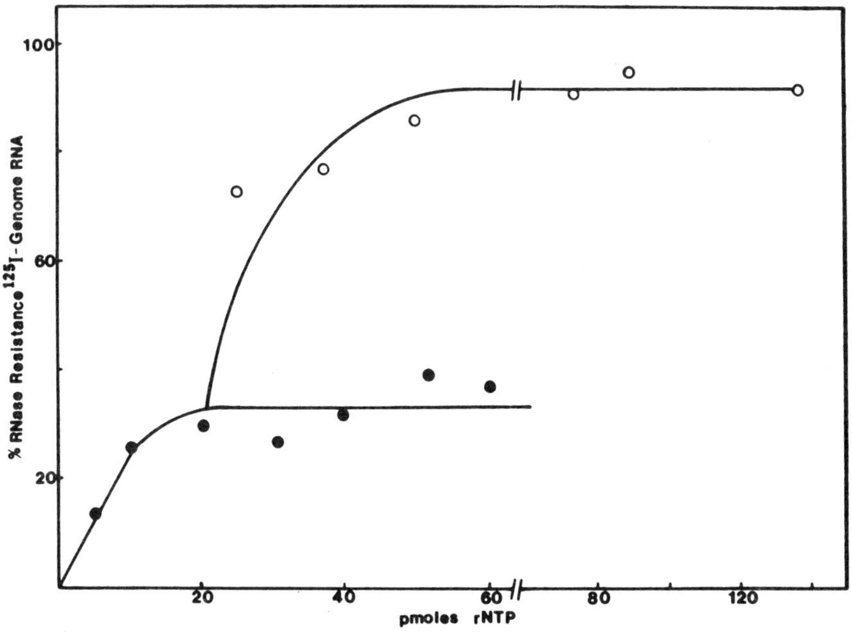

Figure 4. Annealing of virion RNA and mixtures of two size classes *in vitro* polymerase product RNA. Saturating amounts of 28–33S RNA (●----●) were mixed with increasing amounts of 12–24S RNA (○----○) and the mixtures were hybridized to ^{125}I-labeled genomes. The total products synthesized *in vitro* in the Hepes system were separated into 12–24S and 28–33S fractions as described in Figures 1 and 2. The amount of (28–33S) RNA which saturates all genome sites available (32%) (20 pmol rNTP) was mixed with increasing amounts of 12–24S RNA and then hybridized to ^{125}I-labeled genomes (1 ng, 1.6×10^{4} DPM) as described in Materials and Methods.

inhibits H-bonding.[8,9] The second method (Figure 5b) utilized glyoxal which binds to amine functions of nucleotides and inhibits H-bonding.[6] Although the sedimentation profiles differ somewhat, both methods of denaturation yield major products near 18S and additional products sedimenting at 22, 33, 42 and 50S. Thus, among the products of the RNA synthesized *in vitro* by Newcastle disease virus are RNA molecules which sediment at 50S relative to 18S ribosomal RNA under two different conditions of denaturation. The amount of radioactivity incorporated into the large RNA products is quite small. In this regard, *in vitro* RNA synthesis mimics viral RNA in synthesis in infected cells because with the Beaudette C strain no distinct peak of 50S RNA is observed in infected cells.

Transcripts co-sediment with genome RNA. To determine if the 50S RNA molecules synthesized *in vitro* were precisely the same size as genome RNA, the RNA sedimenting at 50S (fractions 27–29, Figure 5b) was pooled, denatured with glyoxal and resedimented on a glyoxal gradient. Genome RNA

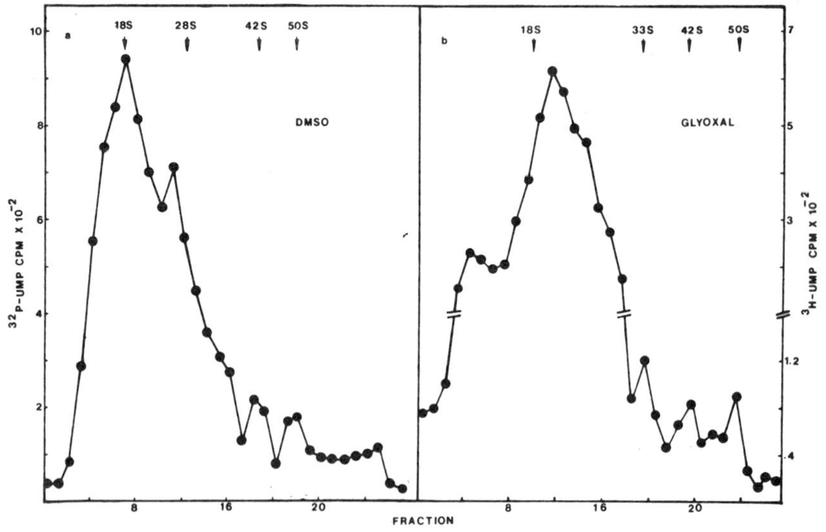

Figure 5. Velocity sedimentation analysis of the RNA products synthesized in an NDV polymerase reaction *in vitro*. The position of ^{3}H-uridine labeled 18S ribosomal RNA from chicken embryo fibroblasts was used as an internal sedimentation marker in all gradients. Viral RNA was synthesized in a 0.1 ml volume under standard *in vitro* polymerase reaction conditions as described under Materials and Methods. The *in vitro* RNA product was phenol extracted, chromatographed on a Sephadex G-50 (Pharmacia Fine Chemicals) column to remove unincorporated triphosphates, and ethanol precipitated prior to analysis by velocity sedimentation. Conditions for glyoxal and DMSO denaturation, as well as the conditions for centrifugation, are described in Materials and Methods. (a) Dimethyl sulfoxide, and (b) glyoxal denaturation of RNA synthesized *in vitro*. Sedimentation is from left to right.

was extracted from virus particles, labeled with ^{125}I and purified on an initial gradient after DMSO treatment. The ^{125}I-labeled genome RNA was then denatured with glyoxal and resedimented on a glyoxal gradient. Both the genome RNA and the 50S RNA synthesized *in vitro* migrate with identical sedimentation coefficients (data not shown).

Hybridization of in vitro labeled RNA. To conclusively establish that the 50S RNA synthesized *in vitro* was a plus strand and to determine the extent of the genome which had been transcribed into the 50S RNA molecules, hybridization experiments were performed (Figure 6). The 50S RNA molecules synthesized *in vitro* annealed to the ^{125}I-labeled genome, proving that the 50S *in vitro* product is complementary in base sequence to genome RNA and, therefore, contains plus strand RNA. Quantitation of the extent of the genome transcribed into the 50S molecules is dependent upon the self annealing values of the reactants. Self annealing values for ^{125}I-labeled genome RNA, 50S *in vitro* polymerase product and the total *in vitro* polymerase products are

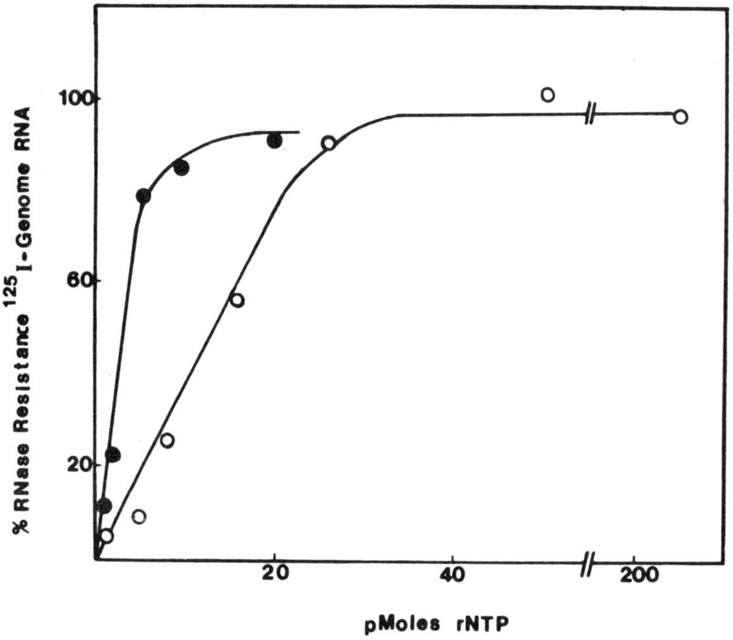

Figure 6. Annealing of Newcastle disease virus (NDV) polymerase products synthesized *in vitro* with NDV genome RNA. The total *in vitro* synthesized RNA products (12–50S) which had been labeled with ^{3}H-UMP and denatured in DMSO were pooled from fractions of a gradient and then were ethanol precipitated. The sample was resuspended in 0.03M sodium citrate pH 7.0 and increasing amounts of the ^{3}H-UMP labeled RNA were then annealed to approximately 1 ng of ^{125}I-labeled virus genome as described under Materials and Methods (○----○) The *in vitro* synthesized RNA products were denatured in DMSO and centrifuged on a 15–30% sucrose gradient as described in Figure 5. The fractions containing 50S RNA molecules were pooled and the RNA was ethanol precipitated. The 50S RNA was resuspended in 0.03 M sodium citrate pH 7.0 and increasing amounts of the ^{3}H-labeled 50S RNA were annealed to 1 ng of ^{125}I-labelled virus genomes as described under Materials and Methods (●----●).

4.6, 4.7 and 6.4%, respectively. The low self annealing values indicate that the majority of the RNA molecules in each sample are of a single polarity. A 4.6% self annealing value for genome RNA limits the maximum amount of the genome which can be proven to be transcribed into the 50S RNA synthesized *in vitro* to 95.4% because 4.6% of the genome will be protected from ribonuclease even in the absence of added plus strand RNA. After annealing and subtraction of self annealing, the 50s plus strand rendered the genome 92–95% resistant to ribonuclease (Figure 6). Thus, at least 95% of the genome is transcribed into the 50S *in vitro* RNA. Hybridization of 50S *in vitro* RNA to genome RNA was more complete at lower RNA concentrations than was the hybridization of the total *in vitro* polymerase products to genome RNA. This is to be expected since in the total product the transcripts from some portions of the genome are more abundant than from other portions,[2,10] whereas the 50S RNA product appears to be a single RNA molecule containing an essentially complete genome transcript.

Summary

Complete transcription of the Newcastle disease virus (NDV) genome was achieved *in vitro* using a Hepes buffered system. The RNA products sedimenting between 12 and 35S contained transcripts from at least 92–95% of the genome. The RNAs sedimenting at 12–24S contained sequences from approximately 68% of the genome, while RNAs sedimenting at 28–33S contained sequences from 32% of the genome. The 12–24S and the 28–33S RNAs are transcribed from different portions of the genome since they do not share common sequences. A small amount of the *in vitro* polymerase products sedimented at 50S under two different denaturing conditions. The 50S RNA molecule appears to be a full length plus strand, presumably representing the first step in genome replication.

References

1. Buetti, E. and Choppin, P.W. (1977): *Virology* 82:493–508.
2. Collins, P.L., Hightower, L.E., and Ball, L.A. (1978): *J. Virol.* 28:324–336.
3. Colonno, R.J. and Stone, H.O. (1975: *Proc. Natl. Acad. Sci. U.S.A.* 72:2611–2615.
4. Huang, A.J., Baltimore, D., and Bratt, M.A. (1971): *J. Virol.* 7:386–394.
5. Chang, S.H., Hefti, E., Obijeiski, J.F., and Bishop, D.H.L. (1974): *J. Virol.* 13:652–661.
6. Hsu, M., Jung, H., and Davidson, N. (1973): *Cold Spr. Harb. Sym. Quant. Biol.* 38:943–950.
7. Commerford, S.L. (1971): *Biochem.* 10:19993–1999.
8. Kolakofsky, D. and Buschi, A. (1975): *Virology* 66:185–191.
9. Strauss, J.H., Jr., Kelley, R.B., and Sinsheimer, R.L. (1968): *Biopolymers* 6:793–807.
10. Morrison, T., Weiss, S., Hightower, L., Spanier-Collins, B., and Bratt, M.A. (1975): In: *In Vitro Transcription and Translation of Viral Genomes.* Haenni, A.L. and Beaud, G. (Eds.), INSERM, Paris, pp. 281–290.

David H. L. Bishop and Richard W. Compans, eds.
The Replication of Negative Strand Viruses

Virus-Host Cell Interaction During the Adsorption-Penetration Phase of Paramyxovirus Infection

M.A.K. Markwell,[a] C.A. Kruse,[a] J.C. Paulson,[a] and L. Svennerholm [b]

During the initial phase of infection paramyxoviruses adsorb to the host cell surface and penetrate the plasma membrane releasing their genetic material into the cytoplasm. The interactions which occur during this phase to permit infection of the host require the participation of specific elements of the viral envelope and of the cell surface membrane. Although the general sequence of the early events of paramyxovirus infection have been known for years, the details of the adsorption-penetration process are only currently being elucidated.

Participation of the Viral Glycoproteins in the Adsorption-Penetration Phase

The envelope of paramyxoviruses is studded with hundreds of surface projections in the form of spikes about 8 nm long and spaced 8–10 nm from each other.[1] Two types of spikes, HN and F, have been identified by differences in their biological, serological, morphological and chemical properties.[2,3] The HN spike mediates attachment of the virus to the cell surface and also contains the neuraminidase activity of the virus.[3-5] The F spike is associated with the viral hemolytic and fusing activities.[6,7] The viral spikes are composed of the glycoproteins HN and F whose structure and host-dependent processing have been recently reviewed.[8] However, the actual conforma-

[a] Department of Microbiology and Department of Biological Chemistry, School of Medicine, and Molecular Biology Institute, University of California, Los Angeles, California.
[b] Department of Neurochemistry, Psychiatric Research Centre, University of Göteborg, S-422 03 Hisings Backa, Sweden.

tional grouping of the glycoproteins within their spikes which confer the abilities to bind specifically to membrane receptors and to mediate fusion of the viral envelope with the host cell membrane is still largely unknown, Our recent studies indicate that the isolated glycoproteins of Sendai virus retain the hemagglutinating and fusing capabilities of the whole virus [9] and thus do not require the lipid environment of the viral envclope for expression of these activities or to maintain the protein-protein interactions within the functional subviral unit, the spike.

The size of detergent-solubilized spikes [2,3] indicate that each consists of several of its respective HN or F proteins. The HN protein of Sendai virus forms disulfide-bonded tetramers and dimers in both egg-grown and MDBK cell-grown viruses.[10] Because only one size population of this spike has been observed by electron microscopy, these data strongly suggest that the HN spike is tetravalent. Covalent homooligomers of F containing up to 4 proteins are formed by by chemical cross-linking of egg-grown Sendai virus.[10] Although this has not yet been confirmed with virus grown in other hosts, it supports the idea that the F spike is also tetravalent. Moreover, since the size of the HN and F glycoproteins and the corresponding spikes of Sendai virus are similar to those of other paramyxoviruses, multivalent hemagglutinin and fusion spikes seem likely to be a common characteristic of this family of viruses.

The multivalency of paramyxoviruses may play an important role in their interaction with host cells during adsorption and penetration. Not only do the spikes consist of more than one protein molecule, but neighboring spikes on the same virion are in a topographically favorable arrangement to interact simultaneously with the same region of the host cell membrane to facilitate the adsorption-penetration process. Furthermore, because of its relatively large size and pleomorphic shape, a single paramyxovirus could bridge two or more discrete membrane areas. This phenomenon has been observed during the fusion of Sendai virus with the chorioallantoic membrane.[11]

Paramyxovirus Receptors on Host Cells

It is generally accepted that the HN spikes of paramyxoviruses adsorb to sialic acid-containing macromolecules on the cell surface as the first step in infection. To complete its function as a receptor the host cell component must not only specifically bind the virus but must also promote the infectious process. The question of specificity of binding is a complicated one for multivalent particles such as viruses which are theoretically capable of cooperatively binding to low affinity sites in addition to high affinity ones. In fact it has been only very recently shown by McDonald et al. in their study of receptor-purified virus that the binding of Sendai virus to continuous cell lines is a specific saturable reaction.[12] The problems of interpreting beinding data because of high backgrounds of non-specific or low affinity binding in complex cell systems led us to choose an approach to defining paramyxovirus receptors which relies on the ultimate result of infection, production of progeny virus. Extensive treatment of our host cells, Madin-Darby bovine kidney (MDBK) cells, with *Vibrio cholerae* neuraminidase (sialidase) makes them resistant to infection by Sendai virus.[13] Susceptibility to infection can be fully restored to neuraminidase-treated cells (asialo cells) by endogenous replacement of receptors or by incubation with purified sialyltransferases which elaborate specific carbohydrate sequences on glycoproteins and glycolipids. The observation that the sequence NeuAcα2,3Galβ1,3GalNAc functions as a receptor sequence for Sendai virus but the sequence NeuAcα2,6Galβ1,4GlcNAc does not,[13] indicates that the specificity of the paramyxoviruses for their host cell receptors is defined by more than just the presence of a terminal sialic acid.

The ability of liposomes containing sialic acid in the form of glycoproteins [14,15] or glycolipids [16,17] to absorb paramyxovirus and to compete with erythrocytes for virus as measured by hemagglutination inhibition suggests that both types of sialoglycoconjugates may act as host cell receptors. However, the actual functioning of cell surface gangliosides or sialoglycoproteins during infection has not been previously demonstrated. Therefore, the asialo MDBK cells previously described [13] were incubated with individual highly purified gangliosides [18] before inoculation with Sendai virus (Table 1). Incubation of the cells with GD1a, GT1b, and GQ1b produced a successful infection. All of these gangliosides contain the sequence NeuAcα2,3Galβ1,3GalNAc which had been previously shown to function as a receptor sequence for Sendai virus on host cells [13] and erythrocytes.[19] Incubation with gangliosides lacking this sequence, i.e., GM1 and GD1b, produced no detectable virus. By varying the amount of ganglioside incubated with the cells it was determined that the ganglioside GQ1b which contains the terminal sequence NeuAcα2,8NeuAcα2,3Galβ1,3GalNAc was 100-fold more effective in imparting susceptibility to infection than any of the other gangliosides tested (Table 1). These data obtained with host cells correlate closely to the specificity of binding of Sendai virus observed by Svennerholm and coworkers using polystyrene-absorbed gangliosides [20] and confirm the validity of using this simple system to describe the interactions between paramyxoviruses and their hosts.

Is adsorption of specific gangliosides sufficient to impart susceptibility to infection? Asialo MDBK cells that were incubated with 2.5 μg of GD1a at 0°C instead of 37°C produced no detectable virus (<2 HAU/ml) by 48 hr after inoculation. In a parallel set of experiments, asialo cells were incubated with GD1a for 20 min at 37°C, then half of the cells were trypsinized to remove gangliosides adsorbing to proteins on the cell surface. Both sets of cells produced a titer of 64 HAU/ml. These results are consistent with the premise

Table 1. Sendai Virus Infection of Asialo Cells after Incubation with Specific Gangliosides.

Cell preparation[a]	Added ganglioside[b] Type, Amount (μg)	Virus produced[c] (HAU/ml)
Native	None	256
Asialo	None	< 2
Asialo	GM1, 2.5	< 2
Asialo	GD1a, 2.5	64
Asialo	0.25	4
Asialo	0.025	< 2
Asialo	GD1b, 2.5	< 2
Asialo	GT1b, 2.5	64
Asialo	0.25	4
Asialo	0.025	< 2
Asialo	GQ1b, 0.25	128
Asialo	0.025	64
Asialo	0.0025	4

[a]The susceptibility to Sendai virus infection of MDBK cells which had been treated with neuraminidase (asialo cells) and then incubated with enxogenous gangliosides for 20 min at 37°C was compared to that of untreated (native) cells.

Abbreviations for gangliosides follow the nomenclature system of Svennerholm.

[c]Sendai virus production was measured by HA titer of the infected culture media 48 hr after inoculation.

that it is not only necessary for gangliosides to contain the proper sequence, but they must be an integral part of the cell membrane to function as paramyxovirus receptors.

ACKNOWLEDGMENTS
We thank Dr. Hugh C. McDonald for providing us his manuscript prior to publication. This work was supported by research Grants from the National Institutes of Health, AI-15629, AI-16165, and RR-07009 and from the Swedish Medical Research Council, 3X-627 and 16X-3382.

References

1. Kingsbury, D.W., Bratt, M.A., Choppin, P.W., Hanson, R.P., Hosaka, Y., Ter Meulen, V., Norby, E., Plowright, W., Rott, R., and Wunner, W.H. (1978): *Intervirology* 10:137–152.
2. Shimizu, K. Shimizu, Y.K., Kohama, T., and Ishida, N. (1974): *Virology* 69:90–101.
3. Scheid, A. Caliguiri, L.A. Compans, R.W., and Choppin, P.W. (1972): *Virology* 50:640–652.
4. Towaza, H., Watanabe, M., and Ishida, N. (1973): *Virology* 55:242–253.
5. Seto, J.T., Becht, H., and Rott, R. (1974): *Virology* 61:354–360.
6. Homma, M. and Ohuchi, M. (1973): *J. Virol.* 12:1457–1465.
7. Scheid, A. and Choppin, P.W. (1974): *Virology* 57:475–490.
8. Choppin, P.W. and Scheid, A. (1980): *Rev. Infect. Dis.* 2:40–61.

9. Kruse, C.A., Markwell, M.A.K., and Spector, E.B. (1980): *Fed. Proc.* 39:2050.
10. Markwell, M.A.K. and Fox, C.F. (1980): *J. Virol.* 33:152–166.
11. Morgan, C. and Howe, C. (1968): *J. Virol.* 2:1122–1132.
12. McDonald, H.C., Luderer, A.A., Hess, D.M., and Credle, L. (1980): *Anal. Biochem.* 106:127–133.
13. Markwell, M.A.K. and Paulson, J.C. (1980): *Proc. Nat. Acad. Sci. U.S.A.* 77:5693–5697.
14. Tiffany, J.M. and Blough, H.A. (1971): *Virology* 44:18–28.
15. Wu, P.-S., Ledeen, R.W., Udem, S., and Isaacson, Y.A. (1980): *J. Virol* 33:304–310.
16. Haywood, A.M. (1974): *J. Mol. Biol.* 83:427–436.
17. Sharom, F.J., Barratt, D.G., Thede, A.E., and Grant, C.W.M. (1976): *Biochem. Biophys. Acta* 455:485–492.
18. Svennerholm, L. and Fredman, P. (1980): *Biochim. Biophys. Acta.* 617:97–109.
19. Paulson, J.C., Sadler, J.E., and Hill, R.L. (1979): *J. Biol. Chem.* 254:2120–2124.
20. Holmgren, J. Svennerholm, L., Elwing, H., Fredman, P., and Strannegard, O. (1980): *Proc. Nat. Acad. Sci. U.S.A.* 77:1947–1950.

David H. L. Bishop and Richard W. Compans, eds.
The Replication of Negative Strand Viruses

Specific Inhibition of Paramyxovirus and Myxovirus Replication by Hydrophobic Oligopeptides

Christopher D. Richardson, Andreas Scheid, and Purnell W. Choppin [a]

The paramyxoviruses possess two membrane glycoproteins. The first, designated HN, is responsible for receptor-binding (hemagglutinating) and neuraminidase activities,[1–3] and the second glycoprotein, F, is involved in virus penetration, hemolysis, and virus-mediated cell fusion. The fusion protein is activated by a proteolytic cleavage by a host enzyme to yield two disulfide-linked polypeptides, F_1 and F_2.[4–10] This cleavage generates a new hydrophobic N-terminus on the F_1 polypeptide.[8,11–13] The first 20 amino acids of the F_1 polypeptides of 3 paramyxoviruses with different natural hosts have been determined[13] and are shown below.

Sendai Phe- Phe- Gly- Ala- Val- Ile- Gly- Thr- Ile- Ala- Leu- Gly- Val- Ala- Thr- Ala- Ala- Gln- Ile- Thr

SV5 Phe- Ala- Gly- Val- Val- Ile- Gly- Leu- Ala- Ala- Leu- Gly- Val- Ala- Thr- Ala- Ala- Gln- Val- Thr

NDV Phe- Ile- Gly- Ala- Ile- Ile- Gly- Gly- Val- Ala- Leu- Gly- Val- Ala- Thr- Ala- Ala- Gln- Ile- Thr

Several factors suggested that this N-terminal region was involved in the biological activities of the F-protein: the activities appear when this N-terminus is generated by cleavage; it is highly hydrophobic and could interact with membranes; the sequence is highly conserved among different paramyxoviruses, including mutants activated by different proteases;[6,7] the sequence is similar to that of an oligopeptide which inhibited cell fusion

[a] The Rockefeller University, New York, New York.

mediated by measles virus.[14–18] We therefore hypothesized that it might be possible to inhibit specifically the activities of the F-protein and thus virus replication by competitively interfering with this region of the protein. We therefore synthesized a number of oligopeptides which resembled the amino terminus of the F_1 polypeptide and tested their ability to inhibit the growth of several paramyxoviruses.[13] The results of these studies are summarized below.

Proteolytic cleavage of the hemagglutinin protein of myxoviruses also yields two disulfide-linked polypeptides, HA_1 and HA_2, and this cleavage activates infectivity.[19–22] A new N-terminus is generated on the HA_2 subunit and the amino acid sequence of this region resembles that of the F_1 polypeptide of paramyxoviruses. These initial sequences determined for three strains of influenza virus by Skehel and Waterfield [23] are shown below.

Influenza B/Lee	Gly-Phe-Phe-Gly-Ala-Ile-Ala
Influenza A_0/Bel	Gly-Leu-Phe-Gly-Ala-Ile-Ala
Influenza A_2/Singapore	Gly-Leu-Phe-Gly-Ala-Ile-Ala

The structural and functional similarities between the HA and F-proteins suggested that oligopeptides might specifically inhibit influenza virus also. Therefore oligopeptides that resembled the N-terminus of the influenza HA_2 polypeptide were synthesized and tested for their ability to inhibit influenza virus replication.

Inhibition of Paramyxoviruses by Synthetic Oligopeptides

Oligopeptides that resembled the N-terminal region of the Sendai F_1 protein [13] were synthesized, and also oligopeptides which varied in sequence, steric conformation, and the presence of amino-terminal or carboxy-terminal additions. These peptides were tested for inhibitory activity against Sendai, SV5, measles, and canine distemper (CDV) viruses and the results are summarized in Table 1.

Sendai and SV5 viruses were specifically inhibited by oligopeptides that resemble their F_1 N-termini; Z-D-Phe-L-Phe-Gly and Z-L-Phe-L-Phe were the most effective. Measles virus, however, was found to be much more sensitive to the action of these oligopeptides. Figure 1 illustrates that a 10 μM concentration of Z-D-Phe-L-Phe-Gly completely abolished plaque formation. The amino acid sequence of the N-terminus of the F_1 polypeptide of measles virus has not yet been determined because of the difficulty in obtaining enough pure polypeptide, however because of the highly conserved nature of this sequence in the other paramyxoviruses which have been examined, it seems reasonable to assume that the structure of the measles virus F_1 will not be greatly different. The peptides Z- D- Phe- L- Phe- Gly- D- Ala- D- Val- D- Ile- Gly, Z- D- Phe- L- Phe- Gly, and Z-D-Phe-L-Phe-L-(NO_2) Arg

Table 1. Inhibition by Oligopeptides of Plaque Formation by Paramyxoviruses and Myxoviruses.

Virus	Peptide[a]	50% effective concentration (*m*M)
Sendai	Z-D-Phe-L-Phe-Gly	320
	Z-D-Phe-L-Phe-L-(NO_2)Arg	540
	Z-Gly-L-Phe-L-Phe-Gly	> 1000
SV5	Z-D-Phe-L-Phe	320
	Z-D-Phe-L-Phe-Gly	320
	Z-D-Phe-L-Phe-L-(NO_2)Arg	500
Measles	Z-D-Phe-L-Phe-Gly-D-Ala-D-Val-D-Ile-Gly	0.02
	Z-D-Phe-L-Phe-Gly	0.20
	Z-D-Phe-L-Phe-L-(NO_2)Arg	0.20
	DNS-D-Phe-L-Phe-Gly	0.34
	Z-D-Phe-L-(pBr)-Phe-Gly	0.52
	Z-D-Phe-L-Tyr-Gly	9.30
	Z-D-Phe-D-Phe-Gly	10
	Z-D-Phe-L-Phe-Gly (methyl ester)	20
	Z-D-Phe-L-(Benzyl)Tyr-Gly	20
	Z-L-Phe-L-Phe-Gly	23
	Z-D-Phe-L-Phe-Gly-D-Ala-D-Val-D-Ile-Gly	130
	Z-Gly-L-Phe-L-Phe-Gly	870
Measles mutant R93	Z-D-Phe-L-Phe-Gly	20
	Z-D-Phe-L-Phe-L-(NO_2)Arg	>1000
Canine distemper	Z-D-Phe-L-Phe-Gly	1.50
	Z-D-Phe-L-Phe-L-(NO_2)Arg	>1000
Influenza A (WSN)	Z-Gly-L-Leu-L-Phe-Gly	20
	Z-Gly-L-Phe-L-Phe-Gly	53
	Z-D-Phe-L-Phe-Gly	290

[a]Z denotes a carbobenzoxy group; DNS, a dansyl group.

were the most active inhibitors of measles virus, as shown in Figure 2 and Table 1. Substitutions on the second phenylalanine with a para-OH group (to yield tyrosine), or bromine groups, radically reduced inhibition. A free carboxyl group was important for maximum inhibitory activity; Z-D-Phe-L-Phe-Gly methyl ester, was much less effective than non-esterified peptide. The presence of a carbobenzoxy (Z) or a dansyl (DNS) group on the N-terminal amino acid greatly increased activity; the hepta- and tripeptides with the same amino acid sequence but lacking these groups were far less active.

Steric configuration of the N-terminal phenylalanine was also a factor in determinine the optimum level of inhibitory activity (Table 1). Peptides containing a terminal D-Phe were much more effective than those with the L-enantiomer. Substitution of the second Phe with the D-isomer to yield Z-D-Phe-D-Phe-Gly reduced activity and Z-L-Phe-D-Phe-Gly was even less effective.

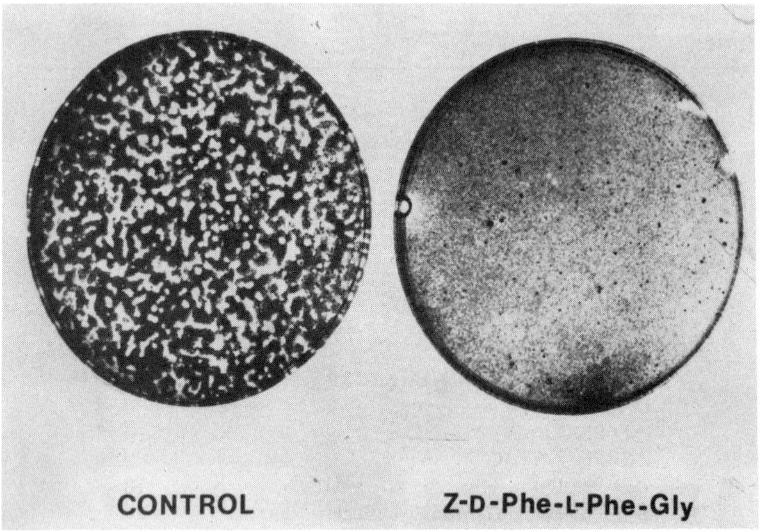

Figure 1. Plaque inhibition of measles virus by Z-D-Phe-L-Phe Gly. The peptide was present in the inoculum during virus adsorption and also in the agar overlay.

Inhibition of a Measles Virus Mutant and Canine Distemper Virus

A drug-resistant mutant of measles virus, R93 was isolated by passing the virus in the presence of Z-D-Phe-L-Phe-L-(NO_2)Arg. As shown in Table 1, the mutant was almost completely resistant to this tripeptide but remained relatively sensitive to Z-D-Phe-L-Phe-Gly. Canine distemper virus (CSV) is a member of the morbillivirus subgroup of paramyxoviruses, and the CDV F protein is immunologically related to that of measles virus.[24] CDV is sensitive to Z-D-Phe-L-Phe-Gly, but is resistant to Z-D-Phe-L-Phe-L-(NO_2)Arg (Table 1), resembling the R93 measles virus mutant. These results with CDV and a resistant measles virus mutant emphasize the amino acid sequence-specific nature of the inhibition, and indicate that the third amino acid can be an important determinant of activity.

Inhibition of Influenza Virus

Because of the structural and functional similarities between the F and HA glycoproteins of paramyxoviruses and myxoviruses discussed above, the inhibitory activity of oligopeptides that resembled the amino termini of the HA_2 polypeptides of influenza virus was investigated. Z-Gly-L-Leu-L-Phe-Gly and, to a slightly lesser extent, Z-Gly-L-Phe-L- Phe-Gly were effective inhibitors of the WSN strain of influenza virus (Table 1). These tetrapeptides mimic the N-termini of the HA_2 polypeptides of influenza A and B

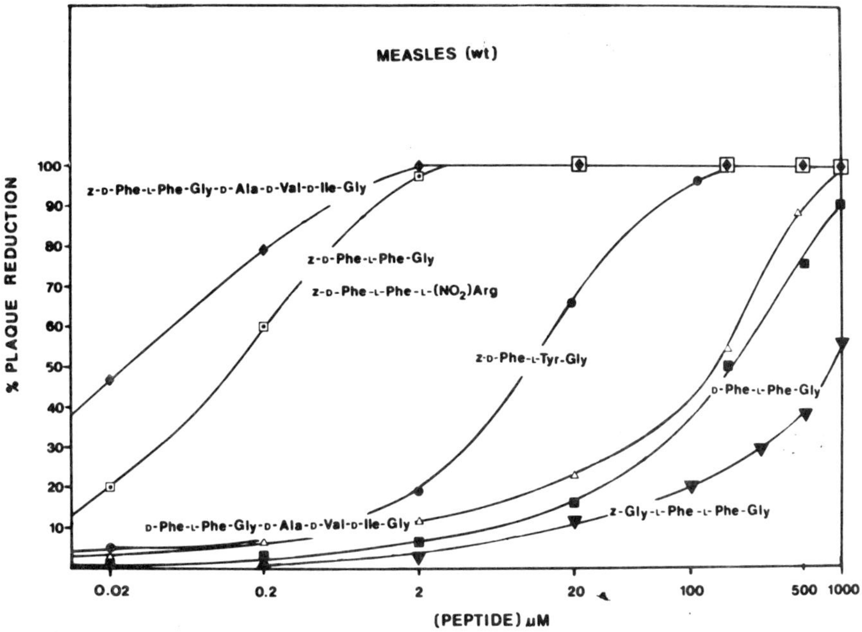

Figure 2. Inhibition of measles virus plaque formation by various oligopeptides at different concentrations. The peptides were present in the virus inoculum during adsorption and in the agar overlay. Each monolayer was inoculated with 100–200 PFU and the percentage plaque reduction was calculated relative to the number of plaques formed in control monolayers in the absence of peptides. Plaque reduction is plotted versus peptide concentration on a $\log_{10}$ scale.

viruses, respectively. In contrast, Z-D-Phe- L-Phe-Gly which resembles the N-terminus of the F_1 polypeptide of paramyxoviruses was less effective against influenza virus. These results indicate that although the effective doses were higher than those required for measles virus, there was amino acid sequence-specific inhibition of influenza virus replication by the appropriate oligopeptides.

We are continuing studies on the site and mechanism of action of these oligopeptide inhibitors which should provide further information on this possible new approach to chemical inhibition of virus multiplication, as well as additional knowledge of the mechanisms involved in the penetration of paramyxoviruses and myxoviruses into cells, in virus-induced cell fusion and hemolysis, and membrane fusion in general. The site of action of these inhibitors is currently being investigated in binding studies using labeled oligopeptides, including Z-D-Phe-L-Phe-L- (^{125}I)Tyr, Z-D-Phe-L-(^{3}H) Phe-Gly, and DNS-D-Phe-L-Phe-Gly. The parameters of the association of peptides with the host cell and/or virions are being determined.

Summary

Previous evidence suggested that the N-terminal region of the F_1 polypeptide of paramyxoviruses might be involved in virus penetration, virus-induced hemolysis and cell fusion. Oligopeptides which were structurally similar to this region and to the analogous N-terminal region of the HA_2 polypeptide of influenza viruses, were synthesized with the hypothesis that they might inhibit the biological activities of these polypeptides and viral replication. Specific oligopeptides were found to be highly effective inhibitors. Several characteristics of the inhibitory activity were defined: (1) amino acid sequences similar to those of the F_1 or HA_2 amino termini were required, indicating that inhibition was sequence specific; (2) the presence of a carbobenzoxy group on the N-terminal amino acid increased inhibitory activity; (3) esterification of the carboxy-terminus decreased activity; (4) the steric configuration of the first two amino acids affected activity. The precise mechanism of action of these oligopeptides remains to be elucidated; however, the data suggest that they act by interfering with the N-terminal region of the F_1 polypeptide of paramyxoviruses or the HA_2 polypeptide of influenza virus.

References

1. Scheid, A. Caliguiri, L.A., Compans, R.W., and Choppin, P.W. (1972): *Virology* 50:640–652.
2. Scheid, A. and Choppin, P.W. (1973): *J. Virol.* 11:263–271.
3. Tozawa, H., Watanabe, M., and Ishida, N. (1973): *Virology* 55:242–253.
4. Homma, M. and Ohuchi, M. (1973): *J. Virol* 12:1457–1465.
5. Scheid, A. and Choppin, P.W. (1974): *Virology* 57:475–490.
6. Scheid, A. and Choppin, P.W. (1975): In: *Negative Strand Viruses.* Mahy, B.W.J. and Barry. R.D. (Eds.), Academic Press, London, pp. 177–192.
7. Scheid, A. and Choppin, P.W. (1976): *Virology* 69:265–277.
8. Scheid, A. and Choppin, P.W. (1977): *Virology* 80:54–66.
9. Homma, M. (1975): In: *Negative Strand Viruses.* Mahy, B.W.J. and Barry, R.D. (Eds.), Academic Press, London, pp. 685–697.
10. Nagai, Y., Ogura, H., and Klenk, H.-D. (1979): *Virology* 69:523–538.
11. Scheid, A., Graves, M.C., Silver, S.M., and Choppin, P.W. (1978): In: *Negative Strand Viruses and the Host Cell.* Mahy, B.W.J., and Barry, R.D. (Eds.), Academic Press, London, pp. 181–193.
12. Gething, M.J., White, J.M., and Waterfield, M.D. (1978): *Proc. Nat. Acad. Sci. U.S.A.* 75:2737–2740.
13. Richardson, C.D., Scheid, A., and Choppin, P.W. (1980): *Virology* 105:205–222.
14. Nicolaides, E., DeWald, H, Westand, R., Lipnik, M., and Posler, J. (1968): *J. Med. Chem.* 11:74–79.
15. Miller, F.A., Dixon, G.J., Arnett, G., Dice, J.R., Rightsel, W.A., Schabel, F.M., and McLean, J.W. (1968): *App. Microbiol.* 16:1489–1496.
16. Norrby, E. (1977): *Virology* 44:599–608.

17. Norrby, E. and Sievertsson, H. (1974): *Antimicrobial Agents Chemother.* 5:426–430.
18. Graves, M., Silver, S.M., and Choppin, P.W. (1978): *Virology* 86:254–263.
19. Lazarowitz, S.G., Compans, R.W., and Choppin, P.W. (1971): *Virology* 46:830–843.
20. Lazarowitz, S.G., Goldberg, A.R., and Choppin, P.W. (1973): *Virology* 56:172–180.
21. Lazarowitz, S.G. and Choppin, P.W. (1975): *Virology* 68:440–454.
22. Klenk, H.-D., Rott, R., Orlich, M., and Blödorn, J. (1975): *Virology* 68:426–439.
23. Skehel, J.J. and Waterfield, M.D. (1975): *Proc. Nat. Acad. Sci. U.S.A.* 72:93–97.
24. Hall, W.W., Lamb, R.A., and Choppin, P.W. (1980): *Virology* 100:433–449.

David H. L. Bishop and Richard W. Compans, eds.
The Replication of Negative Strand Viruses

Enveloped Viruses-Cell Interactions

Frank R. Landsberger,[a] Noa Greenberg,[b] and Larry D. Altstiel[b]

Enveloped viruses such as influenza, parainfluenza, and rhabdoviruses have a membrane which is acquired during the assembly process at the plasma membrane of the host cell.[1-3] The lipid composition of the virion reflects that of the host cell plasma membrane while the protein composition is determined by the viral genome. The outer surface of an enveloped virion contains glycoproteins which form spike-like projections and are involved in attachment of the virion to the cell surface and in penetration of the host cell plasma membrane.

The initial step in virus infection involves attachment to the cell surface.[4,5] The process by which enveloped viruses penetrate the cell may involve either phagocytosis or fusion of the viral envelope with the plasma membrane. The view that for parainfluenza viruses fusion of the envelope with the plasma membrane and infectivity are related is supported by the observation that in these viruses both fusion and infectivity are activated by proteolytic cleavage of a single glycoprotein (F_0).[6,7] It has been suggested that Semliki Forest virus [8] and vesicular stomatitis virus (VSV) [9] penetrate the host cell plasma membrane by phagocytosis.

We have been investigating membrane-membrane interactions involved in enveloped virus penetration of the host cell using several approaches includ-

[a] The Rockefeller University, New York, New York.
[b] Present address: (N.G.), Department of Membranes and Ultrastructure, The Hebrew University-Hadassah Medical School, Jerusalem, Israel; and (L.D.A.), Department of Cell and Developmental Biology, The Biological Laboratories, Harvard University, Cambridge, Massachusetts.

ing spin label electron spin resonance (ESR) techniques. Two types of spin labels have been predominantly used in these studies:

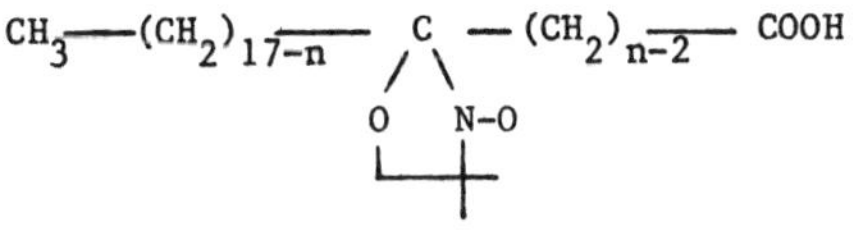

(C_n)

which is a nitroxide derivative of stearic acid and

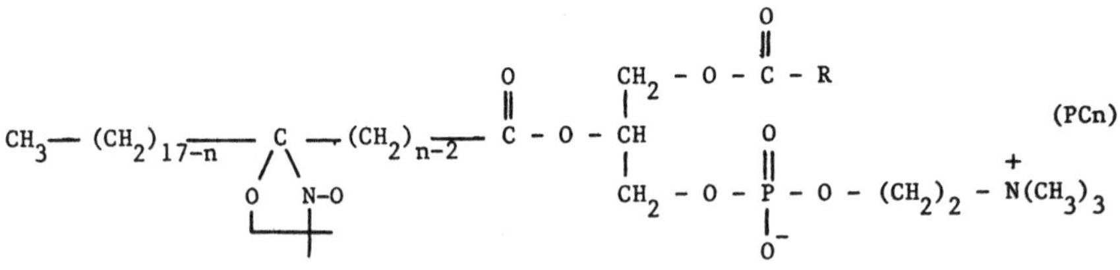

(PCn)

which is a nitroxide derivative of phosphatidylcholine. These spin labels can be incorporated into the lipid bilayer of either an enveloped virus particle or of a plasma membrane. The observed ESR spectrum of C_5 spin labeled intact tissue culture cells appears to be due to spin label in the plasma membrane.[10] The carboxyl group of the stearic acid probes incorporated into a lipid bilayer is in the vicinity of the phosphate moiety of the lipid bilayer phospholipids.[11]

Figure 1 shows the ESR spectrum of C_5 spin labeled BHK cells. The distance between the outermost peaks of the spectrum, $2A'_{zz}$, reflects the motion of the probe in the local environment in which it is inserted. $2A'_{zz}$ decreases with increasing fluidity of the environment and increases with decreasing fluidity of the local environment. (For a more detailed discussion of the relationship between $2A'_{zz}$ and the average or effective fluidity of a lipid bilayer of a biological membrane, refer to reference 12.)

Adsorption of VSV to the Cell Surface

Adsorption of VSV to BHK cells causes a detectable increase in the effective rigidity of the plasma membrane as determined by the observed increase in $2A'_{zz}$ of the ESR spectrum obtained from C_5 spin labeled BHK cells (cf Table 1). Colchicine inhibits the VSV adsorption induced change in plasma membrane bilayer structure.

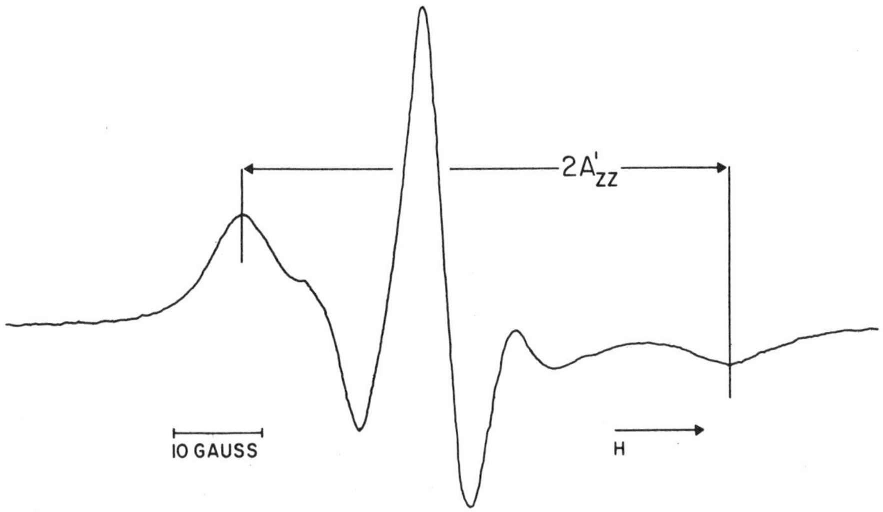

Figure 1. C_5 spin labeled BHK cells. The vertical lines are drawn through the extrema of the outermost peaks. The distance between these two lines is defined as $2A'_{zz}$

Cross-linking of virus receptors in the plane of the plasma membrane appears to be involved in the change in plasma membrane structure upon virus adsorption. VSV-infected cells release a water soluble form (G_s) of the viral glycoprotein (G)[13–15] which appears to be monovalent. Addition of G_s to BHK cells causes no detectable change in the ESR spectrum of C_5 incorporated into the BHK cell plasma membrane. However, if anti-G IgG is added to cells pretreated with G_s, a similar change in the cell surface lipid bilayer is observed as shown in Table 1. These results are in qualitative agreement with those previously obtained upon addition of Sendai or influenza virions to avian erythrocytes.[16,17] Using ^{35}S-methionine labeled VSV, the binding of VSV to L cells and BHK cells appears to show positive cooperativity. It is possible that binding cooperativity and the structural change in the plasma membrane observed upon virus binding share similar underlying mechanisms.

Parainfluenza Virus Envelope Fusion with Membranes

Spin label ESR techniques can be used to measure the rate and extent of fusion of Sendai virus envelopes with a membrane.[18] If the appropriate spin label (e.g., a particular member of the PCn group) and target membrane are chosen, the observed $2A'_{zz}$ of spin labeled virus and target bilayer will be

Table 1. VSV Binding Induced Structural Change in the Plasma Membrane of BHK Cells.

Experiment	$2A'_{zz}$ (experiment) − $2A'_{zz}$ (control) (Gauss)
BHK (control)	0.0
BHK + VSV	1.3
BHK + G_s	0.0
BHK + G_s + IgG	1.0
BHK + colchicine + VSV	0.3

The order of the reagents, e.g., BHK + colchicine + VSV, is the order in which they were added before harvesting and spin labeling with C_5. BHK cells were treated with 100 pfu/cell of VSV at 37°C for 30 min. A change greater than 0.3 gauss in the difference $2A'_{zz}$ (experiment) − $2A'_{zz}$ (control) is considered significant.

significantly different. Fusion of virus with a membrane results in the mixing of viral lipids with those of the target membrane. When PCn labeled virus is added to unlabeled membranes, the observed $2A'_{zz}$ of the ESR spectrum will change upon fusion from that characteristic of the virus envelope to that of the target membrane.

Using electron microscopy methods, Haywood[19,20] showed that Sendai virus could fuse with protein-free lipid bilayers (liposomes) containing gangliosides as virus attachment sites. Spin label methods were used to investigate the rate and extent of Sendai virus with liposomes of different compositions.

As shown in Table 2, a significant amount of fusion between egg-grown Sendai virions which has the F glycoprotein in the active form and liposomes

Table 2. Kinetics of Egg-Grown Sendai Virus Fusion with Liposomes.

Liposome Composition[a]						
Sph	PE	Chol	PC	Gang	$\alpha_f(\infty)$[b] (%)	$t_{1/2}$[c] (min)
+	+	+	+	+	60	53
	+	+	+	+	45	33
		+	+	+	59	65
			+	+	40	30
+	+	+	+	−	30	
−	−	+	+	−	30	

[a]Multilamellar liposomes were prepared by vortexing Tris buffered saline over a dried thin film containing as indicated 0.33 mg Sphingomyelin (Sph), 0.17 mg phosphatidylethanolamine (PE), 0.5 mg cholersterol (Chol), 0.73 mg phosphatidylcholine (PC), and 22 wt% beef brain gangliosides.

[b]$\alpha_f(\infty)$ is the fraction, expressed in percent, of total virus fused.

[c]$t_{1/2}$ is the reaction half-time.

can be detected. No fusion was observed between liposomes and in MDBK-grown Sendai virions which lacks fusion activity. The extensive fusion between egg-grown Sendai virions and liposomes indicates that the fusion reaction does not absolutely require the presence of host cell proteins. Ths fusion process appears to follow the form of a first-order reaction which is independent of virus concentration. The qualitative similarity in the kinetics of Sendai virus fusion with liposomes and erythrocytes [18] suggests that liposomes are appropriate models for the study of viral and cell plasma membranes. The rate and extent of fusion of Sendai virus with liposomes is dependent on the composition of the liposomes (cf Table 2). No apparent correlation between liposomal bilayer fluidity (i.e., $2A'_{zz}$ measured with PCn spin labels) and the fusion kinetics with Sendai virus has been observed.

The presence of gangliosides as virus xinding sites in the liposomes does not appear to be a prerequisite for fusion with Sendai virus. In the ESR experiments, the fusion reaction occurs in the small volume ($\sim 50\ \mu L$) of the sample holder filled with liposomes. It is possible that in these experiments, the binding step is replaced by the comparatively close contact between the virions and the liposomal surfaces.

ACKNOWLEDGMENTS
This work was supported by United States Public Health Service Grant AI-14040 from the National Institute of Health and National Science Foundation Grants PCM78-09346 and PCM79-22956. F.R.L. is an Andrew W. Mellon Foundation Fellow.

References

1. Choppin, P.W. and Compans, R.W. (1975): In: *Comprehensive Virology*. Fraenkel-Conrat, H. and Wagner, R.R. (Eds.), Plenum Press, New York, 4:95–178.
2. Compans, R.W. and Choppin, P.W. (1975): In: *Comprehensive Virology*. Fraenkel-Conrat, H. and Wagner, R.R. (Eds.), Plenum Press, New York, 4:179–252.
3. Wagner, R.R. (1975): In: *Comprehensive Virology*. Fraenkel-Conrat, H. and Wanger, R.R. (Eds.), Plenum Press, New York, 4:1–93.
4. Dales, S. (1973): *Bact. Rev.* 37:103–135.
5. Lonberg-Holm, K. and Phillipson, L. (1974): *Monogr. Virol.* 5.
6. Scheid, A. and Choppin, P.W. (1974): *Virology* 57:475–490.
7. Homma, M. and Ohuchi, M. (1973): *J. Virol.* 12:1457–1465.
8. Helenius, A., Kartenbeck, J., Simons, K., and Fries, E. (1980): *J. Cell Biol.* 84:404–420.
9. Miller, D.K. and Lenard, J. (1980): *J. Cell Biol.* 84:430–437.
10. Landsberger, F.R. and Compans, R.W. (1976): *Biochemistry* 15:2356–2360.
11. Godici, P.E. and Landsberger, F.R. (1975): *Biochemistry* 14:3927–3933.
12. Landsberger, F.R. and Altstiel, L.D. (1980): *Annals N.Y. Acad. Sci.* 348:419–425.
13. Kang, C.Y. and Prevec, L. (1971): *Virology* 46:678–690.
14. Little, S.P. and Huang, A.S. (1977): *Virology* 81:37–47.
15. Little, S.P. and Huang, A.S. (1978): *J. Virol.* 27:330–339.
16. Lyles, D.S. and Landsberger, F.R. (1976): *Proc. Natl. Acad. Sci. U.S.A.* 73:3497–3501.

17. Lyles, D.S. and Landsberger, F.R. (1978): *Virology* 88:25–32.
18. Lyles, D.S. and Landsberger, F.R. (1979): *Biochemistry* 18:5088–5095.
19. Haywood, A.M. (1974): *J. Mol. Biol.* 87:625–628.
20. Haywood, A.M. (1978): *Annals N.Y. Acad. Sci.* 308:275–280.

David H. L. Bishop and Richard W. Compans, eds.
The Replication of Negative Strand Viruses

Respiratory Syncytial Virus-Specific RNA

Dennis M. Lambert, Marcel W. Pons and Gustave N. Mbuy[a]

Introduction

Respiratory Syncytial (RS) virus, a pleomorphic enveloped virus containing a single-stranded RNA genome, is regarded as the most important etiological agent associated with lower respiratory disease in human newborns.[1] Because of a few morphological dissimilarities with parainfluenza viruses, RS virus as well as its close relative pneumonia virus of mice have been placed in a separate genus *(Pneumovirus)* in the family Paramyxoviridae.[2] A number of studies designed to characterize the genome of RS virions have been carried out and several have reported the presence of an RNA component of approximately 50S.[3,4] However, little if any information has been published concerning the strand sense of this virus. Consequently, very little information exists on the mRNAs made by this virus during the infectious cycle. We report here an investigation of the RS virus (strain Long) genome and the isolation of viral mRNAs from infected cells.

Virus Purification

Extensive purification of labeled virus, harvested at 48–72 hr postinfection when cytopathology was complete, was achieved by centrifuging successively through two discontinuous sucrose gradients followed by a linear velocity gradient centrifugation. All sucrose solutions were made in STE buffer (0.1 M NaC1, 0.01 M Tris-HC1, pH 7.0, 0.001 M EDTA) containing 1

[a] The Christ Hospital Institute of Medical Research, 2141 Auburn Avenue, Cincinnati, Ohio.

M urea. After the first discontinuous sucrose gradient (composed of 1 ml 60% sucrose and 10 ml 30% sucrose), the virus was homogenized and briefly sonicated to release virus trapped in large vesicles co-purifying with virions. This treatment increased recovery of infectious virus by about four-fold. Figure 1 shows the sedimentation profile of 5,6- ^{3}H-uridine labeled RS virions in a 20–60% linear sucrose gradient after 30 min centrifugation at 25,000

Figure 1. Sedimentation of 5,6- ^{3}H-uridine labeled RS virions in a linear sucrose gradient. Virus was grown in HEp2 cells treated at 4 hr p.i. for 1 hr, with 0.5 μg/ml actinomycin D. At 5 hr p.i., the media was replaced with drug-free media containing 20 μCi/ml 5,6- ^{3}H-uridine. At 48 hr p.i., the media was harvested and cells and debris removed by centrifuging at 1500 x g for 10 min. Virus which had been centrifuged twice through 30% to 60% discontinuous sucrose gradients was layered through 5 ml mineral oil onto a 20 ml 20% to 60% linear sucrose gradient containing STEU buffer (0.1 M NaC1, 0.1 M Tris-HC1, pH 7.2, 0.001 M EDTA, 1.0 M urea). Centrifugation was at 25,000 rpm for 0.5 hr in a SW27 rotor.

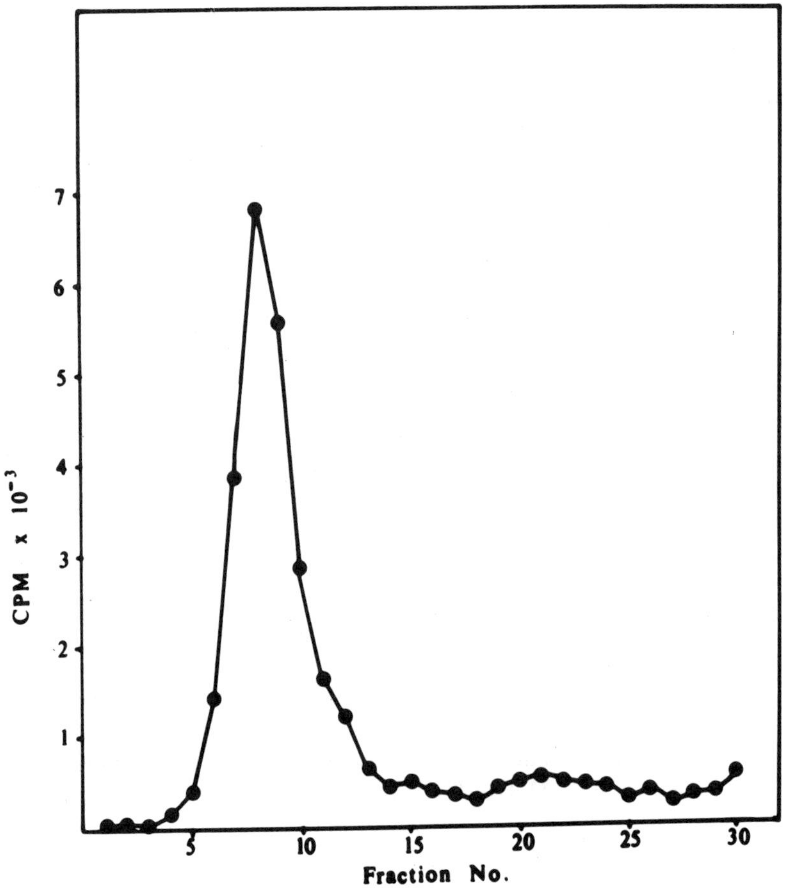

rpm in the SW27 rotor. A small amount of slower-sedimenting material was present in the gradients, but the bulk of the labeled material was in the virus peak. The purification procedure yielded 20–25% recovery of infectious virus and resulted in an overall purification of approximately 300-fold on the basis of PFU/mg protein.

Virus was grown in HEp2 cells treated with actinomycin D (0.5 μg/ml for 1 hr) because a 4–5-fold stimulation of infectious virus was observed as compared to virus from untreated cultures. In addition, labeled virus from untreated cultures contained host cell 28S and 18S RNA species which were not present as labeled components of virus preparations from actinomycin D-treated cell cultures. The precise mechanism of actinomycin D stimulation of infectious virus is unclear, but it could be due either to the inhibition of synthesis of low levels of interferon or to less competition for precursor molecules as a result of the inhibition of host RNA synthesis.

Isolation of Virion 50S RNA

Fractions containing purified RS virus labeled with 5,6- ^{3}H-uridine were pooled from linear sucrose gradients and sedimented onto a 1 ml cushion of 60% sucrose. After diluting to 4 ml in STE buffer, virus was treated with 500 μg proteinase K per ml at 37°C for 20 min; then 0.1 vol of 10X SLA buffer (5% SDS, 1.4 M $LiCl_2$, 5×10^{-2} M sodium acetate pH 4.9) was added, and the incubation was continued for 10 min. Virion RNA was extracted with phenol and chloroform: isoamyl alcohol and then precipitated at −20°C overnight with 2.5 vol 95% ethanol after adding 0.5 mg yeast tRNA.

Virion RNA, dissolved in STE buffer, was centrifuged in 5–20% linear sucrose gradients. Figure 2 demonstrates the sedimentation profile of virion RNA. The RNA sedimented predominantly as 50S with a small amount of 4S RNA present at the top of the gradient. Sedimentation values were determined by centrifugation of 18S and 28S chick embryo fibroblast ribosomal RNAs in identical gradients. Isolation of 50S RNA was also accomplished by centrifuging non-ionic detergent-lysed virions through discontinuous glycerol gradients (data not shown). Under the conditions of centrifugation, the viral ribonucleoprotein (RNP) pelleted through a 70% cushion of glycerol. RNA extracted from the RNP sedimented at 50S and no 4S material was recovered from the RNP pellet. This data suggests that the 4S RNA is not an integral part of the viral RNP and may represent packaged host cell tRNA. 50S RNA extracted from virions was 99% sensitive to ribonuclease A and T_1 digestion whereas the 4S peak was 48% resistant.

Infected Cell Cytoplasmic RNA

An investigation of cytoplasmic extracts of RS virus infected cells was undertaken to isolate and identify viral-specific RNA species. HEp2 cell

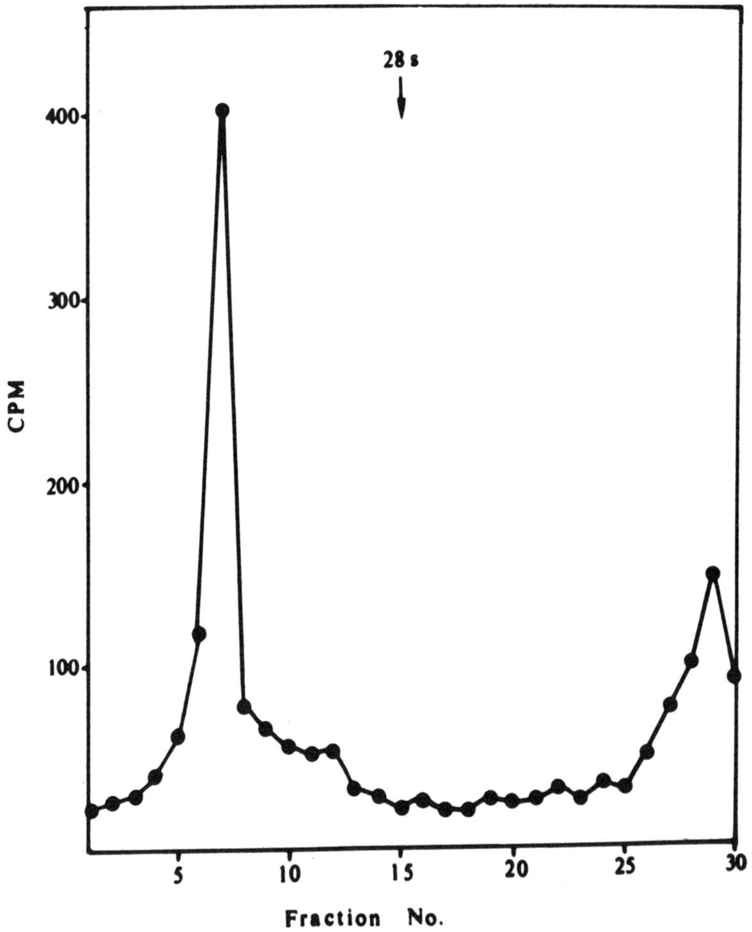

Figure 2. Sedimentation analysis of RS virion RNA. After treatment of the virus sample with 500 μg/ml proteinase K at 37°C for 20 min followed by addition of 0.5% SDS, 5,6- ^{3}H-uridine labeled virion RNA was extracted from purified virus with equal volumes of phenol and chloroform: isoamyl alcohol (24:1). RNA was precipitated overnight at −20°C with 2.5 vol 95% ethanol after adding 500 μg yeast tRNA as carrier. RNA dissolved in STE buffer was layered over 5–20% (w/v) linear sucrose gradients containing 0.1 M NaC1, 0.001 M EDTA, and 0.01 M Tris-HC1, pH 4.8. Samples were centrifuged in a SW27 rotor at 17,000 rpm for 17 hr.

monolayers were infected with RS virus at a multiplicity of 2 PFU per cell, and mock infected cultures were used as uninfected controls. At 4 hr postinfection the cells were treated with 5 μg/ml actinomycin D, and one hr later, 100 μCi of 5,6- ^{3}H-uridine per ml was added without removing the drug. At 23 hr postinfection, all cultures were harvested and the RNA extracted with phenol, buffered at pH 9.0, and then precipitated with 2 volumes of 95% ethanol. Figure 3 shows the sedimentation profile of cytoplasmic RNA species after centrifugation on a 5–20% sucrose gradient. Infected HEp2 cell

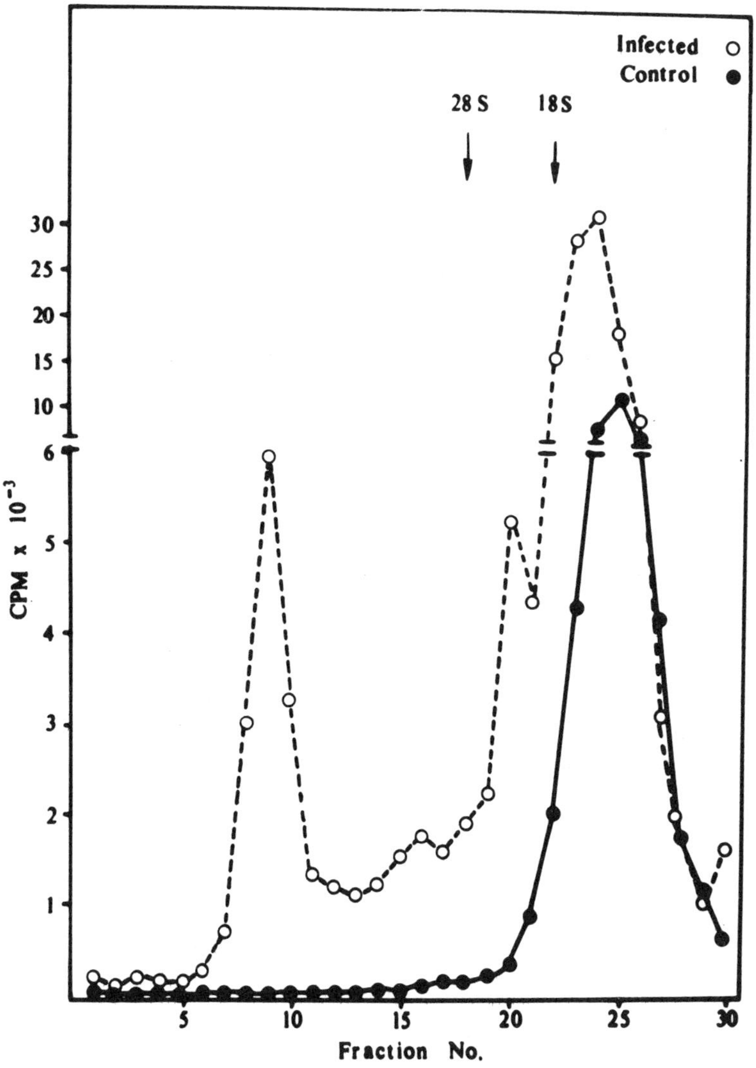

Figure 3. Sedimentation analysis of the cytoplasmic RNA of RS virus infected and mock-infected HEp2 cells at 23 hr p.i. in 5–20% sucrose gradients. The mock-infected and infected cell cultures were treated with 5 μg/ml actinomycin D at 4 hr p.i. and at 5 hr p.i. 100 μCi/ml 5,6- ^{3}H-uridine was added. RNA was extracted as described in the text. The conditions of centrifugation were the same as described for Figure 2.

cytoplasm at 23 hr postinfection contained a major peak of 12S to 24S RNA, as well as 32S and 50S RNA species. The control cell cytoplasmic extracts labeled under identical conditions contained small amounts of 12S material.

In similar experiments infected and control (labeled and unlabeled) cultures were harvested, the cytoplasmic RNA extracted and chromatographed on oligo (dT)-cellulose columns. The polyadenylated RNA was collected and then concentrated by alcohol precipitation. The final recovery ratio of polyadenylated RNA from infected versus control cells was 80:1.

Unlabeled polyadenylated RNA was dissolved in water and various amounts were annealed with 5 μl ^{3}H-uridine labeled 50S virion RNA. The RNA mixture was adjusted to 2X SSC (1X is 0.15 M NaCl and 0.015 sodium citrate) and 50% formamide and annealed at 50°C. The amount of hybridization was determined by treating one half of each sample with RNase A and T_1. The results are given in Table 1. The 11% self-annealing exhibited by 50S vRNA indicated the presence of a small amount (5.5%) of complementary RNA in the 50S RNA population. Only about 60% of the labeled vRNA was driven into double stranded molecules by increasing amounts of mRNA. This amount of annealing was not increased by longer annealing times. It should be noted that if the annealing volumes of the samples were decreased, the amount of annealing also decreased. The reasons for this unusual phenomenon are unclear; however, based on these results and those of other experiments, we feel that it may be due to intermolecular aggregation.

PAGE of Viral mRNAs

Polyacrylamide gel electrophoresis (PAGE) of RS mRNAs was carried out in order to determine both the number of messengers and their molecular weights. Messenger RNAs from infected cells were isolated on oligo (dT) cellulose as described above. After ethanol precipitation, the RNA was annealed to poly-dT and treated with ribonuclease H to remove poly A tracts.[5] The mRNA preparation was then subjected to PAGE in gels containing 6 M urea.[6]

The results of two different mRNA preparations are shown in Figure 4. Eight discrete bands having molecular weights of 8.5×10^5 and 4.8×10^5, 3.55×10^5, 3.4×10^5, 3.2×10^5, 1.05×10^5, 0.87×10^5 and 0.54×10^5

Table 1. Annealing of Unlabeled RS mRNA to ^{3}H-Uridine Labeled 50S Virion RNA.[a]

		CPM[b]		
	mRNA vol (μul)	**With RNase**	**Without RNase**	**% resistant**
Infected-cell mRNA	(0)	114	970	11.8
	(0.1)	198	937	21.1
	(1.0)	472	940	50.2
	(10)	546	964	56.6
	(20)	574	986	58.2
Uninfected-cell mRNA	(20)	126	930	13.6

[a] Approximately 2000 cpm of 50S vRNA and unlabeled RS mRNA in a total volume of 500 μl of 2X SSC and 50% formamide were annealed at 50°C for 46 hr.

[b] Values are the means of four determinations.

were resolved (Figure 4A,B). The molecular weight determinations are an average of seven different mRNA preparations using in each case influenza WSN RNA as a marker (Figure 4C). These eight mRNAs have a total mol wt of 2.59×10^{6} daltons which represents approximately 45% of the total RS genome (assuming 50S = 5.73×10^{6} daltons), leaving approximately 3.14×10^{6} daltons unaccounted for. If one assumes that this amount represents one piece of RNA, it would have a sedimentation coefficient of about 37.5S. We have repeatedly seen a 30–33S RNA in sucrose gradients of infected cell cytoplasmic extracts (Figure 3). It is probably that the 33S RNA does not enter 2.5% gels under the conditions at which they were run, and this would account for the apparent discrepancy in estimating the coding capacity of the RS virus genome.

Summary

The results presented here confirm the reports of others that RS virus has a 50S genome (approximately 5.7×10^{6} daltons). In addition, we have established that RS virus is a negative strand RNA virus, the genome of which

Figure 4. Polyacrylamide gel electrophoresis of 5,6- ^{3}H-uridine labeled RS virus mRNAs extracted from infected cells. Lanes A and B are the mRNA patterns obtained from two different experiments. Lane C is Influenza (strain WSN) marker RNA. mRNAs were prepared as described in the text and subjected to electrophoresis in 2.5% polyacrylamide gels containing 6 M urea as previously described.[6]

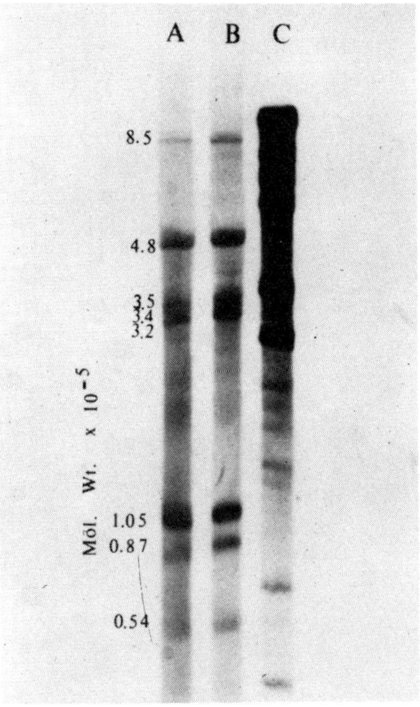

encodes for at least eight messenger RNAs having molecular weights ranging from 8.5×10^5 to 0.54×10^5 daltons. These data suggest that RS virus should contain a viral-specific polymerase. Although the mRNAs are complementary to the virion RNA, they account for only 45% of the theoretical coding capacity of an RNA of 5.7×10^6 daltons. These properties in conjunction with the general morphological features of the virus support its classification in the paramyxovirus family.

ACKNOWLEDGMENTS
This work was supported by U.S. Public Health Service Grant No. AI-14133 of the National Institute of Allergy and Infectious Diseases. The authors wish to express their appreciation to Dr. Olga Rochovansky for her many helpful suggestions.

References

1. Chanock, R.M. (1970): *Science* 169:248–256.
2. Kingsbury, D.W., Bratt, M.A., Choppin, P.W., Hanson, R.P., Hosaka, V., Ter Meulen, V., Norrby, E., Plowright, W., Rott, R., and Wunner, W.H. (1978): *Intervirology* 10:137–152.
3. Wunner, W.H., Faulkner, G.P., and Pringle, C.R. (1975): In: *Negative Strand Viruses,* Vol. 1. Mahy, B.W.J. and Barry, R.D. (Eds.), Academic Press, New York, pp. 193–201.
4. Zhdanov, V.M., Dreizin, R.S., Yankevici, O.D., and Astakhova, A.K. (1974): *Rev. Roum. Virol.* 23:277–280.
5. Etkind, P.R., Buchhagen, D.L., Herz, C., Broni, B.B., and Krug, R.M. (1977): *J. Virology* 22:346–352.
6. Pons, M.W. (1976): *Virology* 69:789–792.

Copyright 1981 by Elsevier North Holland, Inc.
David H. L. Bishop and Richard W. Compans, eds.
The Replication of Negative Strand Viruses

Separation and Characterization of the RNAs of Human Respiratory Syncytial Virus

Yung Huang, Nancy Davis and Gail W. Wertz[a]

Respiratory syncytial (RS) virus is the most important cause of serious lower respiratory tract illness during infancy and early childhood. Analysis of the proteins of RS virus and the proteins synthesized in RS virus infected cells have been reported by several groups.[1-3] The analysis of RS virion RNA and the viral-specific messenter mRNAs synthesized in infected cells, however, has been initiated [4-7] but is far from complete. In particular, other than the observation that seven genetic complementation groups exist,[8] there is no firm biochemical data describing the number of genes on the viral genome or the number and identification of the gene products specified. We have used analytical and preparative gel electrophoretic techniques under denaturing conditions which allow high resolution separaration and molecular weight estimation of a broad range of RNA sizes and poly(U) Sepharose chromatography to analyze the RNA of RS virions and the RNA from RS virus infected HEp2 cells.

Kinetics of RNA Synthesis

Monolayers of HEp2 cells were mock infected or infected with RS virus, strain A2, at a multiplicity of one plaque forming unit (PFU) per cell. At 11 hours postinfection, cells were treated with 5 μg/ml of actinomycin D and two hours later, cells were exposed to ^{3}H-uridine. Infected cells were harvested at two-hour intervals thereafter and assayed for total actinomycin

[a]Department of Bacteriology and Immunology, School of Medicine, University of North Carolina, Chapel Hill, North Carolina.

D-resistant acid precipitable radioactivity. Incorporation increased above the uninfected control levels beginning at 15 hours postinfection, reached maximum levels at 19 to 21 hours and declined to almost background levels at 24 hours (Figure 1).

Analysis of Intracellular RNA Species

The RNAs synthesized in mock or RS virus infected actinomycin D-treated cells exposed to ^{3}H-uridine from 16 to 20 hours postinfection were analyzed. Infected cells were harvested, cytoplasmic or nuclear extracts were prepared, the RNA was separated from proteins by phenol extraction and analyzed by electrophoresis in 1.5% agarose, 6 M urea gels at pH 3.0. No RNA synthesis was detectable in the nuclei of either infected or uninfected actinomycin D-treated cells. A fluorogram of the dried gel displaying RNAs present in the cytoplasm of RS virus infected or uninfected cells is shown in Figure 2. Eight major actinomycin D-resistant species of RNA could be detected in the cytoplasm of RS virus infected cells (Figure 2, lanes 2, 3). No RNA was labeled during this period in uninfected actinomycin D-treated control cultures.

Figure 1. Kinetics of RNA synthesis in RS virus infected cells. HEp2 cells were mock or RS infected at multiplicity of infection (moi) of 1 PFU/cell. At 11 hours postinfection (p.i.), the cells were treated with 5 μg/ml actinomycin D and two hours later labeled with ^{3}H-uridine. Cultures were harvested at the times indicated and radioactivity in trichloroacetic acid (TCA) precipitable material assayed. The radioactivity in mock infected cultures has been subtracted.

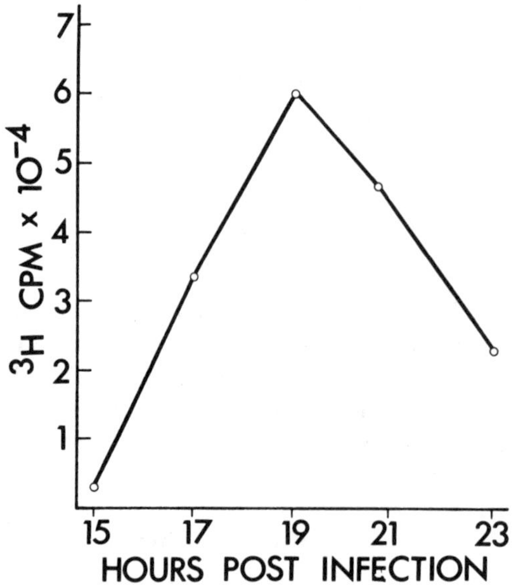

Wunner et al.[7] and Lambert et al.[6] have presented data which suggests that RS virus is a negative strand virus. One of the characteristics of a negative strand virus is that the virion contains an RNA dependent RNA polymerase that is capable of symthesizing messenger RNA in the presence of inhibitors of protein synthesis. The synthesis of the genome size RNAs, however, is sensitive to inhibition of protein synthesis.

The effect of inhibition of synthesis by cycloheximide (100 μg/ml) on RNA synthesis in RS virus infected cells was examined. The data presented in Figure 2, lane 1 show that the seven smaller RNAs synthesized in untreated cells (Figure 2, lane 2) are made in the presence of cycloheximide in undiminished amounts. The synthesis of the largest RNA, band 8, however, it sensitive to inhibition by cycloheximide. This finding makes the large RNA (band 8) a likely candidate for the genome sized RNA and suggests that the seven smaller RNAs are mRNAs. Further studies to confirm these findings are reported below.

Figure 2. Electrophoresis of RS virus cytoplasmic RNAs. HEp2 cells were mock or RS virus infected at moi of 1 PFU/cell. At 14 hours p.i., the cells were treated with 5 μg/ml of actinomycin D or as otherwise stated and labeled two hours later with ^{3}H-uridine. The cells were harvested at 20 hours p.i. RNA was extracted as described by Wertz and Davis[9] and analyzed by electrophoresis in 1.5% agarose, 6 M urea gels. A fluorogram of the dried gel is shown. Lane 1, RS virus infected cells treated with 100 μg/ml cycloheximide in addition to actinomycin D; lane 2, RS virus infected cells, actinomycin D-treated only; lane 3, RS virus actinomycin D-treated infected cells, RNA sample heated at 65°C for one minute prior to electrophoresis; lane 4, mock infected, actinomycin D-treated control culture.

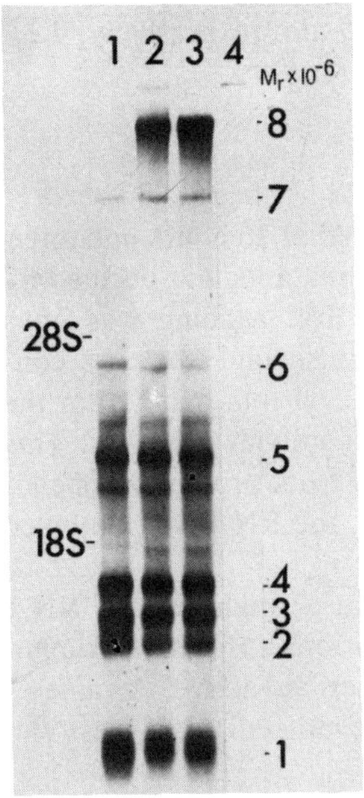

Analysis of Polyadenylated RNAs

A signal feature of messenger RNAs of eukaryotic cells is the presence of 3′-terminal polyadenylate residues. The presence of this structure on a molecule allows one to select the polyadenylated mRNA molecules out of total cytoplasmic RNA by chromatography on poly(U)Sepharose columns. When total cytoplasmic RNA from RS virus infected cells was chromatographed on poly(U)Sepharose, approximately 70% of the total ^{3}H-uridine labeled material specifically bound to the resin. This material was eluted from the column with 90% formamide and analyzed by electrophoresis in agarose-urea gels. As shown by the data in Figure 3, lane 4, all seven of the major small RNA species were retained by the poly(U)Sepharose. This result shows that these RNAs have poly(A) residues and suggests that they may be messenger RNAs.

The non-binding material (Figure 3, lane 5) was rechromatographed on a second poly(U)Sepharose column. Only 3% of the material that did not bind in the first cycle of chromatography bound on the second cycle. This indicates that approximately 97% of the polyadenylated RNAs were bound in a single cycle of chromatography.

The approximate molecular weights of the seven polyadenylated RNAs were calculated based on their relative electrophoretic mobility in agarose-urea gels, using BHK cell 18S and 28S ribosomal RNAs and the VSV mRNAs electrophoresed in adjacent lanes of the same gel as internal standards. Values for the VSV mRNAs were based on data provided by Dr. J. Rose (personal communication) from sequence analysis of cloned cDNA of the messages. A value of 125 poly(A) residues was added in order to use the mRNAs as standards. By this method approximate molecular weight values for the polyadenylated RNAs identified as bands 1–7 are, respectively, 0.21, 0.33, 0.37, 0.41, 0.67, 1.1 and 2.56×10^{6} daltons.

Analysis of Virion RNA

RS virions grown in the presence of ^{3}H-uridine added at 16 hours postinfection were harvested from infected cell culture supernatants and purified by two cycles of banding in sucrose gradients. The first banding was on a discontinuous 30%, 35%, 60% step gradient, and the second was on a continuous 30% to 60% sucrose velocity gradient. The visible band from the second gradient comigrated with the single peak of radioactive label. This material was collected and the RNA was separated from protein by phenol extraction. After recovery by ethanol precipitation, the RNA was analyzed by electrophoresis on agarose-urea gels.

The data presented in Figure 3, lane 1 show that a single labeled RNA species (band 8) was isolated from the banded virions. The approximate molecular weight of this RNA was calculated as described above. A value of 4.88×10^{6} daltons was obtained for the approximate molecular weight of the RNA extracted from the RS virus.

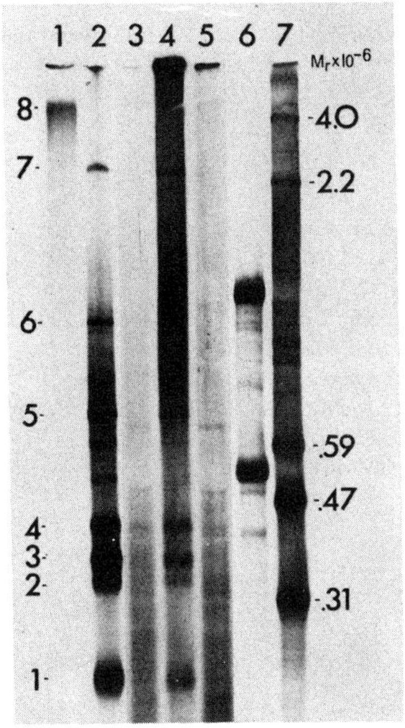

Figure 3. Electrophoretic analysis of RS virion RNA and RS virus cytoplasmic RNAs separated by poly(U)Sepharose chromatography. RS virions were grown HEp2 cells infected at a moi of one in the presence of ^{3}H-uridine added at 16 hours postinfection. The supernatant fluids from the infected cultures were harvested at 48 hours when the cells showed extensive syncytial cytopathic effect. Cell debris was removed by centrifugation at 11,000 x g for 10 minutes and virus was pelleted by centrifugation at 65,000 x g for 90 minutes. The virus pellet was resuspended in Tris buffer containing 10% sucrose and banded on a discontinuous 30%, 35%, 60% sucrose gradient for one hour at 150,000 x g. The viral band at the 35%, 60% interface was collected, diluted with Tris buffer, sonticated and centrifuged in a 30% to 60% continuous sucrose velocity gradient at 150,000 x g for one hour. The visible virus band was collected, the RNA separated by phenol extraction, recovered by ethanol precipitation and analyzed by electrophoresis in a 1.5% agarose, 6 M urea gel (lane 1).

RNA from HEp2 infected, actinomycin D-treated, cytoplasmic extracts was prepared as described in the legend to Figure 2 and fractionated by chromatography in poly(U)Sepharose. The RNA was then analyzed by electrophoresis on 1.5% agarose, 6 M urea gels. The RNAs were visualized by fluorography. Lane 2, total cytoplasmic RNA from RS virus infected HEp2 cells treated with actinomycin and cycloheximide; lane 3, total cytoplasmic RNA from mock infected actinomycin D-treated HEp2 cells; lane 4, RS infected cell cytoplasmic RNA which bound to and then was eluted from poly(U)Sepharose by 90% formamide; lane 5, RS infected cell cytoplasmic RNA which did not bind to poly(U)Sepharose; lane 6, BHK cell 18 and 28S ribosomal RNA markers; lane 7, VSV RNAs co-electrophoresed in marker RNAs.

ACKNOWLEDGMENTS
This work was supported by Public Health Service Grants AI-12464 and AI-15134 from the National Institute of Allergy and Infectious Disease.

References

1. Levine, S., Peeples, M., and Hamilton, R. (1977): *J. Gen. Virol.* 37:53–63.
2. Wunner, W. and Pringle, C. (1976): *Virology* 73:228–243.
3. Cash, P., Pringle, C., and Preston, C. (1979): *Virology* 92:375–384.
4. Hodes, D., Schauf, V., and Chanock, R. (1974): *Proc. Soc. Exp. Biol. Med.* 146:287–290.
5. Baldridge, P. and Senterfit, L. (1976): *Proc. Soc. Exp. Biol. Med.* 151:684–688.
6. Lambert, D., Pons, M., and Mbuy, G. (1980): (*Abstr.*) *Amer. Soc. Micro.*, p.255.
7. Wunner, W., Faulkner, G., and Pringle, C. (1975): In: *Negative Strand Viruses.* Mahy, B. and Barry, R. (Eds.), Academic Press, New York, pp. 193–201.
8. Gimenez, H.B. and Pringle, C.R. (1978): *J. Virol.* 27:459–464.
9. Wertz, G. and Davis, N. (1979): *J. Virol.* 30:108–115.

David H.L. Bishop and Richard W. Compans, eds.
The Replication of Negative Strand Viruses

Translation of the Separated Messenger RNAs of Newcastle Disease Virus

Peter L. Collins,[a] Gail T.W. Wertz,[b]
L. Andrew Ball,[c] and Lawrence E. Hightower[a]

Introduction

The polypeptide coding capacities of Newcastle disease virus (NDV) and other paramyxoviruses have not been established. However, in hybridization experiments, the single-stranded RNA genome of Sendai virus was annealed completely by its 18S and 33S mRNA size classes.[1] Several lines of evidence indicate that an identical situation exists for NDV.[2–5] Therefore, determination of all of the mRNA coding assignments of these two size classes would identify all of the unique viral proteins. This approach was used in the present work to investigate the virus-specific products of NDV strain Australia Victoria (*AV*).

Results and Discussion

1. *mRNA fractionation.* Radiolabeled, poly(A)-selected NDV mRNA was extracted from infected, actinomycin D-treated secondary cultures of chicken embryo (CE) cells and was fractionated by electrophoresis in a preparative agarose gel under denaturing conditions, as described in the legend to Figure 1. During electrophoresis, fractions were collected from a buffer-filled chamber cast in the gel. The fractions were analyzed by electrophoresis using the same gel system on an analytical scale (Figure 1).

[a] Microbiology Section, Biological Sciences Group, University of Connecticut, Storrs, Connecticut.
[b] Department of Bacteriology and Immunology, University of North Carolina, School of Medicine, Chapel Hill, North Carolina.
[c] Biophysics Laboratory, University of Wisconsin, Madison, Wisconsin.

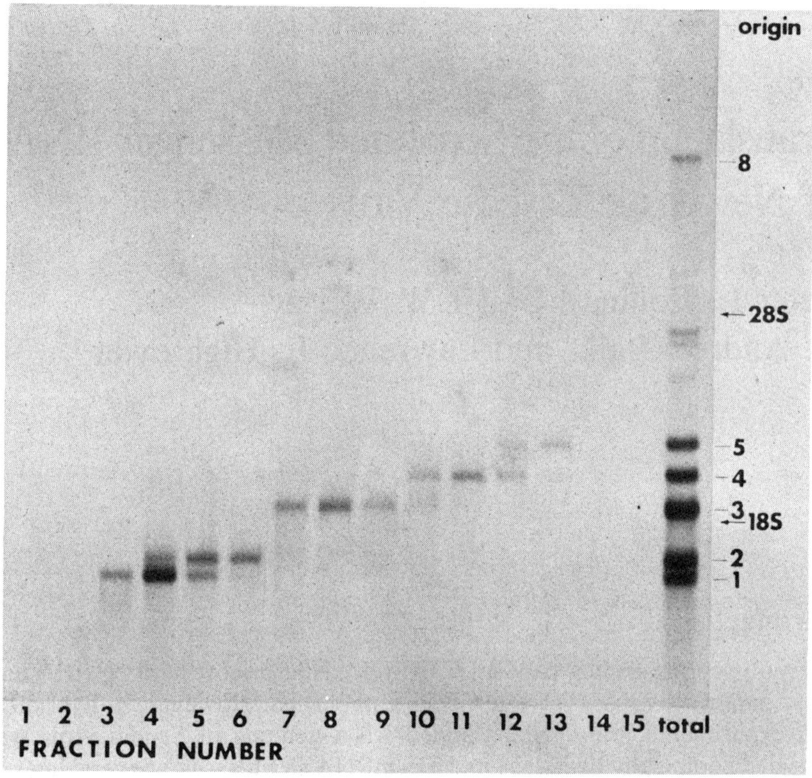

Figure 1. Analytical agarose gel electrophoresis of fractions of NDV mRNA that were partially purified by preparative electrophoresis. Secondary cultures of CE cells, infected with strain *AV* at an input multiplicity of 10, were treated with actinomycin D (2 μg/ml) beginning at 3.25 hr postinfection and with ^{3}H-uridine (100 μCi/ml) at 4 hr.[8] At 10 hr postinfection, the cultures were solubilized and mRNA was collected using oligodeoxythymidylic acid-cellulose.[8] Electrophoresis on both the analytical and preparative scales was performed using 1.5% agarose gels containing 6 M urea and 0.025 M Na citrate pH 3.5.[14] Fractions (1–15) of eluate from the preparative gel were collected, ethanol precipitated, and analyzed in an analytical gel along with a sample of unfractionated (total) NDV mRNA. A fluorogram[15] of the analytical gel is shown with the individual viral electrophoretic bands labeled 1–5 (18S size class) and 8 (35S size class) in order of decreasing mobility. The positions of 18S and 28S ribosomal RNAs in an adjacent channel are marked.

Substantial purification of the 5 electrophoretic bands of the 18S mRNA size class was obtained.

2. *Translation of the separated mRNAs.* The samples of fractionated mRNAs were also analyzed by individual translation in reticulocyte lysates, and the products were resolved using sodium dodecyl sulfate polyacrylamide gel electrophoresis (SDS-PAGE) as shown in Figure 2. Comparison of the results shown in Figures 1 and 2 provided the following coding assignments:

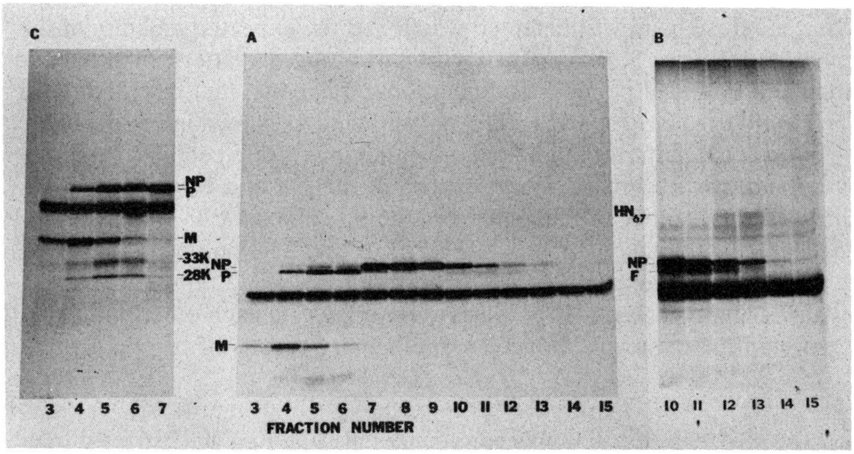

Figure 2. SDS-PAGE of polypeptides synthesized *in vitro* in response to fractions (3–15) containing partially purified NDV mRNAs. The mRNA fractionation is described in the legend to Figure 1. Fractions were individually translated in nuclease-treated rabbit reticulocyte lysates[16] using conditions reported previously[9] and analyzed by SDS-PAGE.[17] (A) shows a fluorogram of an 11.5% gel run at 20 mA for 16 hr; (B) shows a longer exposure of a portion of the same gel, and (C) shows some of the identical translation products analyzed using an 11.5% gel run for 11 hr. Positions of the viral polypeptides are shown; HN_{67} is a 67 kilodalton unglycosylated form of the HN glycoprotein.[5,7,8] The translation products of fractions 1 and 2 are not shown because they contained only endogenous reticulocyte products. The abundant product migrating between the P and M-proteins in all of the channels is an endogenous product.

mRNA band 1 (M-protein), band 2 (the P [6], 33K, and 28K polypeptides), band 3 (NP), band 4 (F-protein), and band 5 (HN-protein). Some of the virus-specific cell-free system products have been described.[5,7,8,9] However, 33K, 28K, and the cell-free system form of the F glycoprotein have not been reported previously.

33K and 28K were of particular interest because mRNA band 2 is known to contain two components that differ in sensitivity to UV inactivation, evidence that they are transcribed from different regions of the genome.[10] In addition, Sendai virus and other paramyxoviruses code for a nonstructural protein C,[11] but no analogous polypeptide has been identified for NDV.

3. *Peptide mapping*. For the fractionation experiments, relatively small amounts of message were used and the incorporation of ^{35}S-methionine into viral polypeptides using the separated mRNAs was insufficient for tryptic peptide analysis. However, partial digest peptide mapping[12] was possible and established that the polypeptide migrating slightly faster than NP in fractions 10–13 (Figures 2A,B) was unrelated to the NP and P-proteins (not shown). Therefore, this protein was a candidate for the cell-free system form of F. In SDS-PAGE separations of viral polypeptides made *in vitro* using total

mRNA, a diffuse band was detected which had the expected mobility of the F candidate (Figure 3, channel 3). The partial digest pattern of this band contained NP-related peptides and additional fragments that matched those of the putative *in vitro* F polypeptide (not shown). As shown in Figure 4, the two-dimensional pattern of methionine-containing tryptic peptides from this diffuse band was a composite of the maps of the NP and F_0 proteins. This established the identity of the cell-free system form of the F-protein.

Partial digest peptide mapping of the P, 33K and 28K polypeptides, using several proteases over a range of concentrations, yielded reproducible patterns that suggested that 33K and 28K were related to each other but not to the P-protein (not shown). However, preliminary results of tryptic peptide

Figure 3. SDS-PAGE of radioactive polypeptides made *in vitro* and extracted from cultured cells. CE cells were (a) mock-infected and (b) infected with NDV at an input multiplicity of 5, incubated for 9.5 hr, and exposed to ^{35}S-methionine for 0.5 hr.[9] At 10 hr postinfection, the cultures were solubilized and radioactive proteins were analyzed by SDS-PAGE as described before.[8] An adjacent channel (c) contained ^{35}S-methionine labeled polypeptides synthesized in a reticulocyte lysate in response to mRNA made *in vitro* by detergent-activated virions.[8] Another gel was used to analyze polypeptides made *in vitro* using poly(A)-selected mRNA extracted [8] at 6 hr postinfection from CE cells that were either (d) mock-infected or (e) infected at an input multiplicity of 10. Actinomycin D (2 μg/ml) was present continuously in the cultures beginning 0.5 hr before infection. Autoradiograms of 11.5% gels run at 20 mA for 11(a–c) and 9 (d, e) hr are shown. The viral and inducible cellular [9] polypeptides (p88, 72, 71 and 23) are marked.

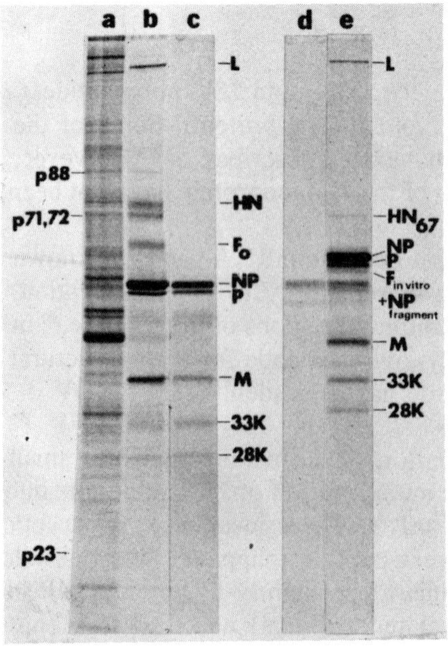

mapping showed a partial match of 33K, 28K and the P-protein. This remains to be confirmed by mixing experiments. More extensive tryptic peptide analysis will be necessary to establish the uniqueness of 33K and 28K.

4. *33K and 28K are virus-specific.* Extracts of NDV-infected cells (Figure 3b) contained ^{35}S-methionine labeled polypeptides that comigrated on SDS-polyacrylamide gels with the 33K and 28K cell-free system products (Figure 3c). The 33K and 28K polypeptides from infected cells had the same

Figure 4. Two-dimensional chromatography of ^{35}S-methionine labeled tryptic peptides of the F_0 glycoprotein from infected cells, NP made *in vitro,* and the cell-free system form of F made by unfractionationed mRNA. The latter pattern contained peptides contributed by a fragment of NP that comigrated with *in vitro* F. The F-related peptides in the pattern were numbered. To prepare the maps, ^{35}S-methionine labeled polypeptides were separated by SDS-PAGE, excised from the dried gel, and digested exhaustively with trypsin as described before.[13] The digests were analyzed on cellulose thin layer plates as described previously,[18] except that electrophoresis was the first dimension and ascending chromatography was the second.

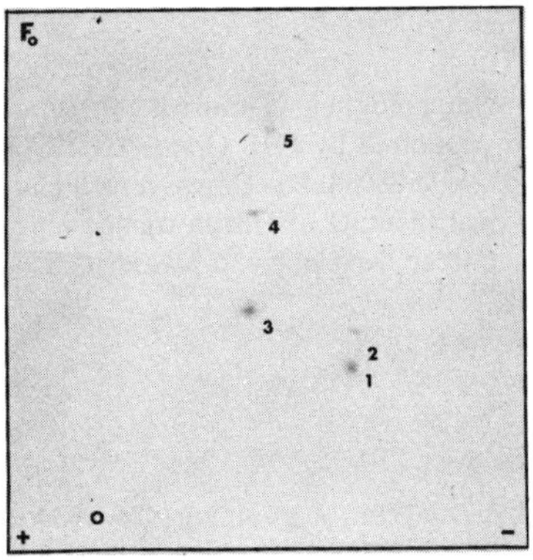

partial digest peptide maps as their cell-free synthetic counterparts. Therefore, these proteins were not simply cell-free system artifacts. In the experiment shown in Figure 3c, 33K and 28K were both synthesized in cell-free systems programmed with mRNA made *in vitro* by the detergent-activated virion-associated transcriptase. Therefore, these small polypeptides were both virus-specific.

Summary

The 5 electrophoretic species of the 18S size class of NDV mRNA were fractionated by preparative gel electrophoresis and identified by individual translation. Each mRNA band coded for a unique [5–9,13] viral protein. In addition to the P-protein, mRNA band 2 encoded 33K and 28K, two related polypeptides that may be either unique or P-related.

The 35S size class of NDV mRNA contains a single messenger that codes for the L-protein.[2,5] The 5 electrophoretic species of the 18S size class used in our study account for the remainder of the genome coding capacity.[1–3] Thus, the polypeptides detected by translation of the fractionated mRNAs probably account for all of the unique viral proteins.

ACKNOWLEDGMENTS

We thank Liz Jean and Jean Winters for help in preparing this manuscript, and Glenn Smith for photography. This work was supported by PHS Grants HL23588 and CA14733 (Cell Culture Facility) and by NSF Grant PCM78-08088. P.L.C. was a recipient of a National Science Foundation fellowship and a National Institute of Health traineeship. L.A.B. was a recipient of a National Institute of Health Career Development Award. G.T.W.W. was supported by PHS Grants AI12464 and AI1513.

References

1. Roux, L. and Kolakofsky, D. (1975): *J. Virol.* 16:1426–1434.
2. Spanier, B.B. and Bratt, M.A. (1977): *J. Gen. Virol.* 35:439–453.
3. Kaverin, N.W. and Varich, N.L. (1974): *J. Virol.* 13:253–260.
4. Varich, N.L., Lukashevich, I.S., and Kaverin, N.V. (1979): *Acta. Virol.* 23:273–283.
5. Morrison, T.G., Weiss, S., Hightower, L.E., Spanier-Collins, B., and Bratt, M.A. (1975): In: *In Vitro Transcription and Translation of Viral Genomes*. Haenni, A.L. and Beard, G. (Eds.), INSERM, Paris, pp. 281–290.
6. Smith, G.W. and Hightower, L.E. (1981): *J. Virol.* 37, in press.
7. Clinkscales, C.W., Bratt, M.A., and Morrison, T.G. (1977): *J. Virol.* 22:97–101.
8. Collins, P.L., Hightower, L.E., and Ball, L.A. (1978): *J. Virol.* 28:324–336.
9. Collins, P.L., Hightower, L.E., and Ball, L.A. (1980): *J. Virol.* 35:682–693.
10. Ball, L.A., Collins, P.L., and Hightower, L.E. (1978): In: *Negative Strand Viruses and the Host Cell.* Mahy, B.W.J. and Barry, R.D. (Eds), Academic Press, London, pp. 367–382.
11. Etkind, P.R., Cross, R.K., Lamb, R.A., Merz, D.C., and Choppin, P.W. (1980): *Virology* 100:22–23.
12. Cleveland, D.W., Fischer, S.G., Kirschner, W.M., and Laemmli, U.K. (1977): *J. Biol. Chem.* 252:1102–1106.

13. Hightower, L.E., Morrison, T.G., and Bratt, M.A. (1975): *J. Virol.* 16:1599–1607.
14. Wertz, G.W., Davis, N.L., and Edgell, M.H. (1980): *Anal. Biochem.* 106:148–155.
15. Laskey, R.A. and Mills, A.D. (1975): *Eur. J. Biochem.* 56:335–341.
16. Pelham, H.R.B. and Jackson, R.J. (1976): *Eur. J. Biochem.* 67:247–256.
17. Laemmli, U.K. (1970): *Nature* 277:680–685.
18. Lamb, R.A. and Choppin, P.W. (1977): *Virology* 81:382–397.

David H.L. Bishop and Richard W. Compans, eds.
The Replication of Negative Strand Viruses

Nucleocapsid-Associated RNA Species from Cells Acutely or Persistently Infected by Mumps Virus

Micheline McCarthy[a]

Introduction

Mumps virus is a paramyxovirus that causes acute inflammatory disease of the respiratory tract, salivary glands, gonads, pancreas, and the central nervous system of man.[1,2] Fresh isolates of mumps virus typically grow in primate cells, causing varied cytopathic effect (cpe). Both neuroadapted and attenuated, chick embryo adapted strains were developed in the laboratory for research or vaccine purposes.[3–5]

There is relatively little data on the structure and replication of mumps virus. Mumps virus is very difficult to grow in comparison with other paramyxoviruses such as Sendai or Newcastle Disease Virus (NDV). Previously East and Kingsbury[6] identified negative sense 50S RNA as the predominant RNA species in virions of the chick embryo adapted Enders strain, an attenuated, prototype vaccine strain. The genomic RNA, like that of NDV and Sendai, is partially self-complementary. Recently I have isolated and purified nucleocapsids from cultured cells which were lytically or persistently infected with mumps virus. Viral RNA can be obtained in good yields from these nucleocapsids, and RNA species can be characterized and compared.

Results

Productive, acutely cytopathic infections in culture may result from infection of either chick embryo fibroblasts (CEF) with the Enders strain of

[a] Department of Neurology, Johns Hopkins University School of Medicine, Baltimore, Maryland and Laboratory of Molecular Genetics, NINCDS, NIH, Bethesda, Maryland.

mumps virus or from infection of Vero cells with nonadapted or neuroadapted strains.[7] The Enders strain causes a productive but nonlethal infection in the chick embryo. I have used 5-^{3}H-uridine or ^{32}P-orthophosphoric acid to label infected cells between 24 to 72 hr postinfection; ^{3}H-uridine was also used to label viral RNA *in ovo*. Nucleocapsids were isolated from egg-grown virions[8] or from cell lysates and purified through a discontinuous CsC1 gradient[9] followed by a continuous CsC1 gradient.[10] Nucleocapsids isolated from virions or from infected CEF or Vero cells have a density of 1.30 g/ml and contain predominantly 50S RNA (Figure 1). However, variable amounts of 28S RNA frequently occur as a shoulder or second, smaller peak in the RNA profile, and the relative amount of this RNA appears to depend on whether nucleocapsids are phenol-extracted or dissociated with SDS. The 50S RNA will self-anneal to a maximum of 30% ribonuclease resistance, while the 28S RNA exhibits 60% to 90% maximum ribonuclease resistance. The 50S RNA will hybridize with unfractionated mRNA from mumps virus infected cells up to a maximum of 85% ribonuclease resistance. No hybridization occurs with a similar RNA fraction from uninfected cells. Hybridization between 28S RNA and mRNA is difficult to demonstrate because of the high degree of self-annealing in the

Figure 1. Sucrose velocity gradient centrifugation of nucleocapsid-associated RNA from the Enders strain of mumps virus. Nucleocapsids were isolated from lysates of ^{32}P-labeled, infected chick embryo fibroblasts (left panel) or from ^{3}H-uridine labeled, egg-grown virions (right panel). Sedimentation is from right to left. Ribosomal 28S, 18S, and 4S RNA species were used as size markers in sucrose velocity gradients.

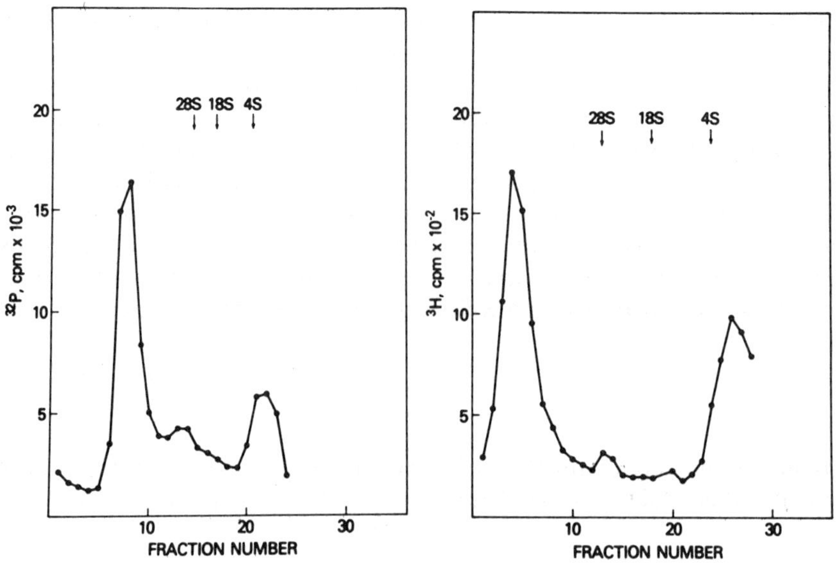

28S RNA. However, when labeled 28S RNA is denatured at low concentration, it will hybridize with excess unlabeled 50S RNA. A maximum of 50% of the label is ribonuclease resistant, suggesting that 50% of the 28S RNA is plus sense sequences. Thus, both the 50S and 28S nucleocapsid-associated RNAs are virus-specific.

The 50S RNA fraction appears to contain a single size RNA as determined from electrophoresis on composite agarose/acrylamide gels (Figure 2). The 28S RNA does not migrate readily into such gels unless it is first heat-treated or digested with ribonuclease. Then a single major species migrates into the gel with a higher mobility than 50S RNA. This species of RNA is double-stranded and hybridizes with excess 50S RNA. From this evidence, it would appear that 28S RNA contains predominantly duplexes of full-length plus and minus sense 50S RNA, since the duplex 50S would be expected to have a 28S sedimentation value. Thus, nucleocapsids isolated from culture cells acutely infected with various strains of mumps virus contain predominantly a single size class of RNA—50S in sedimentation value. Both plus sense and minus sense RNA species are encapsidated, with plus sense constituting 15% to 30% of the total sequences.

A persistent, relatively noncytopathic infection results when the Enders strain of mumps virus infects Vero cells. Although the Enders strain grows acutely and productively in chick cells in culture,[7] in Vero cells the infection is initially non-productive, lacking even the release of non-infectious particles, although infection can be demonstrated by hemadsorption (HAD) or immunofluorescence. When infected cells are trypsin treated and passed several times, scattered cpe is observed. Mumps virus is released, with a titer varying from 3 to 5 logs $TCID_{50}$. This is two to four logs lower than titers obtained from acute infections. Persistently infected cells contain the normal spectrum of viral proteins, including glycoproteins and M-protein. There is no temperature-sensitive behavior either of this infection or of the virus released from persistently infected cells. The mumps virus released by the cells will interfere with the growth of nonadapted or neuroadapted mumps virus strains in normal Vero cell culture (Figure 3).

The amount of viral RNA replication in persistently infected cells, measured by ^{3}H-uridine or ^{32}P label incorporation into nucleocapsids, is 20% to 50% that in acutely infected cells. Nucleocapsids purified from persistently infected cells contain 50S RNA plus a spectrum of subgenomic size RNAs (Figure 4). Both genomic and subgenomic size RNA fractions contain relatively higher amounts of plus sense RNA as judged by self-annealing. The 50S RNA may contain up to 30% plus sequences, the subgenomic RNA up to 45% plus sequences. Electrophoretic analysis of this RNA confirms that 50S RNA is homogeneous, while the subgenomic RNA is very heterogeneous (data not shown). With increasing passage level of the persistently infected cells, the amount and size heterogeneity of subgenomic RNA increases relative to 50S RNA (Figure 5). Correspondingly, the titer of released mumps

Figure 2. Gel electrophoresis of viral 50S and 28S RNA species obtained from intracellular nucleocapsids. Appropriate fractions from sucrose velocity gradients of nucleocapsid RNA were pooled, ethanol precipitated, and subjected to electrophoresis on 2% acrylamide—0.5% agarose gels. The 28S RNA in the center lane was heated to 100°C, then cooled and incubated at 60°C prior to electrophoresis. The 28S RNA in the right lane was exhaustively digested with pancreatic and T_1 ribonucleases prior to electrophoresis.

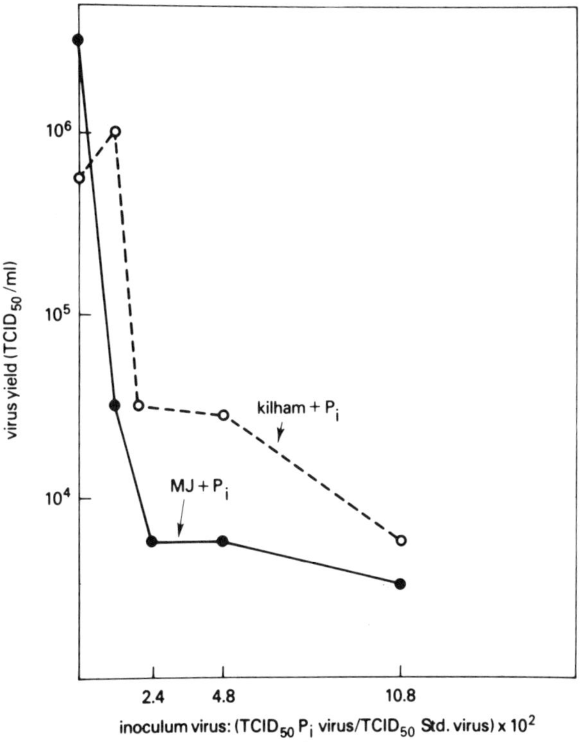

Figure 3. Infection of Vero cells with mumps virus plus virus from persistently infected cells (P_i virus). Normal Vero cells were inoculated with P_i virus mixed in the indicated ratios with either the Kilham neuroadapted or the MJ nonadapted strain of mumps virus. The multiplicity of infection was 0.1 $TCID_{50}$ Kilham or MJ virus per cell. Infectious virus yields from these infections were determined by hemadsorption titration of Vero cells.

virus diminishes, as does the extent of HAD and cpe. Thus, the persistent infection tends to be self-limiting.

Fingerprint analysis of both 50S and subgenomic RNA fractions demonstrates that the small RNAs are viral-specific (Figure 6). There is correspondence between pancreatic ribonuclease-derived and T_1-derived oligonucleotides from both fractions. The pancreatic fingerprint of the subgenomic RNA fraction lacks four of the 43 major oligonucleotides present in the 50S RNA pattern and contains twelve additional oligonucleotides, which may derive from the plus sense sequences constituting almost half the subgenomic RNA fractions. In addition, matching oligonucleotides in subgenomic and 50S RNA fingerprints often differ dramatically in intensity. Thus, the population of subgenomic RNA species must include sequences from most of the genomic 50S RNA; however, oligonucleotide intensity differences suggest that the subgenomic RNA is relatively enriched in some

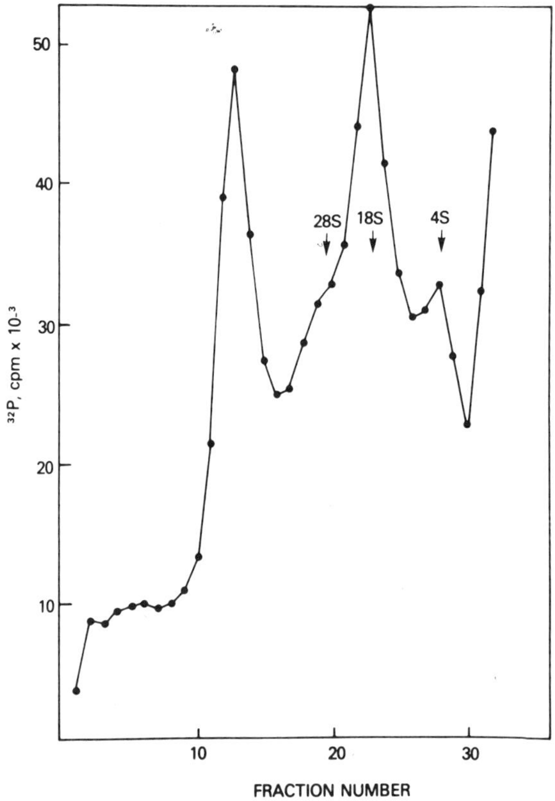

Figure 4. Velocity gradient centrifugation of nucleocapsid RNA from ^{32}P-labeled, persistently infected Vero cells at the ninth passage.

genomic sequences. It is unlikely that the subgenomic RNA species arise via random degradation of intracellular nucleocapsids by soluble host cell factors. When lysates from persistently infected cells and acutely infected cells are mixed and nucleocapsids subsequently co-purified, intact 50S RNA is the only major RNA species obtained from acutely infected cells.

Discussion

Persistent, relatively noncytopathic infection must derive from a unique host-virus interaction between the chick embryo adapted Enders virus and the Vero cell line. Why unique? Neither the virus strain nor the host cell line has any particular predisposition to establishing persistent infection. A major population of subgenomic nucleocapsid-associated viral RNA species such as I have described appears only when the attenuated Enders strain infects Vero cells. I have demonstrated that 50S RNA is the single major RNA species synthesized by Enders virus acutely infecting chick cells or by

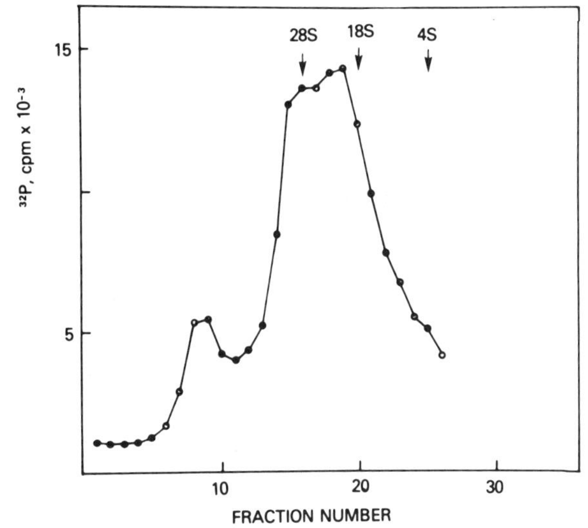

Figure 5. Velocity gradient centrifugation of nucleocapsid RNA from ^{32}P-labeled, persistently infected Vero cells at the seventeenth passage.

Figure 6. Fingerprints of oligonucleotides derived from exhaustive pancreatic ribonuclease digestion of 50S RNA (left panel) and subgenomic RNA (right panel) from persistently infected cells. The tracing of major oligonucleotides includes common oligonucleotides (open spots), subgenomic RNA oligonucleotides (filled spots), and 50S RNA oligonucleotides (hatched spots). Positions of bromophenol blue and xylene cyanol dyes are indicated by Xs.

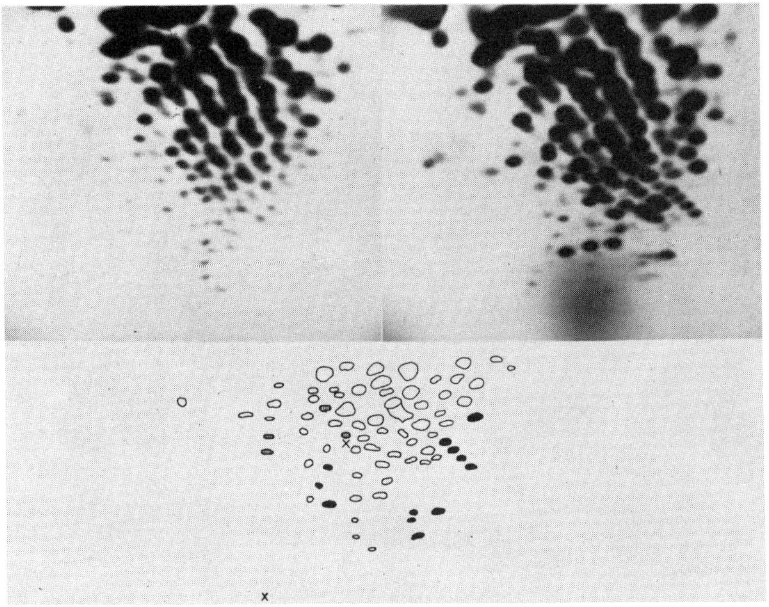

other mumps virus strains acutely infecting Vero cells. Viral RNA synthesis must be aberrant in Enders-infected Vero cells such that defective viral RNA species appear and evolve through the course of the persistent infection. Moreover, the gradual decrease in genomic length 50S RNA relative to increased accumulation of subgenomic viral RNA correlates with loss of infectious virus production. Thus, defective interfering RNA species apparently modulate viral replication during this persistent infection.[11-13] Defective RNA species evolve with such diversity and heterogeneity that the complexity of the DI population approximates the complexity of the genome. Moreover, the replication of DI RNAs is favored over the replication of genome RNA, so that the infection is self-limiting.

References

1. Bang, H.O. and Bang, J. (1943): *Acta Medica Scandinavica* 113:487–505.
2. Azimi, P.H., Cramblett, H.G., and Haynes, R.E. (1969): *J. Amer. Med. Assoc.* 207:509–512.
3. Kilham, L. and Overman, J.R. (1953): *J. Immun.* 70:147–151.
4. Enders, J.F., Levens, J.H., Stokes, J. Jr., Maris, E.P., and Berenberg, W. (1946): *J. Immun.* 54:283–291.
5. Buynak, E.B. and Hilleman, M.R. (1966): *Proc. Soc. Exp. Biol. Med.* 123:768–775.
6. East, J.L. and Kingsbury, D.W. (1971): *J. Virol.* 8:161–173.
7. McCarthy, M., Jubelt, B., Fay, D.B., and Johnson, R.T. (1980): *J. Med. Virol.* 5:1–15.
8. McCarthy, M. and Johnson, R.T. (1980): *J. Gen. Virol.* 46:15–27.
9. Mountcastle, W.E., Compans, R.W., Caliguiri, L.A., and Choppin, P.W. (1970): *J. Virol.* 6:677–684.
10. Hall, W.W. and Martin, S.J. (1973): *J. Gen. Virol.* 19:175–188.
11. Kiley, M.P., Gray, R.H., and Payne, F.E. (1974): *J. Virol.* 13:721–728.
12. Rima, B.K., Davidson, W.B., and Martin, S.J. (1977): *J. Gen. Virol.* 35:89–97.
13. Hall, W.W., Martin, S.J., and Gould, E. (1974): *Med. Microbiol. Immunol.* 160:155–164.

Copyright 1981 by Elsevier North Holland, Inc.
David H.L. Bishop and Richard W. Compans, eds.
The Replication of Negative Strand Viruses

Internal Structural Differentiation of the Plasma Membrane in Sendai Virus Maturation

Thomas Bächi and Martin Büechi[a]

Morphogenesis of Sendai virus and other members of the paramyxovirus family appears in the electron microscope as a process of budding at the host cell surface.[1] Lateral and transmembranal interactions of viral molecules at the plasma membrane required for the viral particles are, however, poorly understood. To further analyze the structural relationship of viral elements with the cell surface during its conversion into a viral envelope, we have studied four different domains of the plasma membrane (Figure 1). Freeze-etching and freeze-fracturing techniques [2] were employed to reveal the external surface (ES), and the hydrophobic exoplasmatic (EF) and protoplasmatic (PF) fracture faces of cell membranes and viral envelopes, respectively. The plasmatic surface (PS) has for the first time become accessible to direct observations by freeze-drying electron microscopy of isolated membranes from virus infected cells (Büechi and Bächi, in preparation).[3] A comparison of the virus-specific elements present in these different membrane domains allowed an analysis of structural relationships between the glycoprotein spikes (HANA, F) projecting from the external surface and proteins lining the internal surface, i.e., the viral membrane protein (M) and the nucleoprotein (NP).

Localization of Viral Structures in the Plasma Membrane

The mouse lymphoma cell line EL-4 was infected with egg-grown Sendai virus and processed for electron microscopy as described previously.[4] The

[a] Institute for Immunology and Virology, University of Zürich, CH-8028 Zürich, Switzerland.

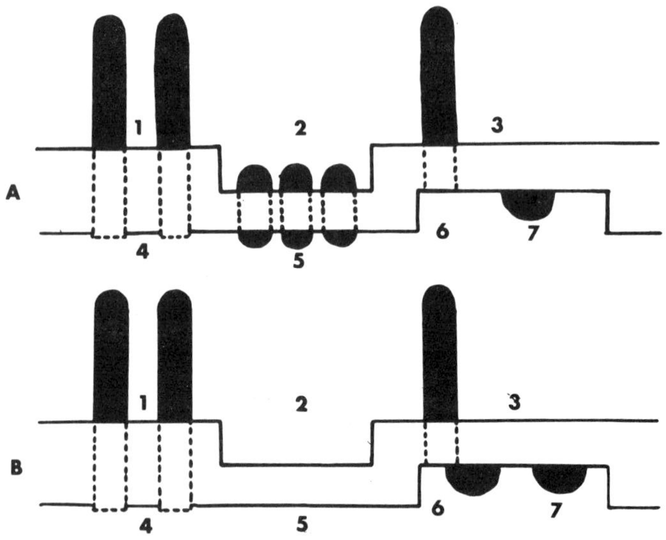

Figure 1. Model of the structure of host plasma membranes after virus-specific conversion and of young viral envelopes (A), or of old viral envelopes (B). The glycoprotein spikes (1) on the ES (3) are seen in superposition to areas of the PF (2) containing virus-specific membrane particles. The latter are arranged in a crystalline pattern in A exclusively and cannot be detected in B.[4] In envelopes of old virions (B) the PF (2) reveals a relatively smooth surface, whereas the EF (6) shows more large membrane particles (7).[9] Crystalline areas are also seen on the PS (5) of plasma membranes of infected cells (A), suggesting the spanning of the inner lipid leaflet by an element which is thought to be related to the viral protein M.

most prominent structural change seen 24 hr to 48 hr after infection was the appearance on the PF of patches of virus-specific intramembrane particles (IMP) arranged in an orthogonal crystalline array (Figure 2). These were often coated over bulging areas of the plasma membrane representing different stages of viral budding. Membranes containing crystalline structures could be detected also in the corresponding fracture face of the viral envelope and were thus incorporated into the virus particles (Figure 3). This structural feature, however, did not persist in the viral envelope: in populations of virions aged *in vitro* or *in vivo,* a diminishing portion was seen to contain crystalline elements (Figure 1B).[4]

By immunoferritin labeling of virus glycoproteins, a superposition of surface antigens (glycoprotein spikes) with hydrophobic domains of the plasma membranes containing crystalline patches was seen. Moreover, nucleoprotein strands appeared to align underneath the crystalline areas.[4]

To study the plasmatic surface (PS) of the plasma membrane of infected cells, covalent attachment of cells to glass cover slips was followed by removal of the cell body, except for the immobilized membrane (Büechi and

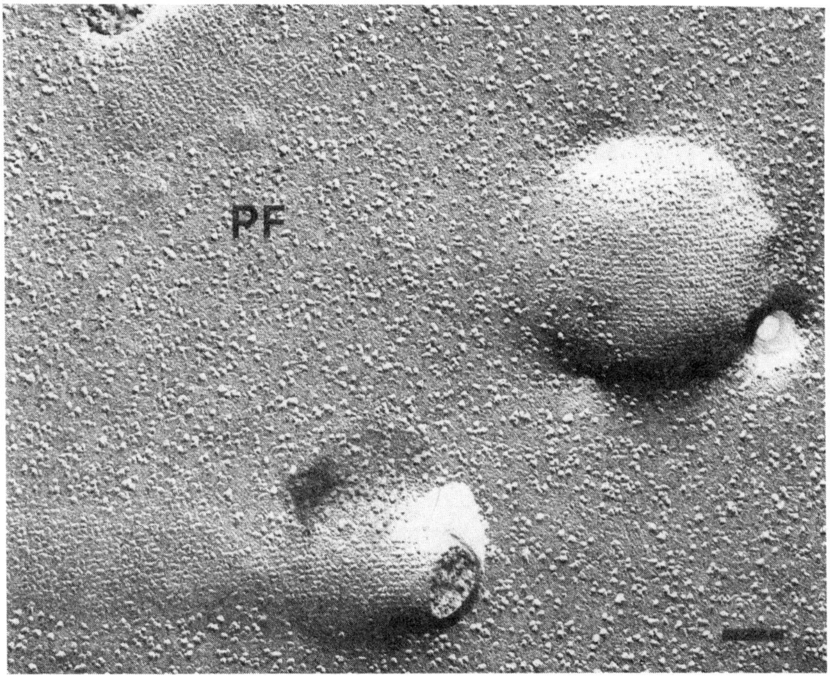

Figure 2. Freeze-fractured plasma membrane of EL-4 cell 48 hr after infection with Sendai virus. The PF is covered with host-specific intramembrane particles and with crystalline arrays of virus-specific particles. Bar: 100 nm.

Bächi, in preparation).[3] Freeze-drying electron microscopy of membranes isolated by this technique reveals virus-specific structures at the PS. Strands of NP were often associated with crystalline patches similar to the ones observed on the PF (Figure 4). When membrane preparations were reacted with an IgG fraction of hyperimmune serum to purified Sendai virus (labeled in a following step with a protein A ferritin conjugate),[5] the marker molecules could be detected only in certain areas of the PS. The ES which was occasionally exposed by curling of the edges of the membrane generally revealed ferritin molecules arranged in various densities (Figure 5). The discrepancy of the two labeling patterns indicate a highly asymmetric disposition of viral antigens with respect to the ES and PS.

The expression of similar virus-induced crystalline patterns on both the ES and PF suggests the presence of a structural element spanning the inner lipid leaflet of the host plasma membrane and of young viral envelopes (Figure 1A). Moreover, our observations demonstrate an association of crystalline areas of the PS with strands of NP on one side and with the surface glycoprotein spikes on the other side (Figure 1A). A relationship of

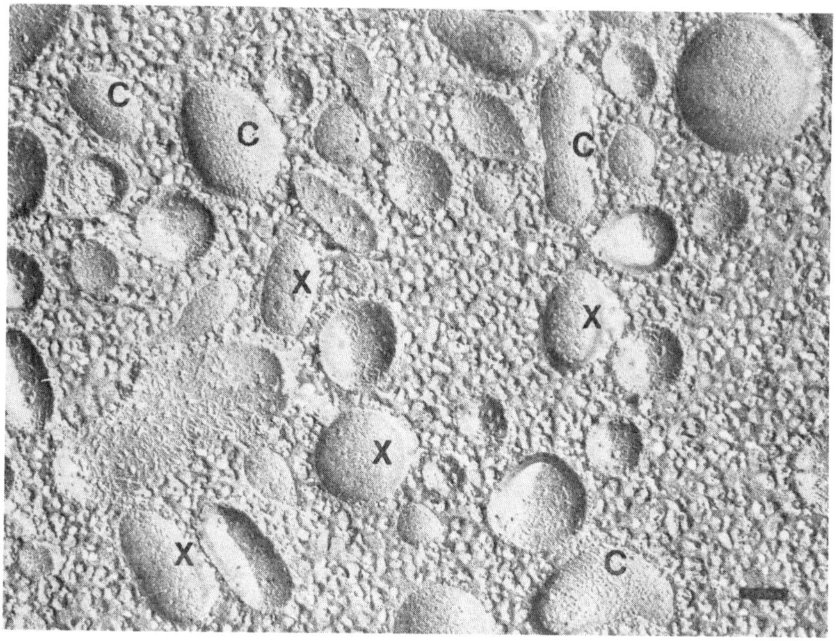

Figure 3. Sendai virus particles harvested from the allantoic fluid 48 hr after infection of chicken eggs. The envelopes of a large portion of these virions possess crystalline structure (C) identical to the one seen on the PF of plasma membranes. The crystalline structures disappear from the fracture faces of virus particles aged *in vivo* or *in vitro* (X). Compare also model of Figure 1. Bar: 100 nm.

these crystalline elements with the viral M-protein is therefore strongly suggested. These observations allow the conclusion that, during the production of the virus particle and in the young viral envelope, the viral M-protein is interacting with the inner lipid leaflet in a fashion which results in the formation of crystalline arrays.

Correlation of Structure and Function

A discussion of mechanisms of Sendai virus maturation and of the acquisition of cell fusing and hemolytic activities as a function of age[6] should take into account the observation that these phenomena are paralleled by a change in crystalline organization. The recruiting of proteins to the site of viral budding could greatly profit from an element possessing the ability to undergo lateral aggregation. The essential role for the triggering of viral budding which has been assigned to the M-protein[7,8] is concordant with a

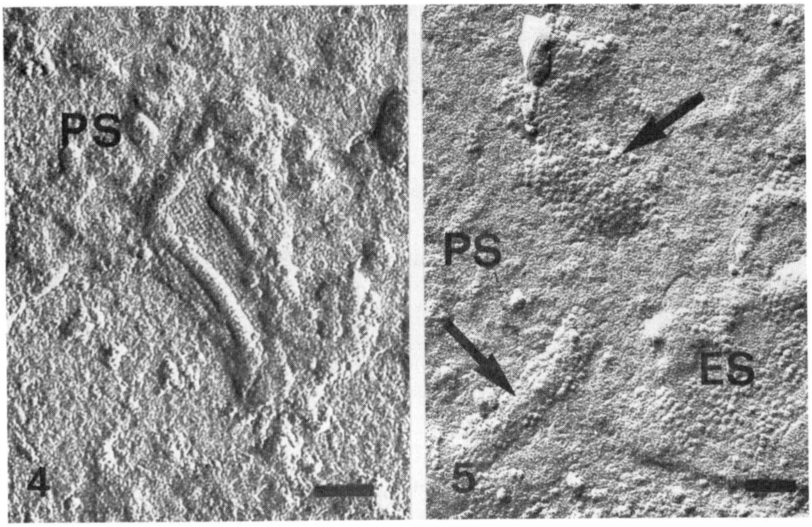

Figures 4 and 5. Plasmatic surface (PS) of membrane preparations of cells 48 hr after infection of cells with Sendai virus. Strands of nucleoproteins are seen to be associated with areas of the internal surface exhibiting a crystalline structure (Figure 4). Ferritin labeling of membranes with virus-specific antibodies reveals the virus-specific nature of certain structures (arrows) seen on the PS and a more generalized occurrence of viral antigens on the ES which is exposed by folding of membrane (Figure 5). Bar: 100 mn.

concept involving the formation of a crystalline scaffolding element. Furthermore, the interaction of this element with the lipid bilayer and its association with NP points to a key role for the establishment of transmembranal interactions of viral proteins.

The presence of these crystalline elements in envelope of young but not of old virus particles[4] adds a further parameter to other structural alterations accompanying the aging of Sendai virus.[9] The loss of crystallinity is undoubtedly ensued by an increase in flexibility which is characteristic of old pleomorphic virus particles.[10] The rigidity of the viral envelope indeed could be an important factor for the development of biological activities residing in this part of the virus, namely hemolysis and cell fusion. These two phenomena require fusion of the viral envelope with the target cell plasma membrane. While both young and old virus particles possess this ability,[6,7] cell lysis and cell-cell fusion are an exclusive capacity of old (flexible) virus (Table 1). Flexibility is required for the lateral diffusion of viral elements in the plasma membrane which follows virus-cell fusion, and this in turn is important for hemolysis and cell-cell coalescence.[2,12,13] Thus, the possibility

Table 1. Correlation of Structure and Function of Sendai Virus Activities.

Event of virus-cell interaction	Virus type		Functional component of virus
	Young[a]	Old[b]	
Adsorption	+	+	HANA
Virus-cell fusion	+	+	F
Lysis[c] cell-cell fusion	−	+	M

[a]Young virus containing active protein F and M-protein in a *crystalline* state. This type is capable of virus-cell fusion; its intrinsic rigidity, however, does not allow a lateral diffusion of envelope proteins in the plasma membrane.

[b]Old virus containing active protein F and M-protein in an *amorphous* state (M), resulting in a flexible envelope structure allowing diffusion of viral components in the membrane.

[c]Interrelation of lysis and cell-cell fusion may depend on osmotic swelling.[11]

emerges that the viral M-protein not only plays an important role in viral morphogenesis but that, by undergoing secondary structural alterations, also influences physical and biological parameters of Sendai virus-cell interactions.
pKelly

References

1. Compans, R.W. and Klenk, H.D. (1979): In: *Comprehensive Virology*. Fraenkel-Conrat, H. and Wagner, R.R. (Eds.), Plenum, New York, pp. 293–407.
2. Bächi, T., Aguet, M., and Howe, C. (1973): *J. Virol.* 11:1004–1012.
3. Büechi, M. and Bächi, T. (1979): *J. Cell Biol.* 83:338–347.
4. Bächi, T. (1980): *Virology* 106:41–49.
5. Bächi, T., Dorval, G., Wigzell, H., and Binz, H. (1977): *Scand. J. Immunol.* 6:241–246.
6. Homma, M., Shimizu, K., Shimizu, Y.K., and Ishida, N. (1976): *Virology* 71:41–47.
7. Yoshida, T., Nagai, Y., Yoshii, S., Maeno, K., and Matsumoto, T. (1976): *Virology* 71:143–161.
8. Yoshida, T. Nagai, Y., Maeno, K., Iinuma, M., Hamaguchi, M., Matsumoto, T., Nagayoshi, S., and Hoshini, M. (1979): *Virology* 92:139–154.
9. Kim, J., Hama, K., Miyake, Y., and Okada, Y. (1979): *Virology* 95:523–535.
10. Hosaka, Y, Kitano, H., and Ikeguchi, S. (1966): *Virology* 29:205–221.
11. Knutton, S. and Bächi, T. (1980): *J. Cell Sci.* 42:153–167.
12. Bächi, T., Deas, J.E., and Howe, C. (1977): In: *Virus Infection and the Cell Surface*. Poste, G. and Nicolson, G.L. (Eds.), North-Holland, Amsterdam, New York, Oxford, pp. 83–127.
13. Bächi, T., Eichenberger, G., and Hauri, H.P. (1978): *Virology* 85:518–530.

David H.L. Bishop and Richard W. Compans, eds.
The Replication of Negative Strand Viruses

Interaction of Sendai Viral Proteins with the Cytoplasmic Surface of Cellular Membranes

Douglas S. Lyles, Henry A. Bowen
and Susan E. Caldwell[a]

Paramyxovirus proteins are incorporated into the host cell plasma membrane at two stages in the virus life cycle. Virion proteins are incorporated into the host plasma membrane biosynthetically prior to virus maturation by budding.[1] The cytoplasmic surface of the membrane is of considerable interest in studying the budding process, since that is where the viral nucleocapsid interacts with the viral envelope proteins. Parental virion proteins are also incorporated into the host plasma membrane as a result of fusion of the virus envelope with the cell membrane,[2,3] which is probably the mechanism of virus penetration.[4,5] The structure of viral proteins on the cytoplasmic surface of cell membranes following envelope fusion probably reflects the interactions occurring on the interior surface of the virus envelope. In addition, these interactions may influence the course of the early events in virus infection.

The cytoplasmic surface of cell membranes is relatively inaccessible to experimental manipulation, because the membrane itself presents an effective permeability barrier. As described here, two systems have been developed for studying the cytoplasmic surface of the plasma membrane of virus infected cells. In the first, the erythrocyte membrane is used as a model for the membrane of virus infected cells. The advantage to using the erythrocyte membrane is that it can be made into inside-out (IO) vesicles, with the cytoplasmic surface of the membrane selectively exposed.[6] The proteins of Sendai virus are incorporated into the erythrocyte membrane by fusion of the virus envelope with the erythrocyte membrane. IO vesicles are prepared

[a] The Bowman Gray School of Medicine of Wake Forest University, Winston-Salem, North Carolina.

from these membranes, which contain the internal virion proteins on their external (cyto-plasmic) surface.[7]

We have also purified plasma membranes from virus infected cells using a technique described by Cohen et al.,[8] which selectively exposes the cytoplasmic surface of the membrane. The preparation involves the attachment of virus infected cells to polycationic polyacrylamide beads with subsequent lysis of the cells. The plasma membranes are retained with the external surface of the membrane attached to the bead and the cytoplasmic surface exposed and thus are functionally inside-out. To identify which viral proteins are exposed on the cytoplasmic surface, inside-out membranes from uninfected and Sendai virus infected BHK cells labeled with ^{35}S-methionine were treated with trypsin, which only digests proteins on the exposed surface. The proteolysis products were analyzed by sodium dodecyl sulfate (SDS)-polyacrylamide gel electrophoresis (Figure 1). By comparison with proteins from purified Sendai virions, each viral structural protein is present in the plasma membranes of infected cells, as previously described.[9] The internal virion proteins L, P, NP, and M are largely digested by trypsin treatment, indicating that they are present on the exposed (i.e., cytoplasmic) surface of the inside-out membrane. In contrast, the glycoproteins of the virus, HN and F_0, were partially protected from digestion. However, trypsin digestion decreased their molecular weights by approximately 1000–2000. Thus, a small segment of these glycoproteins is exposed on the cytoplasmic surface of the membrane. Since a major portion of these molecules is known to be exposed on the external surface of the membrane, these results imply that the Sendai virus glycoproteins span the membrane lipid bilayer. A similar pattern of protection from proteolysis is obtained when the glycoproteins are labeled with ^{3}H-glucosamine (Figure 2). Figure 2 shows a clearer resolution of the F_1 region of the gel (arrow in Figure 2) than in gels of membranes from ^{35}S-methionine-labeled cells. Thus, it can be noted, there is little cleavage of F_0 to F_1 upon treatment of membranes from infected cells with trypsin, indicating that the external surface is, in fact, largely protected from trypsin digestion. Similar results have been obtained with Sendai virus infection of other cell types. The finding of a cytoplasmically exposed segment of the viral glycoproteins has implications for the transmembrane communication among virion components, e.g., in the budding process. For instance, such a segment could serve as a recognition site for assembly of viral nucleocapsid and M-proteins on the membrane.

It has been shown previously that HN and F_1 are in a transmembrane configuration following fusion of Sendai virus with erythrocyte membranes.[7] Thus, a transmembrane structure appears to be a general feature of these proteins, independent of the state of virus maturation. It is likely that the same regions of these proteins are present on the cytoplasmic surface of both the erythrocyte membrane after envelope fusion and the BHK plasma membrane before budding. Since F_1 is derived from the carboxyl terminus of F_0,[10] the fact that both F_1 and F_0 have cytoplasmic regions of similar size suggests

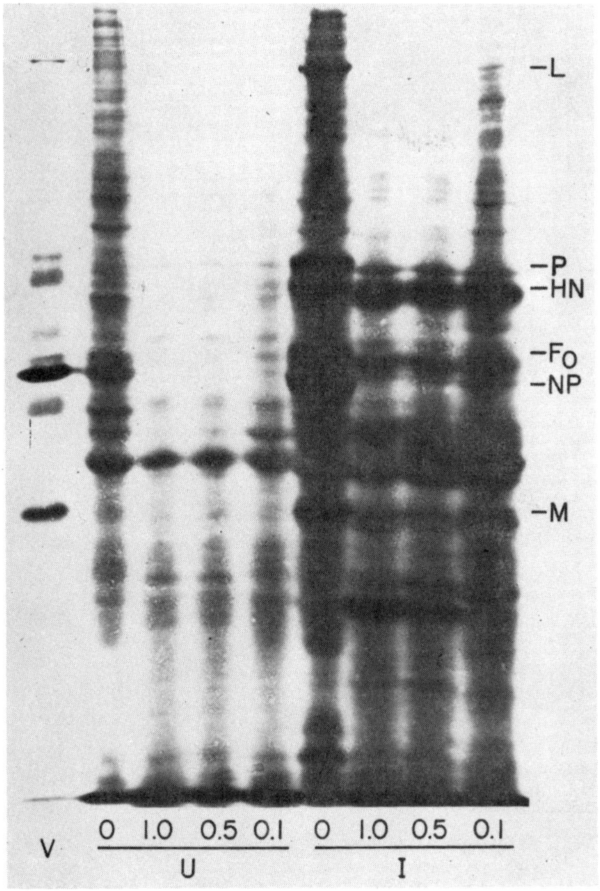

Figure 1. Proteolysis of plasma membranes of Sendai virus infected BHK cells. Cells were labeled with ^{35}S-methionine (10 μCi/ml) for 2 hr at 18 hr postinfection. Plasma membranes were isolated,[8] and suspended in 100 μl of buffer. A 10 μl aliquot of trypsin at the indicated concentrations (mg/ml) was added and incubated at room temperature for 15 min. Digestion mixtures were subjected to SDS gel-electrophoresis and fluorography. V: purified virus labeled with ^{35}S-methionine; U: plasma membranes of uninfected cells; I: plasma membranes of Sendai virus infected cells.

that the carboxyl terminus of both proteins is exposed on the cytoplasmic membrane surface. A similar structure is also likely for HN, as predicted from the solubility properties of proteolytic fragments of HN.[11] The presence of a small cytoplasmic carboxyl segment appears to be characteristic of a large class of viral and cellular glycoproteins.

The viral nucleocapsid appears to associate specifically with the host plasma membrane in contrast to intracellular membranes.[9] The interaction of internal virion components, the nucleocapsid and the M-protein, with the

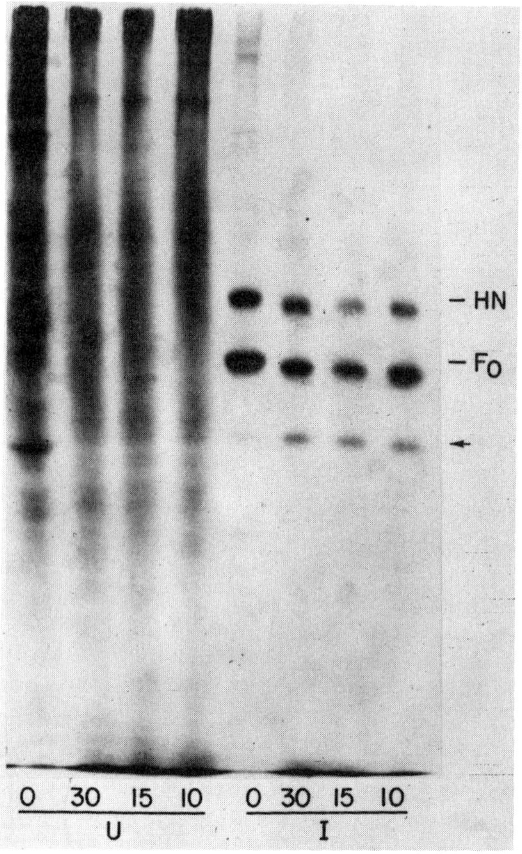

Figure 2. Proteolysis of ^3H-glucosamine-labeled plasma membranes from Sendai virus infected BHK cells. Plasma membranes isolated from labeled cells were incubated with trypsin (final concentration of 50 μg/ml) for the indicated times (min). Fluorographs of the dried gels are shown. U: plasma membranes from uninfected cells; I: plasma membranes of Sendai virus infected cells.

cytoplasmic surface of the plasma membrane appears to be preserved in the preparation described here. Nucleocapsids purified from infected cells and added to the membrane preparation do not associate non-specifically with the isolated plasma membranes or beads (data not shown). Thus, it may be feasible to use this membrane preparation to characterize the interaction among the viral nucleocapsid and envelope components in the budding process.

Sendai viral nucleocapsids are also associated with the cytoplasmic surface of erythrocyte membranes following fusion of the virus envelope with the cell membrane.[3,7] Nucleocapsids purified from virions do not associate non-specifically with IO vesicles derived from erythrocyte membranes (data not shown). The forces binding Sendai viral nucleocapsid and M-proteins to the erythrocyte membrane have been characterized by treating IO vesicles containing viral proteins with agents that perturb protein structure while leaving the membrane lipid bilayer intact. Such agents selectively remove peripheral proteins from membranes while integral membrane proteins remain associated with the lipid bilayer.[12] Furthermore, a gradient of avidities of proteins for the membrane can be established, depending on the strength of treatment required for removal from the membrane. IO vesicles containing Sendai viral proteins labeled with ^{35}S-methionine were treated with various disrupting agents. The disruption products were analyzed by sedimentation in sucrose gradients. Slowly sedimenting peripheral membrane proteins remained at the top of the gradient, while proteins remaining associated with the vesicles sedimented to density equilibrium in the middle of the gradient. The presence of Sendai viral proteins in these two fractions was assessed by SDS-gel electrophoresis and fluorography to detect the labeled viral proteins. Figure 3 shows the results of three representative experiments. Treatment of IO vesicles containing Sendai viral proteins with 6 M guanidine removes the nucleocapsid and M-proteins, suggesting that they are peripheral membrane proteins, while the integral membrane glycoproteins HN and F_1 remain associated with the vesicles. Treatment with 6 M urea or 5 mM p-chloromercuribenzenesulfonate removes the nucleocapsid proteins but not the M-protein, suggesting that the M-protein interacts more extensively with the membrane. Treatment with 40 mM lithium diiodosalicylate (LIS), 4 M potassium thiocyanate, or high pH (12–13) removed both the nucleocapsid and M-proteins, similar to treatment with guanidine. Some treatments that remove peripheral proteins from erythrocyte membranes resulted in no elution of viral proteins. These include extremes of ionic strength (1 M KC1-0.5 mM sodium phosphate), cation chelation, freezing and thawing, and lower concentrations of other eluting agents (1 M urea, 10 protein mM LIS pH 10). Under no conditions tested were intact nucleocapsids found to be released from the membrane without disruption of the membrane lipid bilayer (e.g., with Triton X-100). This suggests that the forces binding nucleocapsids to membranes are similar in magnitude to those responsible for nucleocapsid integrity. It is not known with what membrane components viral proteins are associated on the cytoplasmic surface of erythrocyte membranes. An association between NP and M of Sendai virus has been identified by treatment of intact virions with chemical cross-linking reagents.[13] Such an association could account for some of the membrane interactions observed here.

Since nucleocapsids remain associated with the membrane after fusion of the viral envelope with the cell membrane, the initial transcription events in virus replication (primary transcription) may occur on membrane-bound nucleocapsids, or, alternatively, an additional step after viral envelope fu-

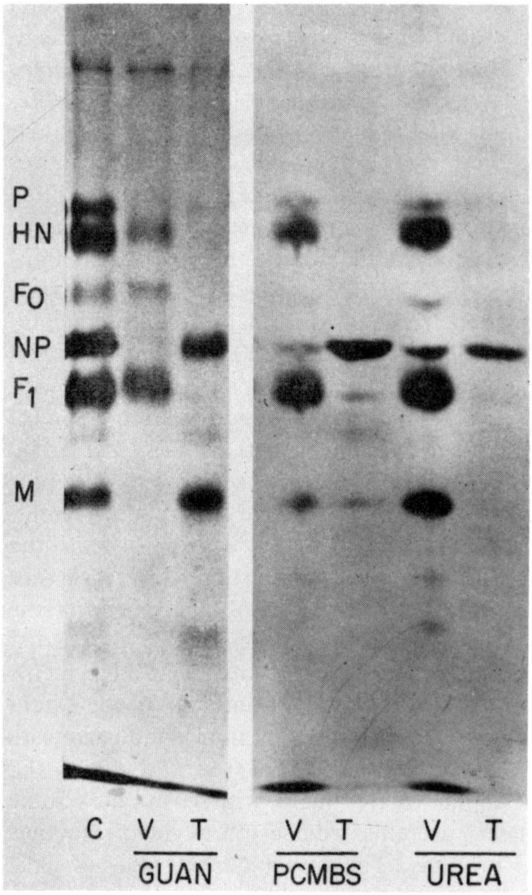

Figure 3. Elution of Sendai virus proteins from IO vesicles. ^{35}S-methionine-labeled virions were allowed to fuse with erythrocytes and IO vesicles were prepared from these membranes.[7] Vesicles were resuspended in the indicated reagents in 5 mM phosphate buffer, pH 8. Disruption products were centrifuged on a 15–45% sucrose gradient at 120,000 × g for 2 hr. The gradient was fractionated, peak fractions containing radioactivity were pooled, acid-precipitated, and subjected to SDS-gel electrophoresis and fluorography. C: control vesicles treated with 5 mM phosphate; GUAN, PCMBS, UREA: vesicles treated with 6 M guanidine, 5 mM p-chloromercuribenzenesulfonate, or 6 M urea, respectively; V: gradient fractions containing vesicles; T: fractions at the top of the gradients.

sion may be necessary for releasing nucleocapsids into the cytoplasm. Preliminary data, obtained using the plasma membrane purification procedure described above, suggest that nucleocapsids remain membrane-bound following penetration of Sendai virus into cell types other than erythrocytes. Membrane interactions similar to those described here may also occur on the internal surface of the virus envelope. It has been suggested that M is a peripheral membrane protein in virus envelopes, based upon disruption of virions with LIS,[14] and in electron micrographs nucleocapsids are frequently observed to be peripherally associated with the inside surface of the viral envelope.[15] Finally, similar membrane interactions may also be involved in virus maturation by budding from host cell membranes. This hypothesis should be testable using inside-out membranes from virus-producing cells.

ACKNOWLEDGMENTS
This research was supported by Grant AI 15892 from the National Institute for Allergy and Infectious Diseases, by a Grant from the North Carolina United Way and Forsyth Cancer Service, and by pilot funds from the Oncology Research Center Support Grant CA 12197, Bowman Gray School of Medicine from the National Cancer Institute.

References

1. Choppin, P.W. and Compans, R.W. (1975): In: *Comprehensive Virology,* Vol. 4. Fraenkel-Conrat, H. and Wagner, R.R. (Eds.), Plenum Press, New York, pp. 95–178.
2. Bächi, T., Aguet, M., and Howe, C. (1973): *J. Virol.* 11:1004–1012.
3. Büechi, M. and Bächi, T. (1979): *J. Cell Biol.* 83:338–347.
4. Dales, S. (1973): *Bacteriol. Rev.* 37:103–135.
5. Scheid, A.S. and Choppin, P.W. (1976): *Virology* 69:265–277.
6. Steck, T.L. (1974): In: *Methods in Membrane Biology,* Vol. 2. Korn, E. (Ed.), Plenum Press, New York, pp. 245–281.
7. Lyles, D.S. (1979): *Proc. Natl. Acad. Sci.* 76:5621–5625.
8. Cohen, C.M., Kalish, D.I., Jacobson, B.S., and Branton, D. (1977): *J. Cell Biol.* 75:119–134.
9. Lamb, R.A. and Choppin, P.W. (1977): *Virology* 81:371–381.
10. Scheid, A.S. and Choppin, P.W. (1977): *Virology* 80:54–66.
11. Scheid, A.S., Graves, M.C., Silver, S.M., and Choppin, P.W. (1978): In: *Negative Strand Viruses and the Host Cell.* Mahy, B.W.J. and Barry, R.D. (Eds.), Academic Press, New York, pp. 181–193.
12. Steck, T.L. (1974): *J. Cell Biol.* 62:1–19.
13. Markwell, M.A.K. and Fox, C.F. (1980): *J. Virol.* 33:152–166.
14. Li, J.K-K., Miyakawa, T., and Fox, C.F. (1980): *J. Virol.* 34:268–271.
15. Kim, J., Hama, K., Miyake, Y., and Okada, Y. (1979): *Virology* 95:523–535.

David H.L. Bishop and Richard W. Compans, eds.
The Replication of Negative Strand Viruses

Permissive Temperature Analysis of RNA$^+$ Temperature-Sensitive Mutants of Newcastle Disease Virus

Mark E. Peeples, James P. Gallagher, and Michael A. Bratt [a]

As in the case for many paramyxoviruses, the virion envelope of Newcastle disease virus (NDV) contains three protein species, HN, F, and M. HN is a glyco-protein which functions in attachment and possesses both hemagglutinating and neuraminidase activities (HA and NA). F is a glycoprotein involved in fusion and hemolysis, and seems to be required for entry into the cell. A third protein, M, is not glycosylated and is associated with the inner surface of the envelope. M is thought to interact with the internal nucleocapsid protein (NP) in virion maturation, and might function in organization of the HN and/or F.[1,2]

We have been studying the independently isolated temperature-sensitive (*ts*) mutants[3,4] of the Australia-Victoria strain of NDV (AV-WT) to determine: (1) the biological effects of each mutation; (2) the altered protein in each mutant; and (3) the interactions among these proteins. Complementation studies placed these *ts*-mutants in 5 groups, A, B, C, D and E, and a sixth group, BC which complemented neither B nor C. Mutants in groups B, C, D and BC were RNA$^+$ (synthesized RNA at non-permissive temperature), and exhibited a variety of biological effects suggesting they might represent envelope protein mutations. Mutants in group D were *ts* for fusion from within (FFWI), a result suggesting mutations in the gene for the F-protein. Some mutants in groups B, C, and BC were variously affected in FFWI or hemadsorption, consistent with mutations in the HN gene. The low levels of complementation between B and C mutants suggested possible

[a] Department of Molecular Genetics and Microbiology, University of Massachusetts Medical School, Worcester, Massachusetts.

intracistronic complementation of mutants with lesions in the same gene. One of several possible explanations for the existence of group BC, was that mutants in this group have mutations in the same gene represented by groups B and C. The evidence described herein strongly suggests that mutants in groups B, C and BC all contain lesions in the HN gene.

In the experiments described here we have ignored the temperature-sensitivity of these mutants, using them simply as a source of mutants. Virus stocks were grown and experiments performed at permissive temperature only.

Biological and Physical Properties

Table 1 compares infectivity, HA, and NA of virions of mutants in groups B, C, BC and D. Group B, C and BC virions are, on average respectively, 85, 110 and 99% as infectious as AV-WT. Group D mutants consistently show

Table 1. Biological Activities and HN Amount in Virions.

	% AV-WT			
Virus[a]	**pfu**[b] / **μg protein**[c]	**HA**[d] / **μg protein**	**NA**[e] / **μg protein**	**HN**[f] **in virion**
AV-WT	100	100	100	100
B1	106	32	8.0	4.1
B2	76	61	42	34
B4	90	52	34	20
B5	76	24	26	6.8
B6	126	43	34	32
B7	33	29	14	13
C1	83	37	34	27
C2	137	73	75	46
BC1	67	34	28	28
BC2	115	54	11	28
BC3	115	58	9.2	15
D1	22	100	123	89
D2	19	126	118	126
D3	7.2	138	109	95

[a]Virus stocks were grown in embryonated hen eggs between 36°C and 37.5°C, concentrated and purified by centrifugation.[5]

[b]Infectivity determined by plaque assay at 37.5°C.

[c]Determined by method of Lowry et al.[6]

[d]Determined by the fractional dilution method at 37.5°C.

[e]Determined by sialic acid released from fetuin[7] in 1 hr at 37.5°C.

[f]HN protein was determined by electrophoresing purified virions on a 10% polyacrylamide gel,[8] staining the gel with Coomassie blue, scanning the wet gel with an Ortec 4310 Densitometer, measuring the area under the HN and (P + NP) peaks with a Wang System 2200 Digitizer, and calculating HN/(P + NP) ratio. P (or P55) is a newly described phosphoprotein of NDV,[9-11] which migrated too close to NP to be distinguished in the tracing.

lower infectivity (on average 16% AV-WT). For HA the opposite is true: B, C, and BC mutants average, respectively, 40, 50 and 49% as much HA as AV-WT, while D mutants average 121%. Similar results are found for NA with average values of 26, 54, and 16% for B, C, and BC mutants respectively, and 117% for D mutants. Thus, B, C and BC mutants are defective in both HA and NA, while D mutants are not. Only D mutants are consistently defective in infectivity.

Reduced HN activities could reflect either lower amount or lower activity of HN. To distinguish between these possibilities, virions were electrophoresed, and the amounts of HN relative to P + NP determined. The results in the last column of Table 1 indicate that virions of B, C, and BC mutants contain less HN than AV-WT, while group D virions do not (averaging, respectively, 18, 36, 24, and 103%). Similar results have been obtained with mutant C1 by Smith and Hightower.[12] Portner et al.[13] also reported that the single RNA^+ mutant of Sendai virus (*ts*271), when grown at permissive temperature, contains lower NA and less HN protein.

Examination of Table 1 also reveals that the amount of HN correlates with NA, indicating that the mutations have not affected the NA site on the HN molecule. The amount of HN does not correlate as well with HA, probably because HA is a more complex virion function. Finally, there is no apparent correlation between the amount of HN and infectivity.

Stability of HA

Tsipis and Bratt[4] have previously shown that the infectivity of several of the B, C and BC mutants are more thermolabile than that of AV-WT. In Figure 1, the thermostability of HA is presented. The HA of AV-WT (and D1, D2, D3, C2, B1, B2 and B7, now shown) is stable under these conditions. In order of increasing sensitivity were: C1 (BC3, not shown), B5, BC1, B6, B4 (not shown), and BC2. Clearly, the thermolabile HA phenotype is associated with mutants in the B, C, and BC groups, but not with the D group, again suggesting that these mutants are defective in the HN gene; the D group represents mutants with lesions in another gene. Comparison of these results with the last column in Table 1 shows no correlation between thermolability of HA and amount of HN in virions. Thermolability of HA, therefore, appears to be an intrinsic property of HN.

Polypeptide Migration Differences

In infected cells, four of the mutants produce proteins with altered migration patterns: BC3 has a more slowly migrating HN (Figure 2A); B2 and C1 have more rapidly migrating HNs (Figure 2B), and D1 has a more rapidly migrating M (Figure 2B). The differences in migration rates of the HN of B2 and C1 reflect differences in the HN polypeptide rather than in their carbohydrate content, since in the presence of tunicamycin (TM), which prevents glycosy-

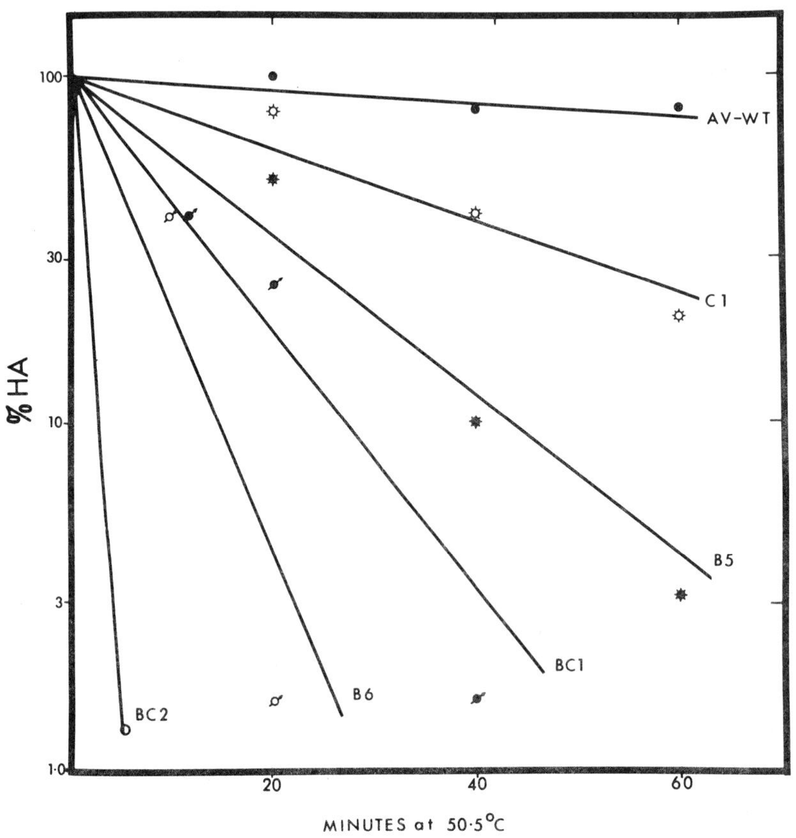

Figure 1. Thermolability of HA. Stocks were diluted to 0.6 mg/ml in 0.01 M HEPES (pH8.0), 0.03 M NH_4Cl, 0.00012 M EDTA, and 10% sucrose, and heated at 50.5°C. Samples were taken at times indicated and assayed by the fractional dilution method for remaining HA. Values are expressed as percentage of 0 time value.

lation of both HN and F,[9] their unglycosylated HNs still migrate faster than those of AV-WT or D1 (Figure 2C). A similar experiment with BC3 is not yet complete.

Conclusions

We have presented evidence that Groups B, C, and BC all reflect lesions in the HN gene: (1) Most members of all 3 groups show significantly reduced HA and NA which appear to be a direct reflection of reduced HN content (rather than reduced activity). (2) Some members of each group show decreased thermostability of HA. (3) A member of each group shows an altered HN migration.

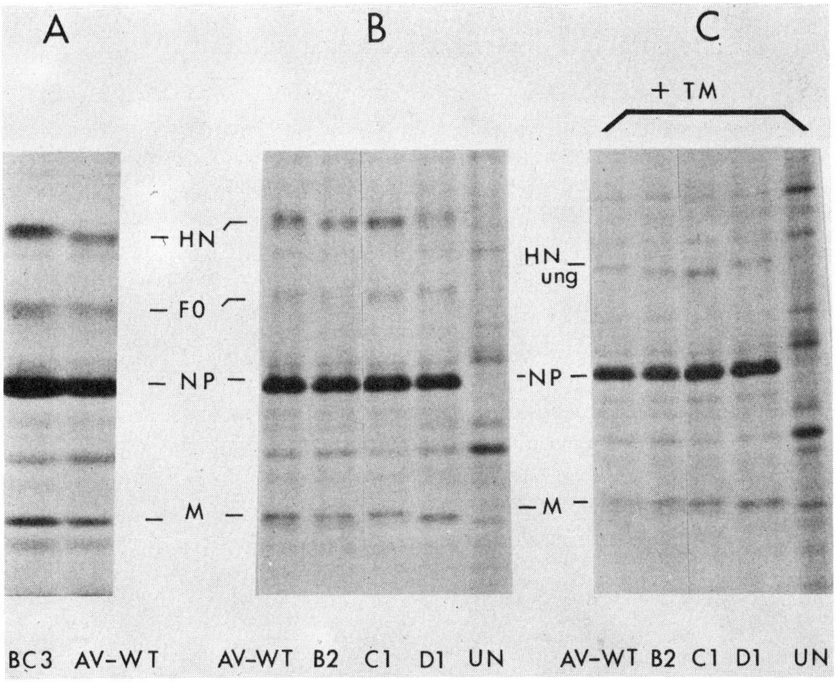

Figure 2. Differences in migration of *ts*-mutant intracellular proteins. Chicken embryo cells pulsed with ^{35}S-methionine (10 μ Ci/ml) for 1 hr at 5-hr postinfection, were lysed and electrophoresed on 7% (A), or 5% to 12.5% gradient (B and C) SDS polyacrylamide gels.[14] Only the bottom half of the gels are shown. Cells in C were labeled in the presence of 1 μg/ml TM. HN_{ung} is the unglycosylated HN.

In addition, these studies indicate that virions need not possess a full complement of HN to be infectious since mutants in groups B, C, and BC all have infectivity/HN ratios higher than AV-WT. As expected, HA, which in the extreme might only require a minimum of two HN spikes/virion, is less sensitive to reduced HN content than is the enzymatic NA activity which closely reflects HN content.

The D group of mutants are more puzzling and their assignment to a specific gene is by no means complete. The three D group mutants are 5–15 × less infectious than AV-WT and most of the B, C, and BC mutants, as well. It is clear that the decreased infectivity is not due to decreased HA or NA, or to decreased HN protein. Nor is it due to increased thermolability of HA, or infectivity (unpublished results). This group is uniformly temperature-sensitive for FFWI (Tsipis and Bratt[4] and unpublished results). This was suspected to reflect alterations of the F protein. But mutant D1 contains an M-protein with an increased migration rate. Whether the decreased specific infectivity or the temperature-sensitivity of infectivity and FFWI reflect altered M or F-proteins, awaits analysis of revertants.

ACKNOWLEDGMENTS
We would like to thank Rhona Glickman, Michael Glass and Judy Brackett for their excellent technical assistance, and Joan Nelson and Pam Chatis for help in manuscript preparation. These studies were supported by a Grant to MAB (AI 12467) and a fellowship to MEP (AI 05874) from the National Institute of Allergy and Infectious Disease.

References

1. Choppin, P.W. and Compans, R.W. (1975): In: *Comprehensive Virology,* 4. Fraenkel-Conrat, H. and Wagner, R.R. (Eds.), Plenum Press, New York, pp. 95–178.
2. Bratt, M.A. and Hightower, L.E. (1977): In: *Comprehensive Virology,* 9. Fraenkel-Conrat, H. and Wagner, R.R. (Eds.), Plenum Press, New York, pp. 457–533.
3. Tsipis, J.E. and Bratt, M.A. (1975): In: *Negative Strand Viruses.* Mahy, B.W.J. and Barry, R.D. (Eds.), Academic Press, New York, pp. 777–784.
4. Tsipis, J.E. and Bratt, M.A. (1976): *J. Virol.* 18:848–855.
5. Weiss, S.R. and Bratt, M.A. (1974): *J. Virol.* 13:1220–1230.
6. Lowry, O.H., Rosebrough, N.J., Farr, A.L., and Randall, R.J. (1951): *J. Biol. Chem.* 193:265–275.
7. Webster, R.G. and Campbell, C.H. (1972): *Avian Diseases* 16:1057–1066.
8. Blattler, D.P., Garner, F., Van Slyke, K., and Bradley, A. (1972): *J. Chromatography* 64:147–155.
9. Morrison, T.G. and Simpson, D. (1980): *J. Virol.* 36:171–180.
10. Smith, G.W. and Hightower, L.E.: *J. Virol.,* in press.
11. Madansky, C.H. and Bratt, M.A. (1980): *Animal Virus Genetics–ICN-UCLA Symposia on Molecular and Cellular Biology, Vol. XVIII.* Fields, B., Jaenisch, R., and Fox, C.F. (Eds.), Academic Press, New York, in press.
12. Smith, G.W. and Hightower, L.E. (1980): In: *Animal Virus Genetics–ICN-UCLA Symposia on Molecular and Cellular Biology, Vol. XVIII.* Fields, B., Jaenisch, R., and Fox, C.F. (Eds.), Academic Press, New York, in press.
13. Portner, A., Scroggs, R.A., Marx, P.A., and Kingsbury, D.W. (1975): *Virology* 67:179–187.
14. Laemmli, U.K. (1970): *Nature* 227:680–685.

David H.L. Bishop and Richard W. Compans, eds.
The Replication of Negative Strand Viruses

Comparison of Lytic and Persistent Measles Virus Infections by Analysis of the Synthesis, Structure and Antigenicity of Intracellular Virus-Specific Polypeptides

John R. Stephenson, Stuart G. Siddell and V. ter Meulen[a]

Introduction

Measles virus is a common pathogen of man, causing an acute childhood disease.[1] This infection is normally controlled and the virus eliminated by the different defense mechanisms of the host. However, under conditions which are still unknown, a persistent measles virus infection of the central nervous system (CNS) is established as a complication of acute measles and later develops into a chronic disease process, known as subacute sclerosing panencephalitis (SSPE).[2] In this rare disorder, brain cells are persistently infected by measles virus without infectious virus being detected. Moreover, indirect evidence suggest that persistency and lack of infectious virus may be linked to the failure of the infected brain cells to synthesize the M polypeptide of measles virus.[3,9] These observations require further investigation of the mechanism of measles virus persistency and, in the present study, some aspects of the replication events of measles virus in a lytic and a persistent infection in tissue culture were analyzed. VERO cells were lytically infected with the SSPE virus "LEC", derived from brain cells of a patient with SSPE.[4] A persistent infection with the same virus was obtained by explanting brain cells from a hamster which revealed a subacute encephalitis after ic inoculation of SSPE LEC virus infected tissue culture cells. The cells were cocultivated with VERO cells to establish a persistent infection and have been passaged over 250 times. The culture contains measles virus antigens but does not produce infectious virus. This reflects

[a] Institut für Virologie and Immunbiologie, Universität Würzburg, 8700 Würzburg, Federal German Republic.

the situation described in brain cells derived from SSPE patients.[2] In these two virus cell systems the synthesis of virus-specific polypeptides *in vivo* was compared by immunoprecipitation of radiolabeled cell lysates and a nuclease treated cell-free system from rabbit reticulocytes was used to analyze the function of the messenger RNA (mRNA) *in vitro*. In addition, the antigenicity of the hemagglutinin (HA) was characterized by monoclonal antibodies raised against purified measles virus in an indirect immunofluorescence test in order to detect any antigenic changes in measles HA during the passage of virus in a persistent state.

Synthesis *in Vivo* and *in Vitro* of Measles Polypeptides From Persistent and Lytic Infections

When VERO cells are lytically infected with measles virus and labeled with ^{35}S-methionine, at least four structural polypeptides can be precipitated with hyperimmune rabbit sera raised against purified virus. Polypeptides corresponding in mobility to the N (major nucleocapsid protein), F (fusion protein) and M polypeptides are precipitated. The polypeptide labeled H is thought to be the molecular species responsible for the hemagglutinin properties of the virus, as it can be precipitated with a monoclonal antibody raised against purified virus which contains high hemagglutination inhibition (HI) activity. In addition, the polypeptide of 50,000 molecular weight is thought to correspond to the uncleaved precursor of the fusion protein (F_0). The "NC" polypeptide is interpreted, by analogy with Sendai virus,[5] to be a cleavage product of the N polypeptide.

When polypeptides from a persistent infection are analyzed, only the H, N, F_0 and NC virus-specific polypeptides can be precipitated. Similar results in brain cultures from a case of SSPE have been found by other workers,[3] which suggest that the synthesis of M-protein is blocked, either at the level of transcription or at the level of translation. However, as virus-specific polypeptides can only be labeled inefficiently in persistent infections, it is possible that less abundant antigens such as F and M would not be labeled sufficiently well to detect under these conditions.

In order to detect virus-specific messengers, poly (A) containing RNA from uninfected, lytically infected and persistently infected cells were translated in a cell-free system and products analyzed. When mRNA from a lytic infection is translated *in vitro* and the products precipitated by hyperimmune serum, the following polypeptides are detected. Species which correspond in mobility to the major nucleocapsid protein (N) and M-protein of the virus are seen. The minor species of molecular weight 44,000 could be a cleavage produce of the N polypeptide as described above. The species of molecular weight 50,000 correspond with the putative uncleaved precursor to the fusion protein seen in infected cells labeled *in vivo*, and thus could represent its nonglycosylated precursor. Three other polypeptides of molecular weight

34,000, 30,000 and 18,000 of unknown function are also seen. As these polypeptides have not been observed in purified virus particles they may represent non-structural viral proteins. In addition to these species, two polypeptides of molecular weight 76,000 and 74,000 are observed, which, as they are precipitated by monoclonal antibody, are thought to be nonglycosylated precursors to the hemagglutinin (H).

When mRNA from a persistent infection is translated *in vitro,* and analyzed directly or after immunoprecipitation, polypeptides corresponding to the N, M 34K, 30K and 18K can be detected. When equal amounts of total cell mRNA are compared directly, the messengers from persistent cultures appear to contain a lower proportion of viral-specific messengers, although it is not possible at present to distinguish whether this observation is due to the translational efficiencies of the mRNAs, or their absolute amount. In addition, the cell-free products of the sample from persistent infections do not contain the slower moving component of the "pre-H" doublet, but the faster moving band is present and is precipitable with hyperimmune serum or with a HA specific monoclonal antibody.

Antigenic Changes in the Hemagglutinin Arising During Persistent Infection

The observation that mutations appear to arise during persistency by VSV,[7,8] led to a search for changes in the antigenicity of virus structural proteins in cells persistently infected with measles virus. For this approach, monoclonal antibodies were raised against Edmonston virus. As the HA is the major surface antigen of the virus and differences could be detected at the mRNA level, clones were initially selected which contained antibody activities directed against hemagglutinin. As shown in Table 1, all seven clones neutralized the homologous virus and reacted with this strain in a radioimmune assay. All seven clones also immunoprecipitated the hemagglutinin of Edmonston strain, whereas only six clones revealed HI activity (Table 1), thus indicating that this structural protein of measles virus carries at least two different epitopes. By using these antibodies in indirect immunofluorescent studies, distinct differences could be noted in cells lytically or persistently infected with LEC virus (Table 1). Only one out of four monoclonal antibodies reacting with the LEC virus in the lytic infection recognize the HA in the persistent infection, suggesting antigenic change has arisen during the persistent infection. When these antibody clones were reacted with various isolates of measles and SSPE viruses in the same assay, differences between the wild type Woodfolk isolate of measles and the attenuated vaccine isolate of measles (Edmonston), as well as between the two SSPE isolated (LEC and Mantooth), could also be detected. Thus, not only do antigenic changes of the HA occur during persistency but also different measles virus isolates show distinct antigenic differences.

Table 1. Analyses of the Antigenic Relationships Between the HA of Various Strains of Measles Viruses by Monoclonal Antibodies.

		34	13/1	26	32A	15/1	15/2	17/3
Biological	**HI**	**64**	**2048**	**2048**	**2048**	**2048**	**2048**	**Negative**
activities	**NT**	**320**	**1280**	**1280**	**2560**	**1280**	**2560**	**320**
(homologous virus)	**RIA**	$10^{-3.5}$	$10^{-6.88}$	$10^{-7.5}$	$10^{-7.5}$	$10^{-8.0}$	$10^{-8.0}$	10^{-5}
Edmonston		+	+	+	+	+	+	+
Woodfolk		−	+	−	+	−	+	+
LEC		−	+	−	+	−	+	+
Mantooth		+	+	+	+	+	+	+
LEC (persistent)		+	−	−	−	−	−	−

Conclusion

Comparative analyses of certain replicative steps of measles virus in both lytic and persistent infections has revealed differences in the synthesis of viral-specific proteins. Also differences in the structure and antigenicity of the hemagglutinin have been detected. Whereas in the lytic infection all major structural proteins are easily detected, in the persistent infection only H, N and F can be found readily. This would suggest that the expression of particular viral proteins is blocked either at the level of transcription or at the level of translation. However, functional mRNA can be detected for all major viral proteins, including M, in a persistent infection, albeit at a lower level than that found in lytic infections. Although the data suggest a block at the level of translation, the methods used to detect protein synthesis in persistent infections *in vivo* may not be sensitive enough to permit such an interpretation.

Using monoclonal antibodies raised against purified virus, the polypeptides responsible for the hemagglutinin reaction have been identified from persistent and lytic infections, both *in vivo* and *in vitro*.

In addition, when mRNA from a lytic infection is translated *in vitro*, 2 polypeptides are precipitated with monoclonal antibodies directed against the hemagglutinin; but only 1 species can be seen in samples from persistent infections. Also antigenic differences are observed when samples from lytic and persistent infections are assayed with a family or different monoclonal antibodies. These differences can be distinguished from those between naturally arising measles variants and are thought to arise from mutations occurring during the persistent infection.

References

1. Fraser, K.B. and Martin, S.J. (1978): *Measles Virus and Its Biology*. Academic Press, New York.
2. ter Meulen, V., Katz, M., and Müller, D. (1972): *Curr. Top. Microbiol. Immunol.* 57:1–38.
3. Hall, W.W. and Choppin, P.W. (1979): *J. Virol.* 99:443–447.
4. Barbanti-Brodano, G., Oyanagi, S., Katz, M., and Koprowski, H. (1970): *Proc. Soc. Exp. Biol.* 134:230–236.
5. Lamb, R.W. and Choppin, P.W. (1977): *Virology* 81:382–397.
6. Carrasco, L. and Smith, A.E. (1976): *Nature* 264:807–809.
7. Holland, J.J. and Villarreal, L.P. (1974): *PNAS* 71:2956–2960.
8. Rowlands, D., Grabau, E., Spindler, K., Jones, C., Semler, B., and Holland, J. (1980): *Cell* 19: 4:871–880.
9. Stephenson, J.R. and ter Meulen, V. (1979): *Proc. Natl. Acad. Sci. U.S.A.* 76:12:6601–6605.
10. Stephenson, J.R., Hay, A.J., and Skehel, J.J. (1977): *J. Gen. Virol.* 36:237–248.
11. Siddell, S.G., Wege, H., Barthel, A., and ter Meulen, V. (1980): *J. Virol.* 33:10–17.
12. Koprowski, H., Gerhard, W., and Croce, C.M. (1977): *Proc. Natl. Acad. Sci. U.S.A.* 74:7:2985–2988.

David H.L. Bishop and Richard W. Compans, eds.
The Replication of Negative Strand Viruses

Acute St Plus DI Infection of BHK Cells leading to Persistent Infection: Accumulation of Intracellular Nucleocapsids [a]

Laurent Roux, Pascale Beffy and
Francis A. Waldvogel [b]

Abstract

We have followed the accumulation of intracellular nucleocapsids after infection of BHK cells with standard (St) and defective interfering (DI) particles of Sendai virus (mixed virus infection). We have found that in mixed infections, DI particles, which allowed complete survival of infected cells, did not cause a dampening of viral metabolism. In fact, early after infection (30–80 hours), the amount of nucleocapsids found in mixed virus infected cells even exceeded the one found in St virus infected cells. Later on (5 to 20 days), in mixed virus infected cells, the amount of nucleocapsids (mainly of DI size) decreased to a minimum, increased during the next 10 days reaching a level corresponding to the one found in persistently infected cells.

Introduction

Two lines of data have been published dealing with the understanding of the mechanism of RNA virus persistency. On the one hand, various authors have shown involvement of temperature-sensitive (*ts*) mutants in establishment and maintenance of reovirus, NDV and VSV persistent infections in mouse cells.[1–3] On the other hand, numerous reports have demonstrated the role that DI particles play in establishing and maintaining persistency of various RNA viruses in BHK cells. These reports concern VSV,[4] reovirus,[5]

[a] These results are part of data which have been submitted to publication in *Virology*.

[b] Infectious Disease Division, Department of Medicine, University of Geneva Medical School, University Hospital, 1211 Geneva 4, Switzerland.

rabies virus,[6] LCM virus,[7,8] measles virus,[9] Japanese encephalitis virus,[10] Sendai virus,[11] Sindbis virus,[12] and Semliki forest virus.[13]

Recently, Semler and Holland,[14] Holland et al.,[15] and Rowlands et al.[16] have described the evolution of VSV genomes during long-term DI mediated infection of BHK cells. They showed that both non-defective (ND) and DI genomes were undergoing numerous mutations during the course of persistency and that finally, after 69 months, an extremely mutated virus was recovered which was able to re-establish persistent infection without the aid of DI particles. These recent data demonstrate that DI particles are not required to establish persistency, when the infecting virus has lost its cytopathogenicity, due to accumulation of debilitating mutations.

The fact still remains that DI particles are an absolute requirement for establishing a persistent infection with a highly infectious virus in cells, especially when the cells are not protected by interferon. We were therefore interested in precising the role of DI particles during an acute infection leading to survival of infected cells and establishment of persistency. To do so, we infected BHK cells with St plus DI particles of Sendai virus (mixed virus infection) and measured the accumulation of intracellular nucleocapsids throughout the period of establishment of persistency.

Results

Follow-up of Nucleocapsids Accumulation After Acute Mixed Virus Infection

One way to follow the effect of DI particles upon mixed virus infection was to measure the amount of nucleocapsids present in the cells. Since nucleocapsids are composed of viral genomic RNA and protein, their amount is somehow a reflection of viral genome replication, transcription and translation in infected cells.

Therefore, St. virus or mixed virus infected BHK cells were harvested at various times after infection. Cytoplasmic extracts were prepared[11] and the nucleocapsids were purified by sedimentation on CsC1 gradients.[11] The nucleocapsids were pelleted and resuspended in polyacrylamide gel electrophoresis sample buffer[17] and electrophoresed on discontinuous polyacrylamide slab gels.[17] After the electrophoresis, the proteins were stained with Coomassie brillant blue and scanned in a Joyce-Loebl densitometer. The amount of N-protein was estimated and this served as a measure of nucleocapsid amount. To be able to compare samples at different times of harvest, ribosomal RNA present in the pellet of the CsC1 gradient tube was measured (OD_{259}), and served as a mean to standardize the amount of nucleocapsids per number of cells.

Figure 1 shows the results of such an experiment. Early after infection (30–80 hours), the nucleocapsid accumulation in mixed virus infected cells reached a peak which exceeded the one found in St virus infected cells.

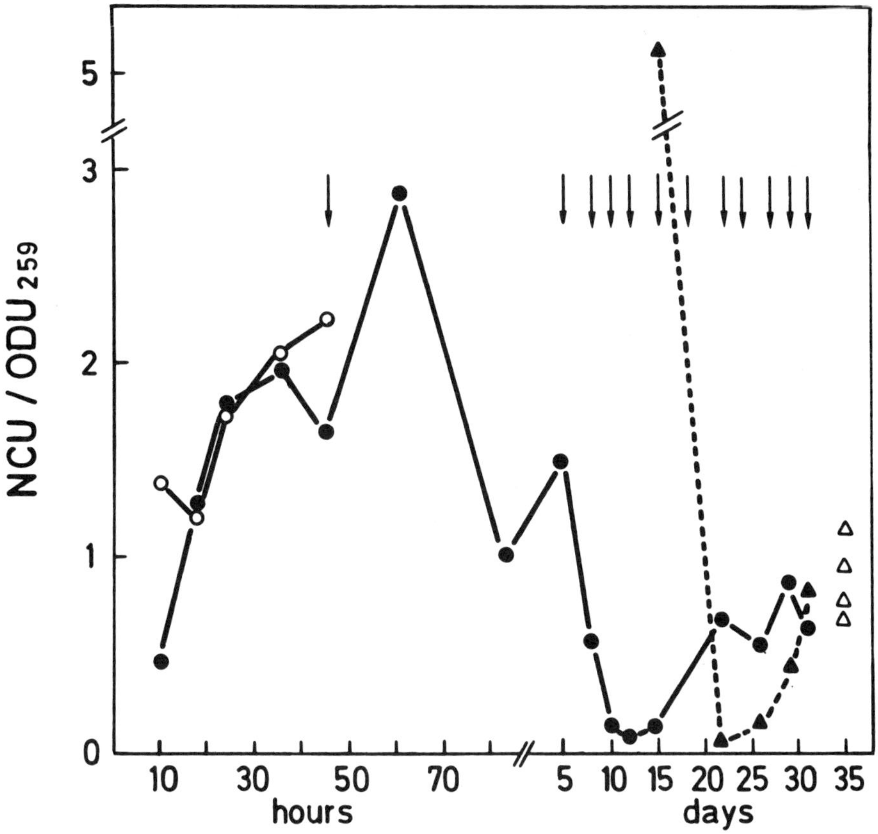

Figure 1. Long-term accumulation of nucleocapsids in St plus DI virus infected BHK cells. Samples of 2×10^7 St virus or mixed virus infected BHK cells were harvested at various times after infection and intracellular nucleocapsids were isolated as described previously.[11] Their amount was estimated as described in the text. At day 14, a sample of cells was challenged with 10–20 PFU/cell of St virus and further on the challenged culture was analyzed in parallel with the non-challenged culture. (○—○) St virus infection. (●—●) Mixed virus infection. (▲---▲) Mixed virus infected culture challenged with St virus at day 14. (△) Nucleocapsid amount in 4 different isolates from a Sendai persistently infected BHK culture (6 months old). Arrows indicate times at which the cultures were passaged. NCU/OD_{259}: relative nucleocapsid unit expressed as mg of N-protein per unit of absorbance at 259 μm of the ribosomal RNA found in the pellet of the CsC1 gradient used to isolate nucleocapsids.[11]

Later on, this amount decreased to become barely detectable around 10 to 15 days It then slowly increased during the next 15 days to reach a level similar to that found in various isolates from persistently infected (pi) cells (open triangle). From this experiment, we concluded that survival of the cells after mixed virus infection cannot be correlated with dampening of infection by DI particles.

Characterization of the Mixed Virus Infected Cells at Later Phases After Infection

It was of interest to characterize further the mixed infected cells during the intermediate phase of the infection (5 to 20 days) to see if the cells were still infected, even though the amount of viral macromolecules were barely detectable.

To do so, the mixed virus infected cells were challenged at day 14 with St Sendai virus (moi 10–20) and the fate of nucleocapsids was followed. The cells fully resisted to challenge and as shown in Figure 1 (day 15, closed triangle), the challenge provoked a burst of intracellular nucleocapsids. The nature of the nucleocapsids present at day 15 in the challenged and the non-challenged culture was determined by sizing the ^{35}S-methionine pulse-labeled nucleocapsids by sedimentation on sucrose gradients. As shown in Figure 2, panel 3E, the nucleocapsids synthesized at day 15 in the non-challenged culture is of DI size. Interestingly, the majority of the newly synthesized nucleocapsids after challenge was also of DI size (Figure 2, panel C).

Another way to characterize the mixed virus infected cells was to stain them with anti-Sendai fluorescent antibodies at day 15 prior or after St virus challenge. As shown in Figure 3, if only a small percentage of the cells showed a positive reaction prior to the challenge (Figure 3A), all the cells, however, were infected after challenge (Figure 3B). These results indicate that at day 15, all the cells contained at least a small amount of DI genomes capable of interference with the incoming St virus. During the late phase of the infection (20 to 30 days), while the amount of nucleocapsids in the unchallenged cells was progressively reaching the level found in persistently infected cells, the amount of cellular viral antigens also increased spontaneously, so that by day 31, all the cells showed a positive reaction to antibody

Figure 2. Sucrose gradient analysis of nucleocapsid sizes. In parallel with the follow-up of NC accumulation presented in Figure 1, samples of 10^7 cells were labeled with ^{35}S-methionine (20 μCi/ml) for 3 hours before harvest of the cells at the time indicated in Figure 1. ^{35}S-methionine labeled cells were disrupted and ^{35}S-methionine NC isolated on CsC1 gradients as described previously.[11] CsC1 solution containing NC (0.5 ml) were diluted 10 times with TNE and 1 ml samples were loaded onto linear (15–65%) (w/v) sucrose gradients in TNE and centrifuged for 2½ hr at 36,000 rpm, 10°C, in a SW 41 Beckman rotor. Fractions were collected and position of NC determined by liquid scintillation counting. Similar analyses were performed with ^{3}H-uridine labeled NC to etermine the position of ND or DI-NC on this type of gradient. (A) ND-^{3}H-uridine labeled NC. (B) ND and DI ^{3}H-uridine labeled NC. (C) ^{35}S-labeled NC isolated from challenged culture at day 15. (D) ^{35}S-labeled NC isolated at 45 hours after initial mixed virus infection. (E) ^{35}S-labeled NC isolated from the non-challenged culture at day 15. (F) Mock infected sample. Sedimentation is from right to left.

fluorescent staining (Figure 3C). At that time, the synthesized nucleocapsids were again mostly of DI size (data not shown). In this late phase, the culture was found fully resistant to challenge by St Sendai virus, but was completely destroyed by VSV infection (data not shown).

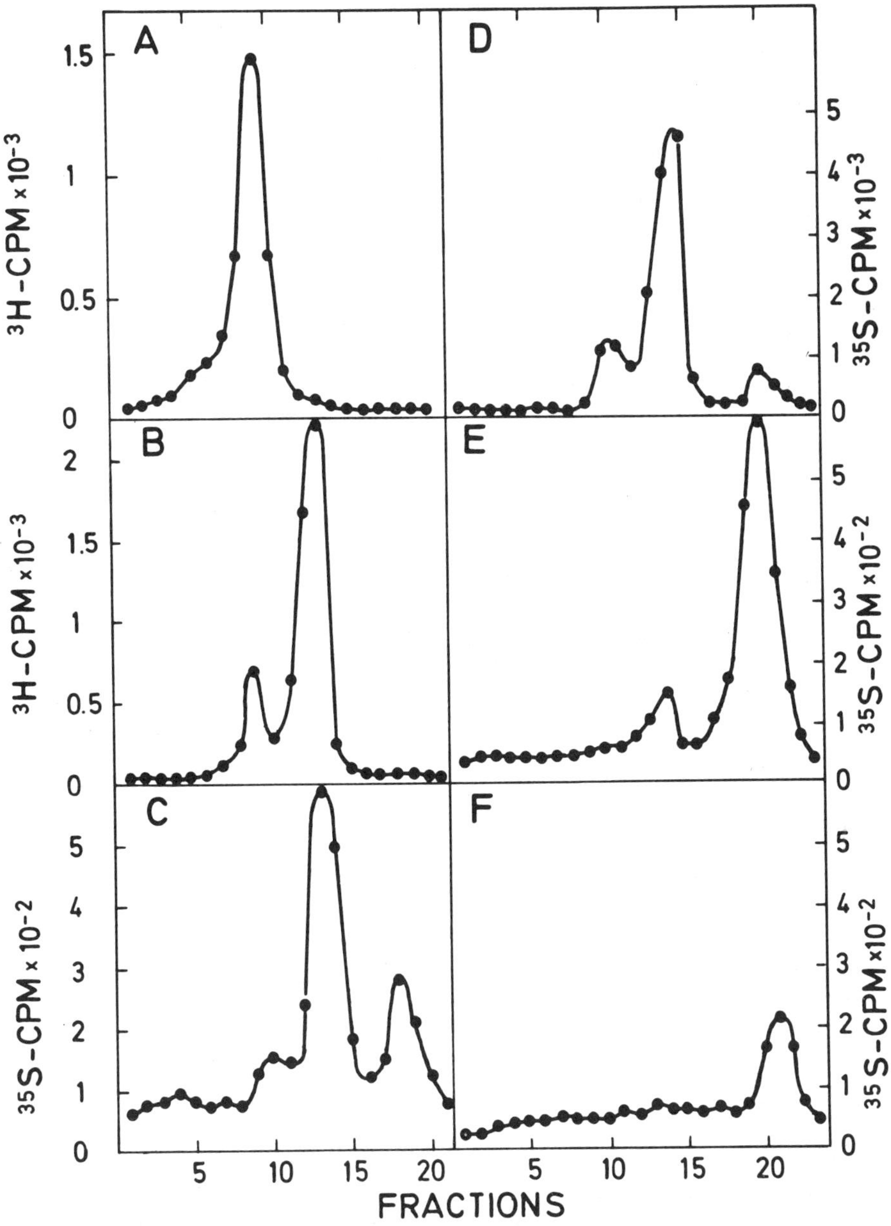

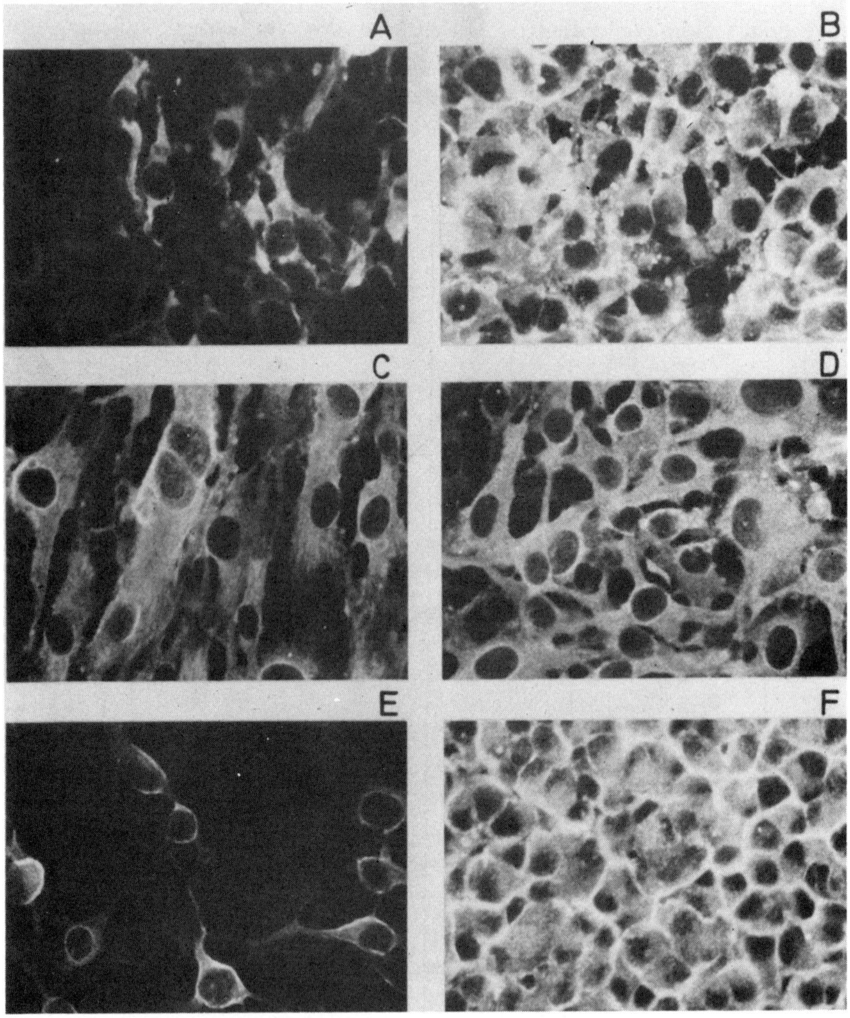

Figure 3. Presence of viral antigen in mixed virus infected cells. Infected St virus BHK cells (moi 0.1 or 20), mixed virus infected BHK cells challenged or non-challenged with St virus (see Figure 1 and text) are stained with fluoroscein labeled anti-Sendai antibodies as described previously.[11] (A) Mixed virus infected cells at day 15. (B) Mixed virus infected cells at day 15 but challenged at day 14. (C) and (D) Mixed virus infected cells at day 31, non-challenged and challenged. (E) St virus infected cells (moi 0.1). (F) St virus infected cells (moi 20).

Discussion

The data presented here demonstrate that survival of mixed virus infected BHK cells leading to persistent infection is not due to dampening of the initial infection by DI particles. They also show how the amount of intracellular nucleocapsids initially varies to reach the level observed in persistently infected cells.

Since survival of cells upon mixed virus infection occurs without dampening of the infectious process by DI particles, the cells must therefore respond to the mixed virus infection by making the decision of survival. This response must be triggered by presence of DI genomes or DI products in the cell. As it is not yet understood why cells are dying upon St virus infection, we have no good hint to explain why survival takes place. However, we can certainly conclude that cell death cannot be uniquely due to large intracellular accumulation of viral products.

The following explanation can be offered to account for the fluctuating level of viral NC inside the mixed virus infected cells. Due to the replicating advantage of the DI genomes relative to the ND genomes, the DI genomes are replicated preferentially using the helper functions provided by the ND genomes, i.e., the viral polymerase and N-proteins. With time, the relative ratio of the DI genomes has increased to the level that the remaining ND genomes can no longer support this level of genome replication. This point represents the initial phase of the mixed virus infection (0–90 hours) and leads to a large accumulation of mostly DI genomes. This accumulation can easily exceed the amount of viral genomes found at the height of a St virus infection (Figure 1). During this time, however, cell division continues to take place at near normal rate and this now leads to a dilution of the viral nucleocapsids present per infected cell in the culture, which brings the NC level to a minimum level at approximately 15 days after infection (intermediate phase). At this point in time, our results indicate that all the cells present in the culture are still infected, at least with DI genomes, since: (a) these cells completely survive a challenge of the culture with St virus; (b) the challenge of the culture leads to a large increase in both the rate of NC synthesis and the total amount of viral NC (mostly DI) in the challenged culture and (c) all the cells are actually infected with the challenging virus.

To explain the late phase in the establishment of the persistent infection, i.e., the increase to the level of NC which we found in a long-term pi culture, two possibilities come to mind. The first is that cell division has re-established a lower ratio of DI to ND genomes in a fraction of the infected cells, and that in these cells the replication of the ND genomes can now escape the interference due to the large molar excess of the DI genomes which limits genome accumulation during the initial phase of the infection. It is also possible that a small portion of cells (positive to antibody fluorescent staining without challenge ?) eventually appear which, due to cell division, now contain only ND genomes. These provide the necessary helper function to allow an increase in the level of DI genomes found at 30 days after infection. In this way, a new equilibrium between DI and ND genomes is reached which now leads to a different level of NC (again mostly DI) present in pi culture.

In our description of the mixed virus infection leading to persistency, we were only concerned with the amount of intracellular NC and with the relative ratios of DI to ND genomes in the infected cells. However, while the mixed infection was evolving, another process could have taken place, which could also modulate the new balance of intracellular NC seen in the late phase, i.e., the appearance of mutations in the viral genomes. Holland and coworkers have recently shown that DI and ND genomes undergo multiple mutations during the course of a VSV-BHK persistent infection. These authors found that the ends of the DI genomes in particular contained a surprisingly high percentage of base substitution in the first 46 nucleotides[14] as well as mutations throughout the entire ND genomes.[15] After 69 months of persistency, they eventually isolated a ND virus capable of establishing persistent infection without the aid of DI particles, presumably due to the multiple mutations that had taken place.[16] The appearance of these mutations are understood as a process of selection towards a more suitable virus for prolonged non-cytocidal infection.

It is therefore possible that in the late phase (20 to 30 days) of the mixed infection we have described, mutations have already occurred in the viral genomes, which now modulate the relationship between DI and ND genomes differently. Similarly, mutations in the ND genome alone could have already altered the cytopathogenicity of the St virus in such a way that the St virus by itself could now maintain the carrier state. However, in view of the period of time necessary to generate the highly non-cytocidal VSV capable of establishing persistent infection by itself (69 months), we consider this possibility to be unlikely. But characterization of the viral genomes during the establishment of our persistent infection is necessary before we can eliminate the above possibility.

References

1. Fields, B.N. and Raine, C.S. (1974): In: *Mechanism of Viral Diseases*. Fox, C.F. (Ed.), Menlo Park, California, W.A. Benjamin, pp. 161–167.
2. Preble, O.T. and Youngner, J.S. (1973): *J. Virol.* 5:559–567.
3. Youngner, J., Dubovi, E.J., Quagliana, D.O., Kelly, M., and Preble O. T. (1976): *J. Virol.* 19:90–101.
4. Holland, J.J. and Villarreal, L.P. (1974): *Proc. Nat. Acad. Sci.* 71:2956–2960.
5. Ahmed, R. and Graham, A.F. (1977): *J. Virol.* 23:250–262.
6. Kawai, A., Matsumoto, S., and Tanabe, K. (1975): *Virology* 67:520–533.
7. Welsh, R.M., Burner, P.A., Holland, J.J., Oldstone, M.B.A., Thompson, H.A., and Villarreal, L.P. (1975): *Bull. World Health Organization* 52:403–408.
8. Popescu, M. and Lehmann-Grube, E. (1977): *Virology* 77:78–83.
9. Rima, B.K., Davidson, W.B., and Martin, S.J. (1977): *J. Gen. Virol.* 35:89–97.
10. Schmaljohn, C. and Blair, C.D. (1977): *J. Virol.* 24:580–589.
11. Roux, L. and Holland, J.J. (1979): *Virology* 93:91–103.
12. Weiss, B., Rosenthal, R., and Schlesinger, S. (1980): *J. Virol.* 33:463–474.

13. Meinkoth, J. and Kennedy, S.I.T. (1980): *Virology* 100:141–155.
14. Semler, B.L. and Holland, J.J. (1979): *J. Virol.* 32:420–428.
15. Holland, J.J., Grabau, E.A., Jones, C.L., and Semler, B.L. (1979): *Cell* 16:495–504.
16. Rowlands, D., Grabau, E.A., Spindler, K., Jones, C.L., Semler, B.L., and Holland, J.J. (1980): *Cell* 19:871–880.
17. Roux, L. and Holland, J.J. (1980): *Virology* 100:53–64.

David H. L. Bishop and Richard W. Compans, eds.
The Replication of Negative Strand Viruses

Evidence of Antigenic Variation in a Persistent *in Vitro* Measles Virus Infection

Steven J. Robbins and Fred Rapp[a]

Two viruses isolated shortly after the establishment of a persistent measles virus infection in AV_3 cells were characterized by various immunological methods. The isolated viruses were different from the parental Edmonston strain of measles virus when examined by neutralization, hemagglutination inhibition (HAI), and analysis of virus proteins by sodium dodecyl sulfate polyacrylamide gel electrophoresis (SDS-PAGE) of specific immunoprecipitates. These findings indicate that antigenic variants of measles virus are generated early in the establishment of persistent infection of AV_3 cells in the absence of immunological selection pressures.

Introduction

A variety of factors have been theorized to play a role in the establishment and maintenance of persistent measles virus infections *in vivo*. Recent molecular and immunological studies suggest that SSPE involves genetic variants of measles virus.[1-3] However, it has not been determined whether such virus variants are involved in the primary infection of SSPE patients or arise during the course of virus replication within the patient.

Previous studies by Gould and Linton [4] and others [5-8] have shown that virus variants can be isolated from measles virus carrier cultures, following long-term passage of the cells in the laboratory. In this report, we describe

[a] The Pennsylvania State University College of Medicine, Hershey, Pennsylvania.

the appearance of virus antigenic variants in a measles virus carrier culture shortly after establishment of the persistent state. These findings suggest that variants of measles virus arise in the absence of immunological selection pressures and that virus genetic drift occurs early in the evolution of persistent measles virus infections.

Materials and Methods

Virus, cells, immunological methods, and the procedures used for radiolabeling proteins and polyacrylamide gel electrophoresis are described elsewhere.[9-12]

Results

General characteristics of AV_3 cell cultures persistently infected with measles virus. A measles virus carrier cell line (AV_3/MV) was established from AV_3 cells surviving a primary infection by the Edmonston strain of measles virus. Measles virus-specific antiserum was not used in the establishment or maintenance of this cell line. The carrier cultures shed infectious virus, demonstrated measles virus-specific antigens in 90–100% of the cells when assayed by a fluorescent antibody technique, and hemadsorbed green monkey erythrocytes.

*Isolation and partial characterization of virus isolates from AV_3/*MV cells. In an attempt to determine whether temperature-sensitive (*ts*) virus variants were generated in the AV_3/MV cells, plating efficiency (P.E.) studies were performed on representative virus isolates selected from early passages of the AV_3/MV cell cultures; passage 10 (P10) virus, isolated 78 days following primary infection; and passage 19 (P19) virus, isolated 58 days later.

Table 1 shows the results of P.E. studies performed with plaque-purified Edmonston, P10, and P19 viruses. No significant differences in titer were observed for the three viruses when stocks were titered at 39.5 or 33.5° (P.E. ratios $> 10^{-1}$). In addition, none of the other early passage virus isolates we examined to date were more temperature-sensitive than the P10 and P19 isolates.

Further characterization of the three viruses by other virological methods (fluorescent antibody, plaque size, and hemadsorption) did not reveal any clear differences between the viruses. Neither the P10 nor P19 virus isolates seemed more predisposed to establishing persistent infections than parental virus. These observations suggest that the isolated virus were simply reisolated parental virus. To test this hypothesis, we measured the relatedness of the viruses to the parental Edmonston strain of measles virus by certain immunological criteria.

Table 1. Plating Efficiency of Measles Virus and Virus Variants.

Virus	Titer[a]		Plating efficiency ratio
	39.5°	33.5°	$Titer_{(39.5°)}/Titer_{(33.5°)}$
Edmonston	0.90	1.55	0.58
P10 isolate	0.51	0.84	0.61
P19 isolate	0.36	1.34	0.27

[a]Titers shown indicate the number of plaque forming units (PFU) per cell in virus stocks derived from low multiplicity infections (0.01–0.05 PFU/cell) in Vero cells incubated at 37°. Titration of virus stocks was conducted at the temperatures indicated on Vero cell monolayers as described previously.[13]

Immunological studies. Neutralization assays performed on P10, P19, and parental measles viruses are shown in Figure 1. While the P10 virus isolate was not substantially more resistant to neutralization with parental virus-specific antiserum than the parental virus, the P19 virus clearly was. Simi-

Figure 1. Neutralization titration curves of Edmonston measles virus (●—●) and two virus variants [P10 ■—■) and P19 (▲—▲)] derived from AV_3/MV cells. Units listed on the ordinate refer to the percent of plaques present on experimental plates as compared to control plates (where virus aliquots were incubated with non-immune animal antiserum).

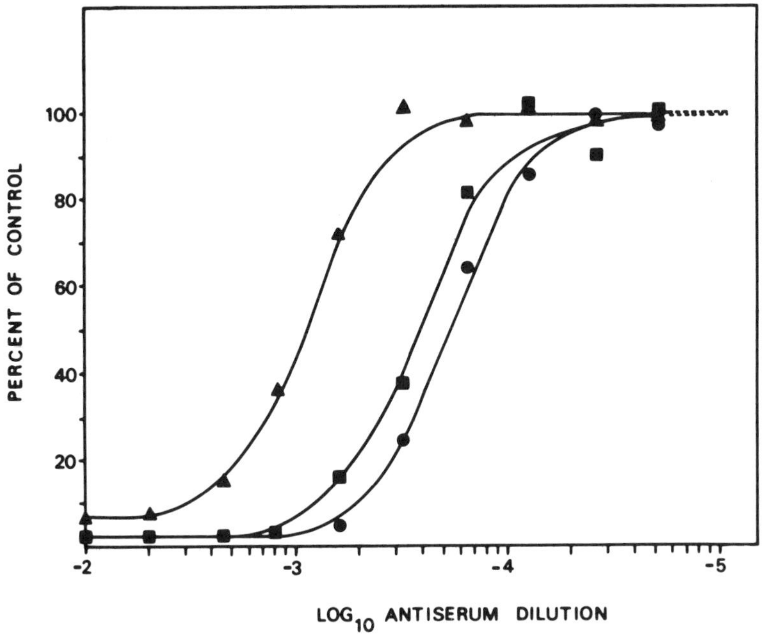

larly, the specificity of HAI by parental virus-specific antiserum was different for the three viruses (see Table 2).

Analysis of the virus-specific polypeptides immunoprecipitated from infected Vero cells and analyzed by SDS-PAGE are shown in Figure 2. While most of the major virus-specified structural proteins were detected in each immunoprecipitate, subtle differences between the patterns were apparent. All immunoprecipitates showed the presence of a fragment of the NP protein, designated NP_1, but the P10 isolate generated substantially more of this fragment. The increase in the P10 NP_1 band also corresponded to a substantial decrease in the P10 NP protein band. Although this might have been the result of proteolytic degradation of the NP band in the immunoprecipitation reaction, the absence of equivalent degradation of the parental and P19 NP bands argued against this. The prevalence of the NP_1 band in the P10 immunoprecipitate was similar, however, to that reported previously by Ramsey et al.[14] for a measles-like virus (IP-3) isolated from an SSPE patient.

The polypeptides precipitated from P19 virus infected cells showed a decrease in the upper band of the two H-proteins observed in parental virus infected cell immunoprecipitates. While this might also be attributable to proteolysis, the reproducibility of this difference and the similarity of this observation to that reported previously by Wechsler et al.[8] suggest that it represented a real difference between the parental and P19 viruses. Both the P10 and P19 viruses showed a decrease in the relative amount of F_1 protein precipitated and the corresponding appearance of a second protein band immediately underneath the F_1 band.

Discussion

The present study has demonstrated that subtle antigenic variants of measles virus are produced shortly after the establishment of persistent infection in AV_3 cells. The variants produced were virologically indistinguishable from parental virus and were not temperature-sensitive. The antigenic variation we observed in the two virus isolates was retained after plaque-purification, indicating that it was genetic in origin and not due to a temporary phenotypic change.

Table 2. HAI Titrations of Edmonston Measles Virus-Specific Animal Antiserum and Gradient-Purified Viruses.

Virus	HAI titer[a]
Edmonston	1536
P10 isolate	512
P129 isolate	1024

[a]The HAI titer given is the average titer based on three separate microtiter assays and is expressed as HAI units/0.025 ml.

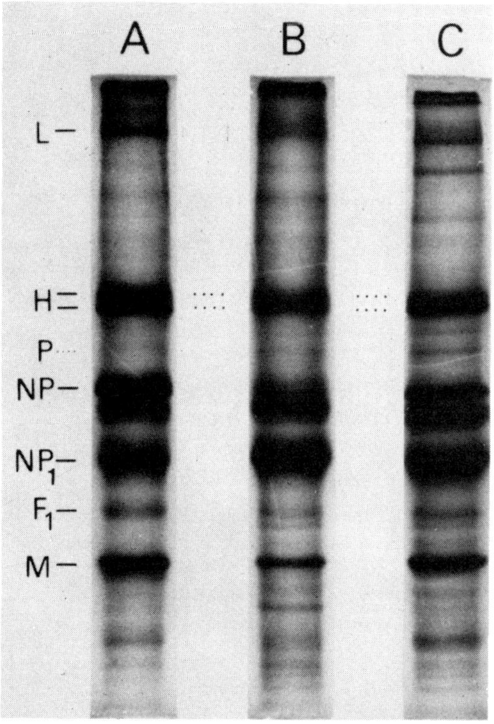

Figure 2. SDS-PAGE polypeptide profiles of ^{35}S-methionine-labeled virus polypeptides immunoprecipitated with Edmonston measles virus-specific rabbit antiserum from Vero cells productively infected with (A) Edmonston measles virus, (B) P10 isolate, and (C) P19 isolate. Designations on the left of the figure indicate the measles virus structural proteins: L, H, P, NP, F_1, and M. The NP_1 designation refers to the migration of a major NP polypeptide fragment.[14]

Since immune sera were not used in the establishment or maintenance of the AV_3/MV cells, the evolution of antigenic variants within the cultures seems paradoxical. While it is possible that the antigenic variations arose as side effects of functional mutations which conferred survival advantages on the viruses, it seems equally possible that the antigenic variations occurred as the result of simple virus genetic drift in an infection where other variables were regulating the virus-carrier state. The former of these two prospects, adaptation to persistence, fits the observations made by others concerning the evolution of *ts*-virus mutants from certain measles virus carrier cell lines.[4,5,7,8] However, this does not seem to fit our present observations. On the basis of this discrepancy, simple virus drift seems a more likely explanation for the antigenic variation we have observed. This suggests that measles virus genetic drift is a concurrent phenomenon with *in vitro* persistence, but the two processes are not necessarily connected.

References

1. Payne, F.E. and Baublis, J.V. (1973): *J. Infect. Dis.* 127:505–511.
2. Wechsler, S.L. and Fields, B.N. (1978): *Nature* 272:458–460.
3. Hall, W.W., Kiessling, W., and ter Meulen, V. (1978): *Nature* 272:460–462.
4. Gould, E.A. and Linton, P.E. (1975): *J. Gen. Virol.* 28:21–28.
5. Chiarini, A., Ammatuna, P., di Stefano, R., and Sinatra, A. (1978): *Arch. Virol.* 56:263–268.
6. Fisher, L. and Rapp, F. (1979): *J. Virol.* 30:64–68.
7. Ju, G., Udem, S., Rager-Zisman, B., and Bloom, B.R. (1978): *J. Exp. Med.* 147:1637–1652.
8. Wechsler, S.L., Rustigian, R., Stallcup, K.C., Byers, K.B., Winston, S.H., and Fields, B.N. (1979): *J. Virol.* 37:677–684.
9. Robbins, S.J. and Bussell, R.H. (1979): *Intervirol.* 12:96–102.
10. Robbins, S.J. and Rapp, F. (1980): *Virology* 106:317–326.
11. Lamb, R.A., Etkind, P.R., and Choppin, P.W. (1978): *Virology* 91:60–78.
12. Laemmli, U.K. (1970): *Nature* 227:680–685.
13. Portner, A. and Bussell, R.H. (1973): *J. Virol.* 11:46–53.
14. Ramsey, J.C., Eron, L., Sprague, J., Bensky, P., Jones, D., Dunlap, R., and Albrecht, P. (1978): In: *Persistent Viruses.* Stevens, J.G., Todaro, G.J., and Fox, C.F. (Eds.), Academic Press, New York, pp. 635–641.

David H. L. Bishop and Richard W. Compans, eds.
The Replication of Negative Strand Viruses

Chronic Measles Virus Infection of Mouse Nerve Cells *in Vitro*

Bernard Rentier, Anne Claysmith,
William J. Bellini, and Monique Dubois-Dalcq[a]

Summary

Mouse embryo spinal cord cultures were infected at 10 days with a high multiplicity of Edmonston measles virus. Two days after infection, measles virus antigens were detected in cell bodies of 5 to 10% of the neurons. After 7 days and up to 40 days, 30 to 40% of the neurons were infected as shown by immunofluorescence, immunoperoxidase, and hemadsorption while neither cell fusion, lysis, nor other cytopathic effects could be seen. Viral antigens were found in neuron soma as well as on the intricate neurite network, but not on other cell types. Infected culture medium failed to infect Vero cells or other neuron cultures and did not interfere with infection of Vero cells by parental virus. Polypeptide analysis showed that all measles proteins were present and that the HA and P polypeptides were slightly modified. Cocultivation of infected neuron cultures with Vero cells allowed rescue of infectious measles virus.

Introduction

Measles virus causes persistent infections of the central nervous system such as subacute sclerosing panencephalitis (SSPE). However, the etiological mechanism by which measles virus persists in nerve cells is still poorly understood. Chronic infection of animals or cells in culture has been obtained with a variety of strains such as neuroadapted or species-adapted

[a]Infectious Diseases and Neuroimmunology Branches, National Institute of Neurological and Communicative Disorders and Stroke, National Institutes of Health, Bethesda, Maryland.

measles virus, or SSPE isolates.[1] Our aim was to investigate if inoculation of dissociated mouse neuron cultures with a wild type strain of measles virus could result in persistent infection.

Results

Dissociated neuron cultures from 13-day-old mouse embryo spinal cord were grown on glass coverslips or in Petri dishes.[2] After 10 days in culture, they were inoculated with a high multiplicity (20 PFU/cell) of a plaque-purified measles virus, Edmonston B strain.[3] The course of infection was followed by immunofluorescence (IF) [4] using human anti-measles IgG.[5] Two days postinfection (p.i.), measles virus antigens were detected in neurons somas but not in non-neuronal cell (Figure 1). Long neuronal processes rarely contained viral antigens. Seven days p.i., 30 to 40% of the neuron population contained measles virus antigens in their processes as well as in the somas (Figure 2). From 2 to 4 weeks after infection, the extensive neurite network that had formed in the cultures was heavily labeled by IF (Figure 3). Two months p.i., the percentage of infected neurons had remained the same although a subtle loss of neurons occurred with time in both infected and uninfected cultures. No cytopathic effect was detected by light microscopy at any time. No cell fusion was observed, nor any alteration in shape or neurite outgrowth. Infected neurons and their processes, which contained viral antigens as detected by IF, were also hemadsorbant [5] (Figures 2, inset, and 4). The membrane of infected neurons and processes could also be labeled by a monoclonal antibody specific for measles virus hemagglutinin (HA).[6]

Infectious virus could not be recovered in medium which had been collected from infected cultures every other day from 2 to 32 days p.i. and inoculated into Vero cells. No syncytium formed and no viral antigen appeared in Vero cells even after 7 days at 37°C or at 32°C. Infected neuron culture medium was also inoculated into other 10-day-old neuron cultures in five successive blind passages. No viral antigen was found after any passage. However, 2 days after seeding Vero cells over neuron cultures infected for 7, 14, 18 and 25 days respectively, syncytia appeared which were brightly labeled by IF for measles virus antigens. Infectious virus was recovered in coculture medium. Its titer progressively increased from 10 PFU/ml at 7 days, to 10^3 PFU/ml at 14 days and 2×10^4 PFU/ml at 18 days. It subsequently decreased to 2×10^2 PFU/ml at 25 days.

Electron microscopic observation of infected neurons after immunoperoxidase labeling [5,7,8] revealed the presence of viral antigen on the plasma membranes of neurons identified by synaptic contacts on their surface (Figures 5 and 6). Neuron cytoplasm contained large inclusions consisting of "fuzzy" nucleocapsids. However, no nucleocapsids were detected in close association with the plasma membrane. Neurites as well contained

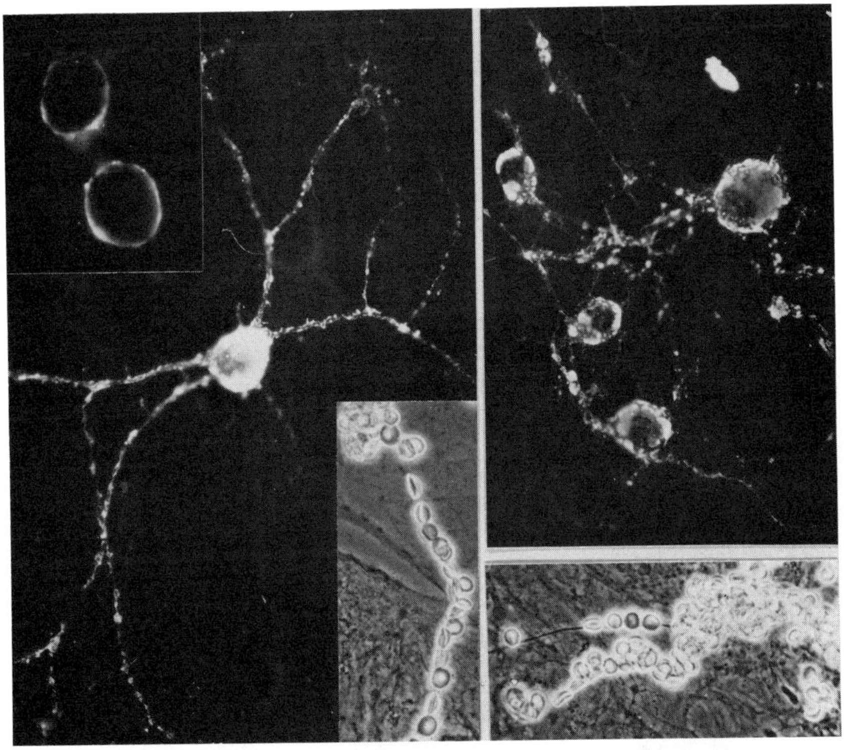

Figure 1–4. Labeling of measles virus antigens in neuron cultures. In Figure 1, at 2 days p.i., neuron soma, not neurites, displays virus-specific fluorescence. At 7 days p.i., neuron soma contains large viral inclusions, processes are stained as well (Figure 2). At 17 days p.i., several neurons and their intricate neurite network contain viral antigens (Figure 3). Figure 2, inset, and 4: hemadsorption on measles virus infected neuron cultures, at 7 days p.i. Erythrocytes adhere to neuron soma, and neurites.

viral nucleocapsids, sometimes of the "smooth" type, and had diffuse viral antigen on their membranes. Neuron plasma membrane formed numerous protrusions 300 to 600 nm in diameter, often filamentous, which resembled viral buds, except that they rarely contained nucleocapsids. Instead, they were usually filled with vesicles or tubules, approximately 30 nm in diameter, of unknown nature (Figures 6 and 7).

Measles virus polypeptides in the infected neuron cultures were analyzed by polyacrylamide gel electrophoresis [9] after a 16-hour pulse with ^{35}S-methionine (25 μCi/ml) at 13 days p.i., followed by lysis in RIPA buffer [10] and immunoprecipitation with hyperimmune antibody to measles virus [3] or monoclonal anti-hemagglutinin antibody.[6] Autoradiograms of the gels showed the presence of all measles virus polypeptides in the neuron culture lysates. When compared to HEp 2 cells persistently infected with measles

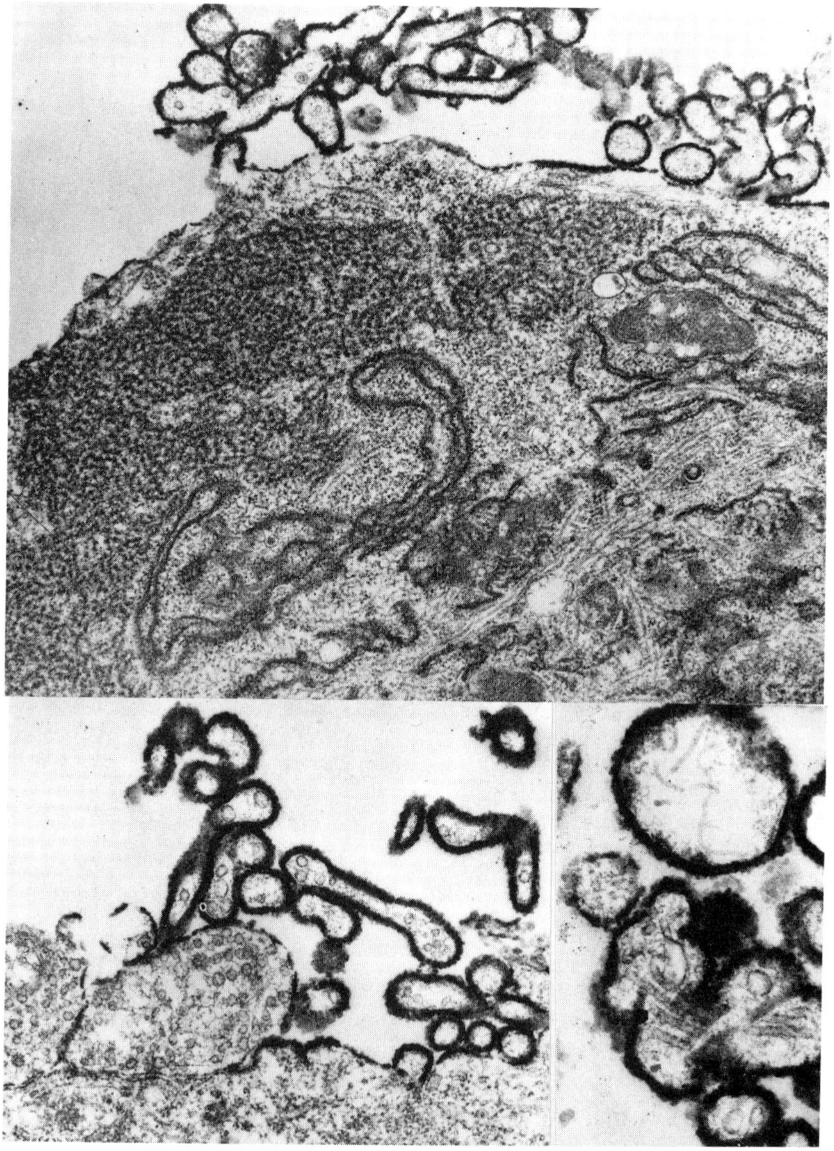

Figure 5–7. Electron micrographs of measles virus infected neurons, at 7 days p.i., after immunoperoxidase staining for surface virus antigens. Viral nucleocapsids (N) accumulate in cytoplasm. Areas of cell surface are labeled with black peroxidase reaction product, but label is heavier on bud-like particles. These are round or elongated and often contain vesicles (long arrows in Figures 5,6). Figure 6 shows detail of infected neuron to which synaptic terminal (S) is apposed. Virus-like particles rarely enclose nucleocapsids (short arrows) and sometimes display 30 nm wide tubules (arrowheads) (Figure 7).

virus (HEp2 P.I.), some differences were noted. Nucleoprotein (NP) was proportionally less abundant than the other polypeptides. HA migrated slightly faster and P polypeptide slightly slower than the corresponding polypeptides in HEp2 P.I. cells. M polypeptide was normal (Figure 8). In the infected neuron cultures, monoclonal anti-HA antibody precipitated an HA polypeptide which had the same modified electrophoretic mobility as when precipitated by anti-measles virus serum. Neuron-Vero cocultures showed normal synthesis of measles polypeptides with the normal electrophoretic mobilities (Figure 8).

Discussion

A selective and persistent infection of cultures spinal cord neurons was obtained *in vitro* with the Edmonston strain of measles virus. Several factors may have contributed to the establishment of viral persistence in neurons. The age of the cultures was important. A period of 8 to 12 days of *in vitro* maturation was necessary for successful infection, whereas neuron cultures more mature at the time of inoculation showed very few infected cells. This is reminiscent of the age dependency observed for establishment of persis-

Figure 8. Autoradiogram of a polyacrylamide gel after electrophoresis of immunoprecipitated cell lysates. In lanes *a* and *g*, lysate of HEp2 cells persistently infected with measles virus (HEp2PI) with normal rabbit serum (NRS). In *c*, HEp2PI lysate with measles convalescent human serum (CHS). In *d*, neuron culture lysate with HRS. In *e*, neuron culture lysate with NRS. In *f*, neuron culture lysate with CHS. In *h*, neuron-Vero coculture lysate with HRS. M-protein is present in neuron cultures (*d,f*) and is not modified. HA and P bands are closer to each other in neuron culture lysates than in HEp2PI (*a, c, g*) or in cocultures (*h*).

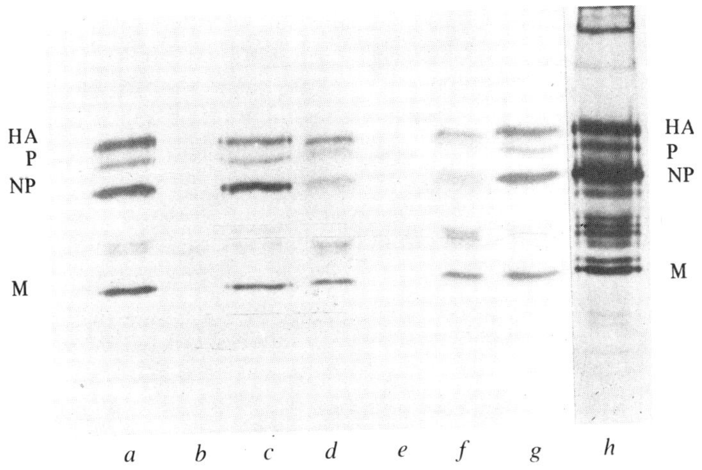

tent measles infection *in vivo*.[11,12] Neuronal infection was reproducible only when a high multiplicity was used and even then, only a fraction of the neurons present in the cultures could ever be infected. It is not known whether infected and non-infected neurons represented different classes of neurons. Specific markers should be used to investigate this question.

Virtually no infectious virus was ever released by the persistently infected neurons. The only way measles virus could be rescued was by cocultivation of permissive cells with infected neurons. Infection seemed then to occur by cell-to-cell contact or by fusion, since primary syncytia in cocultures always formed in the vicinity of infected neurons identified by IF. All these observations pointed to a defect in viral maturation which was confirmed by electron microscopy. Many viral buds were present at the neuron surface but they apparently failed to incorporate nucleocapsids and were instead, filled with vesicles and tubules of undetermined nature, larger than the cellular microtubules. Failure to incorporate nucleocapsids has been observed in experimental conditions with SSPE strains of measles virus [7,13,14] and has usually been associated with an absence of or a defect in M-protein.[15,16] In our system this was apparently not the case since M was present and its electrophoretic mobility was not modified. Only a small amount of NP was found even though nucleocapsids were seen in large numbers in neuron cytoplasm. However, the ^{35}S-methionine pulse of 16 hours was carried out on the 13th day of infection and at this time the synthesis of NP might have been slower because of a feedback regulation following early accumulation of nucleocapsids in the infected cell.

If the defective assembly cannot be attributed to any qualitative or quantitative defect in M or N-proteins, it might be related to the slight modification of electrophoretic mobility of HA and perhaps P. These changes suggest a possible alteration of glycosylation and phosphorylation which are important steps in virus maturation. For instance, cytochalasin B which at high doses blocks glycosylation, also blocks production of infectious measles virus and prevents synthesis of a normal HA polypeptide (Stallcup, Raine and Fields, in preparation). In neuron cultures, even though the electrophoretic mobility of HA was altered, its biological function and antigenicity were preserved since infected neurons were hemadsorbant and were labeled by a monoclonal antibody to measles virus HA.

In conclusion, a selective infection of mouse spinal cord neurons was obtained *in vitro* with the Edmonston strain of measles virus. This infection was persistent and no infectious virus was produced except after cocultivation. The defect in viral maturation was not related to an absence of M-protein synthesis by the host cell. It is possible that the HA molecule was not properly glycosylated and that the P-protein was also modified. The virus used was neither an SSPE isolate nor a neuroadapted strain and the success of this infection was more frequent at an early stage of *in vitro* neuronal maturation. Establishment of chronic measles infection in these neurons thus appears to be a host-dependent phenomenon.

ACKNOWLEDGMENTS

We wish to thank Gayl D. Silver and Raymond T. Rusten for their skillful help during this study and Lin Aspinall for the typing. Part of this study was presented at the 56th Annual Meeting of the American Association of Neuropathologists, New Orleans, Louisianna.[17]

References

1. Dubois-Dalcq, M. (1979): *Internat. Rev. Exp. Pathol.* 19:101.
2. Faulkner, G., Dubois-Dalcq, M., Hooghe-Peters, E.L., McFarland, H.F., and Lazzarini, R.A. (1979): *Cell.* 17:979.
3. Bellini, W.T., Trudgett, A., and McFarlin, D.E. (1979): *J. Gen. Virol.* 43:633.
4. Ecob-Johnston, M. (1977): *Neuropath. and Appl. Neurobiol.* 3:103.
5. Rentier, B., Hooghe-Peters, E.L., and Dubois-Dalcq, M. (1978): *J. Virol.* 28:567.
6. McFarlin, D.E., Bellini, W.J., Mingioli, E.S., Behar, T.N., and Trudgett, A. (1980): *J. Gen. Virol.* 48:425.
7. Dubois-Dalcq, M., Barbosa, L.H., Hamilton, R., and Sever, J.L. (1974): *Lab. Invest.* 30:241.
8. Hooghe-Peters, E.L., Rentier, B., and Dubois-Dalcq, M. (1979): *J. Virol.* 29:666.
9. Laemmli, U.K. (1970): *Nature, Lond.* 227:680.
10. Lamb, R.A., Etkind, P.R., and Choppin, P.W. (1978): *Virology* 91:60.
11. Griffin, E.E., Mullinix, J., Narayan, O., and Johnson, R.T. (1974): *Infect. Immun.* 9:690.
12. Herndon, R.M., Rena-Descalzi, L., Griffin, D.E., and Coyle, P.K. (1975): *Lab. Invest.* 33:544.
13. Raine, C.S., Feldman, L.A., Sheppard, R.D., Barbosa, L.H., and Bornstein, M.B. (1974): *Lab. Invest.* 31:42.
14. Dubois-Dalcq, M., Reese, T.S., Murphy, M., and Fucillo, D. (1976): *J. Virol.* 19:579.
15. Hall, W.W., Lamb, R.A., and Choppin, P.W. (1979): *Proc. Nat. Acad. Sci. U.S.A.* 76:2047.
16. Hall, W.W. and Choppin, P.W. (1979): *Virology* 99:443.
17. Rentier, B., Claysmith, A.P., Dubois-Dalcq, M., and Bellini, W.J. (1980): *Proc. 56th Ann. Meet. Am. Assoc. Neuropathol.*, p. 386, abstr. 147.

David H. L. Bishop and Richard W. Compans, eds.
The Replication of Negative Strand Viruses

Functional Analysis of Anti-HN Hybridoma Antibodies

Jonathan Yewdell and Walter U. Gerhard[a]

Introduction

The hemagglutinin-neuraminidase (HN) of paramyxoviruses if a dimeric glycoprotein (MW 140,000) which mediates adsorption of virus to target cell surfaces.[1] The HN molecule is responsible for the viral hemagglutinin and neuraminidase activities observed *in vitro*.[1] Immune mechanisms directed at the HN molecule appear to play an important role in the prevention and limitation of paramyxovirus infections. Thus, antisera to HN have been shown *in vitro* to inhibit hemagglutinating and neuraminidase activities and to neutralize the virus.[2] *In vivo* such antisera have been shown to protect animals from paramyxovirus infections.[2]

Due to the polyclonality inherent to antisera, previous inquests into the interaction between antibodies and the HN-molecule have been of limited resolution. In the present study, we have used monoclonal hybridoma antibodies specific for the HN of 6/94 virus (a parainfluenza type 1 virus closely related to Sendai virus)[3] to investigate the interaction of antibodies with individual structures on the HN molecule.

Results

Frequency of Antigenic Variants Selected with Anti-HN Hybridoma Antibodies

Using the allantois on shell culture system, antigenic variants of 6/94 virus were selected by growing the virus in the presence of an overneutralizing dose of anti-HN hybridoma antibody. Under these conditions only variant

[a] The Wistar Institute, Philadelphia, Pennsylvania.

viruses which are no longer neutralized by the selecting hybridoma antibody are able to propagate.[4] Variant viruses selected in this manner exhibit decreased avidity for the selecting antibody when assayed in an indirect RIA. By dividing the titer of the virus grown in the presence of overneutralizing doses of hybridoma antibody (i.e., the variant virus titer) by the titer of virus grown in the presence of a control hybridoma antibody not able to bind 6/94 virus, the frequency of antigenic variants present in egg-grown 6/94 virus could be calculated.[4] As shown in Table 1, the frequency of 6/94 virus antigenic variants ranged from $10^{-5.3}$ to $10^{-7.1}$. It should be noted that these frequencies fall within the range of frequencies of A/PR/8/34 influenza virus antigenic variants observed using the same experimental system (Table 1).[5] Similar variant frequencies have been observed in preparations of other strains of influenza viruses, paramyxoviruses, and rhabdoviruses [6-9] propagated in other cell systems. Thus, as noted in previous reports, [6,9] it appears that these three families of negative stranded enveloped RNA viruses produce antigenic variants with similar frequencies *in vitro*. This is in distinct contrast to the situation observed in nature where influenza viruses undergo considerable antigenic variation while paramyxoviruses and rhabdoviruses are antigenically stable.

At present, it is not possible to reconcile this discrepancy except to note that this finding strongly suggests that antigenic variation in the influenza HA-molecule is not due to an intrinsic hypermutability of the HA gene.

Creation of an Operational Antigenic Map of the HN-Molecule

The binding of 6 anti-HN hybridoma antibodies to 6/94 virus antigenic variants selected with 3 of these antibodies was studies in an indirect RIA in

Table 1. Frequency of Parainfluenza and Influenza Virus Antigenic Variants Selected with Anti-HN or Anti-HA Hybridoma Antibodies.

Hybridoma antibody used for selection (specificity)		Frequency of antigenic variants (Log_{10})
A1	(6/94 HN)	−7.1
B1	(6/94 HN)	−5.3
B2	(6/94 HN)	−5.4
Sa4	(A/PR/8/34 HA)	−6.5
Sb6	(A/PR/8/34 HA)	−5.3
Sb7	(A/PR/8/34 HA)	−5.6

Hybridoma anti-HN and anti-HA antibodies were produced as described[14] by fusing P3/x63 myeloma cells with splenocytes from BALB/c mice previously immunized with either 6/94 or A/PR/8/34 virus, respectively. The frequency of antigenic variants present in egg-grown virus stocks was determined from the ratio of virus titers obtained in the presence of selecting and control hybridoma antibodies using the allantois on shell culture system.

which binding of mouse hybridoma antibodies to virus immobilized on polystyrene wells was quantitated by the addition of ^{125}I-labeled rabbit anti-mouse $F(ab')_2$ (Table 2).[10] Based on the results, anti-HN antibodies could be assembled into 3 groups designated A, B and C respectively: those which recognized (i.e., did not bind) variants selected with antibody A1; those recognizing variants selected with antibodies B1 and B2 and those which did not recognize any of the variant viruses. The HN antigenic structures corresponding to the antibody specificities A and B are characterized by the variants selected in the presence of these antibodies. The antigenic structures delineated by group C are less well characterized, since they were unable to select antigenic variants in the presence of the corresponding antibodies.

Correlation Between Epitope Specificity and Anti-Viral Activity Mediated by Anti-HN Antibodies

The potency (activity on a molar basis) of anti-6/94 hybridoma antibodies in virus neutralization and hemagglutination inhibition (HI) tests was assessed by functional assays and to antibody concentration in a radioimmunoassay. With regard to HI potency (Table 3), antibodies segregated into 2 groups: antibodies A1, C1, C2 and C3 with HI potencies ranging from roughly 50 to 100% of maximal activity (defined by the potency of antibody C1); and antibodies B1 and B2 with potencies of less than 30% of the maximum activity. More striking differences were observed in neutralization assays (Table 3), in which antibodies segregated into 3 groups paralleling epitope specificity. Thus, antibody A1 was roughly 25 times more potent in virus neutralization than antibodies B1 and B2, which, in turn, were 3 to 8 times more potent than antibodies C1, C2 and C3. Furthermore, calculation of the ratio of

Table 2. Reactivity of Anit-HN Hybridoma Antibodies with 6/94 Antigenic Variants Selected in Vitro with Hybridoma Antibodies.

	Reactivity with antigenic variants selected with antibody		
Anti-HN antibody	**A1**	**B1**	**B2**
A1	−	+	+
B1	+	−	−
B2	+	−	−
C1	+	+	+
C2	+	+	+
C3	+	+	+

The binding of anti-HN hybridoma antibodies to 6/94 virus antigenic variants selected with either antibody A1, B1 or B2 was compared to binding to parental 6/94 virus as determined by an indirect RIA. −, <25% of binding to parental virus; +, >25% of binding to parental virus.

Table 3. Potency of Anti-HN Hybridoma Antibodies in Hemagglutination Inhibition (HI) and Virus Neutralization (VN) Assays.

Antibody	Isotype	HI	VN	VN/HI × 100
A1	γ 2a	.83	1.00	120
B1	γ 1	.20	.04	20
B2	γ 1	.32	.03	9
C1	γ 1	1.00	.01	1
C2	γ 2a	.73	.01	1
C3	γ 2a	.51	.005	1

The potency (activity on a molar basis) of anti-HN hybridoma antibodies in HI and VN assays was calculated by dividing observed antibody titers in functional assays by antibody concentration as determined by an indirect RIA. Values are given relative to the most potent antibody in HI and VN assays (C1 and A1 respectively). Antibody isotyping was performed as described.[14]

neutralization to HI potencies (Table 3) clearly segregates the hybridoma antibodies according to their epitope specificity.

The correlation between epitope specificity and antibody activity was particularly striking when anti-HN hybridoma antibodies were tested in neuraminidase inhibition assays (Table 4). Thus, antibody A1 inhibited, antibodies B1 and B2 slightly enhanced, and antibodies C1, C2 and C3 strongly enhanced viral neuraminidase activity. These contrasting effects on the neurminidase activity were observed whether fetuin (MW 48,000) or sialal lactose (MW 633) was used as substrate. It should be noted that these differences in antibody activity are not due to differences in antibody isotype since antibodies A1, C2 and C3 are of the gamma 2a isotype and antibodies B1, B2, and C1 are of the gamma 1 isotype (Table 3).

Discussion

Based on comparative antigenic analyses of *in vitro* selected HN-mutants, we have been able to arrange 6 anti-HN hybridoma antibodies into 3 specificity groups. That this grouping is not based on differences in antibody avidity is evident from the reciprocal relationship between the effect of mutations on the binding of homologous (selecting) and heterologous antibodies. The finding that antibodies of a distinct specificity group exhibit distinct anti-viral activities suggests that the anti-viral activity of an anti-HN antibody depends largely upon the structure of the HN-molecule to which the given antibody binds. A similar conclusion has been drawn from experiments performed with a large panel (more than 50 unique antibodies) of anti-influenza virus A/PR/8 hybridoma antibodies. In those studies, it was found that the potency of anti-HA antibodies in HI assays, the ability of the antibodies to bind virus under various conditions and the effect of anti-HA antibodies on viral neuraminidase activity are all related to the antibody epitope specificity.[11]

Table 4. Effector Function of Anti-HN Hybridoma Antibodies in Neuraminidase Assays.

Antibody	Neuraminidase activity
A1	Complete inhibition
B1	2-fold enhancement
B2	2-fold enhancement
C1	10-fold enhancement
C2	10-fold enhancement
C3	15-fold enhancement

The effect of anti-HN hybridoma antibodies on 6/94 virus neuramindase activity was determined essentially according to Russ et al.[15] Degree of neuraminidase enhancement was calculated by titrating virus in the presence of a saturating, constant amount of hybridoma antibody.

In general, antibodies may interfere with the function of a protein by allosteric modification or by steric hindrance of the interaction between the protein's active site(s) and the corresponding ligand.[12] In the present study, possible examples of steric and allosteric antibody mediated alteration of HN function are offered by the action of antibodies A1 and C1 respectively. Thus, the observation that antibody A1 inhibits hydrolysis of both large and small MW substrates [and does so in the presence of large amounts of enhancing antibody C1 (data not shown)] suggests that antibody A1 binds in close proximity to the enzyme active site and sterically prevents substrate access. In contrast, antibody C1 probably enhances enzyme activity by allosterically altering the HN to provide optimal conditions for substrate cleavage. Although these proposals are speculative, at present, they are readily amenable to experiment verification.[13]

ACKNOWLEDGMENTS
This work has been supported in part by grants from the National Institute of Allergy and Infectious Diseases, the National Multiple Sclerosis Society and the Medical Scientist Training Program.

References

1. Choppin, P.W. and Compans, R.W. (1975): In: *Comprehensive Virology*. Fraenkel, H. and Wagner, R. (Eds.), Plenum Publishing Company, New York, 4:95–178.
2. Orvell, C. and Norrby, E. (1977): *J. Immun.* 119:1882–1887.
3. Koprowski, H. and ter Meulen, V. (1975): *J. Neurol.* 208:175–190.
4. Yewdell, J.W., Webster, R.G., and Gerhard, W.U. (1979): *Nature* 279:246–248.
5. Gerhard, W., Yewdell, J.W., and Frankel, M. (1980): Submitted for publication.
6. Portner, A., Webster, R.G., and Bean, W.G. (1980): *Virology* 104:235–238.
7. Lubeck, M., Schulman, J.L., and Palese, P. (1980): *Virology* 102:458–462.

8. Webster, R.G. and Laver, W.G. (1980): *Virology* 104:139–148.
9. Wiktor, T.J. and Koprowski, H. (1980): *J. Exp. Med.* 152:99–112.
10. Frankel, M.E. and Gerhard, W. (1979): *Molecular Immun.* 16:101–116.
11. Gerhard, W.U., Yewdell, J.W., and Frankel, M.E. (1980): In: *Structure and Variation in Influenza Virus.* Laver, G. and Air, G.M. (Eds.), Elsevier/North Holland Publishing Company, New York, p. 273–282.
12. Arnon, R. (1973): In: *The Antigens.* Sela, M. (Ed.), Academic Press, New York, I, 89–159.
13. Frackelton, A.R. and Rotman, B. (1980): *J. Biol. Chem.* 255:5286–5290.
14. Gerhard, W., Yewdell, J, Frankel, M, Lopes, D., and Staudt, L. (1980): In: *Monoclonal Antibodies.* Kennett, R., McKearn, T., and Bechtol, K. (Eds.), Plenum Publishing Company, New York, 317–333.
15. Russ, G., Vareckova, and Styk, B. (1974): *Acta Virol.* 299–306.

David H.L. Bishop and Richard W. Compans, eds.
The Replication of Negative Strand Viruses

Immunocytochemical Localization of Mumps Virus Antigens *In Vivo* by Light and Electron Microscopy

Jerry S. Wolinsky, George Hatzidimitriou, Melvin N. Waxham, and Susan Burke [a]

Mumps virus causes an acute meningoencephalitis in newborn Syrian hamsters. During the early phase (days 3–10) of infection following intraperitoneal inoculation with a neuroadapted mumps virus strain (Kilham), virus can be isolated from clarified brain homogenates and budding virus detected at the plasma membranes of infected cells.[1,2] Animals which survive the acute infection continue to harbor virus in brain for several months. This has been shown by recovery of virus using explant and cocultivation techniques and demonstration of cells containing viral nucleocapsids by electron microscopy, however, cell-free virus can no longer be isolated and budding virus is no longer ultrastructurally detectable.[2,3] This finding has suggested that defective production or insertion of viral-specific proteins may underlie the chronic phase of the infection. A similar pattern of infection has been seen in several other model paramyxovirus infections including parainfluenza I virus infection of mice,[4,5] and measles virus infection of hamsters.[6,7] Further, recent evidence from several laboratories suggests that there may be altered presentation of viral antigens in the course of the human persistent paramyxovirus infection, subacute sclerosing panencephalitis.[8–10] In order to approach this problem directly and also to attempt to determine the sites of viral polypeptide synthesis and assembly in infected cells *in vivo,* we had adapted immunohistochemical techniques to study viral antigens in mumps virus infected hamster brain tissue. This report details the feasibility of such studies.

[a] Department of Neurology, John Hopkins University, School of Medicine, Baltimore, Maryland.

Antisera used in these studies were generated by hyperimmunizing rabbits with gradient purified egg-grown Enders virus or ion exchange chromatography purified hemagglutinin of Enders virus; [11] preimmunization sera served as control reagents for each. The spectra of antibody activity was determined by immunoprecipitation with labeled neuroadapted mumps virus infected cell lysates (Figure 1). For light microscopy, formaldehyde perfused

Figure 1. Characterization of rabbit antisera by immunoprecipitation. Cell lysate targets were prepared from neuroadapted mumps virus infected Vero cell monolayers metabolically labeled by a four-hr pulse of either ^{35}S-methionine (lanes 1, 3, 5, 7, 8) or ^{3}H-glucosamine (lanes 2, 4, 6) beginning about 48 hr after infection. Scraped cells were lysed in 1% zwittergent 3–14, 0.5% sodium deoxycholate, 0.1% sodium dodecylsulfate (SDS), 0.5 50μl M NaCl, 25mM Tris pH 7.6 and equal aliquots reacted with 50 ul of preimmune rabbit sera (lanes 1, 2), hyperimmune antisera to ion exchange purified hemagglutinin purified from Enders virus (lanes 5, 6), a rabbit antisera with activity primarily directed to the fusion protein of Enders virus kindly provided by Dr. Erling Norrby (lane 7),[15] or an antisera generated against a "nucleocapsid" preparation of Enders virus using previously described technics.[16] Antigen-antibody complexes were precipitated with protein A Sepharose and the washed gel disrupted in 5% SDS and 2% dithiotheritol and subjected to discontinuous electrophoresis under reducing conditions using a 1% polyacrylamide resolving gel.[17] The gel was fixed, impregnated with 2, 5-diphenoloxyzole, dried and autoradiographs prepared.[18] The origin of the resolving gel (0) and position of molecular weight markers in kilodaltons are shown to the left. The form of the hemagglutinin which appears on the surface of infected cells (HNs) and its probable intracellular precursor (HNp), fusion protein (F), tentative designations for the intact nucleocapsid (NP), intracellular form of the nucleocapsid (NPi) and matrix protein (M) are given in parenthesis on the left.

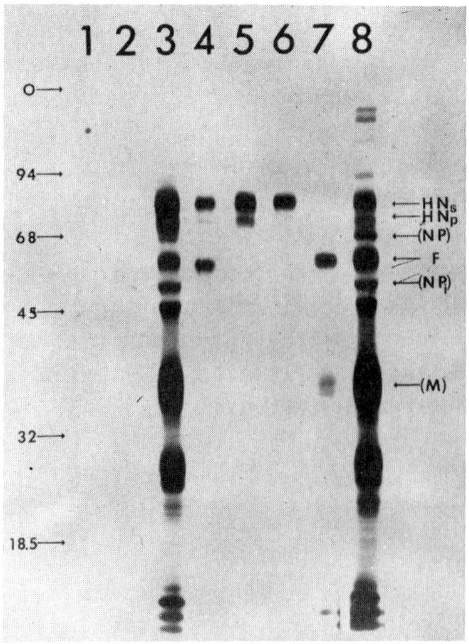

hamster brains from animals infected by peripheral inoculation with either neuroadapted mumps virus or saline solution were embedded in paraffin, sectioned and deparaffinized. The sections were then treated with acid citrate solution followed by trypsin,[12] reacted for 2 hr with an appropriate dilution of the primary reagent followed by swine anti-rabbit antibody, rabbit peroxidase-anti-peroxidase immune complexes and a reaction product generated with diaminobenzidine and hydrogen peroxide using modifications of established methods.[13] Specific staining of infected cells was seen, using

Figure 2. Light microscopic localization of viral antigens in epoxy embedded tissue from the parietal region of a suckling hamster sacrificed nine days after intracerebral inoculation with neuroadapted mumps virus. The ependymal (A, C, E) and meningeal surfaces (B, D, F) of adajcent sections were reacted with preimmune rabbit immunoglobulins diluted 1:80 (A, B), broadly reactive hyperimmune rabbit antisera diluted 1:400 (C, D) or monospecific rabbit antiphemagglutinin immunoglobulins diluted 1:80 (E, F). The specific reaction product appears black against the unstained brain tissue. The regions of lateral ventricle (V), ependymal surface (E), subependymal parenchyma (P), neuronal cell bodies (N), dendritic arborizations (D) and meninges (M) are delineated.

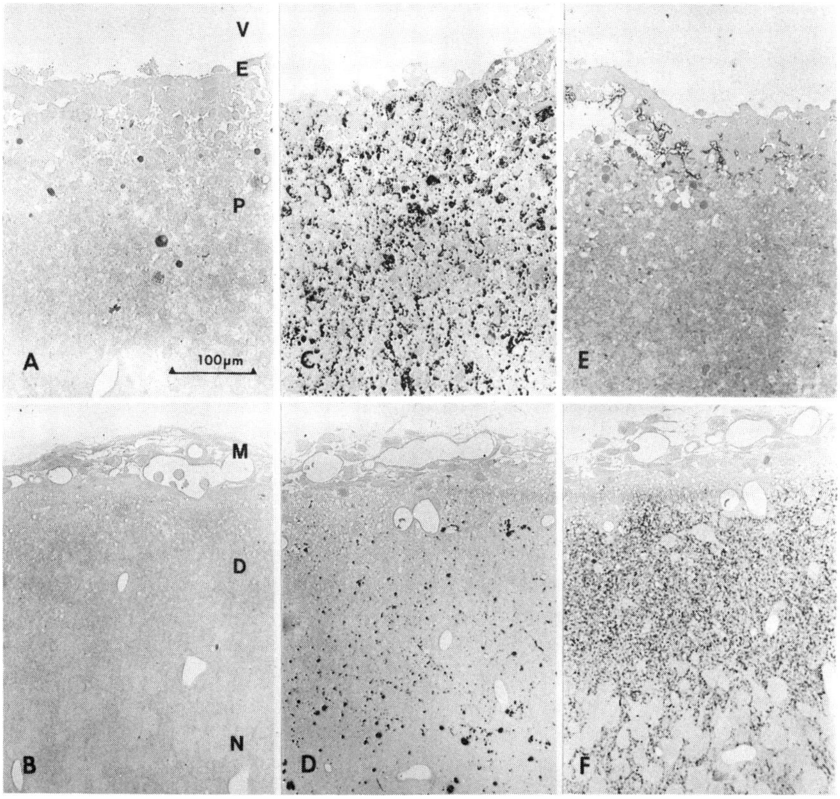

dilutions of the primary antibody as high as 1:10,000 and infected cells could be found for up to 21 days postinoculation (data not shown). These staining patterns were similar to those found using conventional immunofluorescent techniques.[1]

For localization of antigens in resin embedded material intracerebrally or parenterally infected and mock inoculated animals were perfused with 1% paraformaldehyde, 1.25% glutaraldehyde and selected regions of brain processed for embedding in epon or epon-araldite after post fixation in osmium tetroxide as previously detailed.[2,3] Thick sections (0.5–1 um) were then etched with sodium ethoxide,[14] and reacted using the peroxidase-anti-peroxidase technic described above. Primary antibody dilutions up to 1:2,000 successfully located antigen. Thin sections were cut at approximately 90 nm thickness, etched by flotation on either 10% hydrogen peroxide or sodium ethoxide, reacted with antibody, conjugates and diaminobenzidine as above and the reaction product rendered electron dense by osmication.[13] Localization of structures in etched tissues was made by comparison to conventionally stained adjacent 60 nm sections.

DEAE-Sephadex A-50 purified preimmune rabbit immunoglobulins failed to stain either infected or control tissues at dilutions as low as 1:20. Ion exchange purified hyperimmune rabbit immunoglobulin with activity to the hemagglutinin-neuraminidase stained at dilutions up to 1:160 and the broad spectrum antibody stained at dilutions up to 1:2,000. Uninfected material was unstained by these antisera (Figure 2).

By electron microscopy, specific staining or intracytoplasmic nucleocapsid inclusions was achieved in infected neurons using the broad spectrum primary immunoglobulins (Figure 3). Nucleocapsid structures undergoing

Figure 3. Electron microscopic localization of antigen to an intraneuronal nucleocapsid inclusion in a hamster six days after infection. Conventional staining delineates a typical intracytoplasmic inclusion (arrows) (A) for comparison with the localization of reaction product (arrows) on a hydrogen peroxide etched section (B). The plasmalema (P), nucleus (N), Golgi apparatus (G) and mitochondria (M) are noted.

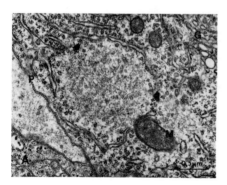

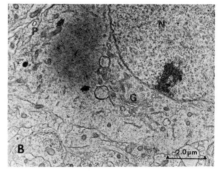

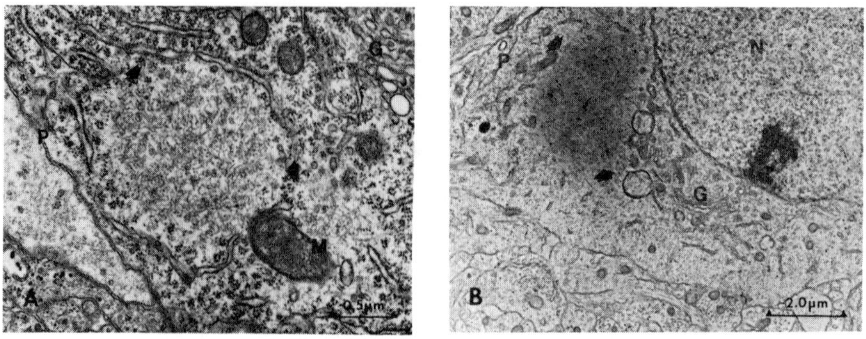

apparent proteolytic digestion within phagolysosomes stained intensely (Figure 4). Staining of thin sections with the monospecific rabbit anti-hemagglutinin and analysis of tissues from different phases of the infection is in progress.

These preliminary studies support the usefulness of this technique for the fine localization of viral polypeptides in infected tissues. Utilization of monospecific reagents with specificity for the other polypeptides of mumps

Figure 4. Ultrastructure of partially digested nucleocapsids in phagolysosomes of a macrophage from the subependymal region of same animal illustrated in Figure 2. Conventionally stained section (A) is contrasted with the same cell from adjacent hydrogen peroxide etched section using broadly reactive rabbit antisera diluted 1:1000 (B). At higher magnification (C), the typical pentagonal structure of the peroxidase-anti-peroxidase molecule can be appreciated (arrows). An adjacent section reacted with a 1:1,000 dilution of preimmune rabbit sera shows no reaction product (D). The cell nucleus (N) and two phagolysosomes (P_1, P_2) containing typical nucleocapsid structures are delineated.

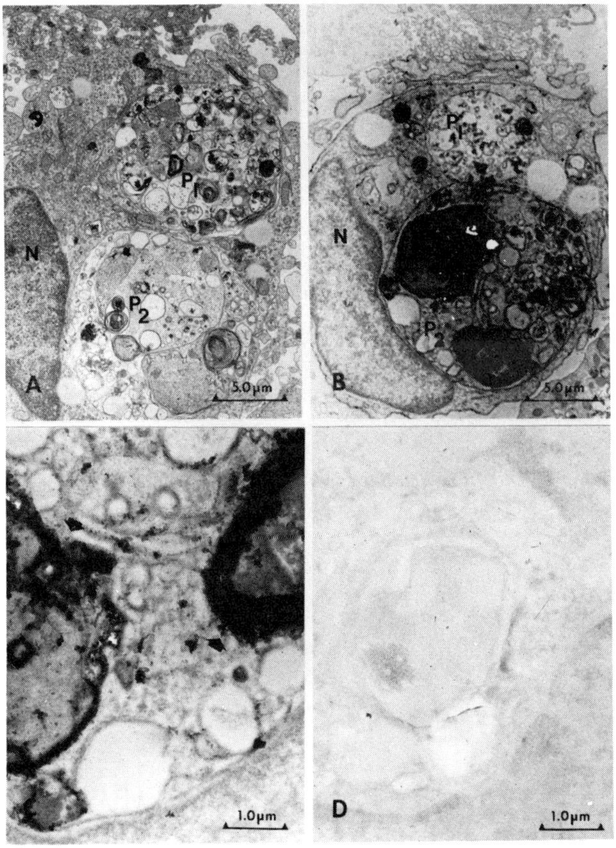

virus and adaption of the technique to utilize mouse monoclonal immunoglobulins of defined specificity should enable us to understand the topography of individual polypeptide synthesis and assembly in cells during the acute phase of the infection and contrast this with patterns seen during the chronic phase of infection as a means of better understanding both the replication of the paramyxoviruses and mechanisms of viral persistence *in vivo*.

ACKNOWLEDGMENTS
Supported by Public Health Service Grants AI 115721 from the National Institute of Allergy and Infectious Diseases and a Research Career Development Award (JSW) NS 00443 from the National Institute of Neurological and Communicative Disorders and Stroke.

References

1. Wolinsky, J.S., Klassen, T., and Baringer, J.R. (1976): *J. Infect. Dis.* 133:260–267.
2. Wolinsky, J.S., Baringer, J.R., Margolis, G., and Kilham, L. (1974): *Lab Invest.* 31:403–412.
3. Wolinsky, J.S. (1977): *Lab. Invest.* 37:229–236.
4. Rorke, L., Gilden, D.H., Wroblewska, Z., and Wolinsky, J.S. (1976): *J. Neuropath, Exp. Neurol.* 35:247–258.
5. Wolinsky, J.S., Gilden, D.H., and Rorke, L. (1976): *J. Neuropath, Exp. Neurol.* 35:271–286.
6. Raine, C.S., Byington, D.P., and Johnson, K.P. (1975): *Lab Invest.* 33:108–116.
7. Raine, C.S., Byington, D.P., and Johnson, K.P. (1975): *Lab. Invest.* 31:355–368.
8. Wechsler, S.L., Weiner, H., and Fields, B.N. (1979): *J. Immunol.* 123:884–889.
9. Hall, W.W., Lamb, R.A., and Choppin, P.W. (1979): *Proc. Natl. Acad. Sci. U.S.A.* 76:2047–2051.
10. Hall, W.W. and Choppin, P.W. (1979): *Virol.* 99:443–447.
11. Server, A.C., Merz, D.C., Waxham, M.N., and Wolinsky, J.S.: In preparation.
12. Swoveland, P.T. and Johnson, K.P. (1979): *J. Infect. Dis.* 140:758–764.
13. Sternberger, L.A. (1979): In: *Immunocytochemistry*. John Wiley and Sons, New York, pp. 104–169.
14. Erlandsen, S.L., Parsons, J.A., and Rodnig, C.B. (1979): *J. Histochem. Cytochem.* 27:1286–1289.
15. Örvell, C. (1978): *J. Gen. Virol.* 41:517–526.
16. McCarthy, M. and Johnson, R.T. (1980): *J. Gen. Virol.* 46:15–27.
17. Laemmli, U.K. (1970): *Nature* 227:680–685.
18. Bonner, W.M. and Laskey, R.A. (1974): *Eur. J. Biochem.* 46:83–88.

Copyright 1981 by Elsevier North Holland, Inc.
David H.L. Bishop and Richard W. Compans, eds.
The Replication of Negative Strand Viruses

Immune Response in Subacute Sclerosing Panencephalitis and Multiple Sclerosis: Antibody Response to Measles Virus Proteins

Steven L. Wechsler,[a,g] H. Cody Meissner,[b]
Usha R. Ray,[c] Howard L. Weiner,[d]
Robert Rustigian,[e] and Bernard N. Fields[b,f]

It is generally believed that the slowly progressive neurological disease subacute sclerosing panencephalitis (SSPE) represents a persistent and progressive infection of the central nervous system by measles virus (MV), or a varient of MV.[1-4] Persistent infection with MV has also been tentatively linked to multiple sclerosis (MS).[4-7] In this report, we will describe some of our findings concerning (1) viral proteins in cells persistently infected by MV (Pi cells) and (2) the host humoral immune response to MV proteins in individuals with SSPE and in individuals with MS.

Virion Proteins

Measles virions contain 6 viral proteins. The proteins and their molecular weights are: L, a nucleocapsid associated protein, 200,000; H, the hemagglutination protein, 80,000; P, a nucleocapsid associated protein, 70,000; NP, the major nucleocapsid protein, 60,000; F_0, the fusion protein,

[a] Department of Molecular Virology, Christ Hospital Institute of Medical Research, Cincinnati, Ohio.
[b] Department of Microbiology and Molecular Genetics, Harvard Medical School, Boston, Massachusetts.
[c] NIDR, National Institute of Health, Bethesda, Maryland.
[d] Department of Medicine, Division of Neurology, Peter Bent Brigham Hospital, Boston, Massachusetts.
[e] Veterans Administration Hospital, Brockton, Massachusetts.
[f] Department of Medicine, Division of Infectious Diseases, Peter Bent Brigham Hospital, Boston, Massachusetts.
[g] Author to whom reprint requests should be addressed.

62,000, composed of polypeptides F_1 and F_2 with molecular weights of 41,000 and 12–20,000 respectively; and M, the matrix protein, 37,000.[8–10] Tryptic peptide analysis reveals that each of these polypeptides is a unique species (data not shown). The NP-protein sometimes appears to be composed of two bands on SDS-PAGE (see Figures 5 and 6). Partial proteolytic digestion reveals only one difference between the digestion products of these two bands (Figure 1, arrow). Thus, these bands represent different forms of the NP-protein (the smaller possibly being a breakdown product of the larger).

An additional major protein component that comigrates with cellular actin is seen in purified measles virions. The assumption that this protein is cellular actin[9,11,12] is confirmed by 2-dimensional peptide analysis in Figure 2.

Polypeptides of Cells Persistently Infected with MV

K11 cells were isolated as a persistently infected clonal cell line following acute infection of HeLa cells with the Edmonston strain of wild type (wt)

Figure 1. Partial protease digests of measles virus NP-protein. ^{35}S-methionine-labeled measles virus was grown, harvested, purified, run on SDS-PAGE, and processed for autoradiography as previously described.[9] Two bands (A and B) present in the NP-protein region of the gel were individually cut out of the gel and digested according to the method described by Cleveland et al.[19] Digestion was for 30 minutes with 3 micrograms of *S. aureus* protease V8. The arrow points to the only observed difference between the two NP-protein digests.

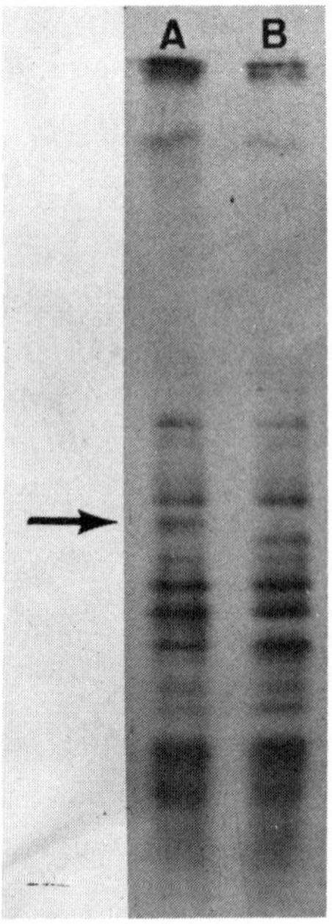

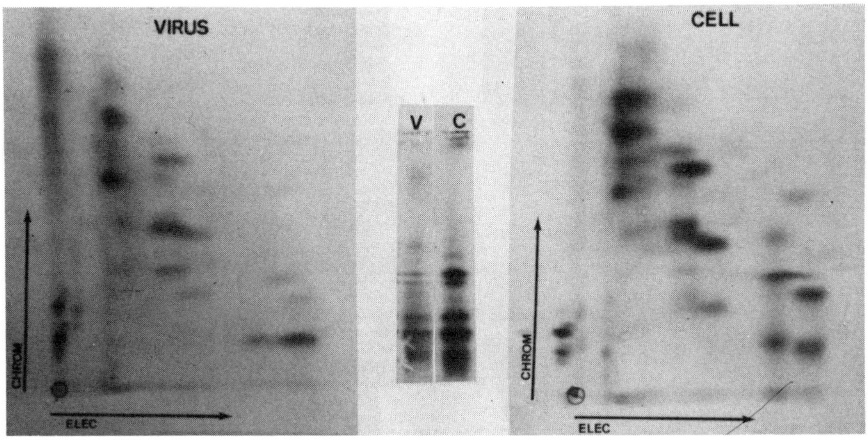

Figure 2. Comparative proteolytic digests of cellular actin and actin recovered from purified measles virions. ^{35}S-methionine-labeled measles virus was grown in CV-1 cells, harvested, purified and run on SDS-PAGE as previously described.[9] Uninfected CV-1 cells were also labeled with ^{35}S-methionine and cell extracts run on SDS-PAGE. The 43,000 molecular weight band (actin) from each preparation were cut out of the gels and subjected to either partial protease digests with *S. aureus* protease V8 as described in Figure 1: (V: virus derived actin; C: cell derived actin) or to two dimensional tryptic digests [20] (Virus: virus derived actin; Cell: cell derived actin).

MV.[13] A second clonal cell line, K11A, was derived from the K11 population.[14] All four of the intracellular viral proteins (H, P, NP, and M) detected in these Pi cells by SDS-PAGE (Figure 3, Pi cells) differ from those of the parental wt virus.[15] The M-proteins from both Pi cells migrate more slowly than the wt M-protein. The wt H-protein appears as an intense band with a diffuse, less intense, trailing band, possibly representing a more highly glycosylated form. This trailing band is not seen in either Pi cell, even if the gel lanes are intentionally overloaded to produce H bands of greater intensity than the wt H band (not shown). The K11 cell H band also appears wider than the wt H band, while the K11A cell H band appears to be relatively reduced in amount compared to the K11A cell P and NP-proteins. The K11 cell NP-protein band is broader than the wt NP band, and both Pi cell NP-proteins migrate more slowly than the wt NP-protein. Although no difference is seen in the ^{35}S-methionine-labeled P-proteins in Figure 3, ^{32}P-labeling does reveal a migrational difference between the Pi and wt phosphorylated forms of the P-protein (data not shown). Thus, in these Pi cells, all four of the detected viral proteins differ from those of the parental wt virus.

The aberrant viral proteins in the Pi cells are due to mutations in the virus rather than alterations in the host cell (Figure 3).[15] K11 and K11A cells cured of their persistent infections do not contain any detectable viral proteins

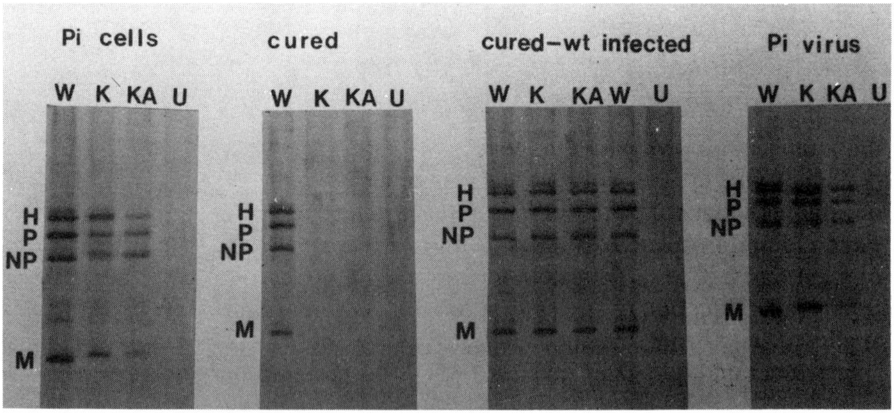

Figure 3. Intracellular measles virus proteins from two cell lines (K11 and K11A) persistently infected with MV. Cells were grown, infected, labeled with ^{35}S-methionine, harvested, immune precipitated with measles antisera, and subjected to SDS-PAGE as previously described.[15] W; Hela cells acutely infected with wt MV; U: uninfected HeLa cells. Pi cells: K; K11 cells; KA: K11A cells. Cured: K: cured K11 cells; KA: cured K11A cells. Cured-wt infected: K: cured K11 cells acutely infected with wt MV; KA: cured K11A cells acutely infected with wt MV. Pi virus: K; HeLa cells acutely infected with virus recovered from K11 cells; KA: HeLa cells acutely infected with virus recovered from K11A cells.

(Figure 3, "cured"). Infecting these cured cells with wt MV, results in the synthesis of viral proteins that are indistinguishable from the wt proteins made in normal HeLa cells (Figure 3, "cured-wt infected"). Finally, when virus recovered from the Pi cells is used to infect normal HeLa cells, the viral proteins synthesized are identical to those found in the original Pi cells.

Reduced Levels of Anti-M-Protein Antibody in the Sera of Individuals with SSPE

The host humoral immune response in individuals with SSPE was studied by means of immune precipitation of ^{35}S-methionine-labeled MV proteins followed by SDS-Page. 10 "normal," 10 convalescent, and 10 SSPE sera were examined for their ability to immune precipitate MV proteins. We found a striking reduction in the ability of sera from patients with SSPE to precipitate the M-protein as compared to the precipitation of M-protein by sera from "normal" adults who had natural measles in childhood, or by convalescent sera obtained 3 to 5 weeks after a naturally occurring measles infection (Figure 4).[16]

The amounts of wt MV-protein immune precipitated by each of the 30 sera were quantitated by scanning autoradiograms with a densitometer and calculating the areas under the peaks. A statistical analysis using a Student's t-test showed that the decrease in precipitation of M-protein by SSPE sera

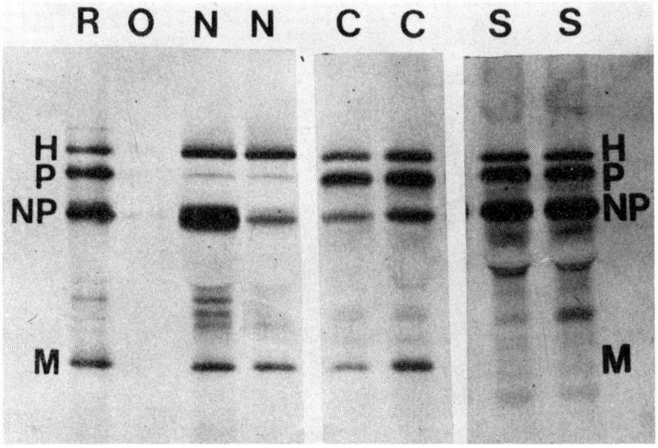

Figure 4. Immune precipitation of ^{35}S-methionine-labeled wt measles virus proteins with "normal," convalescent, and SSPE sera. Experimental protocol was as described previously.[16] Briefly, CV-1 cells were infected with wt MV, labeled with ^{35}S-methionine, and lysed. Equal aliquots of the cell lysate, containing labeled MV-proteins, were incubated with various antisera (10 "normal," 10 convalescent, and 10 SSPE sera) and immune complexes were precipitated with Protein A-Sepharose-CL-4B. The precipitates were dissolved in gel sample buffer and subjected to SDS-PAGE, followed by autoradiography. R: rabbit anti-measles antiserum; O: control-no antiserum; N: 2 representative "normal" sera from adults with exposure to natural measles as children; C: 2 representatiave convalescent sera from otherwise normal individuals 3 to 5 weeks after the onset of natural measles infection; S: 2 representative SSPE sera from patients with SSPE.

was highly significant ("normal" vs SSPE, $p = 0.001$; convalescent vs SSPE, $p = 0.003$).[16] To test if the lack of precipitation of the M-protein could be due to antibody excess, representative sera were serially diluted and immune precipitations performed (data not shown). No antibody excess against the M-protein could be detected. The reduced precipitation of M-protein by SSPE sera was also not due to a lack of immunologic cross-reactivity between wt and SSPE M-proteins.[16] Analysis of the other proteins showed the following results: (1) Undiluted SSPE sera precipitated less H-protein than "normal" or convalescent sera. However, this was due to anti-H-protein antibody excess.[16] (2) "Normal" sera precipitated less P-protein than the other sera. (3) SSPE sera precipitated more NP-protein than convalescent sera.

Humoral Immune Response to Individual MV-Proteins in Patients with MS

Sera from 24 MS patients and sera from 24 "normal" with similar HI titers were analyzed for their ability to immune precipitate MV-proteins (representative results are shown in Figure 5 and 6). The amount of each viral protein

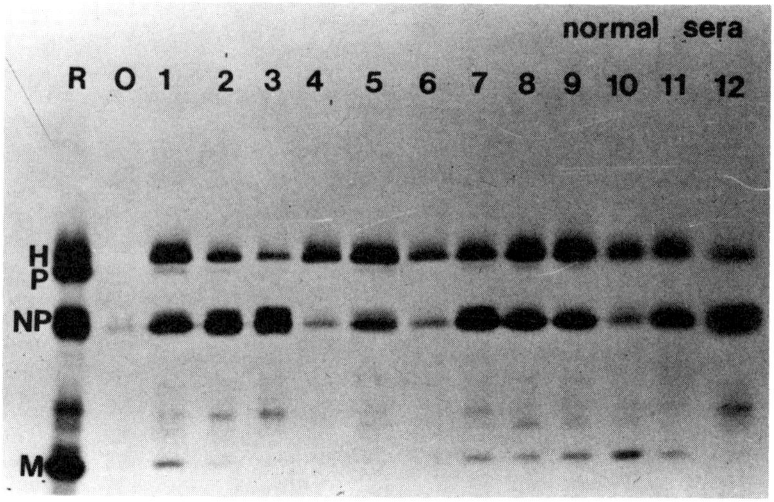

Figure 5. Immune precipitation of ^{35}S-methionine-labeled wt measles virus proteins with "normal" sera. Immune precipitations were done with 24 "normal" sera. Experimental protocol was as described in the legend to Figure 4. R: rabbit anti-measles antiserum; O: control-no antiserum; 1–12: 12 representative "normal" sera from adults with exposure to natural measles as children.

Figure 6. Immune precipitation of ^{35}S-methionine-labeled wt measles virus proteins with multiple sclerosis sera. Immune precipitations were done with 24 MS sera. Experimental protocol was as described in the legend to Figure 4. R: rabbit anti-measles antiserum; O: control-no antiserum; 1–12: 12 representative MS sera from patients with MS.

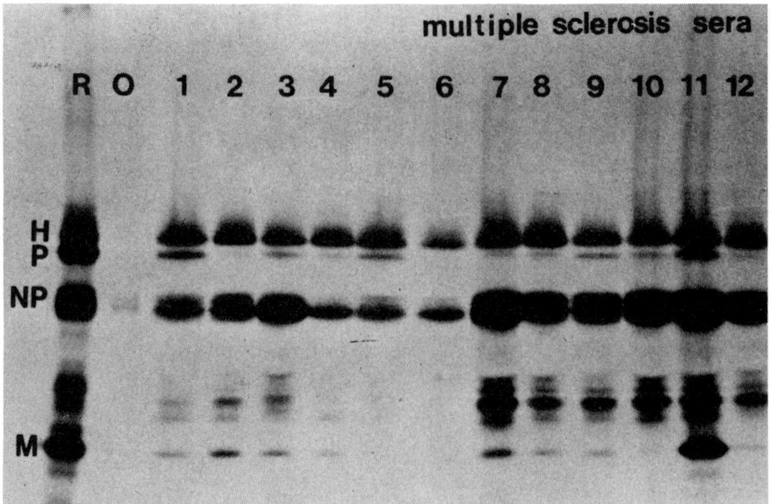

immune precipitated was quantitated by scanning the autoradiograms and integrating the areas under each peak. Representative sera were serially diluted and immune precipitations performed to test for possible antibody excess. The combined analysis showed the following results: (1) For the H and M-proteins, there was no statistically significant difference between the "normal" and MS sera. (2) The MS sera precipitated slightly more NP-protein. However, because of antibody excess to NP in both the MS and "normal" sera, it was difficult to determine if this difference was statistically significant. (3) The MS sera precipitated significantly more P-protein ($p < 0.01$) than the "normal" sera. Identical experiments were also done comparing sera from individuals during an MS exacerbation (5 sera) to sera from individuals with MS who were not undergoing an exacerbation at the time sera was collected (the 24 MS sera used above). Although the exacerbation sera had generally higher levels of antibodies to all the MV-proteins, no statistical difference was detected between the MS and MS-exacerbation sera (data not shown). Thus, the only statistically significant difference between "normal" and MS sera was that MS sera precipitated more MV P-protein.

Discussion

Point mutations usually do not result in detectable alterations of protein mobilities on SDS-PAGE. Thus the finding that all four of the viral proteins seen in K11 and K11A cells differ from the parental wt virus proteins suggests that the virus in these Pi cells has undergone extensive genetic change. In cells persistently infected with VSV, it has been shown that viral mutations appear to accumulate over an extended period of time.[17] The same may be true of cells persistently infected with MV. It is likely that in Pi cells a wide variety of mutations can be tolerated because the persistent virus does not require all of the viral functions that are needed in a productive lytic infection. In fact, in Pi cells, multiple viral mutations may be selected for, since they would greatly reduce the possibility of the virus reverting to a lytic form.[15]

Despite a vigorous antibody response to the other MV-proteins in sera from patients with SSPE, there is a dramatically reduced antibody response to the M-protein of MV, suggesting that only small amounts of the M-protein are present in SSPE brain cells.[16,21] This hypothesis is supported by a recent report in which no M-protein was detected in a single SSPE brain explant, although several other MV-proteins were seen.[18] It is currently not known whether the apparent decrease in the amount of M-protein in SSPE brains is a cause of, or a result of, the persistent MV infection involved with SSPE.

The only statistically significant difference detected between MS sera and "normal" sera is that MS sera has higher anti-P-protein antibody titers. Because the relative amount of anti-P-protein antibody appears to be highest shortly after measles infection, we have previously suggested that antibody against the P-protein may be indicative of a recent or ongoing MV infection.[16] Although the increased anti-P-protein antibody titer in individuals with MS seems to mimic an immune response to an ongoing MV infection, the increased anti-P titer is more probably due to the general elevated humoral immune response that is known to occur in individuals with MS.

ACKNOWLEDGMENTS
We thank Karen B. Byers and Elaine Freimont for their expert technical assistance. This work was supported by the National Multiple Sclerosis Society Grant RC991-B-4 and by the Veterans Administration Research Program.

References

1. ter Meulen, V., Katz, M., and Muller, D. (1972): *Curr. Top. Microbiol. Immunol.* 57:1–38.
2. Horta-Barbosa, L., Fuccillo, D.A., Zeman, W., and Sever, J.L. (1969): *Nature* 211:974.
3. Payne, F.E., Baublis, V.V., and Itabashi, H.H. (1969): *N. Engl. J. Med.* 281:585–589.
4. Morgan, E.M. and Rapp, F. (1977): *Bacteriol. Rev.* 41:636–666.
5. Cathala, F. and Brown, P. (1972): *J. Clin. Pathol.* 25:141–151.
6. Lancet editorial. (1974):*Lancet* i:247–249.
7. Levy, N.L., Auerback, P.S., and Hayes, E.C. (1976): *N. Engl. J. Med.* 294:1423–1427.
8. Wechsler, S.L. and Fields, B.N. (1978): *J. Virol.* 25:285–297.
9. Stallcup, K.C., Wechsler, S.L., and Fields, B.N. (1979): *J. Virol.* 23:166–176.
10. Mountcastle, W.E. and Choppin, P.W. (1977): *Virology* 78:463–474.
11. Wang, E., Wolf, B.A., Lamb, R.A., Choppin, P.W., and Goldberg, A.R. (1976): In: *Cell Motility,* Book B, Vol. 3. Goldman, R.D., Pollard, T., and Rosenbaum, J. (eds.), Cold Spring Harbor Laboratory, pp. 589–599.
12. Tyrrell, D.L.J. and Norrby, E. (1978): *J. Gen. Virol.* 39:219–229.
13. Rustigian, R. (1966): *J. Bacteriol.* 92:1972–1804.
14. Rustigian, R. (1966): *J. Bacteriol.* 92:1804–1811.
15. Wechsler, S.L., Rustigian, R., Stallcup, K.C., Byers, K.B., Winston, S.H., and Fields, B.N. (1979): *J. Virol.* 31:677–684.
16. Wechsler, S.L., Weiner, H.L., and Fields, B.N. (1979): *J. Immunol.* 123:884–889.
17. Holland, J.J., Grabau, E.A., Jones, C.L., and Semlar, B.L. (1979): *Cell* 16:495–504.
18. Hall, W.M. and Choppin, P.W. (1979): *Virology* 99:443–447.
19. Cleveland, D.W., Fisher, S.G., Kirschner, M.W., and Laemmli, U.K. (1979): *J. Biol. Chem.* 252:1102–1106.
20. Brown, E. and Prevec, L. (1978): *Virology* 89:7–21.
21. Hall, W.W., Lamb, R.A., and Choppin, P.W. (1979): *Proc. Nat. Acad. Sci.* 76:2047–2051.

RHABDOVIRUSES

David H. L. Bishop and Richard W. Compans, eds.
The Replication of Negative Strand Viruses

Structural Characteristics of Spring Viremia of Carp Virus

Polly Roy[a]

Analyses of members of the *Vesiculovirus* genus of rhabdoviruses (for example, vesicular stomatitis virus, VSV, Indiana serotype) have established that they have three major polypeptides, a glycoprotein, G, a nucleocapsid protein, N and a matrix (or membrane) protein, M. In addition, there are two minor virion polypeptides, a large polypeptide, L, and a phosphoprotein, NS, which are transcriptase components.[1] Recent studies have indicated that the VSV NS phosphoprotein can be separated into two species (NS1 and NS2) by DEAE cellulose column chromatography,[2] or by acid-urea polyacrylamide gel electrophoresis.[3] It has been proposed that the NS polypeptides may be involved in a regulatory mechanism for RNA synthesis [3] and, in part, this has been supported by recent studies which have shown that different degrees of stimulation of *in vitro* RNA synthesis occur in reconstituted systems involving VSV nucleocapsids and L polypeptide, together with one, the other, or both NS phosphoproteins.[2]

The fish rhabdovirus SVCV has at least two, easily distinguished virion phosphoproteins,[4] both of which can be additionally phosphorylated by the resident virion protein kinase activity. The two phosphoproteins of SVCV are readily separated from each other on the basis of their respective electrophoretic mobilities. In 8%,pH 7 phosphate buffered, polyacrylamide gels, one of the phosphoproteins has an electrophoretic mobility similar to that of N (or another polypeptide), or an alternate form of the other phosphoprotein, but this has not been proven.[4]

[a]Department of Public Health, University of Alabama in Birmingham, Birmingham, Alabama.

The present studies were initiated in an attempt to characterize the two phosphoproteins and provide evidence that the two viral phosphoproteins are related to each other, although it appears that they differ with respect to their sites of phosphorylation. From studies involving hydrolysis of *in vivo* or *in vitro* labeled SVCV NS1 and NS2 species, the major phosphorylated amino acid appears to be phosphoserine.

Identification of the SVCV Virion Phosphoproteins

In previous analyses, we have shown that preparations of SVCV labeled with radioactive amino acids and ^{32}P-phosphate (either *in vivo* or *in vitro* by the action of the resident protein kinase) yield two principal phosphoproteins as well as another minor phosphoprotein, in addition to the non-phosphorylated structural polypeptides L, G, N and M.[4] Using a phosphate buffered, pH 7.0, 8% polyacrylamide gel electrophoretic system to resolve the viral polypeptides, the slower moving phosphoprotein (NS1) migrates on the leading edge of the viral N polypeptide, making it difficult to purify free of N, while the faster migrating phosphoprotein (NS2) is located almost midway between the viral N and M polypeptides.[4] The third (minor) phosphorylated protein present in SVCV virus preparations migrates faster than the NS1 phosphoprotein.[4] In discontinuous 10% slab gels of polyacrylamide,[5] as shown in Figure 1, the one viral principal phosphoprotein migrates slower relative to the other viral polypeptides and is well resolved from them. The third, minor phosphoprotein in 10% discontinuous polyacrylamide gel systems migrates slightly ahead of the viral N polypeptide (Figure 1B). The major SVCV virion polypeptides in a discontinuous gel system are readily distinguished from those of VSV Indiana, as shown in Figure 1A. By reference to the published mol wt values of VSV polypeptides in a discontinuous gel system,[6] (L, 149×10^3; G, 63×10^3; NS, 54×10^3; N, 46×10^3; M, 27×10^3) the SVCV polypeptides have apparent molecular weight values of: L, 149×10^3; G, 73×10^3; NS1, 48×10^3; N, 42×10^3; NS2, 34×10^3; and M, 19×10^3.

Tryptic Peptide Analysis of SVCV NS1 and NS2

In order to compare the structural polypeptides of NS1 and NS2, we have prepared either ^{3}H or ^{14}C-leucine and ^{35}S or ^{3}H-methionine-labeled SVCV, resolved the viral polypeptides by 10%, discontinuous slab gel electrophoresis after fluorescamine treatment [7,8] and recovered them for tryptic peptide analysis. Shown in top panel of Figure 2 are the tryptic peptide analyses of a codigest ^{35}S-methionine NS1 and ^{3}H-methionine NS2. For the methionine labeled tryptic peptide profiles of NS1 and NS2, apart from the minor differences in relative proportions of the peaks (frequently observed in this type of analysis),[7] the profiles were found to be quite compar-

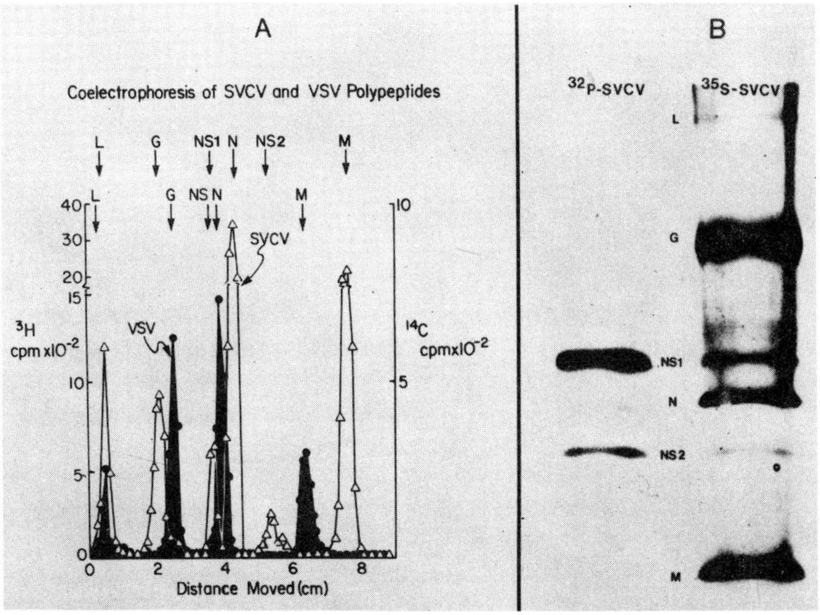

Figure 1. Identification of SVCV viral phosphoproteins. A. A preparation of ^{14}C-leucine-labeled VSV and ^{3}H-leucine-labeled SVCV were mixed and resolved in a 10% Laemmli gel.[5] B. A preparation of an *in vitro* ^{32}P-labeled SVCV protein kinase reaction product, and a purified ^{35}S-methionine-labeled SVCV proteins were resolved in 10% Laemmli gel.[5] On the left is shown the autoradiogram of the ^{32}P-labeled phosphorylated polypeptides, and on the right, is shown the flourograph of the ^{35}S-labeled viral polypeptides.

able to each other, except at the front of the column (fractions 5–10), between fractions 80 and 90, and in particular, a one fraction difference in a peak around fractions 180. This last minor difference has only been observed once and probably represents an artifact, however, the difference observed at the front of the column has been seen in two other similar analyses and may represent tryptic peptide(s) in NS1, not found in NS2. The peak at fraction 80 of the NS2 digest is always present and has been observed one out of three times in an NS2 digest. The reason for this variation in NS1 is not known. Because of this problem, a codigest of ^{14}C-leucine NS1 and ^{3}H-leucine NS2 were analyzed, which are shown in the bottom panel of Figure 2. The two profiles were essentially identical both in the positions and relative amounts of the 14 major peaks. It was concluded from the results obtained that the viral NS1 and NS2 phorphoproteins share a common polypeptide backbone. When ^{32}P-labeled NS1 and NS2 preparations were similarly analyzed, all the ^{32}P eluted in the void volume of the ion exchange column.

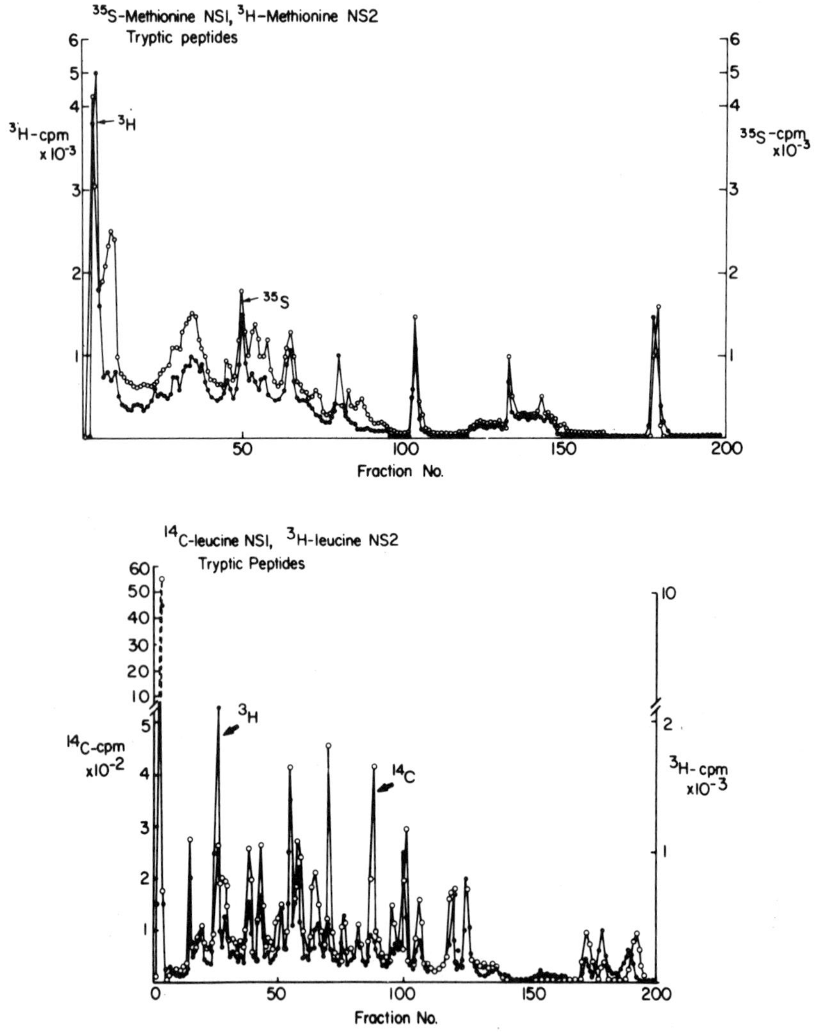

Figure 2. Tryptic peptide analyses of the NS1 and NS2 phosphoproteins of SVCV. A purified preparation of ^{35}S-methionine labeled NS1 and ^{3}H-methionine labeled NS2 mixture (top panel) and a mixture of ^{14}C-leucine NS1 and ^{3}H-leucine NS2 preparation (lower panel) were digested by TPCK-trypsin, and the peptides resolved by high pressure ion exchange resin column chromatography.

Analyses of the Phosphorylated Components of SVCV NS1 and NS2

It is possible that the two SVCV NS phosphoproteins differ in terms of their sites of phosphorylation. In order to analyze this possibility, preparations of *in vivo* or *in vitro,* ^{32}P-γ ATP labeled SVCV, NS1 and NS2 phosphoproteins were purified by polyacrylamide gel electrophoresis, recovered and subjected to partial proteolysis using *Staphylococcus aureus* V8 protease.[9] As shown by the single and codigests of the *in vitro* or *in vivo* labeled NS1 and NS2 species (Figure 3), the results indicate that the phosphorylated oligopeptide fragments of the two NS polypeptides were different. When the *in vivo* or *in vitro* phosphorylated NS1 and NS2 species were hydrolyzed by acid and the phosphorylated amino acids separated by paper electrophoresis, together with optical quantities of marker phosphoserine, phosphothreonine and phosphotyrosine, only labeled phosphoserine was identified.

Discussion

The results obtained in this investigation have established that SVCV has at least two different phosphorylated forms of the virion and interacellular NS polypeptide, NS1 and NS2. Tryptic peptide analyses have shown that the

Figure 3. Limited proteolysis of SVCV NS1 and NS2 phosphoproteins. Preparations of *in vivo* (left) or *in vitro* (right) ^{32}P-labeled NS1 and NS2 bands were digested either separately or together with 2μg of *S. aureus* V8 protease and analysed by electrophoresis in slab gel. An autoradiograph of the product is shown.

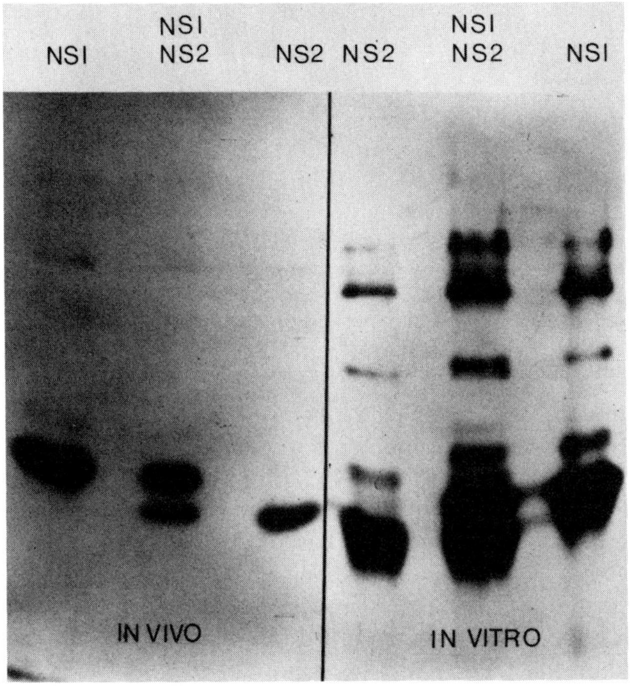

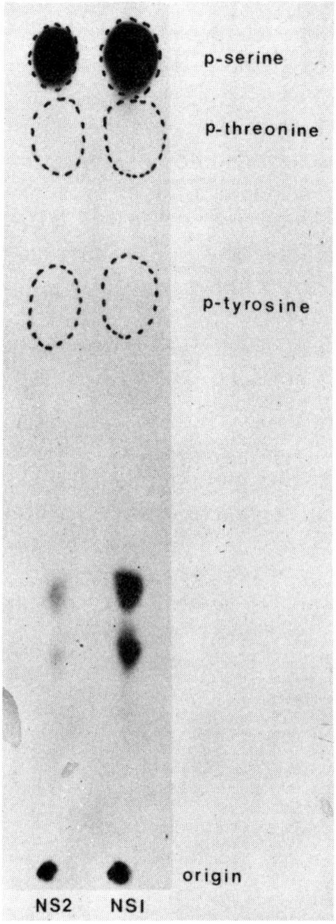

Figure 4. Identification of phosphoaminoacids of NS1 and NS2. The *in vitro* phosphorylated NS1 and NS2 were hydrolysed by 6N HCl for 2 hours and the phosphorylated amino acids separated by high voltage paper electrophoresis, together with optical quantities of marker phosphoserine, phosphothreonine and phosphotyrosine.

two NS phosphoproteins can be distinguished from the four nonglycosylated structural polypeptides (not shown). Amino acid sequence analyses will be needed to determine to what extent NS1 and NS2 are identical and whether they differ only in the sites of phosphorylation or whether there are more fundamental differences, such as their amino or carboxy terminal sequences. The question of the functions of the two phosphoproteins also needs to be determined. If, as suggested by Clinton and associates,[3] and Kingsford and Emerson,[2] the two phosphoproteins are involved in different facets of the regulation of RNA synthesis, then it will be important to know to what extent the different NS forms are involved in RNA transcription signals (initiation, polyadenylation, termination), processing (capping, methylation), progression (chain elongation) or complete RNA replication. These are questions which warrant further investigation into the structure and function of rhabdovirus phosphoproteins.

ACKNOWLEDGMENTS
This research was supported by a Grant from the USPHS National Institute of Allergy and Infectious Diseases A 13686.

References

1. Bishop, D.H.L. (Ed.) (1979): *Rhabdoviruses,* Vol. 1, CRC Press, Florida.
2. Kingsford, L. and Emerson, S.U. (1980): *J. Virol.* 33:1097–1105.
3. Clinton, G.M., Burge, B.W., and Huang, A.S. (1978): *J. Virol.* 27:340–346.
4. Roy, P. and Clewley, J.P. (1978): In: *Negative Strand Viruses and the Host Cell.* Mahy, B.W.J. and Barry, R.D. (Eds.), Academic Press, New York, pp. 116–125.
5. Laemmli, U.K. (1970): *Nature (London)* 227:680–685.
6. Obijeski, J.F., Marchenko, A.T., Bishop, D.H.L., Cann, B.W., and Murphy, F.A. (1974): *J. Gen. Virol.* 22:21–33.
7. Gentsch, J.R. and Bishop, D.H.L. (1978): *J. Virol.* 28:417–419.
8. Udenfriend, S., Stein, S., Bohlen, P., Dairman, W., Leimbruber, W., and Weigele, M. (1972): *Science* 178:871–872.
9. Cleveland, D.W., Fisher, S.G., Kirschner, M.W., and Laemmli, U.K. (1977): *J. Biol. Chem.* 252:1102–1106.

David H. L. Bishop and Richard W. Compans, eds.
The Replication of Negative Strand Viruses

Structural Differences in the Glycoproteins of Rabies Virus Strains

Bernhard Dietzschold [a]

Analysis of several rabies virus strains revealed a great diversity in the reactivity of the viruses with a panel of monoclonal antibodies directed against the glycoprotein antigen.[1] In order to determine the structural basis of the antigenic variability of the rabies virus glycoprotein, a more detailed chemical analysis of the structure of this protein is necessary.

The present report describes the fine structure analysis and comparision of the glycoproteins from four rabies virus strains. The results from this analysis revealed strain-specific variations in the number of sugar side chains and strain-specific differences in the primary structure of the glycoprotein. We report also the isolation and characterization of immunogenic fragments of the rabies virus glycoprotein.

We have shown that two size classes of the envelope glycoprotein can be demonstrated in rabies viruses.[2] The larger glycoprotein which is present in strain ERA is also present in Flury LEP (data not shown), the smaller glycoprotein is found in Flury HEP virus and both size classes are present in CVS and PM strains. The two glycoproteins found in CVS, however, have identical tryptic peptides. The smaller glycoprotein contains less carbohydrate than the larger glycoprotein, indicating the existance of differences in the glycosylation among these two glycoproteins. Figure 1 demonstrates the characteristic differences between the two forms of glycoprotein of CVS by analysis of the desialated sugar containing tryptic peptides. Three glycosylated tryptic peptides were obtained from the larger glycoprotein and only

[a]The Wistar Institute, 36th Street at Spruce, Pheladelphia, Pennsylvania.

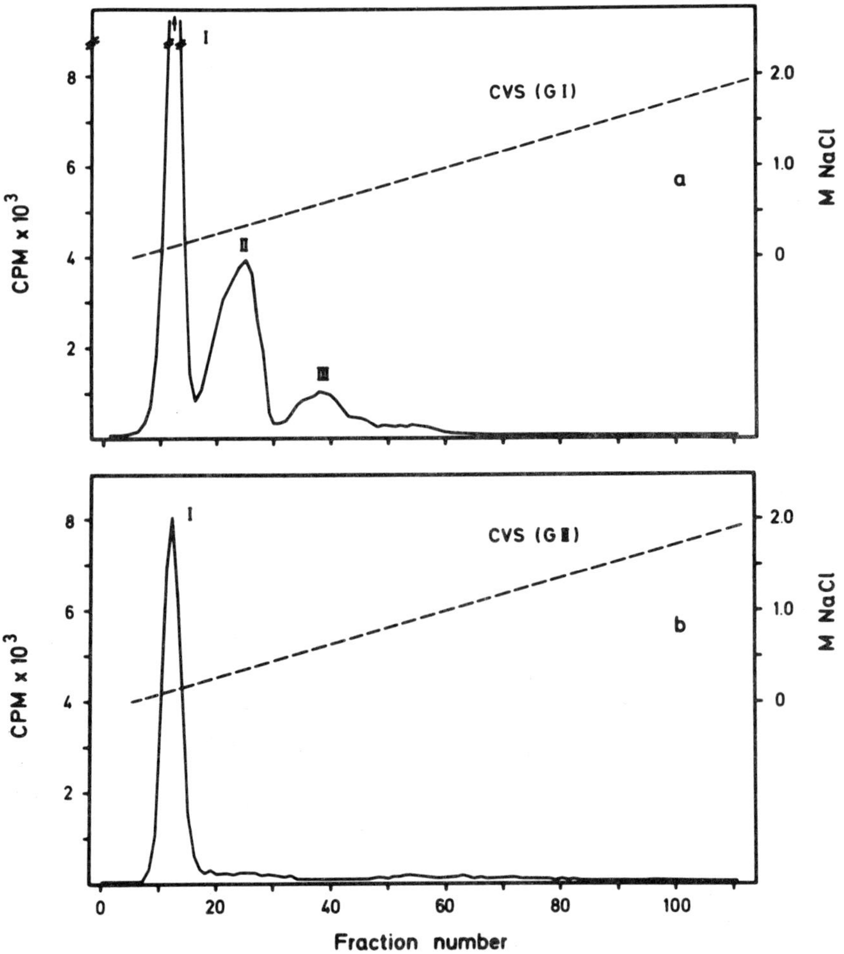

Figure 1. Tryptic glycopeptides of CVS virus glycoproteins GI (a) and G II (b) on DEAE cellulose. The G proteins were purified from ^{3}H-glucosamine-labeled CVS, treated by mild acid hydrolysis and digested with trypsin.[5] The tryptic peptides were analyzed on DEAE cellulose as described.[5]

one glycosylated peptide was obtained from the smaller glycoprotein. Analysis of the desialated glucosamine-containing tryptic glycopeptides in the other virus strains (Figure 2) reveals that the Flury HEP virus glycoprotein, like the small type glycoprotein of CVS, produces only one tryptic glycopeptide, whereas the large type glycoprotein in ERA and Flury LEP viruses produces three glycopeptides as in the larger glycoprotein of CVS. In order to characterize and compare the sugar chains of rabies virus glycoproteins the adjacent amino acids of the glycopeptides were removed by hy-

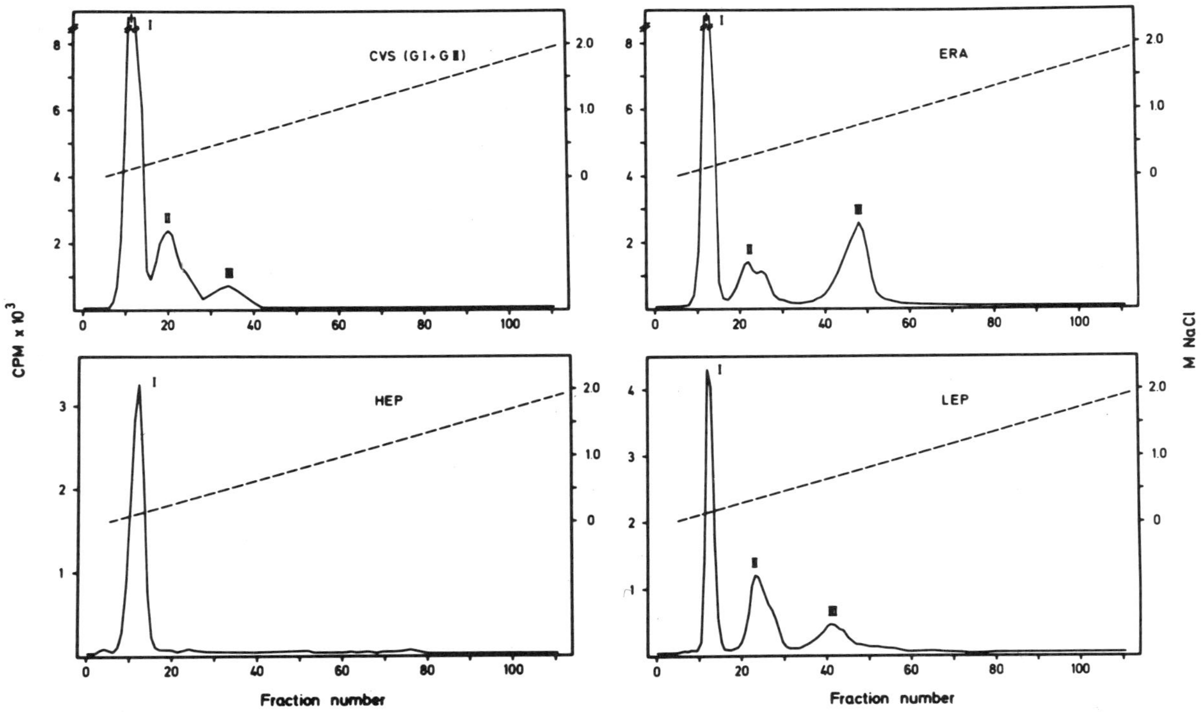

Figure 2. Desialated tryptic glycopeptides of CVS, ERA, Flury HEP and Flury LEP glycoprotein. The tryptic peptides were prepared and analyzed on DEAE cellulose as described.[5]

drazinolysis.[3] In Figure 3, the ^{14}C-glucosamine-labeled amino acid-free sugar chains of Flury LEP were compared with ^{3}H-glucosamine-labeled free sugar chains of (a) ERA, (b) Flury HEP and (c) CVS. These results revealed that the size of the glycoprotein side chains of different rabies virus strains is identical. The apparent molecular weight of this sugar chain is 2,200 as estimated by the method of Burge and Huang.[4]

The sugar chains of the glycoprotein of Flury LEP, Flury HEP and CVS contain similar amounts of D-mannose, D-glucosamine, D-galactose and L-fucose. No change in the elution profile of these sugar chains was observed after endo-β-N-acetylglucosaminidase H digestion. From these results, we conclude that both small and large type glyproteins have complex sugar chains of identical size and composition.

However, the small and large type glycoproteins differ in the number of sugar side chains per molecule. Analysis of the N-terminal amino acid of the larger and smaller forms revealed that both glycoproteins have identical N-terminal amino acids indicating that the smaller glycoprotein is not a likely proteolytic cleavage product of the larger one. In addition, from more recent amino acid sequence determinations, the first four amino acids after the N-terminal lysine shows that the N-terminal region of ERA virus glycoprotein is hydrophobic (Table 1). From an analysis of the leucine-containing tryptic peptides, it is evident that ERA glycoprotein shares only 40 to 50% of its peptides with CVS, Flury HEP or PM glycoprotein, whereas Flury HEP glycoprotein and Flury LEP glycoprotein have 75% common peptides (data not shown).

In contrast to the G-protein, remarkable (80%) sequence conservation was observed among the N, M_1, M_2 and L-proteins of different rabies virus strains.[2] The higher degree of sequence conservation among these proteins is possibly explained by the special functions of these proteins such as interactions with RNA, other proteins and lipids.

In order to identify the regions of the rabies virus glycoprotein to which antibodies bind, we have produced peptide fragments of the rabies virus glycoprotein by cyanogen bromide cleavage. Seven methionine-containing BrCN peptides of ERA glycoprotein can be resolved by SDS-PAGE (Figure 4a). Two of these BrCN peptide fragments (7K and 11K) are glycosylated. Furthermore only three BrCN peptide fragments (11K, 21K and 28K) contain leucine (data not shown), indicating a polar distribution of the hydrophobic amino acids. A strain comparison of the BrCN peptide fragments reveals substantial differences in methionine and glucNH$_2$-labeled peptides (data not shown). Flury HEP produces only one glycopeptide fragment, and Flury LEP and CVS strains each produce two glycopeptide fragments which are different from the ERA strain.

A mixture of BrCN peptide fragments of ERA glycoprotein induced virus neutralizing antibodies when injected into mice. Two of the fragments (11K

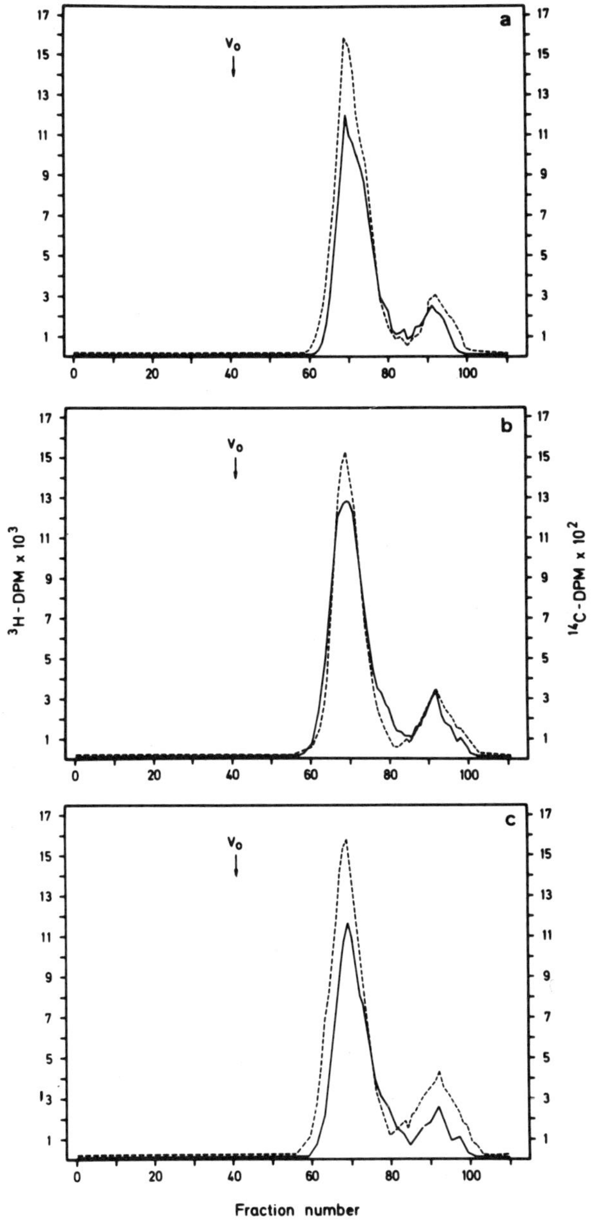

Figure 3. Cochromatography on BioGel P6 of ^{14}C-labeled amino acid-free sugar chains of Flury LEP(——) with ^3H-glucosamine-labeled free sugar chains of ERA (a,---), HEP (b,---) and CVS (c,---). The desialated glycoproteins were digested with trypsin. After dialysis and lyophylization the tryptic peptides were subjected to hydrazinolysis and the hydrazinolysis products were N-acetylated as described.[3]

Table 1. Analysis of N-Terminal Amino Acids.

CVS GI:	H_2N-Lys
CVS GII:	H_2N-Lys
ERA G:	H_2N Lys-Ala-Ala-Leu-Ala

The N-terminal amino acids of CVS GI, CVS GII, ERA G and of shortened proteins derived from ERA G by stepwise Edman degradation were determined by dansylation and chromatography on polyamide sheets as described.[8]

and 16K) were specifically immunoprecipitated with a hyperimmune anti-ERA glycoprotein serum (Figure 4b). To determine whether the immunogenic fragments of ERA G are present in all rabies virus strains, the BrCN fragments of Flury LEP, Flury HEP and CVS were immunonoprecipitated with antiserum directed against ERA G protein. Figure 5 shows that most of the immunoprecipitable fragments are present in all strains but quantitative differences exist.

Figure 4. Polyacrylamide gel electrophoresis of (a) BrCN cleavage fragments of ERA glycoprotein and (b) immunoprecipitates of BrCN fragments of ERA G. One nanomole of glycoprotein, labeled with ^{3}H-methionine, was precipitated with ethanol. The precipitate was dried under vacuum and dissolved in 200 μl 70% formic acid containing 50 mg per ml BrCN and 0.1% 2-mercaptoethanol. The mixture was left 24 hr at 20° in the dark and then lyophilized. For immunoprecipitation, BrCN fragments were dissolved in 0.5 ml of 0.05 M Tris-HCl pH 7.4, 0.1 M NaCl, 1% Triton X-100, 0.5% deoxycholate and incubated with 5 μl of either rabbit antiserum directed against ERA glycoprotein (b) or normal rabbit serum (c) at 4°C for 15 hr. The antigen-antibody complexes were precipitated by addition of 50μl of a 10% suspension of *S. aureus* as described.[6] The BrCN fragments of ERA G and the immunoprecipitable fragments were solubilized with a small volume of 1% SDS and 1% 2-mercaptoethanol and electrophoresed in a 15 to 20% gradient polyacrylamide slab gel. The peptide bands were visualized by fluorography.[7]

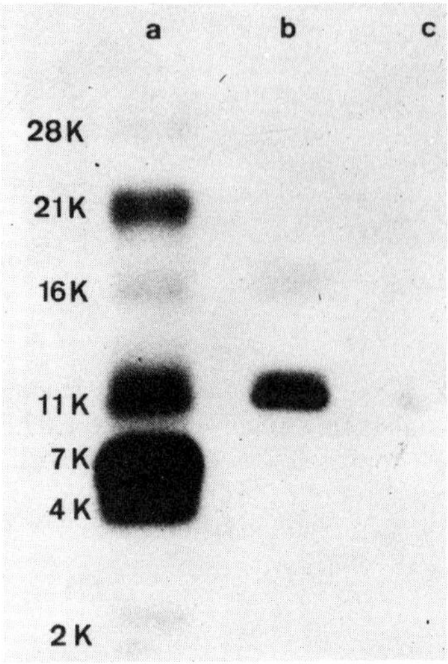

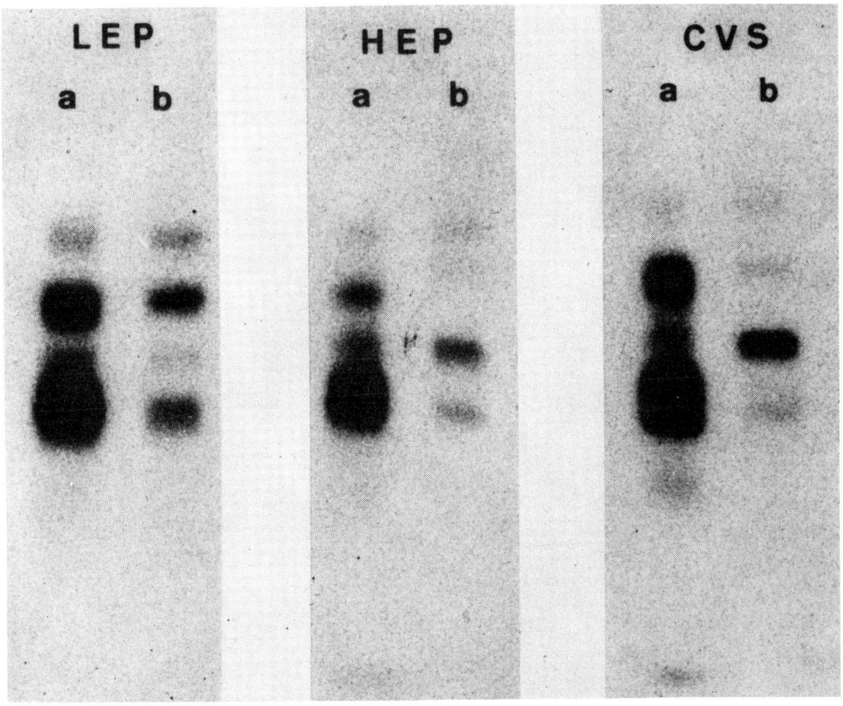

Figure 5. Comparison of the immunoprecipitable BrCN fragments of the glycoproteins of Flury LEP, Flury HEP and CVS virus. BrCN peptides were prepared from purified ^{3}H-methionine-labeled glycoproteins and immunoprecipitated with an antiserum directed against ERA glycoprotein as described in Figure 4. The BrCN fragments (a) and the immunoprecipitates were analyzed in a 15 to 20% gradient polyacrylamide gel.

We expect that the isolated BrCN peptide fragments will provide powerful tools to study structural and functional relationships of the rabies virus glycoprotein. Structural fragments of the glycoprotein will be correlated with the epitope regions as determined by monoclonal antibodies. Using specific modifications of the immunogenic BrCN peptides such as carboxymethylation of the SH-groups or amidination of the α and ϵ amino groups, we could determine whether conformational structures are preserved in the immunogenic peptides.

References

1. Wiktor, T.J., Flamand, A., and Koprowski, H. (1980): *J. Virol. Meth.* 1:33–46.
2. Dietzschold, B., Cox, J.H., and Schneider, L.G. (1979): *Virology* 98:63–75.
3. Yosizawa, Z. and Sato, T. (1966: *Biochim. Biophys. Acta* 121:417–420.
4. Burge, B.W. and Huang, A.S. (1980): *J. Virol.* 6:176–182.

5. Dietzschold, B. (1977): *J. Virol.* 23:286–293.
6. Shih, T.Y., Weeks, M.O., Young, H.A., and Scolnick, E.M. (1979): *J. Virol.* 31:546–556.
7. Bonner, W.M. and Laskey, R.A. (1974): *Eur. J. Biochem.* 46:83–88.
8. Matheka, M.D. and Bachrach, H.L. (1975): *J. Virol.* 16:1248–1253.

David H. L. Bishop and Richard W. Compans, eds.
The Replication of Negative Strand Viruses

Reevaluation of the Structural Proteins M_1 and M_2 of Rabies Virus

James H. Cox, Frank Weiland,
Bernard Dietzschold, and Lothar G. Schneider[a]

Treatment of rabies virus with deoxycholate resulted in the solubilization of the structural proteins M_1, M_2 and G, suggesting that these three proteins are associated with the viral envelope.[1] However, of the three major proteins associated with the virus membrane, only the glycoprotein was found to be located on the external surface.[2] Whether the two nonglycosylated proteins M_1 and M_2 are essential constituents of the virus membrane remained to be clarified.

In this report, we have studied the localization and structural interaction of the rabies virus structural proteins M_1, M_2, G, and N by immunoprecipitation, immunofluorescence and immunoelectronmicroscopy procedures. Our results indicate that the M_1 protein is associated with the N-protein forming the nucleocapsid complex. However, in addition to the G-protein we found that also the M_2 protein of rabies virus is localized on the surface of infected cells.

Immunoprecipitation with antisera directed against purified viral structural proteins was carried out in order to investigate the interrelations of the rabies virus structural proteins. For the preparation of specific antisera, M_1 and M_2 protein were purified by sodium dodecyl sulfate (SDS) polyacrylamide gel electrophoresis. The elution of the proteins from the gel has been reported.[2] In order to remove the majority of SDS, the eluted proteins were precipitated with 5 vol ethanol, dissolved in 8 M urea 0.05 M Tris-HC1 (pH 2.5) and reprecipitated with ethanol. The purification of nucleocapsids

[a]Federal Research Institute for Animal Virus Diseases, P.O. Box 1149, D7400 Tübingen, Federal Republic of Germany.

and G-protein and the production of antisera was recently described.[2,3] The specificity of the antisera employed is depicted in Figure 1. Precipitation experiments were carried out with SDS lysed rabies virus. Purified rabies virus was boiled in 1% SDS and 1% 2-mercaptoethanol. The lysate was then diluted 20-fold with lysis buffer and immunoprecipitated. Columns b and c in Figure 1 show that the anti-M_2 as well as the anti-M_1 serum react monospecifically with their corresponding antigens. The fact that anti-N serum precipitates traces of M_1 could be due to the preparation procedure of the antigen. Anti-N serum was raised against purified nucleocapsids which were possibly not completely free of M_1 protein. Figure 2 shows an autoradiography of a SDS-gel electrophoresis of immunoprecipitates from lysates of uninfected (f) and rabies infected BHK cells (a, b, c, d, e, g). Uninfected or ERA infected cells (24 hpi) were labeled with ^{35}S-methionine (10 μCi/ml) for 12 hr and were lysed by three passages through a 22 guage needle in lysis buffer containing 1% Triton X-100, 0.5% deoxycholate, 0.02 M Tris-hydrochloride (pH 7.2), 0.1 M NaC1, and 200 U/ml Trasylol (Bayer, Germany). The cell lysates were centrifuged at 5000 $\times$ g for 10 min and the supernatant was precleaned by incubation with formaldehyde-fixed *Staphylococcus aureus* suspension followed by immunoprecipitation as previously reported by Shih et al.[4] Lane d in Figure 2 demonstrates that in addition to the M_1 protein significant amounts of the N-protein were coprecipitated by the anti-M_1 serum. Furthermore, the anti-N serum also (lane c) precipitates a mixture of M_1 and N-protein whereas the anti-M_2 serum predominantly precipitates the M_2 protein.

The observation that the anti-M_1 as well as anti-N serum precipitate a mixture of M_1 and N-protein from lysates of rabies virus infected cells may indicate that the M_1 protein can be found in association with the intracellular nucleocapsids. In contrast, the M_2 protein is found mainly free in cell extracts.

That each protein is an independent antigen was also confirmed by indirect immunofluorescence of infected and non-infected cells. In order to determine which of the four antigens are located on the cell surface or within the cell, infected BHK cells were harvested and the suspended cells stained by indirect immunofluorescence. Briefly, the suspended cells were washed and incubated with the specific rabbit antisera at the following dilution: anti-G and anti-RNA: 1:600; anti-M_1 and anti-M_2: 1:50. After further washings, the cells were incubated with fluorescein conjugated goat anti-rabbit immunoglobulins. In parallel, infected cell monolayers were first permeabilized by fixation in -20°C acetone and then treated as above.

In Figure 3 it can be seen that only the G-protein (upper left) and the M_2 protein (upper right) show membrane fluorescence with infected cells which are non-permeable. No fluorescence was observed with the RNP-protein (lower left) and the M_1 protein (lower right). Infected cells which have been

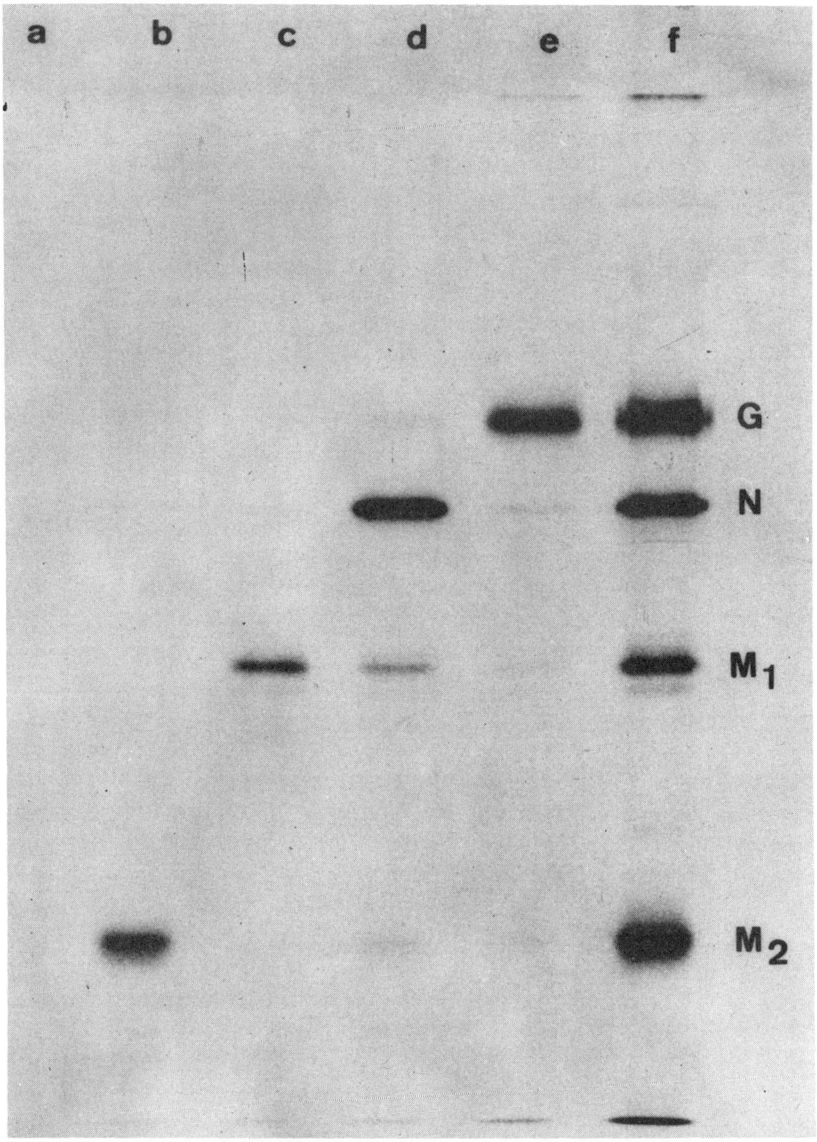

Figure 1. Autoradiogram of immunoprecipitates. Rabies virus labeled with ^{35}S-methionine was dissolved in 1% SDS, 1% 2-mercaptoethanol and boiled for 2 minutes. The disrupted virus was diluted 20-fold with 0.02 M Tris-HC1 (pH 7.2), 0.1 M NaC1, 1% Triton X-100, immunoprecipitated and run on a 10% polyacrylamide slab gel as previously reported.[6] Antisera used were: lane a, non-immune serum; lane b, anti-M_2; lane c, anti-M_1; lane d, anti-nucleocapsid serum; lane e, anti-G. Lane f represents the polypeptide pattern of ^{35}S-methionine labeled rabies virus.

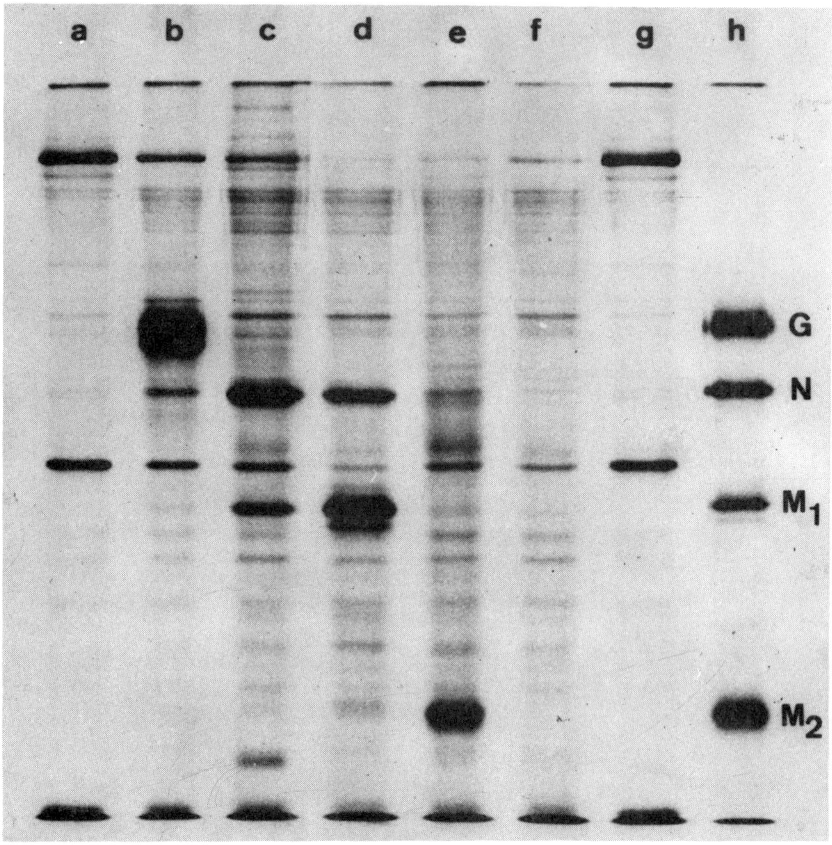

Figure 2. Autoradiogram of immunoprecipitates. ^{35}S-labeled cell extracts of uninfected (f) and rabies infected (a, b, c, d, e, g) BHK cells were labeled for 12 hr with 10 μCi of ^{35}S-methionine per ml, immunoprecipitated with the indicated antisera as described,[4] and run on a 10% polyacrylamide slab gel as indicated in Figure 1. Antisera used were: lanes a and g, non-immune serum; lane b, anti-ERA G; lane c, anti-nucleocapsid serum; lane d, anti-M_1; lanes e and f, anti-M_2. Lane h represents the polypeptide pattern of ^{35}S-methionine labeled rabies virus.

made permeable by acetone fixation showed fluorescence with all four antisera (not shown). No fluorescence was observed in non-infected cells. Therefore, each antibody identified its respective antigen within the infected cell.

A further study to determine whether the specific antibodies are bound at random on the cell surface or localized on the virus particles was carried out using indirect immunoelectronmicroscopy. Confluent monolayers of BHK cells grown in 5 cm Petri dishes were infected with ERA virus. 48 hr after

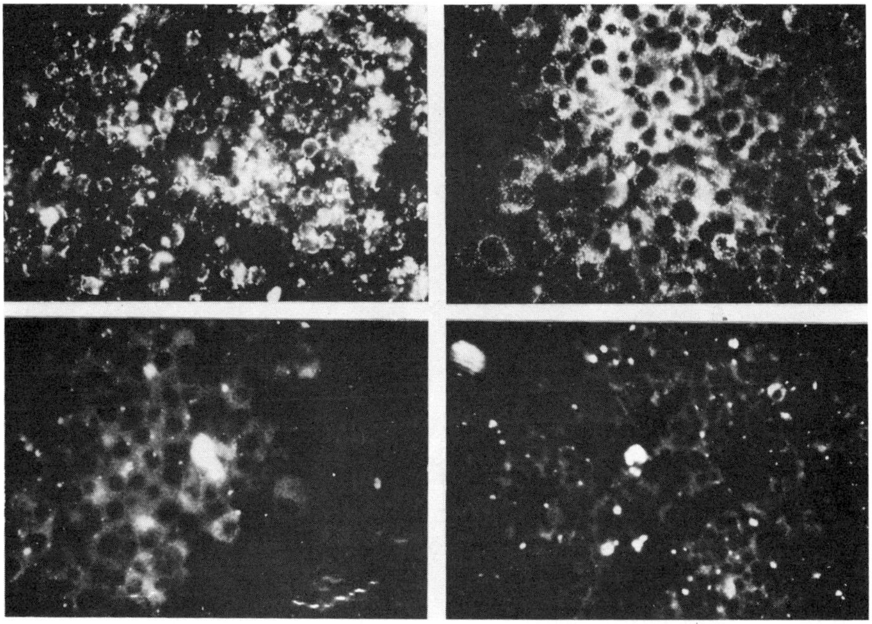

Figure 3. Immunofluorescent localization of rabies virus proteins in virus infected BHK cells. Infected cells were incubated with the indicates specific rabbit antisera and processed for indirect immunofluorescence as described in the text. ERA virus infected cells pretreated with anti-ERA G (upper left), anti-ERA M_2 (upper right), anti-ERA N (lower left), and anti-ERA M_1 (lower right).

infection, the cells were fixed with 0.5% glutaraldehyde in phosphate buffered saline (PBS), pH 7.4, for 4 min at 4°C. Then the cell monolayer was washed with PBS and preincubated with bovine serum albumin in PBS to eliminate any residual non-specific binding. Thereafter, the cells were incubated with the specific rabbit antiserum at the following dilutions: anti-G: 1:50; anti-RNA: 1:50; anti-M_1: 1:10; anti-M_2: 1:10. A further washing procedure was followed by incubation with ferritin conjugated sheep anti-rabbit immunoglobulins. Subsequently the cell monolayer was washed again, fixed with 2.5% glutaraldehyde in 0.1 M cacodylate buffer, pH 7.2, for 1 hr. Cells were then scraped from the Petri dishes, centrifuged and postfixed with 1% osmium tetroxide. En bloc staining was performed with saturated uranyl acetate. Thin sections were mounted on 100 mesh copper grids and examined without further staining.

Figure 4 demonstrates that only the G-protein (d) and the M_2 protein (a) of rabies virus can be visualized on the surface of infected cells by indirect immunoelectronmicroscopy. No ferritin binding could be detected on rabies virus infected cells incubated with either anti-N (not shown) or anti-M_1 (b).

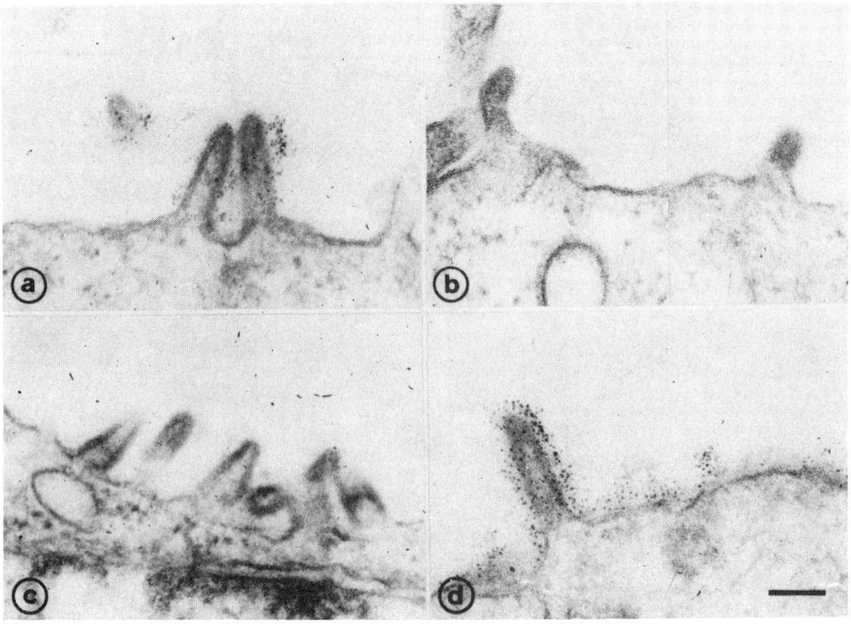

Figure 4. Immunoelectronmicroscopic localization of rabies virus proteins in virus infected BHK cells. Rabies virus or VSV infected cells were incubated with the indicated specific rabbit antisera and processed for indirect immunoelectronmicroscopy as described in the text. After reaction with ferritin conjugated anti-rabbit antibodies, the cells were rinsed extensively with PBS, fixed in 2.5% glutaraldehyde, pH 7.2, and stained with saturated uranyl acetate. Sections were cut and examined without further staining. (a) rabies virus infected BHK cells pretreated with anti-ERA M_2, (b) rabies virus infected BHK cells pretreated with anti-ERA M_1; (c) VSV infected BHK cells pretreated with anti-ERA M_2; (d) ERA infected cells pretreated with anti-ERA G.

Infected monolayers pretreated with rabbit normal serum also showed no ferritin binding (not shown). The specificity of ferritin binding for anti-G and M_2 sera was demonstrated by the absence of ferritin binding in uninfected monolayers (not shown) and in vesicular stomatitis virus (VSV) infected cells (c).

There are obvious differences in ferritin distribution patterns between rabies infected BHK cells pretreated with anti-M_2 (a) and anti-G (d) serum. After incubation of cells with anti-G serum, ferritin bound not only to budding particles but also to patches of G-protein in the cell membrane. In cells pretreated with anti-M_2 the ferritin label was restricted to the area of budding or to released virus particles. From the results of immunofluorescence and immunoelectronmicroscopy, it can be suggested that the M_2 protein of rabies virus represents a membrane protein. The fact that the M_2 protein can be detected only on the surface of budding particles but not on other areas of the cell membrane may suggest a possible function of the M_2 protein during viral budding. It may serve as a bridge between the G-protein and the nucleocapsid complex.[5] The finding that the M_2 protein is exposed on the external surface of the virus membrane is apparently in contrast to recent observations which indicated that the M_2 protein does not traverse the viral membrane.[2] However, only a small portion of the M_2 protein may be exposed on the external surface of the virus membrane. This could explain the failure of labeling of the M_2 protein by the $Na_3B^3H_3$-pyridoxalphosphate technique.[2]

Unlike the M_2 protein, the M_1 protein seems to be an internal virus protein. In rabies infected cells an appreciable amount of the M_1 protein is found in association with the N-protein. Recently, we have found that the M_1 protein of rabies virus is phosphorylated.[6] Analogous to the NS protein of VSV, a highly and a less phosphorylated form of the M_1 protein was resolved.[7] These findings imply that the M_1 protein of rabies virus is functionally related to the NS protein of VSV.

Summary

The structural proteins M_1 and M_2 of rabies virus were studied with respect to their association with other viral proteins and their topographical distribution in infected cell membranes. By immunoprecipitation, immunofluorescence and immunoelectronmicroscopy, it was found that the M_1 protein is associated with the N-protein forming the nucleocapsid complex whereas the M_2 protein is a membrane protein with a small portion exposed on the external surface of the viral membrane.

References

1. Sokol, F., Stancek, D., and Koprowski, H. (1971): *J. Virol.* 7:241–249.
2. Dietzschold, B., Cox, J.H., Schneider, L.G., Wiktor, T.J., and Koprowski, H. (1978): *J. Gen. Virol.* 40:131–139.
3. Schneider, L.G., Dietzschold, B., Dierks, R.E., Matthaeus, W., Enzmann, P., and Strohmaier, K. (1973): *J. Virol.* 11:748–755.
4. Shih, Y., Weeks, M.O., Young, H.A., and Scolnick, E.M. (1979): *J. Virol.* 31:546–556.
5. Dubovi, E.J. and Wagner, R.R. (1977): *J. Virol.* 22:500–509.
6. Dietzschold, B., Cox, J.H., and Schneider, L.G. (1979): *Virology* 98:63–75.
7. Clinton, G.M., Burge, B.W., and Huang, A.S. (1978): *J. Virol.* 27:340–346.

David H. L. Bishop and Richard W. Compans, eds.
The Replication of Negative Strand Viruses

Identification of Intermediates in the Branch Glycosylation of the VSV Glycoprotein

James R. Etchison[a]

Introduction

Elucidation of the complex intracellular pathway for the synthesis and processing of asparagine-linked oligosaccharide moieties of glycoproteins is important for a comprehensive understanding of the biogenesis of enveloped viruses and the interplay of viral gene products and host cell macromolecular synthesis. Delineation of the individual reactions and their collation into a descriptive pathway provides a conceptualization of the cellular processes leading to the formation of plasma membranes and the envelopes of animal viruses which mature by budding through the plasma membrane. Recent studies on the synthesis of viral envelope glycoproteins have clearly shown that the mechanisms employed by the host cell for the synthesis of the complex oligosaccharide moieties of membrane glycoproteins are intricate and do not necessarily follow linear extrapolations from known mechanisms to known structures.[1-3]

Following the en bloc transfer of a glycosylated oligomannosyl precursor oligosaccharide ($Glc_3Man_9GlcNAc_2$) from a polyisoprenylpyrophosphate donor to the nascent polypeptide, a series of processing reactions occur during the transit of the VSV glycoprotein from its site of synthesis in the rough endoplasmic reticulum to its final destination in the plasma membrane or viral envelope.[1,2,4-6] One aspect of this processing involves the removal of the terminal glucose residues from the precursor oligosaccharide and the subsequent trimming of the oligomannosyl core to yield a structure which

[a] California Primate Research Center and Departments of Internal Medicine and Biological Chemistry, University of California, Davis, California.

contains only three mannoses in the mature oligosaccharide. In lectin-resistant cell mutants which lack a specific N-acetylglucosaminyl-transferase, the trimming is incomplete and an oligosaccharide containing a pentamannosyl core is found on the VSV G protein.[2,7,10] From these and other studies, it has been proposed that the final phase of the trimming occurs after the initiation of the synthesis of one of the branches found in the mature oligosaccharides.[8–10]

The results described below represent initial studies in the isolation and characterization of intermediates in the branch glycosylation of the VSV G protein. An intermediate containing a terminal N-acetylglucosamine attached to a pentamannosyl core has been identified after pulse-labeling of VSV-infected cells. In addition, the data suggest that transfer of galactose to this initiated branch may also occur prior to the final trimming to a trimannosyl core.

Results and Discussion

In order to analyze the initial steps in the branch glycosylation of newly synthesized glycoproteins, it is desirable to be able to selectively introduce label into the branch chains and not into other portions of the oligosaccharide. Since glucosamine occurs both in the branch chains and in the core and since glucosamine is metabolized into and incorporated as sialic acid as well, it was necessary to devise special conditions for the pulse-labeling of the branch glycosylation intermediates with glucosamine. Since transfer of the core region of the oligosaccharide occurs on the nascent polypeptide and is, therefore, closely coupled to protein synthesis,[11] inhibition of protein synthesis will result in a rapid cessation of incorporation into the core glucosamine residues. Figure 1 shows that there is an almost immediate reduction in glucosamine incorporation following inhibition of protein synthesis with puromycin. On the other hand, glucosamine incorporation does continue at a reduced rate for at least 30 to 40 minutes following protein synthesis inhibition. In order to maximize the inhibition of glucosamine incorporation into the core region and minimize any untoward effects due to protein synthesis inhibition, we chose to carry out the pulse-labeling of VSV-infected cells between 10 and 20 minutes after protein synthesis inhibition.

Puromycin-treated, VSV-infected cells were pulse-labeled for 10 min with 100 μCi/ml ^{3}H-glucosamine, immediately frozen on a block of dry ice, scraped off the plate into 1% NP40 in 0.01 M Tris-HC1, pH 7.5, and stored on an ice bath. When all plates had been harvested and pooled, the samples were extracted at 80° for 5 min with occasional vortexing. Total glycoprotein was precipitated by extraction with n-butanol and washing the precipitate

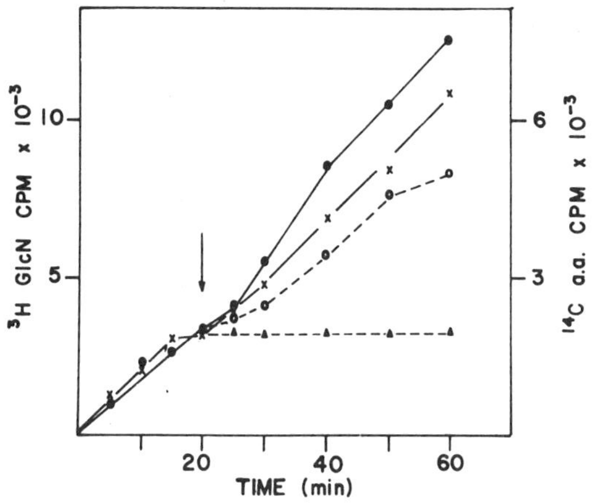

Figure 1. Inhibition of glucosamine and amino acid incorporation into VSV-infected BHK cells by puromycin. At 3.5 hr postinfection labeling medium containing 1 μCi/ml ^{3}H-glucosamine and 0.02 μCi/ml ^{14}C-lysine was added to matched monolayers of BHK–21 cells which had been infected with VSV at an moi of 10 pfu/cell. Incorporation was determined by TCA precipitation after dissolving rinsed monolayers with 2% SDS. After 20 min μg/ml puromycin was added to half the remaining monolayers. (●——●) ^{3}H-GlcN, no puromycin; (X——X) ^{14}C-lys, no puromycin; (○----○) ^{3}H-GlcN, plus puromycin; and (△——△) ^{14}C-lys plus puromycin.

with 95% ethanol. The precipitated glycoproteins were digested with 10 mg/ml Pronase in 0.1 M Tris-HC1, pH 7.5, at 37° for 48 hr . Pronase was inactivated at 100° for 5 min after the incubation. The sample was chromatographed on a 1.5 × 150 cm column of BioGel P4 and the "precursor glycopeptide" fractions (S_3 and smaller, i.e., between 1500 and 2300 daltons) were pooled, lyophilized, and desalted on Sephadex G10. These precursor glycopeptides were digested with neuraminidase and endo-β-N-acetylglucosaminidase H for 24 hr and the neutral products isolated by chromatography on AG1-X2 (formate). The neutral products were analyzed by gel filtration through a 1.5 × 175 cm column of BioGel P2 (200–400 mesh) eluted with 1 mM sodium azide.

Figure 2A shows the gel filtration analysis of the neutral oligosaccharides released by endo H digestion of the pulse-labeled precursor glycopeptides. Of particular interest were the peak and its prominent shoulder eluting slightly larger than a pentamannosyl-N-acetylglucosamine marker (M_5G). The difference in size of the major peak and its shoulder is approximately one hexose unit (~160 daltons). The fractions comprising these peaks were pooled, lyophilized, desalted on G10, and digested with either α-

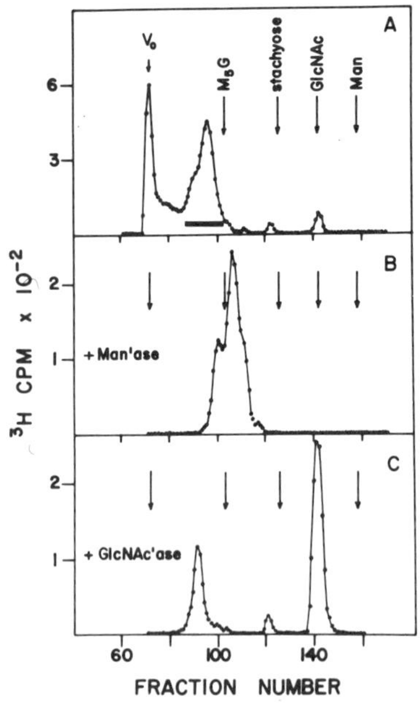

Figure 2. BioGel P2 analysis of neutral oligosaccharides from Endo H digested precursor glycopeptides. Analyses of the neutral products prepared as described in the text were carried out using previously described methods.[1,7,12] The fractions indicated by the solid bar in panel A were pooled and the oligosaccharides were reanalyzed after α-mannosidase (B) or β-N-acetylglucosaminidase (C) digestion.

mannosidase or β-N-acetylglucosaminidase and rechromatographed on BioGel P2. Figure 2B shows that α-mannosidase treatment results in a decrease in molecular weight which corresponds to the loss of 2 to 3 mannose residues. Figure 2C shows that treatment with β-N-acetylglucosaminidase has a differential effect on the major peak and the shoulder: a peak identical in size and proportion of label as the shoulder seen in Figure 2A remains while the lower molecular weight peak was digested and the label released as free N-acetylglucosamine.

In a subsequent experiment the peak remaining after the glucosaminidase digestion was pooled as shown by the bar in Figure 3A. The pooled fractions were lyophilized, desalted, and digested with either β-galactosidase or a mixture of β-galactosidase and β-N-acetylglucosaminidase. Figure 3B shows that the β-galactosidase altered the size of this structure by approximately one hexose unit. Figure 3C shows that most of the glucosamine label becomes sensitive to glucosaminidase after removal of the galactose.

Figure 4 shows a schematic interpretation of the sequential degradation data given in Figures 2 and 3 with the most likely structures based on the protocol used and the elution properties. It should be emphasized that the

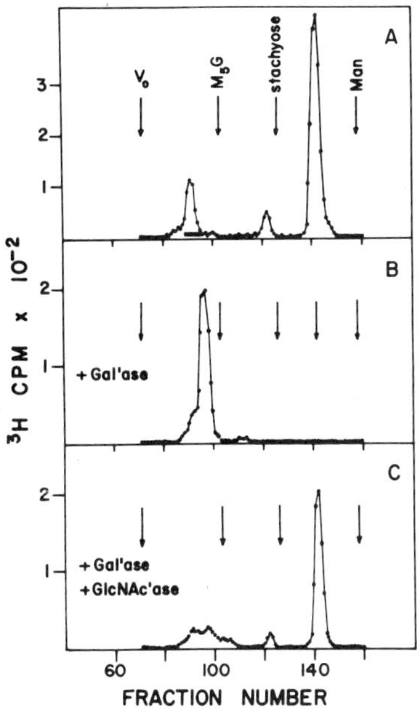

Figure 3. BioGel P2 analysis of glucosaminidase resistant oligosaccharide fraction. Oligosaccharides resistant to glucosaminidase digestion were isolated by gel filtration (A). Pooled fractions indicated by the solid bar were re-analyzed after β-galactosidase digestion (B) or after digestion with a mixture of β-galactosidase and β-N-acetylglucosaminidase (C).

use of endo H in the isolation of these precursor oligosaccharides selects for oligomannosyl structures and selects against trimannosyl structures and probably half of the theoretical tetramannosyl structures. The rationale for using this selection was to obtain a population of structures at the earliest possible stage of branch glycosylation, e.g., containing a pentamannosyl or larger core.

Figure 5 depicts a portion of the processing scheme, based on the above result, for the synthesis of the VSV G oligosaccharides. In this scheme, it is indicated that the galactose may be transferred prior to the final trimming to the trimannosyl core. The alternative pathway, final trimming prior to galactose transfer would not have been seen since the product would be endo H resistant. It should be noted that the synthesis of branched oligosaccharide structures may theoretically be quite complex as a result of the possibility of synthesizing the same structure by more than one pathway or sequence of reactions. The trimming of the oligomannosyl cores further compounds this problem. The possibility of dead end pathway exists. Further studies will be necessary to ascertain if the portion of the glycosylation pathway suggested above is in fact a viable one in the overall synthetic sequence.

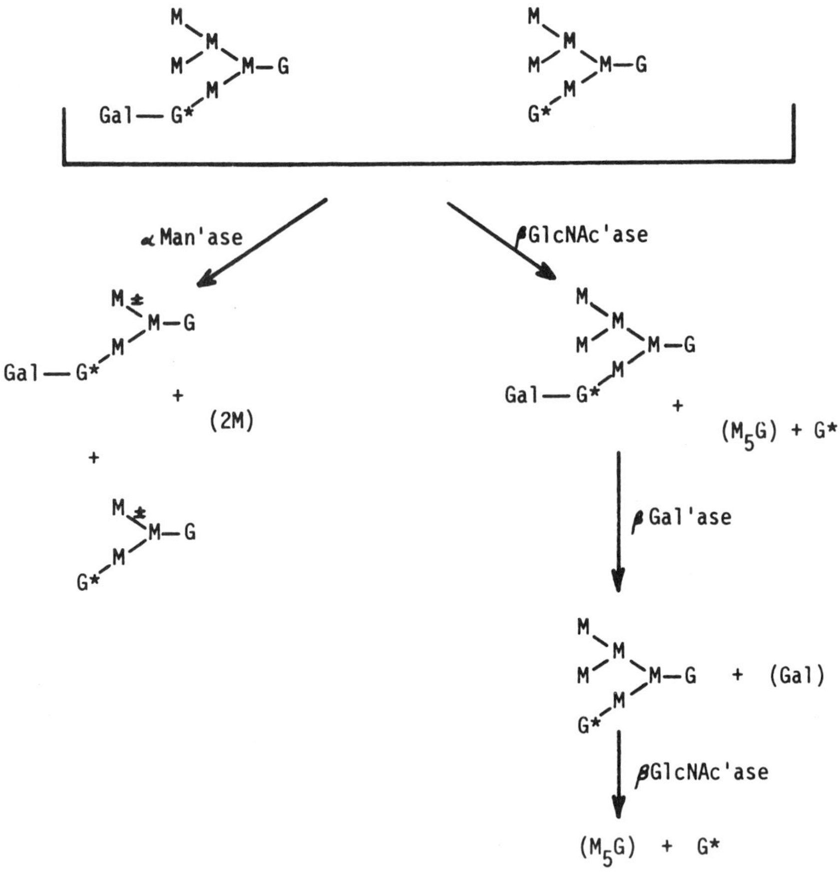

Figure 4. Schematic representation of sequential degradation of proposed structures isolated from the pulse-labeled precursor glycopeptides. M is mannose, G is N-acetylglucosamine, and G * is the incorporated radiolabeled N-acetylglucosamine.

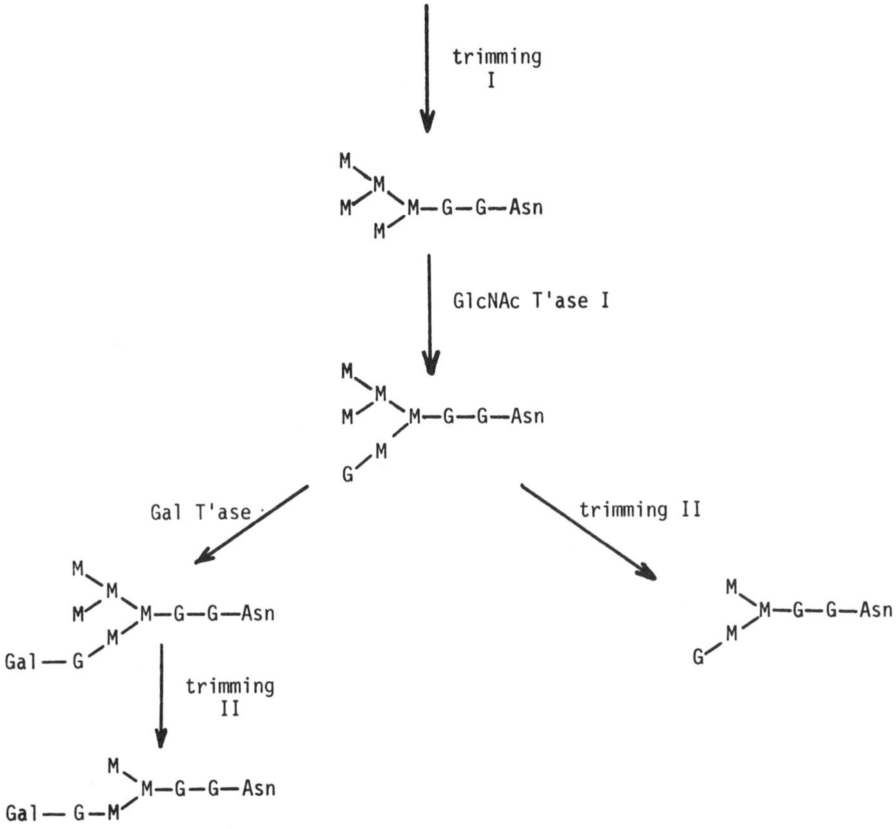

Figure 5. Sequence of processing reaction which would be consistent with intermediates identified in these analyses.

ACKNOWLEDGMENTS

This work was supported in part by Grant number AI 17267 from the National Institutes of Health.

References

1. Hunt, L.A., Etchison, J.R., and Summers, D.F. (1978): *Proc. Nat. Acad. Sci. U.S.A.* 75:754–758.
2. Tabas, I., Schlesinger, S., and Kornfeld, S. (1978): *J. Biol. Chem.* 253:716–722.
3. Nakamura, K. and Compans, R.W. (1979): *Virology* 93:31–47.
4. Li, E., Tabas, I., and Kornfeld, S. (1978): *J. Biol. Chem.* 253:7762–7770.
5. Hunt, L.A. and Summers, D.F. (1976): *J. Virol.* 20:646–657.
6. Hunt, L.A. (1979): *J. Supramol. Structure* 12:209–226.

7. Robertson, M.A., Etchison, J.R., Robertson, J.S., Summers, D.F., and Stanley, P. (1978): *Cell* 13:515–526.
8. Tabas, I. and Kornfeld, S. (1978): *J. Biol. Chem.* 253:7779–7786.
9. Harpaz, N. and Schachter, H. (1980): *J. Biol. Chem.* 255:4894–4902.
10. Li, E. and Kornfeld, S. (1978): *J. Biol. Chem.* 253:6426–6431.
11. Rothman, J.E. and Lodish, H.F. (1977): *Nature* 269:775–780.
12. Etchison, J.R., Robertson, J.S., and Summers, D.F. (1977): *Virology* 78:375–392.

David H. L. Bishop and Richard W. Compans, eds.
The Replication of Negative Strand Viruses

Viral Membrane Glycoproteins: Assembly and Structure

H.P. Ghosh,[a] J. Capone,[a] R. Irving,[a] G. Kotwal,[a] T. Hofmann,[b] G. Levine,[c] R. Rachubinski,[a] G. Shore,[c] and J. Bergeron[c]

The envelope glycoprotein G of vesicular stomatitis virus has been successfully used as a model system to study the sequence of events involved in the synthesis, insertion into membranes, and intracellular transport of membrane proteins.[1] G-protein is synthesized on membrane bound ribosomes and is inserted into the RER in a cotranslational event involving the processing of an NH_2-terminal signal peptide containing 16 amino acids.[3] In the RER, the newly synthesized polypeptide maintains a transmembrane asymetry with the bulk of the protein inside the lumen, while a peptide of 25 amino acids at the COOH terminus is exposed to the cytoplasm.[2,4] The transport of newly synthesized G from RER to the Golgi complex and finally to the plasma membrane is mediated by clathrin coated vesicles.[5] The maturation of G-protein involves processing of high mannose oligosaccharides, addition of terminal sugar residues, and covalent attachment of fatty acids.[1,6] The sequence of events in the biogenesis of G-protein are summarized in Figure 1.

This report will focus mainly on the following aspects of glycoprotein assembly and structure: migration of G through the intracellular membrane systems, NH_2-terminal sequence analyses of *in vitro* synthesized precursor of G and the virion G from different serotypes of VSV, orientation of G-protein in the viral envelope and role of fatty acids attached to G.

[a] Department Biochemistry, McMaster University, Hamilton, Ontario.
[b] Department Biochemistry, University of Toronto, Ontario.
[c] Departments Anatomy and Biochemistry, McGill University, Montreal, Province of Quebec, Canada.

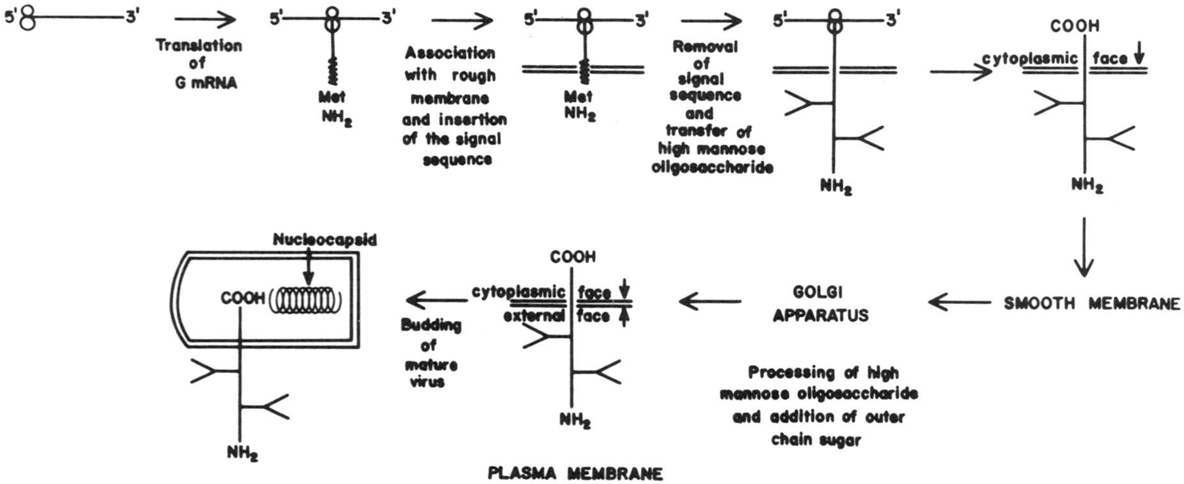

Figure 1. The sequence of events involved in the synthesis and maturation of the VSV glycoprotein.

Intracellular Transport of VSV Glycoprotein

Previous studies examining the transport of G-protein from its site of synthesis to its ultimate destination in the plasma membrane have relied primarily on cell fractionation procedures and subsequent detection of G-protein in the various fractions.[7] In order to visualize this process directly, VSV infected BHK cells were pulse-labeled for a very short period of 2 min with ^{3}H-mannose and chased with nonradioactive mannose for 10, 30 and 50 min. After the chase periods, cells were fixed and sections were visualized by electron microscopy and the grains of ^{3}H were quantitated by autoadiography (Figure 2). The results indicated that at 10 min, radioactivity appears in the RER showing that this is the site of the en bloc transfer of high mannose chains. After 30 min, radioactivity is chased into the Golgi complex. Quantitation of the number of grains present in RER and in Golgi complex shows a decrease in the number of grains in the Golgi bodies indicating that the high mannose oligosaccharides are processed there. Finally, at 50 min, the grains migrate to the plasma membrane which is also the site of virus budding. The EM autoradiographic data thus provide direct evidence that during its maturation G-protein migrates through the intracellular membrane systems as depicted in Figure 1.

NH_2-Terminal Sequence Analyses

The existence of a signal peptide of 15–30 amino acids at the amino terminus has been established for many secretory and membrane proteins.[12] This sequence, which is highly hydrophobic, is believed to be required for the initial ribosome-membrane junction and is generally cleaved soon after insertion into the RER. In an *in vitro* protein synthesizing system in the absence of membrane, G is synthesized as a nonglycosylated precursor G_1 (63,000 D) which contains a 16 amino acid NH_2-terminal extension that is absent in the mature form of G.[3] Figure 3a shows the amino terminal sequences of G_1 from Indiana and New Jersey serotypes of VSV as well as from Cocal virus. The data shows that the signal peptide in each case contained 16 amino acids with a high proportion of hydrophobic amino acids. It was interesting to note that there was very little homology in the signal peptide present in the various serotypes. Analysis of the amino terminal sequences from the mature glycoprotein obtained from the virions of the respective serotypes (Figure 3b) shows a high degree of conservation of sequence between the Indiana and the Cocal strains. As expected, the partial sequence of New Jersey G did not show any homology with Indiana or Cocal G-protein.

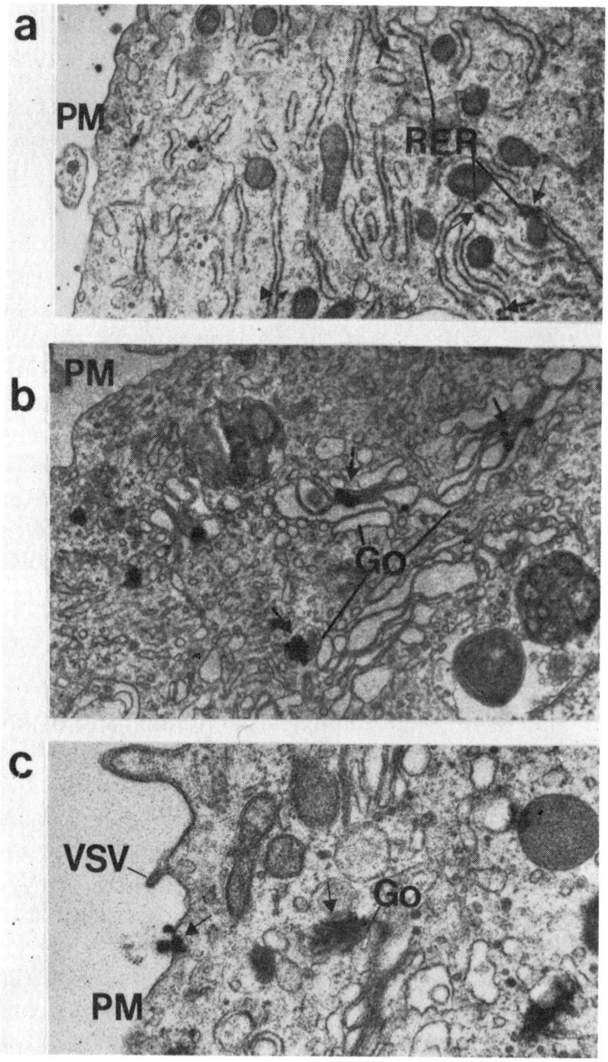

Figure 2. EM-autoradiography of sections of VSV infected BHK cells. Panels a, b, and c represent sections from cells pulsed for 2 min with ^{3}H-mannose and chased with cold mannose for an additional 8 min, 28 min, and 48 min, respectively. Arrow indicates location of black silver grains; RER, Go, and PM: rough endoplasmic reticulum, Golgi complex and plasma membrane, respectively; VSV: budding virion.

Topology of G-Protein and Fatty Acid Attachment

Previous studies have shown that intracellular G is a transmembrane protein that has a COOH terminal region containing 25 amino acids exposed to the cytoplasm.[2,4] One would thus expect that the COOH terminus of G would be on the inside of the viral envelope and the NH_2-terminus on the outside. To

Figure 3a. Signal sequences of viral glycoproteins.

INDIANA		
	1 10 16 ↓	
VSV San Juan	MetLysCysLeuLeuTyrLeuAlaPheLeuPheIleGlyValAsnCysLys	(Ref. 10)
	↓	
VSV Toronto	MetLysCysLeuLeuTyrLeu-PheLeuPhePro-Val-GlyLys	(Ref. 2)
	↓	
VSV New Jersey	MetLeu-ProLeu-Phe-Leu-Val-Pro-Leu- -	
	↓	
Cocal	Met-PheLeuLeu-Leu- - -Leu-Leu- - - -	

examine the topology of G on the viral envelope, VSV was digested with either a mixture of carboxypeptidase A and B, or aminopeptidase M. These exopeptidases specifically remove the COOH or NH_2-terminal ends of proteins. The termini would be insensitive to these enzymes if they were protected by the viral envelope. As seen in Figure 4, when purified VSV was digested with either carboxypeptidase or aminopeptidase (lanes b and d), there is no effect on G-protein. However, if Triton X-100 is added to the reaction mixture to dissolve the membrane, there is a definite reduction in the size of G in both the cases (lane c and e). This indicated that both termini are protected by the viral envelope.

In order to characterize which portion of the G-protein is buried in the membrane, purified VSV was digested with thermolysin or trypsin. As seen in Figure 5 (lanes b, c, e), treatment of VSV with protease results in complete removal of G with concomitant appearance of two small molecular weight peptides of sizes 9,000 and 6,500–7,500 D. These peptides, designated A and B respectively, are protected from digestion presumably by being buried in the membrane. Other workers have reported only one protected fragment [8–10] but we consistently observe two bands irrespective of serotype of virus used or the cell line used (data not shown). That the fragments were membrane associated was further demonstrated by the fact that they are extracted with Triton X-100 under conditions where only the G-protein is released (Figure 5, lane d).

Figure 3b. NH_2 terminal sequences of glycoproteins.

INDIANA		
	1 10	
VSV San Juan	LysPheThrIleValPheProHisAsnGlnLysGlyAsnTrpLysAsnVal	
	20	
	ProSerAsnTyrHisTyrCys	(Ref. 10)
Toronto	LysPheThrIleValPheProTyrAsnGlnLysGlyAsnTrpLysAsnVal	
	ProSerAsnTyrHisTyrCys	
Cocal	LysPheSerIleValPheProGlnSerGlnLysGlyAsxTrpLysAsxVal	
	ProSerSerTyrTyrTyr-	

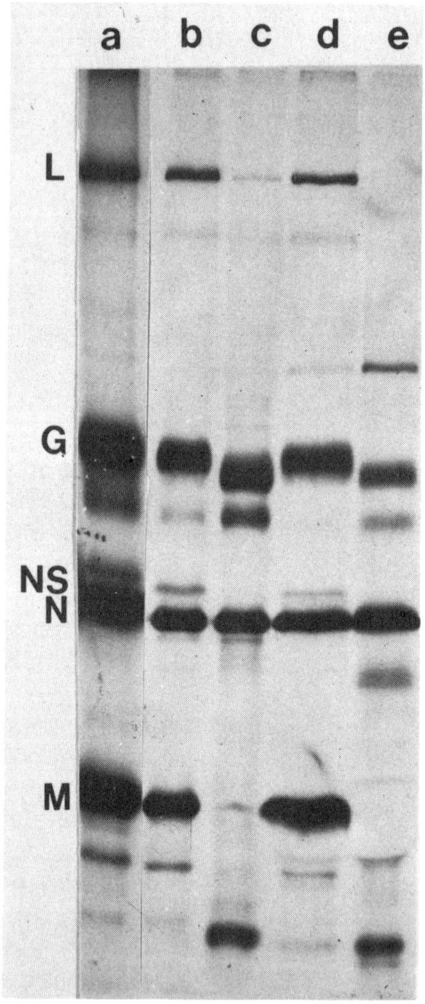

Figure 4. Autoradiogram of VSV digested with aminopeptidase and carboxypeptidase. Well a ^{35}S VSV treated with 1% Triton X-100. Wells b and c, ^{35}S VSV incubated with aminopeptidase M in the absence and in the presence of 1% Triton X-100, respectively; wells d and e, ^{35}S VSV incubated with a mixture of carboxypeptidase A and B in the absence and in the presence of 1% Triton X-100, respectively.

In order to characterize the membrane bound fragments, the tryptic peptides of these fragments were compared with those generated from G-protein that had been partially digested with carboxypeptidase. It was found that fragment A contained tryptic peptides that were present in G but absent in carboxypeptidase treated G (data not shown). Thus, G-protein is anchored in the membrane by a region at the COOH terminus. Furthermore, it was demonstrated that fragment A contains the same tryptic peptides that were

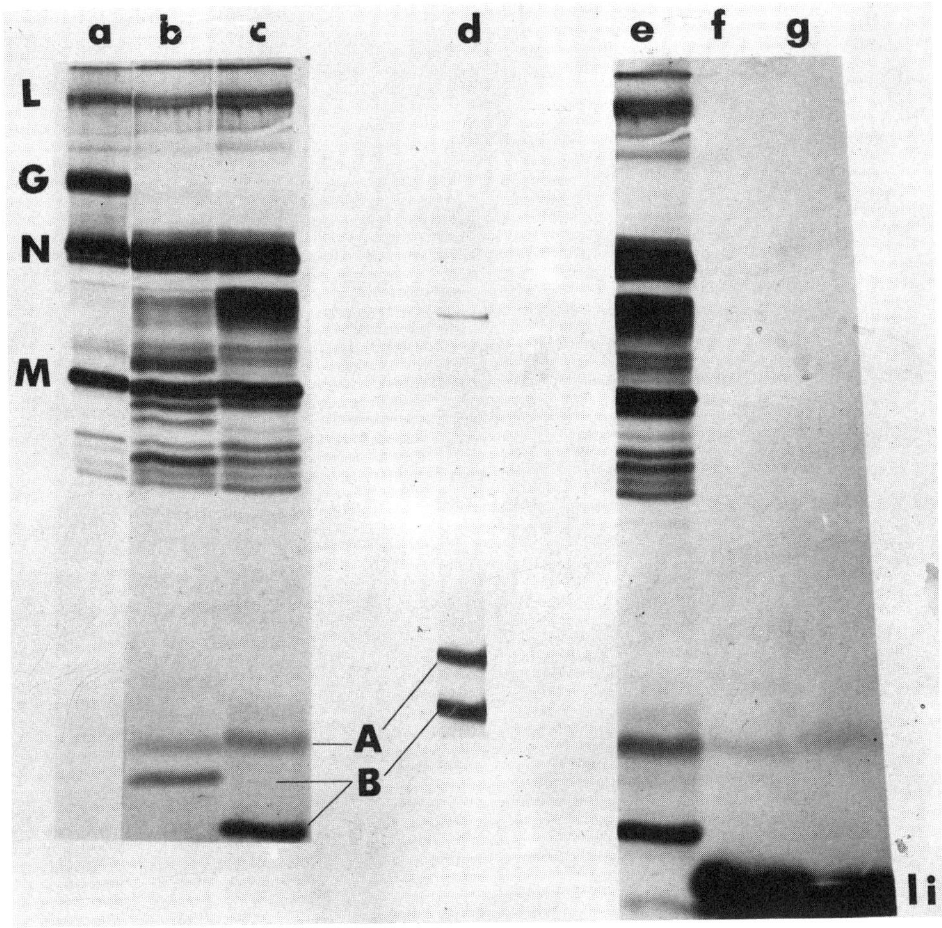

Figure 5. Autoradiogram of membrane associated fragments obtained from VSV digested with protease and analyzed on an SDS containing 17.5% polyacrylamide gel. Lane a, VSV; lane b, VSV digested with trypsin; lane c and e, VSV digested with thermolysin; lane d, extraction of membrane bound fragments from thermolysin treated VSV with Triton X-100; lanes f and g, VSV labeled with ^{3}H-palmitate and digested with trypsin and thermolysin, respectively. L, G, N, M indicate viral proteins, A and B designate peptides generated by proteolysis, Li indicates lipid front.

present in the exposed COOH terminus of intracellular G. Fragment B contains peptides in common with A, but is possibly contaminated with membrane peptides of similar size or is an artifact of our isolation procedures.

The recent report of fatty acid attachment to G-protein [6] suggested that the fatty acids may be involved in the insertion of G-protein in the membrane. To examine this, VSV was labeled with ^{3}H palmitic acid and purified virus was digested with either thermolysin or trypsin. As shown in Figure 5 (lanes f and g), only the fragment A is labeled with ^{3}H palmitic acid.Quantitation of the radioactivity present in palmitate labeled G and in fragment A revealed that all the label originally present in G-protein was also present in fragment A. Thus, the fatty acids are attached to the region of G-protein that anchors it to the membrane.

Discussion

The EM autoradiographic data show for the first time the pathway of migration of a membrane glycoprotein molecule during its maturation along the cellular network in a cultured fibroblast cell. The loss of grain counts in the Golgi complex also provides direct demonstration of the trimming of the precursor oligomannose core in the Golgi complex.

The presence of a predominant amount of hydrophobic amino acids in the signal peptides of the different serotypes of VSV agree with the hydrophobicity of signal sequences analyzed.[12] However, the signal sequences are not conserved in these viruses even though the NH_2-terminal sequences of G-protein of Indiana and Cocal serotypes show a very high degree of homology. Similar lack of homology in the signal sequence of the envelope glycoproteins of various strains of influenza viruses has also been reported (Air et al. and Goething, this volume).

The observation that the G-protein is anchored in the viral membrane by the COOH terminus is in agreement with the transmembrane topology of intracellular G. Sequence analysis of cDNA corresponding to the 3′-end of G mRNA showed the presence of a highly hydrophobic domain at the COOH terminus of G-protein which is also protected by the lipid bilayer from digestion by proteases.[10] The highly hydrophobic sequence of 20 amino acids is presumably the domain spanning the membrane and interacting with the lipophilic core of the membrane.[10]

The observation that fatty acids are attached to the membrane anchoring region of G suggest that they may augment the forces involved in the transmembrane disposition and stabilization of the glycoprotein. Recent studies with *ts*-mutants of VSV that are deficient in fatty acid attachment indicate that fatty acids are required for the proper migration of G-protein from the Golgi complex to the plasma membrane.[11] These results suggest that the fatty acids along with the hydrophobic COOH terminal domain of G, may play an essential role in the process whereby G-protein is recognized by a specific class of clathrin coated vesicles that are responsible for the intracellular transportation of G from the Golgi complex to the plasma membrane.[5]

ACKNOWLEDGMENTS

This research was supported by grants from Medical Research Council of Canada.

References

1. Ghosh, H.P. (1980): *Reviews of Infectious Diseases* 2:26–39.
2. Toneguzzo, F. and Ghosh, H.P. (1978): *Proc. Natl. Acad. Sci.* 74:1516–1520.
3. Irving, R.A., Toneguzzo, F., Rhee, S.H., Hofmann, T., and Ghosh, H.P. (1979): *Proc. Natl. Acad. Sci.* 76:570–574.
4. Katz, F.N. and Lodish, H.F. (1979): *J. Cell Biol.* 80:416–426.
5. Rothman, J.E. and Fine, R. (1980): *Proc. Natl. Acad. Sci.* 77:780–784.
6. Schmidt, M.F.G. and Schlesinger, M.J. (1980): *J. Biol Chem.* 255:3334–3339.
7. Knipe, D.M., Baltimore, D., and Lodish, H.F. (1977): *J. Virology* 21:1128–1139.
8. Mudd, J.A. (1974): *Virology* 62:573–577.
9. Schloemer, R.H. and Wagner, R.R. (1975): *J. Virology* 16:237–249.
10. Rose, J., Welch, W.J., Sefton, B.M., Esch, F.S., and Ling, N.C. (1980): *Proc. Natl. Acad. Sci.* 77:3884–3888.
11. Zilberstein, A., Snider, M.D., Porter, M., and Lodish, H.F. (1980): *Cell* 21:417–427.
12. Blobel, G. (1980): *Proc. Natl. Acad. Sci.* 77:1496–1500.

David H. L. Bishop and Richard W. Compans, eds.
The Replication of Negative Strand Viruses

A Role for Oligosaccharides in the Synthesis of the G-Protein of Vesicular Stomatitis Virus

Sondra Schlesinger and Ron Gibson[a]

Introduction

Viruses such as vesicular stomatitis virus (VSV) have provided a valuable probe for determining the steps in the synthesis of oligosaccharide chains covalently attached to asparagine residues in glycoproteins.[1–4] The glycoprotein (G) of VSV contains two complex-type oligosaccharide chains.[5,6] Glycosylation occurs during the synthesis of G on membrane bound ribosomes[7] and is initiated by the transfer of the oligosaccharide, $Glc_3Man_9GlcNAc_2$ (Figure 1, structure I) from the lipid carrier to the nascent polypeptide. After completion of translation, the G-protein migrates from the rough endoplasmic reticulum, through the Golgi membranes to the plasma membrane.[8] During this transport the glucose and six of the mannose residues are removed and N-acetylglucosamine, galactose, fucose and sialic acid residues are added in a stepwise fashion directly from nucleotide sugars (Figure 1A).

The discovery of oligosaccharide processing raises the question of why the initial precursor is large. We set out to test the hypothesis that, for some glycoproteins, the large oligosaccharide chains are required during folding of the newly synthesized polypeptide for the protein to achieve the correct conformation. Our work with nonglycosylated forms of G-proteins provided the experimental basis for testing this hypothesis.

[a] Department of Microbilogy and Immunology, Washington University School of Medicine, St. Louis, Missouri.

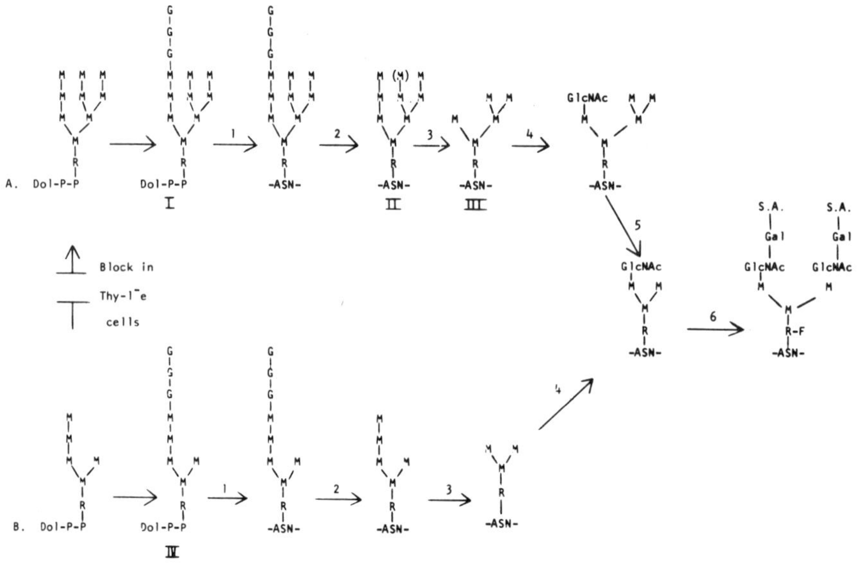

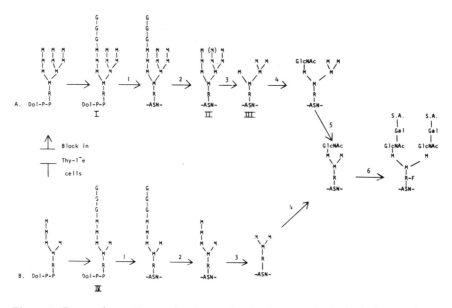

Figure 1. Processing pathways for the synthesis of asparagine-linked oligosaccharide chains on protein. (A) The major pathway used by exponentially-growing wild type cells. (B) The alternative pathway described initially for the mutant Thy-1⁻e cells. For each pathway the oligosaccharide attached to the lipid carrier, dolicholpyrophosphate, is transferred to an asparagine residue in the nascent polypeptide (reaction1). After completion of protein synthesis stepwise removal of glucose residues (reaction 2) and αl, 2 linked mannose residues (reaction 3) occurs. In pathway A, the addition of N-acetylglucosamine (reaction 4) precedes the action of the late stage processing α mannosidase (reaction 5). The addition of N-acetylglucosamine, galactose, fucose and sialic acid from sugar nucleotides (reaction 6) completes the synthesis of a complex-type oligosaccharide chain. Abbreviations used: Dol-P-P: dolicholpyrophosphate; M: mannose; G: glucose; GlcNAc: N-acetylglucosamine; Gal: galactose; S.A.: sialic acid; F: fucose; R; GlcNAc $\beta 1 \rightarrow 4$ GlcNAc; ASN: asparagine.

The Synthesis and Properties of Nonglycosylated G-proteins

We had studied the effect of tunicamycin (TM), an antibiotic which prevents the glycosylation of nascent proteins, on the replication of VSV. We compared two strains of VSV containing different but related G-proteins and found that TM inhibited virus production almost completely with VSV (San Juan); with VSV (Orsay) inhibition was very strong at 38°C, but only two-fold or less at 30°C.[9] In those cases in which virion production was strongly inhibited, the G-protein was synthesized in normal amounts but did not localize to the outer surface of the infected cell plasma membrane—a step essential for infectious virion formation. Furthermore, in these cases, the G-protein was extracted from the cells in a highly aggregated form. In contrast, when inhibition by tunicamycin was weak the nonglycosylated G-protein localized to the cell surface and, like the glycosylated molecules, was not isolated from the infected cell in an aggregated form. The aggregated, nonglycosylated G-protein could be solubilized in guanidine hydrochloride. For VSV (Orsay), the extent of renaturation *in vitro* by dialysis showed a temperature sensitivity analogous to that seen during synthesis; dialysis at 38°C to remove guanidine hydrochloride led to aggregation as measured by sedimentation at 100,000 × g, but dialysis at 30°C did not. Based on these data we proposed that for the G-protein of VSV, oligosaccharides play a crucial role during the folding of the polypeptide and that the San Juan G-protein has a more stringent requirement for oligosaccharides during folding than the Orsay G-protein.

The Synthesis and Properties of G-Protein Containing Oligosaccharides of Different Sizes

To obtain G-protein wth different oligosaccharides, we grew VSV in cell lines altered in the processing pathway.[10] The G-protein obtained from VSV grown in one cell line (clone 6) contains primarily the oligosaccharide structure $Man_8GlcNAc_2$ (Figure 1, structure II minus the mannose in par-

Table 1. Temperature-Sensitive Renaturation of G-Protein as Measured by Aggregation.

G-protein from	Structure of oligosaccharide	Temp. (°C) of dialysis to remove GdmCl	% CPM S-100	pellet
VSV (San Juan)	$Man_8GlcNAc_2$	30	95	5
		40	89	11
	$Man_5GlcNAc_2$	30	94	6
		40	52	48
	$Man_1GlcNAc_2$	30	40	60
		40	15	85
VSV (Orsay)	$Man_8GlcNAc_2$	30	95	5
		40	93	7
	$Man_5GlcNAc_2$	30	97	6
		40	90	10
	$Man_1GlcNAc_2$	30	88	12
		40	70	30

Purified G-proteins (~ 5 μg) were diluted into 1 ml of 6 M guanidine hydrochloride (GdmCl) in PBS containing 0.2% Triton X-100. The samples were incubated for 15 min at either 30° or 40° and were then dialyzed against 200-fold volume excess of PBS containing 0.2% Triton X-100 for 8 hr at 30° or 40° respectively. After dialysis, the samples were centrifuged at 40,000 rpm for 90 min and the supernatant and pellet fractions were assayed for radioactivity.

enthesis); the oligosaccharide chains on the G-protein of VSV grown in another cell line (15B) have the structure $Man_5GlcNAc_2$ (Figure 1, structure III). In our initial studies, we treated purified G-proteins containing either $Man_8GlcNAc_2$ or $Man_5GLcNAc_2$ oligosaccharide chains with guanidine hydrochloride and measured their ability to refold using an *in vitro* aggregation assay (Table 1). The San Juan G-protein with $Man_5GlcNAc_2$ oligosaccharides aggregated to a much greater extent when dialysis to remove guanidine was carried out at 40°C rather than at 30°C. The Orsay G-protein with $Man_8GlcNAc_2$ oligosaccharides and both proteins containing Man_8 $GlcNAc_2$ oligosaccharides did not aggregate upon dialysis at either temperature. When the G-proteins containing $Man_8GlcNAc_2$ oligosaccharides were treated with α-mannosidase to produce the oligosaccharide structure Man_1 $GlcNAc_2$, an increase in aggregation was observed.

We are beginning to analyze the denaturation and renaturation of G-proteins containing oligosaccharides of different sizes using the physical parameters of intrinsic fluorescence and circular dichroism. These spectrophotometric measurements should provide us with a more precise decription of the folding and unfolding of G molecules. The intrinsic fluorescence spectra of proteins usually measures the local environment of tryptophan residues.[11] In the native protein tryptophan residues may be in a non-polar environment as evidence by a decrease in λ_{max} and an increase in intensity. Upon denaturation, these tryptophans become exposed to the polar solvent resulting in a λ_{max} and a decrease in intensity. Our data obtained with the San Juan G-protein show that the ability of the protein to recover the appropriate fluorescence intensity can be affected by the oligosaccharide chains covalently bound to asparagine residues (Figure 2 and Table 2). Only the G-protein with $Man_8GlcNAc_2$ oligosaccharide chains recovered the initial quantum yield (Table 2). Thus, our initial results corroborate the date based on aggregation.

There are several aspects of the spectral studies which require further investigation. First, it is essential to establish that in all cases the protein molecules are equally unfolded. The only criterion we have at present that the proteins are all denatured to the same extent in GdmC1 is that they all show the same shift in λ_{max} in their fluorescence spectra. Second, we are assuming that recovery of physical properties is a measure of renaturation, but this assumption would be strengthened, if we could measure a biological function. Third, we don't know what effects the detergent have on protein folding. We do know, however, that the extent of aggregation can be influenced by the choice of detergent. We used Brij 58 in the fluorescence studies instead of Triton X-100 because of the high UV absorbance of the latter. The extent of aggregation of G in Brij 58 is much less than in Triton X-100 (Table 3). We were able to increase the percent of aggregation in Brij 58 by increasing the concentration of G. Further studies on the effects of different detergents are underway.

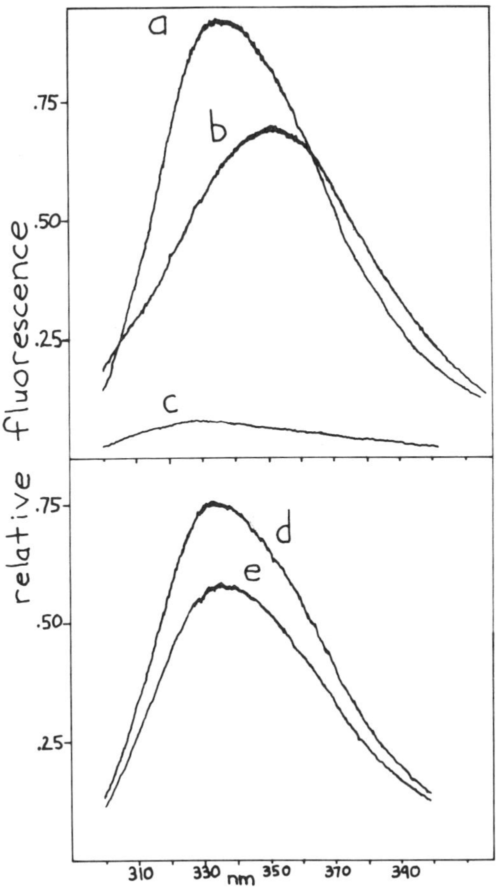

Figure 2. Fluorescence spectra of the San Juan G-protein which has the oligosaccharide structure $Man_5GlcNAc_2$. Uncorrected fluroescence spectra were recorded in the single beam mode on an Aminco SPF-500 spectrophotofluorometer. Excitation was at 280 nm. (a) Native San Juan G-protein (8 μg/ml) in PBS containing 0.2% Brij 58. (b) San Juan G-protein (8 μg/ml) in 7.2M GdmC1 in PBS containing 0.2% Brij 58. (c) 7.2 M GdmC1 in PBS containing 0.2% Brij 58. (d) San Juan G-protein treated with 7.2 M GdmC1 and the denaturant removed by dialysis at 30° against PBS containing 0.2% Brij 58. (e) San Juan G-protein as in (d) but the dialysis to remove GdmC1 was at 40°.

The Synthesis of VSV in Two Mouse Lymphoma Cell Lines which Synthesize Lipid-Linked Oligosaccharides of Different Sizes

An alternative processing pathway has been described in which the oligosaccharide chains transferred from lipid have the structure

Table 2. Temperature-Sensitive Renaturation of the San Juan G-Protein as Measured by Intrinsic Fluorescence Spectra.[a]

Structure of oligosaccharide on G	Treatment of G	λ_{max}[b] (nm)	$\Delta\lambda$[c] (nm)	$RF\lambda_{max}$[d] (corrected)
$Man_8GlcNAc_2$	—	333	61	1.0
	7.6M GdmCl	352	89	0.68
	7.6M GdmCl dialyzed at 30°	334	64	1.0
	7.6M GdmCl dialyzed at 40°	334	64	0.98
$Man_5GlcNAc_2$	—	334	60	1.0
	7.2M GdmCl	351	74	0.76
	7.2M GdmCl and dialyzed at 30°	334	60	1.07
	7.2M GdmCl and dialyzed at 40°	335	62	0.86
$GlcNAc_1$	—	333	66	1.0
	7.2M GdmCl	351	88	0.43
	7.2M GdmCl and dialyzed at 30°	334	70	0.59
	7.2M GdCl and dialyzed at 40°	335	70	0.45

[a]Uncorrected fluorescence spectra were recorded in the single beam mode on an Aminco SPF-500 spectrophotofluorometer. Excitation was at 280 nm using an excitation band pass of 5 nm and an emission band pass of 10 nm. The denaturation and dialysis are described in Table 1.

[b]λ_{max}: emission maximum of fluorescence upon excitation at 280 nm.

[c]$\Delta\lambda$: band width at half height.

[d]$RF\lambda_{max}$ (corrected): fluorescence amplitude at λ_{max} relative to the λ_{max} of the native molecules. The value has been corrected for any losses based on the recovery of protein as assayed by radioactivity.

Table 3. Temperature-Sensitive Renaturation of San Juan G-$Man_5GlcNAc_2$ Protein in Triton X-100 and Brij 58.

Detergent present during *in vitro* aggregation assay	Temp. at which GdmCl was removed by dialysis °C	Distribution of protein (% cpm) S-100	Pellet
Triton X-100	30	88	12
	40	24	76
Brij 58	30	98	2
	40	87	13

G-protein was extracted from virions with β-D octylglucoside and this detergent was replaced with either Triton X-100 or Brij 58 by dialysis at 4°C. The G-protein samples were treated with 7.6 GdmCl at the appropriate temperature for 15 min before dialysis. After dialysis, the samples were centrifuged at 100,000 × g for 90 min and the supernatant (S-100) and pellet fractions assayed for radioactivity.

$Glc_3Man_5GlcNAc_2$ (Figure 1B, structure IV.) A recently characterized cell mutant, Thy-1⁻e, is blocked in the synthesis of dolicholphosphorylmannose and transfers only the truncated oligosaccharide to nascent polypeptides.[12] We used these cells to determine if there is an *in vivo* parallel for the *in vitro* results. If large oligosaccharide chains were important for the folding of G during its synthesis, then the polypeptide with $Glc_3Man_5GlcNAc_2$ oligosaccharide chains should not fold as well as one with the larger oligosaccharide chains. The oligosaccharide chains on G molecules synthesized in Thy-1⁻e cells and in the wild type Thy⁺ cells differ only during the initial steps in the processing pathway (Figure 1). We observed that (1) VSV assembly is relatively temperature-sensitive in Thy-1⁻e cells; and (2) VSV (San Juan) is more sensitive than VSV (Orsay) (Table 4). The differences between the two cell lines are not observed in the presence of TM as would be expected if the differences depended on glycosylation.[10] These results provide evidence that the size of the oligosaccharide chains transferred to the nascent G polypeptide can determine how well the mature G-protein functions in virion assembly.

Discussion

The viability of the Thy-1⁻e cells demonstrates that many cellular glycoproteins are functional subsequent to glycosylation with the truncated oligosaccharide. In exponentially growing wild type cells, however, the larger $Glc_3Man_9GlcNAc_2$ oligosaccharide appears to be the only precursor transferred to nascent polypeptides. Our results with the G-proteins of VSV suggest a possible answer for why the initial high mannose oligosaccharide is large. An oligosaccharide which serves as the only precursor for the glycosylation of asparagine residues must meet the structural demands of the most sensitive glycoproteins. High mannose chains may be the more primitive

Table 4. The Effect of Temperature on the Production of VSV in Thy-1⁺ and Thy-1⁻e cells.[a]

VSV	Cell Line	TM (1.0 μg/ml)	40°/30° yield	Index[b]
San Juan	Thy-1⁺	−	0.74	0.24
	Thy-1⁻e	−	0.18	
Orsay	Thy-1⁺	−	0.89	0.52
	Thy-1⁻e	−	0.46	
Orsay	Thy-1⁺	+	0.10	0.90
	Thy-1⁻e	+	0.09	

[a]The procedures have been described.[10]
[b]Index = The 40°/30° yield in Thy-1⁻e cells ÷ the 40°/30° yield in Thy-1⁺ cells.

form of oligosaccharides and some glycoproteins may have evolved with the large high mannose oligosaccharide chains assuming an important role during the initial folding. The evolution of an oligosaccharide processing pathway can be considered as a means to expand the repertoire of carbohydrates and, thus, the biological roles of oligosaccharide chains. The retention of the large high mannose precursor would be essential for certain glycoproteins to achieve a correct tertiary structure. Once the initial folding has occurred, the large oligosaccharide structure may no longer be necessary and processing begins.

ACNOWLEDGMENTS
This research was supported by Public Health Service Grant R01 AI 11377 to S.S. R.G. is a recipient of Public Health Serice Grant 5 T32 GM 07200.

References

1. Tabas, I., Schlesinger, S., and Kornfeld, S. (1978): *J. Biol. Chem.* 253:716–722.
2. Hunt, L.A., Etchison, J.R., and Summers, D.F. (1978): *Proc. Natl. Acad. Sci. U.S.A.* 75:754–758.
3. Robbins, P.W., Hubbard, S.C., Turco, S.J., and Wirth, D.F. (1977): *Cell* 12:895–900.
4. Kornfeld, S., Li, E., and Tabas, I. (1978): *J. Biol. Chem.* 253:7771–7778.
5. Etchison, J.R., Robertson, J.S., and Summers, D.F. (1977): *Virology* 78:375–392.
6. Reading, C.L., Penholt, E.E., and Ballou, C.E. (1978): *J. Biol. Chem.* 253:5600–5612.
7. Rothman, J.E. and Lodish, H.F. (1977): *Nature* 269:775–780.
8. Knipe, D.M., Baltimore, D., and Lodish, H.F. (1977): *J. Virol.* 21:1149–1158.
9. Gibson, R., Schlesinger, S., and Kornfeld, S. (1979): *J. Biol. Chem.* 254:3600–3607.
10. Gibson, R., Kornfeld, S., and Schlesinger, S. (1981): *J. Biol. Chem.*, in press.
11. Freifelder, D. (1976): *Physical Biochemistry Applications to Biochemistry and Molecular Biology*. W.H. Freeman and Co., San Francisco, pp. 415–417.
12. Chapman, A., Fujimoto, K., and Kornfeld, S. (1980): *J. Biol. Chem.* 255:4441–4446.

David H. L. Bishop and Richard W. Compans, eds.
The Replication of Negative Strand Viruses

Fatty Acid Acylation of VSV Glycoprotein

Milton J. Schlesinger,[a] Anthony I. Magee,[a]
and Michael F.G. Schmidt[b]

The glycoproteins of enveloped viruses undergo a variety of modifications as they move from their site of synthesis in the cells' rough endoplasmic reticulum to their final insertion into the virus particle membrane. One of the posttranslational processing events consists of the attachment of fatty acid to the glycoprotein, leading to a stable covalent ester-type bond that appears to link fatty acid directly to the polypeptide chain.[1–3] In a recent report,[3] we showed that this acylation occurs about midway in the transit pathway from synthesis to arrival at the cell surface; thus, placing the acylation reaction in the Golgi apparatus of the cell. In this communication we describe some additional properties of the fatty acid-bound glycoprotein, suggest a site on the polypeptide chain where attachment occurs, and offer speculation on a function for the fatty acid residues in glycoprotein.

Site of Fatty Acid Acylation on VSV-G-Protein

Our previously published data showed that G-protein, isolated from virions and purified free of lipid, contained 1 to 2 moles of fatty acid per mole polypeptide.[2] We were able to generate fatty acid-containing proteolytic fragments by treating this delipidated G with pepsin and thermolysin and we isolated several fatty acid labeled peptides by two dimensional chromatography on silicic acid thin layer plates.[2] Yields were low (~1 to 10%) and the

[a] Department of Microbiology and Immunology, Washington University School of Medicine, St. Louis, Missouri.
[b] Present address: Institut fur Virologie, Justus Liebig Universitat, 63 Giessen, Federal Republic of Germany.

most intensely labeled radioactive fragment showed a high content of serine. In attempting to define more precisely the amino acid residue in G that has been acylated with fatty acid, we have performed two additional experiments. We treated the intact VSV particle (labeled with either ^{35}S-methionine or ^{3}H-palmitate) with chymotrypsin under conditions that would "shave off" external portions of G but retain the particle intact.[4,5] These particles were separated from chymotrypsin by centrifugation in the presence of TPCK, the potent anti-chymotrypsin inhibitor, and the pelleted material analyzed by fluorography of SDS-polyacrylamide gel electropherograms. A significant amount of a ^{3}H-palmitate labeled peptide of low molecular weight was recovered in the pellet fraction (Figure 1) indicat-

Figure 1. SDS-polyacrylamide gel analysis of ^{35}S-methionine and ^{3}H-palmitate labeled VSV particles after treatment with various proteases. "Con" refers to untreated sample of virus.

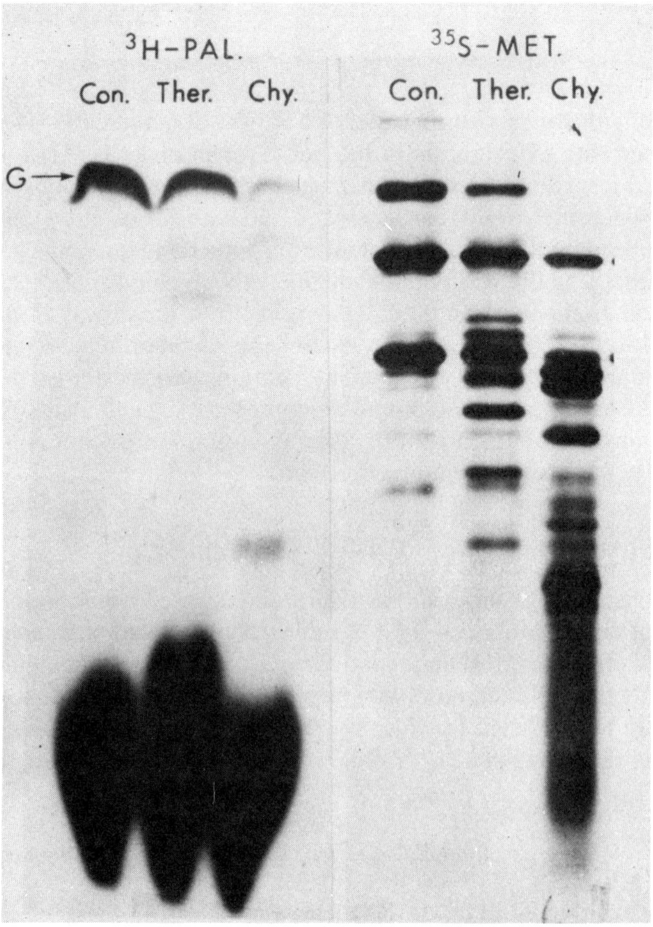

ing that the fatty acid was bound to that fraction of G retained in the membrane of the virion.

Rose et al.[6] have recently published a portion of VSV-G primary sequence, based on DNA sequences from cloned recombinant molecules of the VSV-G gene. In the carboxy-terminus of the protein, a sequence of 20 very hydrophobic amino acids has been tentatively assigned as the portion of the protein spanning the lipid bilayer. On the extra-cytoplasmic side of this G sequence, there are 4 serine residues; two of them lie between two tryptophanes. To determine if fatty acid might be acylated to one of these serines, we extracted a ^{3}H-palmitate labeled preparation of G from 100 mg of VSV by stirring virus at 4°C for 1 hr in 10 mM Tris HCl, pH 8.0, with 30 mM octyl-β-D-glucopyranoside. The protein was separated from nucleoprotein by centrifugation and precipitated with acetone. Unbound lipid was removed by extracting this precipitate once with acetone, three times with chloroform-methanol (2:1) and twice with diethyl ether. The delipidated protein was reduced and carboxymethylated with iodoacetic acid in 8 M urea and acetone precipitated. Protein was solubilized with 1.5 ml of 4 M guanidine hydrochloride in 80% (v/v) acetic acid and treated with iodosobenzoic acid for 24 hr at 22°C in the dark. β-Mercaptoethanol (12 μl) was added to stop the reaction which cleaved the polypeptide at tryptophan residues. A 5 ml solution of ethanol 88% formic acid (4:1) was added and the mixture applied to a column of Sephadex LH60 equilibrated with the same solvent. Four peaks of fluorescamine-positive material were detected and the most retarded fraction contained 90% of the ^{3}H label recovered from the column (this accounted for about 20% of the radioactivity in the sample treated with the cleavage reagent). The peak was pooled, dried under N_2, redissolved in 1 ml 88% formic acid followed by 4 ml ethanol, and rechromatographed on Sephadex LH20. The ^{3}H peak eluted at a position corresponding to an apparent molecular weight of ~600. A 6 N HCl, 24-hr hydrolysate was analyzed for amino acid composition, yielding phenylalanine and serine as well as smaller amounts of glycine, alanine, leucine, aspartic acid, glutamic acid, and threonine. Dansylation of a sample showed a strong spot in the region of dansyl-phenylalanine. These data suggest that a fatty acid residue is attached to G close to the "extra-cytoplasmic" part of the polypeptide region embedded in the membrane, although further purification of this peptide and sequence data are necessary before the site of acylation can be established and assigned with certainty.

In Vivo Acylation

The addition of ^{3}H-palmitic acid to VSV-infected tissue culture cells showed a selective labeling of the VSV-G-protein.[2] We used this kind of label to determine when G synthesis commenced during the infection cycle.

Labeling of G with ^{3}H-palmitate was clearly detectable 3 hours postinfection (Figure 2, lane 5) in chicken embryo fibroblasts and could be seen with longer exposures of the autoradiogram as early as two hours postinfection. ^{35}S-methionine labeled G was also seen clearly at 3 hours postinfection (Figure 2, lane 6), but synthesis of host cell proteins obscured the G label at earlier times. Many host cell proteins were labeled with fatty acid (Figure 2, lane 1) and the pattern of ^{3}H-labeled proteins was distinct from that of the ^{35}S-label (Figure 2, lane 2). These host cell fatty acid-labeled proteins had properties similar to those found for the VSV-G-protein; they were located in cellular membranes, the fatty acid residues were added only to newly synthesized polypeptide chain; and the lipid appeared to be linked covalently directly to the polypeptide chain.[7] These cellular lipid-bound proteins were not extracted into organic solvent and some of them appeared to be glycoproteins. In order to detect host cell proteins with bound lipid, it was

Figure 2. SDS-polyacrylamide gel analysis of ^{35}S-methionine and ^{3}H-palmitate labeled chicken embryo fibroblast cells infected with VSV. The odd numbered lanes contained ^{3}H; the even numbered lanes contained ^{35}S. 1, 2—uninfected cells; 3, 4—infected for 2 hours; 5, 6—infected for 3 hours; 7, 8—infected for 4 hours. Label was present for 30 minutes before cells were harvested. See reference 7 for experimental details.

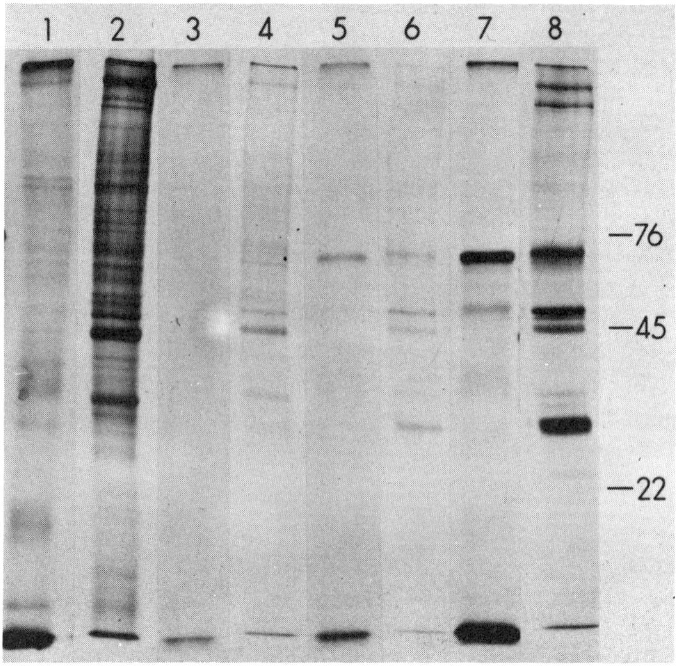

necessary to increase by ten-fold the amount of ^3H-sample added to the SDS-gels.

A Speculative Model for the Role of Fatty Acid in Glycoprotein Function

The VSV-G-protein becomes acylated with fatty acid after the polypeptide chain has been completely formed, inserted into the intracellular membranes, and moved into the Golgi apparatus. The precise location for acylation has not been established but was inferred from the relation of fatty acid acylation to the oligosaccharide trimming reactions.[3] Several VSV *ts*-mutants altered in G have been examined for their ability to be acylated with fatty acids.[2] Most of these fail to bind fatty acid at the non-permissive temperature and their oligosaccharides are not converted from the high mannose form. The latter suggests that, at the non-permissive temperature, the glycoproteins cannot move into the Golgi apparatus. In one of these *ts* VSV mutants, however, the oligosaccharides of the G-protein do get processed but no fatty acid is attached and the protein fails to reach the cell surface membrane.[8] Tunicamycin-treated cells infected with the San Juan strain of VSV make G-proteins that aggregate, do not reach the cell surface, and do not become acylated with fatty acid. The Orsay strain of VSV, however, makes a G-protein—in cells incubated with tunicamycin at 30°—that does move to the cell surface, is incorporated into virions, and acylated with fatty acids. Recent studies with monensin, a sodium ionophore which allows proteins to move into the Golgi apparatus but not to the cell's surface, showed that VSV G could be both acylated with fatty acids and its oligosaccharides processed in monensin-treated cells.[9] From these data, one can make a case for the essentiality of fatty acid acylation in the post-Golgi transport of glycoprotein.

What role might fatty acids play in the transport of proteins from Golgi to surface membrane? One possibility is that the fatty acids provide an "anchor" that keeps the protein attached to membranes during the movement of protein from Golgi to cell surface. For this stage of transport, intracellular vesicles are postulated to bud from the Golgi membranes and then fuse with the plasma membrane. Another role may be to select for specific kinds of "boundary" lipid that associate with the embedded protein. If the acylated fatty acids were asymmetrically disposed in the polypeptide chain, then membrane lipid asymmetry would result, as seen in Figure 3. This kind of lipid selection and asymmetry could be important in various membranal activities, including virus budding. These hypotheses are testable utilizing current methods for reconstituting glycoprotein into membrane vesicles and it will be interesting to see if fatty acids do, in fact, play a role in glycoprotein-membrane interactions.

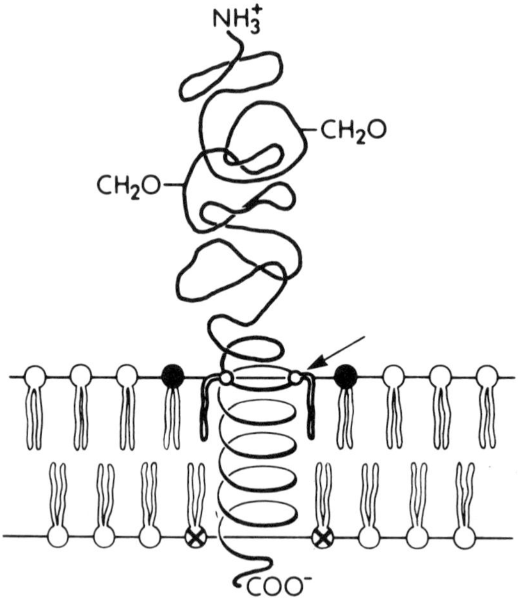

Figure 3. Postulated site on G for fatty acid acylation and the effect on membrane lipid asymmetry. ● and ○ refers to different phospholipids; arrow indicates protein-bound fatty acid.

ACKNOWLEDGMENTS
This research was supported by Grants 2 R01 CA 14311 and 5P30 CA 16217 from the National Cancer Institute.

References

1. Schmidt, M.F.G., Bracha, M., and Schlesinger, M.J. (1979): *Proc. Natl. Acad. Sci. U.S.A.* 76:1687–1691
2. Schmidt, M.F.G. and Schlesinger, M.J. (1979): *Cell* 17:813–819.
3. Schmidt, M.F.G. and Schlesinger, M.J. (1980): *J. Biol. Chem.* 255:3334–3339.
4. Mudd, J.A. (1974): *Virology* 62:573–577.
5. Schloemer, R.H. and Wagner, R.R. (1975): *J. Virol.* 16:237–249.
6. Rose, J.K., Welch, W.J., Sefton, B.M., Esch, F.S., and Ling, N.L. (1980): *Proc. Natl. Acad. Sci. U.S.A.*, in press.
7. Schlesinger, M.J., Magee, A.I., and Schmidt, M.F.G. (1980): *J. Biol. Chem.*, in press.
8. Zilberstein, A., Schneider, M., Porter, M., and Lodish, H.F. (1980): *Cell* 21:417–427.
9. Johnson, D.C. and Schlesinger, M.J. (1980): *Virology* 103:407–424.

David H. L. Bishop and Richard W. Compans, eds.
The Replication of Negative Strand Viruses

Vesicular Stomatitis Virus Glycoprotein-Phospholipid Interactions

William A. Petri, Jr., Ranajit Pal,
Yechezkel Barenholz, and Robert R. Wagner[a]

Vesicular stomatitis (VS) virus is unique amongst the enveloped viruses in that its membrane contains only a single intrinsic glycoprotein (G).[1] The G-protein forms the spikes seen on the viral surface, has a molecular weight of 69,000 [1] and is anchored in the viral membrane by ~5 K mol wt, thermolysin-resistant tail fragment. [2,3] Recent sequencing of a cDNA clone of the 3′ end of G mRNA has identified a region of the G-protein 29 amino acids from the COOH-terminus that contains 20 consecutive hydrophobic amino acids, by which the G-protein presumably spans the viral membrane.[4] We present here evidence for the reconstitution of the G-protein with phosphatidylcholine into liposomes and a characterization of the G-protein-phosphatidylcholine interaction in the reconstituted membrane.

Isolation and Reconstitution of the G-Protein

VS virus of the Indiana serotype was purified from infected baby hamster kidney-21 cells as previously described.[5] The G-protein and viral lipids were solubilized from VS virus with the nonionic detergent β-D-octylglucoside, and the G-protein was separated from the viral lipids by centrifuging the G-protein into a sucrose gradient containing octylglucoside (Figure 1A). The G-protein isolated from the sucrose gradient was 97% pure as determined by polyacrylamide gel electrophoresis and contained no detectable cholesterol and about 1 molecule of residual phospholipid/molecule G-protein.[6] Upon

[a] Department of Microbiology, University of Virginia School of Medicine, Charlottesville, Virginia, and Department of Biochemistry, School of Medicine, Hebrew University, Jerusalem, Israel.

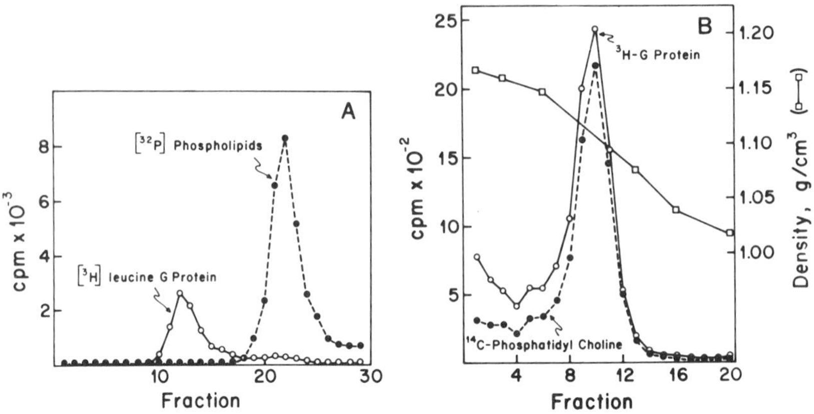

Figure 1. Isolation and reconstitution of VS virus G-Protein. (A) The G-protein was isolated from purified VS virus grown in the presence of ^{3}H-leucine and ^{32}P-orthophosphate. VS virus at a protein concentration of 0.9 mg/ml was treated with 30 mM β-D-octylglucoside in 10 mM Tris (pH 7.5) for 1 hr at room temperature. The nucleocapsids were removed by centrifugation at 150,000 $\times$ g for 90 min. The supernatant, which contained 90% of the viral glycoprotein, was layered onto a 15 to 30% sucrose gradient containing 60 mM octylglucoside, 0.5 M NaC1, and 50 mM Tris (pH 7.5). After centrifugation in an SW 60 rotor at 55,000 rpm for 16 hr, fractions were collected and radioactivity was counted by scintillation spectrometry. (B) 150 μg of octylglucoside-extracted ^{3}H-G-protein was reconstituted by detergent dialysis with 100 μg of ^{14}C-phosphatidylcholine, A 0 to 30% sucrose gradient containing 0.5 M NaCl and 50 mM Tris (pH 7.5) was layered over the reconstituted vesicles made 40% in sucrose and then centrifuged at 190,000 $\times$ g for 12 hr at 4°C. Each fraction was analyzed for radioactivity by scintillation spectrometry and for density by refractometry.

SOURCE: *Reprinted from Petri and Wagner, 1979.*

removal of the octylglucoside by dialysis the G-protein formed micelles [6] which would spontaneously partition into preformed sonicated egg phosphatidylcholine vesicles.[7]

The G-protein was reconstituted with phosphatidylcholine by the detergent dialysis method.[6] 150 μg of ^{3}H-leucine-labeled G-protein was mixed with 100 μg of ^{14}C-phosphatidylcholine in 60 mM octylglucoside in 50 mM Tris, pH 7.5, and the detergent removed over 36 hr by dialysis against two 1-liter changes of either 10 mM Tris (pH 7.5) or 50 mM KC1. Equilibrium centrifugation of the reconstituted vesicles on a sucrose flotation gradient demonstrated superimposable peaks of ^{3}H-leucine-labeled G-protein and ^{14}C-phosphatidylcholine at a density of 1.11 g/cm ^{3}. (Figure 1B), whereas vesicles formed in the absence of G-protein had a density of 1.02 g/cm^{3}. Equivalent results were obtained when dimyristoylphosphatidylcholine (DMPC) and dipalmitoylphosphatidylcholine (DPPC) were used in place of egg phosphatidylcholine. Negative stain electron microscopy of DMPC-G-protein vesicles showed the vesicles to vary in diameter from 300–1200Å

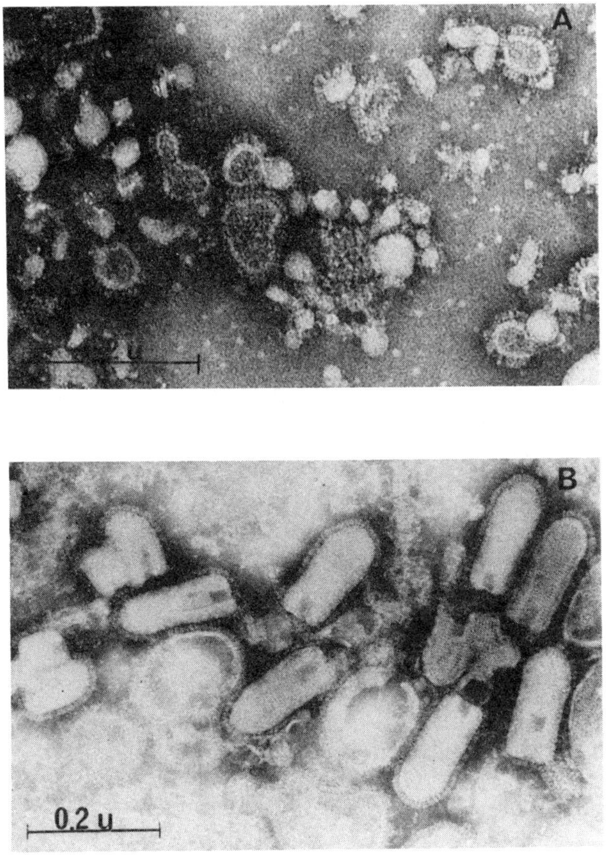

Figure 2. Negative stain electron microscopy of G-protein-dimyristoyllecithin detergent dialysis vesicles and VS virus. Negative stain electron microscopy of (A) reconstituted vesicles of G-protein and dimyristoylphosphatidylcholine (DMPC) and (B) VS virus. 100 μg of G-protein in 60 mM octylglucoside was added to 100 μg DMPC and dialyzed at room temperature against 50 mM KCl for 48 hr. A drop of the reconstituted vesicles or of purified VS virus was placed on a Formvar-coated grid for 30 s and the excess liquid was blotted off before adding 1 to 2% phosphotungstic acid for 15 s. After removing excess phosphotungstic acid, the grids were examined with a Siemens 1A electron microscope.

(Figure 2A) and to have oriented externally from their surface spike-like projections that were indistinguishable from the glycoprotein spikes protruding from the VS virus membrane (Figure 2A and B).

Thermolysin Digestion of Reconstituted G Protein

In order to determine whether the G-protein was inserted in the reconstituted membrane in the same manner as in the virus membrane, the reconstituted vesicles and native virus were treated with thermolysin. The electropherograms shown in Figure 3 demonstrate that the vast majority of the G-protein in the reconstituted vesicles was digested by thermolysin and was, therefore, presumably oriented externally on the vesicles, thus confirming the results of the electron microscopic negative staining (see Figure 2A). Integrated gel scans of lanes C and D revealed that thermolysin degraded 97% of lipid-associated G-protein. Remaining in association with the re-

Figure 3. Polyacrylamide slab-gel electrophoresis fluorogram of (A) proteins of whole VS virus, (B) proteins of whole VS virus treated with thermolysin, (C) reconstituted with glycoprotein-phosphatidylcholine vesicles, (D) reconstituted viral glycoprotein-phosphatidylcholine vesicles treated with thermolysin, (E) viral glycoprotein in octylglucoside treated with thermolysin, and (F) viral glycoprotein rosettes treated with thermolysin. VS viral proteins were labeled by growing the virus in the presence of ^{14}C-amino acids. Whole VS virus, reconstituted glycoprotein vesicles, glycoprotein rosettes, and phospholipid-free glycoprotein in 60 mM octylglucoside were treated with 50 units of thermolysin/mg of viral protein for 30 min at 37°C. The virus and vesicles were then repurified by equilibrium sedimentation, and electrophoresed on a 17.5% acrylamide, 7 M urea, sodium dodecylsulfate slab gel. L, N, M are viral proteins, 2G is the glycoprotein dimer and T is the thermolysin-resistant tail fragment of G-protein. To demonstrate better the thermolysin-resistant tail fragment (T), lane D was loaded with 7 times the number of vesicles than was lane C; this accounts for the apparent greater density of the vesicle T compared with the virion T fragment.

SOURCE: *Reprinted from Petri and Wagner, 1979.*

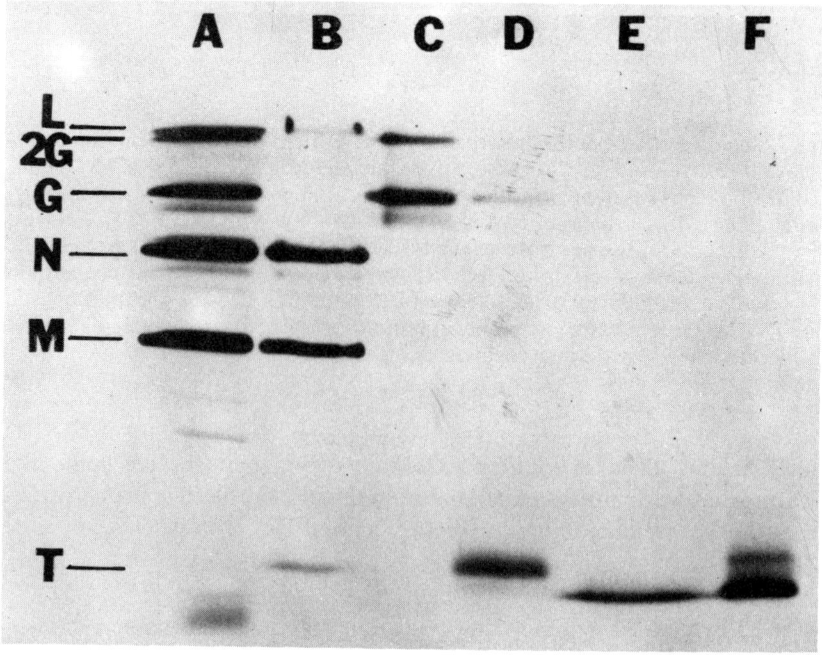

purified protease-treated vesicles was a fragment of apparent $M_w \approx 5000$ which comigrated with the hydrophobic fragment (T) of thermolysin-treated VS virus (Figure 3 lanes B and D). This finding indicates that the G-protein in the reconstituted vesicles was inserted in the bilayer by means of the same hydrophobic segment inserted in the intact virus membrane. In contrast, the isolated phospholipid-free G-protein treated with thermolysin in the presence of octylglucoside did not yield the same hydrophobic segment found in the reconstituted glycoprotein vesicles and virus. Treatment with thermolysin of G-protein rosettes in the absence of octylglucoside gave rise to two fast migrating bands (Figure 3, lane F); a faint band comigrated with the viral hydrophobic piece and was likely due to the presence of residual viral phospholipid that formed glycoprotein vesicles occasionally seen with the glycoprotein rosettes on negative staining.

Differential Scanning Calorimetry of G-Protein-DPPC Vesicles

To probe the nature of the G-protein interaction with the phosphatidylcholine in the reconstituted vesicles, differential scanning calorimetry was used to monitor the phase transition of dipalmitoylphosphatidylcholine (DPPC) in DPPC-G-protein vesicles which formed on detergent dialysis. Increasing the mol % G-protein veiscles which formed on detergent dialysis. Increasing the mol % G-protein in the DPPC vesicles resulted in a decrease of the phase transition temperature and enthalpy change (Figure 4). The enthalpy change as a function of the mol % G-protein could be fit to a straight line by a least-squares procedure. Extrapolation of the results to the glycoprotein concentration at which the enthalpy change was zero indicated one G-protein molecule removed 270 ± 150 molecules of DPPC from the phase transition.[8]

Fluorescence Depolarization Studies of G-Protein-DPPC Vesicles

To gain insight into the mechanism by which the G-protein affects the DPPC phase transition, the phase transition was also monitored by the fluorescence anisotropy of four fluorescent probes: *trans*-paranaric acid; 16-(9-anthroyloxy) palmitoylglucocerebroside; 1,6-diphenyl-1,3,5-hexatriene; and 4-heptadecyl-7-OH-coumarin. The phase transition temperature measured by all four probes was decreased in the presence of G-protein (Figure 5), in agreement with the calorimetry data (Figure 4). However, it was possible to demonstrate the presence of a phase transition by fluorescence anisotropy at protein to lipid ratios where the scanning calorimetry data predicted there would be no observable phase transition. There are two possible explanations for this discrepancy between the fluorescence and scanning calorimetry data. First, at high protein concentrations the G-protein may self-aggregate so that it interacts with a fewer number of DPPC molecules per molecule of G-protein. The transition temperature lowering caused by the G-protein became progressively smaller as the G-protein concentration

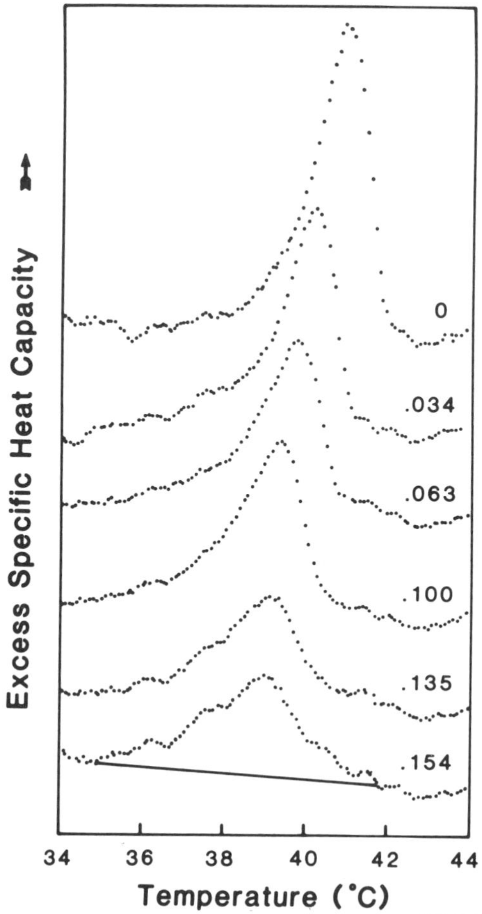

Figure 4. Differential scanning calorimetry of DPPC vesicles reconstiuted with varying mole percentages of G-protein. Phospholipid concentrations for the calorimetry were between 3 to 5 mM, and the scanning rate was approximately 15°C/hr. Heat capacity was calculated per mole of phospholipid. A 20 point least-squares fit was used to smooth the curves. The mole percentage of G-protein in the reconstituted vesicles was determined by phosphate and protein assays after each scan, and is listed to the right of each curve.

SOURCE: *Reprinted with permission from Biochemistry 19:3088. Copyright 1980 American Chemical Society.*

increased, which may be a reflection of G-protein aggregation.[8] If the G-protein did aggregate at the high protein-to-lipid ratios studied by fluorescence anisotropy, the scanning calorimetry data of 270 DPPC molecules removed from the transition per G-protein would still be valid as a measure of the effect of the G-protein monomer on the DPPC phase transition.

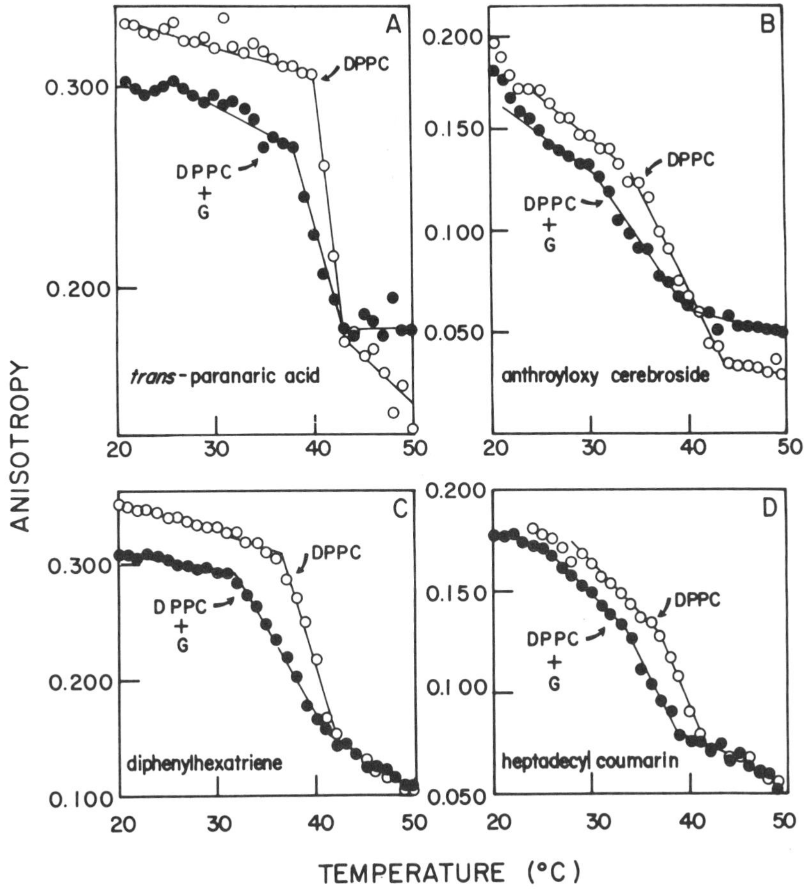

Figure 5. Fluorescence anisotropy of (A) *trans*-paranaric acid, (B) 16-(9-anthroyloxy)palmitoylglucocerebroside, (C) 1,6-diphenyl-1,3,5-hexatriene and (D) 4-heptadecyl-7-OH-coumarin in G-protein-DPPC vesicles as a function of temperature. The fluorescent probes were incorporated into the reconstituted vesicles and the fluorescence anisotropy measured as described in the text. The DPPC vesicles contained either no glycoprotein (O) or (A) 0.45 mol%, (B) 0.34 mol%, (C) 0.56 mol%, (D) 0.40 mol *4* glycoprotein (●).

Alternatively, the fluorescence probes may not be accurately reporting the thermotropic behavior of all the DPPC in the reconstituted vesicles but may be preferentially located in relatively undisturbed DPPC domains containing little G-protein. The use of four different fluorescent probes should minimize such a possibility.

The decrease in the fluorescence anisotropy of all four probes in the DPPC gel state in the presence of G-protein is evidence that the G-protein disrupts the ordering of the fatty acyl chains in the gel state.[9] This disordering of the gel phase is at least partly responsible for the decreased DPPC transition enthalpy change caused by the G-protein. A similar disordering of the gel phase of dimyristoylphosphatidylcholine by the M13 coat protein has been seen using *cis*-paranaric acid fluorescence intensity.[10] The increase in anisotropy of the liquid-crystalline state of G-protein-DPPC vesicles seen with *trans*-paranaric acid and anthroyloxy cerebroside suggests that the G-protein may additionally depress the transition enthalpy change by ordering the fatty acyl chains in the liquid-crystalline state. Diphenylhexatriene may be too deep in the membrane interior and heptadecyl coumarin too superficial to monitor this ordering effect.

ACKNOWLEDGMENTS
This research was supported by Grants MV-9D from the American Cancer Society, PCM-77-00494 from the National Science Foundation, and AI-11112 from the National Institute of Allergy and Infectious Diseases. W.A. Petri, Jr. has been a predoctoral and postdoctoral trainee supported by National Cancer Institute Training Grant CA-09109.

References

1. Wagner, R.R. (1975): In: *Comprehensive Virology,* Vol. 4. Fraenkel-Conrat, H. and Wagner, R.R. (Eds.), Plenum Press, New York, pp. 1–93.
2. Mudd, J.A. (1974): *Virology* 62:573–577.
3. Schloemer, R.H. and Wagner, R.R. (1975): *J. Virol.* 16:237–249.
4. Rose, J.K., Welch, W.J., Sefton, B.M., Esch, F.S., and Ling, N.C. (1980): *Proc. Natl. Acad. Sci. U.S.A.* 77:3884–3888.
5. McSharry, J.J. and Wagner, R.R. (1971): *J. Virol* 7:59–70.
6. Petri, W.A., Jr., and Wagner, R.R. (1979): *J. Biol. Chem.* 254:4313–4316.
7. Petri, W.A., Jr. and Wagner, R.R. (1980): *Virology,* in press.
8. Petri, W.A., Jr., Estep, T.N., Pal, R., Thompson, T.E., Biltonen, R.L., and Wagner, R.R. (1980): *Biochemistry* 19:3088–3091.
9. Petri, W.A., Jr., Pal, R., Barenholz, Y., and Wagner, R.R. (1980): *Biochemistry,* submitted for publication.
10. Kimelman, D., Tecoma, E.S., Wolber, P.K., Hudson, B.S., Wickner, W.T., and Simoni, R.D. (1979): *Biochemistry* 18:5874–5880.

Copyright 1981 by Elsevier Norhh Holland, Inc.
David H. L. Bishop and Richard W. Compans, eds.
The Replication of Negative Strand Viruses

The Role of VSV Proteins and Lysosomes in Viral Uncoating

Douglas K. Miller and John Lenard[a]

Introduction

Two different pathways have been suggested for the entry of enveloped viruses into cells (penetration) and the removal of their envelopes (uncoating): fusion with the plasma membrane and endocytosis, or viropexis. Paramyxoviruses are thought to fuse with the plasma membrane within minutes after binding, simultaneously completing both penetration and uncoating.[1,2] Semliki Forest virus (SFV) and vesicular stomatitis virus (VSV), on the other hand, are thought to penetrate into the cell by endocytosis with uncoating occurring later in the lysosomes.[3-6] The endocytosis of SFV follows a sequence very similar to that outlined by Goldstein et al.[7] for the internalization of hormones and low density lipoproteins. Endocytosis occurs by the invagination of coated pits bearing SFV particles, leading to coated vesicles in the cytoplasm each containing a single SFV particle. These coalesce to form larger uncoated vesicles containing several viral particles that eventually fuse with primary lysosomes to become secondary lysosomes. SFV first appears in secondary lysosomes within 5 min after its adsorption to the cell surface. The low lysosomal pH [8] (< 4.5) permits fusion of the viral envelope with the lysosomal membrane with resulting release of an uncoated nucleocapsid into the cytoplasm. Consistent with this uncoating scheme, infection by SFV could be completely inhibited by the addition of weakly basic lipophilic amines (lysosomotropic agents) [9] which immediately diffuse to the lysosomes and raise the lysosomal pH.[3,8]

[a] Department of Physiology and Biophysics, College of Medicine and Dentistry of New Jersey, Rutgers Medical School, Piscataway, New Jersey.

Both Sendai and SFV uncoat by a fusion event brought about by an activated viral spike: in the case of Sendai, there is a proteolytic activation prior to adsorption;[10] for SFV, the activation is provided by the low lysosomal pH.[11] A precedent for a young lysosomal role in viral uncoating has been provided by the experiments of Silverstein et al.[12,13] with reovirus. In this case, reovirus uncoating was facilitated by the lysosomally induced proteolytic cleavage of the viral protein envelope.

In this paper, we report a series of experiments that were begun to determine the role of the VSV proteins in the uncoating of VSV. In these experiments, we made use of a number of temperature-sensitive (ts) mutants of VSV that were thermolabile.[15] Such mutants are distinguished by the inactivation by moderate heat treatment (1 hr at 45°) of the mutated viral protein in virions grown at the permissive temperature. Wild type (wt) virions are only slightly inactivated by this treatment, but thermolabile mutants become non-infectious because of denaturation of the mutated protein. The other proteins of the virion are identical to those of wt VSV and consequently are little affected. The thermolabile property implies that the mutated protein is essential in the viral infection at a stage at or preceding primary transcription; following primary transcription, a functional mutated protein can be synthesized by the cell. In order to define the role of each protein in these early events more precisely, the mutant virions, inactivated by ultraviolet irradiation were used as inhibitors of wt VSV infection before and after thermal inactivation of the mutated protein. Infection was assayed by determination of the amount of ^{3}H-uridine labeled primary or secondary viral RNA formed after infection by a standard amount of VSV. Irradiation of the inhibitor virions was sufficient to prevent them from synthesizing their own RNA.

Experiments were also performed to determine the role of the lysosomes in VSV uncoating. Additional experiments with Sendai, SFV, and influenza, done in parallel to those with VSV, suggest an obligatory lysosomal role in the uncoating of all these viruses.

Importance of G-Protein in VSV Infection

Since we believed that the G-protein should provide the initial viral contact with the cell, we compared the inhibitory activity or excess G-protein, presented to the cell prior to infection, in two different forms: as a reconstituted vesicle containing only G-protein and viral lipids and as an intact virion. To assess the specificity of the G-protein effect, both UV-irradiated virions and vesicles were prepared from wt VSV and from *ts*(V)045, a thermolabile G-protein mutant. The UV-VSV and UV-*ts*45 virions inhibited viral RNA production at a concentration equivalent to 0.3–0.4 μg G/10^6 BHK cells. After heat treatment to inactivate the mutant G-protein, inhibition by UV-*ts*45 was completely lost (Figure 1B), while inhibition by UV-VSV was only

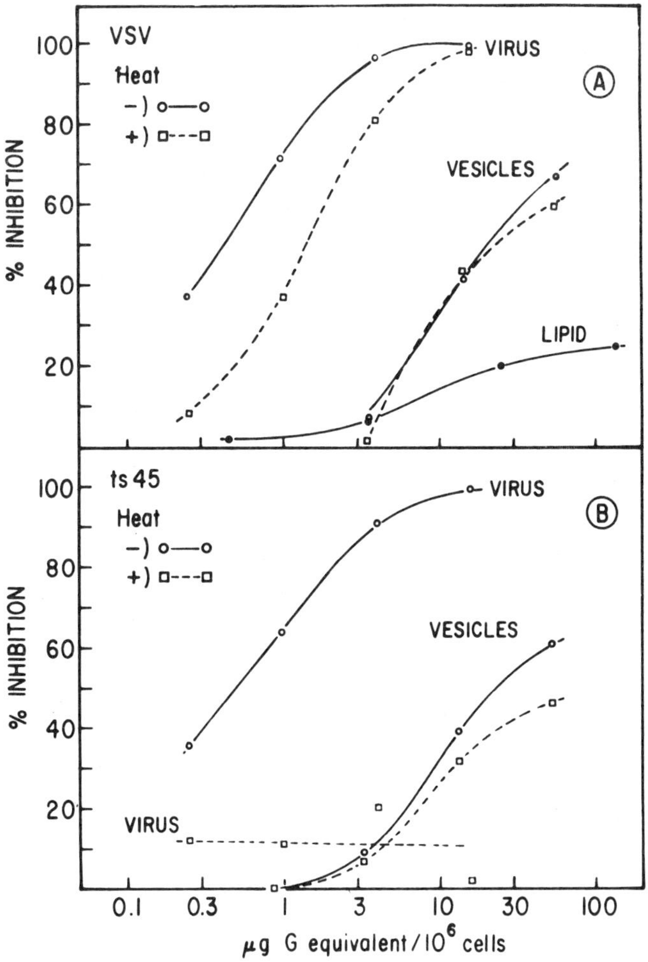

Figure 1. Inhibition of VSV transciption by UV-irradiated intact virus or G-protein vesicles from VSV (A) or *ts*045 (B). Virus was grown in BHK cells and purified as previously described.[15] Vesicles were prepared from octyl glucoside extracts of the purified virions.[15] Samples of virus and vesicles were UV-irradiated with 12,500 ergs/cm ^{2} at 254 nm. Heat-treated samples were then incubated 1 hr at 45°. For the transcription assay, confluent 35 mm monolayers of BHK cells were preincubated for 1 hr with the indicated inhibitor at 37° before addition of 10 pfu/cell VSV in presence of actinomycin D. After incubation for 1 hr at 37°, the plates were washed, and incubated at the presence of ^{3}H-uridine for an additional 4 hr. The viral RNA was precipitated in TCA, and the measured cpm were compared with those of a standard curve prepared from serial dilutions of infectious VSV in the absence of any G-protein inhibitor.[15] □, samples treated with heat; ○, samples not heated. In A, protein-free lipid vesicles (●) were tested at the same lipid concentration (based on cholesterol) as found in the G vesicles.[15]

slightly reduced (Figure 1A). In addition to *ts*45, four thermolabile and one thermostable UV-irradiated G-protein mutants were tested, and in each case the loss of inhibitory activity after heating was as predicted from the thermolability of the mutant.[6]

The vesicles prepared from either VSV or *ts*45 were only 2% as effective as UV-VSV per μg of G-Protein (Figure 1). After heat treatment, the vesicles prepared from *ts*45 were inactivated no more than those prepared from VSV. Since vesicles prepared from viral lipids alone were much less effective than those containing G-protein (Figure 1A), the G-protein in the vesicles did have some inhibitory effect.[15]

Since *a priori* the UV-VSV should be processed by the cell similarly to the infectious VSV, these results implied that functional G-protein was necessary for infection by the virus. Furthermore, since the G-protein vesicles from the VSV and *ts*45 were both much less effective inhibitors than the intact virions, the vesicles lacked some element of the virus that was a strong inhibitor of infection. The alternative explanation that the G-protein became inactivated in the preparation of the vesicles seemed unlikely considering (a) the mildness of the procedure, (b) the equal inhibitory activity of the *ts*45 protein relative to wt G, despite the former's lower inherent stability, and (c) the more effective inhibition by the heat treated *ts*45 vesicles relative to the heat inactivated *ts*45 virions. Thus, while the G-protein was necessary for inhibition by UV-VSV, it was not sufficient.

Importance of Nucleocapsid Proteins

Similar experiments carried out with UV-irradiated thermolabile mutants of the L, M, and N proteins (complementation groups I, III, and IV, respectively) indicated that the nucleocapsid proteins N and L were necessary for the inhibition but that the M-protein was not. Only thermolabile M-protein mutants were still able to inhibit RNA production after heat treatment.[16] Thus, the presence in the UV-VSV of nucleocapsid proteins that were strong inhibitors of VSV infection was apparently responsible for the added inhibitory activity of the UV-irradiated viruses beyond that found with the G-protein vesicles (Figure 1).

The inhibitory properties of defective interfering (DI) particles were used to assess the relative role of the nucleocapsid proteins and RNA in UV-VSV inhibition. Purified wt DI particles were compared with UV-VSV in their ability to inhibit secondary and primary VSV transcription, both before and after UV-irradiation of the DI particles. Unirradiated DI particles were potent inhibitors of secondary transcription, reflecting their classical viral interference (Figure 2). While the particles were much less effective after UV-irradiation, they still inhibited secondary transcription as well as did UV-VSV. When unirradiated DI particles, UV-DI, and UV-VSV were tested for their ability to inhibit primary transcription, all three were equally effec-

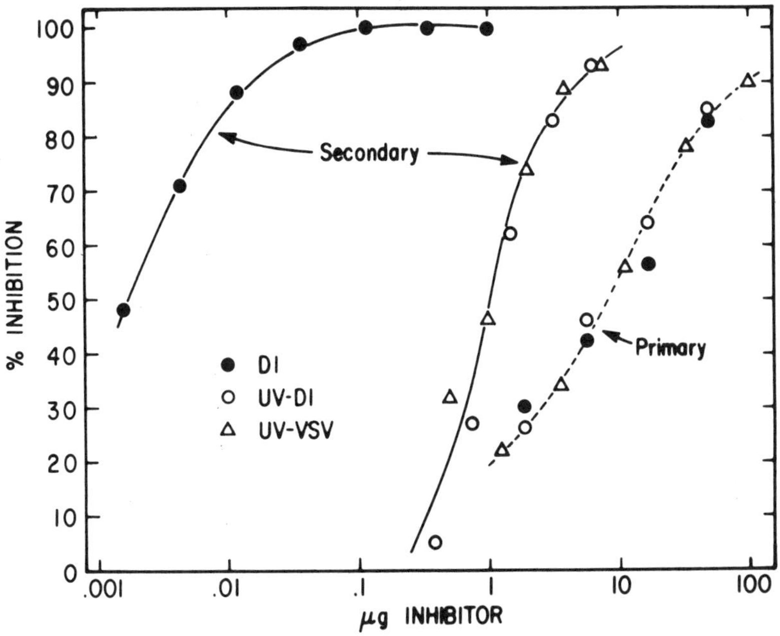

Figure 2. Inhibition of primary (dashed line) and secondary (solid lines) VSV transcription by UV-VSV and DI particles. DI particles and VSV grown in the same BHK flasks were purified,[16] UV-irradiated, and assayed as described in Figure 1. For primary transcription assays, VSV infection was performed at 1000 pfu/ml in 10 μg/ml DEAE-dextran in media that also contained 100 μg/ml cycloheximide.[17] For secondary transcription assays 4 pfu/cell was used.[16]

tive (Figure 2). This indicates that inhibition of infection by UV-VSV depends upon the presence of functional proteins, but does not require the presence of functional genomic RNA.[16]

G-Protein Performs Its Function At an Internal Site

Experiments were performed to show that UV-VSV does not inhibit by competition with infectious virus for attachment to the cell surface or internalization into the cell. Neither binding nor internalization (measured by the acquisition of protease resistance) differed significantly between wt VSV and *ts*45, whether or not it had been heat inactivated. Binding of all these different virion preparations increased linearly with time and with virus concentration, indicating no saturation of putative receptor sites. In addition, the binding and internalization of a small amount of highly labeled infectious VSV (< 1 pfu/cell) was unaffected by the presence of a 350-fold excess of unlabeled UV-VSV or UV-*ts*45.[6] This observation rules out the possibility that infection proceeds through the binding of a small number of specific receptors on the cell surface. Furthermore, since UV-VSV does not inhibit at the cell surface or during internalization, it must act at a subsequent intracellular location.

Importance of the Lysosome in VSV Infection

The importance of the lysosome in VSV uncoating was determined by the use of a number of agents known to inhibit lysosomal function, in particular the antimalarial chloroquine. At concentrations generally used to inhibit lysosomal function, chloroquine completely inhibited VSV infection at all multiplicities tested (2-350 pfu/cell), indicating that the chloroquine sensitive step was required for infection and could not be bypassed.[6]

A wide variety of chemically dissimilar agents also are lipophilic weakly basic amines and, like chloroquine, were expected to inhibit lysosomal function. VSV infection was inhibited by all such agents tested (Table 1). As is illustrated in Figure 3 using chloroquine, the inhibition by all those lipophilic bases tested occurred early in infection, at a point preceding both primary and secondary transcription. Since inhibition could only be seen when chloroquine was added prior to RNA synthesis (up to 3 hr for secondary transcription, Figure 3A, and 1 hr for the primary transcription, Figure 3B), the action of chloroquine was not on transcription *per se,* but had to be on a

Table 1. Inhibitors of VSV Transcription.

Group	Drug	mM concentration for 50% inhibition
Antimalarial	Chloroquine	0.02
Local anaesthetics	Dibucaine	0.025
	Tetracaine	0.20
	Lidocaine	0.22
	Procaine	3.3
Antihistamines	Pyrilamine maleate	0.05
	Chlorpheniramine	0.4
	Promethazine HCl	0.5
Antipyretic	Aminopyrine	2.3
Miscellaneous amines	Dansylcadaverine	0.4
	Ethylene diamine	1.5
	l-propylamine	4
	Imidazole	4
	Methylamine	5.5

The compounds were incubated with the cells for 30 min at 37° prior to infection with VSV (4 pfu/cell), and continuously thereafter until harvest. The cells were incubated with ^3H-uridine and harvested as described under Figure 1.[17]

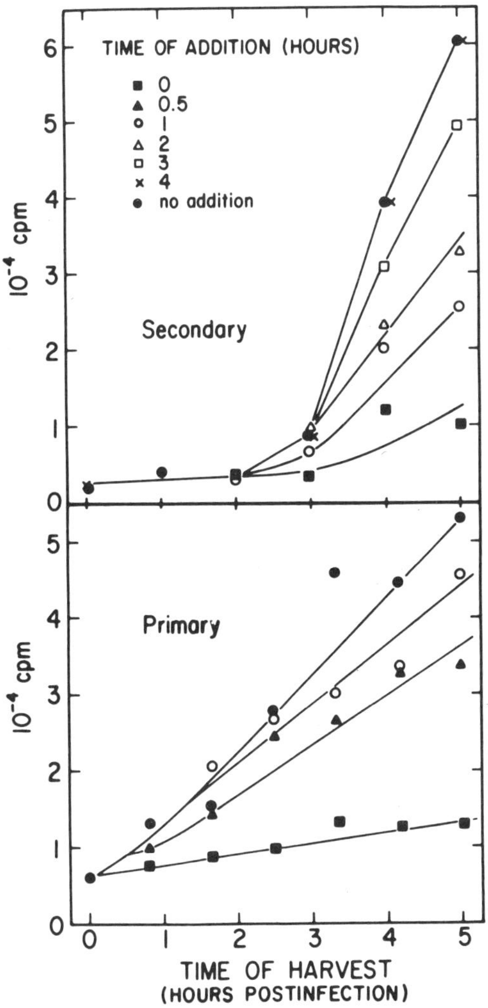

Figure 3. Inhibition of VSV secondary (top) and primary (bottom) transcription by chloroquine. To BHK cells, 1000 pfu/cell (primary) or 15 pfu/cell (secondary) of infectious VSV was added for 1 hr at 5°. The cells were washed, and 1 ml of the appropriate medium containing ^{3}H-uridine was added to each. At times ranging from 0–4 hr, 50 μl of chloroquine at 2 mM was added to each plate. The plates were incubated from 1–5 hr at 37° and then harvested and assayed as described in Figure 1.[17]

cellular event preceding primary transcription.[17] This is consistent with its site of action being at the lysosomes. Other experiments showed that binding, internalization, and *in vitro* transcription were all unaffected by lipophilic bases in concentrations sufficient to inhibit viral infection (Table 1).

Temporal Order of Lysosomal and UV-VSV Inhibition

Experiments with UV-VSV similar to those shown for chloroquine (Figure 3) indicated that UV-VSV also inhibited prior to primary transcription (data not shown).[16] To compare directly the rate of movement of infectious virus through the UV-VSV sensitive step relative to its movement through the lysosomes, a high concentration of inhibitor was required.[16] Inhibition by chloroquine or pyrilamine maleate decreased as a function of time of addition of the drugs postinfection so that by 30 min, 50% of the inhibition was lost as measured in a primary transcription assay. In contrast, the decrease of inhibition by UV-VSV occurred with a halftime of 60 min postinfection (Figure 4). Thus, the UV-VSV inhibition site was reached after the lysosomal site.[16,17]

It is attractive to consider the lysosomes as the site of the action of the G-protein, where, like the spike proteins of SFV, it acts as a fusogenic agent for the viral and lysosomal membranes. That the UV-VSV inhibition occurs soon after the lysomal step (relative to the onset of transcription, Figure 4),

Figure 4. Proposed sequence for the passage of VSV through the lysosomes, UV-VSV inhibitory site, and primary and secondary transcription. Lines represent that proportion of the total effect as measured 5 hr p.i. "Lysosomes" refers to the loss in primary transcription assays of inhibition by 100 μM chloroquine or 1.25 mM pyrilamine maleate; after 2 hr little inhibition can be found indicating that those infectious virions in the assay have moved beyond that point of inhibition.[17] "UV-VSV inhibition" refers to the loss in primary transcription assays of inhibition by 50 μg UV-VSV.[16] "Primary" and "secondary transcription" refer to the rate of formation of each type of RNA when infected at a multiplicity of infection of 1000.

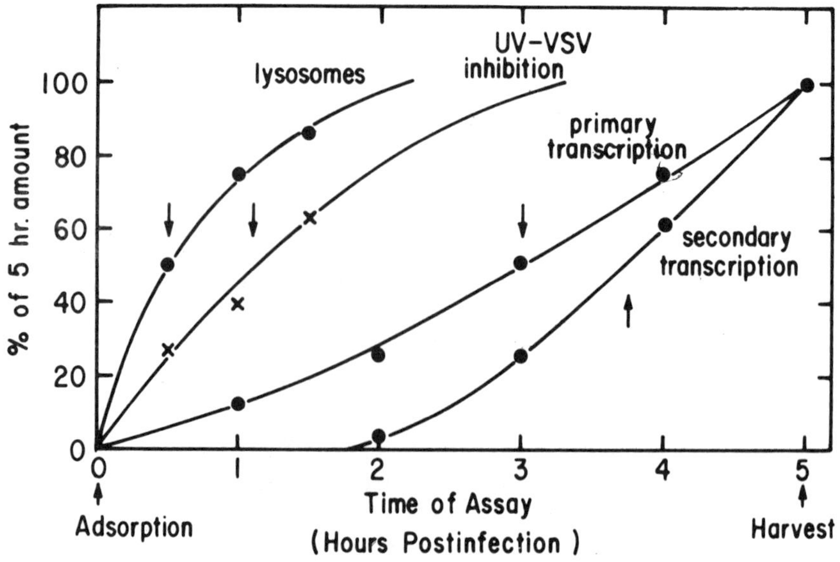

when the G-protein is presumably absent, is consistent with the observation that much, if not most, of the inhibition is related to the presence of functional nucleocapsid proteins. Hence, the nucleocapsid inhibition may indicate an interaction with the cell at a separate intracellular site. An alternative conclusion consistent with the data is that the nucleocapsid actually interacts with the G-protein, either after exit from the lysosome or while still in the viral and lysosomal membranes.

Inhibition of Sendai, Influenza and SFV Infection by Chloroquine and UV-VSV

Infection by a number of enveloped viruses was reported in the earlier literature to be inhibited by agents that were subsequently shown to be lysosomotropic.[18,19] We have found that chloroquine inhibits infection by both high and low fusing strains of Sendai (Z and Obayashi), by influenza (WSN) as well as by SFV and VSV. Similar concentrations of chloroquine are required, and inhibition of Sendai infection resembles the data reported above for VSV (Figure 3), occurring at a step prior to primary transcription.[17] Thus, an early lysosomal step appears obligatory in the infectious cycle of all these viruses.

This requirement is quite unexpected in the case of Sendai, which is known to fuse with plasma membranes under normal conditions. Although the fusogenic capacity of Sendai must be activated by the F_0 to F-protein conversion in order for Sendai to be infective, [1] we know of no evidence that fusion *at the plasma membrane* is either necessary or sufficient to initiate a Sendai infection. Two alternative possibilities are (a) fusion at the plasma membrane is necessary but not sufficient, an additional lysosomal step being necessary to complete uncoating; (2) fusion at the plasma membrane is neither necessary nor sufficient and represents an unproductive dead end. Instead, endocytosis and subsequent fusion with the lysosomal membrane is required for productive uncoating to occur. Recent data on Sendai-erythrocyte fusion reported by Lyles has shown that the Sendai nucleocapsid remains tightly associated with the inner surface of the plasma membrane after fusion.[20,21] If a similar situation occurs after fusion with other plasma membranes, then one of the above alternatives might be required. Sendai retains appreciable hemolytic activity at pH 5.5, showing that it can fuse at the lysosomal pH.[22]

Sendai, influenza, Sindbis, and SFV, as well as VSV, are all inhibited by similar concentrations of UV-VSV (Figure 5). In addition, inhibition of Sendai, SFV, and VSV all showed a similar temporal relationship between chloroquine inhibition, UV-VSV inhibition and transcription.[16,17] In view of the markedly different replication strategies employed by these different enveloped viruses, it appears that UV-VSV is acting on the cell rather than on a specific step in the viral infectious cycle. This inhibition may therefore

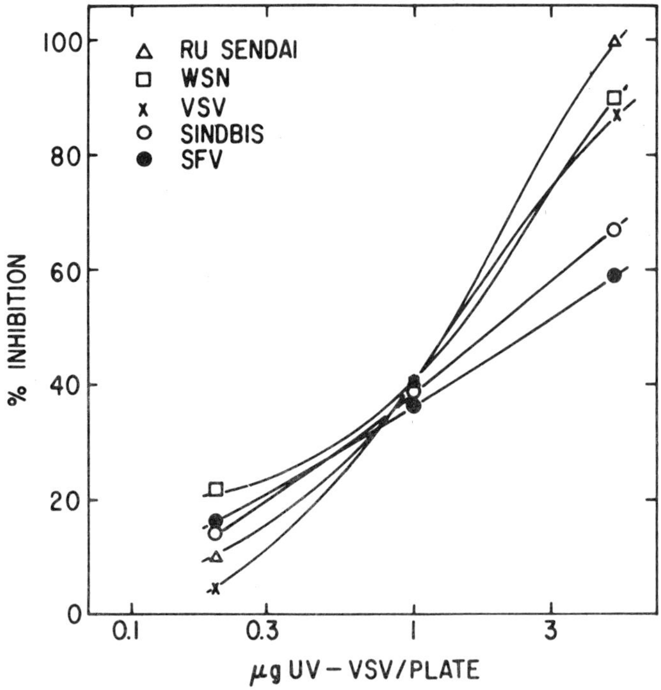

Figure 5. Inhibition by UV-VSV of viral infection by Sendai, (RU strain), influenza, (WSN strain), VSV, Sindbis, and SFV. Plates were incubated with the indicated concentrations of UV-VSV for 1 hr at 37° followed by addition of the virus for 1 hr at 37°. Incubations were for an additional 4 hr for VSV, Sindbis, and SFV; 10 hr for Sendai; and 22 hr for influenza. Concentrations of virus added: 20 hemagglutinating units/plate, Sendai; 2 pfu/cell, influenza; 4 pfu/cell, VSV; 2 pfu/cell, Sindbis; 1 pfu/cell, SFV.[16]

reflect the general cytopathic effect shown to occur with UV-VSV or DI particles.[23–30] That such cytopathic effects have been related to specific changes in lysosomal permeability [18] underscores the importance of a lysosomal interaction in viral infection.

Summary

Lysosomotropic agents, including chloroquine, local anaesthetics, antihistamines, and primary aliphatic amines all inhibit infection by VSV at a step following internalization of the virus. This step is thought to be uncoating, and the effect of these agents is interpreted to mean that uncoating must occur in the lysosomes. SFV, Sendai and influenza are all inhibited at a similar early step by the same drugs at identical concentrations, suggesting a lysosomal role in the uncoating of all of them.

UV-VSV inhibits VSV infection at a later time then the inhibition by lysosomotropic agents. A functional G-protein and nucleocapsid proteins, but not functional genomic RNA or M-protein, are required for this inhibition. The G-protein is thought to be required for uncoating of UV-VSV in the lysosomes, while the major inhibitory effect may be caused by the uncoated nucleocapsid. In addition to VSV, SFV, Sindbis, Sendai, and influenza are all inhibited at a similar stage in their viral cycle by similar concentrations of UV-VSV. Inhibition thus occurs as a result of a cytopathic effect on the cells, rather than by specific interference with a step in the viral cycle.

References

1. Choppin, P.W. and Compans, R.W. (1975): *Comprehensive Virology* 5:95–178.
2. Poste, G. and Pasternak, C.A. (1978): *Cell Surface Reviews* 5:306–367.
3. Helenius, A., Kartenbeck, J., Simons, K., and Fries, E. (1980): *J. Cell Biol.* 84:404–420.
4. Simpson, R.W., Hauser, R.G., and Dales, S. (1969): *Virology* 37:285–290.
5. Dahlberg, J.E. (1974): *Virology* 58:250–262.
6. Miller, D.K. and Lenard, J. (1980): *J. Cell Biol.* 84:430–437.
7. Goldstein, J.L., Anderson, R.G.W., and Brown, M.S. (1979): *Nature* 279:679–685.
8. Ohkuma, S. and Poole, B. (1978): *Proc. Natl. Acad. Sci. U.S.A.* 75:3327–3331.
9. deDuve, C., deBarsy, T., Poole, B., Trouet, A., Tulkens, P., and van Hoof, F. (1974): *Biochem. Parmacol.* 23:2495–2531.
10. Scheid, A. and Choppin, R.W. (1974): *Virology* 23:370–380.
11. White, J. and Helenius, A. (1980): *Proc. Natl. Acad. Sci. U.S.A.* 77:3273–3277.
12. Silverstein, S.C. and Dales, S. (1968): *J. Cell Biol.* 36:197–230.
13. Silverstein, S.C., Astell, C., Levin D.H., Schonberg, M., and Acs, G. (1972): *Virology* 47:797–806.
14. Keller, P.M., Uzgiris, E.E., Cluxton, D.H., and Lenard, J. (1978): *Virology* 87:66–72.
15. Miller, D.K., Feuer, B.I., VanDeroef, R., and Lenard, J. (1978): *Virology* 87:66–72.
16. Miller, D.K. and Lenard, J. (1981a): *J. Cell Biol.* submitted for publication.
17. Miller, D.K. and Lenard, J. (1981b): *J. Cell Biol.* submitted for publication.
18. Allison, A.C. (1967): *Perspectives in Virology* 5:29–61.
19. Poste, G. (1972): *Intern. Rev. Cytol.* 33:157–252.
20. Lyles, D.S. (1979): *Proc. Nat. Acad. Sci. U.S.A.* 76:5621–5625.
21. Roach, S. and Lyles, D. (1980): *Amer. Soc. Microbiol. Abstracts,* p. 265.
22. Lenard, J. and Miller, D.K. *Virology,* submitted for publication.
23. Huang, A.S. and Wagner, R.R. (1965): *Proc. Natl. Acad. Sci. U.S.A.* 54:1579–1584.
24. Wagner, R.R. and Huang, A.A. (1966): *Virology* 28:1–10.
25. Bablanian, R. (1975): *Prog. Med. Virol.* 19:40–83.
26. Baxt, B. and Bablanian, R. (1976): *Virology* 72:370–382.
27. Baxt, B. and Bablanian, R. (1976): *Virology* 72:383–392.
28. Dubovi, E.J. and Youngner, J.S. (1976): *J. Virol.* 18:526–533.
29. Dubovi, E.J. and Youngner, J.S. (1976): *J. Virol.* 18:534–541.
30. Marcus, P.I. (1977): Adv. Pathobiol. 6:192–213.

David H. L. Bishop and Richard W. Compans, eds.
The Replication of Negative Strand Viruses

Origin and Properties of a Tyrosine Kinase Activity in Virions of Vesicular Stomatitis Virus

Gail M. Clinton and Nicholas G. Guerina[a]

Vesicular stomatitis virus (VSV) is one of the best characterized animal viruses and therefore provides a system for studying the regulatory events surrounding protein phosphorylation. VSV virions contain host derived protein kinases which phosphorylate viral proteins *in vitro* and presumably *in vivo* during the infectious cycle [1] (and Clinton, G.M., unpublished observations). A study of phosphorylation of VSV proteins should, therefore, help define how host enzymes interact with viral components during virus replication.

One mechanism of regulation by protein phosphorylation has been suggested by studies with VSV. In this case, the extent of phosphorylation of a protein results in conformational changes of that protein and affects the interaction with other macromolecules. The NS-protein-has been found in two phosphorylated forms.[2–5] Only one of the phosphorylated forms is found in association with viral nucleocapsids.[2,5] Because nucleocapsids are the site of transcription and replication, it has been suggested that interconversion of NS phosphorylated forms may be regulatory in viral RNA synthesis.[2,4] The other VSV phosphoprotein, the membrane associated M-protein, also, is found in two phosphorylated forms.[2] The M-protein, in association with the membrane, functions in viral morphogenesis.[6] When the M-protein is in the soluble fraction of the cytoplasm or with nucleocapsids, it regulates the rate of viral transcription.[7–9] Interconversion of the phosphorylated

[a] Department of Microbiology and Molecular Genetics, Harvard Medical School, and Division of Infectious Diseases, Children's Hospital Medical Center, 300 Longwood Avenue, Boston, Massachusetts.

forms is an attractive explanation for the multifunctional nature of the M-protein.

Another hypothesized mechanism of regulation is by control of the separate kinases which phosphorylate these proteins. This mechanism was suggested by the finding that separate kinases interact differently with the two VSV phosphoproteins. The NS and M-proteins which have different functions and locations in virions and in the cytoplasm have a different distribution of phosphoamino acids, while the two phosphorylated forms of the NS-proteins have the same spectrum of phosphoamino acids. The NS-protein are greatly enriched in their phosphothreonine content, while the M-protein is the only VSV-protein containing phosphotyrosine.[10,11] Separate regulation of the tyrosine and threonine kinases could control the extent of phosphorylation and location of the NS and M-proteins during the infectious cycle.

While phosphothreonine and phosphoserine are common and represent most of the phosphoamino acids in cellular proteins, phosphotyrosine is relatively rare, representing less than 0.5% of the phosphoamino acids.[11,12] An amplification in phosphorylation of tyrosine in cells has been found to occur as a result of transformation with at least some of the RNA tumor viruses.[13] It has been suggested that increased phosphorylation of tyrosine in specific cell proteins leads to changes in cytoskeletal elements resulting in altered cell morphology and growth properties. The proteins so far known to contain phosphotyrosine are the transformation specific protein of the avian sarcoma virus pp60src, a 50,000 dalton protein associated with src in immune precipitates [11] and the membrane associated M-protein of VSV.[10,11]

The M-protein is known to interact with the transmembrane glycoprotein G and to form the submembrane skeleton of the virus. It is possible that phosphorylation of tyrosine in these proteins may be regulatory in the interaction of skeletal elements with the membrane.

Kinase Activity in Virions of VSV

Phosphoamino Acid Analysis

Because phosphotyrosine was found in the M-protein,[10,11] it was expected that a tyrosine specific kinase would be in association with VSV-proteins in virions. Purified virions were disrupted with a non-ionic detergent and incubated with (γ^{32}P)ATP in a kinase reaction mixture. The products of the kinase reaction, after partial acid hydrolysis, were separated by high voltage paper electrophoresis. Electrophoresis in the first dimension from top to bottom was at pH 3.5 and in the second dimension from right to left was at pH 1.9 (Figure 1). A two-dimensional analysis insured that no contaminants comigrated with the three phosphoamino acids. Figure 1 is an autoradiogram of the paper following electrophoresis. Of the amino acids phosphorylated in

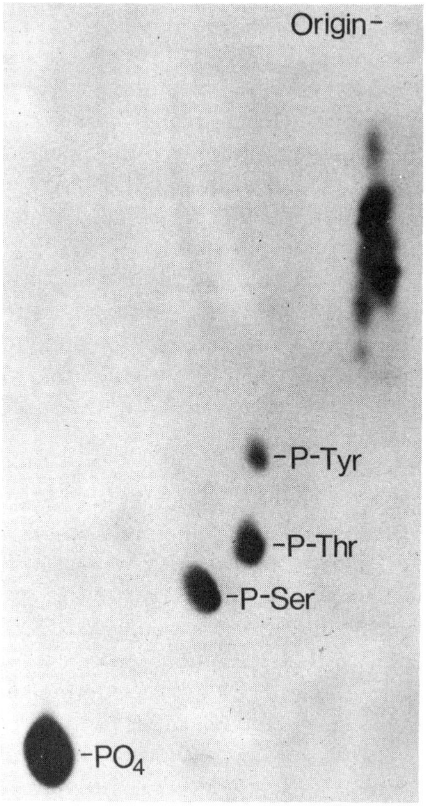

Figure 1. Amino acids phosphorylated by the vesicular stomatitis virus-associated protein kinases. VSV (Indiana serotype) was grown in BHK-21 cells. Purified virions (20 μg protein) were incubated in a kinase reaction mixture (300 μl) containing 10 mM Tris pH 8.0, 10 mM $MnC1_2$, 5 mM dithiothreitol, 50 μM ATP, 1% NP40, 250 μg casein and 5 μCi (γ ^{32}P)ATP at 2000 Ci/mmol (New England Nuclear Corp.). The reaction was incubated at 31° for 30 min and precipitated in 25% TCA and 100 μM ATP. The pellet was washed once with 25% TCA containing 100 μM ATP and twice with 100% acetone. The precipitated proteins were suspended in 6 N HC1 and hydrolyzed in a sealed capillary at 110°C for 1.5 hours. After drying *in vacuo*, the sample was suspended in 5 μl of water containing 1 mg/ml phosphoserine, phosphothreonine and phosphotyrosine as markers. The sample was spotted on 3 MM paper and electrophoresed from top to bottom at 2400 volts for 1 hour in a buffer of 5% acetic acid and 0.5% pyridine at pH 3.5. Electrophoresis in the second dimension, from right to left was for 1 hr at 2000 volts in a buffer of 7.8% acetic acid and 2.5% formic acid at pH 1.9. ^{32}P-labeled material was detected by autoradiography. The position of each of the phosphoamino markers was determined by ninhydrin staining.

this reaction tyrosine was about 20%, threonine represented 25%, and serine was 55%. VSV virions must therefore contain a tyrosine specific kinase activity in addition to kinases which recognize threonine and serine.

Assay for src-Related Tyrosine Kinase Activity

The M-protein of VSV and pp60src have similar modes of synthesis and locations in cells. Both are synthesized on soluble polysomes, are not glycosylated, and after synthesis are associated with the plasma membrane beneath the surface.[6,14–19] Because the normal cell homolog to src has a similar location in cells, and like pp60src, has tyrosine kinase activity,[17,20–22] we decided to assay for a src-related activity in the virions of VSV by the method developed by Collett and Erikson.[23]

Immunoprecipitates were made with highly purified virions of VSV and serum from tumor bearing rabbits infected with the avian sarcoma virus. This antiserum cross-reacts with and specifically precipitates pp60src and its normal cell homolog. To detect a src-like activity, the washed immunoprecipitate was assayed for the transfer of ^{32}P from (γ ^{32}P)ATP to the IgG heavy chain of the tumor bearing rabbit serum. Figure 2 (panel A, lanes 1, 2, and 3) shows the Coomassie blue stained IgG heavy chain in a Laemmli gel [24] after electrophoresis of the immune complex. Panel B is the autoradiogram of the gel. A comparison of lanes 1, 2 and 3 (Figure 2) demonstrates that phosphate transfer occurred only when VSV was immunoprecipitated with serum from tumor bearing rabbits which recognized src but not serum from normal rabbits or antiserum to VSV nucleocapsids. This analysis demonstrates the presence of a src-related enzyme in VSV virions. The same assay was carried out on virions collected early, 8 hours after infection, before extensive cell damage, and on defective interfering particles purified by sedimentation twice in sucrose gradients (data not shown). Because a src-like activity was always found, we concluded that the enzyme was incorporated into virions and not a contaminant from cells. Analyses are underway to determine if this enzyme interacts with viral components and is concentrated in virions, or whether it represents a component of the cell membrane which VSV picks up when budding out of the cell.

Properties of the Tyrosine Compared to the Serine and Threonine Kinases

Because separate regulation of the amino acid-specific kinases was suggested by us to control the level of phosphorylation of the NS compared to the M-proteins,[10,11] The properties of the VSV-associated kinases were compared. This was done by varying the components of the kinase reaction mixture and analyzing the products by high voltage paper electrophoresis in one dimension at pH 3.5. The standard reaction mixture contained 10 mM Tris pH 8.0, 5 mM dithiothreitol, 10 mM $MnCl_2$, 50 μM ATP and 1% NP40 and was incubated at 31° for 30 min. When the following parameters were changed, little or no effect was observed on the proportion of phosphotyrosine to phosphothreonine and phosphoserine: sonication to disrupt virions (without NP40); NP40 concentrations of 0.02%, 0.5%, 1%, 2%; temperature of incubation of 0°, 31°, 37°, 42°C; time of incubation for 10 min, 30 min, 60 min, 120 min (data not shown). However a large effect on the proportions of the three phosphoamino acids was observed when 10 mM Mn^{++}

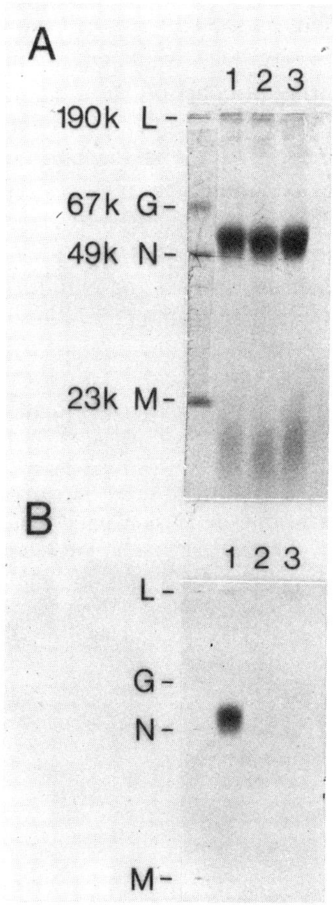

Figure 2. Detection of src-related protein kinase activity in immunoprecipitates of vesicular stomatitis virions. Purified virions (20 μg protein) were incubated with 5 μl serum on ice for 2 hr in 200 μl or RIPA buffer (1.0% Triton X-100, 1.0% sodium deoxycholate, 0.1% sodium dodecyl sulfate, 0.15 M NaC1, 0.05 M Tris pH 7.2). To absorb the immune complexes, 50 μl of a 10% solution of *Staph. aureus* in RIPA buffer was added and the mixture was incubated on ice for 60 min. The bacteria were washed three times with 1 ml of RIPA buffer, and twice with 1 ml of 0.15 M NaC1 and 0.05 M Tris pH 8.0. For the kinase reaction, 10 mM Tris pH 8.0, 10 mM $MnC1_2$ and 10 μCi (γ ^{32}P)ATP (2000 Ci/mmol from New England Nuclear Corp.) were added. After incubation at 31°C for 10 min, the reaction was made 1× in RIPA buffer. The pelleted bacteria were suspended in 50 μl of Laemmli sample buffer and incubated in a boiling water bath for 2 min. The supernatant from the pelleted bacteria was electrophoresed in a 10% polyacrylamide gel as described by Laemmli.[24] Electrophoresis was at 100 volts for 14 hours. Panel A is a photograph of the Coomassie blue stained gel. Panel B is an autoradiogram of the gel. Proteins from virions of VSV were used as molecular weight markers. Lane 1 contains immunoprecipitates of VSV with 5 μl antiserum from tumor-bearing rabbits infected with the Schmidt-Ruppin strain of the avian sarcoma virus. Lane 2 contains immunoprecipitates of VSV with 5 μl normal rabbit serum. Lane 3 contains immunoprecipitates of VSV with 5 μl of antiserum to VSV nucleocapsids.

compared to 10 mM Mg^{++} was used as the divalent cation (Figure 3). When Mg^{++} was present, the amount of phosphotyrosine was reduced about three-fold, phosphothreonine was increased slightly, and phosphoserine was increased about four-fold. An increased activity with Mn^{++} for the tyrosine kinase of the transformation specific protein, p120, of the Abelson murine leukemia virus has also been observed.[25] Little or no amino acid phosphorylation was observed in the absence of divalent cations or when 10 mM Ca^{++} was added (data not shown). These observations suggest that the VSV-associated kinases, particularly the tyrosine and serine kinases, have different cation requirements and may be under separate and opposing regulation.

Figure 3. The effects of $MgCl_2$ or $MnCl_2$ on the amino acids phosphorylated by the vesicular stomatitis virus-associated protein kinases. Purified virions were incubated in a protein kinase reaction mixture without casein and the reaction was processed for phosphoamino acid analysis as described in Figure 1. The hydrolyzed proteins were electrophoresed at pH 3.5 from top to bottom in one dimension only. The phosphoamino acids were detected by autoradiography. The kinase reaction in sample 1 contained 10 mM $MgCl_2$ as the divalent cation. Sample 2 contained 10 mM $MnCl_2$ as the divalent cation.

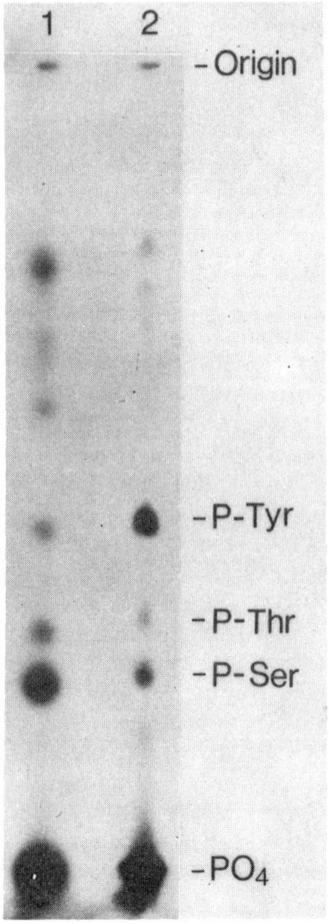

Conclusion and Summary

The presence of a src-related tyrosine kinase in virions of vesicular stomatitis virus raises the possibility that a lytic virus may use this host enzyme during the infectious cycle. The conservation of this gene throughout the vertebrate species in addition to its role in transformation by Rous sarcoma virus implies that the enzyme functions as a vital regulatory element in cell growth.[26] An investigation of the interaction of the src-like tyrosine kinase with VSV-proteins will provide insight into the properties of this enzyme and into the use that VSV might have for such an enzyme. It is possible that VSV does not use this src-related tyrosine kinase during replication, but only incorporates the activity into virions in the process of budding out through the host plasma membrane. Apparently, the sites on the membrane through which VSV buds out are not random. For two different cell lines it has been shown that VSV incorporates a non-random selection of host membrane-associated proteins.[27] In addition, there is selection for the sites where the transmembrane glycoprotein, G, is embedded.[6,28] VSV may provide a vehicle to enrich for and capture this enzyme in its native, membrane associated state.

ACKNOWLEDGMENTS
This work was supported by research Grants from the American Cancer Society MV-54G and from the National Institutes of Health AI 16625. Trudy Lanman provided excellent technical support and Suzanne Ress prepared the manuscript. We thank Dr. Joan Brugge for helpful advice and for supplying serum from tumor-bearing rabbits. We thank Dr. Ray Erikson for the generous supply of serum from tumor-bearing rabbits. We thank Dr. Alice S. Huang for support and encouragement during the course of these studies.

References

1. Imblum, R.L. and Wagner, R.R. (1974): *J. Virol.* :113–124.
2. Clinton, G.M., Burge, B.W., and Huang, A.S. (1978): *J. Virol.* 27:340–346.
3. Clinton, G.M., Burge, B.W., and Huang, A.S. (1979): *Virology* 99:84–94.
4. Kingsford, K. and Emerson, S.U. (1980): *J. Virol.* 33:1097–1105.
5. Hsu, C.-H. and Kingsbury, D.W. (1980): In: *Animal Virus Genetics,* ICN-UCLA Symp., Vol. XVIII. B. Fields, R. Jaenisch, and C.F. Fox (Eds.), Academic Press, New York, in press.
6. Wagner, R.R., Snyder, R.M., and Yamazaki, S. (1970): *J. Virol.* 5:548–588.
7. Clinton, G.M., Little, S.P., Hagen, F.S., and Huang, A.S. (1978): *Cell* 15:1455–1462.
8. Martinet, C., Combard, A., Printze-Ané, C., and Printz, P. (1979): *J. Virol.* 29:123–134.
9. Carroll, A.R. and Wagner, R.R. (1979): *J. Virol.* 29:134–142.
10. Clinton, G.M. and Huang, A.S. (1980): Proc. Juselius Symp., *Expression of Eukaryotic Viral and Cellular Genes,* Helsinki, in press.
11. Clinton, G.M. and Huang, A.S. (1981): *Virology,* in press.
12. Hunter, T. and Sefton, B.M. (1980): *Proc. Natl. Acad. Sci.* 77:1311–1315.
13. Sefton, B.M., Hunter, T., Beemon, K., and Eckhart, W. (1980): *Cell* 20:807–816.
14. David, A.E. (1977): *Virology* 76:98–108.
15. Knipe, D., Baltimore, D., and Lodish, H.F. (1977): *J. Virol.* 21:1128–1139.

16. Willingham, M.C., Jay, G., and Pastan, I. (1979): *Cell* 18:125–134.
17. Courtneidge, S.A., Levinson, A.D., and Bishop, J.M. (1980): *Proc. Natl. Acad. Sci.* 77:3783–3787.
18. Lee, J.S., Varmus, H.E., and Bishop, J.M. (1979): *J. Biol. Chem.* 254:8015–8022.
19. Sefton, B.M., Beemon, K., and Hunter, T. (1978): *J. Virol.* 28:957–971.
20. Collett, M.S., Erikson, E., Purchio, A.F., Brugge, J.S., and Erikson, R.L. (1979): *Proc. Natl. Acad. Sci.* 76:3159–3163.
21. Karess, R.E., Hayward, W.S., and Hanufusa, H. (1979): *Proc. Natl. Acad. Sci.* 76:3154–3158.
22. Oppermann, H., Levinson, A.D., Varmus, H.E., Levintow, L., and Bishop, J.M. (1979): *Proc. Natl. Acad. Sci.* 76:1804–1808.
23. Collett, M.S. and Erikson, R.L. (1978): *Proc. Natl. Acad. Sci.* 75:2021–2024.
24. Laemmli, U.K. (1970): *Nature* 227:680–682.
25. Witte, O.N., Dasgupta, A., and Baltimore, D. (1980): *Nature* 283:826–831.
26. Spector, D., Varmus, H.E., and Bishop, J.M. (1978): *Proc. Natl. Acad. Sci.* 75:4102–4106.
27. Lodish, H.F. and Porter, M. (1980): *Cell* 19:161–169.
28. David, A.E. (1973): *J. Mol. Biol.* 76:135–142.

David H. L. Bishop and Richard W. Compans, eds.
The Replication of Negative Strand Viruses

In Vitro Modification of the 3′ End of Vesicular Stomatitis Virus Nucleocapsid RNA

David M. Coates,[a] Elizabeth A. Grabau,[b] and David J. Rowlands[a]

The RNA extracted from vesicular stomatitis virus (VSV) particles can be readily modified *in vitro* at both the 5′ and 3′-termini. The 5′ end can be dephosphorylated with alkaline phosphatase and uniquely labeled in the 5′ nucleotide using (^{32}P) γ-ATP and polynucleotide kinase, thus providing a means of sequencing the 5′ portion of the genome.[1] The 3′-terminal nucleotide can be specifically labeled by convalently linking radioactive pCp to the available 3′OH group using T_4 RNA ligase.[2,3] The 3′ end can also be polyadenylated *in vitro* by polynucleotide phosphorylase or poly(A) polymerase to provide a template for transcription by reverse transcriptase using an oligo(dT) primer.[4,5]

The RNA within nucleocapsids extracted from infected cells or prepared from purified virions by mild detergent disruption is, in contrast to naked RNA, resistant to RNase digestion.[6] However, Chanda and Banerjee [7] have recently shown that high levels of micrococcal nuclease can introduce nicks into the central region of the RNA of approximately half of the nucleocapsid population. The 5′-terminus of the RNA is protected from RNase digestion within the nucleocapsid [8] and Chanda and Banerjee [7] concluded that the 3′-terminal sequence is similarly protected since the RNA from RNase treated nucleocapsids retained the ability to hybridize to leader sequence, a small RNA transcript representing the complement of the extreme 3′ region of virus RNA.[9]

[a] Animal Virus Research Institute, Pirbright, Surrey, England.
[b] Department of Biology, University of California, San Diego, La Jolla, California.

We have investigated the reactivity of the 3′ end of the nucleocapsid RNA to determine whether modifications are possible in this region and with the ultimate intention of studying the biochemical and biological effects of such alterations, since nucleocapsids are infectious and transcriptionally active *in vitro*. We first examined the accessibility of the 3′OH terminus of nucleocapsid RNA for extension with poly(A) chains using ADP and polynucleotide phosphorylase. Polyadenylation was then assayed by extracting the RNA from the nucleocapsids and measuring the extent of binding to oligo(dT) cellulose.[5] The results of such an experiment are shown in Figure 1 in which

Figure 1. Binding of ^3H VSV RNA to oligo(dT) cellulose. The RNA was loaded on to a short column of oligo(dT) cellulose in binding buffer containing M KC1, 10 mM Tris HC1 and 0.5% SDS, pH 7.5. The column was washed with one ml aliquots of the same buffer. The arrow indicates a change from binding buffer to elution buffer (10 mM Tris, 0.5% SDS, pH 7.5). Fractions were counted in Triton/toluene scintillant cocktail. The elution of control RNA is shown in a and RNA from nucleocapsids which had been polyadenylated in b.

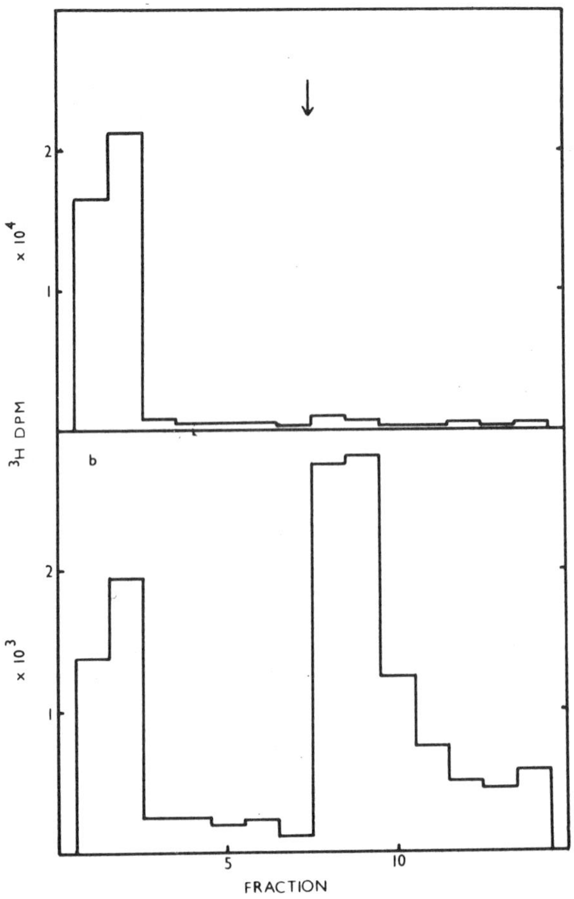

nucleocapsids released from ^{3}H-uridine labeled virus with 0.1% DOC and purified by sucrose density gradient centrifugation followed by pelleting were incubated in polyadenylation buffer [5] with or without polynucleotide phosphorylase for 6 min. Negligible binding was observed with RNA from mock treated nucleocapsids whereas up to 70% of the RNA from polyadenylated nucleocapsids bound to the oligo(dT) cellulose. This figure was similar to the results obtained with naked RNA and shows that the 3′-terminus of RNA within the ribonucleoprotein complex is fully accessible for the addition of ribonucleotide bases. Nucleocapsids prepared by Nonidet NP40 disruption [10] which, in contrast to DOC extracted nucleocapsids, retain the L and NS-protein in addition to N-protein, could also be polyadenylated by this technique.

We next examined the possibility of joining nucleotides to the 3′-terminus of nucleocapsid RNA using T_4 RNA ligase. Nucleocapsids were released from purified virus by treatment at 0°C with either 0.1% DOC [6] or 0.1% Triton N101 [11] and purified as above. The DOC extracted material was shown by polyacrylamide gel electrophoresis to contain only N-protein and RNA whereas the Triton extracted nucleocapsids retained the L and NS-proteins. Both types of particle were incubated with 5′-(^{32}P)pCp and RNA ligase under the conditions described by England and Uhlenbeck [2] except that DMSO was omitted from the mixture. Nucleocapsids were then separated from unincorporated pCp on a Sephadex G100 column and analyzed by sucrose gradient centrifugation. The results (Figure 2) showed that ^{32}P counts now sedimented with the ^{3}H-labeled nucleocapsids. When the RNA was extracted from these nucleocapsids and analyzed on further sucrose gradients it was found that the ^{3}H and ^{32}P counts co-sedimented to the position of undegraded virus RNA, showing that the (^{32}P)pCp was not simply adsorbed to nucleocapsids.

The specificity of attachment of the radioactive pCp was checked by analysis of RNase digest of the RNA by high voltage DEAE paper electrophoresis. The 3′ sequence of VSV RNA is ---pGpU$_{OH}$ [3] which would be expected to give ---pGpU ^{32}pCp on ligation with ^{32}pCp. Digestion of this product with T_1 nuclease should give U ^{32}pCp as the single radioactive product and further digestion with pancreatic RNase would be expected to produce U ^{32}p. The analysis shown in Figure 3 is consistent with this scheme and shows that the pCp is added at a unique site. Figure 3 also shows that ligation of pCp to free, purified VSV RNA gave greater heterogenity than when nucleocapsids were used as acceptors.

Having shown that pCp can be ligated to at least a portion of the nucleocapsid RNA, we examined the susceptibility of the added ^{32}P label to removal with RNase. Nucleocapsids labeled with pCp were incubated with T_1 or pancreatic RNase under conditions which were shown not to alter the sedimentation behavior of RNA extracted from such enzyme-treated nucleocapsids (1 U/ml T_1, or 0.5 μg/ml pancreatic RNase for 1 hr at 20°C in the presence of 20 μg of tRNA) and then analyzed by sucrose gradient centrifugation. In both cases the ^{32}pCp derived label was completely removed from

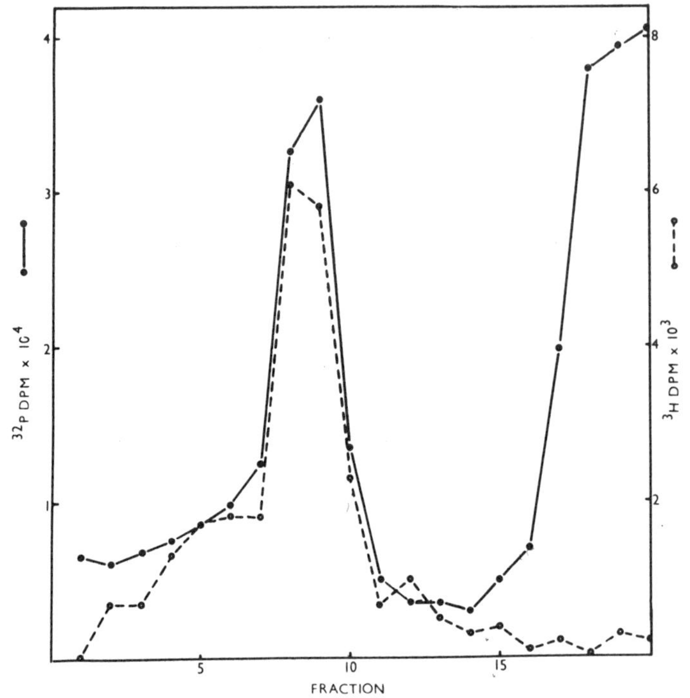

Figure 2. Sucrose gradient centrifugation of ^3H VSV nucleocapsids after incubation with ^{32}pCp and RNA ligase. The nucleocapsids were separated from unincorporated ^{32}pCp on a short column of Sephadex G100 before sedimenting through a 15–45% sucrose gradient at 30,000 rev/min for 2 hrs. Fractions were counted in Triton/toluene scintillant.

the nucleocapsids, from which it was concluded that at least two bases are accessible to RNase cleavage.

In conclusion we have shown that ribonucleotide base addition can be made at the 3′ end of VSV ribonucleoprotein with polynucleotide phosphorylase and with RNA ligase. In the case of polyadenylation with polynucleotide phosphorylase, the proportion of nucleocapsids modified was as high as with free RNA, suggesting that all nucleocapsid particles may be potential acceptors of added bases. The proportion of nucleocapsids labeled with pCp and RNA ligase is lower but this may be due to steric hindrance by the protein molecules at the 3′ end since it is known that RNA ligase requires an acceptor RNA of at least 3 bases.[12]

It will now be of interest to examine the effect of the base additions on the transcriptional activity of nucleocapsids to determine whether the added

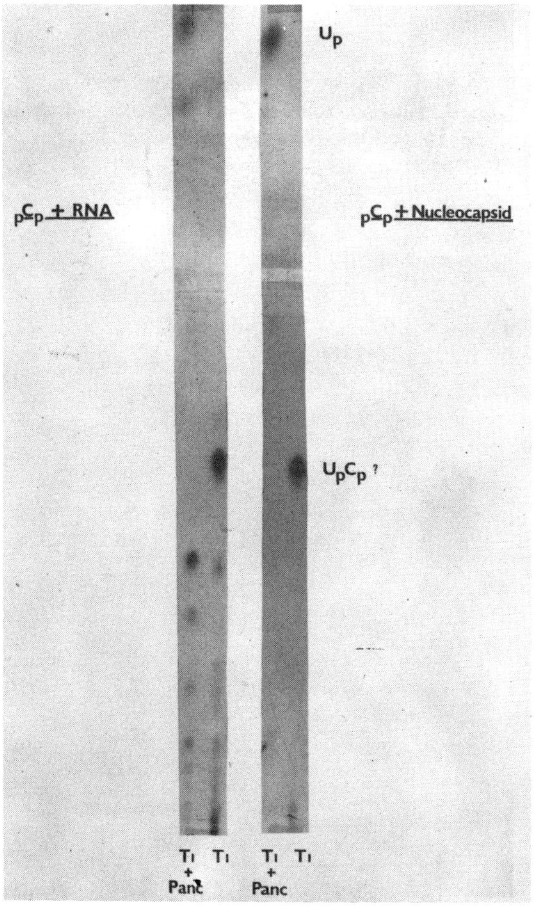

Figure 3. Autoradiograph of high voltage electrophoresis on DEAE paper of T_1 or pancreatic RNase digests of VSV RNA incubated with ^{32}pCp and RNA ligase either free or as nucleocapsid. The RNAs were purified by passing through Sephadex G100 and sedimenting on sucrose gradients before digestion. The position of nucleoside monophosphates were determined by electrophoresing markers on a parallel track.

bases are inhibitory and, if not, whether they are transcribed. A second possible extention will be to ligate the unique 3′ end "stem" sequences characteristic of most VSV DI particle RNAs [13] to nucleocapsids to determine whether this sequence alone is sufficient to confer the property of interference. Preliminary studies have suggested that such stem structures can in fact be ligated to nucleocapsids.

References

1. Semler, B.L., Perrault, J., and Holland, J.J. (1979): *Nucleic Acids Res.* 6:3923–3931.
2. England, T.E. and Uhlenbeck, O.C. (1978): *Nature (London)* 275:560–561.
3. Keene, J.D., Schubert, M., Lazzarini, R.A., and Rosenberg, M. (1978): *Proc. Natl. Acad. Sci. U.S.A.* 75:3225–3229.
4. McGeoch, D.J. and Dolan A. (1979): *Nucleic Acids Res.* 6:3199–3211.
5. Rowlands, D.J. (1979): *Proc. Natl. Acad. Sci. U.S.A.* 76:4793–4797.
6. Cartwright, B., Smale, C.J., and Brown, F. (1969): *J. Gen. Virol.* 7:19–32.
7. Chanda, P.K. and Banerjee, A.K. (1979): *Biophys. Biochem. Res. Comm.* 91:1337–1345.
8. Bishop, D.H.L. and Smith, M.S. (1977): In: *The Molecular Biology of Animal Viruses.* Nayak, D.P. (Ed.), Marcel Dekker Inc., New York, pp. 167–281.
9. Banerjee, A.K., Abraham, G., and Colonno, R.C. (1977): *J. Gen. Virol.* 34:1–8.
10. Breindl, M. and Holland, J.J. (1975): *Proc. Nat. Acad. Sci. U.S.A.* 72:2545–2549.
11. Szilagyi, J.F. and Uryvayev, L. (1973): *J. Virol.* 11:279–286.
12. England, T.E. and Uhlenbeck, O.C. (1978): *Biochemistry* 17:2069–2076.
13. Perrault, J., Semler, B.L., Leavitt, R.W., and Holland, J.J. (1978): In: *Negative Strand Viruses and the Host Cell.* Mahy, B.W.J. and Barry, R.D. (Eds.) London, Academic Press, pp. 527–538.

Copyright 1981 by Elsevier North Holland, Inc.
David H. L. Bishop and Richard W. Compans, eds.
The Replication of Negative Strand Viruses

Vesicular Stomatitis Virus Gene Structure and Transcription Attenuation

J.K. Rose,[a] L.E. Iverson,[a,b]
C.J. Gallione,[a] and J. R. Greene[b]

We have determined the nucleotide sequences of the vesicular stomatitis virus (VSV) mRNAs encoding the N, NS, M, and G-proteins from the sequences of cloned cDNAs. Each mRNA contains an open reading frame for translation which extends from the 5′-proximal AUG codon to within 47–139 nucleotides from the poly(A). Differences between predicted and apparent molecular weights for VSV-proteins are discussed. Only two potential sites of VSV G-protein glycosylation were identified from the predicted amino acid sequence. Comparison of the position of these sites with the known timing of G glycosylation during synthesis indicates that glycosylation occurs on the appropriate asparagine residues as they traverse the membrane of the rough endoplasmic reticulum.

We have analyzed the process of partial transcription termination (attenuation) which results in non-equimolar synthesis of VSV mRNAs during sequential transcription. Comparison of the level of transcription of defined regions of the VSV genome by DNA/RNA hybridization shows that attenuation occurs at or near the intergenic regions rather than non-specifically throughout the genome. Transcription decreases 29–33% across the junctions of the N-NS, NS-M, and M-G genes resulting in a cumulative effect on gene expression. Analysis of the kinetics of transcription *in vitro* shows that transcription appears to be discontinuous with significant pauses (2.5–5.7 min) occurring at or near the intergenic regions.

[a] Tumor Virology Laboratory, The Salk Institute, Post Office Box 85800, San Diego, California.
[b] Department of Biology, University of California, San Diego, La Jolla, California.

Introduction

We have previously described the isolation and characterization of cDNA plasmids containing sequences derived from the vesicular stomatitis virus (VSV) mRNAs.[1,2] Using these as well as additional plasmids which contain nearly full-length copies of each mRNA sequence, we have been able to obtain complete sequences of the VSV mRNAs encoding the N, NS, M, and G-proteins. These sequences will be reported elsewhere; however, some important features of the mRNA and predicted protein sequences are discussed here.

Experiments measuring the sensitivity of VSV transcription to UV irradiation have indicated that transcription begins at the 3′-end of the vRNA followed by sequential transcription of the viral genes in the order (3′)N-NS-M-G-L(5′).[3,4] Despite the fact that transcription is sequential, the five mRNAs are not present in equimolar ratios in infected cells. Instead, there is a gradient of decreasing mRNA abundance that follows the gene order with N > (NS and M) > G > L.[5] Because the rates of degradation of VSV mRNAs are the same,[6] the differences in abundance of viral mRNAs must be due to differential rates of synthesis.

The non-equimolar synthesis of VSV mRNAs must result from partial transcription termination (attenuation) which occurs either non-specifically throughout the genome or at a few specific sites. We describe here experiments which have allowed us to distinguish between these two models. We also present data on the kinetics of VSV transcription *in vitro* which provide direct evidence that transcription is sequential. These experiments also suggest that transcription proceeds discontinuously with pauses in transcription occurring at or near the intergenic boundaries.

Results and Discussion

Some basic conclusions regarding VSV mRNA and protein sizes derived from our nucleotide sequence data are summarized in Table 1. The lengths of the N, NS, M, and G mRNAs [without poly(A)] were determined directly from the DNA sequences of nearly full-length cDNA clones using the chemical method.[7] The precise length of the N mRNA as well as the number of 3′ non-coding nucleotides should be regarded as tentative because we have not yet sequenced both DNA strands completely and because there is an uncertainty in the sequence (due to a sequence compression) approximately 200 nucleotides from the 3′-end. The sequence of the L mRNA has not been determined except for a few nucleotides at both ends.[2,8,9] Difficulties related both to obtaining a cDNA clone the size of L, and to lack of sufficient intellectual motivation are likely to prevent the determination of the L mRNA sequence in the near future. The length of the L mRNA was esti-

Table 1. VSV mRNA and Protein Sizes.

mRNA	Length without poly (A)	Non-coding nucleotides (5′)	Non-coding nucleotides (3′)	Protein molecular weights (Calculated)	Protein molecular weights (Gel mobility)
N	1311	13	47	47,000 (417 a.a.)	52,500
NS	815	10	139	25,100 (222 a.a.)	43,000–55,000
M	832	41	98	26,100 (312 a.a.)	27,000
G	1665	29	103	57,700 (511 a.a.)	63,500 (nonglycosylated)
L	6732 ± 1000	10	n.d.	n.d.	200,000

Lengths of mRNAs are expressed without the poly(A) or 5′-terminal cap. Abbreviations used are a.a.: amino acids and n.d.: not determined.

mated by subtracting the total lengths of the N, NS, M, and G mRNAs as well as the lengths of the leader RNA, the intergenic regions and the 5′-non-coding region of the genome,[2,9–11] (4,768 nucleotides) from an estimated length of 11,500 nucleotides for the VSV genome.[12]

The molecular weights predicted for the VSV-proteins from the amino acid sequences agree reasonably well with the apparent molecular weights determined by gel electrophoresis,[13] except in the case of the NS-protein. The apparent molecular weight of NS is approximately twice that which is encoded by the NS mRNA. Inspection of the amino acid sequence did not reveal obvious features that might contribute to this anomalous migration. Possibly phosphorylation contributes to this anomaly, or perhaps the protein is a dimer which is not disrupted by SDS polyacrylamide gel electrophoresis under reducing conditions.

Glycosylation Sites in the VSV G-Protein

The VSV G-protein is known to contain two asparagine-linked complex oligosaccharides.[14,15] Insertion of the G-protein into the rough endoplasmic reticulum (RER) and glycosylation occur when the protein is a nascent chain.[16,17] Because the precise timing of the two glycosylation events had been determined relative to the fraction of the protein synthesized,[16] it was of interest to examine the predicted G-protein sequence for potential glycosylation sites of the form Asn × Ser or Thr. The complete sequence of G contains 511 amino acids including the 16 amino acid leader peptide. There are 18 Asn residues in the sequence, but only two (Figure 1) occur in possible glycosylation sequences. Therefore, we presume that these are the actual sites. The nearby exact correspondence of the positions of these sites with

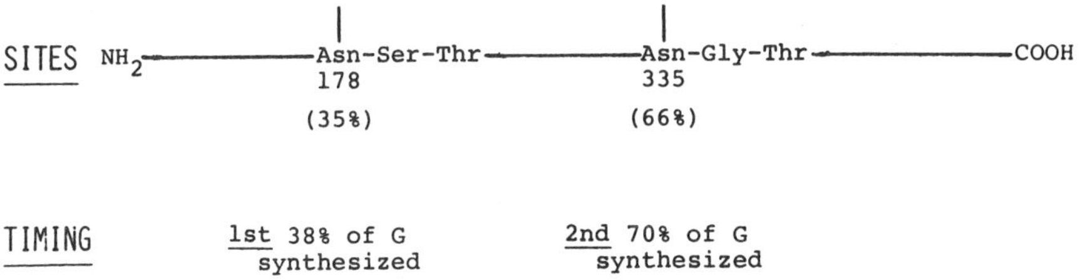

Figure 1. Sites and timing of glycosylation in VSV G. The data on the timing of glycosylation is from reference 16.

the fractions of G synthesized when each glycosylation event occurs (Figure 1) indicates that the oligosaccharide groups are transferred to the nascent chain just as the appropriate Asn residues traverse the RER.

Localization of Transcription Attentuation Sites

If attenuation of VSV transcription occurs non-specifically throughout the genome, then, for any given gene the level of synthesis of 5′-mRNA sequences should be greater than the level of synthesis of 3′-mRNA sequences. If transcription attenuation were localized in the gene termini, however, a gradient of transcription should be seen only when the level of transcription of two different genes is compared. In order to distinguish between these two models, we first identified a set of eight recombinant plasmid DNAs containing sequences derived from the 5′- and 3′-ends of four of the five VSV mRNAs (N, NS, M, and G). The positions of the sequences present in these cloned DNAs relative to the VSV genome are shown in Figure 2.

Transcription was measured by hybridization of ^{3}H-uridine labeled mRNA purified from VSV-infected BHK cells to excess plasmid DNAs immobilized on nitrocellulose filters.[18] A more detailed account of these experiments will be published elsewhere.[19] In all experiments, plasmid DNA was in excess as shown by the linear relationship between the level of hybridization and the RNA concentration. Because ^{3}H-uridine was used for labeling viral RNA, we have corrected for the uridine content of the mRNA sequences hybridizing. This value was obtained directly from the sequences of the insert DNAs. The data are expressed as the molar ratios of mRNA sequences hybridizing, where hybridization to pN161 (5′-proximal N mRNA sequences) is set at one. Table 2 summarizes the results obtained in such experiments using either mRNA labeled for 1 hour or pulse-labeled for 15 minutes. Essentially identical results were obtained in each case. It is clear that the molar ratios of 5′-proximal and 3′-terminal mRNA sequences were nearly equal for the four genes tested. Attenuation of transcription was apparent only when separate genes were compared and it occurred to a similar extent across each gene junction, with transcription decreasing 32%

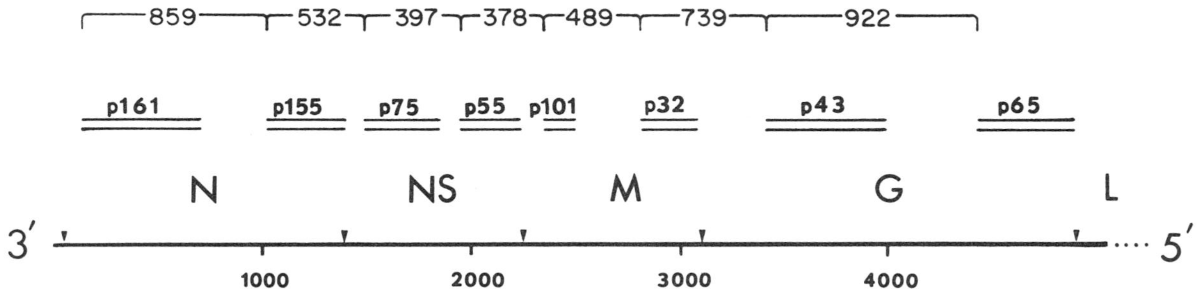

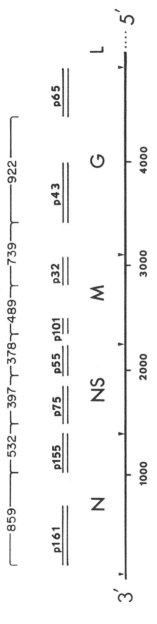

Figure 2. Positions of insert DNA sequences relative to the VSV genome. The single line represents the 3′ half of the VSV genome with lengths given in thousands of nucleotides. Double lines represent regions of the mRNA sequence present as inserts in the indicated plasmids. Solid triangles represent gene boundaries. Bracketed regions with lengths given in nucleotides were used in the calculation of transcription rates.

Table 2. Molar Ratios of Total VSV RNA, Pulse-Labeled RNA and RNA Synthesized by UV-Irradiated VSV.

Plasmid DNA	Total VSV RNA (1-hour label)	Pulse-labeled RNA (15-minute label)	mRNA synthesized by UV-irradiated VSV (2-hour label)
pN161 (5′ N)	1.00	1.00	1.00
pN155 (3′ N)	0.90	1.01	0.12
pNS75 (5′ NS)	0.57	0.69	0.00
pNS55 (3′ NS)	0.54	0.64	n.d.
pM101 (5′ M)	0.42	0.52	n.d.
pM32 (3′ M)	0.46	0.56	n.d.
pG43 (5′ G)	n.d.	0.35	n.d.
pG65 (3′ G)	0.36	0.35	n.d.

Molar ratios of mRNA sequences hybridizing using total VSV RNA (not purified by gradient sedimentation) labeled from 2.5–3.5 hr p.i.; pulse-labeled RNA (3.0–3.25 hr p.i.) and RNA synthesized in BHK cells infected with UV-irradiated VSV (labeled 0.5–2.5 hr p.i.).

at the N-NS junction, 29% at the NS-M junction, and 33% at the M-G junction. The experiment using UV irradiated VSV virions shows that it is possible to detect prematurely terminated VSV transcripts under artificial conditions. This control experiment rules out the possibility that attenuation is due to polymerase termination within genes followed by rapid degradation of incomplete transcripts.

Transcription Attenuation and Discontinuous mRNA Synthesis In Vitro

To analyze sequential transcription of the VSV genome in detail and to determine the extent of transcription attenuation *in vitro,* we analyzed the kinetics of mRNA synthesis by hybridization to plasmid DNAs used in the previous experiments. In these experiments, we used the transcriptase activity present in VSV virions under conditions which yield full-length mRNA transcripts.[20]

VSV transcription was initiated *in vitro* by the addition of $MgCl_2$ to an otherwise complete and pre-warmed reaction mixture. Samples were taken at various times and the RNA from each sample was extracted with phenol, precipitated with ethanol, and then hybridized to DNA from the eight recombinant plasmids. The results of this experiment (Figure 3) demonstrate the sequential nature of VSV transcription *in vitro,* with transcription occurring in the order N, NS, M, and G. The data are expressed as the molarities (arbitrary units) of the sequences hybridizing from a fixed volume of reaction taken at the indicated times. The slope of each line in Figure 3 is proportional to the level of transcription from the particular region of the VSV genome. As *in vitro,* the level of transcription of 5′-proximal and 3′-terminal sequences of each mRNA are the same and the transcription of each gene is decreased relative to the preceding gene. The degree of attenuation observed

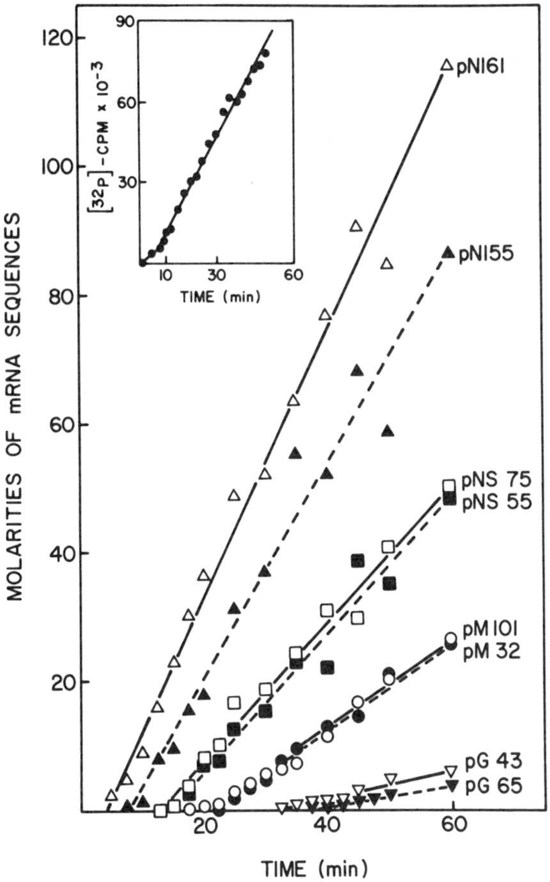

Figure 3. Kinetic analysis of sequential transcription of DNA-RNA hybridiation. RNA samples taken at the indicated times after initiation of the *in vitro* transcription reaction (30°C) were hybridized to each of the eight recombinant plasmid DNAs shown in Figure 2. Molarities of mRNA sequences are in arbitrary units calculated as the cpm [^{32}P] UMP hybridized divided by the number of UMP residues in the sequence hybridizing. Decreasing sample volumes were taken during the reaction to compensate for increasing [^{32}P] UMP incorporation. Therefore, all values are correct to the cpm hybridizing from a constant volume (1 μl) of the reaction mixture. Examples of actual cpm hybridized were 14,853 cpm (pN161, 50 min) and 514 cpm (pG65, 50 min). The background level of hybridization to pBR322 DNA at 50 min was 65 cpm. The inset shows total acid precipitable [^{32}P] UMP incorporation with time. The following symbols are used: pN161 (△), pN155(▲), pNS75 (□), pNS55 (■), pM101 (○), pM32(●), pG43 ( ), pG65( ).

in vitro is greater than that observed in infected cells (29–33%) and appears to increase with distance from the 3′-end of the genome, with 36% attenuation observed at the N-NS juction, 38% at the NS-M junction and 70% at the M-G junction.

An interesting feature of the kinetics of *in vitro* transcription is evident from calculations made from the data in Figure 3. Extrapolation of each set of data points to the X-axis gives the time of appearance of transcripts from each region. Because the distance in nucleotides along the genome between the regions contained in each cloned DNA segment is known (Figure 1), we could calculate the rate of elongation to be 3 to 4 nucleotides per second within the N, NS, M, and G genes. In contrast, calculation of the elongation rate in regions which span the intergenic regions gives the slower rate of approximately 1.5 nucleotides per second. We suggest that this apparently slow transcription is due to transcription pauses of 3–5 minutes occurring at the intergenic regions. Such pauses may occur during polyadenylation by a slippage mechanism at the U_7 sequences present at each gene junction, or may be due to some other process (such as capping or methylation) which is slow relative to transcription.

References

1. Rose, J.K. and Iverson, L.E. (1979): *J. Virol.* 32:404–411.
2. Rose, J.K. (1980): *Cell* 19:415–421.
3. Ball, L.A. and White, C.N. (1976): *Proc. Nat. Acad. Sci.* 73:442–446.
4. Abraham, G. and Banerjee, A.K. (1976): *Proc. Nat. Acad. Sci.* 73:1504–1508.
5. Villarreal, L.P., Breindl, M., and Holland, J.J. (1976): *Biochemistry* 15:1663–1667.
6. Pennica, D., Lynch, K.R., Cohen, P.S., and Ennis, H.L. (1979): *Virology* 94:484–487.
7. Maxam, A.M. and Gilbert, W. (1979): *Proc. Nat. Acad. Sci.* 74:560–564.
8. Rose, J.K. (1978): *Cell* 14:345–353.
9. Schubert, M., Keene, J.D., Herman, R.C., and Lazzarini, R.A. (1980): *J. Virol.* 34:550–559.
10. Colonno, R.J. and Banerjee, A.K. (1976): *Cell* 8:191–204.
11. McGeoch, D.J. (1979): *Cell* 17:673–681.
12. Repik, P. and Bishop, D.H.L. (1973): *J. Virol.* 12:969–983.
13. Knipe, D., Rose, J.K., and Lodish, H.F. (1975): *J. Virol.* 15:1004–1011.
14. Etchison, J.R., Robertson, J.S., and Summus, D.F. (1977): *Virology* 78:375–392.
15. Reading, C.L., Penhoet, E.E., and Ballou, C.E. (1978): *J. Biol. Chem.* 253:5600–5612.
16. Rothman, J.E. and Lodish, H.F. (1977): *Nature* 269:775–780.
17. Toneguzzo, F. and Ghosh, H.P. (1977): *Proc. Nat. Acad. Sci.* 74:1516–1520.
18. Gillespie, D. and Spiegelman, S. (1965): *J. Mol. Biol.* 12:829–842.
19. Iverson, L.E. and Rose, J.K. (1981): *Cell,* in press.
20. Rose, J.K., Lodish, H.F., and Brock, M.L. (1977): *J. Virol.* 21:683–693.

David H.L. Bishop and Richard W. Compans, eds.
The Replication of Negative Strand Viruses

Cloning of Full Length cDNA from the Rabies Virus Glycoprotein Gene

Peter J. Curtis, Algis Anilionis, and William H. Wunner[a]

The negative-strand RNA genome of rabies virus is transcribed upon infection to produce complementary monocistronic mRNAs which code for each of the five structural proteins, L, G, N, M_1 and M_2 of the rabies virion. The envelope glycoprotein (G) which appears as a spike projecting from the virion surface is mainly responsible for the induction of virus neutralizing antibody *in vivo*. Studies by Cox et al.[1,2] and Dietzschold[3] have shown that purified virion glycoprotein or cyanogen bromide fragments of the glycoprotein induce virus neutralizing antibody when injected into mice. Therefore, cloning of the glycoprotein mRNA and expression of the clonal cDNA sequence in *E. coli* might provide a protein capable of acting as an immunogen.

The present report describes the cloning of the rabies virus glycoprotein gene in plasmid pBR-322 and an analysis of the cloned cDNA sequence. The results of this analysis show that the entire coding capacity of the glycoprotein is contained in this cDNA clone.

Earlier work demonstrated that the glycoprotein mRNA sedimented at 18S, one fraction ahead of the 16S nucleocapsid (N) protein mRNA.[4] The respective mRNAs were identified by their ability to direct the synthesis of glycoprotein and nucleocapsid protein which were detected by specific monoclonal antibodies after microinjection into *X. laevis* oocytes. To clone glycoprotein mRNA, double-stranded complementary DNA was synthesized from 18S poly(A) RNA purified from cells infected with the ERA strain of rabies virus using AMV reverse transcriptase and *E. coli* DNA

[a] The Wistar Institute, 36th Street at Spruce, Philadelphia, Pennsylvania.

polymerase I.[5] High molecular weight DNA fractionated by sucrose gradient centrifugation was then inserted into the Pst1 site of pBR-322 by dG:dC tailing.[6] Tetracycline resistant transformants of *E. coli* X1776 were screened using rabies virion RNA labeled with ^{32}P by polynucleotide kinase. Approximately 1% of the colonies bound labeled probe to varying degrees.

Screening of cloned plasmid DNAs with restriction enzyme $BamH_1$ indicated the presence of two different cloned sequences. Restriction with Pst1 identified those clones containing the largest inserts. To assign the clones, DNA from clones A344 and B333, selected as having the largest insert of each group, was labeled by nick translation. The labeled probes were then hybridized to poly(A) RNA which had been fractionated by sucrose density gradient centrifugation. Formation of cDNA:mRNA hybrid was assayed using S_1 nuclease. Figure 1 shows that the probe derived from A344 detected a mRNA sedimenting at 18S, while that from B333 detected a slightly smaller mRNA. We conclude from this that clone A344 contained rabies glycoprotein mRNA sequences (denoted as pRG), while clone B333 contained rabies nucleocapsid protein mRNA sequences (denoted as pRN).

Restriction mapping established the size of the inserted DNA of pRG as 1.75 kilobase pairs. Preliminary DNA sequencing identified the location of a poly(dA) tract which enabled us to determine that the orientation of the glycoprotein mRNA sequence is the same as that of the β-lactamase gene of pBR-322. To determine how many nucleotides might be missing from the glycoprotein cDNA insert, a small labeled fragment bounded by PvuII and HindIII, which corresponds to a region close to the 5′ end of the glycoprotein mRNA sequence (Figure 2a) was hybridized to 18S poly(A) RNA containing rabies virus-specific mRNA, and the primer was elongated by AMV reverse transcriptase with the glycoprotein mRNA as template. Products from the reaction were resolved on a denaturing gel and identified by autoradiography (Figure 2b). The observed major products of the elongation reaction were approximately 42 and 165 nucleotides in length. The smaller fragment presumably arose by filling in the recessed HindIII site formed by reannealing of the PvuII-HindIII strands, while the larger fragment is the elongated primer directed by the glycoprotein mRNA. Since the PvuII site is located 130 base pairs from the 5′-terminus of the cloned glycoprotein gene sequence, our result indicates that 35 additional nucleotides were added as directed by the glycoprotein mRNA template.

Discussion

Characterization of DNA clones complementary to the rabies surface glycoprotein was based firstly on hybridization of clones with labeled purified rabies negative-strand virion RNA and secondly by hybridization to glycoprotein mRNA. The specificity of the latter was demonstrated by the coincidence of translational capacity of the 18S glycoprotein mRNA and 18S

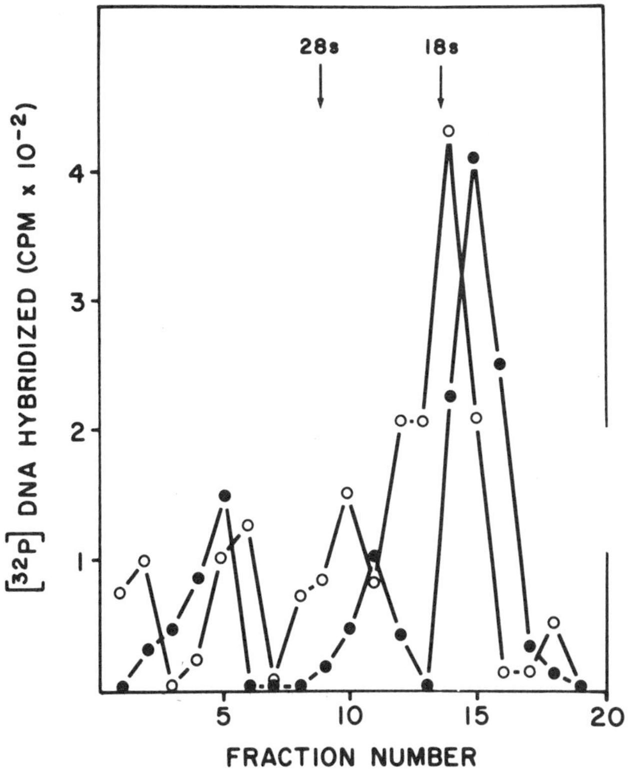

Figure 1. Detection of mRNA for glycoprotein and nucleocapsid protein by cloned DNAs. Poly(A) RNA from cells infected with ERA strain of rabies virus was fractionated in a 5–23% sucrose density gradient centrifuged at 25,000 rpm for 16 hr at 10° C in a Beckman SW41 rotor. Each fraction was diluted 1:60 with water and 1 μl was hybridized in 10 μl 50% formamide, 0.75 M NaC1, 10 mM Hepes pH 6.8 with nick-translated ^{32}P-labeled insert DNA, 3,000 cpm, (prepared from plasmids pBR322/A344 and pBR322/B333, sp. act. 10^8 cpm/μg) at 45° C for 16 hr, under paraffin oil. To terminate hybridization, 100 μl ice cold 0.3 M NaCl, 0.3 M Na acetate pH 4.5, 1 mM $ZnSO_4$ was added and after removal of paraffin oil hybrids were incubated with 1,000 units of S_1 nuclease at 37° C for 45 min. Acid insoluble counts were determined by TCA precipitation; -●- A344, -○- B333.

poly(A) RNA as detected by hybridization. L-protein has a much larger mRNA (> 28S), while M_1 and M_2 proteins are synthesized from significantly smaller mRNAs.[7]

One form of the surface glycoprotein which has a molecular weight of approximately 69,000, is present in the ERA strain of rabies virus,[8] while a smaller form of the same protein which is not fully glycosylated exists intracellularly and has a molecular weight of approximately 65,000 (Wunner, unpublished observations). The latter form comprising an estimated 540

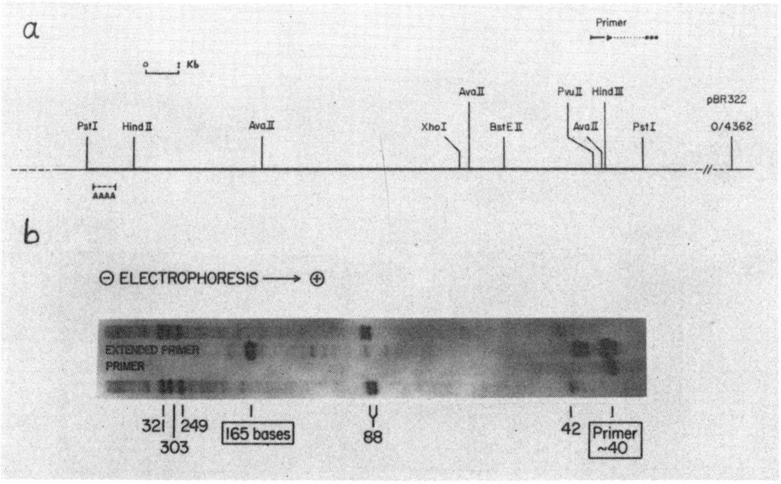

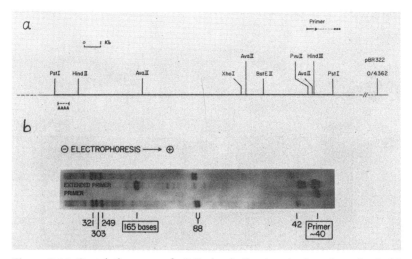

Figure 2.(a) Restriction map of pRG also indicating the location of polydA tract as determined by DNA sequencing. (b) Elongation of a DNA primer by AMV reverse transcriptase. pRG was restricted with PvuII, labeled by γ- ^{32}P-ATP ($\geqslant$ 5000 Ci/mmol, Amersham) with polynucleotide kinase, and subsequently digested with HindII. A 40 base pairs labeled fragment was resolved on a 5% polyacrylamide (0.1% bisacrylamide) gel from which it was eluted. The 5′-labeled fragment was dried, dissolved in 20 μl 80% (w/v) formamide and then denatured by heating at 100° C for 3 min. A sample of 18S poly(A) RNA (10 μg) containing rabies virus-specific mRNA was dried down with 8 μl 5 M NaC1, 0.5 μl 1 M Hepes buffer pH 6.8 and 0.2 μl 0.5 M EDTA pH 7.5. The solution of denatured ^{32}P-labeled DNA was added to the dried RNA and incubated at 60° C, 58° C, 56°C, and 54° C for 1 hr at each temperature. After annealing, the mixture was transferred into 100 μl cold 0.3 M sodium acetate buffer, pH 5.5, and precipitated by 2.5 vol. ethanol at −20° C. The precipitated nucleic acids were dissolved in 40 μl of 50 mM Tris-HC1, pH 8.3, 10 mM $MgC1_2$, 35 mM KC1 30 mM β-mercaptoethanol, 0.5 mM each of the four deoxynucleotide triphosphates, 200 μg/ml BSA, 12.5 μg/ml actinomycin D and AMV reverse transcriptase, 12 units (kindly supplied by J. Beard) and incubated at 37° C for 90 min. Nucleic acids were precipitated from the reaction mixture with ethanol, heat denatured, resolved on a 8% polyacrylamide gel containing 50% (w/v) urea, and detected by autoradiography.

amino acids would require 1620 nucleotides for its mRNA coding length. The size of the cloned glycoprotein cDNA insert of 1.75 kilobase pairs is consistent with the presence of the entire glycoprotein coding capacity.

Elongation of a small primer derived from the 5′ end of the cloned cDNA indicates that only 35 nucleotides are missing from the cloned DNA sequence. Since the 5′ non-coding region of VSV glycoprotein mRNA contains 30 nucleotides followed by 48 nucleotides for the signal peptide,[9,10] we expect our clone to contain all of the sequences required for the mature glycoprotein.

ACKNOWLEDGMENTS
The authors gratefully acknowledge the excellent technical assistance of Erik Whitehorn and Chris Brewer.

This work was supported by U.S. Public Health Service research Grants AI-09706 from the National Institute of Allergy and Infectious Diseases and RR-00540 from the Division of Research Sources.

References

1. Cox, J.H., Dietzschold, B., and Schneider, L.G. (1977): *Infect. Immun.* 16:754–759.
2. Cox, J.H., Dietzschold, B., Weiland, R., and Schneider, L.G. (1980): *Infect. Immun.*, in press.
3. Dietzschold, B. (1980): In: *Proceedings of the 4th International Symposia of Negative Strand Viruses*. Bishop, D.H.L. and Compans, R.W. (Eds.), Elsevier, Amsterdam.
4. Wunner, W.H., Curtis, P.J., and Wiktor, T.J. (1980): *J. Virol.* 36:133–142.
5. Wickens, M.P., Buell, G.N., and Schimke, R.T. (1978): *J. Biol. Chem.* 253:2483–2495.
6. Villa-Komaroff, L., Efstratiadis, A., Broome, S., Lomedico, P., Tizard, R., Naber, S.P., Chick, W.L., and Gilbert, W. (1978): *Proc. Nat. Acad. Sci.* 75:3727–3731.
7. Pennica, D., Holloway, B.P., Heyward, J.T., and Obijeski, J.F. (1980): *Virology* 103:517–521.
8. Dietzschold, B., Cox, J.H., and Schneider, L.G. (1979): *Virology* 98:63–75.
9. Rose, J.K. (1980): *Cell* 19:415–421.
10. Rose, J.K., Welch, W.J., Sefton, B.M., Esch, F.S., and Ling, N.C. (1980): *Proc. Nat. Acad. Sci.* 77:3884–3888.

David H. L. Bishop and Richard W. Compans, eds.
The Replication of Negative Strand Viruses

Analysis of Structure in VSV Virion RNA

Gail W. Wertz and Nancy Davis [a]

A pivotal question in the biology of the negative strand virus, vesicular stomatitis virus (VS), concerns the mechanism by which the two RNA synthetic processes, that of transcription of the five discrete mRNAs and leader RNA and that of the synthesis of the genome-sized virion complementary (VC) RNA is controlled. At present, we can distinguish only that replication of genome size RNAs, in contrast to transcription of individual mRNAs, requires concomitant protein synthesis.[1]

Two models for the mechanism of mRNA synthesis exist. The first model is based on UV inactivation studies which show that synthesis of the five messages is processive.[2,3] One interpretation of these results is that a single initiation event occurs. It was proposed [2] that the individual messages could be generated by a cleavage process, and that according to this model, failure to process nascent RNA into mature mRNAs could yield the full length VC RNA. The second model is that individual initiation and termination events occur for each message. In this case, synthesis of full length VC RNA might arise via a process that involved failure to terminate and reinitiate. For either model, one must postulate signals for either processing events or for individual initiations and terminations. Data obtained by sequencing cDNA to the VSV RNAs show that extensive homologies exist at the junctions of the five VSV genes.[4,5]

[a] Department of Bacteriology and Immunology, School of Medicine, University of North Carolina, Chapel Hill, North Carolina.

It is possible that secondary structure in VSV RNA, alone or in concert with the binding of specific proteins may, in addition to primary sequence, play a role in signaling and controling several molecular events. We examined VSV virion RNA for the presence of secondary structure using various single or double-strand specific nucleases as probes. We previously reported that the procaryotic RNA processing enzyme, RNase III, recognizes and cleaves at sites on the VSV virion RNA to yield 20–24 size classes of fragments which range from 3.5 to 0.1 × 10 ^{6} daltons.[6] The specificity of cleavage by RNase III appears to involve a complex recognition by the enzyme of both primary sequence and secondary structural features.

This paper reports the characterization of the sites of RNase III cleavage of VSV virion RNA and it describes the mapping of the cleavage sites relative to the 3′ end of the RNA molecule.

Structure Associated with Cleavage Sites

Two lines of evidence indicate that the RNase III cleavage sites in VSV virion RNA involve secondary structure. Cleavage of the virion RNA by RNase III can be detected only if the RNA is denatured after enzyme treatment and prior to analysis. Analysis of the sedimentation coefficients of undenatured RNase III treated and untreated RNAs by centrifugation in non-denaturing sucrose-SDS gradients, shows that both sediment at 40S. No change in the sedimentation rate of the treated RNA is observed. If, however, the RNAs are first denatured by heating to 100°C, then the untreated RNA again sediments at 40S, while the enzyme-treated RNA sediments in a broad range from 6–36S. This result indicates that the cleavage sites in the virion RNA are within areas of secondary structure that are sufficient to hold the molecule intact after enzyme digestion. Only after the cleaved RNA has been subjected to denaturing conditions are the resultant cleavage fragments visualized.

The second finding indicative of secondary structure associated with the sites of cleavage comes from analysis of the amount of VSV RNA resistant to solubilization by single-strand specific nucleases before and after treatment with RNase III. We find that 2% of the VSV virion RNA is resistant to digestion by the single-strand specific ribonucleases A, T_1, or T_2 alone or in combinations. This resistance is dependent on salt concentration. If, however, the virion RNA is treated with RNase III prior to treatment with the single-strand specific nuclease, then the 2% of the RNA resistant to solubilization is reduced to 0.6%. Treatment with RNase III alone renders none of the RNA soluble. This result indicates that RNase III may cleave in areas of structure normally resistant to nucleases T_1, T_2, and A and that in so doing it renders these areas susceptible to digestion by the single-strand specific nucleases.

Mapping of RNase III Cleavage Sites

The sites of cleavage of VSV virion RNA by RNase III were mapped relative to the 3′ end of the virion RNA. The location of the cleavage sites was accomplished by digestion of ^{32}P-3′ end-labeled RNA followed by analysis of the size of the fragments generated which retained a 3′ end label.

Virion RNA uniformly labeled with ^{3}H-uridine was prepared, deproteinized and purified as described by Wertz and Davis.[6] One-half of this RNA was end-labeled *in vitro* by addition of ^{32}pCp to the 3′ end of the molecule using RNA ligase according to the method of Keene et al.[7] Identical quantities of RNA labeled uniformly with ^{3}H-uridine alone or with ^{3}H-uridine plus a ^{32}P-3′ end label were digested with RNase III. The fragments generated by cleavage of the RNA by RNase III were analyzed by electrophoresis in denaturing agarose-urea gels. The ^{3}H-uridine labeled RNAs were visualized by fluorography of the dried gels and the ^{32}P-labeled RNAs by autoradiography.

RNA fragments that extend from the 3′ end of the molecule were identified by comparing digests of the uniformly labeled RNA with that of the 3′ end-labeled RNA as shown in Figure 1. Fragments designated A′, B, D, G, H, I, J, K, and L contain a 3′ end label. (Figure 1, lane D.) Fragments A, C, E, and F, for example, do not. It was found that digestion with lower concentrations of enzyme or for very short times allowed discernment of 3′ end label in higher molecular weight fragments. The high molecular weight fragments have been identified previously as partial digestion products. With increased time of digestion, label present in the high molecular weight partials disappears and increased amounts of 3′ end label are observed in the smaller molecular weight fragments.

The sizes of the end-labeled molecules were determined by electrophoresis in denaturing gels and were calculated from a standard curve based on the molecular weight values of the sequenced RNAs of Qβ and 16S and 23S *E. Coli* ribosomal RNAs and the VSV mRNAs, all of which had been co-electrophoresed in the same gel as the RNase III generated fragments.

The sites of RNase III cleavage were determined using a "ruler method." That is, the size of an end-labeled fragment generated by RNase III cleavage was used to indicate the distance of the RNase III cleavage site which had generated that fragment from the 3′ end of the molecule. A map of the cleavage sites discernable by this 3′ end-labeling technique is drawn to scale in Figure 2. The values for the size of the coding regions in the map for the four VSV genes, N, NS, M, and G, are based on sequence analysis data provided by Dr. J. Rose (personal communication).

The map as shown is based on reactions that were approaching a limit digest, that is, few of the high molecular weight partial cleavage products are shown. For this reason this technique and this map identify predominantly

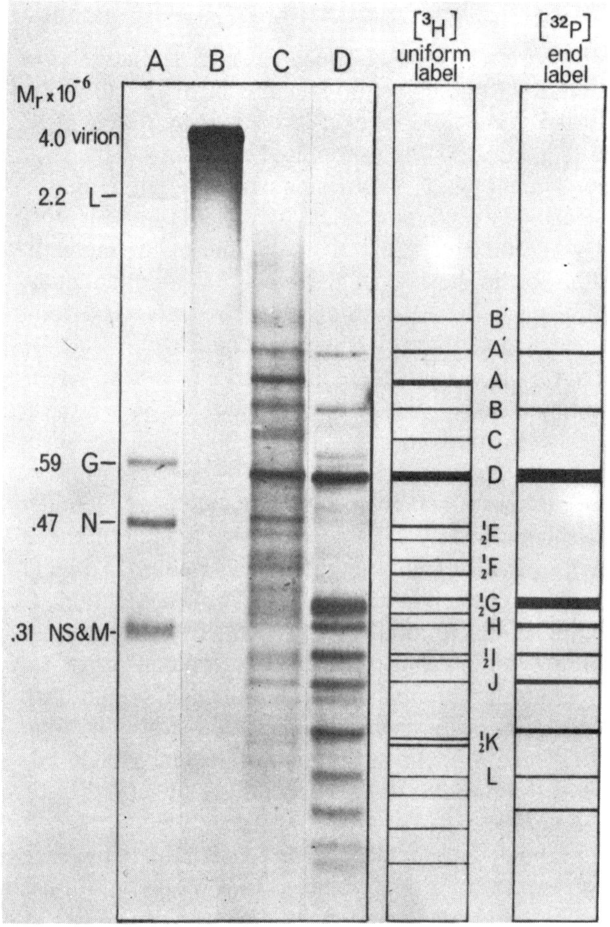

Figure 1. Identification of fragments generated by RNase III digestion of VSV virion RNA that extend from the 3′ end of RNA. VSV RNAs either uniformly labeled with ^{3}H-uridine or labeled only at the 3′ end (using ^{32}pCp and RNA ligase) were digested with three untits of RNase III and the digestion products were visualized by electrophoresis in denaturing agarose-urea gels. (A) VSV intracellular marker RNAs; (B) VSV virion RNA, no enzyme; (C) ^{3}H-uridine uniformly labeled RNA digested with three units of RNase III (fluorogram); (D) ^{32}P-3′ end-labeled VSV RNA digested with three units of RNase III (autoradiogram).

cleavage sites in the 3′ half of the genome. These sites include six cleavage sites in the N gene, two sites in the NS gene, one site in the M gene, two sites in the G gene and at least five sites in the L gene. Other mapping techniques including 5′ end-label analysis will be required to map the sites in the extreme 5′ end of the genome.

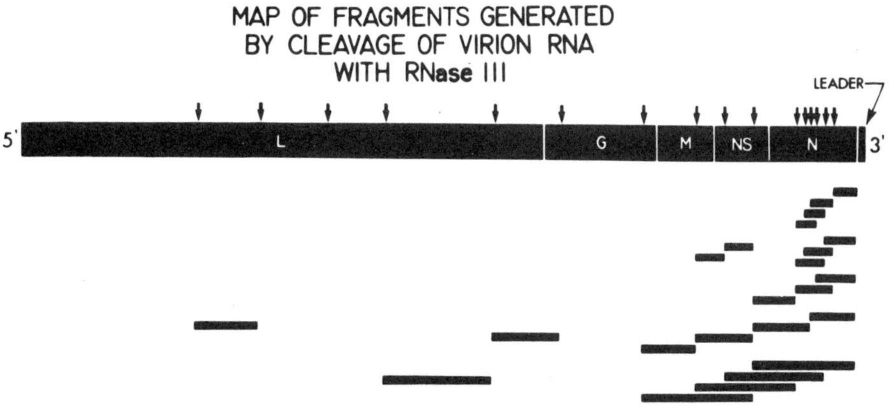

Figure 2. Based on the location of the RNase III cleavage sites (arrows) as determined by digestion of 3′ end-labeled RNA the indicated array of possible partial digestion fragments was generated (solid lines).

Using the cleavage sites indicated in Figure 2 (arrows), it is possible to catalogue possible partial cleavage fragments arising from the 3′ half of the molecule. When this is done, it is possible to predict that numerous fragments having the same size can be generated from several regions of the genome. (See Figure 2 as indicated by solid bars.) This observation in turn predicts that the various size classes of fragments visualized in the agarose-urea gels are not unique but contain similarly sized fragments from several locations on the genome.

To test the hypothesis that fragments having the same size could arise from several locations on the genome, we used the technique of Northern blotting and hybridization using individual VSV mRNAs as probes. The fragments generated by RNase III digestion of unlabeled virion RNA were separated by electrophoresis in agarose-urea gels. The RNA fragments were then transferred to diazobenzyloxymethyl (DBM) paper using the method of Alwine et al.[8] Several replicate transfers were made. Each identical lane of unlabeled virion RNA fragments immobilized on DBM paper was then hybridized to a specific ^{32}P-labeled mRNA probe. The four VSV mRNAs, N, NS, M, and G, that had been separated to a high degree of purity using the preparative gel electrophoresis system described by Wertz et al.[8] were used as probes.

When a mixture of all five of te VSV mRNAs was used as probe, it was found that they hybridized to all of the size classes of RNase III generated fragments. When each of the individual VSV mRNAs was used separately as a probe, it was found that each of the messages also annealed to all of the size classes of fragments. Appropriate negative controls using virion RNA and extremely stringent hybridization and washing conditions all indicated that these hybridizations were well matched.

Discussion

The results presented here show that fragments having the same size can be generated by RNase III cleavage from within each coding region of the VSV genome. This data, that identically sized fragments arise from within each coding region, taken together with the repeating fragment sizes predicted by the 3′ end-label map of the RNase III cleavage sites, indicate that RNase III is able to recognize and cleave the VSV genome at sites that occur in a regular pattern. Based on knowledge of the specificity of cleavage by RNase III and the data presented here indicating involvement of secondary structure in RNase III cleavage sites of VSV genome RNA, we propose that VSV virion RNA can assume a regular structure. We can currently testing whether or not the structures we have probed result from long-range interactions in the RNA or are due to interactions between proximal sequences. The possible roles of regular structure in the RNA in the binding of proteins, in the maturation of the virion nucleocapsid or in any other regulatory processes are being considered.

ACKNOWLEDGMENTS

We thank John Glass for his valuable assistance. This work was supported by Public Health Service Grants AI-12464 and AI-15134 from the National Institute of Allergy and Infectious Disease and Grant CA 19014 from the National Cancer Institute, DHEW.

References

1. Wertz, G.W. and Levine, M. (1973): *J. Virol.* 12:253–264.
2. Ball, L.A. and White, C.N. (1976): *Proc. Nat. Acad. Sci.* 73:422–446.
3. Abraham, G. and Banerjee, A. (1976): *Proc. Nat. Acad. Sci.* 73:1504–1508.
4. McGeoch, D. (1979): *Cell* 17:673–681.
5. Rose, J. (1980): *Cell* 19:415–421.
6. Wertz, G.W. and Davis, N. (1979): *J. Virol.* 30:108–115.
7. Keene, J., Schubert, M., and Lazzarini, R.A. (1980): *J. Virol.* 33:789–794.
8. Alwine, J.C., Kemp, D.J., and Stark, G. (1977): *Proc. Nat. Acad Sci.* 74:5350–5354.

Copyright 1981 by Elsevier North Holland, Inc.
David H. L. Bishop and Richard W. Compans, eds.
The Replication of Negative Strand Viruses

Structure and Origin of Terminal Complementarity in the RNA of DI-LT (HR) and Sequence Arrangements at the 5′ end of VSV RNA

Jack D. Keene, Helen Piwnica-Worms, and Cheryl L. Isaac [a]

Introduction

The defective interfering particles derived from the heat-resistant (HR) strain of vesicular stomatitis virus contains RNAs that are internal deletions of the VSV genome.[1,2] Both DI-LT RNAs have conserved predominantly the 3′ half of the VSV genome and differ in length by about 10%. The larger of the two DI-LT RNAs (DI-LT2) has conserved 1000 or more bases from the 5′ end of VSV RNA and possesses complementary termini (panhandles) similar to those described for the class of DI particle RNAs derived from the 5′ half of the VSV genome.[3] The smaller of the DI-LT RNAs (DI-LT1) is a simple internal deletion mutant and has conserved about 350 bases from the 5′ end of VSV RNA. Thus, the DI-LT1 RNA has a 3′-terminal sequence identical to VSV RNA and contains the smallest portion of the 5′ end of the VSV genome of any previously described DI RNA. Since all DI particles of VSV have conserved various portions of the 5′ end of the genome, this region probably plays an important role in RNA packaging.

In this rcport, we describe sequence arrangements in the region of terminal complementarity of DI-LT2 RNA and present a model for the origin of the complementary termini. We also describe sequence relationships at the 5′ end of VSV Indiana and VSV New Jersey RNAs as an approach to understanding the role of the minimally conserved portion of the VSV genome in virus assembly and packaging.

[a] Department of Microbiology and Immunology, Duke University Medical Center, Durham, North Carolina.

Materials and Methods

Cells and viruses were grown as described previously.[4] RNAs were purified, labeled and sequenced as described.[5,6]

Results

Sequence of 3' and 5'-Termini of the RNA of DI-LT2

We have sequenced the 3′ end of the RNA derived from a stock of DI-LT that contains predominantly the larger, circular DI-LT2 molecules with complementary termini. Figure 1 shows a sequencing gel starting 43 nuc-

Figure 1. Chemical RNA sequencing on 12% urea-acrylamide gel of 3′ labeled DI-LT2 RNA. Following the sequence that is complementary to the polyadenylation site on the 5′ end at base 70, the nondefective VSV leader region begins at position 72.

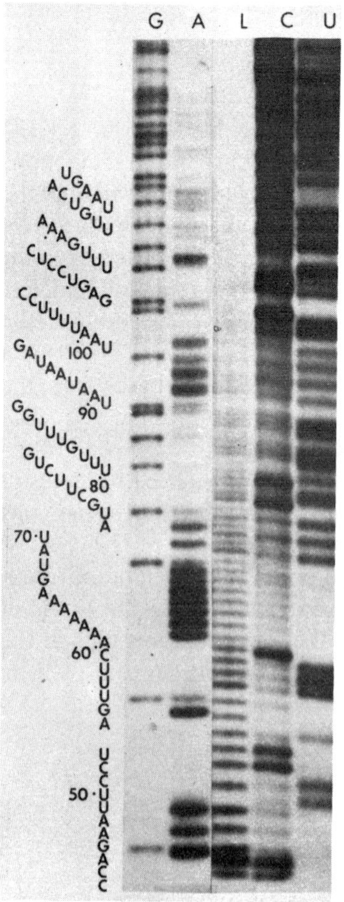

leotides from the 3′ end of this RNA. For a full 70 nucleotides this sequence is the precise complement of the 5′ end of VSV RNA.[7] Beginning at position 71 from the 3′ end, however, the sequence is identical to the 3′-terminus of VSV RNA.[4–6] Thus, the DI-LT2 RNA contains covalently linked defective and nondefective 3′ ends and 3′–5′ base sequence complementarity for 70 nucleotides. Surprisingly, neither the internal VSV leader gene nor the N, NS, M, or G genes of DI-LT2 produce transcription products *in vitro* (See Lazzarini et al., this volume). However, evidence from our laboratory indicates that polymerase is bound in the internal leader region. Thus, this site is transcriptionally inactive because of the failure of polymerase to initiate synthesis at the start of the internal leader gene.

In Figure 2, we propose a model to explain the origin of the terminal complementarity of DI-LT2 RNA. During the production of progeny minus strands, a premature termination of synthesis takes place at the sequence . . . GAA . . . located 71 to 73 nucleotides from the 3′ end of VSV plus strand.[7] This purine rich sequence is characteristically found at other sites of termination or cleavage in VSV and DI particle RNAs.[6] According to the

Figure 2. Model for the origin of the complementary termini in DI-LT2 RNA. A nascent minus strand, terminated at position 70 from the 3′ end of VSV or DI-LT plus strand, resumes synthesis at the 3′-terminus of a completed VSV or DI-LT minus strand. The new plus strand generated by this transcriptional leap gives rise to an RNA with covalently linked defective (70 bases) and nondefective 3′ ends when converted to a minus strand. The sequence for 130 bases at the 5′ end is shown in Figures 3A and 4 and was described previously.[3,7]

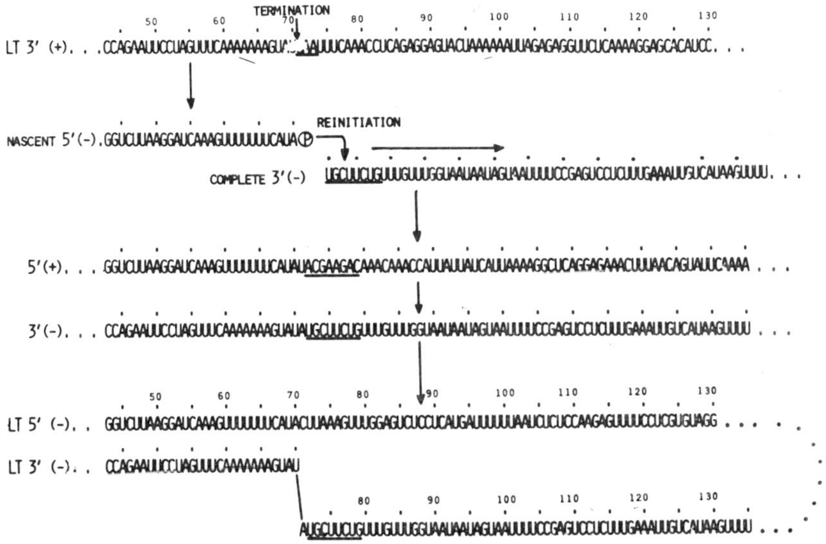

ORIGIN OF THE DI-LT RNA WITH COMPLEMENTARY TERMINI

model, the nascent minus strand with polymerase attached resumes synthesis at the 3′ end of another complete minus strand. The completed product of this transcriptional leap is a new plus strand with covalently linked complements of the 3′ ends of defective and nondefective VSV RNAs. When this plus strand is itself transcribed into a new minus strand, the DI-LT2 RNA with covalently linked defective and nondefective 3′ ends and 70 base terminal complementarity results. The only feature that is unexplained by the model is the origin of the A residue at position 71. However, the VSV polymerase is known to occasionally insert A residues (e.g., at the end of the leader RNA) and this may account for the extra A in position 71.[6] This model has features in common with that proposed by Leppert et al.,[8] for the origin of DI particle RNAs derived from the 5′ end of VSV RNA, in that premature termination followed by resumption of synthesis by the VSV polymerase results in the production of complementary 3′ and 5′-termini. In the latter model, the polymerase resumes synthesis at an internal site near the 5′ end of the nascent RNA molecule. In our model for the origin of DI-LT2, on the other hand, polymerase resumes synthesis at the 3′ end of another molecule. Both models share the properties of sequence specific termination and resumption by the VSV RNA polymerase and explain how DI particle RNAs with terminal base sequence complementarity can originate on RNAs derived from the 5′ as well as the 3′ half of the VSV genome.[3] Since both kinds of DI particles have identical 3′ sequences for up to 47 nucleotides, they both produce the same DI transcription product *in vitro* (Lazzarini et al., this volume). The presence of the internal, transcriptionally muted VSV leader gene makes the DI-LT2 particles useful tools for the study of VSV transcription and replication.

Our model does not address the mechanisms that gave rise to the internal deletions of DI-LT2 RNA. We conclude that DI-LT2 arose as a two-step event and our model is proposed to explain only the origin of the complementary termini. We do not know which of the two steps was primary and which was secondary in the origin of DI-LT2.

Sequences at the 5′ End of VSV New Jersey and VSV Indiana

During virus assembly, VSV proteins distinguish only minus strands for packaging into progeny infectious particles. The minimally conserved base sequences among the RNAs of all the strains of VSV Indiana and their DI particles is about 350 nucleotides at the 5′ end of the VSV genome.[1,2] The majority of this 350 base region falls within the domain of the L gene since the polyadenylation site is present 60 bases from the 5′ end of VSV Indiana.[7] However, the termination codons for L-protein have not been reported and thus, the non-coding region at the 5′ end remains undefined. We have determined the RNA sequences at the 5′ ends of VSV Indiana and VSV New Jersey by extrapolation from sequences of the 3′ ends of hairpin DIs as described previously.[3,7] Figure 3 shows the sequence of the 3′ ends of a

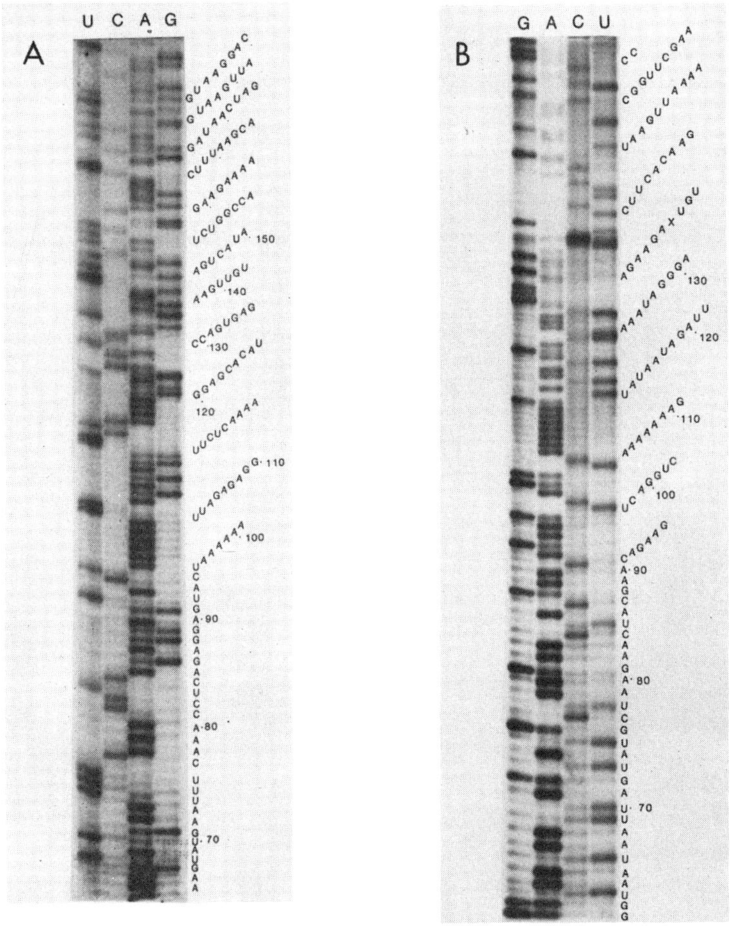

Figure 3. Chemical RNA sequencing of the 3′ end of the hairpin RNAs of (A) VSV Indiana and (B) VSV New Jersey. Those sequences are complementary to the 5′ end of the VSV genome.[3,7] The X, in Figure 3B, denotes an undenatured hairpin between positions 133 and 145.

Hazelhurst-New Jersey DI RNA and DI-011 Indiana from positions 65 to 200. The sequence for up to 130 bases of DI-011 agrees with that reported previously using other methods.[7] Near the top of Figure 3A; translation termination codons are evident suggesting that a region of about 120 nucleotides at the 3′ end of the L messenger RNA may not serve a coding function. Thus, there may be as many as 180 nucleotides at the 5′ end of VSV Indiana RNA whose sequences are not constrained for the synthesis of protein. Figure 4 compares the sequence at the 5′ ends of RNAs of the two serotypes for 130 nucleotides. Interestingly, the mRNA polyadenylation site

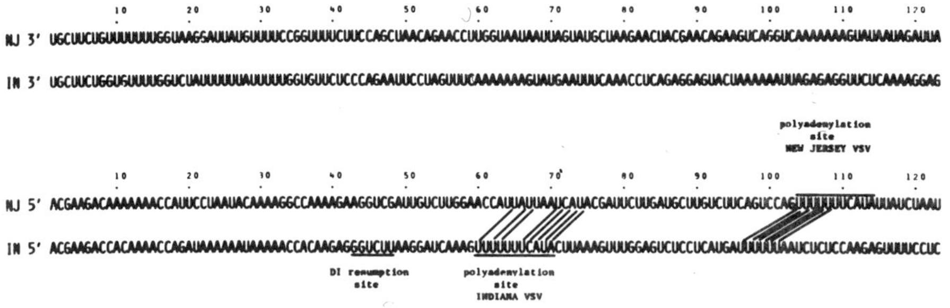

Figure 4. Comparison of the 5′ ends of VSV New Jersey and VSV Indiana serotype RNAs by extrapolation from the sequences of the 3′ ends of hairpin RNAs. The shifted polyadenylation sites and the vestiges of these sites in the RNA of the other serotype are noted. The internal polymerase resumption sites that may be involved in DI RNA production are underlined.

in the VSV New Jersey genome is about 40 nucleotides farther from its 5′ end than that of VSV Indiana. In addition, each serotype has a vestigial sequence of the other's 11 base polyadenylation signal (shown in Figure 4 as connecting lines). Thus, it is not readily possible to determine whether one serotype evolved from the other, even though major sequence arrangements have occurred at their 5′ ends. The L gene polyadenylation sites for both serotype RNAs are preceded by the sequence . . . UAA . . . and followed by the sequence . . . AG. . . .

The 5′ end of the New Jersey genome contains two short sequences which resemble the internal polymerase resumption site (. . GGUCUU . .) that we have proposed is involved in the origin of DI particles whose RNAs were derived from the 5′ half of VSV Indiana RNA.[3] In one case 4 of the 6 bases . . . GGUC . . . are present at positions 44 to 47 and in the other case, 5 of the 6 bases . . . GUCUU . . . are present 53 to 56 nucleotides from the 5′ end of VSV New Jersey. DI particle RNAs from the New Jersey serotype which contains short complementary termini of known length have not been described; thus, it is not known whether these regions can function as competent polymerase resumption sites.

The 5′ sequences shown in Figure 4 are highly conserved in VSV and DI RNAs of both serotypes. It is likely that specific-viral proteins interact with this region in selecting VSV minus strands for packaging. We are presently using chemical probes of RNA structure to investigate such interactions.

References

1. Epstein, D.A., Herman, R.C., Chien, I., and Lazzarini, R.A. (1980): *J. Virol.* 33:818–829.
2. Perrault, J. and Semler, B.L. (1979): *Proc. Natl. Acad. Sci. U.S.A.* 76:6191–6195.
3. Schubert, M., Keene, J.D., and Lazzarini, R.A. (1979): *Cell* 18:749–757.

4. Keene, J.D., Schubert, M., Lazzarini, R.A., and Rosenberg, M. (1978): *Proc. Natl. Acad. Sci. U.S.A.* 75:3225–3229.
5. Keene, J.D., Schubert, M., and Lazzarini, R.A. (1979): *J. Virol.* 32:167–174.
6. Keene, J.D., Schubert, M., and Lazzarini, R.A. (1980): *J. Virol.* 33:789–794.
7. Schubert, M., Keene, J.D., Herman, R.C., and Lazzarini, R.A. (1980): *J. Virol.* 34:550–559.
8. Leppert, M., Kort, L., and Kolakofsky, D. (1977): *Cell* 12:539–552.

David H. L. Bishop and Ricard W. Compans, eds.
The Replication of Negative Strand Viruses

Mechanisms of mRNA Capping and Methylation in Spring Viremia of Carp Virus

Kailash C. Gupta [a] and Polly Roy [b]

We have investigated mechanisms of mRNA capping and methylation by spring viremia of carp virus (SVCV), to test whether mRNA synthesis in this virus occurs by independent initiation of each message [1] or by cleavage of a precursor RNA molecule.[2] Our approach has been to synthesize oligonucleotides in transcription reactions lacking one or more nucleoside triphosphates (NTPs). Heré we report that both capped and uncapped methylated oligonucleotides were synthesized in these partial transcription reactions.

Methylation of Partial Transcripts

No incorporation of radioactive methyl groups was obtained when ATP alone was used in partial reactions (Table 1). However, a reaction containing ATP and CTP gave 1.5% of the incorporation obtained in the complete reaction. If ATP or UTP was added to this reaction as a third nucleotide, incorporation increased to about 4 to 6% of the control. No incorporation was detected in the other combinations of NTPs that were tested (Table 1). These results show that both ATP and CTP are essential for the methylation of partial transcripts.

[a] Present address: Division of Virology, St. Jude Children's Research Hospital, Memphis, Tennessee.
[b] Department of Public Health, University of Alabama Medical Center, Birmingham, Alabama.

Table 1. Incorporation of (^{3}H-Methyl)-S-Adenosyl-L-Methionine into the Partial Reaction Products.

Reaction NTP (s)	Net CPM	% incorporation
A	N.D.	0
A+ C	103	1.5
A + G	N.D.	0
A + U	N.D.	0
A + C +U	344	4.5
A + C + G	422	5.6
A + G + U	N.D.	0
A + C + G + U	7,549	100

The reactions were incubated at 22°C for two hours. 50 μl samples of the products were spotted on DE-81 cellulose filter paper squares (2 × 2 cm) at two hours. The filter squares were washed immediately with several changes of deionized water, dried and counted in toluene-Omnifluor scintillation cocktail. N.D.: Not detected.

Characterization of Methylated Products

The total reaction products from the partial reactions ATP + CTP + GTP (ACG) and ATP + CTP + UTP (ACU) were analyzed by DEAE-cellulose column chromatography. In both cases, three peaks of radioactivity eluted in the region of negative charges 5 to 7 (Figure 1). A similar elution profile was obtained for a reaction containing ATP and CTP, except the incorporation of label was less than in the ACG or ACU reactions (data not shown). Oligonucleotides carrying net negative charges greater than 7 were not synthesized in these reactions, because no radioactivity eluted when columns were stripped at the end of the gradient with 1.6 M LiC1. From these results, we surmised that the products of both partial reactions were three to six nucleotides in length.

Each peak was pooled separately, a portion was digested with alkaline phosphatase and analyzed again on DEAE-cellulose. About 40 to 60% of the material in the pools was not labile to phosphatase and was probably capped (Figure 2). In addition, minor peaks appeared, ranging in negative charges from 1 to 5. These results suggested that the oligonucleotides had capped and uncapped termini in about equal proportions.

To learn whether the oligonucleotides had the same terminal structures as *in vitro* transcripts of SVCV,[3] the oligonucleotides were digested with RNase T2 and alkaline phosphatase and the digests were chromatographed as described above. ACG and ACU pool II yielded a major peak slightly after charge −3, which corresponds to m7GpppAmpA, whereas, ACG and ACU pool III yielded a major peak slightly before charge −4, which corresponds to GpppAmpA (Figure 3). These results suggest that the capped and methylated termini of the oligonucleotides are in fact the same as those of *in vitro* transcripts.

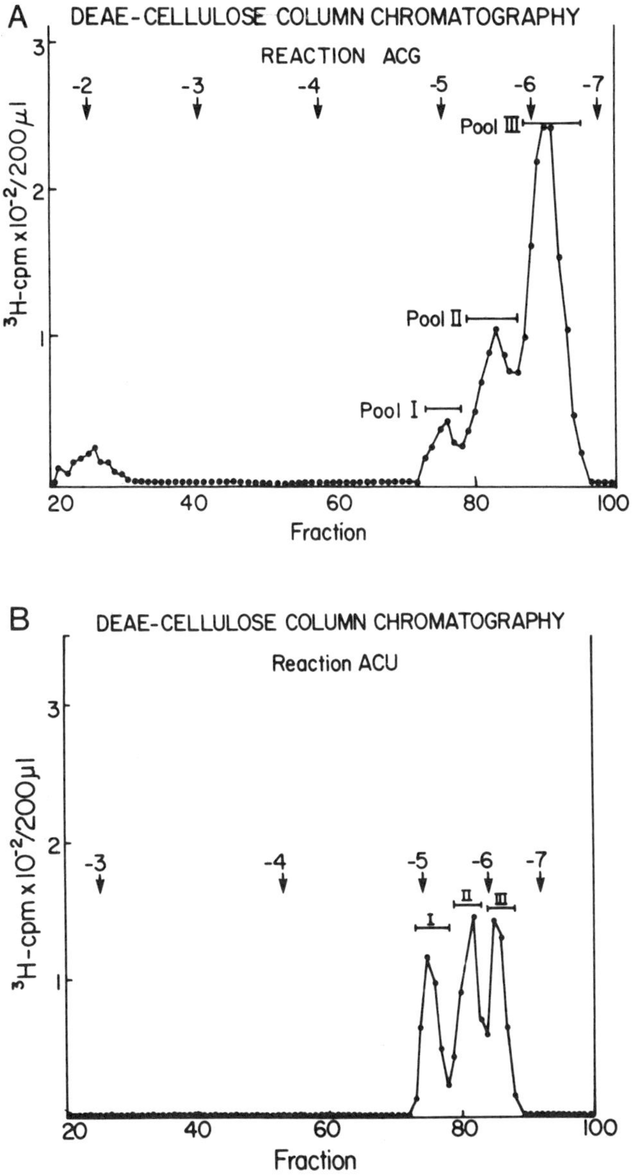

Figure 1. DEAE-cellulose column chromatography of partial transcripts synthesized in the presence of ^{3}H-SAM. Products were analyzed from reactions in which UTP (reaction ACG) or GTP (reaction ACU) was omitted. After the elution of unincorporated ^{3}H-SAM from the column, the bound oligonucleotides were eluted with a 130 ml linear gradient of 0-250 mM LiC1 in 10 mM tris-HC1, pH 7.2, 2 mM EDTA and 7 M urea. Each fraction was measured at 260 nm for the position of marker oligonucleotides. Peaks I, II and III were pooled and analyzed independently.

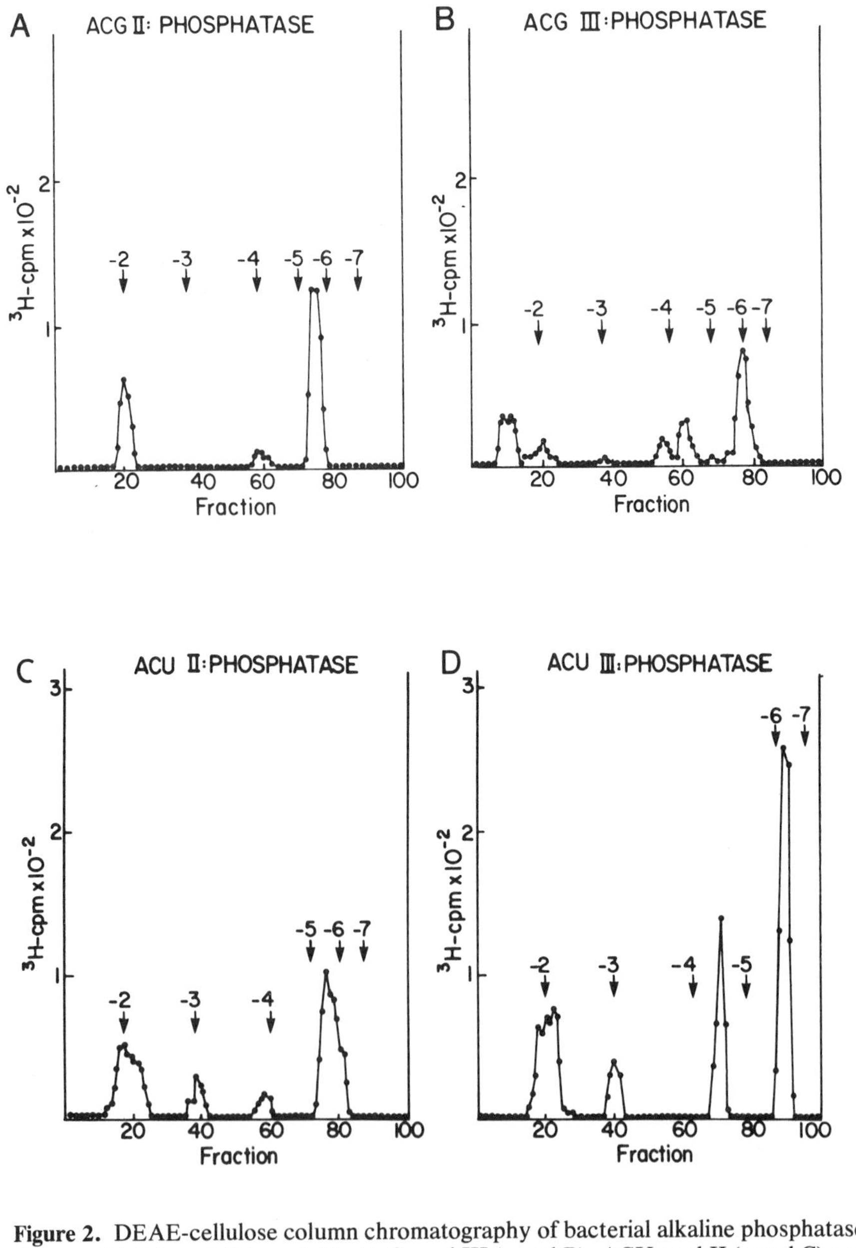

Figure 2. DEAE-cellulose column chromatography of bacterial alkaline phosphatase digests of ACG pool II (panel A), ACG pool III (panel B), ACU pool II (panel C) and ACU pool III (panel D).

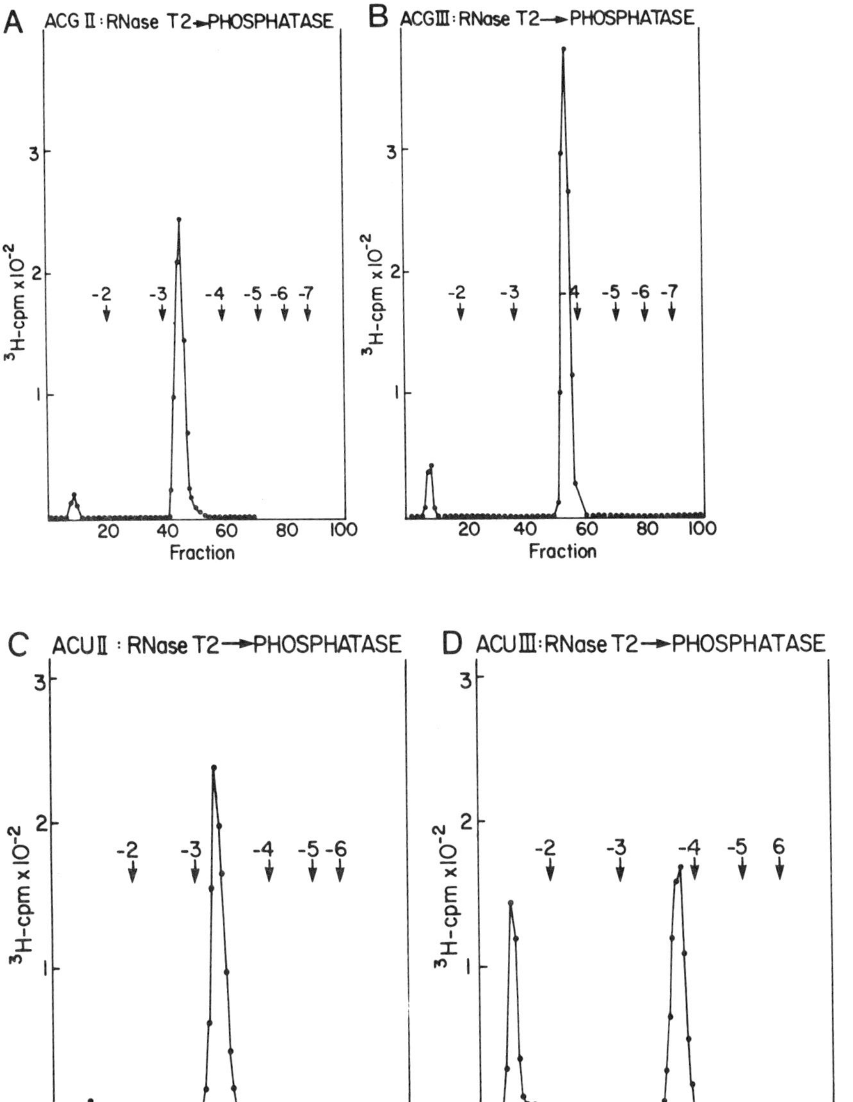

Figure 3. DEAE-cellulose column chromatography of RNase T2 and alkaline phosphatase digests of ACG pool II (panel A), ACG pool III (panel B), ACU pool II (panel C) and ACU pool III (panel D).

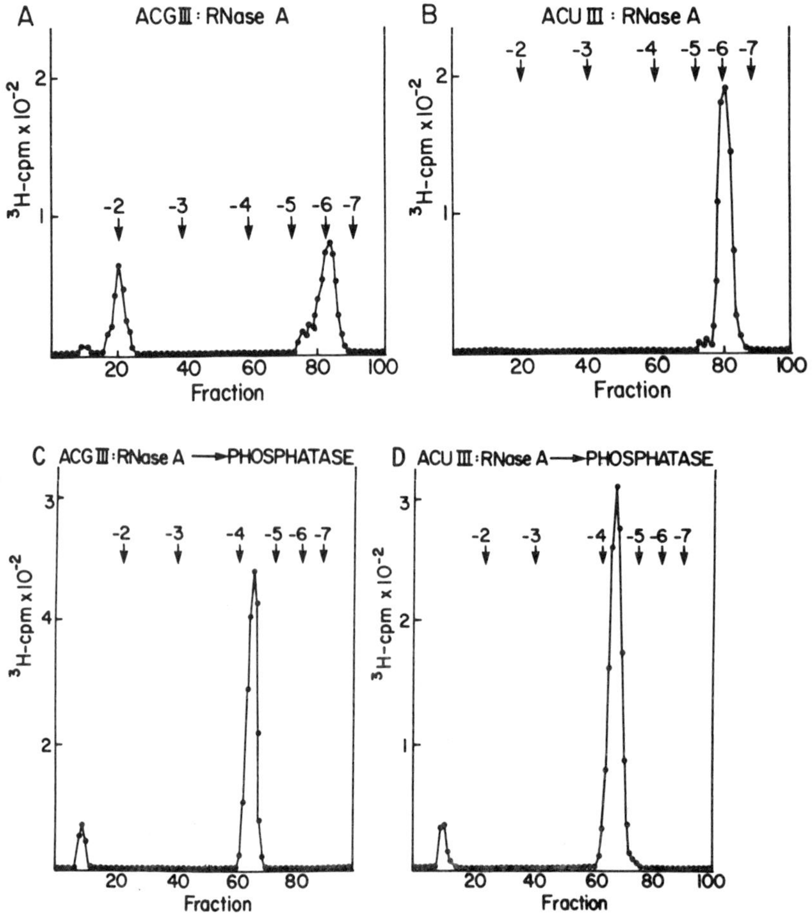

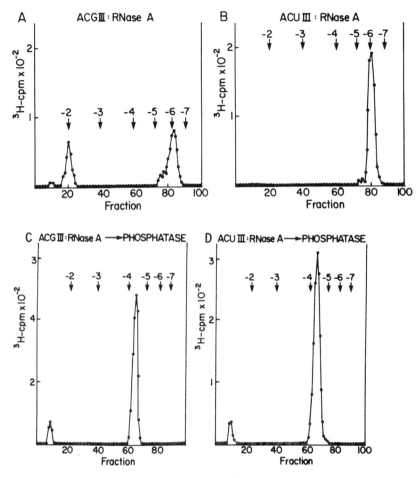

Figure 4. DEAE-cellulose column chromatography of RNase A and alkaline phosphatase digests (A) ACG pool III, RNase A; (B) ACU pool III, RNase A; (C) ACG pool III, RNase A; (B) ACU pool III, RNase A; (C) ACG pool III, RNase A plus alkaline phosphatase; (D) ACU pool III, RNase A plus alkaline phosphatase.

We have shown [3] that cytidine occupies the fourth position from the capping residue of *in vitro* transcripts of SVCV. Since all partial reactions had an absolute requirement for ATP and CTP to synthesize methylated oligonucleotides, it was of interest to find out the position of cytidine in the oligonucleotides. Therefore, pools III of reactions ACG and ACU were digested with RNase A or RNase A and alkaline phosphatase and analyzed on DEAE-cellulose columns. RNase A alone did not change the elution patterns of the oligonucleotides, but when RNase A digestion was followed by phosphatase a peak between charges −4 and −5 appeared (Figure 4). This is the expected result if cytidine occupies the fourth position in the capped and methylated oligonucleotides.

The synthesis of capped oligonucleotides in the absence of GTP (the ACU reaction) is puzzling. The possible presence of GTP in each of the other NTP stocks was ruled out by phosphototransferase reactions [4] and the possibility of endogenous GTP in virions was eliminated by using nucleocapsids instead of virions (data not shown). It is conceivable that these tests failed to detect very small amounts of contaminating GTP. It is also possible that a nucleoside other than guanosine was incorporated in the cap position.

Conclusions

Based on the charges carried by the oligonucleotides and their digestibility with various enzymes, as described above, the following structures have been assigned to each pool.

Pool I: pppAmpNpC, ppAmpNpCpN, m7GpppAmpNpCpN
Pool II: pppAmpNpCpN, ppAmpNpCpNpN, m7GpppAmpNpCpNpN
Pool III: pppAmpNpCpNpN, ppAmpNpCpNpNpN, GpppAmpNpCpNpN

The results presented here demonstrate that methylation and capping can occur at the onset of transcription by SVCV, a feature that has also been observed with other viral [5–7] and eukaryotic [8] systems. These results favor a model of transcription wherein mRNAs are initiated independently.

A scheme of capping and methylation is proposed, based on our present and previous results (Figure 5).[3,9] However, a more detailed analysis of capping and methylation will be needed to confirm this hypothesis.

Figure 5. Proposed mechanisms of capping and methylation in SVCV.

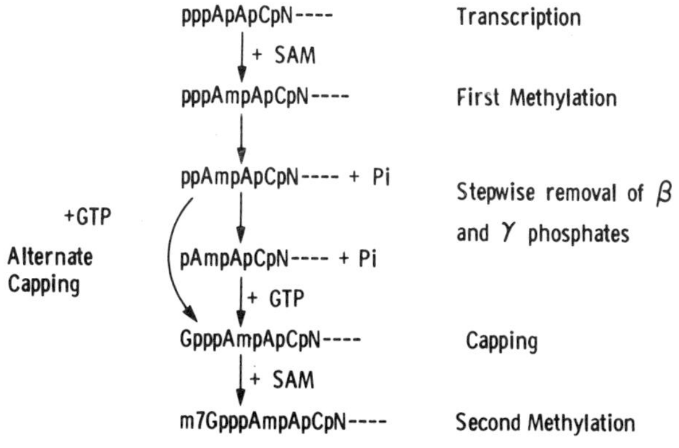

References

1. Bishop, D.H.L. (1977): *Comprehensive Virol.* 10:117–278.
2. Banerjee, A.K., Abraham, G., and Colonno, R.J. (1977): *J. Gen. Virol.* 34:1–8.
3. Gupta, K.C., Bishop, D.H.L., and Roy, P. (1979): *J. Virol.* 30:735–745.
4. Roy, P. and Bishop, D.H.L. (1971): Biochem. Biophys. *Acta* 235:191–206.
5. Paoletti, E., Lipinskas, B.R., and Panicali, D. (1980): *J. Virol* 33:208–219.
6. Furuichi, Y. (1978): *Proc. Natl. Acad. Sci. U.S.A.* 75:1086–1090.
7. Furuichi, Y., Muthukrishnan, S., Tomasz, J., and Shatkin, A.J. (1976): *Progr. Nucl. Acid Res. Mol. Biol.* 19:3–20.
8. Salditt-Georgieff, M., Harpold, M., Chen-Kiang, S., and Darnell J.F., Jr. (1980): *Cell* 19:69–78.
9. Gupta, K.C. and Roy, P. (1980): *J. Virol.* 33:735–745.

Published 1981 by Elsevier North Holland, Inc.
David H. L. Bishop and Richard W. Compans, eds.
The Replication of Negative ftrand Viruses

Studies of the Mechanism of VSV Transcription

R.A. Lazzarini, M. Schubert, and I. Ming Chien[a]

Considerable attention has been focused on the mechanism of VSV mRNA synthesis and to date three distinct models have been proposed for the process, each supported by a different constellation of experimental results. In the *Processing Model,*[1] the VSV polymerase initiates synthesis at the extreme 3′ end of the genome and has the potential to transcribe the genome in a single, uninterrupted synthetic event. The polycistronic transcript, perhaps while it is still incomplete, is processed into the monocistronic capped and polyadenylated mRNAs. The second model, the *3′ Entry, Start-Stop Model,*[1] is similar to the first in that the polymerase can only enter the template at the precise 3′-terminus, but initiates transcription at the beginning of each cistron and terminates at the end of each cistron as it moves down the genome. Thus, the potential to synthesize a polycistronic mRNA is absent in this model. The newest model, the *Multiple Entry with Cascade Synthesis,*[2] specifies that the polymerases enter the template at the beginning of each cistron but can only perform a limited transcription. The further extension of the short mRNA transcripts (mRNA leaders) requires the complete transcription of the 3′ proximal cistron. Thus, a cascade of synthetic activity is envisaged which begins with the completion of the leader RNA and sweeps down the entire genome.

Many of the experimental results that support one or another of these models are not compatible in a single scheme. Consequently. it seems likely

[a] Laboratory of Molecular Genetics, National Institute of Neurological and Communicative Disorders and Stroke, National Institutes of Health, Bethesda, Maryland.

that some of the observations may have little biological significance and may be manifestations of aberrant transcriptional events (mistakes) that are not directly related to mRNA synthesis.

In this communication, we re-examine two central phenomena: the synthesis of the short triphosphate terminating mRNA sequences [2,3] and the constraint that the polymerases enter the template at the 3′-terminus.[4,5] For these experiments we have made use of γ-thio ATP, an analog in which one of the oxygens of the γ-phosphate is replaced by a sulfhydryl group.

Identification of Triphosphate Terminating Transcripts With γ-thio ATP

Transcripts which bear thio-triphosphate termini are easily separated from other RNAs by affinity chromatography using mercury-agarose.[6,7] Figure 1 (lane A) shows a polyacrylamide gel profile of ^{32}P-VSV transcripts that were synthesized with thio-ATP, bound to mercury-sepharose and eluted with dithiothreitol. The major component is a family of leader RNAs that migrate slower than xylene cyanol dye.[8] The precise size of the RNA at the top of the gel is unknown but it is single-stranded and larger than 600 bases. It is clear from the work of others [2,3] that some of the transcripts shown in Figure 1 contain mRNA sequences. The aggregate of all mRNA transcripts can be estimated by digesting the mixed transcripts with RNase T_1 prior to the affinity column chromatography. Under these conditions only the 5′-terminal oligonucleotides will be purified since they alone carry the thio-triphosphate groups. since all VSV mRNAs begin with the pentamer AACAG and leader RNA begins ACG,[9,10] the relative contributions of mRNA and leader RNA can be determined by an analysis of these oligonucleotides. Lane B of Figure 1 shows an analysis of the thio-triphosphate bearing RNase T_1 oligonucleotides. Only the expected two bands corresponding to HSpppApApCpApGp and HSpppApCpGp are apparent. Since both oligonucleotides contain a single C residue, the radioactivity in each oligonucleotide can be directly compared. Identical results to those shown in lane B were obtained when transcripts were labeled with ^{32}P-GTP. Since under the latter conditions, all T_1 oligonucleotides (except those from the extreme 3′-terminus of the transcript) must be labeled, these results show that there are only two initiating sequences that begin with pppA . . . .

Transcripts Prepared With Thio-ATP Are Capped and Methylated

Since our purpose was to explore the possibility that triphosphate terminating mRNA sequences are the direct precursors of capped and methylated mRNAs, it was important to demonstrate that capping and methylation pro-

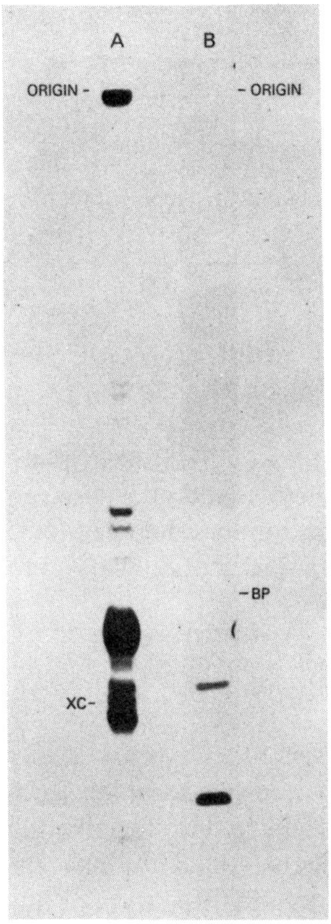

Figure 1. Gel electrophoresis of thio-triphosphate terminating transcripts and thio-triphosphate terminating T_1 oligonucleotides derived from them.

ceeded normally when thio-ATP replaced ATP in the transcription mixture. Table 1 shows the incorporation of ^{3}H-SAM and ^{32}P-GTP into RNA. That the ^{3}H-methyl groups of SAM were incorporated into cap I or cap II structures was demonstrated by electrophoresis of nuclease P_1 and calf intestine phosphatase digests of the transcripts (data not shown). These syntheses do not depend upon prior conversion of thio-ATP to ATP during transcription. Control experiments to estimate the conversion of thio-ATP to ATP by enzymatic exchange by desulfurization during transcription showed this conversion to be negligible. We conclude that VSV transcripts are synthesized, capped and methylated as efficiently in the presence of thio-ATP as with ATP. If triphosphate terminating transcripts are precursors to capped and methylated mRNAs, then thio-triphosphate terminating transcripts are capped and methylated efficiently.

Table 1. Incorporation of ^{3}H-SAM and ^{32}P-CTP into VSV mRNAs in the Presence of Thio-ATP.

	Time	^{3}H CPM	^{32}P CPM
Thio-ATP	2 hours	24,660	51,165
	4 hours	40,995	95,657
ATP	2 hours	11,452	35,752
	4 hours	29,302	105,750

Triphosphate Terminating Transcripts Accumulate During *In Vitro* Transcription of VSV

The appearance of triphosphate terminating transcripts in VSV transcription reactions was followed as a function of time using the method shown in Figure 1, lane B. Both leader and mRNA termini accumulate linearly for 5 hours (Figure 2). In several such experiments the pentanucleotide was 1/5 to 1/2 the concentration of the trinucleotide.

Triphosphate Terminating Transcripts Are Not Associated With The Transcriptive Complex

We have used two methods of separating the transcriptive complex from released transcripts to determine whether the triphosphate terminating transcripts are bound to the template as would be expected for nascent transcripts. Transcription complexes were removed by centrifugation or filtration through nitrocellulose filters under conditions where proteins and nucleic acids bound to them adhered to the filter while free nucleic acids passed through the filter. RNA prepared from both the particulate and soluble or filterable fractions were analyzed for the content of the trimer and pentamer. We estimate from both types of experiments that approximately 95% of the triphosphate bearing mRNAs are found in the supernatant or filtrate and therefore are not associated with the transcriptive complex.

Triphosphate Terminating Transcripts Are Metabolically Inert

If thio-triphosphate terminating mRNAs are capped by either of the capping reactions proposed for rhabdovirus mRNAs,[11,12] the thiophosphate group would be lost and the capped RNAs not recoverable by mercury-agarose chromatography. Thus, there would be a continuous flux of material through the mercury-agarose bindable fraction that would be easily visualized by a radiochemical pulse-chase experiment. A VSV transcription reaction using thio-ATP and α- ^{32}P-CTP was allowed to proceed for 2 hours. A sample was withdrawn and 20-fold more unlabeled CTP was added to the remainder of

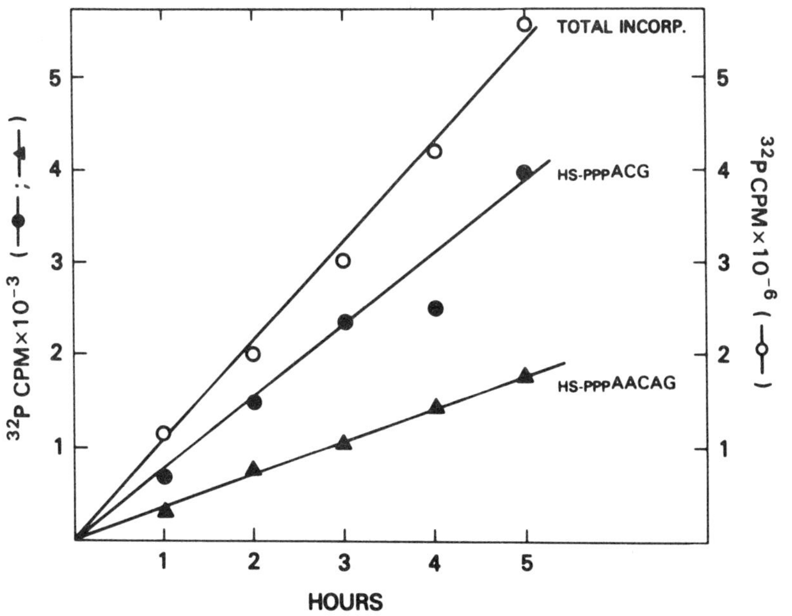

Figure 2. Time course of ^{32}P incorporation into total transcripts and triphosphate terminating RNase T_1 oligonucleotides derived from them.

the incubation mixture. After a 1-hour chase period, the remainder of the reaction was withdrawn. Both the prechase and chased samples were separated into template bound and unbound fractions by filtration and RNA prepared from each. The analysis of the mercury-agarose purified T_1 oligonucleotides is shown in Figure 3. The radiolabel in the thio-triphosphate bearing oligonucleotides of neither the template nor the filtrate fraction was diminished by the 1-hour chase with unlabeled CTP. We conclude that most of the triphosphate bearing RNAs are metabolically inert and no more than 15% of the label in these materials (the limit of resolution of our experiments) is "chasable."

Thus, the triphosphate terminating mRNA transcripts accumulate for at least 5 hours during the transcription reaction, are smaller than functional message, have been released from the transcriptive complex, and are metabolically stable. Taken together, these results suggest that the uncapped transcripts are products of aborted transcription that accumulate during the transcriptive process and are not precursors of functional mRNAs. Our results however, *do not* lead to the conclusion that independent initiation of functional mRNA is unlikely. Firstly, we observe only two types of triphosphate terminating transcripts, those beginning with pppACG and ppp AACAG—although our method is capable of detecting all types of trans-

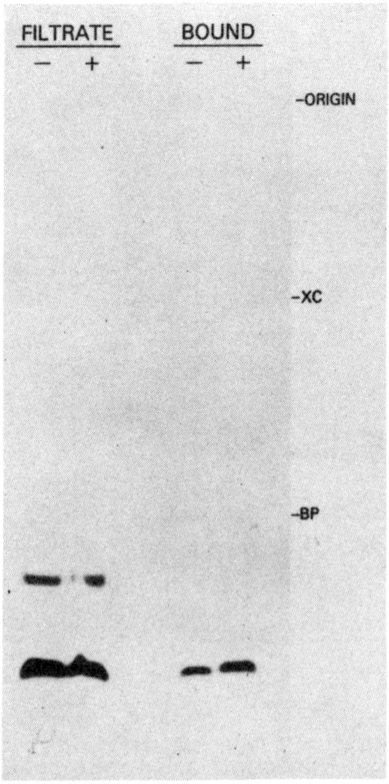

Figure 3. Metabolic stability of radiolabel in 5′-terminal RNase T_1 oligonucleotides. The − and + refer to samples drawn before and after the "chase", respectively.

cripts that begin with pppAp . . . . This restriction of the initiation to the precise 3′-terminus of leader and the beginning of the viral mRNA cistrons suggest a sequence specificity that is most easily reconciled with the independent initiation models. Secondly, initiation of transcription at the beginning of mRNA cistrons is quite frequent—1/2 to 1/6 as frequent as initiations at the leader region. This level of promoter activity must be a minimum since only the metabolically stable, aborted transcription products have been scored—capped RNAs initiated at these sites would go undetected by our methods. Thirdly, our procedures would not detect the capping of very short transcripts since only material that is excluded by Sephadex G-50 was analyzed. These results suggest that if capped and methylated RNAs are independently initiated, the capping and methylation must take place when the transcripts are very short. It is conceivable that the capping and methylation form a integral part of the initiating event.

The Dysfunction Of An Internal VSV Leader RNA Gene

The one, undisputed initiation site for VSV RNA synthesis is the 3′ terminus of the genome, the proposed unique entry site for the polymerase in two of the models of VSV transcription. Until now, it has not been possible to determine the relative importance of the base sequences of this region and the 3′-terminal position. The recent structural analysis of a DI particle which we refer to as DI-LT_2 has made that determination possible. DI-LT_2, like DI-LT, is derived from the 3′ portion of the parental genome and contains the four smaller mRNA cistrons.[13,14] The sequence analysis by Keene has shown that the 3′-terminus of the DI-LT_2 RNA is complementary to the 5′-terminus for 70 nucleotides. A complete leader RNA gene begins at position 72 from the 3′ end and is followed by the intercistronic sequence and the beginning of the N cistron (Figure 4). The presence of the defective particle terminal sequence, the internal leader RNA gene and the mRNA genes raised the possibility that DI-LT_2 particles might be capable of initiating RNA synthesis at its 3′-terminus as well as at the beginning of the leader RNA and mRNA genes. To distinguish between initiation at the beginning of the leader and the DI particle product genes, we have made use of the size of the larger RNase T_1 oligonucleotides that are characteristic of each RNA. Leader RNA yields a single large RNase T_1 oligonucleotide, a 28′mer,[10] while DI particle product yields two unique oligonucleotides, a 21′mer and 13′mer.[15] To minimize the contribution of large RNase T_1 oligonucleotides derived from other RNAs, the transcripts to be analysed were prepared with γ-thio-ATP and purified by affinity chromatography with mercury-agarose, digested with RNase T_1 and analysed by polyacrylamide gel electrophoresis (Figure 5). Labeled transcripts were prepared using approximately equal numbers of VSV, DI-LT_2 and DI-T particles. The VSV and DI-T served as a source of leader RNA and DI particle product RNA, respectively, for comparison. From the autoradiograms shown, it is clear that the leader and DI particle product RNAs are readily distinguishable, that DI-LT_2 synthesizes approximately as much DI particle product as the DI-T particles and that DI-LT_2 *does not* synthesize significant amounts of leader RNA.

We have analyzed DI-LT_2 and VSV transcripts prepared with normal ATP but labeled with ^{3}H-SAM and α- ^{32}P-CTP for the presence of methylated cap structures. Nuclease P_1 and calf intestinal phosphatase digests of such transcripts were separated by ionophoresis on thin layer sheets of PEI cellulose. VSV transcripts contained 1940 cpm ^{3}H in authentic cap I and II structures while transcripts from an equal number of DI-LT_2 particles contained only 65 cpm of ^{3}H label in caps. This comparison shows that the DI-LT_2 preparation synthesized only 3.5% as many methylated caps as did an equal number of VSV particles. Since the DI-LT_2 preparation has trace contaminants of both VSV (0.8%) and the linear DI-LT (3–4%), a small

Figure 4. Sequence of the 3′-terminal region of the DI-LT_2 genome.

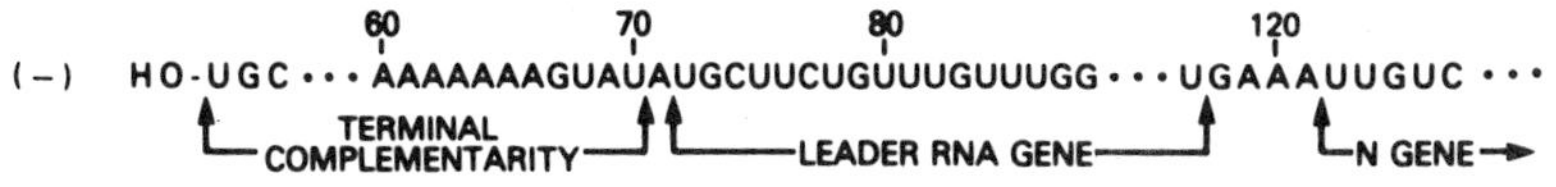

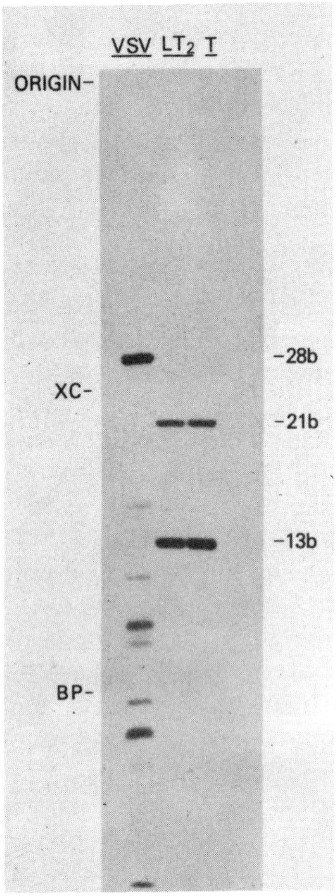

Figure 5. Acrylamide gel electrophoresis of the 5′-terminal RNase T_1 oligonucleotides from transcription products from DI-LT_2, DI-T and VSV.

amount of capped synthesis by the mixture is expected. We conclude from these experimetns that DI-LT_2 particles do not efficiently synthesize leader RNA or capped mRNAs *in vitro,* although its genome contains the genes for both types of transcripts. These results suggest that base sequence by itself does not specificy efficient initiation of the polymerase—there is an additional requirement that the sequence be located at the precise 3′-terminus of the template.

There are no clear indications as to why the polymerase efficiently initiates only at termini. However, it is possible that some structural feature of the terminus (i.e., a specific protein or a short stretch of RNA protruding beyond the first N-protein) may be indispensable. Alternatively, topological considerations may also be important. If transcription requires a perturbation of the nucleocapsid structure, such as partial displacement of the N-protein, it is possible that the perturbation may only be initiated at the 3′-terminus.

References

1. Banerjee, A.K., Colonno, R.J., Testa, D., and Franze-Fernandz, M.T. (1978): *J. Gen. Virol.* 34:1–8.
2. Testa, D., Chanda, P.K., and Banerjee, A.K. (1980): *Cell* 21:267–275.
3. Roy, P. and Bishop, D.H.L. (1973): *J. Virol.* 11:487–501.
4. Ball, L.A. and White, C.N. (1976): *Proc. Natl. Acad. Sci. U.S.A.* 73:442–446.
5. Abraham, G. and Banerjee, A.K. (1976): *Proc. Natl. Acad. Sci. U.S.A.* 73:1504–1508.
6. Reeve, A.E., Smith, M.M., Pigiet, V., and Huang, R.C.C. (1977): *Biochemistry* 16:4464–4469.
7. Smith, M.M., Reeve, A.E., and Huang, R.C.C. (1978): *Cell* 15:615–626.
8. Carrol, A.R. and Wagner, R.R. (1979): *J. Biol. Chem.* 254:9339–9341.
9. Rhodes, D.P. and Banerjee, A.K. (1976): *J. Virol.* 17:33–42.
10. Colonno, R.J. and Banerjee, A.K. (1978): *Cell* 15:93–101.
11. Abraham, G., Rhodes, D.F., and Banerjee, A.K. (1975): *Nature (London)* 255:37–40.
12. Gupta, K.C. and Roy, P. (1980): *J. Virol.* 33:292–303.
13. Epstein, D.A., Herman, R.C., Chien, I., and Lazzarini, R.A. (1980): *J. Virol.* 33:818–829.
14. Perrault, J. and Semler, B.L. (1979): *Proc. Natl. Acad. Sci. U.S.A.* 76:6191–6195.
15. Schubert, M., Keene, J.D., Lazzarini, R.A., and Emerson, S.U. (1978): *Cell* 15:103–112.

David H. L. Bishop and Richard W. Compans, eds.
The Replication of Negative Strand Viruses

Mode of Transcription and Replication of Vesicular Stomatitis Virus Genome RNA *in Vitro*

Amiya K. Banerjee, Pranab K. Chanda, and Sohel Talib [a]

Vesicular stomatitis virus (VSV)—a prototype of negative strand viruses [1] —serves as a model system to study transcription and replication of viral genome *in vitro*. The virion-associated RNA polymerase transcribes the linear single-stranded genome RNA *in vitro* into five mRNA species and a 47 nucleotide leader RNA representing the complementary sequences of the 3′-terminal end of the genome RNA.[2] The sum of the molecular weights of these RNA species correspond closely to the molecular weight of the genome RNA indicating almost complete transcription of the genetic material *in vitro*.[3] However, under these conditions the full-length plus strand of the genome RNA, the required intermediate of replication, is not detected although it has been identified in infected cells.[4] Thus, it appears that transcription of the full-length VSV genome RNA *in vitro* possibly requires modified experimental conditions or mediation by host or intracellular viral proteins.

The Transcription Process *in Vitro*

In order to understand better the regulatory mechanisms involved in the transcription-replication processes *in vitro*, it was important to study first the mechanism by which the mRNAs and the leader RNA were synthesized *in vitro* by the virion-associated RNA polymerase. In recent years, the following basic observations with regard to the *in vitro* transcription processes were

[a] Department of Cell Biology, Roche Institute of Molecular Biology, Nutley, New Jersey.

made: (a) the leader RNA is always produced as a unique length,[5] (b) all of the mRNA species are capped and only the α-phosphate of the penultimate adenosine base of the mRNAs is retained in the capped structure, $G_{(5')}\overset{\alpha\beta}{pp}\text{-}\overset{\alpha}{p}_{(5')}AACAG$, [6] (c) there is a differential molar abundance of the mRNA species synthesized *in vivo* [7] and *in vitro,*[8] in the order N>NS>M>G>L, (d) transcription mapping experiments involving inactivation by UV irradiation indicate that the VSV genome RNA behaves as a single transcription unit with individual mRNA species transcribed sequentially in the order 3′N-NS-M-G-L 5′.[7,9] Based essentially on these observations, a processing model for mRNA biosynthesis was proposed[2,10] in which initiation of RNA synthesis occurs at a single site at the 3′-end of the genome RNA. All five mRNA species and the leader RNA are cleaved from the growing RNA chain in a polar fashion, i.e., 3′ leader RNA-N-NS-M-G-L 5′.

Recent data however, indicate that the transcription process *in vitro* may be more complex than originally contemplated. In addition to the leader RNA, at least three distinct small RNA species were detected in the transcription reaction.[11] Two of the RNA species were identified at the 5′-terminal portions of N-mRNA (40 bases) and NS-mRNA (28 bases). These small RNA species contain an uncapped polyphosphorylated 5′-terminus and are continuously synthesized during transcription. In contrast, the polyadenylated mRNA species are synthesized sequentially in the order N-NS-M-G. Based on these observations a tentative model for VSV transcription has been proposed (Figure 1),[11] in which the termination of leader

Figure 1. A model for the biosynthesis of VSV mRNAs *in vitro.* (R) Regulatory element; (R *) modified regulatory element; (□) Cap; (—) Poly(A). For detailed discussion see text and reference 11.
Source: *Figure taken from reference 11 with permission.*

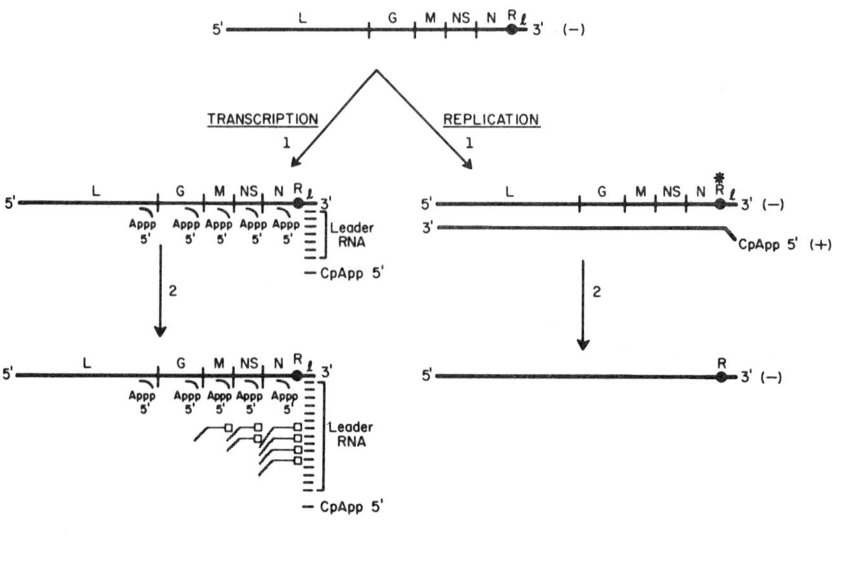

RNA at base 47 occurs by attenuation by a putative regulatory element (R) rather than by cleavage. The individual mRNA species are initiated independently at different promoter sites but the elongation and completion of the individual mRNAs depends on prior transcription of the 3′-proximal gene resulting in an overall polar effect.

Analyses of the Reaction Products of Incomplete Reaction *in Vitro*

In order to determine whether the transcription process is mediated by multiple initiation events, we analyzed the reaction products formed in incomplete reaction mixtures *in vitro*. Reactions were carried out in the presence of ATP, α- ^{32}P CTP, and GTP but without UTP. Based on sequence data of the leader RNA [5] and the mRNA species,[12] it is expected, according to the model, that the reaction products should include a leader fragment of 18 residues (5′-ACGAAGACAAACAAACCA), 5′-AACAG for N-mRNA, 5′-AACAGA for NS-mRNA and so on. The incomplete reaction products were treated with alkaline phosphatase (to release ^{32}Pi from α- ^{32}P CTP) and analyzed by polyacrylamide gel electrophoresis (Figure 2). The bands migrating faster than the XC dye were analyzed by various enzymatic digestions and confirmed to be 5′-fragments of leader RNA of sizes ranging from 18 to 8 bases (data not shown). The prominent band slightly above the BPB dye (Figure 2, +VSV lane) was confirmed to be uncapped 5′-AACAG and the band below it uncapped 5′-AACAGA (data not shown). These experiments indicate that the leader RNA and the mRNAs are synthesized by independent initiations at separate promoter sites. Moreover, the mRNA fragments do not contain the 5′-cap indicating that the capping event occurs subsequent to initiation and possibly after elongation of mRNAs to specified lengths.[11] Similar incomplete reactions using only ATP and α- ^{32}P CTP have shown that the only oligomers synthesized *in vitro* are polyphosphorylated AC and AACA (data not shown) indicating multiple initiations (see Figure 5).

Synthesis of Genome-Length Plus Strand *in Vitro*

The function of the leader RNA in relation to the transcription process is unclear. Is the synthesis of leader RNA obligatory for sequential synthesis of the mRNAs? Is its synthesis attenuated at template residue 47 by a regulatory element as suggested in the model (Figure 1)? It is apparent that suppression of this attenuation process is essential for the synthesis of full-length plus strand, the intermediate required for replication. How does the putative regulatory element control termination and antitermination of transcription (R to R*, Figure 1)? Phosphorylation seems to play a role in this process. This notion was suggested from our finding that the β, γ,

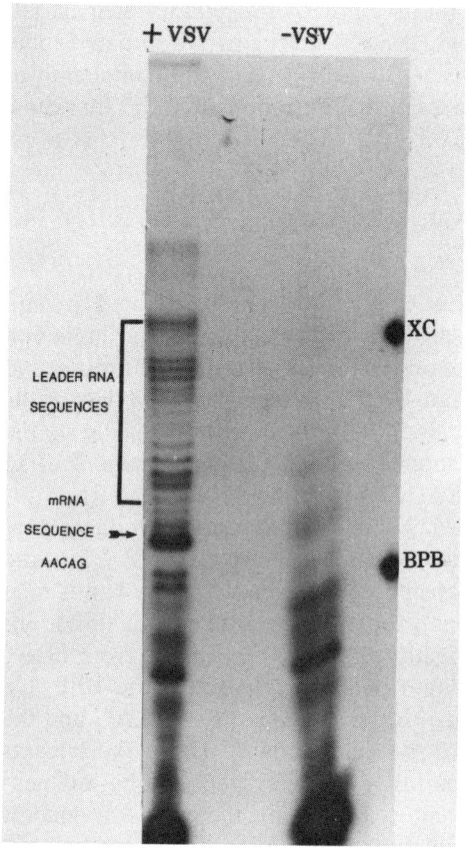

Figure 2. Analysis of the incomplete reaction products *in vitro*. RNA synthesis by Triton disrupted VSV was carried out in 100 μl standard reaction [5] containing 1 mM ATP, 30 μM CTP and GTP, 200 μCi CTP (400 Ci/mmole) and incubated at 30° C for two hours. The reaction products were purified by phenol extraction followed by ether extraction and treated with 50 units/ml of calf intestine alkaline phosphatase for one hour at 37° C. The digestion was analyzed by 20% polyacrylamide sequencing gel [5] and autoradiographed. SC and BPB represent the migration positions of xylene cyanole FF and bromophenol blue dyes, respectively. A similar reaction without VSV (−VSV) served as control.

imidoanalogue of ATP (AMPPNHP) failed to support transcription when substituted for ATP in the reaction mixture.[13] This analogue does not serve as a substrate for phosphatases and protein kinases. In contrast, the ribonucleoprotein (RNP) complex purified following transcription preinitiation with ATP and CTP, did support subsequent transcription in the presence of AMPPNHP (Figure 3). The radioactivity sedimenting at 42S region (Peak I) was confirmed to be 42S plus strand by various criteria.[14] These results,

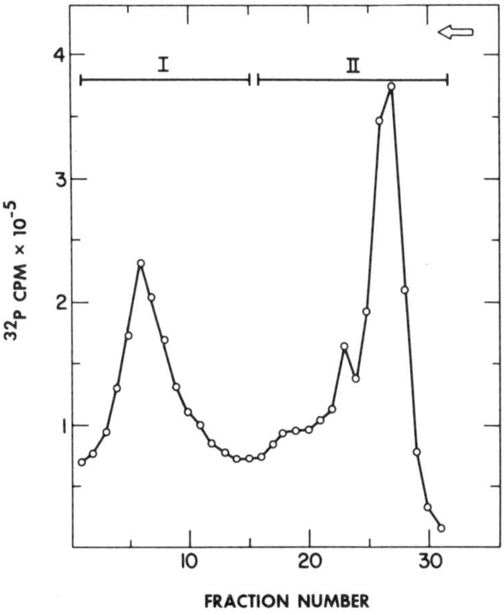

Figure 3. Synthesis of 42S plus strand *in vitro*. ^{32}P-UMP labeled RNA products synthesized by preinitiated RNP in the presence of AMPPNHP were analyzed by velocity sedimentation through a linear 15–30% sucrose gradient as described.[14] The RNA contained in Peak I represents the 42S plus strand.

while not confirmatory, lend support to the hypothesis that the phosphorylated state of a protein may be involved in the termination and antitermination process. In this respect, it is interesting to note that a different analogue of ATP, ATP-SH, is as effective as ATP in supporting VSV transcription *in vitro*.[15] This analogue, although not a substrate for alkaline phosphatase, is a donor for protein kinases.[16]

Effect of Protamine on *in Vitro* RNA Synthesis

The only phosphorylated protein in purified VSV is the NS-protein. Interestingly, it exists in different phosphorylated forms in purified virus and the degree of phosphorylation apparently determines its affinity for the RNA; the less phosphorylated form of NSI binds poorly to the RNP complex compared to its hyperphosphorylated counterpart (NS2).[17,18] The NS-protein is also a substrate for the protein kinase activity of unknown function that is associated with the RNP.[19,20] In an effort to understand the role of phosphorylation, if any, in the transcription/replication processes of VSV, we have studied the effect of adding proteins that are excellent phosphate acceptors to the *in vitro* transcription reaction mixture. Protamine, a low molecular weight highly basic protein of trout sperm,[21] has been found to be a highly–potent inhibitor of the virion-associated transcriptase. As shown in Figure 4A, at a concentration of 1 μg/ml (2×10^{-7}M) of protamine, RNA synthesis was virtually abolished. The same degree of inhibition was observed when protamine was added at different times after the onset of transcription (Figure 4B). The sizes of the mRNA products remained full-length during inhibition (data not shown), indicating that the inhibition occurred at or near the initiation site of the polymerase. As shown in Figure 5, in a partial reaction containing ATP and α- ^{32}P CTP, the formation of dinucleotides ApC (representing leader RNA) and AACA (representing the mRNAs) (Figure 5B) was inhibited more than 80% in the presence of protamine (Figure 5A). These results indicate that the effect of protamine is probably at the level of initiation of RNA synthesis. Next, we studied the effect of protamine on the endogenous protein kinase. As shown in Figure 6, the phosphorylation of the NS-protein was inhibited more than 80% in the presence of protamine (Figure 5A). These results indicate that the effect of protamine is probably at the level of initiation of RNA synthesis. Next, we studied the effect of protamine on the endogenous protein kinase. As shown in Figure 6, the phosphorylation of the NS-protein was inhibited more than 80% in the presence of protamine. It is interesting to note that a separate phosphorylated protein appears in the presence of protamine migrating

Figure 4. Inhibition of *in vitro* transcription by protamine. (A) Standard transcription reaction mixtures were set up [5] using ^{3}H-UTP as the labeled precursor. Protamine was added at different concentrations and acid-insoluble ^{3}H radioactivity was measured. (B) In separate experiments 1 μg/ml of protamine was added to standard reaction tubes at different times as indicated by arrows and RNA synthesis measured.

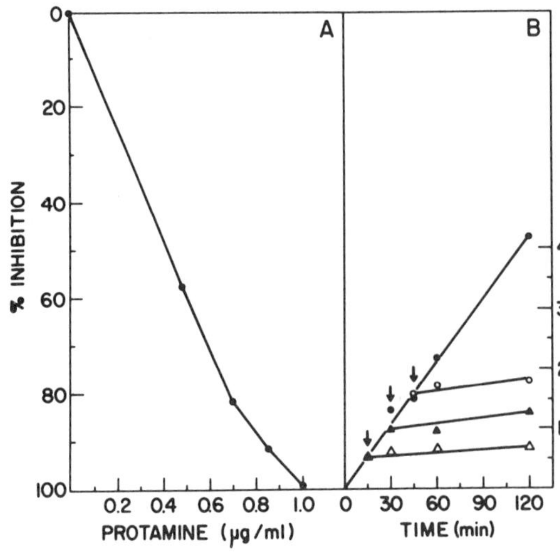

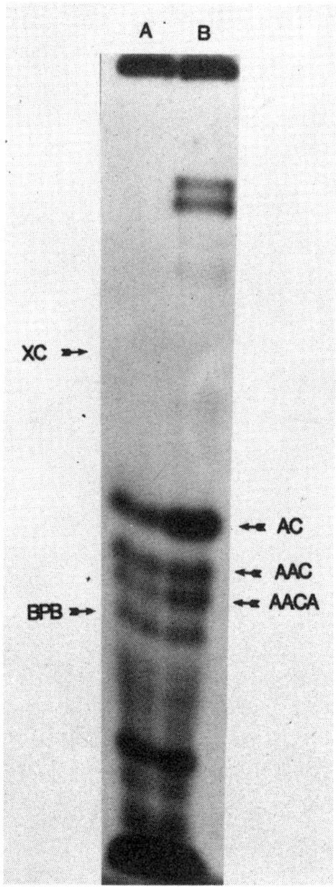

Figure 5. Effect of protamine on the initiation of RNA synthesis. RNA synthesis by Triton-disrupted virus was carried out in the presence of ATP (1 mM) and α- ^{32}P CTP (final specific activity = 1.4 × 10 ^{5} cpm/pmole) for two hours at 30° C similar to Figure 2 but containing 1 μg/ml of protamine (A) and in the absence of protamine (B). The reaction products were analyzed by polyacrylamide gel electrophoresis as described in Figure 2. The oligonucleotide sequence of the bands was determined after elution from the gel.

slightly faster than the major NS-protein band (Figure 6B, C, and D). It is unclear whether this band represents the less phosphorylated NS1 as described by other workers.[17,19] The above results suggest that inhibition of phosphorylation may be related to transcription. In order to study whether protamine has also an inhibitory effect on the replicative process *in vitro,* the RNP complex preinitiated with ATP and CTP was used to synthesize the full-length plus strand in the presence of protamine. As shown in Figure 7B, the synthesis of the 42S plus strand (chain length 10 to 12,000 bases) was inhibited by only 50% at a protamine concentration of 1 μg/ml, whereas the 12–18S mRNA (average chain length 1200 bases) synthesis was inhibited by more than 95% under the identical condition (Figure 7A). These results

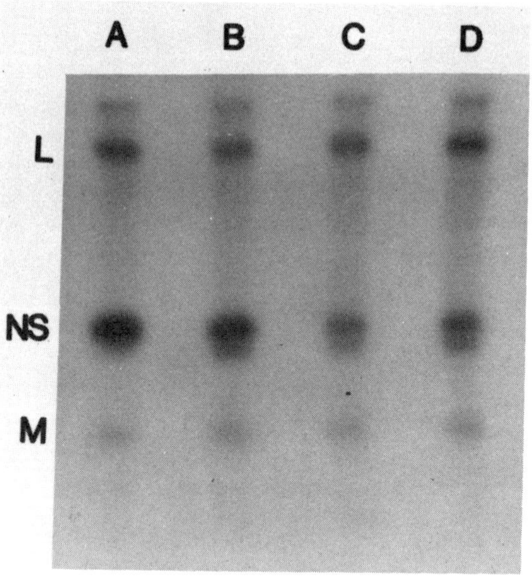

Figure 6. Effect of protamine on the endogenous protein kinase activity. VSV RNP (50 μg) was incubated in the complete reaction mixture (0.2 ml) in the absence (A) and in the presence of 0.25 μg/ml (B), 0.5 μg/ml (C); and 1 μg/ml (D) of protamine with 50 μM ATP and γ ^{32}P-ATP (final specific activity = 4000 cpm/pmole) for 15 minutes at 37° C. The phosphorylated proteins were analyzed by 10% polyacrylamide slab gel electrophoresis as described.[18] L, NS, M denote the migration positions of the VSV structural proteins.

indicate that the enzyme(s) involved in the replicative process may be less sensitive to protamine action than the enzyme(s) that constitutes the transcriptase.

Conclusions

The results presented in this paper indicate that the transcription of the VSV genome RNA *in vitro* occurs in a unique manner. Although there appear to be multiple sites on the genome RNA where initiation or RNA synthesis occurs, the elongation and completion of the RNA chains take place in a polar fashion yielding a sequential synthesis of mRNAs in the order N-NS-M-G-L. The mechanism by which the full-length plus strand is synthesized appears to be mediated by suppression of termination of the leader RNA which eliminates the polarity of transcription. This regulation may be mediated by a phosphorylation/dephosphorylation of a regulatory element (possibly a protein) since in the absence of a β-γ hydrolyzable bond of ATP in the reaction mixture, the full-length plus strand of the genome RNA can be

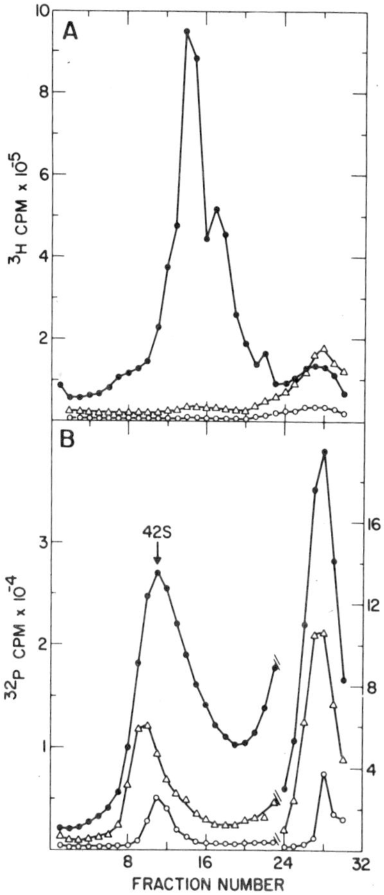

Figure 7. Effect of protamine on the synthesis of 42S plus strand *in vitro*. (A) RNA synthesis in the presence of protamine using ^{3}H-UTP as the labeled substrate. After incubation at 30° C for two hours, the mRNA synthesis was measured by velocity sedimentation in sucrose gradients under conditions as described previously.[3] (B) RNA synthesis was carried out in the presence of AMPPNHP using α- ^{32}P UTP as the labeled precursor as described previously.[14] The 42S plus strand synthesis in the presence of protamine was measured by velocity sedimentation in sucrose gradients as described.[14] No protamine, ●—●; 0.5 μg/ml protamine, Δ—Δ; 1 μg/ml protamine, ○—○.

synthesized. It was also shown that since protamine—a highly basic phosphate acceptor protein—inhibits transcription and protein kinase activity *in vitro*, the effect on the replicative process is less. It is not clear which of the viral proteins interacts with protamine during transcription and whether the dephosphorylated state of the protein(s) regulates transcription and replication. Further studies along this line will shed light into the precise role of phosphorylation, if any, in the VSV replicative processes.

References

1. Wagner, R.R. (1975): In: *Comprehensive Virology,* Vol. 4. Fraenkel-Conrat, H. and Wagner, R.R. (Eds.), Plenum Press, New York, pp. 1–80.
2. Banerjee, A.K., Abraham, G., and Colonno, R.J. (1977): *J. Gen. Virol.* 34:1–8.
3. Moyer, S.A. and Banerjee, A.K. (1975): *Cell* 4:37–43.
4. Wertz, G.W. (1978): *Virology* 85:274–285.
5. Colonno, R.J. and Banerjee, A.K. (1978): *Cell* 15:93–101.
6. Abraham, G., Rhodes, D.P., and Banerjee, A.K. (1975): *Cell* 5:51–58.
7. Abraham, G. and Banerjee, A.K. (1976): *Proc. Nat. Acad. Sci. U.S.A.* 73:1504–1508.
8. Villarreal, L.P., Breindl, M., and Holland, J.J. (1976): *Biochemistry* 15:1663–1667.
9. Ball, L.A. and White, C.N. (1976): *Proc. Nat. Acad. Sci. U.S.A.* 73:442–446.
10. Ball, L.A., White, C.N., and Collins, P.L. (1976): In: *Animal Virology, ICN-UCLA Symposia on Molecular and Cellular Biology,* Vol. 4. Baltimore, D., Huang, A.S., and Fox, C.F. (Eds.), Plenum Press, New York, pp. 419–438.
11. Testa, D., Chanda, P.K., and Banerjee, A.K. (1980): *Cell* 21:267–275.
12. Rhodes, D.P. and Banerjee, A.K. (1976): *J. Virol.* 17:33–42.
13. Testa, D. and Banerjee, A.K. (1979): *J. Biol, Chem.* 254:2053–2058.
14. Testa, D., Chanda, P.K., and Banerjee, A.K. (1980): *Proc. Nat. Acad. Sci. U.S.A.* 77:294–298.
15. Carroll, A.R. and Wagner, R.R. (1979): *J. Biol Chem.* 254:9339–9341.
16. Gratecos, D. and Fischer, E.H. (1974): *Biochem. Biophys. Res. Commun.* 58:960–967.
17. Clinton, G.M., Burge, B.W., and Huang, A.S. (1979): *Virology* 99:84–94.
18. Kingsford, L. and Emerson, S.U. (1980): *J. Virol.* 33:1097–1105.
19. Moyer, S.A., and Summers, D.F. (1974): *J. Virol.* 13:455–465.
20. Imblum, R.L. and Wagner R.R. (1974): *J. Virol.* 13:113–124.
21. Ando, T. and Watanabe, S. (1969): *Int. J. Protein Res.* 1:221–224.

David H.L. Bishop and Richard W. Compans, eds.
The Replication of Negative Strand Viruses

Initiation of Transcription by Vesicular Stomatitis Virus Occurs at Multiple Sites

Clayton E. Naeve and Donald F. Summers[a]

Several models have been proposed to account for the transcription of five mRNAs and a leader RNA from the 11,000 nucleotide genome of vesicular stomatitis virus (VSV). Transcription may proceed by (1) multiple simultaneous initiations at all possible sites, (2) multiple sequential initiations, and (3) single-site initiation with processing.[1] The first two models are similar in that both would require initiation at multiple sites; the last two are similar in that both would require initiation at the 3′-terminus of the genome prior to subsequent initiation or processing. The published information regarding multiple versus single site initiation is often contradictory. Recently, a fourth proposal has been offered by Testa et al.[2] in which initiation of transcription occurs simultaneously at multiple sites but that elongation occurs at variable rates such that completed mRNAs appear in a sequential order.

Our approach to this question was based on the recently available RNA sequence information of the leader RNA/N mRNA junction region[3–5] and all other mRNA/mRNA junction regions.[6] From this information (Figure 1), one can see that the first UTP requirement for synthesis of leader is 19 nucleotides in from the 5′ end of the transcribed RNA whereas the first UTP requirement for synthesis of all mRNAs is 6–9 nucleotides in from the 5′ end of uncapped mRNAs. Thus, in an *in vitro* transcription system lacking UTP, one might expect premature terminations to occur resulting in the synthesis of an 18 base oligonucleotide for molecules initiated at the 3′ end of the genome and 5–8 base oligonucleotides for internally initiated molecules.

[a] Department of Cellular, Viral and Molecular Biology, University of Utah College of Medicine, Salt Lake City, Utah.

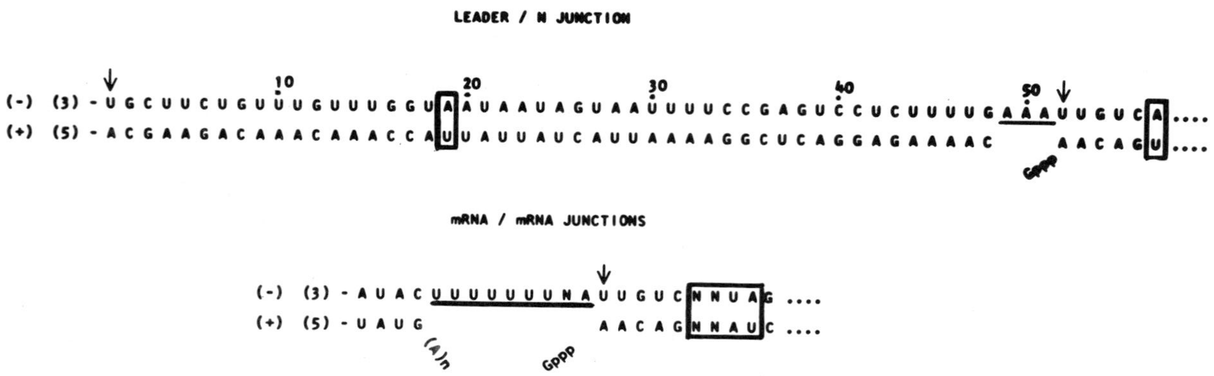

Figure 1. RNA sequence of the leader/N mRNA junction and all other mRNA/mRNA junction regions for VSV Indiana. The leader and leader/N mRNA junction sequences are from Colonno and Banerjee[3] and Rowlands.[5] The mRNA/mRNA junction sequences are from McGeoch[4] and Rose.[6] The boxed nucleotides represent the first UTP requirements during transcription from assumed initiation sites marked with arrows. The underlined nucleotides represent those present in the template but not in the product (intergenic regions).

Results

The RNAs made in an *in vitro* transcription reaction lacking UTP (-U) were examined by electrophoresis on 40 cm 20% acrylamide-7M urea gels[7] after pronase digestion (Figure 2A). Oligonucleotides 18–21 bases long were observed (compared to poly(A)-pCp ladders) in addition to a large amount of material remaining at the top of the gel. The multiple bands are not surprising since the VSV polymerase makes multiple leaders.[3] We suspected the material at the top of the gel to be 18–21 base oligonucleotides (prematurely terminated transcription product, PTTP) annealed to the vast excess of template. To test this suspicion, we phenol extracted the RNAs made in a parallel reaction and denatured with glyoxal-formamide[8] before applying the sample to the gel (Figure 2B). Virtually all of the material at the top of the gel disappeared. As a control, RNA synthesized in a standard reaction mixture was extracted and denatured in the same fashion and electrophoresed in the adjacent lane. In this case, leader and full length mRNA were made while the PTTP clearly was not made.

These results suggested to us that only PTTP is made in a -U reaction and not oligonucleotides 5–8 bases long. However, each preparation of -U RNA examined had activity at the bottom of the gel (Figure 2B). We initially assumed this activity was unincorporated (α-^{32}P) ATP remaining after several ethanol precipitations. This material, in fact, was shown to be 5–8 base oligonucleotides by removal of their 5′-terminal triphosphates with bacterial alkaline phosphatase (A.K. Banerjee, personal communication; unpublished observations). The 5–8 base oligonucleotides with 5′ triphosphates migrate faster than dinucleotides on these gels whereas the 18–21 base oligonucleotides with 5′ triphosphates migrate in the expected position. Thus the −U reaction produces oligonucleotides 18 bases long and 5–8 bases long suggesting initiation occurs at the 3′ end of the genome as well as internally.

Four different approaches were taken to determine the identity of PTTP (the 5–8 base oligonucleotides have not yet been examined), i.e., does the PTTP correspond to the first 18–21 bases of the leader RNA.

The first approach was to determine the base composition of the PTTP. A −U reaction was carried out in the presence of all four (α-^{32}P) NTPs. The concentration of labeled UTP (2.5×10^{-4} μmol) was well below the 30 μmol K_M for elongation so that only PTTP and 5–8 base oligonucleotides were made. The reaction mixture was phenol extracted, ethanol precipitated and placed on a 20% acrylamide-7 urea gel. All of the multiple PTTP bands were excised and eluted from the gel in a single pool, digested to completion with ribonuclease T2 and the resulting ribonucleoside monophosphates resolved on PEI TLC plates. The mean of three determinations is presented in Table 1. The measured base composition is very close to the predicted values for the first 18 bases of the leader.

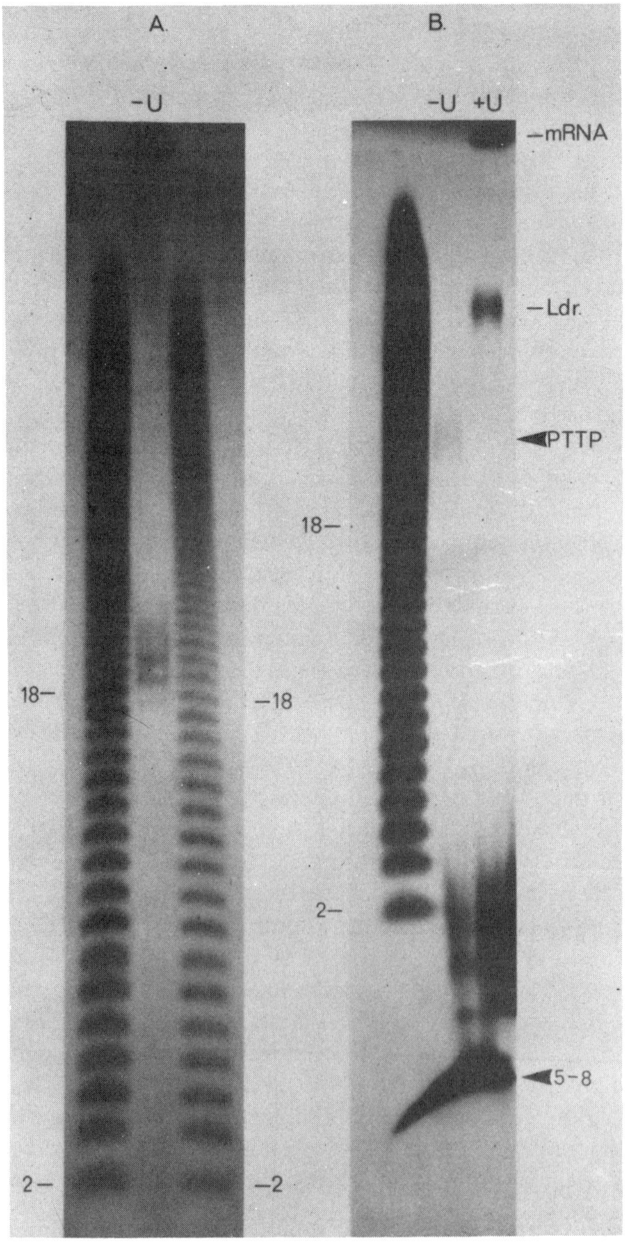
A.
B.
-U
-U +U
–mRNA
–Ldr.
PTTP
18–
–18
2–
–2
5–8

The second approach was to examine the ribonuclease T1 and A digestion products of (α-^{32}P) ATP labeled PTTP (Figure 3A). On the basis of the known sequence (Figure 1), one would predict RNase A digestion of the first 18 bases of the leader to produce fragments 1, 2, 4, and 6 bases long. The single base fragments would be obscured by unincorporated label on these gels and the position to which the dinucleotide would migrate in these gels is not clear due to its 5′ triphosphates. Therefore, only the 4 and 6 base fragments are of diagnostic value. The RNase A fragments of the PTTP are in the predicted size range (Figure 3A, lane A), i.e., 4 and 6 bases long. Likewise, one would predict RNase T1 to produce two fragments 3 bases long, one of which includes the 5′ triphosphates, and a fragment 12 bases long. If the variability in the PTTP length is at the 3′ end of the PTTP the 12 base fragment should also exhibit variability. As can be seen in Figure 3A (lane T1), RNase T1 generates a fragment 12–15 bases long showing variable length and a fragment which seems to comigrate with the putative 4 base RNase A fragment. The essential point of this experiment is that RNase A and T1 generate fragments in the approximate size range of those predicted for the first 18 bases of the leader and more importantly the fragments generated are discrete. If the PTTP were in fact a variety of initiated *in vitro* transcription products, one might expect a heterogeneous population of RNase generated fragments. This data is consistent with the PTTP being the first 18 bases of the leader with variability at the 3′-termini.

Figure 2. Polyacrylamide gel of oligonucleotides made in transcription reactions lacking UTP. Reaction mixtures of 200 μl contained RNP (0.75–1.0 mg protein); 50 mM Tris-HCl, pH 8.0; 5 mM $MgCl_2$; 100 mM NaCl; 10 mM S-adenosyl-L-methionine; 5 mM dithiothreitol; 0.05% Triton N101; 1 mM CTP, GTP and 500 μM unlabeled ATP and 250 μCi of labeled ATP. The transcription mixtures were incubated at 32°C for various times.

Nucleoside triphosphates (Sigma Chemical Co.) were assayed for purity by thin layer chromatography on Polygram CEL 300 PEI plates (Brinkman Instruments, Inc.) as described by Randerath and Randerath.[9]

The entire 200 μl volume was treated with pronase (panel A) and applied to a 20% acrylamide-7M urea gel[7] center lane). Carbonate hydrolyzed 3′ pCp end-labeled[10] poly(A) (10,000 Cerenkov cpm) was used as oligonucleotide size markers (A, outside lanes). The RNA made in a parallel −U reaction was phenol extracted and denatured with glyoxal-formamide[8] and applied to a 20% gel (B, center lane). A standard complete reaction (+U) was processed in the same fashion and coelectrophoresed (B, right lane) along with poly(A)-pCp marker (B, left lane). mRNA refers to the material remaining at the top of the 20% gel (shown to be nascent and completed mRNA), Ldr refers to leader RNA, PTTP refers to the 18 base oligonucleotide called the "prematurely terminated transcription product" and 5–8 refers to the length of small oligonucleotides migrating aberrantly on these gels. The remaining numbers refer to poly(A) marker nucleotide length.

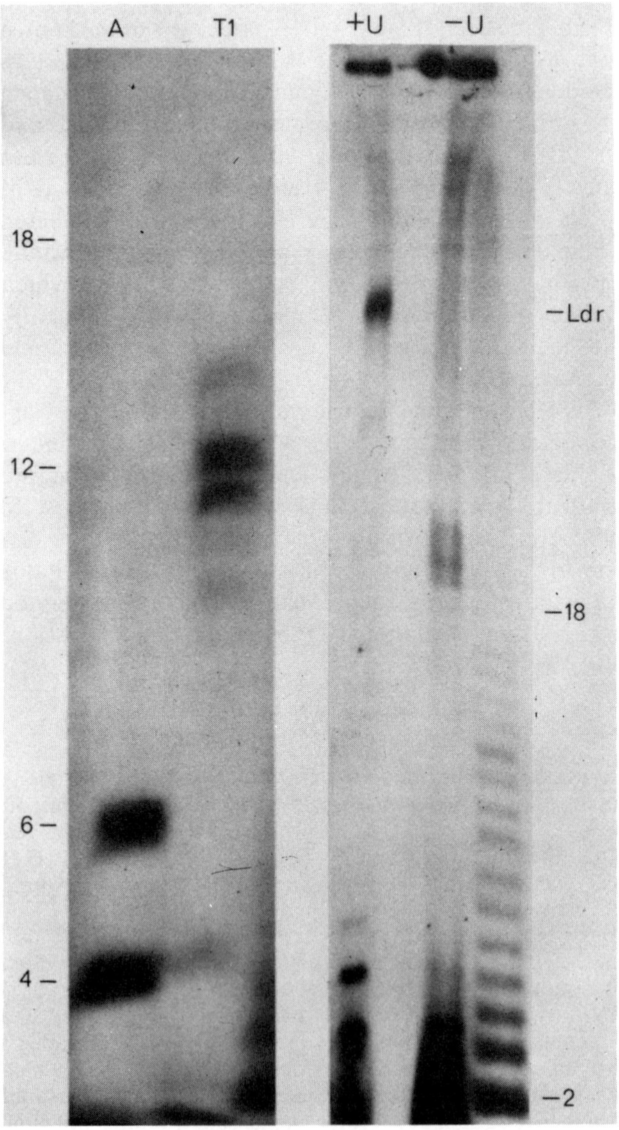

Figure 3. Polyacrylamide gel of RNase A and T1 treated PTTP and (β-^{32}P) ATP labeled products. For ribonuclease A digestion, ethanol precipitated RNA (containing 40 μg yeast RNA carrier) was collected by centrifugation for 5 min in an Eppendorf microfuge and brought to a final concentration of 12 mM Tris-HC1 (ph 7.4), 25 mM NaC1, 1mM EDTA, and 5 μg ribonuclease A (Calbiochem-Behring Corp.) in a volume of 15 μl. The sample was incubated at 37°C for 30 min then diluted with an equal volume of sample buffer, heated to 50°C for 3 min and applied directly on a 20% acrylamide gel (A, left lane). For ribonuclease T1, RNA samples were brought to a

Table 1. PTTP Base Composition.

	Calculated		Measured	
	#	%	#	%
G	2	12	2.5	14
A	10	58	9.7	54
C	5	29	5.5	31
U	0	0	0	0

The base composition of the PTTP was determined by phenol extracting the RNA made in a −U reaction and resolving the RNAs made on a 20% acrylamide-7M urea gel. The PTTP region of the gel was excised and the multiple bands eluted in 0.5M ammonium acetate, 10 mM $MgCl_2$, 0.1 mM EDTA and 0.5% SDS at 37°C overnight. The eluted RNA was precipitated with ETOH, resuspended in 0.1 M ammonium acetate pH 4.5, and digested to completion with RNase T2 (200 U/ml) for 2 hr at 37°C. Samples were applied directly to PEI cellulose thin layer chromatography plates (Brinkman Instruments, Inc.), along with nucleoside 5′-monophosphate markers, washed in methanol to remove buffer salts and developed to 4 cm above the origin with 1 N acetic acid then to 15 cm above the origin with 0.33 M LiCl.[11] The spots were cut out and assayed by Cerenkov counting.

The number of nucleotides calculated and measured is presented along with the percentage of the total counts each comprises.

A third experiment designed to identify the origin of the PTTP was based on the use of (β-^{32}P) ATP as label. Colonno and Banerjee[3] had shown that (β-^{32}P) ATP labels only the leader RNA since all mRNAs lose the β and γ phosphate of the 5′ ATP during the capping reaction. The incorporation of the β-^{32}P label into PTTP should reveal whether or not the PTTP is an initiated molecule. The results of such an experiment are shown in Figure 3B. In the presence of all four nucleoside triphosphates (+U lane) the major labeled species is the leader RNA. In the absence of UTP (Figure 3B, −U lane), leader is not made while the PTTP is made and is clearly labeled with (β-^{32}P) ATP. The material at the top of the gel was shown to be leader and PTTP annealed to excess genome by glyoxal-formamide denaturation and electrophoresis on a 20% acrylamide-7M urea gel (data not shown). Since the leader and the PTTP are both labeled with (β-^{32}P) ATP, we suggest both are initiated molecules.

On the basis of size, base composition, ribonuclease A and T1 products and (β-^{32}P) ATP incorporation, we conclude the PTTP corresponds to the first 18 nucleotides of the leader RNA.

---

final concentration as above but containing 0.5U ribonuclease T1 (Calbiochem-Behring Corp.) in a volume of 15 μl. The sample was incubated qt 37°C for 30 min, diluted with sample buffer, heated, and applied to the acrylamide gel (A, right lane).

PTTP RNA synthesized in a −U reaction in the presence of (β-^{32}P) ATP (B, −U) was compared to a standard complete control reaction (B, +U) with poly(A)-pCp ladders as markers (B, right lane) by electrophoresis on 20% gels.

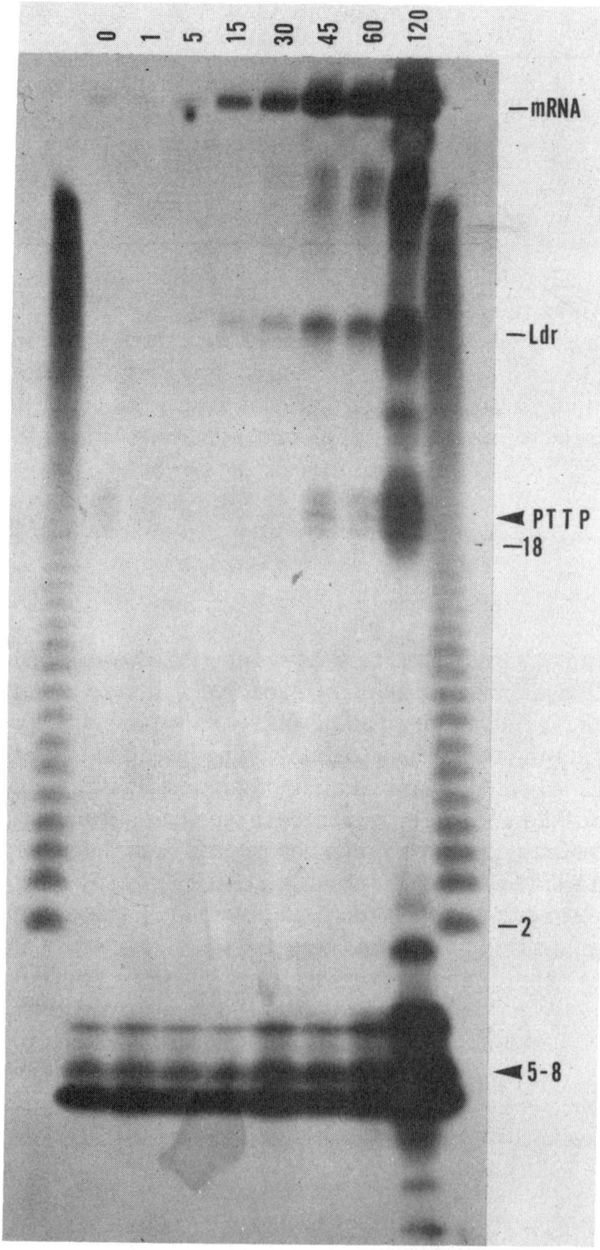

Figure 4. Polyacrylamide gel of oligonucleotides made in a −U reaction at various times after the addition of U. *In vitro* transcription was carried out −U for 30 min at which time U was added and samples taken at intervals (0, 1, 5, 15, 30, 45, 60, and 120 min). The samples were phenol extracted and electrophoresed on a 20% acrylamide gel along with poly(A)-pCp ladders (outside lanes). The labels are as described in the legend to Figure 2.

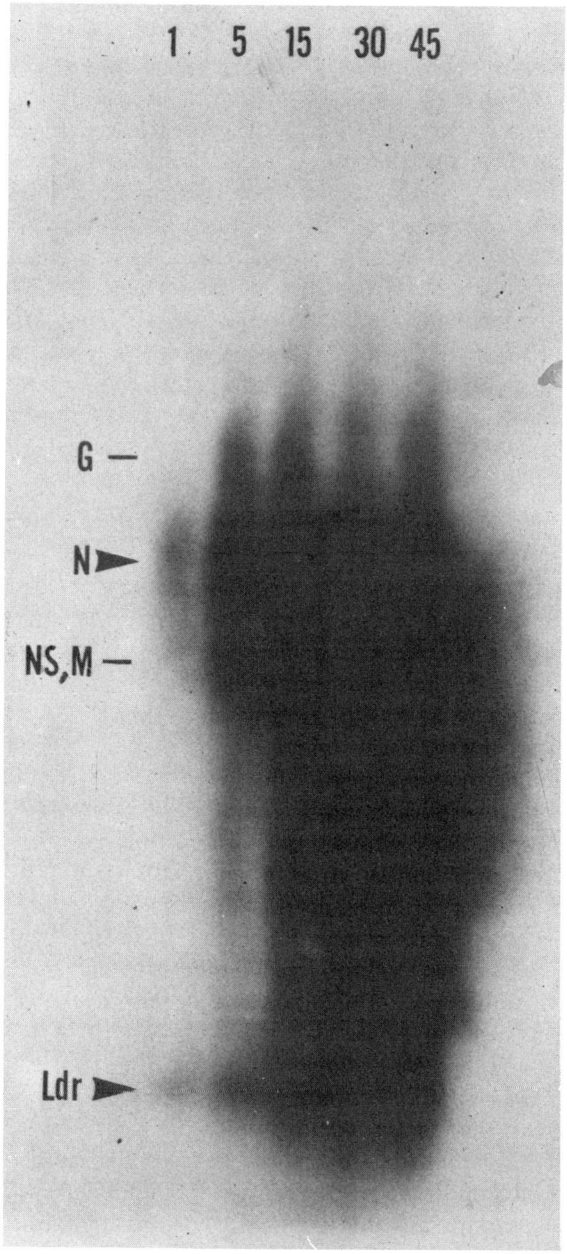

Figure 5. Agarose-formaldehyde gel[12] of aliquots of the same samples analysed in Figure 4 (1, 5, 15, 30, and 45 min after U addition to a −U reaction). L mRNA was not made in the reaction while G, N, NS, and M were and are labeled as such. The leader RNA (Ldr) runs near the gel front.

The fourth approach to determine the identity of the PTTP was an attempt to chase the PTTP into leader by the addition of either UTP alone or UTP plus unlabeled ATP to "preinitiated" ribonucleoproteins. This approach might be expected to demonstrate elongation of the PTTP into leader as well as a sequential or synchronous appearance of subsequent mRNAs. Transcription reactions (−U) were carried out as previously described, after 30 min at 32°C (time zero) U was added and samples removed at various intervals. The samples were phenol extracted, ethanol precipitated and applied to a 20% acrylamide-7M urea gel (Figure 4). It is immediately apparent that neither the PTTP nor the 5–8 base oligonucleotides decline with time after addition of U nor after addition of U and excess unlabeled A (data not shown); thus the PTTP and 5–8 base oligonucleotides are not elongated. Examination of the RNAs found in a reaction mixture from which template had been removed by centrifugation revealed that all the products made are found in the supernatant and are therefore released from template (data not shown). The larger RNAs made in this reaction were analyzed on agarose-formaldehyde gels (Figure 5); the leader and N mRNA clearly precedes synthesis of the remaining mRNAs. Thus, the oligonucleotides made during "preinitiation" are not elongated when U is supplied to the reaction mixture, yet the leader and mRNAs appear in a sequential manner.

Discussion

Since the products made in a −U reaction include the predicted 18 base oligonucleotide and 5–8 base oligonucleotides, we suggest initiation occurs at the 3′ end of the genome template and internally as well. The products of a −U reaction are released from the template after synthesis, so that when U is supplied to the reaction synthesis of the leader RNA and mRNAs proceeds normally, i.e., sequentially. These results may support the model of Testa et al.,[2] which suggests initiation occurs simultaneously at multiple sites but that elongation of each RNA proceeds at varying rates so that the order of completion is sequential. However, it is not yet clear whether the 5–8 base oligonucleotides represent internal initiations at all possible sites or only at the start of N mRNA.

References

1. Banerjee, A.K., Abraham, G., and Colonno, R.J. (1977): *J. Gen. Virol.* 34:1–8.
2. Testa, D., Chanda, P.K., and Banerjee, A.K. (1980): *Cell* 21:267–275.
3. Colonno, R.J. and Banerjee, A.K. (1978): *Cell* 15:98–101.
4. McGeoch, D.J. (1979): *Cell* 17:673–681.
5. Rowlands, D.J. (1979): *Proc. Natl. Acad. Sci. U.S.A.* 76:4793–4797.
6. Rose, J.K. (1980): *Cell* 19:415–421.

7. Donis-Keller, H., Maxam, A.M., and Gilbert, W. (1977): *Nuc. Acids Res.* 4:2527–2538.

8. Cech, T.R. and Pardue, M. (1976): *Proc. Natl. Acad. Sci. U.S.A.* 73:2644–2648.

9. Randerath, K. and Randerath, E. (1964): *J. Chromatog.* 16:111–125.

10. England, T.E., Gumport, R.I., and Uhlenbeck, O.C. (1977): *Proc. Natl. Acad. Sci. U.S.A.* 74:4839–4842.

11. Randerath, K. and Randerath, E. (1967): In: *Methods in Enzymology*. Grossman, L. and Moldave, K. (Eds.), Academic Press, New York, Vol. XII, pp. 331–347.

12. Lehrach, H., Diamond, D., Wozney, J.M., and Boedtker, H. (1977): *Biochemistry* 16:4743–4751.

David H.L. Bishop and Richard W. Compans, eds.
The Replication of Negative Strand Viruses

Replication and Assembly of Vesicular Stomatitis Virus Nucleocapsids *in Vitro*

V.M. Hill, L.L. Marnell, and D.F. Summers [a]

VSV RNA synthesis is known to be comprised of transcription, which results in the synthesis of five VSV specific mRNAs, and a short, leader sequence, and replication which results in the synthesis of virion-length positive and negative sense RNA molecules which have been identified in infected cells as RNP particles or complexes comprised of the RNA replication products plus VSV specific proteins, L, NS, and N.[1,2] The virion-associated transcriptase, which has been shown in numerous studies to function very efficiently *in vitro,* has been found to require both NS and L for enzyme activity,[3,4] and utilizes RNA in a RNP complex as template. It is assumed that the VSV specific replicase is also composed of VSV proteins, L and NS, and likewise utilizes a ribonucleoprotein (RNP) complex containing positive or negative strand RNA as template.[5,6] Replication has been shown to require ongoing protein synthesis.[5] In addition, host factors also appear to be required for replication because of the occurrence of viral host range mutants[7,8] and replication-restricted host cell lines.[9]

There have been only two reports of the reproducible synthesis of full-length RNA *in vitro.*[10,11] The *in vitro* system of Testa et al.[11] used the β, γ imido analogue of ATP to synthesize full-length positive-strand RNA. It is likely that this synthesis is an aberrant transcription reaction since only the transcriptase is present and it is not coupled with protein synthesis or nucleocapsid assembly. The *in vitro* system of Batt-Humphries et al.[10] more closely resembled *in vivo* conditions. Using a VSV-infected cell supernatant

[a] Department of Cellular, Viral and Molecular Biology, University of Utah College of Medicine, Salt Lake City, Utah.

fraction (S-10) they were able to show the production of 42S RNA in addition to the 5 VSV mRNAs. This system had the additional *in vivo* characteristic of being dependent on protein synthesis for replication.[10]

In this paper, we show that a VSV-infected S-10 optimized for VSV RNA synthesis and translation not only synthesizes virion-length VSV RNA, but that this RNA is assembled into nucleocapsids.

Results

We first examined the kinetics of synthesis of total VSV specific RNA and proteins and the assembly of 42S RNA and proteins into nucleocapsids. The *in vitro* system used was an S-10 fraction from VSV-infected HeLa cells harvested at 4–5 hr postinfection. This system was derived from the transcriptional-translational system of Batt-Humphries et al.[10] and modified to optimize for RNA and protein synthesis. Figure 1A shows the incorpora-

Figure 1. Total RNA and protein synthesis in a VSV-infected HeLa cell supernatant. The cytoplasmic supernatant fraction (S-10) was prepared as described previously[10] except the cell pellet was resuspended in an equal volume of PS buffer (10 Mm KC1, 1.5 mM magnesium acetate, 20 mM Hepes, pH 7.4, 0.5 mM dithiothreitol). Aliquots of the S-10 were added to an equal volume of reaction mixture containing 60 mM Hepes (pH 7.4), 120 mM KC1, 6 mM magnesium acetate, 2 mM dithiothreitol, 50 mM mixed amino acids, 2 mM spermidine, 2 mM GTP and ATP, 0.2 mM UTP (labeling with ^{35}S-methionine) or no cold UTP (labeling with α-^{32}P UTP), 20 mM creatine phosphate, and creatine phosphokinase (80 μg/ml). TCA-precipitable counts were determined at intervals throughout the reaction. (—●—) average. (——) range. (A) RNA labeled with α-^{32}P UTP; (B) proteins labeled with ^{35}S-methionine.

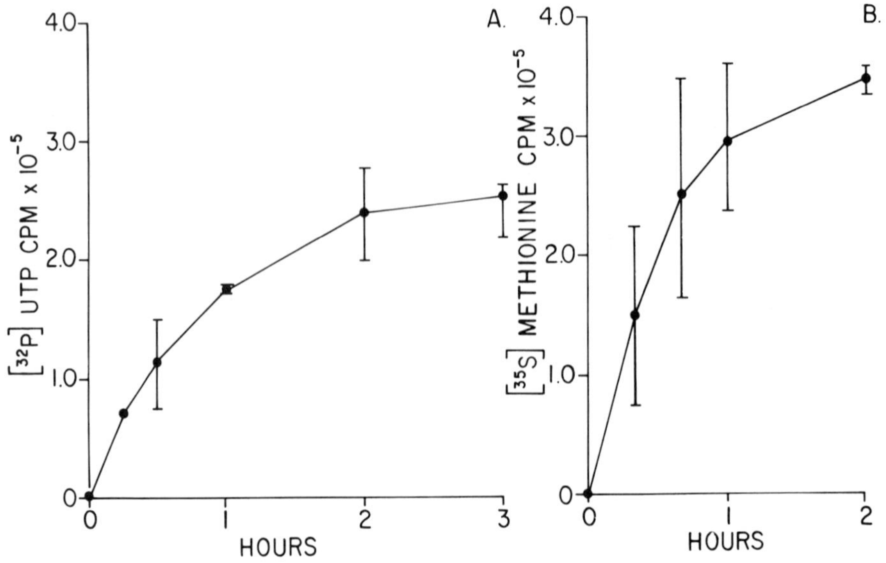

tion of α-^{32}P UTP into total VSV specific RNA over a 3-hr period. Transcription represents the majority (98%) of RNA synthesis in this system and therefore the kinetics of replication cannot be deduced from this figure. Figure 1B shows the *in vitro* incorporation of ^{35}S-methionine into VSV specific proteins. The data show that proteins are synthesized very rapidly during the first hour of the reaction after which synthesis declines considerably.

Since the examination of total RNA synthesis does not provide any information about replication, we examined the kinetics of assembly of newly synthesized proteins and 42S RNA into nucleocapsids. Full-length RNA is found only in nucleocapsids *in vivo* [12,13] and so nucleocapsid synthesis is a direct reflection of replication. The nucleocapsids, synthesized *in vitro* were labeled with either α-^{32}P UTP (Figure 2A) or ^{35}S-methionine (Figure 2B), treated with pancreatic RNase and isolated twice in CsC1 gradients. The nucleocapsids were RNAase resistant and had the same density as that of native virion nucleocapsids, approximately 1.3 gm/cc (data not shown). The

Figure 2. Assembly of nucleocapsids *in vitro*. Nucleocapsids were synthesized in the *in vitro* S-10 described in Figure 1. Aliquots taken at time intervals during the reaction were treated with pancreatic RNase (30 μg/ml) and the nucleocapsids isolated by CsC1 equilibrium centrifugation, as described previously.[12] The amount of radioactivity in α-^{32}P UTP-labeled nucleocapsids was determined directly by Cerenkov counts (A). TCA precipitable counts in ^{35}S-methionine-labeled nucleocapsids were determined (B).

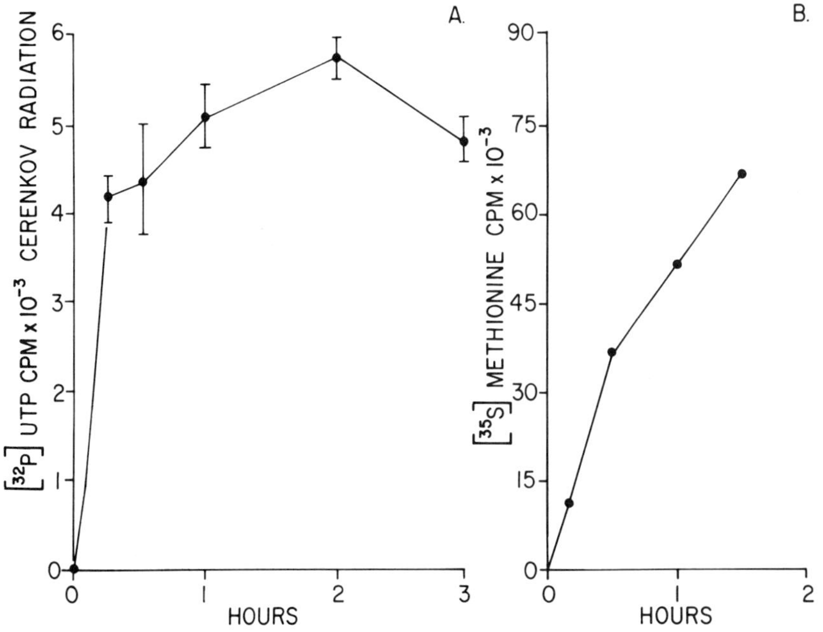

data show that the newly replicated RNA molecules were assembled into nucleocapsids very rapidly in the first 15 min of the reaction, after which assembly occurred at a slower rate. This was in contrast to total RNA synthesis (Figure 1A) which was linear for 1 hr and continued at a reduced rate for 1 and 2 hr more.

Assembly of proteins into nucleocapsids (Figure 2B) occurred at a slower rate for a longer period of time. Recent indirect experiments discussed below suggest that nucleocapsid assembly can occur using pools containing both presynthesized unlabeled proteins and newly synthesized labeled proteins, while RNA is assembled very shortly after it is synthesized. This may account for the apparent slower rate of protein incorporation into nucleocapsids (Figure 2B).

The *in vitro* assembled nucleocapsids were further characterized and compared to virion nucleocapsids. Figure 3 shows the profile of a CsCl gradient containing α-^{32}P UTP-labeled nucleocapsids synthesized *in vitro* and ^{3}H-uridine-labeled virion nucleocapsids. The data show that they both band approximately at the same density (1.3gm/cc).

To establish that the RNA product of *in vitro* replication was similar to that previously characterized following *in vivo* synthesis,[2,13] the CsCl isolated nucleocapsids were collected and the RNA contained therein was analyzed on denaturing gels. Figure 4 shows that the predominant species isolated was 42S RNA. There was a small amount of message size RNA which banded in CsCl, however the small RNA was sensitive to RNase (lane B). Therefore *in vitro* the only RNA assembled into nucleocapsids was 42S RNA. This is identical to the results of previous *in vivo* studies.[12,13] Preliminary results have shown that the RNA assembled into nucleocapsids is predominantly negative strand RNA (data not shown).

Our previous *in vivo* studies of the VSV replicative process had shown that newly synthesized 42S RNA becomes rapidly associated with all 3 nucleocapsid proteins (L, NS, and N). This association seems to occur during the process of RNA chain elongation, since all newly synthesized replicative RNA is RNase resistant and can be shown to be in association with all 3 nucleocapsid proteins.[5] Therefore, we examined the *in vitro* product to see if it is assembled into nucleocapsids, as is found *in vivo*. Figure 5A shows a CsCl gradient of the proteins radiolabeled *in vitro* with ^{35}S-methionine cosedimented with ^{3}H-uridine-labeled virion nucleocapsids. This gradient separates the nucleocapsids (fractions 11–15) from a broad band of VSV specific proteins (fractions 15–27). Fractions from each peak were pooled and the proteins analyzed on 10% polyacrylamide gels (Figure 5B). The only VSV proteins cosedimenting with the nucleocapsids was N-protein. Since the other VSV-proteins are dissociated from nucleocapsids in CsCl,[14] this technique could not be used to determine the initial composition of *in vitro* assembled nucleocapsids. The lighter heterogeneous peak contained four VSV-proteins and no RNA.

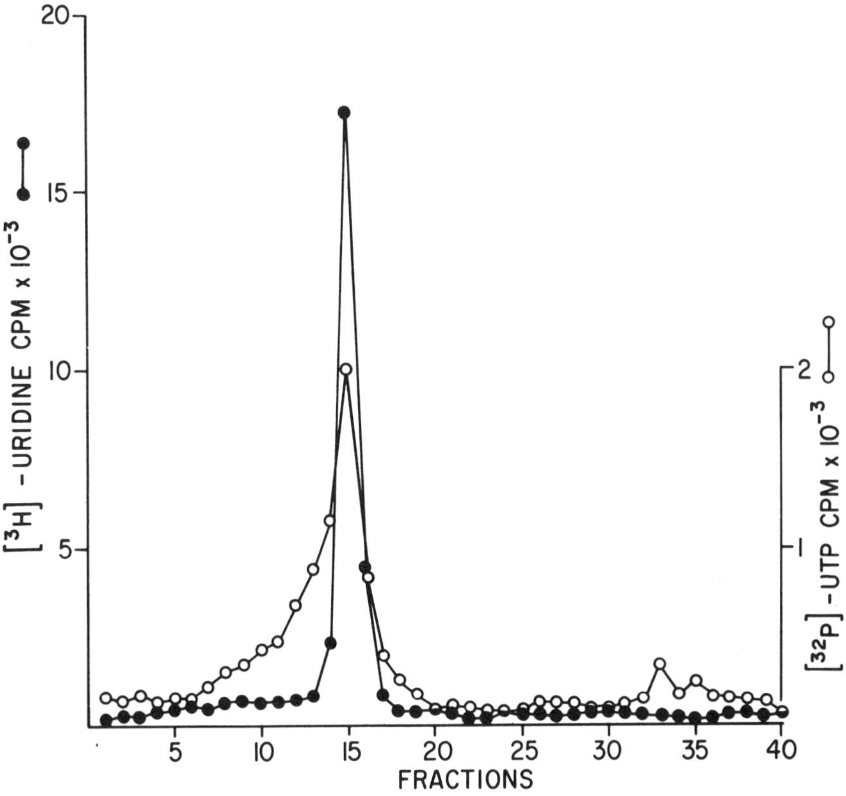

Figure 3. RNA products of *in vitro* synthesis separated by CsCl centrifugation. Nucleocapsids, made *in vitro,* were treated with 1% DOC and isolated on a preformed 20–40% CsCl gradient, as described in Figure 2, except they were not treated with RNase, mixed with marker virion RNPs, and cosedimented on a 20–35% CsCl gradient. TCA-precipitable counts of aliquots of the gradient fractions. (○--○) α-^{32}P UTP-labeled *in vitro* nucleocapsids. (●--●) ^{3}H-uridine-labeled virion RNPs.

Since Wertz and Levine[15] and Rubio et al.[5] have shown that inhibition of protein synthesis in infected cells caused the cessation of replication, we utilized pactamycin to inhibit VSV specific protein synthesis *in vitro*. These studies (data not shown) have shown that continued synthesis of VSV-proteins is not required for the assembly of newly synthesized RNA into RNase resistant complexes with the density of VSV RNP particles. This indicates that the S-10 preparation used in these studies contained pools of presynthesized VSV nucleocapsid proteins sufficient to permit synthesis and assembly of VSV RNA into nucleocapsids for about 1 hr. Our previous studies have indicated that in infected cells protein pools are sufficient for replication to occur for about 30 min.[5] Because the S-10 is a very concen-

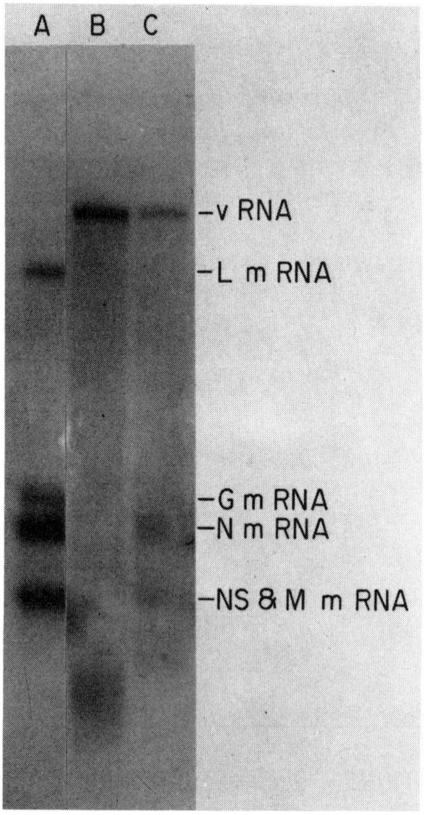

Figure 4. RNA assembled into nucleocapsids *in vitro*. An *in vitro* reaction mixture was either treated with micrococcal nuclease for 15 min at 18°C, and then inactivated with EGTA or untreated. The nucleocapsids were then isolated, by twice banding in CsCl and then were phenol-extracted. The ^{32}P UTP labeled RNA was analyzed on 1% agarose gels in 1.2 M formaldehyde, 0.05 M borate buffer.[16] Lane A—VSV mRNA markers isolated from the pellet of a CsCl gradient; lane B—micrococcal nuclease treated; lane C—untreated.

trated cell supernatant, it apparently has sufficient protein pools that allow replication to proceed at least 1 hr in the absence of protein synthesis.

Discussion

In this paper, we have described an *in vitro* system, optimized for VSV RNA and protein synthesis, which replicates VSV 42S RNA. We have shown that replication in this system is very similar to *in vivo* replication in that it makes full-length RNA and this RNA is predominantly (70%) negative strand. In

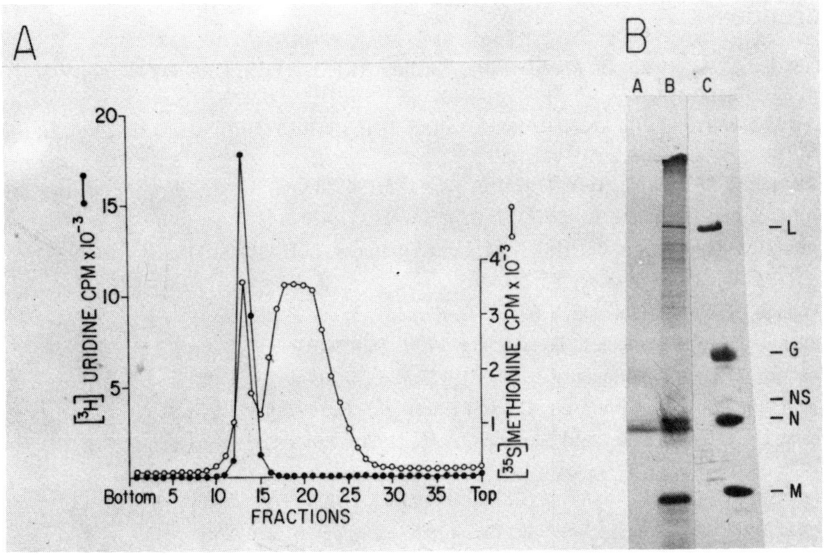

Figure 5. Protein products of *in vitro* synthesis as separated by CsC1 equilibrium centrifugation. The heterogeneous peak, isolated on a 20–40% CsC1 gradient, as described in Figure 3, was mixed with marker virion RNPs and reisolated on a 20–35% CsC1 gradient. (A) TCA precipitable counts of aliquots of the gradient fractions; (○--○) ^{35}S-methionine *in vitro* labeled proteins; (●--●) ^{3}H-uridine labeled virion nucleocapsids. (B) 10% polyacrylamide gel[5] of proteins pooled from the CsC1 gradient. Lane A—fractions 11–15; lane B—fractions 15–27; lane C—VSV virion markers.

addition, we have shown that the 42S RNA is assembled into nucleocapsids containing N-protein and these *in vitro* assembled nucleocapsids are the same density as *in vivo* nucleocapsids. This assembly occurs mostly within the first 30 min *in vitro,* after which there is more protracted synthesis and assembly of RNP-proteins into nucleocapsids.

Ongoing viral protein synthesis has been shown to be necessary for replication *in vivo.* [5,15] We have previously shown, by using either pulse-chase experiments or an inhibitor of protein synthesis, that there is a 30-min pool of proteins available for replication *in vivo.*[5] By using pactomycin to inhibit protein synthesis in the infected S-10, we have found that it also has a large protein pool available for replication.

Now that we have done the preliminary characterization of an *in vitro* replication system, we will use it to investigate replication further. Some of the areas we would like to study are the following: the requirement of NS and L for assembly, the initiation of replicative RNA strands—does it occur and what are the requirements; the host factors involved in replication and further definition of the system for replication.

References

1. Banerjee, A.A. (1980): In: *Rhabdoviruses.* Bishop, D.H.L. (Ed.), CRC Press, Inc., Vol. II, pp. 35–50.
2. Wertz, G.W. (1980): In: *Rhabdoviruses.* Bishop, D.H.L. (Ed.), CRC Press, Inc., Vol. II, pp. 75–94.
3. Emerson, S.U. and Yu, H-Y. (1975): *J. Virol.* 15:1348–1356.
4. Naito, S. and Ishihama, A. (1976): *J. Biol. Chem.* 251:4307–4314.
5. Rubio, C., Kolakofsky, C., Hill, V.M., and Summers, D.F. (1980): *Virology* 105:123–135.
6. Wertz, G.W. (1978): *Virology* 85:271–285.
7. Simpson, R.W. and Obijeski, J.F. (1974): *Virology,* 357–368.
8. Szilagyi, J.F. and Pringle, C.R. (1975): *J. Virol.* 16:927–936.
9. Thacore, H.R. and Younger, J.S. (1973): *Virology* 56:505–511.
10. Batt-Humphries, S., Simonsen, C., and Ehrenfeld, E. (1979): *Virology* 96:88–99.
11. Testa, D., Chanda, P.K., and Banerjee, A.K. (1980): *Proc. Natl. Acad. Sci. U.S.A.* 77:294–298.
12. Simonsen, C., Batt-Humphries, S., and Summers, D.F. (1979): *J. Virol.* 31:124–132.
13. Soria, M., Little, S.P., and Huang, A.S. (1974): *Virology* 61:270–280.
14. Naeve, C.W., Kolakofsky, C.M., and Summers, D.F. (1980): *J. Virol.* 33:856–865.
15. Wertz, G.W. and Levine, M. (1973): *J. Virol.* 12:253–264.
16. Lehrach, H., Diamond, D., Wozney, J.M., and Boedtker, H. (1977): *Biochem.* 16:4743–4751.

David H.L. Bishop and Richard W. Compans, eds.
The Replication of Negative Strand Viruses

Preliminary Characterization of a Cell-Free System for Vesicular Stomatitis Virus Negative-Strand RNA Synthesis

Nancy L. Davis and Gail W. Wertz[a]

Transcription of the vesicular stomatitis virus (VSV) mRNAs has been studied extensively in well-defined *in vitro* systems, while replication of genome-length RNA remains poorly understood, due to the lack of an efficient *in vitro* system. The construction of an *in vitro* system for replication must take into consideration the information gained from studies of this process as it occurs within the infected cell. First, the template for progeny genome synthesis is probably a full-length positive-strand RNA associated with viral proteins in a nucleocapsid structure.[1–3] Second, newly synthesized genomes are incorporated into nucleocapsid structures either during or immediately following their synthesis.[1,4] Third, the initiation and continuation of synthesis of genome-length RNA (both positive and negative polarities) require concurrent protein synthesis.[5] Fourth, it is possible that host factors may be involved in the replication of VSV RNA.[6,7]

In this paper, we describe the preliminary characterization of a cell-free system which produces VSV negative-strand RNA. The system is composed of (1) a high speed pellet fraction prepared from VSV-infected BHK cells to be the source of intracellular nucleocapsids which serve as templates, (2) a micrococcal nuclease-treated mRNA dependent rabbit reticulocyte lysate[8] primed with purified VSV mRNA to be the source of newly synthesized viral proteins, and (3) the precursors and cofactors needed both for protein synthesis and RNA synthesis combined to give maximum protein synthesis while allowing concurrent RNA synthesis.[9] The goal of this work is to reconstitute *in vitro* the factors required for replication in as well-defined a form as possible.

[a] Department of Bacteriology and Immunology, School of Medicine, University of North Carolina, Chapel Hill, North Carolina.

Proteins Produced *in Vitro*

Figure 1A shows the ability of components of the *in vitro* system, that is, the reticulocyte lysate, the template fraction and the purified mRNAs, alone and in combination, to stimulate protein synthesis (measured by ^{35}S-methionine incorporation). VSV mRNA, when added to this reticulocyte lysate system, promotes a level of ^{35}S-methionine incorporation which reaches 8 to 20 times the control level. The nucleocapsid-containing pellet fraction alone is also capable of stimulating protein synthesis, due to the translation of polysome-associated viral mRNAs which are sedimented with the nucleocapsid templates. The combination of purified mRNA and template fraction with the reticulocyte lysate leads to a constant rate of protein synthesis for 60 to 90 minutes which is higher than for either component alone.

The analysis of ^{35}S-methionine-labeled products by polyacrylamide gel electrophoresis indicates that all five VSV-proteins are synthesized. Purified VSV mRNA in the absence of the template fraction promotes the synthesis of the NS, N, and M-proteins, a precursor of G-protein and a small amount of a polypeptide which coelectrophoreses with L-protein (Figure 2A, lanes 1 and 5).

Reaction mixtures which contain the template fraction and no added purified mRNA, synthesize, in addition to NS, N, and M-proteins, an increased amount of a polypeptide of the size to be L-protein (Figure 2A, lanes 2 and 6). These reactions also produce a protein which comigrates with fully glycosylated G-protein (Figure 2A, lane 2). A template fraction from which most of the cellular membranes have been removed stimulates the synthesis of much smaller amounts of the mature size G-protein (Figure 2A, lane 6).

The products of the combined template and mRNA fractions incubated with the reticulocyte lysate include the VSV-proteins NS, N, M, fully glycosylated G-protein, and a protein which comigrates with L (Figure 2A, lanes 3 and 7). The greatly increased production of the L-protein in the presence of the template fraction may reflect the activity of "stabilizing factors" which promote complete translation of this large mRNA, and/or the presence of a disproportionate amount of L mRNA in the polysomal aggregates which copurify with the template fraction.

RNAs Produced *in Vitro*

RNA synthesis *in vitro,* as measured by incorporation of ^{3}H UTP into acid-precipitable material, is template-dependent and continues at a constant rate for 60 to 90 minutes (Figure 1B). The addition of purified VSV mRNA, which stimulates an increased rate of protein synthesis, reproducibly results in a decreased rate of total RNA synthesis (Figure 1B).

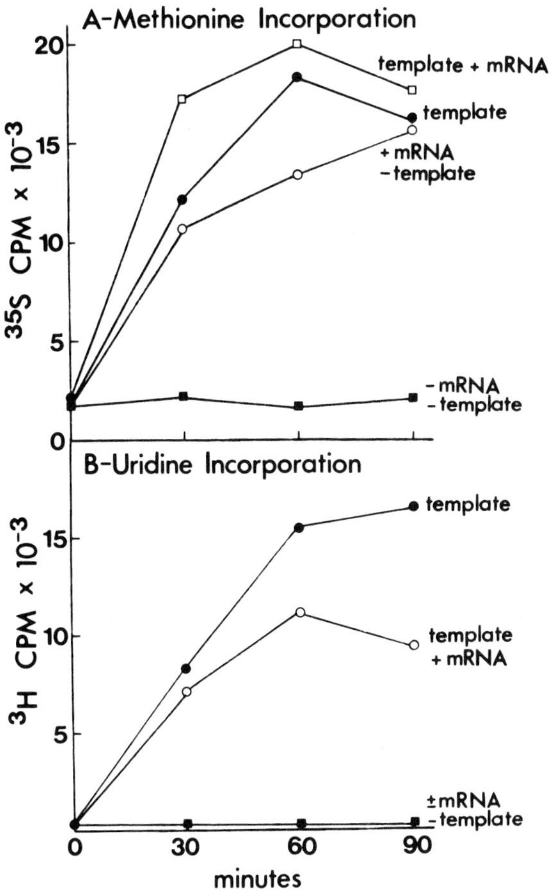

Figure 1. Time course of protein and RNA synthesis *in vitro*. The components of the *in vitro* system were prepared as follows. (1) The template fraction was prepared from actinomycin D-treated virus infected cells lysed by dounce homogenization at five to six hours postinfection and freed of nuclei by low-speed centrifugation. The cytoplasm was diluted two-fold with the swelling buffer (3 mM Tris-C1, pH 8.5, 3 mM NaC1, 0.5 mM $MgC1_2$) and centrifuged at 35,000 rpm for one hour in a Beckman SW50.1 rotor. (This treatment quantitatively pellets structures which are 120S or larger.) The pellet was gently resuspended in 10 mM Tris, pH 8, 2 mM DTE, 10% glycerol and stored frozen at −20°C. (2) The reticulocyte lysate was purchased from Green Hectares and made mRNA dependent as described by Pelham and Jackson.[8] (3) VSV mRNA was isolated from actinomycin D-treated BHK cells at five hours postinfection by phenol extraction and poly(U) sepharose chromatography. A typical reaction mixture contained 40% (by volume) reticulocyte lysate, 30% pellet fraction or buffer control, 4% purified mRNA or buffer control, plus cofactors and precursors as suggested by Ball and White[9] in a total volume of 50 μl. At the indicated times during incubation at 30°C, two μl samples were assayed for acid-precipitable radioactivity either after heating at 95°C (^{35}S) or without heating (^{3}H).

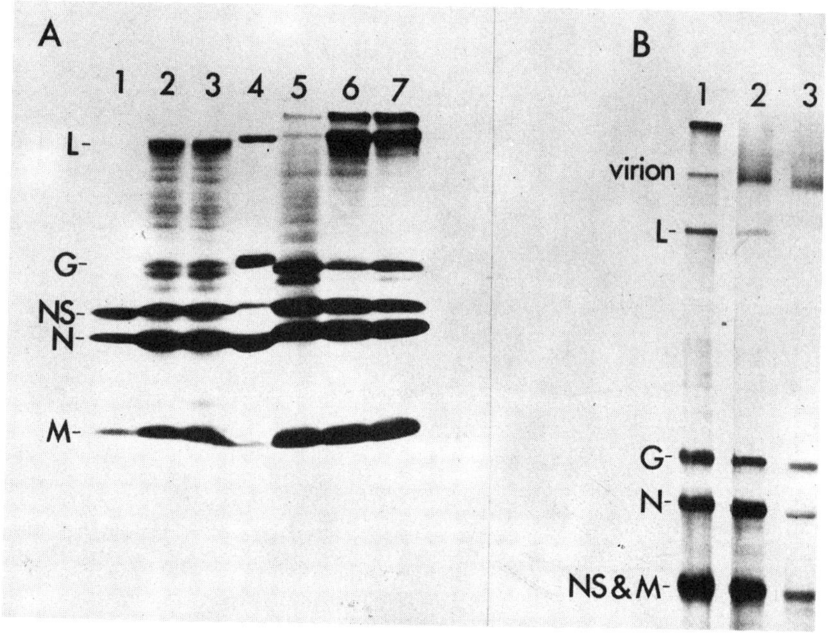

Figure 2. Electrophoretic analysis of protein and RNA products. (A) Reaction mixtures containing ^{35}S-methionine were incubated as described in Figure 1 for 90 minutes. Two to five μl portions were heated to 100°C for two minutes in sample buffer containing 2% SDS and 0.1 M DTE and electrophoresed in a Laemmli SDS-containing 7.5 to 18% gradient polyacrylamide gel cross-linked with DATD.[11] The gel was assayed by fluorography. Lanes 1 and 5, reticulocyte lysate + purified mRNA alone; lane 2, reticulocyte lysate + template fraction alone; lane 3, reticulocyte lysate + template + mRNA; lane 4, purified ^{14}C amino acid-labeled VSV virions; lane 6, reticulocyte lysate + a template fraction isolated from a 10,000 x g supernatant; lane 7, reticulocyte lysate + same template fraction as in lane 6 + mRNA. (B) Reaction mixtures were prepared and incubated in the presence of ^{3}H-UTP as described in Figure 1. After 90 minutes, RNA was purified by phenol extraction and ethanol precipitation. Samples dissolved in 6 M urea-containing sample buffer were electrophoresed in a 6 M urea-agarose gel.[12] Lane 1, *in vivo* ^{3}H-uridine-labeled VSV RNA isolated from actinomycin D-treated BHK cells at five hours postinfection; lane 2, *in vitro* ^{3}H-labeled RNAs purified from a reaction mixture containing reticulocyte lysate + template fraction alone; lane 3, *in vitro* products of a reaction with reticulocyte lysate, template fraction and purified mRNA.

The RNAs produced during 90 minutes of incubation in reaction mixtures containing ^{3}H-UTP, reticulocyte lysate, and template fraction either with or without added mRNA were purified by phenol extraction and electrophoresed in urea-containing agarose gels. The RNA products of the complete *in vitro* system are resolved into four major bands (Figure 2B, lane 3), three of which coelectrophorese with the mRNA-containing bands (G, N, NS, and M) present in the *in vivo* ^{3}H-labeled marker (Figure 2B, lane 1). In addition to the *in vitro* products which comigrate with the mRNAs, a band of product RNA also migrates at the position of genome-length RNA (Figure 2B, lane 3, "virion"). In order to test whether the radioactivity in this high molecular weight band represents genome-length RNA or small nascent RNAs hybridized to unlabeled genome-length template molecules, a similar sample was analyzed following denaturation by treatment with glyoxal.[10] Glyoxal treatment reduced but did not eliminate the radioactive band at the position of genome-length RNA (data not shown).

The omission of mRNA from the reaction reproducibly leads to increased transcription of all five species of viral mRNA (Figure 2B, lane 2), which accounts for the increase in total RNA synthesis (Figure 1B). Thus, the synthesis of L mRNA, which was below the level of detection in the reaction with added mRNA, is detectable in this reaction. In the experiment shown in Figure 2, this reaction also produced RNA which migrates at the position of genome-length RNA. However, similar experiments, with different template fractions with no added purified mRNA, have produced either no labeled product with the migration rate of full-length RNA, or reduced amounts compared to the parallel reaction with added mRNA. One interpretation of these combined results is that different template fractions contain different amounts of viral mRNA and therefore are able, without added mRNA, to program different levels of protein synthesis. The amount of viral protein synthesized by a particular template fraction in a mixture with the reticulocyte lysate may determine the amount of genome-length RNA which is synthesized.

Equilibrium centrifugation in CsCl gradients separates VSV genome-length RNAs, which band as nucleocapsids ($\rho = 1.3$ gm/cm^3), from mRNAs, which sediment to the bottom of the gradient. In order to determine whether the newly-synthesized RNA would band at the density characteristic of VSV nucleocapsids, *in vitro* reaction mixtures containing ^{3}H-uridine-labeled template RNA and ^{32}P-UTP-labeled product RNA were centrifuged in preformed CsCl gradients. Gradient profiles of a reaction mixture incubated for 60 minutes with no added mRNA (Figure 3A) and a mixture containing added mRNA (Figure 3A) both show ^{32}P-labeled product RNA which cobands with the ^{3}H-uridine label present in template nucleocapsids (fractions 16–18).

Nucleocapsid-associated RNAs from reactions with or without added mRNA (Figure 3A and 3B, fractions 16–18) were purified by treatment with 0.1% SDS followed by sucrose density gradient centrifugation in the presence of 0.1%SDS. The 40S regions of the sucrose density gradients, marked by a peak of ^{3}H-labeled template RNA, also contained a peak of ^{32}P-labeled *in vitro* product representing 11–15% of the ^{32}P cpm from the nucleocapsid region of the CsCl gradient.

Product and template RNAs, labeled with ^{32}P and ^{3}H respectively, were hybridized to a large excess of specific VSV RNA probes.[14] Hybridization conditions were used under which 92 to 94% of the ^{3}H-labeled template

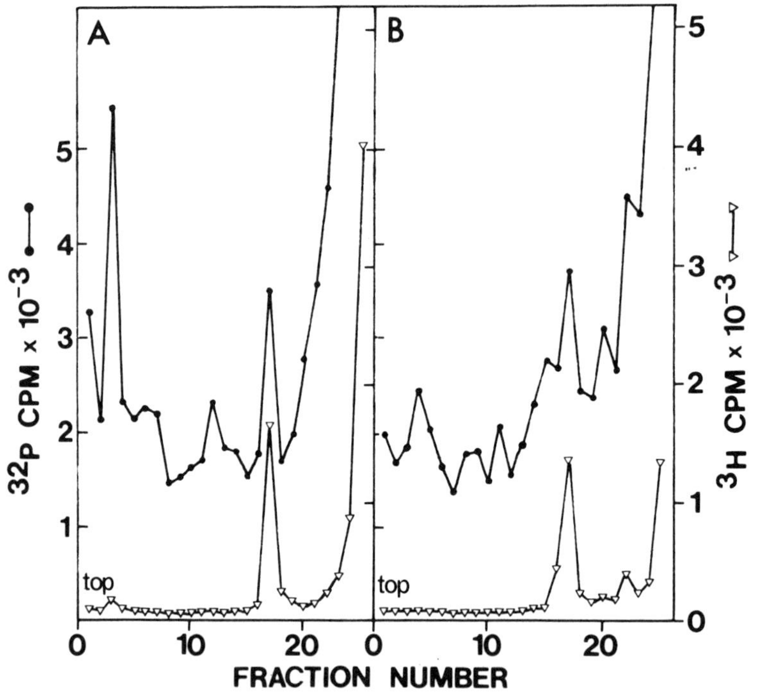

Figure 3. CsCl density gradient centrifugation of *in vitro* reaction mixtures. Reaction mixtures containing a ^{3}H-uridine-labeled template fraction and ^{32}P-αUTP were prepared and incubated for 60 minutes as described in Figure 1 either without (A) or with (B) added purified VSV mRNA. The reactions were stopped by dilution into 2.5 ml of cold 3 mM Tris, pH 8.5, 3 mM NaCl, 0.5 mM $MgCl_2$ and 0.5% NP-40 and layered onto an 8-ml 20 to 40% (w/w) preformed CsCl gradient.[13] Following 16 to 18 hours of centrifugation at 225,000 x g, the gradients were fractionated from the top and a portion of each fraction was assayed for acid-precipitable radioactivity.

RNA was protected by the positive-strand probe and 11–12% was protected by the negative-strand probe. Therefore, the template RNAs were, under these conditions, completely hybridized by the particular RNA probe and could not protect RNA product. Under these conditions, 40% of the ^{32}P-labeled *in vitro* product RNA was protected by mRNA (negative-stranded) and 60% was protected by virion RNA (positive-stranded).

Discussion

We have described a cell-free system capable of producing VSV RNA products which include RNAs which comigrate with genome-length RNA in agarose-urea gels, band in a structure with the buoyant density of nucleocapsids, and are of genome (negative-strand) polarity.

The further development of this *in vitro* system will provide a tool to examine the nature of the protein synthesis requirement for VSV negative-strand synthesis. Of first priority is the isolation of a template fraction which is completely dependent on added viral mRNAs for viral protein synthesis. This refinement will introduce the possibility of programming the system with individual purified VSV mRNAs and thereby identifying specific viral proteins involved in VSV negative-strand RNA synthesis.

ACKNOWLEDGMENTS
We appreciate the able assistance of John Glass. This work was supported by Public Health Service Grants AI-12464 and AI-15134 from the National Institute of Allergy and Infectious Disease and Grant CA19014 from the National Cancer Institute, DHEW.

References

1. Soria, M., Little, S.P., and Huang, A.S. (1974): *Virology* 61:270–280.
2. Wertz, G.W. (1978): *Virology* 85:271–285.
3. Wertz, G.W. (1980): In: *Rhabdoviruses,* Vol. II. Bishop, D.H.L. (Ed.), CRC Press, Inc., Boca Raton, pp. 75–93.
4. Hill, V.M., Simonsen, C.C., and Summers, D.F. (1979): *Virology* 99:75–83.
5. Wertz, G.W. and Levine, M. (1973): *J. Virol.* 12:253–264.
6. Nowakowski, M., Bloom, B.R., Ehrenfeld, E., and Summers, D.F. (1973): *J. Virol.* 12:1272–1278.
7. Obijeski, J.F. and Simpson, R.W. (1974): *Virology* 57:369–377.
8. Pelham, R.B. and Jackson, R.J. (1976): *Eur. J. Biochem.* 67:247–256.
9. Ball, L.A. and White, C.N. (1978): *Virology* 84:479–495.
10. McMaster, G.K. and Carmichael, G.C. (1977): *Biochemistry* 74:4835–4838.
11. Anker, H.S. (1970): FEBS Letters 7:293.
12. Wertz, G.W. and Davis, N.L. (1979): *J. Virol.* 30:108–115.
13. Simonsen, C.C., Batt-Humphries, S., and Summers, D.F. (1979): *J. Virol.* 31:124–132.
14. Wertz, G.W. (1978): *Virology* 85:271–285.

David H.L. Bishop and Richard W. Compans, eds.
The Replication of Negative Strand Viruses

An Unusual Messenger RNA Synthesized by VSV DI-LT

Ronald C. Herman[a] and Robert A. Lazzarini[b]

Defective interfering (DI) particles of vesicular stomatitis virus (VSV) contain only a portion of the parental viral genome and consequently are unable to replicate in the absence of wild type helper virus.[1] The vast majority of VSV DI particles are derived from the 5′ half of the viral genome and have a short nonparental sequence at their 3′ end which is the complement of the normal 5′-terminus.[2] A few DI particles, such as DI-LT, which was generated by the heat-resistant strain of VSV (VSV-HR),[3] contain genetic information primarily from the 3′ half of the genome (Figure 1).

Data from several laboratories have shown that, unlike most VSV DI particles, DI-LT arose by a true internal deletion of the parental genome and thus retains both parental termini.[4-7] Because DI-LT contains genetic information from the 3′ half of the genome and retains the parental 3′-terminal sequence, it can synthesize the leader RNA and biologically functional messages for the nucleocapsid (N), nonstructural (NS), matrix (M), and glyco- (G) proteins.[8,9] Hybridization and electron microscopic analyses show that the DI-LT genome contains only 320-350 nucleotides of the normal 5′-terminal sequence (Figure 1).[6] The electron microscopic analysis further suggests that the deletion probably occurred entirely within the polymerase (L) gene, and up to several hundred nucleotides of L gene sequence may remain in the DI-LT genome on each side of the deletion point. These observations suggest that transcription of DI-LT could proceed beyond the end of

[a] Infectious Disease Institute, Division of Laboratories and Research, New York State Department of Health, Albany, New York.
[b] Laboratory of Molecular Biology, National Institute of Neurological and Communicative Disorders and Stroke, Bethesda, Maryland.

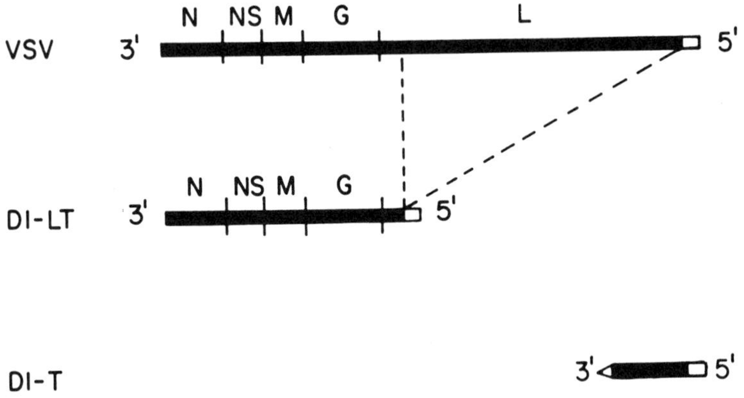

Figure 1. Map of VSV and DI particle genomes. The relative lengths of the VSV and DI-LT genomes, as well as the locations of the intercistronic boundaries, were directly determined by electron microscopic measurements.[16] The length of the DI-T genome was estimated from its sedimentation (19S) relative to the wild type viral RNA (42S). The maps of the two DI particles have been aligned with that of the wild type to reflect the location of the parental sequences contained in each. The dashed lines indicate the approximate location of the deletion which gave rise to DI-LT. The open box at the 5′ end of each map represents the 320-350 nucleotides of parental 5′-terminal sequence retained by DI-LT and is the region of homology shared with DI-T. The tapered end of the DI-T map represents the short nonparental 3′-terminal sequence present in DI particles derived from the 5′ end of the genome.[2]

the G cistron and prompted us to look for a novel small transcript of the remnant polymerase gene containing only the 5′ and 3′-terminal sequences of the L message.

Polyadenylated messenger RNA was isolated from suspension BHK cells 4.5 hr after infection with either VSV-HR or DI-LT. The mRNAs, labeled with ^{32}P-orthophosphate, were resolved in a 1.5% agarose gel containing 6 M urea at pH 3.0 (Figure 2).[10] The messages for the NS, M, and N proteins synthesized by DI-LT comigrate with the corresponding mRNAs synthesized by VSV-HR, and no other small transcript is detectable. Surprisingly, only very little of the polyadenylated DI-LT mRNA comigrates with authentic G message. Instead, a new and larger message is observed (G*).

This unusual poly(A)$^+$ RNA nevertheless contains G message sequences. Polyadenylated VSV-HR and DI-LT mRNA, labeled *in vivo* the ^{3}H-uridine, was resolved in a 1.5% agarose gel and then transferred to a nitrocellulose filter by blotting.[11] The filter-bound RNA was subsequently hybridized to a ^{32}P-labeled probe which had been prepared by nick-translation of plasmid pG65 containing a cloned sequence from the 3′ end of G message.[12-14] As shown in Figure 3, the probe anneals both to the authentic glycoprotein message synthesized by VSV-HR and to the larger mRNA synthesized by DI-LT. This clearly shows that G* does contain G message sequences.

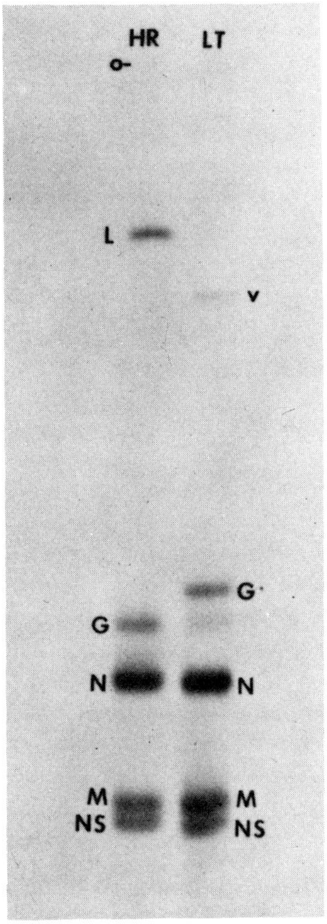

Figure 2. Polyadenylated messenger RNA. Suspension BHK cells were infected with VSV-HR or DI-LT in the presence of actinomycin D (5 μg/ml). ^{32}P-orthophosphate (50 μCi/ml) was added at 30 min postinfection, and the cells were harvested 4 hr later. Cells were disrupted with 0.5% NP40 in hypotonic buffer and the nuclei pelleted. Cytoplasmic RNA was extracted with phenol/chloroform. Polyadenylated mRNA was selected by oligo(dT)-cellulose chromatography and resolved by electrophoresis in a 1.5% agarose gel (pH 3.0) containing 6 M urea at 120 V for 18 hr.[10] The wet gel was autoradiographed against Kodak X-omat R film with an intensifying screen. **o** indicates the origin. The band **v** had the mobility of a RNA the size of the DI-LT genome but has not been characterized further. The bromophenol blue tracking dye has run off the bottom of the gel.

The decreased mobility of G* suggests that it is perhaps 500-600 nucleotides longer than G. This increased size is not due to an increase in the length of the poly(A). Since the poly(A) is not directly encoded in the VSV genome, it remains single-stranded after the messages are annealed to the

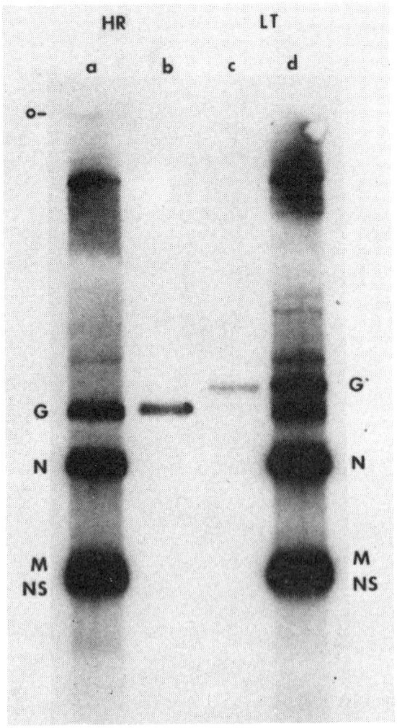

Figure 3. Nitrocellulose filter hybridization. ^{3}H-labeled VSV-HR (lane b) and DI-LT (lane c) poly(A)$^+$ *in vivo* mRNAs were resolved in an agarose gel as described in the legend to Figure 2. Uniformly ^{32}P-labeled VSV-HR (lane a) and DI-LT (lane d) *in vivo* mRNAs were used as markers. The mRNA was transferred to a nitrocellulose filter and hybridized as outlined in the text.

42S viral RNA and can be removed by trimming such duplexes with RNase T_2.[15] All of the messages deadenylated in this manner migrate slightly more rapidly than their adenylated forms (Figure 4). Even though deadenylated G* migrates more rapidly than poly(A)$^+$ G*, it still does not comigrate with deadenylated G message. This shows that excessive polyadenylation does not cause the altered mobility of G*. In addition, this result implies that the poly(A) is located at the 3′ end of G*, which therefore cannot be interrupted by a (large) intervening poly(A) of the type that we previously described.[16] Identical results were also obtained with *in vitro* transcripts of VSV-HR and DI-LT (unpublished observations).

The results of these experiments suggest that the viral polymerase does not terminate transcription at the end of the G cistron but continues without apparent interruption into the remnant polymerase gene, perhaps all the way to the 5′ end of the DI-LT genome. We have used DI-T as a probe to test for transcription of the 5′ end of the DI-LT genome. As illustrated in Figure 1, DI-T is derived exclusively from within the polymerase gene and shares, at most, only the last 320-350 nucleotides of the 5′-terminal parental sequence

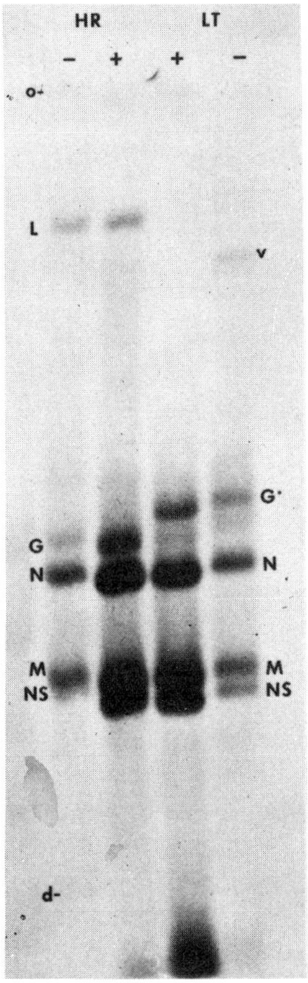

Figure 4. Deadenylated mRNA. Polyadenylated mRNA (50,000 cpm) was annealed to 240 ng of the 42S VSV-HR genome for 1 hr at 60°C in 0.4 M NaCl, 0.01 M Tris, pH 7.5. The resultant duplexes were trimmed with RNase T_2 (5 units/ml) at 37°C for 30 min. After precipitation with ethanol, the RNase-resistant material was resuspended in sample buffer, heated in a boiling water bath for 1.5 min, and electrophoresed in a 1.5% agarose gel as described in the legend to Figure 2 (+). Control, untreated mRNA (25,000 cpm) was resolved in adjacent lanes (−). **o** indicates the origin and **d** the location of the bromophenol blue tracking dye.

with DI-LT. Gel-purified ^{32}P-labeled VSV-HR G and DI-LT G* were separately annealed to DI-T RNA and the resulting partial duplexes trimmed with RNase T_2. After heat-denaturation, the RNase-resistant material was resolved in a 12% polyacrylamide gel containing 8 M urea (Figure 5).[17] As expected, no nuclease-resistant material was detected when either G or G* was self-annealed in the absence of the DI-T RNA. A protected sequence was detected only when G* was annealed to DI-T. The size of this fragment is

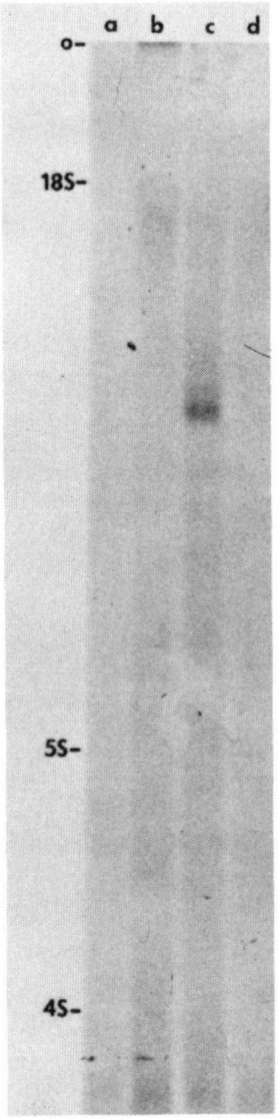

Figure 5. Hybridization of mRNA to DI-T. Equal amounts of gel-purified ^{32}P-labeled G and G* mRNA were incubated at 60°C for 1 hr in 0.4 M NaCl, 0.01 M Tris, pH 7.5, either alone or in the presence of 1.5 μg of DI-T RNA. RNase T_2 was added to each sample (10 units/ml) and incubation continued for 30 min at 37°C. Resistant material was precipitated with ethanol, resuspended in sample buffer, heated for 1.5 min in a boiling water bath, and electrophoresed in a 30-cm 12% polyacrylamide gel. Lanes: (a) G alone, (b) G × DI-T, (c) G* × DI-T, (d)G* alone. The locations of ^{32}P-labeled HeLa cell cytoplasmic RNAs run in a parallel lane are indicated. **o** indicates the origin of the gel.

estimated to be approximately 250 nucleotides by comparison to cellular 4S and 5S RNA. Since the L gene stops at position 60 from the 5′ end of the genome,[18] it seems likely that transcription to yield G* terminates at or very near to the 5′ end of the remnant polymerase gene. In fact, the region of G* which anneals to the DI-T genome is contiguous with the poly(A) tail (unpublished observations), and this suggests that the L gene polyadenylation signal encodes polyadenylation of G*.

The results presented here characterize an unusual messenger RNA synthesized by DI-LT. This RNA contains the glycoprotein message covalently linked to a transcript of the remnant polymerase gene. The events which lead to the synthesis of G* are not currently known. Any model which is formulated to explain this process must take into consideration the various alternative mechanisms that have been proposed for normal VSV transcription[19] and the fact that the deletion which gave rise to DI-LT appears to have occurred entirely within the polymerase gene (unpublished observations). Thus it seems likely, regardless of which mechanism for transcription is ultimately proven, that sequences in the genome (or in the transcript itself) distinct from those at the intergenic region play a role in the synthesis of normal VSV mRNA. Experiments are in progress to sequence the G/L intercistronic boundary in DI-LT to confirm the location of the deletion.

ACKNOWLEDGMENTS
We thank J.K. Rose for providing plasmid pG65 and D.L. Panicali for his expert assistance with filter hybridizations. R.C.H. was supported by a NIH Biomedical Research Support Grant to the New York State Department of Health.

References

1. Huang, A.S. (1973): *Ann. Rev. Microbiol.* 27:101–117.
2. Perrault, J. and Leavitt, R.W. (1977): *J. Gen. Virol.* 38:35–50.
3. Petric, M. and Prevec, L. (1970): *Virology* 41:615–630.
4. Perrault, J. and Semler, B.L. (1979): *Proc. Natl. Acad. Sci.* 76:6191–6195.
5. Clerx-Van Haaster, C., Clewley, J.P., and Bishop, D.H.L. (1980): *J. Virol.* 33:07–17.
6. Epstein, D.A., Herman, R.C., Chien, I., and Lazzarini, R.A. (1980): *J. Virol.* 33:818–829.
7. Chanda, P.K., Kang, C.Y., and Banerjee, A.K. (1980): *Proc. Natl. Acad. Sci.* 77:3927–3931.
8. Colonno, R.J., Lazzarini, R.A., Keene, J.D., and Banerjee, A.K. (1977): *Proc. Natl. Acad. Sci.* 74:1884–1888.
9. Johnson, L.D., Binder, M., and Lazzarini, R.A. (1979): *Virology* 99:203–206.
10. Lehrach, H., Diamond, D., Wozney, J.M., and Boedtker, H. (1977): *Biochemistry* 16:4743–4751.
11. Southern, E.M. (1975): *J. Mol. Biol.* 98:503–517.
12. Wahl, G.M., Stern, M., and Stark, G.R. (1979): *Proc. Natl. Acad. Sci.* 76:3683–3687.
13. Rigby, P.W.J., Dieckmann, M., Rhodes, C., and Berg, P. (1977): *J. Mol. Biol.* 113:237–251.
14. Rose, J.K. (1980): *Cell* 19:415–421.
15. Freeman, G.J., Rose, J.K., Clinton, G.M., and Huang, A.S. (1977): *J. Virol.* 21:1094–1104.
16. Herman, R.C., Adler, S., Colonno, R.J., Banerjee, A.K., Lazzarini, R.A., and Westphal, H. (1978): *Cell* 15:587–596.
17. Peacock, A.C. and Dingman, C.W. (1967): *Biochemistry* 6:1818–1827.
18. Schubert, M., Keene, J.D., Herman, R.C., and Lazzarini, R.A. (1980): *J. Virol.* 34:550–559.
19. Banerjee, A.K., Abraham, G., and Colonno, R.J. (1977): *J. Gen. Virol.* 34:1–8.

David H.L. Bishop and Richard W. Compans, eds.
The Replication of Negative Strand Viruses

Synthesis of the Complete Plus-Strand RNA from the Endogenous RNA Polymerase Activity of Defective Interfering Particles of Vesicular Stomatitis Virus

C. Yong Kang,[a] Pranab K. Chanda,[b] and Amiya K. Banerjee[a]

Defective interfering (DI) particles can be generated by serial undiluted passages of most, if not all, animal viruses. DI particles are a form of deletion mutants, which contain a part of the standard viral genome and the same structural proteins as the standard virus including virion associated viral-specific enzymes. Replication of DI particles requires co-infection with the standard virus, and during this process the DI particles interfere with the standard virus replication. This suppressive effect on viral infection has been implicated in persistent infection and is also viewed as a possible important determinant in the outcome of natural viral infection.[1-4] Such DI particles are produced by vesicular stomatitis virus (VSV),[5] a rhabdovirus that contains 4 $\times 10^6$ molecular weight, linear single-stranded RNA of negative polarity and five structural proteins designated as N, NS, M, G, and L.[6] The standard VS virions contain an RNA polymerase that transcribes the genomic RNA to synthesize a leader RNA (48 nucleotides long)[7] and all five monocistronic mRNAs. Virtually all of the DI particles isolated from VSV infected cells contain variable lengths of the standard virus genome RNA from the 5′-end[8,9] (referred to as 5′-DI particles) with only one exception, which retains almost the entire 3′-terminal half of the standard genome.[9-12] The exception DI particles have been isolated from a heat-resistant strain of VSV (designated as DI-LT) and are capable of transcription *in vitro* identical to that of the standard virus, with the synthesis of the leader RNA and the monocistronic mRNAs coding for N, NS, M, and G-proteins, except the

[a] Department of Microbiology, The University of Texas Health Science Center, Dallas, Texas.
[b] Department of Cell Biology, Roche Institute of Molecular Biology, Nutley, New Jersey.

L-protein mRNA.[13] DI-LT associated virion transcriptase can also synthesize limited amounts of the full-length plus-strand RNA *in vivo* in the absence of new L-protein synthesis.[14] The 5′-DI particles, on the other hand, synthesize *in vitro* a small RNA product, 46 nucleotides (referred to as DI-46mer), from its 3′-end.[15-18] The base sequence of the small RNA is different from the leader RNA synthesized *in vitro* but is identical to the 5′-terminal 46 bases of the standard virion RNA.[19]

The termination of RNA synthesis at the 46 nucleotides and accumulation of the DI-46mer in infected cells has been implicated for the mechanism of viral interference by DI particles. The molecular mechanism of DI-particle-mediated viral interference can be investigated, if we have a system to separate the transcriptional process from the replicational event *in vitro*. The switch-over mechanism from the synthesis of monocistronic mRNAs (transcription) to synthesis of the full-length complement (replication) of the standard genome RNA in virus infected cells is not understood.

Testa et al.[20] have shown that under certain conditions the full-length complement of the standard genome RNA can be synthesized *in vitro*. In this communication, we demonstrate that, by using the same procedure, it is possible to synthesize *in vitro* the full-length complement of the DI particle RNA.

In Vitro Transcription of the DI Particles

The 5′-DI particle synthesizes *in vitro* a small RNA product, DI-46mer, from its 3′-end. To determine whether DI-46mer synthesis is common in all 5′DI-particles, *in vitro* transcription reactions were carried out using partially purified DI particles as shown in Figure 1. The electrophoresis analysis of RNA products from the RNA polymerase reaction of DI particles in a 20% polyacrylamide slab gel shows the product is small and migrated close to the marker leader RNA synthesized by standard virus (Figure 2). It can be seen that there are distinct bands of radioactivity migrating approximately two-thirds the way down in all three DI particles. These small RNAs hybridized specifically to the 5′-DI particle RNAs. Little radioactivity was seen at the top of the gel, indicating that virtually no large RNA species were synthesized *in vitro* by DI-1, DI-2, or DI-3. The faint radioactive band at the top of the gel and slightly above the heavy DI-46mer band (arrow in Figure 2) indicates that the DI preparation has a low level of contamination by the standard virus because these RNAs hybridize to the standard virus genome, but not to their respective DI RNAs (data not shown).

From these experiments we conclude that all three 5′-DI particles which we have isolated transcribe *in vitro* a small RNA product similar in length to the leader RNA. These small RNA products presumably represent the 46-base-long product RNA synthesized by other 5′-DI particles *in vitro* as reported previously.[15-18] Therefore, it appears that all 5′-DI particles transcribe 46 nucleotides *in vitro*.

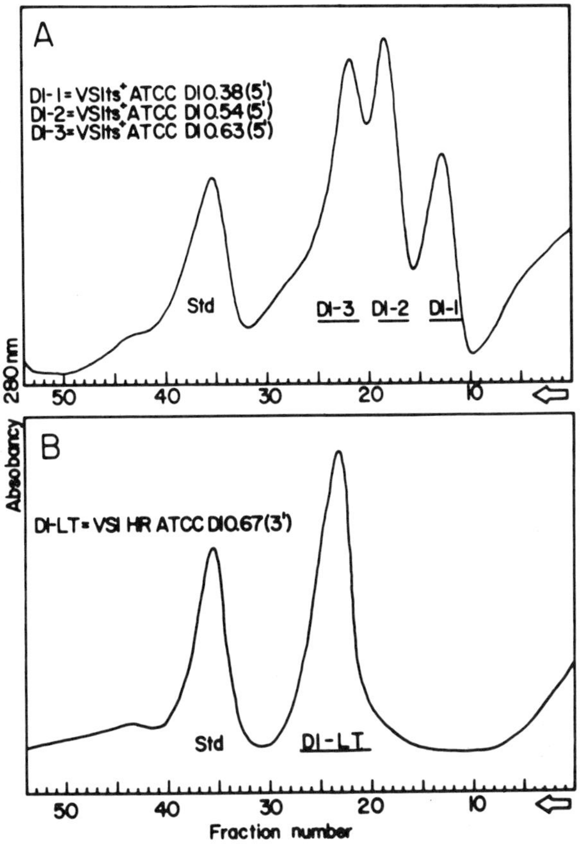

Figure 1. Sucrose gradient purification of DI particles. Approximately 2 × 10^7 R(B77) cells per 100 mm dish were infected with either 0.2 ml of VSV_{IND} stock virus which will yield maximum amounts of DI-1, DI-2, and DI-3 (A) or 0.2 ml of VSV_{IND-HR} stock virus which will yield maximum amounts of HR-LT (B). Samples of lysates from 20 culture dishes each were collected after 15 hours of infection and centrifuged at 600 × g for 15 minutes. The virus particles were pelleted from the supernatant by centrifugation at 81,000 × g for 90 minutes in a Spinco SW 27 rotor, resuspended in phosphate-buffered saline, and layered on a linear 5 to 80% sucrose gradient made in phosphate-buffered saline as described previously.[21] After centrifugation at 110,000 × g at 5°C for 35 minutes in a Spinco SW 41 rotor, the gradients were collected from the top using a Buchler Auto-Densi-Flow IIC. The presence and distribution of virus in the gradient was monitored continuously by passing the effluent through on LKB Uvicord II, with absorbance at 280 nm being continuously recorded on a Fisher Recordall 500. Positions of the standard virion and DI particles are indicated in the figure. Appropriate fractions were pooled and recentrifuged at 81,000 × g for 120 minutes to concentrate the purified DI particles. The virus pellet was resuspended in Tris-HC1 buffer (pH8.0) containing 0.1 M NaCl and 10% DMSO. The protein concentration of purified DI particles were approximately 500-1000 μg/ml. The recent nomenclature for DI particles are given for all DI particles as shown in the figure. For convenience we will use designation as DI-1, DI-2, DI-3, and DI-LT in this paper.

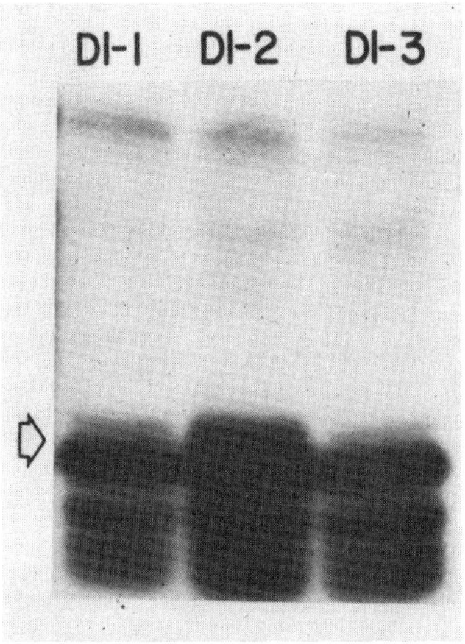

Figure 2. Analysis of the *in vitro* product RNAs of 5′-DI particles by polyacrylamide gel electrophoresis. The product RNAs were synthesized *in vitro* by using DI-1, DI-2 or DI-3 in reaction mixtures that contained (0.2ml): 50 mM Tris HCl (pH 8.0), 0.1 M NaCl, 5 mM $MgCl_2$, 4 mM dithiothreitol, 0.05% Triton-N 101, 0.1 mM UTP, 1 mM ATP, CTP and GTP, 40 μCi of α-^{32}P UTP (24.9 Ci/mmol), and approximately 30 μg of purified DI particles. The incubation was at 30°C for 3 hours. The product RNAs were extracted with phenol, purified by Sephadex G-50 chromatography and electrophoresed on a 20% polyacrylamide slab gel containing 7 M urea.[20] Electrophoresis was at 200V for 17 hours and autoradiography was with Kodak SB5 X-ray film. The arrow indicates the migration position of marker leader RNA.

Synthesis of the Full-Length Plus-Strand RNA of the 5′-DI Particles

We have used the preparation of DI particles as shown in Figure 1 to synthesize the full-length complement of the DI RNA, using the method employed for the synthesis of the full-length plus-strand of the standard genome RNA.[20] The detergent-disrupted DI particles were "preinitiated" with ATP and CTP, followed by isolation of the ribonucleoprotein (RNP) core particles by centrifugation onto the glycerol cushion. The RNA core particles were then incubated in a standard RNA transcription reaction mixture containing GTP, α-^{32}P UTP, and CTP with AMPPNP (1 mM) replacing ATP. The purified RNA products were then analyzed by velocity sedimentation. The results obtained by DI-1, DI-2, and DI-3 particles are shown in Figure 3. It can be seen that distinct peaks of ^{32}P radioactivity sedimented with their corresponding particle RNA according to their sizes (Figure 3A, B, C). The RNA was then isolated from the pooled fractions and denatured with glyoxal

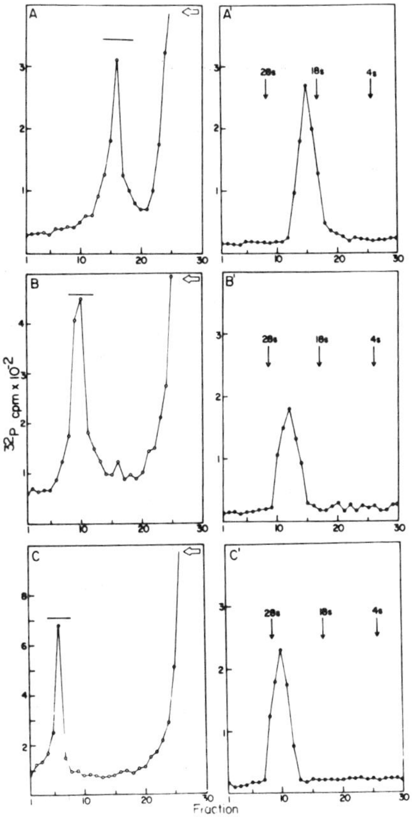

Figure 3. Sucrose gradient sedimentation of AMPPNP product RNAs of DI-1, DI-2, and DI-3. ^{32}P UMP-labeled RNA product was synthesized by preininitiated DI-1, DI-2, and DI-3 cores in the presence of AMPPNP as described previously. Briefly, purified DI particles were disrupted with Triton N-101 (0.05%) in the presence of 1 mM ATP and 0.5 mM CTP. After incubation for 30 minutes at 30°C, the preinitiated cores were purified by centrifugation and the RNA syntheses were carried out under standard transcription conditions including α-^{32}P UTP as the labeled precursor, except ATP was replaced by adenosine 5′(β,γ-imido) triphosphate (AMPPNP). The incubation was at 30°C for 2 hours. The product RNAs were then sedimented through linear 15-30% sucrose gradients at 23°C for 6 hours at 45,000 rpm in a Spinco SW 60 rotor. Fractions were collected from the bottom of the tubes and radioactivities were determined (A, B, C). The product RNAs sedimented at the same positions of corresponding DI particle RNAs in separate gradients and are shown by the bars. Fractions indicated by the bars in panel A, B, and C were then pooled separately for each DI particle RNA and precipitated with ethanol. The RNA samples were then dissolved in 10 mM sodium phosphate buffer, pH 7.0, containing 500 mM glyoxal and heated at 39°C for 60 minutes. The denatured RNA was layered onto a linear 5-30% sucrose gradient containing 100 mM glyoxal, 10 mM sodium phosphate and 0.05% SDS and centrifuged at 23°C for 6 hours at 40,000 rpm in a Spinco SW 60 rotor. The radioactivity in each fraction was determined. ^3H-uridine-labeled HeLa cell ribosomal RNA was used as marker (A, A′) DI-1; (B, B′) DI-2; (C, C′) DI-3.

and sedimented in glyoxal gradients (Figure 3A′, B′, C′). All three newly synthesized DI particle RNAs sedimented characteristically according to their sizes. The final proof that these RNA products were indeed complementary to their respective particle RNA is shown in Table 1. All of the *in vitro* RNAs were hybridized to their respective particle RNAs to the extent of nearly 100%. The large amounts of radioactive material that sedimented at the top of the gradient (Figure 3A, B, C) contained almost exclusively a small RNA product similar to that shown in Figure 2. The proportions of the full-length RNA synthesized *in vitro* by each of the DI particles varied and were approximately 7%, 5%, and 4% for DI-1, DI-2, and DI-3, respectively. From these experiments we conclude that the 5′-DI particles, under appropriate *in vitro* transcription conditions, synthesize a full-length complement of their respective particles RNAs in addition to the characteristic small RNA that is synthesized under normal transcription conditions.

Synthesis of the Full-Length Plus-Strand RNA of 3′-DI Particle

A similar experiment was done with DI-LT for the synthesis of full-length complementary RNAs *in vitro*. As shown in Figure 4A, the *in vitro* product RNAs of the DI-LT sedimented at the same rate as the authentic DI-LT ^{3}HRNA. The RNA was then isolated from the pooled fractions (Figure 4A) and denatured with glyoxal and sedimented in a glyoxal gradient. As shown in Figure 4B, the radioactivity peak sedimented close to the 28s ribosomal

Table 1. Hybridization of ^{32}P UMP-Labeled Complementary Full-Length DI Particle Product RNAs with Their Corresponding Particle RNAs.

DI RNA	Particle RNA μg/ml	Treatment	Acid insoluble cpm	RNase resistance, %
DI-1	None	None	740	—
	None	RNase	50	7
	2.0	Hybridized, RNase	730	99
DI-2	None	None	700	—
	None	RNase	40	6
	2.0	Hybridized, RNase	700	100
DI-3	None	None	990	—
	None	RNase	60	6
	2.0	Hybridized, RNase	960	97

^{32}P UMP-labeled full-length RNAs of DI particles were synthesized *in vitro* and purified by velocity sedimentation as in Figure 3. Aliquots of each product RNA were hybridized with the standard genome RNA and then the hybrids were treated with RNases A and T1 as described.[22] Trichloracetic acid-precipitable radioactivity in each reaction mixture was determined.

RNA marker, indicating that the *in vitro* RNA products were indeed linear single-stranded RNA identical in size to the corresponding DI-LT particle RNA. All of the *in vitro* RNA products were hybridized to DI-LT genomic RNA to the extent of nearly 100% indicating that the product RNA is indeed the complementary strand of the particle RNA (data not shown). The ^{32}P radioactivity at the top of the gradient (Figure 4A), upon further analysis, was found to contain predominantly RNA molecules migrating as leader RNA and other small RNA molecules found previously with the standard virus.

Discussion

The 5′-DI particles of VSV transcribe, by the endogenous RNA polymerase reaction *in vitro,* a 46 bases long RNA which is complementary to the 3′-terminal of the particle RNA.[15-18] In contrast, the 3′-DI particle (only one isolated so far which is designated as DI-LT) transcribes *in vitro* the leader RNA and mRNAs coding for N, NS, M, and G-proteins.[13] In this communication, we have shown that under appropriate *in vitro* reaction conditions all of these DI particles synthesize their full-length complementary RNA—the

Figure 4. Sucrose gradient sedimentation of AMPPNP product RNA of DI-LT particle. ^{32}P UMP-labeled RNA product was synthesized by purified DI-LT cores in the presence of AMPPNP and purified as described in Figure 3. The ^{3}H-uridine-labeled radioactivity represents the DI-LT virion RNA (A). Fractions indicated by bar in panel A were pooled, denatured with glyoxal, and analyzed in the glyoxal gradient as described in Figure 3 (B). The sedimentation positions of ^{3}H-uridine-labeled HeLa cell ribosomal RNA markers are shown by arrows.

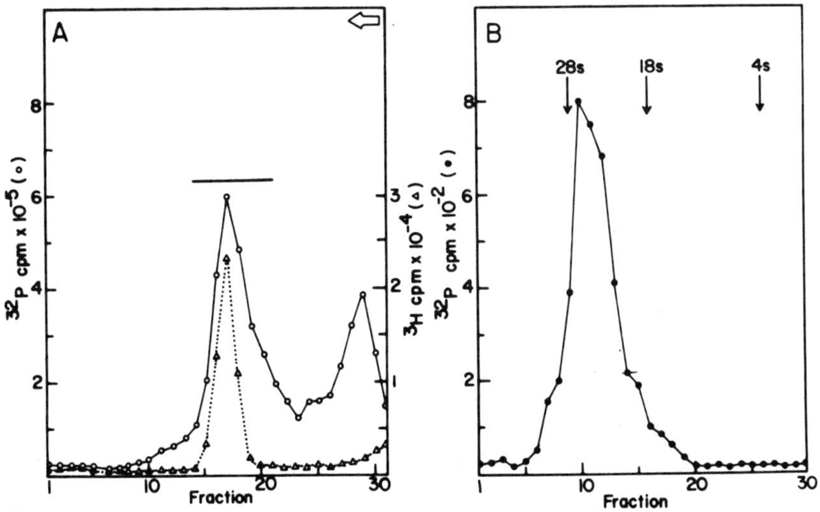

required intermediate of replication. The reactions involved preinitiation of the DI-particle cores with ATP and CTP, followed by RNA chain elongation with AMPPNP replacing ATP. Using this method, Testa et al.[20] have recently been able to synthesize the full-length complement of the standard genome RNA *in vitro*.

The above results indicate that the mechanism of complementary full-length RNA synthesis *in vitro* by the DI particles is similar, if not identical, to that operative in the standard virus. The results also indicate that the enzyme(s) involved in this replicative process is associated with and used by both the DI particles and the standard virus. Again during transcription, the 5′-DI particles synthesize a 46-base-long RNA from their 3′-ends,[15-18] in a manner similar to that of the standard virus, which synthesizes the leader RNA (48 bases long) and in addition synthesizes five mRNAs.[23] These results, taken together suggest that the modes of transcription and replication of the RNAs of 5′-DI particles and the standard virus genome RNA contains additional genes for transcription that the 5′-DI particles lack. The DI-LT, on the other hand, behaves similarly to the standard virus, because both contain identical 3′-terminal sequences.

From studies demonstrating differential inhibition by dATP of standard virus transcription and full-length plus-strand synthesis *in vitro*,[20] it was suggested that two enzymes may be involved in these processes: one involved in the synthesis of mRNAs and the other in the synthesis of the leader RNA, which under appropriate conditions extends it to the full-length complement of the genome RNA. These results, coupled with those obtained in the present studies, suggest that in the DI particles only the latter enzyme is operative, synthesizing the small RNA product attenuated at base 46 and extending it to full-length particle RNA only under conditions of presumptive dephosphorylation (in the presence of AMPPNP). It is yet unclear which proteins (L, NS, or N) associated with the transcribing ribonucleoproteins of the DI particles and the standard virus are involved in these processes. It is important to note that the leader RNA and the small RNA of the DI particles share identical eight-base sequences from the 5′ end—i.e., ppA-C-G-A-A-G-A-C.[15-18] Thus, the corresponding eight bases from the 3′-end of their particle RNAs (. . . G-U-C-U-U-C-G-U_{OH}) appear to be the common binding sequence of the putative replicase shared by both of the transcribing particles. The presence of a possible common binding sequence for replicase in both the particle RNAs will argue against any replicative advantage of the DI particle over the standard virus.

The synthesis of the full-length complement of the DI and the standard virion RNAs *in vitro* has permitted us to test *in vitro* whether DI particles interfere with the production of the standard virus at the level of replication or transcription. Experiments are in progress to learn more about the molecular mechanism of DI-particle-mediated viral interference using this system.

References

1. Huang, A.S. and Baltimore, D. (1977): In: *Comprehensive Virology*. Fraenkel-Conrat, H. and Wagner, R.R. (Eds.), Plenum, New York, Vol. 10, pp. 73–116.
2. Holland, J.J. and Villarreal, L.P. (1974): *Proc. Natl Acad. Sci. U.S.A.* 71:2956–2960.
3. Sekellick, M.J. and Marcus, P.I. (1978): *Virology* 85:175–186.
4. Younger, J.S., Dubovi, E.J., Quagliana, D.O., Kelly, M., and Preble, O.T. (1976): *J. Virol.* 19:90–101.
5. Clewley, J.P., Bishop, D.H.L., Kang, C.Y., Coffin, J., Schnitzlein, W.M., Reichmann, M.E., and Shope, R.E. (1977): *J. Virol.* 23:152–166.
6. Wagner, R.R. (1975): In: *Comprehensive Virology*. Fraenkel-Conrat, H. and Wagner, R.R. (Eds.), Plenum, New York, Vol. 4, pp. 1–93.
7. Banerjee, A.K., Abraham, G., and Colonno, R.J. (1977): *J. Gen. Virol.* 34:1–8.
8. Leamnson, R.N. and Reichmann, M.E. (1974): *J. Mol. Biol.* 85:551–568.
9. Stamminger, G. and Lazzarini, R.A. (1974): *Cell* 3:85–93.
10. Perrault, J. and Semler, B. (1979): *Proc. Natl. Acad. Sci. U.S.A.* 76:6191–6195.
11. Petric, M. and Prevec, L. (1970): *Virology* 41:615–630.
12. Schnitzlein, W.M. and Reichmann, M.E. (1976): *J. Mol. Biol.* 101:307–325.
13. Colonno, R.J., Lazzarini, R.A., Keene, J.D., and Banerjee, A.K. (1977): *Proc. Natl. Acad. Sci. U.S.A.* 74:1884–1888.
14. Johnson, L.D. and Lazzarini, R.A. (1977): *Proc. Natl. Acad. Sci. U.S.A.* 74:4387–4391.
15. Emerson, S.U., Dierks, P.M., and Parsons, J.T. (1977): *J. Virol.* 23:708–716.
16. Riechmann, M.E., Villarreal, L.P., Kohne, D., Lesnaw, J., and Holland, J.J. (1974): *Virology* 58:240–249.
17. Schubert, M., Keene, J.D., Lazzarini, R.A., and Emerson, S.U. (1978): *Cell* 15:103–112.
18. Semler, B.L., Perrault, J., Abelson, J., and Holland, J.J. (1978): *Proc. Natl. Acad. Sci. U.S.A.* 75:4704–4708.
19. Perrault, J. and Leavitt, R.W. (1977): *J. Gen. Virol.* 38:35–50.
20. Testa, D., Chanda, P.K., and Banerjee, A.K. (1980): *Proc. Natl. Acad. Sci. U.S.A.* 77:294–298.
21. Kang, C.Y., Glimp, T., Clewley, J.P., and Bishop, D.H.L. (1978): *Virology* 84:142–152.
22. Banerjee, A.K. and Rhodes, D.P. (1973): *Proc. Natl. Acad. Sci. U.S.A.* 70:3566–3570.
23. Colonno, R.J. and Banerjee, A.K. (1978): *Cell* 8:197–204.

Copyright 1981 by Elsevier North Holland, Inc.
David H.L. Bishop and Richard W. Compans, eds.
The Replication of Negative Strand Viruses

Association of the Transcriptase and RNA Methyltransferase Activities of Vesicular Stomatitis Virus with the L-Protein

Exeen M. Morgan and David W. Kingsbury[a]

The NS and L-proteins of the vesicular stomatitis virus (VSV) nucleocapsid possess the enzymatic activities responsible for mRNA transcription[1-3] and steps in its posttranscriptional modification, including capping and methylation.[4,5] However, it is not known which enzymatic activities reside in L and which reside in NS.

One approach to this question is selective covalent modification of the viral protein that carries a given enzymatic activity. The ideal reagent would be an affinity label,[6] a compound with enough resemblance to a substrate to bind specifically to an active site, but also containing a chemically reactive group capable of forming a covalent bond at that site. Alternatively, a compound that is not an active site reagent might still interact specifically with a single enzyme and thereby identify the relevant protein species.[7] Previously, we found that pyridoxal 5-phosphate (PLP) was a selective label of the mRNA-synthesizing enzymes of reovirus.[7] PLP is structurally similar to nucleoside triphosphates and it contains an aldehyde group that is reactive with exposed ϵ-amino groups of lysine residues in proteins. Here, we report the effects of PLP on the transcriptase and RNA methyltransferase activities of VSV.

Noncompetitive Inhibition of Transcriptase

PLP markedly inhibited VSV transcriptase (Figure 1A). The aldehyde and phosphate moieties were both essential for inhibition, as shown by the rela-

[a] Division of Virology, St. Jude Children's Research Hospital, Memphis, Tennessee.

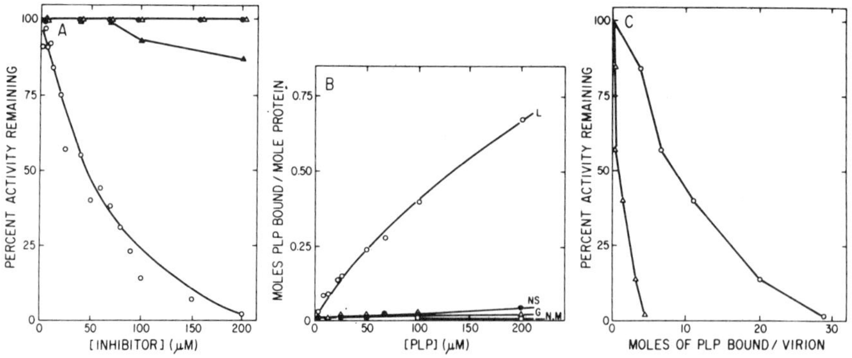

Figure 1. Inhibition of VSV transcriptase activity and labeling of VSV polypeptides as a function of PLP concentration. Panel A: The indicated concentrations of PLP or one of its congeners were added to transcriptase reactions (see Table 1), which were incubated for 30 minutes at 30°C. The untreated controls incorporated 7.8 picomoles of ^{3}H UMP per μg of viral protein per hr ○, PLP; ●, pyridoxal; ▲, pyridoxamine 5-phosphate, △, pyridoxamine. Panel B: Triton N-101-treated virions were mixed with PLP and ^{3}H-borohydride (see Figure 3). The stoichiometry of PLP bound to each polypeptide was based on the numbers of protein molecules per virion given by Bishop and Smith.[8] Panel C: The inhibition of transcriptase activity by PLP compared with the binding of PLP to polypeptides NS (△) and L (○).

tive inactivity of pyridoxamine, pyridoxamine 5-phosphate, and pyridoxal (Figure 1A). Transcriptase inhibition by PLP was readily reversible upon dilution, whereas reduction with KBH_4, which produces a covalent PLP-protein complex, made the inhibition irreversible (data not shown). All of these results are characteristic of PLP interactions with protein ϵ-amino groups *via* Schiff base formation.[7]

Kinetic analyses, in which each of the four nucleoside triphosphate substrates of transcriptase was varied independently, gave noncompetitive patterns (data not shown). Therefore, PLP appears to inhibit VSV transcriptase by binding to a site distinct from the nucleoside triphosphate binding site(s). The apparent K_i values, which provide a measure of the affinity of PLP for the enzyme, were similar for all four nucleoside triphosphates, suggesting that binding of PLP to a single site on the enzyme molecule is responsible for transcriptase inactivation.

Competitive Inhibition of RNA Methyltransferase

To measure methyltransferase activity, it was necessary to choose conditions that allowed transcription, since the VSV methyltransferase activities will not accept exogenous unmethylated mRNA as a substrate.[4] Accordingly, PLP concentrations used in methyltransferase analyses were chosen to give no more than 60% inhibition of transcriptase. We found that PLP inhibited VSV methyltransferase independently and that it was competitive with the substrate, S-adenosyl-L-methionine (AdoMet), as shown in Figure 2. Thus, PLP appears to be an active site inhibitor of VSV RNA methyltransferase activity.

The VSV nucleocapsid contains two methyltransferase activities, one specific for the ribosyl moiety of the penultimate nucleotide and the other specific for the N-7 position of the guanine moiety of the cap.[5,9] Since modification by the former activity must occur before the latter can act, we presume that we have been measuring predominantly 2-0-methyltransferase activity in our assays. This view was supported by the similarity of the apparent K_m of 1.2 μM for AdoMet that we obtained (Table 1) to a previously reported value.[9] Table 1 also shows that the apparent K_i of 29 μM for PLP inhibition of methyltransferase is within the range of the apparent K_i values for the interaction of PLP with VSV transcriptase. This is consistent with the idea that the active site of the methyltransferase is the same site that

Figure 2. Kinetic analysis of VSV methyltransferase and its inhibition by PLP. The reaction conditions are described in Table 1. Velocity (V) is expressed as picomoles of methyl groups incorporated into RNA in 30 minutes at 30°C. PLP concentrations (μM): ○, none; △, 20; ●, 60.

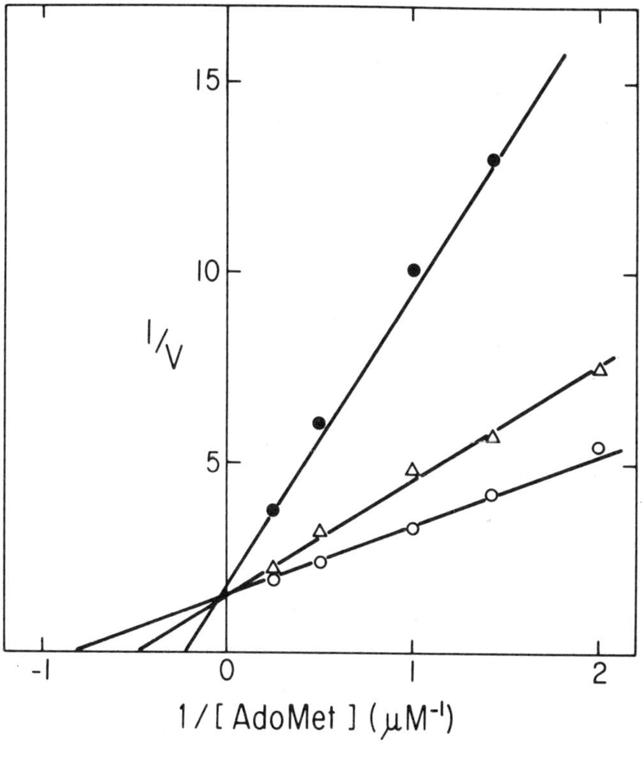

Table 1. Kinetic Constants of VSV Transcriptase and RNA Methyltransferase.

Activity	Substrate	$K_m(M)^c$	$K_i(M)^d$
Transcriptase[a]	ATP	$3.7(\pm1.0)\times10^{-4}$	$3.1(\pm0.2)\times10^{-5}$
	UTP	$1.3(\pm0.4)\times10^{-5}$	$3.2(\pm0.5)\times10^{-5}$
	GTP	$2.2(\pm0.3)\times10^{-5}$	$7.2(\pm1.0)\times10^{-5}$
	CTP	$3.2(\pm0.9)\times10^{-5}$	$3.4(\pm1.6)\times10^{-5}$
Methyltranserase[b]	AdoMet	$1.2(\pm0.07)\times10^{-6}$	$2.9(\pm0.5)\times10^{-5}$

[a]Transcriptase reaction mixtures contained 100 mM NaCl, 50 mM Hepes (pH 8.0), 5 mM $MgCl_2$, 4 mM dithiothreitol, 0.05% Triton N-101, 1 mM each ATP, CTP, GTP, 0.1 mM UTP, 2 μCi of ^{3}H UTP, 0.5 μg of actinomycin D, and 10 μg of VSV per 0.1 ml reaction.[10] The samples were incubated at 30°C for 30 minutes.

[b]Methyltransferase reaction mixtures were the same as the transcriptase reactions, except they contained 1 mM each ATP, CTP, GTP and UTP and 0.2 to 4.0 μM ^{3}H S-adenosyl-L-methionine (AdoMet).

[c]Values (plus or minus standard errors) were derived from assays in which the velocity at several substrate concentrations was determined in triplicate. Nonlinear least squares fit to the Michaelis-Menton equation.[11]

[d]Coefficient of inhibition by PLP. For noncompetitive inhibition (transcriptase), K_i was determined by the formula $K_i = i/(V/V_p - 1)$ where i is the PLP concentration and V and V_p are the V_{max} values in the presence and absence of PLP, respectively.[12] For competitive inhibition (methyltransferase), K_i was determined by $K_i = i/K_p/K - 1)$ where i is the concentration of PLP and K and K_p are the K_m values obtained in the absence or presence of PLP.[12]

is responsible for transcriptase inactivation by PLP, both enzymatic activities being resident on a single protein molecule. Alternatively, different lysine residues in a single protein or in different proteins might coincidently have very similar affinities for PLP.

PLP-Directed Labeling of VSV Polypeptides

In a single step, PLP can be covalently linked to a protein and radioactivity introduced into the complex if ^{3}H-borohydride is used as a reducing agent.[7] When VSV proteins treated in this way were separated by electrophoresis, we saw radioactivity in all viral protein species (Figure 3). However, the stoichiometry of labeling indicated a marked selectivity for the L-protein (Figure 1B). PLP-directed labeling of L and transcriptase inactivation were correlated in a basically linear fashion up to about 50% inhibition of the enzyme (Figure 1C). Over the same range of input PLP concentrations, labeling of NS, the other candidate for enzymatic activity, was negligible (Figure 1). At higher PLP concentrations, labeling of both L and NS increased out of proportion to transcriptase inactivation, indicating that PLP was now binding to sites that had no direct relationship with enzyme activity. Thus, interaction of PLP with a limited number of sites on a subset of the L-protein molecules in a VSV nucleocapsid inactivates both methyltransferase and transcriptase activities. We conclude that the L-protein con-

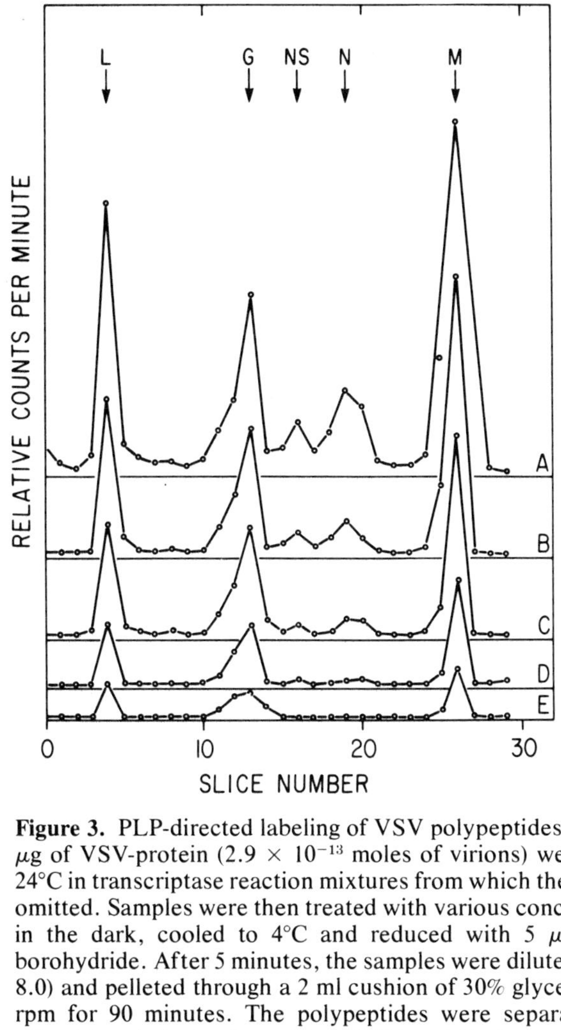

Figure 3. PLP-directed labeling of VSV polypeptides. 0.1 ml samples containing 100 μg of VSV-protein (2.9 × 10^{-13} moles of virions) were incubated for 30 minutes at 24°C in transcriptase reaction mixtures from which the nucleoside triphosphates were omitted. Samples were then treated with various concentrations of PLP for 5 minutes in the dark, cooled to 4°C and reduced with 5 μl (1.5 mCi) of potassium ^{3}H-borohydride. After 5 minutes, the samples were diluted to 3 ml with 5 mM Hepes (pH 8.0) and pelleted through a 2 ml cushion of 30% glycerol in an SW 55 rotor at 45,000 rpm for 90 minutes. The polypeptides were separated by discontinuous polyacrylamide gel electrophoresis in SDS[13] and the incorporation of tritium into the individual polypeptides was determined.[7] PLP concentrations (μM): A, 200; B, 100; C, 50; D, 25; E, 12.5.

tains the methyltransferase active site, and we also suspect that it contains the transcriptase active site.

Summary

We have shown that PLP is a competitive inhibitor of VSV mRNA methyltransferase and a noncompetitive inhibitor of the viral transcriptase. Inhibition of these enzymatic activities was correlated with preferential binding of

PLP to the L-protein species. The L-protein species appears to contain the methyltransferase active site and another site that influences the activity of the transcriptase, perhaps within the same protein molecule.

ACKNOWLEDGMENTS
Karen Rakestraw and Donna Clift provided skillful technical assistance. This work was supported by Research Grant AI-05343 from the National Institute of Allergy and Infectious Diseases and by ALSAC.

References

1. Emerson, S.U. and Wagner, R.R. (1972): *J. Virol.* 10:297–309.
2. Naito, S. and Ishihama, A. (1976): *J. Biol. Chem.* 251:4307–4314.
3. Emerson, S.U. and Yu, Y. (1975): *J. Virol.* 15:1348–1356.
4. Abraham, G. and Banerjee, A.K. (1976): *Virology* 71:230–241.
5. Banerjee, A.K. (1980): *Microbiol. Rev.* 44:175–205.
6. Singer, S.J. (1967): *Adv. Protein Chem.* 22:1–54.
7. Morgan, E.M. and Kingsbury, D.W. (1980): *Biochemistry* 19:484–489.
8. Bishop, D.H.L. and Smith, M.S. (1977): In: *The Molecular Biology of Animal Viruses.* Nayak, D.P. (Ed.), Marcel Dekker, New York, pp. 167–280.
9. Testa, D. and Banerjee, A.K. (1977): *J. Virol.* 24:786–793.
10. Banerjee, A.K. and Rhodes, D.P. (1973): *Proc. Natl. Acad. Sci. U.S.A.* 70:3566–3570.
11. Wilkinson, G.N. (1961): *Biochem. J.* 80:324–332.
12. Dixon, M. and Webb, E.C. (1964): *Enzymes.* Academic Press, New York.
13. Laemmli, U.K. (1970): *Nature (London)* 227:680–685.

David H.L. Bishop and Richard W. Compans, eds.
The Replication of Negative Strand Viruses

A Role for NS-Protein Phosphorylation in Vesicular Stomatitis Virus Transcription

David W. Kingsbury, Chung-H. Hsu, and Exeen M. Morgan[a]

The NS-protein of vesicular stomatitis virus (VSV) is a multifunctional phosphoprotein that participates in viral RNA transcription and genome replication.[1-3] It may also provide a function in virus development unrelated to RNA synthesis.[3] Several phosphorylated species of NS have been identified recently, suggesting that different functions of NS are regulated by the phosphorylation or dephosphorylation of different specific sites in the protein.[4-6] Kingsford and Emerson[6] resolved five phosphorylated species of NS molecules. Only the more highly phosphorylated species were capable of restoring *in vitro* transcription in reconstitution experiments. In the present study, we exploited the ability of bacterial alkaline phosphatase (BAP) to dephosphorylate NS as an independent test of the involvement of protein phosphorylation in VSV transcription.

Enzymatic Dephosphorylation of NS-Protein Species

Alkaline phosphatase can cleave phosphoester bonds in proteins in addition to a variety of low molecular weight metabolites.[7] We used the bacterial (*E. coli*) enzyme, because it has been studied extensively.[7] However, we found it necessary to add the protease inhibitor aprotinin[8] to block a protease that contaminates commercial BAP. Figure 1 displays the action of BAP on the proteins in VSV nucleocapsids. Before enzyme treatment, electrophoresis[6]

[a] Division of Virology, St. Jude Children's Research Hospital, Memphis, Tennessee.

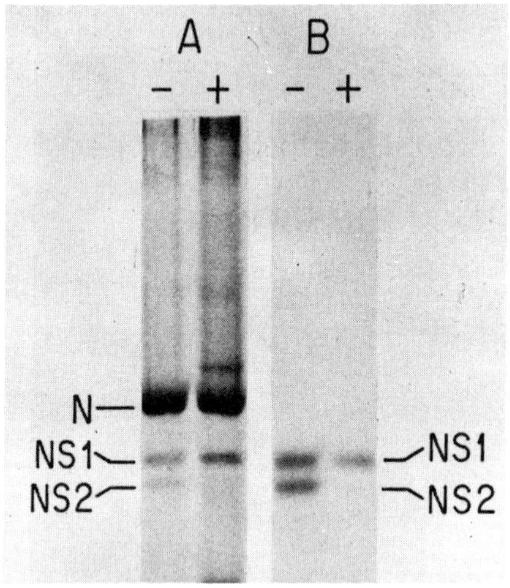

Figure 1. Enzymatic dephosphorylation of NS. Virions were isolated from cells incubated with ^{32}P-phosphate and disrupted 1% Triton X-100 in 0.01 M Tris-HCl, 0.001 M $MgCl_2$, 0.15 M NaCl (pH 7.4) at 24°. BAP (Worthington) was added to a final concentration of 2 mg/ml. The concentration of viral protein was 400 μg/ml and Aprotinin (Sigma) was present at 67 μg/ml. Incubation for 1 hr at 30° was followed by centrifugation on a sucrose step-gradient to isolate nucleocapsids.[9] The nucleocapsid proteins were then examined by polyacrylamide gel electrophoresis.[6] (A) Coomassie blue stain; (B) ^{32}P-radioactivity; (+) BAP-treated; (−) untreated control.

resolved two classes of NS molecules, which we have designated NS1 and NS2. These correspond to components "NS top" and "NS bottom" of Kingsford and Emerson[6] and to "NS1" and "NS2" of Clinton et al.[4] In terms of protein concentration, NS1 is more abundant than NS2, but in terms of the specific activity of radiophosphate, NS2 is the more highly phosphorylated species.[4,6] As BAP removed phosphate, NS2 was converted to NS1 and the level of phosphate label in the NS1 class of molecules was reduced (Figure 1). About 20% of the radiophosphate originally present in the total NS population remained in the NS1 class after exhaustive BAP treatment. Most of the phosphate residues in NS were thus revealed to be accessible on the exterior of the intact nucleocapsid. Presumably, the fraction of phosphate resistant to BAP is in a sterically sheltered position. We saw no proteolytic degradation of NS molecules after BAP treatment, and when we used another electrophoretic procedure,[9] the mobilities of all nucleocapsid protein species were unaltered, confirming the abrogation of proteolysis by the added aprotinin (data not shown).

Loss of Transcriptase Activity on Dephosphorylation

The changes shown in Figure 1 were accompanied by a marked reduction of *in vitro* transcriptase activity (Figure 2). Relative to a control that was preincubated at 30° in the absence of BAP, the loss of activity in this experiment was about 70%, with a range of 50% to 80% in other experiments. Relative to a control nucleocapsid preparation that was not preincubated, the difference was slightly greater (Figure 2). The products of both the control and inhibited reactions were predominantly viral messenger RNA species and there was no change in their relative abundances (data not shown).

Figure 3 shows that the viral genomic RNA template was not altered in its sedimentation rate by the incubation with BAP. Thus, the loss of transcriptase activity cannot be ascribed to cleavage of the template RNA by a contaminating nuclease. Since BAP destroys nucleoside triphosphates, we determined that traces of BAP remaining after nucleocapsid isolation did not interfere with transcription by substrate depletion. We observed no loss of the L-protein from nucleocapsids after enzymatic dephosphorylation of NS. In addition, we have observed no phosphate incorporation into L *in vivo,*

Figure 2. Time course of RNA synthesis by VSV nucleocapsids. Virion nucleocapsids were treated with BAP, isolated by centrifugation as described in Figure 1 and tested for RNA synthesizing ability[10] at 30°. ○, BAP-treated; △, untreated, but incubated and centrifuged in parallel with the treated sample; ●, untreated, unincubated control.

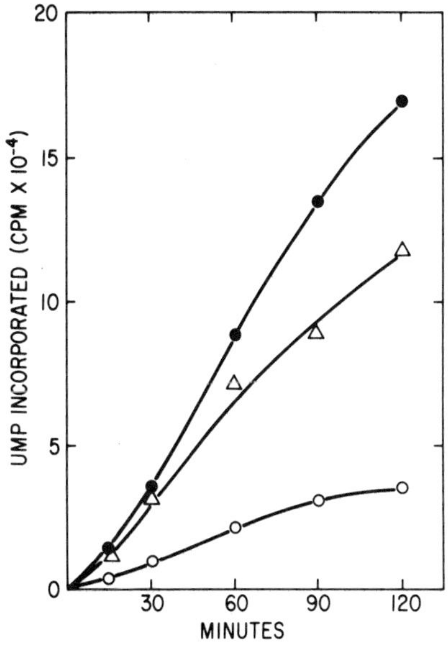

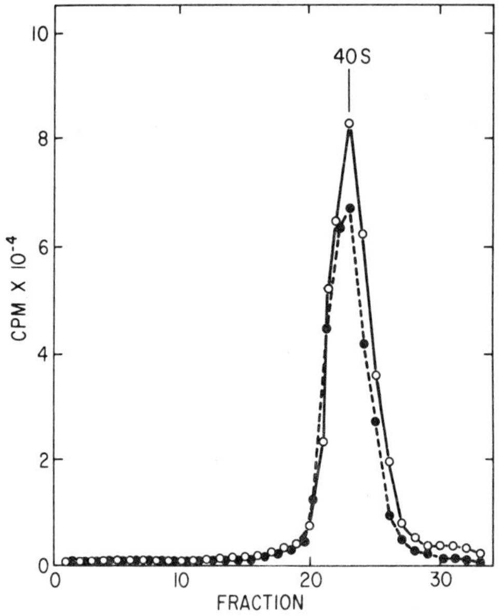

Figure 3. Template RNA after incubation with BAP. ^{3}H-uridine-labeled virions were disrupted, treated with BAP, and centrifuged as described in Figure 1. RNA was then analyzed by sucrose gradient centrifugation and acid insoluble radioactivity was determined. Centrifugation was for 16 hr at 18,000 rpm, 20°C. The direction of sedimentation is shown from left to right. ○, untreated control; ●, BAP-treated.

confirming that L is not a phosphoprotein.[11,13] Therefore, it appears that the enzymatic dephosphorylation of VSV NS-protein is the event responsible for the decline in transcriptase activity that we observed.

Unfaithful Rephosphorylation

VS virions contain a protein kinase which can add phosphate to NS and other virus proteins *in vitro*.[11-13] Since ATP, a transcriptase substrate, is a phosphate donor in the kinase reaction, we expected that rephosphorylation of NS would occur during the transcriptase assay. However, as shown in Figure 4, although NS was rephosphorylated under these conditions, the original pattern was not restored. Lane 1 displays the products of the endogenous protein kinase, when whole virions were disrupted with non-ionic detergent and incubated with γ-^{32}P ATP. NS and M were the most avid phosphate acceptors and more phosphate accumulated in NS2 than in NS1, in agreement with previous reports.[4,11-13] In lane 2, kinase activity remaining associated with isolated nucleocapsids generated lesser amounts of the same NS products. As shown in lane 3, when nucleocapsids were treated with BAP, isolated, and then incubated with γ-^{32}P ATP, NS1 was the principle

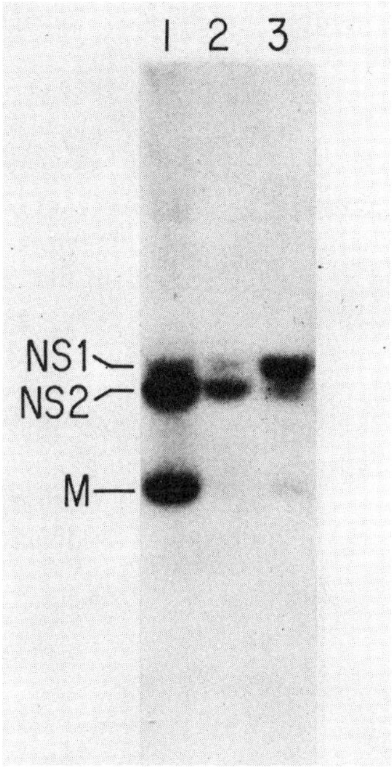

Figure 4. Phosphorylation of viral proteins by virion protein kinase. Samples were incubated in a reaction mix identical to that used in transcription (Figure 2), but γ-^{32}P ATP was used as the radioactive label. Nonradioactive ATP was added to a final concentration of 1 mM, the same concentration used in the transcriptase assay.[10] Incubation was for one hour at 30°. Proteins were analyzed by electrophoresis.[6] (1) Virions treated with 0.1% Triton N-101; (2) nucleocapsids isolated from Triton X-100-disrupted virions by centrifugation on a sucrose step gradient; (3) nucleocapsids treated with BAP and then isolated on a sucrose step gradient, as described in Figure 1.

phosphorylated product and only a small amount of NS2 was produced. Therefore, *in vitro* rephosphorylation under transcriptase assay conditions did not restore the pattern obtained when NS was phosphorylated *in vivo* (compare Figure 1).

Discussion

There are several possible explanations for the failure of the virion-associated protein kinase to restore the *in vivo* phosphorylation pattern of NS. A kinase that has nothing to do with the intracellular phosphorylation of VSV-proteins may be the one that is incorporated into virions when they bud at the cell membrane and this kinase might have a different specificity for sites in NS. Alternatively, dephosphorylation of NS might produce a conformational change in the protein or *in vitro* conditions might alter kinase specificity, preventing accurate rephosphorylation.

Our data provide independent evidence that the state of NS phosphorylation is relevant to the transcriptase activity of VSV. Since the NS2 class of molecules was not restored by *in vitro* rephosphorylation, this class may contain the essential form for transcriptase activity.[6] However, we cannot rule out the participation of molecules that migrated as NS1, since phosphate was also removed from this class. In this regard, it may be important that a significant fraction of transcriptase activity remained when all of NS2 was eliminated by BAP treatment. Either NS2 is not absolutely required for transcription or incomplete dephosphorylation of a critical NS1 species might account for the residual activity. Clearly, a complete separation of all the phosphorylated forms of NS will be necessary to sort out their individual contributions to transcription. When these forms are tested for function, BAP may help to reveal how covalently bound phosphate residues modulate their activities.

Summary

Enzymatic removal of phosphate from NS-proteins in VSV nucleocapsids by BAP reduced *in vitro* transcriptase activity up to 80%. The highly phosphorylated NS2 class of molecules was especially sensitive to BAP, whereas the NS1 class was partially resistant. Rephosphorylation by the protein kinase in virions did not restore the native phosphorylated species, accounting for the irreversibility of transcriptase inactivation. We conclude that phosphate residues located at specific sites in NS potentiate the performance of the protein in transcription.

ACKNOWLEDGMENTS

Sallie Clark and Karen Rakestraw provided skillful technical assistance. Supported by Research Grant AI 05343 from the National Institute of Allergy and Infectious Diseases and by ALSAC.

References

1. Evans, D., Pringle, C.R., and Szilagyi, J.F. (1979): *J. Virol.* 31:325–333.
2. Lesnaw, J.A. and Reichmann, M.E. (1975): *Virology* 63:492–504.
3. Szilagyi, J.F. and Pringle, C.R. (1979): *J. Virol.* 30:692–700.
4. Clinton, G.M., Burge, B.W., and Huang, A.S. (1979): *Virology* 99:84–94.
5. Hsu, C.-H. and Kingsbury, D.W. (1980): In: *Animal Virus Genetics, ICN-UCLA Symposia on Molecular and Cellular Biology,* Vol. 18. Fields, B., Jaenisch, R., and Fox, C.F. (Eds.), Academic Press, New York, in press.
6. Kingsford, L. and Emerson, S.U. (1980): *J. Virol.* 33:1097–1105.
7. Reid, T.W. and Wilson, I.B. (1971): In: *The Enzymes,* Vol. 4. Boyer, P.D. (Ed.), Academic Press, New York, pp. 373–415.

8. Werle, E. (1972): In: *New Aspects of Trasylol Therapy,* Vol. 5. Brendel, W. and Haberland, G.L. (Eds.), Schattauer, Stuttgart, pp. 9–16.
9. Hsu, C.-H., Kingsbury, D.W., and Murti, K.G. (1979): *J. Virol.* 32:304–313.
10. Banerjee, A.K. and Rhodes, D.P. (1973): *Proc. Natl. Acad. Sci. U.S.A.* 70:3566–3570.
11. Sokol, F. and Clark, H.F. (1973): *Virology* 52:246–263.
12. Imblum, R.L. and Wagner, R.R. (1974): *J. Virol.* 13:113–124.
13. Moyer, S.A. and Summers, D.F. (1974): *J. Virol.* 13:455–465.

David H.L. Bishop and Richard W. Compans, eds.
The Replication of Negative Strand Viruses

In Vitro Transcription Alterations in a Vesicular Stomatitis Virus Variant

J. Perrault, J.L. Lane, and M.A. McClure[a]

Introduction

The transcription process of vesicular stomatitis virus (VSV) has been under intensive investigation for several years but several aspects of its regulation are poorly understood. Purified viral cores which contain only three proteins (N, NS, and L) synthesize all five VSV mRNAs *in vitro* and also carry out the posttranscriptional modifications, i.e., capping, methylation, and polyadenylation.[1] The latter reactions appear to be intimately coupled to the processes of initiation and termination of transcription. Conceivably, cleavage of short-lived precursor molecules is also involved but evidence is lacking. A small leader RNA, 48 bases long and complementary to the 3′ end of the template, is synthesized during the transcription reaction. The transcripts appear sequentially and in stepwise decreasing amounts reflecting their location on the template, i.e., 3′-leader-N-NS-M-G-L-5′.

We have recently described a VSV variant, pol R1, which shows important changes in virus particle polymerase activitics. The mutated virus generates defective interfering virus (DI) particles which synthesize nearly full size copies of their templates *in vitro*.[2] DI particles generated from wildtype virus were shown previously to synthesize only a small 46 base-long transcript complementary to the 3′ end of the DI RNA template.[3]

Pol R1 virus was obtained after a selection procedure involving several cycles of heat inactivation of wildtype virus (Mudd-Summers, standard ATCC Indiana strain) and growth of survivors (details to be published

[a] Department of Microbiology and Immunology, Washington University School of Medicine, St. Louis, Missouri.

elsewhere). The variant phenotype has remained stable through several successive clonings and yields are similar to wildtype at 37°C. Curiously, infectious particles of the variant are not heat resistant as compared to wildtype. The *in vitro* polymerase activity and stability as a function of temperature is also similar to wild type (unpublished observations). We have detected minor electrophoretic mobility changes in SDS gels for at least two of the viral proteins (N and M) but the genome T1 oligonucleotide maps are identical (in preparation). We have also shown that it is the mutation(s) in one or more polymerase protein components of pol R1, as opposed to changes in template recognition sites, which is or are responsible for the pol R1 phenotype measured in DI particles.[2]

We report here that infectious pol R1 virions synthesize mRNA transcripts *in vitro* with larger and more heterogeneously-sized poly(A) tracts than wildtype virus. In addition, the variant synthesizes reduced amounts of leader RNA.

Results

Incorporation of labeled precursor by wildtype and pol R1 core particle preparations. Virus preparations were purified as described before[4] and stripped of their envelopes by the method of Breindl and Holland.[5] Transcriptase reactions with cores were carried out in the presence of high salt (0.3 M potassium acetate) as described previously[2] to maximize synthesis of full-size mRNA transcripts. Figure 1 shows a comparison of wildtype vs pol R1 activity under these conditions. Incorporation levels off sooner for the variant than wildtype and accumulates about one-third as much labeled ^{3}H-UTP precursor by two hours of reaction. Similar relative levels of activity (~ six-fold net syntheses for wildtype) were also obtained with α-^{32}P-labeled GTP precursor. However, only about a two-fold difference of activity was observed with ^{3}H-ATP as precursor. Interestingly, little or no difference in ^{3}H-UTP incorporation was observed between wildtype and pol R1 when whole virus preparations rather than core preparations were used.[2]

Analysis of glyoxal-denatured products in agarose gels. The ^{3}H-UTP-labeled products accumulated at two hours of reaction were purified by CsCl centrifugation,[6] followed by proteinase K digestion, phenol-chloroform extraction, and ethanol precipitation.[4] They were then separated into poly(A)$^{+}$ and poly(A)$^{-}$ fractions by fractionation on oligo(dT) columns.[7] In two separate experiments, 81% and 86% of the wildtype products bound to the column, whereas 63% and 83% of the pol R1 products bound. Figure 2 shows an analysis of the poly(A)$^{+}$ and poly(A)$^{-}$ RNA products after denaturation in glyoxal as described previously.[8] The poly(A)$^{+}$ samples were divided into two aliquots, one of which was annealed to oligo(dT) and digested with RNase H to remove poly(A) tails as described by Rowlands et al.,[9] and

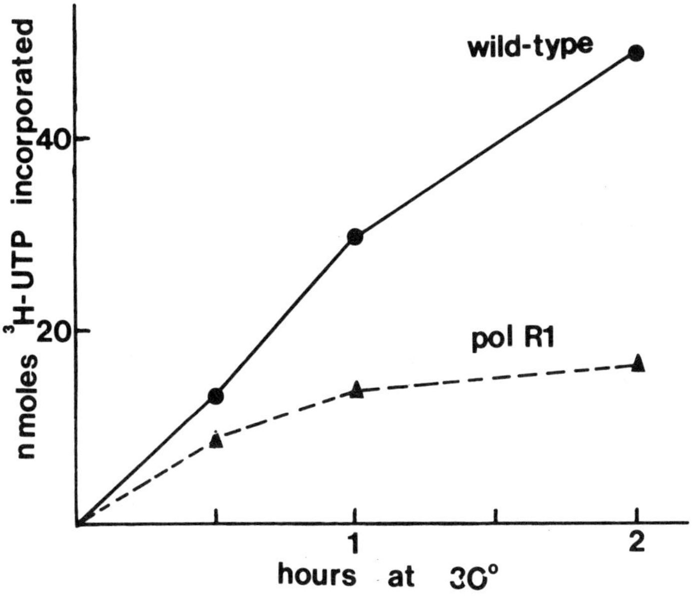

Figure 1. Incorporation of ^{3}H-UTP by purified cores of wildtype and pol R1 virus *in vitro*. The 1 ml reactions contained cores from 1 mg of whole virus.

repurified before glyoxal denaturation and analysis on gels. Figure 2A compares products before and after RNase H digestion. Broad bands corresponding to N, NS + M, and G mRNAs, as well as the sharper high molecular weight L mRNA band, can be clearly seen in the untreated wildtype products (lane a). However, the untreated pol R1 products appear much more heterogeneous and diffuse (lane c). Only the L mRNA can be seen as a distinct band, in addition to a band migrating slightly faster than N mRNA and which is also present in small amounts in wildtype products. Removal of poly(A) tails with RNase H (lanes b and d) results in a dramatic sharpening of bands for both wildtype and pol R1. The individual poly(A)$^+$-selected mRNAs can now be clearly observed and little or no difference is seen between wildtype and pol R1 products. This experiment therefore suggests that the more heterogeneous and diffuse pattern of pol R1 products is due to longer and more heterogeneously-sized poly(A) tails than those found in wildtype products. This conclusion is supported by the results presented below.

We also considered the possibility that pol R1 mRNAs might be linked via intervening poly(A) tracts of variable size.[10] However, analysis of ^{3}H-S-adenosyl-methionine-labeled products as in Figure 2A also show identical patterns for wildtype and pol R1 after removal of poly(A) tracts (not shown). It is thus unlikely that a large fraction of undigested pol R1 mRNAs are

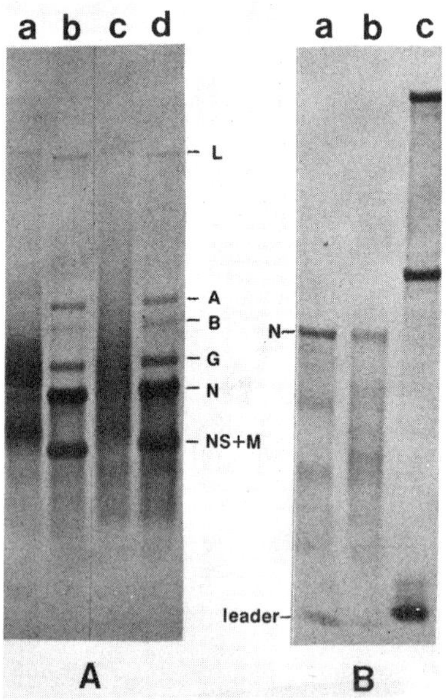

Figure 2. Agarose gel analysis of glyoxal-denatured wildtype and pol R1 *in vitro* transcriptase products. (A) poly(A)$^+$-selected RNA: lanes a and b, wildtype products before and after specific removal of poly(A) tails as described in the text; lanes c and d, pol R1 products before and after removal of poly(A) tails. (B) poly(A)$^-$-selected RNA: lane a, wildtype; lane b, pol R1; lane c, ^{3}H-uridine-labeled marker RNAs from HeLa cells. The upper portion of the gel did not contain any labeled material and is not represented in these fluorographs.

linked by poly(A) since this would lead to a significant reduction in labeled cap structures.

In addition to the expected mRNAs, both wildtype and pol R1 products contain two bands, A and B, larger than G mRNA (Figure 2A, lanes b and d). These products are also found *in vivo* and appear to be unprocessed or read-through, polycistronic transcripts (manuscript in preparation). The band migrating slightly faster than N mRNA in the undigested samples co-migrates with N mRNA lacking poly(A) tails and is enriched in the poly(A)$^-$ fractions (see Figure 2B). Hybridization experiments not shown here suggest that this band is, in fact, N mRNA with little or no poly(A) tails. Under these transcription conditions, some N mRNA (< 10%) is apparently synthesized without the usual homopolymeric extension.

Figure 2B shows the poly(A)$^-$ products which did not bind to the oligo(dT) column. In addition to N mRNA lacking poly(A) tails, we observe two broad bands of smaller transcripts, roughly 1000 and 750 bases long, which also appear somewhat larger and more heterogeneous in pol R1. We have not yet characterized these products. Lastly, we can see the small 48 base-long leader RNA in both virus preparations but in clearly reduced amounts in pol R1. The differences between the variant and wildtype small RNA products are described in more detail below.

Size estimate of poly(A) tracts in wildtype vs pol R1 products. Total products from transcription reaction carried out as above but labeled with α-^{32}P-ATP were digested with RNases A and T1 (10 μg/ml and 50 μg/ml respectively, 37°C, 30 min) in 80 μl of 0.3 M NaCl, 10 mM Tris-Cl, pH 7.6, 1 mM EDTA, containing 50 μg of carrier yeast RNA. The RNase resistant fractions, 32% and 39% of the total cpm for wildtype and pol R1 products respectively, were analyzed in agarose gels as described above. Densitometer tracings of the autoradiographic films were carried out as described previously.[4]

The results shown in Figure 3 indicate that the distribution of poly(A) chain lengths peaks at ~ 310 bases for pol R1 products and ~ 200 bases for wildtype. Furthermore, 35% of pol R1 poly(A) chains are > 500 bases in length as opposed to 12% for wildtype. The accuracy of the size estimate for the homopolymeric poly(A) chain length relative to glyoxal-denatured markers ribosomal and transfer RNAs has not been evaluated in this gel system. However, it is clear that the pol R1 transcripts do contain longer poly(A) tails than wildtype transcripts.

Figure 3. Size distribution of poly(A) chain lengths from wildtype and pol R1 transcriptase products. The RNase resistant fractions (see text) were analyzed as described for Figure 2, and the autoradiographs were quantitated by densitometry. The positions of the ^{3}H-uridine-labeled marker RNAs from HeLa cells (run in a parallel gel slot) are indicated.

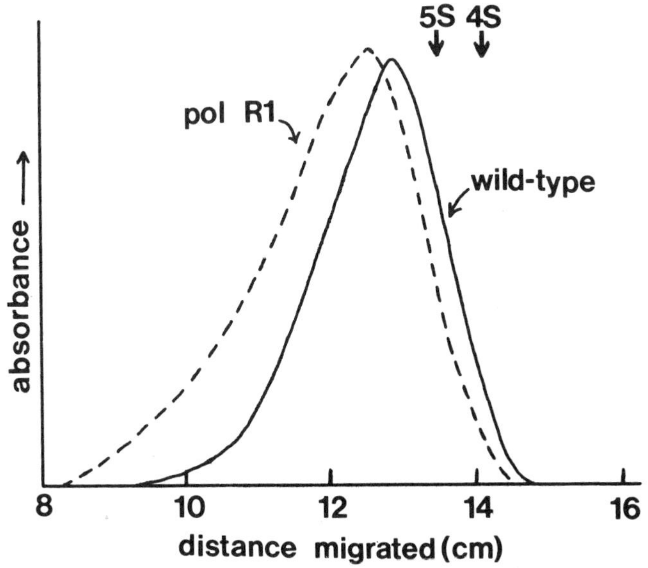

Analysis of small RNA transcripts in urea-acrylamide gels. Poly(A)$^-$ RNA transcripts obtained from reactions identical to the ones described above but labeled with α-^{32}P-GTP were analyzed in 7M urea—12% acrylamide gels. Figure 4 shows the results obtained. Both wildtype and pol R1 products contain a distinct and homogeneous 48 base-long leader RNA but the amount in pol R1 is only ~ 1/3 that of wildtype (quantitated by densitometry). In addition to leader RNA, pol R1 products contain a series of bands somewhat larger than leader not present in wildtype products. We will show elsewhere that some of these larger products correspond to "read-through" or unprocessed transcripts containing leader RNA sequences (in preparation).

Testa et al.,[11] have recently reported that three small distinct RNAs, ~ 28, 42, and 70 bases long can be detected along with leader RNA in the products of *in vitro* transcription by VSV. Two of these small RNAs were

Figure 4. Analysis of small poly(A)$^-$ RNA transcripts from wildtype (lane a) and pol R1 (lane b) by polyacrylamide gel electrophoresis. Poly(A)$^-$ selected RNA samples equivalent to the same fraction of total RNA synthesized by both virus preparations were loaded on the gel in 30 μl of 10 mM Tris-Cl, pH 7.6, 1 mM EDTA, 20% glycerol, and marker dyes (bromphenol blue and xylene cyanol) after heating to 100°C for 1 min and quick cooling. The 12% acrylamide (0.33% bis-acrylamide) −7M urea gel (1.5 mm thick, 40 cm long) and the running buffer contained 0.1 M Tris-borate and 1 mM EDTA at pH 8.3. Electrophoresis was carried out at ~ 250 v for ~ 15 hours and the bands were located by wet gel autoradiography as before.[8]

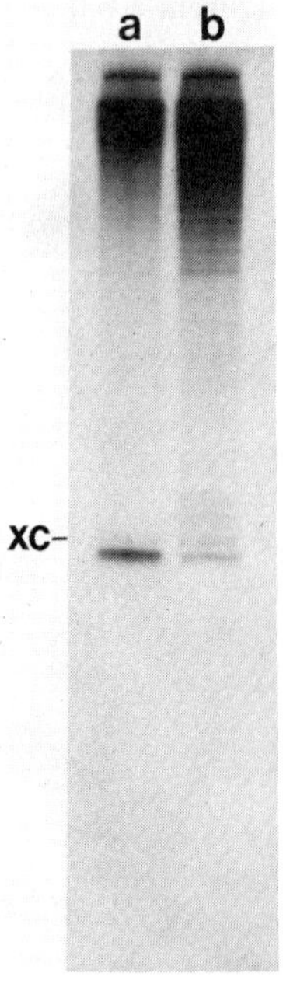

shown to correspond to the 5′-terminal sequences of the N and NS mRNAs and are thought to result from independent initiation of individual mRNAs. We did not detect such RNAs under our transcription reaction conditions (Figure 4) even with longer autoradiographic exposures (not shown).

Discussion

We have previously proposed that the pol R1 mutation(s) affects the regulation of transcription vs replication functions of the VSV-coded polymerase.[2] A central feature of this regulation must necessarily involve a switch from synthesis of individual monocistronic mRNAs to synthesis of a full-size complement of the VSV genome. It is within this context that the pol R1 variant properties, described here and previously, are suggestive of a mutation affecting this regulation.

The increase in poly(A) size and heterogeneity of the pol R1 products is intriguing because it is likely that the VSV polymerase itself is responsible for poly(A) addition.[1] A similar and even more dramatic effect on poly(A) size also occurs when transcription is carried out in the presence of the methylation inhibitor, S-adenosyl homocysteine.[12] These observations suggest a close coupling between methylation and posttranscriptional addition of poly(A) to RNA transcripts. However, in the case of pol R1 no obvious differences in methylation of transcripts were observed.

As far as leader RNA synthesis is concerned, the pol R1 variant polymerase appears to differ from wildtype either by "reading-through" what is normally a termination site 48 bases from the 3′ end of the template, or by being deficient in a hypothetical cleavage activity responsible for producing leader RNA. In either case, this phenotypic change can be considered a partial step towards synthesis of full-size complements of the template. We have previously shown what appears to be "read-through" synthesis of nearly full-size complements of DI RNA templates by pol R1 DI[2] and experiments to be reported elsewhere do show an increase in "read-through" or unprocessed *in vitro* transcripts for pol R1 infectious virus as compared to wildtype. It is not yet clear whether all of the pol R1 properties described above are due to a single mutation in one of the viral proteins or to more than one unrelated mutations. We have, however, isolated an additional independent variant, employing a similar heat selection procedure, which shows a phenotype nearly identical to pol R1 (unpublished). The availability of these novel VSV mutants should prove useful in elucidating the mechanism of VSV polymerase regulation.

ACKNOWLEDGMENTS

We thank Ms. Patricia Schnarr for excellent technical assistance and Donald Rao for helpful comments. This work was supported by National Institutes of Health Grant AI 14365.

References

1. Banerjee, A.K. (1980): In: *Rhabdoviruses,* Vol. II. Bishop, D.H.L. (Ed.), CRC Press Inc., Florida, pp. 36–50.
2. Perrault, J., Lane, J.L., and McClure, M.A. (1980): In: *Animal Virus Genetics, ICN-UCLA Symposia on Molecular and Cellular Biology,* Vol. XVIII. Fields, B., Jaenisch, R., and Fox, C.F. (Eds.), Academic Press, New York, in press.
3. Semler, B.L., Perrault, J., Abelson, J., and Holland, J.J. (1978): *Proc. Nat. Acad. Sci.* 75:4704–4708.
4. Perrault, J. and Leavitt, R.W. (1978): *J. Gen. Virol.* 38:21–34.
5. Breindl, M. and Holland, J.J. (1976): *Virology* 73:106–118.
6. Breindl, M. and Holland, J.J. (1975): *Proc. Nat. Acad. Sci.* 72:2545–2549.
7. Rowlands, D.J. (1979): *Proc. Nat. Acad. Sci.* 76:4793–4797.
8. Perrault, J. and Semler, B.L. (1979): *Proc. Nat. Acad. Sci.* 76:6191–6195.
9. Rowlands, D.J., Harris, T.J.R., and Brown, F. (1978): *J. Virol.* 26:335–343.
10. Hermann, R.C., Adler, S., Lazzarini, R.A., Colonno, R.J., Banerjee, A.K., and Westphal, H. (1978): *Cell* 15:587–596.
11. Testa, D., Chanda, P.K., and Banerjee, A.K. (1980): *Cell* 21:267–275.
12. Rose, J.K., Lodish, H.F., and Brock, M.L. (1977): *J. Virol.* 21:683–693.

David H.L. Bishop and Richard W. Compans, eds.
The Replication of Negative Strand Viruses

The Role of the L and NS Polypeptides in the Transcription by Vesicular Stomatitis Virus New Jersey Serotype

József Ongrádi and József F. Szilágyi[a]

Vesicular stomatitis virus (VSV) New Jersey serotype consists of a single-stranded RNA molecule of approximately 40S and five polypeptides L, G, N, NS, and M.[1,2] The virion contains an RNA transcriptase which synthesizes messenger RNAs both *in vitro* and *in vivo*.[1] Temperature-sensitive (*ts*) mutants of VSV New Jersey have been isolated and classified into six non-overlapping complementation groups.[2,3]

Mutants of complementation group A transcribe RNA *in vitro* similarly to wild type virus but do not synthesize virion RNA in infected cells at the restrictive temperature (39°) indicating that the mutated polypeptide is involved in the replication of the virion RNA.[4-7] A representative mutant (*ts*B1) of complementation group B synthesizes very little RNA *in vitro* at 39°, strongly suggesting that the affected polypeptide in this group is involved in the transcription process. Mutants of complementation groups C and D transcribe and replicate their RNA normally and therefore the mutated polypeptides of these mutants are presumably involved in some late stage of virus development. The two mutants of complementation group F are heterogeneous since *ts*F1 appears to be restricted in transcription and *ts*F2 in replication.[4,5]

Three mutants (*ts*E1, *ts*E2 and *ts*E3) have been classified in complementation group E.[5] The *in vitro* transcriptase activity of *ts*E1 is almost completely inhibited at 39° while *ts*E2 and *ts*E3 transcribe their RNA normally at the restrictive temperature.[4] Mutant *ts*E3 is unable to replicate its virion RNA at

[a] Medical Research Council Virology Unit, Institute of Virology, University of Glasgow, Scotland.

the restrictive temperature in infected cells while the replicase activity of *ts*E2 appears to be unaffected by the mutation.[5] These results indicate that the mutated polypeptide in the group E mutants is multifunctional having some role in transcription, in replication and in some late event in virus development.[4]

Polyacrylamide gel electrophoresis of the virion polypeptides showed that the polypeptides of the mutants of complementation groups A, B, C, and F have electrophoretic mobilities identical to those of wild type virus polypeptides.[8] However, the NS polypeptides of all three mutants of complementation group E were found to have altered electrophoretic mobilities in relation to one another and to wild type virus.[8] Since the electrophoretic mobilities of the NS polypeptides of the two revertant clones *ts*E1/R1 and *ts*E3/R1 are identical to that of the wild type virus, it was concluded that the alteration in the mobilities of the NS polypeptides of the group E mutants is almost certainly the result of the *ts* mutation.[8] The polypeptide compositions of the group E mutants are shown in Figure 1.

Dissociation of the Virion into Subviral Fractions

To determine the role of the viral polypeptides in the transcription process wild type virus, the mutant *ts*E1, and its revertant clone, *ts*E1/R1, were used in dissociation and re-association experiments. Mutant *ts*E1 is especially suitable for such experiments since it is a transcriptase negative mutant, and the *ts* mutation was shown by polyacrylamide gel electrophoresis to affect its NS polypeptide.

The dissociation of the purified virions into subviral fractions is shown in Figure 2 and the polypeptide composition of the fractions in Figure 3. The first step of the fractionation was to obtain transcribing nucleoprotein (TNP) complexes from purified virions by disruption of the virion with Triton N 101 and CsCl and by isolation of the liberated TNP by sucrose gradient centrifugation.[9] The TNPs contained the virion RNA, polypeptides L, NS, and N, and in some preparations traces of G. At the next stage, the TNP was dissociated into a pellet (P) and a supernatant (S) fraction by treatment of the TNP with digitonin and 0.9 M LiCl and subsequent separation of these fractions by centrifugation through a glycerol gradient.[10] The pellet fraction contained the virion RNA and polypeptide N and only minute traces of L and NS. The supernatant fraction contained the L and NS polypeptides and traces of G. Finally, the two polypeptides in the supernatant fraction were separated by phosphocellulose column chromatography.[11] Elution of the column with buffer of low ionic strength gave the so called "NS fraction," which contained the NS polypeptide as well as traces of G and N, but was completely devoid of polypeptide L. Subsequent elution of the column with buffer containing 2 M NaCl gave a fraction which contained virtually pure L polypeptide and, therefore, it was termed the "L fraction."

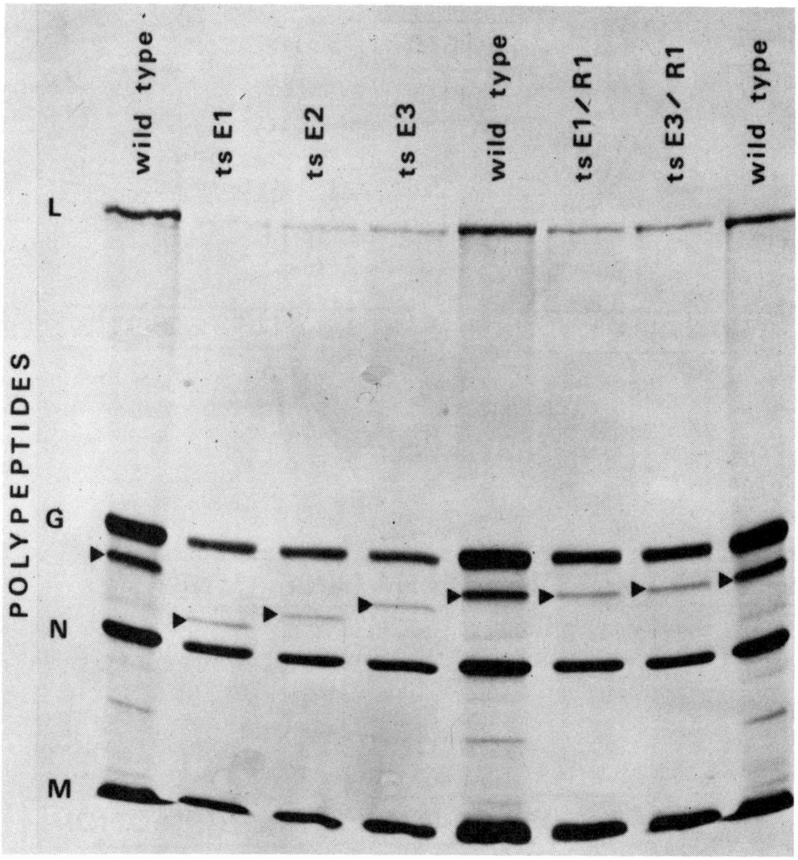

Figure 1. Polypeptide composition of the group E mutants. Purified virion polypeptides of wild type VSV New Jersey, the temperature-sensitive mutants *ts*E1, *ts*E2, and *ts*E3 and the revertant clones *ts*E1/R1 and *ts*E3/R1 were analyzed by SDS polyacrylamide gel electrophoresis. The polypeptides L, G, N, and M are indicated by letters and the NS polypeptides by triangles (▶).

Re-constitution of the Subviral Fractions and Assay of Their Transcriptase Activities

When assayed in isolation, neither the pellet fractions, the supernatant fractions, the L fractions nor the NS fractions contained any transcriptase activity. After re-constitution of the various subviral fractions, transcriptase activities were assayed *in vitro* and the most important results are shown in Figure 4.

Panel A shows that while at 31° the rate of *in vitro* RNA synthesis by the TNP of *ts*E1 was similar to that of wild type TNP and at 39° the transcriptase activity of the mutant was strongly inhibited. This shows that the prepara-

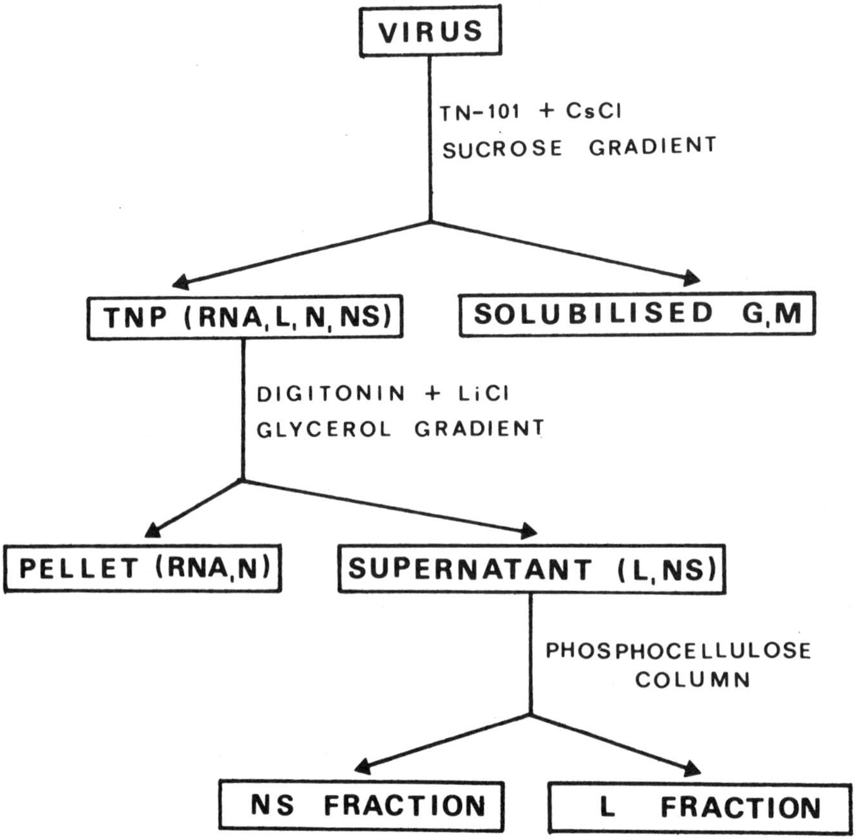

Figure 2. Preparation of the subviral fractions of VSV New Jersey.

tion of *ts*E1 we used for these experiments possessed a temperature-sensitive transcriptase.

Panel B shows that at 39° there was considerable RNA synthesis when the wild type pellet was re-constituted with wild type or revertant supernatants while only small amounts of RNA were synthesized when the supernatant fraction of *ts*E1 was used. Therefore, the mutated polypeptide is in the supernatant fraction.

Panel C shows that RNA was synthesized at 39° at almost identical rates when the wild type pellet was reconstituted with wild type, mutant or revertant L fractions. Since the rates of RNA synthesis were similar to those observed in Panel B using wild type or revertant supernatant fractions, it is unlikely that the trace amounts of polypeptide NS present in the pellet fraction are sufficient for this recovery of the transcriptase activity. Thus, the result in Panel C indicates that polypeptide L alone is capable of acting as

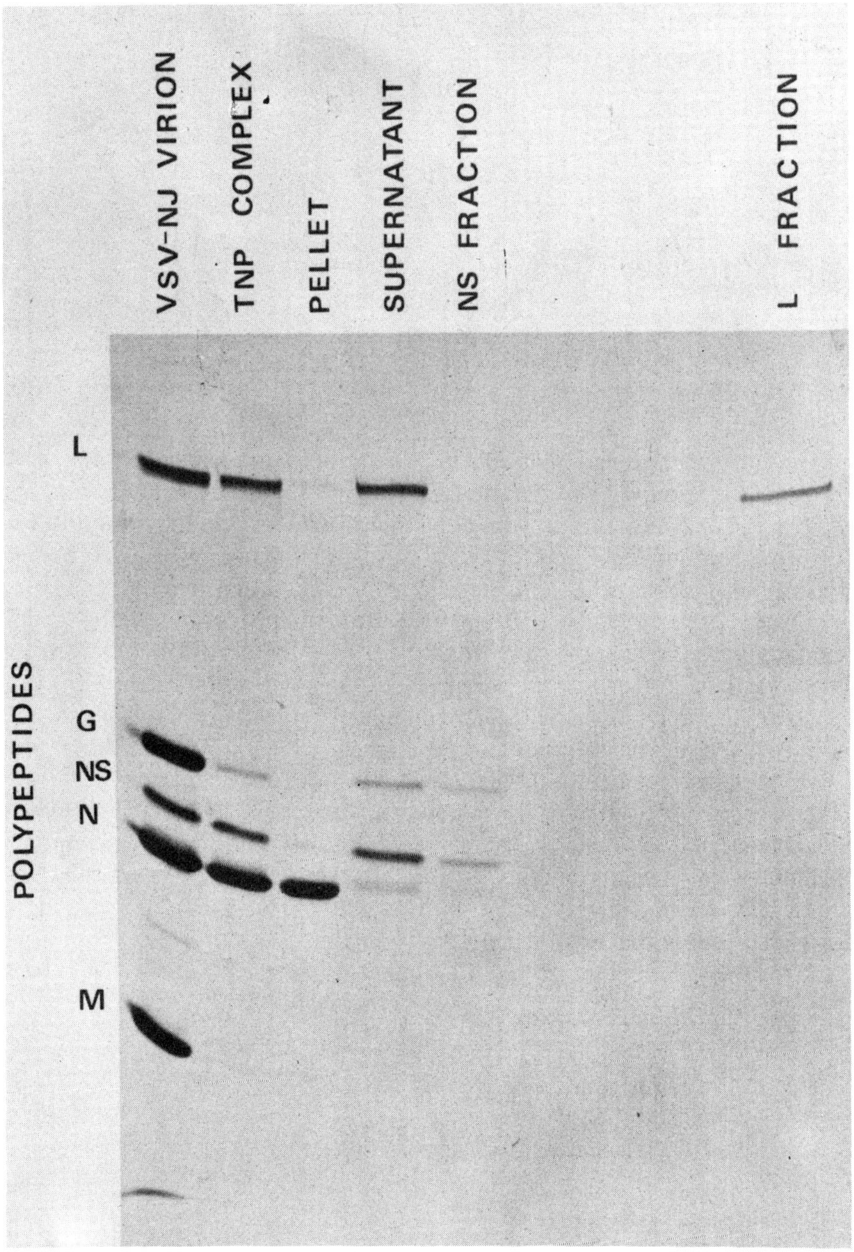

Figure 3. Polypeptide compositions of the subviral fractions of VSV New Jersey. Polypeptides of the subviral fractions of wild type VSV New Jersey were separated by SDS polyacrylamide gel electrophoresis.

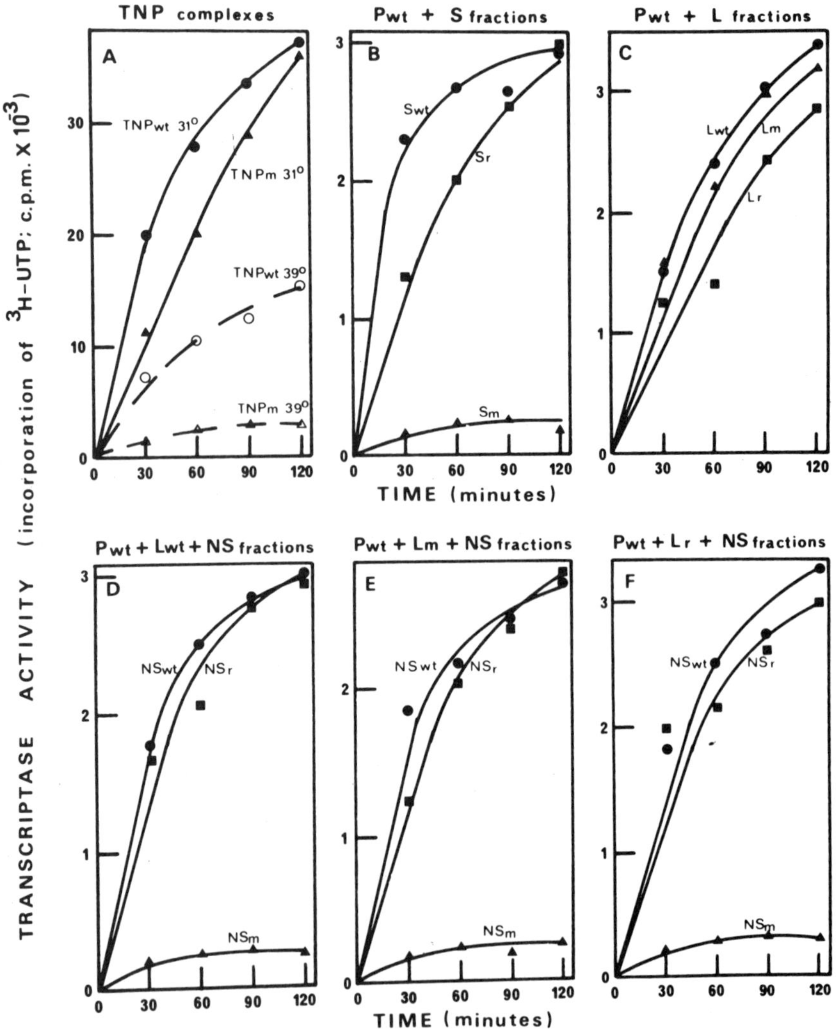

Figure 4. Transcriptase activities associated with the subviral fractions isolated from wild type virus (wt), mutant *ts*E1 (m) and revertant *ts*E1/R1 (r) of VSV New Jersey. (Panel A) Transcriptase activities associated with wild type TNP at 31° (●) and 39° (○) and *ts*E1 TNP at 31° (▲) and 39° (△). (Panel B) Transcriptase activities at 39° after reconstitution of wild type pellet with the supernatant fractions (S) of wild type virus (●), *ts*E1 (▲) and *ts*E1/R1 (■). The pellet and supernatant fractions alone did not have any transcriptase activity (results not shown). (Panel C) Transcriptase activities at 39° after reconstitution of wild type pellet with the L fractions (L) of wild type virus (●), *ts*El (▲) and *ts*E1/R1 (■). (Panels D, E, F) Transcriptase activities at 39° after re-constitution of wild type pellet with the L fractions of wild type virus (Panel D), *ts*El (Panel E) and *ts*E1/R1 (Panel F). The reaction mixtures also contained the NS fractions (NS) of wild type virus (●), *ts*El (▲) and *ts*E1/R1 (■).

the transcriptase. Furthermore, since the mutant L fraction was as active as the other two L fractions, it can be concluded that the mutation did not affect the L polypeptide of *ts*E1.

Panels D, E, and F show that when the wild type pellet was re-constituted with either of the three L fractions in the presence of wild type or revertant NS fractions the rates of RNA synthesis were comparable to those shown in Panel C. However, when the NS fraction of the mutant *ts*E1 was present in the reaction mixtures containing any of the three L fractions, the RNA synthesis in each case was strongly inhibited. Therefore, one can conclude that the *ts*E1 mutation affected the NS polypeptide in such a way that it inhibited the transcription by the L polypeptides.

Conclusions

Re-constitution experiments using the mutant *ts*E1 demonstrated that the L polypeptide synthesized RNA *in vitro* when it is re-constituted with a template which contains only the viral genome and polypeptide N. For this reason polypeptide L can be considered to be the transcriptase itself. These results sharply contrast with the findings of Emerson et al.[11] who found that both L and NS polypeptides are required for transcription, and thus both of these polypeptides were presumed to constitute the transcriptase. At present, experiments are in progress to find out the reasons for these divergent results.

Since the *in vitro* re-constitution experiments also showed that it is the NS polypeptide of *ts*E1 which is affected by the mutation, this confirms the results previously obtained by polyacrylamide gel electrophoresis. However, the most interesting result of our experiments is that the NS polypeptide of *ts*E1 inhibited the RNA synthesis by polypeptide L. This implies that, although it does not take part in the transcription, the NS polypeptide exerts some form of control over the transcription process. Experiments are in progress to elucidate the precise nature of this inhibition. It will be especially important to compare the RNA species which are synthesized by polypeptide L with those synthesized when polypeptide NS is also present.

ACKNOWLEDGMENTS

Dr. J. Ongrádi was a Unilever Fellow of the Biochemical Society for 1979-80 at the Institute of Virology, Glasgow. We thank C. Cunningham for his excellent technical assistance.

References

1. Wagner, R.R. (1975): In: *Comprehensive Virology,* Volume 4. Fraenkel-Conrat, H. and Wagner, R.R. (Eds.), Plenum Press, New York, pp. 1-93.
2. Pringle, C.R. (1977): In: *Comprehensive Virology,* Volume 9. Fraenkel-Conrat, H. and Wagner, R.R. (Eds.), Plenum Press, New York, pp. 239–289.

3. Pringle, C.R. and Szilágyi, J.F. (1980): In: *Rhabdoviruses*, Volume 2. Bishop, D.H.L. (Ed.), CRC Press, Florida, pp. 141–161.
4. Szilágyi, J.F. and Pringle, C.R. (1979): *J. Virol.* 30:692–700.
5. Pringle, C.R., Duncan, I.B., and Stevenson, M. (1971): *J. Virol.* 8:836–841.
6. Lesnaw, J.A. and Dickson, L.R. (1978): *Virology* 91:51–59.
7. Lesnaw, J.A. and Reichmann, M.E. (1975): *Virology* 63:492–504.
8. Evans, D., Pringle, C.R., and Szilágyi, J.R. (1979) *J. Virol.* 31:325–333.
9. Szilágyi, J.F. and Uryvayev, L. (1973): *J. Virol.* 11:279–286.
10. Szilágyi, J.F., Pringle, C.R., and Macpherson, T.M. (1977): *J. Virol.* 22:381–388.
11. Emerson, S.U. and Yu-Hwa Yu (1975): *J. Virol.* 15:1348–1356.

David H.L. Bishop and Richard W. Compans, eds.
The Replication of Negative Strand Viruses

Vesicular Stomatitis Virus Genome Replication and Nucleocapsid Assembly in a Permeable Cell System

Jon H. Condra and Robert A. Lazzarini[a]

The genome of vesicular stomatitis virus (VSV) is a single 42S strand of negative polarity RNA. In the mature virion, this RNA is associated with the viral N, NS, and L-proteins to form an RNAse-resistant nucleocapsid. During infection of susceptible cells, these nucleocapsids serve as templates for synthesis of the five viral mRNAs which encode the five viral proteins.

Replication of the viral RNA is thought to proceed through synthesis of full-length 42S complementary RNA strands which can serve as templates for synthesis of progeny negative-stranded viral genomes. In infected cells, all detectable 42S RNA molecules of both polarities exist as nucleocapsids,[1,2] suggesting that encapsidation may be an essential part of the replication process. This is supported by the finding that cycloheximide inhibits genome replication without blocking mRNA transcription.[3,4]

Although *in vitro* transcription systems have been useful in studies of mRNA synthesis in VSV, very little is known about the details of genome replication and nucleocapsid assembly. This is due in large part to the lack of an easily manipulated *in vitro* replication system. Despite numerous attempts,[5-9] unequivocal genome replication *in vitro* has not been accomplished. In this paper, we report the development of a permeable cell system that supports viral mRNA transcription, translation, genome replication, and nucleocapsid assembly.

Suspension cultures of BHK cells were infected with VSV(HR) at 30 PFU/cell and harvested at 3 hr postinfection at 37°. Cells were washed,

[a] Laboratory of Molecular Genetics, National Institute of Neurological and Communicative Disorders and Stroke, National Institutes of Health, Bethesda, Maryland.

permeabilized with lysolecithin,[10,11] and incubated at 30° in a buffer[5] containing nucleoside triphosphates, phosphoenolypruvate (PEP), pyruvate kinase (PK), 20 amino acids, S-adenosylmethionine, spermine, BHK cytoplasmic RNA, and actinomycin D. A more detailed description of the system will be presented elsewhere.[12]

Figure 1A shows the kinetics of RNA synthesis in these cells as measured by incorporation of ^{3}H-UTP into acid-insoluble material. Permeabilized, VSV-infected cells incorporated ^{3}H-UTP linearly for 90 min in the presence of PEP and PK. When this triphosphate regenerating system was omitted, incorporation ceased after 60 min, suggesting that the triphosphate concentrations became limiting. Very little incorporation was seen in mock-infected permeable cells. As expected, infected, intact cells showed much less incorporation than did their permeable counterparts.

Figure 1B shows the incorporation of ^{35}S-methionine into protein in these cells. VSV-infected permeable cells synthesized protein to the same extent as did intact cells in the presence of glucose. In both preparations, incorporation continued for 60 min after which some turnover was evident. No incorporation was observed when PEP and PK were omitted, or when GTP was substituted with its nonhydrolyzable analog, GPP(CH_2)P, to which intact cells are impermeable.[13] These results show that efficient protein synthesis was occurring in cells freely permeable to nucleoside triphosphates.

The RNAs synthesized by VSV-infected, permeable cells were labeled with α-^{32}P-CTP, pelleted through 20-40% CsCl gradients and analyzed by zone sedimentation in SDS-sucrose gradients (data not shown). The RNAs sedimented in two peaks: 12-18S and 31S, as is characteristic of *in vivo* VSV mRNAs. These RNAs annealed to VSV genomic RNA and bound to oligo(dT)-cellulose (data not shown). Thus, by the criteria of size distribution, annealing properties, and polyadenylation, these appear identical to *in vivo* VSV mRNAs.

In order to examine VSV genome replication in these permeable cells, RNA was labeled with α-^{32}P-CTP and cytoplasmic extracts were centrifuged in CsCl gradients. Figure 2 shows a typical gradient profile. Virtually all the radioactivity in a mock-infected extract remained at the top of the gradient. In contrast, the VSV-infected extract showed a peak of radioactivity at the buoyant density of an ^{35}S-methionine labeled nucleocapsid (NC) marker. Over several experiments, 5 to 10% of the radioactivity of the pellet was found in this peak, consistent with previous observations that genomic RNA synthesis comprises 10-16% of viral RNA synthesized in VSV-infected cells.[4,14]

A similar experiment using ^{35}S-methionine label is shown in Figure 3. Again, a peak of radioactivity at the density of NCs was observed (fractions 17-19). Mock-infected extracts showed no peak at this position (data not shown). Fractions of this gradient were pooled in pairs and examined by SDS-polyacrylamide gel electrophoresis[15] and fluorography. Figure 4 shows that all five viral proteins were present in various regions of the gradient, L

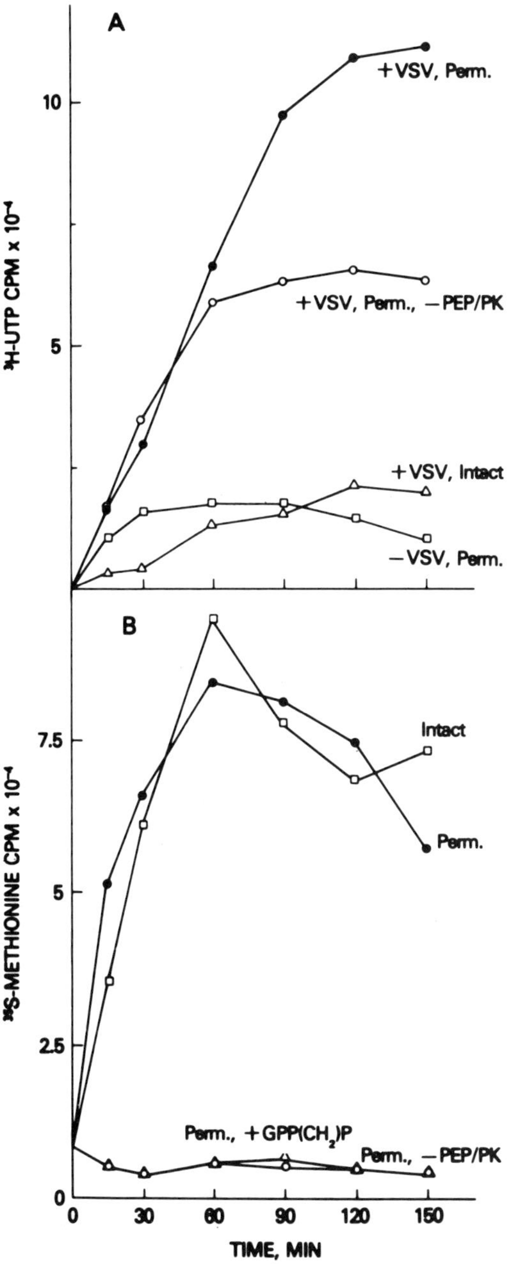

Figure 1. (A) RNA synthesis in permeable cells: Reaction mixtures (1×10^6 cells) were labeled with 5,6 – ^3H – UTP (20 Ci/mmol, 10 μM) at 30°C. At intervals, TCA-precipitable radioactivity was determined. (B) Protein synthesis in VSV-infected cells: Mixtures containing 1×10^6 VSV-infected cells were labeled with ^{35}S-methionine (5 Ci/mmol, 12.5 μM) at 30°C. TCA-precipitable radioactivity was determined at intervals.

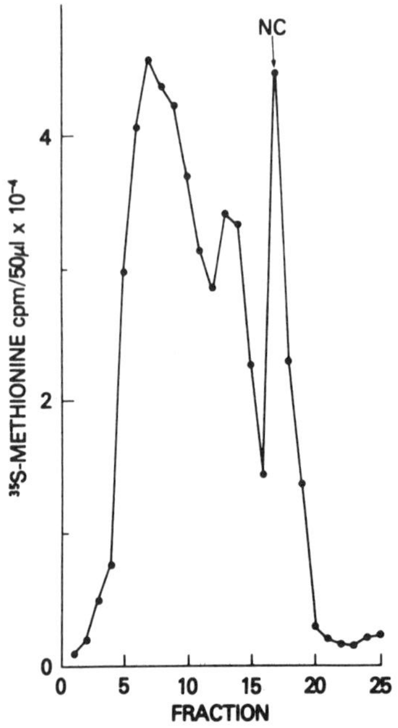

Figure 2. CsCl centrifugation of RNAs synthesized in permeable cells: 2 × 10^7 permeable cells were labeled 60 min with α-^{32}P-CTP (30 Ci/mmol, 10 μM) at 30°. Cytoplasmic extracts were prepared and centrifuged CsCl gradients as described elsewhere.[12] TCA-precipitable radioactivity was determined. The location of an ^{35}S-methionine-labeled nucleocapsid (NC) marker in a parallel gradient is indicated by an arrow. Sedimentation was from left to right. Closed circles: VSV-infected cells; open circles: mock-infected cells.

and G were found between fractions 11 and 16; NS and M were more uniformly distributed from fractions 5-16. N-protein was found in these upper fractions as well, but most appeared in fractions 17-20, corresponding exactly to the peaks of ^{32}P and ^{35}S in Figures 2 and 3, respectively. Moreover, this was the only labeled protein present in significant amounts in these fractions.

These results indicate that newly-synthesized viral RNA and protein appear in CsCl gradients at the buoyant density of VSV nucleocapsids. To demonstrate a direct association between the RNA and protein, immunoprecipitation experiments were performed. 78% of the radioactivity in the ^{32}P-labeled putative nucleocapsids shown in Figure 2 was precipitable by rabbit anti-VSV serum,[16] while only 0.2% of phenol-extracted nucleocapsid RNA was precipitable under the same conditions. Only 0.4% of the radioactivity of either preparation was precipitated by preimmune serum. After RNAse T_1 treatment, 74% of the label in the putative nucleocapsids was immunoprecipitable, showing that they were RNAse-resistant.

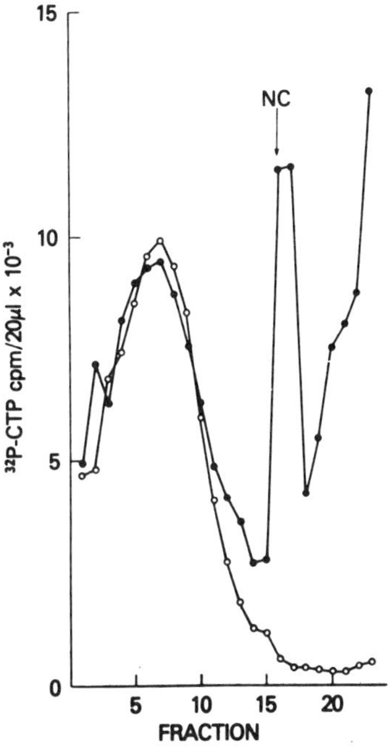

Figure 3. CsCl centrifugation of proteins synthesized in VSV-infected permeable cells: 2×10^7 VSV-infected, permeable cells were labeled with ^{35}S-methionine for 60 min at 30°. A cytoplasmic extract was centrifuged in a CsCl gradient as described.[12] TCA-precipitable radioactivity was determined. Sedimentation was from left to right. The NC peak is shown by an arrow.

The size distribution of the RNA in these nucleocapsids was examined by zone sedimentation in SDS-sucrose gradients. Figure 5A shows that most of the ^{32}P-CTP-labeled RNA sedimented at 42S although a broad shoulder extended down to 18S. This shoulder probably represents nascent chains of replicating RNA. A small peak of radioactivity at 4-7S was observed. This was probably a contaminant in the CsCl nucleocapsid band, since it was present in greater abundance in adjacent regions of the CsCl gradient, and it did not reband with nucleocapsids in a second CsCl gradient (data not shown). In addition, when nucleocapsids were bound to nitrocellulose filters (which retain proteins but not naked RNA),[17] washed, and eluted with SDS, the 4-7S RNA disappeared (Figure 5B).

Finally, the polarity of the nucleocapsid RNA was examined by annealings. 80% of the ^{32}P-labeled RNA was rendered RNAse-resistant by saturating amounts of VSV mRNA, while 20% annealed to virion RNA. Therefore, the labeled RNA in nucleocapsids was composed of viral sequences of both plus and minus strands, and was about 80% minus-stranded. Only 2% of this RNA bound to oligo(dT)-cellulose, indicating an absence of poly(A).

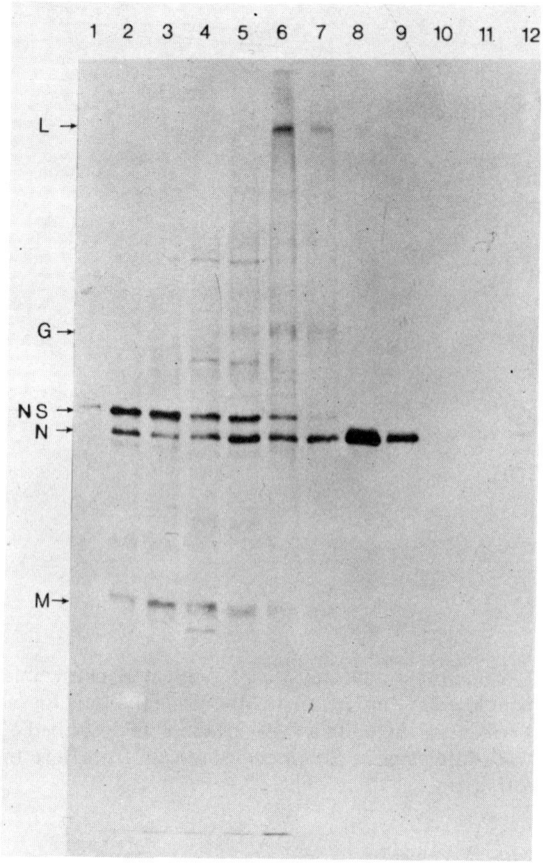

Figure 4. SDS-polyacrylamide gel electrophoresis of proteins synthesized in VSV-infected, permeable cells: Pairs of fractions from the CsCl gradient shown in Figure 3 were pooled, TCA-precipitated, electrophoresed in a 10% polyacrylamide gel,[15] and fluorographed. Migrations of ^{35}S-VSV marker proteins are indicated. Lane 1, fractions 3-4; lane 2, 5-6; lane 3, 7-8; lane 4, 9-10; lane 5, 11-12, lane 6, 13-14; lane 7, 15-16; lane 8, 17-18; lane 9, 19-20; lane 10, 21-22; lane 11, 23-24; lane 12, pellet.

These experiments show that in VSV infected permeable cells, newly-synthesized 42S viral RNA and N-protein are assembled into structures having the buoyant density, RNAse-resistance, immunoprecipitability, and polarity[18] of normal intracellular nucleocapsids. We conclude that genome replication and nucleocapsid assembly take place in this permeable cell system.

Permeabilization of mammalian cells with lysolecithin has proved useful in recent studies of DNA replication and repair.[10,11] Here, we have described conditions under which this method can be used to study VSV gene expression and genome replication.

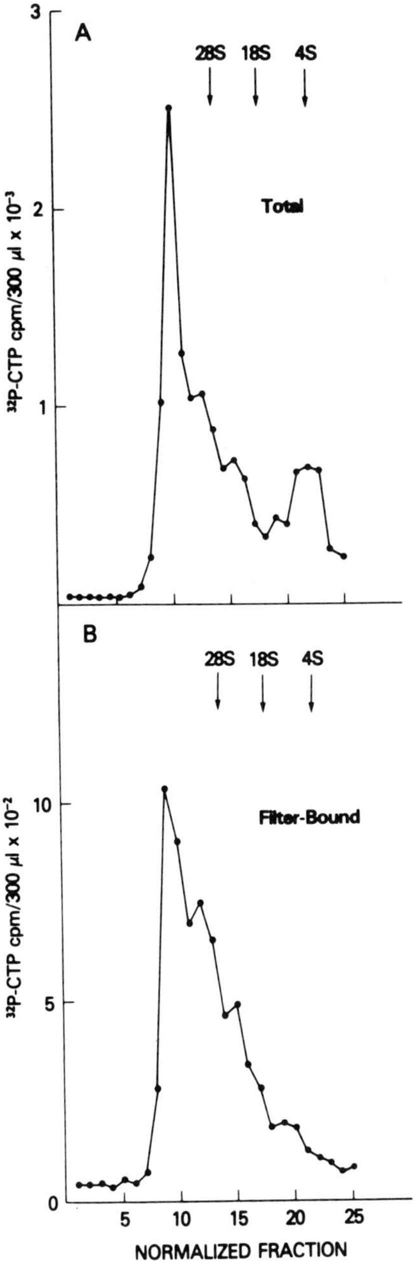

Figure 5. SDS-sucrose zone sedimentation of nucleocapsid RNAs: CsCl-purified ^{32}P-CTP-labeled nucleocapsids were disrupted with 1% SDS and centrifuged in 10-30% sucrose gradients.[12] Sedimentation was from right to left. (A) Total RNA from nucleocapsid band. (B) Filter-bound RNA from nucleocapsid band. TCA-precipitable radioactivity was determined.

One of the greatest advantages of this method is that it closely approximates conditions found in intact cells. Since cells are permeabilized during the middle of the infectious cycle, viral and host cell factors that may influence the course of the infection[19-23] are retained. Moreover, since changes in the phosphorylation state of NS-protein may have regulatory significance[24] and cellular NS differs from virion NS,[25] present *in vitro* transcription systems may lack many components normally involved in RNA synthesis. Our data suggest that in permeable cells many, if not all, of these components are still functional.

Permeable cells represent the only *in vitro* system described so far in which unequivocal genome replication and nucleocapsid assembly have been demonstrated. Previous cell-free RNA synthesizing systems, even when coupled to protein synthesis,[5-8] have not been shown to synthesize nucleocapsids.

The major advantage of permeable cells over intact cells is in the accessibility of the components of the infection to compounds normally unable to penetrate the plasma membrane. Newly-synthesized viral RNAs can be specifically labeled with nucleoside triphosphates. Because permeable cells lose their intracellular nucleotide pools,[10,11] the specific activities of these labels can be controlled and unambiguous pulse-chase experiments with nucleoside triphosphates are possible. Similarly, inhibitors such as AMPP(NH)P, which alters the pattern of VSV RNA synthesis *in vitro,*[9] can be used to study aspects of viral replication not approachable in a purely *in vitro* system.

Finally, permeable cells promise to be useful in studying the mechanisms of viral autointerference. Using a stock of VSV known to contain a DI particle, we have observed preferential replication of DI genomic RNA over the standard virus (data not shown). The processes of autointerference therefore seem to operate in permeable cells. In some instances, DI replication appeared to be favored more strongly in permeable cells than in similarly infected intact cells. The ability to manipulate experimental conditions during autointerfered infections of permeable cells may permit studies of the mechanisms of genesis and replication of defective interfering particles, not only in VSV, but in other virus-host systems as well.

References

1. Soria, M., Little, S.P., and Huang, A.S. (1974): *Virology* 61:270–280.
2. Morrison, T.G., Stampfer, M., Lodish, H.F., and Baltimore, D. (1975): In: *Negative Strand Viruses.* Mahy, B.W.J. and Barry, R.D. (eds.), Academic Press, New York, pp. 293–300.
3. Perlman, S.M. and Huang, A.S. (1973): *J. Virol.* 12:1395–1400.
4. Wertz, G.W. and Levine, M. (1973): *J. Virol.* 12:253–264.
5. Ball, L.A. and White, C.N. (1978): *Virology* 84:479–495.
6. Breindl, M. and Holland, J.J. (1975): *Proc. Nat. Acad. Sci.* 72:2545–2549.

7. Breindl, M. and Holland, J.J. (1976): *Virology* 73:106–118.
8. Batt-Humphries, S., Simonsen, C., and Ehrenfeld, E. (1979): *Virology* 96:88–99.
9. Testa, D., Chanda, P.K., and Banerjee, A.K. (1980): *Proc. Nat. Acad. Sci.* 77:294–298.
10. Miller, M.R., Castellot, J.J., Jr., and Pardee, A.B. (1978): *Biochemistry* 17:1073–1080.
11. Miller, M.R., Castellot, J.J., Jr., and Pardee, A.B. (1979): *Exp. Cell Res.* 120:421–425.
12. Condra, J.H. and Lazzarini, R.A. (1980): *J. Virol.*, in press.
13. Carrasco, L. (1978): *Nature (London)* 272:694–699.
14. Stamminger, G. and Lazzarini, R.A. (1977): *Virology* 77:202–211.
15. Laemmli, U.K. and Favre, M. (1973): *J. Mol. Biol.* 80:575–599.
16. Johnson, L.D., Binder, M., and Lazzarini, R.A. (1979): *Virology* 99:203–206.
17. Yarus, M. and Berg, P. (1967): *J. Mol. Biol.* 28:479–489.
18. Simonsen, C.C., Batt-Humphries, S., and Summers, D.F. (1979): *J. Virol.* 31:124–132.
19. Nowakowski, M., Bloom, B.R., Ehrenfeld, E., and Summers, D.F. (1973): *J. Virol.* 12:1272–1278.
20. Simpson, R.W. and Obijeski, J.F. (1974): *Virology* 57:357–368.
22. Kang, C.Y. and Allen, R. (1978): *J. Virol.* 25:202–206.
23. Kang, C.Y., Glimp, T., and Allen, R. (1978): In: *Negative Strand Viruses and the Host Cell.* Mahy, B.W.J. and Mahy, R.D. (Eds.), Academic Press, New York, pp. 501–513.
24. Clinton, G.M., Burge, B.W., and Huang, A.S. (1978): *J. Virol.* 27:340–346.
25. Kingsford, L. and Emerson, S.U. (1980): *J. Virol.* 33:1097–1105.

David H.L. Bishop and Richard W. Compans, eds.
The Replication of Negative Strand Viruses

Interaction of Mutant and Wild Type M-Protein of Vesicular Stomatitis Virus with Nucleocapsids and Membranes

John Lenard,[a] Tazewell Wilson,[b] Denise Mancarella,[a] Jeffrey Reidler,[c] Paul Keller,[d] and Elliot Elson[e]

Introduction

The vesicular stomatitis virus (VSV) matrix (M) protein, a nonglycosylated polypeptide of mol wt 29,000, is one of the three major structural proteins of the virion.[1] Recent studies have elucidated two functions of M-protein, one structural and one regulatory. Experiments utilizing VSV temperature-sensitive (*ts*) mutants and viral pseudotypes have shown that M-protein is essential for the budding of virions or virus-like particles from the plasma membrane of an infected cell.[2-4] Virus-like particles have been isolated that lack each of the other viral structural proteins (G or N) but that always contain M-protein. It is therefore thought that M-protein organizes the nascent viral bud by specifically interacting with the viral nucleocapsid and with a portion of the host cell plasma membrane containing G-protein. The nature of these interactions remains largely unknown.

In addition to this structural role, M-protein appears to have a regulatory role in virus-directed RNA synthesis in infected cells[5,6] and to act as an inhibitor of VSV *in vitro* transcription.[7-9]

[a] Department of Physiology and Biophysics, CMDNJ-Rutgers Medical School, Piscataway, New Jersey.
[b] Department of Physiology and Biophysics, SUNY Health Science Center, Stony Brook, New York.
[c] Department of Structural Biology, Stanford University School of Medicine, Stanford, California.
[d] Department of Virology and Cell Biology, Merck Research Laboratories, West Point, Pennsylvania.
[e] Department of Biological Chemistry, Washington University School of Medicine, St. Louis, Missouri.

In this paper, we describe the use of group III *ts*-mutants of VSV, which have lesions in the M-protein, to study the interaction of M with each of the other major viral components. The mutants were chiefly studied at the permissive temperature, where M-protein is functional during the infectious cycle. The demonstrable differences between mutant and wild type (wt) M-protein under these conditions permit several conclusions to be drawn regarding the nature of the interactions between M and other viral components, and the kinetics of the budding process. We also demonstrate that the differences in the properties of mutant and wt M-protein at the permissive temperature are direct consequences of the *ts* phenotype, because revertants isolated from each of the *ts* M mutants all show reversion to wt-like M-protein.

Interaction of Wt and Mutant M-Protein with Nucleocapsids *in Vitro* [10]

These experiments used M-protein mediated inhibition of *in vitro* transcriptase activity as an assay to characterize the interactions between M and nucleocapsids. Transcription was assayed essentially as described by Carroll and Wagner,[11,12] except that the NaCl concentration was varied during disruption of the virus.[10] Figure 1 shows a comparison of the NaCl-dependence of *in vitro* RNA synthesis by wt VSV and by several *ts* M-protein mutants. Wt VSV showed no appreciable RNA polymerase activity until the NaCl concentration in the reaction mixture (representing a five-fold dilution of salt from the virus disruption mixture)[10] exceeded about 0.064 M. The *ts*-mutants, on the other hand, all showed significant polymerase activity at significantly lower NaCl concentrations. The differences in salt dependence were even more dramatic in the presence of poly-L-glutamic acid, which significantly enhances transcriptase activity in whole virus but not in ribonucleoprotein cores lacking M-protein[12] (filled circles, Figure 1). In addition, the maximal activity obtained at high NaCl concentrations was substantially higher for all of the mutants than for wt.[10]

The results of these experiments, and of similar ones using temperature stable revertants derived from the mutants, are summarized in Table 1. The following points may be noted: (1) The Glasgow and Orsay strains of wt VSV show similar behavior. Differences in detail were observed between the salt-dependent profiles of the Glasgow, Orsay and Birmingham (not shown) strains, but each showed the general property of inhibition of polymerase up to quite high NaCl concentrations, and generally lower maximum activity than was found with the mutants. (2) All four of the *ts*-mutants tested showed NaCl dependence that was qualitatively similar (Figure 1). This was reflected in the lower NaCl values required to obtain 50% of maximum activity (Table 1). (3) All of the temperature stable revertants tested (two derived from each of the four mutants) exhibited NaCl dependence that resembled

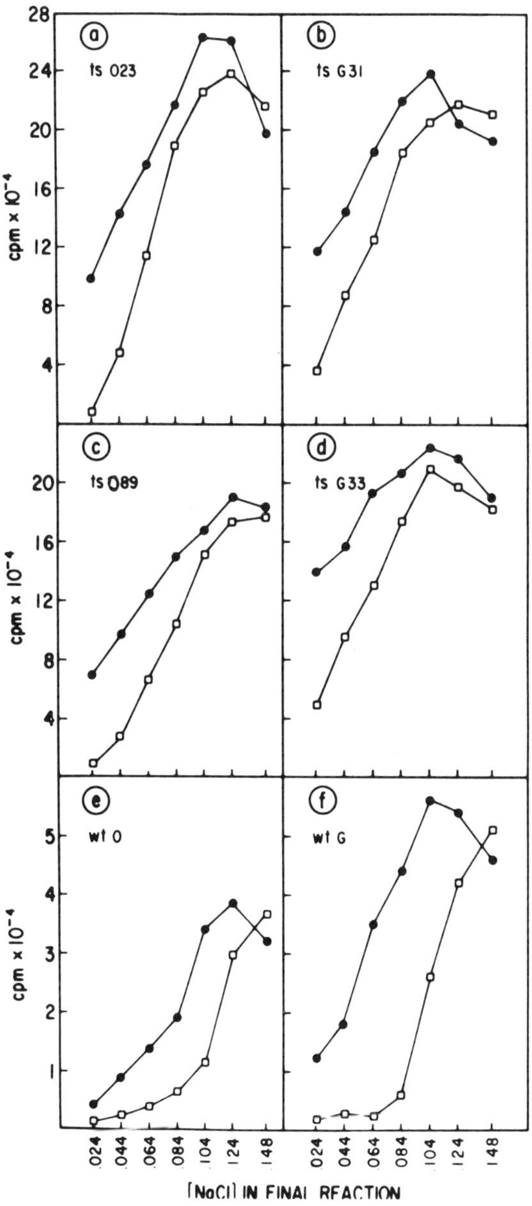

Figure 1. NaCl dependence of *in vitro* transcription: group III mutants and parental wt. Concentrated virus was disrupted in the presence of the appropriate concentration of NaCl and *in vitro* transcriptase activity was assayed either in the presence (circles) or the absence (squares) of poly-Glu (final concentration: 3 mg/ml), as described in reference 10. The final concentration of virus in individual assays was 1 mg/ml; the final NaCl concentrations indicated result from variations in the NaCl present in the disruption mixture.

Table 1. NaCl Dependence of *in vitro* Transcription: VSV wt, Group III Mutants and Revertants.

	NaCl required for 50% of maximum activity	
	− poly Glu	+ poly Glu
wt0	.11	.08
*ts*023	.06	.04
*ts*023r5	.13	.07
*ts*023r7	.11	.09
*ts*089	.08	.04
*ts*089r5	.11	.05
*ts*089r7	.12	.05
wtG	.10	.06
*ts*G31	.06	.02
*ts*G31r4	.09	.04
*ts*G31r8	.10	.05
*ts*G33	.05	<.02
*ts*G33r1	.09	.04
*ts*G33r5	.11	.06

the wt more closely than the mutant from which it was derived. This was true both in the presence and absence of poly-L-glutamic acid (Table 1).[10]

These results suggest that the functional interactions between M-protein and nucleocapsids that result in inhibition of polymerase activity *in vitro* are largely electrostatic in nature. The activity of both wt and *ts*-mutants are independent of detergent concentrations down to levels below the critical micelle concentration, suggesting that hydrophobic effects are not significant in this interaction.[10] It appears that the lesions in all four mutants tested are similar in effecting a decrease of a functional, electrostatically stabilized association between nucleocapsid and M-protein, and that this is an essential aspect of the *ts* phenotype.

Labeling of Wt, Mutant and Revertant Virions with Iodonaphthylazide

The photoreactive probe ^{125}I-iodonaphthylazide (INA) has been used to identify polypeptide chains that are associated with the bilayer component of membranes.[15,16] This assignment is based on the extreme hydrophobicity of INA; it partitions almost quantitatively into hydrophobic regions, most notably the interior of the membrane bilayer and hydrophobic "pockets" of certain proteins. After incubation in the dark, photoactivation by ultraviolet light results in relatively non-specific covalent bond formation with its near neighbors.

When purified wt VS virions were labeled with INA, most of the protein-associated ^{125}I migrated with G-protein in polyacrylamide gels (Table 2). Significant amounts of label were also associated with the other viral proteins (Table 2). The G-protein was labeled quantitatively in its hydrophobic, bilayer-associated peptide. After proteolysis of intact virions and repurification of the spikeless particles, the total amount of label originally associated with G-protein was found in a 5-7,000 mol wt peptide that was absent from the intact virion.[17,18] Proteolysis by trypsin, chymotrypsin, thermolysin, or pronase gave identical results. These results confirm those recently reported by Zakowski and Wagner[19] and indicate that labeling by INA is restricted to the most hydrophobic protein domains.

Consistent differences were observed between the labeling patterns of mutant and wt VSV. Mutant M-protein was 2-3 times more heavily labeled than wt M. Revertants derived from each of the four mutants showed a level of M-protein labeling close to that of wt, indicating that the labeling pattern is related to the temperature-sensitive phenotype of the virus (Table 2).

Interaction of Wt and Mutant M-Protein with Infected Cell Membrane: Fluorescence Photobleaching Studies[13]

In these experiments, we measured the lateral mobility of newly synthesized G-protein that appeared on the surface of BHK cells in response to infection by either wt or group III *ts*-mutants. Comparative measurements were made at the permissive (31°) and the non-permissive (39°) temperatures.

Measurements were made using the technique of fluorescence photobleaching recovery (FPR). The cell surface was specifically labeled with

Table 2. Labeling Ratios of wt VSV, M-Protein Mutants and Revertants.

Virus	Relative ratio INA/Protein (G = 1.0) N	M
A. wt Orsay	.35	.40
wt Glasgow	.12	.37
B. *ts* 0-23	.37	1.19
ts O-89	.28	1.03
ts G-31	.14	.70
ts G-33	.16	.74
C. *ts* O-23r7	.35	.40
ts O-89r5	.40	.56
ts G-31r4	.12	.32
ts G-33r5	.21	.47

fluoresceinated monovalent (Fab) fragments of anti-G antibody. The fluorescence present on a small patch of the cell surface (diameter ca 1 μ) was then bleached by a brief intense pulse of light from a laser beam. The rate and extent of recovery of fluorescence in the bleached spot constitutes the FPR measurement, and provides two parameters of lateral mobility: (1) the *extent* of recovery measures the mobile fraction (R_f) of total surface G-protein, i.e., the fraction that is capable of undergoing lateral diffusion during the time course of the measurement (minutes); (2) the *rate* of recovery of the mobile fraction provides a measure of the lateral diffusion coefficient (D) that characterizes the movement of the mobile fraction of surface G.[14]

G-protein on the surface of wt infected cells had a high R_f value at both permissive and non-permissive temperatures, indicating that most of the molecules were freely diffusible in the plane of the membrane. Significant decreases were found in the R_f of mutant infected cells, but only at the permissive, i.e., virus producing, temperature (Table 3). No significant differences were found in the values of D from the same experiments.[13]

The rates of viral budding in mutant infected cells varied from much slower than wt to about the same rate (Table 4). In no case was a *ts*-mutant observed to bud significantly more rapidly than the wt virus.[13] Experiments that measured the amount of viral protein labeled 6-8 hr postinfection showed that the ratio of total cellular G to M-protein did not differ significantly between wt and mutant infected cells at either temperature.[13] Therefore, the differences in R_f shown on Table 2 did not arise from gross differences in the relative amounts of G and M-protein present in the infected cells.

These findings suggest that the interaction between M-protein and the cell membrane is altered in mutant infected cells. The interaction between mutant M and cell membranes has the effect of immobilizing G-protein on the surface of the cells. The implications for the kinetics of viral budding will be considered below.

Table 3. Mobile Fraction of Cell Surface G-Protein of BHK Cells Infected with wt or *ts* M Mutants of VSV.[a]

	Mobile fraction (%)			
	wt	***ts*23**	***ts*33**	***ts*89**
31°	74 ± 13 (19)[b]	58 ± 12 (18)	48 ± 13 (18)	37 ± 4 (8)
39°	73 ± 13 (52)	74 ± 17 (8)	75 ± 12 (5)	74 ± 11 (14)

[a]Cells infected at a multiplicity of ca 50 pfu/cell. Measurements made 5-10 hr postinfection.
[b]Numbers in parentheses represent the number of measurements.

Table 4. Rate of Virion Production after Infection of BHK Cells with wt or *ts* M-Protein Mutants at 31 °.

	Number of infectious units produced ($\times 10^{-8}$)		
	3-6 hr.	**6-9 hr**	**9-12 hr**
wt	1.5	1.5	10
ts 23	0.05	0.7	8
ts 33	1.2	2	3
ts 89	.001	0.07	0.6

Discussion

The findings reported here demonstrate that all four mutant M-proteins studied possess two coordinate properties in common: an increased association with membranes, and a decreased affinity for nucleocapsids. Although the specific mutations present in each mutant are not known, it seems reasonable that at least some of them represent point mutations, which differ from wt by a single amino acid residue.

How can one envision these coordinate properties, common to all four mutants, arising from such mutations? It is possible that the mutation specifically changes a segment of the polypeptide chain that interacts with nucleocapsids into a segment that interacts with membrane. This is unlikely, however, because such a change might be expected to produce a totally non-functional M-protein. It is also unlikely that every mutation would result in the similar effect on both interactions that is observed. A second possibility is that the mutation causes a conformational change that affects two different regions of the protein in a coordinate manner. It has been reported that M-protein exists in at least two phosphorylated states.[20] Although phosphorylation of M-protein has not yet been shown to have any functional significance, it is possible that a conformational change could arise from a mutation affecting a phosphorylation site.

There is a simpler interpretation of the observed changes in M-protein interactions. The coordinate effect can be explained even if the mutations in M-protein involve a single amino acid change affecting one of two independent binding sites. If, for instance, the interaction of mutant M-protein with nucleocapsids is decreased, then, even if the membrane binding region of the polypeptide chain is completely unaffected by the mutational change, interaction with membranes could increase purely on the basis of mass action:

$$\text{M-NC} \rightleftharpoons \text{M}_{\text{sol}} \rightleftharpoons \text{M-Memb}$$

This interpretation predicts that any mutation affecting one binding site must have an apparent effect on the affinity of the polypeptide for both nucleocapsids and membranes, and provides an explanation for the coordinate effect seen in all the M-protein mutants.[10]

The FPR results reported here are most easily interpreted if it is assumed that mobile G-protein molecules on the surface of cells infected with wt VSV are incorporated into virions.[13] This is consistent with pulse-chase studies, which have shown that G-protein arrives at the cell surface independently of other viral components.[21-23] Subsequent accumulation of G-protein into patches would arise from interaction with other viral components, most notably M-protein.[21] The amount of G-protein immobilized by such interaction in cells infected with wt virus is very low, no greater than the "background" level of immobile G seen on cells infected by mutant virus at the non-permissive temperature (Table 3). This suggests that M-G interaction may be the rate limiting step in the assembly of wt VSV; i.e., once an immobile M-G complex forms, the subsequent steps are relatively rapid, preventing the accumulation of the complexes on the infected cell membrane.

In cells infected with M-protein mutants at 31°, however, the rate of formation of immobile M-G complexes may no longer be the rate-limiting step. The accumulation of immobile G-protein coupled with a decrease (or little change) in the rate of virus production is consistent with this interpretation. If the rate of formation of immobile complexes was still rate-limiting, one would expect that cells with larger amounts of these complexes would produce virus more rapidly.

The defective M-nucleocapsid interaction in these mutants suggests that the rate limiting step in mutant virion formation could be the association of the M-G complex with the nucleocapsid.

Summary

Temperature-sensitive M-protein mutants of VSV (complementation group III) were characterized with regard to their association with nucleocapsids and with membranes using three different types of experiments: (1) Salt dependence of the inhibition of RNA synthesis by disrupted virions shows that the M-protein of all four mutants studied has a decreased functional, electrostatically stabilized association with nucleocapsids; (2) Fluorescence photo-bleaching measurements show that mutant M-protein in infected cells immobilizes G-protein on the cell surface to a much larger extent than wt, suggesting that mutant M-protein interacts to a greater extent with infected cell membranes than does wt M-protein; (3) Labeling of purified virions with the hydrophobic photoactive probe iodonaphthylazide shows that the M-protein of mutant virions is labeled 2-3-fold more strongly than is the M-protein of wt virions, also suggesting greater membrane association for mutant as opposed to wt M-protein. These coordinate effects, exhibited by all the M-protein mutants studied, are interpreted in terms of a single mutational change affecting only one of two independent binding sites. This interpretation provides the basis for a provisional kinetic scheme for budding of VS virions from the plasma membranes of infected cells.

ACKNOWLEDGMENTS
The invaluable technical assistance of Roger VanDeroef is gratefully acknowledged. Supported by NIH Grants AI-13002, GM-21661 and National Foundation March of Dimes Grant I-683.

References

1. Bishop, D.H.L. and Smith, M.S. (1978): In: *Molecular Biology of Animal Viruses*. Nayak, D.P. (Ed.), M. Dekker, Inc., New York, pp. 281–348.
2. Schnitzer, T.J., Dickson, D., and Weiss, R.A. (1979): *J. Virol.* 29:185–195.
3. Schnitzer, T.J. and Lodish, H.F. (1979): *J. Virol.* 29:443–447.
4. Weiss, R.A. and Bennett, P.L.P. (1980): *Virology* 100:252–274.
5. Clinton, G.M., Little, S.P., Hagen, F.S., and Huang, A.S. (1978): *Cell* 15:1455–1462.
6. Martinet, C., Combard, A., Printz-Ané, C., and Printz, P. (1979): *J. Virol.* 29:123–133.
7. Carroll, A.R. and Wagner, R.R. (1979): *J. Virol.* 29:134–142.
8. Combard, A. and Printz-Ané, C. (1979): *Biochem. Biophys. Res. Comm.* 88:117–123.
9. Perrault, J. and Kingsbury, D.T. (1974): *Nature (London)* 248:45–47.
10. Wilson, T. and Lenard, J. (1981): *Biochemistry,* in press.
11. Carroll, A.R. and Wagner, R.R. (1978a): *J. Virol.* 25:675–684.
12. Carroll, A.R. and Wagner, R.R. (1978b): *J. Biol. Chem.* 253:3361–3363.
13. Reidler, J.A., Keller, P.M., Elson, E.L., and Lenard, J. (1981: *Biochemistry,* in press.
14. Schlessinger, J. and Elson, E.L. (1980): In: *Biophysical Methods*. Ehrenstein, G. and Lecar, H. (Eds.), Academic Press, New York, in press.
15. Bercovici, T. and Gitler, C. (1978): *Biochemistry* 17:1484–1489.
16. Cerletti, N. and Schatz, G. (1979): *J. Biol. Chem.* 254:7746–7751.
17. Mudd, J.A. (1974): *Virology* 62:573–577.
18. Schloemer, R.H. and Wagner, R.R. (1975): *J. Virol.* 16:237–249.
19. Zakowski, J.J. and Wagner, R.R. (1980): *J. Virol.* 36:93–102.
20. Clinton, G., Burge, B.W., and Huang, A. (1979): *Virology* 99:84–94.
21. Atkinson, P.H. (1978): *J. Supramol. Struct.* 8:89–109.
22. Knipe, D.M., Baltimore, D., and Lodish, H.F. (1977): *J. Virol.* 21:1128–1139.
23. Rothman, J.E. and Fine, R.E. (1980): *Proc. Natl. Acad. Sci. U.S.A.* 77:780–784.

Copyright 1981 by Elsevier North Holland, Inc.
David H.L. Bishop and Richard W. Compans, eds.
The Replication of Negative Strand Viruses

Formation of Pseudotypes Between Viruses Which Mature at Distinct Plasma Membrane Domains: Implications for Cellular Protein Transport

Michael G. Roth and Richard W. Compans[a]

Introduction

Epithelial cells have two morphologically and functionally distinct surface membrane domains separated by junctional complexes. These domains differ in protein content[1] implying that epithelial cells possess mechanisms for recognizing and differentially transporting at least some surface molecules. In MDCK canine kidney cells, which retain morphological, secretory, and growth characteristics of normal distal tubule epithelia,[2,3] several myxo- and paramyxoviruses mature exclusively at the apical surface, whereas vesicular stomatitis virus (VSV) buds only from the basolateral membrane.[4] The glycoproteins of three of these viruses have been shown to be present exclusively in the membrane domain from which each virus buds, and it has been suggested that the site of insertion of viral glycoproteins determines the site of maturation.[5,6]

It was also reported that the paramyxovirus SV5 assembled at the apical and VSV at the basolateral membrane of MDBK bovine kidney cells.[4] However, it had been observed that co-infection of MDBK cells with SV5 and VSV resulted in the formation of phenotypically mixed virus particles, with a considerable percentage of the virus yield containing a mixture of envelope glycoproteins from each of the parental types.[7] We have investigated the possibility of phenotypic mixing between viruses which mature at distinct sites in MDCK cells, where polarity of maturation appears more sharply defined (Roth, unpublished). Specifically, we wanted to determine whether

[a] Department of Microbiology, University of Alabama in Birmingham, Birmingham, Alabama.

co-infection might result in changes in glycoprotein transport or in virus maturation sites, and whether such changes might occur separately or together.

Methods

A full description of experimental details and the results of double infections of MDCK and BHK-21 cells with influenza virus and VSV are reported elsewhere.[8] Briefly, double infections were performed by primary infection with 10 pfu/cell of A/WSN influenza virus followed by superinfection with 10 pfu/cell of VSV-Indiana some hours later. The production of VSV was measured by plaque assay on BHK-21 cells under conditions where influenza virus does not form plaques. Influenza virus yields were determined by hemagglutination assays. In this report, "VSV(WSN) pseudotype" will refer specifically to particles measured as VSV plaques formed by virus pretreated with antiserum to VSV, which are presumed to contain a VSV core surrounded by an envelope bearing influenza virus glycoproteins and little or no VSV G-protein. VSV particles which contain a mixture of both VSV and influenza glycoproteins in their envelopes are termed "phenotypically mixed," and can be neutralized by antiserum specific for either virus. We measured phenotypically mixed VSV particles by decreases in VSV titers caused by treatment before plaque assay with anti-influenza virus serum.

Results

Initially, double infections were performed with either simultaneous addition of the two viruses, or with a primary infection by influenza virus followed at various intervals by superinfection with VSV. Simultaneous infection with both viruses resulted only in the production of VSV, whereas superinfection later than 6 hr resulted in production of influenza virus alone. Thus it appears that each virus is capable of inhibiting the growth of the other. In experiments where the interval between infections varied from 3 to 5 hr, both viruses were produced and the percentage of VSV(WSN) pseudotypes ranged from 0.5 to 3.4%. In 10 experiments where the interval between infections was 3 hr, the mean percentage of VSV(WSN) pseudotypes was 1.55%, and the mean percentage of phenotypically mixed VSV was 76.6%.

Formation of VSV(WSN) pseudotypes in MDCK cells could occur through a change in glycoprotein transport, or through destruction of the junctional barriers separating distinct sites of viral glycoprotein insertion followed by lateral diffusion of the glycoproteins in the membrane. If directional transport were maintained, it seemed likely that formation of pseudotypes might not occur at early times during infection, although wild type virus would be produced. Figure 1a presents a typical one-step growth curve for VSV and VSV(WSN) pseudotypes in MDCK cells. Both VSV and

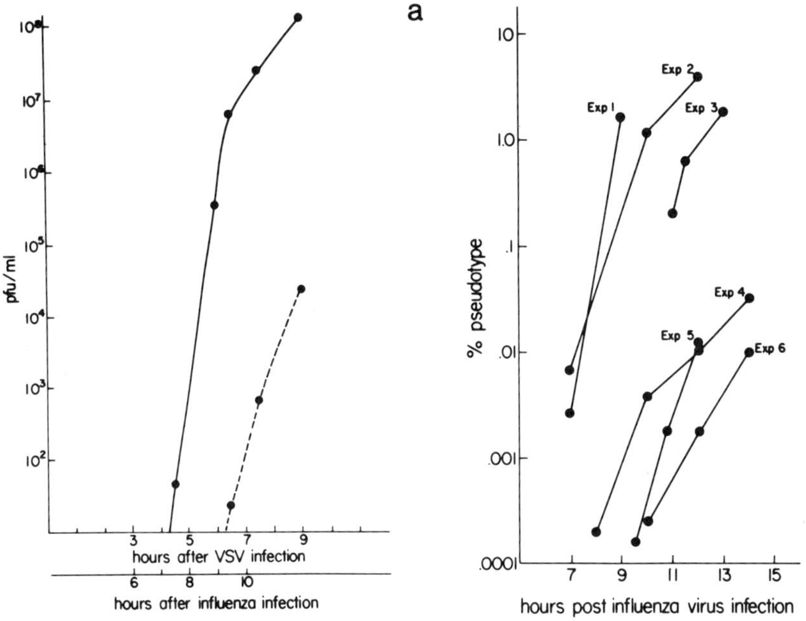

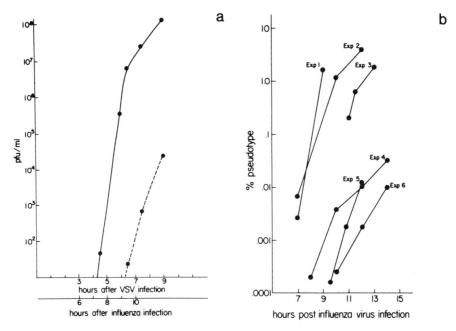

Figure 1. (a) One-step growth curves for VSV and VSV(WSN) pseudotypes in doubly infected MDCK cells. MDCK cells were infected with influenza virus and superinfected with an equal moi of VSV 3 hr later. Residual VSV inoculum was neutralized with anti-VSV serum after virus adsorption. At intervals, the titers of VSV and VSV(WSN) were determined by plaque assay. — VSV; --- VSV(WSN) pseudotypes. (b) Percentages of VSV(WSN) pseudotypes from six separate experiments, plotted as a function of time after infection.

influenza virus were detected in culture supernatants by 9 hr after influenza virus infection (6 hr post VSV infection) and VSV was produced in amounts which allow detection of VSV(WSN) pseudotypes at a level of 0.0001%; however, no pseudotypes were detected.

The exact time of appearance of pseudotypes was observed to vary, as was the onset of visible cytopathic effects in the monolayer. However, in every experiment, VSV(WSN) pseudotypes appeared later than wild type virus, and the percentage of pseudotypes present in culture supernatants increased dramatically at later times p.i. Figure 1b presents data which indicate that the percentage of VSV(WSN) pseudotypes increases relative to wild type VSV at later periods during double infections in MDCK cells. In similar experiments, pseudotypes were not detected as cell-associated virus when doubly infected cells were scraped, frozen, and thawed 9 hr post influenza virus infection. Increasing or decreasing the multiplicity of infection, lengthening the interval between infections with the two viruses from 3 to 5 hr, or infection with a 2:1 ratio of VSV to influenza virus did not affect the result that pseudotypes begin to appear later than wild type viruses. The period when pseudotypes become detectable corresponds to the first visible cytopathic effect in the monolayer.

Under the conditions of this experiment, doubly infected cells begin to hemadsorb strongly 5 to 6 hr after influenza virus infection, indicating that influenza glycoproteins are present in the apical membrane for some time before pseudotypes can be detected, and are certainly present when wild type VSV begins to form. Thus, it is not likely that lack of sufficient concentrations of influenza virus glycoproteins can explain the absence of VSV(WSN) pseudotypes 9 hr after primary infection.

BHK-21 fibroblasts do not form monolayers with cells connected by tight junctions, and polarity of influenza virus or VSV maturation is not observed in these cells. One-step growth curves for VSV and VSV(WSN) pseudotypes in BHK-21 cells are shown in Figure 2a. Doubly infected BHK-21 cells produced a lower percentage of VSV(WSN) pseudotypes than MDCK cells. In contrast to MDCK cells, VSV(WSN) pseudotypes appear to form in BHK-21 cells as soon as wild type virus, at levels between 0.01 and 0.1%. Figure 2b shows that the proportion of pseudotypes does not increase

Figure 2. (a) One-step growth curves for VSV and VSV(WSN) pseudotypes in doubly infected BHK-21 cells. BHK-21 cells infected with influenza virus were superinfected with VSV at 6 hr p.i. Virus titers were determined at various intervals. — VSV; --- VSV(WSN) pseudotypes. (b) Percentages of VSV(WSN) pseudotypes, plotted as a function of time after infection of BHK-21 cells. Data represent the averages from 3 experiments.

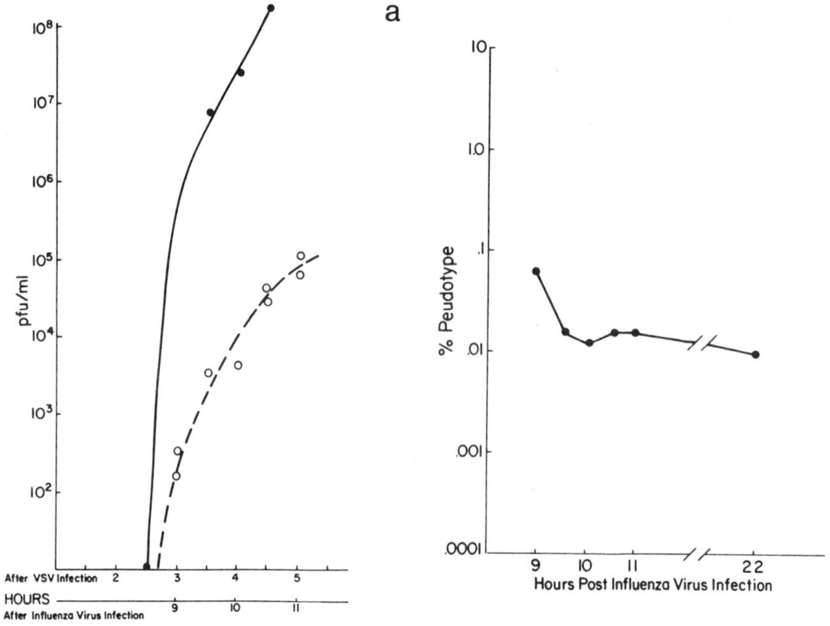

significantly with time p.i. in BHK-21 cells. These data suggest that, in BHK-21 cells, the viral glycoproteins are never segregated in the plasma membrane, and that pseudotype formation can proceed at a constant rate relative to the production of wild type VSV.

Observation of thin sections of doubly infected MDCK cells revealed that polarity of maturation sites for both viruses appeared to be maintained until 10 to 12 hr after the primary infection with influenza virus, a period later than we were able to detect pseudotypes by the more sensitive plaque assays (Figure 3a,b). Where we did observe altered maturation sites in MDCK cells (Figure 3c), we also observed frequent disruption of junctional elements. In cells examined up to 14 hr p.i. with influenza virus (11 hr p.i. with VSV), most influenza virus was associated with apical membranes and most VSV was observed at the basolateral surfaces. In contrast, in doubly infected BHK-21 cells, influenza virus and VSV bud together in the same regions of the cell surface (not shown).

Figure 3. Electron micrographs of doubly infected MDCK cells. MDCK cells were infected with influenza virus and superinfected with VSV 3 hr later. (a, b) Cells were fixed *in situ* 12 hr after primary infection. Influenza virus is found on apical, and VSV on basolateral surfaces. (c) A doubly infected MDCK cell photographed 14 hr after primary infection. VSV and influenza virus are both seen maturing at the apical surface.

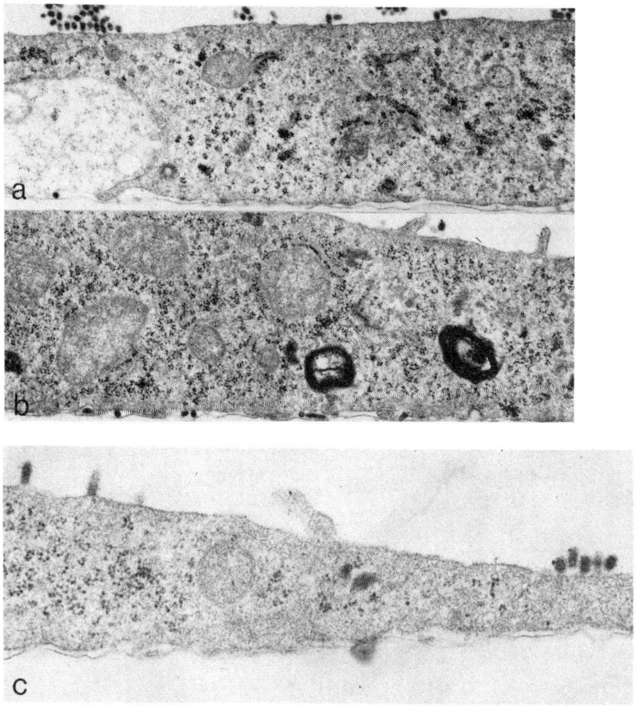

Discussion

The present results confirm reports that the glycoproteins of viruses which show polarity of maturation in MDCK cells are inserted exclusively into the membrane domain from which the virus buds.[5] Since VSV readily incorporates influenza glycoproteins into its envelope, a nondirectional transport of those glycoproteins to basolateral membranes should have resulted in production of a constant percentage of VSV(WSN) pseudotypes in MDCK cells at various times p.i. Exactly such a result is observed in BHK cells, where directional transport of glycoproteins does not occur. It is likely that VSV(WSN) pseudotypes form in MDCK cells after destruction of tight junctions, which allows lateral diffusion and mixing of glycoproteins in the plane of the plasma membrane. Consistent with this idea are observations that the maturation sites of both viruses remain polarized during periods when the monolayer appears intact, and that even at later times a majority of each virus type is found associated with its normal membrane domain. This result suggests that directional transport continues after barriers restricting lateral diffusion of glycoproteins have been removed. These results do not rule out the possibility, which seems less likely, that there is a progressive change in glycoprotein transport that parallels or follows disruption of tight junctions, either through loss of a function dependent on host protein synthesis, or perhaps through virus-induced changes in the membrane.

The present results suggest that directional transport of viral glycoproteins is a stable characteristic of infected MDCK cells, which probably continues after the destruction of junctional elements responsible for maintaining distinct membrane domains. Thus it should be possible to study intracellular transport of viral glycoproteins in doubly infected MDCK cells, and to determine the cellular locations at which sorting of surface molecules occur.

ACKNOWLEDGMENTS
This research was supported by NSF Grant PCM80-06498 and USPHS Grant AI 12680. M.G.R. was supported by a predoctoral traineeship from Grant T32 AI 07150 from the NIAID.

References

1. Louvard, D. (1980): *Proc. Natl. Acad. Sci.* 77:4132–4136.
2. Rindler, M.J., Chuman, M.M., Shaffer, L., and Saier, M.H. (1979): *J. Cell Biol.* 81:636–648.
3. Cereijido, M., Robbins, E.S., Dolan, W.J., Rotunno, C.A., and Sabatini, D.D. (1978): *J. Cell Biol.* 77:853–880.
4. Rodriguez Boulan, E. and Sabatini, D.D. (1978): *Proc. Natl. Acad. Sci.* 75:5071–5075.
5. Rodriguez Boulan, E. and Pendergast, M. (1980): *Cell* 20:45–54.
6. Roth, M.G., Fitzpatrick, J.P., and Compans, R.W. (1979): *Proc. Natl. Acad. Sci.* 76:6430–6434.
7. Choppin, P.W. and Compans, R.W. (1970): *J. Virol.* 5:609–616.
8. Roth, M.G. and Compans, R.W.: Manuscript in preparation.

Copyright 1981 by Elsevier North Holland, Inc.
David H.L. Bishop and Richard W. Compans, eds.
The Replication of Negative Strand Viruses

Phenotypic Mixing Between Vesicular Stomatitis Virus (VSV) and Host Range Variants of Mouse Mammary Tumor Virus (hrMMTV): Expression of MMTV Envelope Glycoprotein GP52 on VSV (hrMMTV) Pseudotypes

James C. Chan,[a] Michael Scanlon,[a]
James M. Bowen,[a] Richard J. Massey,[b]
and Gerald Schochetman[b]

The ability of vesicular stomatitis virus (VSV) to form pseudotypes with unrelated viruses through the incorporation of heterologous viral envelope components was first described by Choppin and Compans.[1] Subsequently, the formation of pseudotypes as a result of phenotypic mixing between VSV and avian,[2-7] murine[8-15] and primate[16] retroviruses has been reported.

Zavada et al.[15] reported the formation of pseudotypes which contained the VSV genome in the envelope of wild type (wt) MMTV. (Following standardized nomenclature pseudotypes are henceforth denoted by a symbol for a virus genome followed by a bracketed symbol for envelope). We also succeeded in the production of VSV(wtMMTV) pseudotypes and were able to detect them by immunoelectronmicroscopy.[17]

In this communication, we report on pseudotype formation between VSV mutants and hrMMTV which had been adapted to propagate in feline kidney cells.[18] In addition, we present results of host range and immunological studies of such VSV(hrMMTV) pseudotypes and the classification of monoclonal antibodies to MMTV-gp52 using such pseudotypes.

[a] University of Texas System Cancer Center M.D. Anderson Hospital and Tumor Institute, Texas Medical Center, Houston, Texas.
[b] Frederick Cancer Research Center, Frederick, Maryland.

Higher Yield of VSV(hrMMTV) Pseudotypes

Two VSV mutants were compared on their ability to yield more hrMMTV pseudotypes of VSV. The mutant t1B17(I and V) is produced with a thermolabile envelope. It has frequently been used because pseudotypes produced by this mutant can readily be distinguished from its genotype on the basis of thermostability.[19] Mutant *ts*045(V) on the other hand has a temperature-sensitive mutation affecting the synthesis of envelope glycoprotein G.

As a source of hrMMTV, three hrMMTV producing feline kidney cell (CrFK) lines were used.[18] These cell lines are designated collectively as CrFK(hrMMTV) and individually as CrFK(C_3H), CrFK(GR) and CrFK(RIII), in which case, the cell line CrFk is followed by a bracketed symbol denoting the strain of hrMMTV it produces. The pseudotypes produced by VSV superinfection of CrFK(C_3H), CrFK(GR) or CrFK(RIII) cells were designated VSV(C_3H), VSV(GR) and VSV(RIII), respectively.

Using the criterion of thermostability (after heating at 45° for 60 min), the percentage yield of VSV(RIII) by t1B17(I and V) was calculated to be 0.0017% of the total progeny (Table 1). In contrast, using *ts*045(V), VSV(hrMMTV) was produced to the extent of 99.21 to 99.37% of the progenies (Table 2). These pseudotypes were neutralizable with monospecific anti-MMTV-gp52 alone (Table 2), suggesting the complete or predominant encapsidation of the VSV genome with MMTV-gp52.

Table 1. Thermoinactivation of VSV Progenies Produced in Cat Kidney Cells and in MMTV-Producing Cat Kidney Cells.

VSV progenies produced in cells[a]	**VSV titers (Log_{10} PFU/ml) after Heating at 45° for min**[b]			
	0	**20**	**40**	**60**
Vero	7.630	NT[c]	NT	<1 (<0.00000023)[d]
CrFK	8.431	4.531 (0.00125)	3.579 (0.0014)	3.086 (0.0004)
CrFK (RIII)	8.262	4.462 (0.0158)	3.698 (0.0027)	3.491 (0.0017)

[a]CrFK and CrFK (RIII) cells were pretreated with dexamethasone (3 μg/ml) and insulin (10 μg/ml) for 4 days prior to infection with VSV-tl B17 at an moi of 3 PFU/cell. Progenies were harvested 24 hr after infection at 37°. Vero cells were infected with a twice-purified plaque suspension of VSV-tlB17 and incubated at 37° until the appearance of maximum cytopathic effects and progeny harvested.

[b]VSV titers were determined by plaque assay in S+L− $MiCl_1$ (Mink) cells.

[c]Not tested.

[d]Expressed as percent of unheated VSV.

Table 2. Enhanced Production of Complete VSV(hrMMTV) Pseudotypes.

Progeny virus produced by VSV-*ts*045 (V) infected cells[a]	Reaction with			Residual VSV titers (Log_{10}pfu/ml)[c]	Percent neutralized[d]
	PRS[b]	Anti-VSV[b]	Anti-MMTV-gp52[b]		
CrFk	+			3.8	—
		+		<1	100
			+	4.0	0
		+	+	<1	100
CrFK (C_3H)	+			5.3	—
		+		5.6	0
			+	3.1	99.37
		+	+	<1	100
CrFK (GR-1)	+			4.2	—
		+		4.2	0
			+	2.1	99.21
		+	+	<1	100

[a]Cells were infected with VSV-*ts*045 (V) mutant at an moi of 3 PFU/cell. Progeny virus was harvested after incubation at 39.5° for 12 hr.

[b]Progeny virus samples were reacted with PRS (preimmune rabbit serum, 1:4 dilution), hyperimmune sheep anti-VSV serum (1:100 dilution), rabbit anti-MMTV-gp52 serum (1:4 dilution) or both. All sera had been decomplemented.

[c]Residual virus titers were determined by plaque assay in NMG (mouse mammary gland) cells.

[d]Percent neutralized = titer after reaction with PRS-titer after reaction with immune sera/titer after reaction with PRS × 100%.

Host Range Studies of VSV(hrMMTV) Pseudotypes

VSV(hrMMTV) pseudotypes were prepared by treatment of progenies of *ts*045(V)-infected CrFK(hrMMTV) cells with hyperimmune sheep anti-VSV serum (decomplemented, 1/640 dilution) and represented VSV virions that were resistant to neutralization by anti-VSV. These pseudotypes were then used in experiments (Tables 3 and 4; Figure1).

The host range of VSV(C_3H) and VSV(RIII) pseudotypes was determined simultaneously in seven cell lines known to be permissive to VSV genotype and their efficiency of plaquing (E.O.P.) indices were calculated. Two interesting patterns were noted (Table 3). First, NRK-2 (rat) cells and bat lung cells were resistant to VSV(C_3H) and VSV(RIII) pseudotypes. Second, whereas VSV genotype had an E.O.P. index of 0.85 in CrFK(C_3H) cells, VSV(C_3H) and VSV(RIII) pseudotypes had unusually high E.O.P. indices of 6.6 and 2.0, respectively in this very same cell line. Since CrFK cells were the parental line of CrFK(C_3H),[18] the productive infection of CrFK cells by C_3H-hrMMTV appeared to account for the differences.

Detection of C_3H (Type) Specific Neutralizing Antibodies in Sera of Mammary Tumor-Bearing C_3H/HeN Mice

VSV(hrMMTV) pseudotypes were utilized to detect MMTV-neutralizing antibodies in sera of mammary tumor-bearing mice (tumored sera). Results (Figure 1) indicated the presence of at least two population of antibodies: one in lower concentration that neutralized VSV(RIII), VSV(GR), and probably VSV(C_3H) also (conceivably group-specific), and the other, in higher concentration that specifically neutralized VSV(C_3H). Since the latter antibody could only neutralize VSV(C_3H) at higher levels (10^4 pfu), it most likely represented C_3H (type) specific antibody reported previously by Schochetman et al.[20] and Teramoto et al.[21] Conversely, the expression of C_3H (type) specific antigenic determinants on the VSV(C_3H) pseudotypes is indicated by the results.

Characterization of Monoclonal Antibodies to MMTV-gp52 by VSV(hrMMTV) Pseudotypes

Monoclonal antibodies to MMTV-gp52 have been prepared by immunization of Balb/c mice with purified C_3H-MMTV-gp52 and subsequent hybridoma techniques.[22] According to their specificity in MMTV-binding assay,[22] these monoclonal antibodies could be grouped into group (C_3H, GR, RIII, C_3Hf) specific, class (C_3H, GR) specific or C_3H (type) specific categories (Table 4). Of the 7 monoclonal antibodies tested, only 4 had neutralizing activities varying from 10^4 to 10^7 neutralizing units per ml (not shown) and all were group specific in MMTV-binding assays. The other 3 (2 class-specific and 1

Table 3. Host Range of VSV and VSV(hrMMTV) Pseudotypes.

		VSV		VSV (C_3H)		VSV (RIII)	
Indicator cells	**Species**	**Titer (pfu/ml $\times 10^4$)[a]**	**E.O.P. Index[b]**	**Titer (pfu/ml $\times 10^4$)**	**E.O.P. Index**	**Titer (pfu/ml $\times 10^4$)**	**E.O.P. Index**
S+L− $MiCl_1$	Mink	4.80	1.00	4.20	1.00	2.40	1.00
S+L− mouse	Mouse	0.48	0.10	0.72	0.17	0.27	0.11
CrFK	Cat	1.78	0.37	4.60	1.10	2.10	0.87
CrFK (C_3H)	Cat	4.10	0.85	27.90	6.64	4.80	2.00
NRK-2	Rat	1.60	0.33	0.26	0.06	0.19	0.08
Cf2th	Dog	1.43	0.30	1.17	0.28	0.56	0.23
Bat lung	Bat	0.46	0.10	0.08	0.02	0.06	0.03
Vero	Monkey	1.67	0.35	0.66	0.16	0.80	0.33

[a]Infectivity titers of VSV or VSV (C_3H) and VSV (RIII) pseudotypes were determined by simultaneous plaque assays in the seven cell lines.

[b]E.O.P.: efficiency of plaquing; E.O.P. index for each virus in S+L− $MiCl_1$ cells was arbitrarily set as 1 and that in other cell lines was prorated.

Table 4. Classification of MMTV-gp52 Monoclonal Antibodies on the Basis of MMTV-Binding Activity and VSV (hrMMTV) Neutralization.

Designation	I_g Subtype	Specificity of MMTV-Binding[a]	Inactivation of VSV (hrMMTV)[b]
79-386-A	IgM	C_3H, GR, RIII, C_3Hf	Group specific
80-500-A	IgG	C_3H, GR, RIII, C_3Hf	Group specific
80-536-A	IgG	C_3H, GR, RIII, C_3Hf	Group specific
80-507S	IgM	C_3H, GR, RIII, C_3Hf	Group specific
79-076-A	IgG_3	C_3H, GR	No activity ± C′[c]
79-298-A	IgM	C_3H, GR	No activity ± C′
79-240-A	IgG_{2a}	C_3H	No activity ± C′

[a]MMTV strain bound by monoclonal antibody.

[b]Group specific inactivation = neutralization of VSV (C_3H), VSV (GR) and VSV (RIII) pseudotypes.

[c]No inactivation of VSV (C_3H), VSV (GR), or VSV (RIII) pseudotypes with or without complement.

Figure 1. Neutralizing antibodies in sera of mammary tumor bearing C_3H/HeN mice. VSV(C_3H), VSV(GR), and VSV(RIII) pseudotypes produced by VSV-*ts*045(V) mutant as described in Table 2 and text were used to detect neutralizing antibodies in serum pools (4 to 5 mice) of normal Balb/c mice, of normal C_3H/HeN mice, and of mammary tumor bearing C_3H/HeN mice. The respective serum pool was decomplemented by heating at 56° for 30 min, and diluted 1:50 in phosphate buffered saline. Increasing amount (from 1 to 4 $\log_{10}$ pfu) of each "strain" of VSV pseudotype was reacted with the 1:50 dilution of each serum pool at 37° for 2 hr. The surviving fraction was determined by plaque assay in S+L- $MiCl_1$ (mink) cells. The percent reduction in VSV(hrMMTV) infectivity was calculated by the formula Vo-Vn/Vo × 100 where Vo = infectivity titer before reaction with serum and Vn = titer after reaction with serum.

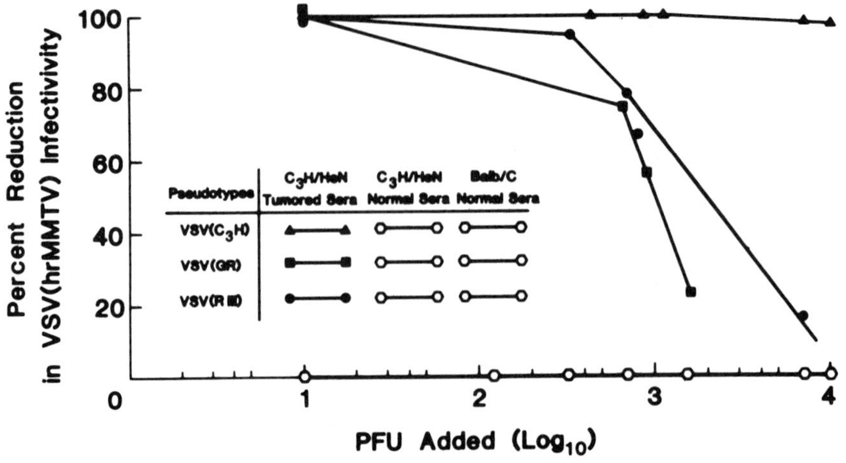

type specific) antibodies were negative in neutralization tests even at 10^{-1} dilution. Addition of mouse or rabbit complement to each of these 3 antibodies (10^{-1} dilution) did not lead to complement mediated virolysis. These results indicate that MMTV-binding monoclonal antibodies to MMTV-gp52 may tentatively be classified into two categories: those that neutralized VSV(hrMMTV) and those that did not.

References

1. Choppin, P.W. and Compans, R.W. (1970): *J. Virol.* 5:609–616.
2. Kang, C.Y. and Lambright, P. (1977): *J. Virol.* 21:1252–1256.
3. Love, D.N. and Weiss, R.A. (1974): *Virology* 57:271–278.
4. Weiss, R.A., Boettiger, D., and Love, D.N. (1975): *Cold Spring Harbor Symp. Quant. Biol.* 39:913–918.
5. Weiss, R.A., Boettiger, D., and Murphy, H. (1977): *Virology* 76:808–825.
6. Zavada, J. (1972): *Nature New Biol.* 210:122–124.
7. Zavada, J. (1976): *Arch. Virol.* 50:1–15.
8. Breitman, M. and Prevec, L. (1977): *Virology* 76:643–652.
9. Chan, J.C., Hixson, D.C., and Bowen, J.M. (1978): *Virology* 88:171–176.
10. Huang, A.S., Besmer, P., Chu, L., and Baltimore, D. (1973): *J. Virol.* 12:659–662.
11. Krontiris, T.G., Soeiro, R., and Fields, B.N. (1973): *Proc. Natl. Acad. Sci.* 70:2549–2553.
12. Livingston, D.M., Howard, T., and Spence, C. (1976): *Virology* 70:432–439.
13. Witte, D.N. and Baltimore, D. (1977): *Cell* 11:505–511.
14. Zavada, J. (1972): *J. Gen. Virol.* 15:183–191.
15. Zavada, J., Dickson, C., and Weiss, R.A. (1977): *Virology* 82:221–231.
16. Schnitzer, T.J., Weiss, R.A., and Zavada, J. (1977): *J. Virol.* 23:444–454.
17. Chan, J.C., Hixson, D.C., Scanlon, M., East, J.L., and Bowen, J.M. (1979): *Proc. Ann. Meeting Am. Soc. for Microbiol.*, p. 297.
18. Howard, D.K. and Schlom, J. (1978): *Proc. Natl. Acad. Sci.* 75:5718–5722.
19. Weiss, R.A. and Bennett, P.L. (1980): *Virology* 100:252–274.
20. Schochetman, G., Arthur, L.O., Long, C.W., and Massey, R.J. (1979): *J. Virol.* 32:131–139.
21. Teramoto, Y.A., Kufe, D., and Schlom, J. (1977): *J. Virol.* 24:525–533.
22. Massey, R.J., Arthur, L.O., Nowinski, R.C., and Schochetman, G. (1980): *J. Virol.* 34:635–643.

David H.L. Bishop and Richard W. Compans, eds.
The Replication of Negative Strand Viruses

In Vivo Inhibition of Primary Transcription of Vesicular Stomatitis Virus by a Defective Interfering Particle

Pauline H.S. Bay and M.E. Reichmann[a]

A large defective interfering (DI) particle isolated from the heat-resistant (HR) mutant of vesicular stomatitis virus (VSV)[1] [VSI HR Toronto *DI 0.5* (3′, 8%)][2] exhibited several unusual properties. Unlike the standard DI particles of the Indiana serotype of VSV, it interfered heterotypically with New Jersey (NJ) serotype (Concan subtype) viral infections,[3,4] it mapped in the 3′ half of the viral genome,[5] and was probably generated by an internal deletion[6-8] rather than by a read-back mechanism.[9,10] We have recently shown that this particle interfered *in vivo* with primary transcription by homotypic or heterotypic virions.[11] We have now constructed HR DI particles containing a temperature-sensitive (*ts*) L-protein derived from either VSV mutant *ts*G11 or *ts*G13. As expected, these DI particles were also *ts* in transcription of their own genome. We demonstrate that interference with viral infection or with primary transcription of the virion requires active transcription of the HR DI particle genome. In contrast, the generation of HR DI particles in mixed infections with NJ virions occurred without either transcription of the former or interference with viral infections.

Results and Discussion

Generation of HR DI Particle Chimeras, ts *in Transcription*

DI particle chimeras[12] were obtained from cells co-infected with *ts*G11 or *ts*G13 virions and a purified HR DI particle preparation using a modified

[a] Department of Microbiology, University of Illinois, Urbana, Illinois.

procedure of Chow et al.[13] The progeny of this infection were assumed to contain the *ts* virion-encoded L-protein,[14] since the HR DI particle genome does not contain the L cistron.[5,7,8] However, as the initial HR DI particle inoculum was contaminated with 10^5 plaque forming units (PFU) of parental virions per optical density (OD) unit (260 nm), HR DI particle preparations were passaged several times in the presence of *ts* virions to ensure a total replacement of the L-protein. HR DI particles were purified after each passage and contaminating virions were monitored for temperature-sensitivity. After the first passage, the difference in the plaque assay at permissive and non-permissive temperatures was already four logs. This indicated that the presence of HR virions in the progeny was negligible. Upon five or more passages, the HR DI particle preparation was found to contain a very high proportion of nontranscribing variants, in agreement with similar observations by other laboratories.[6,8] For this reason, HR DI particle chimeras obtained from the third passage were used in the experiments described below.

In Vivo *Primary Transcription by Wild Type and Chimeric HR DI Particles*

The temperature-sensitivity of primary transcription of wild type and chimeric HR DI particles was tested in the presence of cycloheximide and in the presence or absence of NJ helper virions. Transcription after four hr of infection was monitored by annealing an aliquot of intracellular RNA to an excess amount of radioactive HR DI particle RNA. Since this RNA has very few nucleotide sequence homologies with NJ virion RNA,[15,16] the ribonuclease (RNase) resistant counts obtained after annealing reflect transcription only by HR DI particles. In Table 1, the extent of HR genome transcription is expressed in these terms at the permissive (31°C) and non-permissive (39°C) temperatures. The genomes of the wild type HR DI particle and the HR DI chimera with *ts*G31 proteins (lesion in M-protein) were transcribed at both temperatures. However, transcription by the chimera with *ts*G11 proteins at 39°C was only 6% of the value at 31°C. Similar results were obtained with a HR DI chimera containing *ts*G13 proteins (data not shown). In addition, primary transcription by these various HR DI particles in the absence of helper virus gave essentially the same results as shown in Table 1 (data not shown). Although the same multiplicity of infection (moi) of HR DI particles (based on OD_{260nm} measurements) was used in these experiments, transcription by the chimeras was approximately 1/4 to 1/3 that of the wild type. Whether this was due to a greater instability of the chimeras or to the generation of some nontranscribing DI particle variants during the preparation of the chimeras[6,8] is not known.

Effect of Wild Type and Chimeric HR DI Particles on NJ Virion Primary Transcription

To determine whether the inhibition of virion primary transcription by HR DI particles required active self-transcription by the latter, the *ts*G11

Table 1. Primary Transcription of the HR DI and HR DI Chimera Genomes.[a]

	Ribonuclease resistance after annealing with labeled HR DI RNA (CPM)[b]		Cpm 39°C
DI particle	31°C	39°C	Cpm 31°C
HR DI	18,236	20,170	1.00
HR DI with *ts*G11 proteins	5,999	370	0.06
HR DI with *ts*G31 proteins	6,025	4,500	0.75

[a]BHK cells in 60-mm tissue culture plates (3.5×10^6) cells) were pretreated with 5 μg of actinomycin D per ml for one hr. The cells were then infected with NJ virions (moi, 5) in the presence of the various HR DI particles. After adsorption at room temperature for 30 min, the cells were washed twice with sterile saline and covered with medium containing actinomycin D and 50 μg of cycloheximide per ml and incubated at the appropriate temperature. At four hr postinfection, the cells were washed once with sterile saline and intracellular RNAs were extracted.[5] To determine the amount of HR DI particle transcriptional products, a sample of cytoplasmic RNA was annealed[19] to an excess amount of ^{3}H HR DI RNA probe labeled in all four bases[11] (0.1 μCi/μg of RNA).

[b]The relative concentration of accumulated transcriptional products are expressed as the total ^{3}H CPM of the probe which could be made RNase resistant after annealing. The results were corrected for non-specific annealing by subtracting ^{3}H CPM of HR DI RNA rendered RNase resistant after annealing with intracellular RNA from mock-infected cells.

Note: Viral inocula for the Concan isolate of NJ serotype VSV, for the *ts*G11, *ts*G13, and *ts*G31 isolates of Indiana serotype VSV and for the HR DI particle inocula were prepared in BHK cells using a procedure described previously.[11] DI particle chimeras were obtained from cells co-infected with the appropriate virion and a purified HR DI particle preparation following a procedure described by Chow et al.[13] except that the incubation was at 31°C for 11 hr. Resultant chimeric DI particles were purified using a procedure previously described.[11] DI particles from the third passage were utilized in these experiments.

chimeric HR DI particle was used, since little transcription of this particle was detectable at 39°C (Table 1). Experimental conditions similar to those described in the legend to Table 1 were used except that NJ virion transcription was assayed using a radioactive NJ virion RNA probe. Inhibition of virion primary transcription by the wild type and *ts*G31 chimeric HR DI particles was consistently lower at 31°C (22% and 23%, respectively) than at the higher temperatures (86% and 60%, respectively). In contrast, the *ts*G11 chimera did not inhibit virion primary transcription under conditions non-permissive to its own transcription (39°C), while at 31°C its inhibitory ability was comparable to those of the other two types of HR DI particles (Table 2). These experimental results showed that primary transcription by HR DI particle genomes was required for inhibition of virion primary transcription. One possible basis for this requirement may be the necessity for the transcribing enzymes to move out of the attachment site on the DI particle template so that the freed attachment site can competitively bind a limiting supply of virion transcriptases. *In vitro* experiments, to be published elsewhere, support this conclusion and have demonstrated that the inhibition was strongly dependent on the ratio of HR DI particles to virions.

Table 2. Inhibition of Concan New Jersey VSV Primary Transcription by Various HR DI Particles.[a]

Infecting particles	Ribonuclease resistance after annealing with labeled NJ virion RNA (cpm)		Percent inhibition	
	31°C	39°C	31°C	39°C
NJ virion	17,628	20,239	0	0
NJ virion + HR DI	13,750	2,833	22	86
NJ virion + HR DI with *ts*G11 proteins	13,397	20,410	24	0
NJ virion + HR DI with *ts*G31 proteins	13,574	8,096	23	60

[a]Experimental conditions are similar to those described for Table 1 with the exception that a ^3H NJ RNA probe labeled in all four bases (0.1 μCi/μg of RNA)[11] was used instead of a HR DI RNA probe.

Heterotypic Interference With Virion Replication by Wild Type and Chimeric HR DI Particles

The inhibition of virion primary transcription by HR DI particles may not be an event necessary for interference with viral replication. To investigate whether a correlation existed between these two activities, heterotypic interference with viral yields at 31°C and 39°C was measured in the presence of wild type and chimeric HR DI particles. The results, shown in Table 3, are expressed as $\log_{10}$ reduction in virus yield at 18 hr after infection. A direct correlation between inhibition of virion primary transcription and reduction in virus yield can be seen in the experiments with wild type and *ts*G11 chimeric HR DI particles. The wild type DI particles reduced viral yield more efficiently at the higher temperatures while on the contrary, the *ts*G11 chimeric particle interfered very little with virus generation at the non-permissive temperatures. The parallel between these results and inhibition of virion primary transcription can be seen by comparing the data of Table 2 and Table 3. However, in the case of the *ts*G31 chimera, the interfering ability at the higher temperature was, if anything, smaller than at the lower temperature (Table 3), even though the inhibition with virion primary transcription was approximately 3-fold higher at 39°C than at 31°C (Table 3). The reason for this lack of correlation is unknown. It should be noted, however, that the inhibition of virion transcription by this DI particle at 39°C was not the same as that of the wild type particle. This suggested that some other factors, possibly a greater instability of the particle, may have affected the results.

Table 3. Heterotypic Interfering Ability of the HR DI and HR DI Chimera Particles.[a]

DI particle	Concentration of DI particle solution [OD (260 nm)]	Log_{10} reduction of infectivity (PFU/ml)[b]	
		31°C	39°C
HR DI	0.06	2.53	4.24
	0.03	2.40	3.64
HR DI with *ts*G11 proteins	0.06	2.68	0.93
	0.03	2.15	0.55
HR DI with *ts*G31 proteins	0.06	3.39	3.11
	0.03	2.81	2.14

[a]BHK cell monolayers in 35-mm tissue culture plates (1.5×10^6 cells) were infected with NJ virions (moi, 4) in the presence and absence of various HR DI particles for 30 min at room temperature, rinsed twice with sterile saline and then covered with fresh medium. At 18 hr after infection, samples were removed and assayed at the respective temperatures for plaque-forming ability.[5]

[b]Results are expressed as the $\log_{10}$ difference of infectious particles produced from BHK cells co-infected with virions and DI particles (at various concentrations) as compared to cells infected with virions only. In the absence of DI particles, the yield of virions was 1.7×10^8 PFU/ml.

Generation of HR DI Particles Under Conditions Non-permissive for Interference

Since standard VSV DI particle production may under certain conditions be independent of interference,[12,17,18] the progeny from mixed infections with NJ virion and *ts*G11 chimeric HR DI particles at permissive and non-permissive temperatures were examined. Newly synthesized particles were labeled with ^{3}H-uridine and analyzed by rate zonal centrifugation. A radioactive peak in the position consistent with virion sedimentation properties was obtained when cells were infected with virus alone (Figure 1A, solid circles). In mixed infections with wild type HR DI particles at 31°C (data not shown) or 39°C, a major peak consistent with the sedimentation properties of the HR DI particle was observed (Figure 1A, open circles). Because of interference, the viral progeny was reduced. A similar profile was obtained with the *ts*G11 chimeric HR DI particle at the permissive temperature (Figure 1B, open circles). At the non-permissive temperature, the two peaks corresponding to virion and HR DI particles were considerably larger because of the non-interfering conditions (Figure 1B, solid circles). These data indicated that the *ts*G11 chimera HR DI particle served as a template for its replication, even though interference was not taking place.

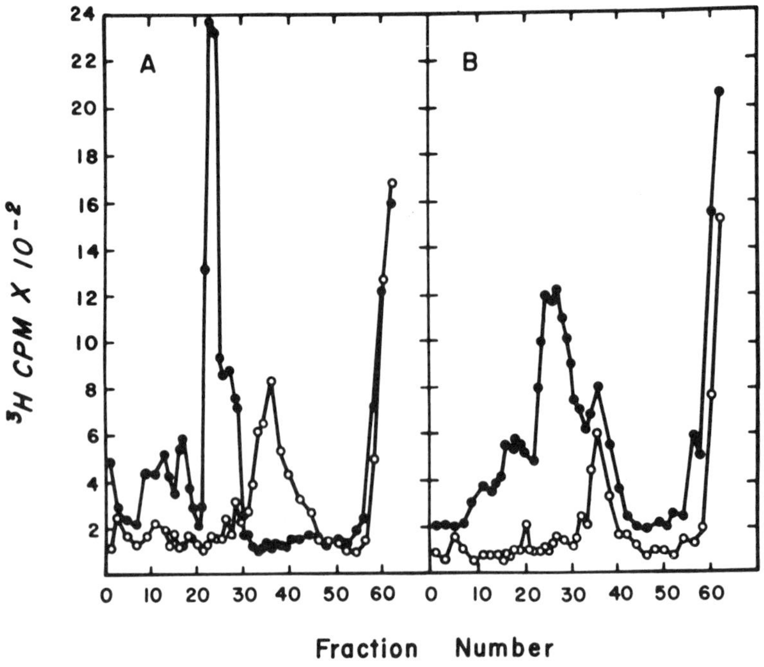

Figure 1. Sucrose gradient profiles of particles produced by BHK cells infected with VSV (NJ) in the presence or absence of various HR DI particles. BHK cell monolayers in 100-mm tissue culture dishes (1×10^7 cells) were infected with NJ virions (moi, 5) in the presence and absence of various HR DI particles for 30 min at room temperature, rinsed twice with sterile saline, and then covered with fresh medium containing 5 μCi/ml of ^{3}H-uridine. At 18 hr after infection at the indicated temperature, the virion progeny was pelleted by ultracentrifugation at $50{,}000 \times g$ and 4°C for 90 min. The pellets were resuspended overnight in 3E buffer,[11] layered on linear 10-35% (wt/vol) sucrose gradients in 3E buffer, and centrifuged at $110{,}000 \times g$ and 4°C for 60 min. Tubes were fractionated after bottom puncture into 10-drop (0.2 ml) samples. A 25 μl aliquot of each sample was spotted on a glass fiber filter and analyzed for radioactivity.[13] The figure shows profiles based on the cpm in the aliquots and obtained from cells infected with the following inocula and incubation temperatures: (A) ●, NJ virions at 39°C; ○, NJ virions and 0.06 OD (260 nm) HR DI particles at 39°C. (B) ●, NJ virions and 0.06 OD (260 nm) *ts*G11 chimeric HR DI particles at 39°C; ○, NJ virions and 0.06 OD (260 nm) *ts*G11 chimeric HR DI particles at 31°C.

Although the *in vitro* stability of the *ts*G11 chimera HR DI particle at 39°C has been observed to be similar to that of the wild type HR DI particle (data not shown), the chimeric DI particle may be unstable *in vivo* at this temperature. The disintegration of the biologically active nucleocapsid would lead to a loss of all its activities and the conclusions about the correlations between these activities would be invalid. The results in Figure 1B indicated that this was not the case. The replication of the *ts*G11 chimeric HR DI particle at the non-permissive temperature clearly demonstrated the presence of a biologically active nucleocapsid of this particle in the cell and that the absence of its primary transcription was due to the lesion in its L-protein rather than to its physical instability.

References

1. Petric, M. and Prevec, L. (1970): *Virology* 41:615–630.
2. Reichmann, M.E., Bishop, D.H.L., Brown, F., Crick, J., Holland, J.J., Kang, C.-Y., Lazzarini, R., Moyer, S., Perrault, J., Prevec, L., Pringle, C., Wagner, R.R., Younger, J.S., and Huang, A.S. (1980): *J. Virol.* 34:792–794.
3. Prevec, L. and Kang, C.Y. (1970): *Nature* 228:25–27.
4. Reichmann, M.E., Schnitzlein, W.M., Bishop, D.H.L., Lazzarini, R.A., Beatrice, S.T., and Wagner, R.R. (1978): *J. Virol.* 25:446–449.
5. Schnitzlein, W.M. and Reichmann, M.E. (1976): *J. Mol. Biol.* 101:307–325.
6. Perrault, J. and Semler, B.L. (1979): *Proc. Natl. Acad. Sci.* 76:6191–6195.
7. Clerx-van Haaster, C., Clewley, J.P., and Bishop, D.H.L. (1980): *J. Virol.* 33:807–817.
8. Epstein, D.A., Herman, R.C., Chien, I., and Lazzarini, R.A. (1980): *J. Virol.* 33:818–829.
9. Huang, A.S. (1977): *Bact. Rev.* 41:811–822.
10. Leppert, M., Kort, L., and Kolakofsky, D. (1977): *Cell* 12:539–552.
11. Bay, P.H.S. and Reichmann, M.E. (1979): *J. Virol.* 32:876–884.
12. Schnitzlein, W.M. and Reichmann, M.E. (1977): *Virology* 80:275–288.
13. Chow, J.M., Schnitzlein, W.M., and Reichmann, M.E. (1977): *Virology* 77:579–588.
14. Hunt, D.M., Emerson, S.U., and Wagner, R.R. (1976): *J. Virol.* 18:596–603.
15. Schnitzlein, W.M. and Reichmann, M.E. (1977): *Virology* 77:490–500.
16. Repik, P., Flamand, A., Clark, H.F., Obijeski, J.R., Roy, P., and Bishop, D.H.L. (1974): *J. Virol.* 13:250–252.
17. Stampfer, M., Baltimore, D., and Huang, A.S. (1969): *J. Virol.* 4:154–161.
18. Adachi, I. and Lazzarini, R.A. (1978): *Virology* 87:152–163.
19. Leamnson, R.N. and Reichmann, M.E. (1974): *J. Mol. Biol.* 85:551–568.

David H.L. Bishop and Richard W. Compans, eds.
The Replication of Negative Strand Viruses

Continuing Evolution of Virus-DI Particle Interaction Resulting During VSV Persistent Infection

Frank M. Horodyski and John J. Holland[a]

Viruses have been isolated from persistent infections in cultured cells which are phenotypically altered from the virus used to establish the persistent infection.[1-8] Many of these viruses are characterized by a temperature-sensitive or small plaque phenotype. The best characterized case of virus evolution during persistent infection is that of vesicular stomatitis virus in BHK-21 cells.[9] Standard virus isolated from this culture has been undergoing extensive and continuous mutational change for 7 years detectable by T_1 oligonucleotide mapping of the viral RNA.[10] These changes are reflected in all five viral proteins as well as in the ends of the viral RNA.[11,12]

Many persistent infections, including the rabies and VSV persistent infections in BHK-21 cells require the addition of homologous defective interfering (DI) particles for the establishment and maintenance of the carrier state.[4,9] Viruses have been isolated from these rabies and VSV persistent infections which are less susceptible to interference mediated by the DI particle used to establish the persistent infection,[8,13] and also from LCM persistent in mice.[14] These virus variants arise by 75 days after the establishment of the persistent infection in the case of BHK-21 CAR4 VSV carrier line described here. The 75-day virus isolate has a small plaque phenotype and no changes from the original *ts*G31 virus are exhibited in the T_1 oligonucleotide map of the viral RNA.[13] Small plaque virus isolated from this persistent infection at 5 years is also less sensitive to interference mediated by the *ts*G31 DI particle, but this virus contains several mutations detectable by T_1

[a] Department of Biology, University of California, San Diego, La Jolla, California.

oligonucleotide fingerprinting.[10,13] It is not known if the DI particles present in the persistent infection promote such a rapid accumulation of mutants by selecting for viruses which are less sensitive to interference mediated by the DI particles.

In this paper, we describe an assay used to screen viruses for their interference properties. We also report the isolation of virus variants from persistent infections established in both BHK-21 and *Aedes albopictus* cells by the 5-year CAR4 virus and DI particle, which are less sensitive to interference mediated by the 5-year DI particle.

DI Particle Assay

A sensitive assay for DI particles, similar to the one used by Khan and Lazzarini,[15] was employed in our previous study quantitating DI particle-mediated interference of viruses isolated from a persistent infection.[13] Briefly, this method involves measuring spectrophotometrically the particle yield of the postinfection supernatant following the separation of standard virus from DI particles on sucrose gradients.

When increasing amounts of the homologous DI particle are included in a high multiplicity infection of VSV, the yield of DI particles is increased while the standard virus yield is decreased relative to the non-interfered infection.[16] However, even viruses which are less sensitive to a particular DI particle may give greater than 90% interference if an extremely high DI particle input (10-100 DI units) is used,[13] where one DI unit is defined as the amount of DI particles in the inoculum which gave 37% of the yield of homologous standard virus relative to the yield using DI particle free inocula.[17] Figure 1 shows that in homologous mixed infection, if the biological activity of DI particles is assayed by measuring the enrichment of DI particles in the yield relative to standard virus (the particle ratio DI/VSV), the response is close to linear with respect to DI particle input multiplicity.

Isolation of Variants from BHK-21 Cells

This assay was used to further evaluate the susceptibility of viruses isolated from CAR4 persistent infection to interference by the *ts*G31 DI particle. The CAR4 cell line was established by co-infection of BHK-21 cells with a high multiplicity of the cloned *ts*G31 mutants of the VSV Indiana serotype and its homologous DI particle.[9] The recovered virus was plaque purified and high titer clonal pools were prepared in BHK-21 cells. The biological activity of the *ts*G31 DI particle against the viruses isolated from CAR4 was tested by co-infecting identical monolayers of BHK-21 cells at a multiplicity of 5 pfu/cell together with increasing amounts of the *ts*G31 DI particle and determining the enrichment of DI particles in the yield (the particle ratio DI/VSV) as described above. Figure 1A extends our previous results,[13] by showing that

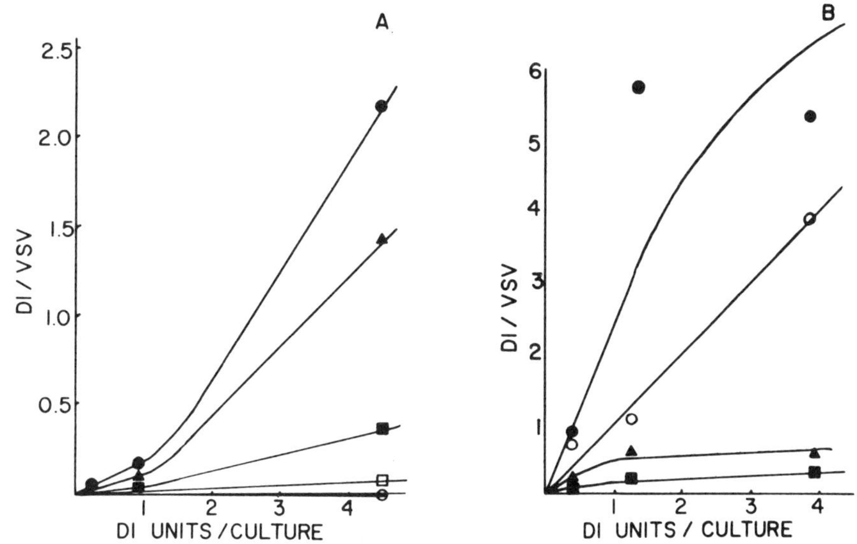

Figure 1. Interference properties of viruses isolated from persistent infections. Virus was isolated from the CAR4 persistent infection established with the *ts*G31 standard virus and the *ts*G31 DI particle, and the CAR51 persistent infection established with the 5-year CAR4 standard virus and the 5-year DI particle. Susceptibility of these viruses to interference by the *ts*G31 DI particle (A) or the 5-year DI particle (B) was tested by co-infecting matched monolayers of 4×10^7 BHK-21 cells with cloned standard virus at a moi = 5 pfu/cell and the purified *ts*G31 DI particle (A) or the 5-year DI particle (B). Since these viruses are temperature-sensitive, all infections were done at 33°C. Culture fluids were harvested at 48 hours postinfection and the particle yield was determined as previously described.[13] DI/VSV is the ratio of DI particles to standard virus in the yield. (A) ●, *ts*G31; □, 75-day SP CAR4; ■, 75-day LP CAR4; ▲, 14-mo CAR4; ○, 5-year CAR4. (B) ●, *ts*G31; ○, 5-year CAR4; ■, 39-day CAR51; ▲, 120-day CAR51.

the small plaque viruses isolated at 75 days and 5 years show little enrichment of DI particles in the yield, even at a DI input multiplicity which gives a DI/VSV particle ratio of greater than 2.0 in the homologous case.

It was of interest to determine whether a virus which has mutated to become less susceptible to a given DI particle can further mutate to become less susceptible to its newly generated homologous DI particle. The 5-year virus isolated from CAR4 has mutated with respect to its susceptibility to *ts*G31 DI particle mediated interference (Figure 1A). A persistent infection (designated CAR51) was established by co-infecting BHK-21 cells with a high multiplicity of the 5-year CAR4 virus and its homologous 5-year DI particle (generated by high multiplicity passage of cloned 5-year virus).[11] Virus was plaque purified directly from the supernatant of the new persistent infection, and high titer clonal pools were prepared in BHK-21 cells. The susceptibility of these viruses to interference mediated by the 5-year DI particle was tested as described above. Figure 1B shows that viruses isolated from this new persistent infection at 39 days and 120 days are less susceptible to 5-year DI particle mediated interference. However, the difference in interference properties between these viruses isolated and the 5-year CAR4 virus is not as extreme as the difference between the *ts*G31 virus and viruses isolated from CAR4 (compare Figures 1A and 1B). Figure 1B also shows that the 5-year DI particle is a potent interfering agent against the *ts*G31 virus.

Isolation of Variants from *Aedes Albopictus* Cells

To determine whether such interference-resistant viruses can be isolated from persistent infections in other cell lines, a persistent infection was established in *Aedes albopictus* cells (designated CAR52) by co-infecting with a high multiplicity of the 5-year CAR4 virus and the 5-year DI particle. Virus was isolated from this persistent infection by cocultivation of carrier cells with uninfected BHK-21 cells. Virus was plaque purified and high titer clonal stocks were prepared in BHK-21 cells. These viruses were tested for their susceptibility to interference mediated by the 5-year DI particle by co-infection of BHK-21 cells with a high multiplicity of the recovered viruses and 6 DI units[17] of the 5-year DI particle. The 3-day virus isolated from CAR51 was included in this experiment as a comparison. The results obtained showed that a variant was isolated after 11 months of persistence in *Aedes albopictus* cells which was less susceptible to interference mediated by the 5-year DI particle. However, the virus isolated after 8 months of persistence shows similar interference properties to the 5-year CAR4 virus with respect to the 5-year DI particle and the *ts*G31 DI particle (data not shown).

Discussion

It is evident from these results that the alterations in virus-DI particle interactions seen with these viruses isolated from persistent infections is unidirectional. The evolved viruses have become less sensitive to interference by the DI particles present at the establishment of the persistent infection, while the DI particles present later in persistent infection retain potent interfering activity against the earlier standard viruses. These phenomena may involve co-selection of mutations in the replicase genes and extensive mutations in the 5′ end sequence of the viral RNA. This in turn would be reflected in the 3′ end of DI particle RNA.[18,19] The mutated replicase gene,[5] might then recognize with lower efficiency the unmutated sequence of the original DI particle RNA and would be much less interfered with in a mixed infection.[13]

The 5′ end sequence of the standard virus RNA has been shown to accumulate extensive mutations during 5 years of VSV persistent infection in BHK-21 cells.[11] Since the 5′ end of the viral RNA is complementary to the 3′ end of both the plus and the minus strand RNAs of most DI particles generated by the virus,[18,19] this sequence has been postulated to play a role in DI particle-mediated interference.[19-20] Comparisons of the 5′ sequence of the viral RNA in a number of virus mutants which have altered interference properties should test this hypothesis.

We are currently examining the role of DI particles in promoting viral genome evolution. Viruses isolated after several serial undiluted passages of VSV in BHK-21 cells are more extensively mutated as seen by T_1 oligonucleotide fingerprinting than viruses isolated after dilute passage (K. Spindler, unpublished results). Since DI particles are only present in appreciable amounts during undiluted passaging of VSV,[21] examination of susceptibility to the homologous DI particle of the original virus following dilute and undilute passage series should allow us to determine if the virus also mutates more rapidly with respect to its interference properties.

ACKNOWLEDGMENTS

We thank Estelle Bussey for excellent technical assistance. This work was supported by Grant Al 14627 from the National Institutes of Health, USPHS. FMH was supported by USPHS Predoctoral Traineeship GM07313.

References

1. Wagner, R.R., Levy, A., Snyder, R., Ratcliff, G., and Hyatt, D. (1963): *J. Immunol.* 91:112–122.
2. Mudd, J.A., Leavitt, R.W., Kingsbury, D.T., and Holland, J.J. (1973): *J. Gen. Virol.* 20:341–351.
3. Preble, O.T. and Youngner, J.S. (1973): *J. Virol.* 12:472–480.
4. Kawai, A., Matsumoto, S., and Tanabe, K. (1975): *Virology* 67:472–480.
5. Holland, J.J., Villarreal, L.P., Welsh, R.M., Oldstone, M.B.A., Kohne, D., Lazzarini, R., and Scolnick, E. (1976): *J. Gen. Virol.* 233:193–211.
6. Ahmed, R. and Graham, A.F. (1977): *J. Virol.* 23:250–262.
7. Igarashi, A., Koo, R., and Stollar, V. (1977): *Virology* 82:69–83.
8. Kawai, A. and Matsumoto, S. (1977): *Virology* 76:60–71.
9. Holland, J.J. and Villarreal, L.P. (1974): *Proc. Nat Acad. Sci.* 71:2956–2960.
10. Holland, J.J., Grabau, E.A., Jones, C.L., and Semler, B.L. (1979): *Cell* 16:495–504.
11. Semler, B.L. and Holland, J.J. (1979): *J. Virol.* 32:420–428.
12. Rowlands, D., Grabau, E., Spindler, K., Jones, C., Semler, B., and Holland, J. (1980): *Cell* 19:871–880.
13. Horodyski, F.M. and Holland, J.J. (1980): *J. Virol.*, in press.
14. Jacobson, S. and Pfau, C. (1980): *Nature* 283:311–313.

15. Khan, S.R. and Lazzarini, R.A. (1977): *Virology* 77:190–201.
16. Huang, A.S. (1973): *Ann. Rev. Microbiol.* 27:101–117.
17. Bellet, A.J.D. and Cooper, P.D. (1959): *J. Gen. Microbiol.* 21:498–509.
18. Leppert, M., Kort, L., and Kolakofsky, D. (1977): *Cell* 12:539–552.
19. Perrault, J., Semler, B.L., Leavitt, R.W., and Holland, J.J. (1973): In: *Negative Strand Viruses and the Host Cell.* Mahy, B.W.J. and Barry, R.D. (Eds.), pp. 527–538.
20. Huang, A.S., Little, S.P., Oldstone, M.B.A., and Rao, D. (1978): In: *Persistent Viruses.* Stevens, J.G., Todaro, G.J., and Fox, C.F. (Eds.), pp. 399–408.
21. Cooper, P.D. and Bellet, A.J.D. (1959): *J. Gen. Microbiol.* 21:485–497.

David H.L. Bishop and Richard W. Compans, eds.
The Replication of Negative Strand Viruses

On the Mechanism of DI Particle Protection Against Lethal VSV Infection in Hamsters

Patricia N. Fultz, John A. Shadduck, C.Y. Kang, and J. Wayne Streilein[a]

Controversy surrounds current ideas about the physiologic role(s) of defective interfering (DI) particles in *in vivo* virus infections. Models for *in vivo* study of the protective effects of DI particles involve diseases of the central nervous system (CNS)[1-6] and most use intracerebral (IC) inoculation as the route of injection of the virus and DI particles. One such model system is the lethal infection of mice following IC injections of vesicular stomatitis virus (VSV). Holland and coworkers[3,7] have shown that simultaneous IC injection of VSV and large numbers of highly purified homologous DI particles can provide complete protection or significant prolongation of life. This protection was shown to require biologically active DI particles,[7] to be specific for homologous virus,[3] and to be unrelated to the action of interferon.[3] In contrast, Crick and Brown,[1] using a similar experimental system, found that biological activity was not a prerequisite and that protection by DI particles was not specific, i.e., they found that inactivated DI particles can protect against both homologous and heterologous VSV challenge. They concluded that DI particles enlist a host mechanism, probably immunological, and that most of the observed protection is not due to true homologous interference. Both groups of investigators, however, agree that induction of and concomitant antiviral activity due to interferon are not important components of DI particle protection *in vivo*.

We have found that VSV, Indiana serotype (VSV_{IND}), causes a lethal systemic infection in certain inbred strains of Syrian hamster following in-

[a] Departments of Cell Biology, Pathology, and Microbiology, University of Texas Health Science Center at Dallas, Dallas, Texas.

traperitoneal (IP) inoculation of small numbers of VSV—10 to 100 plaque-forming units (PFU).[8] Death of infected hamsters occurs within 72 hours and histopathologic studies suggest that cells of lymphoreticular origin are the primary target tissue. More importantly, there appears to be little involvement of the CNS at the time of death.

Systemic infection by VSV in hamsters provides a model for studying the prophylactic effects of DI particles on a lethal disease that is not confined to the restricted environment of the CNS, and that allows for the potential interaction of numerous host defense systems. We found that hamsters highly susceptible to IP infection with VSV could be protected not only by biologically active homologous DI particles, but also by heterologous DI particles and by treatment with poly(I) : poly(C). These findings suggest that protection against a lethal systemic infection by DI particles may not be due to true interference; instead, circumstantial evidence implicates interferon, either directly or indirectly, as being important.

Materials and Methods

The LSH inbred strain of Syrian hamsters was used; hamsters were bred at our facility and were at least three months old. More than 90% of LSH hamsters that receive an IP injection of VSV die of a systemic infection, the majority (80%) within three days; however, if death occurs at later times it is often preceded by hind-limb paralysis. This is true regardless of the number of PFU injected. In all experiments, we injected 10^4 PFU of VSV, a number significantly greater than required to produce the lethal disease (LD_{50} is less than 10). Two homologous VSV_{IND} DI particles were used: DI-2 (VSI ts^+ ATCC *DI* 0.54), a short DI from the 5′ end of the VSV genome, and DI-LT (VSI HR ATCC *DI* 0.67), a long DI from the 3′ end of the genome of a heat resistant variant.[9] Ten-fold dilutions of DI particles were made directly into the VSV inoculum to generate ratios of DI to VSV of from $10^6:1$ to $10^2:1$. The operational ratios are upper limits for the number of DI particles in each inoculum. The actual numbers are probably at least 10-fold less since protein concentration was used to estimate the number of DI particles in stock preparations (made by sucrose gradient banding one time) which contained contaminating standard VSV. DI-2 and standard VSV were inactivated by exposure to ultraviolet (UV) light. This treatment completely nullified the plaque-forming ability of the standard VSV stock (2×10^{10} PFU/ml) and also destroyed the ability of the DI-2 stock to interfere with VSV replication *in vitro*.

Results

Effects of homologous DI particles on lethal VSV disease. We first determined if homologous DI particles, capable of *in vitro* interference, could provide protection from lethal systemic infections caused by IP-injected VSV. Figure 1

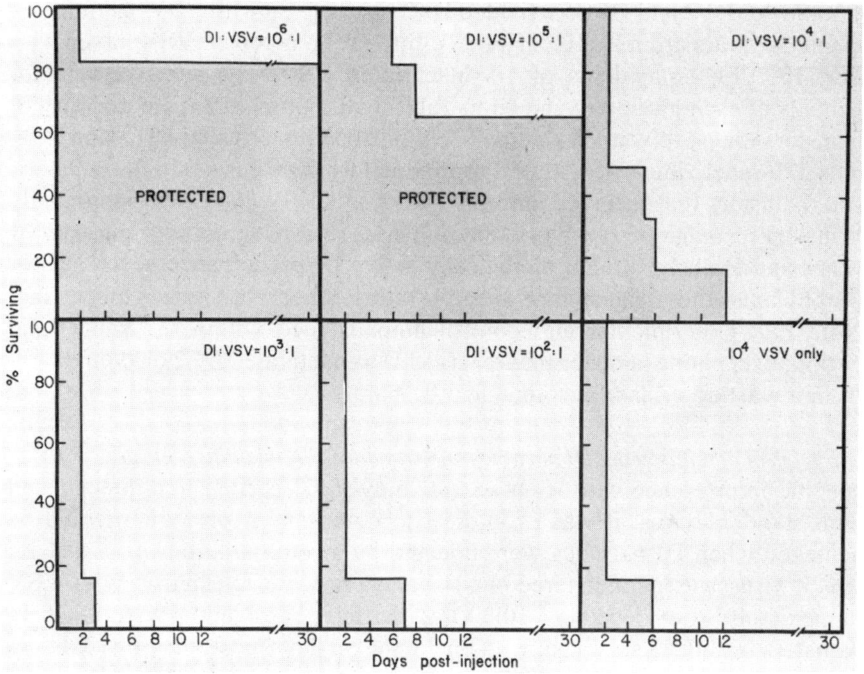

Figure 1. Survival of LSH hamsters after ip injection of VSV and DI-2. In each case 10^4 PFU of VSV were mixed with different numbers of DI-2 particles to generate the ratios of DI : VSV shown in each panel.

illustrates the effect of co-injecting DI-2 and VSV. Significant protection (greater than 65% survival) was achieved at ratios of DI-2 : VSV of $10^5:1$ and $10^6:1$, while at ratios of $10^4:1$ and lower all recipient hamsters died, most within three days. Similarly, when another hemologous DI particle, DI-LT, was co-injected with VSV, 100% of hamsters that received doses of $10^5:1$ and $10^6:1$ survived (data not shown). At ratios of $10^4:1$ and $10^3:1$, 62% and 50% of the recipient hamsters survived, respectively. Thus, both homologous DI particles can protect LSH hamsters from death. The fact that DI-LT appeared to be more effective than DI-2 at lower ratios of DI : VSV may imply that the stock preparation of DI-LT contained more biologically active DI particles. Alternatively, if protection was not due to true interference, then DI-LT may be more efficient at eliciting protective host defense mechanisms.

Effects of UV-inactivated DI particles on lethal VSV disease. Whether inactivated DI particles can protect against ic infection of mice with VSV is controversial since the results of Jones and Holland[7] and Crick and Brown[1] are contradictory. To test this in the VSV-hamster system, comparable amounts of DI-2 and UV-inactivated DI-2 were mixed with VSV and injected ip into separate groups of hamsters. A third group received VSV mixed with 3×10^9 PFU-equivalents of UV-inactivated VSV. The survival data for animals in each group are shown in Table 1, Experiment 1. The majority of hamsters that received VSV alone, VSV plus UV-inactivated DI-2, and VSV plus UV-inactivated VSV died (approximately 20% survival). In comparison, hamsters that received untreated DI-2 plus VSV showed a significantly higher percentage of survivors (80%). These results agree with the data of Jones and Holland[7] in that biologically active DI particles are necessary for prophylaxis and suggest that specific interference was indeed occurring. Note that experimental and control animals died within the same time period, except one hamster that received UV-inactivated DI-2 died on day 11 from a wasting disease accompanied by paralysis.

Effects of administering DI particles and standard VSV via different routes. Since specific interference can occur *in vitro* only if DI particles co-infect a cell with standard virus, it was of interest to determine if protection could be achieved when DI particles were injected by a route different from that by which virus was administered. LSH hamsters were injected ip with VSV immediately after receiving 10^{10} DI-2 particles intravenously (iv). Of 14 hamsters tested, 13 survived, with one dying on day 5 postinfection (Table 1, Experiment 2). Thus, DI particles render hamsters resistant to VSV even if not administered at the same site as infectious virus. This implies (1) that VSV expresses a cellular or tissue tropism in hamsters such that DI and standard virus are transported to or migrate to a common tissue site where interference can occur, or (2) that a protective mechanism other than true interference is being evoked.

Effects of heterologous DI particles on lethal VSV disease. To choose between the above possibilities, we took advantage of the fact that homologous but not heterologous DI particles can interfere with VSV replication in tissue culture cells. A single type of DI particle was generated from VSV of the New Jersey serotype (VSV_{NJ}). This DI particle (DI-NJ) is uncharacterized except for the fact that it does interfere *in vitro* with VSV_{NJ} but not with VSV_{IND} replication. The results of co-infection with VSV_{IND} and DI-NJ are presented in Table 1, Experiment 3, and reveal that 79% of hamsters receiving heterologous DI-NJ survived the infection. We conclude, therefore, that the prophylactic effect of DI particles in systemic infections is very likely not due to true interference.

While protection afforded by heterologous DI particles does not exclude the possibility that true interference is achieved by homologous DI particles *in vivo,* it strongly implicates another mechanism. This mechanism probably does not involve antibody since death within 40 to 72 hours after injection of virus is too early for significant antibody to be secreted to effect protection.

Table 1. Effects of DI Particles, Standard Virus, and Poly(I):Poly(C) on VSV Infections in LSH Hamsters.

Exp.	Inoculum[a] contained 10^4 PFU of VSV and:	No.	Day of death[b]	% survival
		22	11 on day 2; 3, 3, 3, 3, 3, 6, 7, 8	13.6
1	10^{10} DI-2	20	2, 3, 3, 3	80.0
	10^{10} UV-inact DI-2	12	2, 2, 2, 2, 2, 3, 3, 3, 6, 11	16.7
	3×10^9 UV-inact VSV	10	2, 2, 2, 2, 2, 2, 6, 8, 12	10.0
2	10^{10} DI-2 IV[d]	14	5	92.9
3	10^{10} DI-NJ	14	2, 6, 7	78.6
4	poly(I):poly(C)	12	6, 6	83.3

[a]All injections were ip unless otherwise stated.
[b]Day of death of individual hamsters following injection.
[c]Received 10^4 PFU of VSV only.
[d]VSV was administered ip and DI-2, IV.

A second candidate for the protective mechanism is interferon. DI particles are known to be more efficient at inducing interferon than wild type VSV, which is highly cytopathic. Interferon-mediated protection could also explain the observation that fewer DI-LT than DI-2 are required to prevent death. Different DI particles have been shown to have varying capabilities of inducing interferon.[10-12]

Effect of poly(I):poly(C) on lethal VSV disease. The possible role of interferon in attenuation of lethal VSV disease was tested by co-injection of hamsters (ip) with VSV and 100 μg of poly(I):poly(C) per 100 g body weight. Of 12 hamsters that received poly(I):poly(C), two died on day 6 (Table 1, Experiment 4). The approximate 80% survival of hamsters receiving poly(I):poly(C) and of those receiving DI-NJ implicate interferon in the protection of LSH hamsters from systemic VSV infection.

Discussion

To our knowledge, these experiments are the first to examine the protective effects of DI particles on systemic virus infections. In Syrian hamsters, DI particles are very effective at preventing a fatal disease following ip injection of VSV. In fact, the efficiency of protection suggests that host tissues essential for replicating VSV in sufficient quantities to cause death must be limited. This is reasonable if one presumes that DI particles protect by true homologous interference.

Several facts, however, speak against protection being mediated entirely by true interference. The most convincing evidence comes from the ability of heterologous DI particles to protect against death and suggests an alternative mechanism—one that requires biologically active DI particles (not UV-inactivated). Moreover, treatment of hamsters with poly(I):poly(C), an efficient interferon inducer, also protects against lethal VSV infection. These results imply the participation of interferon in DI-induced protection in hamsters. The data, however, do not exclude true homologous interference as playing a contributory role.

It is of interest that Doyle and Holland[3] had to use extremely large ratios of DI:VSV to observe protection in the murine model. Since most investigators doing similar studies have had difficulty in finding interferon in CNS co-infected with virus and DI particles,[1-3,13] it is possible that the capacity of cells of the CNS to produce interferon is limited. If true, then perhaps the only mechanism that can provide early protection after ic infection of mice with VSV is true interference and, to be effective, this requires large numbers of DI particles to insure co-infection of all cells. It is important to note that Gresser et al.[14] have shown that potent interferon preparations administered after onset of VSV replication in brains of infected mice can provide some protection from death.

Much remains to be done in sorting out specific and non-specific effects of DI particles in the VSV-hamster system. We are particularly interested in studying surviving hamsters, protected from the acute lethal disease by DI particles, that subsequently developed lasting paralysis; it appears that persistent, CNS infections have been established. The VSV-hamster model may be the first to provide unequivocal evidence of a role for DI particles in inducing and/or maintaining persistent infections *in vivo*.

ACKNOWLEDGMENT
This work was supported by NIH Grant CA-09082.

References

1. Crick, J. and Brown, F. (1977): *Infec. Immun.* 15:354–359.
2. Dimmock, N.J. and Kennedy, S.I.T. (1978): *J. Gen. Virol.* 39:231–242.
3. Doyle, M. and Holland, J.J. (1973): *Proc. Natl. Acad. Sci.* 70:2105–2108.
4. Popescu, M. and Lehmann-Grube, F. (1977): *Virology* 77:78–83.
5. Spandidos, D.A. and Graham, A.F. (1976): *J. Virol.* 20:234–247.
6. Welsh, R.M., Lampert, P.W., and Oldstone, M.B.A. (1977): *J. Infect. Dis.* 136:391–399.
7. Jones, C.L. and Holland, J.J. (1980): *J. Gen. Virol.* 49:215–220.
8. Fultz, P.N., Streilein, J.W., Shadduck, J.A., and Kang, C.Y. (1980): In: *Hamster Immune Responses and Infectious and Oncologic Diseases.* Streilein, J.W., Hart, D.A., Stein-Streilein, J., Duncan, W.R., and Billingham, R. (Eds.), Plenum Publishing Co., New York, in press.
9. Prevec, L. and Kang, C.Y. (1970): *Nature* 228:25–27.
10. Frey, T.K., Jones, E.V., Cardamore, J.J., and Younger, J.S. (1979): *Virology* 99:95–102.
11. Marcus, P.I. and Sekellick, M.J. (1977): *Nature* 266:815–819.

12. Marcus, P.I. and Sekellick, M.J. (1980): *J. Gen. Virol.* 47:89–96.
13. Faulkner, G., Dubois-Dalcq, M., Hooghe-Peters, E., McFarland, H.F., and Lazzarini, R.A. (1979): *Cell* 17:979–991.
14. Gresser, I., Tovey, M.G., and Bourali-Maury, C. (1975): *J. Gen. Virol.* 27:395–398.

David H.L. Bishop and Richard W. Compans, eds.
The Replication of Negative Strand Viruses

Standard Vesicular Stomatitis Virus is Required for Interferon Induction in L Cells by Defective Interfering Particles

Teryl K. Frey, Debra W. Frielle and Julius S. Youngner[a]

The majority of VSV DI particle RNAs are derived from the 5′ end of the standard virion RNA.[1,2] RNAs of this class have one of two structures: (1) [−] RNA, which is mostly single-stranded except for short complementary sequences at its termini, and (2) [±] or snapback RNA, which is composed of long stretches of complementary sequences joined by a short, single-stranded linker region.[3,4] Marcus and Sekellick reported that a particular [±] RNA DI particle (DI-011) is an excellent inducer of IFN in mouse L cells and primary chicken embryo cells, but that a [−] RNA DI particle (DI-HR) induced no IFN in these cells.[5,6] They postulated that [±] RNA DI particles contain a preformed IFN-inducer, the [±] RNA molecule. Although we confirmed that DI-011 is an excellent interferon-inducer in L cells, we found that two other [±] RNA DI particles and two [−] RNA DI particles also induced IFN, but less efficiently than DI-011.[7] We concluded that the IFN-inducing capacity of a DI particle may be independent of its RNA structure.

To extend these observations, we continued our study using twelve different DI particles provided by three laboratories. We have found that DI particles by themselves cannot induce IFN and that IFN-induction by DI particle preparations is due to co-infection with contaminating standard VSV in the preparation.

Interferon Induction by DI Particles

The DI particles used in this study and some of their characteristics are listed in Table 1; both [±] and [−] RNA DI particles are represented. As can be

[a] Department of Microbiology, School of Medicine, University of Pittsburgh, Pittsburgh, Pennsylvania.

Table 1. DI Particles Used in This Study and Their Capacity for IFN Induction.

DI particle	Reference	DI particles/cell required for IFN induction (8 units or more)	Contamination by standard VSV (PFU/DI particle)
[±] RNA			
DI-wt-U	(7)	20	4.6×10^{-8}
DI-wt-L	(7)	30	5.8×10^{-8}
$DI\text{-}011_Y$[a]	(7)	10	1.9×10^{-6}
$DI\text{-}011_L$[a]	(4, 8)	100	5.5×10^{-9}
DI-C5-ST	(9)	100	6.3×10^{-10}
DI-*ts*G31-ST2	(9)	600	2.9×10^{-9}
DI-3-6 (NJ serotype)	(10)	10	8.0×10^{-8}
[−] RNA			
DI-pi-U	(7)	100	4.5×10^{-9}
DI-pi-L	(7)	30	2.6×10^{-8}
DI-611 Preparation 1	(8)	1,000	$< 1.4 \times 10^{-8}$
Preparation 2		40	1.1×10^{-5}
DI-T Preparation 1	(8)	400	Not done
Preparation 2		1,000	$< 6.7 \times 10^{-9}$
$DI\text{-}wt_{MS}$-ST	(9)	100	9.4×10^{-10}
DI-*ts*G31-ST1	(9)	9,000	$< 2.3 \times 10^{-10}$

[a]$DI\text{-}011_Y$ was grown and purified in our laboratory. $DI\text{-}011_L$ was provided in purified form by Dr. R. Lazzarini.

seen, the different DI particles vary widely in their IFN-inducing capacity in L cells and it is evident that there is no correlation between the ability of a DI particle to induce IFN and its RNA structure (Table 1). The multiplicity of infection required for IFN-induction by both [−] RNA and [±] RNA DI particles varies over the same wide range. Furthermore, with two DI particles, DI-011 ([±] RNA) and DI-611 ([−] RNA), independent preparations of the same DI particle are quite different in their IFN-inducing ability.

Most strikingly, the data in Table 1 suggest a correlation between the IFN-inducing capacity of a DI particle preparation and the degree to which it is contaminated by standard VSV. This correlation was initially observed with the preparations of DI-011 and DI-611. With both of these DI particles, the preparation which is a more efficient IFN-inducer also is significantly more contaminated by standard VSV. Scanning the data in Table 1, it is evident that this correlation holds for most of the DI particles listed. In general, the more contaminated with standard VSV a DI particle is, the fewer the DI particles per cell required for IFN-induction.

In Figure 1, the relationship between IFN-induction and standard virus contamination is shown graphically for four DI particle preparations. The titer of IFN induced and the expected number of PFU per cell (calculated

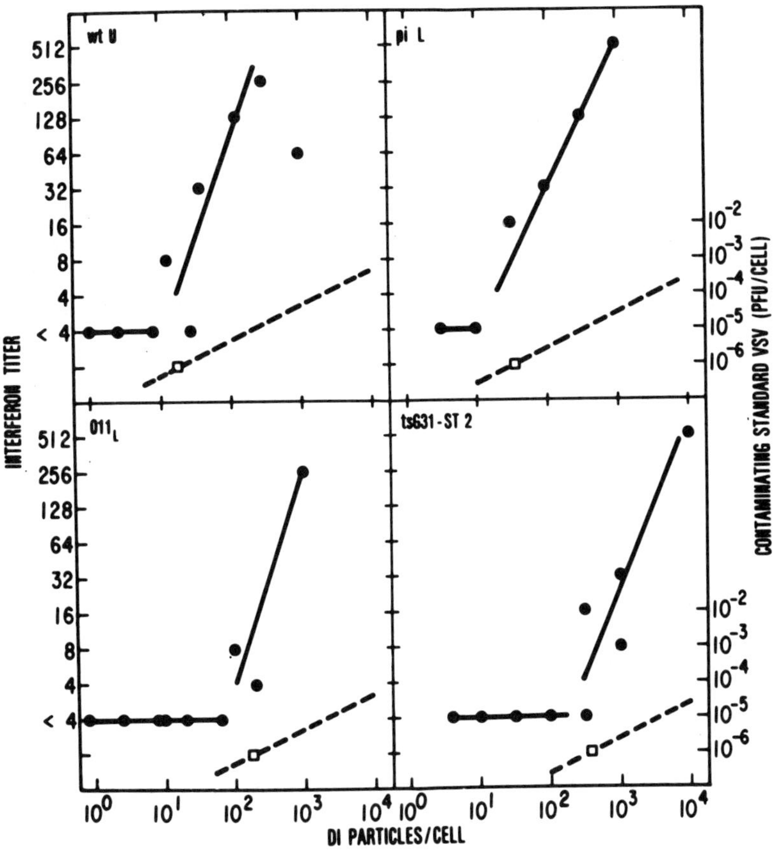

Figure 1. Correlation between IFN induction and standard VSV contamination for four DI particle preparations listed in Table 1. IFN induction in L cells and standard VSV contamination are plotted as a function of the number of DI particles per cell. IFN induction, (●—●); standard VSV contamination (– – –), expressed as PFU/cell, calculated from the contamination ratios in Table 1. The open square (□) marks the multiplicity of DI particles/cell at which the multiplicity of contaminating standard VSV is calculated to be 10^{-6} PFU/cell. At this and higher multiplicities, PFU will be present in the 16 mm wells (10^6 cells) used in this experiment.

from PFU contamination levels given in Table 1) are plotted as functions of the number of DI particles per cell. Since these experiments were conducted in plastic wells which contain approximately 1×10^6 cells, any standard VSV in a well of cells would be present only at multiplicities exceeding 10^{-6} PFU per cell; this value is marked by an open square on each of the four panels in Figure 1. With all four DI particles, IFN is induced only by DI particle multiplicities at which standard VSV is also present in the cell population. Furthermore, an extremely small number of standard VSV (possibly as few as one PFU per well of cells) is sufficient for IFN-induction to occur. When the DI particles contaminated with less than 10^{-7} PFU per DI particle (see Table 1) are analyzed for IFN-induction and standard VSV contamination, the graphs appear similar to those shown in Figure 1. However, the DI particle preparations contaminated with greater than 10^{-6} PFU per DI particle induce IFN at multiplicities of from 10 to 40 DI particles per cell. These multiplicities are 100- to 1,000-times higher than the multiplicity of these DI particles necessary to introduce contaminating standard VSV into the cell population. This finding suggests that, together with small numbers of standard VSV, a minimum number of DI particles per cell is required for IFN induction.

Gradient Purification of DI Particles

A preparation of DI-wt-U was purified by three sequential velocity gradient centrifugations. The original infected culture fluid and the DI particle bands collected after each purification step were assayed for contaminating standard VSV and for IFN-inducing capacity. The three purification steps reduced the amount of contaminating standard VSV by 90-, 6-, and 3-fold, respectively (data not shown). Concomitant with the reduction in standard VSV contamination, the IFN-inducing capacity of the DI-wt-U preparation was reduced at each purification step (Figure 2).

Interferon Induction in Cells Co-infected with DI Particles and Exogenously Added VSV

If our hypothesis is correct, IFN should be induced in L cells by the addition of standard VSV to multiplicities of DI particles which by themselves are incapable of inducing IFN. To test this hypothesis, L cells were mock-infected or infected with DI-011_Y or DI-wt-U. The cells were then co-infected with dilutions of standard VSV calculated to give multiplicities ranging from 10^{-6} to 10^{1} PFU per cell. The wells used in this experiment contained 3×10^4 cells, therefore no standard VSV should be present in the cell population at multiplicities of 10^{-6} and 10^{-5} PFU per cell. As shown in Table 2, with the DI particle multiplicities used, IFN is induced at a multiplicity of 10^{-4} PFU per cell, which corresponds to approximately 3 PFU per well of cells. Interestingly, IFN is not induced when cells are co-infected with DI particles and relatively large numbers of standard VSV per cell. In a similar reconstruction experiment, IFN was induced in cells co-infected with standard VSV and each of the different DI particles listed in Table 1 (data not shown).

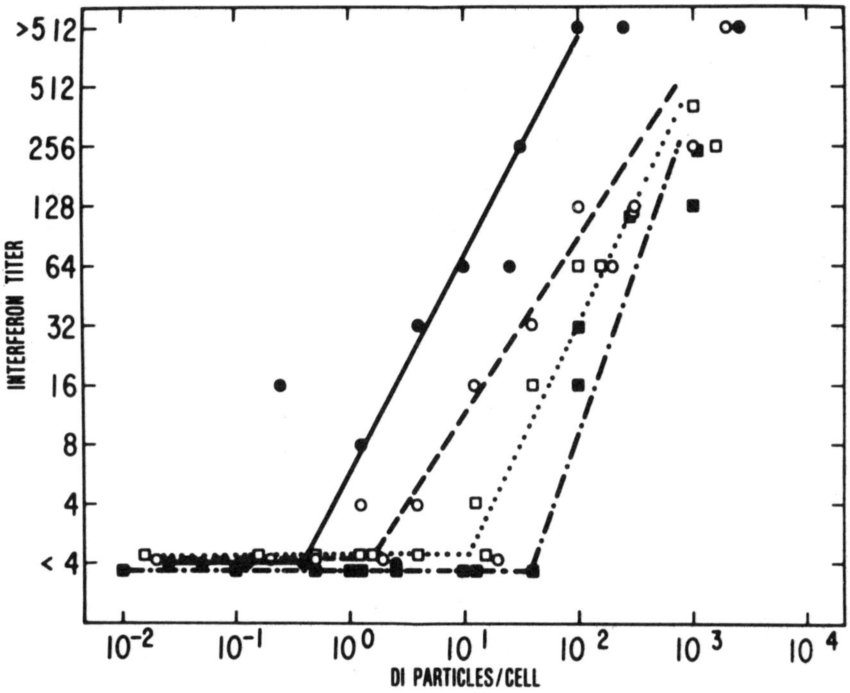

Figure 2. IFN induction by DI-wt-U separated from contaminating standard VSV by three sequential velocity gradient centrifugations. IFN induction in L cells is plotted as a function of the number of DI particles per cell for the original infected culture fluid (●—●) and for the DI particle band after the first (○ – – ○), second (□ . . . □), and third (■ – · – · ■) centrifugation steps.

Discussion

In agreement with our earlier report,[7] this study provides further evidence that the IFN-inducing capacity of VSV DI particles in L cells is not a function of the structure of the DI particle RNA. Instead, a correlation exists between IFN-inducing capacity of a DI particle preparation and its contamination with standard VSV. It seems that while DI particles and standard VSV are incapable of inducing IFN separately, co-infection of L cells with both types of particles leads to an interaction which results in IFN induction. Our results show that IFN-induction in L cells co-infected with DI particles and standard VSV occurs only at certain multiplicities of infection. A minimum number of DI particles per cell is required and, while the presence of standard VSV is necessary, PFU can only be present at relatively low multiplicities of infection. Therefore, it is easy to see how results, obtained

Table 2. Interferon Induction by DI Particles in the Presence of Added Exogenous Standard VSV at Varying Multiplicities.

moi of added Standard VSV (PFU/cell)	No. DI particles	$DI\text{-}011_Y$ 33 DI particles/cell	DI-wt-U 33 DI particles/cell	DI-wt-U 100 DI particles/cell
No PFU added	< 4	< 4	< 4	< 4
10^{-6}	< 4	< 4	< 4	< 4
10^{-5}	< 4	< 4	< 4	< 4
10^{-4}	< 4	16	16	32
10^{-3}	< 4	64	16	64
10^{-2}	< 4	4	64	128
10^{-1}	< 4	< 4	< 4	16
10^{0}	< 4	< 4	< 4	8
10^{1}	< 4	< 4	< 4	< 4

with DI particle preparations heavily contaminated with standard VSV, would lead to the conclusion that infection of a cell with one DI particle was sufficient for IFN-induction.[5,6]

It has been postulated that standard VSV has the capacity to induce IFN, but because of the severe inhibition of host cell macromolecular synthesis which occurs in infected cells, IFN protein is never synthesized.[11] We hypothesize that in cells co-infected with DI particles and standard VSV, the DI particles sufficiently minimize or delay the inhibition of host cell macromolecular synthesis to allow the synthesis of IFN protein. Although it is likely that IFN is induced by some event during the standard VSV replication cycle, we cannot rule out the possibility that IFN-induction is involved with the DI particle replication process.

ACKNOWLEDGMENTS
Expert assistance was provided by John Cardamone, Marion Kelly, Evelyn Ketz, and Joan Winwood. We thank Drs. R. Lazzarini and J. Perrault for providing us with purified DI particles. This work was supported by a research Grant (AI-0624) from the NIAID, NIH. T.K.F. is the recipient of NIAID Postdoctoral Fellowship #1-F32-AI05726.

References

1. Leamnson, R.N. and Reichmann, M.E. (1974): *J. Mol. Biol.* 85:551–568.
2. Stamminger, G. and Lazzarini, R.A. (1974): *Cell* 3:85–93.
3. Perrault, J and Leavitt, R.W. (1977): *J. Gen. Virol.* 38:35–50.
4. Lazzarini, R.A., Weber, G.H., Johnson, L.D., and Stamminger, G. (1975): *J. Mol. Biol.* 97:289–307.
5. Marcus, P.I. and Sekellick, M.J. (1977): *Nature* 266:815–819.
6. Sekellick, M.J. and Marcus, P.I. (1978): *Virology* 85:175–186.

7. Frey, T.K., Jones, E.V., Cardamone, J.J., and Youngner, J.S. (1979): *Virology* 99:95–102.
8. Schubert, M., Keene, J.D., and Lazzarini, R.A. (1979): *Cell* 18:749–757.
9. Perrault, J. and Leavitt, R.W. (1977): *J. Gen. Virol.* 38:21–34.
10. R. Lazzarini: Personal communication.
11. Wertz, G.W. and Youngner, J.S. (1970): *J. Virol.* 6:476–484.

Published 1981 by Elsevier North Holland, Inc.
David H.L. Bishop and Richard W. Compans, Eds.
The Replication of Negative Strand Viruses

Enhanced Mutability Associated With a Temperature-Sensitive Mutant of Vesicular Stomatitis Virus

C.R. Pringle, V. Devine and M. Wilkie[a]

Introduction

Two instances of enhanced mutability have been reported during propagation of vesicular stomatitis virus (VSV) in cultured cells. Mudd et al.[1] observed that mutants with multiple phenotypic changes accumulated during passage of VSV Indiana in insect (*Drosophila melanogaster*) cells. Holland et al.[2] have described progressive mutational change of VSV accompanying a persistent infection of BHK-21 cells initiated by mutant *ts*G31 (III) of VSV Indiana in the presence of its homologous defective interfering (DI) particle. Numerous oligonucleotide map changes accumulated over a period of 5 years, in contrast to the stability of virus maintained by lytic infection.[3] The absence of cytopathogenicity in both instances, and in the first the low temperature of incubation, may have been relevant factors in the accumulation of these variants.

We now describe mutability of a different type associated with an individual temperature-sensitive (*ts*) mutant of VSV New Jersey, and manifested by the appearance of mutants with electrophoretically atypical polypeptides. (A full account of this work is in preparation).

Heterogeneity of Mutant *ts*D1 of VSV New Jersey

Mutants representing 2 of the 6 complementation groups of VSV New Jersey possessed electrophoretically atypical virion proteins. The NS polypeptide

[a] M.R.C. Virology Unit, Institute of Virology, Church Street, Glasgow G11 5JR, Scotland.

of all 3 *ts*-mutants of group E migrated atypically in SDS-polyacrylamide gel, and reversion of *ts* phenotype was invariably accompanied by restoration of normal mobility. Thus complementation group E corresponded to the NS-protein gene.[4]

Previously, Wunner and Pringle[5] had reported that mutant *ts*D1, the solitary member of group D, possessed two altered proteins. Both the G and N polypeptides migrated faster then normal, and only the N polypeptide regained normal mobility on reversion of the *ts* phenotype. Group D thus corresponded to the N-protein gene, an assignment which conflicted with other data.[6] Therefore, the nature of the *ts*D1 lesion was reinvestigated to resolve this paradox.

Analysis of clones isolated from *ts*D1 revealed that the mutant stock was heterogenous with three distinct phenotypes (Figure 1a). Table 1 shows that only 27% of clones exhibited the original double mutant phenotype. (In contrast a similar number of clones from the wild type, or from mutant *ts*E3, showed no similar heterogeneity apart from two wild type revertants of *ts*E3).

Wild type revertants were obtained from each of the three components of the *ts*D1 stock by cloning at 39°C (the restrictive temperature). Table 2 shows that reversion of the *ts* phenotype was not associated with any specific change in the mobility of G or N. (Mutant *ts*E3, on the other hand, showed complete concordance of the *ts* and NS phenotypes). Indeed, additional mobility variants appeared during cloning at 39°C. Figure 1b shows one such clone with atypical mobility of the G, NS, and N polypeptides (track 2).

These results indicated that mutant *ts*D1 was genetically unstable, and that the original observation of co-reversion of the *ts* and N polypeptide phenotypes[5] was fortuitous.

Effect on *ts*E3 of Co-Infection with *ts*D1

Twenty-one clones were isolated from the progeny virus released from BHK-21 cells co-infected with mutants *ts*D1 and *ts*E3. Four clones resem-

Table 1. Clonal Analysis of Wild Type VSV New Jersey and Mutants *ts*D1 and *ts*E3.

Virus stock	Phenotype[a] *ts*	L	G	N	NS	M	Number of clones	Percent of total
Wild type	+	+	+	+	+	+	50	100
*ts*D1	M	+	M	+	+	+	34	71
	M	+	M	M	+	+	13	27
	M	+	+	+	+	+	1	2
*ts*E3	M	+	+	+	M	+	47	96
	+	+	+	+	+	+	2	4

[a]+: wild type electrophoretic mobility, or the *ts*$^+$ phenotype. M: mutant; atypical electrophoretic mobility, or *ts* phenotype.

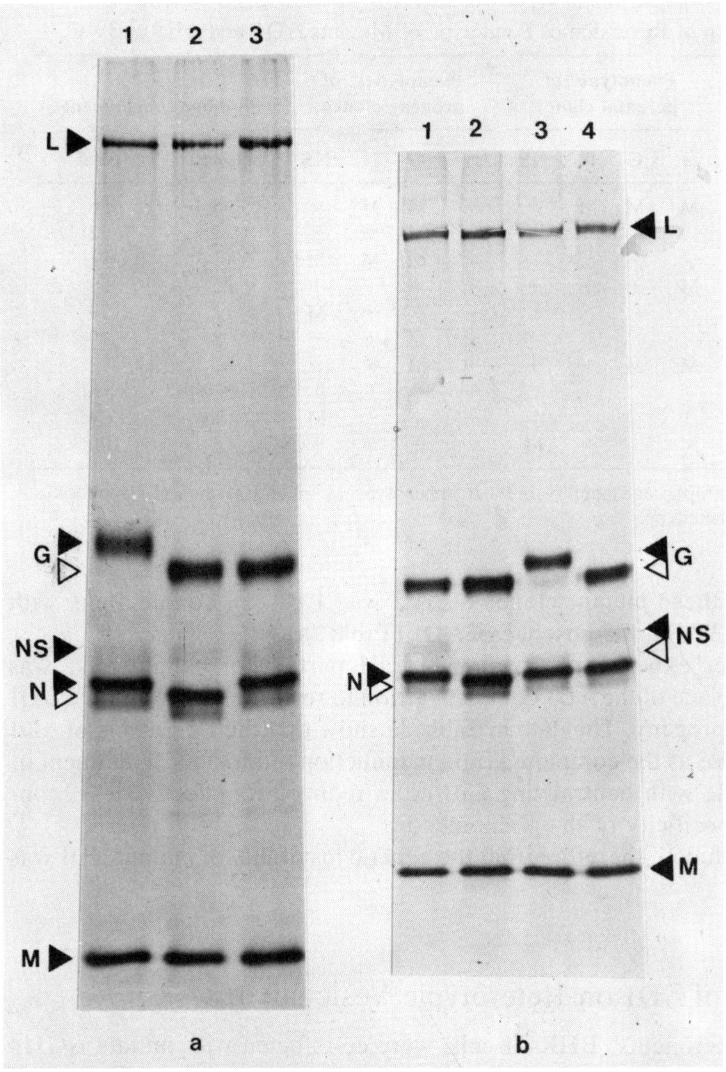

Figure 1. Polyacrylamide gel electrophoresis of the ^{35}S-methionine-labeled virion polypeptides of clones derived from mutant *ts*D1 of VSV New Jersey. (a) The three phenotypes present in the *ts*D1 stock. Track 1, wild type clone; track 2, a clone with mutated G and N polypeptides; and track 3, a clone with mutated G only. (b) Additional phenotypes which appeared during further cloning. Tracks 1 and 4, clones with mutated G; track 2, a clone with mutated G, NS, and N; and track 3, a clone with mutated NS only. The black arrows indicate polypeptides with wild type mobility, and the hollow arrows, polypeptides with atypical mobility.

bled the *ts*D1 parent and 16 the *ts*E3 parent. One clone had an aberrant M polypeptide in addition to the phenotypic characteristics of *ts*E3 (shown in Figure 2, track 2). In other experiments, clones of *ts*E3 with atypical mobilities of the N or G polypeptides were obtained (not shown). Overall the

Table 2. Pattern of Reversion of Phenotype of Mutants *ts*D1 and *ts*E3 at 39°C.

Mutant Stock (VSV New Jersey)	Phenotype[a] of parental clone				Phenotypes[a] of progeny clones				Number of clones	Percent of total
	ts	G	N	NS	*ts*	G	N	NS		
*ts*D1	M	M	M	+	+	M	M	+	5	56
					+	M	+	+	3	33
					+	M	M	M	1	11
	M	+	+	+	+	+	+	+	22	52
					+	+	+	M	12	29
					+	M	+	+	8	19
	M	M	+	+	+	M	+	+	22	92
					+	+	+	+	1	4
					+	+	+	M	1	4
*ts*E3	M	+	+	M	+	+	+	+	16	100

[a]+: wild type electrophoretic mobility, or the ts^+ phenotype. M: mutant; atypical electrophoretic mobility, for *ts* phenotype.

frequency of these mutant clones of *ts*E3 was 1.58% in co-infections with *ts*D1 and ≤0.28% in the absence of *ts*D1 (Table 3a).

In the latter experiments, a standard DI particle of mutant *ts*D1 was employed in place of the *ts*D1 complete virion to restrict recovery of the *ts*D1 parent in the progeny. The data in Table 3a show that the DI particle of *ts*D1 was as effective as the complete virion in induction of mutants. Treatment of the DI particle with neutralizing antiserum reduced its effectiveness, confirming the specificity of the phenomenon.

It was concluded, therefore, that the genetic instability of mutant *ts*D1 was transferable.

The Effect of *ts*D1 on Heterotypic Vesiculoviruses

In similar experiments, BHK-21 cells were co-infected with mutant *ts*G114 (I) of VSV Indiana and DI particles of *ts*D1 (VSV New Jersey). Analysis of the progeny yielded 4 clones (4.7%) with electrophoretically atypical proteins, whereas none (≤1.1%) were isolated in the absence of *ts*D1, or with neutralized DI particles (≤0.8%) (Table 3b).

In another experiment a clone of Chandipura virus with a variant N-protein was obtained from a co-infection with *ts*D1 at a frequency (2.5%) greater than in the absence of *ts*D1 (≤1.5%).

It was concluded, therefore, that co-infection with complete or defective *ts*D1 facilitated isolation of mutants of heterologous vesiculoviruses.

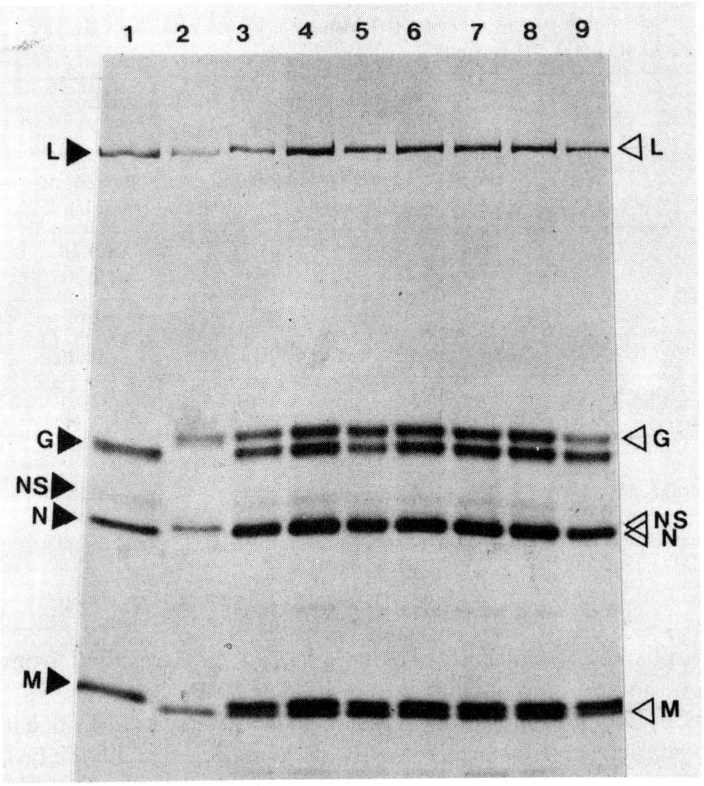

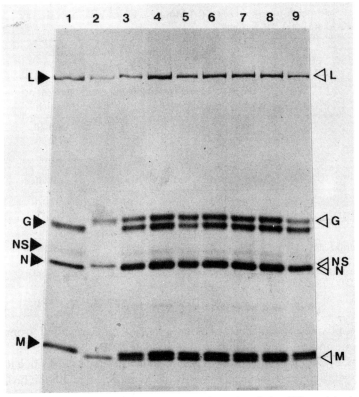

Figure 2. Polyacrylamide gel electrophoresis of the ^{35}S-methionine-labeled virion polypeptides of virus isolated at 39°C from a mixed infection with mutants *ts*E3 and *ts*D1. The black arrows indicate the mobilities characteristic of the *ts*D1 parent, and the hollow arrows mobilities characteristic of the *ts*G3 parent. Track 1, the *ts*D1 parent; track 2, the *ts*E3 parent; tracks 3-9, individual isolates derived from a mixed infection at 39°C, showing presence of G, NS, N, and M polypeptides of both parental mobilities.

Absence of Recombination

The possibility that these variants originated by recombination rather than mutation was examined by co-infection of BHK-21 cells with two clones of VSV New Jersey which were *ts* and had electrophoretically distinct G, N, NS, and M-proteins (Figures 2 and 3, tracks 1 and 2). Potential wild type recombinants were selected by cloning at 39°C, and the isolates screened for the 4 unselected markers. All isolates had polypeptides characteristic of both parents (Figure 2, tracks 3-9). Clones derived from these isolates were *ts* and showed complete segregation of the two parental types (Figure 3, tracks

Table 3. Effect of Co-Infection with *ts*D1 on (a) Mutant *ts*E3 of VSV New Jersey, and (b) Mutant *ts*G114(I) of VSV Indiana.

Inoculum[a]	Number of clones examined	Electrophoretic mobility variants in progeny: Polypeptide affected	Number of clones	Percent of total
(a) *ts*E3 alone	354	nil	0	≤ 0.28
*ts*E3 + *ts*D1/B	62	M	1	1.60
*ts*E3 + *ts*D1/DI	191	N	2	1.60
		G	1	
*ts*E3 + neutralized *ts*D1/DI	156	M	1	0.60
(b) *ts*G114 alone	94	nil	0	≤ 1.10
*ts*G114 + *ts*D1/DI	130	G (slow)	2	4.7
		G (fast)	2	
*ts*G114 + neutralized *ts*D1/DI	131	nil	0	≤ 0.8

[a]BHK-21 cells were infected simultaneously with the target virus (*ts*E3 or *ts*G114) and complete (B) or incomplete (DI) virions of *ts*D1.

3-10). Thus the plaques appearing on monolayers at 39°C resulted from complementation, and no recombinant virus was detected.

The polypeptide "X" in Figure 3 may have been an aggregate of M since it was only observed to migrate faster in extracts of clones of *ts*E3 with the fast M.

Discussion and Conclusions

(1) These results established that mutations were generated at high frequency in a stock of *ts*D1 of VSV New Jersey, and that mutations were also induced in other vesiculoviruses multiplying simultaneously in the same cells.

(2) Complete and incomplete virions of *ts*D1 were equally effective, and the frequency of induction of mutants was reduced by treatment with neutralizing antibody. Hence the mutability was a property of *ts*D1 rather than an adventitious factor. A hypothesis compatible with the data is that mutant *ts*D1 possessed a polymerase with an impaired fidelity of transcription, which generated sequence changes during replication. This hypothesis accommodates both the effectiveness of incomplete virus and its heterologous action, since it has been established that DI particles have a functional, though inactive, polymerase,[7] and that the polymerase of VSV New Jersey can function on a heterologous VSV Indiana template.[8]

(3) Determination of the sequence of nucleotides at the termini of the N-Protein-coding region of the genome revealed that *ts*D1 differed in 2 out of

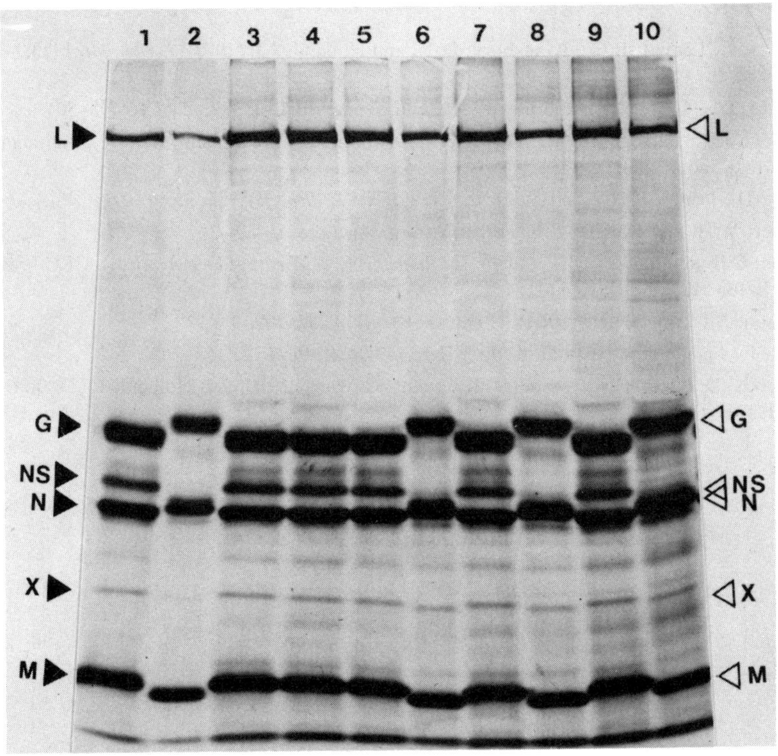

Figure 3. Polyacrylamide gel electrophoresis of the ^{35}S-methionine-labeled virion polypeptides of clones isolated at 31°C from clones of mixed phenotype in the progeny from the mixed infection with mutants *ts*D1 and *ts*E3 (as illustrated in Figure 2, tracks 3-9). The black arrows indicate mobilities characteristic of the *ts*D1 parent, and hollow arrows mobilities characteristic of the *ts*E3 parent. Track 1, the *ts*D1 parent; track 2, the *ts*E3 parent; tracks 3-10, clones obtained from mixed phenotype isolates showing complete segregation of the two parental types.

400 nucleotides from the original VSV New Jersey wild type (D. McGeoch, unpublished data). This suggested that *ts*D1 differed from wild type by approximately 50 base changes over the whole genome. By contrast, Rowlands et al.[9] estimated that the variant of VSV Indiana derived by 5 years propagation as a persistent infection differed from the parental *ts*G31(III) mutant by approximately 200 base changes. It is evident, therefore, that rapid and extensive changes of nucleotide sequence can in certain circumstances occur during lytic infection as well as during non-cytopathic persistent infection. The sequence data also indicated that the polypeptide mobility mutants were probably not unique mutants but merely a subset with a recognizable phenotype.

References

1. Mudd, J.A., Leavitt, R.W., Kingsbury, D.T., and Holland, J.J. (1973): *J. Gen. Virol.* 20:341–351.
2. Holland, J.J., Grabau, E.A., Jones, C.L., and Semler, B.L. (1979): *Cell* 16:495–504.
3. Clewley, J.P., Bishop, D.H.L., Kang, C-Y., Coffin, J., Schnitzlein, W.M., Reichmann, M.E., and Shope, R.E. (1977): *J. Virol.* 23:152–166.
4. Evans, D., Pringle, C.R., and Szilagyi, J.F. (1979): *J. Virol.* 31:325–333.
5. Wunner, W.H. and Pringle, C.R. (1974): *J. Gen. Virol.* 23:97–106.
6. Pringle, C.R. and Szilagyi, J.F. (1980): In: *Rhabdoviruses*. Bishop, D.H.L. (Ed.), CRC Press, Boca Raton, Volume II, pp. 141–161.
7. Emerson, S.U. and Wagner, R.R. (1972): *J. Virol.* 10:297–309.
8. Repik, P., Flamand, A., and Bishop, D.H.L. (1976): *J. Virol.* 20:157–169.
9. Rowlands, D., Grabau, E., Spindler, K., Jones, C., Semler, B., and Holland, J. (1980): *Cell* 19:871–880.

Published 1981 by Elsevier North Holland, Inc.
David H.L. Bishop and Richard W. Compans, eds.
The Replication of Negative Strand Viruses

Temperature-Sensitive Rabies Mutants With an Altered M_1-Protein

Naima Saghi, Florence Lafay, and Anne Flamand[a]

Introduction

Our research is directed towards studying the virulence of rabies virus. We have isolated from the wild type strain of CVS rabies virus 76 temperature-sensitive (*ts*) mutants and characterized them in order to study the relationships between rabies virus and the host cell.[1] Preliminary analyses indicate that the *ts*-mutants have different phenotypes. For example, at non-permissive temperatures, some mutants are blocked after primary transcription, and exhibit low protein synthesis capabilities (F^- mutants); others exhibit an unaltered transcription and replication and protein synthesis capabilities (F^+ mutants); most mutants however have an intermediate phenotype, and are called F^{+-} mutants.[2] Even with mutants of different phenotypes complementation has not been demonstrated.

SDS-Polyacrylamide Gel Electrophoresis of *ts*-Mutants

Several modifications of a protein can alter its electrophoretic mobility in an SDS-polyacrylamide gel system: for example, carbohydrate, phosphate and single amino acid changes.[3,4] The migrations of the viral induced proteins of rabies *ts*-mutants have been studied by SDS polyacrylamide gel electrophoresis. CER cells were infected by each of the mutants and incubated at a permissive temperature. Twenty-four hours after infection, polypeptide labeling was performed using hypertonic conditions in order to lower cellular

[a] Laboratoire de Génétique 2, Bâtiment 400, Université de Paris-Sud, 91405 Orsay Cedex—France.

protein syntheses.[5] Infected cell extracts were prepared.[2] Labeled polypeptides were resolved in 6-14% linear gradient polyacrylamide gels using the discontinuous system of Laemmli.[6] As expected, the polypeptide migrations of most *ts*-mutants were like those induced by wild type rabies virus. The M_1-protein of 2 *ts*-mutants showed an altered electrophoretic mobility. M_1-protein of *ts*0 106 migrated faster than that of wild type rabies (Figure 1) whereas that of *ts*0 52 migrated slower.

SDS-Polyacrylamide Gel Electrophoresis of Revertants

If the variation of the M_1-protein migration was indeed related to the *ts*-defect, we expected that some temperature resistant revertants would exhibit a wild type M_1 phenotype. Twenty revertants were isolated for each mutant. To ensure that most were derived independently, they were isolated

Figure 1. Electrophoretic analysis of *ts* rabies intracellular polypeptides. CER cells infected at a moi of about 1 PFU/cell were incubated at a permissive temperature. After 24 hr, the cells were treated for 30 min with hypertonic medium in order to depress cellular protein synthesis.[5] Labeling was for 2 hr with 25 μCi of ^{35}S-methionine. Cellular extracts were precipitated by 5 volumes of ethanol and the pellet redissolved in Laemmli[6] reducing buffer. Discontinuous, 6-14% linear gradient gels of polycrylamide were used. After electrophoresis the gel was fluorographed. From left to right: uninfected cells; wild type and the following *ts* rabies mutants: *ts*0 16, *ts*0 18, *ts*0 20, *ts*0 22, *ts*0 30, *ts*0 31, *ts*0 34, *ts*0 42, *ts*0 52, *ts*0 93, *ts*0 94, *ts*0 98, *ts*0 102, *ts*0 103, *ts*0 105, *ts*0 106, *ts*0 107, *ts*0 108, *ts*0 109, *ts*0 110, *ts*0 113, *ts*0 114 and *ts*0 115.

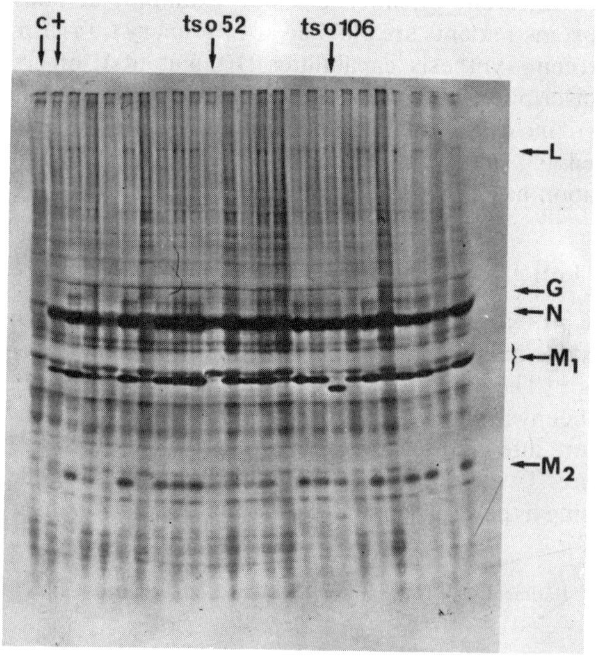

from 9 mutant subclones. Among these revertants, 1 from *ts*0 106 (R_2) and 1 from *ts*0 52 (R_3 and R_4) had M_1-proteins which migrated like that of the wild type M_1 (Figure 2). The 2 *ts*0 52 revertants were isolated from the same mutant clone and may have come from the same mutational event. The presence of several clones revertant for both the *ts* and M_1 characteristics is an indication that the *ts*-defect is correlated with the modified M_1 migration. The other revertants may be phenotypically suppressed by alternate M_1 changes. It remains to be determined if the compensating mutational event for any of the revertants is located in the same structural gene (intragenic suppression), or in another gene (extragenic suppression).

Discussion

Our data suggest that mutants *ts*0 52 and *ts*0 106 are affected in the M_1-protein. Dietzschold et al.[7] have demonstrated that, in addition to the viral N-protein, M_1-protein is phosphorylated. This phosphorylation reduces the electrophoretic mobility of the protein. For *ts*0 52 and *ts*0 106, the relation-

Figure 2. Electrophoretic analysis of the intracellular polypeptides of *ts*0 52, *ts*0 106 and their revertants. Labeled intracellular polypeptides were obtained as described in Figure 1. From left to right: uninfected cells, *ts*0 106 and 2 of its revertants (R_1,R_2), wild type rabies virus, *ts*0 52 and 6 of its revertants (R_1-R_6).

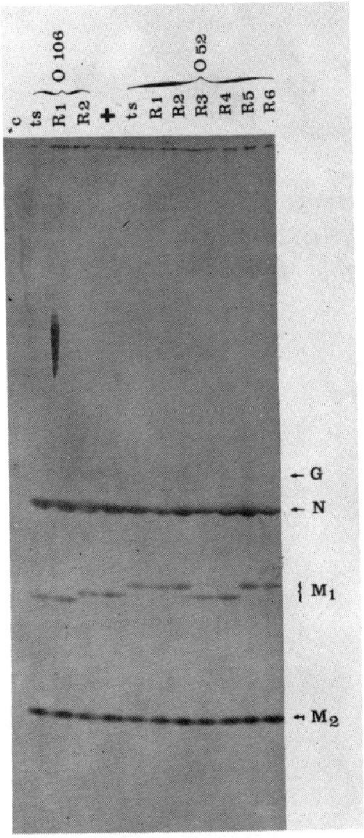

ship between their M_1-protein mobility and its phosphorylation is under investigation. These mutants have been initially classified as F^{+-} mutants on the basis of the fluorescence intensity of infected cells incubated at non-permissive temperatures and treated with anti-nucleocapsid fluorescent antibodies.[2] The RNA synthesis of both mutants at non-permissive temperatures is currently under study. If, as expected, their RNA replication is partially inhibited, the results would indicate that M_1-protein is involved in RNA synthesis. This observation, together with the fact that M_1 is phosphorylated, suggests that M_1 may have a function equivalent to that of the VSV NS-protein. It is interesting to note that some VS New Jersey temperature-sensitive group E mutant have an NS-protein with modified electrophortic mobility.[8,9]

ACKNOWLEDGMENTS

We wish to thank Dr. Philip Vigier for many helpful discussions. The excellent technical assistance of J. Benejean and B. Jaillard is gratefully acknowledged.

This research was supported by the Centre National de la Recherche Scientifique through Grant L.A. 040086.

References

1. Bussereau, F. and Flamand, A. (1978): In: *Negative Strand Viruses and the Host Cells*. Mahy, B.W.J. and Barry, R.D. (Eds.), Academic Press Inc., New York.
2. Saghi, N. and Flamand, A. (1979): *J. Virol.* 31:220–230.
3. DeJohg, W.W., Zweers, A., and Cohen, L.H. (1978): *Biochem. Biophys. Res. Commun.* 82:532–539.
4. Noel, D.K., Nikaido and Ames, G.F. (1979): *Biochemistry* 18:4159–4165.
5. Madore, H.P. and England, J.M. (1977): *J. Virol.* 22:102–112.
6. Laemmli, U.K. (1970): *Nature (London)* 227:680–685.
7. Dietzschold, B., Cox, J.H., and Schneider, L.G. (1979): *Virology* 98:63–75.
8. Evans, D., Pringle, C.R., and Szilagyi, J.F. (1979): *J. Virol.* 31:325–333.
9. Lesnaw, J.A., Dickson, L.R., and Curry, R.H. (1979): *J. Virol.* 31:8–16.

Published 1981 by Elsevier North Holland, Inc.
David H.L. Bishop and Richard W. Compans, eds.
The Replication of Negative Strand Viruses

Are the Drosophila Ref Genes for Piry and Sigma Rhabdoviruses Identical?

Gilbert Brun[a]

Sigma virus is wide-spread in natural populations of Drosophila.[1] But it was classified with difficulty by insect virologists because it is not pathogenic for its host and not contagious, and it is transmitted only by heredity.[2] Since its belonging to rhabdovirus group has been acknowledged[3] and despite the facts (1) that vesiculoviruses are arboviruses transmitted by and multiplying in many diptera, (2) that transovarial transmission of VSV Indiana by phlebotomine sandflies has been demonstrated,[4] and (3) that vesiculoviruses multiply in Drosophila and bring on the same symptom as sigma, a specific CO_2 sensitivity,[5-11] the classification of sigma virus is considered as a problem because it does not multiply in vertebrate cells. So, virus taxonomists have proposed a special rhabdovirus genus for sigma.[12] Our hypothesis is that sigma is directly related to vesiculoviruses. Sigma ancestor would have been analogous to VSV and its accidental isolation (by what accident?) in a non-biting dipteran has bound it down to perfect its hereditary transmission mechanisms for surviving and to save its host any trouble. I shall develop here a new comparison between the biology of sigma and vesiculoviruses in Drosophila.

A set of Drosophila genes has been found which intervene in sigma physiology: ref(1)H, ref(2)M, ref(2)P, ref(3)O, and ref(3)D. Ref is the general name (for refractory), the number in parenthesis refers to the Drosophila chromosome which carries the particular gene designated by the capital letter.[13] Contamine[14,15] and Coulon,[16] by means of temperature-sensitive

[a] Laboratoire de Génétique des Virus, CNRS, 91190 Gif sur Yvette, France.

mutants of sigma and temperature shift experiments, have shown that all the five genes intervene in the early events of sigma cellular cycle. Two early events are distinguishable by their timing and, taking into account the hereditary transmission characteristics of some *ts*-mutants at restrictive temperature and the analogy of sigma and VSV, it is permitted to think that these events are primary transcription and replication.[17] Finally, facts are consistent with the hypothesis that the product of ref(2)P gene plays a role in the efficiency of primary transcription, that of ref(3)O gene in the equilibrium between transcription and replication and that of ref(2)M gene in the lability of viral genome all over sigma cellular cycle.[14]

I have conducted a search of Drosophila ref genes acting on Piry virus in order to draw a parallel between these and those acting on sigma virus.

Mutants of Piry Virus with Affected Growth in Drosophila Flies (ag D Mutants)

Piry has been chosen among vesiculoviruses because it gives the most convenient CO_2 sensitivity symptom, totally identical with sigma symptom, and it is the most adapted to multiplication in Drosophila flies, giving minimum mean incubation time (MIT) for the same number of PFU per injected fly and maximum relative probability of infection (ratio of the probability to infect a fly to the probability to form plaque at 30°C in chick embryo fibroblasts) all over the range of temperatures from 17 to 30°C.

The Paris Drosophila strain has been chosen among laboratory strains because it is the most permissive, giving minimum MIT. This strain is also restrictive for the majority of sigma strains, carrying a restrictive allele of two ref genes, ref(2)P and ref(3)D.[13]

Selection of ag D mutants was conducted using the CO_2 sensitivity symptom: flies were injected in conditions of viral cloning (maximum dilution); the last ones to acquire CO_2 sensitivity symptom were selected and individually crushed; flies which were not CO_2 sensitive in the last CO_2 test were also crushed but collectively; each extract was spread on fibroblast plates; a plaque was picked up and the clone tested for MIT of a lot of inoculated flies at 20 and 28°C (Figure 1).

The great majority of ag D mutants found in two experiments were also temperature-sensitive (*ts*) in chick embryo fibroblasts (Table 1).

Thus, *ts*-mutants have been selected. Among 43 *ts*-mutants, 40 percent were ag D while the 44 ts^+ clones studied in parallel were all ag D^+ (Table 2). We infer from this high ratio of *ts*-ag D mutants that this class contains certainly mutants of L-protein, because of the high ratio of L mutants among spontaneous *ts*-mutants of vesiculoviruses whose genetics has been studied. Mutants of L-protein have also been found among host range mutants of other vesiculoviruses.[18] Other proteins may be touched, but in *ts*-ag D mutants we cannot say yet what they are.

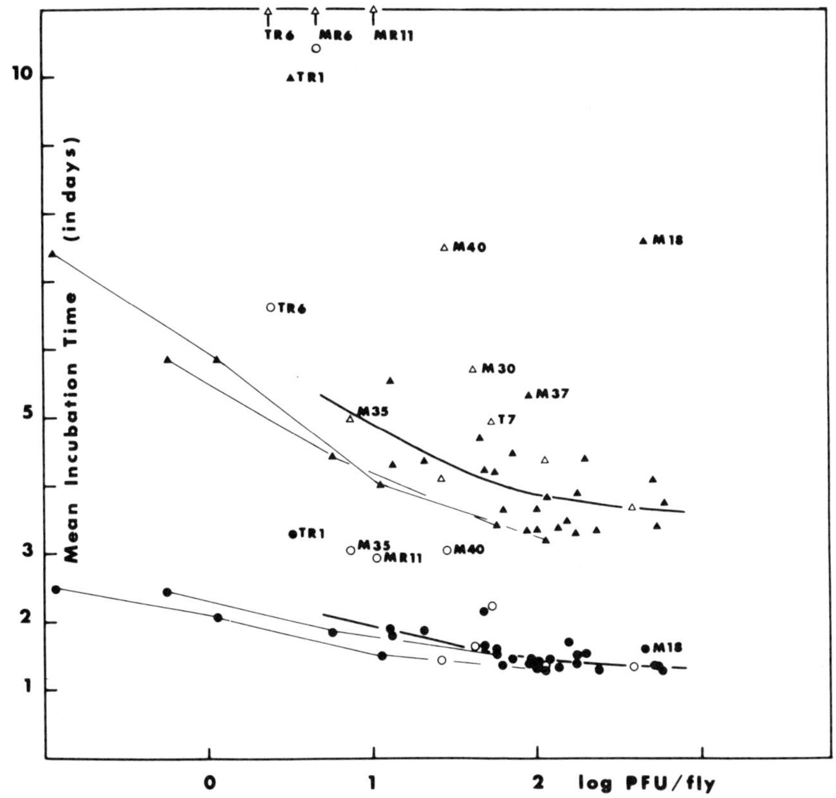

Figure 1. MIT at 20°(▲△) and 28°C (●○) versus PFU per injected fly for clones preselected in the experiment of selection of ag D mutants. Open symbols (△○) refer to *ts* clones. Dark symbols bound by fine lines refer to dilutions of two reference ts^+ clones, systematic divergences with other results are explained by the fact that titrations and fly injections have been made other days (intra-day correlation of MIT versus PFU/fly relationship). Clones refered as "probably ag D" in Table 1 are M30, M37 and T7.

Table 1. Clones Tested in the Experiment of ag D Mutants Selection, Including Two Clones ag D-*ts* and Two Clones ag D^+-ts^+ From Preliminary Experiment.

	ts^+	*ts*	total	%
agD	2	7	9	78
				75
probable agD	1	2	3	
agD^+	18	3	21	14

Table 2. Mutants Selected as *ts* in a Piry Population (One Multiplication After Three Successive Clonings) and in Seven ts^{+} Clones Picked Up From This Population.

tested clones	ts^{+}	*ts*	agD	%
593	551	42		7
tested in Paris flies		43	18	42
	44		0	0

Among ag D mutants, four phenotypic classes can be distinguished from the relation between MIT at 20° and 28°C (Figure 2):

(1) mutants equally affected at the two temperatures,
(2) mutants cold sensitive, more affected at 20°C,
(3) mutants temperature-sensitive in Drosophila, more affected at 28°C,
(4) mutants which do not render the Paris Drosophila flies CO_2 sensitive, neither at 20°, nor at 28°C.

Mutants Restricted Specifically in Some Drosophila Strains

Two main types of *ts*-ag D mutants must exist:

(1) mutants whose affected protein does not work well or cannot be processed in the general Drosophila medium (pH, ionic equilibrium or, for cold sensitive mutants, temperature, since only 30° temperature is common to the two hosts). Such mutants will be restricted in all Drosophila strains.
(2) mutants whose affected protein does not function or is not processed because the defect is expressed in a specific interaction with a definite product of the host. A mutant of this type may be restricted in some Drosophila strains and multiply in others. These two classes of strains allow us to define a ref gene, an allele of which is responsible for restriction of the mutant.

Thus, the most interesting case, the sole one in which we can decide to which type an ag D mutant belongs, is the case of specific restriction. But, because the difference between the two Drosophila strains is shown only by a particular mutant, there is no clue for conducting us in the search of such a situation. And, since testing our 30 ag D mutants against all Drosophila strains was impossible, bets were necessary. The first one has been the choice of the Paris Drosophila strain for selecting mutants.

After a limited number of tests we have found a mutant, referred on Figure 2 as *ts*1 but called also F1 clone, which distinguishes the Paris strain from others. The *ts*1 mutant is one of the fourth-class mutants which do not render CO_2 sensitive Paris Drosophila flies. When it is injected into flies of the V

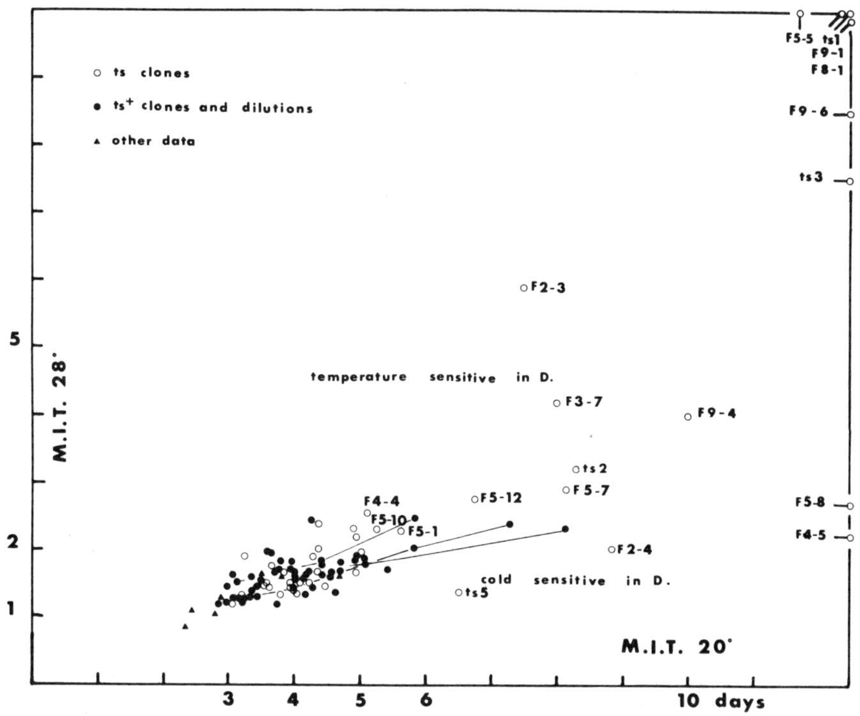

Figure 2. Relationship of MIT at 20° and 28°C: (●) ts^+ clones, (○) the 43 selected *ts* clones. ag D mutants are shown off by their designation.

strain (a strain carrying the two recessive v and bw eye color mutations, the most permissive for Piry after the Paris strain, permissive for all strains of sigma), this mutant renders some of them CO_2 sensitive but very irregularly. The symptom does not show well the difference between the strains. This difference appears in the yield of inoculated flies (Figure 3). For F1 inoculated V flies, a plateau yield is reached more slowly than with wild type (wt) virus, but the yields of individual flies are as homogeneous as for wt. On the contrary, the yields of F1 inoculated Paris flies remain heterogeneous up to 50 days at 20°C. With an inoculum of about 20 PFU per fly, as in the experiments of Figure 3, some Paris flies gave no yield. From the percentage of apparently uninfected flies, it is possible to calculate a titer of the inoculum. The ratio of this titer to the titer in PFU is the ratio of the probability for one virus to induce measureable infection in a Drosophila fly to the probability to form a plaque. This ratio is compared with the relative probability of infecting a fly calculated for wt virus. The two ratios are equal in V Drosophila strain while the ratio for F1 in the Paris strain is two to three hundred times lower than the wild type one.

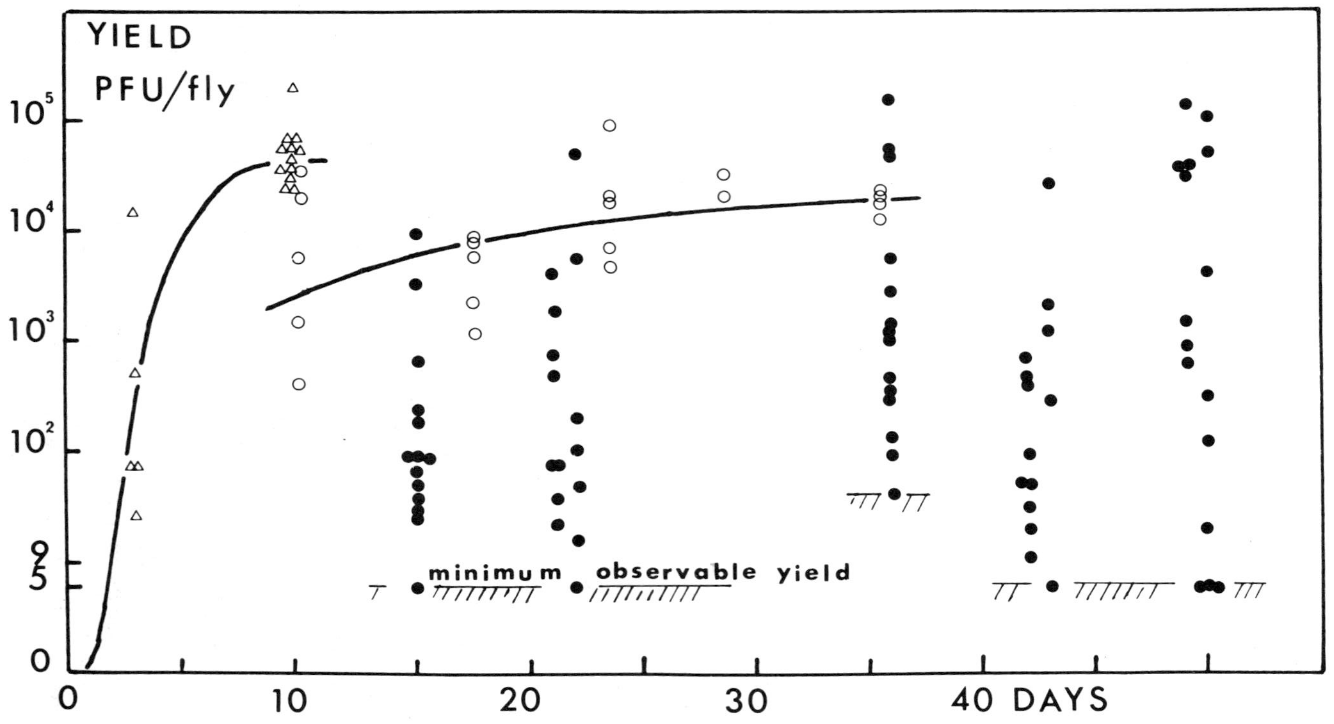

Figure 3. Yield of infected flies versus time after inoculation: (△) Paris flies infected with one infective unit of wt virus (maximum dilution), (○) V flies and (●) Paris flies inoculated with F1 mutant (about 20 PFU/fly). All V flies were infected, while a fraction of Paris flies remained uninfected (about 1/4 with the limits of used detection).

Are these two properties of F1 mutant: incapacity to render CO_2 sensitive flies regarded as permissive and incapacity to grow in Paris strain, consequences of the same mutation? It is very possible, though an "adapted" mutant from F1 has been selected in hybrid Oregon/ebony flies (called Standard flies in experiments with sigma virus) which mutant renders CO_2 sensitive permissive flies and is still restricted in the Paris ones (F1SS4 mutant). Indeed this mutant is one in a broad spectrum of "adapted" or "partially adapted" mutants form F1 and its growing capacities in V and Paris strains have been so changed that its existence cannot affect the conclusion about the number of mutations in F1.

The F1SS4 secondary mutant is useful for locating the corresponding ref gene. This one is located in the left extremity of the third Drosophila chromosome. It is different from the known sigma ref genes and is called ref(3)A.

Discussion

Thus Drosophila ref genes for Piry virus do exist.

Does the fact that the first one found is different from the known sigma ref genes mean that Piry virus works in an other way than sigma virus and requires other host products for its multiplication? I do not believe it. Indeed, study by Contamine of the restrictive characteristics of flies hemizygous for the restrictive allele of ref(2)P (P^P/deficiency) suggests that sigma virus works also with an other host product as a substitute for the P product.[14] Contamine's hypothesis is that there are several host genes whose products can play the same role in the viral physiology (they may be redundancies in the fly genome) and that the spectrum of affinity as the spectrum of efficiency of their products depend on the genotypes in presence. So, if Piry and sigma are related viruses and have the same requirements, we may conceive that, among two host genes which both play the same role for the two viruses, one may be the principal functioning gene and the other the secondary one a the virus, and conversely for the other virus.

The fact that three allelic differences at a ref locus acting in rhabdovirus physiology exist between the Paris and V or Oregon strains must be emphasized. It suggests some interactions or compensations in the physiology of Drosophila between the products of these genes.

However that may be, Piry virus is a better material than sigma for studying with biochemical methods the role of insect host products in the physiology of a rhabdovirus.

References

1. Fleuriet, A. (1972): *C.R. Séance Soc. Biol.* 166:598–601.
2. L'Héritier, Ph (1958): *Adv. Virus Res.* 5:159, 269–283.

3. Berkaloff, A., Bregliano, J.C., and Ohanessian, A. (1965): *C.R. Acad. Sci. (Paris)* 260:5956–5959.
4. Tesh, R.B., Chianolis, B.M., and Johnson, K.M. (1972): *Science* 175:1477–1479.
5. Printz, P. (1967): *C.R. Acad. Sci. (Paris)* 264:169–172.
6. Printz, P. (1968): *C.R. Soc. Biol. (Paris)* 162:372–373.
7. Printz, P. (1973): *Adv. Virus Res.* 18:143–157.
8. Bussereau, F. (1971): *Annales Institut Pasteur* 121:223–239.
9. Bussereau, F. (1972): *Annales Institut Pasteur* 122:1029–1058.
10. Bussereau, F. (1973): *Annales Microbiologie (Inst. Pasteur)* 124A:535–554.
11. Bussereau, F. (1975): *Annales Microbiologie (Inst. Pasteur)* 126B:389–403.
12. Fenner, F. (1976): *Intervirology* 6:1–12.
13. Gay, P. (1978): *Molec. Gen. Genet.* 159:269–283.
14. Contamine, D. (1979): *Thése Paris XI Orsay.*
15. Contamine, D.: Submitted.
16. Coulon, P. (1978): *Thèse de 3ème Cycle, Paris VI.*
17. Contamine, D. (1980): *Annales de Virologie (Inst. Pasteur)* 131E:113–134.
18. Szilagyi, J.F., Pringle, R.C., and Macpherson, (1977): *J. Virol.* 22:381–388.

David H.L. Bishop and Richard W. Compans, eds.
The Replication of Negative Strand Viruses

Isolation and Characterization of the Mitogenic Principle of Vesicular Stomatitis Virus

James J. McSharry and Gail W. Goodman-Snitkoff[a]

The interactions of VSV and murine lymphocytes have been under investigation since the early studies of Edelman and Wheelock.[1] More recent studies by Bloom and coworkers have elucidated the conditions under which VSV will replicate in lymphocytes.[2,3] They showed that VSV will replicate in mitogen-activated T-cells but not in non-activated T-cells or B-cells. Minato and Katsura demonstrated that only a particular subset of T-cells, that consisting of nylon-wool adherent T-cells which are engaged in the suppression of antibody responses, supports the replication of VSV.[4] Additional evidence that B-cells produce little or no virus was recently presented by Creager and Youngner, who showed that infection of activated T-cells results in replication of VSV and cell death, whereas infection of activated B-cells with VSV leads to a persistent infection.[5]

In addition to virus replication in lymphocytes, other virus-lymphocyte interactions lead to stimulation of DNA synthesis and cell division of resting lymphocytes. Thus, in some instances, viruses can act as mitogens. Both infectious and UV-inactivated influenza A viruses (H2N2) activate murine B- and T-cells.[6] Since virus replication is not required for mitogenic activity and mitogenesis is only associated with the H2H2 subtype of influenza A virus, it is likely that one of the virion proteins, particularly H2, could be the mitogen. On the other hand, herpes simplex virus types 1 and 2 are mitogens for B-cells but only infectious virus is mitogenic.[7–9]

[a] Department of Microbiology and Immunology, Albany Medical College of Union University, Albany, New York.

Recently we have demonstrated that VSV is a mitogen for murine B-cells.[10] Both UV-inactivated and defective interfering particles, such as D1011, are B-cell mitogens. The fact that non-infectious virus is mitogenic suggests that no viral replicative functions are required for lymphocyte activation. Therefore, a constituent of the virion, such as a viral protein, could be the mitogen. We report here that the isolated VSV glycoprotein is mitogenic for B-cells, but not T-cells, when assayed in mouse splenic lymphocytes.

Mitogenic Activity of the G-Protein

In order to show that the VSV glycoprotein is a mitogen for murine spleen cells, the G-protein was isolated from purified virus by extraction with Triton X-100 followed by butanol precipitation and extraction with acetone.[11] The purity of the isolated G-protein preparations was determined by SDS-PAGE. Preparations which contained only the G-protein were used in these experiments. Various concentrations of VSV and the G-protein were incubated for 48 hr at 37°C under an atmosphere of 10% CO_2 with splenocytes isolated from CBA/J mice. Twenty-four hours before the end of the incubation period, ^{3}H-thymidine was added to the cultures. At the end of the incubation period, the cells were harvested and incorporation of ^{3}H-thymidine into insoluble material was determined. The results (Figure 1) show that both VSV and the isolated G-protein are mitogenic for murine spleen cells. The optimal response occurs with 10 μg of G-protein and 50 μg of VSV. In previous experiments, we showed that at concentrations of VSV greater than 50 μg, the mitogenic response remained constant or decreased.[10] Since the G-protein represents about 20–30% of protein in the VS virion, 10 μg of G is equivalent to that amount of G represented by 50 μg of virus. Since 50 μg of VSV does not stimulate as much DNA synthesis as 10 μg of G, the data suggest that isolated G is a more efficient mitogen than the virion, possibly because of the aggregated state of the G-protein.

The G-Protein is a B-Cell Mitogen

We have previously shown that VSV is a B-cell mitogen.[10] In order to show that the isolated G-protein is a B-cell mitogen, various concentrations of G-protein were incubated with splenocytes isolated from CBA/J mice or congenitally athymic Balb/c nu/nu mice which have no functional T-cells and the stimulation of DNA synthesis was measured as described above. The data (Figure 2) show that the G-protein was mitogenic for spleen cells isolated from both strains of mice. Maximum stimulation of DNA synthesis occurred with 10 μg of G-protein, a finding similar to previous data (Figure 1). Lipopolysaccharide (LPS), a B-cell mitogen, was mitogenic in both populations of splenocytes, whereas Concanavalin-A (Con-A), a T-cell mitogen,

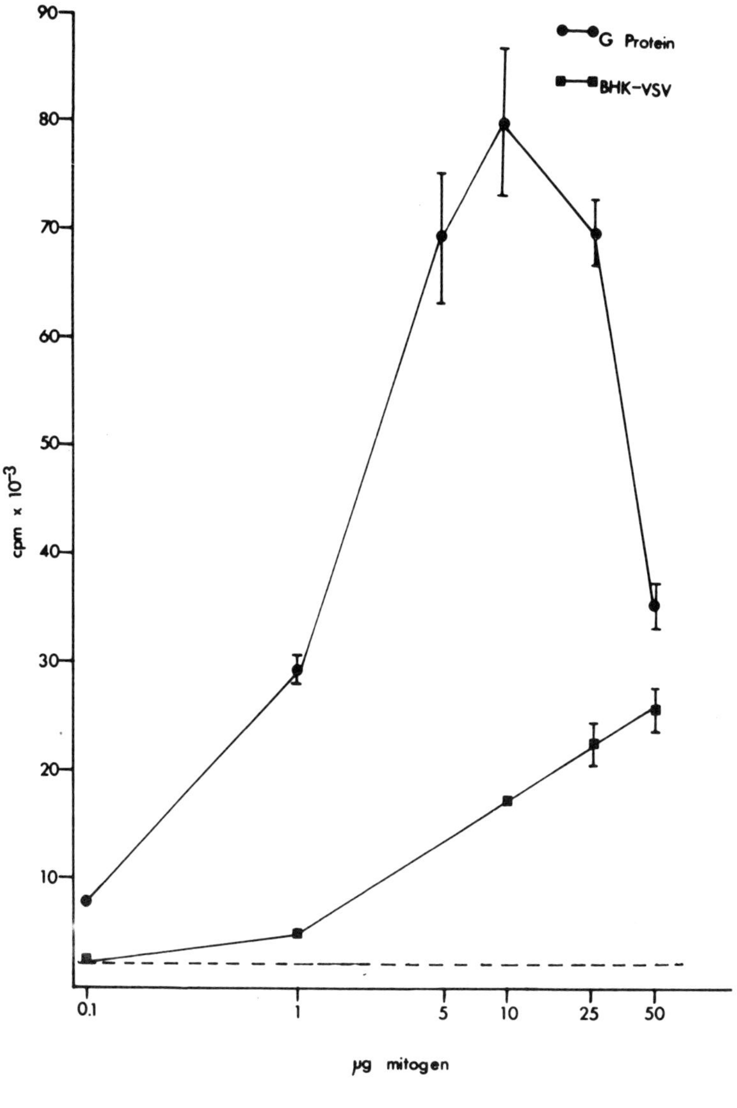

Figure 1. Mitogenic activity of the G-protein. Various concentrations of VSV and the G-protein were incubated with spleen cells isolated from CBA/J mice for 48 hr. Twenty-four hours before the end of the incubation period, ^{3}H-thymidine was added to each culture. At the end of the incubation period, incorporation of ^{3}H-thymidine into insoluble material was determined after extraction of cells onto filters with a multiple automated sample harvester. Radioactivity associated with the filters was determined by liquid scintillation counting.

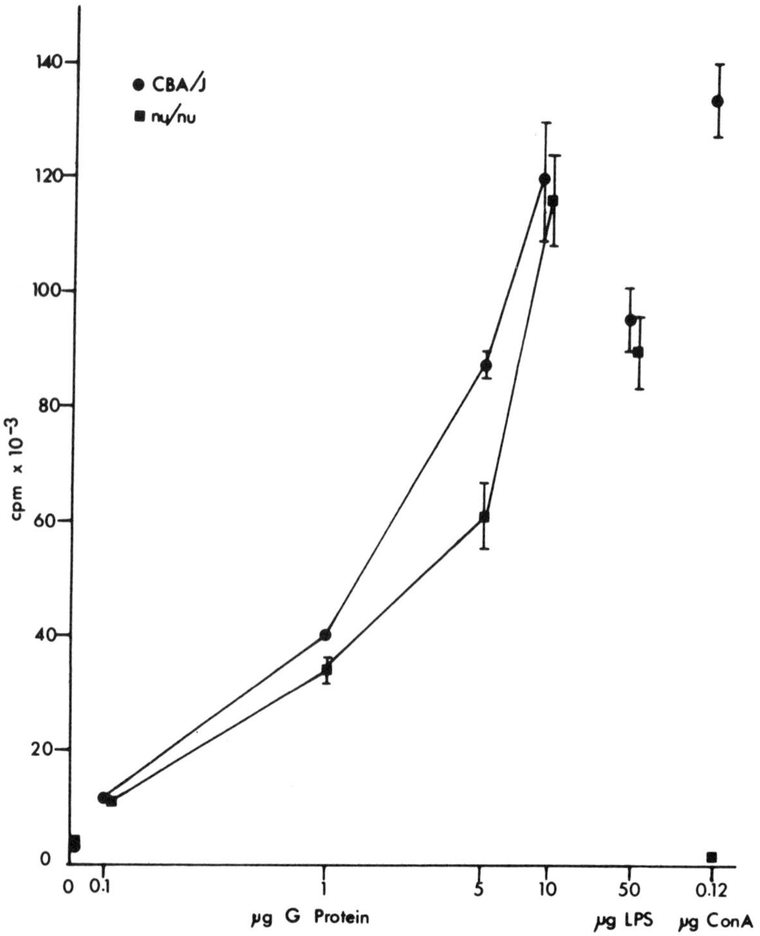

Figure 2. The G-protein is a B-cell mitogen. Various concentrations of the G-protein, 50 μg of LPS (a B-cell mitogen) and 0.12 μg of Con-A (a T-cell mitogen) were incubated with spleen cells isolated from CBA/J or Balb/c nu/nu mice. Mitogenicity was determined as described in the legend of Figure 1.

was mitogenic only for splenocytes isolated from the CBA/J mice, indicating that the Balb/c nu/nu mice lacked functional T-cells. The results of this experiment demonstrate that the G-protein is mitogenic for B-cells in the absence of functional T-lymphocytes.

VSV and the G-Protein are not T-Cell Mitogens

Our previous data, which showed that VSV was a B-cell mitogen, did not address the question of T-cell mitogenicity. Since VSV can replicate in activated T-cells, it was possible that infectious virus activated the T-cell and then proceeded to replicate in the activated cells, resulting in cell death. Since the G-protein stimulates mitogenesis in unseparated spleen cell preparations, as well as in B-cells, and since it is non-infectious, the G-protein is an ideal candidate to determine the mitogenic potential of VSV-proteins for T-cells. Purified T-lymphocytes were recovered from nylon wool columns by the method of Hodes et al.[12] Both adherent and non-adherent cells were washed with RPMI 1640, resuspended and counted. Yields were generally approximately 16% non-adherent cells and approximately 36% adherent cells. VSV and the G-protein were incubated with unseparated splenocytes, nylon-wool column non-adherent (T-cell population) splenocytes, or adherent (B-cell enriched population) splenocytes for 48 and 72 hr, and stimulation was measured as described previously. The data (Figure 3) show that both VSV and the G-protein are mitogenic for unseparated splenocytes and the nylon-wool adherent cells (B-cell enriched preparation), but are not mitogenic for the nylon wool non-adherent cells (T-cell preparation), LPS, a B-cell mitogen, gave a mitogenic response similar to VSV and the G-protein, activating cells in the untreated population and the adherent cell population. Con-A, the T-cell control, was mitogenic for the non-adherent cell population, and to a lesser extent, to the adherent cell population, since this population contains some residual T-cells. In both cases, Con-A gave a maximum response at 72 hr characteristic of this T-cell mitogen. The data presented here demonstrate that VSV and its isolated G-protein are mitogenic for murine B-cells, but not for murine T-cells.

Conclusions and Discussion

The results presented in this report show that the VSV glycoprotein is a B-cell mitogen and that VSV and the G-protein are not T-cell mitogens. Thus the list of viral mitogens has grown to include VSV, influenza virus A (H2N2), herpes simplex virus types 1 and 2, simian virus 5 and Sindbis virus.[6–10] To our knowledge, this report is the first demonstration that an isolated viral glycoprotein is a mitogen, although the report on influenza A virus (H2N2) suggested that the H2 glycoprotein was the mitogen associated with influenza virus.[6] The mitogenic activity of the G-protein parallels that of VSV in that it stimulates DNA synthesis in B-cells but not in T-cells, and it follows the same time course peaking at 48 hr. However, the G-protein is a more potent mitogen, stimulating greater ^{3}H-thymidine incorporation into lymphocyte DNA at lower concentrations of protein than VSV. Fifty μg of VSV contain about 10–15 μg of G-protein; however, optimal VSV activation was one-quarter the optimum obtained with the isolated G-protein. It is possible that the isolated G-protein was more mitogenic than the whole virus because it was aggregated. Alternatively, one or more of the constituents of the virus particle could have inhibited mitogenesis. Further evidence in support of the G-protein being the VSV mitogen comes from experiments in which the G-protein was removed from the virion and the

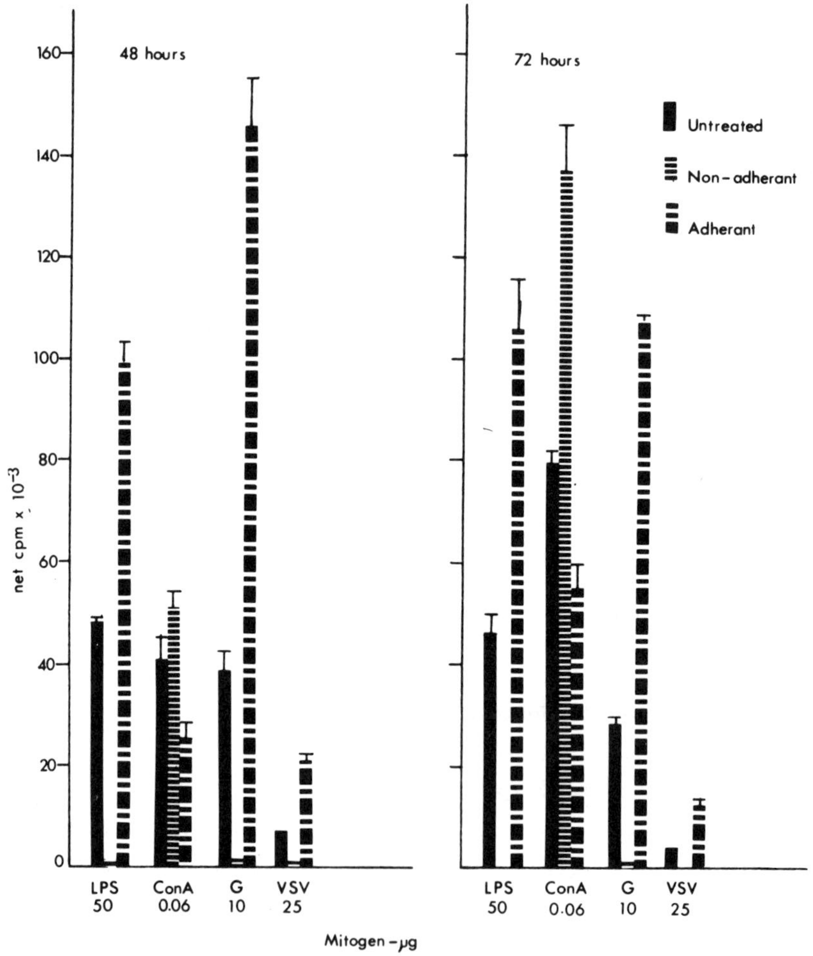

Figure 3. VSV and the G-protein are not T-cell mitogens. Spleen cells were isolated from CBA/J mice and separated into a T-cell population and a B-cell enriched population by absorption to and elution from nylon-wool columns. Optimal concentrations of VSV, G-protein, LPS and Con-A were incubated with untreated, non-adherent (T-cells) or adherent (B-cell enriched) cell populations for 48 and 72 hr. Incorporation of thymidine into DNA was determined as described in the legend of Figure 1.

remaining viral proteins were tested for mitogenesis. Protease-treated virus and the Triton X-100 insoluble material remaining after G-protein isolation, which have greater than 90% of the G-protein removed, were considerably less mitogenic than the virion or the G-protein. The small amount of mitogenic activity associated with the treated material is probably due to residual G-protein.

In this report, we have clearly demonstrated that infectious virus is not required for mitogenesis. This result is in agreement with those of Butchko et al. who showed that UV-inactivated influenza A virus (H2N2) was mitogenic.[6] However, our results are in contrast to those of Mochizuki et al. who showed that heat or UV-inactivated herpes simplex virus types 1 and 2 were not mitogenic, suggesting that infectious virus was required for mitogenesis.[9] It is possible that the heat and UV treatments affected the viral proteins, as well as the genetic material, rendering these antigens non-mitogenic. This can be resolved by isolating herpes virus proteins and determining their mitogenic activity. In addition, it will be of interest to determine the mitogenic activity of the isolated glycoproteins of influenza A virus (H2N2), Simian virus 5 (HN and F), and Sindbis virus (E_1 and E_2).

ACKNOWLEDGMENTS
We thank Ms. Lise Winters for technical assistance and Kathleen Cavanagh for help in preparation of the manuscript. This work was supported in part by the New York State Health Research Council Grant No. 1572, and the National Science Foundation Grant No. PCM-8003126. G.G.S. is supported by a National Institute of Health Postdoctoral Fellowship No. F32A10606001.

References

1. Edelman, R. and Wheelock, E.F. (1966): *Science* 154:1053–1055.
2. Bloom, B.R., Jimenez, L., and Marcus, P.I. (1970): *J. Expt. Med.* 132:16–30.
3. Bloom, B.R., Senik, A., Stoner, G., Ju, M., Nowakowski, M., Kano, S., and Jimenez, L. (1976): *Cold Spring Harbor Symposium on Quantitative Biology* 41:73–84.
4. Minato, M. and Katsura, Y. (1978): *J. Expt. Med.* 148:837–849.
5. Creager, R.S. and Youngner, J. (1980): *Abs. Amer. Soc. Microbiol.* 260.
6. Butchko, G.M., Armstrong, R.B., Martin, W.J., and Ennis, F.A. (1978): *Nature* 271:66–67.
7. Kirchner, H., Darai, G., Hirt, H.M., Keyssner, K., and Munk, K. (1978): *J. Immunol.* 120:641–645.
8. Kirchner, H., Hirt, H.M., Kleiniche, C., and Munk, K. (1976): *J. Immunol.* 117:1753–1756.
9. Mochizuki, D., Hedrich, S., Watson, J., and Kingsburg, D.T. (1977): *J. Expt. Med.* 146:1500–1510.
10. Goodman-Snitkoff, G.W. and McSharry, J.J. (1980): *J. Virol.* 35:757–765.
11. McSharry, J.J. and Choppin, P.W. (1978): *Virology* 84:172–182.
12. Hodes, R.J., Handwerger, B.S., and Terry, W.D. (1974): *J. Expt. Med.* 140:1646–1659.

Published 1981 by Elsevier North Holland, Inc.
David H.L. Bishop and Richard W. Compans, eds.
The Replication of Negative Strand Viruses

Antigenic Variation Between Rabies Virus Strains and Its Relevance in Vaccine Production and Potency Testing

Joan Crick,[a] Fred Brown,[a] A.J. Fearne,[b] and J.H. Razavi,[c]

The *Lyssavirus* genus within the Rhabdoviridae consists of rabies virus and the rabies related viruses.[1,2] A tentative subdivision within the group has placed classical rabies strains within serotype 1, the rabies-related viruses being considered as separate serotypes.[3] The rabies-related viruses appear to be confined to Africa and of no epidemiological significance, whereas rabies virus itself has a worldwide distribution.

Only minor antigenic differences have been reported among the classical rabies virus isolates so that it is still the recommended practice to prepare vaccines and antisera for use throughout the world with a few well characterized strains some of which are listed in Table 1.[4] The Pasteur-derived strains are the most widely used in the preparation of vaccines and antisera, while CVS is the standard challenge strain for vaccine potency testing.[5] The Flury strains[6] are given as live veterinary vaccines for the control of rabies in many parts of the world. Strain 675 is a plaque isolate of HEP virus made by Dr. G. Bijlenga of the School of Veterinary Medicine, Lyon, France.

For some years we have been involved in the development of inactivated vaccines for veterinary use. For this purpose we have used both LEP virus and Strain 675 grown in BHK-21 cells. The vaccines are prepared by inactivating the infected cell supernatants with beta-propiolactone or acetyl ethylene imine.[7–9] We were aware of the small antigenic differences between the Pasteur and Flury strains which had been reported[10] but they were not apparent when we did cross neutralization tests using hyperimmune sera.

[a] Animal Virus Research Institute, Pirbright, Woking, Surrey, United Kingdom.
[b] Mycofarm Ltd., Cut Hedge Farm, London Road, Braintree, Essex, United Kingdom.
[c] State Razi Serum and Vaccine Institute, P.O. Box 656, Teheran, Iran.

Table 1. Rabies Virus Strains Used for Vaccine and Antiserum Production.

Strain	Origin
Pasteur	Bovine-Europe 1882
Pitman-Moore (P-M)	
CVS	
Flury	Human-U.S.A. 1939
LEP	
HEP	
HEP 675	Plaque isolate from HEP

Our interest in vaccine production led us to consider the methods used to measure vaccine potency. Most authorities will license only those vaccines which reach minimum standards in either the Habel[11] or NIH test.[5] In each test, mice are vaccinated and subsequently challenged with CVS and in both the animals receive more than 1 dose of vaccine (Habel—6 doses; NIH—2 doses). Consequently the primary antigenicity of the vaccine is not measured in either test.[12,13] Furthermore, the results depend on the ability of the animals to withstand challenge by the severe and unnatural intracerebral route. For these reasons we suggested an alternative test in which the production of serum neutralizing activty (SNA) after one dose of vaccine was measured.[13,14]

This idea was based on our finding that after a single inoculation the resistance to challenge by mice was related to the SNA developed and that both were dependent (within certain limits) on the dilution of vaccine (amount of antigen) given.[15] In fact, this relationship holds whether one or two doses of vaccines are used and the results of a typical test with a human diploid cell vaccine (HDCV) prepared with P-M virus are given in Table 2. However, as the results show, the single dose test is more stringent: there can be no amplification of potency due to an anamnestic response, so it is to be preferred.

Table 2. Serum Neutralizing Activity and Protection of Mice Against Intracerebral Challenge With CVS After 1 or 2 Intraperitoneal Doses of HDCV.

Vaccine dilution	Protected mice	SNA	Protected mice	SNA
10	20/20[a]	4.2[b]	14/20	3.4
50	13/20	3.7	8/20	2.7
250	4/18	2.3	4/19	1.1
1250	1/20	1.6	1/20	0.9
Inoculation	Day 0 and day 7		Day 0	

[a]Mice challenged with $50ID_{50}$ CVS 14 days after first dose of vaccine.

[b]Depression of virus titer ($\log_{10}$) by 0.015 ml 1/10 pooled sera from groups of 4 mice 14 days after first dose of vaccine. Sera tested in suckling mice against CVS.

We have not been alone in considering new methods for determining vaccine potency as the numerous papers in the WHO/IABS Symposia on Rabies testify[16–18] and as reported in WHO/*Rab.Res.* 80:7.[19] Nevertheless, there is considerable reluctance to discard challenge tests in potency measurement and the most recent WHO recommendations are that for tissue culture vaccines the NIH test should continue to be used.[20] However, an antibody induction test is considered one possible alternative. It has also been suggested that a variation of the NIH test using a single vaccination should be evaluated.[20] Application of the last type of test to our vaccines confirmed the antigenic differences between the Flury and Pasteur-derived strains[10] and indicated that they might be of greater importance than we had previously considered.

Although our LEP vaccines have given very good results in the Habel test (protective indices > 6.0, G. Turner, personal communication), they provide relatively poor protection in the NIH test. Even the vaccines from the HEP 675 strain, which is characterized by massive production of virus particles, give less protection against CVS challenge than vaccine similarly prepared from the P-M strain. Nevertheless, we were somewhat surprised to find that a single dose of LEP vaccine which protected mice well against homologous challenge was much less effective against a corresponding challenge with CVS. The results of one such test are shown in Table 3. The reduced efficacy of the vaccine against the heterologous strain is reflected in the lower level of SNA against it.

We then examined the neutralizing activity in the sera of mice which had received vaccines prepared from P-M, LEP or HEP 675 strains. In this experiment, the mice were given vaccine on day 0 and day 7 as in the classical NIH test[5] and bled on day 14. The results in Table 4 show obvious differences in the neutralizing activities against LEP virus and CVS. In each pooled serum neutralizing activity was highest against the homologous

Table 3. Serum Neutralizing Activity and Protection of Mice Against Intracerebral Challenge With Either LEP or CVS Virus After a Single Intraperitoneal Dose of LEP Vaccine.

Vaccine dilution	Protected mice	SNA	Protected mice	SNA
10	8/8[a]	4.5[b]	5/8	2.7
50	4/8	—	2/8	—
250	3/8	—	1/8	—
1250	1/8	—	0/8	—
Control (no vaccine)	1/8	—	0/8	—
Challenge virus	LEP		CVS	

[a]Mice challenged with either LEP or CVS virus (Ca. 50 ID_{50}) 14 days after vaccination.

[b]Depression of virus titer ($\log_{10}$) by 0.015 ml 1/10 pooled serum from a group of 8 mice 14 days after vaccination. Sera tested in suckling mice against LEP or CVS virus.

Table 4. Serum Neutralizing Activity in Mice After Intraperitoneal Inoculation on Day 0 and Day 7 With Either P-M, LEP or HEP 675 Vaccine.

Vaccine dilution	SNA/LEP	SNA/CVS	SNA/LEP	SNA/CVS	SNA/LEP	SNA/CVS
10	3.5[a]	4.3	3.9	2.3	4.9	4.5
50	2.5	4.3	3.3	0.9	3.9	2.3
250	3.3	4.5	3.5	2.1	3.5	1.7
Vaccine	P-M		LEP		HEP	

[a]Depression of virus titer ($\log_{10}$) by 0.015 ml 1/10 pooled sera from groups of 8 mice killed 14 days after the first dose of vaccine. Sera tested in suckling mice against CVS or LEP virus.

strain, undoubtedly the explanation of the greater ability of P-M vaccines to protect against challenge with CVS.

Our results confirm those of earlier workers concerning antigenic differences between rabies virus strains developed from the original Pasteur isolate and those derived from the Flury strain.[10] Similar differences between these and other laboratory strains have also been reported[10] and are being extensively analyzed with batteries of monoclonal antibodies.[21–24] How significant the strain differences may be in terms of resistance to natural infection is not known, but they may account for vaccine failures and have important implications with regard to laboratory safety and in the selection of future vaccine strains.

Our results also raise the question of the suitability of the Pasteur-derived strains to continue as the absolute standards against which all vaccines and antisera are matched, especially in view of the suggestion that they may differ considerably from certain field strains.[3] Furthermore, the design of currently accepted vaccine potency tests is such that only small antigenic differences, possibly of no significance outside the laboratory, may influence their outcome and hence a manufacturer's ability to meet licensing requirements with his products.

ACKNOWLEDGMENTS

The authors wish to thank Dr. A.C. Allison, International Center for Research on Animal Diseases, Nairobi, Kenya, for the gift of LEP virus; Professor G. Bijlenga, School of Veterinary Medicine, Lyon, France, for HEP 675 virus, Dr. G. Turner, Blood Products Laboratory, Elstree, Herts, United Kingdom, for CVS and for the results of the Habel test and Dr. K.G. Nicholson, Clinical Research Center, Northwick Park Hospital, Harrow, Middlesex, United Kingdom, for P-M virus. Dr. T. McGrath, National Institute for Biological Standards and Control, Hampstead, United Kingdom, kindly provided the HDC vaccine. We also thank WHO, Geneva, Switzerland for permission to reproduce data in WHO/*Rab.Res.*/78:5, WHO/*Rab.Res.*/80:7, and WHO/*Rab.Res.*/80:8, WHO/*Rab.Res.*/80:188.

References

1. Fenner, F. (1976): *Intervirology* 7:42.
2. Brown, F., Bishop, D.H.L., Crick, J., Francki, R.I.B., Holland, J.J., Hull, R., Johnson, K., Martelli, G., Murphy, F.A., Obijeski, J.F., Peters, D., Pringle, C.R., Reichmann, M.E., Schneider, L.G., Shope, R.E., Simpson, D.I.H., Summers, D.F., and Wagner, R.R. (1979): *Intervirology* 12:1–7.
3. Report on Consultation on Rabies Prevention and Control. WHO/*Rab.Res.* 80:188.
4. WHO/Expert Committee on Rabies, Sixth Report (1973). WHO *Technical Report Series* No. 523. WHO, Geneva.
5. Seligmann, E.B., Jr. (1973): In: *Laboratory Techniques in Rabies,* Third Edition. Kaplan, M.M. and Koprowski, H. (Eds.), WHO, Geneva, Ch. 33, pp. 279–286.
6. Koprowski, H. (1954): *Bull. WHO* 10:709–724.
7. Chapman, W.G., Ramshaw, I., and Crick, J. (1973): *Applied Microbiol.* 26:858–862.
8. Crick, J. and Brown, F. (1971): *Res. Vet. Sci.* 12:156–161.
9. Crick, J., Brown, F., Fearne, A.J., Thompson, G., and Razavi, J.H.: Unpublished data.
10. Wiktor, T.J. and Clark, H.F. (1973): *Ann. Microbiol.* (*Inst. Pasteur*) 124A:283–287.
11. Habel, K. (1973): In: *Laboratory Techniques in Rabies,* Third Edition. Kaplan, M.M. and Koprowski, H. (Eds.), WHO, Geneva, Swtizerland, Ch. 31, 276–277.
12. Schneider, W. (1966). In: *International Symposium on Rabies. Symposia Series in Immunobiological Standardization.* R.H. Regamey and others, eds. Karger, Basel, 1:407–410.
13. Crick, J. and Brown, F. (1974): In *International Symposium on Rabies (II). Symposia Series in Immunobiological Standardization.* 1. Regamey, R.H. and others (Eds.), Karger, Basel, 21:316–320.
14. Crick, J. and Brown, F. (1978): In: *International Symposium on Standardization of Rabies Vaccines for Human Use Produced in Tissue Culture (Rabies III). Develop. Biol. Standard.* Karger, Basel, 40:179–182.
15. Crick, J. and Brown, F. (1969): *Nature, London* 222:92.
16. *International Symposium on Rabies. Symposia Series in Immunobiological Standardization 1.* Regamey, R.H. and others (Eds.), Karger, Basel, 1966.
17. *International Symposium on Rabies (II). Symposia Series in Immunobiological Standardization.* Regamey, R.H. and others (Eds.), Karger, Basel, 1974.
18. *International Symposium on Standardization of Rabies Vaccines for Human Use Produced in Tissue Culture (Rabies III). Develop. Biol. Standard.* Karger, Basel, 40, 1978.
19. Report on Cell Culture Rabies Vaccines and their Protective Effect in Man. WHO/*Rab.Res.*80:7.
20. Report of a Consultation on References Preparations and Potency Tests for Rabies Vaccines. WHO/*Rab.Res.*/78:5.
21. Flamand, A., Wiktor, T.J., and Koprowski, H. (1980): *J. Gen. Virol.* 48:97–104.
22. Flamand, A., Wiktor, T.J., and Koprowski, H. (1980): *J. Gen. Virol.* 48:105–109.
23. Wiktor, T.J., Flamand, A., and Koprowski, H. (1980). *J. Virol. Methods* 1:33–46.
24. Wiktor, T.J. and Koprowski, H. (1978): *Proc. Nat. Acad. Sci.* 75:3938–3942.

Copyright 1981 by Elsevier North Holland, Inc.
David H.L. Bishop and Richard W. Compans, eds.
The Replication of Negative Strand Viruses

Antigenic Variations of Rabies Virus

Tadeusz J. Wiktor[a]

Previously, all strains of rabies virus, regardless of their geographic origin or the species of animal from which they were obtained, were considered to be antigenically closely related. Differences in "fixed strains" of virus were separated from field virus on the basis of incubation time or degree of virulence for experimental animals, however antibodies produced by the immunization of animals with any strain of rabies virus were found to be capable of neutralizing all strains of rabies. Only when antibodies produced by animals immunized with whole virions or viral components were assayed by the plaque reduction method were some minor differences detected in the antigenic composition of various rabies strains.[1] The so-called "rabies-related" viruses, Lagos bat, Mokola, and Duvenhage, isolated in Western or South Africa, were found to show some rabies virus characteristics; however, the number of these isolates is limited, and their epidemiological significance and relationship to rabies virus have not been clearly determined.

Nevertheless, antigenic differences among strains of rabies virus can be readily demonstrated through the use of hybridoma monoclonal antibodies. These antibodies are obtained by the fusion of mouse myeloma cells with spleen cells derived from Balb/c mice immunized with inactivated or live rabies vaccine.[2] The specificity of these antibodies has been determined by (a) radioimmunoassay (RIA) for binding to rabies-infected cells, (b) indirect immunofluorescence (FA) with live or fixed virus-infected cells, (c) virus neutralization, (d) immunolysis in the presence of complement, (e) immunoprecipitation, and (f) passive protection from rabies infection.

[a] The Wistar Institute, 36th Street at Spruce, Philadelphia, Pennsylvania

Antibodies directed against nucleocapsid (N) antigens stain intracytoplasmic inclusions of acetone-or formalin-fixed virus infected cells and show no virus-neutralizing or immunolytic activity. Antibodies directed against the glycoprotein (G) antigen stain the membrane of virus infected unfixed cells, lyse these cells in the presence of complement and neutralize the infectivity of the virus. Both types of antibodies bind specifically in RIA to corresponding purified antigens.[2]

The group of hybridomas that secrete antibodies against the M_1 and M_2-proteins is not yet well-defined. No hybridoma has yet been detected with activity directed against the L viral antigen.

Of a large number of hybridomas obtained from several fusion experiments, a panel of 21 hybridomas with an anti-N specificity and 25 with an anti-G specificity were used for the analysis of several laboratory strains of rabies and rabies-related viruses, as well as for the analysis of a number of field rabies isolates.

Antigenic differences were detected on both nucleocapsid and glycoprotein levels. The anti-N antibodies displayed different specificities for different viruses. A panel of three hybridomas was found that can be used for differential diagnosis between rabies and rabies-related viruses by the staining of antigens in impression smears of cells from infected animals or human beings.[3,4]

Similarly, the anti-G monoclonal antibody allows the differentiation of laboratory strains of rabies from each other. The analysis of a limited number of field virus isolates suggests the existence of a pattern of similar reactivity for strains of the same geographical origin.[5]

Treatment of a virus population with an excess of anti-G monoclonal antibodies resulted in the selection of virus clones that resisted neutralization by the antibodies used for their selection. The frequency of variants in a given population was approximately 1:10,000. Animals immunized with variant vaccine were not protected or were only partially protected when challenged with the parent virus or with another variant; however, they were fully protected from challenge with the variant virus used for immunization.[6]

Seven virus isolates from fatal cases of human rabies in the USA were analyzed with a panel of monoclonal antibodies and their pattern of reactivity compared to that of virus strains used for the preparation of antirabies vaccine. These strains can be separated into four groups of reactivity. Animals immunized with a standard rabies vaccine were poorly protected when challenged with field rabies strains, which share only a limited number of antigenic determinants with the virus used for vaccine production. However, these animals were fully protected against homologous challenge and against street viruses, which share several antigenic determinants.[6]

Nearly all antirabies vaccines throughout the world are prepared from a virus strain originally isolated by Pasteur in 1882. The antigenic potency of this vaccine is evaluated by the challenge of immunized mice with another virus derived from the Pasteur strain. This vaccine is used for the protection of men and animals against viruses which may, in different parts of the world, represent variants expressing antigens that do not react with virus strains involved in the vaccine evaluation procedure.

Failures in postexposure rabies treatment have been attributed to either the low potency of the vaccine or to delay in the initiation of treatment. Antigenic differences among strains of rabies have also been suspected as the cause of such failures; however, until now, there has been no way to test this hypothesis. With the availability of monoclonal antibodies, strains of virus can be easily identified and their patterns of reactivity compared with the reactivity of the virus used for vaccine production.

Our results clearly indicate that the selection of vaccine strains and the methods used for the evaluation of the potency of rabies vaccines need to be revised. In the future, vaccines should be tested by the challenge of animals with field virus isolates from those geographical areas in which the vaccine will be used.

References

1. Wiktor, T.J. and Clark, H.F. (1973): *Ann. Microbiol. (Inst. Pasteur)* 124A:283–287.
2. Wiktor, T.J. and Koprowski, H. (1978): *Proc. Nat. Acad. Sci.* 75:3938–3942.
3. Wiktor, T.J., Flamand, A., and Koprowski, H. (1980): *J. Virol. Meth.* 1:33–46.
4. Flamand, A., Wiktor, T.J., and Koprowski, H. (1980): *J. Gen. Virol.* 48:97–104.
5. Flamand, A., Wiktor, T.J., and Koprowski, H. (1980): *J. Gen. Virol.* 48:105–109.
6. Wiktor, T.J. and Koprowski, H. (1980): *J. Exp. Med.* 152:99–112.

David H.L. Bishop and Richard W. Compans, eds.
The Replication of Negative Strans Viruses

Antigenic Determinants of Rabies Virus as Demonstrated by Monoclonal Antibody

Lothar G. Schneider and Susanne Meyer[a,b]

Antirabies monoclonal antibodies were used in a study to determine the degree of antigenic variation occurring among laboratory strains and field isolates of rabies virus.

The antibodies were obtained from cultures of mouse spleen cells fused with murine myeloma cells and cultures *in vitro*.[1,2] Cloning of cultures was obtained by end-point dilutions. Antibodies directed against nucleocapsid (NC) and virus glycoprotein (G) were selected and tested with rabies vaccine and field viruses.

Monoclonal Nucleocapsid Antibodies

Twenty nucleocapsid antibody types (groups W1-W5) were reacted with 16 virus strains by the indirect immunofluorescence method and the strains placed into 10 NC-reaction categories by Dr. Wiktor (Table 1, upper half). Widely cross-reacting viruses are those of groups 1 to 5 representing common vaccine and field viruses. Those of groups 7, 8, and 9 are African viruses previously recognized as different serotypes of the Lyssavirus genus.[3]

With the same antibody panel plus an additional group of 5 monoclonal antibodies from Tübingen (Tü6) we have tested 29 field virus isolates. Of these strains, 9 fitted into 4 of the original 10 reaction groups: strains W238 (squirrel) and E798 (Pakistan dog) into group 1; W797 (cattle) into group 2;

[a] Federal Research Institute for Animal Virus Diseases, P.O. Box 1149, D7400 Tübingen, Federal Republic of Germany.
[b] Part of this study is in collaboration with Dr. Wiktor from the Wistar Institute, Philadelphia.

Table 1. Monoclonal Rabies Nucleocapsid Antibodies.

		W 1				W 2				W 3				W 4				W 5				Tü 6				
		502-2	104-4	111-2	103-7	206	209-1	229-1	590-2	515-3	111-14	389-2	239-10	222-9	377-7	120-2	390-1	102-27	237-3	364-11	422-5	W 239-17	Alo 280-4	Alo 280-5	W 187-5	M 37-3
1	ERA, PM, LEP, Kel.	+	+	+	+	+	+	+	+	+	+	+	+	+	+	+	+	+	+	+	●					
2	AF, SF	+	+	+	+	+	+	+	+	+	+	+	+	+	+	+	+	+	+	●	●					
3	S.Amer. CVS	+	+	+	+	+	+	+	+	+	+	+	+	+	●	+	+	+	+	+	●					
4	Minnesota No.3	+	+	+	+	+	+	+	+	+	+	+	+	+	+	●	+	+	+	+	●					
5	Upper Darby	+	+	+	+	+	+	+	+	+	+	+	+	+	+	+	+	●	+	+	●					
6	N.Amer. No.1, 7, 4	+	+	+	+	+	+	+	+	+	+	+	+	+	+	+	●	●	●	●	●					
7	Lagos Bat	+	+	+	●	●	●	●	●	+	+	+	+	+	●	●	●	●	●	+	+					
8	Duvenhage	+	+	+	●	●	●	●	●	+	+	+	●	●	●	●	●	●	●	+	+					
9	Mokola	+	+	+	●	●	●	●	●	●	●	●	+	+	●	●	●	+	+	+	+					
10	Flury HEP	+	●	●	+	+	+	+	+	+	●	●	●	●	+	+	+	+	●	+	●					
1a	Squirrel	+	+	+	+	+	+	+	+	+	+	+	+	+	+	+	●	+	+	+	●	+	+	+	+	+
2a	Europ. Fox, Rd, S 91	+	+	+	+	+	+	+	●	+	+	+	+	+	+	+	+	+	+	●	●	+	+	+	+	+
2b	Alo 280, W 239	+	+	+	+	+	+	+	+	+	+	+	+	+	+	+	+	●	+	●	●	+	+	+	+	+
2c	3 x fox, 3 x rodent	+	+	+	+	+	+	+	●	+	+	+	+	+	+	+	+	●	+	●	●	+	+	+	+	+
2d	CS 297 BF	+	+	+	+	+	+	+	+	+	+	+	+	+	●	+	+	●	+	●	●	+	+	+	+	+
5a	Apipe	+	+	+	+	+	+	+	+	+	+	+	+	+	●	+	+	●	+	+	●	+	+	+	+	+
8a	SVF 1	+	+	+	●	●	●	●	●	+	+	+	●	●	+	●	●	+	●	+	●	+	●	+	●	●
8b	STADE	+	+	+	●	●	●	●	●	+	+	+	●	●	+	●	●	●	●	●	●	+	●	+	●	●
11	3 x Botsw. jackal	+	+	+	+	●	+	+	●	+	+	+	+	+	+	+	●	●	+	●	●	+	+	●	●	+
11a	1 x Botsw. dog	+	+	+	+	+	+	+	●	+	+	+	+	+	●	+	●	●	+	●	●	+	+	+	●	+

Nucleocapsid reaction groups as obtained from reacting 45 virus strains with 25 types of antibody.
+ = pos. reaction; ● = no reaction.

W56 (rodent) into group 3: and strains CS46 (rodent), E971 (dog), E1034 (dog), SVM 2 (human), and Bobcat (USA) into group 5.

The 20 remaining virus isolates showed different NC-reaction patterns and had to be placed accordingly into another 10 subgroups (Table 1, lower half). Virus strains having similar reaction patterns with only slight antigenic variations were placed into groups 2a through 2d and represent 4 European fox isolates, (W184, W187, W229, W246), 1 cattle (W733), 1 polar fox (Alo 280), 4 rodents (W239, W876, K41, CS297), 1 hedgehog (Igel) and 1 human isolate from Chile (S91).

Closely related to each other and to serotype Duvenhage (group 8) were 2 viruses (groups 8a, 8b) isolated on different occasions from unidentified bats in northern Germany. One bat (SVF1) was picked up by a student from underneath a tree in Hamburg, the other was found dead in the town of Stade, near Hamburg. Originally it was speculated that the unidentified bats belonged to a European species while on their yearly migration. In the light of the close antigenic relatedness to an established exotic serotype, it seems likely that the rabid bats were imported by boat from the African continent.

The NC-reaction patterns of 3 jackal and 1 dog isolates from Botswana Africa are shown in groups 11 and 11a. It is interesting to note that the monoclonal NC antibodies of group Tü6 reacted with all viruses so far tested except those from the African continent (Table 1, last column to the right).

Monclonal Glycoprotein Antibodies

Mouse neutralization tests,[4] were carried out to show the neutralization pattern of 35 rabies virus strains which were reacted with 18 types of selected monoclonal G antibodies. The results are summarized in Table 2. Partial or complete neutralization of a virus strain is depicted by a "plus" sign. Negative reactions, i.e., the virus was not neutralized, are indicated by full black circles. Polyclonal rabbit antirabies sera shown in columns 19 (anti-HEP virion serum) and 20 (anti-ERA virion serum) served as controls.

Viruses were diluted to contain between 50 and 300 mouse LD_{50} per 0.03 ml. Monoclonal antibodies were used undiluted, their homologous titers ranged between 1:50 and 1:1000.

Rabies viruses having similar neutralization patterns are listed in the first two horizontal panels A and B of Table 2. The viruses represent 4 European fox viruses (W184 through W229), followed by 5 rodent isolates and one African virus (Nigerian horse) in panel B. The panel C of Table 2 shows 5 viruses (W156-BF297) of rodent origin exhibiting a wide range of cross-reactions, but no neutralization by antibody LEP132 (column 18).

Panel D shows the neutralization patterns of 7 rabies vaccine virus strains (Pitman Moore through HEP) used throughout the world. On a first glance, the strains Pitman Moore, LEP, and CVS24 on one side and CVS11, S91, ERA, and HEP on the other side, seem to form different reaction blocks.

Table 2. Monoclonal Rabies Glycoprotein Antibodies.

		E 510	E 529	E 559	E 543	C 231	P 75	P 116	P 133	P 4	P 20	P 35	P 33	P 37	P 13	W 131·2	M 37·2	M 37·4	LEP 132	S 53	S 72
A	W 184	+	+	+	+	●	●	●	●	+	+	+	+	+	+	+	●	●	+	+	+
	W 187	+	+	+	+	●	●	●	●	+	+	+	+	+	+	+	●	●	+	+	+
	W 246	+	+	+	+	+	●	●	●	+	+	+	+	+	+	+	●	●	+	+	+
	W 229	●	●	+	+	●	●	●	●	+	●	●	+	+	+	+	●	●	+	+	+
B	K 41	●	●	+	+	●	●	●	●	●	+	+	+	+	+	+	●	●	+	+	+
	W 239	●	+	+	+	●	●	●	●	●	+	+	+	+	+	+	+	●	+	+	+
	W 876	●	+	+	+	●	●	●	●	+	+	+	+	+	+	+	●	●	●	+	+
	W 238	●	●	+	+	●	●	●	●	+	+	+	+	+	+	●	+	●	+	+	+
	W 218	+	+	+	+	●	●	●	●	+	+	+	+	+	+	+	+	●	+	+	+
	Nig. Horse	●	+	+	+	●	●	●	●	+	●	+	+	+	+	+	+	●	+	+	+
C	W 156	●	●	+	+	+	+	+	+	+	+	+	+	+	+	+	+	+	●	+	+
	MIC 37	●	+	+	+	●	+	+	+	+	+	+	+	+	+	+	+	+	●	+	+
	W 56	+	●	+	+	+	+	+	+	+	+	+	+	●	+	+	+	+	●	+	+
	W 131	●	●	+	+	+	+	+	+	+	+	+	+	●	+	+	+	+	●	+	+
	BF 297	●	+	+	+	●	+	●	+	+	+	+	+	●	+	+	+	●	●	+	+
D	Pitm. Moore	●	●	+	+	●	+	+	+	+	+	+	+	+	+	+	+	●	+	+	+
	LEP	●	●	+	+	●	●	+	+	+	+	●	+	+	+	+	+	●	+	+	+
	CVS 24	●	+	+	+	●	+	+	+	+	+	+	+	+	+	+	+	●	●	+	+
	CVS 11	●	●	+	+	+	●	●	●	+	+	+	+	●	+	+	●	●	●	+	+
	S 91	●	+	+	+	+	●	●	●	+	●	+	+	+	+	+	●	●	+	+	+
	ERA	+	+	+	+	+	●	●	●	+	●	+	+	+	●	+	+	●	●	+	+
	HEP	●	●	+	+	+	●	●	●	+	+	+	+	+	+	+	+	●	+	+	+
E	SVM 2	+	+	+	+	+	●	●	●	+	+	+	+	+	+	+	+	●	+	+	+
	SVM 1	●	●	+	+	●	●	●	●	+	●	+	+	●	●	+	●	●	+	+	+
F	KELEV	●	+	●	+	●	●	●	●	●	+	+	+	●	+	+	●	●	+	+	+
	Bobcat	●	●	●	+	●	●	●	●	●	●	+	+	+	+	●	●	●	+	+	+
	CS 46	●	●	●	+	●	●	●	●	●	●	●	+	+	●	●	●	●	●	+	+
	Apipe	●	●	+	●	●	●	●	●	●	●	●	+	+	●	●	●	●	+	+	+
	ALO 280	●	●	+	●	●	●	●	●	●	●	+	●	●	●	●	+	●	●	+	+
	SVF 1	●	●	+	●	●	●	●	●	●	●	●	●	●	●	●	●	●	●	+	+
	STADE	●	+	+	●	●	●	●	●	●	●	●	●	●	●	●	●	●	●	●	+
	DUV	●	●	+	●	●	●	●	●	●	●	●	●	●	●	●	●	●	●	●	+
G	Botsw. 237	●	●	●	●	●	●	●	●	●	●	●	●	●	●	●	●	●	●	+	+
	Mokola	●	●	●	●	●	●	●	●	●	●	●	●	●	●	●	●	●	●	●	●
	LBV	●	●	●	●	●	●	●	●	●	●	●	●	●	●	●	●	●	●	●	+
		1	2	3	4	5	6	7	8	9	10	11	12	13	14	15	16	17	18	19	20

Neutralization patterns of 35 virus strains with 18 types of virus neutralizing antibodies.
+ = virus neutralized; ● = virus not neutralized.

Quite obvious, however, are the significant differences observed between the virus strains CVS24 and CVS11. Both viruses are thought to be derived from the original Pasteur virus strain in Paris having only different passage histories. According to their neutraliziation patterns both CVS strains tested seem to be independent viruses having different antigenic determinants. Future work should inquire about the potentials of antigenic shifts as a consequence of laboratory passage.

Panel E contains two human virus isolates from Turkey. Both viruses, though originating from the same geographical area, show significant differences of their antigenic determinants. A detailed knowledge of the virus strain histories, especially of the biting animal species involved, would certainly help to analyze the existing situation.

The viruses listed in panel F are characterized by a decreasing degree of reactivity with the set of virus neutralizing monoclonal antibodies. The vaccine strain Kelev and the bobcat isolate take an intermediate position since they are still neutralized by 30–40% of the antibodies tested. The other virus isolates, namely from European rodent (CS46), vampire bat (Apipe), polar fox (Alo 280), bats (SVF 1, STADE), and man (Duvenhage from South Africa), reacted with only a fraction (between 1 and 4) of the monoclonal antibodies tested so far. Two of the viruses (STADE and Duvenhage) were not neutralized by the polyclonal anti HEP virion rabbit serum (S53).

Panel G contains 3 viruses which did not react with any of the given monoclonal antibodies: Mokola and Lagos bat viruses are two African viruses previously recognized as independent serotypes of the *lyssavirus* genus. Botswana 237 represents a canine isolate also from Africa. Whereas Mokola was not neutralized by the two anti-virion sera, Lagos bat virus was neutralized by the anti-ERA virion serum and Botswana 237 by both of them (column 19, 20, Table 2).

The neutralization results obtained so far indicate that the several types of monoclonal antibodies induced by vaccine strains showed a good proportion of cross-reactivity with homologous viruses, with old and new world field strains of various origin, and with old world rodent isolates.

The cross-reactivity of antibodies decreased when tested with polar fox virus, a vampire bat, three further bat isolates, and 4 African rabies viruses.

These results indicate that rabies virus is composed of minor and major antigenic determinants or antibody binding sites. The minor determinants apparently are responsible for virus strain differences. In some instances these determinants may even indicate the origin of the virus. For example antibody M37-4 (Table 2, column 17) which was induced by a rodent virus strain, neutralized only rodent-origin viruses.

The major antigenic determinants are found with larger numbers of viruses and may therefore be responsible for determining the serotype. The major determinants apparently occur in most vaccine strains and their respective antibodies cover the majority of existing viruses. For instance, antibodies

E559 and E543 (Table 2, column 3 and 4) together neutralized all viruses except those from panel G.

Protection Study

The poor cross-reactivity observed with some of the viruses of panels F and G caused us to evaluate in a preliminary study the actual protection as effected by the Pitman Moore (PV11) strain, one of the vaccine viruses widely used for the rabies prophylaxis in man and being contained in the antirabies human diploid cell culture vaccine.

Figure 1. Protective activity of Pitman Moore (PM) suckling mouse brain vaccine. Percent of mortality of non-vaccinated (●—●) versus that of vaccinated mice (○--○) after challenge with homologous (PM) and heterologous viruses (rodent viruses W239 and CS46; polar fox virus ALo 280; bat viruses SVF 1 and STADE).

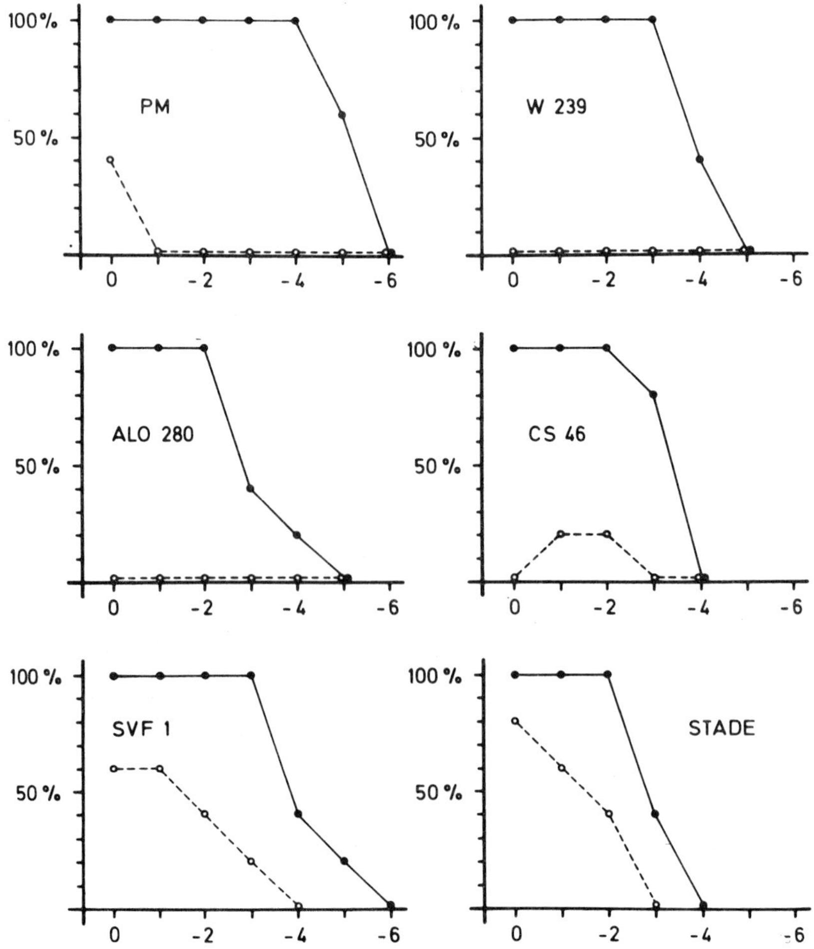

Pitman Moore virus was used to prepare a beta-propiolactone inactivated suckling-mouse brain vaccine.[5] Groups of 30 mice per virus strain to be tested were vaccinated intraperitoneally with 0.5 ml of a 1:30 dilution in PBS of the above vaccine on days 0 and 7. On day 14, the viruses under question were titrated intracerebrally in the vaccinated mice and in non-vaccinated controls. The results are shown in Figure 1.

Complete protection of vaccinated mice was obtained against the challenge infection with the Pitman Moore virus (PM), the rodent isolate W239 and the polar fox virus ALO 280. The rodent strain CS 46 showed some break-through at virus dilutions -1 and -2. Vaccinated mice exposed to SVF1 and STADE bat viruses practically showed no protection.

This indicates that in fact there exist in nature certain variants of the rabies virus against which conventional vaccines do not fully protect.

This study further indicates that a panel of virus neutralizing monoclonal antibodies representing major antigenic determinants may selectively identify such variants of field virus strains, thus being extremely helpful in epidemiological investigations.

Summary

Monoclonal antibodies directed either against the nucleocapsid or the glycoprotein of rabies virus were reacted by indirect immunofluorescence and mouse neutralization tests with vaccine strains and rabies field isolates of various origin. Major antigenic determinants allow the grouping of viruses and minor ones the differentiation of virus strains. A protection study showed the existence in nature of certain virus variants against which conventional vaccines do not fully protect.

References

1. Wiktor, T.J. and Koprowski, H. (1978). *Proc. Nat. Acad. Sci. U.S.A.* 75:3938–3942.
2. Schneider, L.G. (1980): *Proceedings of Cell Culture Rabies Vaccines and Their Protective Effect in Man.* Essen, Germany, March 1980, in press.
3. Schneider, L.G., Dietzschold, B., Dierks, R.E., Matthaeus, W., Enzmann, P., and Strohmaier, K. (1973): *J. Virol.* 11:748–755.
4. Atanasiu, P. (1973): In: *Laboratory Techniques in Rabies, 3rd Ed.* Kaplan, M.M. and Koprowski, H. (Eds.), World Health Organization, Geneva, pp. 314–318.
5. Fuenzalida, E. (1973): In: *Laboratory Techniques in Rabies,* 3rd Ed. Kaplan, M.M. and Kpprowski, H. (Eds.), World Health Organization, Geneva, pp. 216–220.

David H.L. Bishop and Richard W. Compans, eds.
The Replication of Negative Strand Viruses

Host Cell Variation in Response to Vesicular Stomatitis Virus Inhibition of RNA Synthesis

Betty H. Robertson and Robert R. Wagner[a]

Introduction

Vesicular stomatitis virus (VSV) is cytotoxic for a wide variety of vertebrate and invertebrate cell lines. The cytotoxic effects of VSV include inhibition of host cell RNA, DNA, and protein synthesis;[1–8] in addition, a factor responsible for cell killing has also been described, but is not separable from the inhibition of protein synthesis.[9] The mechanism involved in viral shut-off of host cellular RNA and/or protein synthesis is not known, but appears to vary depending upon the cell line infected,[6,10] suggesting that host factors may be involved in the sensitivity to viral inhibition. Unfortunately, the majority of studies thus far have compared cell lines which are relatively permissive for VSV growth, thereby not offering a clear delineation between host susceptibility and viral inhibition.

Lymphoid cells, in contrast to other somatic cells, appear to offer a system which is restrictive not only for VSV growth but also for many other DNA and RNA viruses.[11–16] Nowakowski et al.[16] have presented evidence that lymphoid cells (especially B-cells) which had not undergone antigenic stimulation are restrictive for VSV replication. Previous work by Weck and Wagner[10] demonstrated that MPC-11 cells, a mouse myeloma which secretes immunoglobulins, are extremely sensitive to inhibition of RNA synthesis by VSV compared to mouse L cells or BHK cells infected with VSV.

To explore further the factors which determine susceptibility to infection with VSV, we selected two continuous lines of lymphoid cells which repre-

[a] Department of Microbiology, University of Virginia School of Medicine, Charlottesville, Virginia.

sent the two extremes in B lymphoid cell development as understood at this time. MPC-11 cells were chosen as representative of an end stage cell whose function is to produce immunoglobulins. At the other end of the spectrum, we chose to use 18–81 cells which have been characterized as a pre-B-cell induced by Abelson virus transformation of fetal liver cells; 18–81 cells contain intracellular μ chains[17] and represent a stage in lymphoid cell differentiation prior to acquiring the ability to respond to antigenic stimulation. These cell lines were compared with mouse L cells and human HeLa S3 cells with respect to their ability to support VSV replication, the kinetics of VSV shut-off of host cell RNA synthesis, and the VSV genome segment needed to inhibit cellular RNA synthesis.

Quantitation of Infectious Particles Released from Lymphoid and Non-Lymphoid Cell Lines

The growth curves of VSV in HeLa S3, mouse L, MPC-11, and Abelson virus transformed 18–81 cells are shown in Figure 1. These data indicate that the degree of permissiveness of these cells as determined by the number of infectious particles released per cell, assumes the following rank order: HeLa > L > MPC-11 > 18–81. Although the numbers are not striking, the data do indicate that the two lymphoid cell lines are less permissive for VSV growth than are HeLa S3 or L cells. The pre-B-cell line released 2–3 viral particles per cell, compared to 28–34 for mouse L cells and HeLa S3 cells, respectively; the MPC-11 myeloma cells were intermediate, releasing 10–11 infectious particles per cell. The other interesting feature observed in the growth curves of VSV in these cell lines is the apparent difference in latent period and peak time of release of infectious particles. For HeLa S3, mouse L cells, and the 18–81 cells, the majority of infectious virus released appeared to occur between 4 and 8 hr postinfection with maximal values not apparent until 12 hr postinfection. On the other hand, MPC-11 cells had already begun their release of infectious particles by 4 hr postinfection and maximal values were apparent by 6 hr postinfection.

VSV Shut Off of Host Cell RNA Synthesis

The kinetics of VSV shut-off of host cell RNA synthesis in these four cell lines is depicted in Figure 2. These data confirm observations by other investigators that inhibition of RNA synthesis varies depending upon the cell line. Inhibition of host RNA synthesis in the two lymphoid cell lines was rapid and nearly complete by 3 hr postinfection, while L cell and HeLa S3 cell RNA inhibition was slower and still had not reached a maximum by 6 hr postinfection. The data indicate that the rate of inhibition of host cell RNA

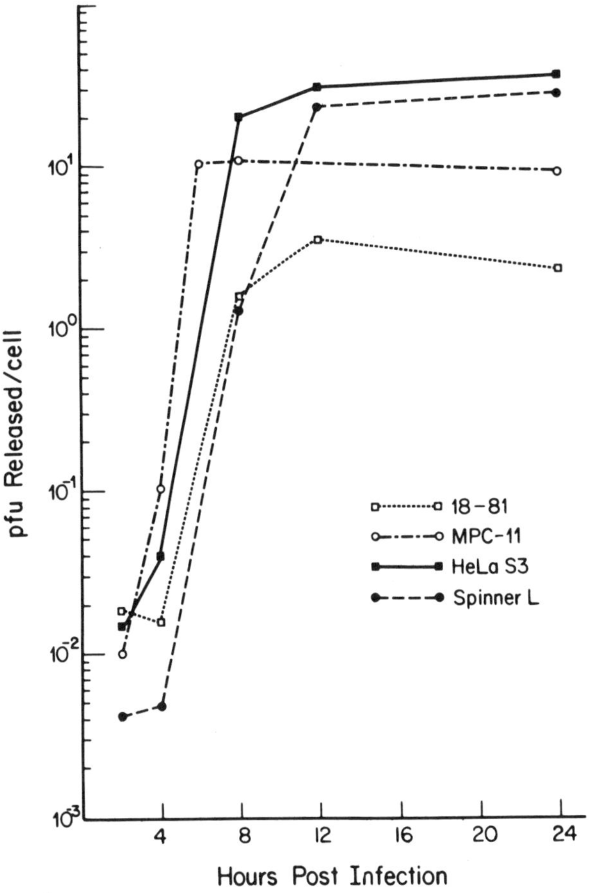

Figure 1. VSV growth curve in HeLa S3, spinner L, MPC-11, and 18–81 cells. Cells in the appropriate media containing no serum (1×10^7/ml) were infected at a moi of 5 with plaque-purified VSV (San Juan strain of Indiana serotype). After one hour absorption at room temperature, cells were pelleted and resuspended at 5×10^5 cells/ml in appropriate media containing serum. Aliquots were removed at the indicated times and after pelleting cells and debris, released virus was assayed by plaques on mouse L cell monolayers. HeLa S3 (obtained from D.F. Summers, University of Utah) and spinner L cells (obtained from J. Brown of our department) were grown in Joklik's minimal essential medium (Flow Labs, McClean, Virginia) containing 2 mM added glutamine and 5% calf serum. MPC-11 and 18–81 cells (obtained from W.M. Kuehl of our department) were grown in Dulbecco's modified Eagle medium (Gibco, Grand Island, New York) with 1% nonessential amino acids and 5% horse serum and RPMI 1640 (Gibco, Grand Island, New York) containing 0.002 M β-mercaptoethanol and 5% fetal calf serum, respectively.

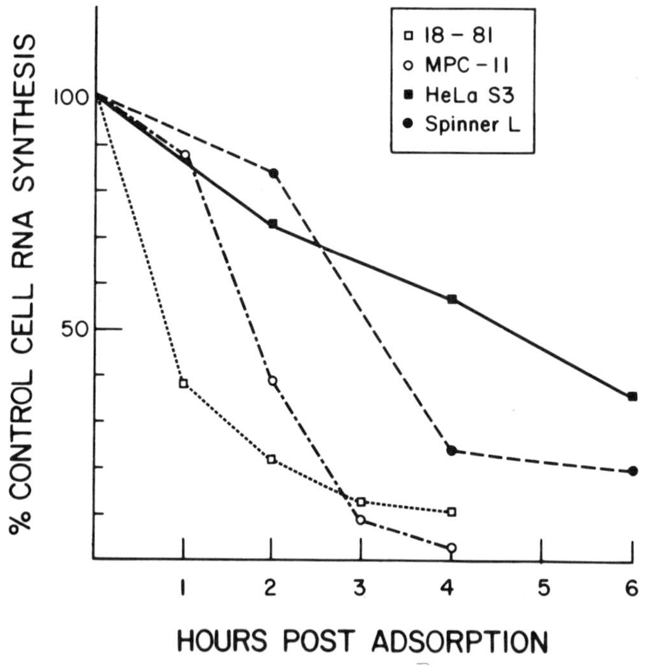

Figure 2. Kinetics of RNA shut-off in HeLa S3, spinner L, MPC-11, and 18–81 cells. Cells were pelleted by centrifugation at 2000 rpm for 10 min and resuspended in the appropriate media containing no serum at a concentration of 1×10^7 cells/ml. Virus was added to provide a final multiplicity of 10 and allowed to absorb for 30 min at room temperature with occasional mixing. The cells were then diluted to $\approx 5 \times 10^5$ cells/ml with media containing no serum (HeLa S3 and MPC-11) or media containing serum (L cells and 18–81), as the latter cells did not remain viable during the time course of the experiment unless serum was present. After dilution, 1 ml of cells were pipetted into triplicate 35 mm tissue culture plates and placed at 37°C in CO_2 incubator. At the appropriate labeling times, 5 μCi of ^{3} H-uridine (New England Nuclear, 1 μCi-ml) was added to each plate and the cells were incubated for 30 min at 37°C. The labeling period was terminated by the addition of 1/10 volume of 10% SDS, followed by an equal volume of 10% TCA. TCA precipitated samples were placed on ice of 30 min, followed by filtration through Whatman GFA glass fiber discs, rinsed thoroughly with 5% TCA and ethanol and dried. Dried filters were counted in Redi-Solve EP (Beckman, Fullerton, California) in a Beckman LS230 liquid scintillation counter with an efficiency of 40% for tritium.

synthesis progresses in the following rank order: 18–81 > MPC-11 > L cells $\geqslant$ HeLa. These data also suggest that the rate of RNA synthesis inhibition is inversely related to the replication and final yield of infectious virus particles in these cell lines.

Viral Genome Segment Responsible for Host Cell RNA Synthesis Inhibition

Since VSV contains a linear genome which is transcribed sequentially, UV inactivation of the viral genome has been used to map the gene sequence[19,20] and to help determine which viral gene products are responsible for various effects on the host cell, such as the inhibition of protein synthesis[18] and RNA synthesis.[21,22] Previous work using UV-inactivated virus to characterize the genome segment responsible for inhibition of RNA synthesis in MPC-11 cells indicated that leader RNA from the 3′ end of the B virion genome might be involved in RNA shut-off,[21] while N and NS cistrons were considered to be implicated in the shut-off of RNA synthesis in mouse L cells.[22] As these results suggest that the measured viral gene product needed for inhibition also varies depending upon the host cell, the four cell lines were compared with respect to the portion of the viral genome needed for inhibition of RNA synthesis.

The results depicted in Figure 3 demonstrate the recovery of uninfected control cell levels of RNA synthesis in each cell line with increasing amounts of UV-irradiation to the virus. Exponentially growing HeLa S3 and MPC-11 cells (5×10^5 cells/ml prior to infection) completely returned to control levels of RNA synthesis at the lower UV dosage (4,000 ergs/mm^2) given to the virus. (For clarity, the MPC-11 line has been omitted). 18–81 and L cells recovered to control levels with intermediate UV dosage to VSV virus, while recovery of stationary phase MPC-11 cells ($\approx 1 \times 10^6$ cells/ml prior to infection) was more resistant with only 65% of control levels of RNA synthesis being attained at 40,000 ergs/mm^2. As the absolute amount of viral shut-off at 4 hr postinfection varies with the cell line, an alternative graphic presentation in terms of the percent of maximal inhibition is shown in Figure 4. Line A depicts the rapid recovery of HeLa S3 and MPC-11 cells (growing to 5×10^5 cells/ml prior to infection) infected with VSV virus exposed to even 4,000 ergs/mm^2. Lines B and C depict the intermediate recovery of 18–81 and L cells, respectively. On the other hand, when MPC-11 cells (line D) approaching stationary phase ($\approx 1 \times 10^6$ cells-ml) are used to measure the effects of UV-irradiated virus on the shut-off of host cell RNA synthesis, a very different slope results compared to that obtained with exponentially growing MPC-11 cells (line A).

Discussion

The data from this comparative study of four cell lines, two B lymphoid and two non-lymphoid, on the growth of VSV and its effect on host cell RNA synthesis confirm that lymphoid cells are relatively non-permissive for VSV growth, yet are sensitive to the cytotoxic effects of the virus. The data

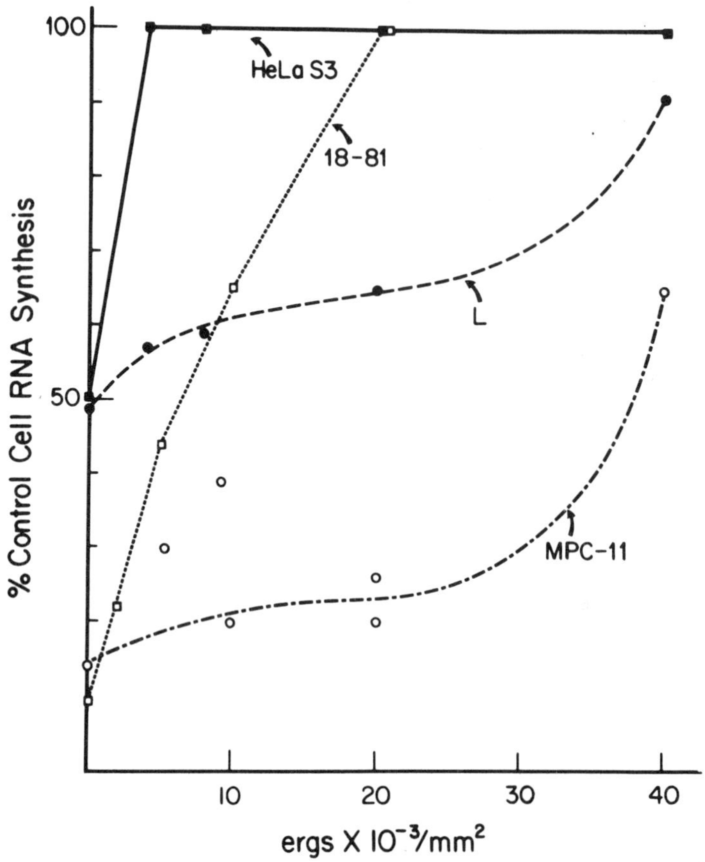

Figure 3. Recovery of host cell RNA synthesis with increasing UV-dosage to VSV. Stock virus was diluted to ≈ 5 × 10^8 pfu/ml with media containing no serum and aliquoted in 0.5 ml volumes into 35 mm tissue culture plates. Irradiation of the virus was performed at a distance of 10 cm from the UV light source at a wave length of 254 nm at a dose of 85 ergs/mm^2 per second. Cells growing at ≈ 5 × 10^5/ml for HeLa S3 (———), 18–81 (---------), L (– – – –), and 1 × 10^6/ml for MPC-11 (— • —) were either mock-infected, infected at a moi of 10 with unirradiated virus, or infected with 10 pfu equivalents with virus subjected to varying amounts of radiation and allowed to absorb for 30 min at room temperature with occasional mixing. After absorption, cells were diluted with the appropriate media containing no serum (HeLa S3 and MPC-11 cells) or with serum (spinner L and 18–81 cells) at a final concentration of 5 × 10^5 cells-ml. After dilution, 1 ml of cells was pipetted in triplicate into 35 mm tissue culture plates and placed at 37°C in a CO_2 incubator. At four hr post-absorption, cells were labeled and processed as described in Figure 2.

presented here demonstrate that, with the lymphoid cell lines we have used, VSV is able to infect and rapidly inhibit host cell RNA synthesis. Unpublished observations by our laboratory indicate that host cell protein synthesis (J.R. Thomas) and DNA synthesis (J. McGowan) is affected in these two

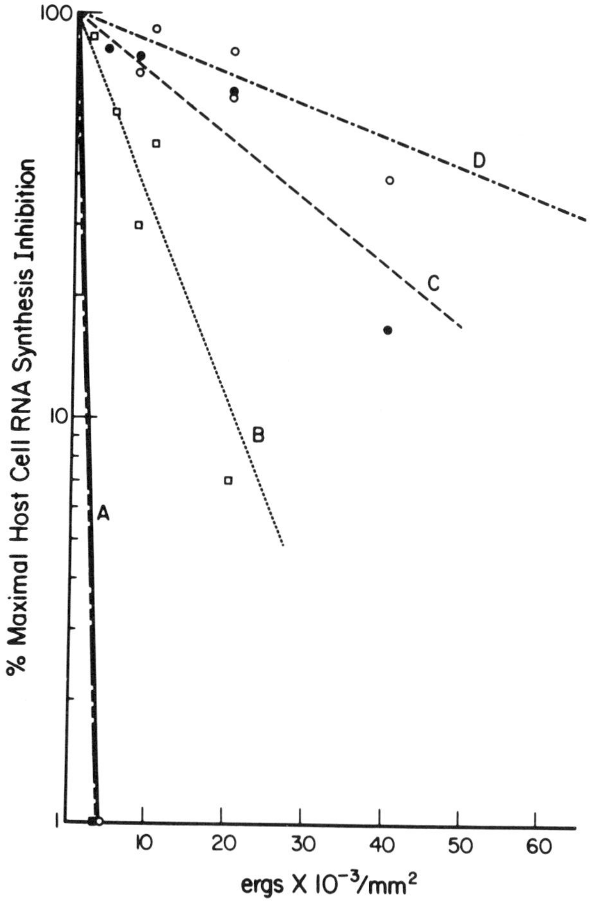

Figure 4. Recovery of host cell RNA synthesis (plotted as percent maximal inhibition) with increasing UV-dosage to VSV. Line A = MPC-11 (——– –) and HeLa S3 (———) growing at 5×10^5 cells/ml prior to infection. B = 81–81. C = spinner L Cells. D = MPC-11 (— – —) growing at 1×10^6 cells/ml prior to infection.

cell lines, and that VSV infection appears to result in cell death of these lymphoid cells. Our observations coupled with previously reported data on VSV infection of lymphoid cells, both *in vitro* cell lines[16,23,24] and cells isolated from animal sources,[11–14] suggest that B-cells undergo some event during differentiation to a cell which is capable of antigenic stimulation which renders them refractive to the cytotoxic components of and infection with VSV. This refractory state is then lost upon antigenic activation and maturation to an immunoglobulin secreting plasma cell.

The delineation between lymphoid and non-lymphoid cells is not observed when the viral gene needed to inhibit host cell RNA synthesis is measured by UV-inactivated VS virus. As variable results were observed not only with all four cell lines, but also within the same cell line, we conclude that the use of UV-inactivated virus in calculating genome equivalents needed for inhibition of host cell functions is subject to multiple parameters not easily controlled *in vivo* as demonstrated by the widely varying slopes of the data presented in Figure 4. Within the MPC-11 cell line, the wide variation between stationary phase cultures, which corresponds to results previously obtained in this laboratory,[21] and exponentially growing cultures most vividly demonstrates this point. Cell line dependent variation in measurement of genome size equivalents was suggested by the results of Wu and Lucas-Lenard using L929 monolayer cells[22] compared to the results using MPC-11 cells in this laboratory.[21] It is obvious that conclusions regarding the genome target size from such experiments are qualified by parameters such as the cell line used and the state of cell growth. Such conclusions are further complicated by the fact that the site affected by UV-irradiation in the virus and the ultimate *in vivo* target of the host cell are widely separated events dependent upon both viral and cellular functions. These studies emphasize that, due to the complex nature of the viral-host interaction within the cell, characterization of the viral gene product responsible for RNA synthesis inhibition and the host cell target affected will necessitate an *in vitro* system for definitive elucidation.

ACKNOWLEDGMENTS

This work was supported by Grants MV-9D from the American Cancer Society, AI-111112 from the National Institutes of Health, and PCM-77-00494 from the National Science Foundation.

We thank R.M. Snyder and O.D. Mason for excellent technical assistance.

References

1. Huang, A.S. and Wagner, R.R. (1965): *Proc. Natl. Acad. Sci. U.S.A.* 54:1579–1584.
2. Wagner, R.R. and Huang, A.S. (1966): *Virology* 28:1–10.
3. Yamazaki, S. and Wagner, R.R. (1970): *J. Virol.* 6:421–429.
4. Wertz, G.W. and Youngner, J.S. (1970): *J. Virol.* 6:476–484.
5. Yaoi, Y., Mitsui, H., and Amano, M. (1970): *J. Gen. Virol.* 8:165–172.
6. Baxt, B. and Bablanian, R. (1976): *Virology* 72:383–392.
7. McAllister, P.E. and Wagner, R.R. (1976): *J. Virol.* 18:550–558.
8. Nuss, D.L. and Koch, G. (1976): *J. Virol.I* 19:572–578.
9. Marcus, P.I., Sekellick, M.J., Johnson, L.D., and Lazzarini, R.A. (1977): *Virology* 82:242–246.
10. Weck, P.K. and Wagner, R.R. (1978): *J. Virol.* 25:770–780.
11. Edelman, R. and Wheelock, E.F. (1968): *J. Virol.* 2:440–448.
12. Miller, G. and Enders, J.F. (1968): *J. Virol.* 2:787–792.
13. Wheelock, E.F. and Edelman, R. (1969): *J. Immunol.* 103:429–436.
14. Jimmenez, L., Bloom, B.R., Blume, M.R., and Oettgen, H.F. (1971): *J. Exp. Med.* 133:740–751.
15. Moehring, J.T. and Moehring, J.M. (1972): *Infect. Immunity* 6:493–500.

16. Nowakowski, M., Felman, J.D., Kano, S., and Bloom, B.R. (1973): *J. Virol.* 3:8–16.
17. Siden, E.J., Baltimore, D., Clark, D., and Rosenberg, N.E. (1979): *Cell* 16:389–396.
18. Marvaldi, J., Sekellick, M.J., Marcus, P.I., and Lucas-Lenard, J. (1978): *Virology* 84:127–133.
19. Ball, L.A. and White, C.N. (1976): *Proc. Natl. Acad. Sci. U.S.A.* 73:442–446.
20. Abraham, G. and Banerjee, A.K. (1976): *Proc. Natl. Acad. Sci. U.S.A.* 73:1504–1508.
21. Weck, P.K. and Wagner, R.R. (1979): *J. Virol.* 30:746–753.
22. Wu, F.S. and Lucas-Lenard, J.M. (1980): *Biochemistry* 19:804–810.
23. Goodman-Snitkoff, G.W. and McSharry, J.J. (1980): *J. Virol.* 35:757–765.
24. Creager, R.S. and Youngner, J.S. (1980): *Abst. Am. Soc. Micro.*, p. 260.

Copyright 1981 by Elsevier North Holland, Inc.
David H.L. Bishop and Richard W. Compans, eds.
The Replication of Negative Strand Viruses

The Effect of the Host Cell and Heterologous Viruses on VSV Production

Sue A. Moyer, Sandra M. Horikami, and Richard W. Moyer[a]

Introduction

Vesicular stomatitis virus (VSV) will grow in an unusually wide range of host cells. The isolation of VSV mutants with a more limited host range by Szilagyi and Pringle[1,2] and Simpson and Obijeski,[3,4] however, provides strong evidence that certain cellular factors are required in order for a productive infection to occur. There are, moreover, some cell lines which are naturally restrictive for wild type VSV. Nowakowski et al.[5] have described a human lymphoblastoid cell line (Raji) which seems to block VSV development at the level of replication. Another interesting system is that described by Thacore and Youngner.[6] Their data suggest that rabbit cornea (RC-60) cells also restrict VSV growth at the level of replication, however the normally abortive infection can be converted into a productive VSV infection by co-infection with vaccinia virus. The complementation of VSV by poxviruses in the doubly infected cells requires only the expression of the prereplicative genes of the poxvirus genome.[6] In a detailed study of the nonpermissive VSV infection of RC-60 cells we have shown[7] that genome replication specifically is suppressed by 80% and the limited genome RNA that is synthesized accumulates as nucleocapsids which fail to mature. These results suggest that there is a block in the maturation process as well as an inhibition of VSV replication in these cells. Poxvirus (and by analogy the host factor(s) in permissive cells) presumably overcomes one or both of the defects. We present here another approach to the study of the role of the

[a] Vanderbilt University School of Medicine, Nashville, Tennessee.

host cell in VSV growth, which is based on studies of the mutagen induced host range (*hr*) mutants of VSV. A collection of these mutants were divided into two groups based on their ability to be rescued under non-permissive conditions by co-infection with a poxvirus. The abortive infections caused by the mutants in the two groups were then characterized biochemically.

Results and Discussion

We have examined 6 VSV *hr*-mutants that were originally isolated by Simpson and Obijeski.[3,4] Each of these six *hr*-mutants grows on the permissive BHK cells at 35°, but is restricted several orders of magnitude on the non-permissive HEp-2 cells at 35°. Each of these *hr*-mutants is also temperature-sensitive (*ts*) on both the permissive and the non-permissive cell lines. We first determined whether any of the mutants can be rescued under various conditions from the non-permissive infection by co-infection with a poxvirus, rabbit poxvirus (RPV), in a manner analogous to the rescue of wild type VSV by RPV in RC-60 cells.

The data in Table 1 show that at the lower temperature RPV can complement the host range defect of mutants *hr*1, *hr*2, *hr*6, and *hr*8. The infections

Table 1. Complementation by Rabbit Poxvirus (RPV) of VSV Host Range (*hr*) Mutants in Non-permissive HEp2 Cells.

Mutant	**Virus yield[a] on HEp2 at 35° (%)**	**Virus yield on BHK (%) 35°**	**39.5°**
wild type VSV	100		
*hr*1	0.61	100	0.78
*hr*1 + RPV	71.5		0.80
*hr*2	1.0		
*hr*2 + RPV	85.4		
*hr*5	3.7		
*hr*5 + RPV	5.5		
*hr*6	0.7		
*hr*6 + RPV	71.3		
*hr*7 + RPV	1.3		
hr + RPV	12.6		
*hr*8	0.6	100	1.7
*hr*8 + RPV	41.9		1.1

[a]HEp2 cells at 35° and 39.5° are non-permissive for the growth of the *hr*-mutants, while BHK cells are permissive at 35°, but non-permissive at 39.5°. The virus yield was determined by quantitating the amount of ^{3}H-uridine incorporated into purified virus under the growth conditions indicated.

caused by the mutants *hr*5 and *hr*7, however, show little or no effect of added RPV. Since all these mutants are also thermolabile, they fail to grow on the normally permissive BHK cells at the non-permissive temperature (39.5°). There is, however, no effect of added RPV on *hr*1 or *hr*8 infected BHK cells at 39.5° (Table 1), even though at 35° RPV can efficiently correct the effects of the host range mutation in both of these mutants. We conclude from these results that poxvirus helper factor(s) can function effectively not only with wild type VSV on RC-60 cells, but also with some, but not all, of the *hr*-mutants on another cell line. It would appear that only some of the host cell functions can be provided in some way by poxvirus. Nevertheless, the ability or inability of the *hr*-mutants of VSV to be rescued by poxvirus allows these mutants to be functionally subdivided into two classes.

Each of the mutants that we have studied has both *hr* and *ts* lesions, however, the relationship between these phenotypic defects is not clear. We have isolated a revertant of *hr*8 (*hr*8R, $hrts^+$) that is no longer temperature-sensitive. The revertant *hr*8R still retains the host range defect and is fully complemented by RPV in the non-permissive HEp2 cells. These observations are consistent with the idea that at least some of the *hr* and *ts*-defects in these viruses might represent independent mutations.

We have begun attempts to relate the ability of a VSV *hr*-mutant to be rescued by RPV to a biochemical phenotype by characterizing the non-permissive infections of the different mutants. All of the following experiments were done on HEp2 cells at 35° to minimize the effects of the host independent *ts* mutations. In the original characterization, Simpson and his coworkers[3,4,8] classified most of the *hr*-mutants on non-permissive HEp2 cells as producing an RNA^- phenotype, since little or no viral RNA could be detected. In order to do more detailed studies of virus-specific RNA synthesis we have analyzed the ^{3}H-uridine-labeled products of non-permissive infections on SDS-sucrose gradients. Mutants *hr*1, *hr*5, *hr*7, and *hr*8 (i.e., mutants representing classes that can and cannot be rescued by RPV) all totally lack the ability to replicate their genome RNA (data not shown). Furthermore, in all these mutants some synthesis of the viral 12-18S mRNAs does occur. The overall level is reduced, however, to 60–80% of the level of primary transcription as determined by measuring mRNA synthesis in wild type VSV infected HEp2 cells in the presence of cycloheximide (Table 2). The level of transcription in the mutant infected non-permissive cells is about half of the level of primary transcription observed with the corresponding mutant infection of permissive BHK cells in the presence of cycloheximide. From these data, the non-permissive infections of all the *hr*-mutants, whether or not they are able to be rescued by poxvirus, are characterized by a limited amount of primary transcription, and an inability to replicate their genome RNA.

We have also attempted to measure the level of VSV-protein synthesis in HEp2 cells infected with the different *hr*-mutants, in order to estimate the

Table 2. Viral RNA Synthesis During the Non-permissive Infections of HEp2 Cells with VSV *hr*-Mutants.

Mutant	12-18S VSV mRNA[a] cpm	%
wild type VSV + cycloheximide	10,713	100
*hr*1	7,719	72.1
*hr*5	8,360	78.0
*hr*7	8,572	80.0
*hr*8	7,238	67.6

[a]The HEp2 cells (2×10^7) at 35° were infected with either wild type VSV in the presence of cycloheximide (100 μg/ml) or the *hr*-mutants, and the viral RNAs were labeled with ^{3}H-uridine (30 μCi/ml) from 1.5 to 5.0 hr postinfection in the presence of actinomycin D (2 μg/ml). Cytoplasmic cell extracts were prepared by detergent lysis as described previously,[9] and the labeled RNA species were separated by centrifugation on 15-30% (w/v) sucrose-SDS gradients[9] in the SW41 rotor at 33,000 rpm for 17 hr at 25°. The total label in the 12-18S VSV mRNAs in each infection was determined and calculated as a percentage of the counts in the wild type VSV infection.

level of functional viral mRNA. We have found that host protein synthesis is not inhibited in HEp2 cells infected with either wild type VSV or the *hr*-mutants, which has made an accurate determination of viral-specific protein synthesis difficult. Our preliminary experiments, however, suggest that only very low levels of any VSV-proteins can be detected in cells infected with *hr*1, *hr*5, *hr*7, and *hr*8. In general these observations correlate with the low level of mRNA synthesis observed under the same conditions, but further work is required to accurately quantitate the viral protein synthesis in mutant infected cells.

One of our initial reasons for characterizing the VSV *hr*-mutants was to try to identify chemically the altered viral protein responsible for the *hr*-mutation. Mutants *hr*1 and *hr*8 and wild type VSV were grown on permissive BHK cells at 35° and labeled with ^{3}H-leucine. The progeny virus was purified and analysis of the virion proteins by discontinuous SDS-10% polyacrylamide slab gel electrophoresis showed no differences in the electrophoretic mobilities of any of the virion proteins of the mutants compared to wild type virus (data not shown).

We have also compared the mutant and wild type virion proteins by 2-dimensional gel electrophoresis.[10] This method resolves multiple forms of the G-protein and 3–4 forms of the N-protein as described previously.[11] We show here that this technique also separates the NS-protein into three and possibly 4 forms based on the apparent differences in their isoelectric points (Figure 1). The virion proteins L and M are not detected in this system. Figure 1, in fact, shows the analysis of a mixture of *hr*8 and wild type VSV. Identical patterns were obtained by 2-dimensional gel analysis for the multiple forms of the G, N, and NS-proteins of the mutants *hr*1 and *hr*8, as were

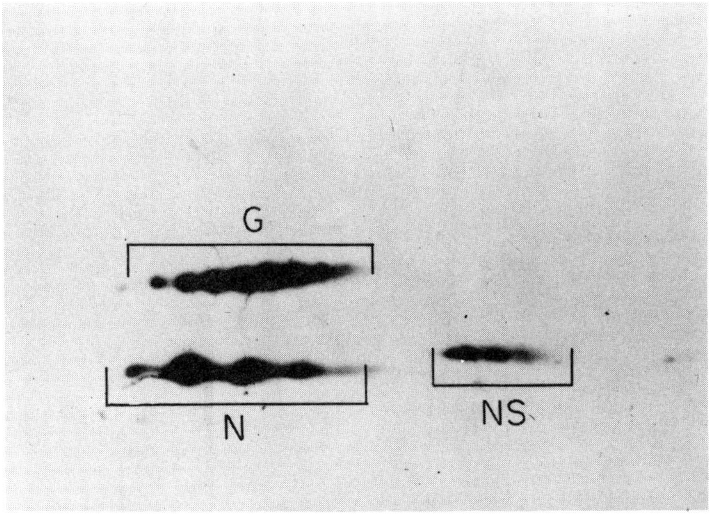

Figure 1. BHK cells at 35° were infected with wild type VSV or *hr*8 and labeled with ^{3}H-leucine (50 μCi/ml) from 1.5 hr postinfection. At 24 hr, the progeny virus was harvested and purified by density gradient centrifugation. Equal amounts of each labeled virus were mixed and analyzed by the method of O'Farrell.[10] Isoelectric-focusing was carried out with an equal mixture of ampholines of pH 4–6.5 and pH 5–8 for 17 hr at 400 V and 1 hr at 800V. The samples were then separated by SDS-10% polyacrylamide gell electrophoresis (from top to bottom) and visualized by fluorography. The pH gradient is from pH 7.5 the left to pH 4.8 on the right.

observed for wild type virus (data not shown). Similar protein analysis of the progeny *hr*-mutant viruses rescued by RPV from the non-permissive infections of HEp2 cells, showed that no changes occurred following RPV complementation in either the number or ratios of the multiple forms of the G, N, and NS-proteins. These data suggest that the *hr,* as well as the *ts*-defects, may reside in one or both of the remaining viral proteins L or M. We tend to favor the idea that the mutations in the *hr*-mutants that we have examined lie in the L-protein. Precedent for this stems from the results of Szilagyi et al.[12] who have shown that the lesion in an independently isolated temperature dependent host range mutant resides in the L-protein.

We have shown that by a relatively simple test, the ability of RPV to complement a non-permissive VSV infection, that we can functionally divide VSV *hr*-mutants into two classes. Analysis of the biochemical phenotype in non-permissive infections with the individual *hr*-mutants indicates that each class enters the cell and initiates primary transcription, but not replication of the genome RNA. This situation is similar to the restriction of wild type VSV in RC-60 cells, an infection which is also rescued by poxvirus, and that shows under non-permissive conditions mainly primary transcription, although in this case some genome replication also occurs.[6,7] Biochemically, it appears clear that host cell factors appear to function in some way in the process of VSV replication. Since co-infection with poxvirus allows us to define two classes of VSV *hr*-mutants, then perhaps two facets of RNA replication require host factor participation. Alternatively, host factors could participate elsewhere in the infectious cycle of VSV, in some process other than RNA replication. In this regard, an examination of changes in the non-permissive phenotypes in the *hr*-mutants co-infected with, but not reduced by, poxvirus might prove to be fruitful.

ACKNOWLEDGMENTS
This work is supported by a grant from the National Institutes of Health and a Faculty Research Award (SAM) from the American Cancer Society.

References

1. Szilagyi, J.F. and Pringle, C.R. (1975): *J. Virol.* 16:927–936.
2. Pringle, C.R. (1978): *Cell* 15:597–606.
3. Simpson, R.W. and Obijeski, J.F. (1974): *Virology* 57:357–368.
4. Obijeski, J.F. and Simpson, R.W. (1974): *Virology* 57:369–377.
5. Nowakowski, M., Bloom, B.R., Ehrenfeld, E., and Summers, D.F. (1973): *J. Virol.* 12:1272–1278.
6. Thacore, H.R. and Youngner, J.S. (1975): *J. Virol.* 16:322–329.
7. Hamilton, D.H., Moyer, R.W., and Moyer, S.A. (1980): *J. Gen. Virology* 49:273–287.
8. Morrongiello, M.P. and Simpson, R.W. (1979): *Virology* 93:506–514.
9. Moyer, S.A. and Gatchell, S.H. (1979): *Virology* 92:168–179.
10. O'Farrell, P.H. (1975): *J. Biol. Chem.* 250:4007–4021.
11. Raghow, R., Portner, A., Hsu, C.H., Clark, S.B., and Kingsbury, D.W. (1978): *Virology* 90:214–225.
12. Szilagyi, J.F., Pringle, C.R., and MacPherson, T.M. (1977): *J. Virol.* 22:381–388.

MARBURG-EBOLA VIRUSES

David H.L. Bishop and Richard W. Compans, eds.
The Replication of Negative Strand Viruses

Marburg and Ebola Viruses: Possible Members of a New Group of Negative Strand Viruses

R.L. Regnery, K.M. Johnson, and M.P. Kiley[a]

Marburg and Ebola viruses are recently isolated agents that can cause severe, often fatal, fevers in man.[1,2] Prior to the completion of the Maximum Containment Laboratory at the Center for Disease Control, most of what was known regarding these viruses concerned the pathology of virus infection. To better understand the agents responsible for these severe human diseases, we initiated studies of the physicochemical properties of these viruses in order to assess their taxonomic position as vertebrate parasites, to understand the biochemistry of infection in cells and to search for possible clues relevant to their still unknown ecology. This paper reviews the characteristics of these viruses which may lead to their inclusion in a new group of negative strand viruses.

Virion Morphology

Upon the isolation of the second of these viruses, Ebola virus, it was immediately recognized that both Marburg and Ebola viruses had similar virus-related structures.[3–5] However, the actual identity of the basic infectious particles for each virus remained unclear. Long flexious particles of varying lengths, toroids, 6-shaped particles, and simple rods have been visualized in electronmicrographs of these viruses. We have shown recently that viral infectivity of both viruses cosediments with discrete particle populations which have sedimentation coefficients of approximately 1300–1400S

[a] Special Pathogens Branch, Virology Division, Center for Disease Control, Atlanta, Georgia.

in rate-zonal gradients.[6] The viral structures in these populations are bacilliform in outline and have average lengths of 790 nm (Marburg virus) and 970 nm (Ebola virus). Particles of both viruses frequently exhibit asymmetries exemplified by one nearly hemispherical end and one blunt or ragged end. It is unclear whether the blunted end is a natural product of viral maturation or is a result of partial disruption during purification and fixation for electron-microscopy. The virion structures have narrow axial cores approximately 20 nm in diameter and have striations perpendicular to the long axis of the particle with a periodicity of approximately 5 nm;[7] both of these structures are visible in only some virus preparations.

Genome Composition

Sensitivity of Ebola genome nucleic acid to ribonuclease established that this moiety is single-stranded RNA. The kinetics of gamma radiation virus inactivation of purified Ebola virus indicated a single genome per particle with an estimated radiosensitive molecular weight of 4.5×10^6.[6] Electrophoresis of glyoxal denatured virion RNA demonstrated that the molecular weights of Ebola and Marburg genomes RNAs were approximately 4.2×10^6 (Figure 1); [8] a size estimate compatible with that determined by

Figure 1. Electrophoresis of the virion RNAs of 2 isolates of Marburg virus. Phenol extracted RNA from purified virions were denatured with glyoxal and coelectrophoresed on a 1.5% agarose slab gel with reference RNAs. (a) VSV and Vero cell ribosomal RNA. (b) Marburg virus (Voge). (c) Marburg virus (Musoke).

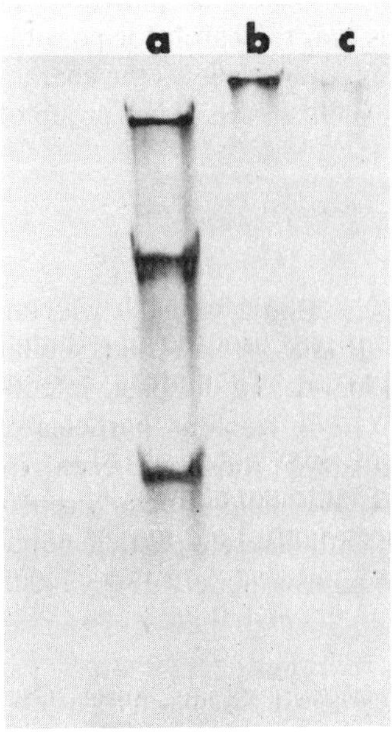

radiosensitive inactivation kinetics. Furthermore, the isolated virion RNA of Ebola virus did not bind to oligo(dT)-cellulose and was non-infectious, suggesting that the genome is a negative strand.

Although the virion RNA of Ebola is the same approximate size as that of VSV or rabies virus and the apparent periodicity of the nucleocapsids is comparable, the virions of Marburg and Ebola are at least twice as long as even the largest plant rhabdovirus particles. If the RNA of the Marburg and Ebola particles traverse the entire length of the virion, it must be concluded that the helix of virion RNA is wound to a smaller diameter within the particle than is the RNA or rhabdoviruses. This is compatible with the observation that the central axial core of Marburg is considerably smaller in diameter than that of known animal rhabdoviruses.[9]

To determine the genetic relatedness between strains of Ebola or Marburg and between the 2 viruses themselves, we are examining the RNA from a variety of isolates by RNA hybridization and analysis of T1 RNase digests. We are also investigating the relationships of intracellular viral specific RNAs to the genome RNA to establish the basic strategy for replication and to further clarify the nature of the strandedness of the genome.

Virion Structural Proteins

Ebola and Marburg virion structural proteins have been analyzed by SDS polyacylamide gel electrophoresis.[10,11] Electrophoresis of isotopically-labeled virion proteins (Figure 2) demonstrates that while the general pattern is the same for both viruses, there are differences in the molecular weights of corresponding proteins. Data on the location and function of the individual proteins demonstrated that when virions are disrupted with non-ionic detergent and high salt 2 proteins, VP2 and VP3, consistently remained associated with the presumptive nucleocapsid (density 1.32 g/cc), whereas only a single N-protein of vesicular stomatitus virus remained nucleocapsid bound with similar treatment. The VP1 proteins of Marburg and Ebola virions are readily labeled with glucosamine and appear to be glycoprotein.[10,11] Their removal from purified infectious samples with bromelain documents a surface position on the virion.

The relationships between relative sizes of proteins and functional roles of proteins are nearly the same for Marburg-Ebola viruses and VSV: L(?) -protein, glycoprotein, nucleocapsid protein, second nucleocapsid protein (Marburg-Ebola viruses) or nonstructural protein (VSV), matrix protein.

Antigenic Properties

Marburg virus has been shown to be serologically unrelated to a variety of viruses it was tested against.[12] Marburg and Ebola viruses showed no serological cross-reactivity by the indirect fluorescent antibody test.[13] In-

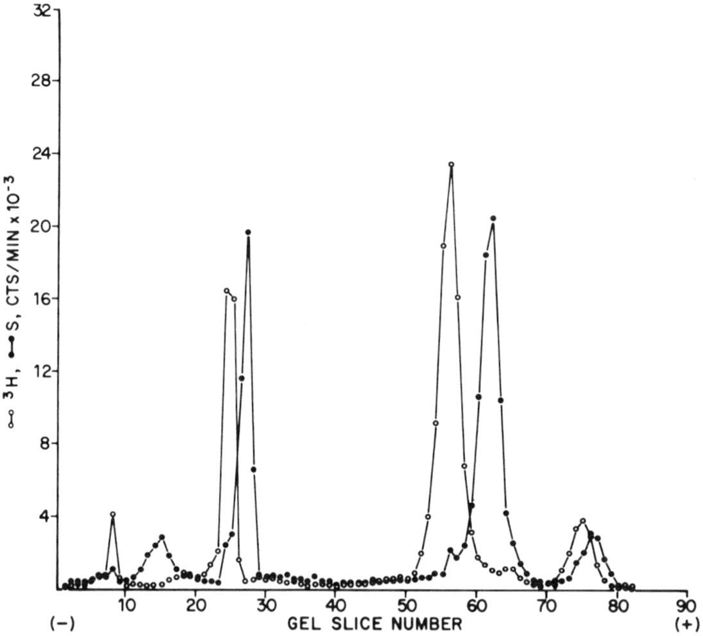

Figure 2. Coelectrophoresis of Marburg and Ebola virus structural proteins. Purified Marburg virus, isotopically labeled with ^{35} S-methionine (○—○), was mixed with purified Ebola virus, isotopically labeled with ^{3}H amino acids (●—●). The mixed sample was disrupted with SDS and mercaptoethanol and boiled for 1 minute prior to electrophoresis as described previously.[10] Coomassie blue staining of the gels confirmed the existence of an "L"-protein for both virions (data not shown).

terestingly, attempts to reduce the infectious titer of unpurified Ebola virus by homologous immune serum *in vitro* have not been successful.[13] Although homologous antiserum to Ebola does not neutralize the virus, we have demonstrated by immunoelectron microscopy that the infectious Ebola virus particles are antigenic and appear to bind large amounts of immunoglobulin (Figure 3). Treatment of Marburg virions with homologous serum yielded similar immunoelectron micrographs.[14]

Summary

Physicochemical data presented here and elsewhere[6,8,10,11] are summarized and compared to the rhabdovirus VSV in Tables 1 and 2. There is general similarity between Marburg-Ebola and rhabdoviruses. However, the length of infectious virions, the diameter of the central core, and the size of the surface glycoprotein of Ebola and Marburg viruses are all different from any known rhabdovirus.

Table 1. Comparative Properties of Marburg-Ebola Viruses and Rhabdoviridae.[15]

I. Properties in common

A. Virion structure

1. Particle symmetry
   bacilliform/bullet shaped
2. Particle density:
   1.14 g/cc (potassium tartrate)
3. Presence of central axial core
4. Presence of striations perpendicular to long axis of particle, 5 nm period
5. Nucleocapsid density:
   1.32/g/cc (CsCl)

B. Genome nucleic acid

1. RNA, single-stranded
2. One piece/genome
3. Molecular weight: $3.5–4.6 \times 10^6$
4. No detectable poly(A)
5. Not infectious

C. Virion proteins

1. 5 proteins including 1 glycoprotein
2. Similar functional/molecular weight order (See Table 2)

II. Properties which differ

A. Virion structure

1. Infectious particle dimensions
   a) Ebola: 970 nm × 80 nm
   b) Marburg: 790 nm × 80 nm
   c) Rhabdoviridae: 130–380 nm × 65–75 nm
2. Central axial core dimensions
   20 nm dia Marburg/Ebola vs 49 nm dia (Rhabdoviridae)
3. Sedimentation Coefficients
   1300s–1400s vs 550s–1000s (Rhabdoviridae)

B. Virion Proteins

1. a) 2 proteins remain nucleocapsid associated with Marburg/Ebola
   b) 1 protein remains nucleocapsid associated with VSV
   } 2.0% Triton X-100 / 1 M KCl
2. Glycosylated protein of Marburg-Ebola ~2 times as large as "G" protein VSV
3. One nucleocapsid protein of Ebola/Marburg ~2 times as large as "N" protein VSV

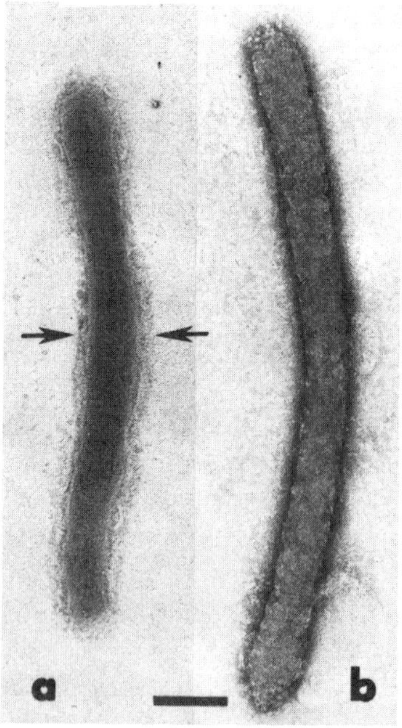

Figure 3. Purified Ebola virions were fixed with 0.5% gluteraldehyde and rinsed twice with PBS by pelleting at high speed. The final, resuspended virion pellet was either treated with convalescent guinea pig antiserum to Ebola virus (a) or with normal guinea pig serum (b) and prepared for electron microscopy by negative staining with 0.5% PTA. Reference bar is equivalent to 100 nanometers.

Table 2. Molecular Weights of Proteins (Descending Order).

Virion proteins	**1**	**2**	**3**	**4**	**5**
Ebola virus	~170,000	125,000	104,000	40,000	26,000
	(L)	(G)	N_1)	(N_2)	(M)
Marburg virus	~170,000	140,000	98,000	38,000	22,000
	(L)	(G)	(N_1)	(N_2)	(M)
Rhabdoviridae[15]	150,000	70-80,000	50-62,000	40-45,000	20,000-30,000
	(VSV)	(VSV)	(VSV)	(VSV)	(VSV)
	(L)	(G)	(N)	(NS)	(M)

References

1. Martini, G.A. and Siegert, R. (1971): *Marburg Virus Disease.* Springer-Verlag, New York.
2. Pattyn, W.R. (1978): *Ebola Virus Hemorrhagic Fever.* Elsevier/North Holland Biomedical Press, New York.
3. Bowen, E.T.W., Platt, G.S., Lloyd, G., Baskerville, A., Harris, W.J., and Vella, E.E. (1977): *Lancet* 1:571–573.
4. Johnson, K.M., Webb, P.A., Lange, J.V., and Murphy, F.A. (1971): *Lancet* 1:569–571.
5. Pattyn, S.R., Jacob, W., VanderGroen, G., Piot, P., and Courteille, G. (1977): *Lancet* 1:573–574.
6. Regnery, R.L., Palmer, E.L., Kiley, M.P., and Johnson, K.M. (1980): In preparation.
7. Murphy, F.A., VanderGroen, G., Whitfield, S.G., and Lange, J.V. (1978): In: *Ebola Virus Hemorrhagic Fever.* Pattyn, W.R. (Ed.), Elsevier/North Holland Biomedical Press, New York, pp. 61–84.
8. Regnery, R.L., Johnson, K.M., and Kiley, M.P. (1980): *J. Virol.* (in press).
9. Murphy, F.A., Simpson, D.I.H., Whitfield, S.G., Zlotnik, I., and Carter, G.B. (1971): *Laboratory Investigation* 24:279–291.
10. Kiley, M.P., Regnery, R.L., and Johnson, K.M. (1980): *J. Gen. Virol.,* in press.
11. Kiley, M.P., Regnery, R.L., Cox, N.J., and Johnson, K.M.: In preparation.
12. Casals, J. (1971): In: *Marburg Virus Disease.* Martini, G.A. and Siegert, R. (Eds.), Springer-Verlag, New York, New York, pp. 98–104.
13. Webb, P.A., Johnson, K.M., Wulff, H., and Lange, J.V. (1978): In: *Ebola Virus Hemorrhagic Fever.* Pattyn, W.E. (Ed.), Elsevier/North Holland Biomedical Press, New York, pp. 91–94.
14. Almeida, J.D., Waterson, A.P., and Simpson, D.I.H. (1971): In: *Marburg Virus Disease.* Martini, G.A. and Siegert, R., (Eds.), Springer/Verlag, New York, pp. 84–97.
15. Brown, F., Bishop, D.H.L., Crick, J., Francki, R.I.B., Holland, J.J., Hull, R., Johnson, K., Martelli, G., Murphy, F.A., Obijeski, J.F., Peters, D., Pringle, C.R., Reichmann, M.E., Schneider, L.G., Shope, R.E., Simpson, D.I.H., Summers, D.F., and Wagner, R.R. (1979): *Intervirology* 12:1–7.

Index

N

Q

R

S

T

U

V